Gallium Arsenide and Related Compounds 1987

Conference Committee

A Christou (Chairman), D W Shaw, G Kiriakidis, L Papadopoulou (Secretary), B Bunnell (Secretary), V G Keramidas, G Kiriakidis, H Beneking

Organising Committee

V G Keramidas, A Christou, J Magarshack, E N Economou, T Nakahara, H Rupprecht, Z Hatzopoulos, G Kiriakidis, C Wood

International Advisory Committee

J V DiLorenzo, C Hilsum, T Nakahara, T Sugano, M Uenohara, T Ikoma, L F Eastman, J Magarshack, D W Shaw, K Thim, K H Zschauer, A Christou

Technical Program Committee

H S Rupprecht (Chairman), J Noblanc, T Ikoma, T N Theis, W Baechtold, K W Benz, M J Coupland, W Harth, L Hollan, W Kellner, J Schneider, W T Tsang, G Weimann

Sponsors

The Symposium was sponsored by the Ministry of Industry of Technology, Greece; Research Centre of Crete; Bell Communications Research, USA; Naval Research Laboratory, USA; USAF European Office of Aerospace Research and Development; US Army European Research Office, London; Fraunhofer Institute for Applied Solid State Physics.

Gallium Arsenide and Related Compounds 1987

Proceedings of the Fourteenth International Symposium on Gallium Arsenide and Related Compounds held in Heraklion, Crete, 28 September–1 October 1987

Edited by A Christou and H S Rupprecht

Institute of Physics Conference Series Number 91
Institute of Physics, Bristol and Philadelphia

CODEN IPHSAC 91 1–834 (1988)

British Library Cataloguing in Publication Data

International Symposium on Gallium Arsenide
and Related Compounds (*14th:1987:Heraklion, Greece*).
1. Gallium arsenide semiconductors
I. Title II. Christou, A. (Aris) III. Rupprecht, H.S. (Hans S)
IV. Series
537.6′22

ISBN 0-85498-182-9

Library of Congress Cataloguing-in-Publication Data are available

ISBN 0-85498-182-9

Published under The Institute of Physics imprint by IOP Publishing Ltd
Techno House, Redcliffe Way, Bristol BS1 6NX, England
242 Cherry Street, Philadelphia, PA 19106, USA

Printed in Great Britain by J W Arrowsmith Ltd, Bristol

Preface

The Fourteenth International Symposium on Gallium Arsenide and Related Compounds was held in Heraklion, Crete, Greece from 28 September to 1 October 1987. Two hundred research papers were presented at the Symposium in three plenary sessions, the regular sessions and poster sessions. The Symposium papers were selected from over two hundred and seventy five abstracts submitted to the Technical Program Committee for evaluation.

The rapid growth of research in gallium arsenide and related compounds over the last three years has resulted in significant progress in the areas of materials growth and characterisation, discrete device physics and processing technology, epitaxial materials growth and ion implantation. However, as was made evident by the Symposium, new breakthroughs and innovations are required as we push forward towards higher levels of integration and nanostructures.

The Symposium was highlighted by eleven invited speakers in the areas of: DX centers, advanced MOCVD optoelectronic devices, electronic states in superlattices, polar–nonpolar epitaxy, delta doping, integrated guided wave optics on III–V semiconductors, wavefunction engineering, rare earth injection lasers, below bandgap photoresponse of undoped GaAs, Raman scattering of superlattices and stable ohmic contacts.

The technical program represents the planning and accomplishments of Dr Hans Rupprecht who organised, with his Technical Program Committee, the reviews of all the abstracts and the formulation of the technical sessions. The technical papers were organised into sessions on materials characterisation, devices, bulk crystal growth, epitaxial layer growth, optical devices, processing technology, heterostructures, quantum wells and superlattices.

The papers in this volume have been selected from the 200 papers presented at the Symposium. Papers from the plenary sessions are in Chapter 1. Submitted papers are arranged by subject in Chapters 2–7 and late news papers are in Chapter 8. Because of the overlap between topics, readers are encouraged to explore chapters on related subject areas.

The Symposium was organised under the auspices of the International Advisory Committee, chaired by John Magarshack. The official sponsors and providers of financial support for the 14th International Symposium on Gallium Arsenide and Related Compounds were: Ministry of Industry and Technology, Greece; Research Centre of Crete; Bell Communications Research, USA; Naval Research Laboratory, USA: Fraunhofer Institute for Applied Solid State Physics. Finally we wish to acknowledge especially the continuing support for this Symposium provided by the US Air Force European Office of Aerospace Research and Development (Dr Eirog Davis) and the US Army Research Office–London.

Aris Christou
General Symposium Chairman
14th International Symposium on
Gallium Arsenide and Related Compounds

During the opening plenary session on 28 September 1987, Dr Hans Rupprecht and Aris Christou are shown above listening to Professor Herbert Kroemer lecture on polar–nonpolar epitaxy.

GaAs Symposium Award and Heinrich Welker Gold Medal

After the discovery of the transistor in 1947 materials other than the classic semiconductor Ge were soon considered for solid state electronics. Besides silicon, which was immediately recognised as important, one of the first new materials to be considered was GaAs, a representative of a new class of semiconductors—the III–V compound semiconductors.

Although progress on III–V compounds was initially slow, bit by bit their performance increased. By 1962 GaAs and other III–V materials were being used in construction of lasers, light emitting diodes, and solid state devices for microwave application. New applications were being proposed at an increasing rate. The upsurge in III–V research led to the proposal, primarily from C Hilsum, that an International III–V semiconductor conference be initiated. This led to the founding in 1966 of the International Symposium on Gallium Arsenide and Related Compounds, commonly called the International GaAs Symposium. By 1976 this conference had grown to the point that its organisers and committee members introduced the Gallium Arsenide Symposium Award, which was given for outstanding III–V research.

In honour of Heinrich Welker, one of the foremost pioneers in the development of III–V semiconductors, his long-term employer Siemens AG of Munich, West Germany, established the Welker Medal, a gold medal, to accompany the GaAs Symposium Award.

The individuals who have received the GaAs Symposium Award and the Welker Medal are:

Nick Holonyak Jr, 1976
Cyril Hilsum, 1978
Hisayoshi Yani, 1980
Gerald L Pearson, 1981
Herbert Kroemer, 1982
Izuo Hayashi, 1984
Heinz Beneking, 1985
Al Cho, 1986.

In 1987 the GaAs Symposium Award and Welker Medal were presented to Dr Zh I Alferov of the A F Ioffe Physical Technical Institute, Leningrad. His pioneering and outstanding contributions in theory, technology and devices since the beginning of the compound semiconductor era are internationally known. Our community recognises Dr Alferov for his work in liquid phase epitaxy, laser diodes, vapour phase epitaxy and for innovative contributions to the technology of compound semiconductors.

Young Scientist Award

The award recognising the outstanding contributions of a young scientist (under the age of forty) was given in 1987 to Dr Naoki Yokoyama. Among his many contributions is the hot electron transistor, the resonant tunnelling hot electron transistor, and the refractory silicide gate metal. The self-aligned structure using refractory silicides is now commonly used throughout the industry. As early as 1981 Dr Yokoyama reported on the self-aligned source-drain planar device for ultrahigh-speed GaAs VLSIs. His recent work on the resonant tunnelling hot electron transistor has opened new possibilities for memory and logic circuits.

We feel honoured to be able to recognise Dr Yokoyama, in a field where there are so many excellent young scientists.

Contents

Chapter 4: Materials characterisation

Chapter 5: Processing

Paper presented at Int. Symp. GaAs and Related Compounds, Heraklion, Greece, 1987

DX centers in GaAs and $Al_xGa_{1-x}As$: device instabilities and defect physics

T.N. Theis

I.B.M. Thomas J. Watson Research Center, P.O. Box 218, Yorktown Heights, NY 10598

ABSTRACT: Recent experimental results relating the unusual electron capture and emission kinetics of the DX center to the band structure of $Al_xGa_{1-x}As$ are reviewed. Perhaps the most remarkable result is the observation of electron trapping by the DX center in heavily doped n-GaAs. A simple model for the kinetics is then presented and the various DX center related device instabilities are explained in terms of the model. Finally, the accumulating evidence that the DX center is just the simple substitutional donor is assessed in the light of these results.

1. INTRODUCTION

Deep donor levels responsible for extremely persistent photoconductivity in n-$GaAs_{1-x}P_x$ were first reported nearly twenty years ago (Craford *et al.* 1968). Similar effects were observed in n-$Al_xGa_{1-x}As$ (Nelson 1977), and studies of the electron capture and emission kinetics led Lang *et al.* (1977, 1979) to conclude that thermal capture and emission occur via a multiphonon process. The photoionization energy was observed to be much larger than the activation energies for thermal capture and emission, suggesting a large energetic relaxation of the lattice upon charge capture. Such a large lattice relaxation was not believed possible for a simple substitutional donor, so a donor complex, the DX center, consisting of a simple donor (D) and an unknown defect (X) was postulated (Lang *et al.* 1979). With the development of GaAs/n-$Al_xGa_{1-x}As$ heterostructure transistors it became apparent that the DX center was the source of dramatic and undesirable device instabilities. The resulting explosion of studies has greatly increased our understanding of the DX center, while generating considerable controversy over the model. Here I present a synthesis of recent experimental results and draw two important conclusions:

1. For $Al_xGa_{1-x}As$ in the direct gap alloy composition range, the electron capture and emission kinetics of the DX center imply that the deep donor level derives its symmetry almost entirely from the L point. This leads to a simple physical description of DX center-related device instabilities including those caused by hot electron trapping.
2. The DX center is intimately linked to the long known (Langer 1980) but poorly understood phenomenon of localized resonant donor levels in compound semiconductors. The deep level in $Al_xGa_{1-x}As$ moves above the band edge with decreasing Al content and becomes a highly localized resonant level in GaAs. Occupation of this state in heavily doped material is a new mechanism, distinct from and acting in addition to self-compensation, which limits the equilibrium free carrier density.

2. THE DX LEVEL AND THE BAND STRUCTURE OF $Al_xGa_{1-x}As$

Both column IV and column VI dopants added to $Al_xGa_{1-x}As$ give rise to a donor level, E_{DX}, which tends to track the L conduction band edge. We shall refer to this level as the DX level to distinguish it from effective-mass-like donor levels which also arise from the same positively charged center (Theis *et al.* 1984, Theis 1986b). A compilation of Hall data from many laboratories (Springthorpe *et al.* 1975, Lifshitz *et al.* 1980, Saxena 1981, Ishikawa *et al.* 1982 Ishibashi *et al.* 1982, Chand *et al.* 1984) is shown in Figure 1a. The Hall activation energy measured with respect to the Γ band edge, $E_\Gamma - E_{DX}$, becomes smaller with decreasing Al mole fraction, x, and vanishes at $x \simeq 0.22$. The activation energies for thermal capture, E_c, and thermal emission, E_e, determined by deep level transient spectroscopy (DLTS), are always found to be larger than $E_\Gamma - E_{DX}$. Thus capture and emission can be viewed as occurring over an effective repulsive barrier as shown in Figure 1b (Lang et al. 1979).

Extrapolation of the Hall data to $x<0.22$ (dashed line in Figure 1a) suggests a resonant level lying above E_Γ. The resonant level may be brought into the fundamental gap by application of hydrostatic pressure (Mizuta *et al.* 1985, Tachikawa *et al.* 1985b). Even when resonant with the band, the DX level is metastable and capable of localizing electrons (Theis *et al.* 1986a). As pointed out by Theis (1986b) the metastability is expected from the behavior of the emission and capture kinetics as a function of alloy composition: While $E_\Gamma - E_{DX}$ goes to zero at $x \simeq 0.22$, E_e is independent of x (Lang *et al.* 1979, Mooney *et al.* 1986) and E_c becomes *larger* as x is reduced (Mooney *et al.* 1986). Extrapolating this behavior to $x<0.22$, one expects the effective barrier of a metastable state as shown in Figure 1c.

For typical free carrier concentrations and temperatures, the equilibrium occupancy of the resonant state in GaAs and $Al_xGa_{1-x}As$ of low Al content is very small. Nevertheless, the state can become heavily occupied under non-equilibrium conditions, such as exist when an electric field heats the free carriers well above the lattice temperature (Theis *et al.* 1986a). The return to thermal equilibrium was studied in samples prepared in this manner, and the thermal emission kinetics at $x=0.14$ were found to be unchanged from those at higher Al mole fractions. Thus the state is unperturbed by its energetic resonance with the conduction band.

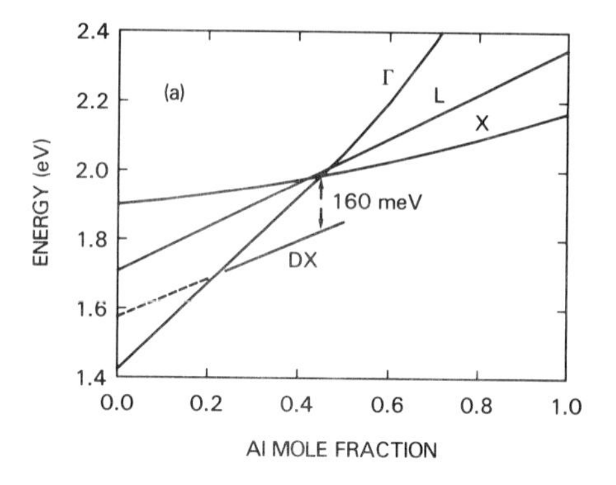

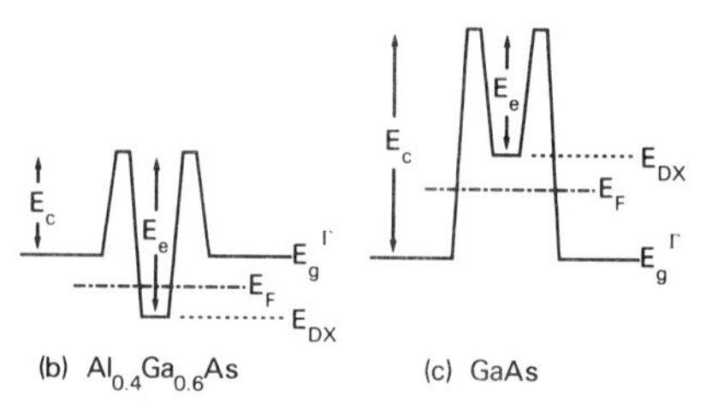

Fig. 1. Lower conduction band edges in $Al_xGa_{1-x}As$ as a function of x. The DX level approximately follows the L band edge and becomes resonant with the Γ band for $x \lesssim 0.22$. (b). Effective potential barrier for thermal capture and emission of electrons by the deep level in $Al_{0.4}Ga_{0.6}As$. (c). Effective potential barrier for the resonant level in GaAs.

Occupancy of the resonant state can also be significant under *equilibrium* conditions in heavily doped degenerate material, as the Fermi level approaches the resonant level. Kirtley *et al.* (1987) were able to show that the DX center is the dominant source of low frequency noise in GaAs/n-$Al_xGa_{1-x}As$ heterojunctions with Al mole fractions as low as 0.14, when the free carrier density, n_o, in the $Al_xGa_{1-x}As$ is $\sim 1 \times 10^{18}$ cm^{-3}. At higher doping levels, thermal occupation of the resonant state becomes observable even in GaAs. Thus the thermal capture and emission were studied by DLTS techniques in MBE grown Si-doped material with $n_o \gtrsim 5 \times 10^{18}$ cm^{-3} (Theis *et al.* 1987a, Mooney *et al.* 1987a). The capture transients displayed a temperature dependence consistent with thermally activated capture from the Fermi level over an effective barrier as shown in Figure 1c. The emission transients were the most nearly exponential ever reported for the DX center, consistent with the absence of alloy broadening (Calleja *et al.* 1986) in GaAs, and explicitly demonstrating the strong localization of the DX state, which is not perturbed by the presence of neighboring Si atoms at an average distance ~35Å. The activation energy for thermal emission in Si-doped GaAs was found to be $E_e=0.33 \pm 0.02$ eV, somewhat lower than the

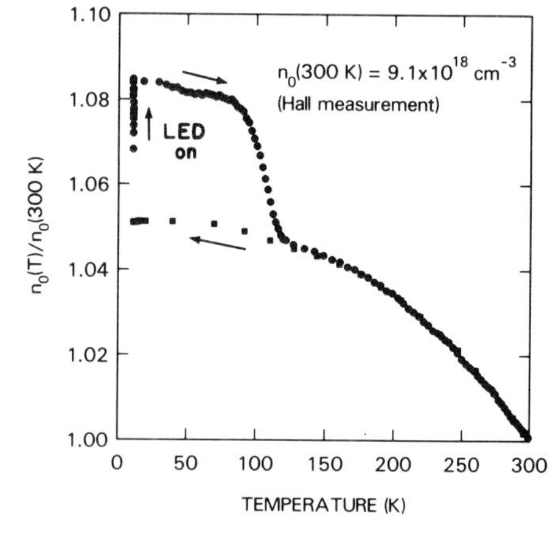

Fig. 2. Hall free carrier concentration in heavily Si-doped GaAs as a function of temperature.

value of 0.44±0.02 eV found in $Al_xGa_{1-x}As$ (0.14<x<0.74), but very close to the values of 0.31 and 0.33 eV reported by Mizuta *et al.* (1985) in GaAs under hydrostatic pressure. Assuming all donors are DX centers (Mizuta *et al.* 1985, Maude *et al.* 1987), and modeling the amplitude of the emission transients, it was found that the occupancy of the resonant level increases rapidly with increasing n_o, so that a crossing of the Fermi level with the resonant level should occur at $n_o \sim 1.5\text{-}2\times10^{19}$ cm^{-3}. As long as the chemical concentration of DX centers exceeds the net uncompensated donor concentration, *this is the highest equilibrium free carrier density obtainable in n-GaAs.* A similar limit is inferred from studies of heavily doped GaAs under hydrostatic pressure (Maude *et al.* 1987).

The counterintuitive influence of the resonant state on transport properties is demonstrated by the previously unpublished Van der Pauw-Hall data of Figure 2. As the heavily Si-doped GaAs is cooled from 300 K, n_o *increases* steadily, as carriers freeze *in* to the conduction band from the higher lying resonant state. Below T~100 K, the emission barrier prevents further electrons from leaving the trap state, except when photoionized by a light source (LED). Upon slow warming the resulting photoconductivity is extremely persistent for T≲120 K, as electrons are unable to surmount the thermal capture barrier.

3. CONFIGURATION COORDINATE MODEL

A probable physical explanation for the phenomenology of the DX center is indicated by the configuration coordinate diagram of Figure 3. U_{DX} is the total energy of the center and a tightly bound electron (charge neutral state), while U_L is the total energy of the ionized center plus an electron in an extended state at the L band edge. (U_L could also represent the total energy of the center with an electron in a loosely bound L donor state (Morgan 1986, Theis 1986b), but, for the sake of brevity, I shall discuss only transitions to and from extended states.) In the simplest picture, electronic transitions between the tightly bound and extended states occur primarily at vibronic levels near the top of the classical potential barriers E_b and E_e. Since there is no confirmed microscopic model of the DX center, there is no *a priori* reason for excluding similar thermal transitions to and from the Γ band. However, the existence of such an optical selection rule is the simplest way to explain the experimental data, as discussed below.

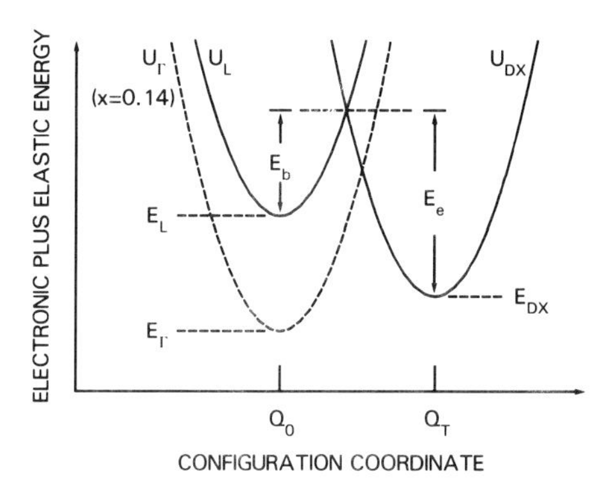

Fig. 3. Configuration coordinate diagram for the DX center. Thermal capture and emission occurs predominantly via L-band states.

If transitions to and from the DX state occur predominantly via L-band states, then the capture rate depends on the electron population of the L band. The capture rate is therefore given by $1/\tau_c = \sigma < v > n$, where σ is the capture cross section, $<v>$ is the average velocity of an L band electron, and n is the concentration of free carriers in the L band. (τ_c is the mean lifetime of the empty state.) For a multiphonon capture process, $\sigma = \sigma_0 e^{-E_b/k_BT}$, where E_b is the capture barrier defined in Figure 3. Under all practical conditions the L minimum is more than a few k_BT above the Fermi level, E_F, so the carrier concentration is $n = 2M_L(2\pi m_L k_BT/h^2)^{3/2} e^{-(E_L - E_F)/k_BT}$ where m_L is the density of states effective mass for an L band electron and M_L is the band state degeneracy. Also the average thermal velocity is $<v> = (3k_BT/m_L)^{1/2}$. Thus the capture rate has a temperature dependence of the form

$$1/\tau_c \sim T^2 e^{-(E_L - E_F + E_b)/k_BT}. \tag{1}$$

From consideration of detail balance, the temperature dependence of the emission rate is found to be

$$1/\tau_e \sim T^2 e^{-E_e/k_BT}, \tag{2}$$

where use has been made of the identity, evident from Figure 3, $E_e = E_L - E_{DX} + E_b$. Both E_e and E_b are independent of Al mole fraction, x, so long as two physically reasonable conditions are satisfied. First, the basis states of the electronic part of the DX wave function must be derived primarily from the vicinity of the L point, so that $E_L - E_{DX}$ is independent of x. Second, the characteristic vibronic energy must also be independent of x. The observed dependence of E_c and lack of dependence ($x \gtrsim 0.14$) of E_e on alloy composition then follows immediately from (1) and (2). Examination of (1) shows that the thermally activated capture barrier measured by experiment is the sum of the activation energy for multiphonon capture, E_b, and the activation energy for population of the L band, $E_L - E_F$. Note that the phenomenological capture barrier of Figure 1 is defined with respect to the Γ band edge, and can be written as $E_c = (E_L - E_\Gamma) + E_b$. Comparison with the exponential factor in (1) shows that E_c can be determined from capture rate data only if the position of the Fermi level is taken into account. The model therefore provides a simple physical explanation for the dependence of capture rate on Fermi level position found by Caswell *et al.* (1986).

We note that Lang (1986) has published a figure similar to Figure 3 except that X point symmetry is assumed. This symmetry might be correct for other alloy systems, but the hydrostatic pressure measurements clearly show that the the DX level in $Al_xGa_{1-x}As$ derives much of its character from the L band (Tachikawa *et al.* 1985a). In addition, the fact that (1) and (2) give a good description of the thermal emission and capture kinetics as a function of alloy composition shows that even in the highly excited vibronic states, the electronic portion of DX wave function retains the L-point symmetry. This is an important constraint on microscopic models of the center.

Some recent models (Hjalmarson and Drummond 1986, Yamaguchi 1986, Henning and Ansems 1987) carry the idea of an L-derived state much further, proposing a much smaller lattice relaxation, with the optical selection rule playing a correspondingly stronger role in suppressing transitions to and from Γ band states. These models also predict exponentially activated capture and emission consistent with (1) and (2), but do not predict the large hole capture cross section (Calleja *et al.* 1985) or the large photoionization energy (~ 1.25 eV for Si-doped $Al_xGa_{1-x}As$) (Lang *et al.* 1979, Legros *et al.* 1987) observed for the DX center. Measurable photoionization at energies as small as 0.32 eV *was* reported by Henning and Ansems (1987) in support of their model of a small energetic relaxation. However, more sensitive measurements by Mooney *et al.* (1987b) showed no measurable photoionization cross section below 0.8 eV, even though a tunable infrared laser source allowed the cross section to be measured eight orders of magnitude below its peak value.

4. DEVICE INSTABILITIES - HOT ELECTRON CAPTURE

As the dominant donor level ($x \gtrsim 0.22$), the DX level controls the free carrier density in bulk n-$Al_xGa_{1-x}As$ and in GaAs/n-$Al_xGa_{1-x}As$ heterojunctions with sufficiently thick $Al_xGa_{1-x}As$ layers. Because of the thermally activated emission and capture kinetics, the time required for the level occupation to approach quasi-equilibrium becomes exceedingly long at low temperatures. In heterostructure field effect transistors (HFET's) the result is undesirable hysteretic changes in operating characteristics induced by the the application of gate and/or source-drain bias voltage. Charge capture and emission in the $Al_xGa_{1-x}As$ layers of large area HFET's can be monitored by capacitance-voltage (C-V) techniques (Mooney *et al.* 1985), and such measurements establish, in turn, the basis for quantitative studies of the thermal emission and capture kinetics (Caswell *et al.* 1986, Mooney *et al.* 1986). The results are entirely consistent with and provide evidence for the model discussed above.

A technologically more troublesome device instability is caused by hot electron capture, observed at high source-drain voltages in short channel HFET's (Rochette *et al.* 1982, Fischer *et al.* 1984, Kinoshita *et al.* 1985, Kastalsky and Kiehl 1986). Capture is most readily observed under gate bias conditions which allow conduction electrons to enter the $Al_xGa_{1-x}As$ beneath the gate (Kastalsky and Kiehl 1986). It occurs primarily in the high field region of the channel between gate and drain. The reduction of fixed positive charge in the $Al_xGa_{1-x}As$ layer reduces the free carrier density and hence the conductance of the high mobility channel in the GaAs layer. As capture proceeds, a portion of the channel may actually become pinched off at low source-drain voltages. In this condition, often referred to as the collapse of the device characteristics, the source-drain I-V relationship becomes diode-like, indicating the development of an asymmetric potential barrier to current flow in the channel (Theis 1987b).

Complexities such as these can be ignored if the hot electron capture only slightly reduces the low field channel conductance. In that case, although the trapping is laterally non-uniform, small changes in channel conductance are approximately proportional to the amount of trapped charge. Theis *et al.* (1986a) were therefore able to study relative capture rates as a function of several variables by monitoring trapping induced transients in the low field conductance of short channel FET's. A threshold source-drain voltage for hot electron capture was observed, and for devices of similar geometry and dimension this threshold voltage increased with decreasing x (increasing Γ-L separation). These observations suggested that the capture involves transfer of electrons to the L-band (Theis 1986a). Additional information is provided by the data of Figure 4 showing relative capture rates at three different source-drain voltages, V_{SD}, as a function of temperature for a sample with $x=0.35$. The relative rates were determined from slopes of capture transients such as those found in Theis *et al.* (1986a), measured at the point where the channel conductance had decreased to 95 per cent of its initial value. The data indicate two competing mechanisms for capture. At temperatures $\lesssim$80 K, the dominant process is purely athermal, that is, independent of lattice temperature. Above this temperature the capture becomes thermally assisted, with an activation energy, E_a~0.1 eV. Neither E_a nor the relative contributions of the thermally assisted and athermal processes to the total hot electron capture rate depended on V_{SD} (*i.e.* average electric field) or on alloy composition ($0.20<x<0.35$).

These observations indicate a two stage hot electron capture process in perfect accord with the configuration coordinate model of Figure 3. Electrons are first transferred to the L band by the action of the electric field, and then captured to the DX level by one of two competing channels. The thermally assisted channel is phonon mediated, and the activation energy is a measure of E_b, the purely multiphonon part of the capture barrier. The athermal capture channel is probably direct radiative capture from the L band (or shallow level associated with the L band) to the DX level. Such a radiative process is expected in addition to the multiphonon process, but has not been observed in thermal capture experiments. It can now be understood that this is simply because radiative capture becomes the dominant capture channel only at temperatures $\lesssim$ 80 K where the capture rate (in the absence of an electric field) is unobservably slow. The effect of the field is to increase the population of the L band, and hence the total capture rate, by orders of magnitude, so that radiative capture can be observed. Note that radiative capture from the Γ band is energetically allowed, but not observed. It must be suppressed according to the same selection rule which modifies the thermal capture and emission kinetics.

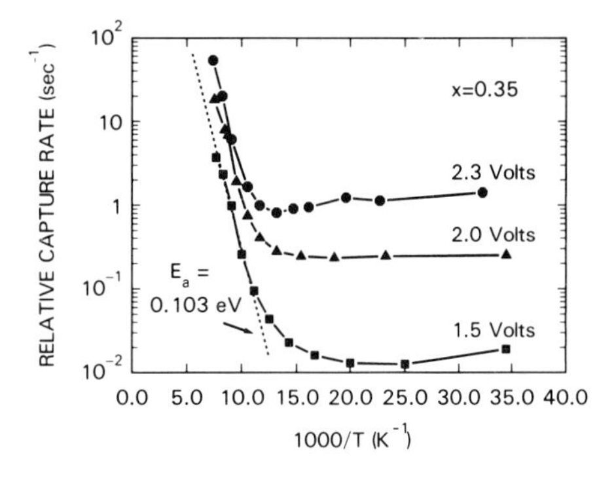

Fig. 4. Relative hot electron capture rates as a function of temperature for three source-drain voltages, V_{SD}, showing thermally assisted (T$\gtrsim$80 K) and athermal (T$\lesssim$80 K) capture.

5. MICROSCOPIC MODELS

The hydrostatic pressure measurements of Mizuta *et al.* (1985) indicate that all or nearly all donors in GaAs act as DX centers. Infrared vibrational spectroscopy indicates the majority of Si donors occupy substitutional sites in GaAs (Maguire, *et al.* 1987), and this has been confirmed on at least one sample where the majority of donors have been shown to act as DX centers (Maude *et al.* 1987). Hydrogen passivation studies also imply the identity of DX centers and isolated Si impurities (Nabity *et al.* 1987). Thus there is growing evidence that the DX center is just the simple substitutional donor. Can a simple donor give rise to the large energetic relaxation of the lattice and strong spatial localization of the electronic charge, for which I have argued? Oshiyama and Ohnishi (1986) and Morgan (1986) have actually considered this possibility, but new ideas may be needed to explain the most recent experimental results. Extended x-ray absorption fine structure (EXAFS) studies suggest a very small bond length change upon charge localization by the donor (Mizuta and Kitano 1987), although significant bond *angle* distortion is not ruled out and is, in fact, consistent with results from Mossbauer spectroscopy (Gibart 1987). A strong localization of the electronic charge is also consistent with the Mossbauer results. Very recent proposals by Khachaturyan and Weber (1987) and Morgan (1987) suggest a small donor displacement with a polaron-like distortion of the surrounding lattice, resulting in a large energetic relaxation of the lattice about a highly localized electronic state. If these ideas should prove correct, they would constitute a revolution in our understanding of donors in III-V semiconductors.

REFERENCES

Calleja E, Munoz E, Jimenez B, Gomez A and Garcia F 1985 *J. Appl. Phys.* **57** 5295
Calleja E, Mooney P M, Wright S L and Heiblum M 1986 *Appl. Phys. Lett.* **49** 657
Caswell N S, Mooney P M, Wright S L and Solomon P M 1986 *Appl. Phys. Lett.* **48** 1093
Chand N, Henderson T, Klem J, Masselink W T, Fischer R, Chang Y C and Morkoç H 1984 *Phys. Rev. B* **30** 4481
Craford M G, Stillman G E, Rossi J A, and Holonyak N Jr. 1968 *Physical Review* **168** 867
Fischer R, Drummond T J, Klem J, Kopp W, Henderson T S, Perrachione D and Morkoç H, 1984 *IEEE Trans. Electron Devices* **ED-31** 1028
Gibart P, Williamson D L, El Jani B and Basmaji P 1987 proc. of this conference
Henning J C M and Ansems J P M 1987 *Semicond. Sci. Technol.* **2** 1
Hjalmarson H P and Drummond T J 1986 *Appl. Phys. Lett.* **48** 656
Ishibashi T, Tarucha S and Okamoto H 1982 *Jpn. J. Appl. Phys.* **21** L476
Ishikawa T, Saito J, Sasa S and Hiyamizu S 1982 *Jpn. J. Appl. Phys.* **21** L675
Kastalsky A and Kiehl R A 1986 *IEEE Trans. Electron Devices* **ED-33** 414
Khachaturyan K A and Weber E R 1987 unpublished
Kinoshita H, Akiyama M, Ishida T, Nishi S, Sano Y and Kaminishi K 1985 *IEEE Electron Device Lett.* **EDL-6** 473
Kirtley J R, Theis T N and Wright S L 1987 *Proceedings of IX International Conference on Noise in Physical Systems, Montreal,* 25-29 May
Lang D V and Logan R A 1977 *Phys. Rev. Lett.* **39** 635
Lang D V, Logan R A and Jaros M 1979 *Phys. Rev. B* **19** 1015
Lang D V 1986 *Deep Centers in Semiconductors* ed S T Pantelides (New York: Gordon and Breach) pp 489-539
Langer J M 1980 *New Developments in Semiconductor Physics* ed F Beleznay, G Ferenczi, and J Giber (Berlin: Springer-Verlag) pp 123-49
Legros R, Mooney P M and Wright S L 1987 *Phys. Rev. B* **35** 7505
Lifshitz N, Jayaraman A, Logan R A and Card H C 1980 *Phys. Rev. B* **21** 670
Maguire J, Murray R and Newman R C 1987 *Appl. Phys. Lett.* **50** 516
Maude D K, Portal J C, Dmowski L, Foster T, Eaves L, Nathan M, Heiblum M, Harris J J and Beall R B 1987 *Phys. Rev. Lett.* **59** 815
Mizuta M, Tachikawa M, Kukimoto H and Minomura S 1985 *Jpn. J. Appl. Phys.* **24** L143
Mizuta M and Kitano T 1987 unpublished
Mooney P M, Solomon P M and Theis T N 1985 *Inst. Phys. Conf. Ser.* **74** 623
Mooney P M, Calleja E, Wright S L and Heiblum M 1986 *Defects in Semiconductors* ed H J von Bardeleben (Switzerland: Trans Tech) pp 417-22
Mooney P M, Theis T N and Wright S L 1987a proc. of this conference
Mooney P M, Northrup G A, Morgan T N and Grimmeiss H G 1987b unpublished
Morgan T N 1986 *Phys. Rev. B.* **34** 2664
Morgan T N 1987 unpublished
Nabity J C, Stavola M, Lopata J, Dautremont-Smith W C, Tu C W and Pearton S J 1987 *Appl. Phys. Lett.* **50** 921
Nelson R J 1977 *Appl. Phys. Lett.* **31** 351
Oshiyama A and Ohnishi S 1986 *Phys. Rev. B* **33** 4320
Rochette J F, Delescluse P, Laviron M, Delagabeaudeuf D, Chevrier J and Linh N T 1982 *Inst. Phys. Conf. Ser.* **65** 385
Saxena A K 1981 *Phys. Stat. Sol. (b)* **105** 777
Springthorpe A J, King F. D. and Becke A 1975 *J. Elec. Mat.* **4** 101
Tachikawa M, Mizuta M, Kukimoto H and Minomura S 1985a *Jpn. J. Appl. Phys.* **24** L821
Tachikawa M, Fujisawa T, Kukimoto H, Shibata A, Oomi G and Minomura S 1985b *Jpn. J. Appl. Phys.* **24** L893
Theis T N, Kuech T F, Palmateer L F and Mooney P M 1984 *Inst. Phys. Conf. Ser.* **74** 241
Theis T N, Parker B D, Solomon P M and Wright S L 1986a *Appl. Phys. Lett.* **49** 1542
Theis T N 1986b *Defects in Semiconductors* ed H J von Bardeleben (Switzerland: Trans Tech) pp 393-8
Theis T N, Mooney P M and Wright S L 1987a *Bulletin of the American Physical Society* **32** 554
Theis T N 1987b *Applied Surface Science* to be published
Yamaguchi E 1986 *Jpn. J. Appl. Phys.* **25** L643

Advanced optoelectronic devices fabricated by MOCVD

P.A. Kirkby

STC Technology Ltd, London Road, Harlow, Essex, CM17 9NA, UK

ABSTRACT: This paper reviews the state-of-the-art of optoelectronic devices made by the metal organic chemical vapour deposition process. The first part reviews recent excellent results on the fabrication yields and reliability of DFB, b.h. lasers and detectors. The second part of the paper reviews the application of MOCVD to quantum well and superlattice based devices. This includes multiquantum well lasers and a comparison of doping and composition superlattices for transmissive detectors and modulators.

1. INTRODUCTION

The past few years have been an important time for the development of the Metal Organic Chemical Vapour Deposition (MOCVD) growth process. Several companies around the world have scaled up the process for 2" wafers and started to transfer the process into production for optoelectronic devices. One key question has been "Will MOCVD live up to its early promise and be able to produce a high yield of high reliability devices for large volume low cost production?" The first half of this paper reviews recent progress towards answering this question particularly for InP-InGaAsP detectors and lasers whilst the second half discusses the properties of some of the new optoelectronic devices based on MOCVD quantum well and superlattice structures. This paper does not cover any detail of the MOCVD growth process itself as this is covered well in other recent papers (Nelson 1985, Dapkus 1984, Thrush 1987).

2. RECENT PROGRESS IN InP-GaInAsP MOCVD

The achievement of high uniformity of layer composition and thickness is not straightforward with MOCVD. A variety of complex gas dynamic, gas diffusion and surface chemistry effects are all important. In order to avoid dislocation formation the lattice matching of GaInAsP to InP requires control of the solid composition to ± 0.7% to give a lattice mismatch $|\Delta a/a|<5\times10^{-4}$. Using a fully automated reactor made by Thomas Swan Ltd., Thrush(1987) has obtained this degree of lattice match for InGaAs over 85% of a 2" InP wafer as shown in Figure 1. For the quaternary composition of GaInAsP suitable for 1.55 μm lasers Nelson(1987) has recently reported composition control sufficiently precise for only a ±5 nm variation in peak wavelength over at least the central 70% of a wafer. Layer thickness control is less well developed for static reactors but the current best results are reported by

Mircea(1986) of ± 1.5% over 2" wafer using a rotating substrate to average out non uniformity.

Obtaining a consistently low background doping level is vital for many devices, particularly detectors and FETs. A process for the purification of trimethyl indium using a diphos adduct has been pioneered by the late Marc Faktor (Moore 1986). Record purity levels for InP have been obtained with this TMI in a low pressure reactor growing at 570°C. Doping levels below $4x10^{13}cm^{-3}$ across 8 µm thick layers with mobilities up to 400,000 $cm^2/V.sec$ at 50 K have been obtained(Baud 1987). This is more than a factor of two higher than previously reported. This improvement, together with careful selection of phosphine bottles has allowed background doping levels in the low $10^{15}cm^{-3}$ range to be routinely obtained, thus allowing detector structures to be manufactured consistently.

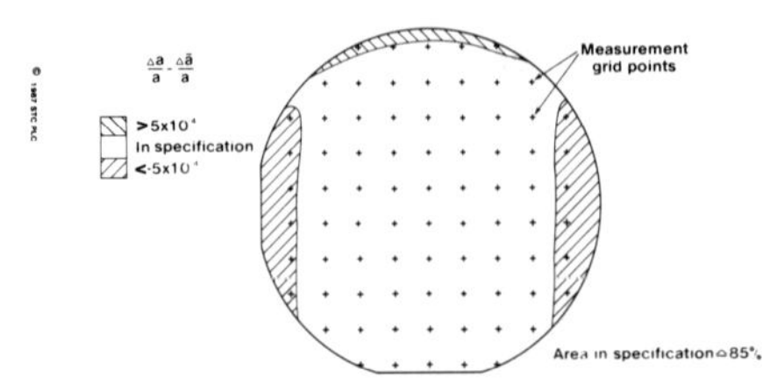

Fig. 1. Contour map of Lattice Mismatch of 2" InGaAs/InP MOCVD Laser

3. DETECTORS

Detectors are being developed (Houghton 1987) using the zinc diffused planar structure(Susa 1981), illustrated in Figure 2. This structure can

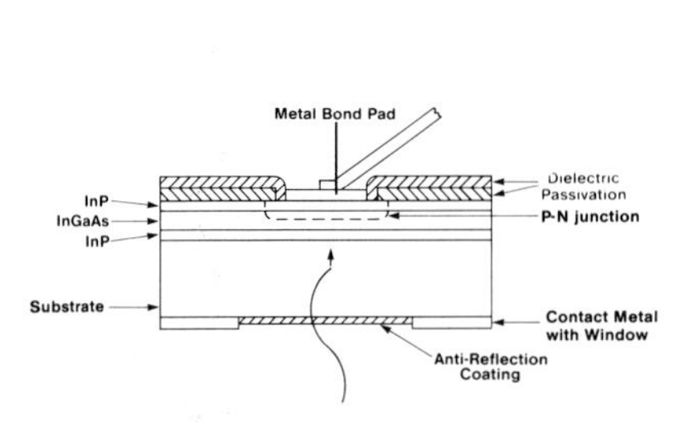

Fig. 2. MOCVD planar nitride-passivated high speed rear-entry pin photodiode

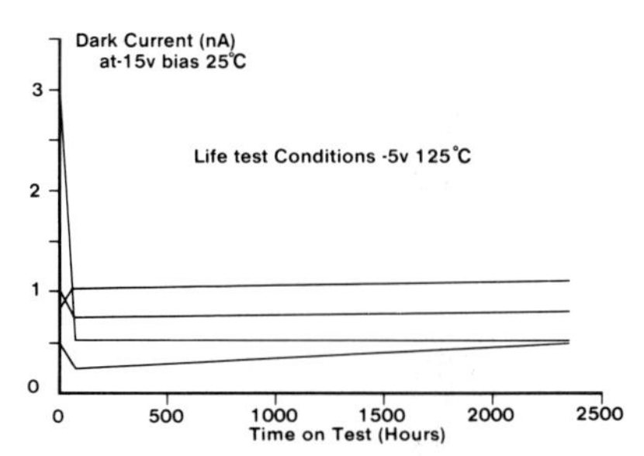

Fig. 3. Life test of MOCVD planar pin diodes

only be made by vapour phase processes because of the dissolution of InGaAs by InP when it is grown by liquid phase epitaxy. The wide bandgap InP surface layer with SiN passivated p n junction virtually eliminates all surface leakage problems prevalent with LPE devices. The total leakage currents are remarkably low being less than 300 pA at -15 V bias at 25°C and still less than 3 nA at -15 V at 85°C. Initial lifetests on these devices are showing excellent stability over 2000 hours at 125°C as illustrated in Figure 3. Gallant(1987) has also reported on MOCVD grown planar pin detectors. He reports a high yield of uniformly low leakage currents across large area wafers. The best wafers exhibit leakage currents down to 10 pA at -10 V for 100 µm diameter devices. His studies indicate that this leakage current is close to the intrinsic bulk leakage limit for GaInAs of this bandgap.

4. LASERS

Several companies around the world are now close to completion of the development of GaInAsP lasers fabricated by MOCVD. The distributed feedback laser which operates at a single frequency at 1.55 µm wavelength is particularly important for the next generation of long haul telecommunications systems. The reliability of these devices is a key parameter. The ridge waveguide DFB laser is giving particularly good results (Rashid 1987, Armistead 1987, Collar 1987). The most critical part of the fabrication process is the MOCVD overgrowth of the DFB grating. Zn doped InP and InGaAs cap layers are grown by MOCVD over the LPE grown GaInAsP grating-waveguide layer (λ_G=1.18 µm) and active layer (λ_G=1.55 µm). The grating comes within 30-60 nm of the active layer so that it is easy for any defects formed at the interface or in the MOCVD InP to diffuse into the active layer and cause performance or reliability problems. In practice the high degree of control of the growth conditions (e.g. III-V ratio and growth rate) and the short exposure time of the wafer to the high temperature reactor environment before growth starts, allows high quality devices to be grown routinely. A set of lifetest results is shown in Figure 4. This shows the drive current required for 5 mW output at 50°C for devices on test for up to six months. The degradation is negligible and extrapolations based on curve fitting indicate that the median change in drive current at 25 years is only 10% with a standard deviation of 21%.

The growth and regrowth quality is even more critical for the all MOCVD buried heterostructure laser because the regrown interface intersects the active filament of the laser (Krakoski 1986, Nelson 1985). The fabrication sequence for such a laser is illustrated in Figure 5. The 1.3 µm MOCVD GaInAsP active filament is first buried by p-InP overgrowth and residual current leakage and stray capacitance are reduced by proton isolation. This structure and the more complex b.h. laser with reverse biased blocking layers (Nelson 1985) can both be manufactured with a very high yield of low threshold devices. This is illustrated in Table 1 which shows the statistics of threshold variation for multiple wafers with many thousands of devices. The lifetest results of these lasers at 50°C to 60°C appear at least as good as their earlier LPE counterparts with degradation rates around 3-4%/1000 hours for the first six months reducing to less than 1%/1000 hours thereafter. This is illustrated in Figure 6.

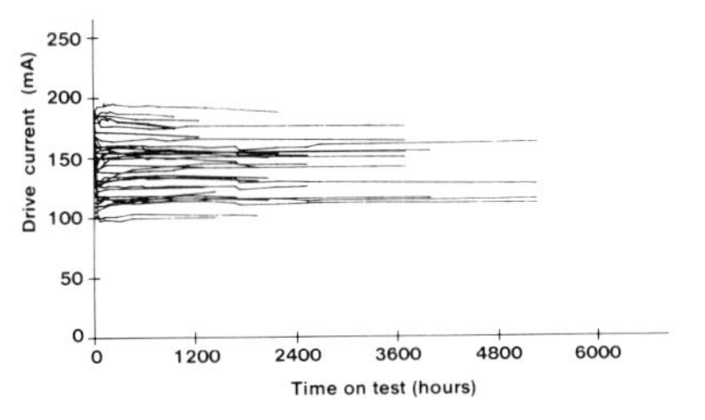

Fig. 4. Longterm reliability of ridge waveguide DFB lasers

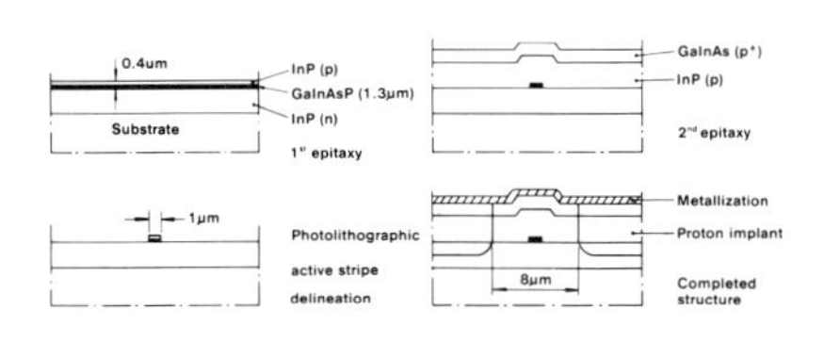

Fig. 5. Fabrication sequence for an all MOVPE b.h. laser (Krakoski)

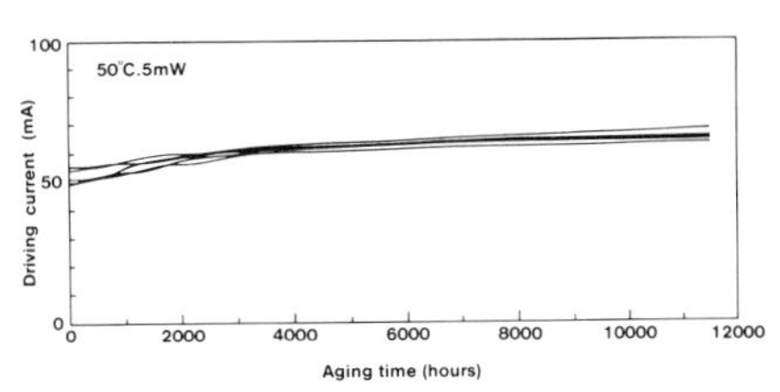

Fig. 6. Lifetest results for all MOVPE b.h. lasers (Krakoski)

Table 1: MOVPE BH-Laser Chip Yields (Nelson 1987)

Slice	Ith<20mA	Mean Ith (mA)	Anded Yield (%) (No. Tested)
A	84%	12.0	73% (547)
B	92%	10.6	58% (562)
C	87%	8.4	65% (581)
D	78%	13.8	55% (569)
E	92%	8.6	53% (677)
F	92%	10.1	46% (730)
G	72%	12.5	56% (430)
H	80%	13.2	70% (747)
I	71%	11.3	56% (590)
J	85%	10.7	64% (619)
K	89%	10.1	77% (1140)
L	94%	9.2	80% (401)

Taken collectively, these results on GaInAsP lasers and detectors show that MOCVD is living up to earlier expectations that it is a good process for large volume production of high performance optoelectronic devices. This is also important for the prospects for developing viable manufacturing processes for much more complex optoelectronic integrated circuits (OEICs). These circuits consist of combinations of optoelectronic devices such as lasers, transistors, detectors and waveguides monolithically integrated on the same chip. The high yield of high reliability InP/GaInAsP discrete devices that has now been demonstrated gives great encouragement that progressively more complex integrated optoelectronic circuits will become manufacturable at high yield in the near future.

5. QUANTUM WELL AND SUPERLATTICE DEVICES

The suitability of MOCVD for growing high quality superlattice and quantum well structures in the GaAs/GaAlAs system and InP/GaInAs system is well known. Recent work by Butler confirms that high quality quantum wells can also be grown in the quaternary GaInAsP/InP system. A test structure of wells ranging from 200 Å down to 35 Å thickness were grown as shown in Figure 7, photoluminescence studies(Nicholas 1987) of these layers carried out at 4.2 K show not only the quantum well shift to higher energies with reducing layer thickness, but also very narrow linewidths ranging from 4 to 12 meV as shown in Figure 8. These narrow linewidths are indicative of very uniform layer thickness and good interface abruptness.

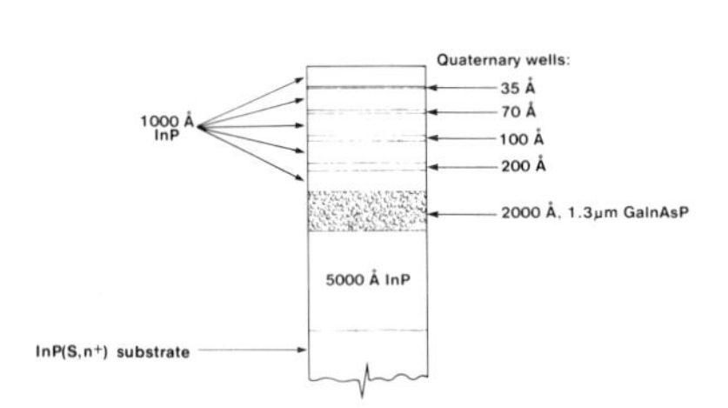

Fig. 7. GaInAsP/InP quantum well test structure

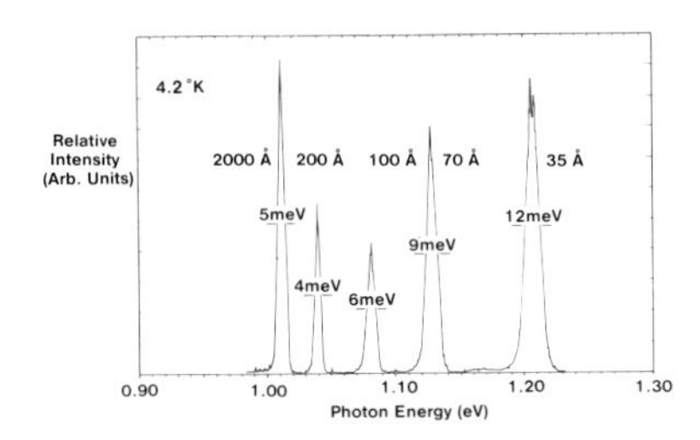

Fig. 8. Photoluminescence spectrum of GaInAs/InP quantum well structure (Nicholas 1987)

Quantum well layers have found their first important application in high power GaAs/GaAlAs lasers. A good example of this type of device is the metal clad ridge waveguide laser reported by B. Garrett(1987). The structure of the device is shown in Figure 9. Electron-hole recombination in the two 100 Å quantum wells provides the optical gain for light guided by the 0.23 µm thick $Al_{0.2}Ga_{0.8}As$ optical guiding layer. Lateral current confinement and optical guidance is provided by the 3 µm wide ridge structure. These devices come closer to the theoretical ideal than any other type of semiconductor laser and may well be the most efficient electrically driven light emitters demonstrated to date. Not only is the threshold and temperature coefficient very low (typically 11 mA at 20°C and 17 mA at 90°C) but also the external differential quantum efficiency above threshold is in the range 85 to 95%. This allows overall power conversion efficiency (including all ohmic losses) to exceed 50% once the drive current exceeds 3.5 times threshold. The very low level of unwanted heat generation in the active layer allows the lasers to be run up to very high output. Recent devices have exceeded 140 mW cw from a 4 µm wide ridge laser (Daniel 1987). There can be few light emitting devices where there is such an intense continuous interaction between electrons, light and matter. The optical generation rate is 5 GW/cm^3 of quantum well active layer. It takes less than one day for the active layer to emit more than its own mass equivalent of light energy.

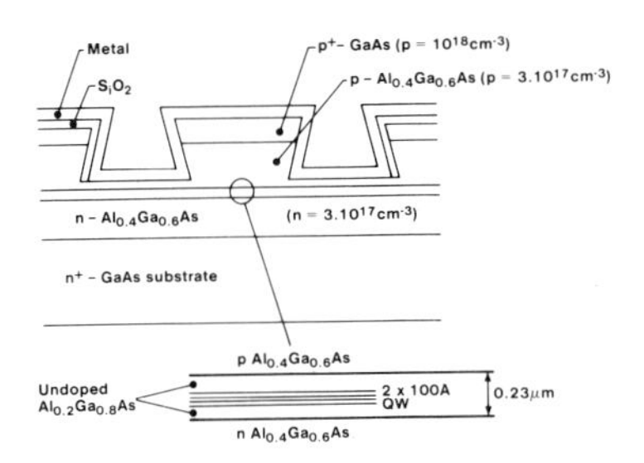

Fig. 9. MOCVD grown GaAs/GaAlAs metal clad ridge waveguide double quantum well separate confinement heterostructure laser

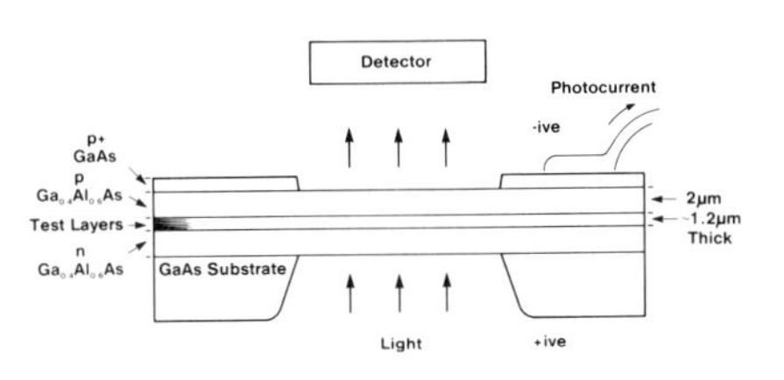

Fig. 10. Test structure for transmissive detector and modulator materials

This remarkable combination of properties for quantum well devices has encouraged the search for new device applications for other related quantum well or superlattice structures. Perhaps the most important new application area is for use in transmissive detector and modulator devices (TDAMs). The structure of a typical TDAM is illustrated in Figure 10. Light from a fibre or other optical system passes through a reverse biased pin diode structure. When zero bias is applied the active layer is transparent at the wavelength of light being used. As reverse bias is increased the effective energy bandgap of the active layers of the device reduces progressively absorbing and detecting more light. In this way the device is at the same time a light modulator and a variable efficiency detector. There are a wide variety of potential applications for such devices ranging from replacing many sources and detectors in serially 'wired' fibre optic networks (Kirkby 1985) through components for VLSI optical interconnect (Hornak 1987) to array devices for optical signal processing (Wood 1987, Optical Engineering 1987). The ideal TDAM would require very high transmission (90–100%) with zero bias, high absorption (50–100%) with a low applied bias of 2 V – 10 V (say) and a wide range of operating wavelength (>50 nm in 900 nm band or >100 nm at 1300 to 1500 nm wavelength). This combination of properties is still far from being achieved but several types of device have been reported that may be sufficiently good to make certain applications viable.

The effect of applied electric field reducing the effective bandgap of a semiconductor is referred to as the electroabsorption effect. In bulk semiconductors this is known as the Franz-Keldysh effect (Wight 1987). Even with applied fields at or close to breakdown (2-4x10^5V/cm) the Franz Keldysh effect is rather weak for use in TDAMs so research has concentrated on finding enhanced effects using multiple quantum wells and doping superlattice structures. Typical examples of the structures have been MOCVD grown and evaluated (Greene 1987). The band diagrams of three such test structures are shown in Figure 11. The first structure uses a bulk GaAs active layer, the second a GaAlAs multiquantum well composition superlattice and the third a p-n-p-n doping superlattice. The photocurrent spectra of Figure 12 clearly show the differing effects of increasing the reverse bias on the absorption edge in the three cases.

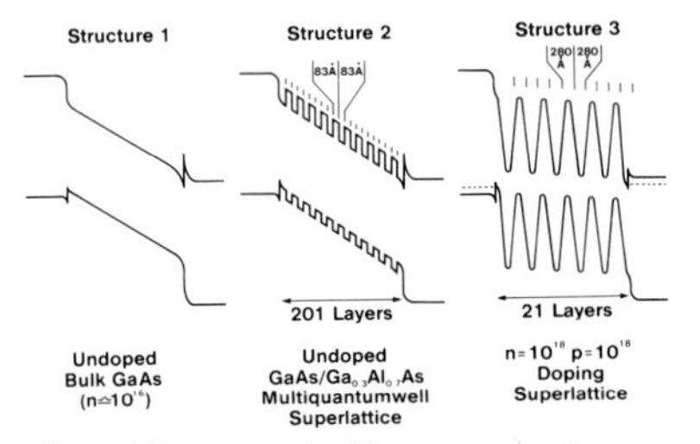

Fig. 11. Band diagrams of alternative transmissive detector and modulator structures (TDAMs)

For the undoped bulk GaAs active layer the applied field increases absorption at long wavelength but reduces it at wavelengths less than the band edge. The effect is fairly small with a useful wavelength range for high transparency and high electroabsorption effect of 16 nm indicated on Figure 12(a). At the optimum wavelength the increase in absorption with applied field for the multiquantum well active layer is much greater as shown in Figure 12(b). The electroabsorption effect is enhanced in MQW structures by the effect of electric field on electron hole pairs (excitons) in the GaAs wells. This effect is known as the quantum confined Stark effect (Miller 1985) and results from the reduction in effective energy gap of the energy levels in the quantum well as electric field increases. A wide variety of novel devices have now been demonstrated (Weatley 1987, Efron 1987, Yaminashi 1987). The best devices have shown over 1.7:1 on to off ratio and impulse response speeds up to 130 ps. All MQW devices, however, have a rather narrow range of operating wavelength. The sample tested in 12(b) has a maximum usable range of only 12 nm unless high absorption is tolerated in the on state.

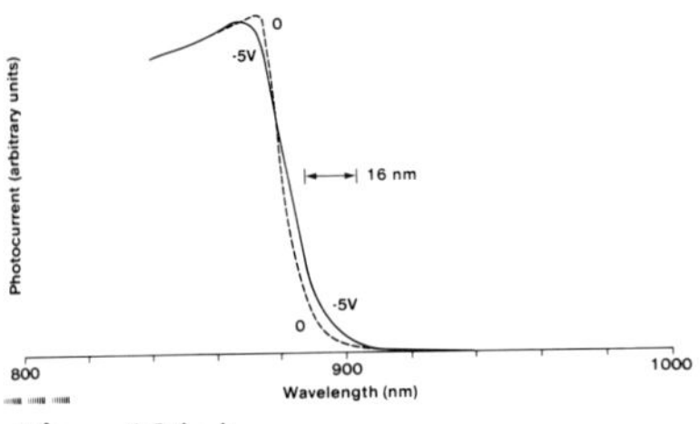

Fig. 12(a)

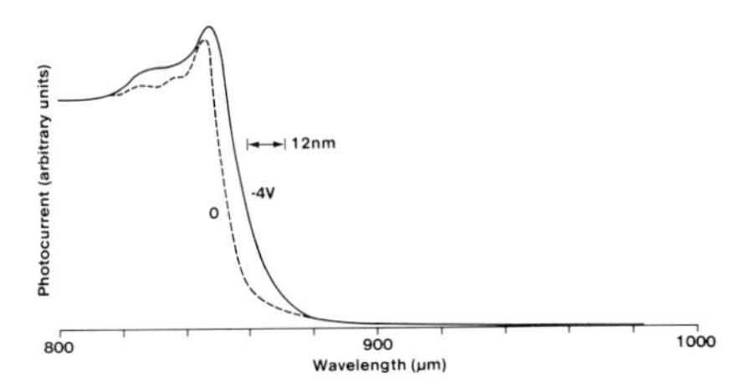

Fig. 12(b)

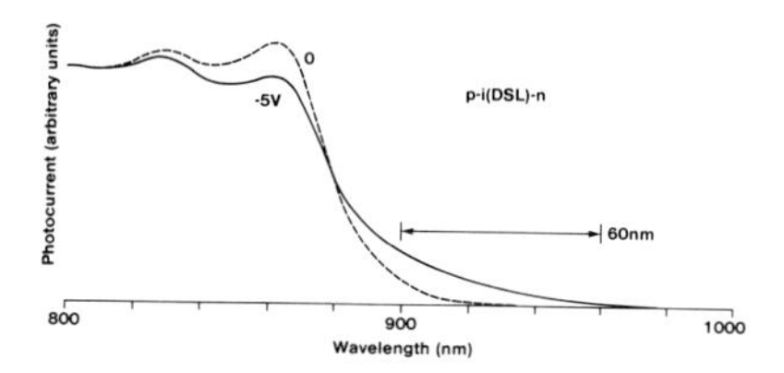

Fig. 12(c)

Fig. 12. Photocurrent spectra of three types of TDAM; a) bulk GaAs active layer, b) multiquantum well, c) doping superlattice (corresponding to structures 1, 2 and 3 respectively of Figure 11).

The doping superlattice structure of Figure 12(c) has a much wider useful range of operating wavelength (60 nm in this case). This wide operating wavelength range was first reported by Horikoshi(1985) and is believed to result from the effect on the absorption edge of the high built in fields of the structure. The wide operating wavelength range of this type of TDAM is expected to be useful in applications where it is undesirable to specify wavelength too precisely and the maximum modulation depth is not required.

CONCLUSIONS

Recent results from groups developing MOCVD for large volume production are confirming earlier expectations that MOCVD is an excellent process for the manufacture of optoelectronic devices. This is increasing confidence that more complex monolithic integrated optoelectronic devices will be manufacturable at high yield in the near future. Research into novel MOCVD grown devices using multiquantum wells and superlattice structures is already demonstrating important new devices of considerable significance for future optoelectronic systems.

ACKNOWLEDGEMENTS

The author is indebted to many colleagues and co-workers for providing up-to-date information for this review, particularly A. Houghton of STC Optical Device Division, A. Nelson of BTRL and B. de Cremoux of Thomson-CSF. The work referred to at STL was carried out with the support of the DTI JOERS programme, ESPRIT Project 263 (Integrated Optoelectronics) and the UK Ministry of Defence (DCVD).

REFERENCES

Armistead C.J., et. al., 'DFB ridge waveguide lasers at λ=1.5 µm ....' Electronics Letts, Vol 23, No 11, 1987, pp 592-3.

Baud J.M., et. al., 'Low temperature electron transport properties of exceptionally high purity InP', Late Newspaper W3 GaAs and Related Compounds, Heraklion 1987.

Butler B.R., STL O-E Lab, London Road, Harlow, UK, private communication.

Collar A.J., et. al, 'DFB lasers fabricated by electron beam lithography....' , Proceedings of European Conference of Optical Communication (ECOC), Sept 1987, Helsinki.

Daniel D., STC Optical Device Division, STC Defence Systems Ltd., Brixham Road, Paignton, Devon, UK, private communication.
Dapkus P.D., 'A critical comparison of MOCVD and MBE for heterojunction devices', Jnl of Crystal Growth 68, 1984, pp 345-355, North Holland.
Efron U., et. al., 'Multiple quantum well based spatial light modulators', Proc SPIE, 1987 p 792.
Gallant M., et. al., 'MOCVD InGaAs Photodiodes ....', Late Newspaper T5, GaAs and Related Compounds, Heraklion, Sept 1987.
Garrett B., et. al., 'Low threshold, high-power .... (AlGa)As/GaAs lasers', Electronics Letts, Vol 23, No 8, pp 371-373, 9th April 1987.
Greene P.D., et. al., STL, London Road, Harlow, UK, private communication.
Horikoshi Y., et. al., 'A new doping superlattice photodetector', Appl. Phys, A37, pp 47-56, 1985 (Springer Verlag).
Hornak L.A. et. al., 'On the feasibility of through wafer optical interconnects ....', IEEE Transactions on electron devices, Vol ED-34, No 7, July 1987.
Houghton A.J.N., Optical Device Division, STC Defence Systems Ltd, Brixham Road, Paignton, Devon, UK, private communication.
Kirkby P.A., 'Fibre optic network component', UK patent application No 8517534, 20th June 1985.
Krakowski M., et. al., 'High yield manufacture of very low threshold high reliability 1.3 µm bh lasers ....', Jnl of Lightwave Technology, Vol LT4, October 1986, p 1470.
Nelson A.W., et. al., 'High performance long wavelength optoelectronic components by atmospheric pressure MOVPE', Jnl of Crystal Growth, Vol 77, pp 579-590, 1986
Nelson A.W., et. al., 'High power low threshold b.h. lasers ....', Electronics Letters, Vol 21, No 20, pp 888-9, 26 Sept 1985.
Nelson A.W., BTRL, Martlesham, Suffolk, UK (to be published).
Nicholas D.J., Dept of Electrical Eng. & Electronics, UMIST, PO Box 88, Manchester, UK, private communication
Miller D.A.B., et. al., 'Electro field dependence of optical absorption....', Phys Rev B 1985, Vol 32, pp 1043-1060.
Mircea A., et. al., 'Instrumental aspects of atmospheric pressure MOVPE growth of InP and InP:GaInAsP heterostructures', Jnl of Crystal Growth, 1986, pp 340-346 North Holland Amsterdam.
Moore A.H., et. al., High mobility InP epitaxial layers prepared by atmospheric pressure MOVPE ....', Jnl Crystal Growth, Vol 77, 1986, pp 19-22, North Holland Amsterdam.
Optical Engineering, special issue on Optical Information Processing, May 1987, 26(5).
Rashid A.A.M., et. al., 'Characteristics of DFB-RWG lasers ....', Proceedings of European Fibre Optic Conference (EFOC), Basle, June 1987.
Susa N., et. al., 'Vapour phase epitaxial growth of InP ....' Jnl of Crystal Growth 51, 1981, pp 518-524, North Holland.
Thrush E.J., et. al., 'Reactor design and operating procedure for InP based MOCVD', Chemtronics, 1987, Vol 2, June (Butterworth Press).
Weatley P., et. al., 'Novel non resonant optoelectronic logic device', Electronics Letters 1987, Vol 23, pp 92-93.
Wight D.R., et. al., 'The limits of electroabsorption in high purity GaAs....'', submitted to Journal of Proceedings of IEE, August 1987.
Wood T.C., et. al., 'High speed spatial light modulator ....', Electronics Letters, 13th August 1987, Vol 23, No 117, pp 916-917.
Yaminashi M., et. al., 'Optical bistability by charge induced self feedback in quantum well structure', O-E Devices and Technologies, Vol 2, No 1., pp 45-51, June 1987.

Inst. Phys. Conf. Ser. No. 91: Chapter 1
Paper presented at Int. Symp. GaAs and Related Compounds, Heraklion, Greece, 1987

Theoretical calculations of the electronic structure of superlattices

M. Altarelli

European Synchrotron Radiation Facility, B.P. 220, F-38043 Grenoble, France
and
Max-Planck-Institut für Festkörperforschung, Hochfeld-Magnetlabor, B.P. 166 X,
F-38042 Grenoble, France

ABSTRACT: A critical overview of envelope-function calculations of electronic states in heterostructures is presented, with emphasis on the specification of input parameters and on the comparison with experiments.

1. INTRODUCTION

Theoretical calculations of electronic subbands in two-dimensional semiconductor systems have been performed in Si MOS structures (see e.g. the excellent review by Ando et al. (1982)) and later in III-V heterostructures. In the latter case, the tight-binding method (Schulman and Chang 1981, 1983), the pseudopotential method (Jaros et al. 1985) or ab initio schemes based on the density functional formalism (Van de Walle and Martin 1986) have been proposed, in addition to the envelope-function or effective-mass method.

In spite of its limitations (most notably, the difficulty in handling states resulting from the admixture of different k-points in the bulk band structure) the envelope-function method has enjoyed a larger popularity, mostly because of its flexibility in incorporating external fields (electric, magnetic and strain fields), the Coulomb potential of impurities and exciton interactions, etc. In this paper, envelope-function calculations are shortly reviewed, with emphasis on the discussion of the input parameters which are necessary, and on comparison with experimental results.

This discussion leads to the conclusion that although these relatively simple calculations are sufficient to reproduce the essential features of many spectroscopic experiments in a remarkable way, detailed quantitative agreement is often beyond the present capabilities.

2. ENVELOPE-FUNCTION DESCRIPTION OF ELECTRONIC STATES

In the envelope-function method (for recent reviews see Bastard and Brum (1986), Altarelli (1986)) the wavefunction in each layer of a superlattice is written in terms of products of a $k=0$ Bloch function of the corresponding bulk semiconductor with a slowly-varying envelope-function (with wavelength of the order of the layer thickness). Suppose for example that the two materials have a parabolic conduction band edge at the Γ point, with effective masses m_A, m_B and energies $E_{\Gamma A}$, $E_{\Gamma B}$ respectively. Then we write the wavefunction of the conduction subbands as:

$$\psi^A(\underline{r}) = F^A(\underline{r})\, u_{\Gamma A}(\underline{r}) \quad \text{in the A layers}$$
$$\psi^B(\underline{r}) = F^B(\underline{r})\, u_{\Gamma B}(\underline{r}) \quad \text{in the B layers} \qquad (1)$$

with F^A, F^B, satisfying the equations:

$$-\frac{\hbar^2}{2m_{A,B}} \nabla^2 F^{A,B} = (E - E_{\Gamma A,B})\, F^{A,B} \tag{2}$$

and joined at an A-B interface, say the z = 0 plane, by the boundary conditions:

$$\begin{aligned} F^A(x, y, z = 0^-) &= F^B(x, y, z = 0^+) \\ \frac{1}{m_A}\frac{\partial}{\partial z} F^A(x, y, z = 0^-) &= \frac{1}{m_B}\frac{\partial}{\partial z} F^B(x, y, z = 0^+) \end{aligned} \tag{3}$$

If the assumption $u_{\Gamma A}(r) \cong u_{\Gamma B}(r)$ is tenable then the boundary conditions (3) ensure the conservation of the total probability current at the interface and are therefore physically justified (Ben Daniel and Duke 1966). This assumption restricts the applicability to pairs of materials with very similar chemical nature and band structure; in practice, it has proven applicable to III-V compounds and alloys with direct band-gap, such as GaAs-$Al_xGa_{1-x}As$, for $x < 0.4$.

The interest of Eqs (1)-(3) is that they are susceptible of far-reaching generalisations. First, any potential varying slowly on the scale of the lattice parameter of the constituents can be included in Eq. (2):

$$\frac{\hbar^2}{2m_{A,B}} (-\nabla^2 + V^{A,B}(\underline{r}))\, F^{A,B}(\underline{r}) = (E - E_{\Gamma A,B})\, F^{A,B} \tag{4}$$

thus allowing consideration of space charge effects in doped superlattices, of charge transfer across the interfaces, of external electric fields (which, however, break the lattice periodicity), etc. A second very important generalisation is the inclusion of band coupling, whenever it is impossible to construct the subband wavefunctions from a single non-degenerate, nearly parabolic band edge as implied by Eq. (1). This happens frequently, because of various factors:

(i) band degeneracy near an extremum, as in the case of the valence band maximum at Γ in all cubic semiconductors.
(ii) deviations from parabolicity, as in the conduction band of direct-gap semiconductors. For narrow-gap materials, like InAs or InSb the non-parabolicity is quite large for energies near the band minimum, but even in GaAs it has a sizeable effect on levels with energy larger than 0.1 eV above the band edge.
(iii) there are situations specific to heterostructures in which the single-band approach fails; if the two materials have a staggered band line-up, then, in a large and interesting energy region, the wavefunction has conduction band character on one side of the heterojunction, and valence band character on the other. InAs-GaSb superlattices provide an example of this situation.

In all these cases it is necessary to treat more than one band at a time, say n bands, on the same footing. This is accomplished via the k.p formalism (Kane, 1982), in which the n band energies $E_l(\underline{k})$, $l = 1, 2, ..., n$ are obtained as eigenvalues of the nxn matrix $H_{lm}(\underline{k})$, expanded, for k near the origin, up to 2nd order in k:

$$H_{lm}(\underline{k}) = E_l(0)\,\delta_{lm} + \sum_{\alpha=1}^{3} P_{lm}^{\alpha} k_\alpha + \sum_{\alpha,\beta=1}^{3} D_{lm}^{\alpha,\beta} k_\alpha k_\beta \tag{5}$$

Here α, β run over the x, y and z directions. The P and D coefficient matrices are written in terms of momentum matrix elements between the Bloch functions at k = 0 of the n bands in question, ant they are, for each material, a set of parameters playing the role of the effective masses.

Following the general ideas of the effective-mass theory in its many-band formulation (as adopted e.g. in the theory of acceptor impurities), we write, instead of Eq. (1)

$$\psi^A(\underline{r}) \cong \sum_{l=1}^{n} F_l^A(\underline{r})\, u_{\Gamma lA}(\underline{r}) + \ldots$$
$$\psi^B(\underline{r}) \cong \sum_{l=1}^{n} F_l^B(\underline{r})\, u_{\Gamma lB}(\underline{r}) + \ldots \tag{6}$$

where the dots replace higher order corrections proportional to ∇F_l which need not be detailed here (Altarelli, 1986). The effective-mass equation, Eq. (4), is now replaced by a set of n differential equations

$$\sum_{m=1}^{n} \left(H_{lm}(-i\underline{\nabla}) + V(\underline{r})\,\delta_{lm} \right) F_m(\underline{r}) = E\, F_l(\underline{r}) \tag{7}$$

for each material, A or B, where, as in Eq. (4), $V(\underline{r})$ is a slowly varying potential. The boundary conditions which must complement Eq. (7) are:

$$F_l^A(x, y, z = 0^-) = F_l^B(x, y, z = 0^+)$$
$$\sum_{m=1}^{n} \left(\sum_{\alpha=x,y} (D_{lm}^{z\alpha} + D_{ml}^{\alpha z})\, k_\alpha - 2i D_{lm}^{zz} \frac{\partial}{\partial z} \right) F_m \quad \text{continuous at } z = 0 \tag{8}$$

The boundary conditions Eq. (8), in analogy to Eq. (3) ensure the continuity of the probability current, provided that one can assume

$$u_{\Gamma lA}(r) \cong u_{\Gamma lB}(r) \qquad l = 1,2,\ldots n \tag{9}$$

They are therefore valid under the condition that the two materials have a set of k = 0 band edges which can be grouped in pairs with similar symmetry and chemical origin. Given the success of the envelope-function method, it appears that Eq. (9) is reasonably appropriate for pairs of lattice-matched III-V (or II-VI) compounds, as far as the lowest conduction band or the upper valence-band edge is concerned. It is important to notice that the equality Eq. (9) implies also the equality of the P coefficient matrix for the two materials A and B (Kane 1982).

The unsatisfactory situation associated with the lack of a rigorous mathematical proof for the boundary conditions Eq. (3,8), is somewhat corrected by the convincing a posteriori evidence offered by detailed comparisons of envelope-function calculations with tight-binding ones, in which no boundary conditions enter. Such evidence was offered for GaAs-AlGaAs by Schuurmans and 't Hooft (1985), for GaAs-AlGaAs and InAs-GaSb by Ando and Mori (1982) and Ando (1987), and by Chang et al. (1985) for the particularly sensitive case of CdTe-HgTe heterostructures.

Some words of caution concerning the boundary conditions are anyway in order. Their adoption is based on the idea of perfect periodicity of both media up to a geometrical plane defining the interface. This is certainly an idealisation and, in the very important case in which one of the components is an alloy (e.g. GaAs-$Al_xGa_{1-x}As$) it is impossible even to define such a plane. It would be more realistic to talk about an interface layer, comprising several atomic planes, separating the two semiconductors, and characterised by given reflection and transmission coefficients. However, these would be parameters of unknown value to be used as input.

3. DISCUSSION OF ENVELOPE-FUNCTION CALCULATIONS

As we already mentioned, envelope-function calculations have been reported for conduction and valence states in GaAs-AlGaAs heterostructures, as well as in other systems, some of which, as InAs-GaSb or CdTe-HgTe, e.g., are characterised by different band line-ups.

If we focus our attention on the GaAs-AlGaAs system, which is the most extensively investigated from the experimental point of view, we find a rather large body of literature. Much theoretical attention has been attracted by the valence subbands, which display a strongly non-parabolic dispersion in the plane parallel to the interfaces (see the reviews of Bastard and Brum (1986) and Altarelli (1986) for a literature survey). While this peculiarities of the valence subbands are detectable in interband optical spectra (see e.g. Sooryakumar et al. 1985) but are complicated by exciton and other many-body effects, they appear most unambiguously in intraband experiments, such as light scattering experiments on p-type quantum wells (Pinczuk 1987 and references therein). The theoretical analysis (Ando 1985), which in these modulation-doped samples includes a self-consistent envelope-function calculation, shows good overall agreement of the energy and lineshape with experiment. It is apparent that the calculations are able to capture the essential features of the subband structure.

A more detailed comparison is achieved by considering spectra in external fields, most notably in magnetic fields which via the Landau quantisation result in spectra characterised by many sharp transitions between completely discrete levels (Fasolino and Altarelli 1984, Broido and Sham 1985, Bangert and Landwehr 1985, 1986, Ekenberg 1986, Iwasa et al. 1986, Heiman et al. 1987, Ancilotto et al. 1987). As the comparison becomes more stringent, discrepancies between envelope-function calculations and experiments are sometimes apparent, especially in the intraband experiments. This is shown for light-scattering experiments by Heiman et al. (1987), while cyclotron resonance experiments are discussed in a short review by Sham (1987).

When it comes to detailed quantitative comparisons, one must not forget that Eq. (2-3) and (5-8) require a large number of input parameters, which we attempt to list here:

1. Electronic structure parameters of bulk compounds: gaps, effective masses, k.p coefficients, as well as dielectric constants, g-factors, deformation potentials, etc., for the case of external fields.
2. Band offsets.
3. Structural and compositional parameters of the heterostructures: layer thicknesses, alloy compositions, doping levels, etc.
4. Additional parameters for special situations: Fermi energy pinning level for heterostructures with free surfaces (see e.g. Altarelli et al. 1987) or $\Gamma \rightarrow X$ reflection and transmission amplitudes if one tries to include different k-points in the calculations, as in Ando 1987.

As for the first group of parameters, it is important to notice that, especially for alloys, they are not known with very good accuracy, and often values for the end compounds are adopted instead.

The band offsets are not easy to measure or calculate and their values are often the subject of controversy (see Duc et al. 1987 for the most recent one). Notice also that sometimes their dependence on an external parameter (e.g. pressure) is required.

The third group of parameters is essential to any comparison of theory with results for a given sample. As far as layer thicknesses are concerned, X-ray diffraction methods often provide much better values than the estimates from growth rates (Fewster and Phillips 1986, Sauvage et al. 1986, Ryan et al. 1987). It would be desirable to have similar techniques for the other parameters of this group.

The last group of parameters is encountered only in special situations and systematic investigation of the relation between envelope-function calculations and experiments can fortunately be carried out without involving them.

4. CONCLUSION

It transpires from the previous considerations that detailed quantitative agreement cannot always be expected from envelope-function calculations, as a consequence of the large number of input parameters, often known with poor accuracy. One should always be cautious when facing claims of excellent quantitative agreement between theory and experiments. Nonetheless, theory has made significant advances in recent years; for example, exciton effects in quantum wells are now better understood theoretically and resolved experimentally (D'Andrea and Del Sole 1987) and very sofisticated calculations including the full complexity of the band structure, as well as external electric and magnetic fields are beginning to emerge (Bauer and Ando, 1987a, b).

REFERENCES

Altarelli M 1986, in: *Heterojunctions and Semiconductor Superlattices, editors* Allan G, Bastard G, Boccara N, Lannoo M and Voos M (Springer: Berlin) p. 12

Altarelli M, Maan J C, Chang L L and Esaki E 1987, *Phys. Rev. B,* in press

Ancilotto F, Fasolino A and Maan J C 1987, *Superlattices and Microstructures* 3, 187

Ando T, Fowler A B and Stern F 1982, *Rev. Mod. Phys.* 54, 437

Ando T and Mori S 1982, *Surf. Sci.* 113, 124

Ando T 1985, *J. Phys. Soc. Japan* 54, 1528

Ando T 1987, in: *Third Brazilian School of Semiconductor Physics*, ed Gonçalves da Silva C E T, Oliveira L E and Leite J R (World Scientific: Singapore) p. 23

Bangert E and Landwehr G 1985, *Superlattices and Microstructures* 1, 363

Bangert E and Landwehr G 1986, *Surf. Sci.* 170, 593

Bastard G and Brum J A 1986, *IEEE J. of Quantum Electronics* QE22, 1696

Bauer G E T and Ando T 1987a, Phys. Rev. Lett. 59, 601

Bauer G E T and Ando T 1987b, in: *Proceedings of the MSS-III Conference,* Montpellier (Les Editions de Physique: Les Ulis) in press

Ben Daniel D J and Duke C B 1966, *Phys. Rev.* 152, 683

Broido A and Sham L J 1985, *Phys. Rev.* B31, 888

Chang Y C, Schulman J N, Bastard G, Guldner Y and Voos M 1985, *Phys. Rev.* B31, 2557

D'Andrea A and Del Sole R 1987, *Excitons in Confined Systems* (Springer: Berlin) in press

Duc T M, Hsu C and Faurie J P 1987, *Phys. Rev. Lett* 58, 1127

Ekenberg U 1986, *Surf. Sci.* 170, 601

Fasolino A and Altarelli M 1984, in: *Two Dimensional Systems, Heterostructures and Superlattices*, editors Bauer G, Kuchar F and Heinrich H (Springer: Berlin)
Fewster P F and Phillips J 1986, Phys. Rev. B34, 268
Heiman D, Pinczuk A, Gossard A C, Fasolino A and Altarelli M 1987 in: *Physics of Semiconductors*, editor Engström E (World Scientific: Singapore)
Iwasa Y, Miura N, Tarucha S, Okamoto H and Ando T 1986, *Surf. Sci.* 170, 587
Jaros M, Wong K B and Gell M A 1985, *Phys. Rev.* B31, 1205
Kane E O, in: *Handbook of Semiconductors, editor* Paul W. (North Holland: Amsterdam) vol. 1, p. 193
Pinczuk A 1987, in: *Third Brazilian School of Semiconductor Physics,* ed Gonçalves da Silva C E T, Oliveira L E and Leite J R (World Scientific, Singapore) p. 114
Ryan T W, Hatton P D, Bates S, Watt M, Sotomayer-Torres C, Claxton P A and Roberts J S 1987, *Semicond. Sci. Technol.* 2, 241
Sauvage M, Delalande C, Voisin P, Etienne P and Delescluse P, 1986, Surf. Sci. 174, 573
Schulman J N and Chang Y C 1981, *Phys. Rev.* B24, 4445
Schulman J N and Chang Y C 1983, *Phys. Rev.* B27, 2346
Schuurmans M F H and 't Hooft G W 1985, *Phys. Rev.* B31, 8041
Sham L J 1987, in: *Application of High Magnetic Fields in Semiconductor Physics,* editor Landwehr G, (Springer: Berlin)
Sooryakumar R, Chemla D S, Pinczuk A, Gossard A, Wiegmann W and Sham L J 1985, in: *Physics of Semiconductors, editors* Chadi J D and Harrison W A (Springer: New York)
Van de Walle C G and Martin R M 1986, *Phys. Rev.* B34, 5621

Paper presented at Int. Symp. GaAs and Related Compounds, Heraklion, Greece, 1987

Polar-on-nonpolar epitaxy: progress and new problems

Herbert Kroemer

Department of Electrical and Computer Engineering
University of California, Santa Barbara, CA 93106

ABSTRACT: Large changes have taken place recently in understanding how the two sublattices of GaAs are allocated to Ga and As during epitaxial growth on a nonpolar substrate such as Si, and how antiphase domain formation is suppressed during that allocation. Either of the two possible sublattice allocations may occur over the entire surface, without breaking up into domains. The antiphase domain suppression does not require the prior formation of a perfectly double-stepped single domain Si starting surface by high-temperature annealing.

1. INTRODUCTION

There has been a rapidly increasing interest over the last few years in research directed towards high-quality epitaxial growth of GaAs on Si substrates, driven by the potentially large practical importance of a successful technology of this kind. Although impressive progress has been made, the goal is still far from having been achieved. In particular, the problem of reducing the huge density of threading misfit dislocations, from a typical present level of around $10^7\,cm^{-2}$ to a desired level of $10^4\,cm^{-2}$ or less, remains largely unsolved. However, large progress has been made in other areas, especially in the area of those problems that arise simply from the fact that GaAs is a *polar* compound semiconductor while Si is a *nonpolar* elemental one. It is by now widely appreciated that this combination is prone to the formation of antiphase domains in the compound layer, which were at one time thought to be a more serious problem than threading dislocations, especially for (100)-oriented growth (Uppal and Kroemer 1985, Kroemer 1986). The problem is by no means unique to the GaAs-on-Si combination, but is common to the growth of all zincblende-type crystals on diamond-type substrate. GaAs-on-Si is simply the most widely studied combination of this kind.

It has become clear since mid-1984 that APD-free growth of GaAs on Si (100) is indeed achievable, by using substrates that are deliberately mis-oriented from the exact (100) plane, by rotation by a few degrees (typically 4°) about the $[0\bar{1}1]$ direction, towards the (011) plane, combined with suitable nucleation conditions (Akiyama et al. 1984, Fischer et al. 1985, Nishi et al. 1985). An essential ingredient in this achievement appeared to be the prior preparation of a Si starting surface in which all terraces on the stepped surface belonged to the same sublattice: It had been reported by Sakamoto and Hashiguchi (1986) that annealing of even a slightly misoriented Si (100) surface for about 20 minutes at temperatures around

1000°C led to such a surface. Inasmuch as a surface anneal has been an essential ingredient in all successful APD-free GaAs-on-Si growths, it was natural to assume that this annealing caused a single-domain Si starting surface to form, on which any subsequent GaAs growth would be inherently APD-free. Actually, the anneal conditions typically employed in the successful APD-free growths of GaAs on Si were significantly milder than those found necessary to achieve perfect single-domain re-arrangement on a clean Si surface. However, it could be argued (Kroemer 1986) that milder anneal conditions were sufficient for surfaces with a larger misorientation, as they were typically employed in GaAs-on-Si epitaxy, and that any residual APDs would become annihilated during the subsequent growth.

The energetic driving mechanism for the step doubling was initially not quite clear, but a model by Aspnes and Ihm (1986) explained the phenomenon in a most natural way via an energy lowering of a particular kind of double step through an atomic reconstruction of that step.

However, at least two major complications arose almost immediately: (a) It was discovered that APD-free GaAs layers could be grown with both possible sublattice allocations, depending on the initial nucleation conditions. (b) It was also demonstrated shortly afterwards that APD-free growth could be achieved even on Si surfaces that are *not* single-domain purely double-stepped surfaces. The present paper deals mostly with the problem raised by these two developments.

2 THE SUBLATTICE ALLOCATION AMBIGUITY

It was first reported by Fischer et al. (1985, 1986) that APB-free growth with *both* sublattice allocations can be achieved. If the growth is initiated by first depositing an arsenic "prelayer" at a sufficiently low temperature, and starting the GaAs formation itself at a low temperature, the sublattice allocation was reported to be the same *as if* the first atomic layer above the Si surface were a simple As layer, as a naive comparison between Si-As and Si-Ga bond strengths would suggest. On the other hand, if the nucleation was started with a Ga prelayer and/or at much higher temperatures — the papers are vague about the exact nucleation conditions — the opposite sublattice allocation was observed. The authors interpreted their observations in terms of a model in which the first atomic plane following an unbroken Si surface is either a complete bulk-like As plane or Ga plane, depending on nucleation conditions, and they suggested that the switch from As to Ga for the first layer is simply a consequence of the loss of As by evaporation at higher temperatures.

However, we consider the formation of a bulk-like Ga plane bonded to a bulk-like Si plane as a result of As loss by selective As evaporation as wholly unlikely (Kroemer 1987a): Although the equilibrium vapor pressure of *bulk* As is much higher than that of *bulk* Ga, the As-Si bond is so much stronger than the Ga-Si bond that As remains bonded to Si up to much higher temperatures than Ga. Furthermore, even if a Ga prelayer is deposited first, the As atoms are likely to displace the Ga atoms initially bonded to the Si surface. In our own work on the MBE growth of GaP on Si(211), which was almost always done with a Ga prelayer, we invariably found that P had displaced Ga from the doubly back-bonded sites, and that upon thermal decomposition of a GaP film the P atoms remained bonded up to much higher temperatures than those at which Ga had disappeared from the surface (Wright 1982). We would expect the behavior of GaAs on Si (100) to be qualitatively the same, and this expectation has in fact ben confirmed in very careful and quantitative work by Bringans et al. (1987).

Clearly, the dependence of the sublattice allocation on nucleation conditions must have a different origin than that proposed by Fischer et al. However, the observation itself is undoubtedly correct, and it has since then been confirmed by Kawabe et al. (1987), and by Pukite and Cohen (1987b). Both groups exposed the Si surface to As_4 fluxes at various temperatures prior to the onset of the Ga beam, and they found that the sublattice allocation of subsequent GaAs growth depends on the temperature of the initial arsenic exposure before turning the Ga beam on.

Kawabe et al. attempt to explain their RHEED data in terms of the very well documented symmetric dimer model of Uhrbach et al. (1986) of the binding of As_2 to Si. They find good agreement for prelayers deposited at high temperatures, but not for prelayers deposited at a low temperature of 450°C. The RHEED pattern of the low-temperature prelayer is not another 2×1 pattern simply rotated in space by 90°, but appears *qualitatively* different, exhibiting two-fold reconstructions in both azimuths, but with different strengths. The authors propose that at low temperatures the cracking of As_4 into As_2 might not take place, and that the sublattice switch is related to this difference. In fact, it appears to this writer that their RHEED pattern is at least qualitatively compatible with a model in which As_4 tetrahedra rather than As_2 dimers are bonded to every second pair of Si surface atoms, with two corner atoms bonded to the Si surface, the other two corners pointing up. The cracking explanation is also compatible with the observation that the surface with the low-temperature atomic arrangement would irreversibly change to the high-temperature surface, and would not return to the low-temperature configuration if the temperature was subsequently lowered again.

For both the low-temperature and high-temperature prelayers, Kawabe et al. report a coverage of roughly one As atom for every Si atom, as evidenced both by the intensity of the As Auger signal, and they find that the As-covered Si surface is well-passivated against oxidation.

On subsequent exposure to a Ga beam, Kawabe et al. found growth of GaAs with a single-domain structure, and with a sublattice allocation that differed for the two starting surfaces, thus confirming the observations of the Illinois group. For the high-temperature surface, the first plane *above* the Si terrace surface became the first As plane, as if the As-As dimer of the Uhrberg model was broken up and the two As atoms in the As-As dimer were re-connected via a Ga atom. This is what one might expect intuitively. However, it is actually opposite to the claim of the Illinois group that nucleation at high temperatures leads to a sublattice allocation that is the same as if the first atomic layer above the Si were a Ga layer. This writer doubts that there is actually a difference in the atomic arrangements between the work of the two groups. A study of the Illinois papers shows that the authors never actually report the orientation of the etch facets relative to the tilt axis of their substrates; they simply seem to *assume* that the sublattice reversal is due to an As starting layer for low-temperature nucleation and to a Ga starting layer for high-temperature nucleation. It should be noted that the data of Kawabe also refute the site interchange model that the present writer had proposed (Kroemer 1987a,b) to explain the high-temperature sublattice reversal of the Illinois group.

If the sublattice allocation for high-temperature nucleation appears plausible in the light of the

Uhrberg dimer model, the real puzzle is the sublattice reversal at low nucleation temperatures. The simplest possibility, that weakly bonded As_4 tetrahedra are simply replaced by Ga atoms, is clearly untenable, for the same reasons that speak against such a configuration at high nucleation temperatures under the Illinois model. Inasmuch as a GaAs layer actually grows, the As_4 tetrahedra are are clearly broken up by the chemical reaction with the incoming Ga atoms, which carry with themselves a high kinetic energy of 2kT, where T is the Ga oven temperature. It appears that the observations call for some sort of site exchange accompanying this breakup, in which As atoms actually interchange sites with part of the Si atoms in the last Si plane, thus leading to a configuration that makes the original last Si plane a partial As plane, and the first plane above the original Si plane a Ga plane, but with the Ga atoms now being much more strongly bonded by having at least part of their bonds going back to As atoms.

This site interchange proposed here, while different in detail, is similar in principle to the site interchange model postulated by the author one year ago, which was refuted in its original form by the data of Kawabe et al. As was pointed out in the earlier paper, the idea of a replacement of between one-quarter and one-half of the Si atoms in the top Si layer is by no means a new ad-hoc idea, but is strongly favored by the electrostatic considerations of Harrison et al. (1978): Purely electrostatically, a model involving suitable site interchanges is always favored over one that leaves the last Si layer intact! The energetically most favorable model has one-quarter of the Si atoms displaced from the top Si layer and moved to the first atomic layer above the original Si surface.

In fact, the electrostatic considerations of Harrison et al. favor a site interchange also in the high-temperature nucleation case, except that now the final sublattice allocation demands that any Si atoms ejected are replaced by Ga atoms rather than As atoms. This seemingly arbitrary reversal of the theoretical argument is not purely ad hoc either: Any electrostatically neutral interface bond configuration necessarily has the same number of As-Si bonds as Ga-Si bonds, hence there is no nearest-neighbor energetic preference between ،wo configurations that differ from one another by a As:Ga sublattice reversal; the actual allocation is determined purely by kinetics!

Another example of the role of kinetics is found in the work by Pukite and Cohen (1987a), mentioned earlier. They report that in the MBE growth of GaAs on Ge the sublattice allocation depends on whether As_2 or As_4 is used as the arsenide species, even though the pre-growth surface reconstructions of the As-covered Ge surface were the same for the two arsenic species, at least in the temperature range used.

This dependence on kinetics rather than final state energetics is one of the important messages that has emerged since mid-1986. The only aspect that remains obscure at this point concerns the exact details of the various evidently quite different kinetic paths for the interface formation. Various observations suggest that the reaction doe not take place over the entire terrace face homogeneously, but that it propagates from the terrace edges sideways.

3. THE STEP DOUBLING PROBLEM

We now turn to the problem of single-domain GaAs epitaxial growth on Si surfaces that are not single domain surfaces, containing steps an odd number of atoms high. The first to give convincing proof of this possibility were Pukite and Cohen (1978a), in the context of a very detailed RHEED study of the nucleation of GaAs on Ge, followed more recently by a shorter second paper by the same authors (1987b), reporting very similar results for GaAs-on-Si, and by a paper by Kawabe and Ueda (1987), who also reported the achievement of single-domain growth of GaAs on Si surfaces that were demonstrably not single domain surfaces prior to growth. We shall concentrate here on the first Pukite and Cohen paper, because of the wealth of information it contains.

The authors investigated Ge starting surfaces that showed clear evidence of being dominated by single-height steps, with only a minority of double-height steps, and with the two surface domains approximately equally represented. Upon exposure to arsenic vapor the RHEED patterns changed drastically, indicating a massive atomic movement along the surface in the presence of arsenic, which changed the arrangement of surface steps. The RHEED pattern strongly suggested "a staircase structure that repeats after four layers." However, the reconstruction pattern still contained a mixture of both 2×1 and 1×2 surface domains, indicating that "there must be some odd-multiple step heights." In fact, the authors propose alternating triple- and single-layer steps adding up to a four-layer periodicity, and they give a model of how such a strange combination might form. The model is not convincing. It involves first a (plausible) replacement of the Ge atoms in certain atomic rows in every second atomic layer, along one of the two kinds of initially present single-height steps. This is followed by a (far less plausible) migration of Ge atoms from one of the terraces to another terrace two steps higher (or lower), yielding the desires alternation in step heights. The trouble with this second assumption is the following: If the tilt rotation axis for the wafer misorientation has *exactly* the intended $[0\bar{1}1]$ direction, then the two terraces between which the Ge atom movement is postulated to take place, should be completely equivalent to each other, and it is hard to see why the atoms should prefer one over the other. Pukite and Cohen do not address themselves to this point.

However, this equivalence argument would break down if the tilt rotation axis had a direction significantly different than the $[0\bar{1}1]$ direction, somewhere in-between $[0\bar{1}1]$ and $[0\bar{1}0]$ or $[001]$. In that case there would be four rather than only two non-equivalent bond configurations at the different single-height step edges, and a three-plus-one step periodicity might indeed arise. Pukite and Cohen give no indication of whether or not such an asymmetry might be present accidentally. Possibly, some other kind of asymmetry in the system might also induce such a lower-symmetry situation even if the tilt axis is not misaligned. Conceivable candidates include the oblique incidence of the As beam or even of the electron beam. It should be self-evident that this point requires further clarification.

Be that as it may, when the Ga beam was turned on following the surface re-arrangement due to As exposure in the work by Pukite and Cohen, the resulting GaAs growth became single-domain growth, *almost from the beginning*, even though the Ge surface was still a two-domain surface after the As exposure. Although the authors do not offer an explanation for this unexpected observation in their first (Ge) paper, in their second (Si) paper they write: "the presence of both sublattices of the Si substrate implies that the GaAs is preferentially nucleating at one type of step, and then overgrowing the other type of step." We believe that this preferential nucleation at one kind of step, followed by overgrowth over the other kind, is indeed the reason for the successful growth of single-domain GaAs on dual-domain Si (or Ge) surfaces, quite independently of the validity of the three-plus-one alternating step model: It is the step *edges* that control what the domain structure the overgrowth will have, not the terraces themselves! In particular, we find the explanation by Pukite and Cohen preferable over the explanation given by Kawabe and Ueda (1987) for the absence of APDs in their GaAs layers. These authors assume that initially both kinds of domains are present, but that one of the two slowly becomes overgrown in a pyramidal fashion. This explanation is not compatible with the observation by Pukite and Cohen that the GaAs growth is single-domain almost rom the beginning.

ACKNOWLEDGEMENTS

This work was supported by the U.S. Army Research Office.

REFERENCES

Akiyama M, Kawarada Y and Kaminishi K 1984 *Jpn. J. Appl. Phys.* **23** L843

Aspnes D E and Ihm J 1986 *Phys. Rev. Lett.* **57** 3054

Bringans R D, Olmstead M A, Uhrberg R I G and Bachrach R Z 1987 *Appl Phys. Lett.* **51** 523-525

Fischer R J, Chand N C, Kopp W F, Morkoç H, Erickson L P and Youngman R 1985 *Appl. Phys. Lett.* **47** 397

Fischer R J, Morkoç H, Neumann D A, Otsuka N, Longerbone M and Erickson L P 1986 *J. Appl. Phys.* **60** 1640

Harrison W A, Kraut E A, Waldrop J R and Grant R W 1978 *Phys. Rev. B* **18** 4402

Kawabe K, Ueda K T and Takasugi H 1987 *Jpn. J. Appl. Phys.* **26** L114

Kawabe K and Ueda 1987 *Jpn. J. Appl. Phys.* **26** L944

Kroemer H 1986 *Heteroepitaxy on Si,* ed J. C. C. Fan and J. M. Poate (Materials Research Society Proceedings **67**) pp. 3-14

Kroemer H 1987a *J. Cryst. Growth* **81** 193

Kroemer H 1987b *J. Vac. Sci. Technol.* B **5** 1150-1154

Nishi S, Inomata H, Akiyama M and Kaminishi K 1985 *Jpn. J. Appl. Phys.* **24** L391

Pukite P R and Cohen P I 1987a *J. Cryst. Growth* **81** 214

Pukite P R and Cohen P I 1987b *Appl. Phys. Lett.* **50** 1739

Sakamoto T and Hashiguchi G 1986 *Jpn. J. Appl. Phys.* **25** L78

Uhrberg R I G, Bringans R D, Bachrach R Z and Northrup J E 1986 *J. Vac. Sci. Technol. A* **4** 1259; also *Phys. Rev. Lett.* **56** 520

Uppal P N and Kroemer H 1985 *J. Appl. Phys.* **58** 2195

Wright S L 1982 Ph.D. Dissertation, University of California, Santa Barbara, unpublished

Paper presented at Int. Symp. GaAs and Related Compounds, Heraklion, Greece, 1987

Delta (δ–) doping in molecular beam epitaxially grown GaAs and $Al_xGa_{1-x}As$/GaAs device structures

K Ploog, M Hauser, and A Fischer

Max-Planck-Institut für Festkörperforschung
D-7000 Stuttgart 80, Federal Republic of Germany

ABSTRACT: Four areas are defined for the application of δ-doping in advanced semiconductor device concepts: (i) nonalloyed ohmic contacts to GaAs, (ii) GaAs field-effect transistors (δ-FET) with a buried 2D channel of high carrier density, (iii) GaAs sawtooth doping superlattices for photonic devices which emit at wavelengths of $0.9<\lambda<1.2$ µm, and (iv) selectively δ-doped n-$Al_xGa_{1-x}As$/GaAs single heterojunctions with very high 2D channel density for application in high electron mobility transistors with improved current driving capabilities. In addition, fundamental studies of δ-doped GaAs and $Al_xGa_{1-x}As$ are discussed.

1. INTRODUCTION

In both AlAs/GaAs ultrathin-layer superlattices (SL) and in monolayer or δ-doping the concept of artificially layered semiconductor structures is scaled down to its ultimate physical limit normal to the crystal surface, as each constituent layer in the $(AlAs)_1(GaAs)_1$ monolayer superlattice as well as the narrow buried δ-doping layer in,e.g., GaAs have a spatial extent of less than the lattice constant of the respective bulk material. The method of δ-doping, also called atomic-plane or sheet doping, which was first proposed by Wood et al (1980) to improve the doping profile in molecular beam epitaxially (MBE) grown GaAs, has recently become important for fundamental studies as well as for application in advanced semiconductor devices. In this paper we give a brief progress report on fundamental and device aspects associated with the deliberate positioning of Si-donors and Be-acceptors in precise numbers and with atomic-layer precision during MBE growth of GaAs and $Al_xGa_{1-x}As$.

2. CONCEPT OF δ-DOPING AND SAMPLE PREPARATION

The term δ-doping stands for the confinement of dopant atoms during crystal growth to a 2D plane one atomic layer thick. The implementation of δ-function like doping profiles by using Si-donors and Be-acceptors is used to generate V-shaped potential wells in GaAs and $Al_xGa_{1-x}As$ with a quasi-2D electron (2DEG) or 2D hole gas (2DHG). The basic concept of Si-δ-doping in GaAs is shown in Fig. 1. The Si-dopant atoms are distributed in a (001)Ga monolayer of the GaAs host crystal, and the ionized Si-donors provide a continuous positive sheet of charge. Due to the electrostatic attraction, the electrons remain close to their parent ionized donors and form a 2DEG in the V-shaped potential well produced by the positive sheet of charge. In the narrow well the electron energies for motion normal to the (001)face are quantized into 2D subbands.

The δ-function like Si and Be doping profiles are obtained through interruption of the growth of GaAs or $Al_xGa_{1-x}As$ by closing the Ga (and Al) shutter, leaving the As shutter open, and opening the shutter of the respective dopant effusion cell. When low growth temperatures in the range $500<T_s<550°C$ are

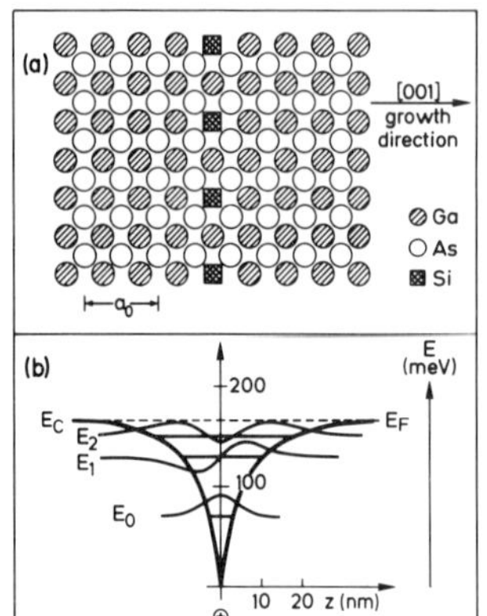

Fig. 1. Schematic illustration of δ-doping in GaAs (a) and of formation of V-shaped potential well (b)

employed, the measured sheet doping density agrees with the number of supplied dopant atoms up to high doping densities which locally exceed the solubility limit of Si and Be in GaAs (Ploog 1987).The Hall mobilities measured for both n- and p-type δ-doping in GaAs compare favorably with homogeneous doping. The concept of δ-doping can also be realized in $Al_xGa_{1-x}As$ layers. However, a certain amount of the Si-donors forms deep levels in the ternary alloy (Schubert et al 1986a). Depending on the alloy composition, the measured Hall electron concentration is thus smaller than the number of incorporated Si atoms, and it depends on the sample temperature and strongly on illumination.

3. SELECTED FUNDAMENTAL AND DEVICE ASPECTS

3.1 Single δ-doping layer structure in homogeneous material

The prototype structure of δ-doping is formed by a single atomic plane of interspersed Si donors in GaAs (Fig. 1). The existence of a 2DEG in δ-doped GaAs was confirmed by the oscillatory magnetoconductivity in a field normal to the doping plane (Zrenner et al 1985). These measurements showed that electrons moving in higher subbands, whose wavefunctions extend far beyond the ionized donors at z=0, have mobilities 5 to 10 times higher than those of the ground-state subband with typical μ_0 of 1500 cm^2/Vs. An oscillatory behavior of the magnetoconductivity was also observed with the field parallel to the doping plane due to the oscillating density of the states at the Fermi level (Zrenner et al 1986). The subband structure in Si-δ-doped GaAs was further investigated by tunneling experiments. Zachau et al (1986) demonstrated tunneling between the subbands of a Si-δ-doping layer embedded in GaAs 20 nm below the surface and a surface metal. In Fig. 2a we show schematically that tunneling occurs in the self-consistent potential profile induced by the δ-doping layer and the potential offset at the metal-semiconductor interface.

The scheme of Fig. 2a is frequently used to illustrate the formation of ohmic metal-semiconductor contacts by heavy doping of the semiconductor. In order to produce low-resistance ohmic contacts, we have to reduce the width of the tunneling barrier in Fig. 2a so that tunneling through the barrier dominates the vertical transport properties. We have examined this idea by placing a δ-doping layer of $N_D^{2D}=1.2\times10^{13}$ cm^{-2} just 2.5 nm below the GaAs surface. The IV-characteristic displayed in Fig. 2b shows strictly linear behavior on all scales of the curve tracer. We have estimated a specific contact resistance below 10^{-6} Ω/cm^2 for these non-alloyed ohmic contacts.

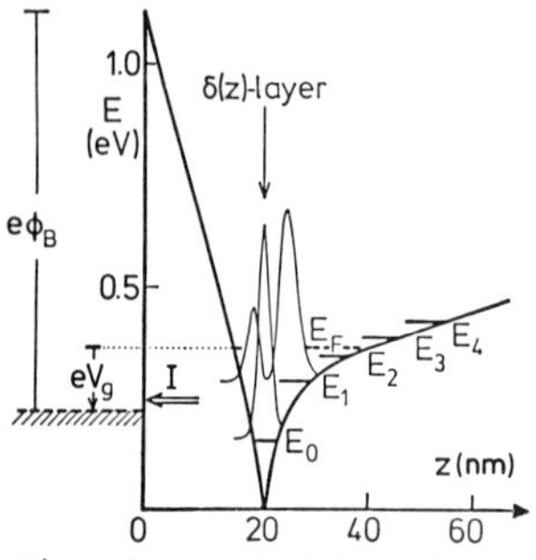

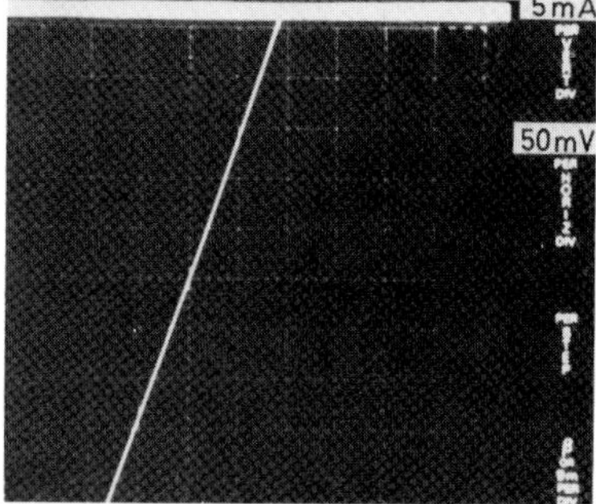

Fig. 2. (a) Schematic illustration of tunneling between the subbands of a Si-δ-doping layer in GaAs and a surface metal; (b) IV-characteristic of a non-alloyed ohmic contact to δ-doped n-type GaAs at 300 K

In the δ-doped field-effect transistor (δ-FET), which is

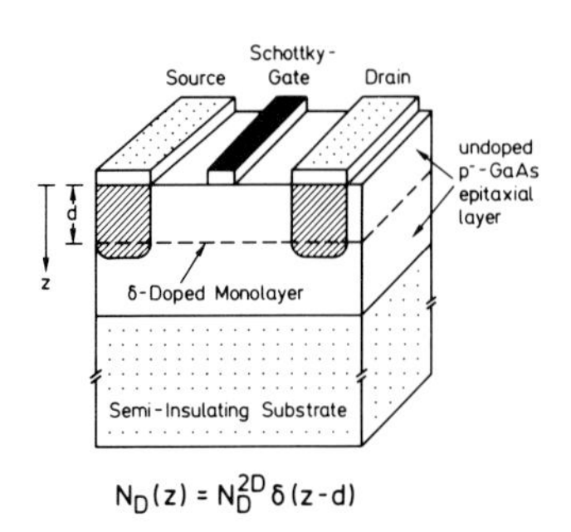

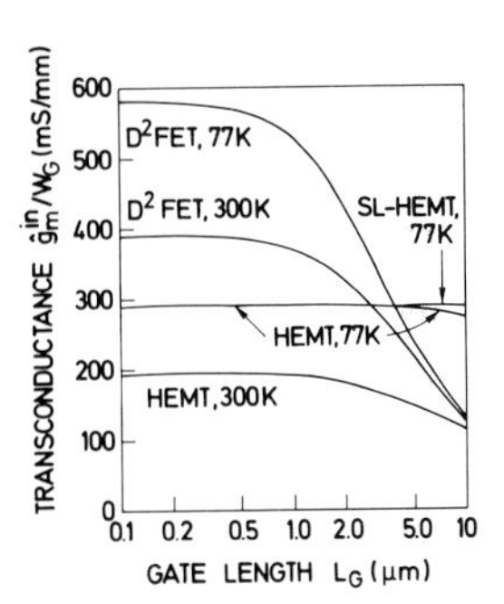

Fig. 3. (a) Schematic illustration of a Schottky-gate δ-doped FET (δ-FET); (b) Calculated transconductance of depletion-mode δ-FET and HEMT assuming a saturated drift velocity of 1×10^7 cm/s at 300 K and 1.5×10^7 cm/s at 77 K, a gate-to-channel distance of 30 nm for the δ-FET and 60 nm for the HEMT, and an electron density of 4×10^{12} cm^{-2} for the δ-FET and 1×10^{12} cm^{-2} for the HEMT

schematically shown in Fig. 2a, the concept of buried-layer transistors (Williams and Shaw 1978) is scaled down to its ultimate physical limit normal to the crystal surface, as the narrow buried δ-doping layer with $N_D^{2D}=(2-4)\times10^{12}$ cm^{-2} has a spatial extent of less than the lattice constant of the host material. In the early attempts to realize FETs having a δ-doping layer (Wood et al 1979), the distance between the gate and the electron channel was too large to fully exploit this concept. We recently fabricated δ-FETs with a narrow distance of 30 nm between the 2DEG and the Schottky gate (Schubert et al 1986b) so that the depletion region below the gate reaches the δ-doping layer at zero bias. In comparison to conventional FETs and to high electron mobility transistors (HEMTs), the advantages of the δ-FET are (i) the high electron concentration of the 2DEG up to 6×10^{12} cm^{-2}, (ii) the large breakdown voltage, and (iii) the high intrinsic transconductance owing to the proximity of the electron channel to the gate. The results of a comparison of the calculated intrinsic transconductance of δ-FET and HEMT are shown in Fig. 3b. For short gate length the saturated velocity model applies for the calculation so that in addition to the saturated velocity only the distance of the 2DEG from the gate determines the transconductance. At high electric fields the ionized impurity scattering becomes negligible. We thus expect δ-doping layers to have the same saturated drift velocity as selectively doped (SD) heterostructures in short-gate FETs. Inspection of Fig. 3b shows that the HEMT has a higher transconductance for L_G>1 μm, while in short-gate devices (L_G < 1 μm) the δ-FET should exceed the performance of the HEMT. In addition, we expect a significant improvement of the high-frequency linearity of the δ-FET because of the linear dependence of the gate-channel capacitance upon the gate bias voltage. The first preliminary experimental data obtained from a 0.5-μm gate δ-FET yielded a normalized external transconductance of 240 mS/mm (Schubert et al 1986c), which will certainly be increased by optimizing the device fabrication process.

At the low and at the high end of the δ-doping scale several phenomena of distinct interest for fundamental research exist. In the low 10^{11} cm^{-2} range the statistical nature of the randomly distributed ionized donors becomes evident when their average distance in the dopant plane is larger than the electron Bohr radius (~10 nm in GaAs). At such low densities, isolated hydrogenic donors exist and the material becomes an insulator. The transition from the degenerate (metallic) 2D sheet, where the individual Bohr radii overlap, to the insulating state with isolated donors was found to occur at a (lower) critical density of 1.35×10^{11} cm^{-2} using the magnetic-field-induced and the density induced transition to an insulating state (Koch et al 1987). At high Si-δ-doping densities beyond 8×10^{12} cm^{-2}, on the other hand, central cell effects at the Γ-point become important, and multi-valley effects at the L- and X-point can lead to a transfer of electrons to higher conduction band minima (Zrenner and Koch 1987). The application of hydrostatic pressure can induce a redistribution of electrons between the various electronic subband levels.

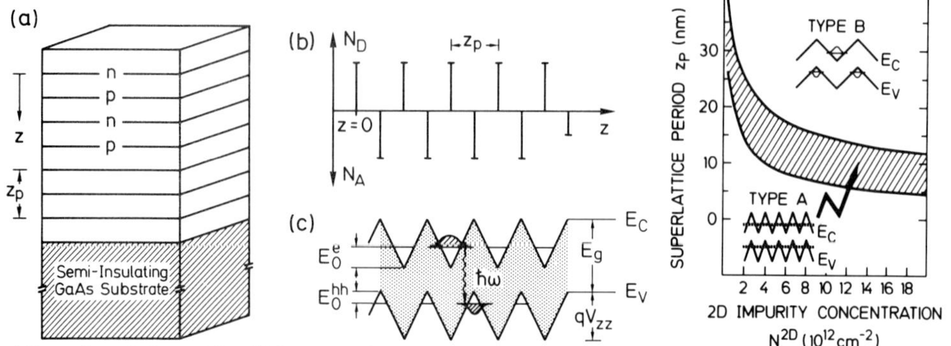

Fig. 4. Schematic illustration of layer sequence (a), doping profile (b), modulation of conduction and valence band edges (c) in a GaAs sawtooth doping superlattice (SDS), and classification of Type A and Type B GaAs SDS(d)

3.2 Periodic δ-doping layer structures in homogeneous material

Two types of periodic δ-doping layer structures were conceived, (i) a periodic sequence of δ-doping layers of the same doping type separated by undoped regions which we call δiδi structure, and (ii) a periodic sequence of alternating n- and p-type δ-doping layers equally spaced by undoped regions which we call sawtooth doping superlattice (SDS). The thickness of the undoped regions between the δ-doping sheets is typically in the range of 2 to 30 nm. The motivation to construct n-type δiδi structures in otherwise intrinsic GaAs comes directly from our experiments on tunneling (Koch et al 1986) and from the possibility to realize extremely high 3D doping concentrations in semiconductors which are several orders of magnitude higher than the solubility limit. For Si-doped GaAs δiδi structures we achieved equivalent 3D electron concentrations as high as 10^{21} cm^{-3} with 300 K mobilities of 1000 cm^2/Vs.

The motivation for developing GaAs SDS was the requirement for both short carrier recombination lifetimes in the ns range and strong modulation of the energy bands in doping superlattices, in order to achieve stimulated (laser) emission (Schubert et al 1985a). The doping profile of GaAs SDS (Fig. 4b) can be described by a periodic train of δ-functions. The periodic variation of the space charge in z-direction leads to a sawtooth-shaped modulation of the energy bands (Fig. 4c). In the narrow V-shaped potential wells size quantization occurs. Radiative recombination involves the lowest electron and heavy-hole subbands. Small period lengths z_p imply a strong overlap of electron and hole wavefunctions from adjacent potential wells, which in turn results in short recombination lifetimes in the ns range. This so-called Type A SDS thus exhibits a stable energy gap smaller than the GaAs gap (Fig. 4c) which does not change upon carrier injection. For a quantitative distinction between Type A and Type B SDS we have related the tunneling probability of carriers between the wells with the amplitude of the respective wavefunctions penetrating the barriers (Ploog et al 1986). In Fig. 4d we show that the Type A GaAs SDS are achievable with material design parameters depicted by the shaded area. Our results of photo- and electroluminescence measurements clearly revealed the differences between Type A and Type B GaAs SDS. The high intensity and the significant redshift of the luminescence also at 300 K makes feasible the application of Type A SDS in photonic devices. We have fabricated light emitting diodes and injection lasers with the superlattice active region sandwiched by n- and p-type $Al_xGa_{1-x}As$ layers for confinement which emit at wavelengths of $0.9 \leq \lambda \leq 1.0$ µm. Finally, it is important to note that in Type A GaAs SDS with strong coupling electron and hole subbands of considerable widths ($\Delta E \simeq 1$ meV) are formed and carrier transport normal to the layers becomes important. We have used IV-measurements to study the vertical transport of holes in a Type A GaAs SDS ($N_D^{2D}=1\text{x}10^{13}$ cm^{-2}, z_p=15 nm) embedded between two p^+-GaAs contact layers (Schubert et al 1986d).

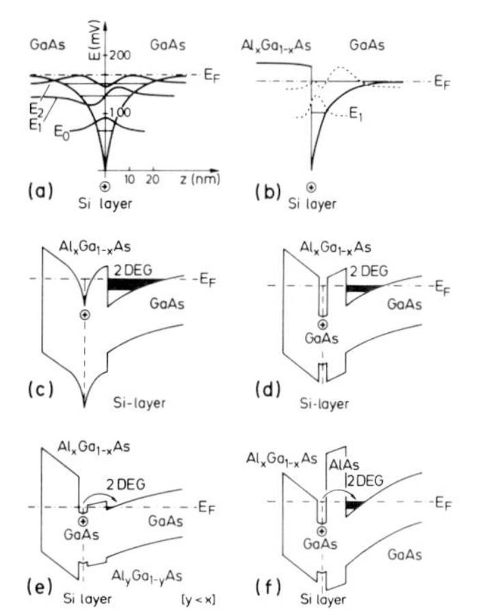

Fig. 5. Real-space energy band diagram of selectively δ-doped n-$Al_xGa_{1-x}As$/GaAs heterostructures with different design concepts

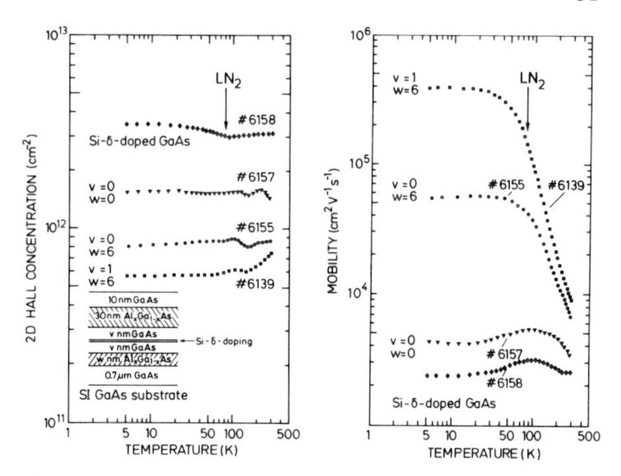

Fig. 6. Hall electron concentration and mobility vs temperature obtained from a single δ-doped GaAs layer (#6158) and from three selectively δ-doped n-$Al_xGa_{1-x}As$/GaAs heterostructures with different design parameters (see inset)

3.3 Confinement of donors (acceptors) to an atomic (001)plane in selectively doped heterostructures

We recently developed the concept of confinement of donors (or acceptors) to atomic (001)planes in SD $Al_xGa_{1-x}As$/GaAs heterostructures for an effective spatial separation of the 2DEG (or 2DHG) from the sheet of ionized impurities, in order to achieve enhanced carrier mobilities. This concept is now used for three purposes, (i) to achieve high mobilities also at high carrier densities in a simple heterostructure, (ii) to reduce the undesired persistent photoconductivity effect, and (iii) to improve the transconductance of HEMTs by minimizing the thickness between the 2DEG and the gate at the crystal surface. We discuss the new concept by means of the real-space energy band diagrams and the results of temperature-dependent Hall measurements shown in Figs. 5 and 6. A conventional single δ-doped GaAs layer with $N_D^{2D}=3x10^{12}$ cm^{-2} having a 300 nm thick cap layer is used as reference. In the heterostructures the cap layer is reduced to 40 nm so that about $1.5x10^{12}$ of the total $3.0x10^{12}$ cm^{-2} electrons have to be transfered to the crystal surface to establish the surface potential. When the δ-doping layer and the heterojunction are both located at z=0 (Fig. 5b), the potential becomes a triangular well and all the electronic screening charge is located on one side of the junction. In the Hall measurements we observe the expected density of $1.4x10^{12}$ cm^{-2}, but no freeze-out of the carriers. The electron mobility is sligthly increased, because the heterojunction leads to a reduced overlap between subband electrons and the ionized impurities located at z=0.

When we move the δ-doping layer away from the heterojunction at z=0, the overlap between subband electrons and the sheet of ionized donors is further reduced. A spatial separation of 6 nm is sufficient to produce a strong enhancement of the 2DEG at the heterointerface (Fig. 6 and Table 1). Although the δ-doping layer and the heterojunction are located at different z, they still form a coupled pair of quantum mechanical systems. As all the donors in these SD heterostructures are confined in the 2D plane, the band bending is linear (Fig. 5), and we can thus expect that HEMTs fabricated from this structure exhibit a constant transconductance and source-to-gate capacitance over a wide bias range. In addition, we easily achieve charge densities of more than $1.3x10^{12}$ cm^{-2} at a single heterointerface having a strongly enhanced mobility even at room temperature (Table 1). The product of mobility and density has thus reached a value never reported before for single $Al_xGa_{1-x}As$/GaAs

Table 1. Material design parameters and transport properties of selectively δ-doped n-$Al_xGa_{1-x}As$/GaAs heterostructures

Sample #	δ-density (cm^{-2})	Configuration and spacer	$n_s(cm^{-2})$ 300 K	$\mu(cm^2/Vs)$ 300 K	$n_s(cm^{-2})$ 77 K	$\mu(cm^2/Vs)$ 77 K	$n_s(cm^{-2})$ 4 K	$\mu(cm^2/Vs)$ 4 K
6158	$3.0x10^{12}$	bulk-type GaAs	$3.1x10^{12}$	$2.5x10^3$	$3.0x10^{12}$	$3.1x10^3$	$3.5x10^{12}$	$2.4x10^3$
6157	$3.0x10^{12}$	δ-doping at heterointerface	$1.4x10^{12}$	$3.5x10^3$	$1.5x10^{12}$	$5.8x10^3$	$1.5x10^{12}$	$4.4x10^3$
6155	$3.0x10^{12}$	6 nm spacer	$8.7x10^{12}$	$6.6x10^3$	$8.6x10^{11}$	$4.4x10^4$	$8.0x10^{11}$	$5.5x10^4$
6170	$6.0x10^{12}$	6 nm spacer	$1.4x10^{12}$	$6.8x10^3$	$1.5x10^{12}$	$5.9x10^4$	$1.3x10^{12}$	$7.3x10^4$
6139	$3.0x10^{12}$	δ-layer embedded in 2 nm GaAs QW,	$7.5x10^{11}$	$9.4x10^3$	$5.9x10^{11}$	$1.6x10^5$	$5.7x10^{11}$	$3.8x10^5$
6163	$6.0x10^{12}$	6 nm spacer	$1.1x10^{12}$	$7.6x10^3$	$9.9x10^{11}$	$9.0x10^5$	$9.7x10^{11}$	$1.6x10^5$

heterojunctions. Combined with the narrow distance between the gate electrode and the high-mobility 2DEG this feature will certainly improve the current driving capabilities of single heterojunction HEMTs.

The transport properties of these new δ-SD heterostructures are further improved when the atomic plane doping in the $Al_xGa_{1-x}As$ is embedded in the center of a 2-nm wide GaAs quantum well (QW) (see Fig. 5). With this modified structure we obtain mobilities of $4x10^5 cm^2/Vs$ with a spacer as narrow as 7 nm In addition, the persistent photoconductivity is strongly reduced, because the $Al_xGa_{1-x}As$ barriers remain essentially undoped. The quantum mechanical coupling between the potential wells from the δ-doping and from the band bending at the heterointerface might give rise to some remaining electron density in the δ-doping channel and thus to a parallel conduction, which is deleterious for some applications. This phenomenon can be eliminated by the design of an asymmetric narrow δ-doped GaAs QW where no bound states are formed if the well width L_z is smaller than a critical value (Horikoshi et al 1987). We obtained very high electron mobilities for δ-doped asymmetric QW heterostructures with AlAs spacer (Fig. 5f). The measured peak mobility of more than $10^6 cm^2/Vs$ is the highest ever reported for a SD n-$Al_xGa_{1-x}As$/GaAs heterostructure with a spacer of only 10 nm (Ploog 1987).

Finally, we present another type of δ-SD $Al_xGa_{1-x}As$/GaAs heterostructures with the confinement of donors (or acceptors) to an atomic (001)plane in the lower gap material close to the heterointerface. In this structure we can accuratel position a plane of repulsive negatively charged Be or of attractive positive charged Si impurities for a controlled interaction with the high-mobility 2DE The influence of repulsive and attractive scattering centers on the magneto-transport properties of the 2DEG is important for a quantitative interpretati of precision measurements of the quantum Hall effect (Haug et al 1987).

Haug R J et al **1987** Phys.Rev.Lett. **59**
Horikoshi Y et al **1987** Jpn.J.Appl.Phys. **26** 263
Koch F et al **1986** Springer Ser.Sol.State Sci. **67** 175; **1987** ibd. **71** 308
Ploog K et al **1986** Surf.Sci. **174** 120; **1987** J.Cryst.Growth **81** 304
Schubert E F et al **1985a** Phys.Rev. **B32** 1085; **1985b** Appl.Phys.Lett. **47** 219
Schubert E F et al **1986a** Solid State Electron. **29** 173
Schubert E F et al **1986b** IEEE Trans.Electron Devices **ED-33** 625
Schubert E F et al **1986c** Appl.Phys.Lett. **49** 1729
Schubert E F et al **1986d** Springer Ser.Sol.State Sci. **67** 260
Williams R E and Shaw D W **1978** IEEE Trans.Electron Devices **ED-25** 600
Wood C E C et al **1979** IEDM 79 Tech.Dig. p 388; **1980** J.Appl.Phys. **51** 383
Zachau M et al **1986** Solid State Commun. **59** 591
Zrenner A et al **1985** Proc.17th ICPS (Springer, Berlin) p 325
Zrenner A et al **1986** Phys.Rev. **B33** 5607; **1987** Proc.18th ICPS (Singapore) p

Inst. Phys. Conf. Ser. No. 91: Chapter 1
Paper presented at Int. Symp. GaAs and Related Compounds, Heraklion, Greece, 1987

Integrated guided-wave optics on III–V semiconductors

M.Erman

Laboratoires d'Electronique et de Physique Appliquée*
3, Avenue Descartes 94451 Limeil Brévannes Cédex (France)

Abstract

Optical telecommunications have stimulated intensive research on semiconductors which offer a number of physical effects of interest in integrated optics. Significant progresses have been achieved in the fabrication of passive and active waveguides. Optical losses have been considerably reduced for both GaAs and InP based waveguides. We will discuss typical recent achievements (laser/external modulator, detector/waveguide and optical amplifier/waveguide integration), which beat great hopes to the field of monolithic integration.

Introduction

Starting in the early 60's with the demonstration of the guiding action of planar p-n junction (1) and the modulation via the electro-optic effect in similar structures (2), guided wave optics on III-V semiconductors has a long history. Nevertheless, it has been faced for a long time to many technological problems which result in devices with high optical losses. However, during last few years, the worldwide research activity on semiconductor integrated optics has been intensified and major breakthroughs have been achieved on both discrete devices and monolithic integration. Though they are many reasons for the revival of interest for integrated optics, the main one is the establishment of optical telecommunications using single mode fiber technology. Together with optoelectronic devices such as lasers and detectors, there is a need for other optical components such as modulators, switches, multiplexers (time, wavelength, space),... With the upcoming coherent transmission systems (3) the need for optical devices will still increase. The monolithic integration is still at the research stage and therefore, cost-effective devices are not expected in the very near future. However, recent progresses achieved in the field of integrated optics indicate that many technological problems can be solved.

Passive waveguides and modulators (phase or intensity) are the basic components of more complex optical circuits. This paper presents various aspects which are related to these devices and is organised as follows. In Section I, the optical properties of III-V semiconductors will be briefly presented, namely the physical phenomena which allows for the guiding and light modulation. Various waveguide and modulator structures will be presented and their performances discussed in Section II. The monolithic integration will be adressed in Section III.

I. Material properties related to guiding and modulation of light in III-V semiconductors

The semiconductor materials commonly used in integrated optics are GaAs and InP as substrate, and the alloys which can be lattice matched to these substrates, i.e. GaAlAs for GaAs and GaInAsP or InGaAlAs for InP. In the case of the GaAs family, all materials are transparent in the near infra-red region of interest for optical telecommunications. On the contrary, the bandgap of GaInAsP lattice matched to InP cover the whole spectral region, allowing for the fabrication of detectors and lasers on same substrate.

Though the materials are used at photon energies below the fundamental bandgap, they still exhibit some residual intrinsic absorption. For doped materials, the dominant effect is due to free carriers (4). Figure 1 gives experimental results for GaAs (5) and InP (6). At wavelengths larger than 3μ, the absorption is proportional to $N\lambda^p$ where N

(*)L.E.P. is a member of the Philips Research Organization

is the carrier concentration, λ the wavelength and $2.5 < p < 3.5$ depending on the absorption mechanisms involved (7). In the 1-2μ range, the absorption is dominated by interband transitions (8) and is proportional to N only for $N > 10^{17} cm^{-3}$ (9). Indeed, at low free carrier concentrations, absorption due to the presence of deep levels can be larger than the free carriers contribution. Thus, the purity of the material is essential to achieve low loss waveguides.

Changes of index of refraction n and absorption k are used for both achieving optical confinement and modulating the guided light. It should be emphasis that in every case n can not be changed without affecting k, and vice versa, due to the Kramers-Kronig relations (10).However, the spectral region where the change occur could differ. For instance, changes of k are very often more localized in energy than the corresponding changes of n.

The consequence on n of the free carriers is to lower the index of refraction (11). A typical value is $\Delta n = -5.10^{-3}$ at $N = 10^{18} cm^{-3}$. This effect is strong enough to provide optical confinement. It is used in homostructure waveguides.

Another way to change n permanently is to change the material chemical composition (heterostructure waveguides). The index change can be as high as 0.3 (12,13). In the case of GaAs family, the ternary has a lower index of refraction and the guiding occurs in the binary GaAs. The situation is opposite for the InP system, where the quaternary has a higher index of refraction than InP. Because the quality of quaternary or ternary epitaxial films (residual doping, compositionnal fluctuations) is generally lower than the binary one, low loss heterostructure waveguides on InP are more difficult to realize than GaAs ones.

Finally, n can be permanently changed by inducing stress into the material. The corresponding elasto-optic effect can change n up to approximatively 10^{-2} (14). This effect is crystal orientation dependent and has been used less frequently than free carrier or compositional change.

Optically properties can be dynamically modulated via electrically, thermally or even optically induced changes. Thermally induced modulation is of small interest because of limited bandwidth. Optically induced changes are promising because they lead to the concept of a fully optical device. Optically induced effect, such as saturable absorption are observed near the fundamental absorption gap, and are particularly strong in the case of quantum well structures (15). However, the number of devices realized to date remains very limited, thus we will only concentrate in this paper on electrically induced modulation.

The general scheme for an active waveguide (i.e. a waveguide in which n or k can be electrically modulated) is a p-i-n structure, the intrinsic region being also the waveguiding region. Also, the p region can be replaced on some materials by a Schottky contact. The device can be operated in two different ways. Direct biasing the p-i-n structure results in injection of electrons through the intrinsic region. The effect on n and k is similar to that obtained by doping.

Reverse biasing the p-i-n structure gives rise to strong electric field in the intrinsic region. This leads to the change of the index of refraction over a wide energy range below the fundamental gap (electro-optic or Pockels effect) and change of the absorption close to the fundamental absorption gap (electro-absorption).

The change of index of refraction due to electro-optic effect is proportionnal to the electric field and is typically in the order of 10^{-4}. The linear contribution (Pockels effect) is crystal orientation dependent and is for (100) substrate maximum for light propagating along 110 and $1\bar{1}0$ directions (16). The sign of Δn is negative for 110 and positive for $1\bar{1}0$. A quadratic contribution (Kerr effect) can add to the linear term, but is generally weaker and rapidly vanishes far from the gap (17). The Kerr contribution has the same sign for 110 and $1\bar{1}0$. The electro-optic modulation can be extremely fast and modulation bandwidth are only limited by the capacitance. The disadvantage of the electro-optic effect is its anisotropy : in the case of (100) substrates, only the TE component of the light can be modulated.

The electroabsorption is caused by the tilting of the conduction and the valence band edges, which results in the reduction of the optical bandgap (18). To be efficient, wavelengths close to the bandgap should be used. Changes in the absorption coefficient can be in the order of 200 cm^{-1} and even higher in the case of quantum well devices, where the electro-absorption is enhanced by the quantum confined Stark effect (19).

2. Waveguides and modulators

A two dimensional optical confinement has to be realized in a waveguide in order to achieve single mode operation. The index change between the substrate and the waveguide is usually obtained by growing a layer with different doping level or composition than the substrate. Several different physical effects and technological approaches exist to achieve lateral confinement. One can locally perturbe the epitaxial planar waveguide in order to create the necessary index step, for instance by depositing a metallic layer - which induces stresses -, or by diffusion. The latter is particularly interesting in the case of superlattices or MQWs where the index difference is due to layer inter-mixing in the superlattice (70). Nevertheless, most of the waveguiding structures are obtained combining etching and epitaxial growth techniques. A number of different structures has been proposed; Figure 2 gives some examples.

The rib structure (Fig. 2a) is commonly used because of its simplicity. For this structure, a planar waveguide is realized first. Thus, all the epitaxial techniques (LPE, VPE, MBE, MOVPE, MOMBE) can be used. The rib is obtained by etching. One of the problems associated with this structure are the scattering losses due to the roughness of the rib etched facets. However, progresses realized in etching techniques have considerably reduced this problem. Indeed, the lowest loss measured to date on a semiconductor waveguide (0.15dB/cm) has been obtained on double heterostructure GaAlAs rib waveguide (39). In the other two structures, i.e. the inverted rib (Fig. 2b) and the buried one (Figure 2c), the etching is performed prior to epitaxy. As a result, the etched interface is embedded and corresponds to a small index difference. The influence of the roughness is thus minimized. In these two structures, an epitaxial growth on a non-planar substrate is required, and, in some cases, localized epitaxy could be advantageous.These two requirements are met by LPE and VPE.

The continuous progress of epitaxy and of the whole III-V technology has been the major factor in the considerable reduction of waveguide losses during the past few years as represented on Figure 3. A definite breakthrough - losses below 1dB/cm - has been realized only recently. In many cases, this corresponds to the use of sophisticated epitaxial techniques such as MBE, VPE and especially MOVPE. The reduction of losses can be explained by the improvement of surface quality, uniformity and purity of the epilayers. Although the best results have been achieved in passive waveguides, they demonstrate the high potentiality of semiconductors for guided wave optics.

An optical modulator (or switch) requires the use of active waveguides. We can distinguished devices based upon one single active waveguide from those using more than one waveguide. A simple active waveguide behave as a phase modulator (electro-optic modulation) for wavelengths far from the gap, and as an intensity modulator for wavelengths near the gap (electro-absorption). Intensity modulator structures using more than one waveguide include the directional coupler (24,25), and the Mach-Zehnder interferometer (57,59). Important modulator parameters are the driving voltage and the optical losses. These two parameters can not be optimized separately (71). Indeed, in an active waveguide, the dominant contribution to the losses is due the absorption of the n^+ and p^+ regions. The driving voltage can be decreased by increasing the electric field in the intrinsic region, which is obtained by decreasing its thickness. By doing this, the optical confinement is reduced, and the portion of the light in the p^+ and n^+ region will increase and so will the losses. Figure 4 giving the absorption loss of the modulator (α in dB/cm) vs the normalized driving voltage (V in V.cm) for several different structures clearly demonstrates this inter-dependence. A general result is that homostructure modulators exhibit higher αV products than the heterostructure ones. This can be understood because in homostructure waveguide the optical confinement is achieved by the n^+ and p^+ regions, and therefore the optical and electrical problems are closely related. This is less the case in heterostructure modulators. In view of the latest achievements on the passive waveguides, it is likely that results on heterostructure waveguides can be still improved. It is also remarquable that the αV product for electro-absorption modulators and the electro-optics devices are similar, though the structures and the physical effects are different. Because the electro-absorption devices operate near the gap, they have higher absorption but lower driving voltage. This separates the plane (α,V) into two regions : electro-optics devices and electro-absorption devices. By changing the wavelength, it is possible to move the device parameters from one region to the other one.

For the application in optical transmission systems, the modulator should be capable of high speed operation. Several results show bandwidths above 1GHz for both electro-absorption modulators (1.6GHz(72), 3.86GHz(73), 3.7GHz for a MQW structure (74)) and electro-optic modulators (4.5Ghz(59), 10GHz(75)). In these examples the bandwidth is limited by the device RC product. Higher frequencies can be in principle obtained by using travelling wave modulators. In this case, the microwave propagates in phase with the optical beam and the bandwidth is no longer capacitance limited. In practice, an ideal travelling wave modulator is difficult to realize on n^+ doped substrate, first because the microstripe line impedance cannot be matched to 50Ω without decreasing the electro-optic interaction, and second because the microwave loss and slowing effect of the n^+ substrate result in a mismatch of microwave and optical refractive index. Bandwidths of 4.1GHz(38) and 9.6GHz(76) have been achieved in this configuration, using GaAs n^+ substrate. These problems can be alleviated by using semi-insulating substrate. The PIN structure can be replaced by two Schottky electrodes. A bandwidth of 20GHz has been measured in this case(40).

3.Monolithic integration

The integration of optoelectronic devices (lasers and detectors) and electronic circuits, leading to the concept of optoelectronic integrated circuits (OEICs) has received much attention (77) because it can improve performances of lasers and detectors by reducing parasitic reactances, by facilitating optoelectronic devices multifunction capabilities and also by facilitating the assembly procedure. Similar arguments can be found for guided wave devices monolithically integrated with OEICs. Because of the numerous technological problems, this issue is a very challenging one.

The integration of OEICs with optical guided wave devices is still in its infancy. However, several attempts have been done. Among them, the one with an immediate practical application is the association of a **laser and an external modulator.** Such a modulator can reduce the wavelength chirping which occurs when laser is directly modulated at high speeds. The simplest structure consists in using the same heterostructure for both the laser and the electro-absorption modulator. The modulator is electrically isolated from the laser by air gap realized by etching. When the laser is operating, the energy of the emitted light is shifted below the absorption edge of the unbiased modulator due to the injected carrier-induced gap-shrinkage. This shift is only significant for MQW structures. Using such laser and modulator structure (GaAlAs/GaAs MQW), the integration of the two elements has been demonstrated(78). A 7dB modulation depth and 0.88GHz cut off frequency has been measured. Laser and modulator performances can be better optimized if different structures are used for each of the two elements. In order to get high modulation efficiency, the coupling between the laser and the modulator should be high. This makes the air gap between the laser and the modulator unsuitable. To achieve a modulator structure monolithically but-coupled to the laser, complex technology, namely multi-step epitaxy is compulsory. Though the difficulties are important, interesting results have been obtained on InP substrates (79,80). In the first example (79), a MQW InGaAs/InAlAs modulator grown by MBE has been integrated with a LPE DFB laser. A good laser-modulator coupling efficiency of 95% has been measured. In the second example (80), a DFB laser has been integrated with guided wave modulator, using bulk electro-absorption. It is worth noting that the same technique (VPE) was used during the whole fabrication process.

Photodiode/waveguide integration on InP substrates has been demonstrated by several authors (49,81,82). In all cases, one step epitaxy has been used: VPE(49), LPE(81) or MOVPE(82). The detector is a PIN photodiode and is delimited by a selective chemical etching. A further etching is necessary in MOVPE(82) to realize the rib structure waveguide, while this could be avoiding in LPE and VPE using inverted ribs. The waveguide/photodiode coupling efficiency can be varied from 0 to 100%, depending on the photodiode/waveguide length. This opens interesting application : end line receiver, monitor photodiode to provide feedback in optical circuit, etc.

The concept of an **optical amplifier** is not only attractive as a separate element, but also from the point of view of integrated optics. Indeed, its integration with an optical circuit can make the device virtually loss free. Further more, integrating the amplifier with the waveguide improve the coupling between these two elements, and reduces the number of anti-reflection coated facets that are needed in the circuit. Attempt has been done by integrating the amplifier and passive waveguide on InP substrate(83). An external cavity gain of 19.5dB has been measured. More complex circuit has been demonstrated that includes an optical switch (total internal reflection induced by carrier injection) and four DFB lasers (84). The laser structure can be used either as a detector or an amplifier. The estimated optical gain is 24dB.

Conclusions

The latest results obtained in the field of III-V semiconductor integrated optics that we have reviewed in this paper demonstrate the significant progresses achieved during the last few years. The low losses, measured on both GaAs and InP substrate waveguides make the III-V semiconductors very competitive materials in the fabrication of optical circuits. However, it has been shown that the waveguide absorption and the modulation efficiency are clearly interdependent; their product gives a figure of merit. From this point of view, heterostructure waveguides appears as a much better solution as compared to homostructure waveguides. The experimental results obtained in different laboratories also show that different growth techniques (VPE, MOVPE and MBE) can be used for the fabrication of high quality waveguides.

Monolithic integration of OEICs and guided wave devices have just started, thus all the advantages of the integration have not yet been demonstrated. Nevertheless, impressive results have already been obtained, and many technological problems have been tackled and solved. In the next years, the technology (improvement of epitaxy, etching techniques and lithography) will certainly continue to be an important issue. How to realize integrated devices cost effectively will remain the main challenge.

References:

/1/ A.Yariv et al., Appl. Phys. Letters **2,** 55 (1963)
/2/ D.F.Nelson et al., Appl. Phys. Letters **5,** 148 (1964)
/3/ T.Kimura, IEEE J. of Light. Techn. **LT-5,** 414 (1987)
/4/ C.R.Pidgeon, Handbook on Semiconductors, T.S.Moss Ed. vol. 2, M.Balkanski Ed.
/5/ W.G.Spitzer et al., Phys. Rev. **114,** 59 (1959)
/6/ O.K.Kim et al., J. Electron. Mat. **12,** 827 (1983)
/7/ W.Walukiewicz et al., J. Appl. Phys. **51,** 2659 (1980)
/8/ F.Fiedler et al., Solid-State Electron. **30,** 73 (1987)
/9/ A.A.Ballman et al., J. Cryst. Growth **62,** 198 (1983)
/10/ L. de Krönig, J. Opt. Soc. Am. **12,** 547 (1926)
/11/ M.A.Mentzer et al., Proc. SPIE, **408,** 38 (1983)
/12/ S.Adachi, J. Appl. Phys. **53,** R1 (1985)
/13/ S.Adachi, J. Appl. Phys. **53,** 5863 (1982)
/14/ N.Suzuki et al., J. Appl. Phys. **22,** 441 (1983)
/15/ J.S.Weiner et al., Appl. Phys. Lett. **47,** 664 (1985)
/16/ S.Namba, J. Opt. Soc. Amer. **19,** 76 (1961)
/17/ H.G.Bach et al., Appl. Phys. Lett. **42,** 692 (1983)
/18/ J.Callaway, Phys. Rev. **134,** A998 (1964)
/19/ D.A.B.Miller et al., Phys. Rev. B **32,** 1043 (1985)
/20/ R.A.Logan et al., J. Appl. Phys. **44,** 4172 (1973)
/21/ F.A.Blum et al., Appl. Phys. Lett. **25,** 116 (1974)
/22/ F.J.Leonberger et al., Appl. Phys. Lett. **29,** 652 (1976)
/23/ J.McKenna et al., J. Appl. Phys. **47,** 2069 (1976)
/24/ H.Kawaguchi, Electron. Lett. **14,** 387 (1978)

/25/ A.Carenco et al., J. Appl. Phys. **50** 5139 (1979)
/26/ F.J.Leonberger et al., Appl. Phys. Lett. **38,** 313 (1981)
/27/ T.M.Benson et al., Proc. ECIO'81, London, (1981)
/28/ A.J.N.Houghton et al., Optics Comm. **46,** 164 (1983)
/29/ M.Erman et al., Appl. Phys. Lett. **43,** 894 (1983)
/30/ R.G.Walker et al., Electron. Lett. **19,** 590 (1983)
/31/ A.J.N.Houghton et al., ibid. **20,** 480 (1984)
/32/ A.Carenco et al., Techn. Digest, Th B4-1 April 24-26, 1984, Kissimme, Florida
/33/ D.J.Jackson et al., Electron. Let. **21,** 44 (1985)
/34/ D.A.Andrews et al., J. Vac. Sci. Technol. **B3,** 813 (1985)
/35/ H.Inoue et al., IEEE J. Light. Technol. **LT-3,** 1270 (1985)
/36/ G.J.Sonek et al., ibid. **LT-3,** 1147 (1985)
/37/ S.Lin et al, Electron. Lett. **21,** 597 (1985)
/38/ S.H.Lin et al., ibid. **22,** 934 (1986)
/39/ E.Kapon et al., Appl. Phys. Lett. **50,** 1628 (1987)
/40/ S.Y.Wang et al., Appl. Phys. Lett. **51,** 83 (1987)
/41/ A.Carenco et al., Appl. Phys. Lett. **40,** 653 (1982)
/42/ C.Bornholdt et al., Electron. Lett. **19,** 81 (1983)
/43/ O.Mikami et al., ibid. **19,** 594 (1983)
/44/ L.M.Johnson et al., Appl. Phys. Lett. **44,** 278 (1984)
/45/ M.Fujiwara et al., Electron. Lett. **20,** 791 (1984)
/46/ K.Okamoto et al., J. Appl. Phys. **56,** 2595 (1984)
/47/ J.P.Donnelly et al., IEEE J. Quant. Electron. **QE-21,** 1147 (1985)
/48/ T.L.Koch et al., Electron. Lett. **23,** 245 (1987)
/49/ M.Erman et al., Proceedings of ECOI'87, Glasgow, May 11-13, 1987
/50/ Y.Bourbin et al., ibid.
/51/ P.W.A.McIlroy et al., Electron. Lett. **23,** 701 (1987)
/52/ P.Cinguino et al., ibid. **23,** 235 (1987)
/53/ S.Ritchie et al., Proc. OFC/IOOC'87, Reno, Nevada, (1987)
/54/ U.Koren et al., Appl. Phys. Lett. **49,** 1602 (1986)
/55/ M.Erman et al., submitted to IEEE J. Light. Technol.
/56/ M.C.Gabriel et al., IEEE J. Light. Technol. **LT-4,** 1482 (1986)
/57/ J.P.Donnelly et al., Appl. Phys. Lett. **45,** 360 (1984)
/58/ M.Erman et al., IEEE J. Light. Technol. **LT-4,** 1524 (1986)
/59/ P.Buchmann et al., Appl. Phys. Lett. **46,** 462 (1985)
/60/ R.G.Walker et al., Proc. ECIO'85, Berlin, 1985
/61/ J.Faist et al., Appl. Phys. Lett. **50,** 68 (1987)
/62/ Y.Noda et al., IEEE J. Light. Technol. **LT-4,** 1445 (1986)
/63/ C.Bornholdt et al., Proc. ECIO'85, Berlin, 1985
/64/ P.M.Rodgers et al., Proc. ECIO'87, Glasgow, 1987
/65/ T.Wood et al., Proc. CLEO'87, Baltimore, Maryland, 1987
/66/ M.D.A.MacBean et al., Proc. ECIO'87, Glasgow, 1987
/67/ T.H.Wood, Appl. Phys. Lett. **48,** 1413 (1986)
/68/ C.M.Gee et al., Proc. IGWO'86, Atlanta, 1986
/69/ U.Koren et al., Appl. Phys. Lett. **50,** 368 (1987)
/70/ F.Julien et al., Appl. Phys. Lett. **50,** 866 (1987)
/71/ A.Carenco, Proc. SPIE **651** Integrated Circuit Engineering III(1986)
/72/ Y.Noda et al., Electron. Lett. **21,** 1183 (1985)
/73/ Y.Noda et al., IEEE J. Light. Technol. **LT-4** 1445 (1986)
/74/ T.H.Wood et al., Electron. Lett. **23,** 540 (1987)
/75/ R.G.Walker et al., Proc. ECIO'87, Glasgow, 1987
/76/ S.H.lin et al., Applied Optics **26,** 1696 (1987)
/77/ O.Wada et al., IEEE J. Quant. Electron. **QE-22,** 805 (1986)
/78/ S.Tarucha et al., Appl. Phys. Lett. **48,** 1 (1986)
/79/ Y.Kawamura et al., IEEE J. Quant. Electron. **QE-23,** 915 (1987)
/80/ M.Suzuki et al., Proc. OFC/IOOC'87, Reno, Nevada, (1987)
/81/ C.Bornholdt et al., Electron. Lett. **23,** 2 (1987)
/82/ S.Chandrasekhar et al., Electron. Lett. **23,** 501 (1987)
/83/ M.G.Öberg et al., IEEE J. Quant. Electron. **QE-23,** 1021 (1987)
/84/ T.Katsuyama et al.,Proc.IGWO'86(Postdeadline paper PDP11),Atlanta,1986

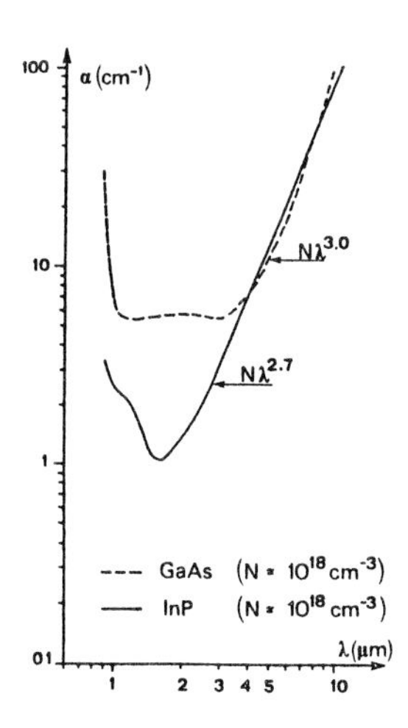

Figure 1 : Absorption curves of N^+ ($\sim 10^{18}$) GaAs (ref.5) and InP (ref.6). The absorption in the 1-2μm range is dependent upon the crystal quality.

$n_g > n_s > n_c$

a: n_c, n_g, n_c, n_s

b: n_c, n_g, n_c, n_s

c: n_c, n_g, n_c, n_c, n_s

Figure 2 : Schematic representation of double heterostructure waveguides : rib (a), inverted rib (b) and burried (c) structure

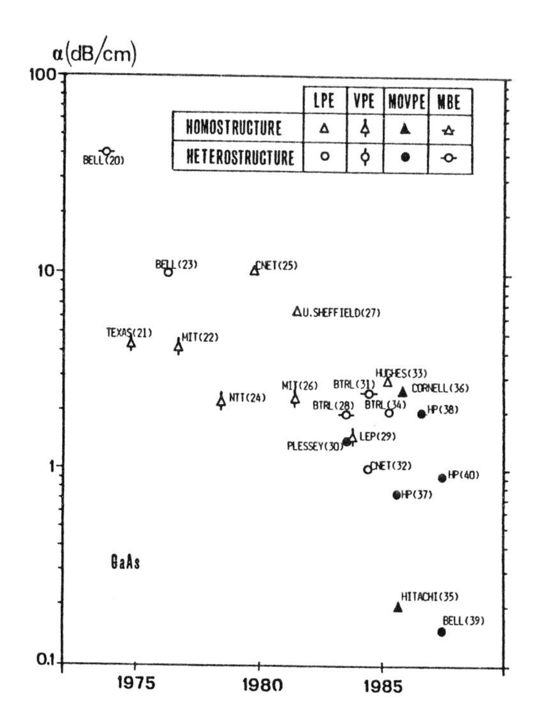

Figure 3-a :Published values of waveguide losses for GaAs based structures.

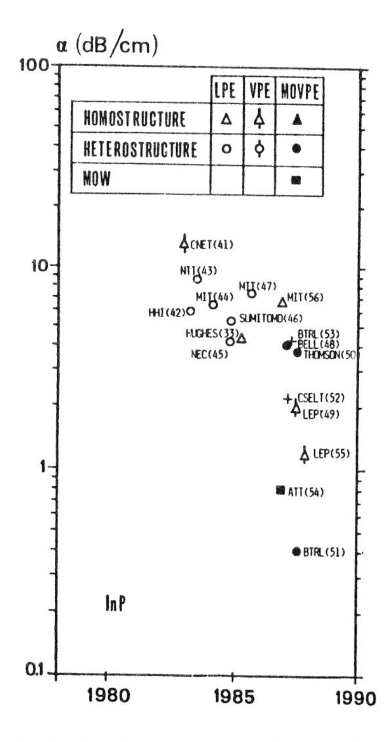

Figure 3-b : Published values of waveguide losses for InP based waveguides.

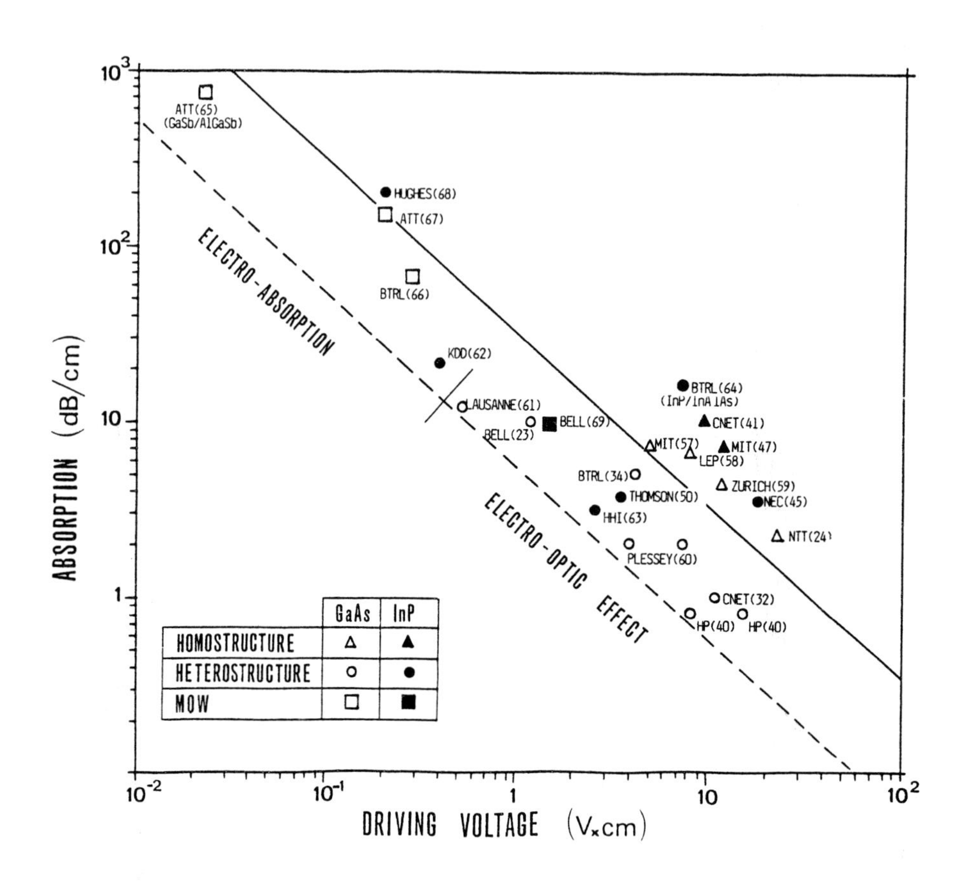

Figure 4 : Absorption versus normalized driving voltage for different active devices, including phase modulators, Mach-Zehnder interferometers and electro-absorption modulators. The full and the broken line calculated for αV = cst correspond to the best values obtained with homostructure and heterostructure devices respectively.

Paper presented at Int. Symp. GaAs and Related Compounds, Heraklion, Greece, 1987

Below band-gap photoresponse of undoped semi-insulating GaAs

U. Kaufmann

Fraunhofer Inst. für Angew. Festkörperphysik, 7800 Freiburg, W. Germany

ABSTRACT Electron-paramagnetic-resonance results for deep defects in undoped as-grown GaAs are summarized. They provide a sound basis for associating EL2 with As_{Ga} and reveal how carriers are exchanged between the various defects under photoexcitation. The As_{Ga}/EL2 mid-gap level is primarily responsible for the photoresponse of the material while other defects are important as trapping centers.

1. INTRODUCTION

If undoped semiinsulating (si) GaAs is illuminated with photons in the range $Eg/2 < h\nu < Eg$ it exhibits a pronounced photoresponse. This is primarily due to the As_{Ga} antisite related mid-gap level EL2. Other deep defects, however, play a role as well since they exchange electrons and holes with As_{Ga}/EL2 under photoexcitation. Depending on the photon energy and the illumination time the charge exchange can be either optically reversible or irreversible (persistent) at low temperatures. The latter effect can be traced back to the metastable properties of the As_{Ga}/EL2 center.

In the following the electron-paramagnetic-resonance (EPR) of defects present in undoped as-grown LEC GaAs is discussed. Their photoresponse is emphasized since it provides a sound basis for the identification of EL2 with As_{Ga} and since it allows to establish a model for the optically induced charge exchange among the various defects. The discussion includes other recent studies in which the optical response of the As_{Ga}/EL2 mid-gap level manifests clearly, too.

2. EPR SPECTRA IN UNDOPED LEC GaAs

If undoped si GaAs is cooled and measured in the dark one observes only the four-line hyperfine pattern of the As_{Ga}^{+} antisite, see Fig. 1. Several authors have confirmed that this defect is omnipresent in standard as-grown material with concentrations around 1×10^{16} cm^{-3}. It has often been claimed, see the references in the review by Dischler and Kaufmann (1987), that several microscopically slightly different As_{Ga} antisite species simultaneously contribute to the spectrum in Fig. 1. For as-grown material this claim is lacking any direct spectroscopic support. Actually, the revised interpretation of electron-nuclear-double-resonance (ENDOR) results for As_{Ga}^{+} (Meyer et al 1987b) suggests that a single As_{Ga} species gives rise to the spectrum in Fig. 1. Its exact structure is still controversial. The ENDOR data have been interpreted as evidence for an As_{Ga}-As_{int} complex. On the other hand the uniaxial stress splittings of the 1.039 eV $(As_{Ga}/EL2)^{\circ}$ zero-phonon-line (Kaminska et al 1985) are indicative of a center with cubic symmetry, i.e. the isolated As_{Ga}. For the sake of brevity we will speak of the As_{Ga} center but the above mentioned ambiguity should be kept in mind. The electronic wavefunction of As_{Ga}^{+} is S-like. Therefore the As_{Ga}^{+} EPR signal strongly saturates below 11 K (Wilkening et al 1987) and is still visible well above 20 K.

Four new EPR signals appear after photoexcitation. They have been labelled FR1, FR2, FR3 (Baeumler et al 1986a, Kaufmann et al 1986) and A_t (Bittebierre et al 1986) since their chemical origin and structure is uncertain. They can have high intensities and thus are likely related to intrinsic defects and/or major impurities like boron or carbon. The new signals are well visible only below 10 K and they do not saturate. This behavior is basically different from that of the S-like As_{Ga}^{+} donor resonance and indicates that the ground states of the new centers are p-like. This is evidence for their acceptor nature, as is further supported by their photoexcitation bands.

The EPR signal FR1, see trace (a) in Fig. 2 consists of several anisotropic lines between ≈3.0 kG and 4.2 kG which strongly overlap. This prevents a proper analysis but there is some evidence that the lines might be due to hyperfine coupling with two As nuclei. Baeumler et al (1986b) have speculated that the FR1 center might be created by a structural rearrangement of As_{Ga}^{o} under 1.2 eV excitation. For instance, one can imagine off-center motion of As_{Ga}^{o} and rebonding such that an off-center As_{Ga}-V_{Ga} vacancy pair is created, in direct analogy with the Si A-center (Bittebierre 1987).

The FR2 EPR signal, see trace (b) in Fig. 2 is composed of the lines FR2 and FR2'. They are isotropic with nominal g-factors of 2.1 and 5.0 respectively which is indicative of a center with cubic symmetry. At present a model cannot be offered. The A_t EPR signal consists of four highly anisotropic lines. Their angular dependence reveals that the corresponding defect has trigonal symmetry. A tentative assignment to a C_{As}-B_{Ga} complex has been made (Bittebierre et al 1986).

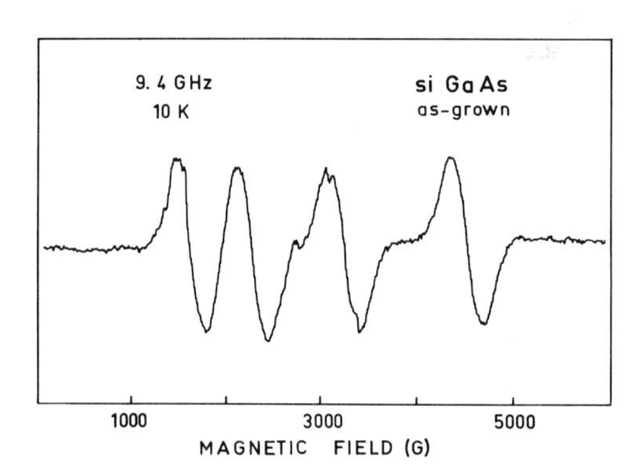

Fig. 1 EPR spectrum of As_{Ga}^{+} in standard as-grown GaAs

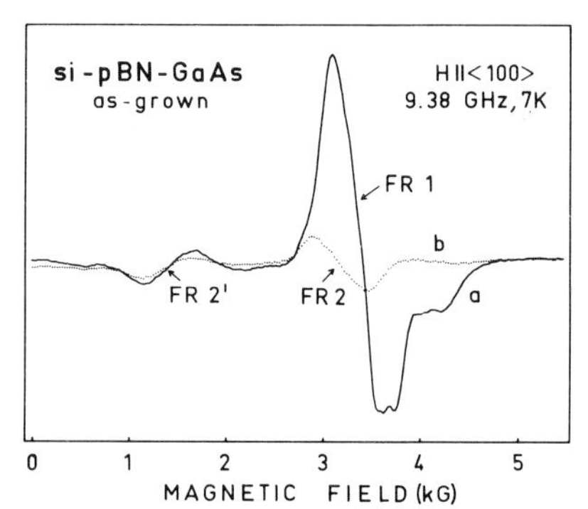

Fig. 2 EPR spectra of undoped GaAs following optical excitation with $h\nu$ = 1.2 eV for 8 min and (b) $h\nu$ = 0.9 eV for 30 s. (Baeumler et al 1986a)

Fig. 3 EPR spectra of the FR3 defect in LEC GaAs for three magnetic field orientations

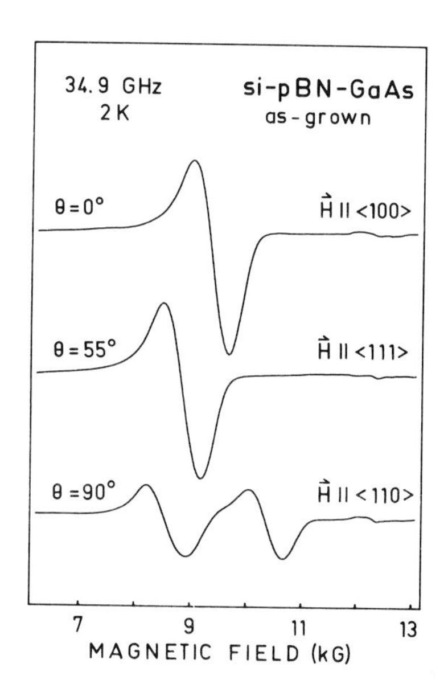

Spectra of the FR3 defect are shown in Fig. 3. It is omnipresent in LEC GaAs and very likely contains B. The angular dependence conclusively demonstrates that the FR3 center has trigonal symmetry. It has been tentatively assigned to a Ga_{As}-B_{Ga} antisite complex (Kaufmann et al 1986)

The structure of the new defects is poorly understood primarily because of the large EPR linewidths (500G - 1200 G) which mask hyperfine signatures. On the other hand, the new defects are definitely important for the compensation mechanism of undoped si GaAs whence their structure should be elucidated.

3. PHOTORESPONSE OF THE EPR SPECTRA

Optically induced EPR signal intensity changes result from removal of carriers or capture at the defect in question. The As_{Ga}^{+} photo EPR data have been reviewed by Dischler and Kaufmann (1987) and we recall only those results which pertain to the correlation between As_{Ga} and EL2. To establish this correlation the knowledge of the EL2 optical properties from photocapacitance studies (Chantre et al 1981, Vincent et al 1982) has been essential.

The first As_{Ga} donor level lies at mid-gap. Changes in the As_{Ga}^{+} EPR intensity result from the release of electrons and holes out of this level (Baeumler et al 1985, Tsukada et al 1985). The spectral dependence of the As_{Ga}^{+} enhancement (process $As_{Ga}^{o} + h\nu \rightarrow As_{Ga}^{+} + e_c$) shows an almost monotonic increase up to the band edge , see Fig. 4a, and resembles that of the EL2 electron ionization cross section $\sigma_n^{\circ}(h\nu)$. On the other hand the As_{Ga}^{+} quenching (process $As_{Ga}^{+} + h\nu \rightarrow As_{Ga}^{o} + h_v$) curve, see Fig. 4b, exhibits two bands peaked near 0.9 eV and 1.2 eV. Within the 0.9 eV band the quenching is optically reversible. Note that the shape of the band up to $\approx$1.0 eV looks very similar to the spectral shape of the EL2 hole ionization cross section $\sigma_p^{\circ}(h\nu)$. Within the 1.2 eV band the quenching is persistent, i.e. the As_{Ga}^{+} signal cannot be restored with light. This persistent As^+_{Ga} quenching band coincides with the EL2 persistent photocapacitance quenching band. Furthermore, within this band the As^+_{Ga} EPR signal intensity varies nonmonotonically with illumination time as it is the case for the EL2 photocapacitance transients. Thus the optical properties of As_{Ga} and EL2 are virtually the same. At present this is the most direct evidence that the EL2 mid-gap level and the As^+_{Ga} EPR originate from the same defect. This view is also supported by the intensity anticorrelation of the 1.039 eV EL2° zero-phonon-line with the As^+_{Ga} signal as observed by Tsukada et al (1986) and by studies of the optically detected EPR of As^+_{Ga} (Meyer et al 1987a).

While the intensity changes of the As^+_{Ga} EPR result from ionization processes at As_{Ga} those of the new EPR signals result primarily from trapping of carriers at the correponding defects. More specifically, the photo-EPR data indicate that the excitation of the new signals proceeds via trapping of holes, released by As^+_{Ga}, at negatively charged acceptors, compare the hole transfer in Fig. 6 (Baeumler et al 1986a, b, Bittebierre et al 1986, Tsukada et al 1985, 1986). For instance the FR1 and FR2 excitation bands coincide with the 0.9 eV and the 1.2 eV As^+_{Ga} quenching bands in Fig. 4b, respectively. No excitation curve is available for the A_t EPR signal but it is presumably similar to that of FR1.

Detailed photo-EPR data have been reported for the FR3 center and its deep acceptor nature has been confirmed (Kaufmann et al 1987). Excitation and quenching bands are shown in Fig. 5. Comparison with Fig. 4 reveals a very close correspondence between FR3 excitation and As^+_{Ga} quenching on one hand and between FR3 quenching and As^+_{Ga} excitation on the other hand. These data suggest the model sketched in Fig. 6. Photoneutralization of As^+_{Ga}, process σ_p°, and trapping of the hole at $FR3_{dia}$ quenches As^+_{Ga} and excites the FR3 EPR. Subsequent photoionization of As_{Ga}, process σ_n°, and trapping of the electron at $FR3_{para}$ excites As^+_{Ga} and quenches the FR3 EPR. Long-time

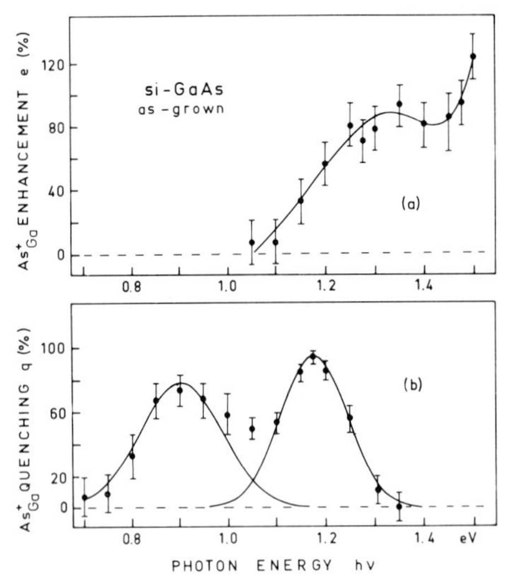

Fig. 4 Spectral dependence of the As^+_{Ga} signal (a) enhancement and (b) quenching. After Baeumler et al (1985)

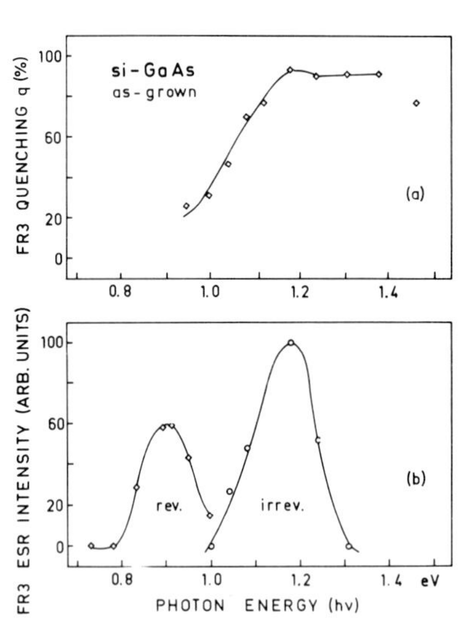

Fig. 5 Spectral dependence of the FR3 EPR signal (a) quenching and (b) enhancement. After Kaufmann et al (1987)

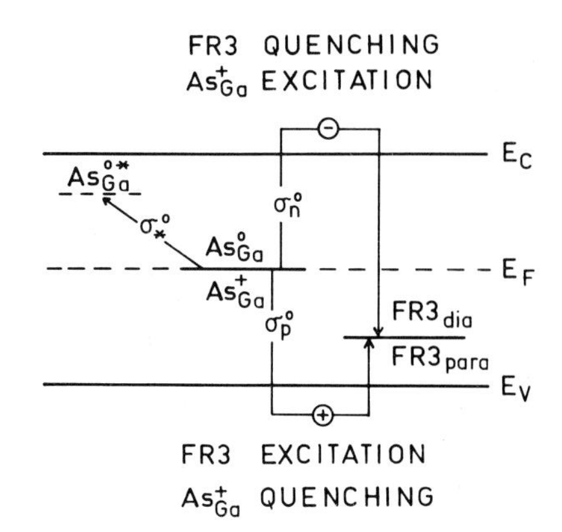

Fig. 6 Model explaining the optically induced intensity changes of the As_{Ga}^{+} and the FR3 EPR signals, see text, in si GaAs. The position of the deep FR3 acceptor level is not excatly known but is shallower than 0.52 eV. The position of the electrically and optically inactive As_{Ga}^{o*} state has no absolute meaning. Photoexcitation of the FR1, FR2 and A_t EPR signals very likely occurs via an analogous hole transfer. After Kaufmann et al (1987).

illumination within the 1.2 eV band of Fig. 5b leads to nonmonotonic FR3 EPR signal transients and to a persistent excitation of the FR3 EPR. These effects have been explained as a direct consequence of the $(As_{Ga}/EL2)^{\circ}$ normal state $\longrightarrow$ metastable state transition, process σ_*° in Fig. 6. Since this process leads to the disappearance of the As_{Ga} mid-gap level the electron source for process σ_n° ceases to exist. The $FR3_{para}$ state is then persistently excited.

4. OTHER OPTICAL MANIFESTATIONS OF THE As_{Ga}/EL2 MID-GAP LEVEL

A well known manifestation of the As_{Ga}/EL2 mid-gap level is its optical absorption band for energies between Eg/2 and Eg. Three optical transitions contribute to this band corresponding to the processes σ_n°, σ_p° and σ_*° in Fig. 6. Its most spectacular feature is its persistent bleaching under illumination with $h\nu$ around 1.2 eV. The photoresponse of the As_{Ga}/EL2 absorption band to secondary light has been reviewed by Dischler and Kaufmann (1987).

A variant of the conventional optical absorption is the magnetic circular dichroism (MCD) absorption technique. For as-grown GaAs the MCD observed by Meyer et al (1984) extends from $h\nu \approx Eg/2$ to Eg and the integrated spectrum consists of two bands peaked at 1.05 eV and 1.29 eV. It has been confirmed that this spectrum is entirely due to As_{Ga}^{+}. The nature of the optical transition giving rise to the As_{Ga} MCD is controversial. Originally, Meyer et al (1984) postulated two internal transitions of As_{Ga}^{+}. Alternatively, Kaufmann (1985) has suggested that process σ_p° in Fig. 6 gives rise to the As_{Ga}^{+} MCD spectrum.

Two photoluminescence (PL) bands peaked near 0.63 eV and 0.68 eV have often been reported for undoped GaAs but their origin remained controversial. The issue has been clarified recently by PL excitation spectroscopy between 0.75 eV and 2.0 eV (Tajima 1987, Tajima et al 1987). The excitation spectra of the 0.63 eV and the 0.68 eV band have the same shapes as the As_{Ga}/EL2 cross sections, $\sigma_n^{\circ}(h\nu)$ and $\sigma_p^{\circ}(h\nu)$ respectively. This strongly indicates that the 0.63 eV and the 0.68 eV band result from electron capture at $(As_{Ga}/EL2)^{+}$ and from hole capture at $(As_{Ga}/EL2)^{\circ}$, respectively.

It is finally noted that shallow acceptor neutralization in si GaAs, as observed by Wan and Bray (1985) in Raman scattering and by Wagner et al (1986) in absorption, is very likely related to a hole transfer process analogous to that in Fig. 6.

5. SUMMARY

Our present knowledge about the correlation of EL2 with As_{Ga} is based on magnetic resonance data (EPR, MCD-EPR, ENDOR). Other experimental techniques have not provided direct support for this correlation but much circumstantial evidence for its correctness has been accumulated. Among the many EL2 models that have been proposed only two alternatives namely the isolated As_{Ga} and the As_{Ga}- As_{int} complex are actually supported directly by experimental results. Ionization processes at the As_{Ga}/EL2 mid-gap level and the structural rearrangement of the (As_{Ga}/EL2)° state appear to be the principal sources for the below-gap photoresponse of undoped si GaAs, the second As_{Ga}/EL2 donor level (Dischler and Kaufmann 1987) apparently being of little importance in such material. Under photoexcitation the mid-gap level exchanges carriers with other defects, in particular with the deep acceptors revealed by EPR

ACKNOWLEDGEMENTS The author thanks M. Baeumler and W. Wilkening for active cooperation and J. Hornung, W. Jantz, and J. Schneider for a careful reading of the manuscript.

REFERENCES

Baeumler M, Kaufmann U, and Windscheif J 1985 Appl. Phys. Lett. 46 781-3
Baeumler M, Kaufmann U, and Windscheif J 1986a Semiinsulating III-V Materials (Tokyo, Ohmsha) pp 361-6 and Refs. therein
Baeumler M, Kaufmann U, Windscheif J, and Wilkening W 1986b Advanced Materials for Telecommunication (Les Ulis, Les Editions de Physique) pp 111-6
Bittebierre J, Cox RT, and Molva E 1986 Defects in Semiconductors (Switzerland, Trans Tech Publications) pp 365-370
Bittebierre J 1987 Thesis, Université de Grenoble
Chantre A, Vincent G, and Bois D 1981 Phys. Rev. B 23 5335-59
Dischler B, and Kaufmann U 1987 Revue Physique Appliquée in press and Refs. therein
Kaminska M, Skowronski M, and Kuszko W 1985 Phys. Rev. Lett. 55 2204-7
Kaufmann U 1985 Phys. Rev. Lett. 54 1332
Kaufmann U, Baeumler M, Windscheif J, and Wilkening W 1986 Appl. Phys. Lett. 49 1254-6
Kaufmann U, Wilkening W, and Baeumler M 1987 Phys. Rev. B in press
Meyer BK, Spaeth JM, and Scheffler M 1984 Phys. Rev. Lett. 52 851-4
Meyer BK, Hofmann DM, and Spaeth JM 1987a J. Phys. C 20 2445-51
Meyer BK, Hofmann DM, Niklas RJ, and Spaeth JM 1987b Phys. Rev. 36, 1332-35
Tajima M 1987 Japanese J. Appl. Phys. 26 L855-8
Tajima M, Iino T, and Ishida K 1987 Japanese J. Appl. Phys. 26 L1060-3
Tsukada N, Kikuta T, and Ishida K 1985 Japanese J. Appl. Phys. 24 L689-92
Tsukada N, Kikuta T, and Ishida K 1986 Phys. Rev. 33 8859-62
Vincent G, Bois D, and Chantre A 1982 J. Appl. Phys.53 3643-9
Wagner J, Seelewind H, and Koidl P 1986 Appl. Phys. Lett. 49 1080-2
Wan K, and Bray R 1985 Phys. Rev. B 32 5265-72
Wilkening W, Kaufmann U, and Baeumler M 1987 Phys. Rev. B submitted

Raman scattering investigations of GaAs/AlAs superlattices structure and interface broadening

Bernard Jusserand, Francois Alexandre and Rosette Azoulay
CNET, Laboratoire de Bagneux,
196 Avenue Henri Ravera, 92220 Bagneux, FRANCE.

Abstract: We review the structural information which can be extracted from Raman scattering on both folded acoustic and quantized optic vibrations in superlattices.

Introduction.

There has been in the past few years an increasing interest on Raman scattering studies of phonons in superlattices. Whereas the peculiar vibrational properties of these periodic structures have been first enlighted on the GaAs/AlAs system (see the review papers by Jusserand and Paquet (1986a), Klein (1986) and Jusserand and Cardona (1988)), similar studies have been recently devoted to Si/GeSi (Brugger et al 1988) and GaSb/AlSb (Schwartz et al 1988) superlattices and to amorphous Si/SiN_x (Santos et al 1987a) and Si/Ge (Santos et al 1987b) multilayers. As a fairly good understanding of their lattice dynamics is now obtained, at least along the growth axis, we will attempt in this paper to review the possible applications of vibrational scattering in characterizing the structure of the superlattices.

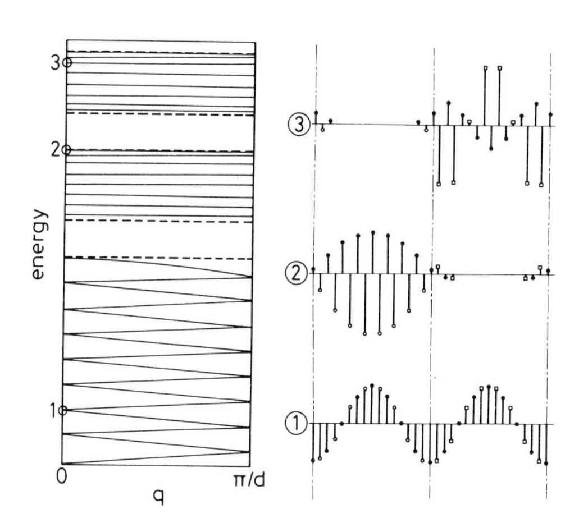

Fig.1: Axial dispersion curve and three typical eigendisplacements for a $(GaAs)_8/(AlAs)_8$ superlattice.

The main feature in the superlattice lattice dynamics is the coexistence of two types of new modes appearing at zone center and displaying very different behaviours. Typical eigendisplacements at zone center and the corresponding dispersion curve along the growth axis are illustrated on Fig.1. Low frequency new modes, build up from the acoustic phonon branches of the bulk constituents, propagate along the modulation axis due to the similarity in the acoustic properties of the bulk constituents. Their dispersion curves along the superlattice axis are understood as coming from the folding of an average bulk dispersion curve. High frequency new modes, build up from the optic phonon branches of the bulk constituants, can also propagate when the optic dispersion curves of the bulk constituants overlap. However such overlap does not often exist and the superlattice high frequency vibrations are usually confined in the layers of one of the bulk constituants. Their frequencies are quantized and display no axial dispersion.

Such a coexistence of confined and propagative vibrations makes of Raman scattering a powerful tool to characterize the structure of the superlattices. While propagative folded acoustic vibrations probe the long range order in the system and thus provide information on its periodicity in a rather similar way to X-Ray diffraction, confined optic phonons probe the local properties and thus provide original information on the individual layer nature, thickness and shape. We will consider successively these two probes in the rest of the paper.

Folded acoustic vibrations.

As first demonstrated by Colvard et al (1980), the folded acoustic vibrations can be very well described using an elastic approximation: they behave as elastic waves propagating in a periodically modulated continuum. Their dispersion curves are obtained first by folding an average bulk dispersion curve and second by opening small gaps at zone center and zone edge. Due to the great similarity in the elastic properties of the III-V compounds, the acoustic modulation is usually very small and can be neglected when describing the folded dispersion curves, excepted very close to zone center and zone edge. As first emphasized by Jusserand et al (1983), the wavevector of the phonon created by backscattering in not negligible with respect to the superlattice Brillouin zone extension, contrary to the case of bulk compounds. The backscattering frequencies, appearing as doublets (Fig.2), thus are unsensitive to the gaps openings and then directly reflect in a simple way the period d. As can be seen on Fig.2, the average frequency of the m-th doublet simply reads:

$$\omega_m = 2\pi m \frac{v}{d}$$

Thanks to the small difference between the sound velocities of the bulk constituants, which bracket the effective one v in the superlattice, a good estimation of d can be obtained from the measured ω_m's, though v is usually only poorly known.

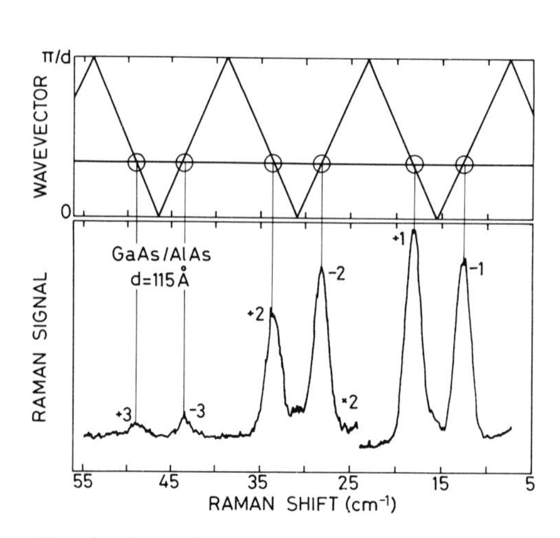

Fig.2: Low frequency Raman spectrum compared with the corresponding axial dispersion curve.

On the other hand, the backscattering frequencies are unable to provide any information on the inner structure of the supercell, i. e. for simple $(GaAs)_{d_1}/(AlAs)_{d_2}$ structures on the thickness ratio $\alpha = d_2/d$. A similar conclusion applies to X-Ray simple diffraction profiles on nearly lattice matched superlattices which only provide information on the periodicity from the splitting between the satellite peaks. Similarly again in both experiments, the inner structure of the supercell can be probed from the intensities of respectively the folded lines and the satellite peaks. In both cases, the modulation of the effect involved in the observation: the photoelastic one for Raman scattering (as first introduced by Klein et al (1984)) and the atomic scattering for X-Ray diffraction, is much larger than the intrinsic modulation of the corresponding sound velocity and lattice parameter.

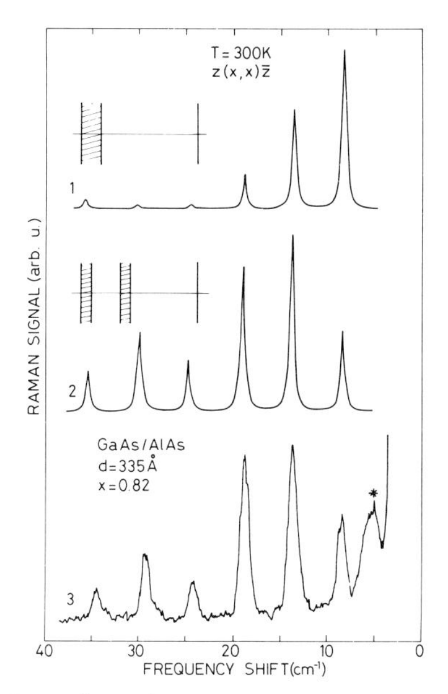

Fig.3 Low frequency Raman spectrum (3) compared with two calculated profiles (1 and 2) of same d and x and two different inner structures shown in inset (GaAs: shaded layer).

The high sensitivity of the folded acoustic intensities to the supercell structure is illustrated on Fig.3. The theoretical profiles (1 and 2) have been calculated taking into account both acoustic and photoelastic modulations (Jusserand et al 1987) for two different inner structures corresponding to identical period and average Aluminium concentration x. They both display a series of peaks of similar frequencies but different intensities. A good description of the Raman spectrum (3) obtained on a sample with the complex structure used for calculation 2 can be only well described when taking into account this structure.

Such sensitivity of the Raman intensities to the detail of the structures can be used for instance to determine the thickness ratio α in any given simple GaAs/AlAs superlattice. We recently showed (Jusserand et al 1987) that the relative intensity of the two components of the first doublet (lines -1 and +1 of Fig.2) is particularly sensitive to this parameter and that a quantitative determination of α can be obtained from this measurement.

The intensities we analysed up to now assuming piecewise constant profiles and abrupt interfaces are also sensitive to interface broadening. The intensity of the folded lines decreases with increasing interface broadening as reported by Colvard et al (1985) on a thermally annealed sample. We introduced a quantitative description (Jusserand et al 1985a) of the folded line intensities as a function of the interface broadening, using an error function profile and we thereby analysed we the line intensities on a series of identical GaAs/AlAs superlattices grown by Molecular Beam Epitaxy at different substrate temperatures. We got evidence of an increasing broadening with increasing growth temperature, in quantitative agreement with the analysis of the quantized frequencies of the optic vibrations, as will be presented in what follows. A similar decrease of intensity was also observed for the same series on the X-Ray diffraction satellites.

To summarize, as compared to X-Ray diffraction, Raman scattering provides similar but sometimes less direct and accurate information on the structure of the superlattices, through the analysis of the folded acoustic lines frequency and intensity. We will now show that it also provides new original informations through the analysis of the confined optic modes frequencies.

Quantized optic vibrations.

Contrary to the case of GaAs/GaAlAs structures where some overlap exist between the optic phonon bands (Jusserand et al 1984), the optic vibrations in GaAs/AlAs superlattices are strongly confined in either the GaAs or the AlAs layers. Their penetration depth in the barrier is usually less than one monolayer and real superlattice effect are unlikely to be observed on their frequencies. This feature justifies the approximate description of the confined optic modes as perfectly confined in an isolated thin slab. Their frequencies then are only dependent on the vibrationnal properties of the corresponding bulk compound and not at all on those of the barrier constituant. From a careful analysis of the boundary conditions at the interfaces, we demonstrated (Jusserand et al 1986b) that the frequency of the m-th confined mode is equal to the bulk frequency at the following wavevector:

$$k_1 = \frac{m\pi}{(n_1+1)a}$$

where n_1 is the number of monolayers (of thickness a_1) in the considered layer. These effective wavevectors correspond to the successive harmonics of a vibrating string of length d_1+a_1. The small difference between the length of the string and the thickness of the layer can be easily understood by considering the atomic displacements around the interface. This is actually the first cation in the barrier and not the interface Arsenic which is pinned. This difference which arises from the microscopic nature of the problem, becomes only important for thin layers.

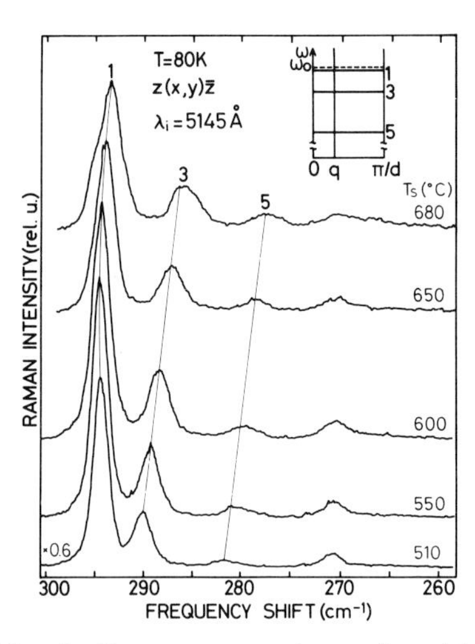

Fig.4 Raman spectra in the GaAs-type optic frequency range on GaAs(25A)/AlAs(15A) superlattices grown by MBE at different substrate temperature T_S.

The slab approximation has been checked by different groups (see Wang et al (1987) and references therein) for GaAs layers, using the bulk GaAs dispersion curve determined by neutron scattering (Waugh and Dolling 1963). In consideration of the moderate accuracy in the layer thickness determinations, a correct agreement has been obtained. It is thus now possible to determine conversely from the frequency of the confined modes the thickness of the corresponding GaAs layers. Moreover, one disposes of several different probes associated to the different quantized levels. In the case of abrupt wells, the deduced thickness must be the same for each of them whereas for a graded well the effective thickness seen by the m-th mode increases with increasing m. These features are illustrated on Fig.4 for the series of MBE samples whose folded line intensities have been previously considered. The Raman spectrum obtained on the low temperature sample can be well fitted assuming a rectangular well and thus no significant interface broadening. As the growth temperature is increased, the confined modes shift towards lower frequencies as a result of the shrinkage of the effective well thickness they probe. Moreover, the high temperature spectra cannot be described using a single thickness but assuming a graded profile. An error function profile allows to correctly fit the three quantized frequencies with the same broadening parameter. The fitted parameter is found to increase with increasing substrate temperature. Its values (for instance 6A at 650C) cannot be explained by a bare thermal interdiffusion during the growth as the diffusion coefficients are extremely low at these temperatures. They thus must be related to growth kinetics. On the other hand, a similar analysis has been applied recently to structures annealed at 850C (Shu-lin Zhang et al 1986). The shift of the confined modes could be very well explained using our model and the broadening parameters deduced from the X-Ray diffraction intensities.

The interface broadening effect on the quantized frequencies can be described qualitatively from the comparison of the energy splitting between the successive confined modes. In a rectangular well, the successive quantized levels ω_m follow a quadratic progression and for instance:

$$\frac{\omega_5-\omega_3}{\omega_3-\omega_1} \equiv \rho = 2$$

In a parabolic well, they follow a linear progression and $\rho=1$. As clearly appears on Fig.4, ρ decreases from nearly 2 to nearly 1 with increasing temperature as a result of the increasing interface broadening. This simple analysis does not require the knowledge of the corresponding bulk dispersion curve which is only assumed to be parabolic. It can then be easily applied to any given superlattice, provided three different quantized levels are observed and indexed.

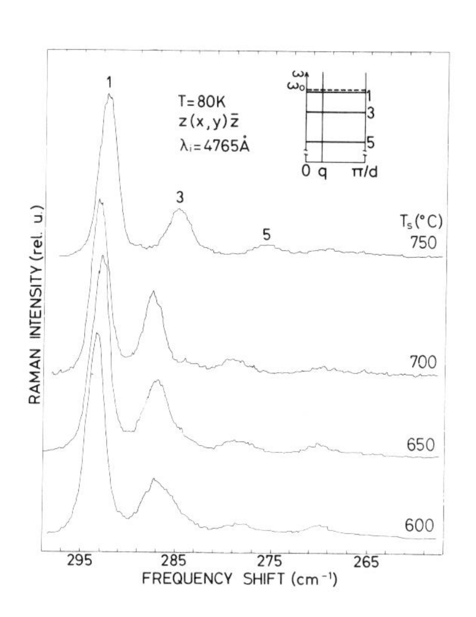

Fig.5: Same as Fig.4 on structures grown by OMVPE.

Such an analysis has been for instance applied (Azoulay et 1986) to GaAs/AlAs superlattices grown by Organometallic Vapor Phase Epitaxy in various conditions. As illustrated on Fig.5, low interface broadening (2 monolayers) can be achieved and no significant dependence on the substrate temperature is observed between 600C and 750C.

The major limitation in such individual layer profile characterization is related to the fact that confinement effects on optic phonons become difficult to observe for thicknesses larger than 70A. Thicker layers behave for optic phonons as bulk samples. This limitation however allows to analyse the intrinsic properties of such layers: their composition in the case of alloys or their strain state in the case of strained layer superlattices.

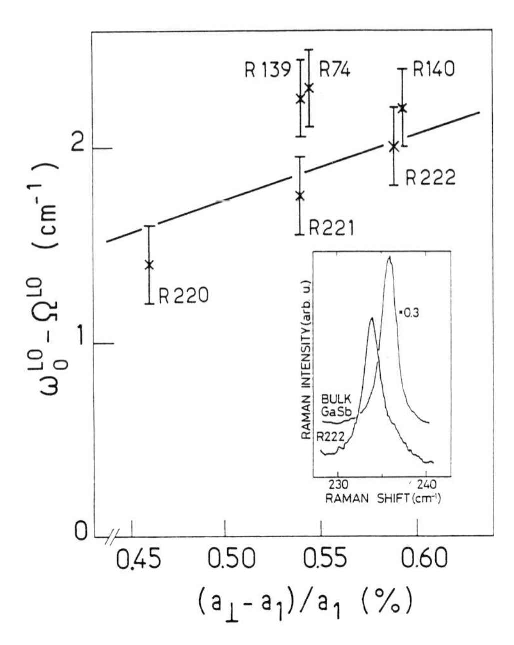

Fig.6: Shift of the GaSb-type LO line in GaSb/AlSb superlattices plotted as a function of the axial strain and compared with the predictions of the elastic theory.

Such quantitative determination of the strain has been applied to various systems: Si/SiGe (Cerdeira et al 1984), GaSb/AlSb (Jusserand et al 1985b), ZnTe/ZnSe (Nakashima et al 1986), CdTe/ZnTe (Menendez et al 1987), ZnTe/ZnS (Le Hong Shon et al 1987). We illustrate these studies on Fig.6 in the case of GaSb/AlSb superlattices. These structures, grown by MBE on a GaAs substrate and a thick AlSb buffer layer, are assumed to be relaxed as a whole with respect to the substrate. The strain state of the layers thus depends on the relative thickness of the GaSb and AlSb constituting layers. In the samples considered on Fig.6, the AlSb layers are always much thicker than the GaSb ones. The lattice mismatch is then mainly accomodated in GaSb, thus inducing a significant shift of the LO line from its bare position. Moreover, some dependence on the sample parameter has been demonstrated and is in good agreement with the theory based on the bulk GaSb deformation coefficients.

Conclusion.

We have shown through various examples that Raman scattering on vibrations in superlattices is a powerful tool to probe the structure of the studied samples.

Laboratoire de Bagneux is 'Unité associée au CNRS' (UA 250).

References.

Azoulay R, Jusserand B, Le Roux G, Ossart P and Dugrand L 1986 J.Cryst.Growth77 546

Brugger H and Abstreiter G 1988 to appear in J.Phys(Paris)

Cerdeira F, Pinczuk A, Bean JC, Batlogg B and Wilson BA 1984 Appl.Phys.Lett.45 1138

Colvard C, Merlin R, Klein MV and Gossard AC 1980 Phys.Rev.Lett.45 298

Colvard C, Gant TA, Klein MV, Merlin R, Fischer R, Morkoc H and Gossard AC 1985 Phys.Rev.B31 2080

Jusserand B, Paquet D, Regreny A and Kervarec J 1983 Solid State Comm.48 499

Jusserand B, Paquet D and Regreny A 1984 Phys.Rev.B30 6245

Jusserand B, Alexandre F, Paquet D and Le Roux G 1985a Appl.Phys.Lett.47 301

Jusserand B, Voisin P, Voos M, Chang LL, Mendez EE and Esaki L 1985b Appl.Phys.Lett.46 678

Jusserand B and Paquet D 1986a Heterojunctions and Semiconductor Superlattices edited by Allan G et al (Springer Heidelberg) 108

Jusserand B and Paquet D 1986b Phys.Rev.Lett.56 1753
Jusserand B, Paquet D, Mollot F, Alexandre F and Le Roux G 1987 Phys.Rev.B35 2808
Jusserand B and Cardona M 1988 to appear in Light Scattering in Solids V edited by Cardona M and Guntherodt G (Springer Heidelberg)
Klein MV, Colvard C, Fischer R and Morkoc H 1984 J.Phys.(Paris)45 C5 131
Klein MV 1986 IEEE J.Quant.Elec.22 1760
Le Hong Shon, Inoue K and Murase K 1987 Solid State Comm.62 621
Menendez J, Pinczuk A, Valladares JP, Feldman RD and Austin RF 1987 Appl.Phys.Lett.50 1101
Nakashima S, Nakakura Y, Fujiyasu H and Mochikuzi K 1986 Appl.Phys.Lett.48 236
Santos PV, Ley L, Mebert J and Koblinger O 1987 to appear in Phys.Rev.B36
Santos PV and Ley L to appear in Phys.Rev.B36
Schwartz GP, Gualtieri GJ, Sunder WA and Farrow LA 1988 to appear in Superlattices and Microstructures
Shu lin Zhang, Levi DH, Gant TA, Klein MV, Klem J and Morkoc H 1986 in Proceedings of the Tenth International Conference on Raman Spectroscopy edited by Peticolas W and Hudson B (University of Oregon Printing Department Eugene) 9-4
Wang ZP, Jiang DS and Ploog K 1987 to appear in Solid State Comm.
Waugh JLT and Dolling G 1963 Phys.Rev.132 2410

Paper presented at Int. Symp. GaAs and Related Compounds, Heraklion, Greece, 1987

Thermally stable, low resistance ohmic contacts to n-type GaAs

Masanori Murakami, Yih-Cheng Shih, W. H. Price, N. Braslau
IBM Thomas J. Watson Research Center, Yorktown Heights, New York 10598, USA

K. D. Childs and C. C. Parks
IBM East Fishkill Product Assurance, Hopewell Junction, New York 12533, USA

ABSTRACT : A family of thermally stable, low resistance ohmic contacts to n-type GaAs has been developed by depositing an extremely thin In layer with refractory metal layers and annealing at high temperatures. The penetration depth of the contact metals to GaAs was shallow and the surface morphology of these contacts was superior to that of the conventionally used AuNiGe contacts. The low resistances are believed to be due to $In_xGa_{1-x}As$ phases which were observed at the metal/GaAs interfaces. The total area of these ternary phases in contact with the GaAs was found to be a key parameter controlling the contact resistance. An ideal microstructure for thermally stable In-based ohmic contacts is proposed.

1. INTRODUCTION

Thermally stable, low resistance contacts are required for high speed GaAs integrated circuits. Some self-aligned devices require heating above 800°C for a short time to activate implants after contact metal deposition. In addition, the devices are subjected to several temperature cycles between room temperature and 400°C after ohmic contact formation. The AuNiGe system, which is the most extensively used ohmic contact metal to n-type GaAs, cannot withstand such temperature cycles. Deterioration of the lithographic profile at the film edges (Shih et al, 1987) and changes in the microstructure and the R_c value during annealing at 400°C (Callegari et al,1985) were observed, because the β-AuGa phase with a low melting point of 375°C exists in the contact metal (Murakami et al, 1986). Thus, development of a new contact metal with improved thermal stability is required.

This paper reports the development of thermally stable, low resistance ohmic contact metals which can withstand both high and low temperature cycles. First, we analysed the MoGeW contact, originally developed by Tiwari et al (1983), which was the only known refractory ohmic contact. We correlated the electrical properties and the microstructure in order to find a key parameter which controls the electrical properties. This study then led to the development of various thermally stable ohmic contact metals by adding a small amount of In to refractory metals. Development of these In-based contacts will be described below.

2. EXPERIMENTAL PROCEDURES

Conducting channels were formed on (100) oriented semi-insulating GaAs substrates by implanting SiF^+ ions at 100 KeV or by implanting twice at 100 and 50 KeV at a dose of 3.5×10^{13} atoms/cm^2 through a photoresist stencil. The implanted substrates were activated by annealing at 800°C for 10 min in an arsine atmosphere. Before metal deposition, the vacuum system was pumped down to $\sim 8\times10^{-6}$ Pa. Various contact metals of MoGeW, MoGeInW, GeInW and NiInW were evaporated using a RF induction or electron beam heating. Deposition sequences are given in the second column of Table 1 where a sign "/" indicates the sequential deposition and "-" between two elements indicates the coevaporation of the two metals. The thicknesses of the top W layers are ~ 30 nm and the total thicknesses of the contact metals are ~ 50 nm. Some of the samples were covered by 50 nm thick Si_3N_4 layers. These samples were annealed at high temperatures and the annealing methods are given in the third column of Table 1, where FA and RTA mean furnace and rapid thermal annealing, respectively.

Table 1. Deposition sequences and annealing methods for various contact metals.

Contact metal	Deposition sequence	Annealing method
MoGeW	Ge/Mo/Ge/W	InAs overpressure
MoGeInW	Ge/In/Mo/W	FA or RTA
GeInW	Ge/In/W, Ge/Ge-In/W	RTA
NiInW	Ni/Ni-In/Ni/W	RTA

3. EXPERIMENTAL RESULTS

3.1. MoGeW contact metal

The MoGeW ohmic contacts were previously prepared by annealing in an InAs overpressure (Tiwari et al, 1983). I-V measurements were carried out for the MoGeW samples which were annealed at high temperatures by FA with an InAs overpressure, arsine atmosphere or Si_3N_4 caps, or by RTA with the caps. Ohmic behavior was observed only when the samples were annealed in an InAs overpressure.

The contact resistances (R_c) of the MoGeW ohmic contacts, prepared by annealing in an InAs overpressure, were measured by transmission line method (TLM). The R_c values lower than 0.5 Ω-mm were obtained for samples with a Mo/Ge thickness ratio in the range 0.6 to 1.3 and annealed at around 800°C. The lowest mean R_c value obtained in our experiments was 0.3 Ω-mm. No changes in the microstructure and the R_c values were observed after annealing the contacts at 400°C for more than 100 hours (Murakami et al, 1987b).

The distribution of each element normal to the film surface was measured by SIMS for the samples which were annealed by the various methods. The depth profiles of Mo, Ge, W, Ga and As were found to be similar for samples annealed at the same temperature by the various methods. However, a significant difference among these samples is the presence of In at the metal/GaAs interface for samples annealed in an InAs overpressure. A comparison of the electrical behavior with the SIMS analysis shows that ohmic behavior is observed only when a small amount of In was detected at the metal/GaAs interfaces.

3.2. MoGeInW contact metal

As described above a small amount of In impurity was able to convert the MoGeW contacts from Schottky to ohmic behavior. Thus, a thin layer of In was directly added to the MoGeW contacts during deposition and the MoGeInW contacts were prepared (Murakami et al, 1987c).

The R_c values of the MoGeInW metals with single or double implant conducting channels were measured as a function of In layer thickness (1.5, 2 and 3 nm), keeping both the Ge and Mo layers fixed. The lower R_c values were obtained for the contacts with the double implant conducting channels, annealed by the RTA method. The contact resistances of ~ 0.5 Ω-mm were obtained in the temperature range of 900 to 960°C for samples with an In layer thickness of 2 nm, which is the lowest value among all the samples. No change in the R_c values was observed during subsequent annealing at 400°C for 100 hrs.

Auger sputter-depth profiles were obtained for a sample with a 2 nm In layer that had been annealed at 880°C. The In intensity peak was observed at the metal/GaAs interfaces. A comparison between samples annealed by FA or RTA suggests that there is more In remaining at the interface for the sample annealed by RTA. TEM microstructural analysis was carried out for the contact with 2 nm In layer, annealed at 920°C. The $In_xGa_{1-x}As$ phases were observed locally at the metal/GaAs interfaces. The typical size of the $In_xGa_{1-x}As$ phases is 80 nm in length and 35 nm in depth along the interface. This size is larger by a factor of about two than that of the $In_xGa_{1-x}As$ phases observed in the contact annealed by FA. The chemical composition of the $In_xGa_{1-x}As$ phase is close to $In_{0.2}Ga_{0.8}As$. From the present microstructural analysis, it is concluded that lower resistance is obtained by increasing the total area of $In_xGa_{1-x}As$ phases in contact with the GaAs.

3.3. GeInW contact metal

Because of the complexity of the quaternary metallization, attempts were made to obtain similar results with ternary metals. Low contact resistances were obtained in the GeInW contact. The effect of the In deposition method on the R_c values was studied using this contact system. Reduction of the R_c values by a factor of ~ two was obtained for the contacts when In was coevaporated with Ge.

Thermal stability of the Ge/Ge-In/W contacts during isothermal annealings at 400°C was investigated. The R_c values started to increase after 3 hrs annealing at 400°C and increased by ~ 0.1 Ω-mm after 100 hrs annealing. Thermal stability for the GeInW contacts is worse than that of the MoGeInW contacts, which is believed to be lack of the high melting point Mo_5As_4 and $Mo_{13}Ge_{23}$ compounds in the GeInW contact metals.

From the GeInW contact experiment, it was learned that uniform In distribution in the as-deposited sample is needed to obtain low R_c values and that thermal stability after contact formation improves if high melting point compounds exist in the contact metals.

3.4. NiInW contact metal

The studies of the above three contact metals have led to development of a new NiInW contact metal which will be applicable to various GaAs devices (Murakami et al, 1987a). The R_c values of the NiInW contacts are shown in Fig. 1 as a function of annealing temperatures. The average values of ~ 0.3Ω-mm were obtained in the temperature range of 800 to 1000°C. It is noted that the temperature dependence of the R_c values is similar for both samples with single and double implant channels.

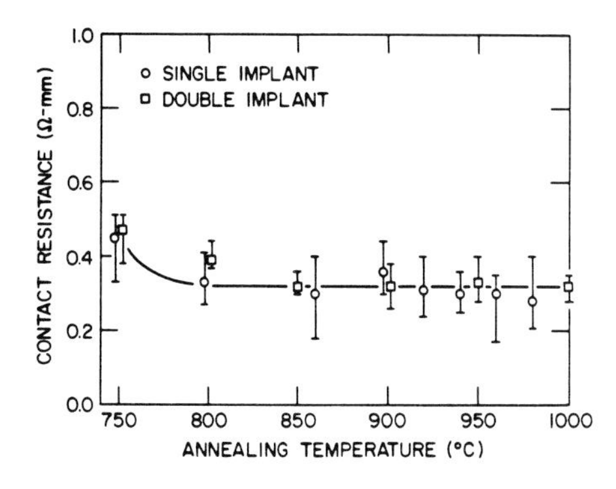

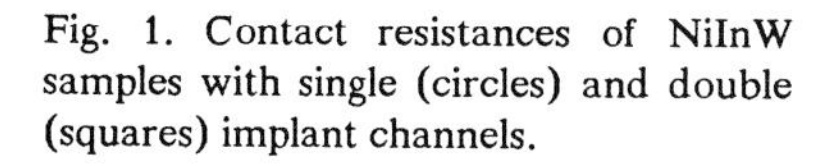
Fig. 1. Contact resistances of NiInW samples with single (circles) and double (squares) implant channels.

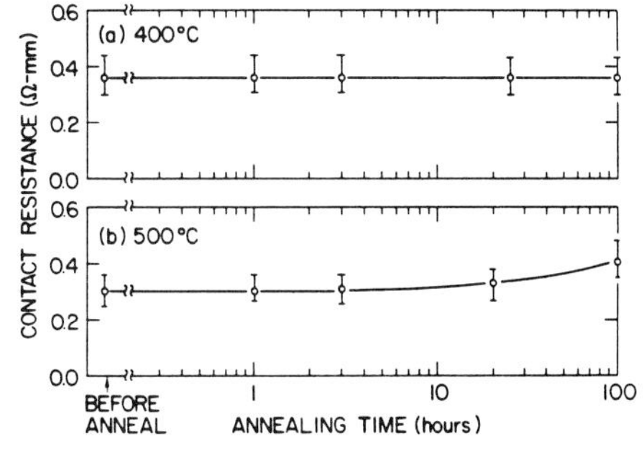

Fig. 2. Contact resistances of NiInW samples during isothermal annealing at 400 (a) and 500°C (b).

Thermal stability of these contacts during subsequent annealings at 400 or 500°C after contact formation was studied. The R_c values during isothermal annealings are shown in Fig. 2 for the contacts, pre-heated at temperatures above 900°C. No change in the R_c values was observed after annealing at 400°C for 100 hrs. During annealing at 500°C the R_c values were constant for 4 hrs and increased by only ~ 0.1Ω -mm after annealing for 100 hrs.

The interfacial microstructure of the NiInW contacts was observed by cross-sectional TEM. After annealing the NiInW contact at 900°C, formation of ternary $In_xGa_{1-x}As$ phases and intermetallic compounds was observed. The typical size of this ternary phase is ~0.3μm in length along the GaAs interface and its depth is ~50 nm. However, the $In_xGa_{1-x}As$ phases are still discontinuous. These ternary phases were found to grow epitaxially on the GaAs substrate. About 2% lattice mismatch was measured by the selective area diffraction. The Ni_3In compounds with typical size of ~40 nm were observed close to the W layer. The TEM micrograph focused on the edge of the NiInW contact showed no deterioration of the edge profile after annealing at 900°C.

4. DISCUSSION

4.1. Comparison between AuNiGe and In-based ohmic contacts

The NiInW contact, which is the best In-based contact among the four contacts studied in the present experiment, is compared with the AuNiGe contact. The results obtained for these contacts are summarized in Table 2. The R_c value of ~ 0.1 Ω-mm is routinely obtained for the AuNiGe contact (Callegari et al, 1985) and this value is lower than that of the NiInW contact by a factor of ~ 3. The R_c value measured by TLM is strongly influenced by the sheet resistance (R_S) of the contact metal (Marlow and Das, 1982) which was 1 and 11 Ω/□ for the AuNiGe and NiInW contacts, respectively. The reduction of R_S of the NiInW contact was achieved by depositing thick (0.2 μm) low-resistance material on top of the contact metal. The R_c values were reduced by ~ 0.1 Ω-mm which was the exact value expected to be reduced in this contact from the theory developed Marlow and Das. This R_c value of the NiInW contact with thick top layer is given in the forth row.

Table 2. Comparison between AuNiGe and NiInW contacts

	AuNiGe	NiInW
Thickness of contact metal	0.2 μm	50 nm
Mean low R_c w/o thick top layer	0.1 Ω-mm	0.3 Ω-mm
Mean low R_c with thick top layer	***	0.2 Ω-mm
Annealing temperature	420-550°C	800-1000°C
R_c after 10 hr at 400°C w/o top layer	0.5 Ω-mm	0.3 Ω-mm
R_c after 10 hr at 400°C with top layer	***	0.2 Ω-mm
Surface morphology	rough	smooth
Edge profile deterioration	0.2-0.5 μm	20 nm
Penetration depth of contact metal	0.2 μm	70 nm

The AuNiGe contact was formed by annealing at ~ 440°C and the range of the annealing temperatures to yield such low R_c values is ΔT ~ 130°C. The NiInW contact is annealed at much higher temperatures. This high temperature annealing is compatible with annealing temperatures required to activate implanted donors and acceptors after ohmic contact formation. The NiInW contact has also a wide annealing temperature range of ΔT ~ 200°C.

The contact resistances measured after annealing for 10 hrs at 400°C after contact formation are given in the sixth row. The AuNiGe contacts are stable after annealing for 4 hrs at this temperature and increased to 0.4(± 0.2) Ω-mm after 5 hrs (Callegari et al, 1985). The R_c values of the NiInW contact with or without the thick top layer did not change after 100 hrs. It is noted that the R_c value of the NiInW contact is lower by a factor of ~ two than that of the AuNiGe contact after 400°C annealing. The NiInW contacts certainly far exceed the requirement of the current device fabrication process for thermal stability after ohmic contact formation. Also, the surface morphology, contact edge profile, and penetration depth of the contact metals to GaAs observed by SEM and TEM after contact formation are given in this table. Comparing between the two contact metals, the microstructure of the NiInW contact is superior to that of the AuNiGe contact.

4.2. Correlation between the contact resistance and the $In_xGa_{1-x}As$ phase

In the present experiment, one of the key parameters which influence the R_c values is believed to be the total area of the $In_xGa_{1-x}As$ phases in contact with the GaAs. The R_c values of the contacts containing a small amount of In are plotted in Fig. 3 as a function of percentage of the GaAs interface covered by the $In_xGa_{1-x}As$ phases. The percentages (in a linear scale) were estimated from the cross-sectional TEM micrographs. Since the thicknesses of these contact metals are thin, their metal sheet resistances (R_S) were large. The contribution of the R_S to the measured R_c was calculated for each contact, substracted from the measured R_c values, and the corrected R_c values are plotted by open symbols in Fig. 3. The R_c value indicated by a closed circle in Fig. 3 is the one measured for the NiInW contact with thick top layer. The contacts were annealed by the RTA method except the MoGeInW contact with R_c = 1.2 Ω-mm. It is noted that the R_c values fall on one curve, although the contact metallugy and annealing method are different. This figure provides conclusive evidence that the total area of the $In_xGa_{1-x}As$ phases in contact with the GaAs is one of the key parameters to control the R_c values. The other parameters which also influence the R_c values will be described elsewhere.

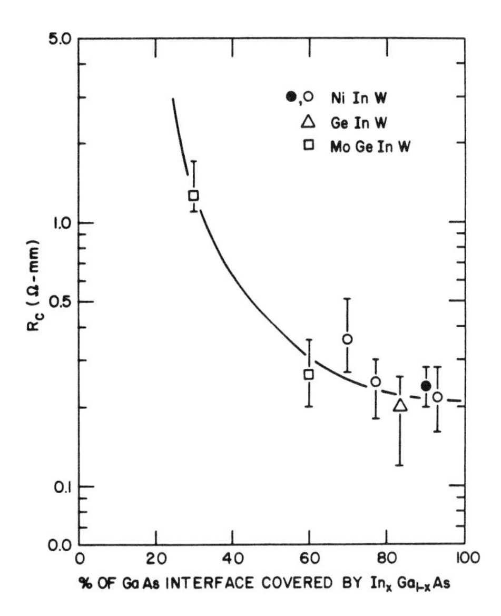

Fig. 3. Contact resistances plotted as a function of percentage of the GaAs interface covered by the $In_xGa_{1-x}As$ phases.

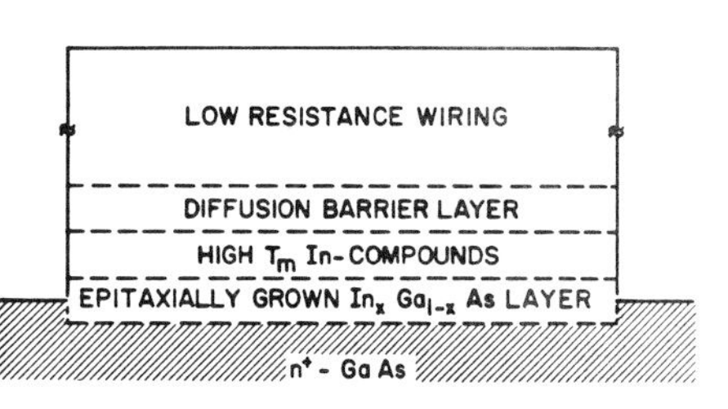

Fig. 4. Cross-section of an ideal, thermally stable In-based ohmic contact.

4.3. Ideal In-based ohmic contact

Cross-section of the microstructure for an "ideal", thermally stable, low resistance In-based ohmic contact is schematically shown in Fig. 4. The $In_xGa_{1-x}As$ layer primarily contributes to reduce the barrier height between the metal and GaAs without forming the *graded* $In_xGa_{1-x}As$ layer as proposed by Woodall et al (1981). Most of the In-based ohmic contacts with thick In layers are thermally unstable, simply because unreacted In which has a low melting point 156°C remains in the contacts. Thermal stability improves significantly when all of the In has formed the $In_xGa_{1-x}As$ phases and high melting point (T_m) compounds with the added metal. In order to avoid unreacted In left in the contact after annealing, amounts of In added to the contact metal should not exceed those needed to form $In_xGa_{1-x}As$ phases and the In compounds. Metals, such as Pd, Pt, Mn, or Co, which also form high melting point intermetallic compounds with In are good candidates for the contact metal to form the thermally stable In-based ohmic contacts. The refractory layer on top of the In compound layer as a diffusion barrier allows flexibility in choosing low resistance material for the wiring.

5. SUMMARY

Several thermally stable, low resistance ohmic contacts have been developed by depositing a small amount of In. The reduction of the R_c values was found to be due to formation of the $In_xGa_{1-x}As$ phases at the metal/GaAs interfaces. The R_c value decreased with increasing total area of the $In_xGa_{1-x}As$ phases in contact with the GaAs. Thermal stability after contact formation was found to improve significantly by adding a metal which forms high melting point compounds with In. The NiInW contact demonstrated ability to withstand high temperature cycles up to 1000°C for a short time and subsequent annealing at 400°C for 100 hrs without deteriorating surface morphology and contact edge profiles. This thermal stability is more than adequate for the current device fabrication process steps.

Acknowledgement

The authors would like to acknowledge valuable discussions with T. N. Theis, T. N. Jackson, and S. L. Wright and the technical assistance of H. J. Hovel, E. L. Willkie, A. T. Pomerene, P. D. Hoh, M. Albert, J. F. DeGelormo, J. W. Mitchell, V. Tom and M. Su.

References

Callegari A, Pan E T S, and Murakami M, 1985 Appl. Phys. Lett. **46** 1141.
Marlow G S and Das M B, 1982 Solid-State Electron. **25** 91.
Murakami M, Childs K D, Baker J M, and Callegari A, 1986 J. Vac. Sci. Technol. **B4** 903.
Murakami M and Price W H, 1987a Appl. Phys. Lett. **51** 664.
Murakami M, Price W H, Shih Y C, Child K D, Furman B K, and Tiwari S, 1987b J. Appl. Phys. (in press).
Murakami M, Price W H, Shih Y C, Braslau N, Childs K D, and Parks C C, 1987c J. Appl. Phys. (in press).
Shih Y C, Murakami M, and Callegari A, 1987, J. Appl. Phys. **62** 582.
Tiwari S, Kuan T S, and Tierney E, 1983 Proc. of IEDM.
Woodall J M, Freeouf J L, Pettit G D, Jackson T and Kirchner P, 1981 J. Vac. Sci. Technol. **19** 626.

ESR study of annealing behaviour of AS_{Ga} in LEC-GaAs

T.Kazuno, Y.Takatsuka, K.Satoh, K.Chino and Y.Chiba*

Electronics Materials Laboratory, Sumitomo Metal Mining
1-6-1 Suehiro-cho, Ohme-shi, Tokyo 198, JAPAN
*Department of Applied Physics, Science University of Tokyo
Kagurazaka, Shinjuku-ku, Tokyo 162, JAPAN

Abstract: Behavior of As_{Ga} antisite defect in semi-insulating LEC-GaAs upon annealing at 1150-1200 °C under various As pressures has been investigated, using the ESR technique. All the samples exhibit p-type conversion with resistivities of the order of 10^4 to 10^7 ohm-cm. Hall measurements indicate generation of around 0.4 eV acceptor level during annealing. Quantitative analyses of Fermi level relocation and ESR signal intensity reveal that EL2 or As_{Ga} concentration ranges from 0.6 to 2.0 $x10^{15}$ cm^{-3} after annealing, depending on both annealing temperature and As pressure. The results are successfully discussed by introduction of the deep acceptor level.

1. INTRODUCTION

Ingot annealing technique of semi-insulating(SI) GaAs crystal has widely been applied to improve uniformity of electrical characteristics of the crystals for ICs. The uniformity is strongly related with the distribution of EL2 (K.Lohnert et al., 1986). Thermal behavior of EL2 has been investigated using DLTS and near infrared absorption measurements (K.Lohnert et al.,1986, J.Lagowski et al.,1986).

This paper describes the direct observation of the annealing behavior of As_{Ga} antisite defect, using the electron spin resonance (ESR) technique. Analyses are carried out based on identification of EL2 with As_{Ga} or its complex. An additional deep acceptor is found to be generated during annealing and play an important role in determining carrier densities in GaAs after annealing.

2. EXPERIMENTAL

The present study of heat-treatment was carried out using

undoped SI(lightly n-type) GaAs crystals with resistivities of about $5x10^7$ ohm-cm grown by the liquid encapsulated Czochralski (LEC) method. The melt composition As/(Ga+As) were in the range from 0.50 to 0.51.

The samples for annealing had dimentions of $5x5x25mm^3$. The samples and the As metal were placed in a high grade quartz ampoule, which was evacuated to about $3x10^{-7}$ Torr and sealed off and set in a two-zone furnace. The samples were annealed for 16 hours at 1150 or 1200°C. The lower temperature zone (500-610°C) was used to control the As vapour pressure. As pressure range was from 50 to 1000 Torr. Each sample was quenched to room temperature within several seconds immediately after annealing by inserting the ampoule into the water. The samples were rapped and etched to dimension of $4x4x24$ mm^3 after annealing.

ESR measurements were performed at 9.2GHz and 4.2K in a cylindrical TE 110 mode cavity. The ESR signal intensity was corrected with the sample weight and Q factor. Resistivities and carrier concentrations were measured over the temperature range 300-400 K using the conventional Van der Pauw method.

3. RESULTS and DISCUSSION

Typical ESR spectra of SI GaAs samples annealed at 1150°C with **different** As pressures are shown in Fig.1 together with a spectrum of an as grown sample. The latter consists of a set of four closely spaced lines and is consistent with a spectrum identified with As_{Ga}(R.J.Wagner et al.,1980, K.Elliot et al., 1984, N.Tsukada et al., 1985). The annealed samples also display As_{Ga} ESR spectra with some additional signals at around g factor = 2 (or 3000 gauss). As_{Ga} signal intensities are higher after annealing than before annealing and depends on As pressure, increasing with increasing As pressure after annealing at both 1150°C and 1200°C. On the contrary, the additional signals observed after annealing decrease with increasing As pressure.

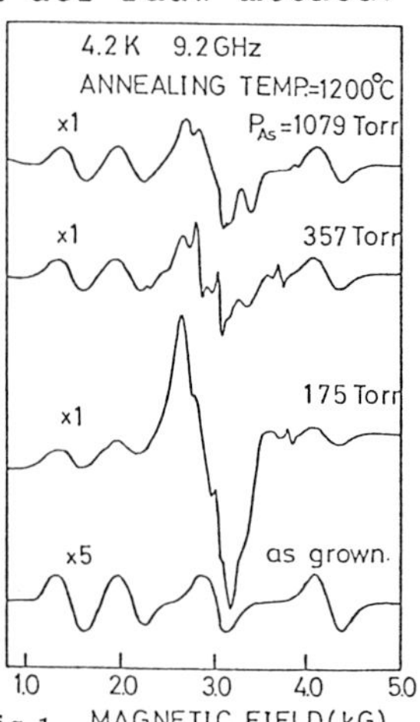

Fig.1 ESR spectra of SI LEC-GaAs before and after annealing at 1200°C.

All the samples investigated(15 pieces) showed p-type conversion after annealing with resistivities of the order of 10^4 to 10^7 ohm-cm. Variable temperature Hall measurements revealed that the samples have around 0.4 eV activation energy for carrier concentration after annealing. This value is close to the theoretically estimated activation energy for Ga vacancy V_{Ga} (P.J.Lin-Chung and T.L.Reinecke, 1983). The additional ESR signals the annealed samples display are suggestive of generation of V_{Ga}(T.A.Kennedy and N.D.Wilsey 1978). p-type conversion was also confirmed in the initially conductive n-type GaAs material with carrier concentration of the order of 10^{16} cm^{-3}. From this the concentration of annealing induced deep acceptor is roughly estimated to be not lower than 10^{16} cm^{-3}. The possibility of contamination with Cu, Mn and Fe can be rejected by both photoluminescence and ESR measurements.

p-type conversion means that Fermi level shifts toward the top of the valence band across the EL2 level, ionizing almost all the EL2 centers that were neutral before annealing and increasing As_{Ga} ESR signal intensity. Both quantitative ESR analysis based on comparison with a calibrated standard and estimation of ionization ratio of the EL2 centers or Fermi level location using carrier concentration obtained through Hall measurements enable to evaluate the total EL2 or As_{Ga} defect concentrations.

In Figs.2 and 3 are plotted EL2 concentrations and hole concentrations at room temperature, respectively, as a function of As pressure for 1150 °C and 1200 °C annealing. The EL2 concentrations after annealing are lower by one order or more as compared with those of the as grown GaAs. The EL2 concentration increases as As pressure increases. This is consistent with the model for As_{Ga} formation which is based on reaction between As atom and V_{Ga}(J.Lagowski et al.,1981) since this reaction is enhanced by increasing As pressure. It is noted that the samples annealed at 1150°C have EL2 concentrations higher than those annealed at 1200 °C do, suggesting formation of EL2 or As_{Ga} is an exothermic reaction. This is also consistent with the literature where 800 °C annealing generates $1.5x10^{16}$ cm^{-3} EL2 centers after annealing(J.Lagowski et al.,1986).

Variation of hole concentration with As pressure as well as p-type conversion with the relatively high

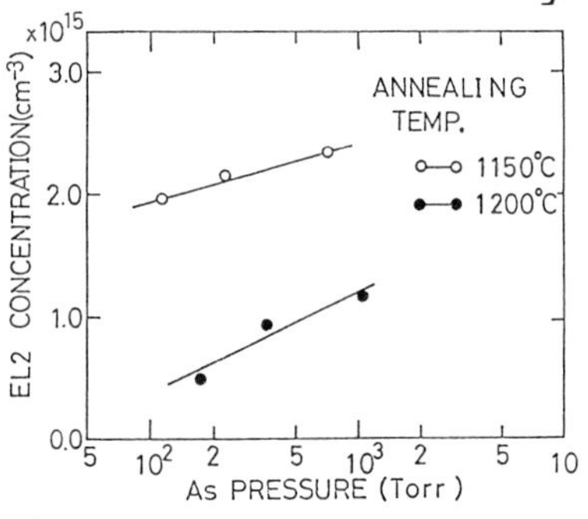

Fig.2 As pressure dependence of EL2 concentrations after annealing.

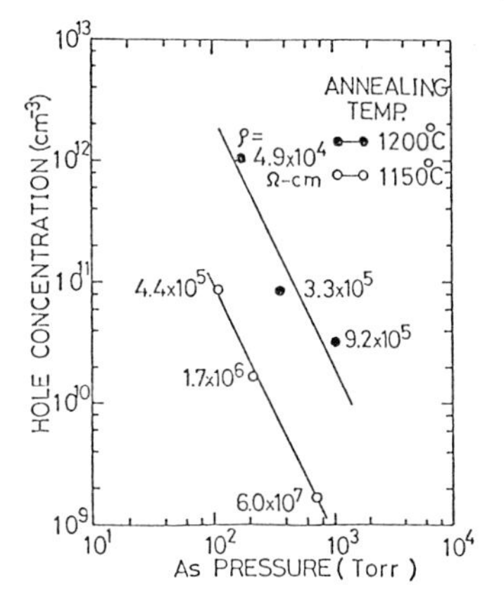

Fig.3 As pressure dependence of hole concentrations after annealing. Resistivities indicated beside data points are those of the corresponding samples after annealing.

resistivities is well explained by introduction of a deep acceptor level in addition to residual carbon acceptor and EL2. Figure 4 illustrates a graphical method (W.Shockley, 1950) to explain behavior of hole concentration and consequently resistivity obsreved in the present study. As EL2 concentration increases with increasing As pressure, Fermi level shifts toward the bottom of the conduction band and hole concentration decreases and resistivity increases along p and ρ lines, respectively. The deep acceptor makes possible a gradual variation of hole concentration and resistivity. Variation of EL2 concentration alone leads to p-type conversion but abrupt change to low resistivity, say 10 ohm-cm or lower.

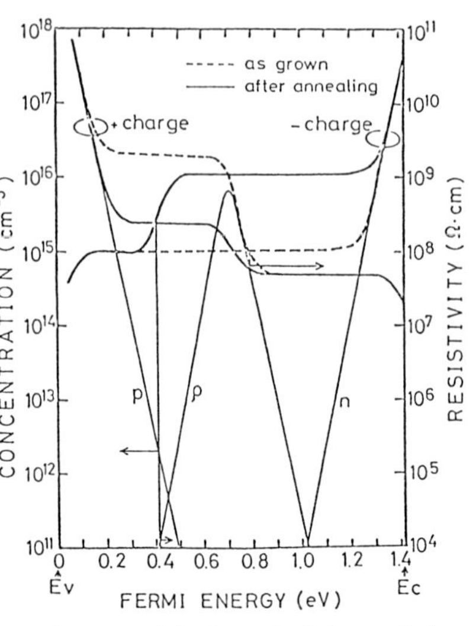

Fig.4 Graphical method to explain annealing behavior of the carrier concentration and resistivity on the basis of introduction of acceptor level.

4. SUMMARY

SI LEC GaAs was subjected to high temperatures annealing at 1150°C or 1200°C under various As pressure conditions. We have found that As_{Ga} or EL2 concentration is higher after annealing at 1150°C than at 1200°C and increases with increasing As pressure. It is also indicated that annealing induces a deep acceptor level.

The authors are grateful to Y.Kadota and T.Abe for supplying GaAs crystals. The continued technical assistance of Y.Sawada is also acknowledged.

References

K.Lohnert et al.,1986 Semi-Insulating III-V Materials, Hakone 267.

J.Lagowski et al.,1986 Appl.Phys.Lett.49 892.

R.J.Wagner et al.,1980 Solid State Commun.36 15.

K.Elliot et al.,1984 Appl.Phys.Lett.44 907.

N.Tsukada et al.,1985 Proceedings of Int.Symp.GaAs and Related Compounds,Karuizawa 205.

P.J.Lin-Chung and T.L.Reinecke,1983 Phys.Rev.B27 1101.

T.A.Kennedy and N.D.Wilsey,1978 Phys.Rev.Lett.41 977.

J.Lagowski et al.,1981 Appl.Phys.Lett.40 342.

W.Shockley,1950 Electrons and Holes in Semiconductors,D.Van Nostrand,Princeton,N.J.

Paper presented at Int. Symp. GaAs and Related Compounds, Heraklion, Greece, 1987

Annealing of bulk-insulating GaAs: A magnetic resonance assessment

S Benakki, E Christoffel, A Goltzene, C Schwab

Laboratoire de Spectroscopie et d'Optique du Corps Solide, Unité Associée au C.N.R.S. n° 232 - Université Louis Pasteur - 5, rue de l'Université, 67084 STRASBOURG Cedex France.

Wang Guangyu, Zou Yuanxi

Shanghai Institute of Metallurgy, Academia Sinica, 865 Chang Ning Road, SHANGHAI 200 050 China.

ABSTRACT : Magnetic resonance data under photoexcitation are obtained on as-grown and plastically deformed semi-insulating GaAs during isochronal anneals in the 20-850°C range, which support the existence of inhomogeneities likely to be potentially detrimental to device performance. We establish also that the antisite-related electron paramagnetic resonance signal generated as well by annealing at 700°C as by a plastic deformation is totally photoquenchable.

1. INTRODUCTION

Microscopic inhomogeneities in a semiconductor substrate material are potentially detrimental to further device performances. Their existence, related to growth and sampling conditions, can be hinted in semi-insulating as grown GaAs by the observation of phase-inverted signals during EPR measurements (Goltzené et al. 1986a, Blazey et al. 1986). Furthermore, the large variations of the values of the Hall R factor (Goltzené et al. 1986b), which translates the homogeneity of a sample, during annealing of fast neutron-irradiated material suggests that intrinsic defects are likely involved.

We bring here further support to these ideas by reporting new magnetic resonance data before and after photoexcitation obtained on as-grown and plastically deformed semi-insulating GaAs during isochronal anneals in the 20-850° range.

2. RESULTS

2.1 As-grown GaAs

EPR spectra of as-grown undoped semi-insulating GaAs have been taken at 4 K after successive isochronal annealing at increasing temperatures T_A. The material is identical to the unirradiated starting material used previously (Goltzené et al. 1985), and the annealing procedure is the same.

First we evidence an increase of the anion antisite As_{Ga}^{4+} signal Q, initially as a non-photosensitive one until T_A = 600°C, but becoming photosensitive above 700°C ; this is shown on **Figure 1** by the intensity variation of the signal before and after illumination, the latter corresponding to the nonphotosensitive fraction of the signal.

A broad signal is superrimposed, which becomes strong for T_A > 300°C ; it increases strongly during a photoexcitation at λ = 1 μm, a variation which not totally reversible when switching off the light.

Finally, a Fe^{3+} signal appears at T_A > 750°C.

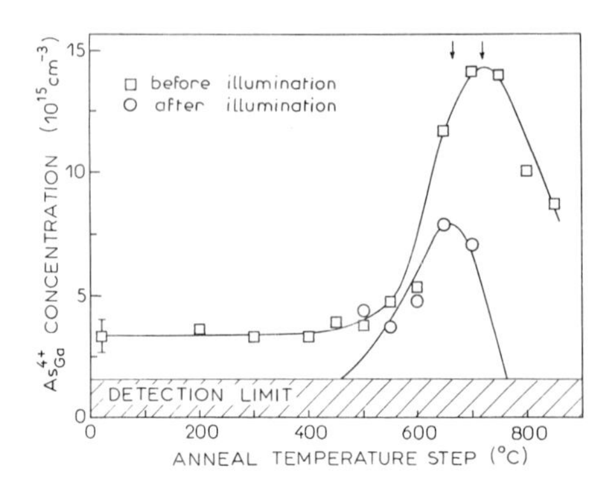

Fig. 1 Mean concentration of the paramagnetic As_{Ga} centers deduced from the quadruplet intensity, versus annealing temperature.

2.2 Stressed GaAs

Figure 2 shows the spectra obtained on stressed undoped GaAs after the same annealing steps, from 450°C on ; the sample corresponds to A6 in our previous report (Benakki et al. 1986). Again a similar quadruplet is generated whose intensity decreases with T_A. It is totally photoquenchable.

A narrower structure may be separated around 3 kG at the highest microwave power ; its intensity decreases also with T_A ; as in the previous case, Fe^{3+} appears only in the upper T_A range.

A broad line is again observed ; its intensity is maximum for $T_A \approx$ 550°C.

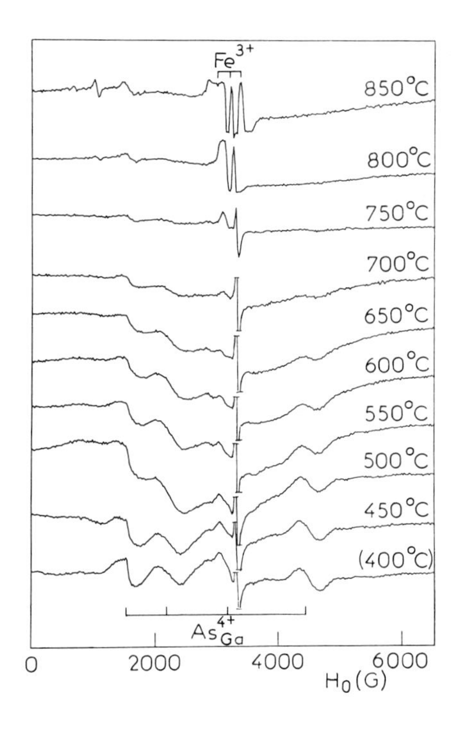

Fig. 2 Effect of successive isochronal annealings on the EPR spectrum of GaAs plastically deformed at 400° C, as recorded at 9 GHz and 4.2 K.

3. DISCUSSION

As well for stressed as for as-grown material, broad signals are observed which may be ascribed to resonance associated to free carriers in local domains of low resistivity ; such an attribution is supported by the strong photoexcitation effects observed by us in as-grown material, to be paralleled with a previous interpretation of the IR absorption data after photoexcitation by

Dischler et al. (1986). However, as the samples remain resistive on the macroscopic scale, the domains should be of small dimensions, as expected if the charges are generated by a local carrier separation induced by micron-sized inhomogeneities (Queisser 1985), which have actually been evidenced by different tomography techniques (Fillard 1985).

On the other hand, we observe that the quadruplet Q generated as well by a strain as by a thermal anneal at $T_A > 700°C$, is totally photoquenchable ; if we associate Q to EL2, this result is in contradiction with a previous observation that the quenchable selective EL2 optical absorption in the deformed material is the same as in the as-grown material before deformation (Omling et al. 1986). For the as-grown GaAs, the peculiar behaviour of Q, being first non photosensitive, is a confirmation of the complex structure of the As_{Ga}-related defects.

Finally, one may stress that as well the generation of Q in as-grown GaAs as the strong intensity variation of the broad line occur within minutes, not hours, at temperatures much lower than those necessary to modify the distribution of the EL2-related optical absorption in the GaAs wafers (Duseaux et al. 1986). Such a difference on the onset temperatures may translate a difference in the required diffusion lengths of the defects when a precipitate is dissolving : in the latter case one needs defects mobile over microns, while in the former one, we observe the activation of traps along an interface, which would only require ion displacements on an atomic scale.

4. REFERENCES

Benakki S, Goltzené A, Schwab C, Wang Guangyu, Zou Yuanxi 1986 Phys. Stat. Sol. (b) **138** 143-149

Blazey K W, Schneider J 1986 Appl. Phys. Lett. **48** 855-857

Dischler B, Fuchs F, Kaufmann U 1986 in Defects in Semiconductors ed von Bardeleben H J. 1986 Materials Science Forum Volumes **10-12** pp 359-364

Duseaux M, Martin S, Chevalier J P 1986 in Semi-Insulating III-V Materials, Hakone Japan, eds Kukimoto H and Miyazawa S (Tokyo : Omsha Ltd) pp 221-226

Fillard J P 1985 Proceedings of the International Symposium on Defect Recognition and Image Processing in III-V Compounds, Montpellier France July 2-4 1985 (Amsterdam : Elsevier)

Goltzené A, Meyer B, Schwab C 1985 J. Appl. Phys. **57** 1332-1335

Goltzené A, Meyer B, Schwab C 1986a J. Appl. Phys. **59** 2812-2816

Goltzené A, Schwab C, David J P, Roizes A 1986b Appl. Phys. Lett. **49** 862-864

Omling P, Weber E R, Samuelson L 1986 Phys. Rev. B **33** 5880-5883

Queisser H J 1985 Appl. Phys. Lett. **46** 757-759

Defect characterization in semi-insulating GaAs crystals by means of recovery characteristics of bleached absorption (RCBA) method

Y Mita*

Opto-Electronics Research Laboratories, NEC Corporation
4-1-1 Miyazaki, Miyamaeku, Kawasaki, Japan 213
* Present address: Tokyo Engineering University
1404 Katakura, Hachioji, Tokyo 192

ABSTRACT: Recovery characteristics of the near infrared absorption band, after bleaching at low temperatures, have been investigated under a slow and controled warm up schedule. It has been shown that the thermal recovery of the bleached absorption takes place rather abruptly near 130K. It has also been shown that the recovery profile is dependent upon the crystal composition and thermal prehistory . Quantitative results have been obtained for In-containing and thermally-treated crystals. These results demonstrate that the RCBA method may present valuable information on defect structures as well as crystal characteristics.

1. INTRODUCTION

One of the representative features of As-rich, deep-lying defects in semi-insulating GaAs crystals is transition to metastable state at lower temperatures. The transition to the metastable state is verified with such methods as bleaching characteristics. Nearly complete bleaching can take place under light irradiation at low temperaure (Martin 1985). There are several detailed reports on the bleaching characteristics (e.g. Mita 1986). However, since the absorption spectrum is broad and structureless, it is usually difficult to obtain further information on the As-rich defects from optical absorption characteristics only.

This report covers investigation on the recovery characteristics of bleached absorption. Thermal recovery processes after bleaching have been studied, especially for GaAs crystals having different compositions or thermal prehistory. It is shown that the method is an effective measure in investigating defect characteristics as well as in crystal evaluation.

2. EXPERIMENTAL PROCEDURES

Optical absorption characteristics were measured for semi-insulating GaAs crystals obtained from several different suppliers. Not a small part of the crystals had specified growing conditions. Crystals were irradiated with 1.13eV light for a definite time at liquid nitrogen temperature(LNT) or liquid helium temperature(LHeT). After bleaching, the entire absorption spectra were measured successively by microprocessor-controlled measuring equipments, while the crystal temperature was varied with a slow and nearly constant (~0.5°C/min) warm up rate. Some of the crystals contained

In at varying concentrations. Some other undoped crystals were thermally treated at 850°C in an inert atmosphere for several hours and subsequently slowly cooled or annealed. After heat treatment, suface layers were eliminated by grinding and polishing. In several experiments photocurrent temperature dependence was measured simultaneously with the optical absorption measurements.

3. EXPERIMENTAL RESULTS

Absorption change during bleaching was measured for many crystals. Most of the crystals measured showed nearly complete bleaching after 1.13 eV light irradiation. However, some crytals showed only partial bleaching, even after a long irradiation time. Along with bleaching near the infrared absorption band, a broad, nearly flat absorption band appeared at a longer wavelength. The absorption band extends toward lower photon energy as far as 0.6eV and grows in reciprocal relation to bleaching of the near infrared absorption. Details regarding this bleaching and the longer wavelength absorption band will be reported elsewhere.

Thermal recovery of the near infrared absorption takes place rather abruptly at temperature range between 110 and 130K, depending upon warm up rate. Representative results are shown in Fig. 1. Similar results obtained in a different crystal are plotted in Fig. 2, taking the temperature as the abscissa. It can be seen that the recovery takes place in a very limited temperature range for all wavelengths almost simultaneously. Activation energies obtained from the optical absorption recovery were considerably different for the crystals adopted and ranged between 0.30 to 0.40 eV. These values seem in fair agreement to the value reported recently (Fuchs 1987). Results for different series of measurements are plotted in Figs. 3 and 4, after taking the temperature derivative numerically. These curves seem to present more direct information and are called "recovery profile". In most undoped, as-grown crystals, recovery profile assumes sharp, nearly symmetric peak, usually having a tail extending toward the higher temperature side. Hoewever, more precise investigation reveals minor difference between crystals. Recovery profile measurements have been carried out for crystals containing varying amounts of In atoms up to 10^{20} cm^{-3} concentrations. Typical results are shown in curves of Fig. 3 for crystals having different In concentration. There is

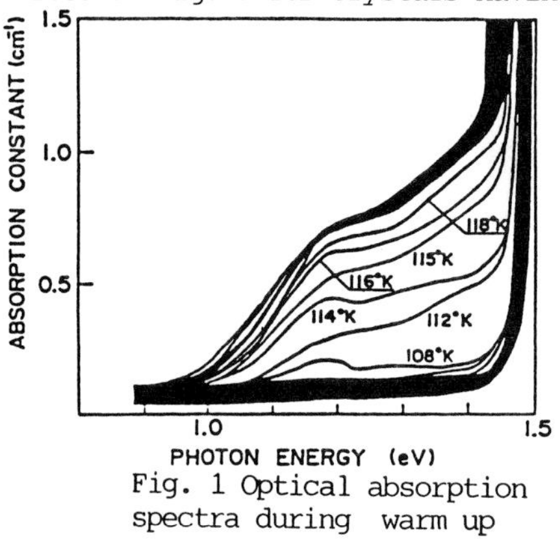

Fig. 1 Optical absorption spectra during warm up

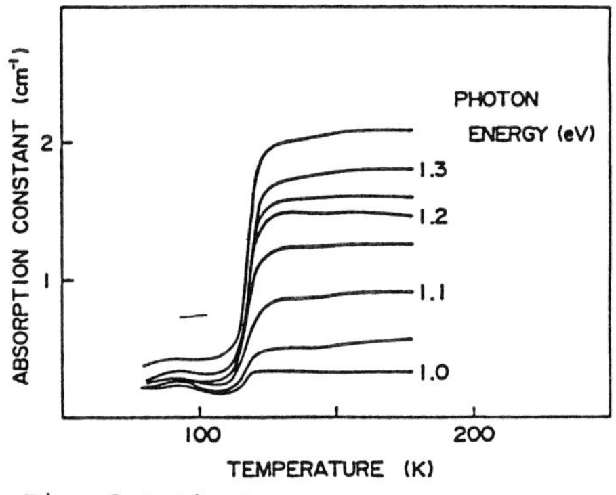

Fig. 2 Optical absorption change

a distinct relation that the width of the recovery profile increases with increasing In concentration. The half-width of the recovery profile peak, however, is not always proportional to the In concentration. It can be seen that a clear increase in width can be observed at an In concentration as low as 10^{19} cm^{-3}. Recovery profile changes after thermal treatment were measured for at least 5 samples. Representative results are shown as curves in Fig. 4. A conspicuous change is observed. After heat treatment the width becomes broader and the peak temperature shifts toward the higher temperature side by approximately 5°. The difference between the results obtained for different cooling conditions seems not essential and seems partly due to difference in effective heat treatment time. Although the profiles for In doped and for heat treated samples resemble each other, no clear shifts in peak temperature have been observed in In doped crystals. It can be seen that the recovery process cannot be specified with single activation energy and that the recovery profiles are due to defects having slightly different recovery temperatures. Simultaneous measurements on extrinsic photoconductivity, along with optical absorption, were carried out for several crystals. Details on photocurrent enhancement phenomena were reported in a previous paper (Mita 1987) and further discussion will be made elsewhere. It was shown that the recovery takes place at the same temperature range for both optical absorption and extrinsic photocurrent. However, simultaneous measurements at bleaching stage showed complicated behavior. Relative spectral efficiencies for transition to and from the metastable state were measured both for optical absorption and photocurrent, except recovery spectrum of optical absorption. The results are compared with available data.

4. DISCUSSIONS

It has long been known that the near infrared absorption intensity is proportional to the As-rich defect concentration. However, since the absorption is broad and no recognizable difference exists in shapes for different crystals, it has been difficult to obtain further information on the defect structures. Thermal recovery characteristics seem to detect a more delicate change in defect structures. Recovery temperature of bleached absorption is in good agreement with that for recovery from photocapacitance quenching effect (Vincent 1985) and extrinsic photoconductivity (Jimenez et al. 1987, Mita 1987). These results are expected from the fact that same defects are relevant to these effects.

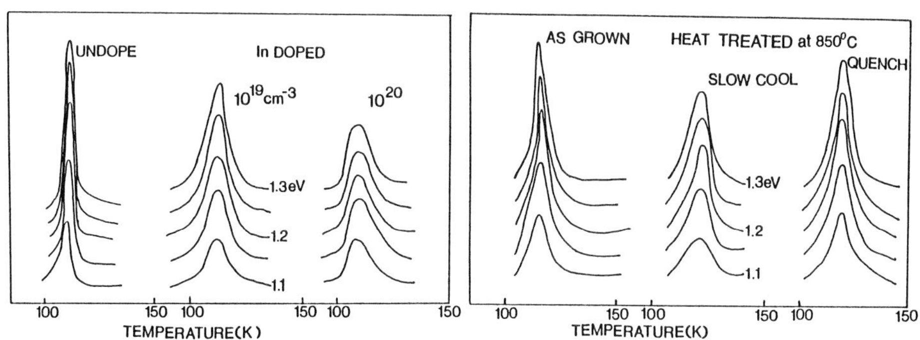

Fig. 3 Recovery profiles for undoped and In doped crystals

Fig. 4 Recovery profiles for untreated and heat treated crystals

According to one model for As-rich defect, the defect is composed of one As antisite atom accompanied with one As interstitial atom (Bardeleben et al. 1985). Transition to and recovery from the metastable state is then corresponding to a flip of As interstitial from next nearest to and from nearest neighbor interstitial position. As has been revealed with bleaching and recovery characteristics, the transition to and from metastable state is apt to be influenced with minor changes in chemical composition or thermal treatment. The In concentration needed to significantly change the recovery profile is on the order of 10^{-3}, which is far smaller than the value expected if In atoms nearest to the transition path can have a substantial effect to the profile. This effect, combined with related thermal treatment effect, may lead to a tentative conclusion that the As-rich defects are composed of a little larger number of defects or, at least, apt to be influenced with distant In atoms. It is fairly accepted that certain kind of native defects, particularly As interstitial atoms, can migrate above moderate temperatures. Considering these situations, it seems plausible that the thermal treatment effect is due to As interstitial diffusion and subsequent association with existing deep level defects. The presence of nearby As interstitial atom then changes the barrier height for atomic displacement during transition, to and from the metastable state, without modifying absorption characteristics appreciably. It seems possible that the defects may be composed of a little larger number of elementary defects than are considered in a simple association model, as proposed by several previous investigators (Ikoma et al. 1984, Wada et al. 1985). A tentative model suggested here is based on the similarity between trigonal As lattice and GaAs lattice. Atomic positions of As atoms in two lattices are closely corresponding to each other with approximately 7% misfit, except central Ga or As_{Ga} atom. Although it is still not possible to assign individual electronic processes relevant to the transitions to metastable state, the defects are presumably composed of several such units and may grow by associating with interstitial As atoms. There seem plenty grounds to consider that the p-type extrinsic photoconductivity is related to the As-rich defects.

The present method, called the RCBA method, can be applied quantitatively and can be nearly non-destructively carried out using liquid nitrogen. Most of the experimental results have demonstrated that the near infrared absorption band shape is not sensitive to In doping or thermal treatment, while the recovery characteristics or sometimes bleaching characteristics are quite sensitive to these effcts. These facts seem to make the present method a convenient tool for investigating defect characteristics.

The author wishes to express thanks to his former colleagues in NEC Corporation for their support and cooperation in the research studies.

REFERENCES

Bardeleben H. J. von, Stievemard D, Deresmes D, Huber A and Bourgoin J. C, 1985, Phys. Rev. B 34,7192.
Fuchs F and Dischler B, 1987, Appl. Phys. Lett. 51, 679.
Ikoma T and Mochizuki Y, 1985, Japan. J. Appl. Phys. 24, 1935.
Mita Y, 1986, Appl. Phys. Lett. 39, 747.
Mita Y, 1987, J. Appl. Phys. 61, 5327.
Vincent G, Bois D, and Chantre A, 1982, J. Appl. Phys. 53, 3643.
Wada K and Inoue N,1985, Appl. Phys. Lett. 47, 945.

Inst. Phys. Conf. Ser. No. 91: Chapter 2
Paper presented at Int. Symp. GaAs and Related Compounds, Heraklion, Greece, 1987

Identity of the EL6 center in GaAs

M. Levinson

GTE Laboratories, Inc., 40 Sylvan Rd., Waltham MA 02254, USA

ABSTRACT: The EL6 center is a defect often found in melt-grown GaAs with concentrations of the same order of magnitude as the much-studied EL2 center. Site symmetry information on EL6 has been obtained by means of both deep level capacitance transient spectroscopy (DLTS) and photoionization measurements under uniaxial stress. Its stress response and photoionization properties are nearly identical to those of the "normal" configuration of the EL2 center. This result leads to a model where EL6, like EL2, is an As_{Ga} antisite-related complex. Furthermore, the relatively large difference between the optical and thermal electron emission energies of this center is interpreted as resulting from a multi-step thermal emission via nearby shallow centers, rather than the conventional large lattice relaxation mechanism.

1. INTRODUCTION

The properties of point defects in GaAs are of great importance to the microelectronic technologies based on these materials. The EL6 center (Martin *et al* 1977) is often found in As-rich, melt-grown GaAs with concentrations of the same order of magnitude as the much-studied EL2 center. EL6, or a defect similar to it, was also observed in neutron-irradiated or ion-implanted GaAs by Martin, *et al* (1984) and Samitier, *et al* (1986). However, this defect has received much less attention than EL2, and its identity is unknown.

The thermal activation energy for the electron emission which gives rise to the characteristic EL6 DLTS peak is 0.35 eV (Martin *et al* 1977). However, the photoionization threshold is ~0.8 eV (Chantre *et al* 1981). This ~0.46 eV difference between the electron thermal activation energy and the photoionization threshold would normally be taken as an indication of a large lattice relaxation which occurs upon change of charge state. Such a lattice relaxation is of interest with regard to the identity of EL6, and also in light of the "metastable" state of EL2, which is generally believed to involve a large lattice relaxation (Vincent *et al* 1982). In fact, the photoionization cross-section of EL6 is very similar to that of the EL2 center in the range 0.7 - 1.3 eV (Chantre *et al* 1981).

Here we have used uniaxial stress, together with both deep-level capacitance transient spectroscopy (DLTS) and polarized excitation photocapacitance measurements as a means of obtaining site symmetry information on this center. These techniques have been previously applied to defects in Si by Meese *et al* (1983) and Stavola and Kimerling (1983), as well as to the EL2 center in GaAs by Levinson and Kafalas (1987). They should also be useful for detecting evidence of lattice relaxations which lower the site symmetry of the defect.

Three kinds of experiments were performed. First, DLTS spectra were recorded both with and without applied stress. Second, photocapacitance transients were recorded under stress at low temperature for light polarization orientations parallel and perpendicular to the stress axis. Third, in an attempt to induce preferred orientations of the defects, samples were annealed under stress at 150 C. (This kind of stress anneal has been shown by Levinson and Kafalas (1987) to induce preferential alignments of the EL2 center). Polarized excitation photocapacitance measurements were then made after the stress was removed at low temperature.

2. EXPERIMENTAL

Samples were prepared from two Se-doped liquid encapsulated Czochralski-grown boules with electron concentration $n = 1 \times 10^{16}$. They were cut to dimensions 1.3 x 1.5 x 7.0 mm^3 with the long axis parallel to the <001>, <110>, or <111> direction. Ti/Au Schottky diodes were fabricated near the center of one face of each sample, and Au wire bonds were made to them. EL6 concentrations, measured by DLTS, were $2\text{-}5 \times 10^{14}$ cm^{-3}.

Uniaxial stress up to 300 MPa was applied parallel to the long axis of the sample using an air cylinder and piston apparatus. The stress apparatus containing the sample was attached to the cold finger of a liquid He dewar for low temperature measurements, but it could be removed and placed in an annealing furnace without interruption of the applied stress.

For DLTS and photocapacitance under stress, the samples were cooled to low temperature before stress was applied. The stress anneals were performed by holding samples at 150 C for 1 hr under 200 MPa of stress. The samples were then cooled to $T < 100$ K before the stress was removed and the photocapacitance measurements were made at zero stress.

For photocapacitance measurements, a Nd:yttrium-aluminum-garnet (Nd:YAG) laser operating at 1.06 μm was used to illuminate the diode through the opposite face of the

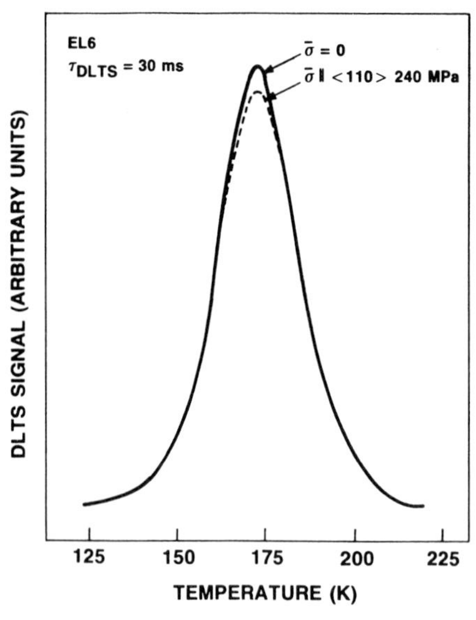

Fig. 1. DLTS spectra showing the EL6 peak at zero applied stress and at 240 MPa with σ ∥ <110>.

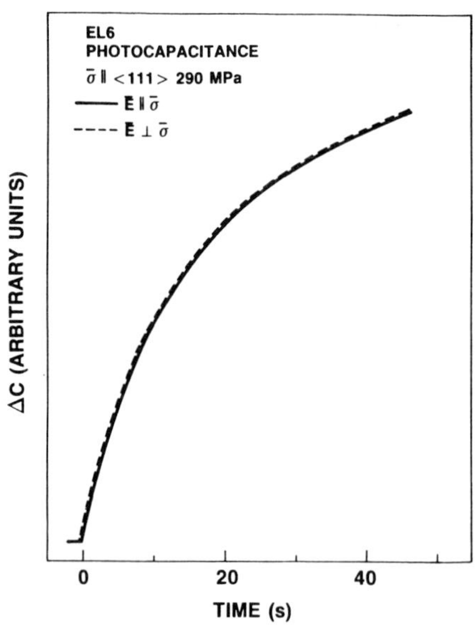

Fig. 2. Photocapacitance transients at σ = 290 MPa with the polarization vector E ∥ σ and E ⊥ σ.

sample. The incident light direction was in all cases <1$\bar{1}$0>. Light polarization was controlled by means of a Glan-Thompson prism and 1/2 wave plate. Verification of equal light intensities for polarization orientations both parallel and perpendicular to the stress axis was provided because identical photocapacitance transients were obtained for both orientations at zero stress.

The samples also contained EL2 concentrations of $\sim 2 \times 10^{15}$ cm^{-3}. These centers did not interfere with the EL6 photocapacitance data because the measurements were made at T $<$ 40 K where the EL2 was made electrically and optically inactive by photoquenching.

3. RESULTS

No stress-induced splitting or broadening of the DLTS peak was observed for any of the three stress directions. Spectra are shown in Fig. 1 for zero stress and for 240 MPa with $\sigma \parallel$ <110>. Similarly, no dichroism was seen in the photocapacitance transients for any stress direction, either under low temperature stress or at zero stress after the 150 C stress anneals. Photocapacitance transients for both parallel and perpendicular polarization orientations are shown in Fig. 2 for 290 MPa with $\sigma \parallel$ <111>.

4. DISCUSSION

The lack of stress effects on the EL6 electron emission would be consistent with a point defect having the full site symmetry of the lattice. However, this interpretation is not conclusive because the defect electronic state involved in the transition may have a higher symmetry than the defect itself. This latter situation is similar to that of the of the EL2 center where the initial trap-emptying photocapacitance transient (corresponding to the "normal" state) shows no stress effects, but the subsequent photoquenching transient (indicative of a transition to the "metastable" state) shows pronounced effects indicative of C_{3v} symmetry (Levinson and Kafalas 1987).

On the other hand, the lack of stress effects for EL6 is surprising because of the large difference between thermal and optical electron emission energies. This difference would normally be taken as evidence of a large lattice relaxation. No stress effects would result for such a relaxation if it involved a breathing mode distortion of an isolated point defect, as the symmetry would be maintained. However, there is evidence (Martin *et al* 1984 and Samitier *et al* 1986) that EL6-related defects due to neutron irradiation in GaAs undergo a multi-step electron emission process involving thermally-assisted tunneling to nearby shallow levels. This process gives an effective thermal activation energy which is significantly smaller than the optical threshold, but without a large lattice relaxation mechanism. Therefore, the optical and thermal energy difference, together with the absence of a uniaxial stress response can also be explained by a defect complex which consists of both a center responsible for the deep level, and one or more shallower defects at distances of 5-20 A. The deep center is relatively isolated as far as optical transitions are concerned, and the random orientations of the shallow defects relative to it prevents observable stress effects in the DLTS measurement.

The dominant electron spin resonance signal obtained from melt-grown GaAs is that of the As_{Ga} antisite defect, and there is now very good evidence (von Bardeleben *et al* 1986 and Meyer *et al* 1987) that the EL2 center is a complex involving As_{Ga}. The very similar photoionization cross-sections of EL6 and EL2 (Chantre *et al* 1982) imply that EL6 also contains the As_{Ga} antisite. It is unlikely to be an isolated antisite because, as discussed above, the difference between the thermal and photoionization energies implies either a large lattice relaxation, which theoretically should not exist for the isolated antisite (Bachelet and Scheffler 1985), or a multistep thermal emission process which requires nearby shallow centers. This information, together with the similar lack of stress response in photocapacitance for EL6 and the "normal" state of EL2, leads to a model of EL6 as an

As_{Ga}-related complex with the As_{Ga} surrounded at some distance by electronically shallower centers.

EL2 and EL6 may in fact be electronically active parts of more extended complexes involving As antisites and interstitials. Such extended complexes have been proposed by Wada and Inoue (1985) as the origin of EL2 in analogy with formation of thermal donors by the aggregation of oxygen in Si. It may be that one local configuration of As atoms within these complexes gives rise to EL2, while another might produce EL6. Tests of these ideas await further experimental information on the formation kinetics and structure of these defects.

5. REFERENCES

Bachelet G B and Scheffler M 1985 *Proc. 17th Intl. Conf. on Phys. of Semiconductors, San Francisco, 1984* ed J D Chandi and W A Harrison (Springer, New York) p 755.
von Bardeleben H J, Stievenard D, Deresmes D, Huber A, and Bourgoin J C 1986 Phys. Rev. B *34* 7192
Chantre A, Vincent G, and Bois D 1981 Phys. Rev. B*23*, 5335
Levinson M and Kafalas J A 1987 Phys. Rev. B *35* 9383
Martin G M, Mitonneau A, and Mircea A 1977 Electron. Lett. *13* 191
Martin G M, Esteve E, Langlade P, and Makram Ebeid S 1984 J. Appl. Phys. *56* 2655
Meese J M, Farmer J W, and Lamp C D 1983 Phys. Rev. Lett. *51* 1286
Meyer B K, Hofmann D M, Niklas J R, and Spaeth J-M 1987 Phys. Rev. B *36* 1332
Samitier J, Morante J R, Giraudet L, and Gourrier S 1986 Appl. Phys. Lett. *48* 1138
Stavola M and Kimerling L C 1983 J. Appl. Phys. *54* 3897
Vincent G, Bois D, and Chantre A 1982 J. Appl. Phys. *53* 3643
Wada K and Inoue N 1985 Appl. Phys. Lett. *47* 945

Inst. Phys. Conf. Ser. No. 91: Chapter 2
Paper presented at Int. Symp. GaAs and Related Compounds, Heraklion, Greece, 1987

P_{Ga}-Antisite complexes in GaP studied with optical detection of magnetic resonance

W M Chen, M Godlewski*), B Monemar and H P Gislason**)

Department of Physics and Measurement Technology, Linköping University, S-581 83 Linköping, Sweden

*) Permanent address: Institute of Physics, Polish Academy of Sciences 02-668 Warsaw, Al. Lotnikow 32/46, Poland
**) Permanent address: Science Institute, University of Iceland, Dunhaga 3, 107 Reykjavik, Iceland

ABSTRACT: Several deep P_{Ga}-related complex defects in Cu,Li-doped GaP are studied with optical detection of magnetic resonance. The electronic structure of the defects is consistent with a strongly localized electron-hole pair ("deep" bound exciton) at a neutral defect, believed to be due to differently separated P_{Ga}-Cu_{Ga} pairs. The role of antisite-related defects in non-radiative recombination processes in GaP is investigated. The experimental procedure to obtain a high concentration of these P_{Ga}-Cu_{Ga} neutral defects is also discussed.

1. INTRODUCTION

Anion antisite defects in III-V semiconductors belong to the most important intrinsic defects. The occurrence of stable isolated antisite defects has been confirmed experimentally in several III-V materials such as GaP (Kaufmann et al 1976), GaAs (Wagner et al 1980) and InP (Kennedy and Wilsey 1984). It has been proved, from studies of optically detected magnetic resonance (ODMR), that such defects act as efficient centers for non-radiative recombination in GaP (Killoran et al 1982, O'Donnell et al 1982) and GaAs (Gislason and Watkins 1986).

In this paper we present the results of an extensive study of the electronic properties of P_{Ga}-antisite related complexes in bulk GaP. In addition to the previously described isolated P_{Ga} centers and associates involving P_{Ga}, we report here the detection of at least three different P_{Ga}-related neutral (isoelectronic) complex defects. Similar defects may be common in III-V materials and play an important role in non-radiative recombination processes.

2. EXPERIMENTAL

The GaP samples studied in this investigation were bulk crystals grown by the liquid encapsulation Czochralski technique. The nominally undoped material was first doped with Cu (diffusion temperature 950°C, 1 h) and then codoped with Li (diffusion temperature 450°C, 4 h).

For the ODMR studies a converted Bruker 200-SRC spectrometer was used. Photoluminescence (PL) was excited with an Ar^+ argon laser and detected with a cooled North Coast EO-817 Ge detector. A Jobin-Yvon 0.25 m grating monochromator was used for the PL measurements, including spectral dependence of the ODMR signals.

3. RESULTS AND DISCUSSION

In Figure 1 we present ODMR spectra of GaP:Cu,Li obtained by monitoring infrared PL. This spectrum is dominated at 4 K by a structured signal, with an additional weak background. The latter is getting relatively stronger at increased temperatures, and dominates the ODMR spectrum for 25 K (P_{Ga}-A) and 50 K (P_{Ga}-B). The relative strength of these signals change from sample to sample, proving that they are related to different defects.

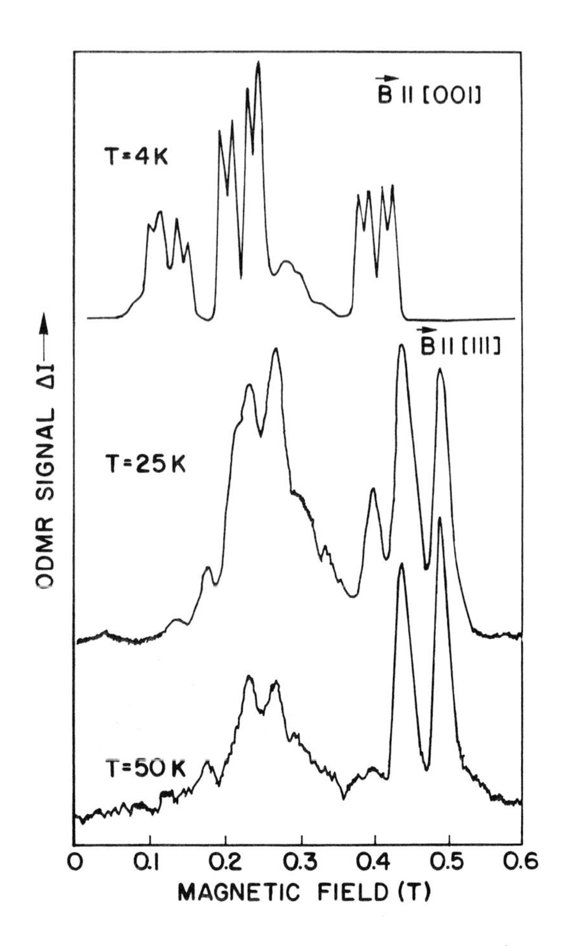

Figure 1. ODMR spectra of GaP:Cu,Li obtained by detecting the infrared emission at different temperatures. The P_{Ga} (nuclear spin I=1/2) central hyperfine structure is resolved for all spectra.

The characteristic feature of all three spectra shown in Figure 1 is a double line splitting of the resonances. This hyperfine structure in the spectra indicates involvement of P_{Ga} as part of the defects studied. Another alternative - an interstitial phosphorus - can be rejected. In the latter case a p-type wavefunction has been predicted (Jaros, 1978), which is inconsistent with the large hyperfine splitting observed for all three defects. This indicates an s-type wavefunction and a strong electron localization, as expected for P_{Ga}-antisite related defects in GaP.

All three spectra were fitted with a spin triplet spin Hamiltonian of the form

$$H_S = \mu_B \bar{B} \cdot \overleftrightarrow{g} \cdot \bar{S} + \bar{S} \cdot \overleftrightarrow{D} \cdot \bar{S} + \bar{I} \cdot \overleftrightarrow{A} \cdot \bar{S}$$

The g-, A- and D-tensor parameters obtained from the fit are summarized in Table 1. The 4 K defect is ⟨110⟩ oriented, whereas the two observed at higher temperatures and labelled P_{Ga}-A and P_{Ga}-B are both of ⟨111⟩ orientation. This suggests a second and third nearest neighbour compensation of the P_{Ga} double donor in GaP, and it is argued that the Cu_{Ga} double acceptor is involved as a compensating acceptor in all three defects discussed here.

Center		P_{Ga}-Cu ⟨110⟩	P_{Ga}-A ⟨111⟩	P_{Ga}-B ⟨111⟩
g-tensor	g_x	1.96	2.00	1.95
	g_y	1.72	2.00	1.95
	g_z	2.27	1.98	2.00
D-tensor (10^{-5} eV)	D_x	0.93	-0.36	-0.53
	D_y	0.75	-0.36	-0.53
	D_z	-1.68	0.72	1.06
A-tensor (10^{-5} eV)	A_x	0.16	0.45	0.55
	A_y	0.11	0.45	0.55
	A_z	0.14	0.66	0.65

Table I. Spin Hamiltonian parameters for three neutral P_{Ga}-Cu_{Ga} defects in GaP. Their z-axes are indicated in the Table.

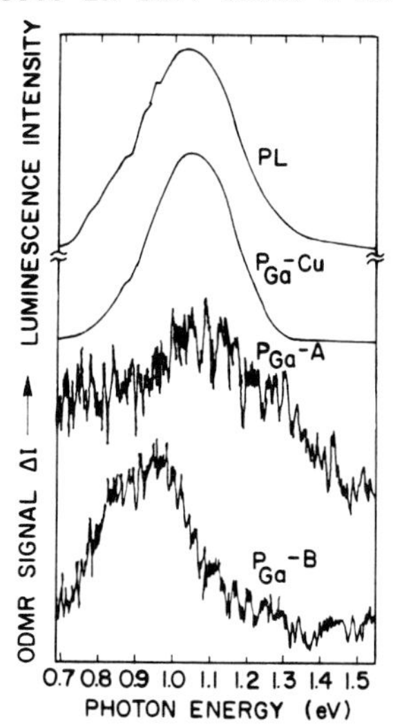

Figure 2. Spectral dependences of ODMR spectra of three P_{Ga}-Cu related neutral defects. The infrared PL is also shown, measured in the same experimental setup as for the ODMR-PL studies.

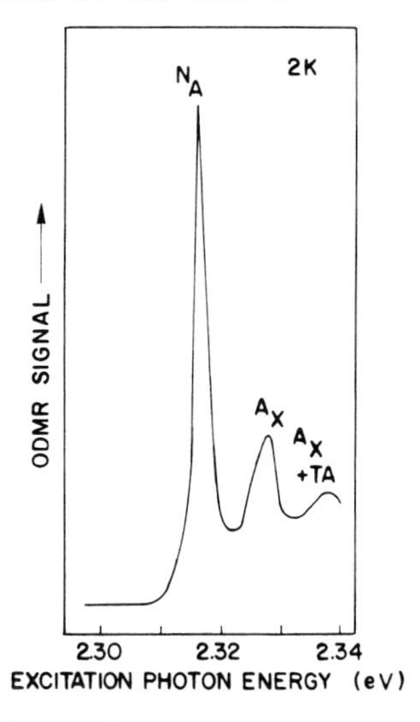

Figure 3. ODMR-PLE spectrum for the P_{Ga}-Cu related neutral complex defects, obtained using a Coherent 590 dye-laser with a Coumarin 540 dye. A fingerprint of the N BE PL excitation is observed, which indicates an efficient energy migration from N_P- to deep P_{Ga}-related complex defects in GaP.

The spectral dependences of the ODMR signals from the P_{Ga}-Cu defects studied are shown in Figure 2. All three centers contribute to the infrared emission, with maxima at about 1 eV. The spin triplet configuration of the initial state of the emission is consistent with the recombination of two localized spin-like particles i.e. an electron and a hole bound at the neutral defects studied. This is an analogue of a closely separated deep donor-acceptor pair, or equivalently a "deep" bound exciton recombination at an isoelectronic center.

The ODMR-PLE experiment, as described elsewhere (Chen et al 1987), shows that the P_{Ga}-related infrared emissions are efficiently excited by an energy migration mechanism. In Figure 3 we show an ODMR-PLE spectrum, proving a very efficient energy transfer from N-related BE:s to P_{Ga}-Cu centers. The energy transfer from N- to P_{Ga}-Cu defects is a phonon assisted process, leading to reduction of GaP visible PL which is replaced by infrared radiation. A simple estimate indicates that energy transfer proceeds at large distances (in the 200 Å range), comparable to the typical inter-center distances in the sample.

4. DEFECT FORMATION

A fundamental question which arises from our studies is whether P_{Ga}-related complexes are common in bulk material. The neutral defects are difficult to detect with electrical or ESR experiments, but as proved here by ODMR-PLE investigations, they may play an important role in recombination processes. A range of samples were studied to answer this question. A most curious result is the necessity of Li codoping to get strong P_{Ga}-Cu related spectra in the ODMR. We believe that Li_{Ga} is not participating in any of the three P_{Ga} defects described here. The Li_{Ga}-related acceptor in GaP is deeper than the Cu_{Ga} one, which would require emission from P_{Ga}-Li defects further out in the infrared than studied here. The role of Li codoping in the formation of the P_{Ga}-Cu neutral defects is either due to a Fermi level effect or due to a replacement of a Cu interstitial in Cu-related complexes by Li, allowing for formation of P_{Ga}-Cu_{Ga} centers. By a Fermi level effect we understand here a change of the charge state of components of a defect, which either helps in defect formation or leads to a defect charge state observed in the experiment. The P_{Ga}-Cu related ODMR spectra reported here are stronger when n-type S-doped GaP was used as starting material. For Zn doped p-type GaP other defects are observed in ODMR, some of them also believed to be P_{Ga}-related.

REFERENCES

Chen W M, Godlewski M and Monemar B 1987 Phys. Rev. B36, to be published
Gislason H P and Watkins G D 1986 Phys. Rev. B33 2957
Jaros M 1978 J. Phys. C11 L213
Kaufmann U, Schneider J and Raüber A 1976 Appl. Phys. Lett. 29 312
Kennedy T A and Wilsey N D 1984 Appl. Phys. Lett. 44 1089
Killoran N, Cavenett B C, Godlewski M, Kennedy T A and Wilsey N D 1982 J. Phys. C15 L723
O'Donnell K P, Lee K M and Watkins G D 1982 Solid State Commun. 44 1015
Wagner R J, Krebs J J, Stauss G H and White A M 1980 Solid State Commun. 36 15

Behaviour of deep levels in GaAs pulled in the controlled As atmosphere

K Sassa, K Tomizawa, Y Shimanuki and J Nishizawa*

Mitsubishi Metal Corporation, 1-297 Kitabukuro Omiya Saitama 330 Japan
*Research Institute of Electrical Communications, Tohoku University, 2-1-1 Katahira, Sendai, 980 Japan

ABSTRACT:The effect of As pressure was studied on the formation of deep levels in GaAs crystals which were grown under various As pressures in a newly developed pulling apparatus. Cusps were found in the As pressure dependence of EL6 and EL3.

1. INTRODUCTION

The composition of a dissociable compound crystal depends on the pressure of the constituent element gas in the crystal puller and the deviation from the stoichiometric composition which leads to the formation of point defects is known to affect strongly the crystal characteristics. Nishizawa etal. (1981) studied As pressure dependence of crystal characteristics on heat treated or LPE-grown GaAs and found that the acceptor density and crystal dilation exibited cusps at specific pressures where stoichiometric composition was thought to be realized. With the aim of applying this idea to Czochralski method, we developed a new pulling apparatus consisting of a double chamber;in an inner chamber a GaAs single crystal was grown under a controlled As pressure. The detail of the apparatus is described elsewhere. In this paper we report the effect of As pressure on the formation of deep levels in crystals grown by this apparatus.

2. EXPERIMENTALS

2.1 Specimens

A set of (100) wafers were sliced longitudinally from crystal ingots of about 50 mm in diameter which were pulled under various As pressures ;these pressures are given by adjusting the temperature at a specified portion of the inner chamber wall to appropriate values (T_{As}). T_{As} ranged from 607 to 625 °C by 3 °C intervals. This temperature range corresponds to the As pressure from 820 to 1330 torr, if Boomgaard's P-T diagram (1957) is adopted. These specimens were conductive and the carrier concentrations measured by Hall Effect were 3-6E14 cm^{-3} . Concentrations of S,Si measured by SIMS and C measured by FTIR are 7E15,3E14 and 3E15 cm^{-3} ,respectively.

2-2 Evaluation of deep levels

Fig.1 is a DLTS spectrum of a peripheral point in a specimen whose T_{As} was 616 C. Specimens became carrierless except for peripheral regions when they were cooled down to -110 C. The spectrum consists of three peaks: The analysis of the first peak (P1) around -120 °C gives 0.34 eV and 1.3E-13 cm^2 of ionization energy and trapping cross section, respectively. These values are close to those of EL6 (Martin et al 1977). These values for the second peak (P2) around -50 C are 0.61eV and 1.2E-12 cm^2, whereas those for the third peak (P3) are 0.82 eV and 1.5E-13cm^2. They correspond to EL3 and EL2, respectively. However DLTS method is not appropriate for the evaluation of these levels because carrier concentration is too low to guarantee exponential electron emission. Therefore, following methods were used,i.e.,TDS-ICTS method (Okumura 1986) and C-V method.

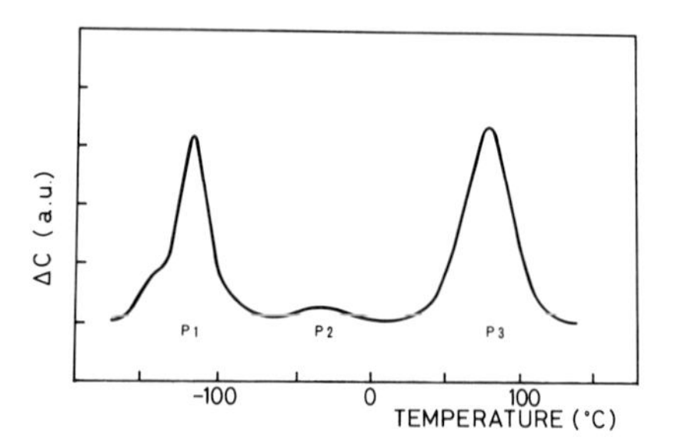

Fig.1 An example of DLTS spectrum of a PCZ crystal

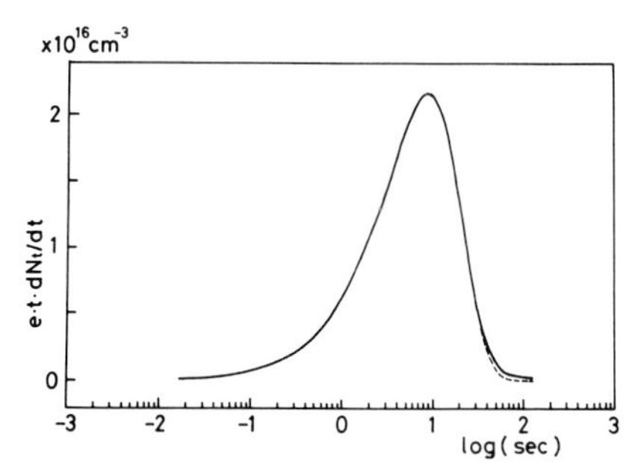

Fig. 2 TDS-ICTS spectrum of EL2
TDS=trap density spectroscopy
ICTS=isothermal capacitance transient spectroscopy

The densities of both EL6 and EL3 were evaluated from the difference of densitiy of ionized centers obtained by C-V method between two temperatures where the ionization of the trap was complete and negligible. For EL6 these temperatures were -70 and -150 C. Because the contact capacitance became negligible at temperatures below -110 C,the density at -70 C was adopted for that of EL6. For EL3 these temperaturtes were -8 and -70 C.

In the case of TDS-ICTS method, the time constant for electron thermal emission (τ) and the density of a deep level (Nt) are correctly evaluated from the capacitance transient even if the carrier concentration is lower than that of the deep level. Generally speaking, both parameters tend to decrease when the measurementrange becomes deeper from the surface. Therefore, in order to evaluate trap densities precisely, this depth profile problem has to be treated correctly. Owing to Okumura (1986), most of the depth profiles observed in the present measurements are mainly due to electron trapping related to the leakage current because log-log plotting of Nt and τ gave straight lines with slopes of about one. Accordingly, measured parameters are affected by the leakage current and their true values are not given independently. But in several cases such plotting gave lumps of points in which electron trapping effect may be ignored. Then τ at 35 °C was found to take a stable value of 6.72 ± 0.04 sec and didn't depend on T_{As}. Using this τ value, Nt was obtained from above mentioned plotting procedure.

3. RESULTS AND DISCUSSION

Fig.3 shows the dependence of trap densities on T_{As}. Both EL6 and EL3 exhibit broad cusps at T_{As} of 616 to 619 C. This temperature range is quite close to that estimated from Nishizawa's relation (1981) and also close to the temperature of the cusps for the EPD and for trap densities (Parsey et al 1982 and referred in Nishizawa et al 1981) Such a phenomenon may be explained by following two models assuming that more than one kind of defects are related to the origin of the levels: 1)the level is associated with two different origins in both sides of the cusp ,but gives similar parameters in the applied measurement. 2)the level is associated with a single state of defects but the formation mechanism is different in the both sides of the cusp. Hurle (1979) proposed a model for the latter case. Because only densities were measured for these levels, it was left unclear if the defects associated with the level were always the same irrespective of As pressure. A precise spectroscopic measurement is desired. In the case of EL2, the trap density appears to change monotonously without a cusp. The time constant for electron emission was quite constant regardless of As pressure and the TDS-ICTS spectrum (fig.2) is always fitted well with an ideal line except for a small deviation in the long time region in agreement with Okumura's result (1986). These facts mean that EL2 level is associated with a single state of defects with a simple structure.

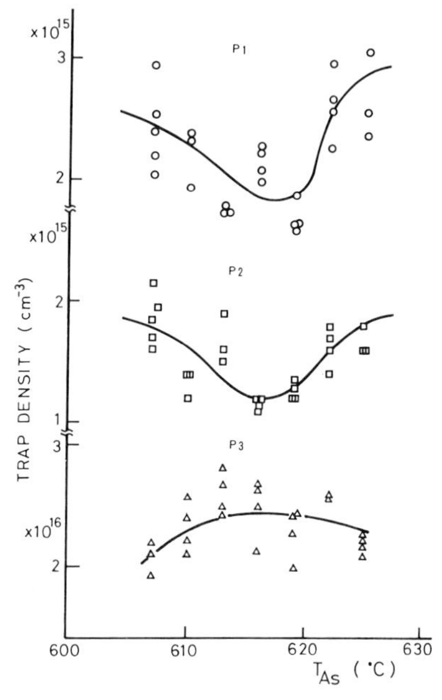

Fig.3 Effect of As pressure on the formation of three deep levels denoted in fig.1.*

The distribution of these levels were measured by the methods descrived above on the (100) wafer corresponding to T_{As} of 616 C. Fig.4 and 5 are the results. The density of EL2 seems to increase with the EPD. On the other hand EL3 is nearly constant. EL6 behaves in a little different way; it increases with the EPD in the radial direction, but it increases toward the tail while the EPD is almost constant. Therefore the formation of EL6 seems to be affected by at least two species of defects or impurities; one interacting with dislocations and another one having a distribution coefficient smaller than one. The latter species seem to be some impurities because

*Data points are recorded from central parts of each wafer.

the defect concentrations have been kept constant under a constant As pressure condition. Sulfur atoms which are main impurity atoms may be the candidates.

4. SUMMARY

The behavior of three deep levels in GaAs ingots grown under various As pressures was studied. Cusps in the As pressuere dependence were seen for the densities of EL6 and EL3, but not for the EL2. Natures of these levels were discussed by taking into account their correlation with the dislocations.

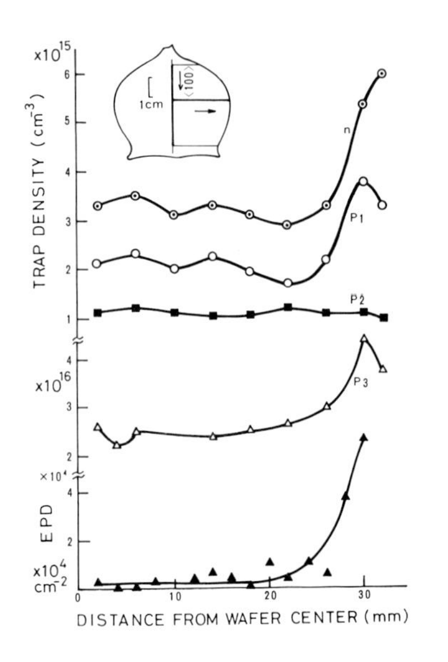

Fig.4 Distribution of deep levels and EPD in the radial direction. n is the density of ionized centers at -8 C.

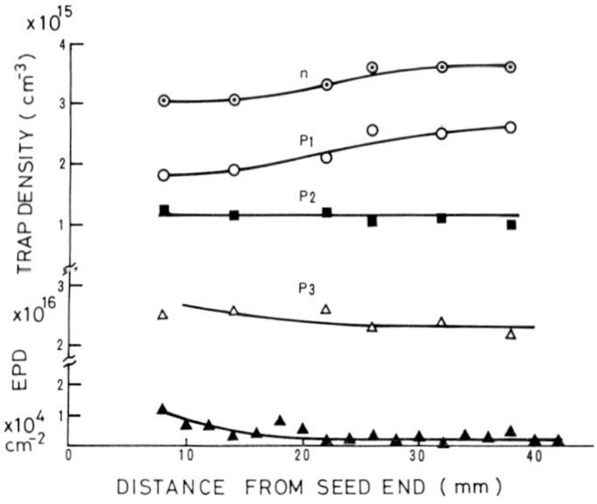

Fig.5 Distributionof deep levels and EPD in the longitudinal direction. n is the density of ionized centers at -8°C.

Acknowledgment

This work was carried out as part of Nishizawa Perfect Crystal Project under the support of the Research Development Corporation of Japan.

References

Boomgaard J V and Schol K 1957 Philips Res. Rep 12 127
Hurle D T J 1979 J Phys Chem Solids 40 627
Nishizawa J, Toyama N and Oyama Y 1981 Proc.3rd Int. School on Semiconductor Optoelectronics, Centniewo, P27
Martin G M, Mitonneau and Mircea A 1977 Electron. Lett.,13 191
Okumura T and Hoshino M 1986 Semi-Insulating III-V Materials Hakone, P409
Parsey J M Nanishi Y Lagowski J and Gatos H C 1982 J. Electrochem. Soc. 129 388

Inst. Phys. Conf. Ser. No. 91: Chapter 2
Paper presented at Int. Symp. GaAs and Related Compounds, Heraklion, Greece, 1987

Dislocation associated defects in GaAs

J.C. Bourgoin, H.J. von Bardeleben
Groupe de Physique des Solides de l'Ecole Normale Supérieure, C.N.R.S. ,
Tour 23, 2 place Jussieu, F-75251 Paris Cedex 05, France

H. Lim
Department of Electronic Engineering, Ajou University, Suwon 170, Korea

D. Stievenard
Département de Physique des Solides, Institut Supérieur d'Électronique du Nord, 41 boulevard Vauban, F-59046 Lille Cedex, France

A. Bonnet
Thomson-CSF, B.P. 10, F-91401 Orsay, France

ABSTRACT: A study of native defects in bulk GaAs by differential thermal analysis shows that thermal annealing at 450° C is accompanied by the recombination of typically 10^{18} cm^{-3} defects. A comparison of these results with those obtained by positron annihilation and electron paramagnetic resonance indicates them to be related to the mobility of α dislocations.

1. INTRODUCTION

Electronic techniques, i.e. electrical, optical techniques and electron paramagnetic resonance (EPR), have detected a number of native defects in GaAs. In particular in unintentionally doped, n-type bulk materials, Deep Level Transient Spectroscopy (DLTS) have shown the presence of electron traps, in the 10^{15} - 10^{16} cm^{-3} concentration range, among which is the well-known EL2 trap. In semi-insulating LEC material the concentration of this EL2 trap dominates and thus compensates the residual acceptor impurities (1). As to EPR, it demonstrated the presence of antisites, As_{Ga}, a defect which is known (2) to be part of the EL2 defect. Absorption and luminescence exhibit, in addition to spectra associated with shallow impurities, bands which have equally been attributed to EL2 (3,4).

Thus the picture deduced from these results obtained by different techniques is that of a material containing a distribution of shallow impurities compensated by deep native defects, the dominant one being EL2 in the 10^{16} cm^{-3} concentration range. The average, macroscopic distributions of both the impurities and the intrinsic defects are approximatively uniform, their concentrations varying by less than one order of magnitude across a wafer.

However, this simple picture is complicated by the presence of dislocations able to getter intrinsic defects, in particular As precipitates (5), and impurities as shown by high resolution scanning photoluminescence

(6,7). The aim of this communication is to describe results obtained by differential thermal analysis (DTA) which provide additional information on native defects apparently associated with dislocations. Combined with positron annihilation (PA) studies, the DTA data demonstrate that the As clusters attached to the dislocations dissociate and the liberated As recombine with vacancies present along the dislocations following a heat treatment at 450° C. The analysis of the recombination kinetics and its comparison with dislocation mobility in similar materials suggest that the dissociation of the As clusters is related to the dislocation mobility. The amplitude of the DTA signal provides an estimation of the concentration of these dislocation associated defects.

2. EXPERIMENTAL RESULTS

In the DTA technique, the difference between the heat powers necessary to maintain the studied sample and a reference sample at the same temperature during a thermal scan is measured. This heat power difference reflects the energy liberated by the defect reaction which takes place in the studied sample, the reference sample having been submitted to the same thermal treatment prior to the measurement. The technique has been applied to two types of materials, highly Si ($1 - 3 \times 10^{18}$ cm^{-3}) doped HB grown n-type and undoped LEC semi-insulating materials; their etch pit dislocation densities are 5×10^2 to 6×10^3 cm^{-2} and $10^4 - 10^5$ cm^{-2}, respectively.

For HB materials the DTA curves exhibit a peak in the range 440 - 470° C corresponding to an energy release of 1 to 2.3×10^{19} eV cm^{-3} (8); no DTA peak is detected in LEC materials (8). The maximum energy which can be released per defect in a reaction involving point defects is the one due to the recombination of a vacancy-interstitial pair, which is known to be 8 - 9 eV in GaAs (9). Therefore, the amount of the energy released implies a minimum defect concentration of the order of $1 - 4 \times 10^{18}$ cm^{-3}; it can be higher if the defect reaction which takes place is less energetic. From the shape of the DTA curve it has been possible to deduce, by a fitting procedure, the activation energy E associated with the reaction. Assuming a first order kinetics, the fit provides E = 2.2 eV; this value is reasonably accurate (± 0.2 eV) and is not much sensitive to the order of the reaction and to the preexponential factor of the annealing rate (a change by 10 modifies E by 5 %).

Such a high defect concentration is surprising. However, P.A. studies confirm the existence of such high defect concentrations : in LEC and HB materials native vacancy concentrations in the 10^{17} to 10^{18} cm^{-3} range (10 - 12) have been reported. They equally exhibit an annealing stage in the same (450° C - 600° C) temperature range.

Native point defect in GaAs have also widely been studied by EPR. However, the application of this technique is limited to the case of semi-insulating samples and thus highly doped n-type samples cannot be studied as such. In semi-insulating LEC materials one dominant defect is observed : the As_{Ga} defect being part of the As_{Ga}-As interstitial complex, the so-called EL2 defect, but only in the 10^{16} cm^{-3} concentration range (13). This defect does not anneal at 450° C. Vacancy related defects have also been observed by EPR, but only in electron irradiated semi-insulating materials (14). From this we can conclude, that vacancy related defects cannot be present as isolated, homogeneously distributed point defects in LEC grown, semi-insulating materials at concentration higher than 10^{15} cm^{-3}. Further, thermal annealing studies of the paramagnetic As vacancy defects have shown them to anneal at the 450° C stage. The recombination is accompanied by the formation of a different paramagnetic defect, tentatively associated with arsenic interstitial ions (15).

The EPR results are confirmed by DLTS studies on undoped LEC and HB n-type GaAs, which detect a variety of electron traps, including EL2, but once again only in the 10^{16} concentration range and whose concentraton does not change significantly under a treatment at 450° C.

3. DISCUSSION

Thus DTA and PA detect the existence of a large amount of defects, but without providing any information on their distribution inside the material. Their non observation by DLTS and EPR techniques, on the other hand, suggests that they are inhomogeneously distributed, i.e. located in regions which are not probed by these techniques. Let us therefore examine the possible correlation between these defects and dislocations. Indeed, dislocations are known to getter impurities (16,17), to be decorated by As precipitates (18) and to modify the distributions of shallow as well as deep levels in their vicinity (19,20). They also should be decorated with vacancy related defects since such defects are revealed by PA; this is confirmed by the fact that the trapping rate of positrons increases linearly with the dislocation density (21).

We are therefore led to the following situation : in a region of radius R around dislocations, the material is heavily perturbed, i.e. contains, in addition to the gettered impurities, As clusters and vacancy defects. The 450° C annealing stage is then related to the recombination of As, originating from the clusters, with vacancies (each recombination liberating an energy of ~ 8 eV) induced whether by the As clusters dissociation or by lattice rearrangement due to the movement of the dislocations. The As interstitials which escape recombination get trapped on impurities, forming a complex detected by EPR. It is actually the dislocation mobility which is the cause of the annealing state for three reasons. The first one is that no energy is released in LEC materials containing a very large dislocation density because these dislocations cannot move appreciably. Second, the activation energy associated with the energy release process is, within experimental accuracy, equal to the one associated with creep rate and the velocity of α dislocations in materials of similar doping level (22). Third, a 500 - 600° C thermal treatment induces a recovery of the luminescence of freshly introduced dislocations (23).

As we shall develop elsewhere, we can obtain an order of magnitude for the radius R (~ 50 μm) of this perturbed region using the variation of the threshold voltage of FETs as a function of their distance to dislocations (24). Then, assuming a uniform defect distribution along the dislocations we get a defect concentration of the order of 10^{19} cm^{-3} in this region.

4. CONCLUSION

The study of the energy release in the 450° C annealing stage, in HB and LEC materials, demonstrate that dislocations are surrounded by a region containing a defect concentration of the order of 10^{19} cm^{-3} whose annealing is triggered by the dislocation mobility.

REFERENCES

1. Martin G M, Farges J P, Jacob G and Hallais J P 1980 *J. Appl. Phys.* **51** 2840
2. von Bardeleben H J, Stievenard D, Deresmes D, Huber A and Bourgoin J C 1986 *Phys. Rev.* **B34** 7192
3. Martin G M 1981 *Appl. Phys. Lett.* **39** 747
4. Tajima M 1986 in *Defects in Semiconductors*, ed. H J von Bardeleben (Aedermannsdorf, Switzerland: Trans Tech Publ.) Material Science Forum **Vol. 10-12** p 493
5. Cullis A G, Augustus P D and Stirland D J 1980 *J. Appl. Phys.* **51** 2556
6. Marek J, Elliot A G, Wilke V and Gein R 1986 *Appl. Phys. Lett.* **49** 1732
7. Heinke W and Queisser H J 1974 *Phys. Rev. Letters* **33** 1082
8. Lim H J, von Bardeleben H J and Bourgoin J C 1987 *J. Appl. Phys.* to be published
9. Lim H J, von Bardeleben H J and Bourgoin J C 1987 *Phys. Rev. Letters* **58** 2315
10. Dlubek G and Brümmer O 1986 *Annalen der Physik* **43** S 178
11. Dannefaer S and Kerr D 1986 *J. Appl. Phys.* **80** 591
12. Stucky M, Corbel C, Geffroy B, Moser P and Hautojarvi P 1986 in *Defects in Semiconductors*, ed. H J von Bardeleben (Aedermannsdorf, Switzerland: Trans Tech Publ.) Materials Science Forum **Vol 10-12** p 265
13. Elliot K, Chen R T, Greenbaum S G and Wagner R J 1984 *Appl. Phys. Lett.* **44** 907
14. von Bardeleben H J and Bourgoin J C 1986 *Phys. Rev.* **B33** 2890
15. von Bardeleben H J and Bourgoin, J C 1986 in *Semi-Insulating III-V Materials*, ed. H Kukimoto and S Miyazawa (Ohmsha, Japan) p. 355
16. Ding J, Chang J S C and Bujatt M 1987 *Appl. Phys. Lett.* **50** 1089
17. Marek J, Elliot A G, Wilke V and Geiss R 1986 *Appl. Phys. Lett.* **49** 1732
18. Cullis A G, Augustus P D and Stirland D J 1980 *J. Appl. Phys.* **51** 2556
19. Kikuta T, Katsumata T, Obokata T and Ishida K 1985 *Inst. Phys. Conf. Ser.* **74** p 47
20. Heinke W and Queisser H J 1974 *Phys. Rev. Letters* **33** 1082
21. Kuramoto E, Takenchi S, Noguchi M, Cheba T and Tsuda N 1973 *J. Phys. Soc. Japan* **34** 103
22. Steinnardt H and Haasen P 1978 *Phys. Stat. Sol. (a)* **49** 93
23. Heinke W 1975 *Inst. Phys. Conf. Series* **23** 380
24. Miyazawa S and Hyngan F 1986 *IEEE Trans., Elect. Dev.* **ED33** 227

Paper presented at Int. Symp. GaAs and Related Compounds, Heraklion, Greece, 1987

Residual dislocations in LEC grown In-alloyed GaAs by photoetching in CrO_3-HF solutions

A Chabli, C Boveyron, E Molva, F Bertin and Ph Bunod

Division d'Electronique, de Technologie et d'Instrumentation - IRDI-CEA, CENG BP 85 X, 38041 Grenoble Cédex, France

ABSTRACT : LEC grown In-alloyed GaAs samples are investigated by microscopic observations after photoetching in CrO_3-HF mixtures. It is evidenced that residual dislocations propagating in the growth direction are generated through the combination of two different remelting steps.

1. INTRODUCTION

Many authors (Jacob et al. 1983, Kobayashi et al. 1986, McGuigan et al. 1986, Ibuka et al. 1986) reported on having successfully grown large diameter GaAs crystals free of dislocations, by heavy In alloying. However, it has not yet been possible to obtain reproducibly controled dislocation density, since In-alloying alone is not sufficient for eliminating all of the dislocations.

We report here on results of In-alloyed GaAs etching in aqueous CrO_3-HF mixtures under illumination (Weyher et al. 1986). The discussion deals with the generation mecanism of the residual dislocations in this material.

2. EXPERIMENTAL

Semi-insulating In-alloyed GaAs crystals with In concentration range of 2×10^{19} - 1×10^{20} at.cm^{-3} were grown along <001> direction in a high pressure Czochralski puller using pBN crucible. They were sliced parallel to the growth axis followed by mechanochemical polishing. The samples have been immersed in aqueous CrO_3-HF mixture - 1,8 mol/l CrO_3 and 2,3 mol/l HF - at room temperature and under white light illumination. In order to support the photo - etching results, transmission X-ray topography was performed using MoK_α radiation.

3. RESULTS AND DISCUSSION

We (1986) have previously reported on that three kinds of dislocations are observed in In-alloyed GaAs, as schematically shown in figure 1. Matsui (1987) confirmed this classification. But, he distinguishes between different kinds of thermal stress induced dislocations (i.e. C type in figure 1) which are drastically reduced by the increase of the In content and the reduction of the thermal stress. The two other types (i.e. A and B in figure 1) are not clearly affected by the In content, and depend strongly on growth conditions.

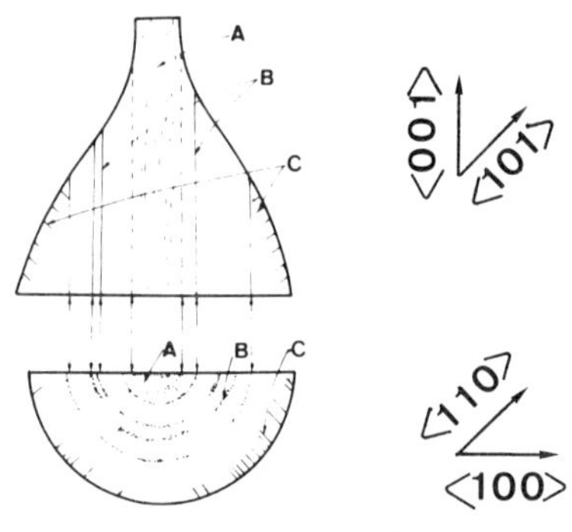

Fig. 1. Schema of configuration of residual dislocations in In-alloyed GaAs. A : generated at the seed-on end, B : generated at the cone, C : slip dislocations on the (III) planes

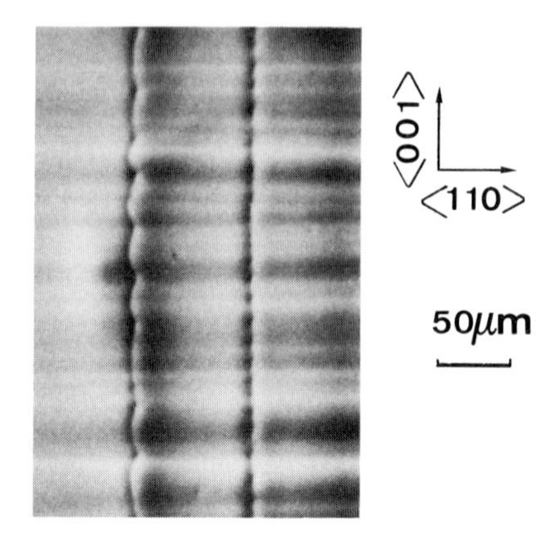

Fig. 2. Optical micrograph of a B type dislocation after photoetching in HF-CrO_3 aqueous solutions

Unlike KOH etching, CrO_3-HF photoetching is sensitive to chemical inhomogeneities and reveals dislocations through their Cottrell atmosphere (Weyher et al. 1986, Bunod et al. 1987). Thus, this method is very convenient for investigations on A and B type dislocations simultaneously with growth striations. The configuration of A and B type dislocations was first described by Scott et al. (1985). They consist of aligned pits -As precipitates- linked by arced grooves -dislocations- (Brozel et al. 1986) as shown on figure 2. The pit distribution seems to be correlated to the growth striations (Matsui 1986). Some of the B type dislocations are associated with a Ga droplet at their beginning as shown on figure 3a. However, the unusual near <001> direction of A and B type dislocations indicates that they propagate in the crystal through the growth. So, they cannot result from the surface decomposition which takes place after the crystal going out of the encapsulant. But the presence of a dislocation may favour the surface decomposition.

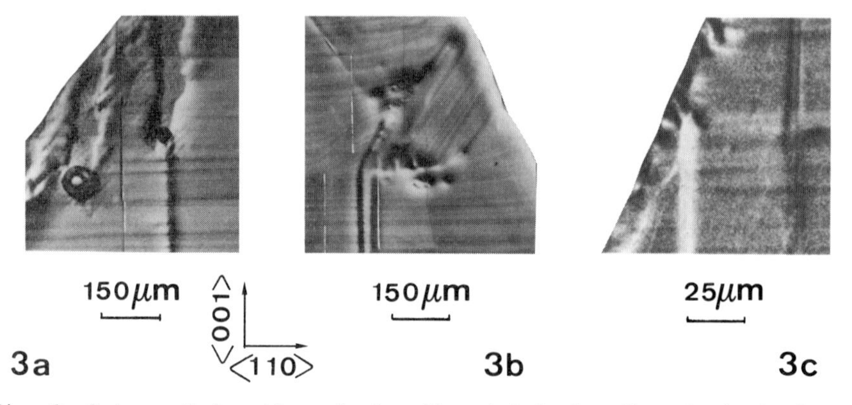

Fig. 3. B type dislocations in In-alloyed GaAs 3a, 3b : Optical micrograph after photoetching in HF-CrO_3 aqueous solutions, 3c : X-ray transmission topograph of (110) slice

On the other hand, when surface decomposition is limited, all B type dislocations are associated with a complex configuration of the growth striations shown on figure 3b. This is confirmed by the transmission X-ray topograph of the figure 3c. This growth striation configuration could be formed in five steps as schematically viewed on figure 4.

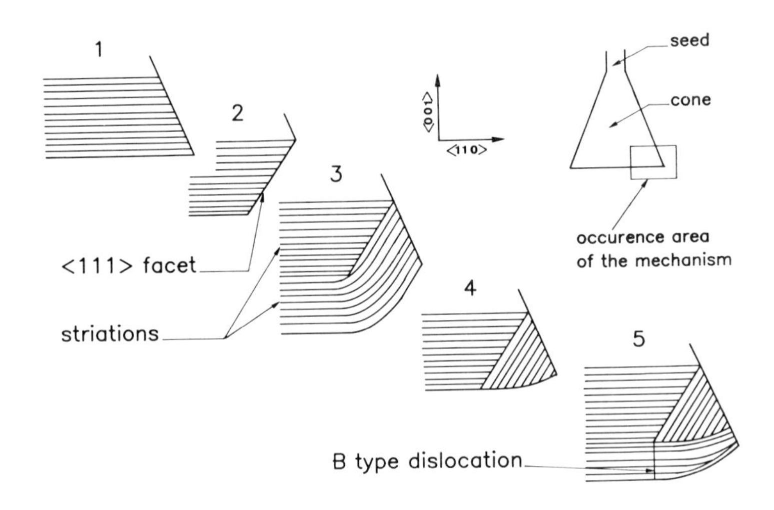

Fig. 4. Generation mechanism of B type dislocations in 5 steps. 1 : growth, 2 : local remelting giving rise to a (111) facet, 3 : growth after generation of lattice mismatch dislocations at the (111) facet 4 : remelting extended to the whole diameter of the crystal, 5 : growth with propagation of a B type dislocation resulting from the mismatch dislocations of step 3.

The step 2 is a remelting at the edge of the solid-liquid interface giving rise to a (111) facet and may be induced by a difference between the thermal axis of the puller and the crystal rotation one. At the step 3, the solid-liquid interface shows rapid fluctuations of the In-content because of the high angle with growth striations. Thus generation of mismatch dislocations may occur at the (111) facet as observed on figure 3b. Then the dislocations are propagated in the growth direction after the step 4 and through the step 5 giving rise to B type dislocations. The step 4 is a more generally observed remelting, induced by temperature or pulling rate fluctuations, and thus extended to the whole diameter. We have never observed dislocation generation by this remelting when it takes place alone, and this may be due to the low angle between growth striations and the solid-liquid interface after the remelting.

The A type dislocations are assumed to be generated by the lattice mismatch between the seed and the growing crystal (Yamada et al. 1986, Matsui 1986). When the remelting takes place at small crystal diameter (i.e. crystal growth beginning), the solid-liquid interface becomes instable and the occurence of lattice mismatch increases giving rise to a severe A type dislocation generation.

The role of the remelting in the A and B type dislocation generation evidences the effect of growth conditions (i.e. temperature and pulling rate fluctuations, puller thermal symmetry) on the residual dislocations in In-alloyed GaAs.

The authors would like to thank their colleagues R. Accomo, A. Basset, C. Faure, J.C. Rival and J. Toffin for their contributions and are gratefull to Y. Faure and G. Rolland for X-ray topography data.
This work was supported by "Ministère de la Recherche et de la Technologie" and by "Ministère de la défense, Direction des Recherches, Etudes et techniques".

REFERENCES

Brozel MR, Clark S and Stirland DJ 1986 *Proc. Conf. Semi-Insulating III-V materials* ed. H Kukimoto and S Miyazawa (Tokyo : Ohmsha) pp 133-138

Bunod P, Molva E, Chabli A, Bertin F and Boveyron C 1987 *Proc. Conf. Defect Recognition and Image processing in III-V compounds* (in press)

Chabli A, Molva E, George A, Bertin F, Bunod P and Bletry J 1986 *Proc. Conf. Advanced materials for telecommunication* ed. PA Glasow (Les Ulis : Les Editions de Physique) pp 27-39

Ibuka T, Seta Y, Tanamura M, Orito F, Okano T, Hyuga F and Osaka J 1986 *Proc. Conf. Semi-Insulating III.V materials* ed. H Kukimoto and S Miyazawa (Tokyo : Ohmsha) pp 77

Jacob G, Duseaux M, Farges JP, van den Boom M M B and Roksnoer PJ 1983 *J. Crystal Growth* 61 417

Kobayashi T, Kohda H, Nakanishi H, Hyuga F and Hoshikawa K 1986 *Proc. Conf. Semi-Insulating III-V materials* ed. H Kukimoto and S Miyazawa (Tokyo : Ohmsha) pp 17-22

McGuigan S, Thomas RN, Hobgood HM, Eldridge GW and Swanson BW 1986 *Proc. Conf. Semi-Insulating III-V materials* ed. H Kukimoto and S Miyazawa (Tokyo : Ohmsha) pp 29-34

Scott MP, Laderman SS and Elliot AG 1985 *Appl. Phys. Lett.* 47 1280

Weyher JL and van de Ven J 1986 *J. Crystal Growth* 78 191

Yamada K, Kohda H, Nakanishi H and Hoshikawa K 1986 *J. Crystal Growth* 78 36

Inst. Phys. Conf. Ser. No. 91: Chapter 2
Paper presented at Int. Symp. GaAs and Related Compounds, Heraklion, Greece, 1987

The effects of plastic deformation on electronic properties of GaAs

M. Skowronski, J. Lagowski, M. Milshtein, C. H. Kang, F. Dabkowski, and H. C. Gatos

Massachusetts Institute of Technology Cambridge, MA 02139 USA

ABSTRACT: Deformation was found to decrease free electron concentration and mobility in n-type GaAs; in semi-insulating material it increased ionization of the midgap EL2 donor, and it had no effect on the transport properties of p-type GaAs. All these effects are associated with a deep acceptor defect introduced at a concentration proportional to the deformation. No evidence was found of any increase of EL2 concentration or other donors. On the contrary, in deformed and annealed GaAs the concentration of EL2 decreased well below that in undeformed sample.

1. INTRODUCTION

The relatively low mechanical strength of GaAs causes melt grown crystals to be very susceptible to plastic deformation induced by thermal gradients (Jordan et al 1981). The magnitude of thermal stress can be particulary large in crystals grown by liquid encapsulated Czochralski (LEC) method. The understanding of structural and electrical consequences of these effects is at present very limited. A number of investigations carried out in recent years were devoted to the effects of plastic deformation on electronic properties of GaAs with focus on the main native deep donor, EL2 and the related antisite defect, As_{Ga} (Ishida et al 1980, Weber et al 1982, Wosinski et al 1983, Hasegawa et al 1984). The present study adresses the question of deformation induced midgap donor, the acceptors, and their effects on deep level occupation. Electronic properties of plastically deformed GaAs were investigated employing a series of n-type, SI, and p-type crystals, which enabled the separation between the creation of new deep levels and the changes in population of the levels already present prior to deformation. Our experimental approach included the measurements of DLTS, Hall effect and optical absorption, all performed on the same sample. Deformed samples were subsequently thermally annealed in order to evaluate the stability of deformation-induced defects.

2. EXPERIMENTAL

The n-type samples, grown by the LEC method, were intentionally doped with selenium. The resulting free carrier concentration was 5 to $10*10^{16}$ cm^{-3} and the mobility was in excess of 3000 cm^2/Vs. Two sets of p-type samples doped with zinc were used. One set with a hole concentration of $5*10^{15}$ cm^{-3} and the other of $5*10^{16}$ cm^{-3} (300K). Each sample used in the experiment had its own reference sample cut in the immediate vicinity from the same crystal. The reference samples were carried through the same thermal treatment as the deformed specimens to precisely determine the effects of deformation.

3. RESULTS AND DISCUSSION

The net change in the majority carrier concentration determined at 300 K from Hall effect measurements is given in Fig. 1 as a function of the deformation.

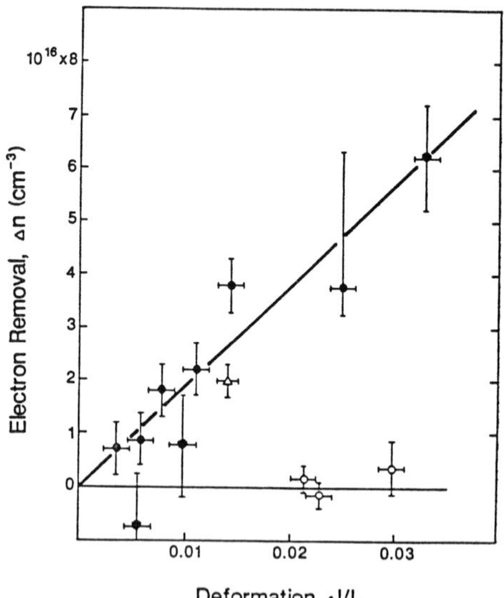

Fig. 1. The electron removal versus the degree of deformation. Full circles correspond to n-type samples; open circles to p-type samples. The open triangle denotes the magnitude of electron removal from EL2 in SI sample.

For n-type samples, the electron concentration decreases linearly with deformation. The same qualtitative trend was observed previously for GaAs plastically deformed or irradiated with electrons. The decrease was tentatively interpreted as the result of the creation of new acceptor type defects (Hasegawa et al 1984, Yamaguchi and Ventura 1985). These stress induced acceptors, N_A, compensate shallow donors, N_D, causing decrease in the electron concentration, $n=N_D-N_A$. In addition, the negatively charged, filled acceptors constitute effective scattering centers decreasing the electron mobility value. Indeed, in n-type samples a large decrease of electron mobility was observed after deformation. For p-type samples (open circles in Fig. 1) no significant change in hole concentration was observed. This behaviour is consistent with the interpretation for n-type samples providing that the stress induced acceptors are sufficiently deep not to be ionized at room temperature. From the Fermi energy in our samples ($E_F-E_V=0.2$ eV) we conclude that the acceptors must be located at an energy exceeding 0.2 eV above valence band. From results for p-type GaAs we also conclude that no donor-type defects are introduced during deformation.
For SI GaAs the concentration of free carriers is as low as 10^7 cm^{-3}. The deformation induced acceptors are compensated by an increased degree of ionization of the midgap EL2 donor without any significant change in the free carrier concentration. The change of EL2 occupation in SI GaAs caused by deformation was measured directly by an optical method described elsewhere (Lagowski et al 1987).
DLTS spectra of deformed and reference n-type samples show EL2 and EL5 electron traps typical for melt grown GaAs. No new electron traps were detected in deformed samples. The concentrations of both traps before and after deformation were identical.
Hole traps located in the lower half of the energy gap were studied employing optical DLTS. Typical ODLTS spectra of deformed and reference n-type samples are shown in Fig. 2.

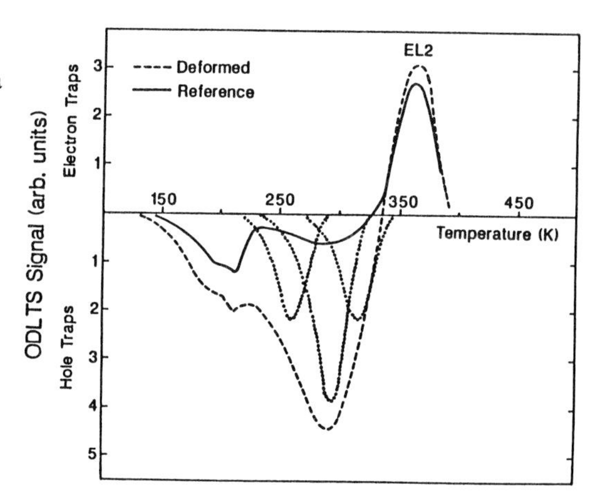

Fig. 2. Optical DLTS spectra of deformed and reference samples of n-type GaAs; t_1/t_2=20ms/40ms.

Electron and hole traps were observed as, respectively positive and negative peaks. The position of the major electron trap coincides with the EL2 peak in the DLTS spectrum. The hole trap spectrum of the deformed crystal is dominated by a large wide band peaked at about 290 K. The corresponding activation energy was estimated to be $0.45 \mp .1$ eV. This trap is deep enough not to affect the free hole concentration in p-type crystals. As shown in Fig. 3, the hole trap concentration increases linearly with deformation.

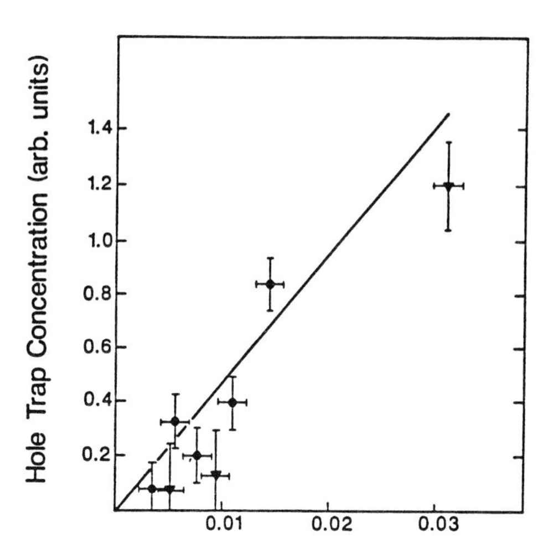

Fig.3. The concentration of the dominant deformation induced hole trap vs. deformation.

ODLTS does not give the absolute value of the trap concentration, but only its lower limit due to incomplete filling by optical pulse. For the 0.45 eV trap this limit is about $1*10^{16}$ cm^{-3} for deformation of 2%. Total hole trap concentration (all components of the broad band) is larger than $2.5*10^{16}$ cm^{-3} at a deformation of 2% in good agreement with the free electron removal data shown in Fig. 1.

Deformed samples were annealed at 850°C in an arsenic overpressure. We found that prolonged annealing at 850°C elliminated deformation induced changes in electronic properties including mobility decrease and electron removal. The most important effect of annealing was the total disappearance of the 0.45 eV hole trap. According to the proposed interpretation this acceptor is responsible for the electron removal and the decrease of mobility. The second important observation was the decrease of EL2 concentration in deformed and subsequently annealed samples.

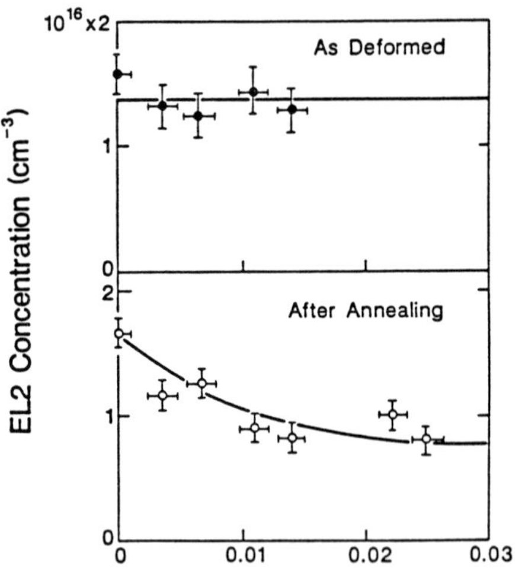

Fig. 4. EL2 concentration vs. degree of deformation for as-deformed samples (upper figure) and after annealing for 10 hours at 850°C (lower figure).

The lower figure shows a consistent trend of decreasing EL2 concentration with increasing deformation. For samples deformed by 1.4%, only half of the original EL2 centers is preserved. The samples used in these measurements were all conducting n-type after annealing. Therefore, all EL2 centers were filled with electrons, and the decrease of the optical absorption was a measure of the decrease in the EL2 concentration rather than the EL2 occupation. Similar results were also obtained with DLTS. It is thus clear that although deformation itself does not change the EL2 concentration, the subsequent annealing of deformed samples leads to partial EL2 annihilation. In undeformed samples the EL2 defect was stable up to 1100°C, therefore this annihilation must result from the interaction of EL2 with deformation induced defects.

REFERENCES

Hasegawa F., Yamamoto N., and Nannichi Y., 1984 Extended Abstracts of the 16th Conference on Solid State Devices and Materials, Kobe Japan p. 169

Ishida T., Maeda K., and Takeuchi S., 1980 Appl. Phys. 21, 257

Jordan A. S., Caruso R., Von Neida A. R., and Nielsen J. W. 1981 J. Appl. Phys. 52, 3331

Lagowski J., Bugajski M., Matsui M., and Gatos H. C., 1987, Appl. Phys. Lett. 51, 511

Weber E., Ennen H., Kaufman U., Windscheif J., Schneider J., and Wosinski T., 1982 J. Appl. Phys. 53, 6140

Yamaguchi F. and Ventura C. 1985 J. Appl. Phys. 57, 604

Influence of residual impurities on the electrical properties and annealing behaviour of s.i. GaAs

M. Baumgartner, K. Löhnert, G. Nagel, H. Rüfer, E. Tomzig

Wacker-Chemitronic, PO Box 11 40, 8263 Burghausen, FRG

ABSTRACT: Resistivity and thermal stability of the electrical properties of undoped s.i. GaAs crystals were studied with respect to EL2 concentration and residual impurity content. Both resistivity and annealing behaviour turn out to be controlled by the main residual shallow acceptor carbon. These results are discussed within the model of compensation of net shallow acceptors by the deep donor EL2. A correlation was also found between the resistivity and the content of boron. Other impurities were only present at very low levels.

1. INTRODUCTION

High resistivity, i. e. $>10^7\ \Omega$ cm, and stability of the electrical properties during thermal processing are essential requirements for semi-insulating (s.i.) GaAs substrates in the fabrication of integrated circuits. According to Martin et al. (1980) the semi-insulating behaviour of undoped GaAs is due to the compensation of the net shallow residual acceptors (N_{SA}-N_{SD}) by the intrinsic deep donor level EL2 which is related to the As_{Ga} antisite defect.
The carrier concentration n and resistivity ρ are then given by

$$n = \left[\frac{N_{EL2}}{N_{SA}-N_{SD}} - 1\right] \times N_C \times \exp[-E'(EL2)/kT]$$

$$\rho = [e\cdot(n\cdot\mu_n + p\cdot\mu_p)]^{-1}$$

where $N_C = 4.7 \times 10^{17}$ cm^{-3} is the effective density of states in the conduction band and E' (EL2) = 0.69 eV is the distance between EL2-level and conduction band. In Fig. 1 the resistivity is plotted as a function of the compensation ratio N_{EL2} /(N_{SA}-N_{SD}).
In this paper we studied the resistivity and thermal annealing behaviour of undoped s.i. GaAs and analyzed the results with respect to the above model.

2. EXPERIMENTAL

The undoped s.i. GaAs crystals (2 and 3 inch diameter) investigated in this study were grown by the LEC technique from pBN crucibles under low pressure (2-3 atm.) conditions. The melt was compounded from 6N Ga and 6N As. All ingots were post-growth annealed (T $>$ 1000° C) before slicing (Löhnert et al., 1986). Thermal annealing of substrate wafers was performed at 850° C, 20 min with the wafers in close contact to each other in a 90% N_2 + 10% H_2 forming gas atmosphere without any additional source of As pressure.
Resistivity and Hall mobility were measured using the van der Pauw technique at room temperature.

The EL2 concentration of the samples was determined by IR absorption measurements at $\lambda = 1.0$ µm (Martin 1981). The carbon content was evaluated by far-infrared local vibrational mode absorption at room temperature using the calibration of Homma et al (1985) with a detection limit of about 5×10^{14} cm^{-3} for 3-4 mm thick slices.
The contents of other impurities in GaAs were assessed by spark source mass spectrometer analyses with specified detection limits of 5 - 10 ppba (Spectrographic Analytical Services, C. Whitehead, UK).
Infrared absorption at 3590 cm^{-1} and a molar extinction coefficient of 141 l mol^{-1} cm^{-1} (Franz, 1965) was used to determine the water content in B_2O_3.

3. RESULTS AND DISCUSSION

In all investigated samples (seed and tail end wafers from more than 70 ingots) a rather constant value of $(1.3 \pm 0.2) \times 10^{16}$ cm^{-3} was measured for the EL2 concentration independent of the sample resistivity ranging from 10^6 to some 10^8 Ωcm. This indicates that the crystals were grown under reproducibly stable stoichiometrical conditions, since the EL2 concentration is strongly influenced by stoichiometry. According to the compensation model where the resistivity is determined by the $N_{EL2}/(N_{SA}-N_{SD})$ ratio (see Fig. 1) this means on the other hand that the individual resistivity values must be controlled by the content of shallow acceptors. As it is shown in Fig. 2 we find indeed a clear correlation of resistivity with carbon concentration, which implies at the same time, that carbon is the dominant residual acceptor. Material with resistivity 10^8 Ωcm showes[C]-values up to 6×10^{15} cm^{-3} while many of the samples with resistivity close above or below 10^7 Ωcm have carbon contents near or below detection limit ($\sim 5 \times 10^{14}$ cm^{-3}). Concentrations of other acceptor impurities (Zn, Mg,...) were found to be well below 10^{15} cm^{-3}, not correlating with the measured resistivity values. A correlation was also found between resistivity and boron concentration, as it is shown in Fig. 3. Low boron contents ($< 10^{16}$ cm^{-3}) are observed for material with resisitivity $< 5 \times 10^7$ Ωcm whereas for high resistivity ($\geq 10^8$ Ωcm) the boron concentrations may rise above 10^{17} cm^{-3}. Since boron is incorporated mainly as an isoelectronic inpuritiy on Ga sites it seems probable that the correlation with resistivity is

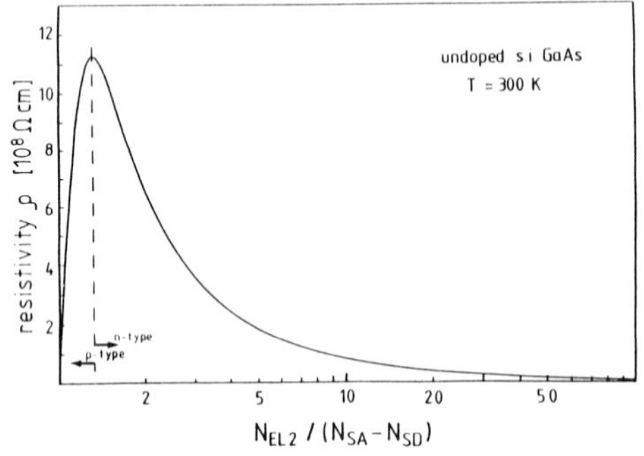

Fig. 1: Resistivity calculated as a function of the compensation ratio

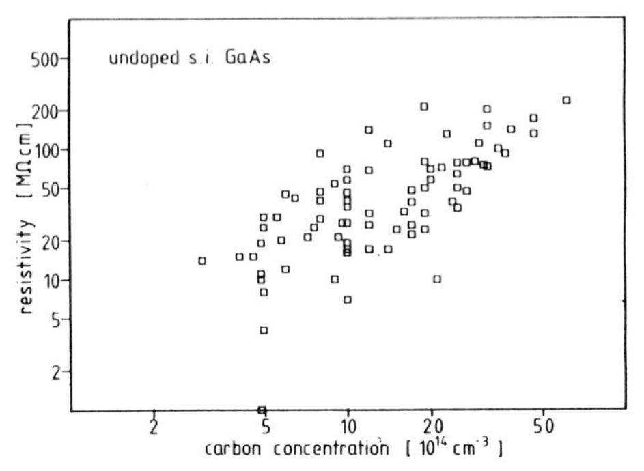

Fig. 2: Resistivity as a function of carbon content

rather indirect and due to a parallel incorporation of boron and carbon. On the other hand there are evidences however, that boron can noticeable affect ion implant results (Osaka et al., 1986).

The incorporation of residual impurities is influenced by the water content of the boric oxide encapsulant (Hunter et al., 1984; Hobgood et al., 1981). Results for carbon concentration for different water contents are shown in Fig. 4. [C] varied by nearly two orders of magnitude for water contents ranging from about 300 ppm to about 1700 ppm. Results on thermal stability are shown in Fig. 5 where the resistivity after a typical post implant annealing is plotted against initial resistivity. It is seen in Fig. 4 that material with high resistivity (i.e. $\geq 10^8$ Ω cm) drops below $10^7 \Omega$ cm with nearly 100 % probability, whereas material with resistivity values close above and also below $10^7 \Omega$ cm generally exhibits a rise in resistivity to the mid $10^7 \Omega$ cm range. A similar dependency holds for the measured Hall mobilities. For the lower initial resistivity material the mobility values after annealing exceed 5000 cm²/Vs. Mobilities drop drastically even to p-type for $\geq 10^8$ Ω cm material. As we have seen that the resistivity before annealing is mainly controlled by carbon, it is clear, that carbon is also the decisive factor for the thermal stability. This can be explained by the compensation model (see Fig. 1) and the outdiffusion of EL2 at the surface during thermal treatment. A loss of EL2 means a reduction of the compensation ratio $N_{EL2}/(N_{SA}-N_{SD})$. This leads to an increase of resistivity when starting from a high ratio which means a low [C] and so a low resistivity before annealing.

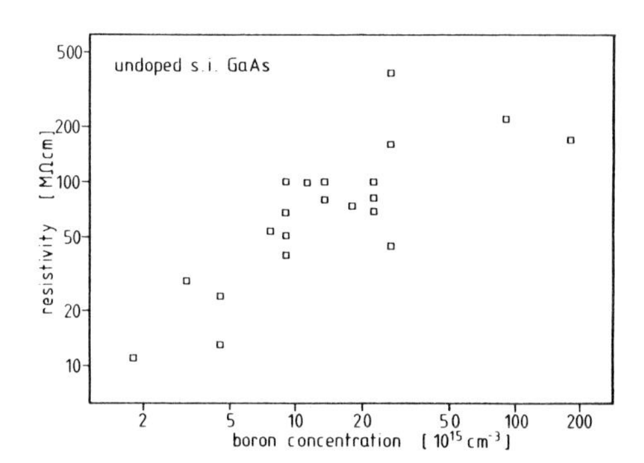

Fig. 3: Resistivity as a function of boron content

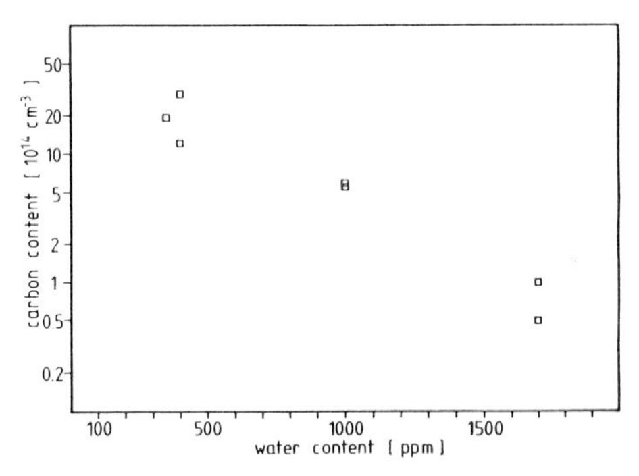

Fig. 4: Carbon content as a function of water content in B_2O_3

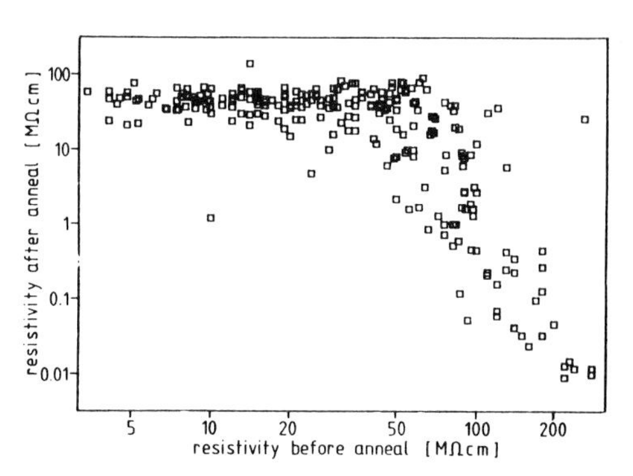

Fig. 5: Effect of annealing on resistivity of s.i. GaAs

In material with high [C] and high initial resistivity where $N_{EL2}/(N_{SA}-N_{SD})$ is lower from the beginning, the loss of EL2 may be sufficient to pass the maximum of the resistivity curve resulting in a drop of resistivity and eventually p-type behaviour. Although the experimental results can be explained qualitatively by this compensation model there are some discrepancies concerning absolute values. On the one side we have never observed resistivity values close below or above 10^9 Ωcm in undoped GaAs as they are calculated from the model. On the other side the observed low resistivity material (e.g.$\sim 10^6$ Ωcm) would require very high compensation ratios (>1000), which would mean unrealistic low values of residual acceptor concentration ($< 10^{13}$ cm^{-3}). One possible explanation could be that s.i. GaAs is not described completely by shallow acceptors and EL2 and also other intrinsic levels have to be taken into account (Kitagawara et al., 1986).

4. CONCLUSIONS

Resistivity in undoped s.i. GaAs is determined by carbon concentration and also shows correlation with boron content, whereas EL2 concentration is very constant throughout all investigated crystals. Carbon is obviously the dominating acceptor down to concentrations in the upper 10^{14} cm^{-3} range. The concentration of other impurities is quite low, not affecting the electrical properties at the reported [C] -levels. Thermal stability is related to initial resistivity and can be explained by the carbon content and the outdiffusion of EL2. High resistivity material ($\geq 10^8$ Ωcm) is thermally unstable. Material with resistivity close to or even below 10^7 Ωcm shows high thermal stability during post-implant annealing and should be favourable for I^2 application.

ACKNOWLEDGEMENTS

Financal support by the Federal Ministry of Research and Technology is gratefully acknowledged.

REFERENCES

Franz H., 1965, Glastechn. Ber. 38, 54
Hobgood H.M., Braggins T.T., Barret D.L. Eldrich G.W. and Thoma R.N, 1981, Am. Conf. Cryst. Growth, 5th, San Diego
Homma Y., Ishii Y., Kabayashi T. and Osaka J., 1985, I. Appl. Phys 57, 293
Hunter A.T., Kimura H., Baukus J.P., Winston H.V. and Marsh O.J., 1984, Appl. Phys. Lett. 44, 74
Kitagawara Y., Noto N., Takahashi T. and Takenaka T., 1986, Semi-insul. III/V materials, Hakone, ed. H. Kukimoto (Ohmsha Ltd.), 273
Loehnert K., Wettling W., Koschek G., 1986, Semi-insul. III-V-materials, Hakone, ed. H. Kukimoto (Ohmsha Ltd.), 267
Martin G.M., Farges J.P., Jacob G. and Hallais J.P., 1980 J. Appl. Phys. 51, 2840
Martin G.M., 1981, Appl. Phys. Lett. 39, 747
Osaka J., Hyuga F., Kobayashi T., Yamada Y., and Orito F., 1986, Appl. Phys. Lett. 50, 191
Ta L.B., Hobgood H.M., Rohatgi A. and Thomas R.N., 1982, J. Appl. Phys. 53, 5771

Inst. Phys. Conf. Ser. No. 91: Chapter 2
Paper presented at Int. Symp. GaAs and Related Compounds, Heraklion, Greece, 1987

Characterization of excess arsenic in GaAs crystals with laser Raman spectroscopy

Takashi Katoda and Koji Yano

Institute of Interdisciplinary Research, Faculty of Engineering,
The University of Tokyo
4-6-1 Komaba Meguro-ku,Tokyo 153, Japan

Abstract: Excess arsenic included in GaAs crystals was characterized by repetition of oxidation, estimation based on Raman spectra and etching. In general the amount of excess arsenic included in LEC GaAs was much larger than those in HB and LPE crystals. Excess arsenic in GaAs grown by MOVPE with $Ga(C_2H_5)_3$ source and $V/III \geq 20$ was comparable to that in LEC GaAs. An example of diffusion constant of the arsenic from LEC GaAs was estimated to be about $7x10^{-11}$ (cm^2/s) at 500°C which is comparable to that of an arsenic vacancy.

1.Introduction

Characterization of deviation from the stoichiometric composition of GaAs is very important to realize high quality crystals for substrates and devices. Although some techniques have been proposed (Fujimoto 1984,Terashima et al. 1985) they are not satisfactory from the stand points of simplicity or theoretical grounds. We tried to characterize excess arsenic in GaAs crystals by repetition of oxidation, estimation based on Raman spectra, and etching. The reason why oxidation is used rather than annealing in a non-oxidizing atmosphere is that a small amount of uncontrollable oxygen lowers reproducibility of arsenic segregation.

2.Experiment

Thermal oxidation of GaAs was done in an atmosphere of $N_2:O_2 = 4:1$ at a temperature between 350 and 600°C. In the estimation of a diffusion coefficient of excess arsenic a relatively large sample is oxidized at first and then Raman spectrum is measured. It is etched with HF solution in order to remove an oxide layer. Then it is cut into seven or eight small chips. Each chip is etched with a solution of $H_3PO_4:H_2O_2:H_2O=10:1:10$ for a different period in order to remove GaAs of a different thickness. The chips are oxidized again (second oxidation) and then Raman spectra are measured. The process is shown in Fig.1. The samples used in the experiment are GaAs crystals grown by liquid encapsulated Czochralski (LEC), horizontal Bridgeman(HB), liquid phase epitaxy(LPE), and metalorganic vapor phase epitaxy (MOVPE). Raman spectra were measured with a back scattering geometry at room temperature.

3.Model used in the calculation of a diffusion constant

Diffusion of particles in a material is presented by the following equation according to Wick's law,

$$\frac{\partial N}{\partial t} = D\nabla^2 N \tag{1}$$

where N is a concentration, t is time, and D is a diffusion constant. If we assume that arsenic which outdiffused to the oxide–GaAs interface does not diffuse into GaAs again, the following equation is obtained.

$$N(x,t) = N_0 \ \mathrm{erf}\left(\frac{x}{2\sqrt{Dt}}\right) \tag{2}$$

N_0 is the initial concentration of excess arsenic. The amount of arsenic N_1 which outdiffused during the first oxidation is expressed by the following equation.

$$N_1(t) = \int_0^{\infty}\{N_0 - N(x,t)\}dx = 2N_0\sqrt{\frac{Dt}{\pi}} \tag{3}$$

If the period of the second oxidation is equal to that of the first oxidation, the amount of arsenic N_2 which outdiffused during the second oxidation is given by the following equation,

$$N_2(t) = N_1(t)\{1 - e^{-\Delta y^2} + \sqrt{\pi}\ \Delta y \ \mathrm{erf}\ (\Delta y)\}$$
$$\Delta y = \frac{\Delta x}{2\sqrt{Dt}} \tag{4}$$

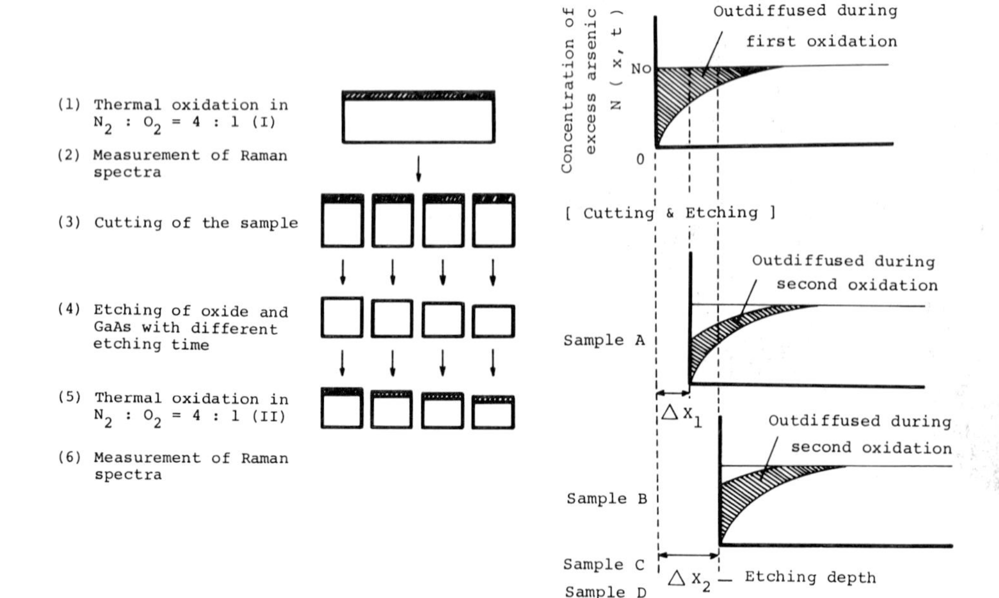

Fig.1. Process of experiment used for estimation of a diffusion constant.

Fig.2. Model used in the calculation of a diffusion constant of arsenic.

where Δx is a depth removed by etching. The model described above is shown schematically in Fig.2.

4. Results and discussion

Figure 3 shows Raman spectra from GaAs oxidized at various temperatures for 60 min. Only a peak corresponding to Eg mode of As crystal is present in the spectra from GaAs oxidized at a temperature lower than 370°C while A_1g and Eg modes are present when GaAs is oxidized higher than 380°C (Lannin et al. 1975). The difference in the spectra shows that there is a stage of oxidation or outdiffusion of arsenic. A relative intensity of Eg mode against GaAs LO mode was used as a measure of amount of arsenic because a part of TO mode is imposed to A_1g mode.

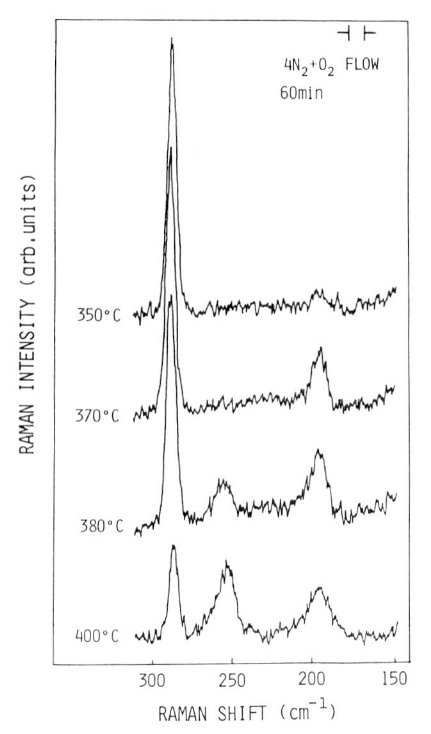

Fig.3. Raman spectra from GaAs after thermal oxidation at various temperatures.

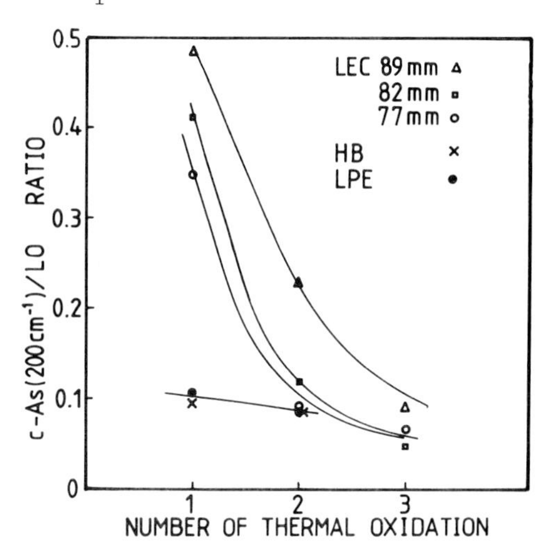

Fig.4. Dependence of amount of As which diffused out during thermal oxidation on the number of oxidation.

Figure 4 shows relative amount of arsenic which outdiffused during oxidation of GaAs grown by LEC, HB or LPE method. In general the amount of arsenic which outdiffused from LEC GaAs is much larger than those from HB or LPE crystals. Although a part of arsenic atoms which was in lattice sites are included in the outdiffused arsenic, they can be neglected because the amount of arsenic which outdiffused from LPE GaAs is very small. That is, most of arsenic atoms which present A_1g or Eg As peak are considered to be excess arsenic atoms which were in interstitial sites before the oxidation.

The amount of arsenic which outdiffused increased with the distance from the seed of a LEC ingot. In general the amount of excess arsenic included in GaAs grown by MOVPE with $Ga(C_2H_5)_3$ source and the condition of $V/III \geq 20$

was comparable to that in LEC crystals. The amount of excess arsenic included in MOVPE GaAs grown with $Ga(C_2H_5)_3$ increased with V/III ratio.

Figure 5 shows dependence of the amount of arsenic which outdiffused during the second oxidation of LEC GaAs on a depth removed by etching after the first oxidation. Dots in the figure are experimental results and the solid curve is one obtained by best fitting based on the model described above. A diffusion constant is obtained by this fitting. Figure 6 shows temperature dependence of diffusion constants of many species in GaAs which have been reported. The result obtained by our work is plotted also in the figure although it was estimated at one temperature. An example of a diffusion constant obtained for LEC GaAs by our work is about $7x10^{-11}(cm^2/s)$ at 500°C which corresponds to that of arsenic vacancy.

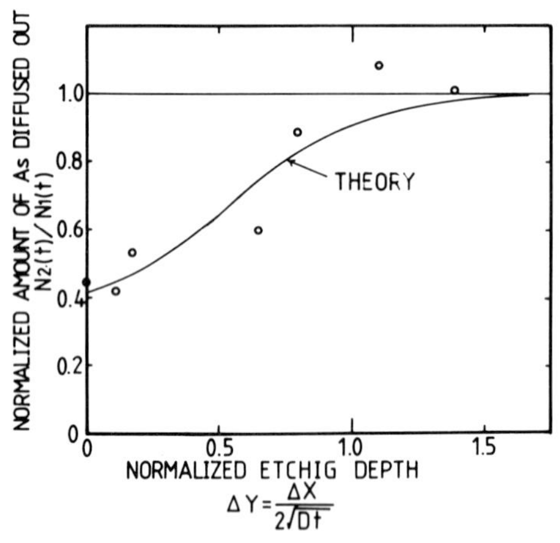

Fig.5. Dependence of the amount of arsenic which outdiffused during the second oxidation on etching depth after the first oxidation.
○ : experimental results
—: best fitted curve based on the model

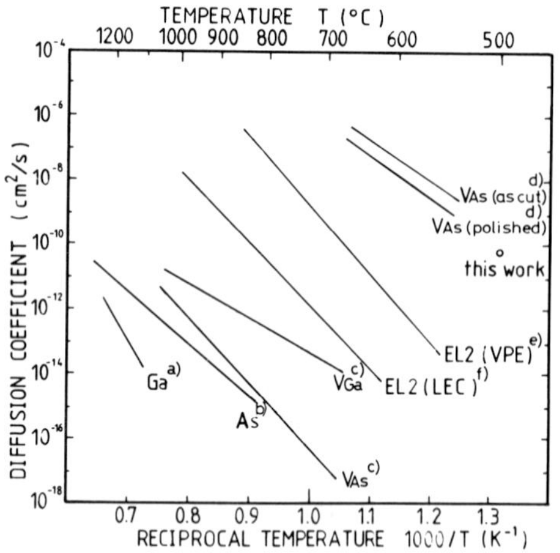

Fig.6. Diffusion constants of various species in GaAs.

5.Summary

A new method to characterize excess arsenic in GaAs crystals with laser Raman spectroscopy was proposed. The amount of arsenic included in LEC GaAs was much larger than those in HB and LPE crystals and increased with the distance from the seed of a LEC ingot. Excess arsenic in GaAs grown by MOVPE with $Ga(C_2H_5)_3$ source and the condition of $V/III \geq 20$ was comparable to that in LEC crystal and increased with V/III ratio. An example of diffusion constant of the arsenic from LEC GaAs corresponded to that of arsenic vacancy.

References

Fujimoto I, 1984 Jpn.J.Appl.Phys. 23, L287.
Launin J.S, Calleja J.M, and Cardona M, 1975 Phys.Rev. B12, 585.
Terashima K, Washizuka S, Nishio J, Okada A, Yasuami S, and Watanabe M, 1985 12th Int.Symp. on Gallium Arsenide and Related Compound, Karuizawa, Japan (1986 Inst. Phys. Conf. Ser. No.79, 37).

Paper presented at Int. Symp. GaAs and Related Compounds, Heraklion, Greece, 1987

Spatial correlation of free-carrier lifetime, near-band-edge and deep-level luminescence across semi-insulating GaAs-wafer

W W Rühle and K Leo

Max-Planck-Institut für Festkörperforschung, Heisenbergstrasse 1, D-7000 Stuttgart 80, Federal Republic of Germany

N M Haegel*

Siemens Research and Development Laboratories, Paul-Gossen-Str.100, D-8520 Erlangen, Federal Republic of Germany

ABSTRACT: Free-carrier lifetime, near-band-edge luminescence, and near-infrared (NIR) deep-level luminescence are mapped across an undoped, semi-insulating (SI) GaAs wafer. The lifetime follows a "W"-shaped profile. The level involved in the emission at 0.8eV could determine the lifetime variation.

1. INTRODUCTION

Photoluminescence (PL) was frequently used to characterize differently grown SI GaAs and inhomogeneities within the wafers (Tajima 1982, Kikuta *et al* 1983, Yu 1984, Wettling and Windscheif 1986, and Noto *et al* 1986). However, the interpretation of the variation of the near-band-edge emission is controversial: Tajima (1982) ascribes this variation to lifetime variations of free carriers, Wettling and Windscheif (1986) evoke a spatial variation of deep-level and shallow acceptor concentrations, i.e. a spatially varying compensation.

We report a direct measurement of free-carrier lifetimes in undoped SI GaAs using time-resolved PL in the picosecond regime. We relate the measured lifetimes to the near-band-edge and deep-level luminescence. Our results are independent of compensation since we use excitation densities larger than $10^{17}cm^{-3}$, i.e. higher than the concentration of any electrical active level in SI GaAs.

2. EXPERIMENTAL

The sample was a diagonal strip along the [100] direction from a 2-in. wafer of SI GaAs grown under B_2O_3 encapsulation from a pyrolitic boron nitride crucible. A synchronously pumped dye laser (pulse width $< 4ps$, $\lambda_{exc} = 790nm$) was used as excitation source. PL at different temperatures ($T_L = 10 - 200K$) was dispersed in a 0.32m monochromator and detected by a two-dimensional Hamamatsu streak camera with S20 cathode ($T_L = 10K$) or S1 cathode ($T_L = 10 \ldots 200K$). The density of excitation was $\approx 10^{17}cm^{-3}$ as calculated from the laser power, focus diameter (FWHM=50μm), and the absorption length at the laser wavelength. Time-integrated

*now: Dep. of Materials Science and Engin., Univ. of California, Los Angeles, CA 90024, USA

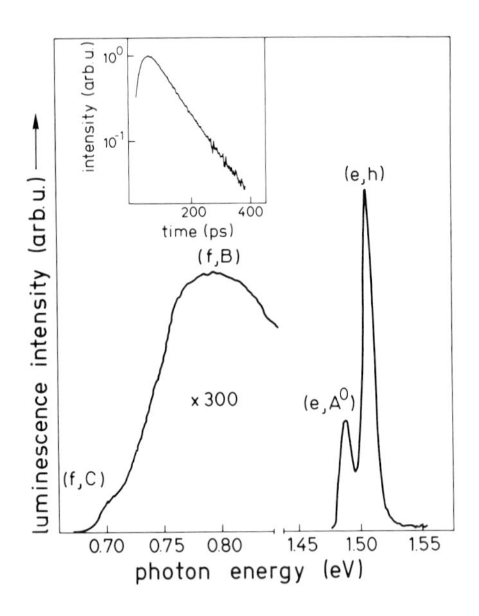

Fig.1 ↑ *Time-integrated PL spectra. The spectra are not corrected for detector sensitivity. The insert shows the luminescence decay of the (e,h) transition measured with the streak camera.*

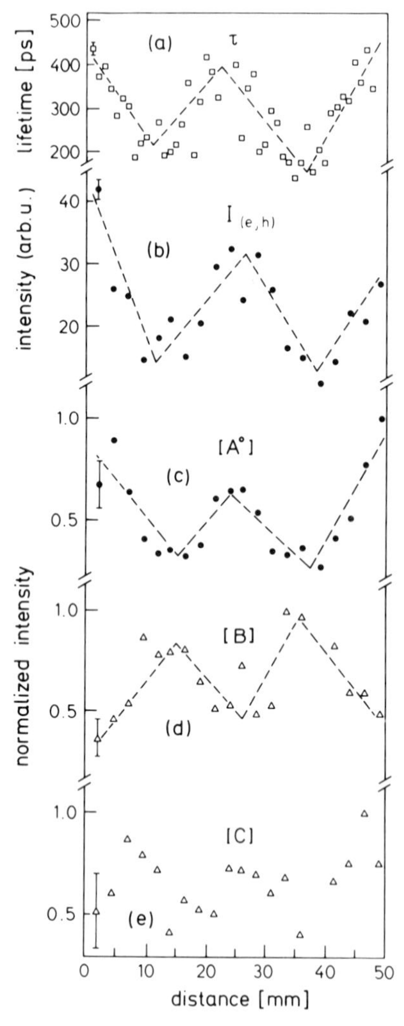

Fig.2 → *Curve a: Distribution of carrier lifetime; curve b: absolute intensity for the band-to-band transition. Curves c-e: The spatial variations of the deep or shallow levels involved.*

luminescence was detected with a cooled (77K) germanium detector using lock-in techniques.

3. RESULTS AND DISCUSSION

Figure 1 shows a time-integrated PL spectrum. The band-to-band (e,h) recombination of free carriers dominates at an excitation density n_{exc} of $10^{17}cm^{-3}$. The close to exponential decay of the (e,h) transition is shown in the insert. The band-to-acceptor recombination is designated with (e,A^0). The NIR emission consists of two bands at 0.8eV (f,B) and at about 0.7eV (f,C). The recombination of a free carrier with a carrier bound at level B is denoted (f,B). The position and intensity of (f,C) is not corrected for the strong sensitivity variation of the Ge-detector in this wavelength region.

The carrier lifetime is limited by recombination via deep levels with concentration N_T. The concentration of the photoexcited electrons (n) and holes (p) is larger than the intrinsic carrier concentration n_i and than N_T. In this case n = p, and the intensity as a function of time is

$$i_{(e,h)} \propto n(t)^2 \tag{1}$$

and the time-integrated intensity is

$$I_{(e,h)} \propto n_{exc}^2 \cdot \tau, \tag{2}$$

where τ is the common lifetime of electrons and holes. The latter relation is experimentally confirmed: We find $I_{(e,h)} \propto n_{exc}^{1.85}$. According to relation (1), the carrier lifetime τ is twice the luminescence decay time.
The time-integrated intensity I_N of the transitions (e,A^0), (f,B), and (f,C) is

$$I_N \propto n_{exc} \cdot N \cdot \tau \tag{3}$$

(N is the concentration of the respective level), and we find experimentally $I_N \propto n_{exc}^{0.75}$, confirming that we really deal with free-to-bound transitions.

Figures 2a and 2b depict the spatial variations of τ and $I_{(e,h)}$ across a [100] diagonal of a 2-in. wafer of SI GaAs. The scattering of the data is not due to the experimental error, but due to a strong short distance variation: the lifetime shows changes as large as a factor of two over distances of 300μm (Leo *et al* 1987a 1987b). On a large scale, both τ and $I_{(e,h)}$ follow a "W"-shape; the dashed lines in Fig.2 are guides for the eye. The close correlation of τ and $I_{(e,h)}$ shows the validity of relation (2).
We have to remove τ from relation (3) in order to obtain information about the distribution of N of A^0, B, and C via the measured I_N (n_{exc} is kept constant). We therefore divide I_N by $I_{(e,h)}$. Figures 2c-e show normalized plots of $I_{(e,A^0)}/I_{(e,h)}$, $I_{(f,B)}/I_{(e,h)}$, and $I_{(f,C)}/I_{(e,h)}$, respectively, and are therefore according to relations (2) and (3) normalized plots of the acceptor, the B-level, and the C-level concentration profiles.
Obviously, the acceptor concentration shows a "W"-shape. The acceptor binding energy is about 26meV, i.e., the acceptor is probably carbon.
The B-level shows an "M"-shape, i.e., its concentration is high where τ is short. We conclude that this B-level is a suitable candidate for the "lifetime-killer" in SI GaAs. This result is consistent with reports by Haegel *et al* (1987) on the annealing behavior of τ and the B-level, and with suggestions made by Tajima (1982) and Kikuta *et al* (1985). The origin of the B-level is not clear: it might be caused by microdefects (Tajima 1982) or by As_{Ga} complexes (Yu 1984, Kikuta *et al* 1983).
The origin of the C-level is also still debated (Samuelson *et al* 1984, Tajima 1987). Using a Ge-detector we are not able to resolve the 0.63eV and 0.68eV emission, and the scattering of the data in Fig.2e is large. A "W"-shape might tentatively be detected.
Finally, we measure the temperature dependence of τ for two sample spots with long and short lifetime at 10K. Using a 50μm and a 20μm pinhole on the sample, we ensure that always the same sample spot is measured. The carrier temperature was determined from the exponential high energy tail of the (e,h) transition. According to the Shockley-Read-Hall statistics, we expect the lifetime to vary as

$$\tau = \frac{1}{v_{th}\sigma N_T} \tag{4}$$

in the case of $n = p > n_i, N_T$, where σ is the smaller of the two capture cross sections for electrons and holes and $v_{th} = \sqrt{3kT/m^*}$ is the thermal velocity (m^* is the effective

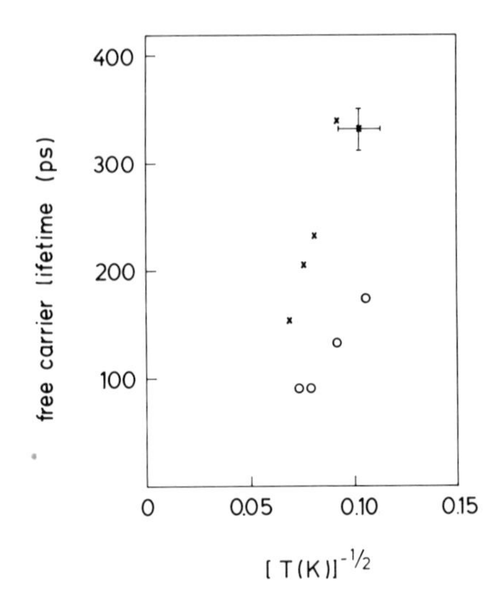

Fig.3 *Temperature dependence of lifetime τ at two spots of the sample. τ is plotted as a function of* $1/\sqrt{T(K)}$.

mass). A plot of τ versus $1/\sqrt{\mathrm{T}}$ as shown in Fig.3 should reveal a straight line through the origin if $\sigma \neq \sigma(\mathrm{T})$. This is obviously not the case; the cross section increases with increasing temperature. Assuming e.g. a smaller cross section σ for holes and typical data of GaAs ($\tau = 200$ps, $\mathrm{N_T} = 2 \cdot 10^{16}\mathrm{cm}^{-3}$, T=80K, $\mathrm{m^*}=0.5\mathrm{m_0}$), one would obtain a value $\sigma_\mathrm{n} = 3 \cdot 10^{-14}\mathrm{cm}^2$.

In conclusion, we have demonstrated that the B-level causing the 0.8eV emission could act as the "lifetime killer" in SI GaAs: Lifetime follows a "W"-shape and the B-level an "M"-shape across the diagonal of a wafer. The spatial variation of the $(\mathrm{e}, \mathrm{A}^0)$ transition is caused by both, a lifetime and an acceptor concentration variation.

4. ACKNOWLEGDMENTS

We thank A.Winnacker for stimulating discussions, H.J.Wolf for providing the samples, L.Ley for a critical reading of the manuscript, and K.Rother and H.Klann for expert technical assistance. The support by the Bundesministerium für Forschung und Technologie is gratefully acknowledged.

5. REFERENCES

Haegel N M, Winnacker A, Leo K, Rühle W W and Gisdakis S 1987 J.Appl.Phys. to be publ.

Kikuta T, Terashima K and Ishida K 1983 Jpn.J.Appl.Phys.**22**, L541-3

Kikuta T, Katsumata T, Obokata T and Ishida K 1985 Inst.Phys.Conf. Ser. No.74 ed B de Cremoux (Bristol and Boston: Adam Hilger) pp 47-52

Leo K, Rühle W W and Haegel N M 1987a J.Appl.Phys., to be published

Leo K, Rühle W W, Haegel N M, Winnacker A and Gisdakis S 1987b Proc. E-MRS Meeting Straßburg 1987, to be publ.

Noto N, Kitagawara Y, Takahashi T and Takanaka T 1986 Jpn.J.Appl.Phys.**25** L394-6

Samuelson L, Omling P and Grimmeis H G 1984 Appl.Phys.Lett.**45** 521-3

Tajima M 1982 Jpn.J.Appl.Phys.**21** L227-9

Tajima M 1987 Jpn.J.Appl.Phys.**26** L885-8

Wettling W and Windscheif J 1986 Appl.Phys.**A40** 191-5

Study of defects in semi-insulating, LEC GaAs by selective photoetching and high spatial resolution photoluminescence

JL Weyher[1], Le Si Dang[2] and EP Visser[1].

[1] Catholic University, Faculty of Science, Solid State Physics III, 6525 ED NIJMEGEN, The Netherlands.
[2] Universitè Scientific et Medicale de Grenoble, Laboratoire de Spectromètrie Physique, B.P. 87-38402, Saint-Martin-d'Heres Cedex, France.

ABSTRACT: Two different techniques have been employed to study crystallographic and chemical defects in undoped liquid encapsulated Czochralski (LEC) GaAs: selective photoetching (the so-called DSL method) and high spatial resolution photoluminescence (PL). The ratio of near band gap PL intensities obtained from grown-in defects and from the dislocation free matrix varies orders of magnitude depending upon growth conditions. In addition, "pure" stress-induced dislocations do not show any recognizable contrast of PL, within spatial resolution of 10 um. Differences in the relation between dislocations and electrically active impurities in different GaAs ingots are discussed in terms of growth parameters, namely constitutional supercooling and thermal stresses.

1. INTRODUCTION

The influence of substrate dislocations on performance of electronic devices, such as FETs', has been reported in a controversial manner (see for instance Winston et al (1984), Miyazawa et al (1983)). These conflicting results are probably due to the use of molten KOH as a tool to locate the defects: this etchant produces well defined pit on the outcrops of dislocations but fails in revealing the impurity atmospheres of the grown-in dislocations (Weyher and Giling (1985)). Recently, new possibilities in this respect were offered by a sensitive DSL photoetching technique (Diluted Sirtl-like solutions used with Light, Weyher and Van de Ven (1983) and Van de Ven et al (1986)). By shallow photoetching different types of defects can be discerned among other "pure" glide and grown-in dislocations (Weyher and Van de Ven 1986). It has also been shown, that in S.I. GaAs, the recombinative impurity atmospheres are distributed non-uniformly around grown-in dislocations (Weyher and Giling (1985) and Weyher and Van de Ven (1986)).
In this paper, the association of different types of dislocations with impurity atmospheres is reported for two semi-insulating, LEC-grown undoped GaAs ingots. The samples were examined by DSL photoetching, KOH etching and high spatial resolution near band gap photoluminescence.

2. EXPERIMENTAL RESULTS

A characteristic radial distribution of dislocations across the wafers

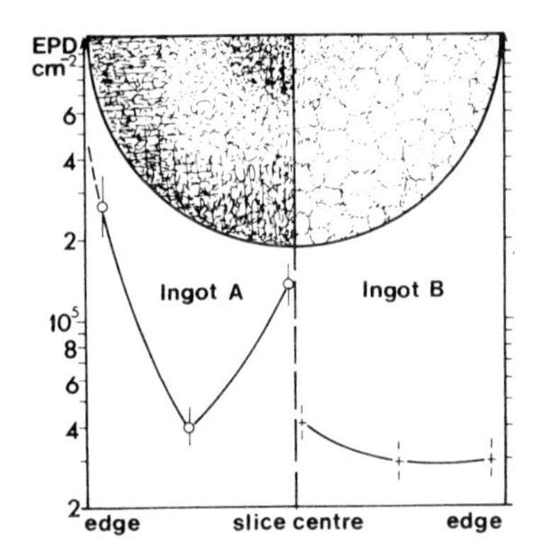

Fig. 1. Radial EPD distributions in LEC GaAs crystals obtained from two producers.

from central parts of both examined GaAs ingots, is shown in Figure 1: crystal "A" shows the W-shape distribution of EPD while the density of dislocations across the wafers from crystal "B" is almost constant and remarkably lower. The DSL photoetching reveals numerous bands of glide dislocations and weakly marked dislocation cells in crystal "A" as is shown in Figure 2a. These bands are arranged along <110> directions and are very dense close to edges of the wafers. On the contrary, crystal "B" contains very uniform dislocation cell structure, without any signs of glide, see drawing in Figure 1 and Figure 3a. The different status of dislocations in the GaAs crystals "A" and "B" appears also when calibration of the DSL photoetched samples with high spatial resolution near band gap PL is done, as shown in Figures 2 and 3. From this calibration and from surface profiling after photoetching it follows: (i) there is a one to one correlation between the DSL-revealed dislocation cells or individual grown-in dislocations and the PL mapping; (ii) "pure" stress-induced dislocations in samples "A" do not show any recognizable contrast of PL although they are clearly visible, together with the glide traces after DSL photoetching, see Figure 2a; (iii) there is essential difference between the relative PL intensities on dislocation cell walls in both samples. In "A" GaAs, the PL intensity on grown-in defects compared with the intensity from the matrix is about 2:1 while in crystal "B" the same relation runs from 50:1 to 100:1, compare Figures 2d and 3d; (iv) the sensitivity of photoetching is essentially higher for crystal "B" (compare Figures 2c and 3c). All these experimental facts lead to the conclusion that the relative concentration of impurities around grown-in dislocations is much higher in crystal "B" than in crystal "A". On the base of the present results it is not possible, however, to distinguish the contribution of deep and shallow impurities for the variations of PL intensities and local photo-etching behaviour.

3. DISCUSSION OF RESULTS

The results of the examination of the structure of both GaAs ingots lead to the definition of their essentially different growth conditions and are summarized in Figure 4. Crystal labelled "A" was grown under condition of a relatively high thermal gradient (high slope of $T_q(A)$). As a result numerous "fresh" glide dislocations are generated at the crystal surface due to the presence of stresses which exceed the critical resolved shear stress (CRSS) of the GaAs lattice. The density of dislocations is therefore the highest in the rim and forms characteristic W-shape distribution across the wafers (Figure 1), according to the theoretical predictions of Jordan et al (1980). As follows from the principles of solidification (Kurz and Fisher 1984) under such growth condtions a small constitutional supercooling occurs (Figure 4, ingot "A"). A stable growth front is then expected and small variations of concentration of impurities should be present. PL results confirm this prediction, as is evident from Figure 2d. On the contrary, crystal "B" was grown under the conditions of

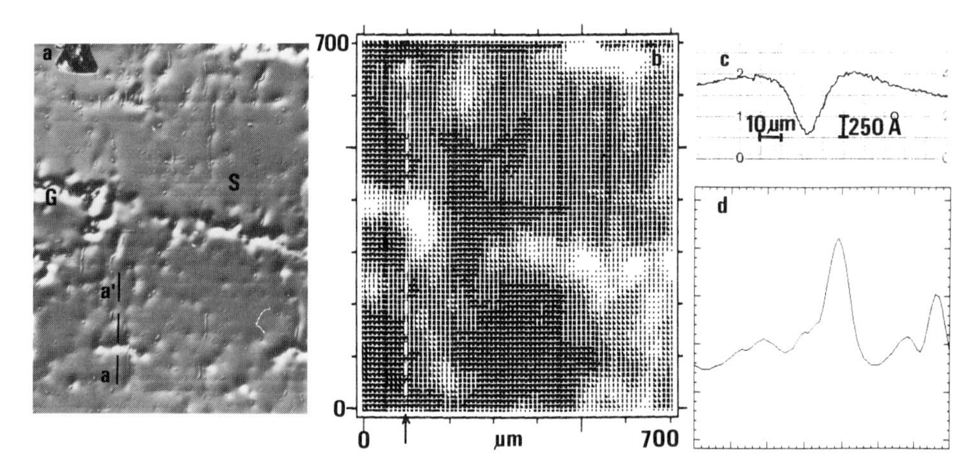

Fig. 2. LEC GaAs, producer "A". (a) surface structure after DSL photoetching: (G) grown-in dislocations, (S) stress-induced glide dislocations; (b) PL mapping for transitions at 820 nm, 1,8°K, the same area as in (a); (c) surface profile along the line a-a' in (a); (d) PL scan along the vertical line 100 um in (b).

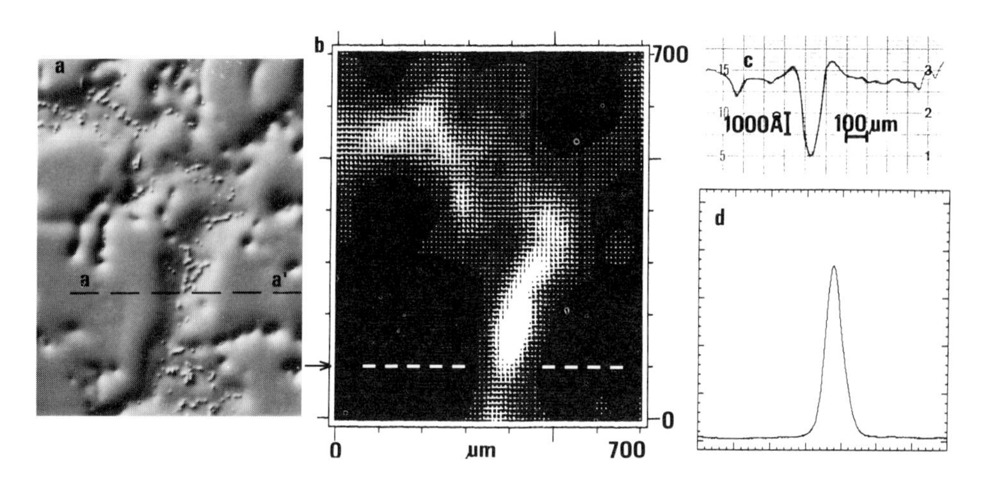

Fig. 3. LEC GaAs, producers "B". (a) surface structure after DSL photoetching; (b) PL mapping for transitions at 820 nm, 1,8°K, the same area as in (a); (c) surface profile along the line a-a' in (a); (d) PL scan along the horizontal line 100 um in (b).

a small thermal gradient (small slope of $T_q(B)$ in Figure 4). As a result no stress-induced dislocations are present but due to high constitutional supercooling of the liquid the growing front becomes locally unstable and dislocation cells are formed. This effect is followed by increasing variations of the concentration of impurities, as is well documented for "B" GaAs by PL examination (Figure 3).

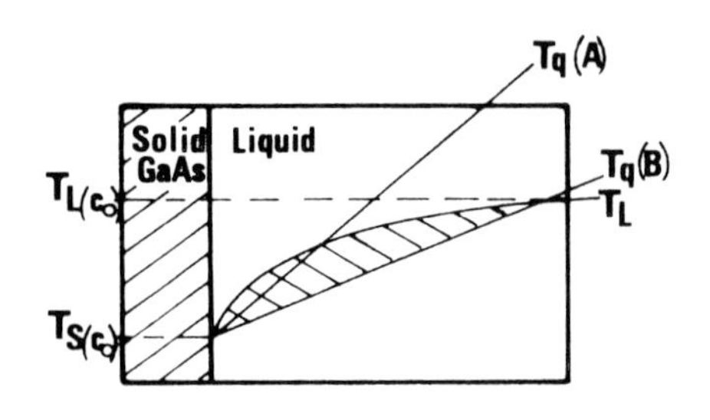

Fig. 4. Illustration of the growth conditions of "A" and "B" GaAs crystals with high (T_q(A)) and low (T_q (B)) gradients of temperature in the liquid.
C_o - initial impurity concentration
T_L - liquidus temperature
T_S - solidus temperature
T_q - real temperature of the liquid dependent upon heat flow.

4. CONCLUSIONS

1. It has been demonstrated by means of DSL photoetching in combination with high spatial resolution photoluminescence, that the dislocations in LEC-grown, undoped GaAs have an essentially different degree of association with impurities. The differences in the degree of decoration result from growth conditions and are explained on the basis of the concepts of constitutional supercooling and thermal stress analysis in LEC grown GaAs.
2. Essential differences in PL efficiency from "pure" stress-induced dislocations and grown-in dislocations is the key to understanding of the influence of individual dislocations and their impurity atmospheres on device performance (V_{th} and dislocation proximity of FETs', see De Raedt et al 1987).

ACKNOWLEDGMENTS

We are pleased to acknowledge Prof. L.J. Giling and drs H. Lochs for the discussions and continuing interest in this work. The authors (J.L.W. and E.P.V.) wish to thank Stichting voor Fundamenteel Onderzoek der Materie (FOM) and NATO (Grants 676/87 and 355/87) for financial support.

REFERENCES

Jordan A S, Caruso R and von Neida A R 1980 Bell Syst. Techn. J. 59 593
Kurz W and Fisher D J 1984 Fundamentals of Solidification (Switzerland: Trans Tech SA) pp 47-95
Miyazawa S, Ishii Y, Ishida S and Nanishi Y 1983 Appl. Phys. Lettr. 43 853
De Raedt W, Van Hove M, de Potter M, van Rossum M and Weyher J L, this Conference
van de Ven J, Weyher J L, van den Meerakker J E A M and Kelly J J 1986 J. Electrochem Soc. 133 799
Weyher J L and van de Ven J 1983 J. Crystal Growth 63 285
Weyher J L and Giling L J 1985 Proc. DRIP-1 Symp. ed J P Fillard (Amsterdam: Elsevier) pp 63-71
Weyher J L and van de Ven J 1986 J. Crystal Growth 78 191
Winston H V, Hunter A T, Olsen H M, Bryan R P and Lee R E 1984 Appl. Phys. Lett. 45 447

The influence of substrates on implanted layer characteristics

J. S. Johannessen* and J. S. Harris

Solid State Electronics Laboratories, Stanford University, Stanford, California.

D. B. Rensch, H. V. Winston, and A. T. Hunter

Hughes Research Laboratories, Malibu, California,

C. Kocot

Hewlett Packard, Palo Alto, California,

A. Bivas

Varian Associates, Palo Alto, California.

ABSTRACT: We have applied a series of diagnostic techniques, including EL2 mapping of as-received semi-insulating GaAs substrates, microwave conductivity mapping (MCM) of the implanted and annealed layers before processing, and mapping of transistor parameters and C-V profiles on diagnostic structures fabricated on the active layer. The substrates were undoped horizontal Bridgman, undoped LEC and In-alloyed LEC. We observe good correlation between microwave conductivity and FET-based characterization, and between FET parameters and EL2 concentration.

1. INTRODUCTION.

Reasonable yields of GaAs integrated circuits made by direct ion implantation depend on achieving tight control of the electrical parameters of the individual FETs. Device parameter uniformity and reproducibility are influenced by both the original choice of GaAs substrate and the nature of the post-implant annealing treatment. Winston et al (1984) found improvements in device uniformity with the use of In-alloyed rather than undoped LEC material that did not appear to be related directly to the low dislocation density of the In-alloyed substrates. Many workers have shown that post growth annealing of undoped material also improves device uniformity. Dobrilla et al (1985) found correlations between local EL2 density and FET parameters, and suggested that EL2 density is a major factor in controlling the activation efficiency and thus determining device parameters. From this point of view, the effect of In-alloying or post-growth annealing is to reduce the local fluctuations of EL2 density, which both are known to do. Here we examine the correlation between EL2 concentration and the transport properties of active implanted layers. We have conducted experiments similar to those of Dobrilla et al (1985), supplemented by microwave conductivity mapping prior to further device processing. We report here on the results for four different kinds of substrates from three different sources. The saturation current of a FET may be expressed by

$$I_{dss} = W\, v_s\, Q_{ms} = W\, v_s\, q\, \eta \int_d^{\infty} N_0(x)dx \qquad 1$$

* Permanent address: ELAB, University of Trondheim, N-7034 Trondheim, Norway.

where W is the channel width, v_s is the average channel saturation velocity, Q_{ms} is the mobile sheet charge density in the channel, and η is the activation efficiency. Here d, the depth of the depletion zone under the gate, depends on the sum of the gate-to-source voltage and the gate Schottky-barrier height, ϕ_b, and $\varepsilon\varepsilon_o$ and q have their usual meanings. The pinch off-voltage is determined by the dopant depth distribution $N_o(x)$, and defined by

$$V_{po} = \frac{q}{\varepsilon\varepsilon_o}\eta \int_0^\infty xN_o(x)dx = \frac{q}{\varepsilon\varepsilon_o}\eta\, D \int_0^\infty xN_o(x)dx \Big/ \int_0^\infty N_o(x)dx = \frac{q}{\varepsilon\varepsilon_o}\eta\, D\, \mathbf{a} \qquad 2$$

Eqn 2 can now be combined with Eqn 1 to define the transfer conductance, $G_T = I_{dss}/V_{po}$, of the channel. In Eqn 2 **a** is the center of gravity of the dopant distribution, and $Q_s = q\eta\, D$ is the sheet charge density, related to the implanted dose D. Notice that the activation efficiency, η, cancels in the expression for G_T.

2. EXPERIMENT.

We have chosen four different semi-insulating GaAs wafers, where A, B, and C are from LEC crystals, and D is from a new HB processed material. Samples B, C, and D are undoped, and A is In- alloyed. Two wafers, C and D, were from outside sources. C was known to be off-specification. All wafers were 2" diameter. They were EL2 mapped as received and then implanted uniformly with 200 keV Si(28) ions of fluence 3 10^{12} per cm^2. After implantation the wafers were capped with 1000 Å plasma-enhanced CVD silicon oxide and furnace annealed at 830 °C for 20 min. The cap was removed and the back side polished for the second EL2 mapping before the wafers were processed to produce an array of FETs over the entire surface. Each cell of the array, 3x3 mm, contains a number of diagnostic devices. We have used a 10 μm wide test FET for I_{dss} and V_{th}, and a 425 μm wide and 125 μm long FATFET for C-V and g_m-V measurements. Based on stepper probing data we have calculated mobile carrier density, N_m, mobile sheet carrier density, N_{ms}, carrier profiles, N(x), mobility profiles, μ(x), by the Pucel-Krumm method (1976), and analyzed these data in terms of the FET state parameters.

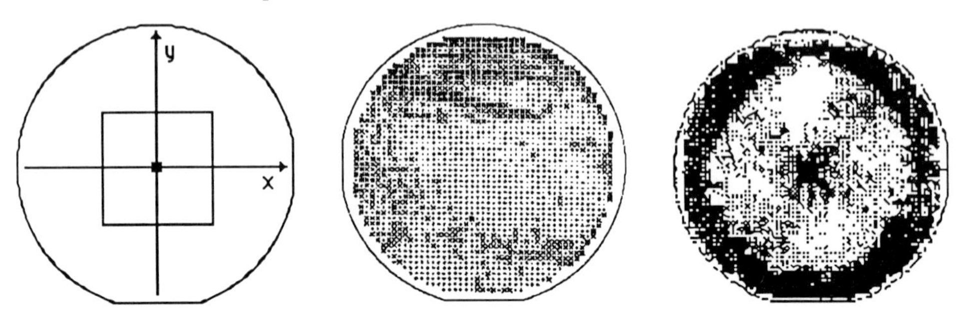

Fig. 1 Fig. 2 Fig. 3

Fig. 1 Orientation of wafers with respect to MCM and FET characterization. Fig. 2 EL2-intensity map of wafer D before implantation. Fig. 3 EL2-intensity map of wafer A after implant and annealing.

Contrast maps of EL2-density on a grey-scale are shown in Figs. 2 and 3 for a HB wafer (D) before implant and for an In-alloyed LEC wafer (A) after implant and anneal, respectively. Notice the lack of radial symmetry in EL2 density of the HB-wafer in Fig. 2. Fig. 1 shows the orientation of our measurements with respect to the wafer. Lines x and y indicate horizontal and vertical line scans for I_{dss} and V_{th}, and the square outlines the area of 18x18 mm mapped by C-V technique on FATFETs. This area is well within the ring of high EL2 density observed after implant anneal, shown in Fig. 3.

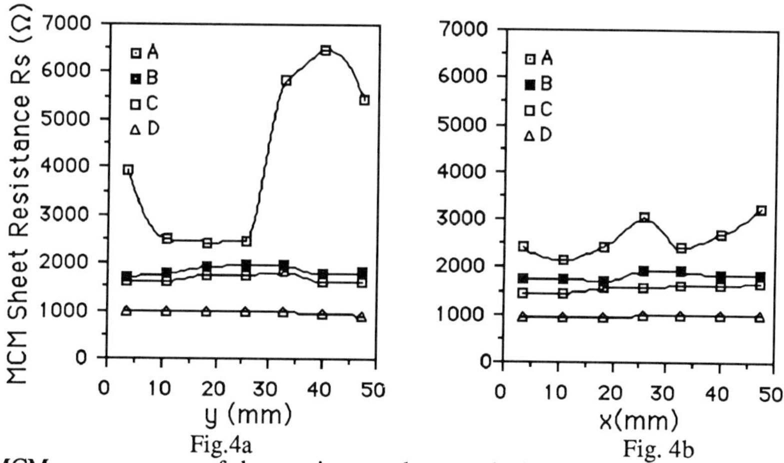

Fig. 4 MCM measurements of sheet resistance along vertical (a) and horizontal (b) diameters of the four wafers investigated.

The implant processing steps add IR absorption in a 5 mm wide ring around the periphery of the wafers. However, the average density of EL2 in the central portion of the wafers does not change significantly after annealing. For this reason we correlate average values of EL2, MCM and FET parameters from this region. Fig. 4 shows how the uniform implant results in a distribution of sheet resistance across the wafers. The standard deviations of R_s are less than 6% for A and B, while for C it is 43%. The uniformity of wafer D is best with $\sigma(R_s)$ less than 3%.The off-specification behavior of wafer C is clearly demonstrated in Fig. 4.

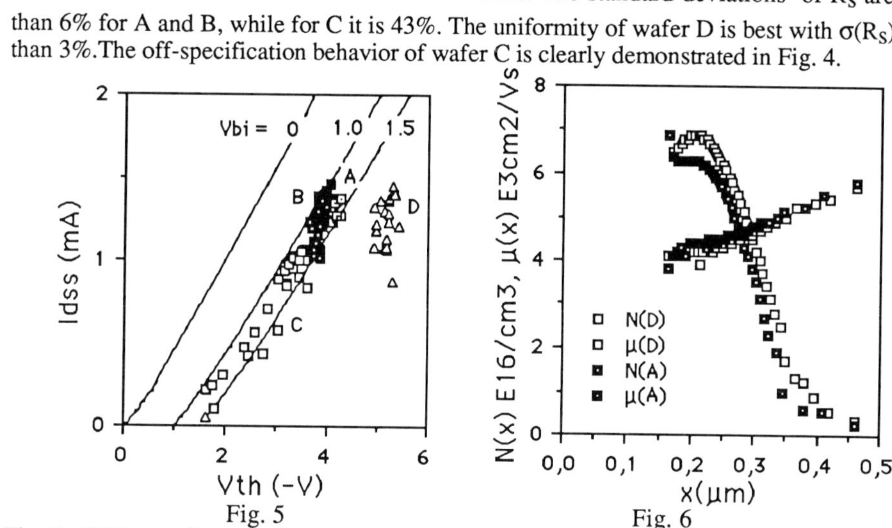

Fig. 5 FET state diagram for wafers A,B,C, and D.

Fig. 6 N(x) and μ(x) profiles from the central regions of wafers A and D.

I_{dss} and V_{th} are largely determined by design and processing ($V_{th} = V_{po} - V_{bi}$). The quality and uniformity of the substrate wafer, in terms of stoichiometry and lattice defects, will be reflected in scattering and deviation from expected values of these FET state parameters. In Fig. 5 we have plotted I_{dss} against V_{th} for all test FETs along x and y lines shown in Fig. 1. Also shown are plots of Eqn.1 for $v_s = 10^7$ cm/s, $\mathbf{a} = 0.2016$ μm (≈ projected range, Pedersen and Johannessen 1985), and the given values of V_{bi}. Here V_{bi} is the sum of $\phi_b =$

0.8 V and the voltage drop between source and gate due to I_{dss}.The data points from LEC wafers are in agreement with expectation for a source-gate voltage drop of about 0.4 V. The HB material deviates considerably by more negative threshold and much lower I_{dss} than expected. More negative threshold means higher carrier activation, and larger I_{dss} unless v_s is significantly reduced. The N(x) profiles of Fig. 6 confirm a higher activation in wafer D than in wafer A. However, the low-field drift mobilities, $\mu(x)$, are almost identical in the two samples! In fact the average low-field mobility of all measurements on all four wafers is 4339 cm^2/Vs with $\sigma = 3.6\%$! Using Eqn. 1 we can now calculate v_s once Q_{ms} is known. By integrating our N(x) profiles we have determined Q_{ms} and find that v_s for wafer D is 0.63 10^7 cm/s, much smaller than 0.9 10^7 cm/s for wafers A and B, in accordance with the low I_{dss} shown in Fig. 5 for wafer D.

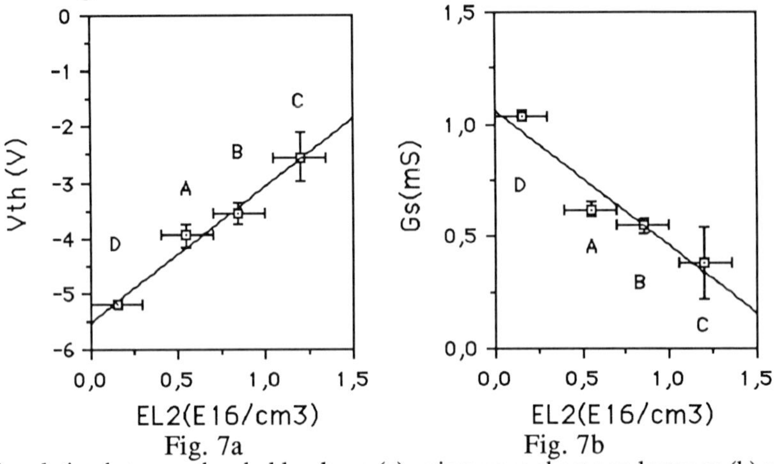

Fig. 7a

Fig. 7b

Fig. 7 Correlation between threshold voltage (a), microwave sheet conductance (b) and the average EL2 density in wafers A,B, C, and D. The error bars indicate the standard deviations in our data.

3. CONCLUDING REMARKS.

Generally $I_{dss}(x,y)$ and $V_{th}(x,y)$ track the $R_s(x,y)$ data shown in Fig. 4. However, process induced scattering and edge effects tend to obscure the uniformity predicted by MCM. Dobrilla et al (1985) correlated local EL2 to I_{dss} and V_{th}. We do not find a consistent local correlation among these parameters. However, average values yield an excellent wafer-to-wafer correlation as shown in Fig. 7 for V_{th} and G_s. Here V_{th} increases with EL2 density. The nature of the cap for post-implant anneal has a profound influence on the manner in which the EL2 and vacancy concentrations interact to influence Si ion activation. Silicon oxide caps are known (Hyuga et al 1986) to prevent the occurrence of a dislocation proximity effect on Si activation, which is observed when a silicon nitride cap is used. We believe that the opposite effects of EL2 concentrations on activation in our work and Dobrilla´s may be related to a difference in post implant anneal caps. We would like to acknowledge contributions from HRL for measurements and data reduction by R. Wong Quen, R. P. Bryan and J. Visher, to H. Kimura for supplying wafers A and B, and to P. T. Greiling for his continuous support. JSJ was supported by NTNF, Elkem A/S and Hughes Aircraft Company.

4. REFERENCES.

Winston H V, Hunter A T, Olsen H M, Bryan R P, Lee R E 1984 *Appl. Phys. Lett.* **45** 447
Dobrilla P, Blakemore J S 1985 *Appl. Phys. Lett.* **47** 602
Pucel R A, Krumm C F 1976 *Electronics Letters* **12** No 10 240
Pedersen U, Johannessen J S 1985 *Proc. 12 Nordic Semiconductor Meeting*
Hyuga F, Watanabe K, Osaka J, Hoshikawa K 1986 *Appl. Phys. Lett.* **48** 1742

Inst. Phys. Conf. Ser. No. 91: Chapter 2
Paper presented at Int. Symp. GaAs and Related Compounds, Heraklion, Greece, 1987

An optical technique for imaging dopant distributions in GaAs and other semiconductors

M.R. Brozel

Department of Electrical Engineering and Electronics and The Centre for Electronic Materials, University of Manchester Institute of Science & Technology, P.O. Box 88, Manchester M60 1QD. England

ABSTRACT: An optical imaging technique which is capable of revealing inhomogeneous distributions of dopants is described. The technique, which relies on the scattering of a light beam which takes place when it traverses a sample of varying refractive index, is capable of revealing growth striations and dislocation networks in GaAs. Applications of this technique to several dopants in GaAs grown by the Liquid Encapsulated Czochralski (LEC) and Horizontal Bridgman (HB) techniques are discussed. Because this technique is not relient on particular properties of GaAs, it will be useful in the assessment of other semiconductor systems.

1. INTRODUCTION

There is still considerable interest in the detection and measurement of non-uniformities in GaAs substrate materials. This results from overwhelming evidence that such non-uniformities give rise to variations in device parameters. For example, it is now well established that GaAs MESFET parameters are related to the presence of nearby dislocations [Ishii et al 1984, Miyazawa & Ishii 1984]. The interaction between dislocations and devices is not fully understood but it is assumed that, in undoped semi-insulating GaAs, variations in the concentration of the deep donor level EL2 are, at least, partially responsible for these effects [Packeiser et al 1986]. It is fortunate that simple infrared absorption imaging techniques have been able to yield valuable information on these non-uniformities of EL2 concentration, [EL2]. Moreover, as high [EL2] is associated with dislocations, it follows that this type of mapping yields considerable data on dislocation distributions [Brozel et al, 1983].

Inhomogenities in doped substrates is also important for devices such as light emitting diodes and lasers. In these cases, variations in dopant concentration can lead to changes in effective series resistance and thus to device lifetimes. It is notionally possible to apply infrared absorption mapping to these doped substrates and observe changes in absorption due to excitation of free carriers. However, free carrier absorption in n-type material, the most common doped substrate, becomes important only at wavelengths longer than about 2 microns and the most useful type of infrared imaging elements, the silicon or PbS/PbO vidicons, are not sensitive at these wavelengths. In this paper we present a new optical technique which is sensitive to changes in dopant concentration and which can reveal inhomogenities in doped GaAs substrates. We also will show that detectable dopant segregation to dislocation cells occurs in this material.

2. EXPERIMENTAL

The technique described here, which relies on the deflection of a light beam as it passes through a sample of varying refractive index, has been reported previously [Brozel 1987].

It is based on the infrared transmission imaging apparatus often used for EL2 imaging, except that the light source is parallel rather than diffused (Figure 1).
In Figure 1, the sample is placed near the focal point of the objective lens of the CCTV camera. Light rays which pass through the sample without being deflected are not focussed, but those which are deflected are focussed onto the detector. If the sample is exactly at the focal point all deflected beams within the acceptance angle of the optics are focussed and the image is nearly uniform. However, if the sample is moved either side of this position, the deflected beams do not recombine exactly and an image of changes in refractive index is obtained.

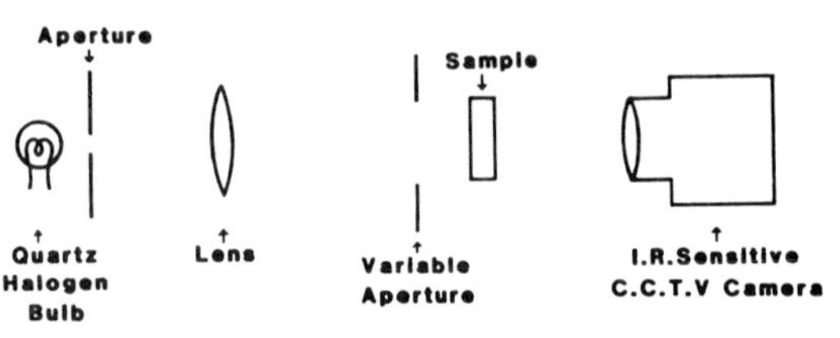

Figure 1: The experimental arrangement of the imaging apparatus. The lamp and lens produce a beam of roughly parallel light.
The camera lens is focussed near to but not on the sample

It has been demonstrated that the angular deflection of such a beam, Θ, is given by: $\Theta \sim t.\partial n/\partial x$, where t is the thickness of the sample and $\partial n/\partial x$ is the rate of change of refractive index in a direction perpendicular to that of the light beam. It follows that if changes in refractive index are due to changes in dopant concentration, then the imaging process can reveal these concentration variations.

3. RESULTS

3.1 Growth Striations

Dopant concentration changes in crystals grown from the melt are often associated with growth striations due to the instability of the crystal/liquid interface shape [See, for example, Brozel et al, 1986)]. Thus, the grown crystal exhibits striae which represents the growth interface shape at every stage in the crystal growth. An image of such striae, produced by the present method, is shown in Figure 2. The {110} sample here is the neck region of an <001> Indium doped GaAs crystal grown by the Liquid Encapsulated Czochralski (LEC) process. The width of the figure represents 1cm. Striae due to growth interface instability and revealed by changes in refractive index produced by variations in indium incorporation are clearly visible. In this image the light beam passes through the sample parallel to the planes of the striae and this offers the greatest sensitivity. Similar results have been obtained by viewing Te [Brozel, 1987] and Sn-doped GaAs crystals in the same way.

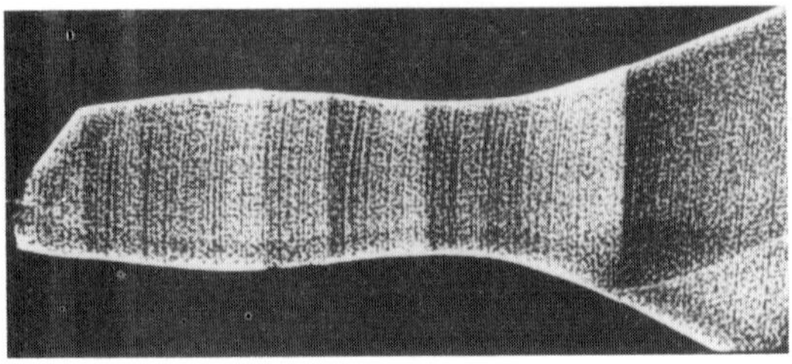

Figure 2: Growth Striations revealed in a {110} section cut diametrically from a <001> In-doped LEC GaAs crystal. Only the neck region is shown. The sample is 5cm in length.

3.2 Non-Uniformities in n-type GaAs Substrates

It is well established that dislocations polygonise in LEC GaAs producing equi-axed cell networks [Brozel et al 1984]. These have been revealed by several methods including etching studies [Stirland 1987], light scattering tomography [Ogawa 1985] and EL2

absorption imaging [Brozel et al 1984]. It has been established that several impurity species segregate to the networks [Kamejima et al 1982] and, in view of this, intentional dopants might behave in the same way. If this is so, it is possible that they can be revealed in a non-destructive manner using the present method.
Figure 3 is a mapping of a conventional Sn-doped {001} substrate ([Sn] ~ $5x10^{17}cm^{-3}$) of 200 microns thickness grown by the LEC method. Cell structure and evidence of lineage is clear. While Figure 3a was obtained with the sample off the focal plane of the camera towards the lens, Figure 3b was obtained with the sample moved off focus away from the lens. It is clear that there is a reversal in contrast, typical of a scattering image. It should be emphasized that the absorption images from all the samples in this paper indicated good uniformity and that the present images are not due to absorption. Additionally, it can be concluded that the absorption images attributed to EL2 in undoped GaAs are not due to the present scattering mechanism as they do not produce images similar to Figures 3a and 3b.

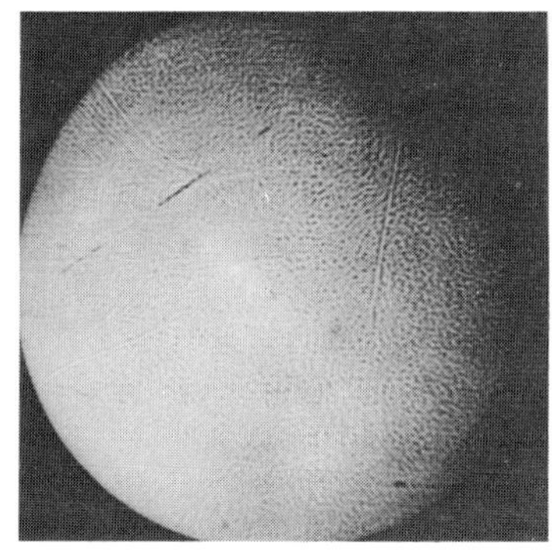

Fig. 3a

Dopant Segregation at dislocation walls revealed in a 200μm thick {001} wafer of Sn doped LEC GaAs
a) off focus away from the camera
b) towards the camera

Fig. 3b

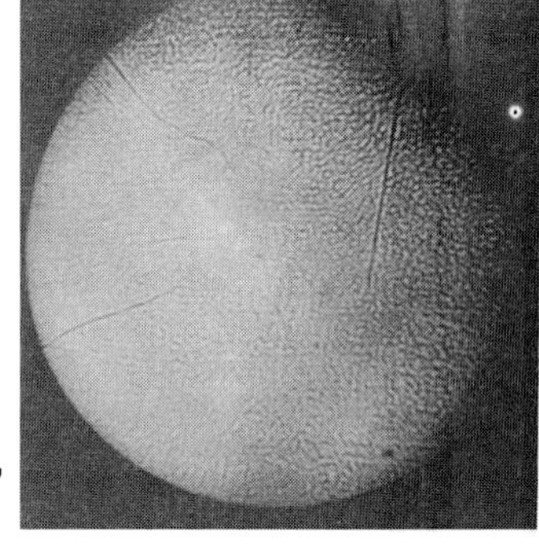

Finally, images of a 2 inch diameter 200 microns thick {001} substrate of silicon doped GaAs grown by the Horizontal Bridgman method are shown in Figures 4a and 4b. These images were obtained in the same way as those images in Figures 3a and 3b. Once again, strong non-uniformity is indicated although details of the image are rather different.

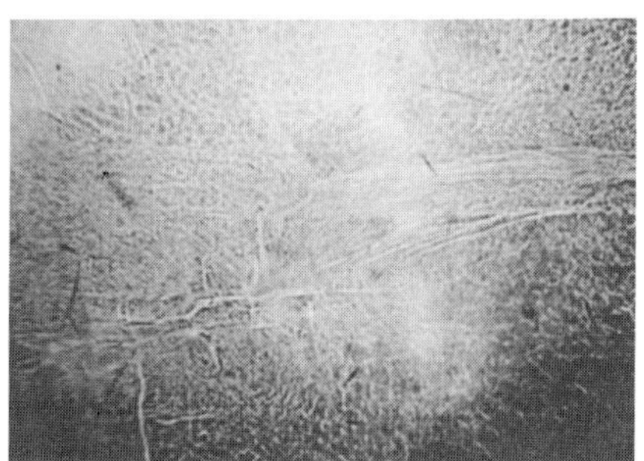

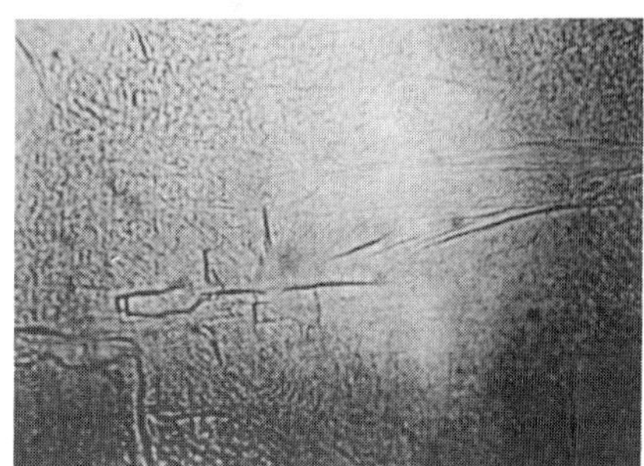

Fig. 4a Fig. 4b
A {001} section of a Si-doped GaAs substrate, thickness 200μm, grown by the Horizontal Bridgman method; a) and b) as defined in Figure 3

4. DISCUSSION

We have presented a novel non-destructive method for assessing the uniformity of doped GaAs substrates. The imaging process is shown to be due to deflection or scattering of below Bandgap light in regions where the refractive index is varying. Changes in the refractive index by dopant incorporation can have several mechanisms. For an isolelectronic dopant such as In in GaAs, this change is due to the simple variation of Bandgap energy. Modelling with this assumption has been shown previously to yield reasonable agreement with observations [Brozel 1987]. However, the concentrations of Te and Sn in our other samples are at least an order of magnitude lower than that of In. In these cases, changes in refractive index will be affected by the presence of free carriers [Pankove et al 1965] and modelling of these systems is now under way and the results will be reported later. The model previously proposed for scattering, requires that the transmitted light beam passes parallel to regions of equal refractive index, that is, that the change in refractive index which gives rise to the scattering mechanism has a strong first derivative with respect to distance in directions orthogonal to the light beam. This is simple to achieve when viewing growth striations through {110} samples cut perpendicular to the growth axis of <001> crystals. However, the images obtained on our {001} wafers must reflect changes in the refractive index in directions perpendicular to (001). This result confirms earlier data obtained on undoped LEC GaAs which concluded that dislocation cell walls were orientated parallel to the <001> growth axis [Brozel et al 1984]. If this were indeed the case in our samples and dopant atoms segregated to these walls, the images of the {001} wafers would correspond to light passing parallel to these walls and in the correct orientiation for the maximum spatial change in dopant concentration. It is surprising perhaps to find such marked non-uniformity effects in HB-grown GaAs. However, it should be pointed out that these effects are probably due to instability in growth, rather than segregation of impurities to dislocations. We expect SIMS and etching experiments which we are presently carrying out to clarify this point.

5. CONCLUSIONS

We have detected considerable non-uniformity of dopant concentration produced by growth instability and segregation in the presence of dislocations, using an optical technique which is both rapid and non-destructive. Images of growth striations are comparable to those produced by X-Ray Topography but are obtained in a matter of a few seconds. Dopant segregation at dislocations allows dislocation networks to be revealed in {001} substrates, also in a non-destructive manner. These effects, due to impurity aggregation which result in modifications of the refractive index, are not present in undoped GaAs where other techniques for revealing dislocation cell structure are already available.

6. ACKNOWLEDGEMENTS

The authors wish to acknowledge ICI Wafer Technology plc for providing many of the samples used in this work and C.A. Graham for her careful help in the preparation of this paper. This work was supported by the Procurement Executive, Ministry of Defence.

7. REFERENCES

Brozel M R 1987 *Proc. Defect Recognition and Image Processing in III-V compounds,* Monterey 1987. To be published.

Brozel M R, Clark S and Stirland D J, 1986, *Semi-Insulating III-V Materials, Hakone.* ed. H.Kukimoto & S. Miyazawa,(Ohmsha and North Holland). p 133

Brozel M R, Foulkes E J and Stirland D J, 1984 *Phys. Conf. Ser. No. 75*,p 59

Brozel M R , Grant I, Ware R M and Stirland D J, 1983, *Appl. Phys. Lett.*42, p 610

Ishii Y, Miyazawa S and Ishida S, 1984, *IEEE.Trans. El.Dev.*,Ed-31, p 1051

Kamejima T, Shimura F, Matsumoto Y, Watanabe H and Matsui J, 1982, *Jpn. J. Appl. Phys.*21, L721

Miyazawa S and Ishii Y 1984, *IEEE Trans. EL.Dev.*, Ed-31, 1057

Ogawa T 1985 *Proc. Defect Recognition and Image Processing in III-V Compounds* ed. J.P. Fillard,(Elsevier) p 1

Packeiser G, Schink H and Kniepkamp ,1986, *Semi-Insulating III-V Materials, Hakone,* ed. H. Kukimoto & S. Miyazawa,(Ohmsha and North Holland), p 561

Pankove J I, Gibson A F, Burgess R E,"*Properties of Heavily Doped Germanium" in Progress in Semiconductors,* 1965, Heywood, 9, p 67

Semi-insulating GaAs:V and its applicability

P S Gladkov and K B Ozanyan

Sofia University, Faculty of Physics, A. Ivanov 5, 1126 Sofia, Bulgaria

ABSTRACT: Bulk properties of SI GaAs:V, subjected to thermal anneal (1 hour in H_2 flow, 550-825°C are studied. Substantial changes in the resistivity, as well as in the PC and PL spectra are observed, probably resulting from point-defect reactions in the bulk. This is evidenced by the 3.2 eV activation energy for the annealing of the 0.8 eV PL band and by the switching of residual Si from Ga to As sites.

1. INTRODUCTION

The idea for technological application of semiinsulating (SI) GaAs:V has become tempting since Kütt et al (1984, 1985) concluded that V-doped high resistivity GaAs substrates showed approximately one order of magnitude less outdiffusion than Cr-doped ones if subjected to various thermal annealing processes or if epitaxial layers were grown upon them. They also observed that the redistribution in originally homogeniously doped SI GaAs:V and GaAs:Cr after Si, Be or B implantation and annealing is much lower for V than for Cr. These results however are relevant to the overall V or Cr concentration, while the electrical parameters of the bulk SI material are known to vary with the changes in the concentration of electrically active centres. It is important to know to what extent the SI state of GaAs:V is preserved after different annealing processes (post-implantation or epitaxial) in order to decide its range of applicability. Investigations on the changes in the compensation upon annealing have not been reported up to now and are expected to improve the understanding of the growth conditions at which SI GaAs:V can be obtained.

2. EXPERIMENTAL

A reference sample ET was cut from the tail part of a monocrystalline GaAs:V ingot, grown by the liquid-encapsulated Czochralski (LEC) method. Other samples were cut from a slice adjacent in position to sample ET and subjected to a 1 hour anneal in a hydrogen atmosphere without encapsulation at temperatures 550, 600, 650, 700, 750, 775, 800 and 825°C (the samples' labels correspond to the annealing temperature). To prepare for photoluminescence (PL) and photoconductivity (PC) measurements, the samples were lapped 50 μm and another 20 μm were additionally etched. For one temperature of annealing - 800°C, additional experiments were performed to reveal the necessary depth of lapping and the effect of the ambient. An additional set of samples was cut from the same slice and consisted of three samples lapped 50, 100 and 200 μm before the final 20 μm etch, and a fourth sample annealed in a sealed ampoule with controlled As-overpressure. The PL spectra of the additional set were in principle equivalent to that of sample 800 and will not be discussed separately. Upon annealing the resistivity decreases from 10^6 Ω.cm for sample ET to 2.10^2 Ω.cm for sample 800. Additional experiments to reveal the mechanisms responsible for the energies observed by DLTS (E_c-0.14 eV and E_c-0.23 eV) were carried out on LPE-grown GaAs:V layers.

3. DISCUSSION

A PL band at 0.8 eV, induced by the V-doping, has been observed only in SI GaAs:V (Gladkov and Ozanyan 1985). Figure 1. presents the semi-logarithmic plot of the intensity of that PL band, related to the band-edge emission, versus the inverse temperature of annealing 1000/T. A linear depcnedence is clearly demonstrated for the temperature range 650-825°C within three orders of magnitude. This allows the estimation of the activation energy for annealing of the 0.8 eV PL band: Ea = 3.2 eV. This value is comparable with the theoretically derived value for the enthalpy change for Ga and As-vacancy formation in GaAs: 2.59 eV (Van Vechten 1975) and the activation energies for a vacancy diffusion in the Ga sublattice - 2.1 eV and As sublattice - 4.0 eV (Chiang and Pearson 1975). The mobilities of As and Ga-vacancies may however become undistinguishable if nearest-neighbour hopping mode prevails (Van Vechten 1984). The observed activation energy implies that for annealing temperatures higher than 650°C, migration of vacancies in both sublattices becomes effective, resulting in point-defect reactions in the bulk. The occurence of such reactions is implied also by the behaviour of near band-edge PL (figure 2.): a definite swithching of residual Si from Ga to As sites is observed upon annealing, as suggested by the induced predomination of the D-A PL band and the appearance of the Ga_{As} antisite acceptor at 1-45 eV (Yu and Reynolds 1982).

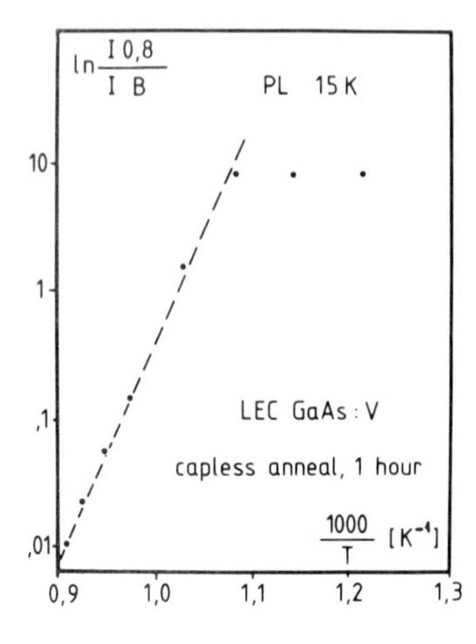

Fig. 1. The dashed line represents activation energy 3.2 eV for the annealing of the 0.8 eV PL band

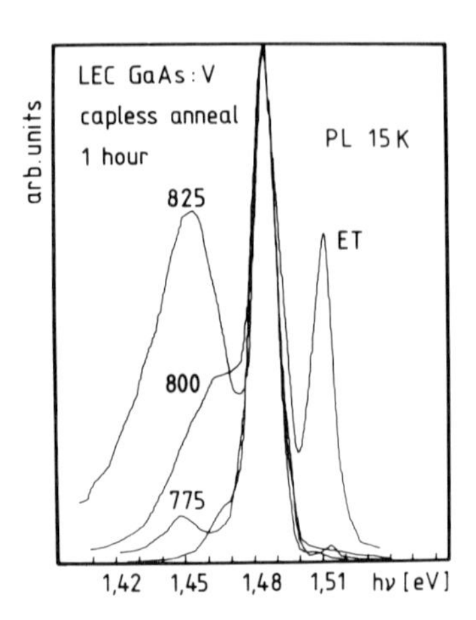

Fig. 2. Changes in the near band-edge PL upon annealing

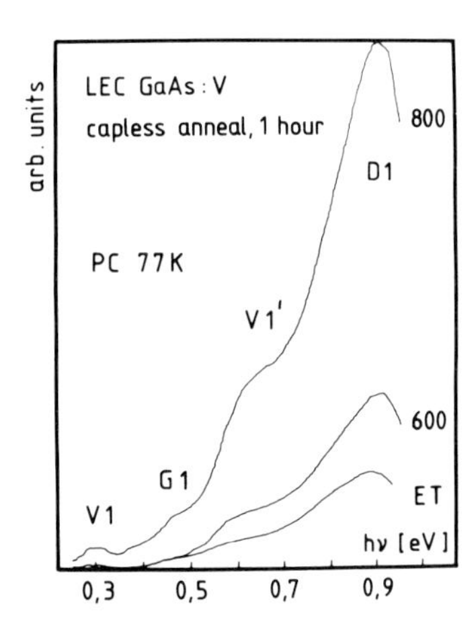

Fig. 3. Influence of annealing on the PC spectra

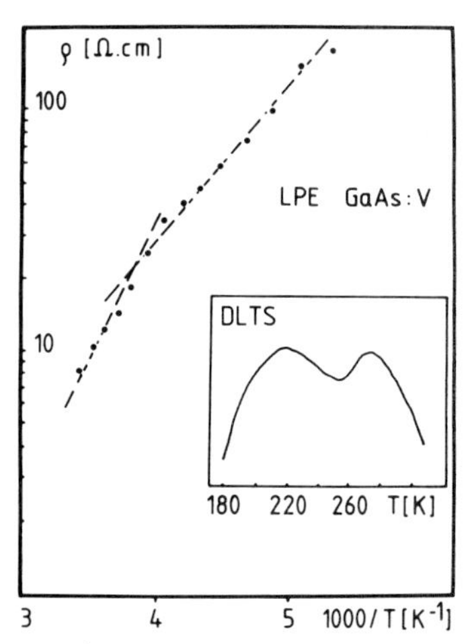

Fig. 4. The dashed lines represent calculated slopes 0.14 and 0.23 eV. The corresponding couple of DLTS peaks are shown on the inset

PC spectra of As-grown SI GaAs:V have been previously investigated (Gladkov and Ozanyan 1987) and PC bands related to isolated V_{Ga} (V1 and V1'), [V_{Ga} - As-vacancy] complex (G1) and valence band-to-deep level transitions (D1) were identified. EL2 induces a PC band peaking at 1.2-1.3 eV (D2) and probably significantly contributes to D1. All those bands monotonically increase with the annealing temperature (figure 3.; for reasons of readability D2 is not shown), since the annealing conditions activate additional point defects and supply conditions for trapping of mobile interstitial As on Ga-vacancies (Weber and Schneider 1983) and effective incorporation of previously inactive V atoms into the Ga sublattice.

The present results, as well as previous work, imply that besides the isolated V_{Ga}, giving a level at E_c-0.14 eV (Clerjaud et al 1985, Brandt et al 1985), V-related complexes can be formed at appropriate conditions of growth:

-[V_{Ga}-Si] at 0.23 eV. In LPE-grown GaAs:V the semi-logarithmic plot of resistivity versus the inverse temperature (figure 4.) shows two slopes corresponding to 0.14 and 0.23 eV - the first one is fairly well defined and the second one is marked only. The same energies are observed by DLTS (figure 4., inset) with comparable amplitudes despite of the much greater cross-section for the 0.23 eV trap. Besides, such a manifestation of the 0.14 eV slope on figure 4. is possible only if the concentration of the 0.23 eV trap is more than an order of magnitude greater than that at 0.14 eV. The 0.23 eV trap is observed in GaAs:V, grown by the horizontal Bridgman (HB) method (Clerjaud et al 1985), but is absent in LEC-grown from a pBN crucible (and presumably Si-free) GaAs:V (Brandt et al 1985). Compared to that, the LPE technique offers conditions, at which the Si contamination is much less than those for quartz-boat grown materials. On this basis, the existence of a [V_{Ga}-Si] complex is proposed for Si-containing GaAs:V crystals.

-[V_{Ga} - As-vacancy] at 0.4 eV. The PC band G1 at 0.4 eV is much less intensive in LEC-grown samples, as compared to HB-grown ones, and very intensive in LPE-grown samples (Gladkov and Ozanyan 1987). In addition, it has been observed to enhance towards the tail of the LEC-grown ingot, probably affected by the reduced As/Ga ratio in the melt. Thus, the serious indications that G1 results from Ga-rich conditions of growth, together with the observation that adding of about 1 at % of Ni, Co or Cr to the Ga-melt for LPE growth results in the appearance of a traps at 0.4 and 0.7 eV (Wang et al 1986), allow an interpretation of G1 as due to a [V_{Ga} - As-vacancy] complex.

-a midgap acceptor, compensating the background shallow donors and responsible for the SI properties of our LEC-grown GaAs:V samples.

4. CONCLUSIONS

In SI LEC-grown GaAs:V, when subjected to thermal treatment typical for post-implantation annealing, LPE/VPE, a strong reduction in resistivity - from 10^6 to 2.10^2 Ω.cm is observed. This is caused by the occurrence of point-defect reactions in the bulk with an activation energy of 3.2 eV. Despite of the lowered resistivity for the anneal material, the advantages of V over Cr as a compensating agent can still be realized in MESFET technology.

5. REFERENCES

Brandt C D, Hennel A M, Pawlowicz L M, Dabkowski F P, Lagowski J and Gatos H C 1985 Appl.Phys.Lett. **47** 607

Chiang S Y and Pearson G L 1974 J.Appl.Phys. **46** 2986

Clerjaud B, Naud C, Deveaud B, Lambert B, Plot-Chan B, Bremond G, Benjeddou C, Guillot G and Nouailhat A 1985 J.Appl.Phys. **58** 4207

Gladkov P S and Ozanyan K B 1985 J.Phys.C:Solid-State Phys. **18** L915

Gladkov P S and Ozanyan K B 1987 submitted to Solid-State Communic.

Kütt W, Bimberg D, Maier M, Kräutle H, Köhl F and Bauser E 1984 Appl.Phys. Lett. **44** 1078

Kütt W, Bimberg D, Maier M, Kräutle H, Köhl F and Tomzig E 1985 Appl.Phys. Lett. **46** 489

Ulrici W, Friedland K, Eaves L and Halliday D P 1985 Phys.Status Solidi b **131** 719

Van Vechten J A 1975 J.Electrochem.Soc. **122** 422

Van Vechten J A 1984 J.Phys.C:Solid-State Phys. **17** L933

Wang Z-G, Ledebo L-Å and Grimmeiss H G 1984 J.Phys.C:Solid-State Phys. **17** 259

Yu P W and Reynolds D C 1982 J.Appl.Phys. **53** 1263

Paper presented at Int. Symp. GaAs and Related Compounds, Heraklion, Greece, 1987

Optical properties and Zeeman spectroscopy of Ti-doped GaP and GaAs

C A Payling*, D P Halliday*, D G Hayes*, M K Saker+, M S Skolnick+, W Ulrici*° and L Eaves*

*Department of Physics, Nottingham University, Nottingham NG7 2RD, U.K.

+Royal Signals and Radar Establishment, Gt. Malvern, Worcs. WR14 3PS, U.K.

ABSTRACT: Zeeman spectroscopy measurements on the sharp line structure observed in the absorption and photoluminescence spectrum of GaP:Ti around 605 meV are interpreted in terms of $^2E \leftrightarrow {}^2T_2$ transitions of Ti^{3+}_{Ga} (d^1). The behaviour of the 566 meV line of GaAs:Ti can also be explained on the basis on the proposed model.

The Ti impurity in III-V semiconductors has attracted considerable interest during the last two years because it offers the technologically important possibility of rendering these materials semi-insulating with improved thermal stability. However, a series of problems concerned with the electronic structure, in particular of the Ti^{3+}_{Ga} charge state, still remain. (Ti^{n+}_{Ga} is referred to as Ti^{n+} in remainder of text).

In both GaAs and GaP the acceptor levels Ti^{2+}/Ti^{3+} (E_C-0.20 eV in GaAs [1,2]: E_C-0.48 eV in GaP [3]) and the donor levels Ti^{3+}/Ti^{4+} (E_V+0.6 eV in GaAs [1,2]; E_V+1.0 eV in GaP [3]) have been detected by capacitance spectroscopy. In both hosts the Ti^{2+} charge state manifests itself by broad absorption bands due to the internal transitions $^3A_2 \rightarrow {}^3T_1(F)$ and $\rightarrow {}^3T_1(P)$ [4,5] as well as by broad absorptions due to the photoionisation transition

$$Ti^{2+} + h\nu \rightarrow Ti^{3+} + e_{cb} \; . \qquad (1)$$

The Ti^{3+} state is characterised by zero-phonon lines (zpl's) observed in absorption and photoluminescence (PL). Figure 1 shows the PL and absorption spectra of GaP:Ti. The zpl's occur at 566.0 meV and 569.3 meV in GaAs and at 604.1 meV, 604.6 meV and 607.9 meV in GaP [4-7]. These lines are assigned to transitions between the ^2E ground state and the $^2T_2$ excited state of Ti^{3+} where the spin-orbit splitting of the $^2T_2$ state ($3\lambda/2$) is reduced by Jahn-Teller coupling to $3K\lambda/2$, with the reduction almost identical for GaAs (K = 0.11) and GaP (K = 0.13).

The investigation of the 566 meV absorption zpl of GaAs:Ti in magnetic fields up to 10 T surprisingly yielded no apparent shift or splitting [4]. In order to clarify whether this result can be understood in terms of the above assignment of the line, we have carried out analogous Zeeman-effect experiments on the PL lines of GaP:Ti.

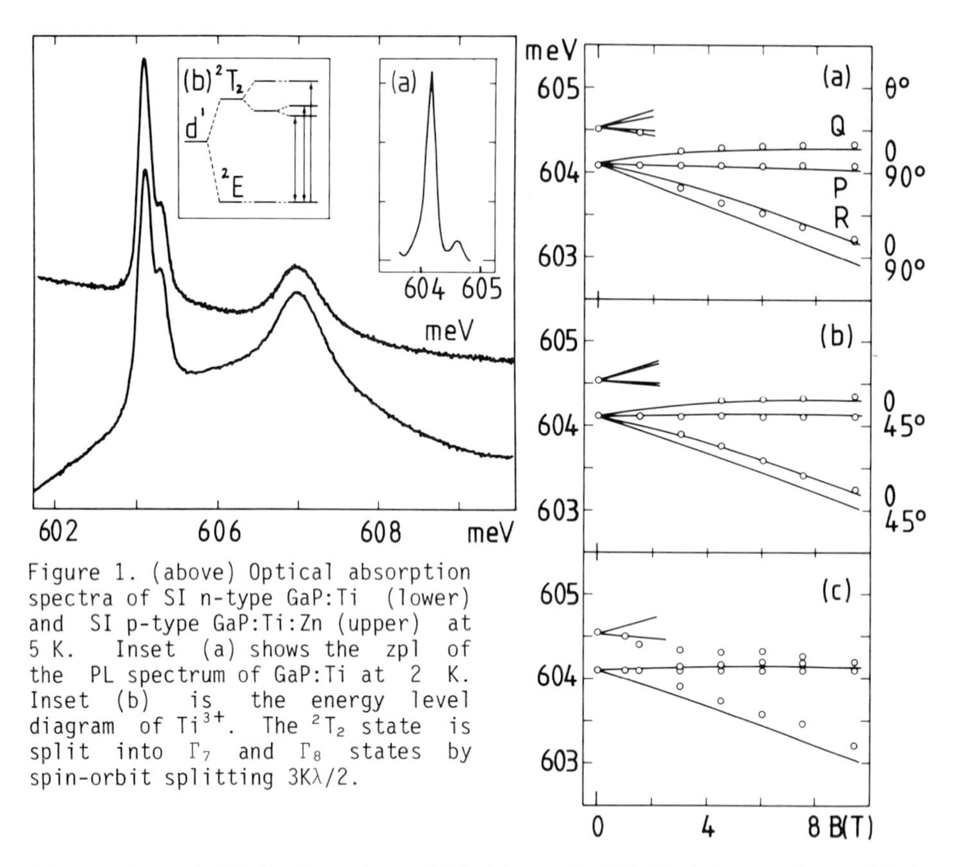

Figure 1. (above) Optical absorption spectra of SI n-type GaP:Ti (lower) and SI p-type GaP:Ti:Zn (upper) at 5 K. Inset (a) shows the zpl of the PL spectrum of GaP:Ti at 2 K. Inset (b) is the energy level diagram of Ti^{3+}. The $^2T_2$ state is split into Γ_7 and Γ_8 states by spin-orbit splitting $3K\lambda/2$.

Figure 2. (right) Energies of PL lines of GaP:Ti (o) vs $\underline{B}$ applied along the three principal crystal axes, (a) [001], (b) $[1\bar{1}0]$ and (c) [111]. Θ is the angle between the defect axes and B. The curves are calculated as described in text.

PL spectra were measured for $\underline{B}||[001]$, [111] and [110] at T = 2 K with Ar-ion laser excitation. The Zeeman splitting patterns for the three directions are shown in Figure 2 (circles = experimental points). The most obvious results are that the Zeeman pattern is only slightly anisotropic and that, with increasing B, most of the PL intensity resides in the main line (P), whose energy remains nearly constant.

We propose the following model to interpret these results. If the zpl's are indeed due to the $^2E \leftrightarrow {}^2T_2$ transitions of Ti^{3+} in cubic symmetry, the Γ_8 (^2E) ground state should exhibit zero orbital angular momentum, i.e. no first order spin-orbit splitting and orbital g-factor, $g_\ell = 0$. Therefore, the Zeeman splitting of the ground state can be described by the Hamiltonian $\mathcal{H}_{gs} = g_s\mu_B m_s B$ with $g_s = 2$, $m_s = \pm\frac{1}{2}$. The splitting of the excited $^2T_2$ state is described by an effective Hamiltonian

$$\mathcal{H}_{ex} = K\lambda\underline{\ell}.\underline{s} + g_s\mu_B\underline{s}.\underline{B} + g_\ell\mu_B\underline{\ell}.\underline{B} + \mathcal{H}_1$$

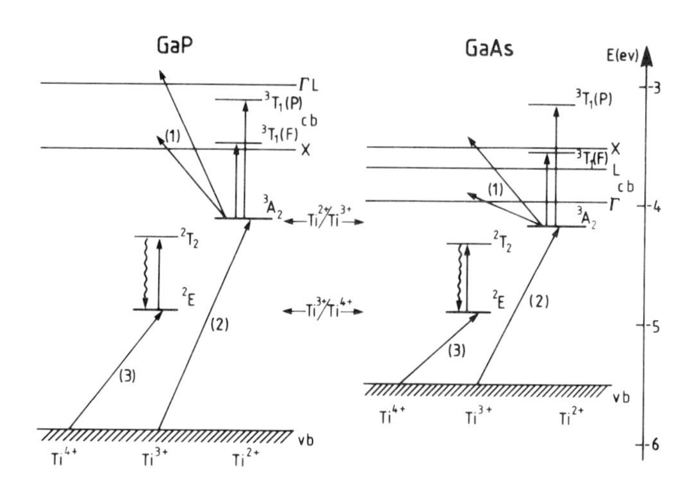

Figure 3. Energy level diagram of Ti in GaAs and GaP.

with $s = \frac{1}{2}$ and $\ell = 1$. The spin-orbit splitting is reduced by Jahn-Teller coupling to $K\lambda = 2.53$ meV. $\mathcal{H}_1$ describes a small axial distortion present only in GaP which causes the 0.5 meV splitting of the Γ_8 ($^2T_2$) state at B = 0. Let us neglect for the moment $\mathcal{H}_1$ and assume complete quenching of ℓ ($g_\ell = 0$, justified by the small value of K). In this case, the j = 3/2 excited state splits with an effective g-factor $g_{eff} = 2/3$ so that in a magnetic field the excited $m_j = -3/2$ state shifts at exactly the same rate as the $m_S = -\frac{1}{2}$ ground state. Therefore, the PL line P can be assigned to the $m_j = -3/2 \rightarrow m_S = -\frac{1}{2}$ transition. It follows from this that the strongly B-dependent line R has to be assigned to $m_j = -3/2 \rightarrow m_S = +\frac{1}{2}$. This transition is forbidden, thus explaining the low intensity of the R line. As only the $m_j = -3/2$ level is occupied in the excited state, no other lines can be observed.

The additional line Q in the Zeeman spectra is assigned to the presence of $\mathcal{H}_1$. A reasonable fit to the Zeeman data can be obtained (solid lines in Figure 2) assuming that $\mathcal{H}_1$ corresponds to a weak [001] axial distortion ($\mathcal{H}_1 = b\ell_z^2$) with b = -0.69 meV and the other parameters as given above. The origin of the weak distortion could be random strain (mainly ε-type) acting on the excited Γ_8 state. Because of the axial symmetry of the distortion, there are magnetically non-equivalent centres, depending on the direction of $\underline{B}$. The P-Q splitting can be fitted well with the above parameters for $\underline{B}||[001]$ and $\underline{B}||[110]$, as can be seen from Figure 2. (Θ is the angle between $\underline{B}$ and the centre axes). However, for $\underline{B}||[111]$ the P-Q splitting should be zero but is found to have a small finite value. This possibly indicates a small orthorhombic component of the axial distortion.

This analysis of the GaP:Ti^{3+} Zeeman results in PL also explains the absence of any Zeeman effect in the zpl absorption of GaAs:Ti^{3+}. The only measured absorption line is due to the $m_S = -\frac{1}{2} \rightarrow m_j = -3/2$ transition, the position of which does not depend on $\underline{B}$ as shown above for GaP:Ti^3 . The transition $m_S = -\frac{1}{2} \rightarrow m_j = -\frac{1}{2}$ is allowed only in π-polarisation. Since the Zeeman experiments [4] were performed in the Faraday configuration, π-polarisation is not observable.

We have also investigated in PL the zpl's of GaP:Ti^{3+} by applying uniaxial stress up to 35 MPa parallel to the three main crystallographic directions. A splitting is found for both lines and the splitting pattern as well as its stress-dependence appear to confirm the proposed assignment of the transitions to Ti^{3+} and to support the assumption that $\mathcal{H}_1$ is due to random strains.

Figure 3 displays the energy level scheme for Ti in GaP and GaAs with the internal transitions of Ti^{2+} and Ti^{3+} and the photoionisation transitions inferred from existing data. The presence of the photoionisation transitions (1) and the transitions

$$Ti^{3+} + h\nu \rightarrow Ti^{2+} + hole_{vb} \quad (2)$$

$$Ti^{4+} + h\nu \rightarrow Ti^{3+} + hole_{vb} \quad (3)$$

are derived from optical absorption spectra [4,6], DLOS spectra [2,3] and common photoinduced rechargings monitored by optical absorption [6] and EPR [8].

°Permanent address: Akademie der Wissenschaften der DDR, 1086 Berlin, DDR.

REFERENCES

[1] Brandt C D, Hennel A M, Pawlowicz L M, Wu Y T, Bryskeiwicz T, Lagowski J and Gatos H C 1986, Appl Phys Letters **48** 1162.

[2] Guillot G, Bremond G, Bencherifa A, Nouailhat A and Ulrici W, 1986 in Semi-Insulating III-V Materials ed H Kukimoto and S Miyazawa (Ohmsa:Tokyo) p 483.

[3] Roura P, Bremond G, Nouailhat A, Guillot G and Ulrici W 1987, Appl Phys Letters (in press).

[4] Ulrici W, Eaves L, Friedland K, Halliday D P, Nash K J and Skolnick M S 1986, J Phys C **19** L525.

[5] Hennel A M, Brandt C D, Wu Y T, Bryskiewicz T, Ko K Y, Lagowski J and Gatos H C 1986, Phys Rev B **33** 7353.

[6] Ulrici W, Eaves L, Friedland K, Halliday D P and Payling C A, Phys Stat Sol (b), to be published.

[7] Halliday D P, Payling C A, Saker M K, Skolnick M S, Ulrici W and Eaves L 1987, Semicond Sci Technol **2** 675.

[8] Kreissl J, Gehloff W and Ulrici W 1987 Phys Stat Sol (b) **143** 207.

Temperature and pressure studies of a deep level in InP:V by frequency resolved capacitance spectroscopy

R P Benyon, K P Homewood*, D J Dunstan, A K Saxena†, A R Adams, B Cockayne+ and K Inabe▲

Department of Physics, University of Surrey, Guildford, Surrey GU2 5XH
*Dept. of Electronic and Electrical Engineering, University of Surrey
†On leave from University of Roorkee, Roorkee, U.P., India
+Royal Signals and Radar Establishment, Great Malvern, Worcs. WR14 3PS, U.K.
▲On leave from University of Kanazawa, Kanazawa, Japan

ABSTRACT: A deep level has been observed in n-type bulk InP:V at a depth of 0.49 +-0.01V. The level displays a hydrostatic pressure shift in the emission rate of 0.16 decades/kbar, while the activation energy is independent of pressure. Frequency Resolved Capacitance Spectroscopy, a new high sensitivity isothermal technique, was used. It employs lock-in methods and a capacitance bridge to detect a quadrature signal due to carrier emission from deep levels in the sample. Pressure measurements, up to 8kbar, were made in a hydraulic piston-cylinder system.

1. INTRODUCTION

Deep levels in semiconductors are states which, unlike shallow donors and acceptors, cannot be described by effective mass theory. They are therefore, not in general well understood, and yet are of great importance in semiconductor devices. In general, the position of the deep level will be determined by contributions from a mixture of conduction and valence band states from the whole Brilloun zone. Hydrostatic pressure provides a unique means of perturbing the band structure of semiconductors. The detailed behaviour under hydrostatic pressure of band states in the important semiconductors is now well documented; in particular, the pressure coefficients of the conduction and valence band extrema are known. Thus, measurements of transition energies and of capture cross sections under hydrostatic pressure can be an invaluable tool for studying deep levels. Of those defects that produce levels deep in the band gap, the $3d^n$ transition metals are probably the simplest and best understood; here we present results on a centre in n-type InP doped with V. Previous authors have reported a deep level associated with V in p-type InP (Deveaud et al, 1986).

2. EXPERIMENT

The standard technique for the measurement of deep levels in semiconductors is Deep Level Transient Spectroscopy (DLTS). Unfortunately, DLTS requires the sample temperature to be ramped over a large temperature range and this can cause a number of difficulties for high pressure measurements. Consequently, most high pressure measurements on deep levels to date have been limited to direct measurements of the capacitance transient (e.g. Zylbersztejn et al, 1978). To avoid these problems we have developed an isothermal deep level measurement technique: Frequency Resolved Capacitance Spectroscopy (FRCS) (Homewood 1987). FRCS is a high sensitivity frequency domain technique that avoids the problems associated with DLTS in high pressure measurements. In the basic form of FRCS, the depletion bias is modulated with a small amplitude sine wave voltage, and the response of the capacitance of the sample to the modulation is measured using quadrature lock-in detection. Sweeping the frequency then generates the deep level spectrum directly. For a more detailed discussion of the technique see Homewood (1987). The hydrostatic pressure measurements were made in a miniature hydraulic pressure cell (Lambkin et al, 1987) in an Oxford Instruments CF1200 flow cryostat. In this system measurements can be made in the range 0-8kbar and 4K to 450K. The single crystal InP:V sample was grown at RSRE by liquid encapsulated Czochralski (Cockayne et al, 1981) and contained $2.10^{16}cm^{-3}$of V.

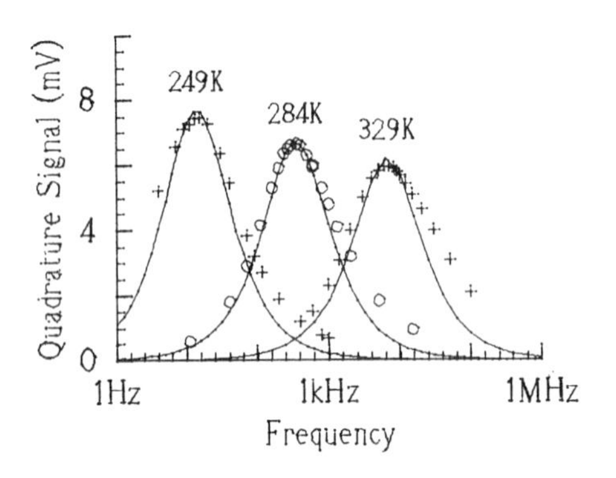

Fig 1. Capacitance spectra at different temperatures measured at ambient pressure. The solid curves are theoretical fits as described in the text.

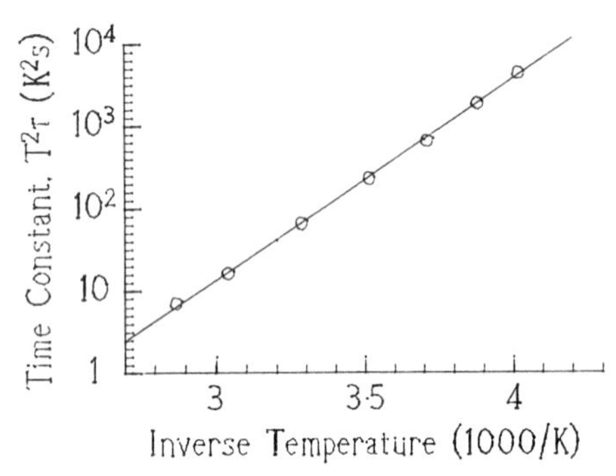

Fig 2. An Arrhenius plot of the data of Fig.1. The solid line is a fit to an activation energy of 490meV.

3. RESULTS

In figure 1 we display typical FRS capacitance spectra from InP:V measured at different temperatures at ambient pressure (0kbar); the solid lines are the theoretical curves (Dunstan and Depinna, 1984) for a single emission frequency, fitted to the peaks of the FRCS spectra. The actual spectra are slightly broader and skewed to higher frequencies; suggesting the presence of an additional level with a faster emission. However, in this paper, we make no attempt to fit to the second minor level. In figure 2 the emission time constants ($T^2\tau$) corresponding to the peak positions of the FRCS spectra are displayed on an Arrhenius plot to obtain an activation energy (490meV) for the dominant level. In figure 3 we show the FRCS spectra for the same level at a fixed temperature (294K) and different hydrostatic pressures in the range 0 to 8kbar. The logarithm of the emission time constant is plotted against hydrostatic pressure in figure 4; this indicates an accurately exponential shift in emission rate with pressure and enables us to obtain a pressure coefficient of 0.16 decades/kbar. In figure 5 we display on an Arrhenius plot the normalised emission rates of the FRCS spectra taken at several pressures with variable temperature. The activation energy of 490meV fits the data at all pressures up to 8kbar.

4. DISCUSSION

The temperature dependence of the emission rate at ambient pressure gives a straight line on the $T^2\tau$ Arrhenius plot and may be interpreted immediately as the activation energy of the centre and identified as its depth at zero temperature from the band to which it is emitting (Lang, 1974). Following previous authors, the pressure dependence of the emission rate would naturally be interpreted in terms of the pressure coefficient of the depth of the centre, and this would give a value of 9.4meV/kbar. This is close to the rate expected for emission to the conduction band gamma-point from a level stationary with pressure with respect to the average of the bands throughout the Brillouin zone, since the gamma point rises at about 10meV/kbar in most of the conventional tetrahedral semiconductors (Paul and Warschauers, 1963) (8.4meV/kbar in InP, Muller et al, 1980). However, this interpretation is ruled out by our measurements of activation energy at high pressure (Fig.5) which show that in fact it is the prefactor, or capture rate, which changes with pressure, and not the depth of the centre. This is a very interesting result. The theory of deep levels (e.g. Ridley, 1978) is not explicit on the relationship between the capture rate and the lattice constant; it appears however that terms which are exponential in phonon frequency will yield an exponential dependence of a multi-phonon capture rate on pressure. Further work is required on this aspect of the problem.

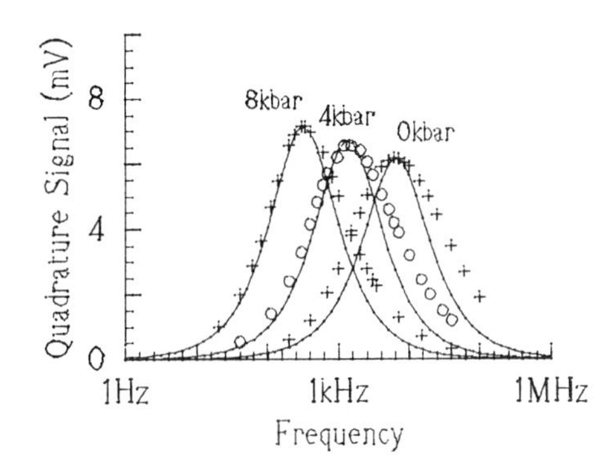

Fig. 3. Capacitance spectra measured at different pressures at 294K. The solid curves are theoretical fits as described in the text.

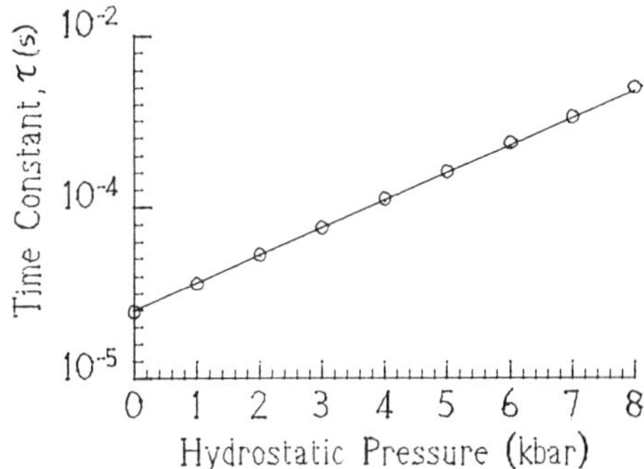

Fig.4. The time constant of the data of Fig.3 is plotted against pressure. The solid line is a guide for the eye, showing the accurately exponential dependence.

The lack of any significant change of the activation energy with pressure is at first sight harder to explain. It is useful first to discuss pressure dependences rather generally. While particular minima in the Brillouin zone may move rapidly in either direction with pressure, the overall effect of pressure on the conduction and valence bands is only to broaden them, eventually narrowing the band-gap. Consequently, a deep level, whose wavefunction is made up of contributions from all the bands, is expected to move slowly, if at all, with pressure. Of the band extrema, the conduction band Γ minimum is anomalous, with its large shift of ~10meV/kbar. This is primarily due to its s-like character. It would be very surprising if a deep level state was made up of only gamma-point conduction band states, and yet this is the only way in which a pressure coefficient of essentially zero with respect to the conduction band

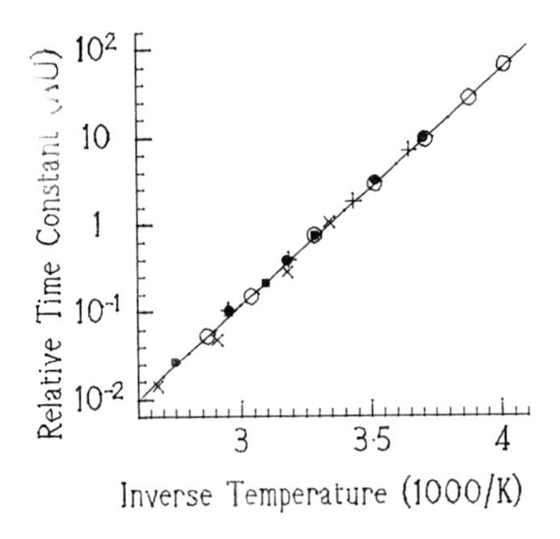

Fig 5. An Arrhenius plot of the time constant at various temperatures at pressures 0kbar (open circles), 2kbar (+), 4kbar (solid circles), 6kbar (squares) and 8kbar (x). For each pressure the time constants have been normalised to unity at 300K. The solid line corresponds to an activation energy of 490meV.

could be obtained. Further experiments are being carried out to clarify this matter. The sample shows photoluminescence at about 1000nm and photoconductivity; the photoconductivity excitation spectrum is a narrow band near 900nm which has previously been observed in the vanadium photoluminescence excitation spectrum (Lambert et al, 1983). Excitation in this band affects the capacitance signal which shows that the centres are related. Details and results from these experiments will be reported in a future paper.

5. CONCLUSIONS

We have presented the pressure and temperature dependence of a deep level in InP:V, and have observed a centre with an activation energy of 0.49eV independent of pressure and a capture cross-section exponentially dependent on pressure. Since our results show that capture rates can change very significantly with pressure, it is important to measure hydrostatic pressure dependences at more than one temperature.

REFERENCES

Cockayne B, Brown G T and MacEwan W R 1981 *J. Cryst. Growth* **54**, 9

Deveaud B, Plot B, Lambert B, Bremond G, Guillot G, Nouailhat A, Clerjaud B and Naud C 1986 *J. Appl. Phys.* **59** 3126

Dunstan D J and Depinna S P 1984 *Phil. Mag.* **B50** 579-597

Homewood K P 1987 (submitted to *Electronics Letters*)

Lambert B, Deveaud B, Toudic Y, Pelous G, Paris J C and Grandpierre G 1983 *Solid State Commun* **47** 337

Lambkin J D, Gunney B J, Dunstan D J and Bristow F G 1987 (submitted to *J. Phys. E*)

Lang D V 1974 *J. Appl. Phys.* **45** 3023

Muller H, Trommer R, Cardona M and Vogl P 1980 *Phys Rev* B **21** 4879

Paul W and Warschauer D M 1963 *Solids at High Pressure*, (McGraw-Hill) chapter 8, p226

Ridley B K 1978 *J. Phys. C: Solid State Phys.*, **11** 2323

Zylbersztejn A, Wallis R H and Besson J M 1978 *Appl. Phys. Lett.* **32** 764

Characteristics of large diameter, undoped SI GaAs grown by the heat exchanger method (HEM)

C. P. Khattak, S. Di Gregorio, F. Schmid

Ghemini Technologies Division, Crystal Systems, Inc., Salem, MA 01970

J. Lagowski

Massachusetts Institute of Technology, Cambridge, MA 02139

ABSTRACT: Undoped, semi-insulating, 7.5 cm diameter GaAs crystals have been grown by the Heat Exchanger Method (HEM™). The *in situ* annealing feature of HEM produces high uniformity of electronic properties. HEM GaAs exhibits high mobilities, low impurity concentration and higher ion-implantation efficiency compared to "state-of-the-art" LEC GaAs.

1. INTRODUCTION

It has been shown that the Heat Exchanger Method (HEM™) is a promising technique for growth of large diameter GaAs crystals with low dislocation density and high resistivity (Khattak et al, 1985). Semi-insulating (SI) GaAs crystals have been produced in silica crucibles by HEM without doping. The deep donor EL2 concentration in HEM GaAs is in the range of $3\text{-}7\times10^{16}$ cm^{-3}, considerably higher than material grown by other melt growth techniques (Wohlgemuth et al, 1986).

This paper describes the scale-up in size of crystals from 5.5 cm diameter to 7.5 cm diameter. The remarkable uniformity and high EL2 concentration have been attributed to the *in situ* annealing feature of HEM. Semi-insulating samples have been characterized for resistivity, mobility and ion-implant efficiency.

2. CRYSTAL GROWTH

The details of HEM growth of 5.5 cm diameter, 5.5 cm high GaAs crystals have been described in a previous paper (Khattak et al, 1985).

In the HEM there is no movement of the crystal, crucible or heat zone; temperature gradients on the solid are dependent on the heat exchanger temperature which is controlled. Under these conditions the temperature gradients in the liquid can be varied during crystal growth. Emphasis is placed on maintaining these gradients to less than 5°C/cm. The temperature gradient on the GaAs crystal is reduced after complete solidification and the boule is *in situ* annealed in the HEM furnace.

3. MATERIAL CHARACTERISTICS

3.1 Structural Characteristics

Single crystal GaAs boules of <100> orientation, 7.5 cm diameter were grown. Twins were observed in the structure with preferential alignment along the <110> orientation, similar to the lineage observed in LEC grown crystals (Thomas et al, 1984).

Samples were mapped for EPD and a W-shaped distribution was observed at the seed end with some flattening of this distribution at the top end. Microscopic examination showed that the cellular structure from the seed had propogated into the HEM-grown material with enlargement of the size of the cells. Figure 1 shows the boundary of the residual seed-grown GaAs; the cell size continued to enlarge with growth. While reduction of EPD in HEM GaAs has been observed, the absolute value and distribution of EPD is being limited by the seed crystal. For 5.5 cm diameter unseeded HEM GaAs 500 cm^{-2} EPD and U-shaped distribution has been achieved (Khattak et al).

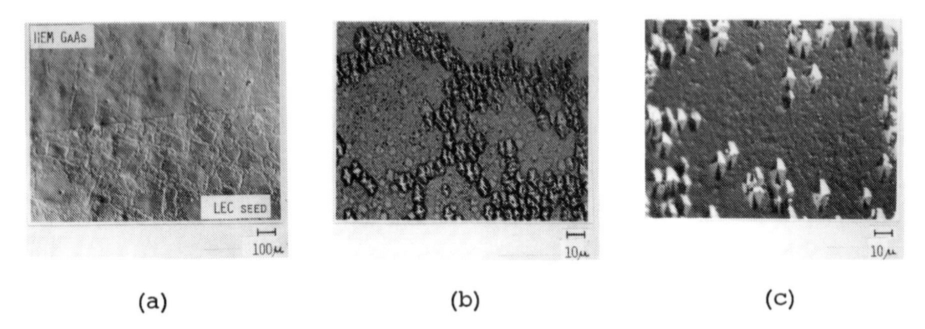

Figure 1. Cellular nature of dislocations at various positions of the boule: (a) LEC seed-HEM GaAs interface; (b) near the seed end and (c) near the top of the boule.

3.2 Electronic Properties

Samples from 7.5 cm diameter HEM GaAs crystals corresponding to the seed and top sections of the boule were evaluated for resistivity and mobility using the Van der Pauw technique. Data showed resistivity of $(4.4 - 5)\times10^{7}$ ohm-cm with mobilities of $(6450 - 6570) cm^{2}V^{-1}s^{-1}$ indicating good quality SI material.

The range of EL2 concentration observed for HEM GaAs is 3 to 7 x 10^{16} cm^{-3} with less than 5% variation within a boule, both along the growth direction and across the wafers. The correlation between EL2 concentration and *in situ* annealing parameters has shown that higher EL2 concentrations are observed for samples annealed at higher temperatures and/or for prolonged periods. This is consistent with the EL2 formation model based on migration of gallium vacancies to neighboring sites at temperatures above 1050K (Lagowski et al, 1982).

In one experiment the furnace power was turned off after solidification and no *in situ* annealing was carried out. The as-grown samples were characterized for EL2 concentration. The samples were then annealed at 850°C for 8 hours. A comparison of the optical absorption spectra at 77K for as-grown and annealed samples from top and seed ends of the boules is shown in Figure 2. An estimate of the EL2 concentration for these samples is shown in Figure 3. The data shows that the as-grown boule exhibited non-uniform EL2 concentration. However, after annealing high uniformity was achieved. In the homogenization by annealing, the EL2 concentration for the top section increased noticeably, whereas it reduced slightly for the bottom section.

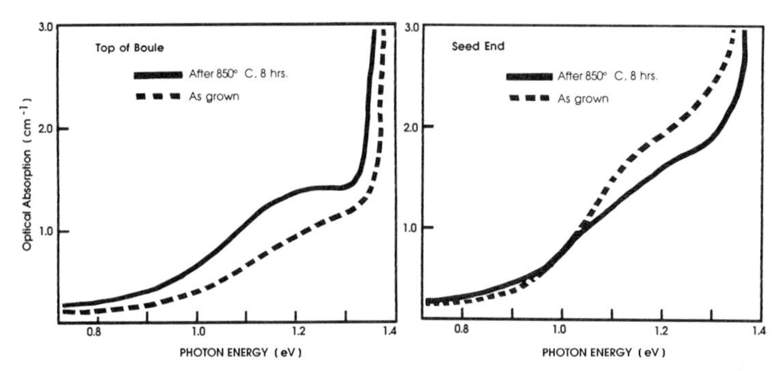

Figure 2. Optical absorption spectra of as-grown unannealed and annealed samples from top and seed sections of HEM GaAs crystal.

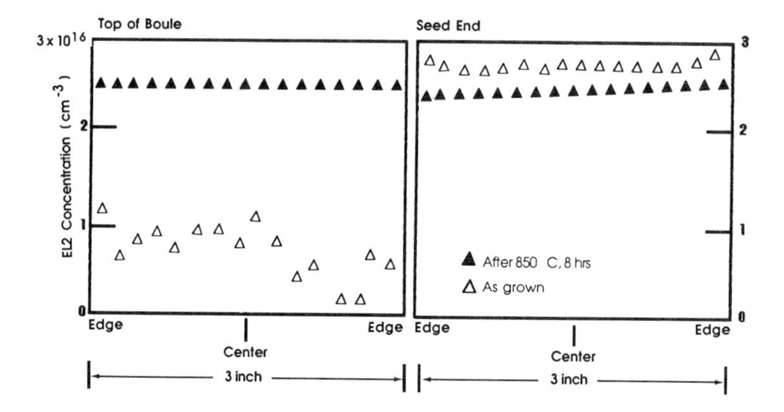

Figure 3. EL2 concentration of as-grown unannealed and annealed samples from top and seed sections of HEM GaAs crystal.

3.3 Shallow Traps

Semi-insulating HEM GaAs samples were characterized using thermally stimulated current (TSC) measurements (Kaminska et al, 1986). Good resolution of the deep levels (Martin, 1980; Martin et al, 1977) was obtained with evidence of thermal emission from shallow traps below 80K. Estimation of the shallow trap concentration yielded a value of about 10^{16} cm^{-3}, similar to high-quality experimental LEC GaAs grown in a multi-zone furnace with post-solidification annealing capability.

HEM GaAs samples evaluated for carbon concentration using FTIR spectroscopy showed no detectable carbon ($<3x10^{14}cm^{-3}$).

3.4 Ion-implantation

Samples of SI HEM GaAs were ion-implanted with silicon at 2 x 10^{12} cm^{-2} along with LEC undoped GaAs and light Cr-doped HB GaAs samples. The data after annealing is shown in Table I. It is clear that ion-implantation efficiency of HEM GaAs is comparable or slightly higher than that of "state-of-the-art" GaAs currently available.

Table I. Electronic properties at 77K after implantation on HEM GaAs and commercially available GaAs crystals.

Material	Implant (cm^{-2})	ρ_s (Ω/□)	n_s (cm^{-2})	μ (cm^2/Vs)	Activation Efficiency (%)
LEC	Si; $2x10^{12}$	1200	$1.2x10^{12}$	4300	60
HB	Si; $2x10^{12}$	1110	$1.2x10^{12}$	4600	60
HEM	Si; $2x10^{12}$	980	$1.35x10^{12}$	4700	67
HEM	Si; $2x10^{12}$	1080	$1.63x10^{12}$	3560	81

4. CONCLUSIONS

HEM growth of undoped GaAs from presynthesized meltstock has been extended to 7.5 cm diameter crystals. SI crystals of <100> orientation have been produced without doping. The dislocation density of the grown material is lower compared to the seed, but some twinning with <110> alignment was observed. The *in situ* annealing feature of HEM allows very high uniformity of EL2 concentration within the crystal and can be controlled with the annealing parameters. HEM GaAs shows lower shallow trap and carbon concentration, high mobilities and higher ion-implantation efficiency compared to "state-of-the-art" LEC GaAs.

5. REFERENCES

Kaminska, M., Parsey, J.M., Lagowski, J. and Gatos, H.C., 1982 Appl. Phys. Lett. 41, 989.

Khattak, C.P., Lagowski, J., Wohlgemuth, J.H., Mil'shtein, S., White, V.E. and Schmid, F., 1985, in Proceedings of Int. Symp. GaAs and Related Compounds, Karuizawa, Japan, 1985, edited by M. Fujimoto (Adam Hilger Ltd., Bristol) p. 31.

Lagowski, J., Gatos, H.C., Parsey, J.M., Wada, K., Kaminska, M. and Walukiewicz, W., 1982, Appl. Phys. Lett. 39, 747.

Martin, G.M., 1980, in Proceedings of Semi-Insulating III-V Materials, Nottingham, 1980, edited by G.J. Rees (Shiva, UK).

Martin, G.M., Mittoneau, A. and Mircea, A., 1977, Electron. Lett. 13, 191.

Thomas, R.N., Hobgood, H.M., Eldridge, G.W., Barrett, D.L., Braggins, T.T., Ta, L.B. and Wang, S.K., 1984, Semiconductors and Semimetals Vol. 20 (Academic, New York) p. 1.

Wohlgemuth, J.H., Khattak, C.P., Lagowski, J. and Skowronski, M., 1986, in Proceedings of Semi-Insulating III-V Materials, Hakone, Japan, 1986, edited by H. Kukimoto and S. Miyazawa (Ohmsha, Tokyo) p. 191.

Paper presented at Int. Symp. GaAs and Related Compounds, Heraklion, Greece, 1987

High quality InP grown by GF method under 27 atm phosphorus vapor pressure

S. Yoshida, N. Nishibe, and T. Kikuta
Yokohama Research Laboratory,
The Furukawa Electric Co.,Ltd.,
2-4-3, Okano, Nishi-ku, Yokohama 220, Japan

Abstract High purity polycrystalline InP crystals have been reproducibly grown by the gradient freeze method under supplying phosphorus vapor pressure of 27 atm.. InP ingots contained no residual elemental In and a large single grain of 4x3.5 cm^2. In more than 90% of the undoped InP ingot grown even from the quartz boat, the carrier concentration was less than $5x10^{15}cm^{-3}$. The best value of Hall mobility reached to 63,000 $cm^2/V{\cdot}sec$ at 77K. The contamination of Si was remarkably reduced by our special treatment of the boat surface.

1. Intoroduction

InP is a promising substrate for optoelectronic devices, high speed integrated circuits and high frequency microwave devices. Especially, InP is required for the substrates of laser diodes with the peak at 1.3 - 1.6μm which correspond to the lowest loss range of the fiber for the optical communication. High purity polycrystalline InP is generally synthesized under the conditions of the low synthesis temperature (<1050 °C) due to prevention of contamination and the low phosphorus pressure(<15atm.) due to conrollability of the temperature. The InP ingots synthesized by this method, (Antypas 1976, Adamski 1983, Kubota et al. 1987), however, have several problems such as the presence of In inclusions, the low yield enough to use industrially and the long synthesis process. InP single crystal is generally grown by the liquid encapsulated Czochralski(LEC) technique using polycrystalline InP as the starting material (Mullin et al. 1968). Direct synthesis of In and P is not available for InP, since the dissociation pressure of InP is very high (27atm.) at the melting point. High quality InP substrates are required more and more for manufacturing high quality devices. Consequently,a cheap and high purity polycrystalline InP is indispensable for InP single crystal growth. In this paper, it is described that the high purity polycrystalline InP has been reproducibly grown by the horizontal gradient freeze (GF) method under supplying 27 atm. phosphorus vapor pressure (Bonner 1981) using a quartz boat.

2. Experimental

Figure 1 shows the apparatus for the growth of polycrystalline InP. Two electric furnaces are installed in the high pressure chamber. In a quartz ampoule, red phosphorus and In with 6N purity are placed in the low and high temperature zone, respectively. The quartz ampoule is sealed in vaccuum less than 1×10^{-7} torr.. The phosphorus vapor pressure during the growth is monitored by the thermocouples which are set outside the phosphorus zone. The temperature is converted into the phosphorus pressure using Bachmann's equation (Bachmann et al. 1974). Corresponding to the phosphorus pressure, the pressure of Ar gas, which is filled in the chamber, should be controlled to prevent the break of the ampoule. Pyrolytic BN boat is generally used to grow the high purity crystal (Henry et al. 1978), since the crystals grown from the quartz boat have high Si contamination due to the sticking between the boat and the InP melt. However, the PBN boat has several problems such as high cost and difficulty for precise manufacture. To solve these problems, the quartz boat was used in this study. The special treatment of the quartz boat has been developed to prevent the sticking, that is, the surface is sand-blasted and thermally finished to have periodic roughness. The horizontal gradient freeze method was used for the growth. The temperature gradient during the growth was 4 - 6 °C/cm. The total growth process was done for 3 days.

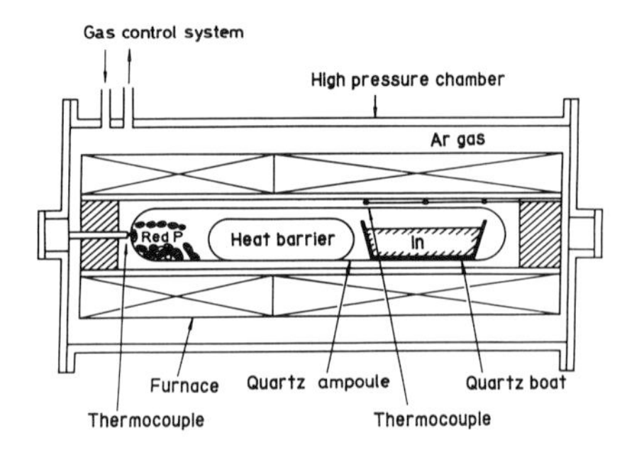

Fig.1 Apparatus for the growth of polycrystalline InP by high pressure GF method.

3. Results and Discussion

The crystals were 2000g in weight and 30 cm in length. Figure 2a shows the bottom view of the InP ingot grown from quartz boat. Note that the bottom view appeared smooth surface with no residual elemental In . The top view of the ingot was also mirrorlike. This suggests that the stoichiometry has been controlled by supplying phosphorus vapor pressure during the growth. In the case of the quartz boat with untreated surface, the crystal was often broken during cooling due to the sticking between the quartz and

InP. However, the crystals with smooth back surface have been reproducibly grown using our special surface treatment of the quartz boats.

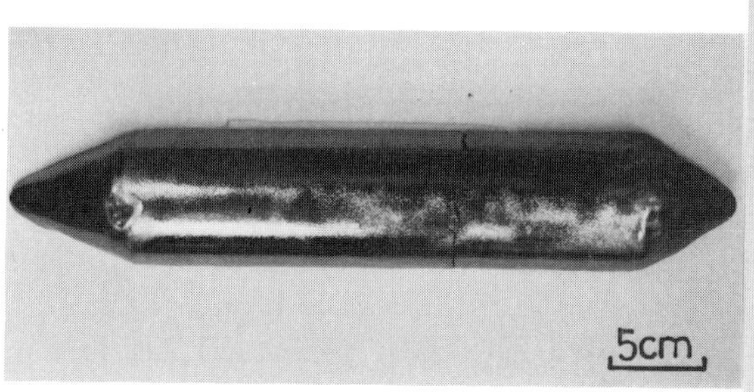

Fig.2a Bottom view of InP ingot.

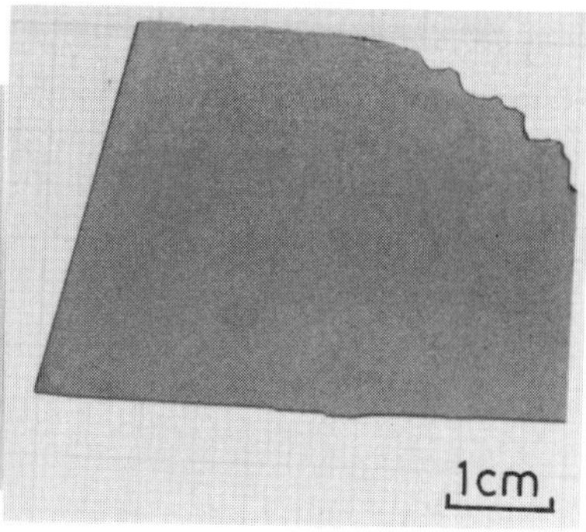

Fig.2b (100) wafer cut from InP polycrystal.

Figure 3 shows the distributions of the carrier concentration and the Hall mobility in undoped InP along the length. All the undoped crystals were n-type conductive. In more than 90% of the ingot, the carrier concentration was less than $5x10^{15}$ cm^{-3}. The best value of Hall mobility reached to 63,000 $cm^2/V{\cdot}sec$ at 77K.

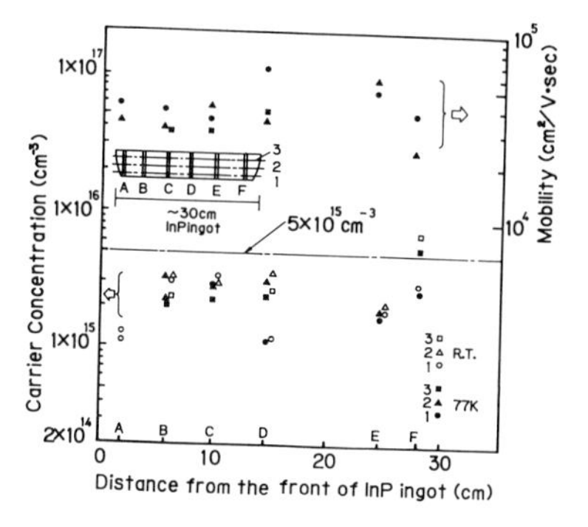

Fig.3 Distribution of the carrier concentration and the Hall mobility in InP ingot.

The impurity concentration were analyzed by SSMS to clarify the effect of the sticking. According to Table 1, the main donor impurity is Si. Note that the Si contamination is remarkably reduced in the case of the quartz boat not stuck to InP melt. This indicates that the Si contamination was mainly based on a quartz boat and our surface treatment was very effective to remove the Si contamination.

Table 1. Results of spark source mass spectroscopy (SSMS) of polycrystalline InP ingot. This shows the comparison between the quartz boat stuck to InP melt and not stuck to InP melt.

Quartz boat stuck to InP melt	N^* (cm^{-3})	impurity concentration (ppma)			
		Si	Mg	Al	Fe
YES	$1.6x10^{16}$	0.342	0.034	0.951	0.014
NO	$3.2x10^{15}$	<0.002	0.004	0.002	0.005

* Carrier concentration (by Van der Pauw method)

Figure2b shows the (100)wafer cut from the polycrystalline InP ingot. The grain size of single crystal was about 4x3.5 cm^2. About 50 wafers with the similar grain size were obtained from the ingot. EPD was partially less than 2,000 cm^{-2}in even undoped InP as shown in Figure 4. This is considered to be due to the growth condition of the low temperature gradient compared to the LEC method. The seed crystal was not used for the growth in this study. It was found that the growth direction was almost (111) and the shape of interface between the melt and the solid was concave toward the melt from the observations of surface morphology, X-ray measurement and the striation. The twin boundary often appeared at the near center of the ingot and the high EPD regions were present along the twin boundary. It was supposed that the crystal was grown from the boat walls to the center of InP melt corresponding to the thermal flow. Then, EPD increased at the center of the ingot, since thermal stress was enhanced there. In order to grow a large InP single crystal with low EPD, the thermal flow pattern should be improved by changing the temperature profile.

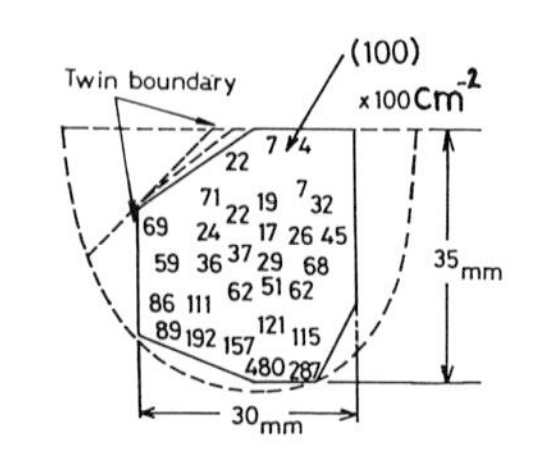

Fig.4 Distribution of EPD in undoped InP(100)wafer.

4. Summary

High purity polycrystalline InP crystals have been reproducibly grown by the GF method under supplying phosphorus vapor pressure of 27atm.. InP ingots contained no residual elemental In and the large single grain of 4x3.5 cm^2. In more than 90% of undoped InP ingot grown even from the quartz boat, the carrier concentration was less than $5x10^{15}$ cm^{-3}. The best value of Hall mobility reached to 63,000 cm^2/V·sec at 77K. The contamination of Si was remarkably reduced by our special treatment of the boat surface.

Acknowledgement

The authors wish to thank Y.Kashiwayanagi and J.Kikawa for usefull discussions. We also wish to thank J.Suzuki and M.Nakayama for preparing the crystal growth.

References

Adamski J.A. 1983 J.Cryst. Growth 64 1
Antypas G.A. 1977 Inst.Phys.Conf.Ser. 33b 55
Bachmann K.J. and Buehler E. 1974 J.Electron. Mater. 3 279
Bonner W.A. 1981 J.Cryst. Growth 54 21
Henry R.L. and Swiggard E.N. 1978 J.Electron. Mater. 7 647
Kubota E., Katsui A. and Omori Y. 1987 J.Cryst. Growth 82 737
Mullin J.B., Heritage R.J., Holliday C.H.and Strughan B. 1968 J.Cryst. Growth 3-4 281

A state of the art LEC growth technique—development of 3-inch diameter, 27 cm long, undoped semi-insulating GaAs single crystal

K.Nambu, R.Nakai, M.Yokogawa, K.Matsumoto, K. Koe and K.Tada

Sumitomo Electric Industries, Ltd., 1-1-1, Koya-kita, Itami-shi 664, Japan

ABSTRACT: We have developed a state-of-the-art LEC growth technique based on a heat flow simulation. Using this technique, a 3-inch diameter undoped semi-insulating GaAs single crystal has been grown to be 27 cm long. The dislocation density shows $3.5\times10^4/cm^2$ at the seed end and monotonically increases towards the tail end up to $5.5\times10^4/cm^2$. The electrical properties exhibit extremely uniform distribution along the growth direction in the crystal.

1. INTRODUCTION

In order to improve the uniformity of the threshold voltage of FET's fabricated on GaAs wafers, many efforts have been devoted such as a reduction of dislocations (Tada et al. 1984), a post-growth annealing (Rumsby et al. 1983) and control of stoichiometry (Holmes et al. 1982) and impurity (Akai et al. 1981). However, since there still exists variations of threshold voltage from ingot to ingot, it has been a key technology to grow a long ingot, which increases a lot in size and therefore, results in a lower qualification test cost per wafer. An LEC growth of a GaAs crystal is carried out in a severe thermal environment, which results in a high dislocation density of 10^4-$10^5/cm^2$. Thus, a lineage structure is apt to occur during the crystal growth, which easily evolves to a polycrystal. In order to grow a longer single crystal, a thermal environment that does not generate this polycrystal, an excellent diameter control technique and stability of the thermal environment during crystal growth are required. In this paper, we specifically describe·how to design a thermal environment which produces an undoped long single crystal and show the results of characterization of this ingot.

2. HEAT FLOW SIMULATION

2-1. One-dimensional model

In the present work, we have used a simple model to understand the variation of the temperature gradient around a solid-liquid (SL) interface in each portion of crystal along the growth direction. A one-dimensional heat transfer equation is given by

$$\rho_S C_S \left[\frac{\partial T}{\partial t} + v\frac{\partial T}{\partial z}\right] = \frac{\partial}{\partial z}\left[K_S\frac{\partial T}{\partial z}\right] + \frac{2Q_r}{r} \quad \text{.............(1)}$$

where ρ_S, C_S, v, K_S, r and Q_r are the density, specific heat, growth rate, thermal conductivity, crystal radius and the heat flow along the crystal radius, respectively. A heat balance at the SL interface can be expressed as

$$-\pi r^2 K_S \left.\frac{dT}{dz}\right|_S = \pi r^2 v C_L - \pi r^2 K_L \left.\frac{dT}{dz}\right|_L \quad \text{..............(2)}$$

Where C_L, K_L are the latent heat and thermal conductivity of melt. As the Péclet number ($=v\rho_S C_S r/K_S$) is negligible in a conventional LEC growth, eq. (1) can be written in the steady-state form. Assuming an appropriate form of Qr, we can estimate the axial variation of growth parameters such as thermal distribution. The second term in the right side of eq.(2) means the heat inflow from the melt into the SL interface. In a one-dimensional model, this term can be written as $dT/dz|_S=(T_m-T_C)/\ell$, where T_m, T_C and ℓ are the melting point of GaAs, the crucible temperature and the melt height, respectively. By solving eq.(1) and eq.(2), we can estimate the change of the temperature gradient around the SL interface with the advance of crystal growth. This model is also appliciable for analysis of the thermal hysterisis of each portion of the crystal after solidification. Fig.1 shows the thermal hysterisis at various points along growth direction. The upper and lower horizontal axis show the duration of growth and the distance from the top of the crystal, respectively. A crystal within 4 cm from the top tends to cool faster after solidification but the remaining part of crystal cools slower and shows the same type of thermal hysterisis. The same thermal hysterisis especially at an elevated temperature, which has a great influence on the formation of native defects, is quite beneficial to a long ingot.

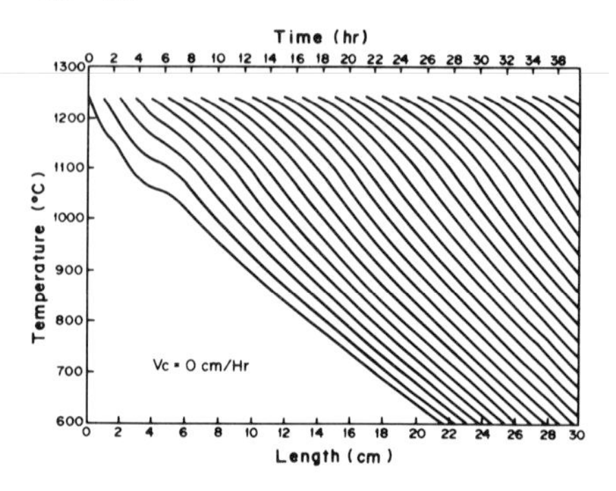

Fig.1 Thermal hysterisis of each portion in crystal. The changing rate of the hot crucible wall height Vc=0 (mm/Hr).

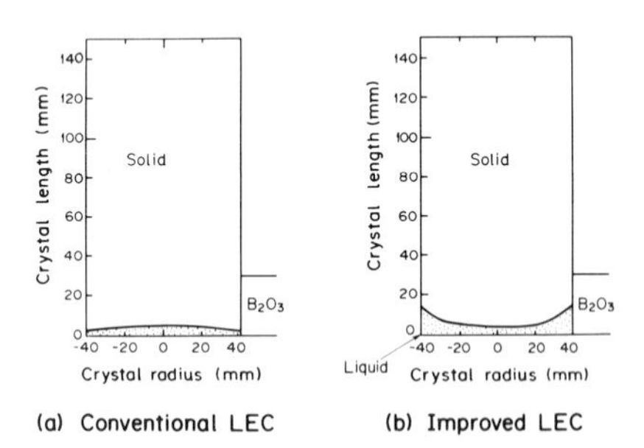

Fig.2 The SL interface shape predicted from the isothermal line in the case of a conventional thermal environment (a) and the improved thermal environment (b).

2-2. Control of SL interface shape

In general, a lineage-originated polycrystal produces a limit to further single crystallization in an LEC-grown undoped GaAs crystal. A dislocation has a natural tendency to propagate perpendicular to the SL interface (Chalmers et al. 1964). Thus in order to avoid the bunching of dislocations, we must make the SL interface shape more convex to the melt. We have tried by improving the thermal environment. Though the observation of SL interface of undoped semi-insulating GaAs crystal was quite difficult because of extremely low carrier concentration, we could design a thermal environment for convexity of SL interface by using a heat flow simulation. Fig.2 shows the SL interface shape predicted from the isothermal line derived from a conventional thermal environment (a), and the improved thermal environment (b). In the case of the improved thermal environment, it can be seen that while the SL interface shape is flat at the center, it becomes more convex at the periphery.

3. EXPERIMENTAL

We used 6-10Kg of GaAs polycrystal pre-synthesized by the HB method as a starting material. A crystal was grown from 6-7.3 inch diameter PBN crucible with a growth rate of 8-12mm/hr under a high pressure N_2 ambient of 15-25atm. A super low water content B_2O_3 (≦100 wtppm) was used as an encapsulant. The puller has three heaters, which enable us to realize the desired thermal environment. The average thermal gradient in B_2O_3 was 80-120 °C/cm. An original ADC system developed in our company (Yamaguchi et al. 1983) was used for a stable diameter control. To characterize the long crystal, the following measurements were performed. The dislocation density was measured as an etch pit density (EPD) revealed by molten KOH etching. The electrical resistivity and mobility were obtained using a Hall effect measurement with Van der Pauw geometry at 298°K. The measurement of Si, S and B concentration were done by SIMS. The concentration of carbon at the arsenic site, $[C_{AS}]$, was estimated by FT-IR measurement at 77°K (Brozel et al. 1986). The neutral EL2 concentration, [EL2], was measured by optical absorption at λ=1μm. The PL measurement was performed at 4.2 °K with an automatic mapping system using Ar laser (λ=5145Å) excitation.

Fig.3 Photograph of conventional and long 3-inch diameter undoped GaAs crystal.

4. RESULTS AND DISCUSSION

A 3-inch diameter GaAs crystal developed in this work is shown in Fig.3. The crystal was almost all single crystal and 27 cm long, which can yield more than 200 wafers. The radial distribution of EPD shows a so-called "W" shape. The axial variation of EPD shows $3.5\times10^4/cm^2$ at the seed end and increases monotonically towards the tail end up to $5.5\times10^4/cm^2$. Though the 1.49 eV PL peak is related to carbon in GaAs, its peak intensity is thought to be strongly affected by the concentration of non-radiative recombination center originating from the native defect. This implies that the variation of PL intensity in a crystal may reflect the thermal history of the crystal, because the formation of the native defect is affected by the thermal history of the crystal. The radial distribution and axial variation of PL intensity at 1.49 eV are shown in Fig.5. A radial distribution of PL intensity shows a strong "W" shape at the seed end and rapidly changes to a weak "W" shape, then maintains the

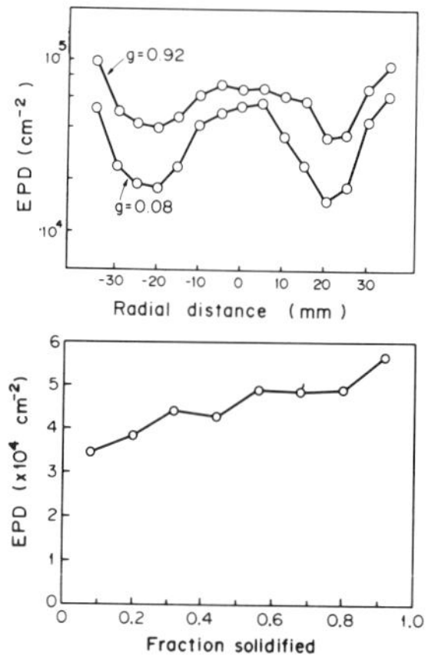

Fig.4 The radial and axial variation of EPD.

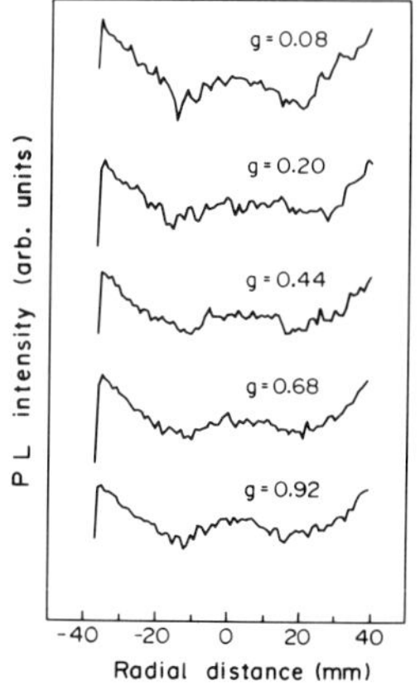

Fig.5 The radial variation of PL intensity at 1.49 eV.

shape towards the tail end. This manner of change of PL intensity along the growth direction totally agrees with the results of the thermal hysterisis analysis mentioned in §2. Fig.6 shows the axial variations of electrical resistivity and mobility. We can see the uniform distributions along the growth direction. According to SIMS analysis, Si and S concentration were under the detection limit ($\leq 1\times10^{14}/cm^3$ and $\leq 6\times10^{14}/cm^3$, respectively). B was in the range of $2\text{-}8\times10^{15}/cm^3$. The axial distributions of [C_{AS}] and [EL2] are shown in Fig.7. Carbon in the GaAs crystal is the dominant impurity which affects the electrical properties of FET's fablicated on a wafer using a direct implantation technique (Chen et al. 1984). As shown in Fig.7, [C_{AS}] exhibits a uniform distribution along the growth axis around $1\times10^{15}/cm^3$. It is well known that a deep donor EL2 is influenced by melt stoichiometry. Since a stoichiometric GaAs polycrystal, grown by the HB method, was used as a starting material, an extremely uniform distribution of EL2 along the growth direction was obtained.

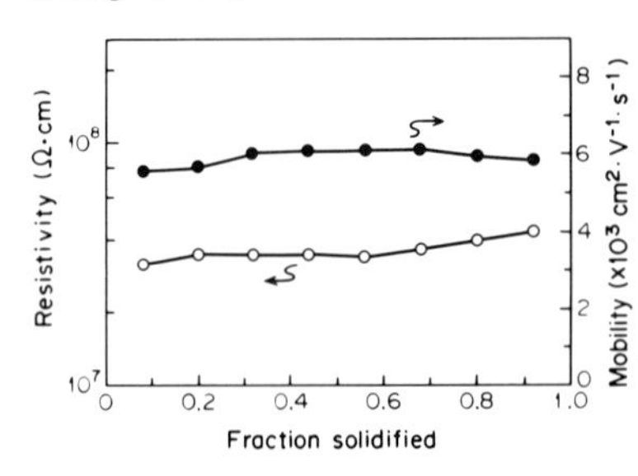

Fig.6 The axial variation of electrical resistivity and mobility.

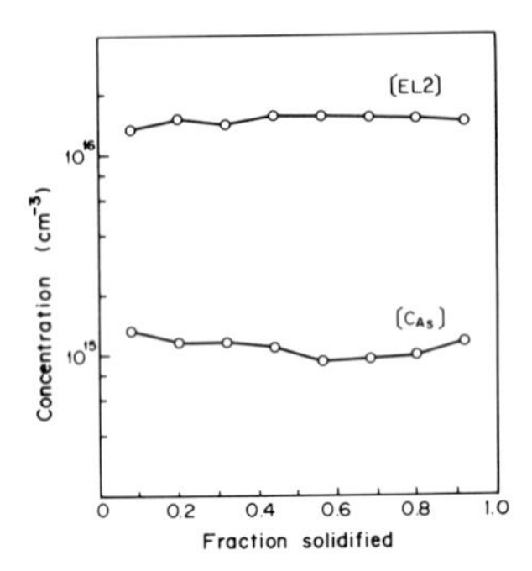

Fig.7 The axial variation of [C_{As}] and [EL2].

5. CONCLUSION

We have developed a 3-inch diameter, 27 cm long, undoped semi-insulating GaAs single crystal by combining a homemade puller equipped with multi-zone heater with a heat flow analysis. The electrical properties show extremely uniform distribution along the growth axis. We are convinced that this technique will be applicable to larger diameter crystals such as 4 or 5 inch ones. This newly developed technique will open a new era not only for the crystal growth of undoped GaAs but also for GaAs IC production.

6. REFERENCES

Tada K, Murai S, Akai S and Suzuki T 1984 Proc. of IEEE GaAs IC Symposium (Boston) p.49.

Rumsby D, Ware R M, Smith B, Tyjberg M, Brozel M R and Foulkes E J 1983 Proc. GaAs IC Conf. (Phoenix).

Holmes D E, Chen R T, Elliot K R and Kirkpatric C G 1982 Appl. Phys. Lett. 40 46.

Akai S, Fujita K, Sasaki M and Tada K 1981 Inst. Phys. Conf. Ser. 63 13.

Chalmers B PRINCIPLES OF SOLIDIFICATION (New York;John Willy&Sons Inc.).

Yamaguch T, Akai S, Tada K, Murai S, Kawasaki A, Nakai R, Takebe T, Shimazu M, Kishi M, and Takenaka K 1983 Denshi Tokyo vol.22.

Brozel M R, Foulkes E J, Series R W and Hurle D T J 1986 Appl. Phys. Lett. 49 11.

Chen R T, Holmes D E and Asbek P M 1984 Appl. Phys. Lett. 45 459.

Theoretical modelling of electrical activation in non-implanted GaAs

N Morris and B J Sealy

Department of Electronic Engineering, University of Surrey, Guildford, Surrey GU2 5XH, U.K.

ABSTRACT: A study of the time dependent behaviour of a number of donor and acceptor ions in GaAs has enabled a theoretical model to be developed which accurately predicts the electrical properties after annealing. The model predicts that, at long times and high temperatures, a saturation level of electrical activity occurs and that, in this region, the change in activation is characterised by an energy, the magnitude of which depends on the ion species. The paper presents the model and discusses the activation mechanisms for various implanted ions in GaAs.

1. INTRODUCTION

There are currently a number of theoretical models available which describe the incorporation of impurities into GaAs. However, in the case of ion-implanted material, the kinetics of electrical activation which occur during the post-implant annealing stage have received remarkably little attention.

In an attempt to obtain a more detailed model, a study of the variation in electrical activity with time and temperature has been performed. A thermodynamic-based analysis of the experimental data has resulted in a single expression which yields two energies, a diffusion energy (Ed) and a saturation energy (Ea), for a range of 'p' and 'n' type impurities in GaAs. This paper presents a discussion of the model and an interpretation of the two energies involved.

2. EXPERIMENTAL TECHNIQUES

Undoped semi-insulating GaAs of (100) orientation was implanted with both donor and acceptor ions at room temperature in a non-channelling direction.The ions, energies and doses studied during this work were (1) the donors: Sn(300 keV, 1E14 cm^{-2}), Se(300 keV, 1E13 cm^{-2}) and S(120 keV, 1E14 cm^{-2}) and (2) the acceptors: Be(75 keV, 5E14 cm^{-2}), Mg(100 keV, 5E14 cm^{-2}) and Zn(100 keV, 1E15 cm^{-2}).

Following implantation, all the donor-implanted samples were encapsulated with 300Å of pyrolytically deposited Si_3N_4 plus 600 Å of reactively evaporated AIN (Bensalem et al, 1983). In the case of the acceptor species, a single encapsulant of 600 Å AIN was used. Annealing of $5mm^2$ samples was performed in a double graphite strip heater (Gwilliam et al,

1985) for times betweeen 1 -200 seconds within a temperature range of 485°C - 1100°C. During the anneal a record was made of the annealing temperature, Ta, the annealing time, ta, and the rise-time, tr. After annealing, samples were cut into a clover-leaf pattern before removing the encapsulant in hydrofluoric acid. Electrical contacts were then made to the implanted samples in preparation for Hall-effect measurements of sheet carrier concentration (Ns, Ps), sheet mobility μs and sheet resistivity ρs.

3. THEORY

Assuming that the dominant process can be modelled by first-order kinetics, then, for any initial concentration of N inactive impurities, the number which become electrically active per unit time is proportional to the number of impurities available for activation at that time, ie:-

$$\frac{d\,N(t)}{dt} = -K\left(N(t)\right) \qquad (1)$$

where K is the rate constant for the process.

However, this equation implies that the reaction will continue at the same rate for all values of time. Since this is not the case for GaAs, equation (1) must be modified so that, at longer annealing times, the activation rate decreases until a saturation level is reached. The required form is therefore:-

$$\frac{d\,N(t)}{dt} = -K\left(N(t) - \Sigma(T)\right) \qquad (2)$$

where $\Sigma(T)$ is the maximum fraction of the implanted dose which can be activated at a given annealing temperature. Hence:-

$$\Sigma(T) \;\alpha\; N_D \exp(-Ea/kT) \qquad (3)$$

where Ea is the characteristic activation energy for the process and N_D is the ion dose. Integration of equation (2) yields:-

$$N(t) = \Sigma(T)\left(1-\exp(-Kt)\right) \qquad (4)$$

where the rate constant K is related to an effective diffusion coefficient, D, for the implanted material. Thus:

$$K = K'D = K'D_o \exp(-Ed/kT) \qquad (5)$$

Hence, by substitution of equations (5) and (3) in equation (4), the time and temperature dependence of activity is given by:-

$$N(t) \;\alpha\; N_D \exp(-Ea/kT) \left[1 - \exp\left\{-K'D_o t \exp(-Ed/kT)\right\}\right] \qquad (6)$$

In order to calculate values for Ea and Ed it is therefore necessary to analyse the data in two distinct regimes. At the longer annealing times equation (6) reduces to:- $N(t) \;\alpha\; N_D \exp(-Ea/kT)$, so that the slope of the graph of $\ell n\, N(t)$ versus reciprocal annealing temperature yields a value for Ea. In the time dependent region, however, it is necessary to plot $\ell n\,(f(t,T)/t)$ versus reciprocal temperature to obtain values for Ed and $K'D_O$, where:-

$$f(t,T) = \ell n \left| \frac{1}{\{1 - (\frac{N(t)}{\Sigma(T)})\}} \right|$$

The resulting slope and intercept of this graph then yield values for Ed and ln $K'D_O$ respectively.

4. EXPERIMENTAL RESULTS

The variations in sheet carrier concentration as a function of annealing time and temperature are plotted as Figure. 1 for the 1E14 cm^{-2} Sn implant and in Figure. 2. for the 5E14 cm^{-2} Be implant[1].

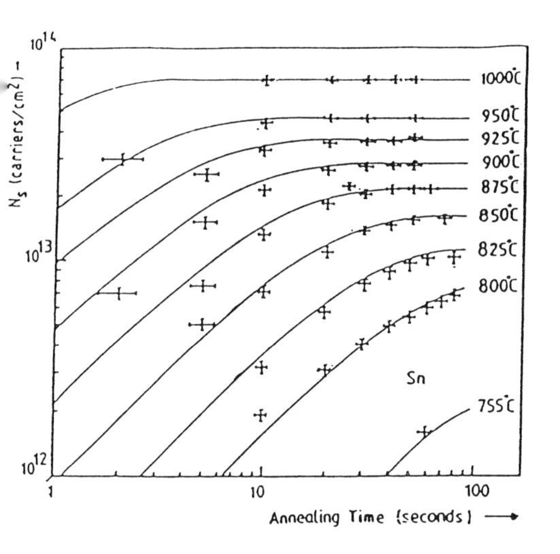

Figure 1. Sheet electron concentration versus annealing time and temperature for 1E14 Sn cm^{-2} 300 keV.

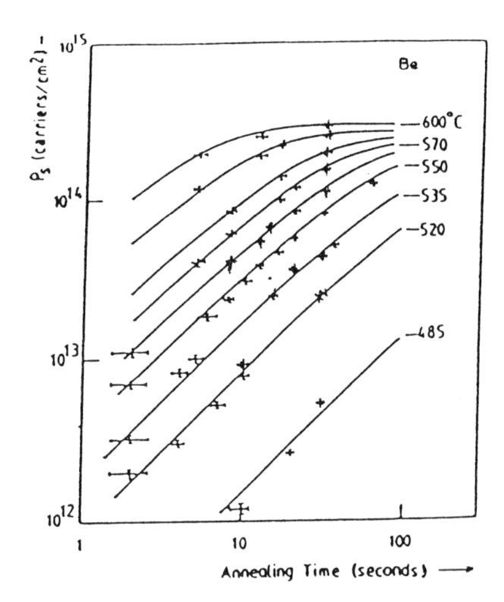

Figure 2. Sheet hole concentration versus annealing time and temperature for 5E14 Be cm^{-2} 75 keV

In each case, the experimental data clearly shows a region where the electrical activity increases with annealing time up to a maximum "saturation" level. The solid lines are predicted by equation (6) using values of Ea, Ed and K'Do calculated from the experimental data in the manner described above. The results of a similar analysis performed for Mg, Zn, Se and S ions are quoted in Table 1.

---

[1] The authors would like to acknowledge N.J. Barrett at GEC for the Be implantation.

TABLE 1. Activation and Diffusion Energies for ion-implanted GaAs

Ion	Ion Dose(cm^{-2})	Ea (eV)	Ed (eV)
Be	5×10^{14}	0.32 ± 0.05	2.3 ± 0.1
Mg	5×10^{14}	0.98 ± 0.1	2.45 ± 0.1
Zn	1×10^{15}	0.37 ± 0.05	-
Sn	1×10^{14}	1.2 ± 0.1	2.5 ± 0.1
Se	1×10^{14}	1.2 ± 0.1	2.5 ± 0.1
S	1×10^{14}	1.1 ± 0.1	-

5. DISCUSSION AND CONCLUSIONS

With reference to Table 1, it is apparent that a distinct separation exists in the range of Ea values obtained for the different ions, ie. (1) values of 0.3 - 0.4eV and (2) values of 1.0 - 1.2eV. We propose that, for ions such as Be and Zn, the inactive fraction of the implanted dose remains in an interstitial state and that, during the annealing cycle, a diffusion of gallium vacancies occurs (with a corresponding diffusion energy, Ed, of approx. 2.5eV) The net activation energy of 0.3 - 0.4 eV is therefore thought to be the total energy required for a reaction of the type ($Be_i + V_{Ga} \rightleftharpoons Be_{Ga}$) to occur. In cases where the activation energies lie in the range 1.0 - 1.2eV, we propose that the inactive fraction exists in the form of impurity-vacancy (I-V) complexes. Thus the model for these ions involves the diffusion of a gallium or arsenic atom to the complex site, thereby allowing the destruction of the (I-V) complex. This reaction may be summarized by:

$$(I_{Ga,As} - V_{Ga,As}) + Ga_i/As_i \rightleftharpoons I_{Ga,As} + Ga_{Ga}/As_{As}$$

This model is also consistent with the values of diffusion energy recorded in Table 1 and with published values of the diffusion energies of gallium, arsenic and vacancies in GaAs which lie in the range 2.5 - 3.0 eV (Kendall et al 1968, Chiang and Pearson, 1975).

In conclusion, we have shown that rapid thermal annealing can be used to study the activation kinetics of ion-implanted GaAs and that, as a result, a comprehensive model for electrical activation has been evolved. This model differs from previous publications in that it accurately describes the activation process over the entire time and temperature regime.

REFERENCES

Bensalem R. Barrett N J and Sealy B J. 1983, Electron Lett 19 112.

Gwilliam R. Bensalem R and Sealy B J. 1985, Physica B 129 440.

Kendall D L. Willardson R K and Beer A C. (eds) 1968, in Semiconductors and Semi-metals, Vol. 4, Academic Press, NY.

Chiang S.Y. and G L Pearson, 1975, J.Appl.Phys. 46 pp2986.

Bond length relaxation around isoelectronic dopants in InP studied by fluorescence-detected EXAFS

H. Oyanagi, Y. Takeda*, T. Matsushita**, T. Ishiguro, T. Yao and A. Sasaki*

Electrotechnical Laboratory, Sakuramura, Niiharigun, Ibaraki 305, Japan
*Department of Electrical Engeneering, Kyoto University, Kyoto 606, Japa
**National Laboratory for High Energy Physics, Tsukubagun, Ibaraki 305, Japan

ABSTRACT: Local structure around isoelectronic dopants (Ga:1x10^{19}/cm^3,As: 7 x 10^{19}/cm^3) in InP has been studied by fluorescence-detected EXAFS. The nearly complete substitution of dopants has been evidenced. The Ga-P and In-As distances in Ga/As-doped InP deviate from the In-P distance and are rather close to those in pure binary compounds (GaP and InAs). The local distortion around impurity atoms causes displacement of second-nearest neighbor atoms along the <111> direction, which may contribute to reduce dislocations by delocalizing strains.

1. INTRODUCTION

Although it is widely recognized that dislocations in bulk crystals can be reduced by In-doping in GaAs (Jacob1982) or co-doping of Ga and As in InP (Shimizu et al. 1986), the operating mechanism of dislocation reduction has not been established yet. The local structure around impurities is a key to understand how these isoelectronic impurities are related with dislocation-free crystal growth. In this study, the local structure around doped Ga and As impurities in InP has been studied *as-grown* by fluorescence-detected EXAFS technique (Sayers et al. 1971) using synchrotron radiation. Efforts have been made to improve sensitivity so that impurities down to 1 x 10^{19}/cm^3 are measured. Obtained bond lengths in Ga/As-doped InP are compared with those in dilute (In,Ga)(As,P) quaternary alloys and discussed in relation with a lattice relaxation around impurity atoms.

2. EXPERIMENTAL

Gas/As-doped InP is a bulk crystal with a low dislocation density prepared by a magnetic field-applied LEC method with a small temperature gradient (Shimizu et al. 1986). The concentration of Ga and As was determined by an ICP emission spectrometry as$1.16 \times 10^{19}/cm^3$ and $7.32 \times 10^{19}/cm^3$, respectively. Lattice-mismatch between the Ga/As-doped InP substrate and the LPE-grown pure InP layer was less than 1×10^{-4}. Dilute (In,Ga)(As,P) alloys of InP-rich composition (Ga- and As-composition less than 0.1), lattice-matched to InP, was grown by LPE at 650 °C. For dilute (In,Ga)(As,P) alloys and Ga/As-doped InP samples, x-ray filters (Zn and GeO_2) were used to reduce inelastic scattering. Further, residual inelastic scattering was removed by a phase height analysis using a Si(Li) solid state detector. Fuorescence yield spectra were obtained by a spectrometer (Oyanagi et al. 1985) using synchrotron radiation from the 2.5 GeV storage ring at the Photon Factory.

3. RESULTS AND DISCUSSION

Ga and As K-edge EXAFS oscillations were extracted from fluorescence yield spectra and normalized. In Fig. 1, (a) and (b) indicate the Fourier transform results of the As K-EXAFS data for Ga/As-doped InP and non-doped InAs, respectively. The magnitude (dotted line) and imaginary part (solid line) are plotted as a function of radial distance R in Å. Peak positions in this figure are shifted to smaller R due to the phase shift effect. A prominent peak located at *ca.* 2.4 Å in Fig. 1 (b) is due to the As-In distance. Using the known In-As distance (2.623 Å) in pure InAs, the empirical phase shift for As-In pair was estimated. The k-dependence of As K-EXAFS envelope indicates that As atoms are coordinated by In cations, i.e., they substitute P atoms in Ga/As-doped InP without forming like-atom bonds or GaAs clusters. The determined As-In distance in Ga/As-doped InP (2.62 Å) is close to that in pure InAs (2.623 Å) rather than the In-P distance (2.541 Å). These results indicate that the nearest In atoms are displaced along the <111> direction by 0.08 Å, which amounts to 3 % bondlength-mismatch.

In Fig. 2, the results of Fourier transform for Ga K-EXAFS data are shown. Curves (a) and (b) indicate the results for Ga/As-doped InP and pure GaP, respectively. The first-nearest peak located at *ca.* 1.9 Å in (b) is due to the Ga-P distance (2.360 Å) in GaP. The k-dependence of Ga K-EXAFS envelope shows that most of Ga atoms in Ga/As-doped InP are coordinated with P atoms, i.e., doped Ga atoms occupy In sites. Therefore the Ga-P distance in (a) can be determined using the Ga-P phase shift extracted from data for pure GaP shown in (b). The determined Ga-P distance (2.39 Å) is much shorter than the In-P distance (2.54 Å) but rather close to the Ga-P distance in pure GaP (2.360 Å). As a result, P atoms are displaced by 0.15 Å toward Ga atoms along the <111> direction.

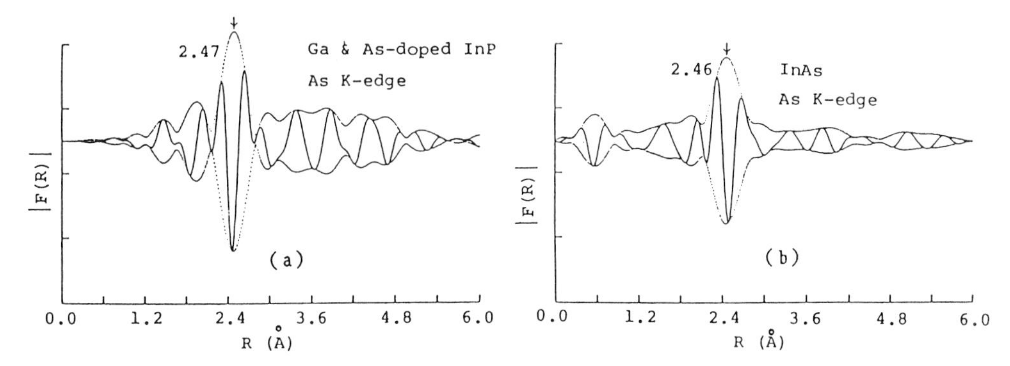

Fig. 1 Fourier transform of As K-EXAFS for Ga/As-doped InP (a) and pure InAs (b).

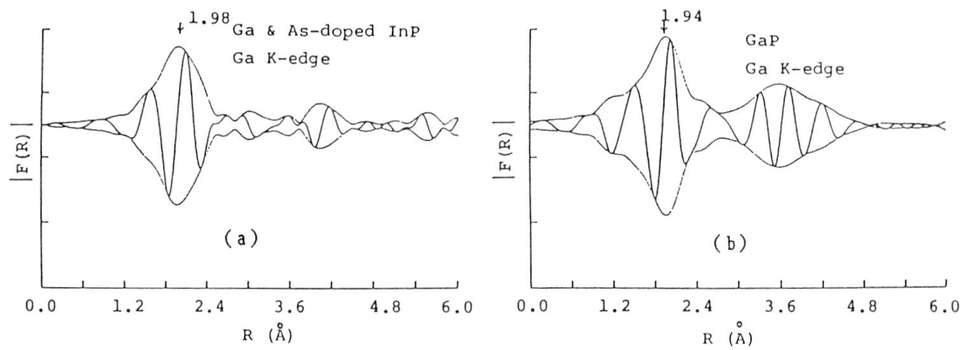

Fig. 2 Fourier transform of Ga K-EXAFS for Ga/As-doped InP (a) and pure GaP (b).

Figure 3 shows the obtained bond lengths as a function of number of bond pairs. First, for the As-In and Ga-P distances in Ga/As-doped InP, the observed values (closed circle) are rather close to those in binary compounds than the interatomic distance of host lattice. The deviations are consistent with those observed for dilute (In,Ga)(As,P) alloys (open circles).

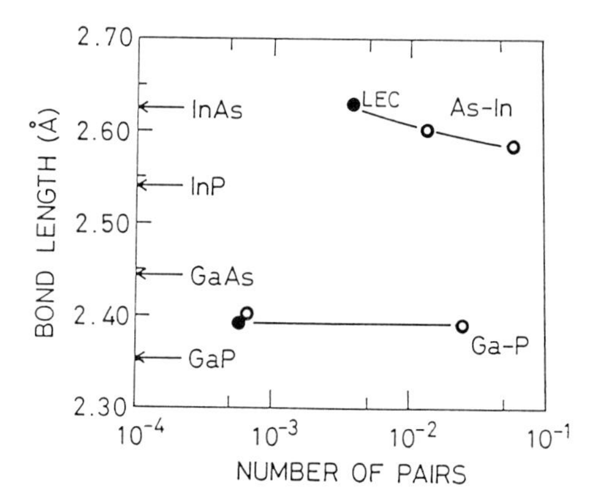

Fig. 3 Bondlength in Ga/As-doped InP and dilute (In,Ga)(As,P) alloys as a function of composition.

Second, the bondleng deviation is larger for the Ga-P than for the In-As pair possibly due to a larger bond length mismatch for the Ga-P pair. We now consider the lattice distortion around impurity atoms. Mikkelsen et al. (1983) have observed the bond length deviation in $In_xGa_{1-x}As$ pseudobinary alloys from the average interatomic distance is a linear function of alloy composition down to a few mol % region. If this can be extrapolated to a dilute limit, one may expect the largest deviation from the value in a pure binary compound. The present result indicates, however, that the *host lattice* relaxes, i.e., the nearest neighbor (host) atoms are displaced along the <111> direction until a bond length mismatch (3-6 %) is accomodated. Although to what extent this local distortion propagates is not clear yet, it is reasonable to assume that the third-nearest neighbors are also influenced. This impurity-originated lattice relaxation may contribute to delocalize strains which otherwise result in dislocations, by averaging out localized strains.

4. SUMMARY

We have shown that, for Ga/As-doped InP, the host lattice relaxes so that bond lengths of impurities are only slightly changed from those in binary compounds. Isoelectronic doping causes tensile or stretching distortions of near-neighbor atoms around impurities which may contribute to reduce dislocations by delocalizing strains.

ACKNOWLEDGEMENT

This work has been performed as a part of a project (Proposal No. 85-017) approved by the Photon Factory Program Advisory Committee and partly supported by the Scientific Research Grant-in-Aid for Special Project Research on "Alloy Semiconductor Physics and Electronics" from the Ministry of Education, Science and Culture, Japan. The authors express their thanks to Sumitomo Electric Industries for providing us the Ga/As-doped InP wafers.

REFERENCES

Jacob G 1982 *Semi-Insulating III-V Materials* (Nantwich: Shiva Publishing) pp. 2-18.

Mikkelsen J C and Boyce J B 1983 *Phys. Rev.* **B28** 7130.

Oyanagi H, Matsushita T, Tanoue H, Ishiguro T, and Kohra K 1985 *Jpn. J. Appl. Phys.* **24** 502.

Oyanagi H, Takeda Y, Matsushita T, Ishiguro T, and Sasaki A 1986 *Inst. Phys. Conf. Ser.* No 79 pp. 295-298.

Sayers D.E., Stern E A, and Lytle, F W 1971 *Phys. Rev. Lett.* **27** 1204.

Shimizu A, Nishine S, Morioka M, Fujita K, and Akai S 1986 *Semi-Insulating III-V Materials* (Tokyo: Ohm-sha) pp.41- 46.

Inst. Phys. Conf. Ser. No. 91: Chapter 2
Paper presented at Int. Symp. GaAs and Related Compounds, Heraklion, Greece, 1987

Extraction of the 'Real' band-gap shifts in heavily p-type doped GaAs from the comparison between experimental and theoretical photoluminescence peaks

B E Sernelius

Department of Physics and Measurement Technology, University of Linköping
S-581 83 Linköping, Sweden
and
Solid State Division, Oak Ridge National Laboratory
Oak Ridge, Tennessee 37831, USA
and
Department of Physics, University of Tennessee
Knoxville, Tennessee 37996, USA

ABSTRACT: We present theoretical photoluminescence peaks for heavily p-type doped GaAs. The complex valued self-energy shifts caused by the doping are taken into account. Their real parts cause band shifts and the imaginary parts give rise to finite lifetimes. The peak shapes agree with experiments to such an extent that a "real" band-gap narrowing can be obtained with the use of the experimental and theoretical peaks.

1. INTRODUCTION

The band-gap, E_g, is a very important parameter in several fields of semiconductor physics and applications. It is, e.g., important in transistor design, crucial for the performance of solar cells and it is one of the parameters to be varied in the tailoring of semiconductor films used as coatings on highly energy-efficient windows.

Experimentally, it is well-established that the value of E_g changes with doping level, and this variation has attracted a hugh attention both experimentally and theoretically during the last two decades. What is most annoying, however, is that the various experimental techniques give results that disagree as to the extent of the band-gap shifts. The underlying reason for this is that the models on which the experimental interpretations rely are too simplified.

The assumption of rigid band shifts, often used, is in general not valid. Different states are shifted different amounts. As a result the density of states in the conduction and valence bands are deformed. In consequence the various experimental techniques should in general give different band-gap shifts, since they involve transitions between different set of states.

One further complication comes from the lifetime broadening effects making the extraction of the experimental values for E_g more difficult. In the luminescence experiments the peaks are broadened due to the short lifetimes of the carriers. The way to go around this problem is to calculate in a consistent manner the energy shifts as well as the lifetimes of the states involved and then to calculate the spectrum taking both the shifts and the lifetimes into account. If the theory is powerful enough the theoretical spectrum is identical to the experimental one, possibly slightly displaced in energy. This displacement gives the correction to the theoretical band-gap shift. We have performed such an analysis of a set of PL-spectra from heavily p-type doped GaAs.

In a photoluminescence experiment on heavily p-type doped GaAs one can gain information about the shifts of the states in both the conduction band and the valence bands in the momentum range from zero up to the heavy-hole Fermi-momentum. In the heavily doped case the spectra are completely dominated by the contribution from the thermalized electrons collected at the bottom of the conduction band. The spectrum consists of one peak from the thermalized electrons and a very weak high energy tail from the hot electrons. The position of the peak determines the shifts of the band eges, while information on the shifts of the states away from the band edges are to be sought in the hot-electron contribution, i.e., in the high energy tail. This tail, however, is very featureless, but the luminescence polarization spectrum from the hot-electron luminescence has a well defined minimum that can be used to gain information about the shifts (*Sernelius, 1986a*). Here we limit ourselves to the contribution from the thermalized electrons.

2. BAND-GAP SHIFTS AND LUMINESCENCE SPECTRA

The shifts of the states are caused by the correlations in the system of valence band holes (coupled heavy and light holes), thermalized conduction band electrons and ionized acceptors. An electron is surrounded by an enhancement in the hole density and it tends to avoid the negatively charged acceptor ions. Both these effects reduce the energy, i.e., the conductron band is shifted downwards in energy. Similarly a valence band hole is surrounded by a depletion in the hole density and tends to stay close to the ions. These effects reduce the energy of the hole, i.e., the valence band is shifted upwards. These shifts are determined in the following way.

The total energy of the system, E, is derived. It consists of the kinetic energy of the electrons and holes and the interaction energy (exchange and correlation energy and the energy from the interactions with the ions). From this energy we obtain the quasiparticle energy, E_p, for a particle in state p according to the following definition

$$E_p = \frac{\delta E}{\delta n_p} = e_p + \Sigma_p \, , \tag{1}$$

where e_p and Σ_p are the kinetic energy and self-energy, respectively, for a particle in state p. Eq. (1) means that the energy of a particle in state p is defined as the change in the total energy of the system when a particle is added to state p, if this state is unoccupied. The energy of a particle in state p is minus the change in the total energy when a particle is removed from state p if this is originally occupied. The self energy Σ contains a real and an imaginary part. The real part gives the energy shifts due to the correlations and the imaginary part gives the particle lifetime. The detailed derivation of Σ in the random phase approximation was performed by Sernelius (1986b). It is rather cumbersome and will not be repeated here.

Our theoretical luminescence peaks are based on the following assumptions. We assume that the electrons taking part in the recombination processes are completely thermalized. We do not know the number of electrons collected at the bottom of the conduction band. We use this number as an adjustable parameter (the only adjustable parameter in the theory) when comparing the results to experiments. We further approximate the matrix elements for the processes with a constant and assume that this constant is the same for both hole types.

The photoluminescence intensity can with these assumptions be expressed as

$$I_{PL}(h\nu) = \sum_{j=lh,hh} \int d\vec{k} \int d\vec{k}' f(e^{e}_{\vec{k}} + \mathrm{Re}\Sigma^{c}_{\vec{k}}) f(e^{j}_{\vec{k}'} + \mathrm{Re}\Sigma^{j}_{\vec{k}'}) \; A(\vec{k},\vec{k}',h\nu) \qquad (2)$$

in its most general form. The functions f are the Fermi-Dirac distribution functions for the two particle types taking part in the process. The energy $h\nu$ is the photon energy after subtraction of the value of the unperturbed band-gap E_g. The value of E_g is temperature dependent while the shifts have negligible temperature dependencies. This means that the experimental band-gap narrowing (BGN) can be plotted as a function of doping level even though the experiments have been performed at different temperatures. We have used the following expression for the function A

$$A(\vec{k},\vec{k}',h\nu) = -\delta(\vec{k}-\vec{k}')\frac{1}{\pi} \frac{\mathrm{Im}(\Sigma^{c}_{\vec{k}}+\Sigma^{j}_{\vec{k}'})}{\mathrm{Im}^{2}(\Sigma^{c}_{\vec{k}}+\Sigma^{j}_{\vec{k}'}) + [e^{e}_{\vec{k}}+e^{j}_{\vec{k}'} + \mathrm{Re}(\Sigma^{c}_{\vec{k}}+\Sigma^{j}_{\vec{k}'}) - h\nu]^{2}} \qquad (3)$$

The first factor is a δ-function in $\vec{k}$, which shows that we have assumed $\vec{k}$-conservation. The rest is the two-particle density of states for the particles involved in the process. This factor transforms into an energy conserving δ-function if the finite lifetimes are neglected.

Fig. 1. displays the comparison between the theoretical, solid curve, and experimental, dash-dotted curve, luminescence spectrum. The experimental spectrum was obtained by Olego and Cardona (1980) at 15 K for a sample with the acceptor density $4\times10^{19}\mathrm{cm}^{-3}$. The dashed and dotted curves are the separate contributions to the theoretical spectrum from the recombination with light and heavy holes, respectively. The theoretical curves have been shifted 10 meV to the right in the figure in order to make the theoretical peak position agree with the experimental one. This means that the filled circle, representing the "real" BGN, in Fig. 2 is, for this particular sample, positioned 10 meV below the solid curve which represents the theoretical BGN. The open circles in Fig. 2 are the "apparent" experimental BGN arrived at by extrapolating the low-energy edges of the peaks to the background. The dotted curve is the theoretical result obtained neglecting the interaction with the acceptor ions. The experiments are from Nathan et al (1963), Olego and Cardona (1980). Titkov et al (1981) and Miller et al (1981).

It is found in Fig. 2 that the agreement between the theoretical and the "real" BGNs are quite good for densities up to around $2\times10^{19}\mathrm{cm}^{-3}$. For higher densities there are deviations. These could be due to effects from self-compensation or acceptor clustering. Both these effects, which are expected to occur to higher degree with increasing doping density, would reduce the ion-contribution to the BGN.

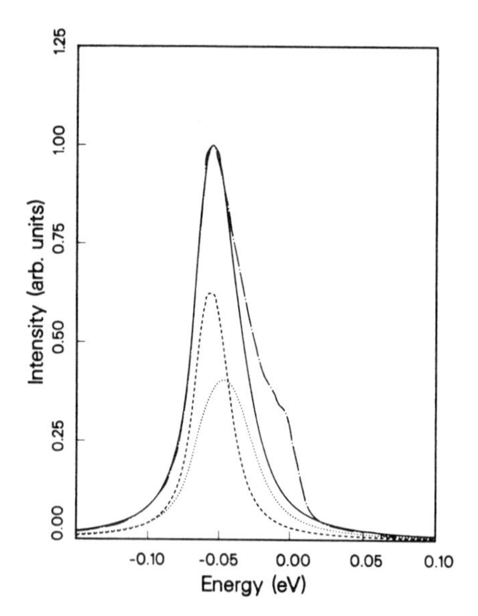

Fig. 1. Comparison between a theoretical, solid curve, and experimental, dash-dotted curve, photoluminescence peak.
See text for details

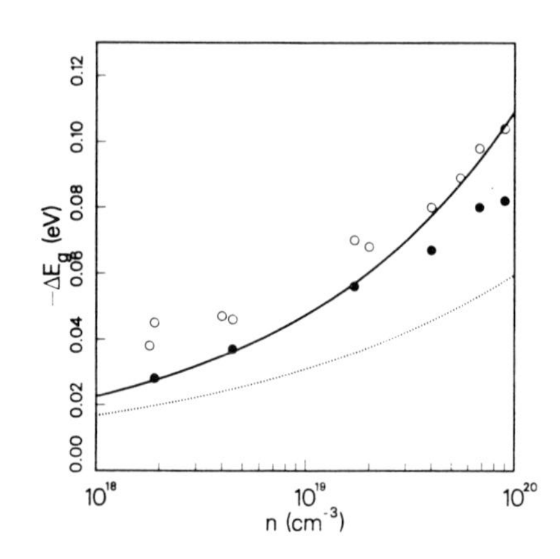

Fig. 2. "Real", solid circles,and "apparent", open circles, experimental band-gap narrowing, compared to the theoretical results, solid curve. See text for details

Two things, in particular, were noted in the comparison between the theoretical and experimental photoluminescence peaks. First of all, the tailing of the low-energy side of the peaks, often believed to be due to band tailing caused by fluctuations in the dopant concentration, was fully reproduced by the lifetime broadening. The peak shapes agreed, for all investigated peaks, to such an extent that a "real" band-gap narrowing could be obtained with the use of the experimental and theoretical peaks. The second thing concerns the high-energy shoulder, which seems to be common to all high-density samples, and which in numerous papers is referred to as being due to non-$\vec{k}$-conserving processes. We found that its position was so far off in energy that this explanation must be wrong. The peak seems to be located at an energy slightly lower than the energy of the band-gap in the pure crystal. One possible explanation for this peak is that it comes from an undoped region or a depletion region somewhere in the sample.

REFERENCES

Miller R C, Kleinman D A, Nordland Jr W A and Logan R A 1981 *Phys. Rev. B* **23** 4399
Nathan M I, Burns G, Blum S E and Marinace J C 1963 *Phys. Rev. B* **132** 1482
Olego D, Cardona M 1980 *Phys. Rev. B* **22** 886
Sernelius B E, 1986a *Phys. Rev. B* **34** 8696
Sernelius B E, 1986b *Phys. Rev. B* **34** 5610
Titkov A N, Chaikina E I, Komova E M and Ermakova N G 1981 *Fiz. Tekh. Poluprovodn.* **15** 345 [*Sov. Phys. Semicond.* **15** 198]

Paper presented at Int. Symp. GaAs and Related Compounds, Heraklion, Greece, 1987

Hot carrier equilibration in degenerate GaAs

H.G. Grimmeiss#, B. Hamilton*, W.T. Masselink and S. T. Pantelides

IBM T.J. Watson Research Center, P.O. Box 218, Yorktown Heights, NY 10598 USA

ABSTRACT: We have used photoluminescence to measure the electron temperature in heavily doped n-type GaAs under steady state excitation. With carefully controlled optical power density we find that much higher electron temperatures are produced in degenerate samples than in lightly doped ones. We propose that a large part of our observations result from a heat capacity reduced by virtue of the Pauli exclusion principle. Other factors such as screening of phonon interactions, short carrier lifetime, and the effective volume for hot carrier interactions must also play an important role

I. Introduction

A knowledge of the non-equilibrium energy distribution function of electrons is useful both as means for better understanding of physics of ballistic transport processes, and for the development of fast devices. It is well known that photoluminescence can be used to probe the distribution function so long as the radiative transition begins with a free particle which has a kinetic energy characteristic (in a statistical sense) of the temperature of a thermalized carrier gas (Dean). The line shape of such transitions includes an high-energy Boltzmann tail from which the temperature can be obtained.

It has been shown that the steady state temperature of electrons heated by optical pumping (Shah and Leite), and by an applied electric field can be measured by this method (Shah). More recently, time resolved measurements have been used to investigate the physics of the early stages of carrier thermalization, especially in highly excited plasmas, and low dimensional systems.

In weakly excited, lightly doped bulk GaAs the mechanisms by which injected hot carriers lose energy to the lattice (the cooling processes) and feed energy into the carrier gas (the heating processes) are reasonably well understood. For example the variation of electron temperature with optical power density can be predicted from power balance considerations for the steady state, (Ulbrich).

Our objective in this work was to evaluate photoluminescence as a technique for the measurement of carrier temperatures in heavily doped GaAs. Much less published work work exists for degenerately doped systems in which scattering rates might be modified. Our results indicate that very high carrier temperatures are produced in degenerately doped epitaxial layers ($n = 2{:}\times 10^{18}$ cm^{-3}). In such samples carrier temperatures of 90 K are achieved with incident optical power densities that produce carrier temperatures of only 15 K in lightly doped epitaxial layers ($n = 10^{15}$ cm^{-3}). We attribute this observation partly to modification of the electronic component of the heat capacity due to the Pauli exclusion principle. However, other factors seem to play a role: possibly screening of the electron-phonon interactions which would reduce the cooling rate, and a short carrier lifetime which might cause recombination heating but will also reduce the effective time available for the carriers to cool.

II. Experimental Technique and Results.

Epitaxial layers grown by both the MOVPE and MBE methods were used in this work. Samples were held in a liquid helium cryostat which allowed variation of the lattice temperature between 4.2 K and 150 K. For most measurements, the 514 nm line of an Argon laser was used to excite the photoluminescence which was dispersed by a one meter grating spectrometer and detected by a cooled GaAs photon counting photomultiplier. The spectral response of the detector is essentially flat over the wavelength range of the measurement. The data reported below were all measured under strictly constant laser power density. It was found that in order to measure consistent and reproducible temperatures in a given sample, all factors which control the incident photon flux, especially the beam focus and the tilt of the sample in the beam, needed to be carefully controlled. Rather weak excitation was used, estimated to be 0.02 Wcm^{-2}. Under these conditions, the photo-injected carrier density will be a small fraction of the thermal equilibrium electron density for all samples.

For the lightly doped epitaxial layers, the free to bound (F-B) transition of the residual carbon acceptor, was used to measure carrier temperature. Figure 1 shows a typical spectrum, measured at 5 K. The dominant feature is the envelope of unresolved bound exciton lines. At lower energy the F-B and donor-acceptor pair (DAP) bands are partially resolved The high energy tail of the F-B band contains the information on the temperature of the electron gas. The inset in Figure 1 illustrates the transition in k space. The thermalized electrons are clustered around the gamma minimum, and the wave function of the holes bound to the carbon acceptor may reasonably be assumed to be uniform in k space near the zone center. The matrix element for the transition is then effectively constant in energy, and the high-energy tail truly reflects the electron energy distribution function. The maximum energy (shortest wavelength) that can be used to track the Boltzmann tail is limited by the low energy tail of the bound exciton signal.

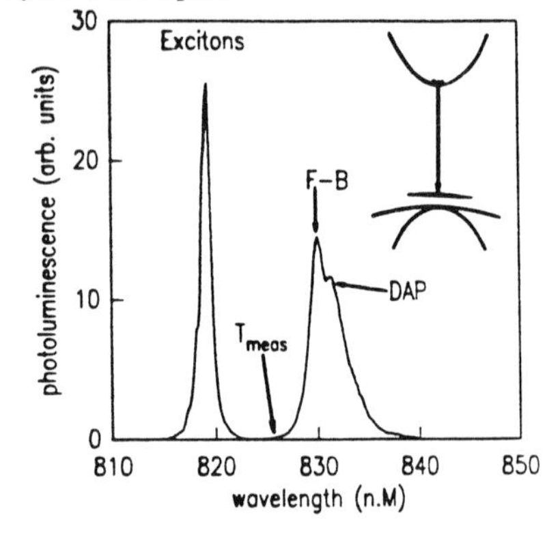

FIG 1. Spectrum for lightly doped material. Electron temperatures were measured from from the tail of the carbon F-B transition

The equivalent data for a degenerate sample, also measured at 5 K. is given in Figure 2. No excitons exist in such heavily doped material. Instead the optical spectrum is very broad and dominated by band to band transitions.

The spectral shape reflects the filling of the conduction band. The inset illustrates the transitions in k space. In this sample, $n=2{:}\times10^{18}(cm.^{-3})$, and the Fermi energy is at about 100 meV above the conduction-band edge. Hot electron transitions commence above E_f and terminate at a photoinjected hole somewhat removed from the zone center. The high hole effective mass ensures that holes are available over the required range of k, and again the matrix element should be constant. We thus expect the Boltzmann tail to reflect the electron temperature, but it is fair to assume that, in a steady state experiment, the electron and hole gases are at the same temperature. Superimposed on the heavy doping spectrum is a sharper emission from the substrate. The epitaxial layer was 1.5 microns thick so the substrate emission probably results from very weak penetration of the laser light, or carrier diffusion.

FIG 2. Spectrum for a heavily doped sample. The electron temperature was measured using the shape of the band to band tail.

Logarithmic plots of the extreme tails of the two free electron related transitions are shown in Figure 3. The onset of the low energy tail of the excitonic emission is clearly visible for the F-B tail of the n-sample. Although the tail of the band to band emission is much weaker than that of the F-B emission it is the shortest wavelength feature.
Measurement of the tail is limited only by noise or stray light. This transition can therefore be tracked down to very low absolute values of intensity. The difference in slope is clear, with the electron temperature in the lightly doped sample being much lower than that in for the degenerate sample. The temperature of 15 K is consistent with previously published values for similar doping levels and excitation densities, (Ulbrich).

The temperature of 85 K for the degenerate sample is at first sight surprising because it is generally assumed that the energy per photo-electron available for heating the whole gas, E^*, is shared amongst all particles, at least in not too dilute systems (Shah and Leite). On this basis we would expect temperature increases of the order of 10^{-2} K, almost four orders of magnitude lower than the observed temperatures.

The electron temperatures in the two sample types were measured as a function of lattice temperature, for constant excitation density. The results are shown in Figure 4. The lightly doped material gives constant electron temperatures until the lattice begins to contribute to carrier heating, at about 15 K, when the lattice and electron temperatures become equal. The degenerate material exhibits a small trend of increasing electron temperature as the lattice cools, and the electron and lattice temperatures only become equal above 100 K.

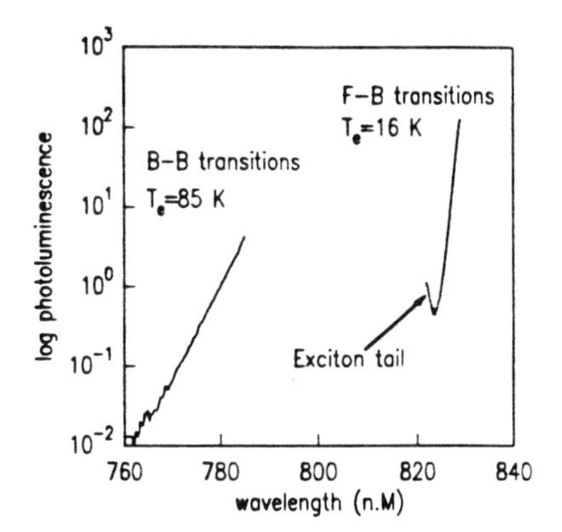

FIG. 3 The Boltzmann tail of the F–B signal from an n– sample, and from the band to band transitions (B–B) in a heavily doped sample. In each case the lattice temperature was 10 K and the incident optical power density, 0.02 (Wcm–2).

III. Discussion

The quantity E^* is approximately the energy in excess of the lowest available state after a hot electron has completed its LO phonon cascade. The lowest available state is roughly the gamma minimum in the n^- samples and E_f in the degenerate samples. In principle, if the laser probe energy creates electrons at an integer number of LO phonons above the lowest available state, one might xpect some "resonantly" low electron temperature. In practice the effect would be somewhat smeared because injected electrons are never monochromatic due to valence band warping.

We now discuss, qualitatively, the difference in electron temperatures produced in the two doping regimes, by the coupling of E^* to the carrier gases.

The simplest effect which might lead to temperature disparity between the two systems is a difference in effective excitation volumes. The 514nm line is absorbed to the same depth in both samples, the extinction depth being about 0.3 microns. It has been suggested (Ulbrich, Dean), that the penetration depth of hot electrons is very great (several microns) because the electrons move ballistically with only small angle scattering for many LO phonon lifetimes.

However, only a modest fraction of hot electrons have k vectors normal to the crystal surface. Although these may move large distances they cannot be matched in speed by holes of the same direction of wavevector. Such "vertically" transported electrons may lead to space charge build up, recombining with holes which have drifted or diffused (both slow processes) the appropriate distance. If we take the ambipolar diffusion length as a guide to the depth distribution of the luminescence, then we might expect the signal from the degenerate layers to originate mainly from the directly excited depth, and that for the n^- samples from about one micron. This would make the excitation volume of the degenerate samples at most five times smaller than that of the n- samples, insufficient to explain the four order disparity.

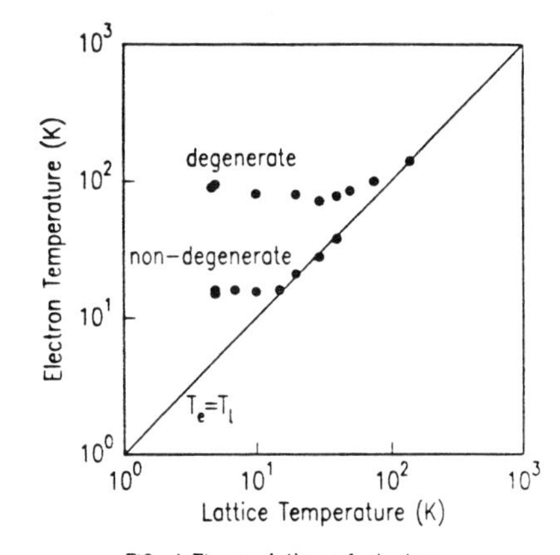

FIG. 4 The variation of electron temperature with lattice temperature for degenerate and non–degenerate n–type GaAs. All data points are for fixed optical excitation rate.

We propose that the a major reason for the high electron temperatures lies in the electron degeneracy. At He temperatures such heavily doped samples are highly degenerate. Electron-electron scattering is excluded for electrons significantly below the Fermi level, by the Pauli principle. The fraction of electrons available for scattering is approximately T_e/T_f times the doping density.

The Fermi temperature of the degenerate samples reported here is about 1200 K. so only seven percent of electrons are available for thermal excitation. The gas is much more dilute the doping density suggests. This factor alone would account for one and a half orders of magnitude of the temperature disparity deduced above from the assumption that E^* is shared amongst all carriers. We are essentially noting that the electronic component of the heat capacity is reduced by the exclusion principle in a metallic fashion. In the completely metallic case we expect that the heat capacity of the electron gas is given by:

$$C_1 = (1/2\pi^2)Nk(\frac{T_l}{T_f})$$

where N is the doping density. In this limit the electron temperature would increase linearly with decreasing lattice temperature, for constant excitation density. We do not see such a linear variation, but note that there is a small trend in the same sense.

The combined effects of reduced heat capacity and reduced excitation volume would make a temperature increase of a few K plausible, but it is clear that other mechanisms are likely to operate to reduce the electron cooling rate. Screening and the fast carrier recombination rate are obvious candidates.

The screening of electron-phonon interactions in degenerate systems is known to be significant (Goebel and Hildebrand). For the present case, with relatively high electron temperatures both polar and acoustic modes probably play a role, even in steady state measurements. Impurity screening may also be relevant. We have not yet made calculations of the effect of screening for our measurements. Recently the role of degeneracy, screening and recombination rate has been discussed in the context of the cooling of highly excited plasmas (Leo and Ruhle). It was demonstrated that fast recombination reduces the plasma cooling rate because the recombination reduces the degeneracy.

We have a fundamentally different situation in which we add an almost negligible number of particles during excitation. The degeneracy is therefore strictly constant during our measurement. The carrier lifetime, though, might play a role in maintaining a hot electron population. At room temperature the radiative lifetime of our degenerate samples is about 3 ns this reduces on cooling, and be is further reduced by Auger and surface recombination. We might then expect subnanosecond lifetimes near 4.2 K. This must increase the probability that hot carriers recombine before completely cooling. We cannot either discount the possibility that recombination feeds energy or hot particles back into the gas.

IV. Conclusions

In conclusion we have measured unexpectedly high electron temperatures in degenerately doped n-type GaAs under weak optical pumping. A major reason for this is that the photo-electrons can only interact with a much reduced fraction of the equilibrium population by virtue of the Pauli exclusion principle. This factor does not completely explain the magnitude of the effect, and it seems likely that screening of scattering processes and short carrier lifetimes are also important.

Acknowledgement — The authors would like to thank T. Kuech and K.E. Singer for the supply of epitaxial layers, and to thank A.Hangleiter and F. Stern for useful discussions.

REFERENCES

Permanent address: #Dept. of Physics University of Lund, Sweden.
Permanent address: * Centre for Electronic Materials University of Manchester Institute of Science and Technology, UK

P.J. Dean *Phys. Stat. Sol. (b)* **98**, 439 (1980)
R. Ulbrich *Phys. Rev. B* **12**, 5719 (1973)
J. Shah and R.C.C. Leite, *Phys. Rev. Letters* **22**, 1034 (1969)
J. Shah and A. Pinczuk, *Appl. Phys. Lett.* **42**, 55 (1983)
E.O. Goebel and O. Hilderbrand, *Phys. Stat. Sol. (b)* **19**, 659 (1986)
K. Leo and W.W. Ruhle, *Solid State Communications*, **62**, 659 (1987)

Paper presented at Int. Symp. GaAs and Related Compounds, Heraklion, Greece, 1987

Direct observation of long range potentials in semi-insulating GaAs and GaAs:In

D. A. Johnson, G. N. Maracas, S. Myhajlenko, J. L. Edwards, R. J. Roedel
Center
Arizona State University
Electrical and Computer Engineering Dept.
Center for Solid State Electronics Research
Tempe, Arizona 85287

H. Goronkin
Compound Semiconductor Tech
Motorola Inc.
5005 E. McDowell Rd.
Phoenix, Arizona 85008

Abstract

We have used the voltage contrast effect to image deep level domains in semi-insulating (SI) GaAs n -i -n resistor structures. Our samples consisted of undoped and indium alloyed SI liquid encapsulated Czochralski material with alloyed AuGe/Ni contacts at spacings from 0.73 mm to 6.3 mm. By viewing the contact side of the samples with a scanning electron microscope (SEM) while the devices were biased in the oscillation region, we observed domain formation and motion from cathode to anode in real time. In a previous paper we described the observation of long range potentials in the GaAs by viewing the polished back surface of the samples (Johnson 1987). That is, the domains which are launched from the front contacts are clearly evident in voltage contrast measurements on the back of the sample. Also, because of the non-uniform charging of the semiconductor surface by the electron beam, we observed interactions between the propagating domains and the cellular dislocation structure in the SI GaAs. These interactions appear as noise in the current oscillation waveforms, and therefore we propose that the current waveforms may provide a simple measure of material uniformity.

1. Introduction

Low frequency oscillations (LFOs) have been observed in semi-insulating (SI) GaAs resistor structures under a constant DC bias (Northrop 1963, Sacks 1970). The LFOs in undoped liquid encapsulated Czochralski (LEC) material are caused by dipole domains which form due to field enhanced trapping by the deep level EL2. In previous work we have investigated the thermionic nature of these oscillations for closely spaced contacts (Maracas 1985, Goronkin 1984).

The first sample consisted of undoped LEC SI material with AuGe/Ni ohmic contacts spaced 1.3 mm apart. The second sample consisted of indium alloyed (= 1.9 atomic %) SI LEC GaAs Hall effect sample with ohmic contacts spaced 6.3 mm apart. Both samples were mounted contact side up in headers and then mounted in a Japanese Electro-optic Laboratories (JEOL) model 840 scanning electron microscope (SEM). A DC voltage was placed across the contacts, and the voltage contrast effect was used to directly view the resulting domain motion in real time. The measurement procedure is described in greater detail in a previous publication (Johnson 1987).

2. Results

Figure 1a shows the results for a typical voltage contrast measurement on the GaAs:In Hall effect sample. For this sample the image was digitized using a Tracor Northern TN-5500 computer system. Figure 1b shows the derivative of the voltage contrast image. This corresponds roughly to a map of the electric field in the sample surface. For the case of 500 V applied bias, the measured domain width is approximately 140 μm. Figure 1c shows a derivative image of the sample with 400 V bias. For this case the domain width is approximately 110 μm.

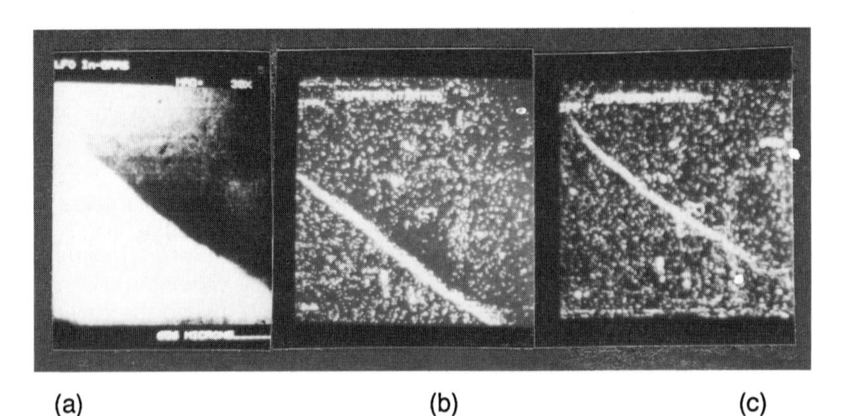

(a) (b) (c)

Figure 1. (a) Digitized voltage contrast images of the 6.3 mm indium alloyed sample (500 V). (b) Differentiated image corresponding to the electric field in the sample. (c) Differentiated image for 400 V applied bias.

This is consistent with the theoretical behavior of a domain in an externally varying electric field. For this sample the bias was applied between contacts which are situated at opposite corners of a Hall pattern with the other two contacts floating. It is interesting to note that as the domain moves across the sample, it appears to spread to fill the volume of the sample. Also, it is interesting to note that the floating contacts can supply enough charge to partially quench the domain.

Figures 2a and 2b are cathodoluminescence images for the indium alloyed and undoped LEC SI GaAs samples respectively. The undoped LEC has a higher dislocation density and exhibits dislocation rings of approximately 100 μm in diameter (Warwick 1985). In comparison, the indium alloyed material has a much lower dislocation density, and the dislocations appear to be uniformly distributed. The bright lines in figure 2a correspond to linear arrays of dislocations which lie along the <110> directions. This type of structure is typical for GaAs:In samples which are cut from near the edge of the wafer (Barrett 1984).

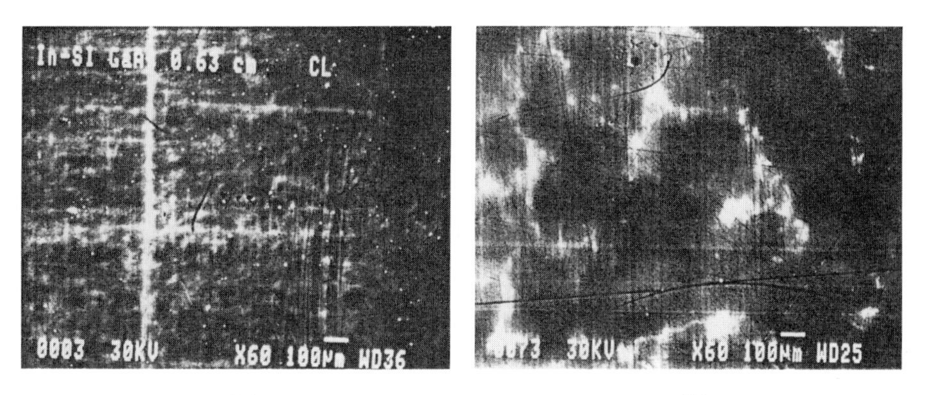

(a) (b)

Figure 2. (a) Band-edge cathodoluminescence image of the indium alloyed material. (b) cathodoluminescence image of the undoped LEC material.

Figure 3a shows a typical current waveform for the GaAs:In material.

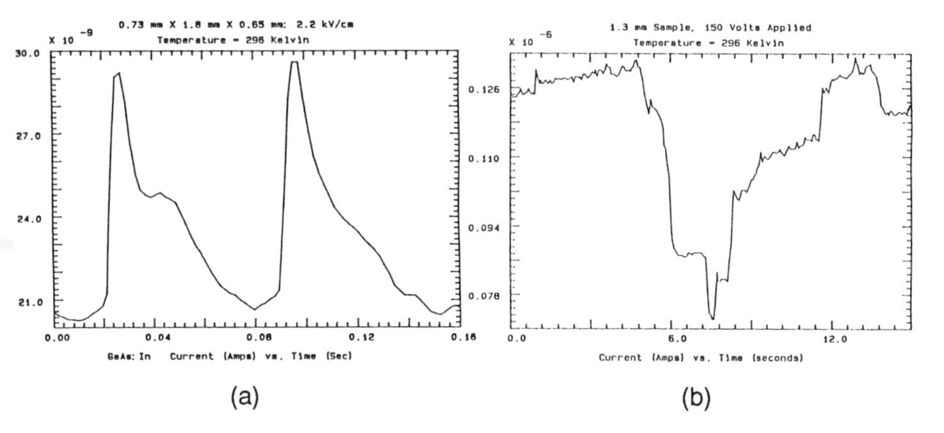

(a) (b)

Figure 3. (a) Current oscillation waveform for an indium alloyed sample with contacts spaced 0.73 mm apart under 150 V bias. (b) current oscillation waveform for the 1.3 mm undoped LEC sample under 150 V bias.

For this case the data was obtained from a sample with contacts spaced 0.73 mm apart under a bias of 150 V. The smaller sample was used for this measurement to facilitate a comparison with the undoped LEC current trace given in figure 3b. The current waveform shown in figure 3b was obtained for 150 V applied bias at room temperature for the 1.3 mm undoped LEC sample.

For reasons which we do not fully understand, the GaAs:In sample did not exhibit steady oscillations for electric field strengths below 1.5 kV/cm, so a direct comparison between figures 3a and 3b is not possible. However, when the samples are biased in the oscillation region, it is clear that the low dislocation indium alloyed material exhibits much less noise in the oscillations then the corresponding signal for undoped LEC material. This information, together with our previous observation of interactions between the moving domains and the cellular dislocation structure in undoped LEC material (Johnson 1987), suggests the possibility of qualitatively determining the uniformity of SI LEC substrate material by observing the shape and magnitude of the current oscillation waveforms for large contact spacings.

3. Conclusions

The voltage contrast mode of an SEM can be used to simply, and directly, view domain motion in SI GaAs resistor structures. We have conducted a comparison between the low frequency oscillations observed in undoped LEC SI GaAs and indium alloyed LEC SI GaAs. The low dislocation density indium alloyed material exhibits much smoother oscillations than the undoped material for similar sample geometries. We attribute this to decreased interaction between the moving domains and the dislocation structure in the material for the indium doped samples.

Acknowledgments: We would like to thank Spectrum Technologies for the SI material and A. Prabakhar from DARPA for the indium alloyed material. This work was supported by the Army Research Office, contract no. DAAL03-86-K-0070. Partial support was also provided by the National Science Foundation.

4. References

Barrett DL McGuigan S Hobgood H Eldridge G and Thomas R 1984 *J. Crystal Growth* 70, 179
Goronkin H, and Maracas GN 1984 *International Electron Devices Symposium Proceedings* (New York: IEEE) pp 182-184
Johnson DA Myhajlenko S Edwards JL Maracas GN Roedel RJ and Goronkin H 1987 *Appl. Phys. Lett.* 12 October
Maracas GN Johnson DA Goronkin H 1985 *Appl. Phys .Lett.* 46 305
Northrop DC Thorton PR and Trezise KE 1963 *Solid State Electronics* 7 17
Sacks HK and Milnes AG 1970 *Int. J. of Elec.* 28 565
Warwick CA and Brown GT 1985 *Appl. Phys. Lett.* 46 574

Low temperature growth of III–V compounds and superlattices by atomic layer molecular beam epitaxy (ALMBE)

F. Briones, D. Golmayo, L. González, M. Recio, A. Ruiz, and J.P. Silveira

CNM - CSIC; Serrano, 144; 28006 MADRID; SPAIN .-

ABSTRACT: A new development of Molecular Beam Epitaxy, the Atomic Layer Molecular Beam Epitaxy (ALMBE), is presented. This promising technique is based on beam alternation synchronized with an atomic layer by layer growth sequence. Demonstrated capabilities of this technique include : growth of good quality GaAs and GaAs/AlAs superlattices and quantum wells al low substrate temperatures ($T_s \leq$ 400°C) with excellent morphology (supressión of oval defects), growth of highly Si doped GaAs ($n=2x10^{19}$ cm^{-3}), modulated doped SL, and for the first time, the possibility of synthesizing alternate monolayer compounds and short period superlattices including different group V elements.

1. INTRODUCTION

Under optimal conditions, MBE growth mechanism for III-V compounds is a competition between terrace propagation and layer by layer (2D) growth, being this last mode the responsible of the oscillations of the RHEED specular beam intensity (Ioo) observed during growth (Neave et al, 1985). Recently Briones et al (1987 a) have shown that a periodic perturbation of the growth front enhances the layer by layer (2D) mechanism. These authors have studied two different perturbations of the growth front: short interruptions of the As_4 flux in phase with the completion of each monolayer under RHEED oscillations feedback, or a complete alternation of group III and group V supply during each monolayer cycle. As experiments have shown (Briones et al, 1987 b) both methods coincide in promoting an homogeneous nucleation simultaneously over the whole surface, even at growth conditions where terrace propagation or island nucleation would dominate the growth process. Using both methods, good quality and excellent morphology layers of GaAs, AlGaAs, QW and SL's have been grown at low substrate temperatures ($T_s \leq$ 400° C). The alternation of beams is also the base of the so called Migration Enhanced Epitaxy (MEE) successfully used by Horikoshi et al (1986) with the aim of improving surface migration of group III elements at low T_s when supplied separately from group V.

2. ATOMIC LAYER MOLECULAR BEAM EPITAXY

Atomic Layer Epitaxy as defined by Suntola (1984) cannot be applied to MBE growth of III-V compounds unless an artificial mechanism be induced to avoid accumulation of group III atoms once one atomic layer of this element is completed. That is possible using RHEED intensity feedback to detect the completion of each atomic layer and consequent operation of beam shutters.

Recently Briones et al (1987, a-b) reported experimental evidence that make Atomic Layer by Molecular Beam Epitaxy (ALMBE) a practical growth method; that is the ability of the layer by layer growth mechanism to adapt its sequence to that imposed by shutters alternation. So, it is possible to continue growing without RHEED feedback (once an optimum sequence has been established by RHEED), and even to use substrate rotation for better uniformity. Fast shutters and fast synchronized substrate rotation need to be used to grow at conventional growth rates (1ml/s). Using standard group V MBE solid sources and shutters for ALMBE is problematic, especially in the case of growths including both As_4 and P_4 alternated beams, because their ON/OFF flux ratio is only about 20, even with extensive cryoshrouds in the sources area. To improve this ratio and avoid memory effects, during the present work we used P_4 pulsed beam generated, as shown on fig.1, by scanned laser evaporation of solid chunks of phosphorous, otherwise kept at room temperature to reduce background pressure in the system.

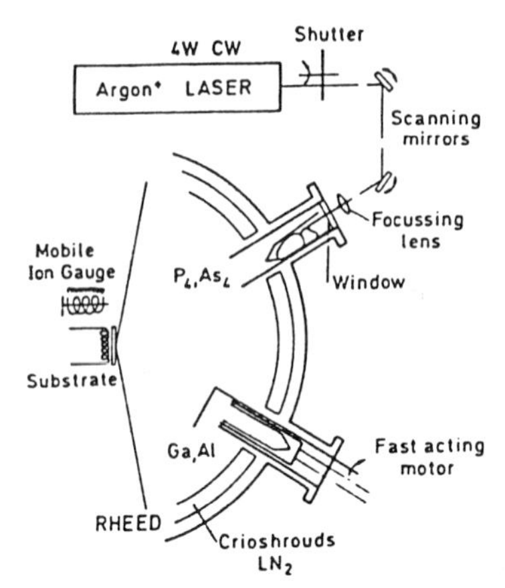

Fig. 1 Schematic diagram of the experimental Atomic Layer MBE set up. P_4 pulsed beams are generated by scanned laser evaporation of solid phosphorous chunks.

The behaviour of RHEED intensities under alternated beam supply must be studied to establish by this technique the adequate growth conditions for ALMBE. Carefully chosen diffraction conditions have to be used in order to minimize multiple scattering effects (Zhang et al 1987). We have recorded specular beam intensity (Ioo) for [110] azimuth and 27 mrad incidence angle, where out-of-phase condition and safe separation of Kikuchi lines are achieved. Under these experimental conditions, we are able to relate Ioo with surface morphology as simply described by a kinematic RHEED model. So it is possible to get some cualitative insight into the growth process from systematic observations of RHEED oscillations during ALMBE growth at various T_s, as shown in fig. 2. A schematic diagram of the expected surface separation from stoichiometry due to the succesive Ga and As_4 supplies at different T_s is also shown in this figure.

For simplicity, Ga ON and As_4 ON intervals were identically set to the deposition time of a monolayer of Ga (6.25×10^{14} at/cm^2). As_4 flux (2.2×10^{15} at/cm^{2}s) was just enough to stabilize the surface during initial oxide desorption and buffer layer growth at T_s=600°C by standard MBE. At T_s=600°C the observed 2x4 surface is in equilibrium with the As_4 beam, which compensates for arsenic evaporation. Surface is flat and ordered, and Ioo is high. A simultaneous cycle As_4 OFF/Ga ON drives the surface out of equilibrium: Ioo drops as expected from a surface disordered by Ga deposition and simultaneous thermal arsenic loss, with more than one atomic layer of Ga accumulated on the surface. As_4 supply causes a very fast initial recovery of the intensity as equilibrium is restablished followed by a slower one as a new monolayer of GaAs is completed. This behaviour is totally in agreement with RHEED observations of Chiu et al (1987) during Chemical Beam Epitaxy growth, and probably involves migration of the excess Ga on the surface simultaneously with the supply of As_4. This complex growth process is mainly 2D and produces good quality material, but differs somewhat from an atomic layer by layer growth process. No RHEED maxima or minima can be used as a reference for an atomic layer completion at 600°C.

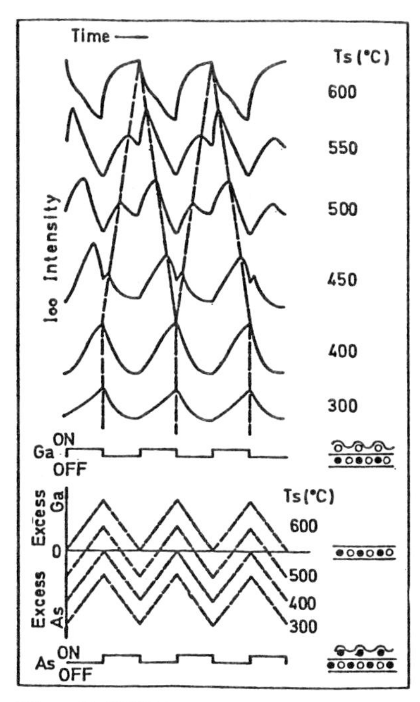

Fig. 2 RHEED specular beam intensity (Ioo) behaviour during Atomic Layer MBE growth of GaAs at different substrate temperatures Ts. Bottom: schematic diagram of the .expected surface separation from stoichiometry during ALMBE growth at different Ts.

At substrate temperatures where no arsenic loss occurs (T_s 500 C) the surface shows a 2x4 structure in equilibrium in vacuum. As_4 supply (As_4ON) causes roughening in the surface as shown by the Ioo drop. Under As_4 overpressure the surface tends to a new equilibrium state, separated from stoichiometry in a fraction of monolayer of As, as pointed out by Briones et al (1985). Ga atoms impinging in that surface react with the As excess reaching Ioo a maximum just after a fractio of atomic layer of Ga is deposited. Ga supply up to a full monolayer causes then a new Ioo decrease due to the excess of this element. A new supply of As_4 will first neutralize the excess of Ga (Ioo will reach a new maximum) and then, supplied in excess, will again drive the surface towards the initial atomically rough and low Ioo state. Accordingly, at $T_s \simeq 500°$ C, Ioo shows two maxima and two sharp minima during each monolayer growth cycle. The phase and relative amplitude of those features strongly depend on T_s and As_4 BEP as would be expected assuming that Ioo during ALMBE would be mainly determined by the surface separation from stoichiometry. According to the schematic diagram at the bottom of fig. 2, the curve representing surface departure from stoichiometry will shift up or down relative to the zero line as a function of T_s and As_4 overpressure, together with the phase of the two zero crossings. Notice that during a monolayer growth cycle, the surface oscillates between a Ga rich and an As rich extrema in a characteristic atomic layer by layer growth mode.

For lower substrate temperatures ($T_s \sim 400°C$) the above two maxima and two minima gradually merge, as shown by the guiding dashed lines, into a single maxima corresponding to the supply of a whole atomic layer of Ga and a minimum for an As saturated surface. Still lower temperatures ($T_s \sim 300°C$) or higher As_4 fluxes, cause a degradation of the RHEED diagram and large decrease of the specular beam oscillations amplitude because, as shown in bottom of fig. 2, surface will not reach stoichiometry at any time during the growth sequence.

In particular, a RHEED specular beam intensity evolution like those for $T_s \simeq 400°C$ on fig. 2, indicates that the surface is never Ga rich, being unexpectable an enhanced migration of Ga as interpreted by Horikoshi et al (1987). Actually our experimental results of low T_s ALMBE growth even at non optimized conditions, show that growth morphology is excellent, with absence of oval deffects. It is well known that for any macroscopic growth front irregularities to be formed, a high surface migration is needed. We think that the uniform high reactivity of each atomic layer deposited during ALMBE causes homogeneous nucleation of the next atomic layer. So, good quality material can be grown at low temperatures without the absolute need for enhanced Ga or As surface migration in order to grow good quality material.

3. GROWTH RESULTS

Exploratory growth runs, under ALMBE mode at low T_s show that very good control of doping and interfaces can be achieved. For example, by growing at T_s=350°C under As excess conditions where low Si autocompensation is expected, n-type doping of GaAs up to $n=2\times10^{19}$ cm^{-3} has been easily obtained together with a defect free, very flat growth morphology. ALMBE

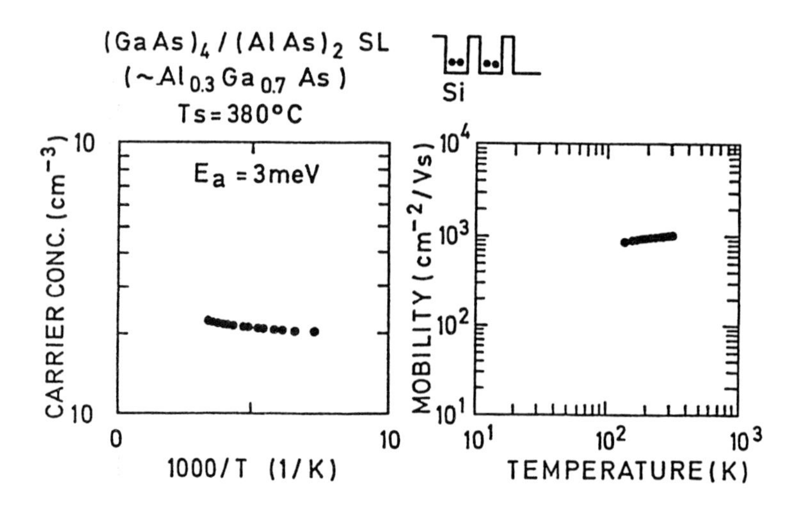

Fig. 3 Hall Carrier concentration and mobility versus temperature measured for selectively doped short period superlattice $(GaAs)_4/(AlAs)_2$ of an average composition equivalent to $Al_{0.3}Ga_{0.7}As$ alloy. This structure was grown by Atomic Layer MBE at Ts=380 C

has also been used to grow GaAs/AlAs SL and QW's of surprisingly good quality at T_s as low as 400°C. PL spectrum (Briones et al, 1987 b) shows FWHM's and intensities comparable to those of similar structures grown by MBE under optimized conditions. In particular, ALMBE growth seems to reduce well width fluctuations quite effectively. A detailed study on the optical properties as a function of well and barrier thickness will be published elsewhere (J.L. Castaño et al, 1987). Figure 3 is another example of what can be achieved by ALMBE modulated doping of short period superlattices (SPS). It shows carrier density and mobility versus temperature for a (GaAs)m/(AlAs)n (m=4,n=2) SPS grown at T_s=380° C, in which only the two central atomic planes of the GaAs wells were heavily doped with Si. High mean carrier densities, small thermal activation energies for donors and good mobilities evidence the ultimate capabilites of the new growth mode to control doping and composition at atomic layer level.

A very promising possibility offered by ALMBE is the synthesis of a new kind of alternate monolayer compounds (AMC), as defined by Petroff (1986), of the general form $(III_A-V_B)/(III_C-V_D)$, and SPS containing also multiple group V elements. The high and homogeneous chemical reactivity of the group III atomic layers during ALMBE is fundamental to reduce the problem of group V molecules competition for incorporation during MBE growth, by allowing fast saturation of each layer under alternate group V pulsed beams. Our preliminary experimental results on the growth of AMCs, involve only the alternation of As_4 and P_4 beams. Figure 4 shows RHEED intensity during ALMBE growth of a $(GaAs)_4(GaP)_1$ AMC, equivalent to $GaAs_{0.8}P_{0.2}$ on a mismatched GaAs substrate. X-Ray diffraction characterization of the grown sample, with excellent morphology, shows

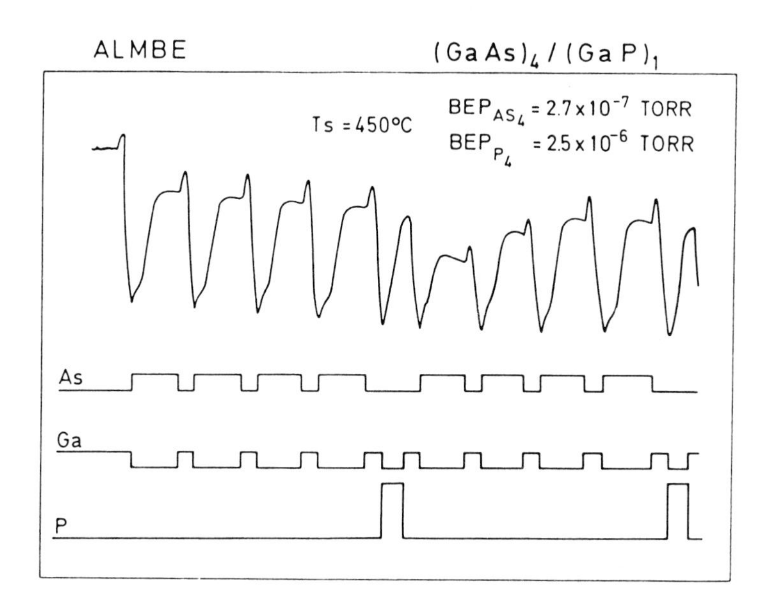

Fig. 4 RHEED specular beam intensity Ioo during Atomic Layer MBE growth on GaAs (100) of a $(GaAs)_4/(GaP)_1$ alternate monolayer compound equivalent to $GaAs_{0.8}P_{0.2}$

satellites for the periodicity of the single plane incorporation of phosphorous.

4. CONCLUSIONS

ALMBE is a promising new development of the MBE technique based on beam alternation synchronized with an atomic layer by layer growth sequence. Demonstrated capabilities of this technique include: growth of good quality GaAs and GaAs/AlAs SL and QW at low temperatures (T_s 400 C) with excellent morphology (suppression of oval defects), growth of highly Si doped GaAs ($n=2x10^{19}$ cm^{-3}), modulated doped SI and, for the first time, the possibility of synthesizing alternate monolayer compounds and short period superlattices including different group V elements.

Acknowledgements : The authors wish to thank J.L. Castaño for PL measurements at Departamento de Física Aplicada, UAM, and to M. Pérez for X-Ray characterization.

REFERENCES

Neave J H, Dobson P J, Joyce B A and Zhang J 1985 Appl. Phys. Lett. 47, 100
Briones F, Golmayo D, González L, Ruiz A 1987a J. Cryst. Growth 81, 19
Briones F, González L, Recio M, Vázquez M 1987b Jpn. J. Appl. Phys, 26, L1125
Briones F, González L, Vela J A 1987c Proc. NATO ARW on RHEED and Reflection Electron Imaging of Surfaces, ed P.K. Larsen and P.J. Dobson. NATO ASI-series (Plenum Press)
Horikoshi Y, Kawashima M, Yamaguchi H 1986 Jpn. J. Appl. Phys. 25, L868
Suntola T 1984 Proc. 16th Conf. on Solid State Devices and Materials, Kobe p. 647
Zhang J, Neave J H, Dobson P J, Joyce B A 1987 Appl. Phys A42, 317
Chiu T H, Tsang W T, Cunningham J E 1987 Robertson A, Jr. J. Appl. Phys. 62, 2302
Briones F, Golmayo D, González L, De Miguel J L 1985 Jpn. J. Appl. Phys. 24 L 478
Castaño J L, Recio M, Briones F 1987, to be published.

Inst. Phys. Conf. Ser. No. 91: Chapter 3
Paper presented at Int. Symp. GaAs and Related Compounds, Heraklion, Greece, 1987

Facts and fancies about the δ-doping layer of Si in MBE-grown GaAs

A. Zrenner and F. Koch
Physik-Department E 16, Technische Universität München, D-8046 Garching

K. Ploog
Max-Planck-Institut für Festkörperforschung, D-7000 Stuttgart

ABSTRACT: We consider the electronic properties of a layer of Si donors embedded in GaAs during MBE growth in order to show that there is spreading of the dopant from the atomic plane into which it was deposited.

1. INTRODUCTION

Molecular-beam-epitaxy growth of GaAs offers the possibility of incorporating Si dopant atoms in a precise and controlled fashion. It has been speculated that Si deposited during an interruption of the growth of the epitaxial layer will result in an atomically sharp, confined layer of donors. The sheet of Si-donor ions together with the electronic screening charge arranged in two-dimensional (2-D) subbands has been referred to as a δ-function layer (Zrenner et al. 1985).

Early work has concentrated on the 2-D properties of the subbands,assuming that the Si ions are contained in a single atomic plane of the substrate (Zrenner et al. 1986, 1987a). Trying to get at the facts of precisely how the Si atoms are built in when growth is continued has been the subject of more recent work (Zrenner et al.1987b). A saturation density of electrically active donors for Si concentrations exceeding 1.2×10^{13} cm^{-2} has been noted. It has been observed, contrary to expectation for genuine δ-confinement, that the sheet conductivity for an identical number of free electrons N_s would vary by ± 10 % or more for different samples. Moreover, changes in the subband occupations were found to relate to the different sheet conductivities in a consistent way.

Samples that we have studied to date all were grown under the same nominal, typical conditions for GaAs MBE growth ($T \sim 580^0$ C). Preliminary results on layers, recently prepared by G. Weimann and for which the growth temperature and the partial pressures of As were deliberately changed to achieve δ-like incorporation of the Si, provide additional evidence that not all that is called δ-doping is atomic-layer-confined Si.

2. POTENTIAL PROFILING USING OSCILLATORY MAGNETOTRANSPORT

In a strong magnetic field applied perpendicular to the layer surface, the sheet resistance is found to oscillate periodically in reciprocal field. A number of distinctly different periods is usually observed. The period measures the number of carriers N_s^i occupying the ith subband. Provided that the field is sufficiently high so that for each of the subband levels $\omega_c\tau > 1$, the sum of the observed periods gives the total free carrier density N_s. In this way the Shubnikov-de Haas experiment provides a complete characterization of the electronic system. A Hall effect measurement of the parallel conducting subbands with their widely differing mobilities, does not provide a reliable sheet carrier density. Moreover, it gives no information at all of the carrier distribution among the subbands.

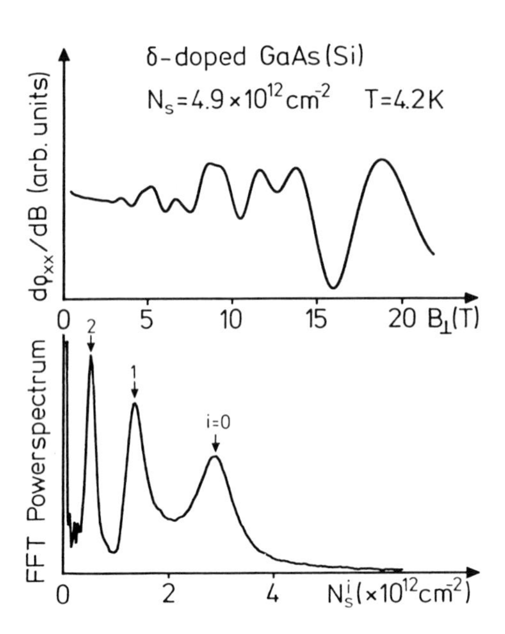

Fig. 1: Shubnikov-de Haas effect and Fourier-spectrum for a typical sample

An example of the resistance oscillations is shown in Fig. 1 for a sample with design doping density $N_D = 5.0 \times 10^{12}$ Si cm^{-2}. This number is known with an accuracy typical of MBE doping procedures. The n^+-layer is embedded in a lightly n-type background. It is placed sufficiently far from the surface in order to have all the electrons in the layer. The periods contained in the resistance oscillation spectrum are shown as peaks in the Fourier-transform plot. The sum of the individual N_s^i is determined as $N_s = 4.7 \times 10^{12}$ cm^{-2}. Together with the estimated $\sim$ 0.2×10^{12} cm^{-2} carriers of the i = 3 level, this gives an experimental N_s value of 4.9×10^{12} cm^{-2}. Such agreement of N_s with the design doping is typical for samples with $N_D \leq 1 \times 10^{13}$ cm^{-2}.

Subbands are usually filled to energies requiring a self-consistent-potential calculation that takes account of the nonparabolicity of the GaAs band structure. Detailed work is to be published elsewhere (Zrenner et al. 1988) and we choose here to discuss only the result pertinent to the example in Fig. 1. In Fig. 2 below we first show how the potential and level energy distribution appear for true δ-doping. There is a distinct and obvious discrepancy with the measured occupations N_s^i.

It is evident that the energy level separation and the relative filling are sensitive functions of the form of the potential. Thus a parabolic potential well would lead to an equidistant spacing of the energies. The sense of deviation between the measured N_s^i and those of the δ-layer calculation in Fig. 2 indicates that the potential is more nearly parabolic at the bottom. To match the data to calculations we choose to model the ion distribution assuming a spreading of the N_D donors over a distance d_z with a constant 3-dimensional density $n_D = N_D/d_z$. In Fig. 2 is also shown the calculation for d_z = 84 Å. The table insert makes clear how much better the N_s^i are described by this distribution of ions. We note that $n_D = 5.8 \times 10^{18}$ cm^{-3}. It is evident from the numbers, in particular the ratio of N_s^0 to N_s^1, that the experiment is quite sensitive to the ion distribution. Spreading of the donors causes a dominant, first-order effect on the occupancy ratio. There is no way to explain the magnetooscillation spectrum using the true δ-potential.

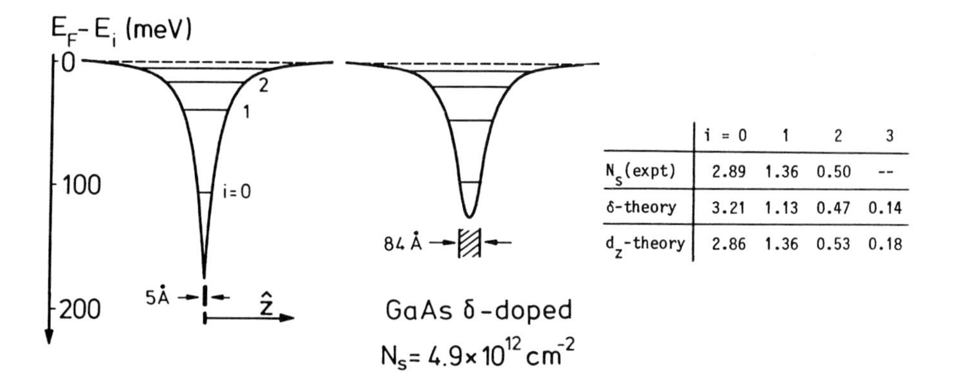

	i = 0	1	2	3
N_s(expt)	2.89	1.36	0.50	--
δ-theory	3.21	1.13	0.47	0.14
d_z-theory	2.86	1.36	0.53	0.18

Fig. 2: Comparison of potentials and occupations for $N_D = 4.9 \times 10^{12}$ cm^{-2}

3. THE SATURATION LIMIT

Increasing N_D in successive steps up to 7×10^{13} cm^{-2} we find that the density of free carriers N_s saturates at 1.2×10^{13} cm^{-2}. Typical of samples with N_D above this number is that with $N_D = 3.4 \times 10^{13}$ cm^{-2} and for which Fig. 3 gives the Fourier-transform spectrum. It shows 7 distinct peaks which sum up to only 1.2×10^{13} free carriers per cm^2. The potential well and the energy level scheme obtained for an assumed spread of 180 Å for the ions is shown as an insert in the figure. The predicted level occupancies are marked by the arrows i = 0, 1, ... etc. We note the nearly equidistant spacings typical of a parabolic potential well. The density n_D of electrically active donors appears as 6.7×10^{18} cm^{-3}.

The calculation also shows that because of the donor spreading the Fermi energy rises to only ∿ 170 meV above the conduction band edge. This is considerably less than the energy required to achieve occupation of the L-point minima of the band structure. It is also less than the expected

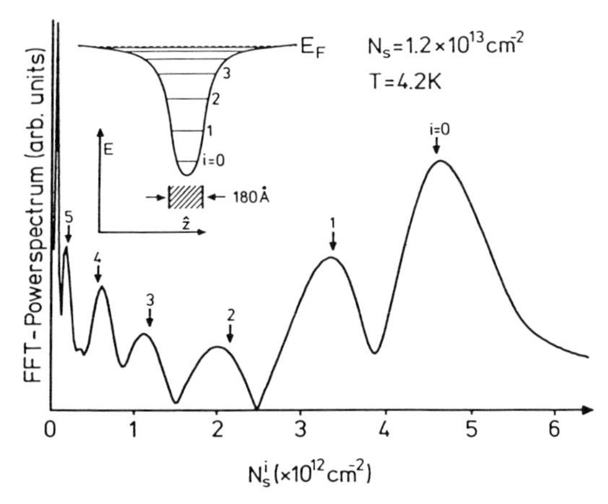

Fig. 3: Fourier-transform spectrum for a sample with N_D above the saturation limit ($N_D = 3.4 \times 10^{13}$ cm^{-2}). The arrows mark the occupations expected for the potential well in the insert.

energy position of the DX-center. In this sense the suggestion made by Zrenner et al. (1987a) that saturation is the result of filling some higher lying states, is not tenable. In that work a δ-potential had been assumed. The Fourier-transform spectrum in Fig. 3 decisively argues against this.

4. CONCLUSIONS. A MODEL HYPOTHESIS. OPEN ENDS.

Having studied a good number of samples with densities N_D ranging over nearly two decades, we are ready to propose a model for the incorporation of the Si ions in the GaAs under the growth conditions that have been employed here. It appears from our observations that the N_D ions deposited on the surface during the growth interruption are incorporated in the newly grown material at a 3-D density which corresponds closely to the known solubility limit for the given growth temperature. The experiments give $n_D \sim 7 \times 10^{18}$ cm^{-3}, with individual samples varying up to $\pm 2 \times 10^{18}$ cm^{-3} about this value. When the number N_D exceeds 1.2×10^{13} cm^{-2}, the excess remains behind on the original surface and is electrically inactive. We have sketched this in Fig. 4. The model explains the gross features of our measurements and is intended to serve as a working hypothesis.

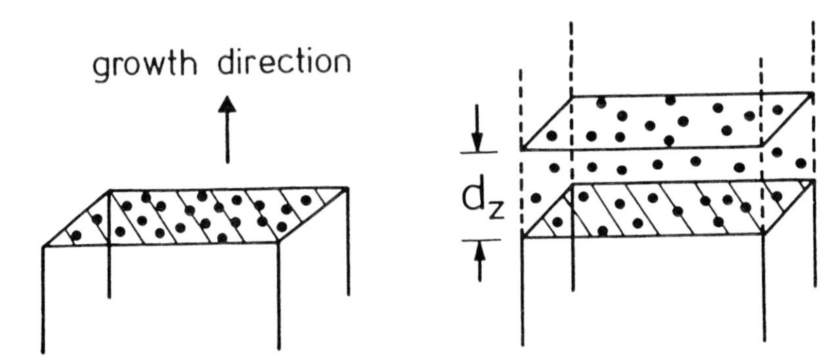

Fig. 4: Incorporation of Si donors (hypothesis). For the given growth conditions, up to 1.2×10^{13} Si cm^{-2} are built into a width d_z of newly grown material with volume density $\sim 7 \times 10^{18}$ cm^{-3}.

The conclusion of this paper is brief - atomic layer confinement for δ-layers grown by the presently employed interrupted growth procedure is a fancy; spreading of the Si ions and incorporation at the 3-D solubility limit near 7×10^{18} cm^{-3} are facts.

REFERENCES

A. Zrenner, H. Reisinger, F. Koch, and K. Ploog, Proc. of the 17th Int. Conf. on the Physics of Semiconductors, San Francisco, 1984, ed. by J.P. Chadi and W.A. Harrison (Springer-Verlag, New York, 1985), p. 325.

A. Zrenner, H.Reisinger, F.Koch, K.Ploog, and J.C.Maan, Phys.Rev. B 33, 5607 (1986).

A.Zrenner and F.Koch, Proc. of the 18th Int.Conf. on the Physics of Semiconductors, Stockholm 1986, ed. by O.Engström (World Scientific, Singapore 1987a), p. 1523.

A.Zrenner and F.Koch, Proc. of the EP2DS VII (1987b).

A.Zrenner et al., to be published (1988)

Paper presented at Int. Symp. GaAs and Related Compounds, Heraklion, Greece, 1987

Substrate temperature dependence of GaAs, GaInAs and GaAlAs growth rates by MOMBE

N. Kobayashi*, J.L. Benchimol, F. Alexandre and Y. Gao

Centre National d'Etudes des Télécommunications, Laboratoire de Bagneux
196 avenue Henri Ravera , 92220 BAGNEUX - FRANCE
* On leave from NTT Electrical Communication Laboratories, Musashino-shi, Tokyo 180, JAPAN

ABSTRACT : The substrate temperature (T_S) dependence (350° to 700°C) of GaAs, InAs and GaInAs growth rates was investigated in metal-organic molecular beam epitaxy (MOMBE), using triethyl gallium (TEG), trimethyl indium (TMI) and solid arsenic (As_4). For GaAs growth, four distinct T_S dependent regions were observed, including a weak desorption process (500° ∿ 650°C) preceding atomic Ga desorption (> 650°C). When adding a TMI flux to grow GaInAs, this desorption process was much enhanced up to 550°C and then diminished due to the occurrence of the atomic In desorption. Consequently, In alloy composition peaks at 550°C. In order to know whether this phenomenon is specific to GaInAs growth or not, we investigated the T_S dependence of GaAlAs growth rate. In contrast with GaInAs, it was found that the weak desorption was strongly minimized.

1. INTRODUCTION

MOMBE using metal-organic compounds as group III sources has been paid much attention because of the possibility of combining some advantages of MOCVD and MBE. With regard to the T_S dependence of GaAs growth rate, several authors (Tokumitsu et al 1984, Tsang 1984 a , Horiguchi et al 1986, Pütz et al 1986) reported that the growth rate increases rapidly with T_S due to the cracking of metal-organic molecules, then reaches a constant value corresponding to mass-transport limited growth rate. However, recently, Tsang et al (1987c) measured the Ts dependence of GaAs growth rate more precisely by RHEED oscillations, and showed that, on the contrary to the results reported so far, growth rate decreases gradually after reaching a maximum value. Furthermore, in GaInAs MOMBE using TEG and TMI, it has been observed that the alloy composition becomes rapidly In-rich as T_S is raised from 500°C (Kawaguchi et al 1986) or 550°C (Tsang 1986b). This phenomenon indicates that growth rates of constituent binary compounds change in the studied T_S range. In order to clarify the mechanism of the phenomena described above, we have investigated, in a wide range (350° to 700°C), the T_S dependence of GaAs, GaInAs growth rates in MOMBE using TEG, TMI and As_4. Moreover, T_S dependence of GaAlAs growth rate was investigated and compared with the results obtained for GaAs and GaInAs MOMBE.

2. EXPERIMENTAL

Metal-organic vapor was evaporated from a bottle immersed in a constant temperature bath, and introduced into vacuum chamber through a variable leak valve. TEG, TMI and triethyl aluminium (TEA) fluxes were measured as beam equivalent pressure by a ionization gauge. (100) oriented semi-insulating GaAs and InP substrates were respectively used for GaAs, $Ga_{1-y}In_yAs$ (y<0,3) and for $Ga_{1-y}In_yAs$ (y>0,3) growth. Growth rate was estimated from secondary ion mass spectroscopy (SIMS) depth profile of a structure composed of several layers grown at different T_s and separated by thin (∿ 50Å) Be doped marker layers. This method can minimize the estimation error on the growth rate caused by both run-to-run fluctuation and position-to-position thickness non-uniformity. Alloy compositions were determined by Auger electron spectroscopy (AES) on bevelled samples or by X-ray diffraction.

3. RESULTS AND DISCUSSIONS

Figure 1 shows the T_s dependence of GaAs growth rate for three different TEG flux values and a constant As_4 flux. Four distinct temperature dependent regions are identified as T_s is increased from 350° to 700°C. The low temperature, decomposition limited region (I in Figure 1) has an apparent activation energy of 15 Kcal/mole. In the T_s range from 400° to 500°C (region II), growth rate is constant. Between 500°C and 650°C (region III), growth rate decreases gradually down to about 80 % of its maximum value. For T_s>650°C (region IV), growth rate decreases rapidly, corresponding to the desorption of Ga atoms, as in conventional MBE (Chika et al 1986). The low decrease in region III is characteristic of GaAs MOMBE. This phenomenon is considered to be caused by the desorption of species containing Ga, which are more volatile than atomic Ga. $Ga-C_2H_5$ molecule seems to be the most probable one, because thermodynamical calculation (Tirtowidjojo et al 1986) shows that this molecule is the most stable among species containing Ga in the temperature range 500° ∿ 800°C.

The weak desorption in region III was much enhanced by the introduction of TMI for $Ga_{1-x}In_xAs$ growth. Figure 2 shows the T_s dependence of alloy composition y and constituent binary growth rates, determined from total growth rate and solid composition. In the T_s dependence of constituent InAs growth rate, no decomposition limited region could be observed in the studied T_s range (> 350°C), indicating that TMI was completely cracked above 350°C. The T_s dependence of constituent GaAs growth rate was quite similar to that obtained in GaAs MOMBE, except in the 500° ∿ 600°C range (region III), where a dip is observed. Constituent GaAs and InAs growth rate behavior can be correlated. That is, desorption of $Ga-C_2H_5$ molecules is much enhanced in region IIIa by the presence of In atoms, then sharply decreases above 550°C (region IIIb) when In desorption occurs. Constituent GaAs growth rate tends to reach that

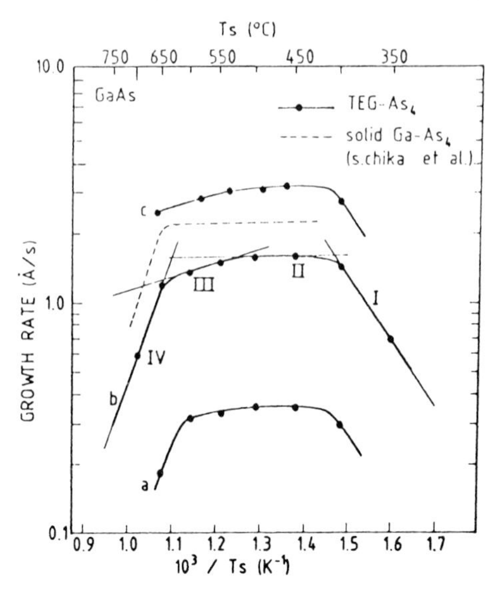

Fig. 1. GaAs growth rate vs.$1/T_s$. TEG fluxes were a) 1.3×10^{-6}, b) 9.0×10^{-6} and c) 1.8×10^{-5} Torr. As_4 flux was $2 \times 10^{-}$ Torr.

of GaAs MOMBE above 600°C, when all In atoms are desorbed. Consequently, In composition peaks at around 550°C. Same phenomenon was observed in $Ga_{1-y}In_yAs$ growth using solid In and TEG (y = 0.16 at T_s = 450°C) and also in higher In content (y = 0.39) GaInAs.

In order to investigate whether this phenomenon is specific to GaInAs or not, GaAlAs was grown using TEG and solid Al or TEA. Figure 3 shows the T_s dependence of constituent binary growth rates and Al solid composition for GaAlAs growth using TEG and solid Al. The weak desorption of region III in GaAs MOMBE was strongly minimized, and as a consequence, the Al solid composition was constant in a large T_s range (400 ∿ 650°C). $Ga_{1-x}Al_xAs$ growth using TEG and TEA led to the same result. Figure 4 shows the T_s dependences of GaAs growth rate in $Ga_{1-x}Al_xAs$ with x = 0.06 and 0.55. For $Ga_{0.94}Al_{0.06}As$ growth, weak desorption with less activation energy than that in GaAs MOMBE was observed, however, at high Al composition of x = 0.55, growth rate was almost constant in a large T_s range (400 ∿ 600°C).

In conventional MBE, minimum As_4 pressure for obtaining As stabilized surface is increased by a factor of 2 for $Ga_{0.95}In_{0.05}As$ and is decreased by a factor of 4 in $Ga_{0.7}Al_{0.3}As$ as compared to GaAs at T_s = 600°C (Harmand et al 1987). This phenomenon indicates that As surface coverage is lower for GaInAs and higher for GaAlAs than for GaAs in the same gorwth conditions. Therefore, both effect (desorption minimization and enhancement) observed in MOMBE may be related to As surface coverage.

4. CONCLUSIONS

In GaAs MOMBE, in addition to Ga atom desorption, two other T_s dependent processes are present, that is, cracking limited and Ga-C_2H_5 desorption regions. The latter process is enhanced or minimized by the presence of In or Al atoms at the surface respectively. As a consequence, T_s range where ternary solid composition is constant decreases from 250°C for GaAlAs to less than 100°C for GaInAs.

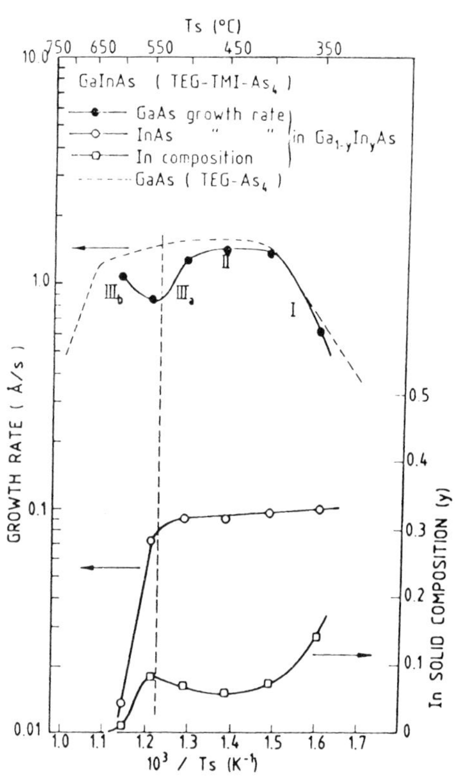

Fig. 2 : Constituent binary GaAs and InAs growth rates and In solid composition vs. $1/T_s$ for GaInAs. TEG, TMI and As_4 fluxes are 9.0x10-6, 1.9x10-6 and 1.9x10-5 Torr.

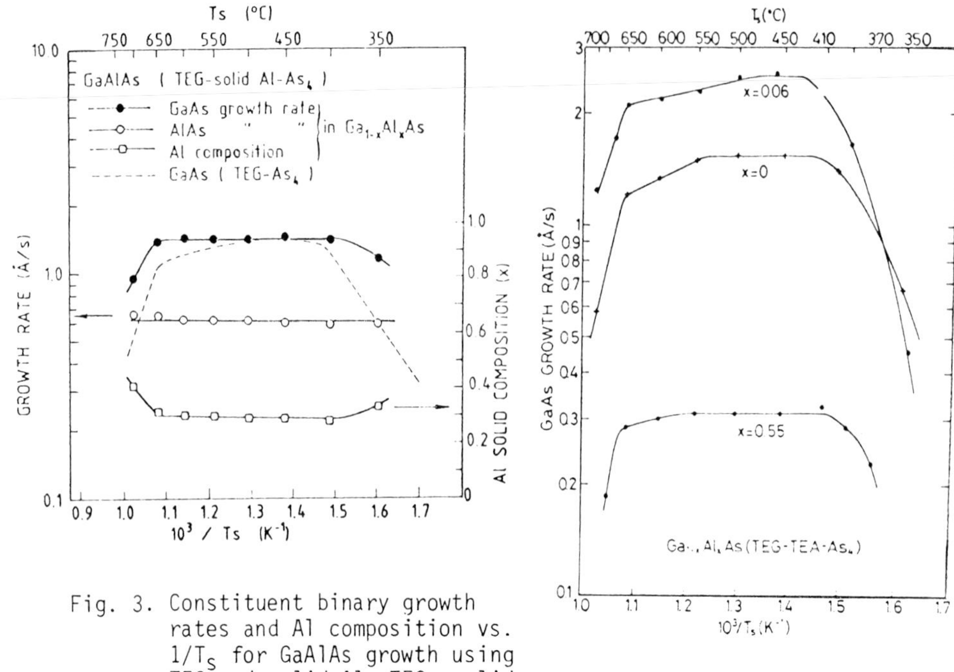

Fig. 3. Constituent binary growth rates and Al composition vs. $1/T_s$ for GaAlAs growth using TEG and solid Al. TEG, solid Al and As_4 fluxes were $1.0x10^{-5}$, $1.5x10^{-7}$ and $2.1x10^{-5}$ Torr.

Fig. 4. GaAs growth rate vs. $1/T_s$ in $Ga_{1-x}Al_xAs$ growth using TEG and TEA. x = 0.06 (TEG $7.3x10^{-6}$, TEA 9.0 x 10^{-7}, As_4 $3.6x10^{-5}$ Torr) x = 0.55 (TEG, TEA $1.4x10^{-6}$ As_4 $3x10^{-5}$ Torr).

ACKNOWLEDGEMENT : The authors would like to thank G. Le Roux for X-ray diffraction and J.F. Bresse for AES measurement.

REFERENCES

Chika S., Kato H., Nakayama M. and Sano N. 1986 Jpn. J. Appl. Phys. 25 1441.

Harmand J.R., Alexandre F. and Beerens J., 1987 Rev. Phys. Appl. 22 821

Horiguchi S., Kimura K., Kamon K., Mashita M., Shimazu M., Mihara M., and Ishii M., 1986 Jpn. J. Appl. Phys. 25 L979.

Kawaguchi Y., Asahi H. and Nagai H., 1986 Extended Abstract of the 18th Conf. of SSDM, Tokyo pp 619-622.

Pütz N., Heinecke H., Heyen M. and Balk P., 1986 J. Cryst. Growth 74 292.

Tirtowidjojo M. and Pollard R., 1986 J. Cryst. Growth 77 200.

Tokumitsu E., Kudou Y., Konagai M. and Takahashi K., 1984 J. Appl. Phys. 55 3163.

Tsang W.T., 1984a, Appl. Phys. Lett. 45 1234.

Tsang W.T., 1986b, J. Electron. Mater. 15 235.

Tsang W.T., Chiu T.H, Cunningham J.E. and Roberston A 1987c, Appl. Phys. Lett. 50 1376.

InAs–GaAs superlattices as a new semiconductor grown by beam separation MBE method

Y.Matsui, N.Nishiyama, H.Hayashi, K.Ono and K.Yoshida

Advanced Semiconductor Devices R&D Department, Sumitomo Electric Industries, Ltd.
1,Taya-cho, Sakae-ku, Yokohama 244, Japan

ABSTRACT : $(InAs)_m(GaAs)_n$ superlattices with the excellent periodicity have been obtained by the Beam-Separation molecular beam epitaxy method. The ratio of the X-ray satellite peak intensity to the 0th peak intensity is 0.3 in the $(InAs)_{12}(GaAs)_1$ superlattices. The misorientation of (001) lattice planes between the epitaxial layers and InP subustrates is extremely small in the $(InAs)_m(GaAs)_n$ superlattices compared with $In_xGa_{1-x}As$ ternary alloys. The effects of selective Si doping to the $(InAs)_m(GaAs)_n$ superlattices have been also discussed.

1. INTRODUCTION

$(InAs)_m(GaAs)_n$ superlattices have attracted attention (Gruntharner 1984,Tamargo et al.1985,Ohno et al.1986) because of the possibility of getting new characteristics. Both the lattice constants and the arrangements of group-III atoms can be variable in the $(InAs)_m(GaAs)_n$ superlattices because the lattice mismatch between InAs and GaAs is large (~ 7%). The large shift of energy band structure can be ,therefore, expected in comparison with the case of lattice-match structures like $(AlAs)_m(GaAs)_n$ superlattices. In this paper, we report the new properties of $(InAs)_m(GaAs)_n$ superlattices. The $(InAs)_m(GaAs)_n$ superlattices presented in this paper have been obtained by the Beam-Separation MBE method (Matsui et al. 1986a). In this method, we can grow the superlattices without using the cell shutters and repeat the growth and the annealing automatically by only rotating the substrate. The effects of annealing on the surface smoothness of the epitaxial layers have been observed recently. It is possible to improve the surface smoothness by the migration or desorption of surface atoms during the annealing (Sakaki et al. 1985, Matsui et al. 1987).We also report,in this paper,the influence of the growth condition in the Beam-Separation method on the crystal quality.

2. EXPERIMENTAL

The epitaxial layers were grown on the Fe-doped (001)InP substrates. The substrate temperatures during the growth were 300~450 °C. The growth rate and the substrate rotation speed were 0.1~1.0 μm/hr and 0.7~6.0 rpm,respectively. The periodicity of superlattices and the crystal quality were observed by the X-ray diffractometer and transmission electron microscope (TEM). The misorientation of crystal lattice planes between the epitaxial layers and the substrates was measured by rotating the sample around the normal of the (001) surface.

3. RESULTS AND DISCUSSIONS

In Fig.1 and Fig.2, the cross-sectional lattice images and the diffraction patterns of TEM are shown for the superlattices grown by the Beam-Separation method. It should be noticed that high order satellite spots can be obviously observed(Fig.1) and the superlattices with the period of 4 monolayers have been also obtained successfully(Fig.2). These results mean the uncertainty of heterointerfaces is less than ±1 monolayer.

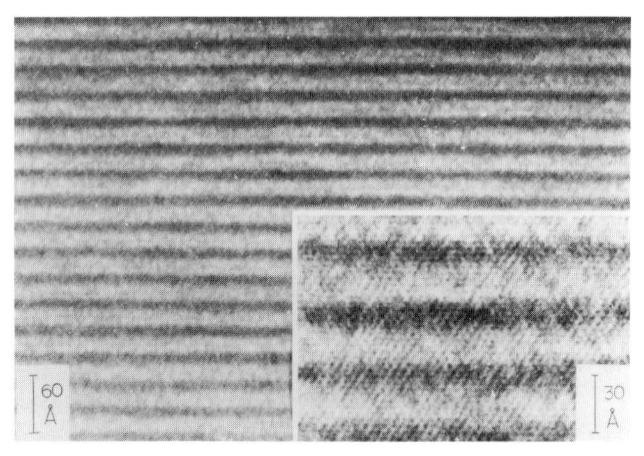

(a)

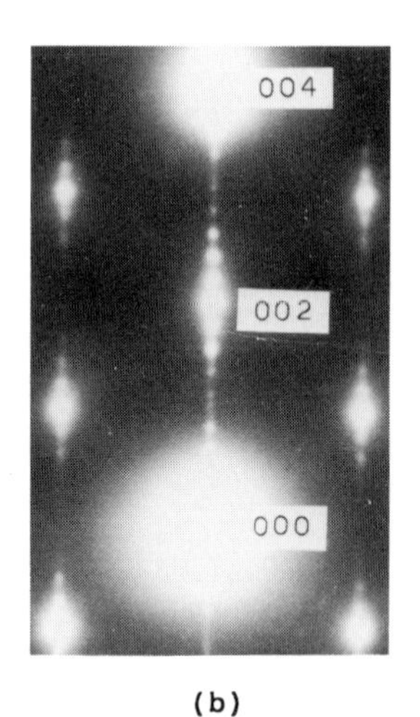

(b)

Fig.1(↖↑). Cross-sectional lattice images (Fig.1(a)) and diffraction pattern (Fig.1(b)) for superlattices with ~30Å period.

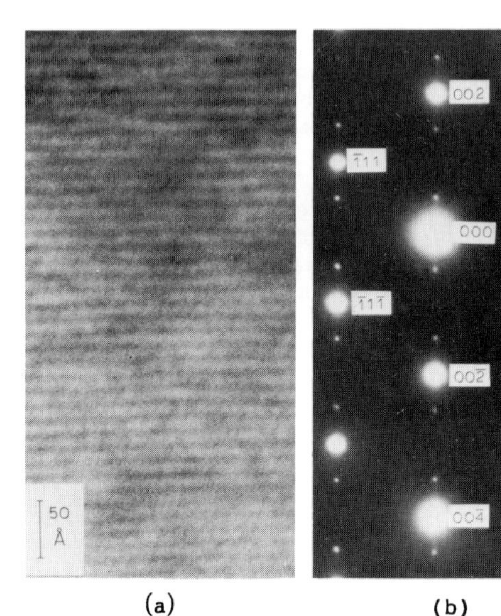

(a) (b)

Fig.2(←). Cross-sectional TEM image (Fig.2(a)) and diffraction pattern (Fig. 2(b)) for $(InAs)_2(GaAs)_2$ superlattices.

In the case of the superlattices with thin period, the periodicity is strongly influenced by the uncertainty of heterointerfaces. It is, therefore, important to enhance the satellite peak intensities, which are attributed to the sharpness of heterointerfaces and the periodicity of superlattices. The ratio of satellite peak intensity to 0th peak intensity is shown in Fig.3. In the case of the samples with the large lattice mismatch between the epitaxial layers and InP substrates, the satellite peak intensities are decreased as the epitaxial layer thickness is decreased. Comparing in the same condition of thickness, the satellite peak intensities are extremely increased by decreasing both the growth rate and the substrate rotation speed. For example, the peak ratio of $(InAs)_{12}(GaAs)_1$ superlattice samples is 0.3, which is 15 times as high as that of superlattices grown in the high growth rate and the high rotation speed. It may be attributed to the difference of time interval (i.e. annealing time) between the InAs growth and the GaAs growth.

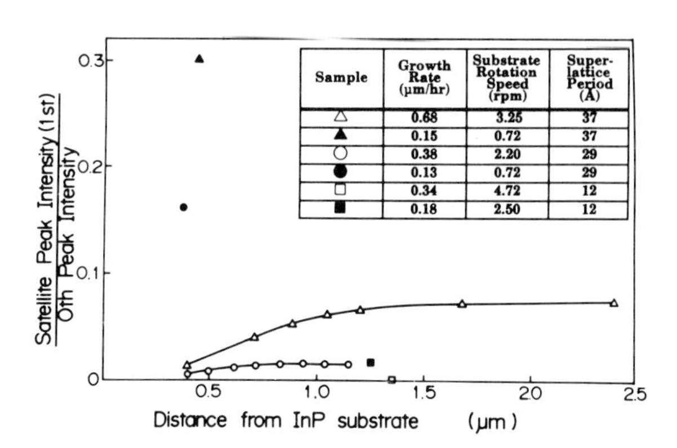

Sample	Growth Rate (μm/hr)	Substrate Rotation Speed (rpm)	Super-lattice Period (Å)
△	0.68	3.25	37
▲	0.15	0.72	37
○	0.38	2.20	29
●	0.13	0.72	29
□	0.34	4.72	12
■	0.18	2.50	12

Fig.3. Satellite peak intensities of $(InAs)_m(GaAs)_n$ superlattices.

The misorientation to InP substrates has been also studied comparing with that in the $In_xGa_{1-x}As$ ternary alloys as shown in Fig.4. The misorientation (α) in the $(InAs)_m(GaAs)_n$ superlattices is extremely small and independent on the lattice mismatch to InP substrates though that in the ternary alloys is increased as the lattice mismatch to InP substrates is increased. The small misorientation in the superlattices is related to the superior surface morphology compared with that of $In_xGa_{1-x}As$ alloys as already reported before (Matsui et al. 1985). The effects of Si doping have been also studied as shown in Table 1. That is, the activation efficiency of Si atoms as n-type dopants in InAs(Si doped)-GaAs(nondoped) superlattices is about 100 times as high as that in InAs(non.)-GaAs(Si) superlattices. This means the diffusion of Si atoms into the adjacent layers dose not occure at all at the growth temperatures (300 ~ 400 °C). That is, the electron

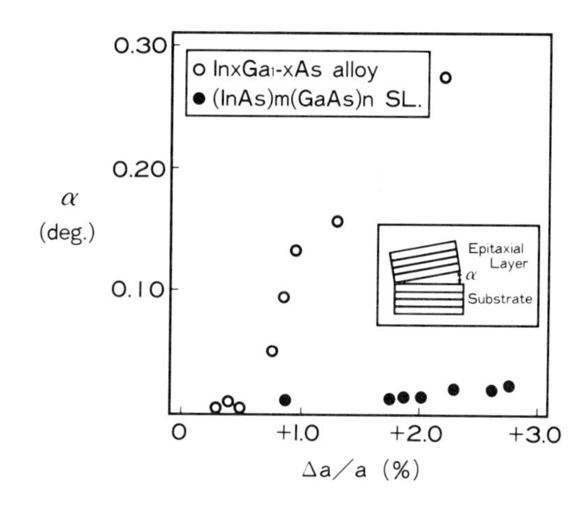

Fig.4. Misorientation(α) of (001) lattice planes between the epitaxial layers and InP substrates

Table 1. Effect of selective Si-doping in $(InAs)_m(GaAs)_n$ superlattices. Samples 1,3 and 5 are nondoped. Sample 2 is the InAs(Si doped)-GaAs(nondoped) superlattice. Samples 4 and 6 are the InAs(nondoped)-GaAs(Si doped) superlattices. Si cell temperatures during the growth for samples 2,4 and 6 are 950°C, 950°C and 1000°C,respectively.

	Si Doping	$\Delta a / a$ (%)	S.L. Period (Å)	Growth Rate (μm/hr.)	Carrier Density (cm^{-3})
Sample 1	no	+0.42	9	0.26	7.65×10^{15}
Sample 2	950°C (InAs)	+0.42	9	0.26	1.18×10^{19}
Sample 3	no	~0	28	0.37	3.80×10^{15}
Sample 4	950°C (GaAs)	~0	28	0.37	2.43×10^{16}
Sample 5	no	+2.73	15	0.42	5.71×10^{16}
Sample 6	1000°C (GaAs)	+2.73	15	0.42	3.23×10^{17}

mobility enhancement in InAs(non.)-GaAs(Si) superlattices (Matsui et al.1986b) should be attributed to the modulation-doped superlattice structure itself as a new semiconductor because each layer is too thin to form the two dimensional electron gas.

4.CONCLUSIONS

$(InAs)_m(GaAs)_n$ superlattices with high periodicity have been obtained successfully. The superior properties of $(InAs)_m(GaAs)_n$ superlattices have been studied.

ACKNOWLEDGEMENTS

The authors are grateful to T.Nakahara,T.Suzuki and S.Akai for their encouragement. This work was performed under the management of the R&D Association for Future Electron Devices as a part of the R&D Project of Basic Tecnology for Future Industries, sponsored by Agency of Industrial Science and Tecnology,MITI.

REFERENCES

Gruntharner F.J. 1984 Inter.Confer. on Superlattices,Microstructures and Microdevices
Matsui Y.,Hayashi H.,Takahashi M.,Kikuchi K. and Yoshida K. 1985 J.Cryst. Growth **71** 280
Matsui Y.,Hayashi H.,Kikuchi K. and Yoshida K. 1986a Surf.Sci.**174** 600
Matsui Y.,Hayashi H. and Yoshida K. 1986b Appl.Phys.Lett.**48** 1060
Matsui Y.,Hayashi H. and Yoshida K. 1987 J.Cryst.Growth **81** 245
Ohno H.,Katsumi R. and Hasegawa H. 1986 Surf. Sci.**174** 598
Sakaki H.,Tanaka M. and Yoshino J. 1985 Japan.J.Appl.Phys.**24** L417
Tamargo M.C.,Hull R.,Greene L.H.,Hayes J.R. and Cho A.Y. 1985 Appl. Phys. Lett.**46** 569

Paper presented at Int. Symp. GaAs and Related Compounds, Heraklion, Greece, 1987

Planar (Al)GaAs structures by selective MOVPE with application to GaAs on Si

P.Demeester, P.Van Daele, A.Ackaert and R.Baets

Univ. of Gent-IMEC, Lab. Electromagnetism and Acoustics, St.Pietersnieuwstraat 41, B-9000 Gent, Belgium

ABSTRACT

The growth behaviour during selective atmospheric pressure MOVPE of GaAs-AlGaAs over masked substrates was studied. Special attention was paid to the problem of irregular polycrystalline deposition on the mask and of non-uniform deposition (in time and space) of monocrystalline material in the channels. Several growth sequences have been investigated to obtain reproducibly planar structures with smooth transitions from mono- to polycrystalline material. The use of a very thin GaAs buffer layer, grown at 450°C, proved to be most successful. Results on optoelectronic devices and on the growth of GaAs on Si using this technique will be discussed.

1. INTRODUCTION

Selective MOVPE growth of GaAs over a masked substrate is made difficult by polycrystalline deposition on the mask, by non-uniform deposition of the monocrystalline material in the channels and by irregular behaviour at the edges. It is known that by growing at low pressure most of these problems can be solved, especially the deposition on the mask can be avoided completely [Kamon 1986]. At atmospheric pressure this can not be achieved. Many studies have been devoted to the use of different mask materials and growth parameters for the selective growth of GaAs and AlGaAs by APMOVPE [Azoulay 1981, Takahashi 1984, Nakai 1984, Ghosh 1984, Yamaguchi 1985, Okamoto 1986]. None of them could avoid the polycrystalline deposition completely and therefore we have to live with that problem. However, in a number of applications a polycrystalline layer can be tolerated and is even useful if it is regular and forms a smooth transition to the monocrystalline regions, that need to be very uniform themselves. In this paper we study a number of growth sequences in order to obtain this result.

2. EXPERIMENTAL PROCEDURE

The growth apparatus consists of a small horizontal reactor working at atmospheric pressure. The growth was carried out at 660°C and the sources were TMG, TMA, and a 5% mixture of AsH_3 in H_2. As doping sources we used DEZ and H_2Se. The selective growth was carried out on patterned substrates by using a special mask design (figure 1). It allowed us to investigate the effect of varying channel (A) and mask (B) width, of channel orientation (C) and of different window shapes and sizes (D). In the experiments we used SiO_2 as masking material that was deposited by PECVD. Prior to growth, a shallow etch was given in a 1 H_2SO_4:1 H_2O_2:18 H_2O solution to remove less than 100 nm of the GaAs. This showed to be very important to obtain reproducible results [Nakai 1984, Heinecke 1986].

3. EXPERIMENTAL RESULTS

It is clear from literature that the polycrystalline deposition on the mask has a strong influence on the growth behaviour in the windows. A high deposition density on the mask will result in uniform growth in the windows and a smooth transistion at the edges [Nakai 1984]. In order to investigate this, we define the nucleation density, namely the number of nucleation points per square millimeter. This number was obtained by counting the polycrystalline dots on the mask when growing a relatively thin layer (nominal thickness in the windows ; 1 µm for GaAs and 0.1 µm for AlGaAs). Only a slight increase of the nucleation density with the TMG molefraction was observed. When adding TMA to the gasphase, we observed an exponentional increase in deposition density due to a high sticking coefficient of the Al on the mask (figure 2) [Azoulay 1981, Takahashi 1984]. The expected uniform growth and smooth transition was indeed observed for a 3µm $Al_{60}Ga_{40}As$ and not for a GaAs layer.

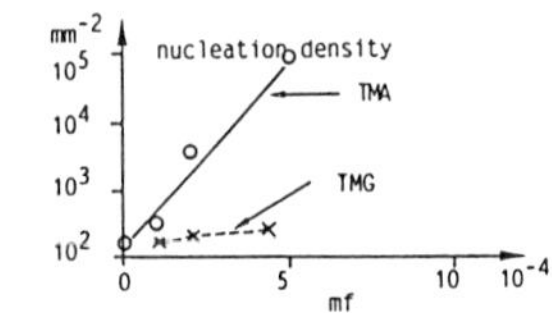

Fig. 2 Influence TMG/TMA mf on nucleation density

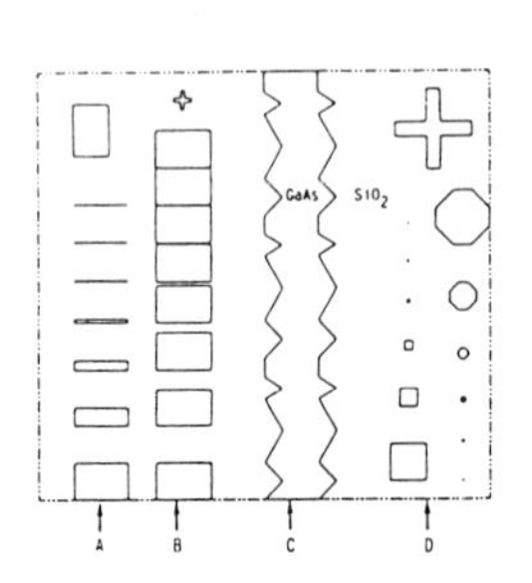

Fig. 1 Mask design

Figure 3 shows the behaviour at the edges of a channel when using a periodic $GaAs/Al_{60}Ga_{40}As$ structure. The nucleation of the polycrystalline material on the mask is clearly seen on the left. On the right we observe the appearance of different crystal planes at the edge of the channel. The planes that appear are strongly dependent on channel orientation and they are difficult to identify because most of the time they interact with the polycrystalline deposition. Although all layers were intentionally grown with the same thickness (100 nm), we clearly observe a higher growth velocity in the channel for the first layers (figure 4). This can be explained qualitatively by considering the arriving Ga species (whatever their exact form is). At the beginning of the growth there will be no nucleation points on the mask and therefore the concentration of the Ga species above the mask will be much larger than above the windows. This results in a lateral diffusion and surface migration towards the window which finally gives a higher growth velocity in the channels than expected for a non masked substrate. When growth proceeds, there will be more nucleation on the mask resulting in a lower lateral supply of Ga species towards the window and finally giving a decreasing growth velocity (figure 5). It is clear that this effect is much stronger for smaller channels.

Most of the problems discussed above are summarized in figure 6 : non uniform growth in the channels, non planar transistion to the edges and non reproducible polycrystalline growth on the mask. In order to solve those problems we tried different growth sequences. By using a thick $Al_{60}Ga_{40}As$ (500 nm) layer, it was possible to obtain a uniform growth velocity over time in the channel. Because the results were still not satisfactory, we tried to grow a thin bufferlayer of GaAs at 450°C and the subsequent layers at normal growth temperature (660°C). This procedure can be expected to lead to a much higher nucleation density [Azoulay 1981, Ghosh 1984, Heinecke 1986]. Moreover it is compatible with the usual growth procedure for GaAs on Si [Akiyama 1986]. The result obtained for a 10 nm bufferlayer is shown in figure 7 where we observe the expected homogeneous nucleation on the mask (no individual nucleation points observable). This results in a constant growth velocity over time in the channel, a smooth transistion from mono- to polycrystalline material and a flat polycrystalline surface.

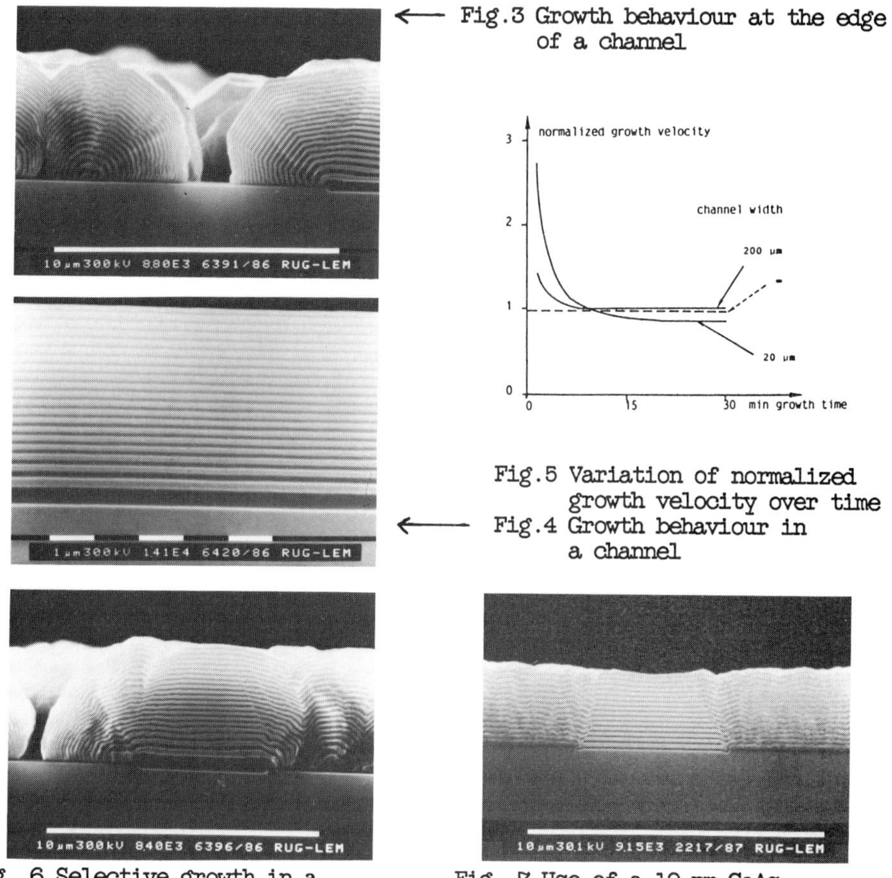

← Fig.3 Growth behaviour at the edge of a channel

Fig.5 Variation of normalized growth velocity over time

← Fig.4 Growth behaviour in a channel

Fig. 6 Selective growth in a 5µm channel

Fig. 7 Use of a 10 nm GaAs layer at 450°C

4. APPLICATIONS

We describe some applications where we use a 10 nm bufferlayer grown at 450°C. High-reflectivity multilayer structures, consisting of a periodic stack of quarter-wavelength GaAs and AlGaAs layers, show a strong reflection peak around a central wavelength λ_c [Baets 1987]. They are very useful for the fabrication of surface-emitting lasers [Gourley 1987]. We compared a selectively grown reflector with a normal one grown during the same run and observed no major difference in reflection characteristic (figure 8). The difference in central wavelength was due to a position-dependent growth velocity in the susceptor.

Double heterostructure IR-LED arrays were selectively grown on a p type substrate. An integration density of 16 LEDs/mm with a power output uniformity of 10% was obtained. A typical output power at 80 mA was 0.8 µW/mA.Sr. Due to the very good isolation of the polycrystalline material between the LEDs, a very simple fabrication scheme without mesa isolation or top dielectric coating was used and resulted in completely planar structures.

For the selective growth of GaAs on Si, we first baked out the substrate at 950°C and then followed the same procedure as described above (10 nm bufferlayer at 450°C). The growth behaviour was very similar to the growth on a GaAs substrate (figure 9). The morphology of the monocrystalline islands is also very similar to that of non-selective GaAs on Si material. The electrical and optical quality is now under investigation.

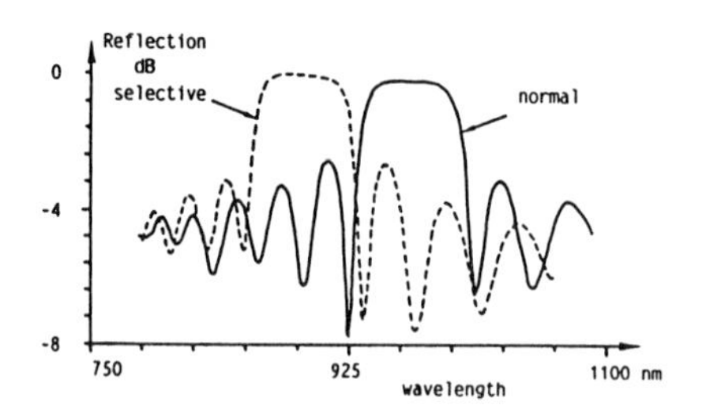

Fig. 8 Reflection plot of selective/ normal reflector

Fig. 9 Selective growth of GaAs on Si

5. CONCLUSION

We developed a growth sequence which is useful for the selective growth of planar structures for optoelectronic devices. This was obtained by using a 10 nm GaAs bufferlayer grown at 450°C. It was also shown that this technique is compatible with the growth of GaAs on Si.

Acknowledgement : The authors thank F.Clauwaert and D.Lootens for device processing and G. Vanden Bossche for optical measurements.

REFERENCES

Akiyama M., Kawarada Y., Ueda T., Nishi S., Kaminishi K. 1986 *J. Cryst. Growth* **77** pp 490–497
Azoulay R., Bouadma N., Bouley J.C., Dugrand L. 1981 *J. Cryst. Growth* **55** pp 229–234
Baets R., Demeester P., Lagasse P.E. 1987 *J. Appl. Phys.* **62** pp 723–726
Ghosh C., Layman R.L. 1984 *Appl. Phys. Lett.* **45** pp 1229–1231
Gourley P.L., Drummond T.J. 1987 *Appl.Phys. Lett.* **50** pp 1225–1227
Heinecke H., Brauers A., Grafahrend F., Plass C., Putz N., Werner K., Weyers M., Luth H., Balk P. 1986 *J. Cryst. Growth* **77** pp 303–309
Kamon K., Shimazu M., Kimura K., Mihara M., Ishii M. 1986 *J. Cryst. Growth* **77** pp 297–302
Nakai K., Ozeki M. 1984 *J. Cryst. Growth* **68** pp 200–205
Okamoto K., Yamaguchi K. 1986 *Appl. Phys. Lett.* **48** pp 849–851
Takahashi Y., Sakai S., Umeno M. 1984 *J. Cryst. Growth* **68** pp 206–213
Yamaguchi K., Okamoto K., Imai T. 1985 *Jap. J. Appl. Phys.* **24** pp 1666–1671

MO-ALE growth of GaAs using $Ga(C_2H_5)_2Cl$ and AsH_3

Kazuo Mori, Masaji Yoshida and Akira Usui

Fundamental Research Laboratories, NEC Corporation
1-1, Miyazaki 4-chome, Miyamae-ku, Kawasaki, Kanagawa 213, Japan

ABSTRACT: Atomic Layer Epitaxy (ALE) for GaAs, using a metal organic (MO) source of diethylgalliumchloride (DEGaCl) and arsine (AsH_3), is presented. Monolayer-unit growth ("digital epitaxy") and extremely uniform growth on a 3-inch GaAs wafer were realized. By Si_2H_6-doping during the period of exposure to DEGaCl, the 1.1×10^{19} cm^{-3} n^{++}-layer growth was carried out.

1. INTRODUCTION

Several reports have been published on Atomic Layer Epitaxy (ALE) for GaAs, using trimethylgallium (TMG) [Nishizawa et al. 1985, Bedair et al. 1985, Doi et al. 1986 and Mori et al. 1986] or triethylgallium (TEG) [Kobayashi et al. 1985] for group III source and AsH_3 for group V source. The other ALE for GaAs uses monolayer adsorption of GaCl [Usui and Sunakawa 1986] and can realize monolayer-unit growth ("digital epitaxy") [Watanabe and Usui 1987] over a wide range of growth conditions. The digital epitaxy is to become selective growth independent from window ratios and uniform growth on large-area wafers. ALE also enables the low temperature growth of III-V semiconductors. These characteristics should be applied to GaAs LSI fabrication. However, a source reaction between Ga metal and HCl gas, to produce GaCl, becomes insufficient as a reactor is scaled up to deal with large-area wafers. This paper presents ALE for GaAs using a metal organic (MO) source of diethylgalliumchloride (DEGaCl) instead of GaCl from Ga/HCl source reaction. Si_2H_6 doping in an ALE process is also carried out.

2. EXPERIMENTAL

A horizontal low pressure MOCVD system [Okamoto et al. 1984] was used for the MO-ALE process. The total pressure in the reactor was 100 Torr, and the total flow rate of H_2 was 9000 SCCM. Cr-O doped GaAs(100), 2° off towards <110> was used as substrates. MO-ALE was conducted by repeating an ALE cycle consisting of AsH_3 purge, MO supply, MO purge and AsH_3 supply.

The DEGaCl bubbler was held at 50°C, and the DEGaCl-line to the reactor was heated up to about 80°C. A DEGaCl vapor pressure is estimated to be 1 Torr at 50°C, using 60°C/2 Torr and the enthalpy of vaporization for DEAlCl [Sumitomo Chemical Co. Ltd.: Tables for properties of MO sources]. H_2 passing through DEGaCl was 400 SCCM, and the partial pressure of DEGaCl was 0.0059 Torr. AsH_3 flow was 15-100 SCCM, which corresponded to the partial pressure of 0.17-1.1 Torr in the reactor.

Thickness of the grown layers was measured by 2° angle-lapping and staining with an optical microscope, or by selective etching of the ALE layer grown on $Al_{0.4}Ga_{0.6}As$ with a surface roughness profiler.

3. RESULTS AND DISCUSSION

The dependence of thickness per ALE cycle on time for the exposure to DEGaCl is shown in Fig. 1, over range of 3 to 15 sec at a growth temperature of 525°C. The exposure to AsH_3 was 0.17 Torr×4 sec and the purging times were 2 sec. The ALE cycle was repeated 636 times. The thickness became constant, which was almost equal to the mono-layer thickness for GaAs(100) (=2.83 Å) at exposure times more than 7 sec.

Figure 2 shows the dependence of thickness per ALE cycle on growth temperature, over a range of 450° to 600°C at a DEGaCl exposure time of 9 sec. The exposure to AsH_3 was 0.44 Torr×2 sec and the purging times were 1 sec. The ALE cycle was repeated 3000 times. In this temperature range, the thickness was almost identical to the monolayer thickness, although it decreased slightly at a growth temperature of 600°C.

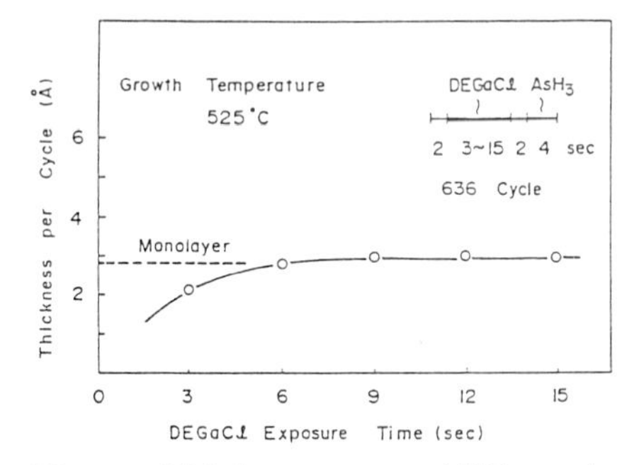

Fig. 1. Thickness per ALE cycle vs. DEGaCl exposure time.

Fig. 2. Thickness per ALE cycle vs. growth temperature.

The digital nature of DEGaCl-ALE for GaAs growth can be well explained by assuming that DEGaCl decomposes to GaCl and that Langmuir monolayer adsorption of GaCl on surface As atoms takes place [Usui and Sunakawa 1987]. On the contrary, MO-ALE using TMG and TEG is controlled by the catalytic TMG/TEG decomposition on surface As atoms and the covering over catalytically active As atoms with Ga-containing decomposition products. When side reactions for TMG/TEG decomposition,for instance, TMG decomposition in vapor phase [Yoshida et al. 1985], are not suppressed, the thickness goes over the monolayer thickness with an increase in the exposure to TMG/TEG or in the growth temperature.

The decrease in the thickness per ALE cycle at a higher growth temperature, 600°C as shown in Fig. 2, is probably due to a decrease in the Langmuir adsorption equilibrium constant or to an increase in the GaCl desorption during the purging time. The thickness at 400°C (not shown) decreased. That is presumably because the DEGaCl decomposition is incomplete or the reaction between adsorbed GaCl and AsH_3 is insufficient at such a low temperature.

Uniform GaAs growth on a 3-inch substrate by DEGaCl-ALE is shown in Fig. 3. ALE conditions were 525°C for the growth temperature, 0.0059 Torr×9 sec for the exposure to DEGaCl, 0.17 Torr × 4 sec for the exposure to AsH_3, 2 sec for

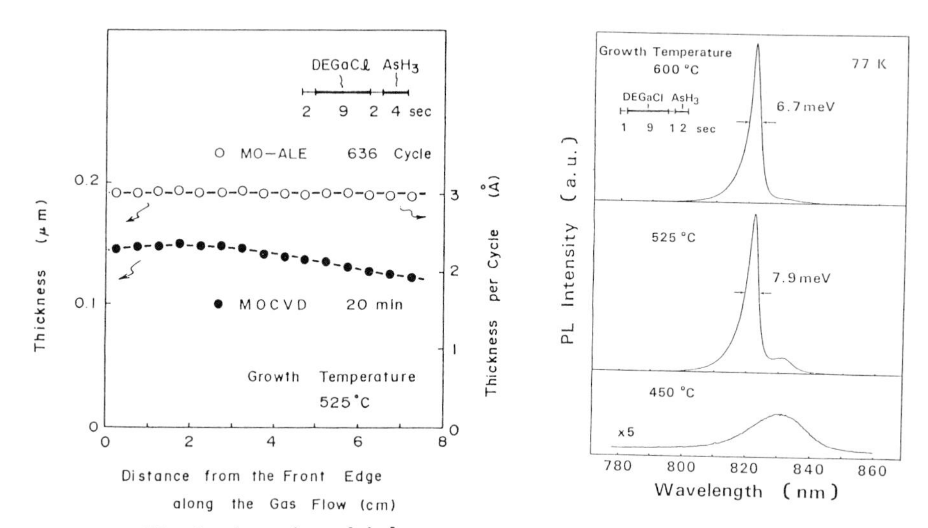

Fig. 3. Variation in thickness along the gas flow.

Fig. 4. 77 K PL from DEGaCl-ALE GaAs layers.

purging times, and 636 for the number of ALE cycles. For comparison, the thickness variation in a GaAs layer grown by MOCVD at 525°C for 20 min is also shown. The variation in the thickness of ALE layer was within the measurement accuracy limits, while the thickness of MOCVD layer decreased by 20 % in the same range along the gas flow. Excellent uniformity in thickness was also obtained at growth temperatures of 450°C and 600°C. It should be mentioned that excellent selective growth on the GaAs substrate partly covered with SiO_2 was also achieved by DEGaCl-ALE.

Hall measurements showed the conversion of DEGaCl-ALE GaAs from p-type to n-type with increasing growth temperature. At 600°C, a 77 K electron mobility of 22400 $cm^2/V{\cdot}sec$ with an electron concentration of 1.1×10^{15} cm^{-3} was obtained. Figure 4 shows 77 K PL spectra for GaAs samples grown at various temperatures. The peak at 832 nm, which is related to carbon acceptor [Mori et al. 1982], decreases as the growth temperature increases. An increase in the growth temperature would seem to contribute to loosening the Ga-C_2H_5 bond.

Finally, Si_2H_6 doping during the period of exposure to DEGaCl was carried out. Growth conditions were 525°C for the growth temperature, 0.0059 Torr × 9 sec for DEGaCl, and 0.44 Torr × 2 sec for AsH_3. A monolayer of Si-doped GaAs was inserted between non-doped p-GaAs layers (p~10^{15} cm^{-3}). The variation in electron concentration with the doped plane ratio, [GaAs(Si)]/{[GaAs]+[GaAs(Si)]}, is shown in Fig. 5. The ratio of Si_2H_6 to DEGaCl, [Si_2H_6]/[DEGaCl], varied from 0.0023 to 0.048. The electron concentration saturates with Si_2H_6 doping due to the self-compensation in GaAs:Si [Druminski et al. 1982]. At a growth temperature of 525°C, a maximum electron concentration for Si-doped GaAs was 8×10^{18} cm^{-3}. At a lower growth temperature, the higher electron concentration should be obtained for GaAs: Si [Ogawa and Baba 1985]. However, in a conventional MOCVD, the decrease in the growth temperature brings a rough surface morphology and a rapid decrease in the growth rate. In contrast, the growth proceeds even at 450°C in the present DEGaCl-ALE system and the mirror-like GaAs layer with the high electron concentration of 1.1×10^{19} cm^{-3} was grown.

Figure 6 shows variation in the electron concentration and mobility in a 3-inch-diameter Si-doped GaAs layer grown by ALE, with the distance along the gas flow. Excellent uniformity was also obtained for the concentration of electron carrier arising from doped Si.

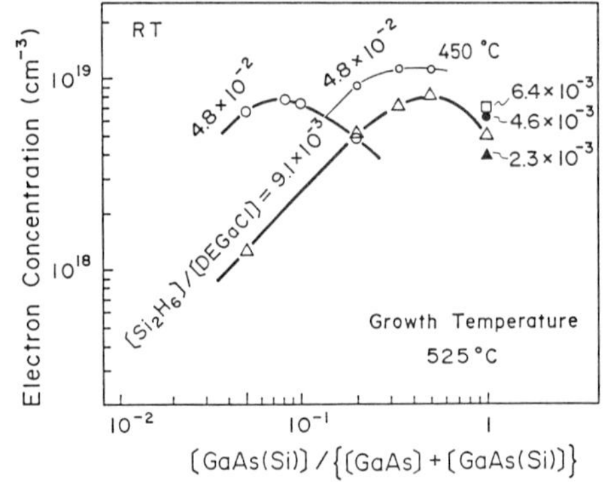

Fig. 5. Electron concentration vs. doped plane ratio.

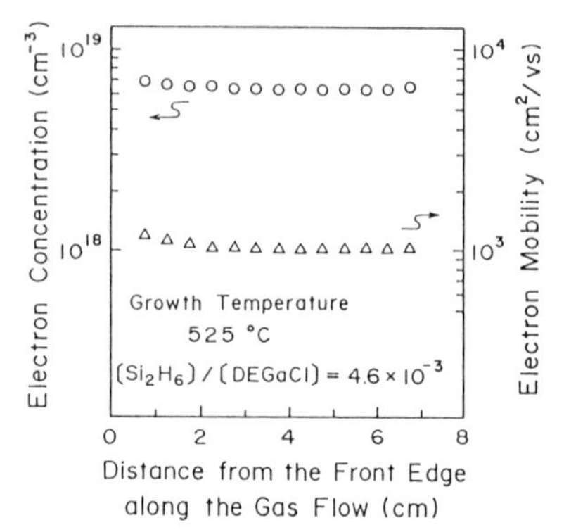

Fig. 6. Variation in electron concentration along the gas flow.

4. CONCLUSIONS

DEGaCl-ALE indicated digital nature in the growth of GaAs; the thickness per ALE cycle corresponding to the growth rate in the conventional MOCVD is independent from the partial pressure of DEGaCl and the growth temperature. Using this characteristic, uniform GaAs growth on a large area, such as on a 3-inch substrate, was achieved. Undoped GaAs grown at 600°C showed n-type conductivity with a carrier concentration as low as 1×10^{15} cm^{-3}. By Si_2H_6 plane-doping at 450°C, n^{++}-GaAs with 1.1×10^{19} cm^{-3} was grown. These results confirm the feasibility of DEGaCl-ALE for the GaAs LSI process technology, such as uniform n^{++}-GaAs selective growth for ohmic contact formation which limits the device performance.

The authors would like to thank H. Terao, H. Watanabe and M. Ogawa for continuing interest and helpful discussions.

REFERENCES

Bedair S M, Tischler M A, Katsuyama T and El-Masry N A 1985, Appl. Phys. Lett. 47, 51.
Doi A, Aoyagi Y and Namba S 1986, Appl. Phys. Lett. 49, 785.
Druminski M, Wolf H -D and Zschauer K -H 1982, J. Cryst. Growth 57, 318.
Kobayashi N, Makimoto T and Horikoshi Y 1985, Jpn. J. Appl. Phys. 24, L962
Mori K, Ogura A, Yoshida M, Terao H 1986, Extended Abstracts of the 18th Conference on Solid State Devices and Materials, Tokyo (1986), p. 743.
Mori Y, Ikeda M, Sato H, Kaneko K and Watanabe N 1982, "GaAs and Related Compounds (Oiso) 1981," Institute of Physics, London (1982), p.95.
Nishizawa J. Abe H and Kurabayashi T 1985, J. Electrochem. Soc. 132, 1197.
Ogawa M and Baba T 1985, Jpn. J. Appl. Phys. 24, L572.
Okamoto A, Sunakawa H, Terao H and Watanabe H 1984, J. Crystal Growth 70, 140.
Usui A and Sunakawa H 1986, Jpn. J. Appl. Phys. 25, L212.
Usui A and Sunakawa H 1987, "GaAs and Related Compounds (Las Vegas) 1986," Institute of Physics, London (1987), p. 129.
Watanabe H and Usui A 1987, ibid., p.1.
Yoshida M, Watanabe H and Uesugi F 1985, J. Electrochem. Soc. 132, 677.

Inst. Phys. Conf. Ser. No. 91: Chapter 3
Paper presented at Int. Symp. GaAs and Related Compounds, Heraklion, Greece, 1987

Monolayer growth of GaAs by switched laser MOVPE

S. Iwai, A. Doi, Y. Aoyagi and S. Namba

The Institute of Physical and Chemical Research
Hirosawa, Wako-shi, Saitama 351-01, Japan

Abstract. Atomic layer epitaxy (ALE) of GaAs is achieved by favoring photochemical decomposition of triethylgallium (TEG) with irradiation and suppressing pyrolytic decomposition. Selective photo-catalytic enhancement of the decomposition of TEG on the As atomic surface is found to occur under irradiation, with no enhancement on the Ga atomic surface. An ideal ALE is realized by this mechanism in which suspension of Ga deposition at 100% coverage on the As atomic layer is achieved.

1. Introduction

Atomic Layer Epitaxy (ALE), which was proposed by Suntola et al (1977), is an attractive method to control the growth thickness in one monolayer unit. Recently, many researchers have interested in ALE since this technique seems to make it possible to produce abrupt interfaces in heterostructures and narrow doped layers at one atomic level. In an ideal ALE, the growth thickness is expected to be determined by the number of growth cycles without fine control of flow rates, pressure and temperature in the growth system. In order to realize an ideal ALE, a mean of arresting the deposition of element at 100% surface coverage is an essential part of the growth mechanism. ALE of GaAs has been reported by metalorganic vapor phase epitaxy (MOVPE) (Nishizawa et al 1985, Bedair et al 1985). However, the ideal growth rate of one monolayer /cycle has been difficult to be achieved. Recently, Usui et al have reported that the ideal ALE has been realized in diethylgallium chloride (DEGaCl)-arsine system with the 100% surface coverage by gallium chloride (GaCl)(1986). Recently, we have found that Ar ion laser irradiation during trimethylgallium (TMG) gas supply largely enhances the growth rate of GaAs and this enhancement is not due to photothermal effect but photochemical effect at the surface (Aoyagi et al 1985). This result suggests that the surface reaction-limited growth will be obtained under the Ar ion laser irradiation.

In this paper, a new ALE method by switched laser MOVPE (SL-MOVPE) using triethylgallium (TEG) is described. It is also shown that this pulse laser technique can be used to study the transient behavior of adsorbed species on a substrate surface.

2. Experimental

In our laser MOVPE system, visible light of 514.5 nm from an Ar ion laser was used in order to enhance the surface reaction of TEG and to suppress the decomposition in gas phase. The growth procedure for SL-MOVPE has been proposed by Doi et al (1986) and is shown schematically in Fig. 1. TEG and

AsH_3 are separately introduced into a growth chamber by switching the valves. A laser beam from a cw Ar laser is switched on by a shutter to irradiate a substrate during desired periods. TEG and AsH_3 are accumulated for 3 sec and are introduced into the chamber for 1 sec alternately. The purging time for 1 sec between TEG and AsH_3 gas pulses was enough to prevent mixing of source gases. The total flow rate of H_2 carrier gas was 2800 sccm and the total pressure in a reactor was 75 Torr. Epitaxial layers were grown on (100) oriented Si doped and Zn doped GaAs substrates. For ALE by laser MOVPE, a substrate was irradiated during the TEG flux pulse, as shown by (1) in Fig. 1. On the other hand, laser irradiation was also performed after the end of the TEG pulse in order to investigate the behavior of TEG molecules adsorbed on a substrate, as shown by (2) in Fig. 1.

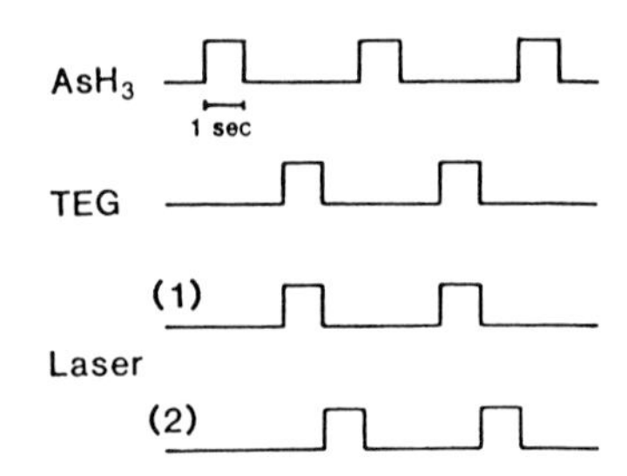

Fig. 1. Sequences of gas flows and laser pulses in SL-MOVPE.

3. Results and Discussions

Typical photographs of the surface grown under the ALE condition at 350 C are shown in Fig. 2. The growth rate of an irradiated area is largely enhanced and the epitaxial layer shows a flat mirror surface in spite of the intensity profile of a laser beam. No deposition is observed in the area without laser irradiation at 350 C. Scanning the laser beam on the substrate, the line pattern of ALE was obtained, as shown in Fig. 2(b). The growth thickness was uniform although the scanning speed was varied.

Figure 3 shows the growth rate per cycle as a function of temperature. The growth rate is enhanced by laser irradiation and is independent of the growth temperature in the range of 340 to 370 C. The growth rate/cycle is equal to the thickness of 1 monolayer on a (100) oriented GaAs surface. On the other hand, the growth rate without irradiation is negligibly small at the temperature around 350 C when TEG and AsH_3 are separately supplied.

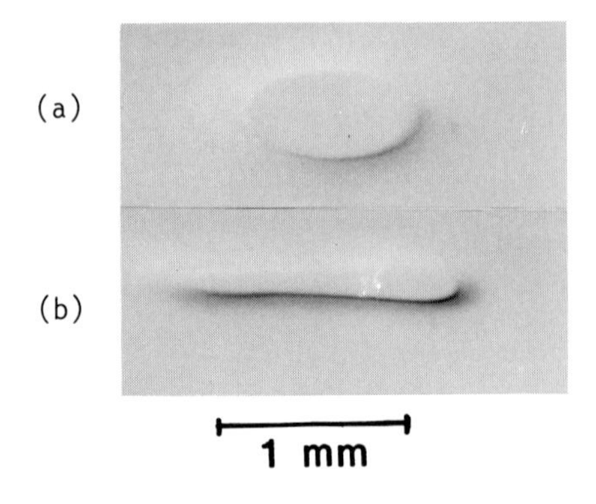

Fig. 2. Nomarski micrographs of the GaAs surface grown under the ALE condition at 350 C. (b) shows a line pattern grown by scanning a laser beam.

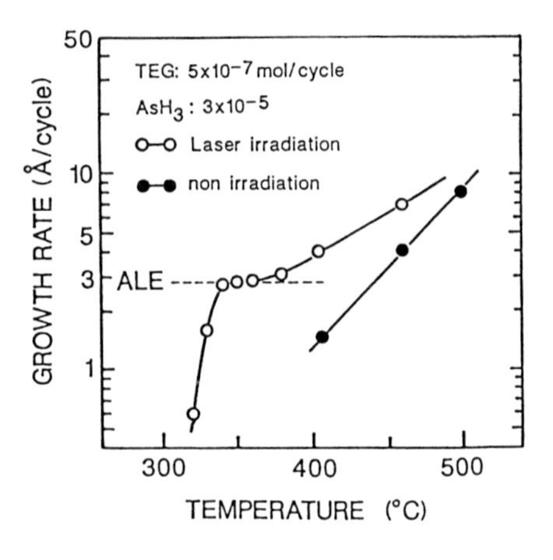

Fig. 3. Temperature dependence of the growth rate of GaAs.

At temperature above 400 C, the growth rate increases with the increase of temperature because the thermal decomposition of TEG is added to the decomposition by photochemical reaction. Above 500 C, the photochemical reaction is almost masked by pyrolytic reaction. ALE by laser MOVPE seems to be realized under the condition where the pyrolytic decomposition of TEG was suppressed.

Figure 4 shows the growth rate/cycle as a function of laser power density at the TEG flux of $5x10^{-7}$ mole/cycle at 350 C. For low power density less than 60 W/cm^2, the grown surface showed a hillock shape similar to the gaussian profile of the laser intensity on the substrate and the height of the hillock increased with the laser power. Above 60 W/cm^2, the growth rate saturates at a constant value of 1 monolayer and the surface becomes flat, as shown in Fig. 2. Increase of the growth rate at high power density above 250 W/cm^2 is due to the local temperature rise caused by laser irradiation. The growth rate for p-GaAs substrates was almost the same as that for n-GaAs substrate.

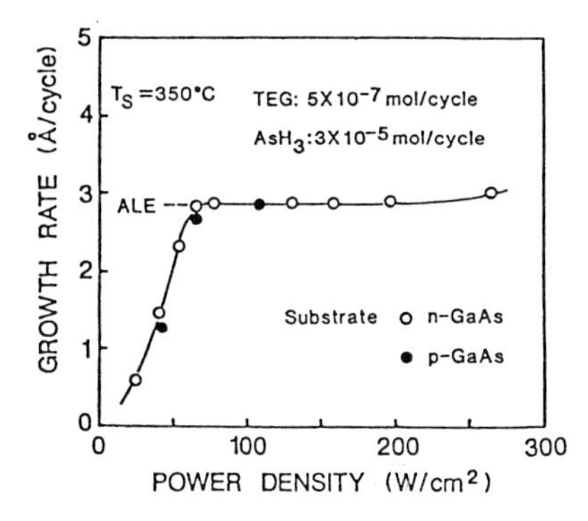

Fig. 4. Growth rate of GaAs as a function of laser power density.

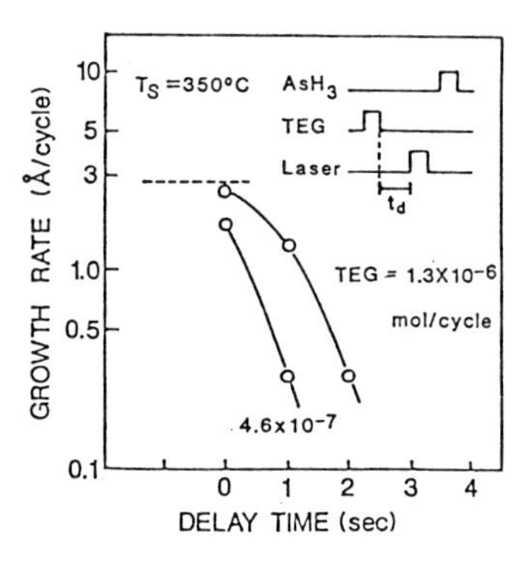

Fig. 5. Growth rate of GaAs as a function of the delay time of a laser pulse after the TEG pulse.

When the laser irradiation is delayed to the TEG pulse, the growth rate decreases rapidly as shown in Fig. 5. In this case, the epitaxial growth is due to the photochemical decomposition of residual TEG adsorbed on the substrate. The adsorbed TEG molecules are easy to desorb from the substrate. The desorption time constant for TEG was obtained to be about 0.6 sec from the slope of curves in Fig. 5. On the other hand, TMG was observed to desorb more rapidly than TEG and the time constant was roughly estimated less than 0.1 sec.

Figure 6 shows the growth rate as a function of TEG flux. The laser power density and the growth temperature are 120 W/cm^2 and 350 C, respectively. Curve (1) in Fig. 6 shows the growth rate of epitaxial layers grown under irradiation during the TEG pulse, as shown by (1) in Fig. 1. Above $1x10^{-7}$ mole/cycle of TEG flux, the growth rate is independent of the TEG flux and remains constant at one monolayer/cycle up to the highest TEG flux in our experiment. On the other hand, curve (2) in Fig. 6 shows the growth rate with irradiation just after the end of TEG pulse. The growth rate of (2) corresponds to the surface coverage by TEG on the substrate and increases with the increase of TEG flux. However, the growth rate saturates at the value less than 1 monolayer/cycle for the TEG flux enough to realize ALE.

This result suggests that the 100% surface coverage by TEG is not necessary to achieve an ideal ALE under this growth condition with irradiation. If TEG flux is supplied with irradiation, the decomposition of TEG occurs continuously and Ga atoms cover the whole surface of an As surface layer. Furthermore, it was found that even if two and more TEG pulses were supplied between two AsH_3 pulses under the ALE condition, the growth rate/cycle remained constant at 1 monolayer. This result suggests that only one Ga atomic layer is formed on the As surface and no deposition of Ga occurs on the Ga surface. Thus, the 100% surface coverage by Ga atoms is realized on the As surface.

These experimental results are explained by the growth model shown in Fig. 7. ALE by laser MOVPE is based on the growth mechanism that the decomposition time constant of TEG (T_{As}) on As atomic layer is much smaller than that (T_{Ga}) on Ga atomic layer under irradiation. Once the whole surface of the As layer is covered with Ga atoms, no decomposition of TEG by irradiation occurs on the Ga layer. The stopping mechanism of Ga deposition at 100% surface coverage is introduced in the decomposition process of TEG with irradiation.

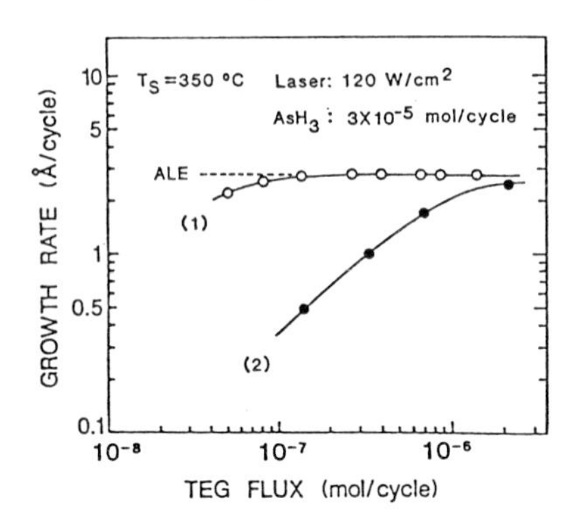

Fig. 6. Growth rate as a function of TEG flow rate under laser irradiation (1) during TEG pulses and (2) just after TEG pulses.

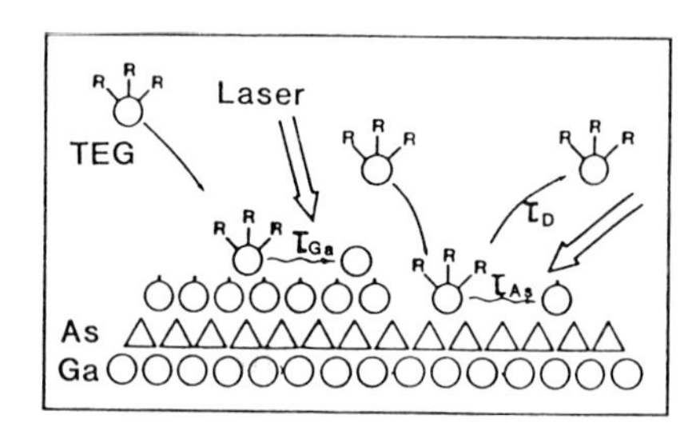

Fig. 7. The growth model for ALE under laser irradiation.

4. Conclusion

ALE was achieved by the enhancement of photochemical decomposition under irradiation and suppressing pyrolytic decomposition of TEG. The selective photochemical decomposition of TEG on the As surface layer seems to be a dominant mechanism to realize ALE in TEG-AsH_3 system by SL-MOVPE.

Acknowledgment

The authors would like to thank M. Mihara for his technical assistance and T. Meguro for useful discussions.

References

Aoyagi Y, Masuda S, Namba S and Doi A 1985 Appl. Phys. Lett. 47 95
Bedair S M,Tischler M A, Katsuyama T and Elemasry N A 1985 Appl.Phys. Lett. 47 51
Doi A, Aoyagi Y and Namba S 1986 Appl. Phys. Lett. 49 785
Nishizawa J, Abe H and Kurabayashi T 1985 J. Electrochem. Soc. 132 1197
Suntola T and Anton M J 1977 US Patent No.4-058-430
Usui A and Sunakawa H 1986 Jpn. J. Appl. Phys. 25 L212

Paper presented at Int. Symp. GaAs and Related Compounds, Heraklion, Greece, 1987

MBE growth and characterization of n-GaAs on InP Substrates and its device application

Kensuke Kasahara, Kazunori Asano, and Tomohiro Itoh

Microelectronics Research Laboratories, NEC Corporation
4-1-1, Miyazaki, Miyamae-ku, Kawasaki, 213 Japan

ABSTRACT: A good-quality n-GaAs on InP substrates (lattice mismatch 3.8%) grown by MBE and successful GaAs MESFETs fabrication on InP are presented. The results indicate that this technology is promising for fabricating InP based long-wavelength OEICs. The n-GaAs epilayers with an undoped GaAs buffer layer (0-1μm) were characterized by RHEED, Photoluminescense, Hall measurement, and DLTS. 1μm-gate GaAs MESFETs on InP exhibited high transconductance of 120mS/mm with a complete pinch-off and good drain current saturation.

1. INTRODUCTION

InP-based, long-wavelength, optical communication systems have conventionally been used with highly transparent optical fibers(Shibata et al 1984). However, InP based electric devices must still be required to overcome many problems before high performance can be achieved. On the other hand, GaAs-based electric devices have been used because of their maturing process technology for high-speed GaAs integrated circuits. Therefore, optoelectric integrated circuits (OEICs), which combine InP-based optical devices with the GaAs-based electric devices, are very attractive because of their matured technology. Recently, GaAs/InGaAs(Chen et al 1985), and GaAs/Si(Choi et al 1984, Fisher et al 1986) have been successfully grown by Molecular Beam Epitaxy (MBE). Also, InP/GaAs has been grown by Vapor Phase Epitaxy(Teng et al 1986). These lattice mismatched epitaxy techniques have proven to be promising for device applications. Using GaAs or AlGaAs as a gate insulator on InP, in spite of a large lattice mismatch (that is about 3.8%), we have already reported the successful fabrication of InP MISFETs with good and stable dc and RF characteristics(Itoh T. et al 1986), and long-wavelength PINFET OEICs on GaAs on InP substrates(Suzuki et al 1987). This paper describes the growth and characterization of n-GaAs on InP substrates, as well as its application to MESFETs.

2. MBE GROWTH

The GaAs epilayers were grown on Fe-doped (100) semi-insulating InP substrates. After wet chemical etching, the substrates were loaded into the MBE chamber. The InP substrates were cleaned by heat treatment, under As pressure to remove the surface oxide(Mizutani et al 1985). Before and during the epitaxial growth, the RHEED patterns were observed. GaAs layers were grown at 500°C with a 0.7μm/h growth rate.

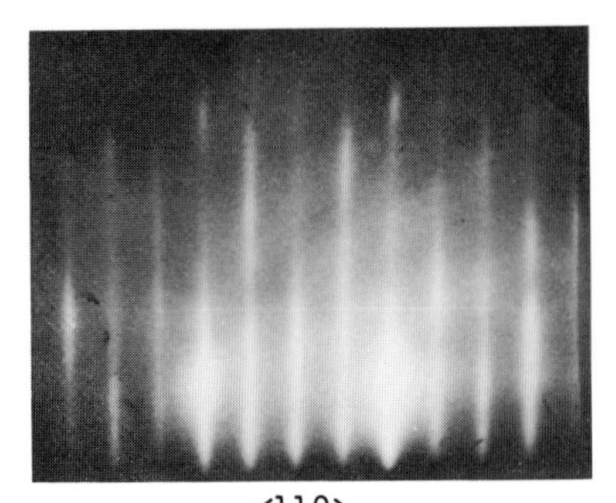

<110>

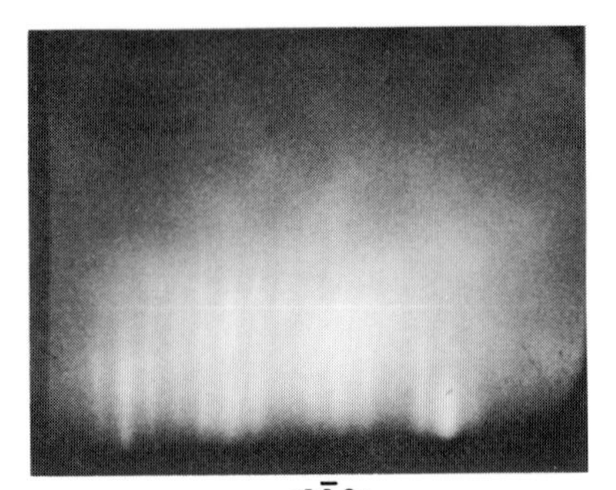

<1$\bar{1}$0>

Fig.1 RHEED Patterns for GaAs surface after 1000Å Growth on InP substrates.

After cleaning the substrate surface, an undoped GaAs buffer layer was directly grown on an InP substrate, followed by growing a 0.5μm thick Si doped n-GaAs layer. To characterize the influence of the lattice mismatch, undoped GaAs buffer layers were grown with different thicknesses, from 0 to 1μm. After the native oxide of the InP substrate was removed, the growth was started by opening the Ga cell shutter. The RHEED pattern changed from a substrate surface streak pattern to a spotty pattern. After the GaAs thickness exceeded about 300Å, the RHEED pattern gradually recovered from the spotty pattern to the streak pattern. Figure 1 shows the RHEED patterns for the GaAs epilayer after the 1000Å growth on the InP substrate. A one-half order of <110> azimuth and one-quarter order reconstruction of <1$\bar{1}$0> azimuth, were observed. These results indicate that, initially, the GaAs was grown three dimensionally, because of the large lattice mismatch. However, by increasing the epilayer thickness, a single crystal with a smooth surface was obtained. Despite the large lattice mismatch, the grown layers had mirror smooth surfaces, and no cross-hatched slip lines were observed using a Nomarski microscope.

3. CHARACTERIZATION

The photoluminescence (PL) spectra for the n-GaAs on the InP substrates with different buffer thicknesses was measured. The PL peak wavelength for the GaAs on the InP substrates was the same as that for the GaAs on the GaAs substrates. The full-width-at-half-maximum (FWHM) value for this spectrum improved by increasing the buffer layer thickness from 0 to 1 μm. Hall measurements were also carried out for these samples. Figure 2 shows the mobility and the carrier concentration as functions of the buffer layer thickness. In spite of the same doping density, as the buffer layer thickness decreased, the carrier concentration of the n-GaAs deteriorated. The same tendency was observed in mobility. For the n-GaAs with a thin buffer layer, in spite of its lower carrier concentration, the mobility was lower than that for a thick buffer layer. These results seem to come from the misfit dislocations. For the n-GaAs on an InP substrate with a 1μm GaAs buffer layer thickness, the mobility was 3,510 $cm^2/V{\cdot}s$, and the carrier concentration was $3.1\times10^{17}cm^{-3}$. These values are close to those for n-GaAs on GaAs, and good enough to fabricate electron devices. By further optimizing the growth condition and the epilayer structure, the quality of n-GaAs on InP substrates will be improved. To understand the influence of lattice mismatch on deep

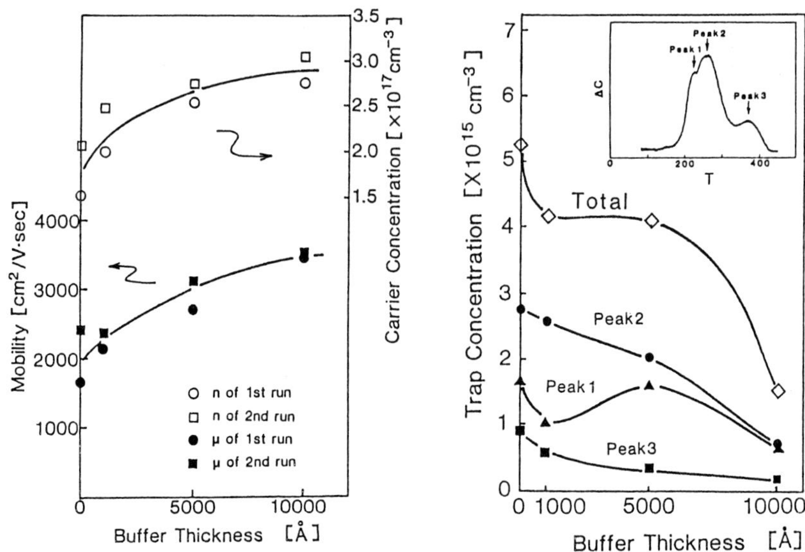

Fig.2 Hall Mobility and Carrier Concentrationofn-GaAs/GaAs/InP at Room Temperature.

Fig.3 Trap Concentrations as a Function of Buffer Layer Thickness.

traps, DLTS measurement were performed. Figure 3 shows a typical DLTS spectrum for the n-GaAs on the InP substrates with a 0.5 μm-thick buffer layer, and the deep trap concentrations as a function of the buffer layer thickness. Three distinct DLTS peaks(Peak 1, 2, 3) were observed, whose activation energies were 0.42eV, 0.34eV, and 1.0eV, respectively. As the GaAs buffer layer thickness increased, the trap concentrations decreased. It is noted in particular that Peak 3 could not be observed in the n-GaAs on GaAs substrates and exhibited monotonic decrease as the GaAs buffer layer thickness increased. From these results, Peak 3 may be the result of the lattice mismatch. However the changes in the trap concentrations as a function of the buffer layer thickness were too small to explain the changes in carrier concentrations. The deterioration of the carrier concentrations may be due to other mechanisms. For example, Si atoms may occupy the interstitial positions or the acceptor sites.

4. MESFET APPLICATION

The recessed gate GaAs MESFETs on InP were fabricated by using conventional photolithography and lift-off technique (Asano et al 1987). The gate length was 1μm, and the gate width was 200μm. Figure 4 shows the drain current-voltage characteristics for GaAs MESFETs. When the buffer layer thickness was 0μm or 0.1μm, non pinch-off characteristics and hysterisis were observed. When the buffer layer thickness was more than 0.5μm, the GaAs MESFET exhibited very good I-V characteristics, with complete pinch-off and drain current saturation. The transconductance of the device with a 1μm-thick buffer layer was about 120mS/mm. The I-V characteristics didn't show any noticeable light sensitivity under intense microscope illumination. These results indicate a good crystal quality for GaAs layers grown directly on InP substrates by MBE .

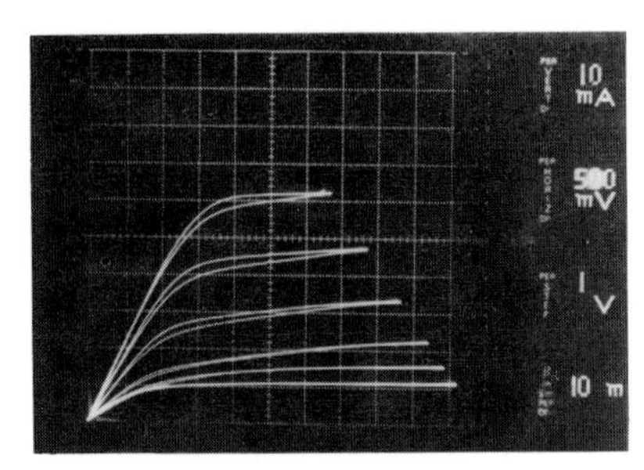

Buffer 0.1μm

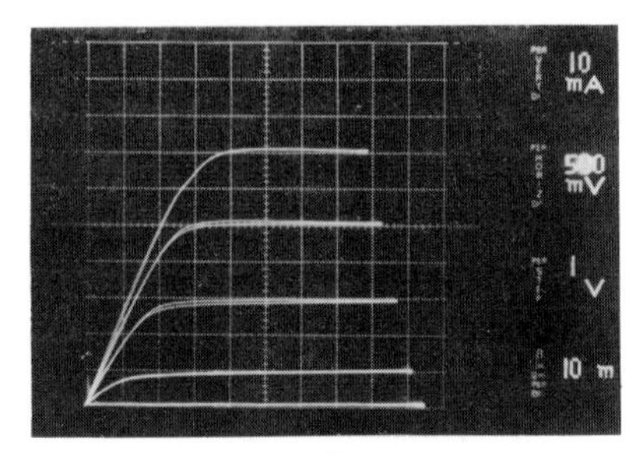

Buffer 1.0μm

Fig.4 Drain I-V Characteristics of GaAs MESFETs on InP substrates.

5. SUMMARY

The successful growth of GaAs on InP semi-insulating substrates, and the fabrication of GaAs MESFETs have been demonstrated. In spite of a large lattice mismatch, single crystal GaAs can be obtained on InP substrates. The n-GaAs has good electric properties, when a 0.5μm or more thick buffer layer is grown on the InP substrate. Deep traps could also be found but their concentrations were very low. The GaAs MESFETs on the InP substrates exhibited good device performances. The results indicate that further development is promising for the monolithic integration of InP, GaAs, and their related compound devices on the same chip.

ACKNOWLEDGMENT

The authors would like to thank T.Mizutani and K.Hirose for their help in MBE growth. They also thanks M.Kuzuhara for helpful discussion on DLTS, and K.Ohata, H.Sakuma for their continuing support.

REFERENCES

Asano K, Kasashra K, and Itoh T, 1987, IEEE Electron Device Lett. EDL-8, p.289-290

Chen C Y, Cho A Y, and Garbinski P A, 1985, IEEE Electron Device Lett. EDL-6, p.20-21

Choi H K, Tsaur B Y, Metze G M, Turner G W, and Fan J C C, 1984, IEEE Electron Device Lett. EDL-5, p.2077-208

Fisher R and Morkoc H, 1986, Solid State Electronics, 29, p269-271

Itoh T, Kasahara K, Ozawa T, and Ohata K, 1986, Int. Conf. Solid State Devices and Materials, p.779-780

Mizutani T, and Hirose K, 1985, Jap. J. Appl. Phys. 24, p119-122

Teng S J J, Ballingall J M, and Rosenbaum F J, 1986, Appl. Phys. Lett. 48, p1217-1219

Shibata J, Nakao I, Sasai Y, Kimura S, Hase N, and Serizawa H, 1984, Appl. Phys. Lett, 45, p191-193

Suzuki A, Itoh T, Terakado T, Kasashara K, Asano K, Inomoto Y, Ishihara H, Torikai T,Fujita S, 1987, Electronics Lett. 23, p954-955

Epitaxial growth and characteristics of Fe-doped InP by MOCVD

K. Nakai, O. Ueda, T. Odagawa. T. Takanohashi, and S. Yamakoshi
FUJITSU LABORATORIES, ATSUGI
10-1 Morinosato-Wakamiya, Atsugi 243-01, Japan

ABSTRACT: Fe-doped InP epitaxial layers were grown by MOCVD, using $Fe(C_5H_5)_2$ as the dopant, and their characteristics were systematically investigated. The maximum concentration of the electrically active Fe atom (deep acceptor level) was found to be 7×10^{16} cm^{-3}, which is corresponds to the Fe solibility at the growth temperature (650°C). The doping condition of a precipitation-free semi-insulating layer with a maximum resistivity of over 10^9 ohm•cm has been obtained.

1. INTRODUCTION

For high speed optical and electronic device applications, epitaxial growth of semi-insulating InP has been a very important technique. Several studies on the Fe-doped epitaxial InP have been reported (Long 1984, Tanaka 1985, Kondo 1985, Speier 1986, Huang 1986, and Kato 1987). The semi-insulating layers with a resistivity of more than 10^9 ohm•cm has been obtained by MOCVD. However, iron-phosphorous precipitates have been observed in this layer (Nakahara 1985 and Chu 1985). There has been a large discrepancy between the Fe concentration determined from SIMS analysis and the deep level density estimated from the current voltage characteristics (Sugawara 1987). The relationship between the characteristics of the Fe-doped layers and their doping condition has never been clear.

In this paper, the characteristics of the Fe-doped layers are systematically investigated and the doping conditions to obtain high quality semi-insulating InP epitaxial layers are clarified.

2. EXPERIMENTAL

InP epitaxial layers were grown by MOCVD using a PH_3/TMI/H_2 system under an atmospheric pressure. Vertical quartz reactor and carbon susceptor were used. The growth temperature was mainly 650 ° C. The molar ratio of PH_3 to TMI was about 20. The electron concentration of nondoped layers was about 2×10^{15} cm^{-3}. Vapor of ferrocine ($Fe(C_5H_5)_2$) was used as the Fe source. The Fe-doping gas, which passed through the container of $Fe(C_5H_5)_2$ at 20 ° C, was first diluted by H_2 and then introduced into the reactor, which makes it possible to perform a very light and precise Fe-doping. The background electron concentration for the Fe-doping was controlled with silicon donor using SiH_4 gas of 20 ppm.

Fig. 1 shows a depth profile of the electron concentration measured with the C-V method, using the Semiconductor Profile Plotter(Ambridge 1980) Keeping the donor concentration at a certain level by Si-doping, Fe was intermittently doped. From the difference of the electron concentrations with and without the Fe-doping, the compensator density due to the Fe-doping was evaluated, as shown with (ΔN) in Fig. 1. For a lot of samples, similarly doped with various doping levels of Si and Fe, the compensator densities were determined.

The deep level in the partially compensated layer was measured with admittance deep level transient spectroscopy (DLTS)(Takanohashi 1984), using p-n junction diode fabricated near the epitaxial surface by Zn diffusion. The total Fe concentration in the epitaxial layer was determined from SIMS analysis, which was performed by Charles & Evance Co Ltd. The observation of the precipitates was carried out with a transmission electron microscope (TEM) of AKASHI EM-002B. The specimens were prepared by chemical etching with Br-C_2H_5OH solusion. The resistivity of the Fe-doped layer was measured with the current-voltage method using an n-SI-n diode structure (Sugawara 1986).

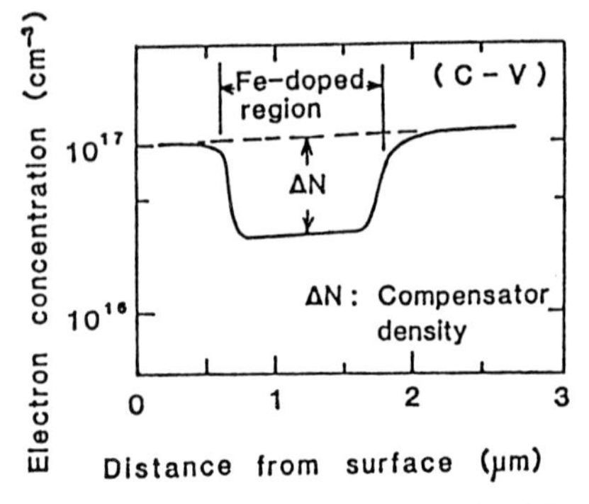

Fig. 1 The carrier concentration depth profile of the Fe-doped layer.

3. RESULTS and DISCUSSION

Fig. 2 shows the concentration of the electrically active Fe atom, which is assumed to be equal to the compensator density, as a function of the flow rate of the Fe-doping gas. The deep level density of Fe are also indicated in this figure.

Fig. 3 shows a typical DLTS spectrum of the layer partially compensated with Fe. On this spectrum, signal from a deep electron trap is observed near 40° C. The activation energy of the deep trap was evaluated to be about 0.63 eV, which almost agrees with that of the Fe deep level (Iseler 1979). No other deep level was observed. As shown in Fig. 2, the deep level density closely matches the Fe concentration determined from compensator density. Therefore, it is concluded that the compesation by the Fe doping is caused by the Fe deep acceptor

In the low doping range, the Fe concentration increases with the doping gas flow rate. However, in the high doping range, the Fe concentration hardly depends on the gas flow rate. The concentration of the electrically active Fe atom is saturated at 7 x 10^{16} cm^{-3}. Therefore, in order to obtain highly resistive layers, the background electron concentration should be

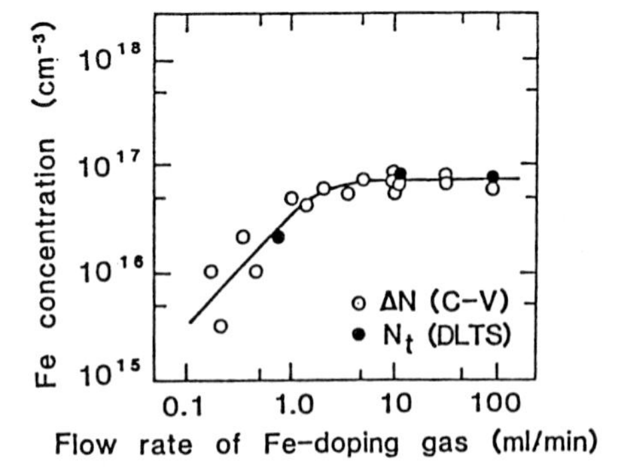

Fig. 2 The electrically active Fe concentration as a function of the Fe-doping gas flow rate.

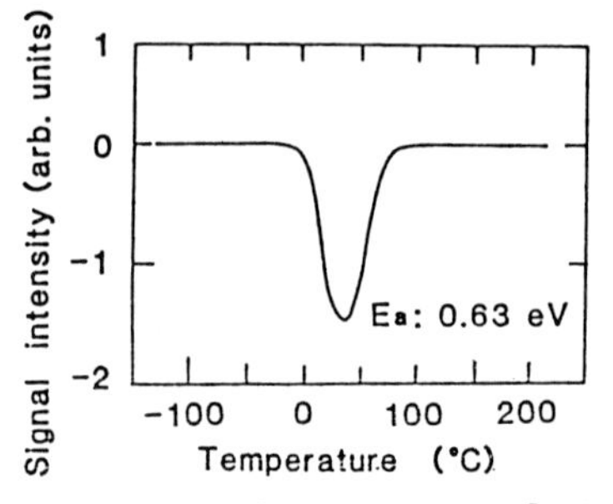

Fig. 3 DLTS spectrum of the partially compensated layer.

less than 7×10^{16} cm^{-3}.

Fig. 4 shows the total Fe concentration determined from SIMS analysis. The total Fe concenration is proportional to the doping gas flow rate up to the Fe concentration over 10^{18} cm^{-3}. It should be mentioned that, in high doping range, there is a discrepancy between the Fe concentation determined from the electrical measurements and from SIMS analysis. This means that Fe atom becomes electrically inactive in heavily doped layers.

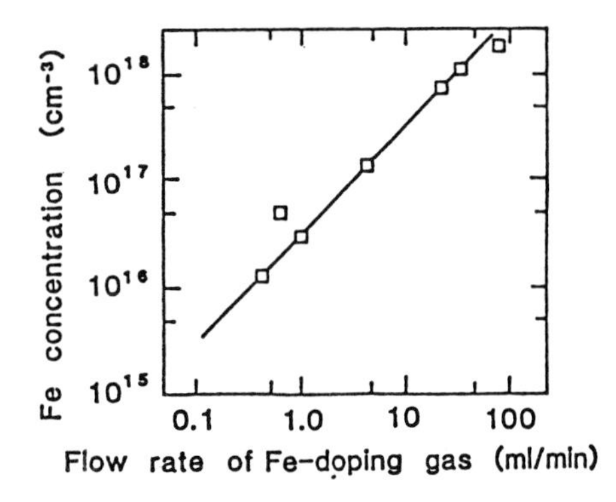

Fig. 4 The total Fe concentration determined from SIMS analysis.

Fig. 5 shows TEM images of epitaxial layers doped with various Fe-doping levels. In the layers doped with Fe-doping gas of more than 20 ml/min, spherical precipitates with a diameter of about 10 nm, which are considered to be iron-phosphorous (Nakahara 1984), are observed. Their density increases from 8×10^{12} to 1×10^{14} cm^{-3}, as the Fe-doping gas flow rate increases from 20 to 100 ml/min, while the size is almost constant. In the layers doped less than 5 ml/min, the precipitates are not observed. The flow rate of 5 ml/min corresponds to the Fe concentration of about 10^{17} cm^{-3}. Therefore, when the doped Fe concentration exceeds 10^{17} cm^{-3}, the excess Fe atoms form the iron-phosphorous precipitates.

Fig. 6 shows the comparison between the saturated concentration of the electrically active Fe atom in this MOCVD and the Fe solubility data obtained from diffusion experiments (Shishiyanu 1976). As shown in this figure, the saturated Fe concentration corresponds to the Fe solubility at the growth temperature. It has been clarified that the concentration of the substitutional Fe atom is determined by the thermal equilibrium level even in the case of the growth by MOCVD.

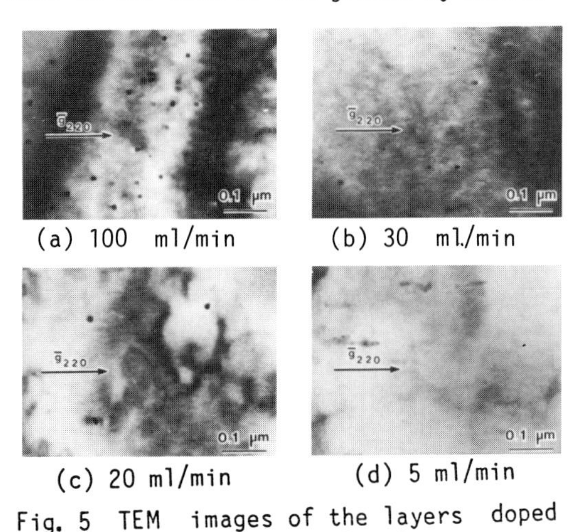

(a) 100 ml/min (b) 30 ml/min

(c) 20 ml/min (d) 5 ml/min

Fig. 5 TEM images of the layers doped with various Fe doping level.

Fe concentration (cm^{-3})
10^{19}
10^{18}
10^{17}
10^{16}
● Diffusion (Shishiyanu)
○ MOCVD (This work)
0.6 0.8 1.0 1.2
$10^3/T$ (K^{-1})

Fig. 6 The comparison between the electrically saturated Fe concentration and the Fe solubility.

Fig. 7 shows the dependence of the resistivities of Fe-doped layers on the Fe-doping gas flow rate, of which the background electron concentration are 2×10^{15} and 1×10^{16} cm^{-3}. Solid and dotted lines are the calculated resistivities under the thermal equilibrium condition, with and without taking account of the Fe solubility, respectively. In this calculation, the activation energy of the Fe deep level, the mobilities of electron and hole are assumed to be 0.63 eV, 3000 and 100 cm^2/Vsec, respectively.

As shown in this figure, the measured resistivities agree with the calculated resistivities with taking account of the Fe solubility. The resistivities become constant in the doping level over the solubility point.

From these results, the optimum doping condition is considered to be near the point indicated with arrow A, where the layer has the background electron concentration of about 2 x 10^{15} cm^{-3} and the highest deep level density of 7 x 10^{16} cm^{-3}. Furthermore, no precipitates could be introduced and a maximum resistivity of over 10^9 ohm•cm was achieved.

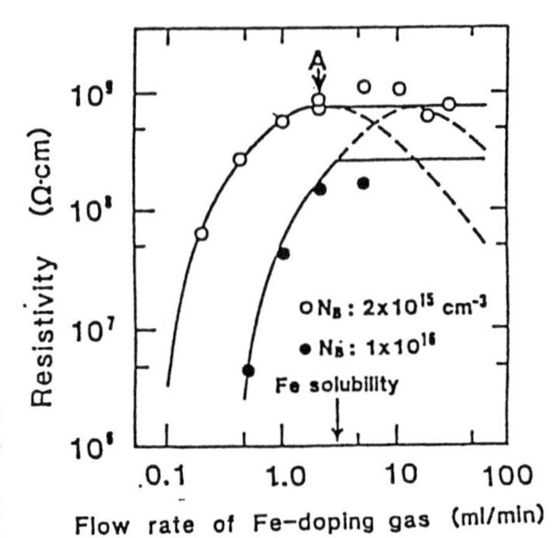

Fig. 7 The dependence the resistivity on the Fe-doping gas flow rate. Solid and dotted lines are the calculated values with and without taking account of the Fe solubility, respectively.

4. SUMMARY

In the Fe-doped InP epitaxial layers grown by MOCVD, the electrically active Fe concentration is saturated at about 7 x 10^{16} cm^{-3}, which corresponds to the Fe solubility at the growth temperature (650°C). The excess Fe atoms doped over the solubility point form the precipitates. The optimum doping condition of a precipitation-free semi-insulating layer with a maximum resistivity of over 10^9 ohm•cm, has been obtained.

Authors thank Mrs. M. Sugawara, K. Tanaka, Drs. K. Kitahara, K. wakao K. Nakajima and T. Sakurai for helpful discussions and for their support.

REFERENCES

Ambridge T, Stevenson J L and Redstall R M 1980 J.Electrochem.Soc.127 222
Chu S N G, Nakahara S, Long J A, Riggs V G, and Johnston W D,Jr, 1985 J. Electrochem. Soc. 132 2795
Huang H and Wessels B W 1986 J. Appl. Phys. 60 4342
Iseler G W 1979 Inst. Phys. Conf. Ser. No 45 pp 144
Kato Y, Kasahara K, Sugou S, Yanase T, and Henmi N 1987 Ext.Abst.of the 19th Conf.on Sol.Stat.Dev. and Mat. Tokyo pp 95
Kondo M, Sugawara M, Yamaguchi A, Tanahashi T, Isozumi S and Nakajima K, 1986 Ext.Abst.of the 18th Conf. on Sol.Stat.Dev. and Mat. Tokyo pp 627
Long J A, Riggs V G and Johnston W D,Jr. 1984 J. Crystal Growth 69 10
Nakahara S, Chu S N G, Long J A, Riggs V G and Jhonston W D,Jr 1985 J. Crystal Growth 72 693
Shishiyanu F S,Gheorghiu V Gh and Palazov S K 1977 Phys.Stat.Sol.(a) 40 29
Speier P, Schemmel G and Kuebart W 1986 Electron. Lett. 22 1216
Sugawara M, Kondo M, Nakai K, Yamaguchi A and Nakajima K 1987 Appl. Phys. Lett. 50 1432
Sugawara M, Aoki O, Nakai K, Tanaka K, Yamaguchi A, and Nakajima K 1986 Semi-Insulating III-V Materials (c) Ohmsha Ltd. pp 597
Tanaka K, Nakai K, Aoki O, Sugawara M, Wakao K, and Yamakoshi S 1987 J. Appl. Phys. 61 4698
Takanohashi T, Komiya S, Yamazaki S and Umebu I 1984 Jpn.J.Appl.Phy. 23 L849

Atmospheric pressure OMCVD growth of high-quality InP-based heterostructures without hydrogen

A. Mircea, B. Rose, Ph. Dasté, D. Robein, B. Couchaux, Y. Gao

Centre National d'Etudes des Télécommunications - Laboratoire de Bagneux
196 avenue Henri Ravera - 92220 Bagneux - FRANCE

C. Carrière

ALCATEL-CIT/CSO, BP. 6, Nozay - 91620 La Ville du Bois - FRANCE

ABSTRACT : Helium was compared to hydrogen in the growth of InP-based heterostructures. The composition range of InGaAsP quaternaries from 1.1 to 1.65µm was explored and in all respects the He-grown structures were equal, if not superior to the H2-grown ones. As a preliminary conclusion, according to our study, helium can be successfully used in place of hydrogen. This makes possible also the realization of integral silica-on-InP structures.

1. INTRODUCTION

The use of inert gases (A, He) as carrier gas in OMCVD is an attractive alternative to hydrogen, since it minimizes the fire hazard. This is particularly interesting in the case of phosphorous-containing materials because phosphorous, as well as phosphine, ignite spontaneously in the presence of oxigen or air. This was the basic motivation of our study. Moreover, there is a scientific interest for this study in as much as the chemical reactions governing the growth are quite imperfectly known. While the global chemical balance does not involve any contribution from the carrier gas, most of the detailed models proposed in the litterature suppose that hydrogen participates to the intermediate phases of the organometallic pyrolysis, while other models emphasize the role of hydrogen as an adsorbate on the substrate surface during growth. It is, therefore, meaningful to obtain experimental data allowing one to appreciate the effective influence of the hydrogen carrier gas during growth, by comparison with growths done under similar experimental conditions, but without hydrogen, e.g. using an intert carrier gas. THe growth oqf GaAs under nitrogen has been studied in recent years by Kuech and Veuhoff (1954). They report essentially no difference on the photoluminescence properties for H2- or N2- grown layers. A more comprehensive recent study of GaAs and GaAlAs grown uder he was made by Azoulay et al (1987). While the do not observe substantial differences for GaAs, in the case of GalAlAs the photoluminescence efficiency of the He-grown layers was weaker and the oxygen content considerably larger than for the H2-grown ones.

In the case of InP and InGaAsP layers, excellent quality growths have been reported by Razeghi (1983) using a mixture of H2 and N2. To our knowledge, the only published information concerning growths of these materials without hydrogen is in our previous report (Mircea et al 1986a).

2. EXPERIMENTAL

The growths were carried out in two different reactors. Reactor no 1 has a conventional horizontal design ; we described it in previous papers (Sacilotti et al 1983, Mircea et al 1986a). Reactor no 2 has a special, T-shaped growth chamber, designed for extremely uniform growth on 2" substrates (Mircea et al 1986b). Both reactors used diffuser-type organometallic sources (Mircea et al 1986a) made from pyrex glass, allowing one to visually observe possible reactions of the organometallics with the carrier gas. We grew layers of InP, InGaAs, InGaAsP at 1.1, 1.2, 1.3 and 1.4μm, InP/InGaAsP double heterostructures, complete laser structures, and insulator/semiconductor heterostructures consisting in a layer of InP and a layer of SiO_2 grown one after another without interruption (Dasté et al 1987). The growth rates for InP were between 1.8 and 4μm/hr. A variety of characterization techniques was applied to these layers, including optical microscopy and defect counting, accurate thickness measurements, simple and double X-ray diffraction, photoluminescence, Hall effect, electrochemical "Polaron" profiling, secondary ion mass spectroscopy (SIMS) and Auger profiling. The complete laser heterostructures were evaluated by making broad-area lasers. The SiO_2/InP heterostructures were evaluated by ellipsometry and by I(V) and C(V) electrical measurements.

3. LAYER CHARACTERIZATION RESULTS

The growth rate under helium was generally smaller than under hydrogen. Keeping constant the organometallic temperatures, in the T-shaped reactor set-up the growth rate ratio was about 0.8/1 in He/H_2. For the horizontal reactor, the ratio was 0.63/1, very nearly the theoretically expected value $(D_{He}/D_{H2})^{3/2}$. The growth rate uniformity in the T-shaped reactor was slightly better with helium : standard deviations measured over 48 mm diameter on 2" substrates were 0.8 percent in the former case (fig. 1) against 1.5 percent in the latter. The morphology was basically featureless in both cases for optimum growth conditions.

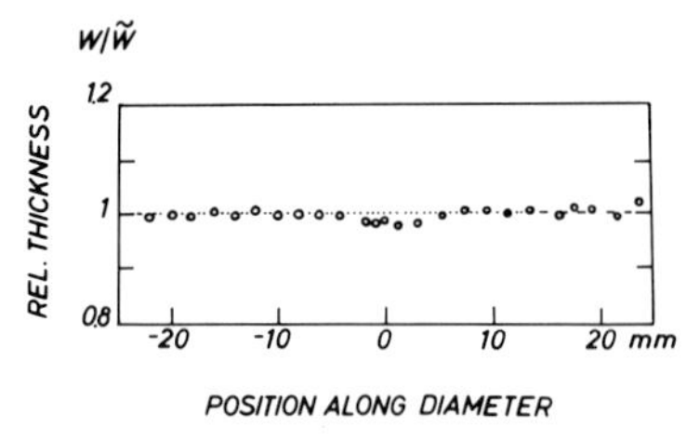

Fig. 1 : Thickness uniformity on 2" substrates.

Double X-ray diffraction data, as shown in fig. 2, for quaternary layers at 1.1 and 1.3μm about 1μm thick, revealed linewidths of 35-45 sec for the layer and 15-20 sec for the substrates. Both these values are 1.5 to 2 times larger than the ideal theoretical ones (Halliwell et al 1984) which have also been closely approached experimentally (Bocchi et al 1987) by LPE, but they are still among the best reported in OMVPE (Nelson et al 1986), GSMBE (Panish et al 1985, Huet and Lambert 1986) or CBE (Tsang et al 1987) at these compositions. Although much more layers were grown under hydrogen than under helium, the best linewidths were observed on the latter ones.

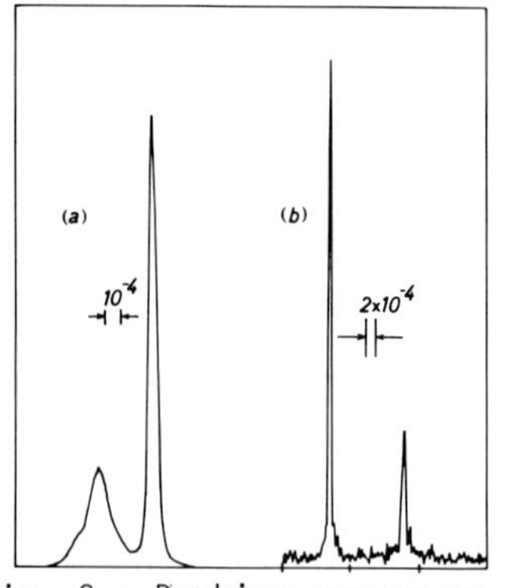

Fig. 2 : Rocking curves on quaternaries at (a) 1.1μm, (b) 1.3μm compositions.

The photoluminescence spectra observed at 300 K on the 1.1, 1.2 and 1.3μm compositions were very intense, stronger than those seen on LPE layers of similar composition. The highest intensities were observed on layers grown under He. The line-widths at 1.1 m were about 40 to 50 meV, at 1.3μm about 50 to 60 meV. The layers purity was adequate for most applications : at 9 K, an undoped InP layer from an integral SiO2/InP heterostructure showed a single, 3.5 meV wide donor-to-valence band line, very little compensation, no carbon. This was confirmed by Hall effect measurements at 77 K on the same layer which yielded a mobility of 85,000 cm^2/Vs with n = $9.10^{14} cm^{-3}$ (Fig.3).

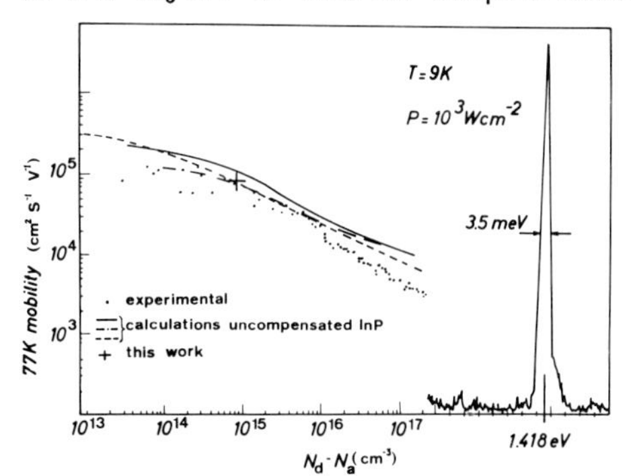

Fig. 3 : Characterization of undoped InP layer (a) Hall effect, (b) photoluminescence.

Secondary ion mass spectroscopy on InP and quaternary layers grown either under hydrogen or helium gave similar results, in that for both, the oxygen content was below the instrumental detection limit of about $10^{17} cm^{-3}$.

4. DEVICE CHARACTERIZATION RESULTS

Preliminary device characterization results for helium-grown structures were obtained for insulator/semiconductor (silica/InP) heterostructures grown in the T-shaped reactor and for 1.3μm laser double heterostructures grown in the horizontal reactor. The I/S structure was obtained in a single thermal cycle without interruption between the InP layer and the silica layer. This kind of process cannot be done with hydrogen due to the explosion hazard of the hydrogen-oxygen mixture. Promising results were obtained concerning the electrical and thermo-mechanical properties of these device structures. Fig. 4 shows the C(V) characteristics of MIS capacitors made with this process and, for comparison, with silica deposited on silicon substrates in the same reactor and operating conditions. The silica/InP structures withstood successfully thermal treatments up to about 800°C.

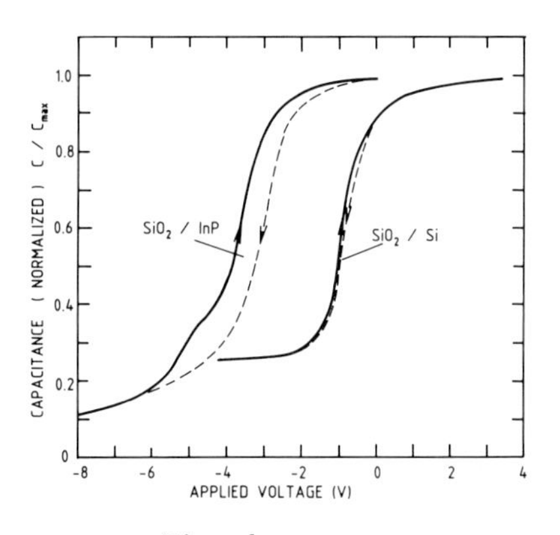

Fig. 4

Many 1.3μm laser heterostructures were grown in the horizontal reactor under hydrogen and evaluated as broad-area 250μm long lasers. The average threshold current density, was 1.7 kA/cm^2 on the best wafer ; the cumulated value, over the last twelve wafers before the helium experiment, was 2.9 kA/cm^2. One of these wafers, having an average broad-area threshold of 2.7 kA/cm^2, was processed on a production line, which normally used LPE/LPE wafers, to obtain buried-heterostructure (BH) lasers, applying the standard second LPE step. The threshold current was 31 ± 7 mA, which compared favorably with the cumulated production values of the line. The lasers withstood successfully the standard high-temperature degradation test of the line.

Some of these lasers have been placed on the lifetest bench. At the time of writing, only one laser wafer grown under helium, with the same reactor and growth conditions, was evaluated as broad-area lasers ; the average threshold was 2.4 kA/cm^2.

5. CONCLUSION

According to our study, helium can be used in place of hydrogen, for growing InP based device structures. The composition range from 1.1 to 1.65 m was explored, albeit not exhaustively and in all respects the He-grown layers were as good, if not better than the H2-grown ones. He-grown lasers had characteristics similar to the H2-grown ones. The chemical stability of helium is a factor which contributes to the overall safety of the process and makes possible some interesting developments, such as the realization of integral silica-on-InP structures.

6. REFERENCES

Anderson DA and Apsley N, 1986 Semicond. Sci. Technol. 1 187.
Azoulay R. Dugrand L, Gao Y and Leroux G, 1987 3rd Biennial OMVPE Workshop Cape Cod (USA), Sept. 1987.
Bocchi C, Ferrari C, Franzosi P, Fornuto G, Pellegrino S and Taiariol F, 1987 J. Electron. Mater 16 245.
Dasté P, Couchaux B, Mircea A and Ossart P, 1987 1st European Workshop on MOVPE, Aachen (Germany) March 29, 1987.
Halliwell MAG, Lyons MH and Hill MJ, 1984 J. Cryst. Growth 68 523.
Huet D and Lambert M, 1986 J. Electron. Mater. 15 37.
Kuech TF and Veuhoff E, 1984 J. Crystal Growth 68 148.
Mircea A, Mellet R, Rose B, Robein D, Thibierge H, Leroux G, Dasté P, Godefroy S, Ossart P and Pougnet AM, 1986a J. Electron. Mater 15 205.
Mircea A, Mellet R, Rose B, Dasté P and Schiavini G, 1986b J. Crystal Growth 77 340.
Nelson AW, Moss RH, Spudens PC, Cole S and Wong S, 1986 Br Telecom Technol. J 4 85.
Panish MB, Temkin H and Sumski S, 1985 J. Vac. Sci. Technol. B3 657.
Razeghi M, 1983 Revue Technique Thomson-CSF, 15.
Rose B, Robein D, Mircea A, Devoldere P, Leroux G and Thibierge H, 1987 1st European Workshop on MOVPE, Aachen (Germany), March 29, 1987.
Sacilotti, M. Mircea A and Azoulay R, 1983 J. Crystal Growth 63 111.
Tsang WT, Schubert EF, Chiu TH, Cunningham JE, Burkhardt EG, Ditzenberger JA and Agyekum E, 1987, Appl. Phys. Lett. 51 761.

Paper presented at Int. Symp. GaAs and Related Compounds, Heraklion, Greece, 1987

MO VPE growth and doping of (AlIn)As on InP for optoelectronic devices

M.Druminski and R.Gessner

Siemens Research Laboratories, Otto Hahn Ring 6,
D8000 Munich, FRG

ABSTRACT: $Al_{1-x}In_xAs$ layers were grown at 630 °C on InP substrates at atmospheric pressure by MO VPE. The composition was studied for x = 0.47 to 0.54. The layers were characterized by photoluminescence (PL), x-ray diffractometry, Hall and SIMS measurements. A linear correlation between the 300K PL peak energy E_{PL} and the In content was found (E_{PL} = 2.606-2.188x). Lattice matched (AlIn)As with $0 \leq a/a \leq 5 \times 10^{-4}$ could be grown reproducibly. As n-type and p-type dopant sources SiH_4 and Cp_2Mg were used. For Si, SIMS measurements indicate negligible diffusion. For Mg a strong dependence of the carrier concentration on the V/III ratio was observed. From SIMS results, the diffusion coefficient of Mg in (AlIn)As was estimated being around $1 \times 10^{-12} cm^2 s^{-1}$ at 630°C.

1. INTRODUCTION

(AlGaIn)As is a promising semiconductor material for optoelectronic applications in the wavelength range between 880nm and 1660nm, because it is an attractive alternative to the "conventional" quaternary system (GaIn)(PAs) (Olego et al 1982). Better control over alloy composition is expected because only one hydride (AsH_3) is used in the case of metal organic vapor phase epitaxy (MO VPE). We have studied the MO VPE growth of (GaIn)As and (AlIn)As for the potential fabrication of lasers on InP substrates. For this purpose n-and p-doped (AlIn)As layers are required as confinement layers. In this paper, we present results on growth and doping of (AlIn)As on InP substrates.

2. EXPERIMENTAL

The experiments were carried out at atmospheric pressure in a reactor described previously (Druminski et al 1986). The trimethyl compounds of the group III elements and arsine were chosen as precursors in a hydrogen carrier gas at a gas flow of 33slm (gas velocity: 0.5 ms^{-1}). SiH_4 and Cp_2Mg were used as n-and p-dopants, respectively. (100) oriented InP:Fe slices served as substrates. The heating cycle for the substrates was carried out without PH_3. The (AlIn)As layers were grown at temperatures of 630 °C with growth rates around 4μmh^{-1} and with an In-fraction 0.47 to 0.54. The samples were evaluated by 300K photoluminescence (PL), multiple crystal diffractometry, Hall measurements and secondary ion mass spectrometry (SIMS). The concentration profiles of silicon and magnesium were determined by means of SIMS using a CAMECA IMS 3f ion microscope. For the evaluation of magnesium and silicon O_2^+ and Cs^+ primary ion beams were used, respectively. The Si and Mg intensities presented below are not calibrated.

3. RESULTS AND DISCUSSION

The quality of Al-containing semiconductor alloys is extremely sensitive to traces of oxygen and water in the gas ambient used for epitaxial growth. Therefore, our epitaxial equipment was checked for the suitability for the growth of (AlIn)As at first by the production of (AlGa)As/GaAs structures. Results on oxide stripe lasers with excellent performances are published elsewhere. (Druminski et al 1986).
Then the MO VPE equipment was used for the growth of (AlIn)As on InP. Lattice matched material could be grown with $0 \leq a/a \leq 5 \times 10^{-4}$ over a distance of about 5 cm along the susceptor (Gessner et al, 1986). Undoped 3μm thick (AlIn)As layers with $a/a \leq 5 \times 10^{-5}$ show a FWHM of 29 arc sec in the x-ray rocking curves. A linear relationship between the room temperature PL peak wavelength and the lattice mismatch of (AlIn)As layers was observed, as shown in fig.1. The lattice mismatch values shown include a 0.47 correction factor for strain (Bartels et al 1983). Assuming Vegards law, the corresponding composition for (AlIn)As can be calculated.
The resulting linear correlation between the PL peak energy E_{PL} and the indium content (x) is found to be $E_{PL} = 2.606 - 2.188x$ for the investigated composition range. For exactly lattice matched material (x= 0.523) $E_{PL} = 1.462$ eV is found. Our results are in very good agreement with those derived from cathodoluminescence measurements for comparable MBE material (Davies et al 1984). The correlation given above applies for undoped as well as for doped (AlIn)As, as shown in fig.1.

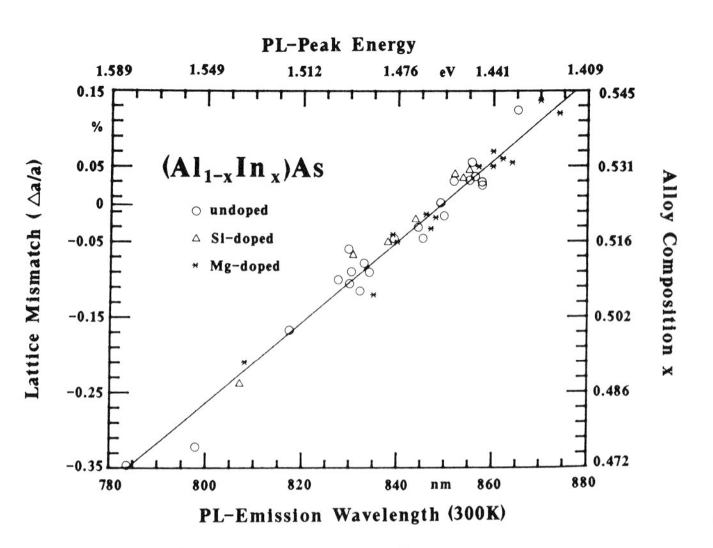

Fig.1 : Correlation between PL peak wavelength/-energy and lattice mismatch/In-content of (AlIn)As

The doping of the layers was investigated only in the range of interest for the potential fabrication of lasers. Two different V/III ratios (20,50) were investigated. For Si doped layers, the electron concentration exhibits a linear dependence on the SiH_4/(TMAl + TMIn) ratio. No influence on the Si incorporation due to the V/III ratio was observed. Unintentionally doped 3μm thick (AlIn)As layers were n-type showing Hall mobilities around $2500cm^2/Vs$ for carrier concentrations of $1-5 \times 10^{15}cm^{-3}$.

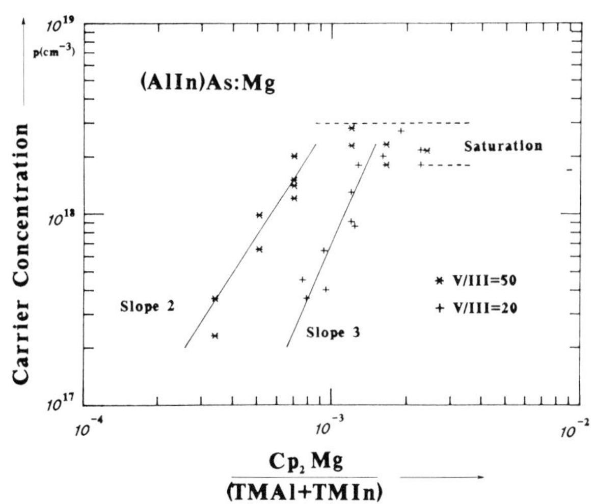

Fig.2:Dependence of carrier concentration on the Mg/(Al+In) flow for two V/III ratios

For the doping of (AlIn)As with Cp_2Mg, a remarkable influence of the V/III ratio was observed. Fig.2 shows for the two investigated V/III ratios the relationships of(N_A - N_D) on the Mg/(Al + In) flow. An increase of the flow leads to a cubic law dependence of the carrier concentration for a V/III = 20 and to a square law dependence for a V/III = 50 ratio. The higher Mg-incorporation efficiency for the higher V/III ratio can be interpreted in terms of an increasing density of unoccupied group III surface sites. The two different superlinear slopes for the different V/III ratios are not understood at this time. The saturation range for electrically active Mg in (AlIn)As appears to be within the range $2-3x10^{18}cm^{-3}$.

In fig.3 SIMS profiles for an (AlIn)As/(GaIn)As DH - structure are shown. The thicknesses for the Si-doped and Mg-doped (AlIn)As confinement layers and the (GaIn)As active and top layer are 1.5μm and 0.2μm, respectively. To identify the position of the interfaces,the ^{27}Al signal was recorded simultaneously. The ^{30}Si intensity drops from a position of about 0.2μm before the active layer, corresponding to an intentional stop of the silane flow during the growth process. The sharp decrease of the ^{30}Si signal indicates a negligible diffusion for silicon in (AlIn)As. The ^{27}Mg signal in the upper confinement layer of the DH Structure

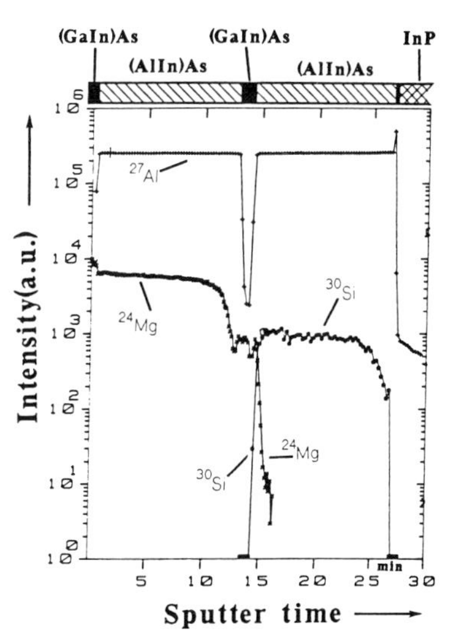

Fig.3:SIMS profiles of an (AlIn)As/(GaIn)As - DH structure

approaches a constant intensity at a position about 0.3μm far from the (GaIn)As layer. This was not intended during the growth, because the Cp_2Mg flow was started together with the growth of the second (AlIn)As layer. The delay of the Mg incorporation may be due to an strong adsorption of Cp_2Mg at the surface of the stainless steel tubes in the epitaxial equipment (Roberts et al 1984). However ,the diffusion of Mg in (AlIn)As appears to be high, because the ^{24}Mg signal is observed even in the first (AlIn)As layer. The increase of the ^{24}Mg intensity in the active (GaIn)As region has to be interpreted as an matrix effect. The diffusion coefficient of Mg in (AlIn)As was roughly estimated assuming the beginning of the diffusion at the position where the Mg intensity has reached a nearly constant value. This resulted for a lot of SIMS profiles in a diffusion coefficient around $1x10^{-12}cm^2s^{-1}$ for Mg in (AlIn)As at 630 °C.

3. CONCLUSION

We have demonstrated the growth of (AlIn)As layers of high crystallographic, electrical and optical quality using MO VPE for the growth process. SiH_4 is an appropriate n-type dopant for (AlIn)As. The use of Cp_2Mg for p-type doping of laser structures might serve satisfactorily. However, for devices requiring steep dopant profiles and reproducible dopant concentrations well below $10^{18}cm^{-3}$, other dopant precursors should be used. Studies for alternative p-type dopant sources are in progress.

4. ACKNOWLEDGEMENT

The autors are indepted to Dr.H.Goebel and Dr.M.Schuster for the x-ray diffractometry measurements, Dr.R.Treichler for the SIMS analysis, Mrs.M.Beschorner for the PL-and Hall-measurements and Mr.O.Schlabbach for his help preparing this manuscript. Stimulating discussions with Dr.E.Veuhoff are highly appreciated. This work has been supported under the technological program of the Federal Department of Research and Technology of the Federal Republic of Germany. The authors alone are responsible for the contents.

REFERENCES

Bartels WJ 1983 J.Vac.Sci. Technol.B1(2)338
Davies GJ, Kerr T, Tuppen CG, Wakefield B and Andrews DA. 1984 J.Vac.Sci.Technol.B2(2)219
Druminski M, Gessner R, Kappeler F, Westermeier H, Wolf HD and Zschauer K-H 1986 Jpn.J.Appl.Phys.25L17
Gessner R and Druminski M 1986 presented at ICMOVPE III April 13-17,Universal City,California USA
Olego D, Chang TY, Silberg E, Caridi EA and Pinczuk A 1982 Appl.Phys.Lett.41 476
Roberts JS and Mason NJ,1984 J.Cryst.Growth. 68 422

Growth of InP and GaInAsP by GSMBE for heterostructure lasers

L. Goldstein, M. Lambert, B. Fernier, D. Bonnevie, C. Starck, M. Boulou

Laboratoires de Marcoussis, CR-CGE, Route de Nozay, 91460 Marcoussis, France

ABSTRACT : We have studied the growth conditions of InP and GaInAsP lattice matched to InP by gas source molecular beam epitaxy (GSMBE). The excellent quality of the layers has been confirmed by the realization of quantum wells and of double heterostructure lasers with low threshold current density ($J_{th} \simeq 1.5$ kA/cm2) from which low threshold (I_{th}=25mA) buried heterostructure lasers have been fabricated. Preliminary results on GSMBE regrowth on engraved substrate will be presented.

1. INTRODUCTION

Quaternary alloys of InGaAsP lattice matched to InP is of particular interest for many optoelectronic devices used in optical communications. Recent development of GSMBE has led to the fabrication of high quality layers of InP and GaInAsP. A wide range of structures has been grown using this technique including double heterostructures for lasers (Panish, 1984) as well as quantum well (Temkin, 1985) and superlattices (Vandenberg, 1986).

2. GSMBE of $Ga_xIn_{1-x}As_yP_{1-y}$ (y = 2.2 x)

The growth has been performed in a RIBER 2300 system equipped with a PBN cracking cell to provide As_2 and P_2 fluxes from thermal decomposition of AsH_3 and PH_3 (Huet, 1986). We have studied the growth conditions of quaternary alloys lattice matched to InP over the whole range of composition ($y = 2.2\ x$, $0 < y < 1$).
The nucleation mechanisms are very similar in GSMBE and in conventional MBE : the sticking coefficients of Ga and In are equal to unity for substrate temperatures lower than the reevaporation limit of III elements, and the growth rate is not dependent of the V elements fluxes. Ga and In fluxes are calibrated using an ion gauge, whereas AsH_3 and PH_3 flows are regulated by mass flow controllers. The variation of y as a function of the arsine to total hydride flux ratio is shown on figure 1. For As rich alloys ($y > 0.8$, $\lambda > 1.5\mu m$) the ratio of the V elements is almost identical in the gas phase and in the solid phase, whatever is the substrate temperature.
For higher phosphorus content alloys ($y < 0.6$, $\lambda < 1.3\mu m$), the relative sticking coefficient of As to P increases as the relative amont of phosphorus in the flux increases. In this case, we have observed a stronger arsenic incorporation when the substrate temperature is reduced :

i.e., for a As/P fluxes ratio = 5.7 %, y = 26 % for T_s = 540°C and y = 35 % for T_s = 480°C.
The uniformity of quaternary layers grown on 2" substrates is better than 5 % in thickness and 10^{-3} in Δa/a. These layers show a low background carrier concentration (n < 2.10^{15} cm^{-3}), a high mobility at 77K, (μ = 25000 cm^2/V.s for λ = 1.55μm), photoluminescence properties comparable to LPE materials and a low defect density.

3. HETEROSTRUCTURES

3.1 GaInAs/InP quantum wells

The abruptness of interfaces of GSMBE heterostructures has been investigated on GaInAs single quantum wells with different well thicknesses (1.5 nm < t < 10 nm) between InP barriers. The figure 2 shows the photoluminescence at 6 K of a structure with 4 different wells. Typical full width at half maximum is 5.5 meV for a well thickness of 10 nm and 10.5 meV for 1.5 nm.
For a well thickness lower than 10 nm, the photoluminenscence line presents two or three peaks. Such a behaviour which has been observed in the GaAs/GaAlAs system has not yet been reported in the GaInAs/InP system and can be interpreted as follows : if the interface exhibits steps with lateral dimensions larger than the exciton radius, then wells with different thicknesses can be observed experimentally. The energy difference between the peaks of figure 2 indicate fluctuations of the well thickness of one atomic monolayer.

3.2 GaInAsP/InP heterostructures for 1.3μm - 1.5μm lasers

Standard double heterostructures (DH) and separate optical and carrier confinement heterostructures (SCH) suitable for laser emission at 1.3μm and 1.5μm have been grown at 540°C. A large number (20) of heterostructures have been characterized by pulse measurements on broad area (100μm wide, 400μm long) lasers. A typical threshold current density J_{th} of 1.7 $kA.cm^{-2}$, with a minimum value of 1.5 $kA.cm^{-2}$ is currently obtained for DH with a 0.15μm thick active layer. A further reduction of J_{th} can be obtained using SCH with a thinner active layer (0.12μm) ; in this case the mean value of J_{th} is 1.5 kA/cm^2 with a minimum value of 1.3 kA/cm^2. The uniformity of the active layer and its reproducible control has been confirmed by the low dispersion of the lasing wavelength on the same wafer (2 nm) and the good reproductibility from run to run (4 nm). Additionnal measurements of J_{th} and external efficiency η_{ext} as functions of the cavity length L give further evidences of the active layer homogeneity and interface quality : J_{th} as low as 1.1 kA/cm^2 for L = 800μm, η_{ext} = 45 % for L = 250μm are currently obtained for 1.5μm DH with a 0.15μm thick active layer.

4. LPE/GSMBE BH LASERS

Buried heterostructure (BH) lasers have been fabricated with GSMBE 1.5μm DH wafers using standard chemical etching and a LPE regrowth processes (Benoit, 1983).
The good uniformity of the GSMBE wafers allows a high fabrication yield and a low dispersion of the characteristics of the BH laser as shown on figure 3. These lasers exhibit a low threshold current (25 mA) and a good linearity up to 15 mW at 20°C (Fernier, 1987). Maximum temperature in excess of 100°C for DC operation is curently obtained.

These lasers are characterized by clean far field patterns and regular spectrum which confirm the high homogeneity of the active layer. Preliminary aging tests have been performed at 70°C under 5 mW constant emitted power conditions. After 1200 h the relative increment of the threshold current at 70°C is less than 15 % for 70 % of the aged lasers. This indicates that no special failure mechanism relevant to GSMBE can be detected.

5. GSMBE REGROWTH ON ENGRAVED SUBSTRATES

The fabrication of advanced laser structures such as low threshold BH lasers and DFB lasers require high quality regrowth on engraved materials. GSMBE InP regrowth on a grating and etched mesa stripes are shown on figure 4. Second order grating for 1.5µm laser with grating pitch of 480 nm and depth of 150 nm have been fabricated on quaternary layers (λ_g = 1.3µm) onto which InP regrowth has been performed. By a proper choice of the regrowth conditions, it is possible to achieve a planar regrowth even for thin layers (0.2µm), without significant alteration of the grating depth.
Preliminary experiments on the regrowth of InP layers onto chemical etched mesa stripes along the [110] direction on (001) substrate indicate very uniform regrown layers and interesting possibilities for all GSMBE BH fabrication.

6. CONCLUSIONS

The high quality and uniformity of $Ga_xIn_{1-x}As_yP_{1-y}$ (y = 2.2 x) materials and heterostructures grown by GSMBE has been demonstrated from the detailed characterization of 2 D and 3 D heterostructures and from the fabrication of simple laser structures with excellent characteristics. GSMBE technique appears very promising either to improve the production yield of standard components, such as 1.3µm - 1.5µm BH lasers, or to fabricate advanced electronic and optoelectronic components and IC's.

7. ACKNOWLEDGMENTS

The authors wish to thank J. Benoit for constant encouragements and helpful discussions. We also wish to thank D. Sigogne, L. Le Gouézigou, B. Bourdon, C. Artigue, Y. Louis, C. Padioleau, F. Poingt and R. Vergnaud for helpful cooperation.

8. REFERENCES

- Benoit J., et al, 9th European Conference on Optical Communication (ECOC'83), Geneve, Switzerland, pp. 35-38 (1983)
- Fernier B., et al, 17th European Solid State Device Research Conference (ESSDERC'87), Bologna, Italy, pp. 873-876 (1987)
- Huet D., Lambert M., J. Electron. Mat., 15, 37 (1986)
- Panish M.B. and Temkin H., Appl. Phys. Lett., 44, 785 (1984)
- Temkin H., Panish M.B., Petroff P.M., Hamm R.A., Vandenberg J.M. and Sumski S., Appl. Phys. Lett., 47, 394 (1985)
- Vandenberg J.M., Hamm R.A., Macrander A.T., Panish M.B., Temkin H., Appl. Phys. Lett., 48, 1153 (1986)

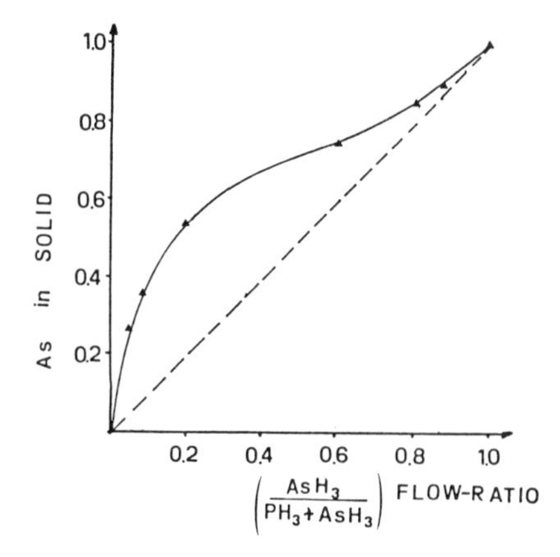

Fig. 1. Arsenic concentration in the solid as a fonction of the relative arsine flow ratio.

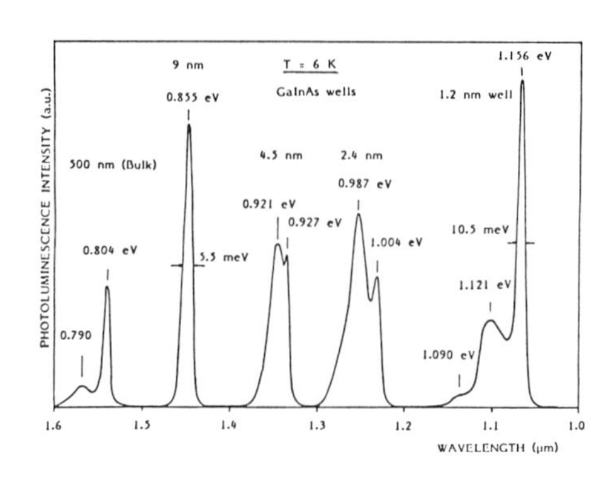

Fig. 2. Photoluminescence spectra at 6K of GaInAs/InP quantum wells.

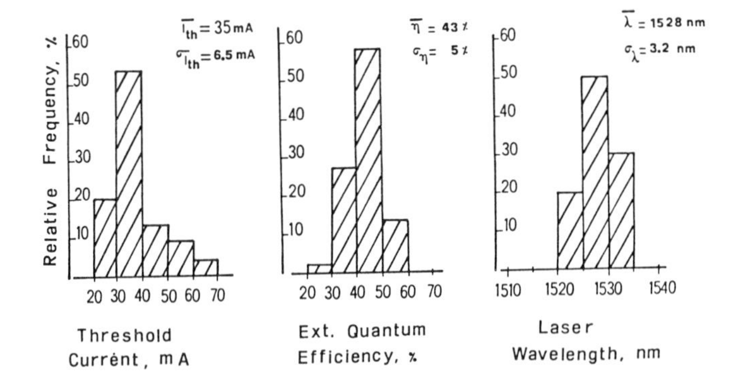

Fig. 3. Histogram of 50 BH lasers from the same wafer.

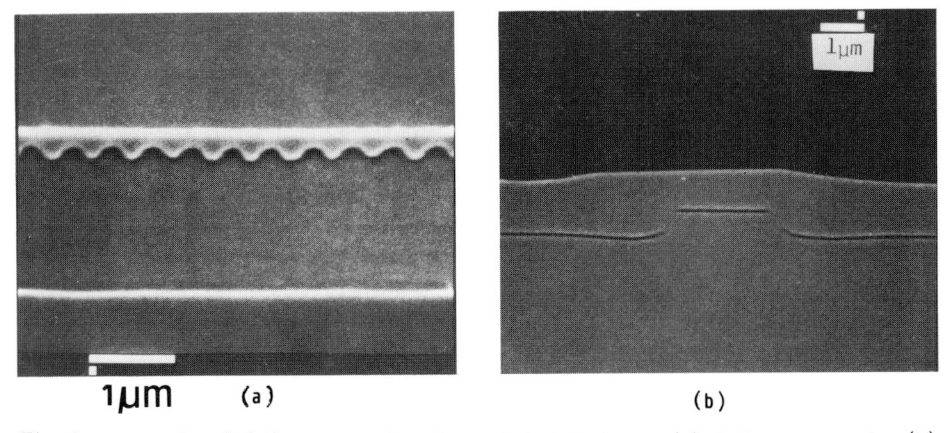

Fig. 4. regrowth of InP: on grating in a GaInAsP layer (a), InP mesa stripe (b)

Paper presented at Int. Symp. GaAs and Related Compounds, Heraklion, Greece, 1987

Theoretical and experimental studies on lattice matched and strained MODFETs

Y. Sekiguchi, Y.J. Chan, M. Jaffe, M. Weiss, G.I. Ng, J. Singh
M. Quillec* and D. Pavlidis

Center for High Frequency Microelectronics
Department of Electrical Engineering and Computer Science
The University of Michigan
Ann Arbor, MI 48109-2122

ABSTRACT: A formalism is presented for the quantum mechanical problem of charge transfer in a 2D strained n- or p-type channel. Bandmixing between heavy and light hole is retained for good understanding of mass decrease in p-MODFETs. Experimental studies confirm the theoretical predictions for n-channel strained InGaAs/InAlAs devices.

1. INTRODUCTION

Strained channel Modulation Doped Field Effect Transistors (MODFETs) have been recently attracting considerable interest as candidates for superior high speed/high frequency devices[1,2]. A much greater degree of bandstructure tailoring is possible in strained systems, potentially leading to higher band discontinuities and lighter carrier masses and thus providing superior device performance. Experimentally, both n- and p-type strained MODFETs have been fabricated and show improved performances[3,4]. The p-type device in particular shows remarkable improvement which appears to be related to the reduction in the mass of the hole gas. In this paper we provide a formalism for addressing the quantum mechanical problem of charge transfer in a 2-D strained n- or p-type channel.

To study the effect of the strained channel, we have carried out a systematic study of device performance in the $In_{0.53+x}Ga_{0.47-x}As$/ $In_{0.52}Al_{0.48}As$ (on InP substrates) system with x values of 0%, 7% and 12%. These experimental studies clearly showed the effect of increased In content on device and material properties. In section 2 we will briefly describe the theoretical formalism. The key results of the theoretical work and experimental studies are shown in section 3 and we conclude in section 4.

2. THEORETICAL ASPECTS

To understand the charge control picture in a strained MODFET, one needs to address the following three problems:

2.1. Strain effects on channel material properties

We consider the effect of strain on bandgap changes and effective mass changes. The tight binding method (TBM) is used to simulate the changes on the bandgap of materials under arbitrary strain. The effects of spin-orbit coupling were included in our tight binding model [5]. By scaling the values of TBM intergrals with the changes in interatomic distance brought on by strain, the effects of strain in the bandstructure was simulated. While the bandgap modulation effect is the same for n- or p-type MODFETs, the effective masses change in very different ways. The electron mass is found to change only about 5% with a 10% excess In concentration in the channel. For the p-type MODFET, the problem is much more complicated, and the changes brought on by strain are much more dramatic. This is due to the fact that the hole band structure is comprised of two bands, the light hole ($\Phi_{3/2,1/2}$) and the heavy hole ($\Phi_{3/2,3/2}$). These two bands couple strongly and the coupling, to a large extent, determines the effective mass of each band. In the absence of strain, the two bands are degenerate at the gamma point. However, strain will split the two hole bands. This splitting causes the energies of the two bands to change in the following manner [6]:

$$HeavyHole : E_{HH} = E_o + \frac{1}{2}\delta_{sh} - \delta_{hy} \qquad LightHole : E_{LH} = E_o - \frac{1}{2}\delta_{sh} - \delta_{hy} \qquad (1)$$

where δ_{sh} and δ_{hy} are the shear and the hydrostatic portions of the change in the band energies from the unstrained band energy position Eo. Evaluation of these equations for $In_xGa_{1-x}As$ yields shifts of $-5.96 \cdot \varepsilon$

$$\delta_{sh} = -2b\,[(c_{11} + 2c_{12})/c_{11}]\,\varepsilon \qquad \delta_{hy} = -2a\,[(c_{11} - c_{12})/c_{11}]\,\varepsilon \qquad (2)$$

for the heavy hole band and $-12.4 \cdot \varepsilon$ for the light hole band [6] where ε is the strain. In order to fully represent

Work supported by Wright-Patterson Air Force Base Contract (No. F33615-87-C-1406) and US Army Office (Grant No.DAAL03-87-U-007).

(*) Centre National d'Etudes des Telecommunications (CNET) 92200, Bagneux, France.

the hole states and the coupling between them, the Kohn Luttinger hamiltonian is used. This is described in the next section.

2.2. The Schrödinger equation

The Schrödinger equation to be solved for the electron case is quite simple since gamma valley conduction band is non degenerate, nearly parabolic, and quite well described by pure Is> type states. To model electrons in the gamma valley of a strained crystal, we can write the Schrödinger equation in the form

$$-\frac{\hbar^2}{2\overline{\overline{m}}^\star}\nabla^2\Psi_n(\mathbf{k}_\parallel, z) + V(z)\Psi_n(\mathbf{k}_\parallel, z) = E_n\Psi_n(\mathbf{k}_\parallel, z). \quad (3)$$

Here, $\overline{\overline{m}}^*$ is a tensor because of the non-spherical nature of the strained gamma valley. This equation can be simplified by separating Ψ into its parallel and perpendicular components. We assume that the electrons are free in the two directions parallel to the interface. Then, the wave function can be written as

$$\Psi_n(\mathbf{k}_\parallel, z) = \Psi_n(z)u_o(r)e^{i\mathbf{k}_\parallel \cdot \rho}. \quad (4)$$

Here,$u_o(r)$are zone center Bloch functions. Eqn. (3), which must be solved for the z dependent envelope function then reduces to

$$-\frac{\hbar^2}{2m_\perp^\star}\nabla^2\Psi_n(z) + V(z)\Psi_n(z) = E_n\Psi_n(z). \quad (5)$$

We solve this equation in order to determine the energy of the bottom of each subband and the wave function in the z direction, using a finite difference numerical eigenvalue technique for our solution. As mentioned earlier, in order to describe the subbands and the band mixing in the p-type MODFET, a more complicated Schrödinger equation is required which uses the Kohn Luttinger hamiltonian which describes the HH and LH states quite accurately. The Kohn Luttinger problem with the strain terms of Eqn.(2) is then solved in presence of the confining potential V(z). This equation is solved analogously to the electron Schrödinger equation, with finite difference techniques. In this case, however, the matrix equation will be four times as big as is the electron case for the same number of spatial steps in the solution.

2.3. The charge control model

Device simulation is obtained by a self-consistent solution of the Poisson equation and the Schrödinger equation. Solution of the Schrödinger equation yields the envelope wavefunction for each subband. The occupation of each subband is obtained by integrating the density of states multiplied by the Fermi distribution function. Taken together, the wave function and the band occupation yield the quantum confined charge distribution in real space. The charge profile in the device is then given as the sum of the doping charge, the free charge and the quantum confined charge i.e.

$$\rho(z) = N_d(z) - N_a(z) - n_{free}(z) + p_{free}(z) - \sum_i p_i\phi_i^\star(z)\phi_i(z) \quad (6)$$

Here, N_a and N_d represent the doping levels, n_{free} and p_{free} are the free electron and hole concentrations which are calculated using Fermi-Dirac statistics, and the sum is over i two dimensionaly confined bands is which the hole concentrations are p_i. The density of free states is calculated in the usual way from conduction and valance band effective masses except where there is quantum confinement. In these regions, the parabolic form of the three dimensional density of states is assumed to begin at the top of the confining potential instead of at the conduction band.

3.1 THEORETICAL RESULTS

We applied out model for a n-type MODFET to two similar structures one with a GaAs substrate, (the other with an InP substrate). The structure of the devices simulated was as follows: a thick buffer of $Al_{.7}Ga_{.3}As$($Al_{.48}In_{.52}As$), 400Å of GaAs($In_{.53}Ga_{.47}As$), a 100Å thick layer of $In_xGa_{1-x}As$($In_{.53+x}Ga_{.47-x}AS$) in which the excess In fraction, x was varied from 0 to 12 percent to simulate different strains, a 100Å spacer of $Al_{.3}Ga_{.7}As$($Al_{.48}In_{.52}As$), a 200Å donor layer of $Al_{.3}Ga_{.7}As$ ($Al_{.48}In_{.52}As$) doped at $2\times10^{18}cm^{-3}$, a 100Å $Al_{.3}Ga_{.7}As$($Al_{.48}In_{.52}As$) undoped layer, and then the gate. In both case, the Schottky barrier height was taken to be 0.8 volt, and the band offsets were modeled at $\Delta Ec/\Delta Ev$=65/35.

The solution of our model for n-type MODFET yields the conduction band profile, the charge profile, and subband energies. The results for n-type structures are shown in Fig.1 and Table 1. Fig. 1 shows the effect of excess In on total sheet charge and ground state occupation in the channel. The increase in sheet charge is modeled for both InP and GaAs based system, but there is a considerable increase in 1st subband occupancy which is expected to lead to better transport properties[7].

As noted earlier, the effect of strain is much more dramatic on p-MODFETs. The hole masses are reduced considerably due to the strain as discussed in section2. In Table 1 we summarize our results for p-type devices. The results are shown for a lattice matched system (GaAs channel) and strained ($In_{.12}Ga_{.88}As$ channel). The sheet charge in the subband levels as well as the density of state masses are shown. At room temperature we expect a overall mass reduction of ~50% and at 77°K a reduction of ~75%.

3.2 EXPERIMENTAL STUDIES

Growth of strained materials:
Strained InGaAs heterostructures were grown by MBE in order to confirm our theoretical prediction. The material composition and thickness was discussed in section 3.1. Growth of the strained region was possible by reducing the temperature of the Ga cell. This was done with no growth interuption and resulted in very small gradients of In-composition between the two channel regions. The growth temperature was 510°C.

Mobility enhancement by strain:
The electron mass reduction due to carrier confinement in the high InAs mole fraction strained layer is translated by mobility enhancement and was confirmed by Hall-measurements. For the lattice matched (53%-In) sample, the mobility varied from $9900cm^2/V\text{-}s$ at 300°K to 36,500 $cm^2/V\text{-}s$ at 16°K. The corresponding change for the 60% system was $11{,}100cm^2/V\text{-}s$ to 55,100 $cm^2/V\text{-}s$. The mobility improvement by increasing the In composition seems to tailor-off for mole fractions above 60%. Our experimental data show no important variation of sheet density with strain. This agrees with the theory which predicts a variation from $1.34x10^{12}cm^{-2}$ to $1.4x10^{12}$ cm^{-2} for an increase of In-composition from 53% to 65%.

Capacitance-Voltage characteristics:
The validity of our charge control model was checked by comparing it to experimentally obtained data. Fig.2 shows C-V measurements on 7% and 12% excess In heterostructures. The effect of strain is very small, especially if one accounts for the slight shift due to different recess and therefore additional fixed capacitance. As discussed earlier the sheet density does not very appreciably with strain and a maximum increase of only 5% is predicted for 12% excess In. Since the capacitance is proportional to n_s[8] one cannot therefore see any important effect of strain on C-V characteristics.

The C-V characteristics were used in order to determine the carrier concentration across the device layer. Fig.3 shows such results for the for the 12% excess In sample. A peak electron concentration is observed at 650Å from the surface which corresponds well to the channel region of our samples. An electron confinement within a region of less than 200Å can also be estimated.

Transconductance improvement by strain:
Devices were processed on materials with 53% and 60% In-composition. The minimum gate length was 1.2μm for the lattice matched system and 2.1μm for the strained device. Their extrinsic transconductance(Fig.4) varies from 157mS/mm to 223mS/mm with 7% excess In. This corresponds to a 211mS/mm to 348mS/mm intrinsic transconductance change. The drain-source current varies from 140 mA/mm to 160mA/mm at Vgs=0V. Transconductance measurements as a function of drain-source voltage confirm the mobility and velocity enhancement in the presence of strain and demonstrate the usefulness of high InAs mole fractions.

Microwave characterization:
Microwave testing was performed on devices with microstrip access lines which were thinned down to 200μm and diced to 1mmx1mm chips. By measuring the S-parameters and extracting the device equivalent circuitr the following changes were observed for 1.5μmx150μm gate size and 7% In increase: (a) gate-source capacitance increase from 0.48pF to 0.53pF, (b) transconductance increase from 28.6mS to 48.1mS, (c) an output resistance which remained almost constant to 480Ω with marginal only increase with strain. The parasitic access-resistance elements were fixed to their DC values during model extraction. A slight gain improvement of about 2dB was observed up to 6GHz in the presence of strain. This was primarily due to the increase of the $f_T=g_m/(2\pi Cgs)$ ratio, since the output and gate resistance are not influenced by the In composition change; $f_T$was in fact found to increase from about 9GHz to 14GHz in the presence of strain.

4. CONCLUSIONS

A self-consistent solution of Poisson and Schrodinger equations was used in order to study the effect of strain in heterostructures. This technique combined with the Kohn-Luttinger formalism demonstrates an important decrease of hole masses making them very promising for high speed applications. N-channel InGaAs/InAlAs strained MODFETs were analyzed theoretically and experimentally. Experimental results confirm the mobility increase for increased In compositions, while demonstrating improved transconductance and cut-off frequency characteristics.

Acknowledgement
The authors would like to thank H.F. Chau, M. Tutt, and J.L. Cazaux for useful discussions and help in the characterizations.

Temp. (°K)	subband #	GaAs channel				$In_{0.12}Ga_{0.88}As$ channel			
		$n_s \cdot 10^{11}$	m^*_{dos}	$E_n - E_F$ (meV)	state	$n_s \cdot 10^{11}$	m^*_{dos}	$E_n - E_F$ (meV)	state
	0	1.19	0.601	-84.8	HH0	1.49	0.257	-56.4	HH0
300	1	0.95	0.481	-84.8	HH0	1.15	0.199	-56.4	HH0
	2	0.70	0.686	-102.	LH0	0.86	0.342	-78.6	HH1
	3	0.64	0.624	-102.	LH0	0.79	0.313	-78.6	HH1
	4	0.45	0.485	-105.	HH1	0.53	0.561	-104.	HH2
	5	0.43	0.463	-105.	HH1	0.39	0.406	-104.	HH2
	6	0.35	0.566	-116.	HH2	0.27	0.569	-122.	LH0
	7	0.33	0.544	-116.	HH2	0.19	0.404	-122.	LH0
	total	5.06	0.564			5.67	0.324		

Table 1 Results of simulations of p-type MODFETs.

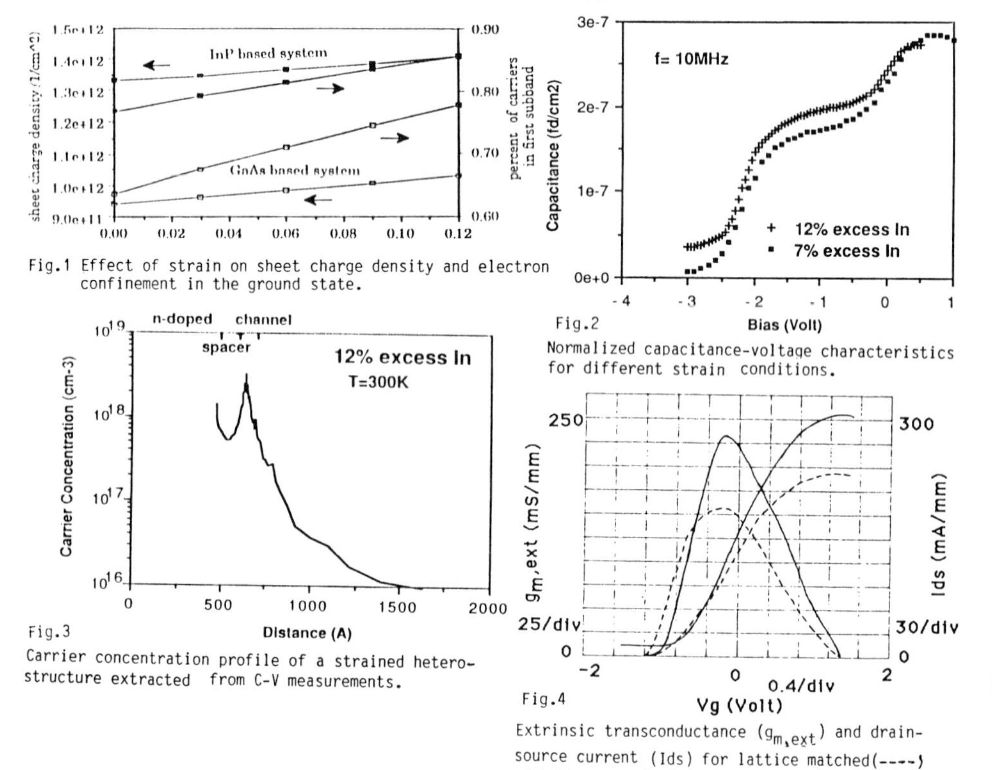

Fig.1 Effect of strain on sheet charge density and electron confinement in the ground state.

Fig.2 Normalized capacitance-voltage characteristics for different strain conditions.

Fig.3 Carrier concentration profile of a strained hetero-structure extracted from C-V measurements.

Fig.4 Extrinsic transconductance ($g_{m,ext}$) and drain-source current (Ids) for lattice matched(----) and 7% excess In (——) MODFETs. (Vds=2.5V).

Reference

[1] A. Swanson, Microwave and RF, p.173, March 1987.
[2] L. Eastman, 1986 IEDM Technical Digest, p.456.
[3] J. Kuo, B. Lalevic, T. Chang, 1986 IEDM Technical Digest, p.460.
[4] C. Lee, H. Wang, G. Sullivan, N. Sheng, D. Miller, IEEE Electron Device Letters, 8(3), p.85, 1987.
[5] M. Jaffe, J. Singh, Solid State Communications, 62(6), p.399, 1987.
[6] C. Kuo, S. Vong, R. Cohen, G. Stringfellow, Journal of Applied Physics, 57(12), p.5428,1985.
[7] S. Mori, T. Ando, Journal of the Physical Society of Japan, 48(3), p.865,1980.
[8] L. Sadwick and K. Wang, IEEE Trans. on Electron. Devices, ED-33(5), p651, 1986.

DLTS study of interface and bulk traps in undoped InP grown by molecular beam epitaxy

A. Iliadis, S. C. Laih, D. E. Ioannou and *E. A. Martin

Joint Program for Advanced Electronic Materials, Electrical Engineering Department, University of Maryland and Laboratory for Physical Sciences ,College Park, Maryland 20742, U.S.A.
*Bendix Aerospace Technology Center Allied Corporation, 9140 Old Annapolis Road, Columbia, Maryland 21045, U.S.A.

ABSTRACT. Deep level transient spectroscopy (DLTS) was used to study the deep traps in the SiO_2/InP interface and the bulk of InP grown by MBE from solid sources for the first time. It was found that deposition of SiO_2 thin films substantially increased the barrier height of the Al/SiO_2/InP contacts , and introduced electron traps at the interface. A study of the characteristics of these traps under different passivating procedures and growth temperature suggested two distinct interface defects of different origin one of which was related to excess phosphorus at the interface. Three electron traps were present in the bulk of the MBE grown InP with activation energies $E_c - E_t = 0.43$ eV, 0.53 eV and 0.57 eV. Trap B1 at 0.43 eV was observed in the high growth temperature samples and was related to a phosphorus vacancy–Fe, Mn complex. The other two traps appeared to be characteristic of the MBE growth.

1.INTRODUCTION

Previous deep level transient spectroscopy (DLTS) studies of InP grown by LEC, VPE, LPE and MOCVD showed a number of deep traps, some of which were found to be related to extrinsic impurities (Wada *et al* 1980) and native defects (Yamazoe *el al* 1981). However very little work has appeared on InP grown by molecular beam epitaxy (MBE). InP grown by MBE from solid sources is normally n-type and it is characterized by a number of unintentional impurities and defects (Iliadis *et al* 1986). Furthermore, the low Schottky barrier height ($\phi_b = 0.45$ eV) of this semiconductor severely limits its device applications, and complicates the use of strong analytical techniques like DLTS. In order to artificially increase the barrier height, we used plasma-enchanced chemical vapor deposition (PECVD) to deposit a thin SiO_2 film on InP and used DLTS on the resultant Al/SiO_2/InP Schottky diodes to study the deep traps in both the bulk of the MBE grown layers and the SiO_2/InP interface. The SiO_2/InP interface may in itself introduce shallow and deep trapping centers and its study is of great importance to MISFET device applications.

2.EXPERIMENTAL PROCEDURE

The epitaxial layers were grown by MBE on (100) n-type and semi-insulating InP substrates, at two different growth temperatures, $T_S = 450°C$ and 530°C, under phosphorus (P_2) stable conditions. The grown layers were unintentionally doped n-type with carrier concentrations N_D between $7 \times 10^{15} cm^{-3}$ and $3 \times 10^{16} cm^{-3}$. A series of bulk LEC grown (100) undoped InP samples were also processed in order to provide information about the PECVD SiO_2 deposition and the electrical properties of the Al/SiO_2/InP system. The samples were etched in HF first and then in a 3:1:1

solution of $H_2SO_4 : H_2O_2 : H_2O$, rinsed in deionized water, blown dry in N_2 and mounted in the SiO_2 chemical vapor deposition reactor (Pande *et al* 1987). The SiO_2 layers had thicknesses ranging between 100Å and 400Å. Aluminum dots for Schottky contacts and Au-Ge pads for ohmic contacts were evaporated in a conventional resistive evaporation vacuum system. Current-voltage (I-V) and capacitance-voltage (C-V) measurements were made at 300°K. The DLTS measurements were made in the temperature range between 77°K and 350°K.

3.RESULTS AND DISCUSSION

3.1 Current-Voltage and Capacitance-Voltage measurements

The deposition of a thin SiO_2 layer prior to metal deposition, resulted in a significant increase of the barrier height of the system. A typical log I versus V plot of an $Al/SiO_2/InP$ diode is shown in Fig.1. The barrier height $\phi_b = 0.66$ eV and the ideality factor n=1.19. Barrier height ranged between 0.55 eV and 0.66 eV and n values between 1.15 and 1.8. Reproducibly good diodes were obtained for SiO_2 thickness d_{ox} between 200Å and 250Å. A typical C-V characteristic of a diode measured at 1MHz gave the thickness of the SiO_2 layer $d_{ox} \simeq 268$Å in good agreement with the SiO_2 deposition data.

3.2 DLTS of interface and bulk traps

The DLTS spectrum of a sample grown at the low substrate temperature of $T_S =$ 450°C, is shown in Fig.2, and the spectrum of a sample grown at high $T_S = 530$°C is shown in Fig.3. Analysis of the spectra using different applied biases (V_R), filling pulse height (V_f) and rate windows, revealed that electron traps I1,I2 and I3 originate from the SiO_2/InP interface, while traps B1, B2 and B3 originate from the bulk of the grown layers. This distinction is made by monitoring the temperature shift of these peaks with V_R and V_f (Yamasaki *et al* 1979). Peak I1 is the main interface trap and has been observed in all our samples. It appears to be independent of surface treatment and/or MBE growth conditions. The energy distribution of this trap is centered around the midgap of the semiconductor and is believed to be due to the intrinsic surface states of InP. Peak I2 is found to be present in the low growth temperature samples ($T_S = 450$°C). The energy distribution E_{ss} of this trap (Table1) is between 0.44 eV and 0.58 eV. Peak I3 has an energy distribution E_{ss} between 0.22 eV and 0.24 eV from the conduction band edge and has been observed only in our MBE samples and not in the bulk LEC grown samples. An experiment was conducted with LEC grown samples whereby one sample was processed normally and another was passivated with phosphorus nitride for 1 min in the CVD reactor prior to SiO_2 deposition. The passivation introduced an additional interface peak with characteristics very similar to those of I3 in the MBE samples. Although the complexity of this interface can be high, there is a clear link here between this relatively shallow electron trap and the excess phosphorus at the interface .

The DLTS data of interface traps clearly indicate that there is not a continuous distribution of a single type of interface trap across the energy band gap, but distinct clusters of interface states the energy distribution of which is given in Table 1 along with the interface density N_{ss}. This is in contrast with the work of Inuishi *et al* (1983), who observed a continuous distribution of interface traps on thermally oxidized bulk InP, but in agreement with the work of Yamaguchi *et al* (1980) who observed two

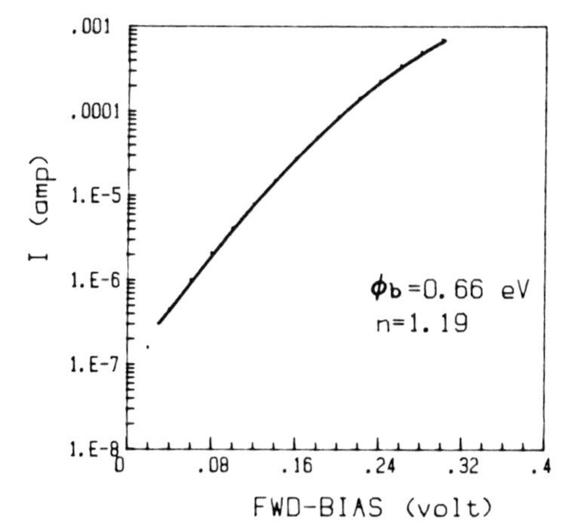

Fig.1 I-V characteristic of an $Al/SiO_2/InP$ diode

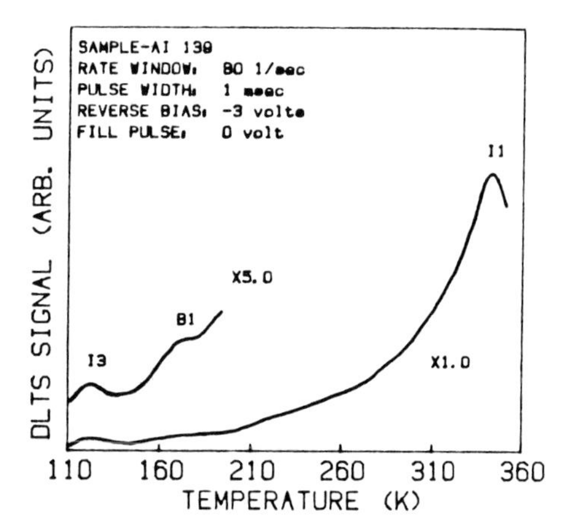

Fig.2 DLTS spectra of MBE grown InP ($T_s = 530°C$)

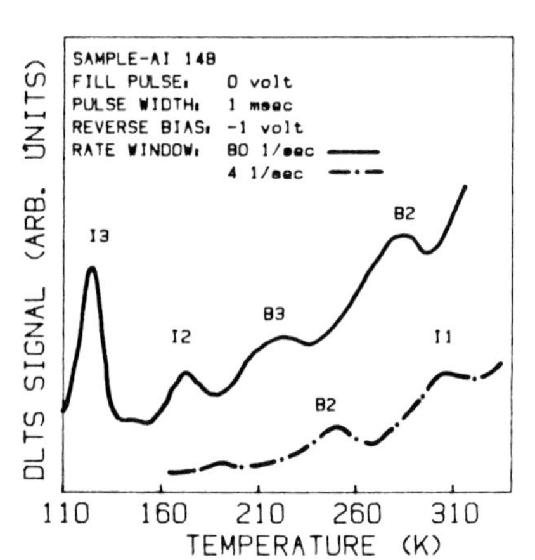

Fig.3 DLTS spectra of MBE grown InP ($T_s = 450°C$)

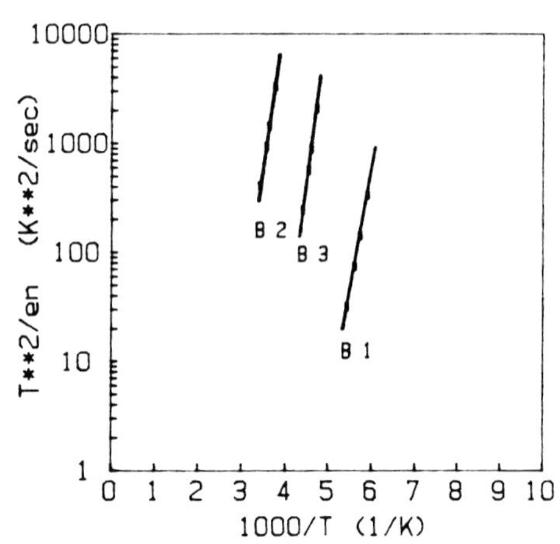

Fig.4 Arrhenius plots of the emission rate T^2/e_n vs temperature

TABLE 1 Interface peaks

Peak	E_{ss} (eV)	N_{ss} ($cm^{-2}eV^{-1}$)	Related to
I1	mid gap	2.2×10^{14} 3.6×10^{14}	Intrinsic
I2	0.44 0.58	2.3×10^{13} 3.4×10^{14}	T_s
I3	0.22 0.24	5.6×10^{13} 9.3×10^{13}	growth

TABLE 2 Bulk peaks

Peak	E_t (eV)	σ_∞ (cm^2)	N_t (cm^{-3})
B1	0.43	2.6×10^{-11}	1.6×10^{14}
B2	0.53	1.3×10^{-14}	2.9×10^{14}
B3	0.576	7.0×10^{-11}	1.3×10^{14}

distinct interface traps at 0.16 eV and 0.075 eV. Furthermore our results indicate that traps I2 and I3 are probably due to different interface defects.
Peak B1 is due to a bulk trap and it is observed in the high growth temperature ($T_S = 530°C$) MBE samples and the LEC bulk samples. The MBE growth of InP at elevated growth temperatures ($T_S > 500°C$), implies an increased concentration of phosphorus vacancies due to the heavy loss of phosphorus from the surface that cannot be replaced by the moderate phosphorus overpressures used during the growth of these layers. This trap with an activation energy $E_c - E_t = 0.43$ eV (Fig.4), has similar characteristics to trap E5 of Yamazoe *et al* (1981) and trap E of Wada *et al* (1980). Yamazoe *et al* (1981) showed that this trap had the same thermal response with the 1.1 eV photoluminescence (PL) band upon heat-treatment and concluded that the trap and the 1.1 eV PL band are related to a phosphorus vacancy-defect complex. However it has been shown by Iliadis *et al* (1985) that the 1.1 eV PL band in these samples is due to Fe and Mn acceptor impurities and secondary ion mass spectroscopy (SIMS) data clearly support this identification. Our data are in support of the phosphorus vacancy-defect complex origin of this trap, where the defect part of the complex is related to Fe and/or Mn impurities. Bulk traps B2 and B3 with activation energies 0.53 eV and 0.57 eV respectively are observed in the low growth temperature MBE sample (Table 2). Trap B2 has similar characteristics to trap MOE2 of Ogura *et al* (1983) and trap B of Wada *et al* (1980). Trap B3 has not been observed before and its origin is currently under investingation.

4.CONCLUSIONS

The DLTS studies of the SiO_2/InP interface and the bulk of InP grown by MBE, showed that traps I1, I2 and I3 are deep electron traps originating from the interface, while traps B1, B2 and B3 are from the bulk of the grown layers. The interface traps form distinct clusters across the energy gap with I1 being the main interface trap at midgap. I2 is related to low substrate temperature during MBE growth, and I3 to excess phosphorus at the interface. Although a significant increase in barrier height is achieved by depositing the SiO_2 films, the presence of these deep interface traps may introduce problems to device fabrication. There is strong indication that bulk trap B1 at 0.43 eV is due to a phosphorus vacancy-defect complex where the defect can be identified with Fe and/or Mn. Traps B2 and B3 are bulk traps observed at low growth temperatures and may be related to native defects. Higher growth temperature samples appear to have a lower number of traps and may be more appropriate in certain device applications.

REFERENCES

Iliadis A, Prior K A,Stanley C R,Martin T and Davies G J 1986 *J.Appl.Phys.* **60** 213
Iliadis A, Stanley C R and Sykes D E 1985 *Semiconductor Quantum Well Structures and Superlattices*,edited by Ploog K and Linh N T Vol.VI,p.167.
Inuishi M and Wessels B W 1983 *Thin Solid Film* **103** 141
Ogura M, Mizuta M, Hase N and Kukimoto H 1983 *Jpn. J. Appl. Phys.* **22** 658
Pande K P, Martin E, Gutierrez D and Aina O 1987 *Solid-State Electronics* **30** 253
Wada O Majerfeld A and Choudhury A N M M 1980 *J. Appl. Phys.* **51** 423
Yamaguchi M and Ando K 1980 *J. Appl. Phys.* **51** 5007
Yamasaki K, Yoshida M and Sugano T 1979 *Jpn. J. Appl. Phys.* **18** 113
Yamazoe Y, Sasai Y and Nishino T 1981 *Jpn. J. Appl. Phys.* **20** 347

Evolution of the electronic properties of highly doped $Ga_{1-x}Al_xAs$ alloy in presence of resonant DX-centers

P. Basmaji[*], J.C. Portal[**], R.L. Aulombard[***], C. Fau[***], D.K. Maude[****], L. Eaves[****] and P. Gibart[*]

[*] Laboratoire de Physique du Solide et Energie Solaire - C.N.R.S., Sophia Antipolis - 06560 Valbonne, France
[**] Service National des Champs Intenses, C.N.R.S.-38042 Grenoble, France.
[***] Groupe d'Etudes des Semiconducteurs, Université des Sciences et Techniques du Languedoc - 34060 Montpellier, France.
[****] Department of Physics, University of Nottingham, NG7 2 RD, England.

Abstract : The electrical transport properties of epitaxial highly doped Sn-$Ga_{0.7}Al_{0.3}As$ grown by MOVPE were studied. Magnetoconductivity was measured at 4.2 K as a function of magnetic field (up to 18 T) and hydrostatic pressure (up to 1 GPa). The Γ_{1c} carrier concentration was deduced from the Shubnikov-de-Haas oscillations. The decrease of free carrier concentration with pressure shows an electron freeze-out from the Γ_{1c} band to the resonant states. The energetic position of the resonant level relative to the bottom of the Γ_{1c} band is found to exhibit a pressure dependent coefficient of -117 meV/GPa. DX character of this level was directly shown through the persistent photoconductivity observation.

1. Introduction

Understanding and controlling deep impurity levels is still a major task for the semiconductor field both from fundamental and technological points of view. It is well known that deep impurities arise from (Jaros 1980) a defect potential which is strong and short ranged. As a consequence such levels can be resonant in the continuum of a band (Ren 1982).

In particular, the alloy system $Ga_{1-x}Al_xAs$ has received considerable attention recently because of its potential application and great interest has been focussed upon the study of metastable localized states (DX centers) introduced by substitutional donors in n-type $Ga_{1-x}Al_xAs$ (Lang 1986.

The aim of the present work was to investigate the behaviour of Sn-related impurity states in heavily doped $Ga_{1-x}Al_xAs$. Magnetoresistivity experiments (up to 18 T) under hydrostatic pressure (up to 1 GPa) have been performed at 4.2 K on layers grown by Metal Organic Vapor Phase Epitaxy (MOVPE). Basmaji (1987) found that the deep donor exhibits a well defined resonance with the Γ_{1c} band. Measurements of the pressure dependence of free electrons trapping from the Γ_{1c} conduction band (at T > 100 K) shows that the energetic position of the donor level varies as $dE_D/dP = -15$ meV/GPa. Moreover, the persistent photoconductivity (PPC), which results from the DX character of this deep level, has been directly observed through the variations of the Shubnikov-de-Haas (SdH) oscillations after illumination with a LED.

2. Experiment

$Ga_{1-x}Al_xAs$ epilayers were grown in a conventional atmospheric MOVPE vertical reactor on semi-insulating Cr-doped GaAs substrates consisting of a 500 A° undoped $Ga_{1-x}Al_xAs$ buffer layer and a 2 μm Sn-doped $Ga_{1-x}Al_xAs$ layer. The Sn-doping level was chosen to be higher than $10^{18} cm^{-3}$.

Magnetoresistivity under pressure was measured at 4.2 K on Hall samples shaped by standard photolithography techniques (Au-Ge-contacts). Higher magnetic fields were produced using a 10 MW Bitter-type Solenoïd allowing continuous changes up to 18 T. The hydrostatic pressure experiments were performed using a Cu-Be clamp cell in the range 0-1 GPa at low temperature.

Typical S.d.H.-measurements at 4.2 K on a Sn-doped sample with initial carrier concentration of $\sim 4 \times 10^{18} cm^{-3}$ are shown in Fig.1 for a range of applied pressure. The pressures quoted are those measured at 4.2 K. As P is increased, the period of the SdH-oscillations increases, corresponding to a decrease in carrier concentration. The Γ carrier density n_Γ at 4.2 K was deduced from the period of the SdH-oscillations :

$$\Delta\left(\frac{1}{B}\right) = \frac{2e}{\hbar} (3\Pi^2 n_\Gamma)^{-2/3}$$

(see table I). This first experimental result shows the emptying of the Γ_{1c} valley under applied pressure.

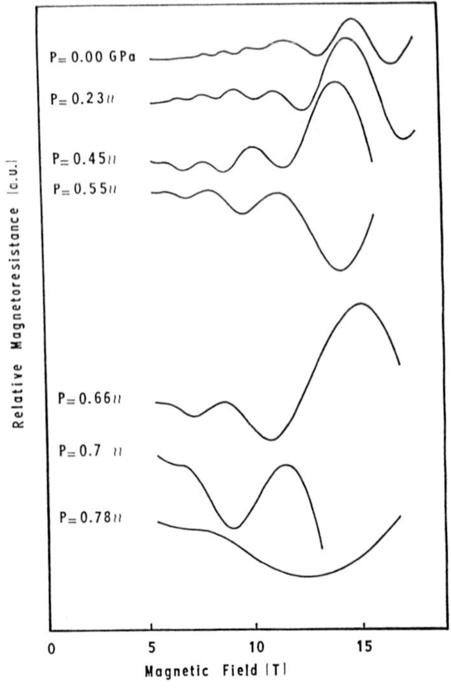

Figure 1 : Shubnikov-de-Haas oscillations for Sn-$Ga_{0.71}Al_{0.29}As$ at 4.2K under different hydrostatic pressure.

At low temperature (4.2 K), after exposure of the sample to light of an energy less than the bandgap of $Ga_{1-x}Al_xAs$, the n_Γ carrier density increased, and this increase persisted after switching off the light. This is evidence of the so-called PPC (the photoionized carriers were prevented from falling back down to their previous state). Here, it is important to note that the pressure was initially applied at room temperature and samples were cooled slowly. This second experimental result shows that the deep level introduced in the band of Sn-doped $Ga_{1-x}Al_xAs$ was characterized by a metastable occupation (DX-center) leading to time dependent effects at T < 100 K similar to silicon in GaAs and GaAlAs as seen by Theis (1987) and Kukimoto (1986).

3. Analysis of the data

At temperature less than 100 K, the free carrier density is not the thermodynamic equilibrium value. Hence the data need to be interpreted with care. Since no "hysteresis" effects were observed at $T \geqslant 100$ K, the use of equilibrium statistics is valid at and above this temperature. Thus, we have analyzed the results in the following way :

i) it is assumed that the total carrier concentration in the conduction band remains constant between 4.2 K and 100 K (only redistribution of carriers between Γ_{1c}, L_{1c} and X_{1c} could occur).

ii) Considering the well known band structure of $Ga_{1-x}Al_xAs$ (Lifshitz 1980), the X- and L-bands are unoccupied at 4.2 K.

The following equation can be written :

$$(n_\Gamma)_{4.2\,K} = (n_\Gamma + n_L + n_X)_{100\,K}$$

The occupation of the Γ_{1c}, L_{1c} and X_{1c} minima of the conduction band and the Fermi energy $(E_F - E_{\Gamma_{1c}})$ were determined at 100 K taking into account non parabolicity as Raymond et al. (1979) did, and the pressure dependence of the effective mass of the Γ_{1c} minimum (three band Kane's model). From the neutrality equation

$$N_A^- + n_\Gamma + n_L + n_X = N_D - \frac{N_D}{1 + 1/2 \exp\left(\frac{E_D - E_F}{kT}\right)}$$

the activation energy $(E_D - E_{\Gamma_{1c}})$ of the donor level was determined. The total density of donors N_D was obtained from a fit of the experimental curve n(P) and N_A was neglected ($N_A \ll N_D$). Typical results we have obtained are listed in table I. The results in Fig.2 clearly show that with increasing pressure the DX level moves toward the Γ_{1c} band edge with the rate slope

$$\frac{d(E_D - \Gamma_{1c})}{dP} = -117\ \text{meV/GPa}$$

However, we observe at the highest investigated pressure an apparently "anomalous" deviation from linearity (at the $\Gamma_{1c} - E_D$ crossover). Taking into account the pressure coefficients given by Lifshitz (1980) or by Chandrasekhar (1986), the pressure coefficient obtained $\frac{dE_D}{dP}$ varies between - 15 and 0 meV/GPa. So, as the pressure increased, the impurity level is driven into the fundamental gap with a variation near that of the X-band.

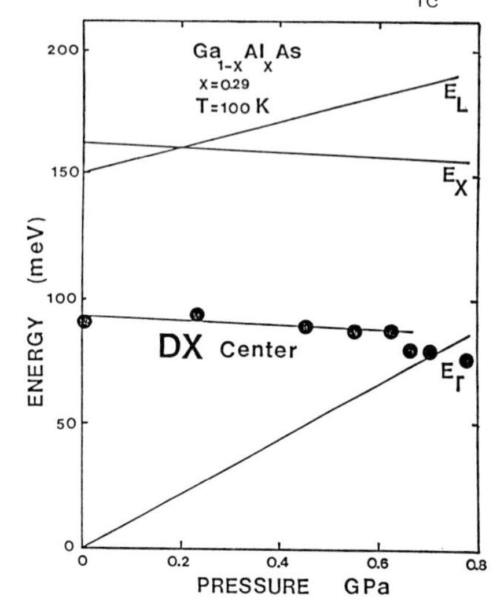

Figure 2 : Energetic position of minima of the band conduction and DX level relative to the bottom of the Γ_{1c} conduction band.

P GPa	n_Γ 4.2 K ($\times 10^{18} cm^{-3}$)	T = 100 K		
		$n_{\Gamma_{1c}}$ ($\times 10^{18} cm^{-3}$)	$E_F - E_{\Gamma_{1c}}$ (meV)	$E_D - E_{\Gamma_{1c}}$ (meV)
0	2.88	2.87	77	91
0.23	2.05	2.05	61	67
0.45	1.16	1.15	41	39
0.55	0.85	0.84	32	27
0.62	0.74	0.74	25	20
0.66	0.44	0.43	18.5	6.5
0.70	0.37	0.36	15.5	2
0.78	0.23	0.22	8.5	- 9.6

Table I : Measured carrier concentration and calculated Fermi energy and donor level energy for the sample $Ga_{0.71}Al_{0.29}As$ ($N_D = 4 \times 10^{18} cm^{-3}$).

4. Conclusion

To summarize, the measurements reported here show that a deep donor of DX character exists in highly Sn-doped $Ga_{0.71}Al_{0.29}As$ ($N_D = 4\times10^{18} cm^{-3}$), in resonance with the Γ_{1c} band continuum. The result indicates that the DX level approximately follows the X minimum in agreement with Wolford's (1987) result on GaAs at high pressure.

References

Basmaji, P., Portal, J.C., Aulombard, R.L., and Gibart, P., 1987, Solid State Comm. Vol. 63 - N° 2.
Chandrasekhar, J.M. et al., 1986, Int. Conf. on the Physics of Semiconductors, Stockholm.
Jaros, M., 1980, Adv. Phys., 89, 409.
Kukimoto, H., 1986, 18th Int. Conf. on the Physics of Semiconductors, Stockholm.
Lang, D.V., 1986, "Deep levels in semiconductors", edited by S.T. Pantelides (Gordon and Breach, New York) p. 489.
Lifshitz, N., Jayaraman, A., Logan, R.A., and Card, H.C., 1980, Phys. Rev. B 21, 670.
Raymond, A., Robert, J.L., and Bernard, C., 1979, J. Phys. C 12, 2289.
Ren, S.Y., Dow, S.D., and Wolford, D.J., 1982, Phys. Rev. B 25, 7661.
Theis, T.N., Mooney, P.M., and Wright, S.L., unpublished 1987.
Wolford, D.J., Kuech, T.F., Steiner, T., Bradley, J.A., Gell, M.A., Ninno, D., and Jaros, M., 1987, Int. Conf. Microstructures Superlattices, Chicago.

The metastable character of Si-impurity in GaAlAs from study of GaAlAs under hydrostatic pressure

R. Piotrzkowski*, J.L. Robert, S. Azema
Groupe d'Etudes des Semiconducteurs, Université des Sciences et Techniques du Languedoc, 34060 Montpellier, Cédex-France.

E. Litwin-Staszewska
High Pressure Research Center of the Polish Academy of Sciences Warsaw, Poland.

J.P. André
Laboratoire d'Electronique et de Physique Appliquée, 94450, Limeil-Brevannes, France.

ABSTRACT: Hydrostatic pressure generated in a gaseous medium is used to study the properties of Si-donor in $Ga_{1-x}Al_xAs$. This technique is particularly convenient since the pressure can be varied at low temperature,i.e. in the region where the effects of metastability can be observed. This is the first time that the transition from metastable to equilibrium behavior has been studied as a function of time, temperature and hydrostatic pressure. The study of the photoconductivity decay kinetics shows that the height of the barrier is pressure dependent.

1. INTRODUCTION

The properties of Si-donor level in GaAlAs are intensively studied, essentially because its metastable character.(Chand et al 1984). The use of hydrostatic pressure is one of the most convenient ways to study such phenomena. This is especially true if high pressures are generated in gaseous medium, making it possible to vary the pressure at low temperature, i.e. in the region where the effects of metastability can be observed. Up to now, transport properties of bulk GaAlAs under such conditions of pressure and temperature have not been studied.
We report for the first time the results of the Hall effect and conductivity measurements under pressure up to 10 kbars, in the temperature range 77-300K.

2. RESULTS

The investigated sample was grown by MOCVD technique. The Al content was chosen equal to 0.26 in order to have reliable results at low temperature, under pressure. For this concentration the conductivity of the sample remains high enough to avoid any difficulty in transport measurements.
Fig. 1 shows variations of the carrier concentration n, measured by Hall effect, when the sample is cooled down from 300K to 77K under different pressures.

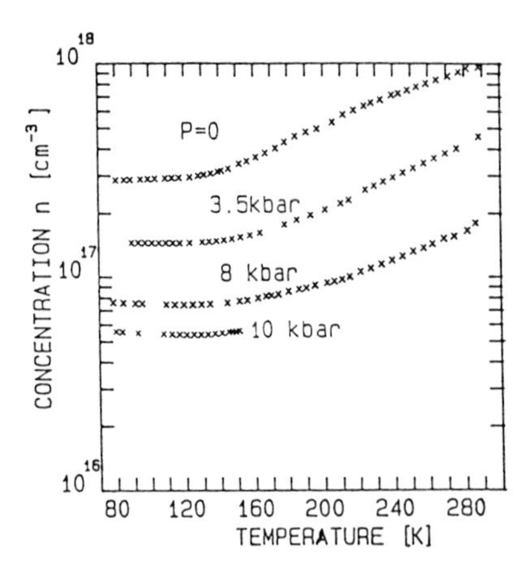

Fig. 1. The electron concentration versus temperature at different pressures. The results were obtained as the sample was cooled.

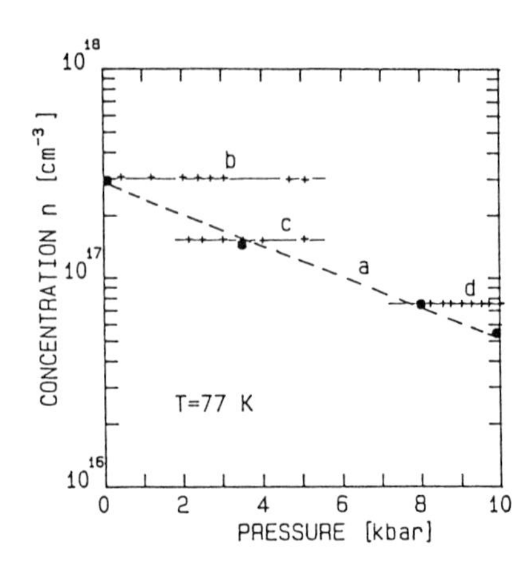

Fig. 2. The electron concentration at 77K versus pressure. a : pressure is varied at T > 160K before cooling the sample. b,c,d : pressure is varied at 77K.

The observed behavior is typical for all the GaAlAs samples with x > 0.22:

a) a thermal activation of n is observed in the temperature range 120K-300K.

b) below 120K, the carrier concentration becomes constant.

The decrease of the carrier density with increasing pressure is due to the increase of the thermal depth of the Si-level.

Curve a) on fig.2 shows the electron concentration at 77K as a function of the pressure P_m at which the sample was cooled. The corresponding experimental points are taken from the curves n versus T at 77K (see fig.1). This curve $n = f(P_m)$ cannot be experimentally reproduced if we vary the pressure at T = 77K: indeed, if we change the pressure at 77K we observe that the concentration does not vary (curves b, c, d). This signifies that, at this temperature, for each value of the pressure, we can have different values of the electron density, depending on the history of the sample : the thermodynamical equilibrium cannot be reached, i.e., the sample is in some metastable state.

The observed behavior is a consequence of the large lattice relaxation character of the Si-level. At low temperature, i.e. when the energy barrier for electron capture cannot be overcome, the occupancy of the Si-states cannot be modified even when the pressure is varied. As long as the temperature is low enough to prevent any passage of electrons above the barrier, the electron gas is in a metastable state and the thermal equilibrium of the sample is not established.

Then, it is clear that, in the low temperature range, we cannot pass from one curve, n versus T, to the other by changing only the pressure. On the other hand, this should be possible above some temperature, when the barrier is ineffective.

This is illustrated by the results shown in fig.3, where two sets of curves are shown. The upper one corresponds to the results obtained at P = 0, the lower to the results obtained at P = 10 kbars. The points marked by circles were taken when the sample was heated, starting from 77K. The initial metastable concentration was obtained by cooling the sample at a pressure P_m different from that at which the measurements were performed.

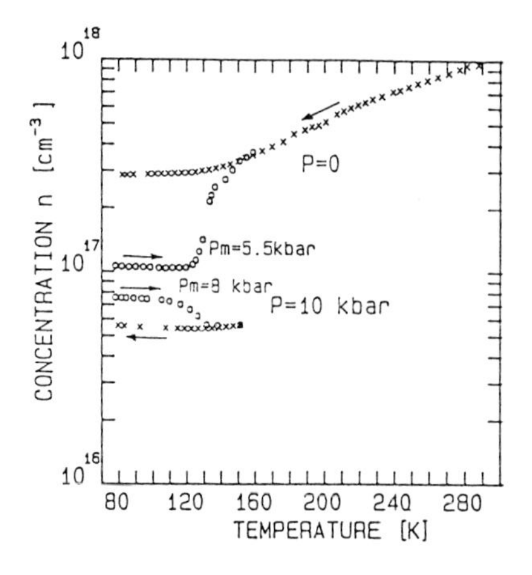

Fig. 3. Variations of electron concentration when the temperature was varied as indicated by arrows. P is the pressure at which the measurements were made. P_m is the pressure at which the initial metastable state was obtained.

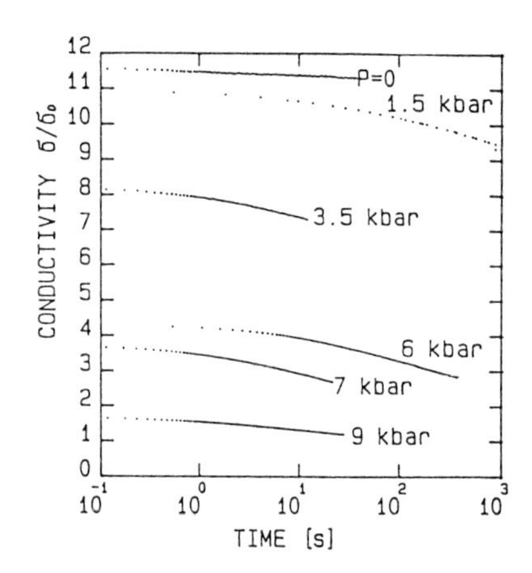

Fig. 4. Decay of the photoconductivity σ under different pressures. The values of σ/σ_o are plotted. σ_o is the conductivity of the sample at 77K when it is cooled at atmospheric pressure.

In the first experiment, the pressure P_m was 5.5 kbars. P_m is higher than the pressure P = 0 at which the measurements were made when the sample was heated.

The pressure P_m corresponding to the second experiment was chosen equal to 8 kbars and was lower than the pressure P = 10 kbars at which the measurements were made.

We observe that the carrier density begins to vary at 110-120K. The n(T) variation observed in the region 110-130K at P = 10 kbars and 120-150K at 0 kbar is due to time dependent processes. Above 150K at P = 0 and 130K at P = 10 kbars the results do not depend on the history of the sample, i.e. the equilibrium conditions are satisfied.

The transition from the metastable conditions to the conditions of thermodynamical equilibrium is thus clearly pointed out.

This is the first time that the transition from metastable to equilibrium behavior has been studied as a function of time, temperature and hydrostatic pressure. By using appropriate gas pressure cycling, it is possible to change the carrier density, which can be lowered as much as 10-fold at zero pressure. This possibility can be used for studying the dependence of GaAlAs properties versus the electron density.

It is worthwhile noticing that the thermal equilibrium does exist only for temperature higher than 150K at P = 0. The attempt to fit the curve n versus T in the whole temperature range would inevitably lead to erroneous results.

The well-known metastability phenomenon resulting from large lattice relaxation character of Si-donor in GaAlAs is the persistent photoconductivity (PPC). Electrons are optically excited from Si-level to the conduction band and cannot be recaptured at low temperature because of the barrier.

To look at the influence of pressure on this phenomenon, we have studied the conductivity when illuminating the sample with a GaAs LED.

We present the results obtained at 77K.
At P = 0, the electron concentration under illumination increases to $3.3.10^{18}$ cm ^{3} (the initial room temperature density was $1.1.10^{18}$ cm ^{3}) independently of light intensity. It remains constant after switching off the light. This is no longer true when the pressure is applied. The stationary value of photoconductivity is lower than that at P = 0 and depends on the light intensity.
For a given intensity, the photoconductivity signal decreases as pressure increases. This indicates that the lifetime of photoelectrons has a finite value.
In fig.4, the photoconductivity decay, measured at 77K under various pressures, is reported in the σ versus log (t) coordinates. After some time, the decay becomes logarithmic with a slope which seems independent of pressure.

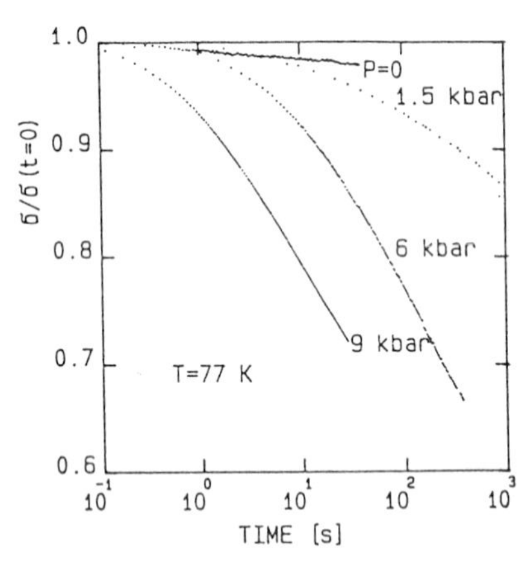

Fig. 5. Decay of the photoconductivity. The values of $\sigma/\sigma(t=0)$ are shown.

In fig.5, the ratio of the photoconductivity signal to its initial value is plotted, making it clear that the increase of the pressure enhances the decay rate.
Our results are very similar to those obtained by Mooney et al (1985) in Si-doped GaAlAs MODFET'S.
This kind of kinetics is explained by the large-lattice relaxation center model with the alloy-broadened barrier for thermal capture. The evolution of the kinetics signifies that the barrier decreases with increasing pressure.

REFERENCES

Chand N, Henderson T, Klem J, Masselink W T, Fischer R, Chang Y C, Morkoç H, 1984 Phys. Rev. B-30-4481.

Mooney P M, Caswell N S, Solomon P N, Wright S L,1985 Material Research Society Symp. Proc. Vol. 46.

Acknowledgements : One of us (E. Litwin-Staszewska) would like to thank the C.N.R.S. for its financial support during her stay in France.

* On leave from High Pressure Research Center. Warsaw. Poland.

Shallow and deep donors in direct-gap Si-doped n-type MBE $Al_xGa_{1-x}As$ with low Al content

P. Debray
Service de Physique du Solide et de Résonance Magnétique, CEN-Saclay, 91191 Gif-sur-Yvette Cedex, France

G. Beuchet, D. Decroix and D. Huet
Thomson-Semiconducteurs, R.D. 128, 91401 Orsay, France

ABSTRACT: Low-temperature (2-300K) Hall and photoluminescence measurements are reported for n-type MBE-grown $Al_xGa_{1-x}As$ doped with Si ($N_{Si} \geq 5\times10^{17}cm^{-3}$) for x = 0.225 and 0.150, respectively. The Hall and photoluminescence measurements both show the presence of a high concentration of deep donor levels in the high x sample. The existence of such deep centres in the low x sample is detected only by the photoluminescence measurements. This is the first observation of deep donor states in $Al_xGa_{1-x}As$ for x < 0.20.

1. INTRODUCTION

It is now generally accepted that the so-called "DX" centre in n-type $Al_xGa_{1-x}As$ is a substitutional deep donor level and is observed in alloys having $x \gtrsim 0.20$ (Lang 1985). The electronic propertie of such alloys are thus controlled by deep donors as well as by hydrogen-like shallow donors. The DX centres are at the origin of the persistent photoconductivity effect and can cause a decrease in carrier concentrations by freezing out at low temperatures. Since their number is of the order of the donor atoms, the presence of such deep levels is highly detrimental to the performance of devices based on selectively doped n-type $GaAs/Al_xGa_{1-x}As$ heterostructures (e.g. TEGFET or HEMT). It is thus important to know the optimum value of x for a given donor atom concentration for which deep donor states are absent. With this objective we have undertaken a systematic study of shallow and deep donors in Si-doped n-type $Al_xGa_{1-x}As$. In this work, we report on our results of Hall electron concentration (Hall n) and photoluminescence (PL) measurements for x = 0.225 (HC) and 0.150 (LC), respectively.

2. EXPERIMENTS

The ternary alloys with a few microns thickness were grown by MBE on semi-insulating GaAs substrates at 630°C as read by a calibrated pyrometer. Buffer layers of thickness equal to a few thousand angströms of the undoped ternary material were used to avoid the formation of a two-dimensional electron gas at the heterointerface. Doping concentrations well in excess of $10^{17}cm^{-3}$ were chosen, since this range is used for the fabrication of high-electron-mobility transistors. Hall n measurements were carried out in the dark on cloverleaf samples using the van der Pauw technique in the temperature range 2-300K in magnetic fields up to 1 Tesla using a fully automated cryomagnetic system. PL measurements were made in the temperature range 4-300K on samples mounted in a continuous-flow cryostat. The excitation source was a chopped 4880Å line of an argon-ion laser with $0.6 wcm^{-2}$ power. The photoluminescence was analyzed by a Jobin-Yvon HR640 spectrometer and detected by a liquid N_2-cooled S1 photomultiplier.

3. RESULTS AND DISCUSSION

Figure 1 shows the temperature dependence of the measured Hall n for both the samples. For the HC sample at temperatures below 150K, the Hall electron concentration remains constant and is equal to $5 \times 10^{17} cm^{-3}$, showing the absence of any carrier freeze-out below this temperature. Above 150K the carrier concentration is thermally activated and increases more or less linearly up to room temperature. The temperature dependence of the resistivity shows two distinctly different regimes corresponding to T below and above 150K, respectively, and thus corroborates the Hall measurements. The Hall concentration of the lower Al-content LC sample, on the contrary, is independent of temperature and remains equal to $8 \times 10^{17} cm^{-3}$ within the accuracy of our measurements in the entire temperature range studied. The thermal activation energy of the shallow donors at these fairly high doping levels is, therefore, practically zero for both the samples.

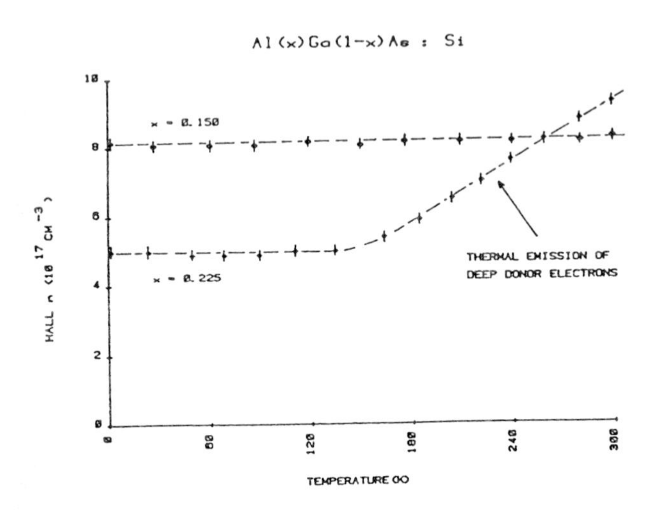

Fig. 1. Temperature dependence of measured Hall electron concentration for Si-doped $Al_xGa_{1-x}As$ for x = 0.150 and 0.225. Dashes are eyeguides.

We interpret the thermally activated increase in Hall n of the HC sample to be due to thermal emission of deep Si donor electrons into the conduction band. The thermal activation energy, E_{dd}, of these deep donors is determined using the simple relations (Schuber and Ploog 1984)

$$n^2 - n N_{SD} \sim \exp(-E_{dd}/kT)$$
$$n = N_{SD} + N_{DD}^{+},$$

where n is the free-electron concentration, N_{SD} is the shallow donor concentration, and N_{DD}^{+} is the concentration of ionized deep donors. The analysis was limited to a temperature range where the deep donor is ionized only weakly. We obtain $E_{dd} = 106 \pm 10$meV. A simple analysis also gives $N_{DD}/N_{SD} \sim 10$. The above observations indicate the presence of a high concentration of deep donors in the sample with the higher Al concentration while that with the lower Al concentration *apparently* has only shallow donors and no deep states.

The low-temperature PL spectra of both the samples show two prominent peaks, a near-band-edge luminescence peak (P_{SD}) and a *deep* luminescence peak (P_{DD}). P_{DD} is located at 130±15meV and 107±20meV, respectively, below P_{SD} for the HC and LC samples. We interpret the P_{SD} peak to be a shallow donor-to-acceptor or band-to-acceptor transition while P_{DD}, we think, is associated with deep states and is the result of radiative recombination of an electron on a deep Si donor with a hole residing on a shallow acceptor. Figure 2 shows the temperature dependence of the *deep* photoluminescence lines. It is interesting to note that the intensity of the P_{DD} peak for both the samples follows the *same* temperature variation and decreases rapidly as the temperature is raised and practically reaches

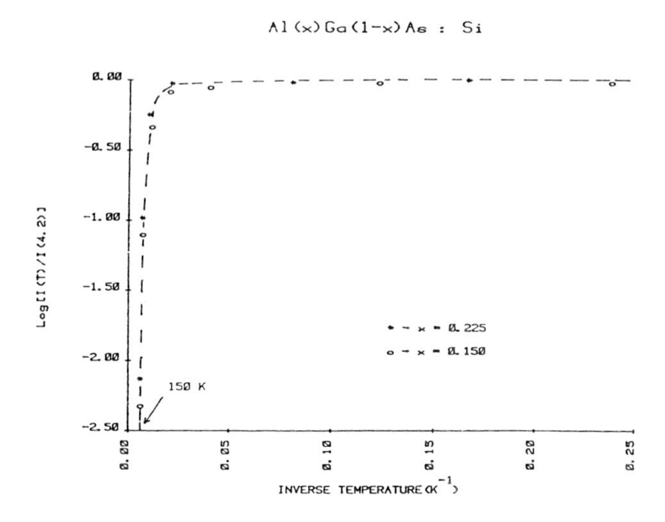

Fig. 2. Temperature dependence of the *deep* photoluminescence lines for n-type Si-doped $Al_xGa_{1-x}As$. I(T) is intensity at TK and I(4.2) that at 4.2K. Dashed curve is an eyeguide.

zero at T ~ 150K. The identical temperature dependence of the P_{DD} peaks for both samples indicates that their *physical origin is the same*. Henning and Ansems (1987) have recently observed *deep* photoluminescence lines similar to P_{DD} in Si-doped n-type $Al_xGa_{1-x}As$ for x ≥ 0.22 and attributed them to donor-acceptor recombinations involving deep donor levels bound to the L minimum of the conduction band. Our PL results for the LC sample indicate that such deep donor levels can also exist in alloys with $x < 0.20$. To the best of our knowledge, this is the first observation of deep donor states in Si-doped $Al_xGa_{1-x}As$ with the value of x well below 0.20. If it is considered that the deep level is tied to the L valley, the optical depth, E_{DD}, as obtained from the line positions turn out to be 268±15meV and 287±20meV for the HC and LC samples, respectively. The deep level for $x = 0.150$ thus lies *below* the Γ valley and is not a metastable trapping level (Theis 1986). The temperature dependence of P_{DD} lines as shown in Figure 2 is consistent with that observed by Henning and Ansems (1987). We have not attempted any analysis by fitting due to lack of sufficient number of data points. The relative intensities of the P_{SD} and P_{DD} peaks give N_{DD}/N_{SD} to be equal to ~ 12 and 0.5, respectively, for the HC and LC samples. The former is consistent with the value obtained from Hall measurements. Finally, it should be mentioned that we have not observed any excitonic emissions in our PL spectra. This is a common observation in heavily doped samples and results due to the screening by dopant impurities of the electrostatic interaction between the hole and electron forming the exciton.

The thermal ionisation energy E_{dd} of the deep donors for the HC sample as obtained from Hall measurements has a value close to the depth of the deep state from the Γ valley minimum. This indicates that the thermal activation of the deep donor occurs to the low-energy Γ conduction band although it is bound to the L minimum. This is not surprising since the electron momentum need not be conserved during thermal ionisation. Why such thermal activation is not observed in the lower Al concentration sample is very likely due to the low concentration of deep donors present, which is perhaps too small to influence the electric transport properties.

The authors would like to thank B. Boucher and M. Bonnet for their constant encouragement during the course of this work.

REFERENCES

Henning J C M and Ansems J P M 1987, Semicond. Sci. Technol. 2, 1
Lang D 1985 *Deep centeres in Semiconductors* (Ed. S T Pantelides. New York: Gordon and Breach) pp. 489-539
Schubert E F and Ploog K 1984, Phys. Rev. B14, 7021
Theis T N, Parker B D, Solomon P M and Wright S L 1986, Appl. Phys. Lett. 49, 1542

On the properties of EL2 related defect in organometallic $Ga_{1-x}Al_xAs$ alloys

A Ben Cherifa, R Azoulay*, A Nouailhat, G Guillot
Laboratoire de Physique de la Matière, INSA de Lyon, 20 Avenue Albert Einstein
69621 Villeurbanne Cédex (France)
* CNET Bagneux, 196 Avenue H. Ravera, 92220 Bagneux (France)

ABSTRACT : Deep electron traps have been studied by means of DLTS and photocapacitance measurements in undoped $Ga_{1-x}Al_xAs$ of n-type grown by MOCVD with $0 \leqslant x \leqslant 0.3$. A dominant deep electron level is present at E_C-0.8 eV in GaAs. Its concentration is found to be independent of x, while its activation energy increases with x. The photocapacitance quenching effect typical for the EL2 defect in GaAs, is observed for the first time for the corresponding defect in the $Ga_{1-x}Al_xAs$ alloys thus confirming clearly that EL2 is also created in MOCVD $Ga_{1-x}Al_xAs$.

1. INTRODUCTION

Metalorganic chemical vapor deposition (MOCVD) has become one of the most attractive techniques for the growth of $Ga_{1-x}Al_xAs$ epitaxial layers for microwave and optoelectronic devices because of its suitability for large scale production, the excellent uniformity of the grown layer and the precise thickness controllability of thin films.

The characterization of deep level properties is an important indicator of the material quality. In MOCVD GaAs, the EL2 near-midgap level with an activation energy of 0.80 eV has been observed in previous studies (Bhattacharya et al. 1980, Samuelson et al. 1981). In MOCVD $Ga_{1-x}Al_xAs$, a serious discrepancy remains in the literature about this level. Wagner et al. (1980) and Matsumoto et al. (1982) reported the presence of a trap at 0.8 eV with an activation energy which remains unchanged with Al content x in the range $0 \leqslant x \leqslant 0.35$. On the other hand, Johnson et al. (1981) have reported a monotonically increasing activation energy with increasing x for the same center as recently observed in $Ga_{1-x}Al_xAs$ grown by molecular beam epitaxy (Yamanaka et al. 1987). Moreover so far there have been no clear evidence that the near midgap level detected in $Ga_{1-x}Al_xAs$ is related to EL2.

The purpose of this paper is to report investigation of deep electron traps in undoped MOCVD $Ga_{1-x}Al_xAs$ and to clearly identify the EL2-related defect in the alloy by the photocapacitance quenching effect typical for EL2 in GaAs (Vincent et al. 1982).

2. EXPERIMENTAL

The layers have been grown in a vertical MOCVD reactor at atmospheric pressure described previously (Azoulay et al. 1986). The substrate is seated on a rotating graphite pedestal. Starting gases are pure arsine, trimethylgallium and trimethylaluminium. They were pyrolyzed in H_2 purified through a Palladium cell at 750°C to form undoped epitaxial layers of $Ga_{1-x}Al_xAs$ ($0 \leq x \leq 0.3$) with thicknesses of 3-5 μm. The layers were grown on (100)-oriented GaAs substrates of n^+-type conductivity doped with Si ($2\ 10^{18}\ cm^{-3}$). The net dopant concentrations obtained by C(V) measurements range between $5\ 10^{14}$ to $3.5\ 10^{15}\ cm^{-3}$ depending on x. Deep level transient spectroscopy (DLTS) and photocapacitance transients were used to investigate electron traps. Schottky diodes were formed on the as-grown layers with Au dots and AuGeNi alloyed ohmic contacts. Deep level concentration was determined taking into account distance between the depletion edge and the position where the deep level crosses the bulk Fermi level.

3. EXPERIMENTAL RESULTS AND DISCUSSION

Only diodes with good I(V) characteristics were used for the DLTS analysis. We found that by increasing x from 0 to 0.3, the barrier height ϕ_{Bn} given by I(V) measurements varied from 0.92 eV to 1.10 eV in good agreement with the results of Best (1979).

The apparent activation energy, capture cross section and concentration of the deep levels detected by DLTS in our samples are given in Table 1 as a function of x. The E_C-0.23 eV which appears for x = 0.25 can be identified to the DX center since it has been generally accepted that the DX center disappears for $x < 0.2$ (Zou et al. 1982). The DLTS spectra at high temperature are dominated by a single electron

x	Energy of electron trap (eV) Capture cross section (cm^2) Trap concentration (cm^{-3})						
0						0.46 $3\ 10^{-15}$ $4\ 10^{11}$	0.80 $5\ 10^{-14}$ $1.7\ 10^{14}$
0.14	0.13 $5\ 10^{-16}$ $1.6\ 10^{13}$	0.16 $5\ 10^{-16}$ $2.4\ 10^{13}$	0.29 $6\ 10^{-15}$ $6\ 10^{13}$		0.36 $5\ 10^{-14}$ $4.5\ 10^{13}$	0.49 $2\ 10^{-14}$ $2.5\ 10^{14}$	0.83 $2\ 10^{-14}$ $1.8\ 10^{14}$
0.25			0.30 $2\ 10^{-15}$ $5\ 10^{12}$	0.23 $2\ 10^{-18}$ $3\ 10^{13}$		0.54 $6\ 10^{-15}$ $7\ 10^{13}$	0.90 $5\ 10^{-14}$ $7\ 10^{13}$

Table 1 - Electron trap detected in undoped MOCVD $Ga_{1-x}Al_xAs$ as a function of x

trap ; its thermal signature for x = 0 is found the same as the one of the EL2 level. Figure 1 shows the shift of the DLTS emission peak of the near-midgap level as a function of x for a constant rate window. The emission peak temperature shifts by about 40 K when x increases from x = 0 to 0.25. The broadening of the DLTS peak which has been observed for the EL2 related level in the $GaAs_{1-x}P_x$ alloy (Samuelson and Omling 1986) is much less pronounced in $Ga_{1-x}Al_xAs$ and appears significant only for x = 0.3. We find that the concentration of this level is nearly independent of x but that its apparent thermal activation energy increases with x (Table 1) confirming results of Johnson et al. (1981) and Yamanaka et al. (1987). This increase exceeds the overall uncertainty in the energies which is estimated to be ± 0.02 eV. The reason why some authors (Wagner et al. 1980, Matsumoto et al.

1982) found an invariant activation energy as a function of x for the corresponding trap is not clear at all.

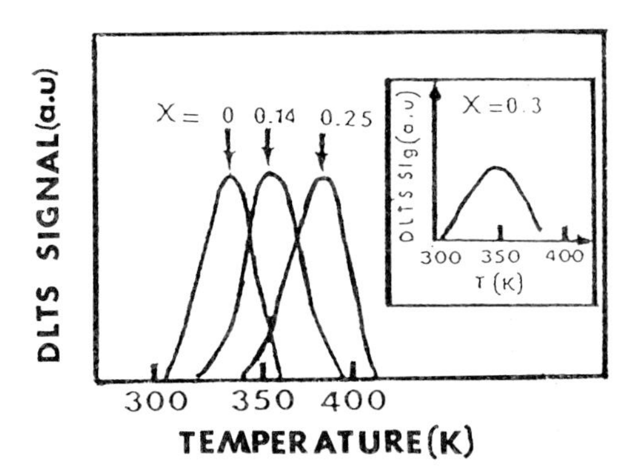

Fig. 1 - Experimental DLTS spectra recorded for various alloy compositions x in the $Ga_{1-x}Al_xAs$: EL2 system. The rate window is 2.8 s^{-1} for x = 0 to x = 0.25 and 0.26 s^{-1} for x = 0.3 (insert). Each spectrum is normalized to its peak intensity.

For the first time, we have shown that this midgap is EL2-related since we observed the photocapacitance quenching effect and the appearance of a metastable state which is regarded as a finger print of EL2 (Vincent et al. 1982) for different values of the aluminium content x (Figure 2). We have checked that the amplitude of the photocapacitance quenching effect corresponds to that of the DLTS signal of the high temperature peak. These results confirm unambiguously the presence of a EL2-related trap in the $Ga_{1-x}Al_xAs$ alloy and that the same type of transformation of the normal state to a metastable state occurs in this alloy for $0 \leqslant x \leqslant 0.3$.

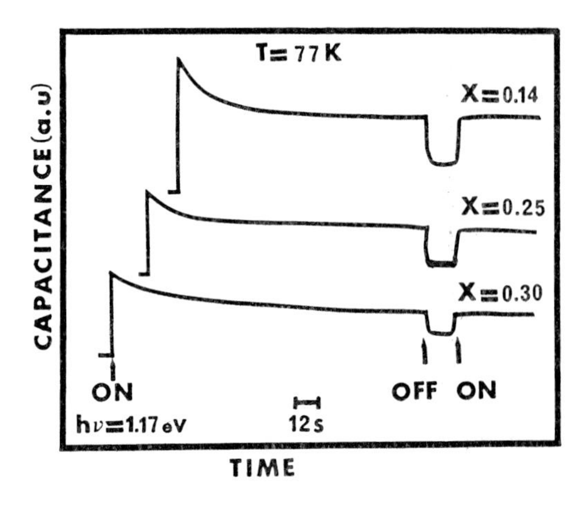

Fig. 2 - Photocapacitance quenching effect typical for EL2 in GaAs as observed for the corresponding EL2 defect in $Ga_{1-x}Al_xAs$.

Bhattacharya et al. (1984) have reported that the concentration of the near midgap level found in MOCVD $Ga_{1-x}Al_xAs$ does not depend on the impurity of the growth ambient suggesting that impurities are not associated with the formation of this trap. In MOCVD GaAs, it has also been found that the incorporation of EL2 varies as the square root of the arsine partial pressure during epitaxial growth hence suggesting a relation between the formation mechanism of EL2 and vacant group III sites (V_{Ga}) or a complex containing a gallium vacancy (Samuelson, et al. 1981). The gallium vacancy is believed to be so unstable so that the arsenic antisite is easily formed from the V_{Ga} (Baraff and Schluter 1985).

These results and our observation of the photoquenching behaviour are confirming that the EL2-related center in $Ga_{1-x}Al_xAs$ has the same microscopic nature as in GaAs namely a complex involving an arsenic antisite As_{Ga} and an As interstitial (Von Bardeleben et al. 1986). This model is in agreement with the observed broadening of the DLTS peak of the EL2-related defect which is much less important in the $Ga_{1-x}Al_xAs$ alloys than in the $GaAs_{1-x}P_x$ alloys (Samuelson et al. 1986). Indeed, in $Ga_{1-x}Al_xAs$, the near environment of the $(As_{Ga}+As_i)$ defect is not changed by varying the alloy composition contrary to the same sort of defect in $GaAs_{1-x}P_x$; therefore one would not expect a very important alloy broadening effect for a highly localized deep level like EL2 in $Ga_{1-x}Al_xAs$ as we observed.

4. CONCLUSION

Deep levels in undoped MOCVD $Ga_{1-x}Al_xAs$ layers were investigated by DLTS. A main trap at E_C-0.80 eV for x = 0 having an apparent activation energy increasing with x was clearly identified as the EL2-related defect on the basis of the photocapacitance quenching effect typical of EL2 observed for the first time in this alloy.

REFERENCES

Azoulay R, Jusserand B, Le Roux G, Arsart P and Dugrand L 1986, J of Cryst Growth 77 546
Baraff GA and Schluter M 1985 Phys.Rev.Lett 55 2340
Best JJ 1979 Appl.Phys.Lett. 34 522
Bhattacharya PK, Ku JW, Owen STJ, Aebi V, Cooper BC and Moon RL 1980 Appl.Phys.Lett. 36 304
Bhattacharya PK, Subramanian S and Ludowise MJ 1984 J.Appl.Phys. 55 3664
Johnson NM, Burnham RD, Fekete D and Yingling D 1981 Defects in Semiconductors ed J. Narayan and TY Tan (Amsterdam ; North Holland) pp. 481-486
Matsumoto T, Bhattacharya PK and Ludowise MJ 1982 Appl.Phys.Lett. 41 662
Samuelson L, Omling P, Titze H and Grimmeiss HG 1981, J.Cryst.Growth 55 164
Samuelson L and Omling P 1986 Phys.Rev.B 34 5603
Vincent G, Bois D and Chantre A 1982 J.Appl.Phys. 53 3643
Von Bardeleben HJ, Stievenard D, Deresmes D, Huber A and Bourgoin JC 1986 Phys.Rev.B 34 7192
Wagner EE, Mars DE, Hom G and Stringfellow GB 1980 J.Appl.Phys. 51 5434
Yamanaka K, Naritsuka S, Kanamoto K, Mihara M and Ishii M 1987 J.Appl.Phys. 61 5062
Zou BL, Ploog K, Gmelin E, Zheng XQ and Schulz M 1982 Appl.Phys. A28 223

Paper presented at Int. Symp. GaAs and Related Compounds, Heraklion, Greece, 1987

Identification of manganese in AlGaAs alloys

F. Bantien and J. Weber

Max-Planck-Institut für Festkörperforschung, Heisenbergstr. 1, 7000 Stuttgart 80, Federal Republic of Germany

ABSTRACT: Low temperature photoluminescence measurements are performed on $Al_xGa_{1-x}As$ samples doped with manganese. The Mn-related $(D°,A°)$-luminescence band shifts linearly to higher energies with increasing Al-content for Al-concentrations up to $x \approx 0.3$. The binding energy $E_i(x)$ of the Mn-acceptor is determined from our measurements to be $E_i(x) = (0.11 + 0.33\ x)$eV. AlGaAs samples with Al-concentrations higher than x=0.3 show a different, broad luminescence band, which is assumed to be the intra-atomic $^4T_1$-$^6A_1$ transition of Mn^{2+}. Time-resolved and temperature-dependent photoluminescence measurements support this assignment.

INTRODUCTION

The photoluminescence technique has now been widely accepted as a nondestructive and highly sensitive method to characterize impurities in semiconductors. In this paper we report on the identification of Mn in AlGaAs layers, using Mn-related photoluminescence lines.

Manganese is an inadvertent impurity in all III-V semiconductors. Increased Mn-concentrations were found after heat treatments and Mn seems to be responsible for the thermal conversion to p-type conductivity in GaAs. (Klein et al. 1980)

Manganese substitutes for the group-III metal site in III-V compound semiconductors and gives rise to an acceptor level. In GaAs this acceptor energy is 113meV, as was determined by Schairer and Schmidt (1974). Recently, Schneider et al. (1987) determined the ground state symmetry of the neutral acceptor level as due to a $Mn^{2+}(d^5)$ negatively charged acceptor core, which binds a more delocalized hole. The optical recombination of the hole bound to the Mn with a donor bound electron gives rise to the well-known $(D°,A°)_{Mn}$ recombination (Schairer and Schmidt 1974).
We study this recombination process in the AlGaAs alloys. Our samples are Mn doped AlGaAs layers (x=0 to 0.9) grown by liquid phase epitaxy (LPE). (Bantien and Weber 1987)

PHOTOLUMINESCENCE SPECTRA

Typical photoluminescence spectra of our $Al_xGa_{1-x}As$ samples are shown in Fig. 1 for five different compositions x. The near bandgap luminescence clearly shows the bound exciton (BE) re-

combination of the unintentionally doped shallow donors and acceptors and the (D°,A°) pair recombination between these impurities. These luminescence lines shift with increasing x to higher energies due to the increase of the energy gap in the alloy. From the position of the BE lines the exact composition x of our layers is determined, according to the data reported in the literature. (Adachi 1985)
The increased linewidth in the alloy is due to the local fluctuations of the Al-Ga composition. The Mn-related $(D°,A°)_{Mn}$ recombination in GaAs shows the zero-phonon line at ~880nm. (Schairer and Schmidt 1974). This band is marked in Fig.1 by an arrow. The $(D°,A°)_{Mn}$ band exhibits a much stronger increase in linewidth compared to the (D°,A°) recombination of the shallow impurities. This observation can be explained by the larger shift of the Mn-binding energy with the composition x compared to the shallow impurities.

In all samples we still detect the near bandgap luminescence of the GaAs substrate. However, the intensity ratio between BE-luminescence and (D°,A°) recombination seems to vary due to absorption processes within the LPE AlGaAs layers.

In Fig. 2 the energetic position of the two Mn-related optical transitions is given for all our samples. For comparison, we also include the shift of the near bandgap BE-luminescence (A°,X). The difference between the $(D°,A°)_{Mn}$ energy position

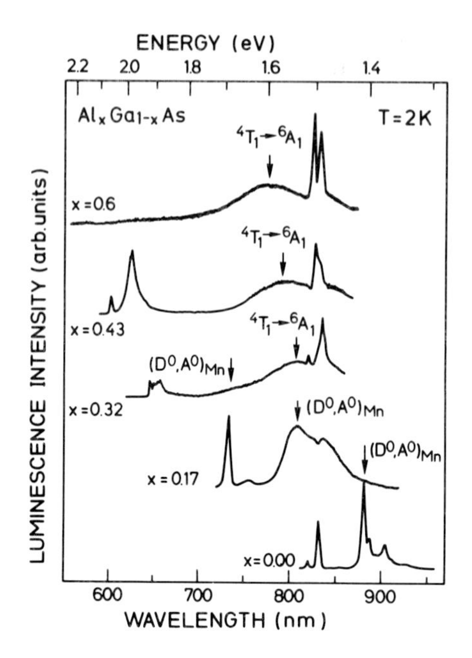

Fig. 1. Low temperature photoluminescence spectra of Mn doped AlGaAs layers.

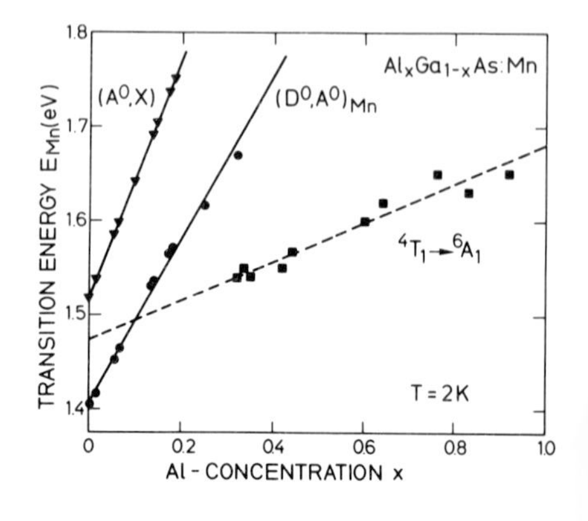

Fig. 2. Energy shift with the composition x for various luminescence lines.

and the bandgap energy gives us directly the increase in binding energy of the Mn acceptor:

$$E_i(x) = (0.11 + 0.33\ x)eV$$

At compositions higher than $x \approx 0.3$ a new Mn related transition appears. (Plot et al. 1986, Bantien and Weber 1987). The linear shift of this band with x is much smaller. We interpret this band as due to the internal optical transition within the $Mn^{2+}(d^5)$ transition metal-ion. Similar to GaP:Mn (Vink and Gorkom 1972) or the II-VI compound semiconductors ZnS (Gumlich et al. 1966), ZnSe or CdS (Langer and Richter 1966) a transition between the $^4T_1$ and $^6A_1$ crystal-field states occurs. The $^4T_1$ level is degenerated with the conduction band in GaAs and in AlGaAs up to $x \sim 0.3$. For $x > 0.3$ however the $^4T_1$ state has a lower energy than the conduction band.

TIME RESOLVED PHOTOLUMINESCENCE MEASUREMENTS

We performed time-resolved measurements on our AlGaAs samples, to verify the suggestion of an internal d-shell transition made in the last chapter. Table I lists our results. The $(D°,A°)_{Mn}$ recombination shows a slightly non exponential decay of the order of a few microseconds, typical for pair-transitions. The internal transition which we see in samples with $x \geqslant 0.32$ has a much longer lifetime of a few hundred microseconds. Such long lifetimes are expected for spin-forbidden optical d-shell transitions. For comparison we give the value for GaP:Mn, where only the internal transition was found.

Table 1. Measured lifetimes τ in microseconds of Mn-related transitions in $Al_xGa_{1-x}As$ alloys

Material	x	τ	Transition
AlGaAs	0	1.5	$(D°,A°)_{Mn}$
	0.06	4.2	
	0.17	5.4	
	0.27	6.8	
	0.32	2.4	
	0.32	350	$^4T_1 \to {}^6A_1$
	0.42	600	
	0.64	600	
GaP	-	1200	

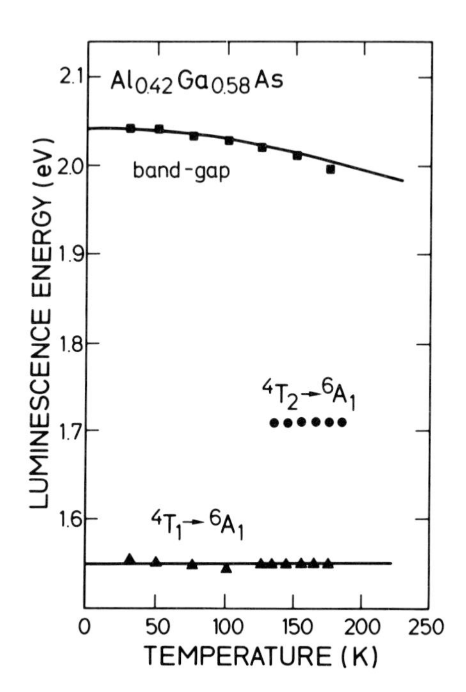

Fig. 3. Temperature dependence of the energy positions of the Mn-related transitions.

TEMPERATURE DEPENDENT PHOTOLUMINESCENCE MEASUREMENTS

Luminescence measurements at higher temperatures were performed. Figure 3 gives the results for a sample with x=0.42. Whereas the bandgap energy shrinks nonlinearly, the internal transition $^4T_1 \rightarrow {}^6A_1$ shows a constant energy.

At temperatures higher than 150K we find another luminescence band at ~1.7eV, which also shows no energy shift with temperature. We assign this transition to another internal transition ($^4T_2 \rightarrow {}^6A_1$) within the $Mn^{2+}(d^5)$ system.

Temperature-dependent time resolved measurements on a sample with x=0.42 give a constant lifetime of the internal transition up to 100K. Increasing the temperature to 200K reduces the lifetime of the transition by 5 orders of magnitude (~10ns).

We determine the same temperature dependent lifetimes for the $^4T_2 \rightarrow {}^6A_1$ transition as we measured for the $^4T_1 \rightarrow {}^6A_1$ transition, thus verifying our suggestion that both belong to the same center.

CONCLUSION

Manganese in AlGaAs can be identified by two different optical transitions. In alloys with $x \lesssim 0.3$ the $(D^\circ,A^\circ)_{Mn}$-luminescence band is found, whereas in alloys with $x \gtrsim 0.3$ an internal transition within the $Mn^{2+}(3d^5)$ ion is observed.

ACKNOWLEDGEMENT

We would like to thank E. Bauser and H.J. Queisser for their support of this work, and W. Heinz for technical assistance during the measurements.

REFERENCES

Adachi S 1985 J. Appl. Phys. 58 R1
Bantien F, and Weber J 1987 Solid State Commun. 55 865
Gumlich H E, Pfrogner R L, Shaffer J C, and Williams F E, 1966 J. of Chem. Phys. 44 3929
Klein P B, Nordquist P E R, and Siebenman P G 1980 J. Appl. Phys. 51 4861
Langer D W, and Richter H J 1966 Phys. Rev. 146 554
Plot B, Devaud B, Lambert B, Chomette A and Regreny A 1986 J. Phys. C 19 4279
Schneider J, Kaufmann U, Wilkening W, Baeumler M, and Köhl F 1987 Phys. Rev. Lett. 59 240
Schairer W, and Schmidt M 1974 Phys. Rev. B 10 2501
Vink A T, and Gorkom G G P. 1972 J. of Lum. 5 379

Properties of DX-center-like traps in AlGaSb

Y. Takeda, Y. Zhu, and A. Sasaki

Department of Electrical Engineering, Kyoto University
Kyoto 606, Japan

ABSTRACT: All of the major characteristics of the DX-center in AlGaAs, such as a strong composition dependence of the concentration, differences among activation energies for thermal electron-emission, electron-capture and photoionization, a negligible capture rate at low temperatures, persistent photoconductivity, a linear dependence of the trap concentration on the donor concentration, and a chemical shift of the trap properties have experimentally been observed for the deep electron-trap in AlGaSb. The origin of the DX-center in AlGaAs and the deep electron-trap in AlGaSb is discussed.

1. INTRODUCTION

In Te-doped AlGaSb, a deep electron-trap which has quite similar characteristics to the so-called DX-center in AlGaAs (Lang 1979) has been detected (Takeda 1987a, 1987b). The observed common characteristics were i) a strong composition dependence of the concentration of the deep electron-trap, ii) differences among activation energies for thermal electron-emission, electron-capture and photoionization, iii) a low capture rate at low temperatures, and iv) a persistent photocapacitance (Takeda 1987b). The DX-center possesses other characteristic features such as v) a persistent photoconductivity, vi) a linear dependence of the trap concentration on the donor concentration, and vii) a chemical shift of the trap properties (Lang 1979).

In the band structure of AlGaSb, Γ-, L- and X-band minima ordering is similar to that in AlGaAs, but the energy separations among them are much smaller and each band crosses at a lower Al composition (Cheng 1976, Alibert 1983). Thus, comparison of the composition dependence of the trap properties in AlGaSb and AlGaAs should be one of the key points for the discussion. Since both Te and Se occupy the Sb-site in AlGaSb, there should be a chemical shift of the trap properties as observed in AlGaAs (Lang 1979). Another interesting point is a possible relation of the Sb-vacancy to the origin of the deep electron-trap in AlGaSb as proposed for the origin of the DX-center in AlGaAs (Lang 1979).

In this paper, we will describe the properties of the deep electron-trap in $Al_xGa_{1-x}Sb$ including the observation of persistent photoconductivity, donor concentration dependence, and donor species (Te and Se) dependence of the trap properties in AlGaSb. Origin of the DX-center will be discussed based on the results of both AlGaAs and AlGaSb.

2. EXPERIMENTS

AlGaSb layers were grown on Te-doped GaSb substrates by LPE at 500°C. The Al composition was covered from 0.0 to 0.77. At a higher Al composition the crystal quality degraded due to a larger lattice-mismatch between the epitaxial layer and the substrate. Au/AlGaSb Schottky barriers were fabricated for DLTS and C-V measurements. Semi-transparent gold film was evaporated and periphery was etched out for photocapacitance measurement. For the Hall effect measurement, the n-type epitaxial layer was electrically isolated from the conductive GaSb substrates by p-n junction.

3. RESULTS

Composition dependence-In Te-doped $Al_xGa_{1-x}Sb$ with x from 0.0 to 0.2 an electron-trap was not detected by DLTS measurements in the temperature range of 77-300K. In Te-$Al_xGa_{1-x}Sb$ with x beyond 0.2 its concentration increased steeply with x in the range of 0.3-0.5. For x>0.5 it saturated and was by more than two order of magnitude greater than shallow donor concentration (Takeda 1987b). This composition dependence is quite similar to that observed in AlGaAs (Lang 1979). However, in Se-doped $Al_xGa_{1-x}Sb$, deep electron-traps were detected even in $Al_{0.1}Ga_{0.9}Sb$ as shown in Fig. 1. Double peaks were also observed in Se-$Al_{0.3}Ga_{0.7}Sb$.
Activation energies-Differences among the activation energies for such as thermal electron-emission, electron-capture, and photoionization are the characteristic features of the DX-center in AlGaAs. In Te-doped $Al_xGa_{1-x}Sb$ those activation energies measured at x, *e.g.*, 0.5 are as follows: E_e for thermal emission is 0.48eV, E_c for electron-capture is 0.35eV, E_H obtained from Hall measurement is 0.12eV, and E_n for photoionization is 0.86eV.
Capture rate-In the thermally stimulated capacitance (TSCAP) measurements at x, *e.g.*, 0.4 irreversible temperature cycles were observed. The capacitance increased by light illumination and it was persistent after removal of light when temperature was below ∿120K, indicating a negligibly small capture rate (Takeda 1987b). These characteristics are exactly the same as those observed in AlGaAs (Lang 1979). In the DLTS measurements a long injection pulse width (>20ms) was necessary to fill up the trap around the temperature at peak due to a very small capture rate.
Hall effect measurements-Increased electron concentration by light illumination in the Hall effect measurement was persistent after removal of light at low temperatures (<∿120K). Electron mobilities also increased under the increased electron concentration as observed in AlGaAs (Chand 1984). The results in Te-AlGaSb are listed in Table 1. Temperature dependences of free electron concentration showed two distinct slopes in the samples with x>0.3, indicating existence of two major levels of donor; the shallow and the deep. The composition dependence of the thermal activation energies for the deep trap is shown in Fig. 2.
Donor concentration dependence-If the deep electron-trap is directly related to the doped donor, its concentration should linearly change with the doped donor concentration. The linearity was confirmed within the experimentally covered range between $\sim 3\times10^{16}$ and $\sim 3\times10^{17}$ cm^{-3}.
Chemical shift-In addition to the Te-doping, Se-doping was tried and trap properties were investigated in a limited number of AlGaSb. As mentioned above, in the DLTS measurement, two peaks were detected in Se-doped $Al_{0.1}Ga_{0.9}Sb$ (E_{e1}=0.24eV and E_{e2}=0.37eV) and $Al_{0.3}Ga_{0.7}Sb$ (E_{e2}=0.35eV. E_{e1} was not calculated since this peak did not shifted with temperature due to a positive peak overlapped). The activation energy, 0.35eV, is lower than that in Te-doped $Al_xGa_{1-x}Sb$ at the same x, 0.3. The capture cross-sections were of the same order (10^{-11}∿10^{-12} cm^2) as those in Te-doped AlGaSb.

4. DISCUSSION

It has been shown that the electron-trap in donor-doped $Al_xGa_{1-x}Sb$ possesses all of the major properties of the DX-center in $Al_xGa_{1-x}As$. Thus, it can be stated that the mechanism to form the deep electron-trap in AlGaSb is the same as that of the DX-center in AlGaAs. Therefore, the model for the origin of the DX-center in AlGaAs should also work for the deep electron-trap in AlGaSb if the model is general. From the composition dependence of

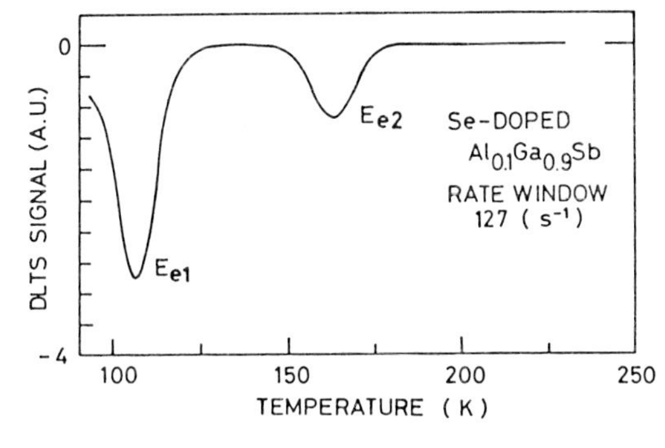

Fig. 1. DLTS spectrum observed in Se-doped $Al_{0.1}Ga_{0.9}Sb$. From the two peaks activation energies for thermal emission were calculated as E_{e1}=0.24eV and E_{e2}= 0.37eV.

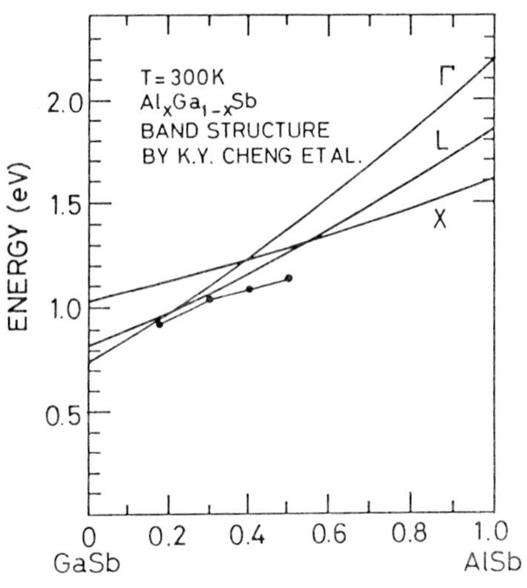

Fig. 2. Composition dependence of Γ-, L-, and X-band minima in $Al_xGa_{1-x}Sb$. Solid circles represent the position of the Te-donor level in the band gap obtained from Hall measurement.

Table 1 Electrical properties of Te-doped $Al_xGa_{1-x}Sb$ in the dark and after illumination.

	300K Dark		77K Dark		After illumination (Persistent)	
x	n (cm^{-3})	μ ($cm^2V^{-1}s^{-1}$)	n (cm^{-3})	μ ($cm^2V^{-1}s^{-1}$)	n (cm^{-3})	μ ($cm^2V^{-1}s^{-1}$)
0.2	5.0×10^{17}	411	2.3×10^{17}	876	2.3×10^{17}	976
0.3	8.7×10^{17}	197	5.1×10^{17}	288	6.8×10^{17}	296
0.4	8.4×10^{17}	182	3.8×10^{16}	91.3	7.0×10^{17}	208
0.5	4.1×10^{17}	119	---	---	4.1×10^{17}	170

the band structure and that of the deep trap concentration, models of a simple association of the donor with the L- or X-band minima and a band-crossing have been denied (Takeda 1987b).

As for the possibility of the donor-V_{Sb} complex formation as the origin for the deep trap, it is not plausible. As investigated in detail by Y. Takeda *et al.* (1985), the acceptor in undoped LPE-grown $Al_{0.1}Ga_{0.9}Sb$ is intrinsic and its concentration has a unique dependence on the growth temperature. At the growth temperature of 500°C employed for the present work, the acceptor concentration is $\sim 1\times10^{17}\ cm^{-3}$. The concentration of the deep electron-trap ($\simeq n_{300K}-n_{77K}$) is greater than that value as can be seen from Table 1. It was shown that the residual acceptor in GaSb is the complex of $Ga_{Sb}V_{Ga}$ (van der Meulen 1967). This is reasonable and a formation of Sb-vacancy requires a higher energy than the Ga-vacancy formation (Van Vechten 1980).

Considering the discussion given above and the very linear dependence of the trap concentration on the donor concentration, it is very likely that the deep trap is formed by the doped donor itself. If it is so, a change of the properties of the deep electron-trap by the donor species can be expected. As shown above, in Se-doped $Al_xGa_{1-x}Sb$, deep electron-traps were detected even in $Al_{0.1}Ga_{0.9}Sb$ at which composition deep traps were not detected in Te-doped AlGaSb. Difference between the central cell potentials of Te and Se atom and the different covalent radii are expected to cause the difference of the activation energies. This is exactly the model proposed very recently by E. Yamaguchi (1987) for the origin of the DX-center. The experimental results in AlGaSb obtained so far fit very well to the numerical values calculated by Yamaguchi (1987).

5. SUMMARY

Properties of the deep electron-trap in Te-doped $Al_xGa_{1-x}Sb$ and Se-doped $Al_xGa_{1-x}Sb$ have been revealed by the DLTS, thermally stimulated capacitance, photocapacitance, and Hall effect measurements. It was shown that the deep electron-traps possess all of the major properties of the DX-center observed in $Al_xGa_{1-x}As$. From the comparison of the composition dependences of the trap properties in AlGaAs and AlGaSb, those simple band-structure-related models for the origin of the deep traps have been denied. Central-cell potential model appeared to be most plausible to explain the experimental results in AlGaSb.

Acknowledgment. This work was supported in part by the Scientific Research Grant-in-Aid for Special Project Research on "Alloy Semiconductor Physics and Electronics" from the Ministry of Education, Science and Culture, Japan.

References

Alibert C, Joullie A, Joullie A M and Ance C 1983 *Phys. Rev.* B 27 4946
Chand H, Henderson T, Klem J, Masselink W T, Fischer R, Chang Y C and Morkoc H 1984 *Phys. Rev.* B 30 4481
Cheng K Y, Pearson G L, Bauer R S and Chadi D J 1976 *Bull. Am. Phys. Soc.* 21 365
Lang D V, Logan R A and Jaros M 1979 *Phys. Rev.* B 19 1015
Takeda Y, Noda S and Sasaki A 1985 *J. Appl. Phys.* 57 1261
Takeda Y, Gong X C, Zhu Y and Sasaki A 1987a *Jpn. J. Appl. Phys.* 26 L273
Takeda Y, Zhu Y and Sasaki A 1987b *Inst. Phys. Conf. Ser. No. 83* (Bristol: Institute of Physics) pp.203-208
van der Meulen Y J 1967 *J. Phys. Chem. Solids*, 28 25
Van Vechten J A 1980 *Handbook on Semiconductors* (Amsterdam: North-Holland) pp. 1-111
Yamaguchi E 1987 *J. Phys. Soc. Jpn.* 56 2835

Paper presented at Int. Symp. GaAs and Related Compounds, Heraklion, Greece, 1987

Behaviour of heavily doped Si atoms in MBE grown GaAs revealed by X-ray quasi-forbidden reflection (XFR) method and photoluminescence measurement

I. Fujimoto, N. Kamata, K. Nakanishi, H. Katahama and Y. Shakuda
ATR Optical and Radio Communications Research Laboratories
Twin 21 MID Tower, 2-1-61 Shiromi, Higashi-ku, Osaka 540, Japan

K. Kobayashi and T. Suzuki
Science and Technical Research Laboratories of NHK
1-10-11 Kinuta, Setagaya-ku 157, Japan

Abstract
By use of both XFR and photoluminescence (PL) measurements, the anomalous behaviour of Si dopants in MBE grown GaAs is investigated at high doping up to $2\times10^{20}cm^{-3}$. Three broad emission bands of PL are observed. The detailed study of these PL spectra including the change of spectra after thermal annealing, together with the XFR results, indicates that autocompensation mechanism assuming Si_{Ga}-Si_{As} pairs does not dominate, but rather that the formation of Si_{Ga} clusters and / or their complexes with native defects plays an important role.

1. Introduction

Heavily doped GaAs is useful for application to high speed electronic and optical devices. Silicon is widely used as a n-type dopant in GaAs grown by molecular beam epitaxy (MBE). However, carrier density saturates or even decreases at Si concentration $[Si]>6\times10^{18}cm^{-3}$. The mechanism of this saturation has been studied by use of various techniques such as photoluminescence (PL) measurement, IR absorption and PIXE. Some mechanisms including the autocompensation (Narusawa et al. 1984, Ogawa 1985) assuming $[Si_{Ga}]\sim[Si_{As}]$ have been proposed, but no definite explanation has been obtained.

The direct determination of the lattice location is particularly important to understand the saturation mechanism because of amphoteric nature of Si. In a previous paper (Fujimoto et al. 1984), the lattice location of Si in GaAs was analyzed by use of X-ray quasi-forbidden reflection (XFR) method. The results in heavily doped GaAs : Si grown by MBE indicate that most of Si atoms occupy Ga site, which is contrary to autocompensation mechanism. On the other hand, autocompensation mechanism dominates in GaAs : Si grown by LPE.

In this work, the PL spectra of heavily doped GaAs : Si are measured in addition to the XFR measurements. Broad bands below gap are observed. Thermal annealing effect on the PL spectra is also investigated. From the results of both XFR and PL measurements, carrier saturation mechanism and thermal behaviour of Si atoms are discussed.

2. Experimental

The GaAs : Si specimens used in this work were grown on (100)substrate by MBE at a substrate temperature Ts=600°C, with flux ratio J_{As} / J_{Ga}=1.4~5.0 and growth rate of 1.3~2.6 μm/h. A face-to-face annealing was carried out at 700~850°C in N_2 gas for 2 hours. PL measurements were performed using a 1 m grating monochromator and a cooled Ge photodetector. The specimens were excited by an Ar ion laser (514.5 nm) or a halogen lamp.The detail of the XFR measurements was described by Fujimoto (1984).

3. Results

The results of the XFR measurements of heavily doped GaAs : Si were shown by Fujimoto et al. (1985). Figure 1 shows the normalized PL spectra of GaAs : Si specimens with various Si concentrations at 11K excited by the Ar ion laser. The band-edge emission at 830 nm is dominant in doping range below $5\times10^{18}cm^{-3}$ but its intensity decreases rapidly with the increase of Si concentration. The specimen with [Si]~$9\times10^{18}cm^{-3}$ shows both band-edge and below-band emission. Three broad bands are observed at 1000, 1060 and 1250 nm depending on Si concentration; The band at 1000 nm which appears in the range of [Si]~$1\times10^{19}cm^{-3}$ corresponds to so called self-activated (SA) center (Williams and Bebb 1972).

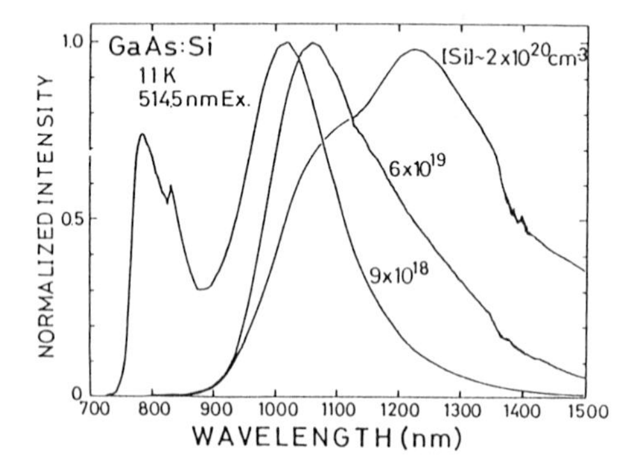

Fig. 1 Normalized photoluminescence spectra of GaAs : Si with Si concentration in [Si] over $9\times10^{18}cm^{-3}$

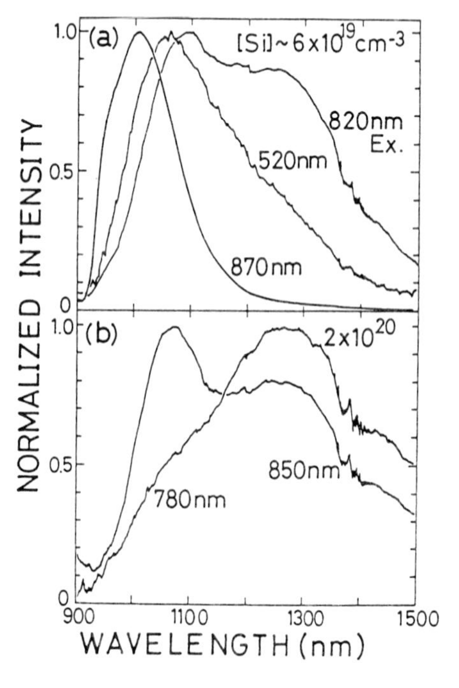

Fig. 2 Excitation wavelength dependence of photoluminescence spectra in heavily doped GaAs : Si
(a)[Si]~$6\times10^{19}cm^{-3}$
(b)[Si]~$2\times10^{20}cm^{-3}$

The PL spectra are shown in Fig. 2 when the halogen lamp is used as a light source. In this case, the excitation power is much lower than the Ar ion laser. The features of the spectra vary by changing excitation wavelength. In specimen (a) ([Si]~6×10^{19}cm^{-3}), the 1060 nm band is dominant with 520 nm exciting source, while the 1000 nm band is observed with 870 nm excitation. More detailed excitation measurements indicate that the origins of the 1000 and 1060 nm bands are different. The 1250 nm band appears as a shoulder with 820 nm excitation source. The observation of the three bands in same specimen implies the coexistence of three kinds of complex centers. In more heavily doped specimen (b) ([Si]~2×10^{20}cm^{-3}), the 1250 nm band becomes dominant with Ar ion laser excitation(Fig. 1). The observation of the 1060 nm band with 850 nm excitation also shows the coexistence of two complex centers.

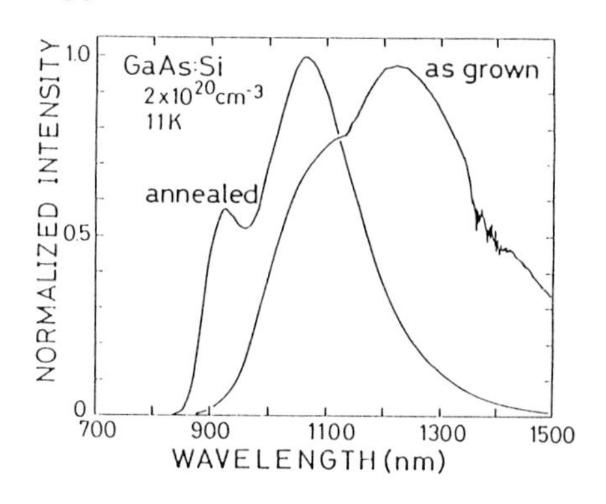

Fig. 3 Change of photoluminescence spectra caused by annealing in GaAs : Si with [Si]~2×10^{20}cm^{-3}

The PL spectra of the specimen (b) after annealing are shown in Fig. 3. The XFR intensity of these specimens was observed to decrease after annealing. Corresponding to the XFR results, PL spectra are also observed to change ; (1) A new peak appears at 930 nm (2) The intensity of 1250 nm band decreases and the 1060 nm band becomes dominant. For lightly doped specimens with [Si] under 6×10^{19}cm^{-3}, no remarkable changes have been observed both in the XFR intensity and in the PL spectra.

4. Discussion

The analysis of the XFR intensity indicates that most Si atoms occupy Ga site ([Si_{Ga}] ≫ [Si_{As}])in MBE-grown specimens contrary to conventional interpretation of autocompensation assuming Si_{Ga}-Si_{As} pair. Possible interpretations of the XFR results are discussed by Fujimoto et al. (1985) ; Si_{Ga} clusters, autocompensation plus high concentration of V_{Ga} and / or As_I, and the interstitial Si atoms in As atomic plane. However, those models including a large number of defects may be physically unreasonable. The most probable model is the formation of Si-clusters. Each cluster consists of several Si atoms on Ga sites as shown in Fig. 4. It does not act as a donor but it forms an inactive center for carrier generation.

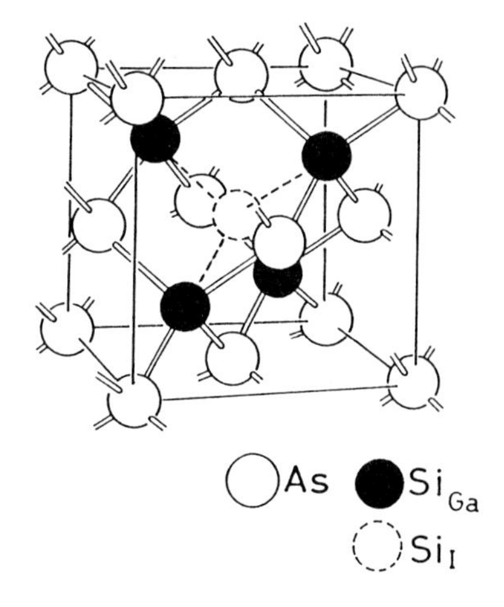

Fig. 4 Schematical view of proposed Si cluster model

The deep emission band of PL is discussed briefly. The 1250nm band is dominant in our most heavily doped GaAs : Si with [Si] over 9×10^{19}cm^{-3}(Fig.1). In this doping range, thermal annealing causes the decrease of both the intensity of XFR(Fujimoto et al. 1985) and that of the 1250nm band of PL(Fig.3). As the decrease of XFR intensity implies the Si site transfer from Ga site to As site, the origin of the 1250nm band is considered to be thermally metastable Si-complex centers. These centers may consist of Si_{Ga} complexes with native defects which allow Si site transfer rather easily. In this point of view, the origin of these centers must be distinguished from that of thermally stable 1060nm band which appears in the doping range $1\times10^{19}<[Si]<9\times10^{19}$cm^{-3}.

Rapid diffusion of Si atoms in heavily doped GaAs has been observed by many investigators (Greiner and Gibbons 1984, Kawabe et al. 1984). The phenomenon has been explained by assuming Si_{Ga}-Si_{As} pairs with vacancies. Since the doping range of Si diffusion enhancement overlaps with that of the appearance of 1250nm band and XFR results rule out Si_{Ga}-Si_{As} pairs, it is likely that the diffusion enhancement is not due to the formation of such Si pairs, but is related to the thermally metastable Si_{Ga}-complex centers with native defects. Heavily doped Si atoms in GaAs grown on a (111)A oriented substrate by MBE, which represents p-type conductivity, also show the behavior similar to that of (100) substrate. The detail will be reported elsewhere.

5. Summary

The XFR intensity and the PL spectra have been measured in order to study the behaviour of heavily doped Si atoms in MBE grown GaAs. The three below-gap bands are observed in the PL spectra. For the doping level over 6×10^{19}cm-3, the band at 1250 nm appears and its intensity decreased after annealing. The thermal annealing effect suggests that the rapid Si diffusion may have a close relation with Si_{Ga}-complex at 1250 nm. Combined with the XFR measurements, the three below gap emission bands may be related to Si_{Ga} clusters and / or their complexes with native defects. It is likely that the saturation or even decrease of carrier density in heavily Si doped GaAs grown by MBE is not due to autocompensation but to the formation of Si_{Ga} cluster and / or their complexes with native defects, in contrast to LPE-grown one.

Acknowledgement

The authors wish to thank Dr. S. Nishine, Dr. H. Okano, Mr. S. Tsuji, Mr. H. Seto and Mr. H. Tsuchiya for their helpful discussions, and Dr. Y. Furuhama for his encouragement through this work.

References

Fujimoto I 1984 Jpn. J. Appl. Phys. **23** L287
Fujimoto I, Kamata N, Kobayashi K and Suzuki T 1985 Inst. Phys. Conf. Ser. No. 79 Gallium Arsenide and Related Compounds, Karuizawa 199
Greiner M E and Gibbons J F 1984 Appl. Phys. Lett. **44** 750
Kawabe M, Matsuura N, Shimizu N, Hasegawa F and Nannichi Y 1984 Jpn. J. Appl. Phys. **23** L623
Narusawa T, Uchida Y, Kobayashi K, Ohta T, Nakajima M and Nakashima H 1984 Inst. Phys. Cont. Ser. No. 74, Gallium Arsenide and Related Compound, Biarritz 127
Ogawa M 1985 Inst. Phys. Cont. Ser. No. 79, Gallium Arsenide and Related Compound, Karuizawa 103
Williams E W and Bebb H B 1972 Semiconductors and Semimetals (New York, Academic) pp 321-392

Non-radiative recombination in aluminium gallium arsenide

J.H.Evans, A.R.Peaker, D.J. Nicholas, M.Missous, K.E.Singer
Department of Electrical Engineering and Electronics and The Centre for Electronic Materials, University of Manchester Institute of Science & Technology, P.O. Box 88, Manchester M60 1QD. England

ABSTRACT: A study of deep states in Molecular Beam Epitaxial layers of $Al_xGa_{1-x}As$ ($0.14 \leq x \leq 0.23$) grown with dimeric arsenic has been carried out. Direct measurement of the electron capture cross-sections and their temperature dependence of all detectable deep electron states has been undertaken and preliminary measurements of hole cross sections made.

1. INTRODUCTION

The ultimate aim of this work is to establish whether the deep levels commonly observed in MBE-grown $Al_xGa_{1-x}As$ act as recombination centres of sufficient significance to adversely affect the performance of bipolar devices such as lasers and heterojunction bipolar transistors (HBTs). Much work has been reported previously on the DX centres and on the activation energies of deep levels in AlGaAs. In addition, several studies of the effects of growth conditions on deep state concentration have been published (e.g. McAfee 1981, Yamanaka 1984, 1987, Mooney 1985, Naritsuka 1985). However, very little information has been reported on the recombination kinetics in AlGaAs, but in one case majority and minority capture processes have been characterised by Deep Level Transient Spectroscopy (DLTS) and Minority Carrier Transient Spectroscopy (MCTS) in material grown by metal organic vapour phase epitaxy (MOVPE) (Wu et al 1982). In a number of cases a negative correlation has been noted between luminescence efficiency and the concentration of particular deep states (Ando 1987, Akimoto 1986) implying that the minority carrier lifetime is deep state limited.

2. EXPERIMENTAL

The $Al_xGa_{1-x}As$ samples used in this study were grown in a Riber 2300 MBE system at a substrate temperature of 690 °C with the minimum V–III flux ratio necessary to maintain As-stable growth as determined from the reflection electron diffraction pattern. This approximates to a net V-III ratio of 1:1. GaAs was used as a source of As_2 molecules. The samples were n-type using silicon doping at the $10^{16}cm^{-3}$ level. Only material with low aluminium content was studied with x = 0.14, 0.21 and 0.23. Epitaxial aluminium deposited by MBE was used to form Schottky barriers, but in some cases semi-transparent nickel (15nm thick) was deposited to facilitate optical excitation of the AlGaAs. Deep level measurements were carried out using a Polaron S4600 DLTS spectrometer at a reverse bias of 2V. To investigate the rate of capture of electrons at the defect the fill pulse length was decreased until the DLTS peak was reduced, and the cross section determined in the usual way (Lang 1974). Minority carrier capture was characterised by replacing the electrical filling pulse by a pulse of near band-gap light which, under appropriate conditions, injects a carrier flux into the depletion region consisting predominantly of holes (Brunwin et al 1979). After the light source is removed, the hole emission transients are analysed by conventional DLTS processing. In some cases the minority carrier cross section of a majority carrier trap was measured using a sequence of a zero bias pulse and light pulse (Wu & Peaker 1982).

3. DLTS AND RECOMBINATION KINETICS IN AlGaAs

A DLTS spectrum for $Al_{0.14}Ga_{0.86}As$ is shown in FIG.1 with the DLTS output scaled so that the amplitude of a peak represents the dopant concentration. This is typical of all samples investigated in this work. Arrhenius plots of the majority carrier traps present in our samples are illustrated in FIG.2.

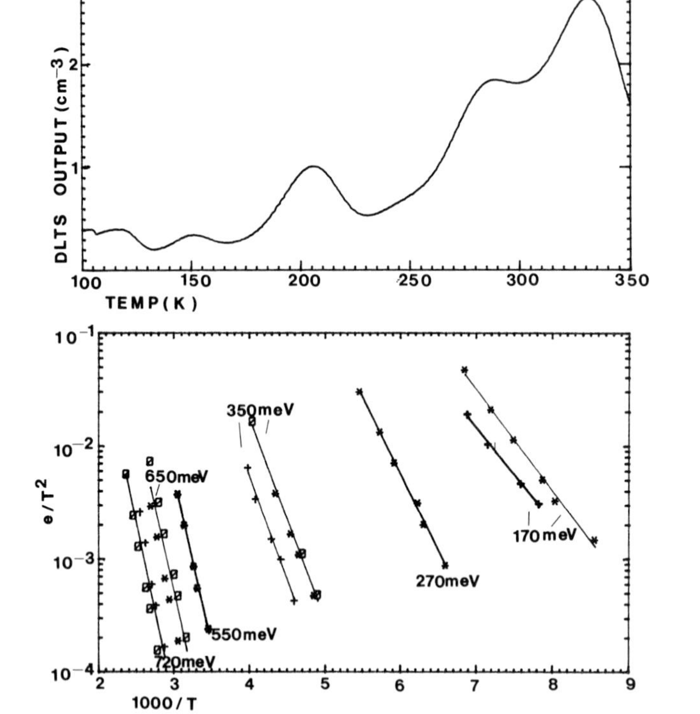

FIG.1 DLTS spectrum for $Al_{0.14}Ga_{0.86}As$ grown by MBE

FIG.2 Arrhenius plots of dominant electron traps in MBE $Al_xGa_{1-x}As$.

* x = 0.14
\+ x = 0.21
0 x = 0.23

The table presents the directly measured capture cross-section results for several electron traps. These show no apparent trend with temperature within the error of measurement except for the traps with activation energies of 170 and 550 meV. These were found to have an electron capture cross-section which increased with increasing temperature.

E_a and A are derived from fitting the temperature dependence of the electron emission transients $e_n = A/T^2 \exp(-E_a/kT)$. σ_n is the directly measured electron capture cross section and T is the temperature. The maximum error in the derivation of σ_n is ± 30%.

The minority carrier trap spectrum consisted of a number of overlapping peaks and a continuous background signal. One deep state with an activation energy of 740meV was apparent at a concentration of ~ $2x10^{13}cm^{-3}$. The presence of the background hole trapping signal complicated the minority carrier capture measurement on the electron traps. However, we can say that in all cases $\sigma_p > \sigma_n$ and for the 350meV state $\sigma_p > 5x10^{-15}cm^2$. In the case of the 550meV trap σ_p appears to be very large ($> 10^{-14}cm^2$).

TABLE OF DEEP STATE PROPERTIES IN $Al_xGa_{1-x}As$

x = 0.14

E_a (meV)	A $(sK^2)^{-1}$	σ_n (cm^2)	Temperature (K)	Typical N_T (cm^{-3})
650	2.26×10^6	8.7×10^{-17}	325-350	2.7×10^{14}
~550	$\sim3.20\times10^6$	2.2×10^{-16}	345	1.8×10^{14}
		1.6×10^{-19}	305	
350	2.13×10^5	1.9×10^{-16}	207-225	1.0×10^{14}
270	7.41×10^5	1.7×10^{-17}	150-161	3.0×10^{13}
170	5.53×10^4	7.8×10^{-20}	133	4.0×10^{13}
		4.9×10^{-20}	119	

x = 0.21

E_a (meV)	A $(sK^2)^{-1}$	σ_n (cm^2)	Temperature (K)	Typical N_T (cm^{-3})
720	3.72×10^6	5.4×10^{-17}	351-365	4.5×10^{14}
~550	$\sim1.04\times10^6$	5.6×10^{-22}	341	3.6×10^{14}
		2.3×10^{-22}	320	
350	1.02×10^5	4.7×10^{-17}	229-243	1.0×10^{14}

4. DISCUSSION

If a deep state is to act as a powerful recombination centre the carrier flux through the state must be comparable with that associated with other mechanisms. The deep state recombination can be limited by either the hole or electron flux whichever is the slowest process. In the simplest case, the minority carrier lifetime is given by $(\sigma N_T <v_{th}>)^{-1}$. σ is the capture cross section for the appropriate carrier and N_T the trap concentration; $<v_{th}>$ is the thermal velocity.

If we take the radiative lifetime of AlGaAs to be ~ 1μs at this doping level, then we conclude that in p-type material the recombination will be limited by the electron flux as $\sigma_p > \sigma_n$ and due to the higher hole concentration the hole flux will be very much larger than the electron flux. Consequently, it seems that only the 350meV state is likely to be important in x = 0.14 material. In n-type MBE the 170meV state can be ignored because of its low capture cross section. At room temperature the electron occupancy of this state will also be low due to thermal re-emission of the captured electron to the conduction band. The 550meV state has a very low electron cross section and irrespective of its hole cross section could only be significant at very low injection levels. The remainder could be important in n-type material if $\sigma_p > 10^{-15}cm^2$.

The deepest trap in our samples has an activation energy of 720 meV and is found in $Al_{0.21}Ga_{0.79}As$ and $Al_{0.23}Ga_{0.77}As$. It has an almost identical emission characteristic to the state named ME6 which has been related to PL efficiency (Yamanaka 1984, Akimoto 1986). It has also been suggested that this level is connected with the outgassing of arsenic oxide during MBE growth (Naritsuka 1985). The trap concentration is reduced by growth at substrate temperatures of 690 ^{0}C and higher.

The next deepest level has an activation energy of 650 meV and is observed in AlGaAs with x = 0.14 and x = 0.23. A defect state with this energy has been reported in MOCVD AlGaAs and is related to the presence of oxygen in the MOCVD growth process (Matsumoto 1982, Bhattacharya 1984). The same trap has also been recently reported in MBE-grown AlGaAs but only at low x-values (Yamanaka 1987).

One of the AlGaAs traps seen in this study has an activation energy of 350meV and is present in all the MBE samples we have grown. However in material with $x > 0.3$ it is difficult to detect because the DX centre dominates this part of the DLTS spectra. The trap is also seen in MOVPE material where it has been proposed that it is an oxygen related defect (Wallis 1981 and Sakamoto 1985). Matsumoto (1982) observed a similar state and measured its electron capture cross section at 140–180K. He reported an increase in cross section with increasing temperature and an electron capture cross section of $3x10^{-17}cm^2$ at 180K compared to our value of $4.7x10^{-17}cm^2$ over the range 229–243K.

Sakamoto has observed that in his material (MOVPE) another state at 270meV is present in identical concentrations. He believed this was a different charge state of the 350meV defect. We also observe a state at 270meV and indeed in our material both the 270meV and 350meV states have identical emission characteristics to those reported by Sakamoto, but the concentrations of the two states are not equal, typically differing by a factor of 3. It is important to note that in MOVPE material, Sakamoto observed a negative correlation between photoluminescence efficiency and the concentration of the two traps. Our measurements on MBE material predict that the 270meV state is expected to be an important recombination centre in n-type material even at moderately high excitation densities.

5. CONCLUSION

Our results indicate that there can be a marked similarity between many of the deep electron states in AlGaAs grown by MBE using dimeric arsenic and MOVPE materials. Some of the states appear to be associated with oxygen contaminants and one in particular with an activation energy of 270meV has characteristics which lead us to believe it could be a very significant recombination centre.

6. REFERENCES

Akimoto K, Kamada M, Taira K, Arai M and Watanabe N 1986 *J. Appl. Phys.* **59** 2833
Ando K, Amano C, Sugiura H, Yamaguchi M and Anne Saletes 1987 *Jap. J. Appl. Phys.* **26** L226
Bhattacharya P K, Subramian S and Ludowise M J 1984 *J. Appl. Phys.* **55** 3664
Brunwin R, Hamilton B, Jordan P and Peaker A R 1979 *Electron. Lett.* **15** 348
Hikosaka K, Mimura T and Hiyamizu S 1982 *Inst. Phys. Conf. Ser.* **63** 233
Lang D V 1974 *J. Appl. Phys.* **45** 3023
Matsumoto T, Bhattacharya P K and Ludowise M J 1982 *Appl. Phys. Lett.* **41** 662
McAfee S R, Tsang W T and Lang D V 1981 *J. Appl. Phys.* **52** 6165
Mooney P M, Fischer R and Morkoc H 1985 *J. Appl. Phys.* **57** 1928
Naritsuka S, Yamanaka K, Mihara M and Ishii M 1984 *Jap. J. Appl. Phys.* **23** L112
Naritsuka S, Yamanaka K, Mannoh M, Mihara M and Ishii M 1985 *Jap. J. Appl. Phys.* **24** 1324
Sakamoto M, Okada T and Mori Y 1985 *J. Appl. Phys.* **58** 337
Wallis R H 1981 *Inst. Phys. Conf. Ser.* **56** 73
Wu R H, Allsopp D and Peaker A R 1982 *Electron. Lett.* **18** 74
Wu R H and Peaker A R 1982 *Solid–St. Electron* **25** 643
Yamanaka K, Naritsuka S, Mannoh M, Yuasa T, Nomura Y, Mihara M and Ishii M 1984 *J. Vac. Sci. Technol.* **B2(2)** 229
Yamanaka K, Naritsuka S, Kanamoto K, Mihara M and Ishii M 1987 *J. Appl. Phys.* **61** 5062

Inst. Phys. Conf. Ser. No. 91: Chapter 3
Paper presented at Int. Symp. GaAs and Related Compounds, Heraklion, Greece, 1987

Optimized MBE procedure and characterization of AlGaInAs grown on InP

J.P. Praseuth, M. Quillec, M. Allovon, M.C. Joncour, J.M. Gérard and P. Henoc

Centre National d'Etudes des Télécommunications
Laboratoire de Bagneux
196 avenue Henri Ravera - 92220 BAGNEUX - FRANCE

ABSTRACT : We first describe a new procedure for the growth by molecular beam epitaxy of AlGaInAs alloys with very easy lattice-matching to InP. Next we present a complete investigation of crystalline, optical and electrical properties for the whole range of composition. Then we show that this material system is an interesting alternative to GaInAsP for microwave and optical devices applications.

INTRODUCTION

The AlGaInAs and GaInAsP quaternary alloys lattice-matched to InP are important materials for heterostructure microwave and optoelectronic devices. However, for molecular beam epitaxy (MBE) growth, it is more convenient to use AlGaInAs, since this material system contains three group III elements, whose sticking coefficients can be easily kept unity, and one group V element, thus avoiding the need of special MBE system due to the problem of P/As ratio control. Despite its unique potentiality, we find very few papers concerning AlGaInAs. Furthermore, alloy properties data they report are rather poor in comparison with other III-V semiconductors. It is often objected that the control of lattice-matching for this alloy system is difficult. Only recently can one find reports about the growth of AlGaInAs lattice-matched to InP with the pulsed molecular beam method (Fujii et al 1986) or with the conventional MBE using two indium cells (Goldstein et al 1987). In this paper, we first describe an improvement in the procedure of the latter method and give a comparison to the former method. Next, we present a complete investigation of structural, crystalline, optical and electrical properties of these alloys for the whole range of compositions. We then compare these characteristics with previous published data for AlGaInAs and also GaInAsP.

EXPERIMENTS

All AlGaInAs layers were grown on (100) oriented Fe-doped semi-insulating InP substrate in a Riber 2300 system with computer monitoring. The substrates were prepared with standard process. Surface oxides were removed by heating the substrate to ∿ 520°C in a As flux. The clean surface showed (2x4) reconstruction as indicated by reflection high energy electron diffraction (RHEED). The substrate temperature (T_S) was monitored with a calibrated infrared pyrometer. An ion gauge monitor measured the beam equivalent pressure (BEP) of the fluxes from the sources. Precise thickness control was made by chemical etching (H_2O_2:H_3PO_4:H_2O, 1:3:40). Growth were carried out at 520°C under As-stabilized surface conditions with a BEP (As) of 2.10^{-5} Torr. All layer were grown as to obtain a thickness of 1,5-2μm. Double crystal X-ray diffraction was used to measure the lattice-mismatch of the films and also to check the structural properties. Transmission electron microscopy (TEM) at 125 keV and photoluminescence (PL) at 10 K under Kr^+ laser excitation (6764 A) were involved in the characterization. The electron mobility and carrier concentration were measured by the Van der Pauw-Hall method in a 6 kG magnetic field at room temperature.

GROWTH PROCEDURE

The two indium cells method was briefly described previously (Goldstein et al 1987). The quaternary is obtained by the superimposition of the two end-ternaries $Al_{0.48}In_{0.52}As$ and $Ga_{0.47}In_{0.53}As$.
So the quaternary can be described as :

$$(Al_{0.48}In_{0.52}As)_z\ (Ga_{0.47}In_{0.53}As)_{1-z}$$

where z is the ratio of the growth rates GR1 and GR2 of the ternaries : $z/(1-z) = GR1/GR2$. If t1 and t2 are the relative lattice mismatches of the two ternaries with the substrate, that of the quaternary, t, can be expressed as : $t = zt1 + (1-z)t2$ (1) (in this expression, we neglect the slight difference in the elastic parameters). Clearly, if the ternaries are lattice matched ($t1 = t2 = 0$), then $t = 0$. Thus the complete control of the quaternary growth, for $0<z<1$, requires a good lattice matching of the two ternaries, and this is true for a wide range of growth rates.

In practice, the usual technique of lattice match control through fluxes measuring and adjusting is not appropriate. We rather use the following sequence :

- Initial cells calibration

After refilling, each cell is calibrated once for all (fig. 1). Besides, the G.R. of each binary (GaAs, InAs, AlAs) is calibrated through ternary layers thickness, combined with lattice mismatch X-rays measurement.

- Everyday, these "initial" data are corrected from a single flux measurement for each cell (see fig. 1). Since the activation energy remains constant for each element, we just keep the slope of the curve Lg|BEP| as a function of 1/T. The binary G.R. can be corrected using the last X-ray diffraction data obtained on a related ternary.

- From the new calibrations, it is easy to compute the cells temperatures required for any desired quaternary (or ternary).

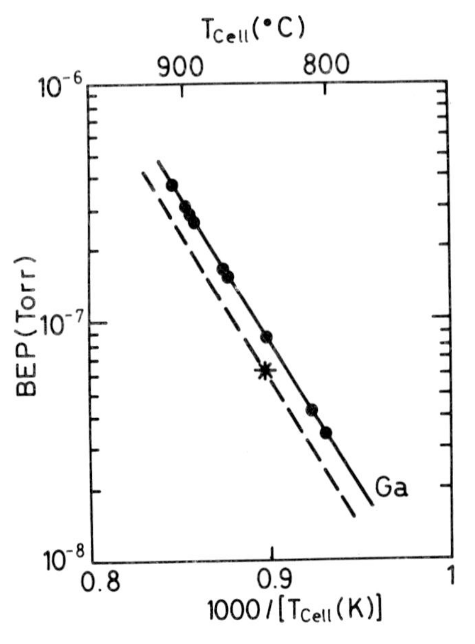

Fig. 1: The beam Equivalent Pressure calibration of the gallium cell is obtained once for all. The subsequent calibrations are deduced from a single (BEP,T) couple.

This procedure leads to routinely closely lattice matched

quaternary epilayers (within 5.10^{-4}) even for different alloys compositions in the same day. Another advantage of this technique is a significant improvement of the lattice matching of the quaternary layers with the substrate in terms of statistics compared to the ternaries. As they are due to many small fluctuations, the probability distributions of t1 and t2 are gaussians, centered on zero, with deviations σ_1 and σ_2. It can be shown that the fluctuation on z leads to second order terms. From (1), we deduce that t follows also a gaussian distribution, with a deviation $\sigma = \sqrt{(z\sigma_1)^2 + |(1-z)\sigma_2|^2}$ if we take $\sigma_1 = \sigma_2 = \sigma'$, $\sigma = \sigma'\sqrt{z^2+(1-z)^2}$
So that $\sigma<\sigma'$ is minimum for $z = 1/2$ ($\sigma_{minimum} = \sigma'/\sqrt{2}$).
It has to be noted that the use of a single indium cell for InGaAlAs growth (Alavi et al, 1987) leads to a degraded statistics compared to the ternaries. The minimum deviation in that case is $\sqrt{3}$ times higher than in our case.

Finally, comparing with the pulsed molecular beam method, we can list some advantages of our method (a) no continuous shutter actions are needed, thus avoiding mechanical failure ; (b) larger growth rates up to 3µm/h comparing to 0.5µm/h, this makes possible growth of heterostructures for devices applications in a reasonable time.

AlGaInAs QUATERNARY ALLOYS PROPERTIES

- Structural properties

The Full Widths at Half Maximum (FWHM) of the X-rays double crystal diffraction profiles are in the range 20 to 30 seconds of arc. They are independent of the composition and larger than the theoretical values. However, FWHM of the corresponding substrate is also superior to theoretical value (9 arc.s). Therefore, we can conclude that we obtained AlGaInAs quaternary alloys of good structural properties for the whole range of composition.

- Crystalline properties

TEM micrographs show excellent crystalline quality for z up to 0,60. However, a granular aspect is observed for the z = 0.75 alloy and this contrast incrases for AlInAs (z=1). This feature is related to a degradation of crystalline quality, we conclude however that quaternary alloys of good crystalline quality can be grown for z lower than 0,60 (Praseuth et al, 1987).

- Optical properties

The energy of the PL emission peak versus alloy composition z and the corresponding FWHM are shown in figure 2. This is an estimation of the energy band gap to within 5 meV. We find that Eg varies linearly from 0,8 to 1,52 eV for z from 0 to 1 without measurable bowing effect. The FWMMs of the PL peaks of quaternary alloys with z $\leq$ 0,60 in the range 7 to 10 meV are much narrower than previous published data : 34 meV at 4 K for z = 0,6 (Stanley et al 1982), 35 meV at 5K for z = 0,5 (Barnard et al 1982) and 23 meV at 4 K for z = 0,4 (Genova et al 1987). They are in the same order as those obtained by the pulsed molecular beam method (Fujii et al 1986).

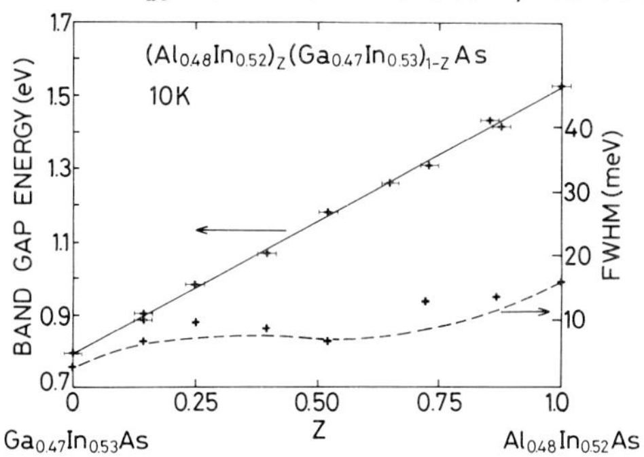

Fig. 2: The band gap energy of AlGaInAs and the FWHM of the PL spectrum as a function of composition

Therefore, we can conclude that our technics lead to quaternary alloys of good optical properties.

- Electrical properties

All unintentionally doped layers were n-type with N_d-N_a values less than $5.10^{15}cm^{-3}$. Room temperature electron mobilities as a function of z are 11000 $cm^2/V.s.$ for GaInAs (z=0), 1200 $cm^2/V.s.$ for AlInAs (z=1) and presents a minimum of 880 $cm^2/V.s.$ for z = 0,88. This minimum is probably due to the contribution of alloy scattering, since the disorder is larger for the quaternary than for the ternaries. For the alloy with z = 0.40, electron mobility up to 5100 $cm^2/V.s.$ is the best result up to now as compared to 1100 $cm^2/V.s.$ reported by Genova et al (1987). Futhermore, when we compare as a function of band gap energy the electron mobility of AlGaInAs with the best published results of LPE GaInAsP (Benchimol et al 1983) (fig. 3). We can notice that for a band gap less than 1.15 eV, electron mobility is of the same order. This is very satisfactory since we made comparison between a MBE material and a LPE one which has generally better characteristics.

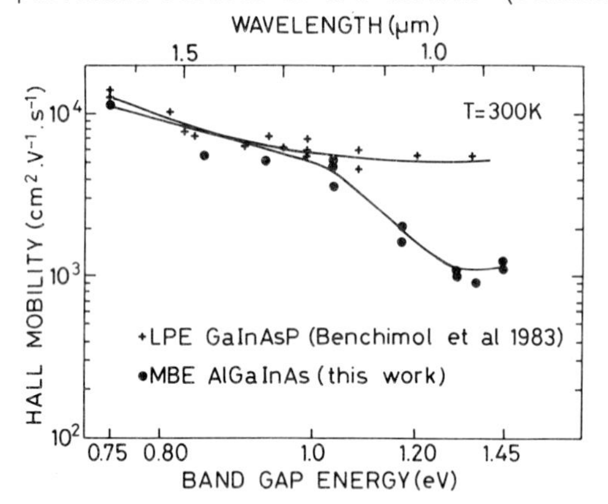

Fig. 3 : Comparison of electron mobility of AlGaInAs with that of GaInAsP as a function of wavelength.

CONCLUSION

AlGaInAs can be easily grown using the two indium cells method for the growth of quaternary AlGaInAs alloys with very easy lattice-matching to InP. An investigation of some of their crystalline optical, and electrical properties shows that we have grown high quality material in this system for z up to 0,60 (0,75<Eg(eV)<1.15). We thus conclude that AlGaInAs is an interesting alternative to GaInAsP for microwave and optical device applications.

REFERENCES

Alavi K, Cho AY, Capasso F and Allam J, J. Vac. Sci. Technol. B5, p 802, 1987
Barnard J, Wood CEC and Eastman LF, 1982 IEEE Electron. Dev. Lett. EDL-3, 318.
Benchimol JL, Quillec M and Slempkes S, 1983, J. Cryst. Growth 64, 96.
Fujii T, Nakata Y, Sugiyama Y, and Hiyamizu S, 1986, Jap. J, App. Phys. 25, L 254.
Genova F, Morello G and Rigo C, 1987, J. Vac. Sci. Technol. B5, 811.
Goldstein L, Praseuth JP, Joncour MC, Primot J, Henoc P, Pelouard JL and Hesto P, 1987, J. Cryst Growth. 81, 396.
Praseuth JP, Joncour MC, Gérard JM, Henoc P and Quillec M, to be published in J. Appl. Phys.
Stanley CR, Welch D, Wicks GW, Wood CEC, Palmstrom C, Pollak FH and Parayanthal P, 1982, Inst. Phys. Conf. Ser. 65, 173.

LPEE growth and characterization of $In_{1-x}Ga_xAs$ bulk crystals

T. Bryskiewicz,* M. Bugajski,** B. Bryskiewicz, J. Lagowski and H.C. Gatos

Massachusetts Institute of Technology, Cambridge, MA 02139, USA

ABSTRACT: A novel procedure for liquid phase electroepitaxial (LPEE) growth of highly uniform multicomponent bulk crystals has been developed and successfully applied to the growth of high quality bulk $In_{1-x}Ga_xAs$ crystals. $In_{1-x}Ga_xAs$ ingots 14 mm in diameter and up to 3 mm thick were grown on (100) InP substrates. In terms of homogeneity, electrical characteristics, and defect structure they are comparable to high quality thin LPE layers.

1. INTRODUCTION

Liquid phase electroepitaxy is a solution growth technique in which the growth process is induced and sustained solely by passing a direct electric current across the solution-substrate interface while the temperature of the overall system is maintained constant (Bryskiewicz 1986 and references therein). It has been found that after initial stages of growth the solute electrotransport towards the interface becomes the dominant driving force for the growth (Bryskiewicz 1978, Jastrzebski et al 1978, Bryskiewicz et al 1980). Therefore, within a few minutes after an electric current is turned on the growth proceeds under isothermal and steady-state conditions. These features of electroepitaxy have proven (both experimentally and theoretically) to be uniquely suited for the growth of ternary and quaternary semiconductor compounds with constant composition (Daniele 1981 Bryskiewicz et al 1980). $Ga_{1-x}Al_xAs$ wafers as thick as 600 μm (Daniele et al 1981), $GaAs_{1-x}Sb_x$ (Biryulin et al 1983), $In_{1-x}Ga_xP$ (Daniele et al 1983), and $Hg_xCd_{1-x}Te$ (Vanier et al 1980) epilayers up to 200 μm, 120 μm, and 500 μm, respectively, grown by LPEE showed a remarkable uniformity of composition, varying by Δx=0.01-0.03 over their entire thickness. However, the growth procedures proposed thus far (Daniele et al 1981, Nakajima 1987) are suitable for the growth of uniform wafers a few hundred microns thick.

In this paper a novel procedure, useful for electroepitaxial growth of bulk crystals (several millimeters thick) of ternary and quaternary semiconductors is proposed. This novel procedure is successfully applied to the growth of high quality $In_{1-x}Ga_xAs$ bulk crystals.

2. NOVEL LPEE GROWTH PROCEDURE

The growth of highly uniform $In_{1-x}Ga_xAs$ bulk crystals was carried out in a novel vertical LPEE apparatus (Bryskiewicz et al 1987a), employed recently

*On leave from Microgravity Research Associates, Inc., Midland, TX 79701, USA.
**On leave from Institute of Electron Technology, 02-668 Warsaw, Poland.

to the growth of epitaxial quality GaAs bulk crystals (Bryskiewicz et al 1987b). A schematic diagram of the growth cell used in our growth experiments is shown in Fig. 1. During electroepitaxial growth solution is contacted by the substrate at the bottom and the source material at the top. Thus, the crystallizing material driven by the electric current is deposited onto the substrate while the solution is being continuously saturated with the source material. A very important characteristic of the growth cell seen in Fig. 1 is the shape of a graphite source holder which allows the current to bypass the source material. This results in a minimization of the Joule heating for an arbitrary form (monocrystal, polycrystal, or chunks) of the source material. The requirements for the source are thus limited to the compositional homogeneity and the chemical composition fitted to the composition of the crystal to be grown. In order to achieve these source characteristics, a procedure for the preparation of a macroscopically homogeneous source material had to be developed.

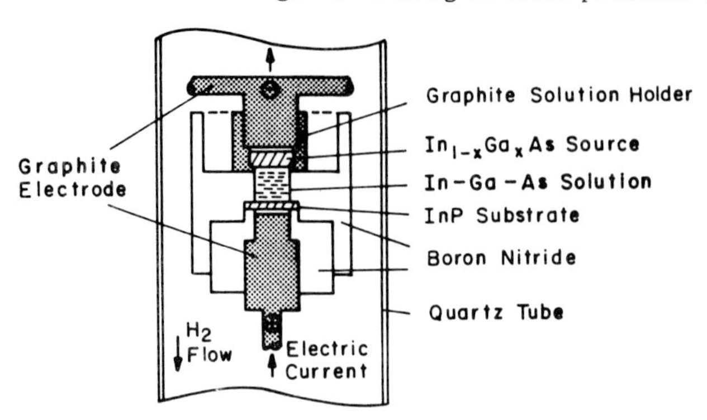

Fig. 1. Schematic diagram of the growth cell used for LPEE growth of $In_{1-x}Ga_xAs$ bulk crystals.

In this study the source material required for the growth of highly uniform $In_{1-x}Ga_xAs$ bulk crystals was prepared in a sealed quartz ampoule evacuated to 10^{-6} Tr. The inner wall of the quartz ampoule was covered with a pyrolytic carbon in order to prevent wetting. A semiconductor grade InAs-GaAs quasibinary mixture and a small amount of a high purity arsenic were sealed in the ampoule, heated up to about 20-30°C above the liquidus temperature and kept molten for one hour. High compositional homogeneity of the $In_{1-x}Ga_xAs$ source material was assured by rapid cooling of the ampoule. $In_{1-x}Ga_xAs$ bulk crystals with compositions between x=0.46 and x=0.48 were grown at 650°C on (100)-oriented, Sn-doped InP substrates. The grown ingots were 14 mm in diameter and up to 3 mm thick, i.e., suitable for slicing up to five wafers.

3. CRYSTAL CHARACTERIZATION

A microphotograph of the etch pits revealed on the (100)-oriented InP substrate and on the $In_{.52}Ga_{.48}As$ bulk crystal is seen in Fig. 2. Although the dislocation loops generation process is very likely to occur in this case near the surface, we did not observe any significant increase in the etch pit density between the InP substrate (EPD∿.5-2x10^5cm^{-2}) and $In_{1-x}Ga_xAs$ crystals.

Electrical parameters of the $In_{1-x}Ga_xAs$ bulk crystals grown from unbaked In-rich solutions are shown in Table I. It is expected that the free electron concentration in the low 10^{16}cm^{-3} range can be reduced considerably by using higher purity solution components and/or by baking the solution prior to each run (Bhattacharya et al 1983). The 300°K mobility as high as 8000 cm^2/Vs and the 77°K mobility of about 13,000 cm^2/Vs is

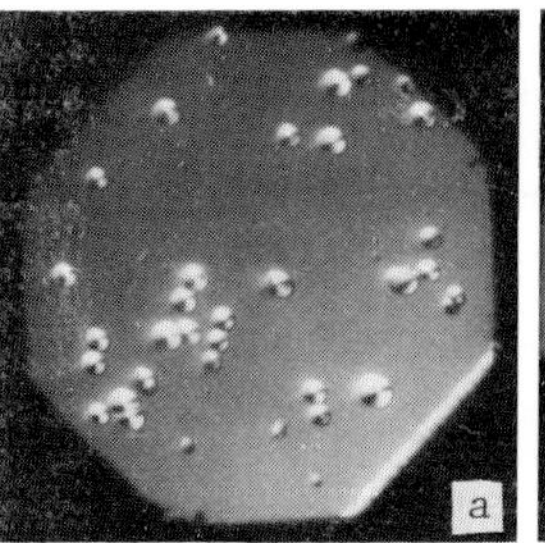

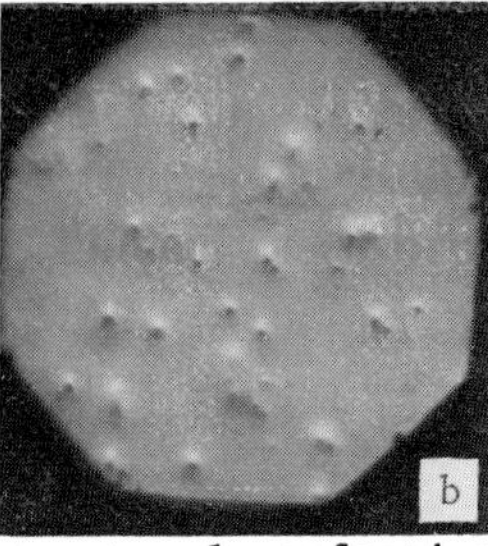

Fig. 2. Optical micrographs showing the typical etch pit pattern observed on (a) (100)-oriented InP substrate; (b) $In_{.52}Ga_{.48}As$ bulk crystal.

quite remarkable for the $10^{16}cm^{-3}$ free carrier concentration. These mobility values suggest low compensation and high homogeneity and structural perfection of the $In_{1-x}Ga_xAs$ ingots.

Table I. Electrical characteristics of LPEE $In_{1-x}Ga_xAs$ bulk crystals.

Composition (at.%)	Conductivity Type	Carrier Concentration (cm^{-3}) 300°K	77°K	Mobility (cm^2/Vs) 300°K	77°K
46	n	$2.7x10^{16}$	$2.4x10^{16}$	6240	13620
47	n	$4.5x10^{16}$	$4.1x10^{16}$	7780	11750
48	n	$4.6x10^{16}$	$4.0x10^{16}$	7640	13140

In addition, DLTS measurements did not reveal any measurable electron traps in these crystals.

Structural perfection as well as high compositional homogeneity of the $In_{1-x}Ga_xAs$ bulk crystals, comparable in quality with thin LPE layers, is documented by a high resolution photoluminescence (PL) spectrum shown in Fig. 3. This spectrum was recorded at 5°K for a nominally undoped n-type

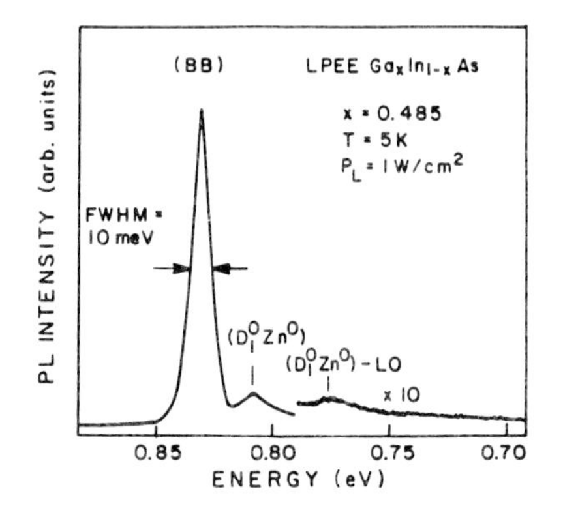

Fig. 3. 5°K PL spectrum of n-type $In_{.52}Ga_{.48}As$ grown on (100) InP substrate; $n = 4x10^{16}cm^{-3}$.

LPEE $Ga_{.52}Ga_{.48}As$ crystal. The dominant line at 0.8303 eV corresponds to the band-to-band transitions (BB). The above assignment was made on the basis of the line shape and the luminescence intensity vs. excitation density dependence (which appeared to be nearly quadratic). As seen in Fig. 3, the full width at half maximum (FWHM) of the (BB) line is equal to 10 meV. This value can be understood in terms of alloy broadening due to the random distribution of the In and Ga cations. From the model developed for $Al_xGa_{1-x}As$ (Schubert et al 1984) the alloy broadening of the BB transitions in $In_{1-x}Ga_xAs$ has been estimated to be in the 9.2-13.7 meV range. The spread in the calculated FWHMs results mainly from the uncertainties of the values of the heavy hole mass and band discontinuity at the $In_{1-x}Ga_xAs$/InP heterointerface. Nevertheless, an overall agreement between theory and experiment is satisfactory.

The second line in Fig. 3 located 20.4 meV below the (BB) peak is due to the presence of the residual Zn accetpor, and it corresponds to the donor-acceptor (DA) type of transitions. The binding energy E_A of the Zn acceptor estimated from the line position in different samples is E_A=20.6-21.5 meV, in close agreement with E_A=22±1 meV reported by Goetz et al (1983).

The measurements of the (BB) peak position were used to determine the compositional variations vs. a distance from the substrate and along the crystal surface. The results are shown in Fig. 4a and b, respectively. An excellent compositional homogeneity of the $In_{1-x}Ga_xAs$ ingots, both perpendicular and parallel to the growth direction, is evident. In both cases the composition fluctuations do not exceed 1%.

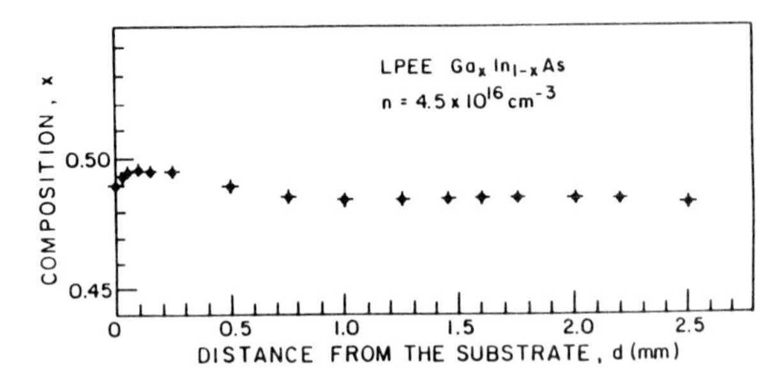

Fig. 4. Composition profile of $In_{.52}Ga_{.48}As$ obtained from PL measurements:

(a) along the growth direction

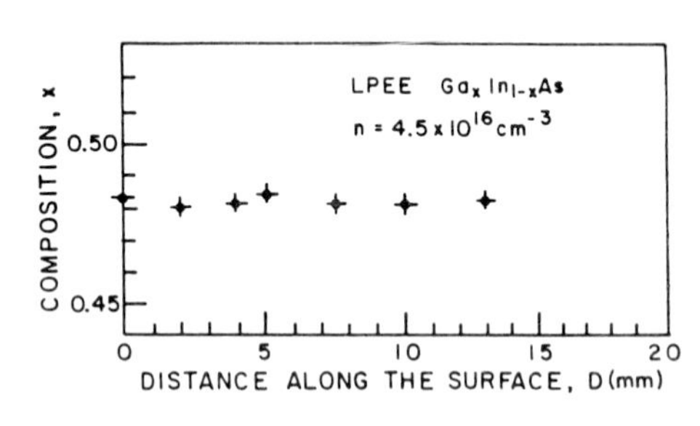

(b) perpendicular to the growth direction.

The authors are grateful to Microgravity Research Associates, Inc., and to the Air Force Office of Scientific Research for financial support.

4. REFERENCES

Bhattacharya P K, Rao M V and Tsai M-J 1983 J. Appl. Phys. 54, 5096

Biryulin Y F, Golubev L V, Novikov S V and Shmartsev Yu V 1983 Sov. Tech. Phys. 9, 68 (Pis'ma Zh. Tekh. Fiz. 9, 155)

Bryskiewicz T 1978 J. Cryst. Growth 43, 567

Bryskiewicz T, Lagowski J and Gatos H C 1980 J. Appl. Phys. 51, 988

Bryskiewicz T 1986 Prog. Crystal Growth and Charact. 12, 29 (Pergamon)

Bryskiewicz T, Boucher Jr C F, Lagowski J and Gatos H C 1987a J. Cryst. Growth 82, 279

Bryskiewicz T, Bugajski M, Lagowski J and Gatos H C 1987b J. Cryst. Growth (to be published)

Daniele J J and Hebling A J 1981 J. Appl. Phys. 52, 4325

Daniele J J and Lewis A 1983 J. Electron. Mat. 12, 1015

Goetz K H, Bimberg D, Jurgensen H, Selders J, Solomonov A V, Glinski G F and Razeghi M 1983 J. Appl. Phys. 54, 4543

Jastrzebski L, Lagowski J, Gatos H C and Witt A F 1978 J. Appl. Phys. 49, 5909; 50, 7269 (1979)

Nakajima K 1987 J. Appl. Phys. 61, 4626

Schubert E F, Gobel E O, Horikoshi Y, Ploog K and Queisser H J 1984 Phys. Rev. B 30, 813

Vanier P A, Pollak F H and Raccah P M 1980 J. Electron. Mat. 9, 153

Inst. Phys. Conf. Ser. No. 91: Chapter 3
Paper presented at Int. Symp. GaAs and Related Compounds, Heraklion, Greece, 1987

Magnetic-field-induced localisation and metal–insulator transition in n-type Si-doped MBE GaAs

P. Debray, M. Sanquer and R. Tourbot
Service de Physique du Solide et de Résonance Magnétique, CEN de Saclay, 91191 Gif-sur-Yvette Cedex, France

G. Beuchet and D. Huet
THOMSON-Semiconducteurs, R.D. 128, 91401 Orsay, France

ABSTRACT: Metal-insulator transitions in barely insulating and barely metallic samples of nominally uncompensated Si-doped n-type GaAs have been induced by the application of a magnetic field. The field is found to first cause a negative magnetoresistance indicating the suppression of weak localisation due to quantum interference. Increasing the magnetic field raises the resistance showing the onset of magnetic freeze-out leading ultimately to an insulating state. The results are in agreement with Shapiro's mobility edge phase diagram.

1. INTRODUCTION

Doped semiconductors have been widely studied to probe the nature of the metal-insulator transition in disordered electronic systems. However, theoretical investigations have produced conflicting results and our understanding of this phenomenon is far from being complete (Mott 1986). More work on well tailored and characterised samples is needed. Experimentally an external perturbation, such as an applied magnetic field, can be used to drive individual samples through the transition. The use of an applied magnetic field has the advantage that it does not require a set of samples and also that all the parameters of the sample, such as compensation, homogeneity etc., except the effective Bohr radius of the impurity, are kept constant. Inspite of this incentive the use of an applied magnetic field to probe the metal-insulator transition has been limited to only a few cases (Pepper 1986). In this paper we report on our results of magnetic-field-induced localisation and metal-insulator transition observed in n-type GaAs.

2. EXPERIMENTS

Two samples of Si-doped n-type GaAs with doping concentrations 10^{16} (S1) and $2\times10^{16}\,cm^{-3}$ (S2) were grown by standard MBE technique on semi-insulating GaAs substrates. These values of the donor concentration are close to the zero-field Mott critical density ($n_c^{1/3}\,a^* = 0.25$ which gives $n_c = 1.6\times10^6\,cm^{-3}$). The samples had thicknesses ~ 4μm which is greater than the relevant length scales at the temperatures used and are thus three-dimensional for electronic conduction mechanisms. The samples are nominally uncompensated. Magnetotransport and Hall measurements were carried out in the dark on cloverleaf samples using the van der Pauw technique in the temperature range 1.5-15K in transverse magnetic fields up to 5T using a fully automated cryomagnetic system.

3. RESULTS AND DISCUSSION

Electrical conductivity measurements in the absence of an applied magnetic field extrapolated to zero temperature gives $\sigma(0)$ equal to 2.58 and $0.063(\Omega cm)^{-1}$, respectively, for the S2 and S1 samples. The two samples can consequently, be considered as barely metallic and barely insulating. The temperature variation of the conductivity in the temperature range 1.5-10K was fitted to the theoretically expected form (Lee and Ramakrishnan 1985)

$$\sigma(T) = \sigma(0) + mT^{\frac{1}{2}} + BT \qquad (1)$$

where the first correction ($\sim T^{\frac{1}{2}}$) arises from electron interactions and the second correction is due to weak localisation effects. In the latter correction ($\sim T^{p/2}$) we have chosen $p = 2$, which is valid for coulomb interactions in the clean limit. The values for m are -1.3 and -2.1, respectively, for the barely metallic and barely insulating samples, while B is equal to 0.5 for both the samples. In both cases, therefore, we observe a negative interaction correction indicating the dominance of the Hartree term in the interaction.

Figure 1 shows the magnetic field dependence of the transverse conductivity σ_{xx} at 2K for both samples S1 and S2. The magnetic field is found to first cause a negative magnetoresistance. This is as expected. An applied magnetic field H destroys or suppresses the quantum interference

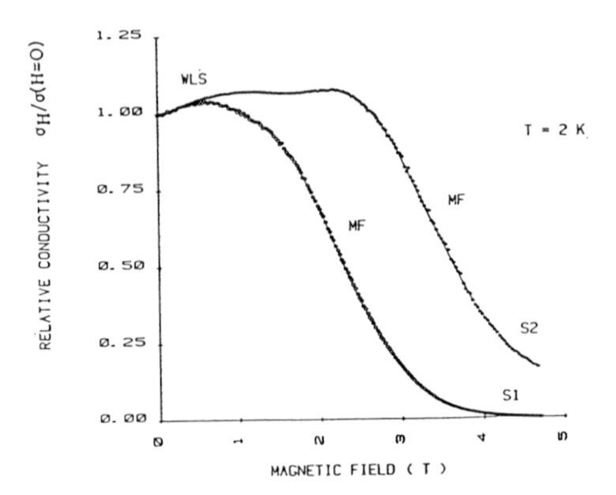

Fig. 1. Field dependence of the transverse relative conductivity of barely metallic S2 ($2\times10^{16}cm^{-3}$) and barely insulating S1 ($1\times10^{16}cm^{-3}$) Si-doped n-type GaAs at 2K. WLS indicates suppression of weak localisation and MF corresponds to magnetic freeze-out.

effect by introducing a phase factor. Hence it reduces the weak localisation effect leading to a positive correction to the conductivity (Kawabata 1980). The effect of an applied field on interaction correction m $T^{1/2}$ of (1) is negative (Lee and Ramakrishnan 1985). If we assume that the two contributions are additive, the overall correction to the conductivity due to the field can take either negative or positive form depending on the relative importance of localisation and interaction corrections. Both, however, predict the same field dependence. For weak fields it is H^2, while for strong fields the dependence is $H^{1/2}$ (Kawabata 1980, Lee and Ramakrishnan 1985). The experimental results shown in Figure 1 indicate a net positive correction to conductivity resulting from a predominance of the positive weak localisation effect in an applied magnetic field. Indeed for small fields the field variation is H^2, consistent with above interpretation. Increasing the field, however, lowers the conductivity, which decreases *continuously* until it reaches practially zero for the barely insulating sample (S1).

In Figure 2 the measured Hall resistance R_{xy} at 2K is plotted as a function of applied magnetic field for both the samples. Initially R_{xy} shows a linear behaviour until at a field H_f it deviates from linearity. There is, therefore, no freezing of free carriers until H reaches H_f. Beyond this field the Hall resistance continues to rise more and more sharply with $\partial R_{xy}/\partial H$ increasing until the latter reaches a value $\sim \infty$ at a field H_c. This indicates a progressive freeze-out of free electron carriers leading ultimately to an insulating state and thus a field-induced metal-insulator transition. Values of H_f are 2.2 and 3.8T, respectively for S1 and S2. H_c = 3.2T for the barely insulating sample and we estimate it to be ~ 5T for the other. Inspection of Figure 2 reveals the absence of any *Hall dip*, which has been observed in n-type $Hg_{0.8}Cd_{0.2}Te$ and InSb (Shayegan et al 1985). We have found that increasing

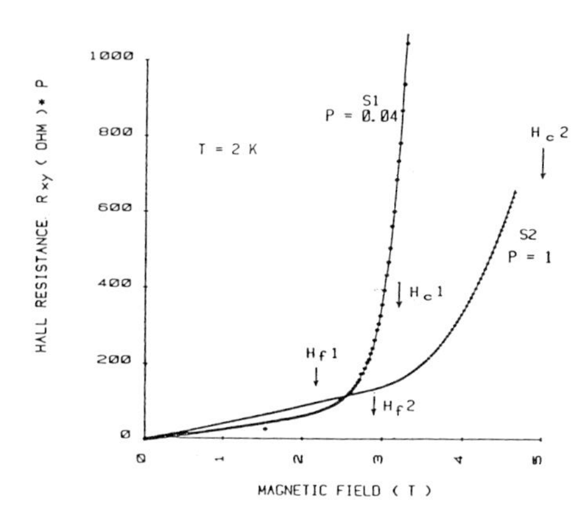

Fig. 2. Field dependence of the Hall resistance R_{xy} of samples of Figure 1 at 2K. H_f indicates the value of magnetic field at which non-linear behaviour sets in. H_c is the field at which an insulating state results.

the temperature shifts both H_f and H_c to higher values until metallic conductivity persists in the entire field range.

We interpret our observations of the decrease of conductivity $\sigma(H)_T$ accompanied by an increase of the Hall resistance $R_{xy}(H)_T$ as depicted in Figures 1 and 2 as due to a magnetic freeze-out of the free electrons such that metallic conduction gives way to *hopping* conduction leading ultimately to an insulating state under the influence of an applied magnetic field. An external magnetic field H introduces into the localisation problem a new length scale $L_H = (\hbar/eH)^{1/2}$, equivalent of the classical cyclotron radius. For weak magnetic fields ($L_H \gg a^*$, where a^* is the effective Bohr radius of the donor electron state in the host), the quantum interference effect is reduced and the eigen states tend to be delocalised. As the field is increased L_H approaches a^* and the magnetic field thus squeezes the electron wave functions, lowers the overlap, giving rise to an important positive contribution to the magnetoresistance. The condition for localisation in a magnetic field is $Na^*L_H^2 \sim 1$, where N is the donor electron concentration. At this point metallic conduction gives way to hopping conduction at a *finite* temperature. The field needed to induce this state will, therefore, be higher as the donor concentration is increased, which is consistent with our results.

Our results are also in agreement with a mobility edge phase diagram proposed by Shapiro (1984). In a disordered electronic system there exists a *mobility edge*, E_c, such that states with $E < E_c$ are localised and those with $E > E_c$ are extended. The mobility edge thus marks the transition between a metal and a non-metal. The initial effect of an external magnetic field is to delocalize the eigen states as explained above which shifts the mobility edge to lower energies. As the field is increased further, the main effect is to reduce overlap between states even for weak disorder. Thus E_c is pushed to higher energies. Thus if one starts with a barely metallic sample then at some critical field a transition to a non-metallic state would occur as we observe for our sample S2. On the other hand, for a barely insulating sample one should expect first an insulator-metal transition followed by a metal-insulator transition when the field is further increased. This has been recently observed in n-type InP (Spriet et al 1986). We have not found this in our barely insulating sample presumably because the temperature is not low enough.

The authors wish to thank B. Boucher and P. Bergé for their constant encouragement and interest for this work.

REFERENCES

Kawabata A 1980 J. Phys. Soc. Japan 49, 628
Lee P A and Ramakrishanan T V 1985, Rev. Mod. Phys 57, 287
Mott N F 1986 *Proceedings of the Thirty First Scottish Universities Summer School in Physics* ed D M Finlayson (Scottish Universities Summer Schools) pp. 29-70
Pepper M ibid., pp. 291-312
Shapiro B 1984 Phil. Mag. B50, 241
Shayegan M, Goldman B J, Drew H D, Nelson D A and Tedrow P M 1985 Phys. Rev. B32, 6952
Spriet J P, Biskupski G, Dubois H and Briggs A 1986, Phil. Mag. B54, L95

The use of X-ray double crystal rocking curves for the evaluation of epitaxial layer thickness

I.C. Bassignana, A.J. SpringThorpe and C.C. Tan

Bell-Northern Research, P.O. Box 3511, Station C,
Ottawa, Ontario, Canada K1Y 4H7

ABSTRACT: Double crystal X-ray diffraction rocking curves can be used to evaluate epitaxial layer thicknesses. In order to assess this technique we have measured the (004) rocking curves for a series of MBE grown $Al_{.3}Ga_{.7}As/GaAs$ epitaxial layers in the regime 0.2 - 6.0μm. The epitaxial layer thicknesses were independently evaluated by SEM measurements. The experimental rocking curves are compared to those generated from computer simulations based on X-ray dynamical scattering theory. In the thickness range 0.5 - 2.0μm Bragg geometery Pendellosung fringes are also observed. The epitaxial layer thicknesses calculated from the integrated intensity ratios and the fringe spacings agree very well with layer thicknesses obtained from SEM sections.

1. INTRODUCTION

X-ray double crystal diffraction (DCD) rocking curves provide a non-destructive method for the characterization of epitaxially grown layers. For single layer structures rocking curve measurements directly provide information on the epitaxial layer composition, quality, and uniformity. For more complex structures, containing multiple, graded or even superlattice layers, this same information can still be obtained by comparing the experimental rocking curves with those calculated by X-ray dynamical scattering theory. It has also been suggested, by Halliwell et al. (1984) that the usefulness of the rocking curve can be extended to the measurement of epitaxial layer thickness. This is significant since the performance of many opto-electronic devices is critically dependent on layer thickness and often it is preferable to non-destructively evaluate the epitaxially grown material before the device is processed.

Thickness information can be obtained from the rocking curve data both from the diffracted intensity and the spacing of the secondary maxima (Pendellosung fringes). In this paper we investigate the potential and limitations of using rocking curves for routine epitaxial layer thickness evaluation. The system chosen for this work was AlGaAs with a composition of 30% Al; this is a simple system but still technologically significant since this composition is important for HEMT devices.

2. EXPERIMENTAL

The rocking curves were recorded for the (004) reflection on a modified

Bede 6" double crystal diffractometer in the parallel (+,-) setting using CuKα radiation. A GaAs single crystal was used as the first crystal; this crystal when rocked with an identical one gave a rocking curve FWHM of 10.5 ± 0.2 arc-sec.

The AlGaAs epitaxial layers were grown by molecular beam epitaxy. The epitaxial layer uniformity parallel to the wafer surface was very good and the DCD rocking curves showed no detectable compositional variation across the 5-7mm cleaved pieces used for this study. The layer uniformity perpendicular to the wafer surface was also very good, as evidenced by the very narrow half widths (~13 arc-sec) of the rocking curves for thick layers. Furthermore, sequentially etched and measured samples showed that the mismatch did not change as the layers became thinner. The epitaxial layer thickness was obtained from SEM measurements of the cleaved and stained cross sections. In order to eliminate errors due to etching of the surface by the staining solution it was protected by a ~1200Å layer of CVD SiO_2. Using this technique epilayer thickness can be measured to better than ±0.05μm. The spatial uniformity of the epitaxial layers is important and in order to insure maximum reproducibility both DCD and SEM measurements were made at exactly the same point on the wafer.

3. RESULTS AND DISCUSSION

A single layer of $Al_{.3}Ga_{.7}As$ on GaAs gives a rocking curve which consists of two peaks. The higher angle peak is the substrate peak and the peak at 98 arc-sec lower angle, corresponding to a larger lattice parameter, is the epilayer peak. A set of rocking curves for the (004) reflection of epitaxial $Al_{.3}Ga_{.7}As$ was simulated using a dynamical scattering theory based on a solution of the Takagi-Taupin equations fig. 1. The method is discussed in detail by M.J. Hill (1985). The thickest epilayer where the substrate peak is no longer visible corresponds to the maximum penetration depth of the X-rays, in this case > 10μm. At the other extreme, the thinnest epilayer which still gives a clearly defined epilayer peak is ~0.2μm.

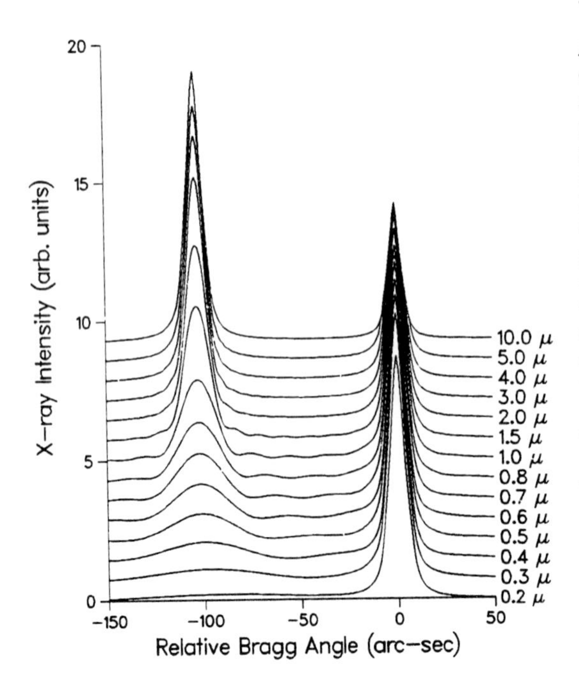

Fig. 1. Double crystal rocking curves for $Al_{.3}Ga_{.7}As$/GaAs, the layer thickness is indicated.

3.1 Integrated Intensities

For AlGaAs layers of constant composition the total or integrated intensity diffracted by the layer is proportional to the scattering volume, i.e. the epitaxial layer thickness. The rocking curve half width is also indicative of the scattering volume; however, this quantity is much more difficult to interpret since it is also a function of the sample curvature, diffractometer alignment (especially tilt angle) and most of all, crystal quality. So, the most significant quantity is the weighted intensity I_{sub}/I_{epi}; i.e. the ratio of the integrated intenisty for the substrate peak I_{sub} to that for the epitaxial layer I_{epi}. The data for rocking curves of 0.2-10μm epilayers are summarized in fig. 2a. This curve is specific for the $Al_{.3}Ga_{.7}As$ composition; AlGaAs layers of lower Al content give a curve of similar shape but lower I_{sub}/I_{epi} ratio, and vice versa for those of higher Al content. In the limit of thin single layers a simple kinematical theory is adequate, but for thicker layers the full dynamical scattering theory must be used in order to correctly predict the scattered intensity. Since the solution to the dynamical theory is a numerical solution it is not possible to give an analytical function for the I_{sub}/I_{epi} curve of fig. 2a.

The experimental data agree well with the theory. The greater experimental scatter for the thinner layers (t<1μm) is twofold. First, SEM cleaved and stained cross sections are at best accurate to within ±0.05μm, and for the thinner layers this becomes significant. The second source of error comes from the integration of the rocking curves. The rocking curves were integrated numerically using a Simpson's Rule algorithm. For thin layers, both the large half width of the epilayer rocking curve peak and the presence of Pendellosung fringes obscure the boundary between the epilayer and substrate peaks. For $Al_{.3}Ga_{.7}As$ this only causes some uncertainty in the integration. However, for AlGaAs layers of lower Al content, where the mismatch is smaller so that layer and substrate peak are closer, this technique is limited in its usefulness. In this case a mathematical deconvolution of the two peaks is not possible since both the rocking curve half widths and the fringe spacings are dependent on the epilayer thickness, which is the unknown parameter.

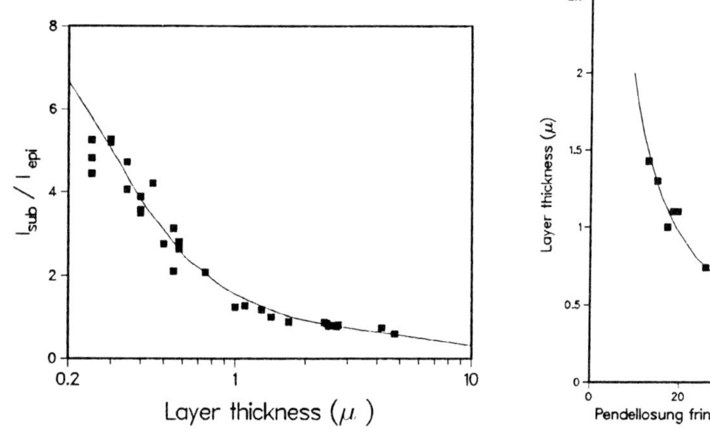

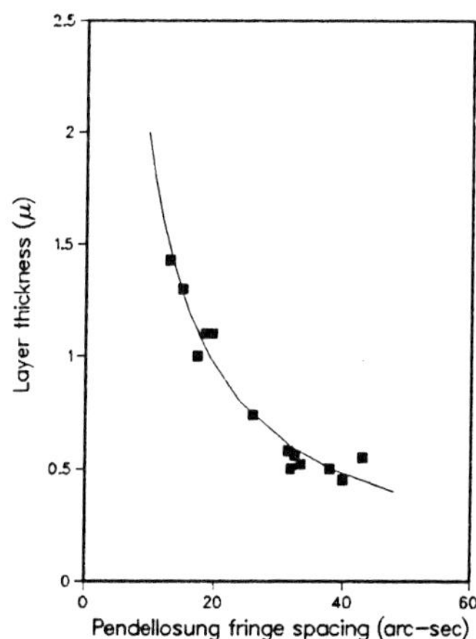

Fig. 1a. Integrated intensity ratios for $Al_{.3}Ga_{.7}As$ epitaxial layers, 1b. Pendellosung fringe spacing dependence on layer thickness, (-) theory, (■) experiment.

3.2 Pendellosung Fringes

Thin layers of high crystal quality can give rise to Pendellosung fringes. The observation of Pendellosung fringes was first reported by Batterman and Hildebrandt (1968). Briefly, the dynamical theory predicts such oscillations for a plane wave incident on a thin crystal in which two tie points on the same branch of the dispersion surface interact and produce interference effects in the diffracted intensities. The observation of Pendellosung fringes is a severe test of epilayer quality since even slight distortions in the crystal will affect fringe contrast. The observation of well-resolved fringes is a good indication of perfection in the epitaxial layer. Independent of layer quality, the fringes are not always present. Their appearance may be obscured by poor morphology of the layer surface or interface. The presence or absence of these alone cannot be used as a test of epilayer quality. However, when the fringes are present, the angular spacing between them ($\Delta\theta$) can be used to give an indication of the layer thickness (t), since $t = (\lambda\gamma_h)/(\Delta\theta \sin 2\theta)$ where γ_h is the direction cosine of the diffracted beam from the inward normal to the sample surface i.e. $\gamma_h = \sin(\theta + \emptyset)$, where for the symmetric (004) reflection of (001) oriented GaAs, $\theta = 33.06$ and $\emptyset = 0.0$. For CuKa radiation ($\lambda = 1.5418$Å), expressing t in microns and $\Delta\theta$ in arc-sec, this expression becomes $t = 19.2/\Delta\theta$.

Pendellosung fringes were observed both on the high and low angle side of the epilayer rocking curve peak. The higher angle fringes were more difficult to observe and were also partly obscured by the substrate peak. This behavior has been noted by Stacy and Janssen (1974). For samples of thickness $t \geqslant 2\mu m$ and $t \leqslant 0.5\mu m$ fringes were difficult to observe. The experimental data is shown in fig. 2b, the solid line is the equation given above.

4. CONCLUSIONS

We have investigated the use of X-ray double crystal rocking curves for the non-destructive evaluation of AlGaAs/GaAs epitaxial layer thickness. The integrated intensity ratio I_{sub}/I_{epi} gives a good measure of the epitaxial layer thickness of 0.2-10μm thick layers. For 0.5-2μm layers the thickness can also be determined by the Pendellosung fringe spacing. The next challenge will be to establish the validity of this method for double and multiple layers.

5. REFERENCES

1) B.W. Batterman and G. Hildebrandt, Acta. Cryst. A 24, 150 (1968)
2) M.A.G. Halliwell, M.H. Lyons and M.J. Hill, J. Crystal Growth 68, 523 (1984)
3) M.J. Hill, Ph.D. thesis, Durham Univ. (1985).
4) W.T. Stacy and M.M. Janssen, J. of Crystal Growth 27, 282 (1974).

Inst. Phys. Conf. Ser. No. 91: Chapter 3
Paper presented at Int. Symp. GaAs and Related Compounds, Heraklion, Greece, 1987

Characterization of n-doped GaSb and of GaSb/InAs heterojunction grown by MOVPE

S.K.Haywood, A.B.Henriques, D.F.Howell, N.J.Mason, R.J.Nicholas and P.J.Walker

Clarendon Laboratory, Parks Road, Oxford, OX1 3PU, U.K.

ABSTRACT: The epitaxial growth of GaSb by MOVPE and n-type doping of the material with selenium have been studied. Undoped layers of $p = 3-4 \times 10^{16}$ cm^{-3} had a hole mobility of $\approx$1000 cm^2/Vs at 295 K rising to $\geqslant$4300 cm^2/Vs at 77 K. Photoluminescence (PL) spectra at 4 K showed a weak residual acceptor peak combined with strong excitonic emissions. N-type material was obtained with Hall mobilities of 3400–850 cm^2/Vs at 295 K for apparent carrier concentrations of 9×10^{15} cm^{-3} – 2×10^{18} cm^{-3}. A Type II heterojunction (p-GaSb/n-InAs) showing a 2D electron gas (2DEG) has also been grown for the first time by MOVPE.

1. INTRODUCTION

In contrast to the amount of work on arsenic and phosphorus based III–V compounds relatively little attention has been paid to the growth of antimony containing layers. However, these materials can be used in a number of physically interesting systems e.g. Type II GaSb/InAs heterostructures [Sakaki et al (1977)]. The 1.6 μm bandgap of GaSb also leads to applications in the 1.2–1.7 μm region important for fibre-optic devices. An understanding of the epitaxy and doping of bulk antimonide layers is a prerequisite to the investigation of more complex systems.

GaSb has been deposited epitaxially by several different techniques including MOVPE [Cooper et al (1979), Manasevit and Hess (1979)], MBE [McLean et al (1984), Lee et al (1985)] and LPE [Miki et al (1974)]. Undoped material is p-type due to a native defect. The highest Hall mobility (μ_H) for undoped GaSb was reported by McLean et al (1984). Heteroepitaxial MBE grown layers of $p = 8 \times 10^{15}$ cm^{-3} gave μ_H = 1400 cm^2/Vs at 295 K The lowest level of n-type doping was obtained by LPE. For $n = 3.8 \times 10^{15}$ cm^{-3}, Miki et al (1974) measured a room temperature mobility of 7,700 cm^2/Vs. At higher levels of n-type doping MBE and LPE give comparable results [μ_H(295K) $\approx$ 3000 cm^2/Vs at $n = 5 \times 10^{16}$ cm^{-3}]. The only electrical data in the literature from MOVPE grown GaSb is μ_H(295K) = 610 cm^2/Vs for undoped material on a GaAs substrate [Manasevit and Hess (1979)]. Our aim here is to examine whether MOVPE can be used to obtain undoped and n-type GaSb comparable to that grown by MBE and LPE.

2. EXPERIMENTAL

A custom-built MOVPE reactor with a horizontal silica cell was used at atmospheric pressure. The reactants were TMGa (Epichem, −9 °C) and TMSb (Alfa, 0°C) with H_2Se (100 ppm in H_2) for n-type doping. Palladium diffused hydrogen was the carrier gas. Typical flow rates through the alkyl bubblers were 25 $ccmin^{-1}$ of TMGa and 30 $ccmin^{-1}$ TMSb. H_2Se flows of 5–200 $ccmin^{-1}$ were used for doping. The substrates (GaSb:Te, GaAs:Si and GaAs:Cr) were positioned on a SiC coated susceptor which was heated by an RF coil. Carrier concentration and mobility were determined from Hall effect heteroepitaxial layers (GaSb/GaAs:Cr) using a field of 0.5 Tesla. The transverse magnetoresistance was measured at 4 K and fields of up to 9 Tesla. PL spectra were taken at 4 K. Excitation was with the 488 nm line of an argon ion laser at a power density of 500 Wcm^{-2}.

2. RESULTS

2.1 Bulk GaSb

Detailed results concerning the growth conditions required to produce undoped material will be published elsewhere [Haywood et al]. In summary, material grown close to the optimum III–V ratio of ≈1.2:1 had $\mu_H(295K) \simeq 1000\ cm^2/Vs$. No variation of μ_H with thickness was found for layers of >1 μm. The mobility rose to 4300 cm^2/Vs at 77 K and reached a peak of 5400 cm^2/Vs at about 50 K. These values are comparable to those obtained from MBE grown material [McLean et al (1984), Lee et al (1986)]. For homoepitaxial samples the PL spectrum was dominated by the narrow bound exciton bands; the acceptor band (777 meV) being much weaker. The most prominent feature in the spectrum was the acceptor bound exciton at 796 meV. This had a FWHM of 1.1 meV, as compared to reported values of 0.3 meV for LPE [Rhule et al (1976)] and 2 meV for MBE material [Lee et al (1986)]. The free exciton was seen at 810 meV.

By varying the H_2Se flow rate, n–type GaSb was obtained with apparent electron concentrations, n_H, of $9.7 \times 10^{15}\ cm^{-3}$ to $2.0 \times 10^{18}\ cm^{-3}$ at 295 K. The corresponding Hall mobilities (μ_H) were 3200 cm^2/Vs and 850 cm^2/Vs. μ_H and n_H as a function of temperature are shown in Figure 1 for typical samples. The increase in n_H with decreasing temperature is due to multi–valley conduction [Kourkoutas et al (1984) and references therein]. The Γ and L valleys are close in energy in GaSb. The population of the L valley (which has a high density of states but low mobility) changes with temperature, as does the separation of the Γ and L valleys. More heavily doped samples will also have a higher L valley population. Other measurements are in progress to investigate the population and separation of the Γ and L valleys. Hall measurements alone are not sufficient to do this since: $n_H = (n_\Gamma\mu_\Gamma + n_L\mu_L)^2/(n_\Gamma\mu_\Gamma^2 + n_L\mu_L^2)$.

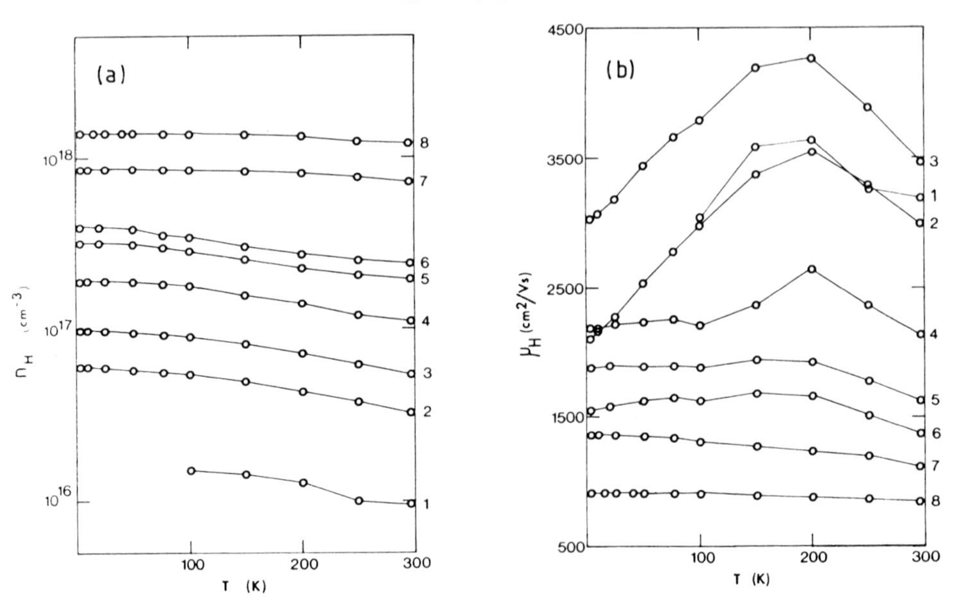

Figure 1: Results of Hall measurements on selenium doped GaSb (a) n_H versus temperature (b) μ_H versus temperature

For a given n_H, μ_H is comparable with material grown by MBE [McLean et al (1985)] and LPE [Miki et al (1974)]. The maximum in μ_H occurs at 180–200 K for most of the samples in Figure 1(b) and a double peak is seen at the higher carrier concentrations, for which the multi-valley conduction is presumably also responsible.

2.2 GaSb/InAs Heterojunction

GaSb/InAs is an example of a Type II heterojunction; the conduction band in InAs lying $\simeq$ 0.15 eV below the GaSb valence band [Sakaki et al (1977)]. Figure 2(a) shows the structure we have grown. The GaSb is undoped and material grown under similar conditions was p-type with a hole concentration of 3 x 10^{16} cm^{-2} at 295 K falling to $\simeq$ 10^{14} cm^{-3} at 4 K. The InAs is n-type with an electron concentration of about 10^{17} cm^{-3}. Figure 2(b) is the postulated band diagram for this heterojunction (see below). Hall measurements on this structure showed the Hall coefficient, R_H, initially rising as the sample was cooled and then levelling out below 30 K. This is a result of two carrier conduction (electrons in the InAs and holes in the GaSb) with R_H increasing as the holes freeze out. At 4 K, R_H yielded a sheet electron concentration, n_H of 5.4 x $10^{12}cm^{-2}$, with μ_H = 2,900 cm^2/Vs.

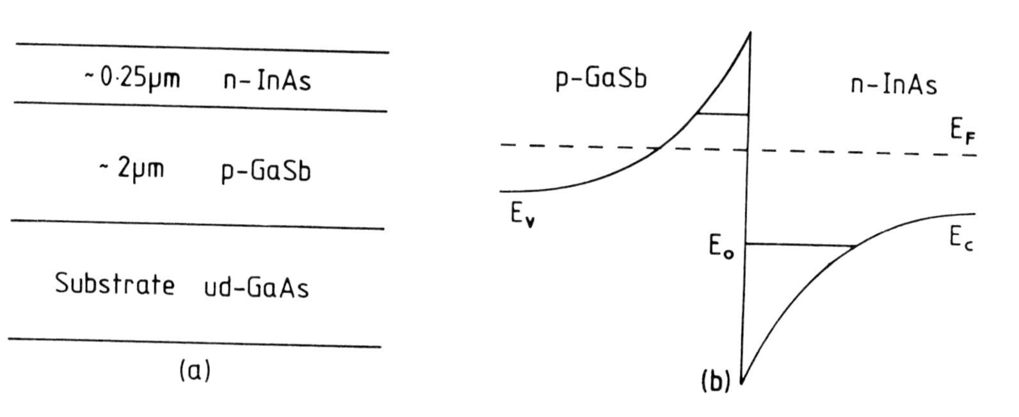

Figure 2: (a) p-GaSb/n-InAs heterojunction grown by MOVPE
(b) Band diagram for this structure: E_v is the GaSb valence band edge, E_c the InAs conduction band edge and E_F the Fermi energy.

Shubnikov–de Haas (SdH) oscillations in the transverse magnetoresistance were observed at 4 K (Figure 3). Rotation of the sample relative to the magnetic field, B, showed these oscillations to be 2D ($B_N\cos\Theta$ remains constant for each minimum).There is therefore a 2DEG in the InAs. The fundamental field, B_F = 8.7 T gives an electron concentration, n_o $\simeq$ 4.2 x 10^{11} cm^{-2}. This is much lower than n_H indicating that there is a significant contribution to conduction from 3D electrons in the degenerate InAs. (N.B. n_H underestimates the sheet electron concentration because the electric field at the heterojunction requires the existence of a 2D hole gas in the GaSb.) Thus, there is effectively an accumulation layer at the interface with several electric subbands occupied. The oscillations observed are a result of electrons in the lowest subband (E_o) only. Oscillations from higher subbands would occur at lower fields but these were not resolved. However, the ratio of the total 2D electron concentration, n_{2D}, to the electron concentration in the lowest subband, n_o is approximately constant for accumulation layers in narrow gap materials [Nicholas (1987)]. Thus, we estimate n_{2D} to be about 6 x 10^{11} cm^{-2}. The 3D conduction in the InAs (together with the occupancy of the higher subbands) is responsible for the small amplitude of the 2D oscillations relative to the magnetoresistance and also for the introduction of the phase factor, φ in Figure 3.

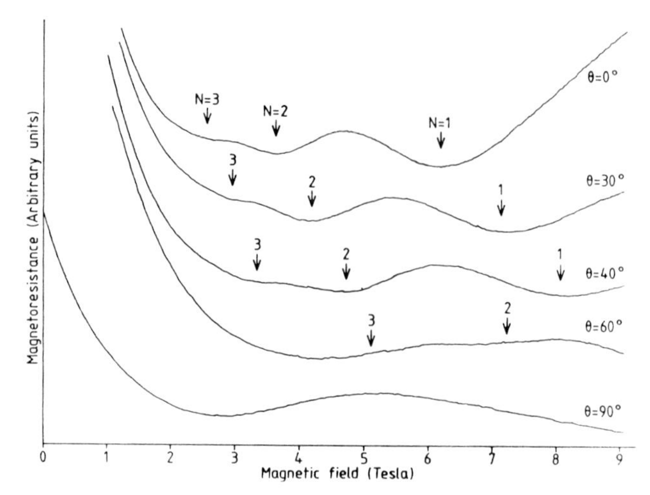

Figure 3: Magnetoresistance of a GaSb/InAs heterojunction for various angles Θ between B & the surface normal. A term proportional to B has been subtracted to enhance the SdH oscillations. The arrows indicate values magnetic field of $B_N(\Theta)$, that satisfy the condition: $(N + \varphi)B_N\cos\Theta = B_F = (h/2e)n_o$ with N integer, $B_F = 8.7$ T & $\varphi = 0.41$

3. CONCLUSIONS

GaSb' of high electrical and optical quality has been deposited epitaxially by MOVPE. N-type doping with selenium allowed control of the Hall electron concentration over more than two orders of magnitude. A GaSb/InAs heterojunction was also grown and shown to have a 2DEG in the InAs.

Acknowledgements:

We thank the Science and Engineering Research Council and GEC Hirst Research Centre, Wembley, U.K. for financial support and Derek Morris for technical assistance. One of us (A.B.H.) acknowledges financial support from CNPq (Brazil).

References:

Cooper C B, Saxena R R and Ludowise M J 1979 *J.Electron.Mat* **11** 1001
Haywood S K, Henriques A B, Mason N J, Nicholas R J, and Walker P J submitted to *Semiconductor Sci and Tech.*
Kourkoutas C D, Bekris P D, Papaioannou G J and Euthymiou P C 1984 *Solid State Comm.* **49** 1071
Lee M, Nicholas D J, Singer K E and Hamilton B 1986 *J.App.Phys.* **59** 2895
Manasevit H M and Hess K L 1979 *J.Electrochem.Soc.* **126** 2031
McLean T D, Kerr T M, Westwood D I, Grange J D and Murgatroyd I J 1984 *Inst.Phys.Conf.Ser.* **74** 145
McLean T D, Kerr T M, Westwood D I, Howell D F and Wood C E C (1985) *Proceedings of Sixth MBE Workshop: University of Minnesota*
Miki H, Segawa K and Fujibayashi K, 1974 *Jpn.J.Appl.Phys.* **13** 203
Nicholas R J, 1987 *Proceedings NATO ASI Erice: Plenum* – to be published
Rühle W, Jakowetz W, Wölk C, Linnebach R and Pilkuhn M 1976 *Phys.Stat.Solidi* **B73** 255
Sakaki H, Chang L L, Ludeke R, Chang C, Sai-Halasz G A, and Esaki L 1977 *Appl.Phys.Lett.* **31** 211

Application of isothermal liquid phase epitaxy to $Al_{0.35}Ga_{0.65}As$ photovoltaic devices

M Gavand*, L Mayet*, B Montégu*, J P Boyeaux*, A Laugier

Laboratoire de Physique de la Matière (U.A.358 CNRS)
Institut National des Sciences Appliquées
20, Avenue A.Einstein 69621 VILLEURBANNE Cédex (FRANCE)
* Permanent address:I.S.I.D.T.-U.C.B. 69622 VILLEURBANNE Cédex (FRANCE)

ABSTRACT: With the object of monolithic multibandgap device realization, high efficiency $Al_{0.35}Ga_{0.65}As$ photovoltaic devices are grown by isothermal liquid phase epitaxy (I.L.P.E.) . They have a thin graded band gap $Al_yGa_{1-y}As$ window layer with a double layer antireflection coating , allowing an improvement of the quantum efficiency for the short wavelengths. The conversion efficiency measured under one sun is 14.6 % (AM 1 , 25°C).

1. INTRODUCTION

To improve the efficiency of photovoltaic devices under concentration , one of the most promising possibilities is the monolithic multibandgap device with several junctions of different energy gaps (Beaumont 1982) . Following theoretical analysis for a two junctions device , the couple GaAs (1.42 eV) - $Al_{0.35}Ga_{0.65}As$ (1.89 eV) represents a very good band gap combination . A simplified LPE technique has been reported to make high efficiency GaAs and $Al_{0.2}Ga_{0.8}As$ solar cells (Kordos 1979 , Gavand 1985 , Zerdoum 1987).In this paper , we describe the application of this technique to the fabrication of the $Al_{0.35}Ga_{0.65}As$ cell which is the upper cell in the cascade system and present the preliminary photovoltaic results.

2. SOLAR CELL PROCESSING

High efficiency $Al_{0.35}Ga_{0.65}As$ solar cells are obtained by growing p: $Al_yGa_{1-y}As$ - p: $Al_{0.35}Ga_{0.65}As$ - n: $Al_{0.35}Ga_{0.65}As$ - n^+ GaAs heterostructures . The upper p: $Al_yGa_{1-y}As$ layer is a thin graded window layer in which y varies from 0.35 at the window layer - p: $Al_{0.35}Ga_{0.65}As$ layer interface to 0.85 at the surface.This reduces drastically the surface recombination velocity by producing a built-in electric field and consequently improves the solar conversion efficiency . The p-n junction is located in the $Al_{0.35}Ga_{0.65}As$ material .The energy band diagram of this structure is shown in Figure 1, with a graded carrier concentration into the p-side of the junction.

These heterostructures are fabricated at 800°C in an horizontal LPE apparatus . The GaAs substrates are <100> oriented and Si doped with $N_D-N_A = 2.10^{18}$ at.cm^{-3} .Two Sn doped and Be doped AlGaAs melts and the n^+ substrate are closed inside a graphite boat with a sliding bottom .Three main steps can be distinguished as shown in Figure 2 corresponding to three relative positions of the sliding bottom / crucible : i) A Sn- doped melt with a suitable amount of Al is saturated with As from an undoped polycrystalline lump for two hours corresponding to an $Al_{0.35}Ga_{0.65}As$ solid according to the phase diagram . ii) A 4 - 8 µm thick $Al_{0.35}Ga_{0.65}As$ layer is grown by equilibrium cooling . The temperature is then kept constant for ten minutes allowing the equilibrium of the Be doped melt with an $Al_{0.85}Ga_{0.15}As$ solid . The passage of the wafer from the first to the second melt is made without its exposure to the furnace atmosphere to prevent the oxidation of the AlGaAs layer . iii) The wafer is pushed under the second melt for one or two minutes . There is no

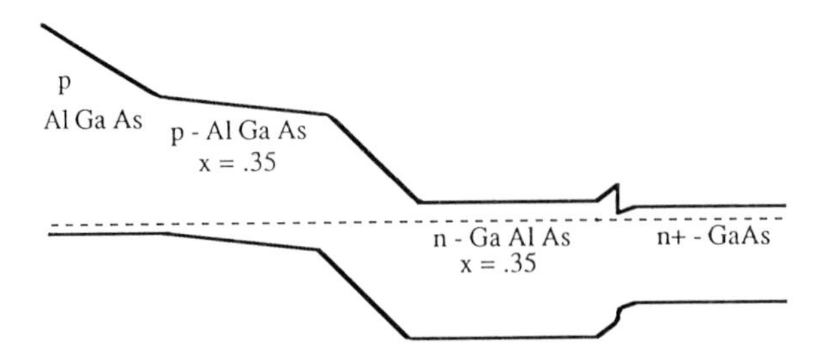

Figure 1: *Schematic energy band diagram of the $Al_{0.35}Ga_{0.65}As$ photovoltaic device under short circuit conditions.*

equilibrium between the ternary $Al_{0.35}Ga_{0.65}As$ solid and the ternary AlGaAs liquid corresponding to the $Al_{0.85}Ga_{0.15}As$ solid . This leads in the first stage to a weak dissolution of $Al_{0.35}Ga_{0.65}As$ and in the second stage , a thin graded AlGaAs layer , about 50 nm thick (Mayet 1987), is formed. A small increase of the temperature ($\Delta T = 0.25°C$) improves the etchback-regrowth mechanism (Woodall 1977).The Al concentration in this window increases from 0.35 at the interface to 0.85 at the surface . During the melt - $Al_{0.35}Ga_{0.65}As$ layer contact , Be diffuses in the low band-gap layer forming an $Al_{0.35}Ga_{0.65}As$ p-n junction at a depth of 1 to 2 μm depending on the contact duration .

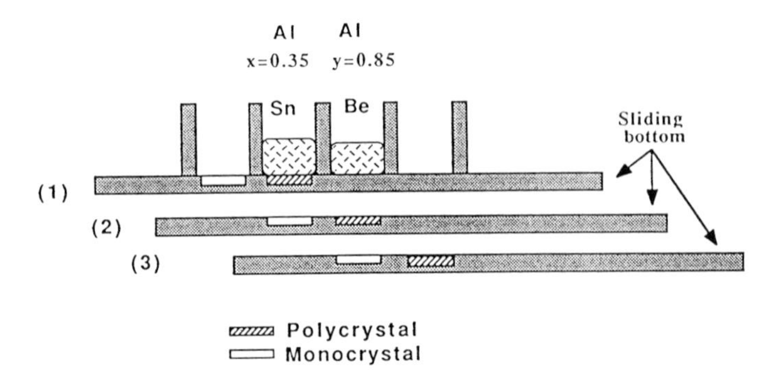

Figure 2: *Growth process for p: $Al_yGa_{1-y}As$ - p: $Al_{0.35}Ga_{0.65}As$ - n: $Al_{0.35}Ga_{0.65}As$ - n^+ GaAs heterostructures .Step 1: 2 hours saturation . Step 2 : $Al_{0.35}Ga_{0.65}As$ layer growth by equilibrium cooling . Step 3 : $Al_yGa_{1-y}As$ graded window layer growth .*

The back contact of the devices is made by classical Au-Ge-Ni vacuum deposition and the front contact grid is obtained by coating a 100 nm Au-Be (99-1)layer.The ohmic contacts are then annealed at 450°C for 3 min in dry H_2. Then the grid thickness is increased up to 500 nm by metallization (unannealed pure silver).Finally an antireflection coating is realized by a double quarter -wave layer structure allowing to adjust the reflectivity by means of thickness and refractive index n . In our case the thermal evaporation of 50 nm of ZnS (n = 2.3) followed by 82 nm of MgF_2 (n = 1.36) gives an average reflectivity for useful photons equal to 2% (Mayet 1987). The cell structure is shown in Figure 3.

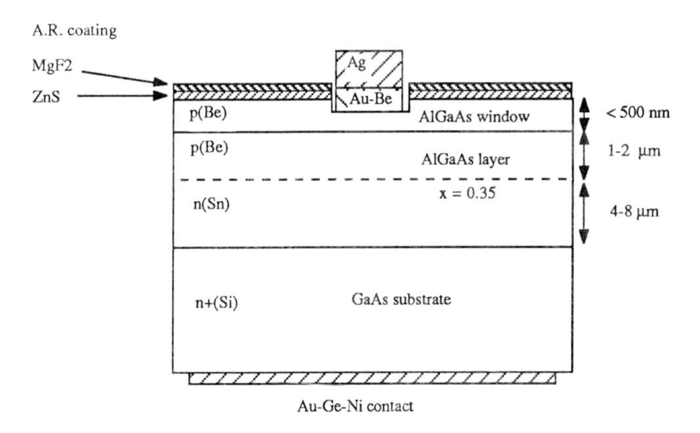

Figure 3: *$Al_{0.35}Ga_{0.65}As$ solar cell structure.*

3. RESULTS AND DISCUSSION

Measurements under illumination using a Xenon lamp simulator give Jsc=12.45 $mA.cm^{-2}$,Voc=1305 mV, FF=0.864 and an efficiency η equal to 14.6 % under one sun illumination (AM1 , 25°C). In a GaAs (1.42 eV) - $Al_{0.35}Ga_{0.65}As$ (1.89 eV) monolithic system , it is possible to obtain an efficiency equal to 25 % .The I(V) characteristic of this cell is plotted in Fig 4 . Figure 5 shows the conversion efficiency of the same cell before and after A.R. deposition. The response improvement in the short wavelength zone can be explained by the drift field created in the thin graded band gap $Al_yGa_{1-y}As$ window which enhances the collection of photogenerated carrier from this layer .

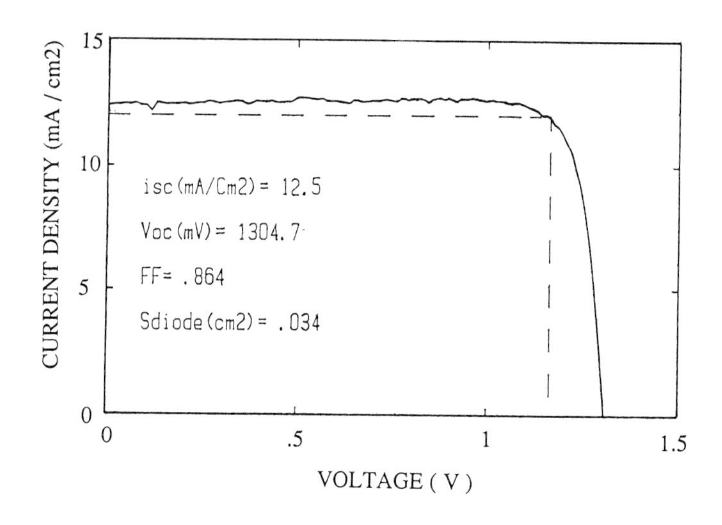

Figure 4: *I (V) characteristic of an $Al_{0.35}Ga_{0.65}As$ cell under AM1 illumination at 25 °C. (η = 14.6 %)*

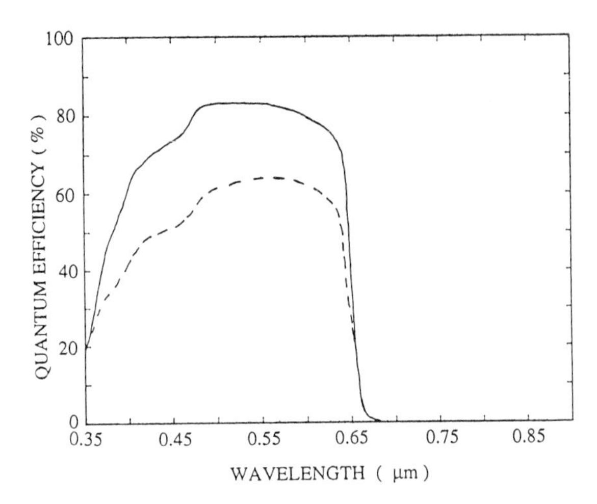

Figure 5: *Collect efficiency of the cell of the Figure 4 before (dashed line) and after (full line) AR coating.*

These results can be compared with the data in the literature . Gale et al.(1984) have obtained η = 16.4 % with x_{AlAs}=0.27 : the part of the solar spectrum useful is more important. Bedair et al.(1981) have also used x_{AlAs}=0.35 by LPE and obtained η = 10 % under AM0 without A.R. coating . Salètes et al.(1987) using MBE and MOVPE techniques were not able to obtain good efficiency for $x_{AlAs} > 0.2$ and have explained this fact due to the existence of deep donors. This hypothesis was pointed by Li (1981) who has shown that the $Al_xGa_{1-x}As$ epitaxial layers (x=0.25 to 0.31) prepared by the LPE technique have a lower defect density than those prepared by the MOCVD and MBE techniques.

ACKNOWLEDGEMENTS

This work was supported by AFME (French Agency for Energy Management) and by the Commission of the European Communities .

REFERENCES

Bedair S M, Timmons M L and Chiang J P C 1981 *Proc. GaAs and Related Compounds 1980* Inst.Phys.Conf.Ser. N° 56 pp 403 -11

Beaumont B, Nataf G , Raymond F and Vèrié C 1982 *Proc.16th IEEE Photovoltaic Specialists Conference* (New York: IEEE) pp 595 -600

Gale R P,Turner G W, Fan John C C, Chapman R L and Pantano J V 1984 *Proc.17th IEEE Photovoltaic Specialists Conference* (New York: IEEE) pp 721-5

Gavand M, Mayet L and Laugier A 1985 *Proc. GaAs and Related Compounds 1984* (Inst.Phys.Conf.Ser.N° 74) pp 469 -74

Kordos P , Pearson G L and Panish M B 1979 *J Appl Phys* 50 pp 6902 -6

Li S S 1981 *Proc. 15th IEEE Photovoltaic Specialists Conference* (New York: IEEE) pp 1283-8

Mayet L,Gavand M, Montégu B, Zerdoum R and Laugier A 1987 *European Materials Research Society Conference* Strasbourg , to be published

Salètes A, El Jani B, N'Guessan K, Rudra A, Leroux M, Contour J P ,Gibart P and Vèrié C 1987 *Proc .7th Eur.Com.Conf.on Photovoltaic Solar Energy* Sevilla (London: Reidel) pp 1117-21

Woodall J M , Hovel H J 1977 *Appl Phys Lett* 21 pp 492-3

Zerdoum R,Mayet L,Gavand M and Laugier A 1987 *Proc.7th Eur.Com.Conf. on Photovoltaic Solar Energy* Sevilla 1986 (London : Reidel) pp 938 - 42

Paper presented at Int. Symp. GaAs and Related Compounds, Heraklion, Greece, 1987

An I/F converter for GaAs defect location using laser probing

S Tsitomeneas, A Arapoyanni, N Theophanous, G Papaioannou

University of Athens, Physics Department, Panepistimioupolis, Ktiria TYPA, Ilissia, Athens 157 71

ABSTRACT: Defect location in semiconductor devices is presented, based on an Intensity-to-Frequency (I/F) converter detecting the photocurrent induced to the wafer nodes by a laser beam. This converter performs repetitive optoelectronic integration which improves the SNR and thus the sensitivity of photocurrent measurements. Multiple scanning combined with additional integration (averaging) provides even better sensitivity and enhances spatial resolution. The particular advantages of our method compared with other techniques using laser beams are also considered.

1. INTRODUCTION

Laser beam probing is being recently developed as a new testing method for integrated circuits, since previous methods suffer from several drawbacks. In fact, needle probing is now pushed to its limits by decreasing IC-feature sizes, while needle probes can damage ICs during probing. Also, electron-beam probing must be realised in a vacuum and continuously execute tight program loops, while it can cause excessive local heating (Scheiber 1986). Inducing currents in semiconductor devices by means of laser beams instead of electron beams has several advantages:Laser photons carry energies much lower than those of e-beams, probing does not take place in vacuo and surface charging effects (induced by e-beam probing) are avoided.

On the other side, laser probes have poorer image resolution than e- beam probes, since visible light wavelengths are longer than those associated with keV-electron beams. Also, until recently, laser probes suffered from serious speed and signal detection problems (Scheiber 1986). Our present system and detection method attempt to cope with the latter drawbacks of the laser probing technique by using optoelectronic integration processing and I/f conversion. First results obtained in practice and expected ameliorations seem rather encouraging.

2. BASIC INFORMATION

We consider the absorption of light in a semiconductor creating electron-hole pairs by the transitions of electrons from the valence band to the conduction band. Electron-hole pairs created by the above mechanism easily recombine, unless an electrical field separates them. This, for example, is the case in semiconductor depletion zones, where the diffusion voltages keep pairs separated and generate the photocurrent that the tester senses. Any defects in the crystalline structure degrade the above current or voltage response. Therefore, scanning a laser beam across a sample and recording induced photocurrents can produce a topological "image" in which pinpoint structural irregularities and grain boundaries of the device will appear as locally reduced values of currents. In this way, a defect-display method is obtained which can be used to investigate p-n junctions, Schottky barriers, and other regions exhibiting reduced recombination of electron-hole pairs (Scheiber 1986).

We note that, more precisely, in a GaAs-MESFET, absorption of optical radiation has two main effects: (a) photovoltaic effects in the gate-Schottky barrier region and in the active channel to substrate (or buffer layer) barrier, and (b) photo-conductive effects on the parasitic resistances in series with the active channel and in the substrate (and buffer layer, if present) (Fig.1).

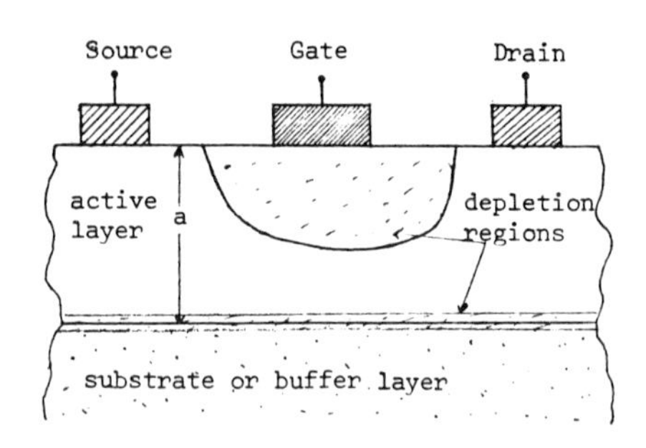

Fig.1. Schematic cross section of a MESFET

Our method exploits the photocurrent due to the photovoltaic properties of the gate-Schottky barrier of the MESFET, and namely the fact that, if V_{DS}=0 Volts and V_{GS}=0 Volts the MESFET performance is analogous to that of a photodiode. The electron-hole pairs created in the depletion region under the gate (Fig.1), or into a diffusion length from it, are separated by the internal field. The corresponding photocurrent flowing to the gate circuit can be written as :

$$I_{ph} = \frac{\eta q}{h \nu} P_{opt} \qquad (1)$$

where η is the quantum efficiency, $h\nu$ the photon energy of a radiation with wavelength λ and P_{opt} the incident optical power. Hence, given that any defect in the MESFET affects the local quantum efficiency η, the defects will appear as local deviations from the expected normal I_{ph} value, only provided that P_{opt} is constant. Consequently a SNR improving detection of I_{ph} is indispensable to reduce the effect of fluctuations in the laser beam power.

3. EXPERIMENTAL TECHNIQUE AND RESULTS

By virtue of the above, we can verify the topology of a MESFET and locate eventual material defects by scanning its surface with a punctual laser-beam and measuring the resulting gate photocurrent, provided that a noise reduction technique is applied.

Rouger J M et al (1984) have already reported an experiment for taking the gate photocurrent profile of a GaAs-MESFET, using a He-Ne laser beam focalized to a spot of 1.5 µm diameter by means of a monomode fiber and microscope optics.

The geometry of the MESFET in question is represented in fig.2, where L=2µm, Z=130µm, L_{DS}=12µm, N_D=1.7 10^{17} cm^{-3}, a=0.16µm.

Since the penetration depth for He-Ne at 632nm is $1/\alpha = 0.3\mu m$, the electron-hole pairs are generated as well in the active region as in the buffer layer.

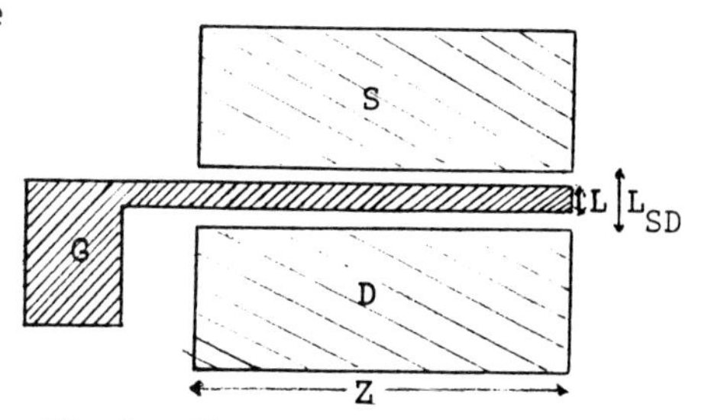

Fig.2. The surface geometry of the MESFET

The distribution of the gate photocurrent along a line perpandicular to the channel is depicted in fig. 3, for an optical power of 0.3 mW. Normally, as seen, two peaks of equal amplitude, symmetrical with respect to the gate (where the light reflectance is maximum) are observed. Any defects into the substrate alter this normal profile. Repeating this scanning all along the entire channel can provide a complete chartography (topological "image") of the device.

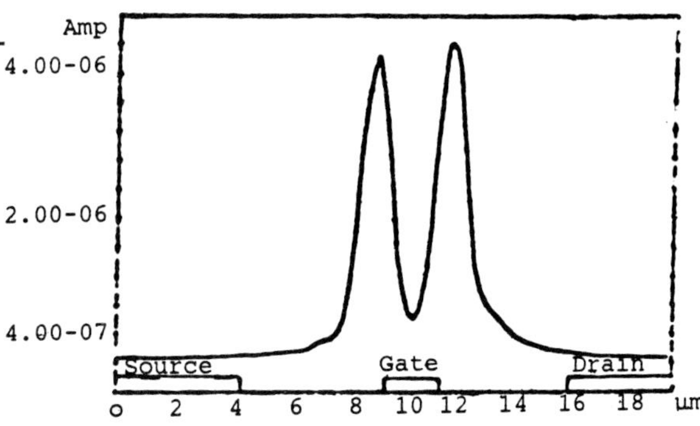

Fig.3. Distribution of the gate photocurrent across the channel of a MESFET

In our method, the induced photocurrent I_{ph} is measured by means of a (current) Intensity to Frequency (I/F) converter (Tsitomeneas et al 1986), whose schematic diagram is shown in fig.4. Here the capacitor C is charged by the photocurrent I_{ph}, which in our case is the gate photocurrent of the MESFET, up to a preset value of reference voltage V_o and then an abrupt discharging follows. Obviously, the charging time of the capacitor will depend on the photocurrent intensity, and -for convenient values of C and V_o - is given (idealy) by: $T=CV_o/I_{ph}$ (2). Thus, the photocurrent intensity can be deduced by measuring the period T of a sawtooth waveform or its frequency f=1/T.

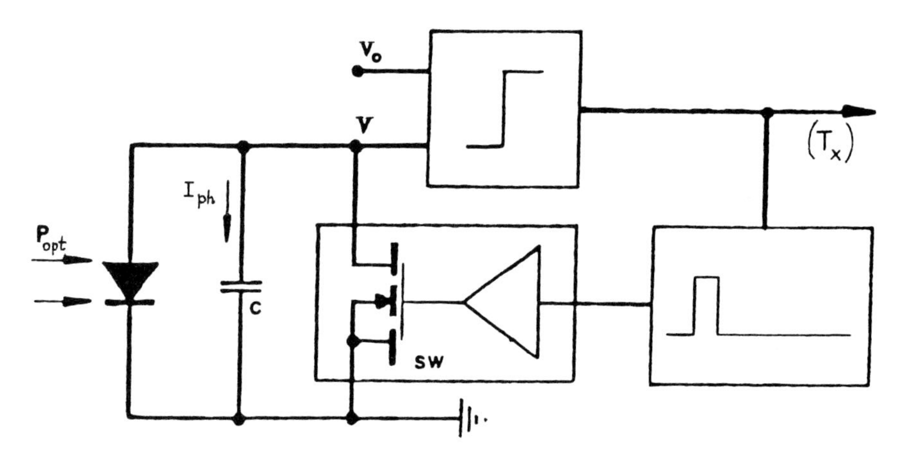

Fig.4. Schematic diagram of the I/F converter

In practice, the output waveform differs slightly from the ideal form, while the actually measured period T_x yields the photocurrent by means of:

$$I_{ph} = \frac{1}{aT_x - b} \quad (3)$$

where a and b are constants depending on the parameters of the circuit.

Testing measurements with our I/F converter in the range of very weak photocurrents have been made with a charging capacitor C=5nF and a reference voltage V_o=1mV. Indicative results are shown in Table I.

Table I

P_{opt} (W)	T_x (μsec)	I_{ph} (A)
10^{-6}	12,6	$5\cdot10^{-7}$
10^{-7}	117	$5\cdot10^{-8}$
10^{-8}	1158	$5\cdot10^{-9}$
10^{-9}	11566	$5\cdot10^{-10}$

The following remarks can be made on the utilizability of our I/F converter for detection of gate photocurrents induced to a MESFET by a laser beam:

1. Although these photocurrents are very weak(fig 3) and their measurement with conventional methodes is very difficult and expensive, with our system no such problems appear, since the measurement of periodes in the range of msec to μsec (Table I), or pulse frequencies from kHz to MHz, has been quite feasible.
2. Our I/F converter is essentially an optoelectronic integrator (Arapoyanni et al 1984), integrating the photocurrent over a period T_x. Thus, the SNR of the measurement set is enhanced, which is very important for the faithful repetitivity of measurements, given that photocurrents in question are very noisy. Moreover, taking the mean value of several periods, T_x, around each point on the wafer, we obtain an additional reduction of noise and so an even better accuracy of measurement.
3. Each integration needs a minimum of time to be performed and this time is as greater the weaker the photocurrent is;this puts a finite lower limit to the time duration of scanning the MESFET's surface. However this does not pose a problem. In fact, e.g.,if we need to take 75 points per scanning line with a time of pause at least 11.5 msec at each point (Table I), the time for one line scanning will be 75x11.5≅0.86 sec, hence for 100 lines, needed to cover the device surface, the total time will be 86 secs, which is not very large. Under these conditions, any method of scanning the MESFET's surface can be accepted (even manual continuous displacement by means of a micropositionner) since the scanning velocity is small enough to permit several integration periods at each point, even for very small photocurrents.

Finally, if the scanning of a line across the MESFET is continuous and uniform, the output of the I/F converter will be a frequency-modulated sawtooth waveform. The additional use of a PLL should allow then to recognize any frequency shift as a DC voltage shift; in this way an illustrative graph of the MESFET's photocurrent defects can be obtained, provided with the additional noise reduction offered by the PLL action.

REFERENCES

Arapoyanni A, Theophanous N, Caroubalos C, Wendel P L 1984,Proc. Opto 84, Paris (ESI publications), pp 116-118

Rouger J M, Perichon R A, Mottet S and Forrest J R 1984,IV J.Nationales Microondes pp 76-77

Scheiber S 1986,Test and Measurements World, May 1986, pp 32-40

Tsitomeneas S, Theophanous N, Arapoyanni A, Caroubalos C 1986,Nat. Conf. on Physics, Athens.

Inst. Phys. Conf. Ser. No. 91: Chapter 3
Paper presented at Int. Symp. GaAs and Related Compounds, Heraklion, Greece, 1987

Ti-Fe Co-doped semi-insulating InP grown by MOVPE

A. G. Dentai and C. H. Joyner
AT&T Bell Laboratories, Crawford Hill Laboratory, Holmdel, NJ 07733

T. W. Weidman
AT&T Bell Laboratories, Murray Hill, NJ 07974

ABSTRACT: A new semi-insulating (s-i) material, InP co-doped with titanium and iron (InP:Ti:Fe), was grown by metal organic vapor phase epitaxy (MOVPE). This material, unlike InP:Fe, remains semi-insulating even in the presence of excess acceptors and therefore it is suitable for the burying or passivation of p-n junction devices. We have shown that as much as 10^{20} cm^{-3} Ti can be incorporated into the epitaxial layers using metal organic Ti sources. The resistivity of these layers is $\rho > 10^7$ ohm−cm. Various low threshold, low leakage buried lasers were made using s-i InP:Ti:Fe for current confinement both on n and p-type InP substrates.

1. *INTRODUCTION*

In order to reduce current leakage, device capacitance and thermal resistance in various InP-based semiconductor devices, it is preferable to use semi-insulating (s-i) InP layers for current confinement or passivation, instead of the more commonly used reverse-biased junctions or dielectrics. Iron-doped s-i InP layers have been successfully grown by MOVPE (1) and used for current confinement in long wavelength lasers (2,3). Iron, however, is a deep acceptor in InP (4) and while it will yield s-i material when $N_{Fe} > N_D - N_A$, the layers will be conductive when the shallow acceptor concentration, N_A, is higher than the shallow donor concentration, N_D. This means that when an Fe-doped s-i layer is grown in contact with a p-n junction, the rapidly diffusing p-type impurities (Zn, Cd, Mg, Be) will render a small volume of the s-i layer conductive (Fig. 1).

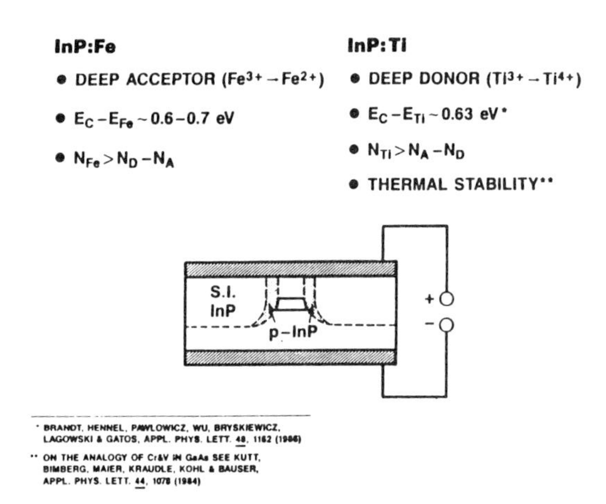

Figure 1 Characteristics of Fe and Ti in InP.

Recently, deep donor levels have been identified in Ti-doped liquid encapsulated Czochralski (LEC) InP (5). The resistivity of this bulk material was $\rho = 3\times10^6$ ohm-cm and the authors predicted higher resistivities (3×10^7 ohm-cm) as well as good thermal stability for this material. Since Ti is a deep donor in InP, it is not expected that acceptor out-diffusion, which makes the use of Fe-doped s-i epilayers difficult in p-n junction devices, will be detrimental for Ti-doped InP. Some of the characteristics of Fe and Ti in InP along with the conditions required for s-i behavior are summarized in Fig. 1.

EXPERIMENTAL

Before the start of the Ti doping experiments we evaluated the effect of p-type dopant out-diffusion from the substrate on the s-i properties of Fe-doped InP. The InP layers were grown by atmospheric pressure MOVPE using ferrocene as the organometallic iron source. In each growth run a p-type and an n-type InP substrate were placed side by side on the susceptor and an $\sim 3\ \mu m$ thick layer of Fe-doped InP was grown at 600°C. The n-type dopants were tin or sulfur, while the p-type dopants were Zn, Cd, Mg and Be. In each case the layers grown on the n-type substrates were s-i, with resistivities in excess of 10^7 ohm-cm and the layers grown on p-type substrates were conductive with resistivities of $\sim$1-10 ohm-cm, for all p-type dopants.

The Ti-doped InP layers were grown in an atmospheric pressure horizontal MOVPE reactor, described previously (6). We evaluated various organometallic Ti sources such as tetrakis(diethylamino)titanium, made by Morton Thiokol Inc./Alfa Products. Since no vapor pressure data were available for these compounds, the source temperatures and the N_2 (or Ar) flow rates through the Ti-source were arrived at empirically.

We have used trimethylindium at +15°C, 250 cc/min and 100% phosphine at 90 cc/min for a growth rate of $\sim 3\mu m$/hour. Dimethylcadmium was used in some of the doping experiments as the p-type dopant source and ferrocene as the iron source. We used secondary ion mass spectroscopy (SIMS) to measure the Ti concentration in the epilayers. Positive secondary ions were monitored under 12kV, $1\mu A$ mass filtered O^+ ion bombardment in a Kratos XSAM 800/SIMS 800 multitechnique surface analysis instrument which employs an Atomika cold cathode oxygen gun. Both $^{48}Ti^+$ and $^{31}P^+$ ions were monitored and the Ti counts were converted into concentrations based on relative sensitivity factors obtained in Ti-implanted InP standards. Depth scales were calibrated based on the crater depth measurements using a Dektax profilometer. The detection limit for Ti under our measurement conditions is $2\times10^{15}\ cm^{-3}$. The accuracy of the concentration values reported here is good to within 30%.

The resistivities of the doped layers were determined by direct resistance measurements on etched mesas with known area and thickness.

RESULTS

Using the commercially available tetrakis(diethylamino)titanium as the organo-metallic titanium source, the net Ti level in the epilayer (after subtracting the SIMS background) was $\sim 4\times10^{15}\ cm^{-3}$. In order to evaluate the effect of donors and acceptors on the resistivity of Ti-doped epilayers a series of wafers were grown on n-type InP. First we grew an n-type InP layer (background carrier concentration $N_D - N_A = 8\times10^{14}\ cm^{-3}$) doped with $4\times10^{15}\ cm^{-3}$ Ti. The I-V curve, measured on an etched mesa, and the resistivity ($\sim$1 ohm-cm) are shown on the top of Fig. 2. The low resistivity indicates that unlike Fe, Ti does not trap electrons. In an other growth run we introduced sufficient amounts of Cd to make the layer slightly p-type. The I-V characteristics of this layer are shown on the center of Fig. 2. Finally, a layer was grown with $N_{Ti} \sim 4\times10^{15}\ cm^{-3}$ and $N_A - N_D = 2\times10^{15}\ cm^{-3}$. The I-V curve measured on this wafer is on the bottom of Fig. 2, and shows that the layer is s-i, with

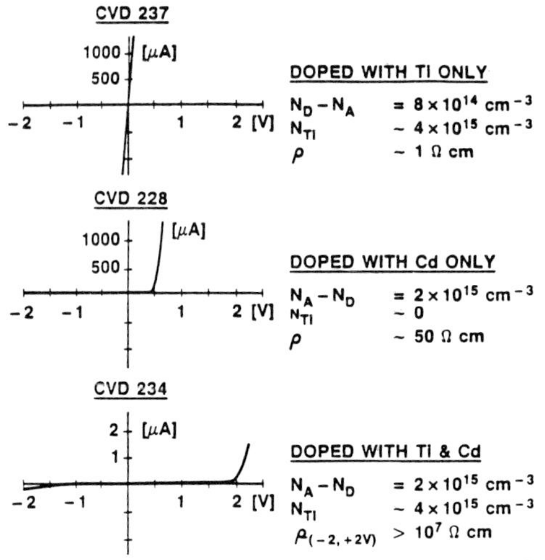

Figure 2 Resistivities of Ti, Cd and Ti-Cd doped layers.

$\rho > 10^7$ ohm−cm. When we tried to grow s-i InP on Zn-doped p-type substrates or on epitaxially grown Zn- or Cd-doped substrates, however, they were low resistivity, as would be expected under conditions when $N_A - N_D > N_{Ti}$.

We synthesized other organometallic compounds which allowed us to incorporate as much as $10^{20}\ cm^{-3}$ Ti into the InP epilayers (Fig. 3). Since not all of this Ti is electrically active, it was not clear what was the maximum net acceptor level that could still be trapped by Ti doping. It is even more important for device fabrication to know the maximum net acceptor level of the p-type substrate in contact with the Ti-doped epilayer that will still yield s-i structures. In order to find this acceptor level we have prepared several p-type InP wafers with the p-type dopant concentration (Zn) ranging from $4 \times 10^{17}\ cm^{-3}$ to $2 \times 10^{18}\ cm^{-3}$, and used them as substrates for the growth of Ti-doped InP. As shown in Fig. 4 we were able to grow s-i material on substrates with

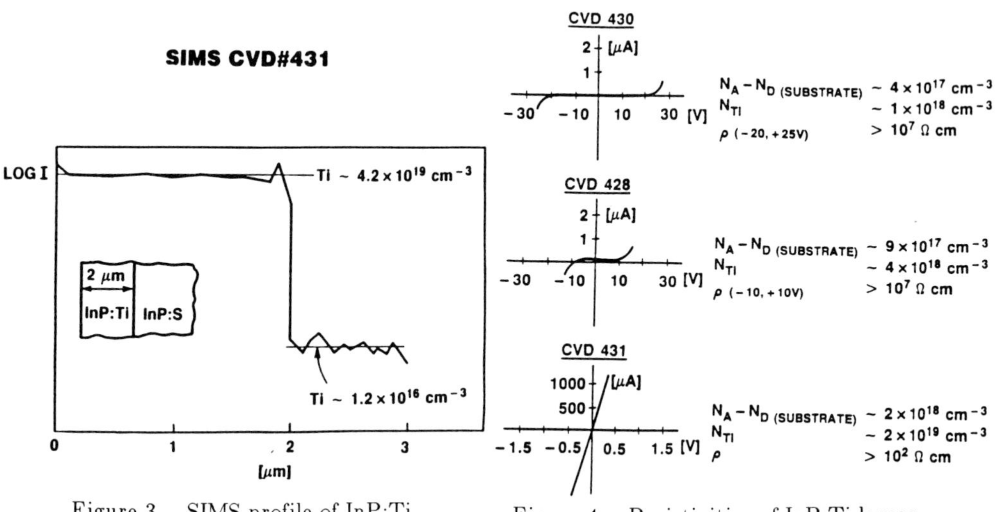

Figure 3 SIMS profile of InP:Ti.

Figure 4 Resistivities of InP:Ti layers grown on p-type substrates.

N_A-N_D up to 9×10^{17} cm^{-3}, while Ti-doped layers grown on more highly doped substrates were conductive.

Since Ti-doped InP epilayers grown on n-type surfaces are not s-i (see top curve of Fig. 2) unless co-doped with either shallow acceptors or iron, we grew InP:Ti:Fe layers with ferrocene as the iron source (at +5°C, 120 cc/min), on both n- and lightly doped ($N_A-N_D \sim 5\times10^{17}$ cm^{-3}) p-type substrates. The resistivities of these layers grown on both types of substrates were similar to those shown on the top two curves of Fig. 4. We used these s-i InP:Ti:Fe blocking layers to make two different laser structures: an inverted (p-substrate) buried cresent laser structure (7) with thresholds as low as 10 mA, which was grown by a MOVPE/LPE hybrid technique, and an all MOVPE planar buried laser with s-i InP:Ti:Fe current confinement which had thresholds of $\sim$25 mA. Neither of these two structures had any additional insulating layers such as SiO_2 or Si_3N_4 applied prior to metallization, thereby eliminating the need for post-growth photolithography, yet maintained excellent current confinement characteristics even when the substrate was p-type Zn-doped InP.

CONCLUSIONS

We have shown that it is possible to grow high resistivity ($\rho > 10^7$ ohm$-$cm) InP on both n- and p-type substrates by atmospheric pressure MOVPE, utilizing Ti and Fe as deep level traps. Our results confirm the deep donor nature of Ti in InP, since only layers with an excess of shallow acceptors became s-i upon Ti doping. We were able to incorporate Ti into the InP epilayer in the range of 4×10^{15} cm^{-3} – 10^{20} cm^{-3}. Buried lasers made with this new s-i material have low thresholds and excellent current confinement characteristics, without the use of additional dielectric layers.

ACKNOWLEDGEMENTS

We would like to thank J. C. Centanni for help with sample preparation, V. Swaminathan for the SIMS measurements and M. A. Pollack and R. C. Alferness for their support and encouragement.

REFERENCES

[1] J. A. Long, V. G. Riggs and W. D. Johnston, Jr., Journal of Crystal Growth, 69, 10-14, (1984).

[2] B. I. Miller, U. Koren and R. J. Capik, Electronics Letters, 22, 947 (1986).

[3] S. Sugou, Y. Kato and K. Kasahara, 10th International Semiconductor Laser Conference, Kanazawa, Japan (1986).

[4] G. Bremond, A. Nouailhat, G. Guillot and B. Cockayne, Electronics Letters, 17, 55, (1981).

[5] C. D. Brandt, A. M. Hennel, L. M. Pawlowitz, Y. T. Wu, T. Bryskiewicz, J. Lagowski and H. C. Gatos, Appl. Phys. Lett., 48, 1162, (1986).

[6] A. G. Dentai, C. H. Joyner, B. Tell, J. L. Zyskind, J. W. Sulhoff, J. F. Ferguson, J. C. Centanni, S. N. G. Chu and C. L. Cheng, Electronics Letters, 22, 1186, (1986).

[7] C. Caneau, C. E. Zah, J. S. Osinski, S. G. Menocal and T. P. Lee, Bell Communications Research, private communication.

Gold growth on GaAs (001)

J. Kanski and T.G. Andersson

Department of Physics, Chalmers University of Technology, S-412 996 Göteborg, Sweden

G. Le Lay

CRMC[2] - CNRS, Campus de Luminy, Case 913, F-13288 Marseille, Cedex 09, France

ABSTRACT: The initial formation of Au overlayers on an MBE-grown GaAs(001) has been studied by means of RHEED, AES, UPS and XPS. Our results show that the metal is first incorporated in a disordered continuous As rich mixture, which grows laminarly during the formation of the first two atomic layers. The laminar growth appears to be stabilized by the presence of As. Above 4Å coverage the Au overlayer grows in epitaxy with the substrate.

1. INTRODUCTION

Detailed understanding of metal-GaAs interfaces is highly desirable, for fundamental physical reasons as well as for practical applications in the increasingly advanced device structures. From earlier studies of Au deposition on sputter-cleaned GaAs(001), a more or less reacted interface region was inferred with segregation of As and Ga (Waldrop and Grant, 1979; Hiraki et al., 1979), while some more recent investigations on MBE- grown GaAs(001) (Narusawa et al., 1984) point towards an atomically abrupt interface with a small amount of As segregated to the overlayer surface.

In this work we present a comprehensive study of the initial formation of an Au overlayer on MBE-grown GaAs(001). The growth process is followed by means of electron diffraction (RHEED) and surface electron spectroscopies (AES, XPS and UPS).

2. EXPERIMENTAL

Our experimental arrangement consisted of a Varian MBE 360 growth chamber, connected via an UHV transfer system to a Vacuum Generators ADES 400 electron spectrometer. The GaAs samples were 0.5 μm thick Si-doped layers (10^{16} cm^{-3}) grown on n-type GaAs substrates. The growth was terminated to produce a c(2x8) surface reconstruction, which can be obtained in a rather wide surface composition range (Svensson et al., 1984). In the present case the relative concentrations of As and Ga were chosen to give similar intensity ratio in AES as for the in situ cleaved (110) surface. Gold films with thicknesses 0.25 to 15Å were grown at room temperature from a stable MBE-type source. The deposition rate was 6.0×10^{-3} Ås^{-1} as calibrated by surface profiling of ~3000 Å thick films. With unity sticking probability, this corresponds to a growth rate of one (001) monolayer during 170 s (Andersson et al., 1987). (The surface atom density on the ideal GaAs(001) surface is 6.26×10^{14} cm^{-2}, while the ideal Au(001) surface is nearly twice as dense (1.22×10^{15} cm^{-2}).

Auger electrons were excited with a 2 keV, 1 μA beam and the spectra were recorded in the first derivative mode using a single pass cylindrical mirror analyzer with 0.3% energy resolution. All UPS spectra reported here were obtained with 21.2 eV radiation at 65° incidence angle, normal emission and energy resolution of 0.1 eV. The XPS spectra were excited with unmonochromatised AlK_{α}- and MgK_{α}- radiation.

3. RESULTS

The development of the Au(69eV), Ga(51eV) and As(32eV) Auger signal intensities was followed as a function of Au coverage. For the Ga signal an exponential decay was observed, corresponding to a mean free path of ~3.5 Å in the overlayer. The As signal was attenuated markedly slower over the whole deposition range, suggesting dissociation and segregation of As. A detailed examination of the low coverage range revealed clear breaks in the intensity slopes, occurring for all three elements near 1 Å Au coverage, see Figure 1. These breaks were interpreted as indications of a completed first monoatomic overlayer. Clearly, this overlayer cannot be pure Au, but must incorporate atoms from the substrate.

The RHEED observations showed a fading intensity of the (2x8) surface reconstruction in the submonolayer range, accompanied by an increased background. At about 0.5 Å coverage only a bulk (1x1) pattern remained and beyond 1 Å the surface appeared disordered. However, after about 5 Å deposition (not shown), a new ordered pattern was observed due to formation of a crystalline Au layer in epitaxy with the substrate. The overlayers were very stable at room temperature and no spectral changes could be detected within several hours after growth.

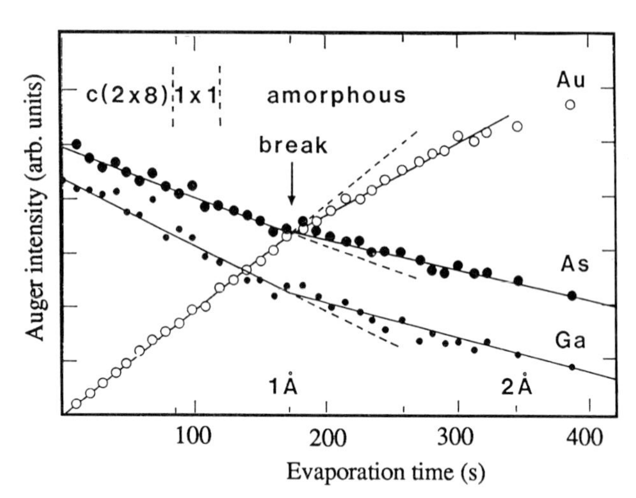

Figure 1. AES and RHEED observations in the low Au coverage range

Similarly to the Auger intensities, the XPS signals from As and Ga showed linear attenuation at low Au coverages, with the Ga signals being markedly more reduced. The attenuation of the As signals actually levelled out towards higher coverages, and even after deposition of a 100 Å thick Au layer the As signals were quite strong, see Figure 2.

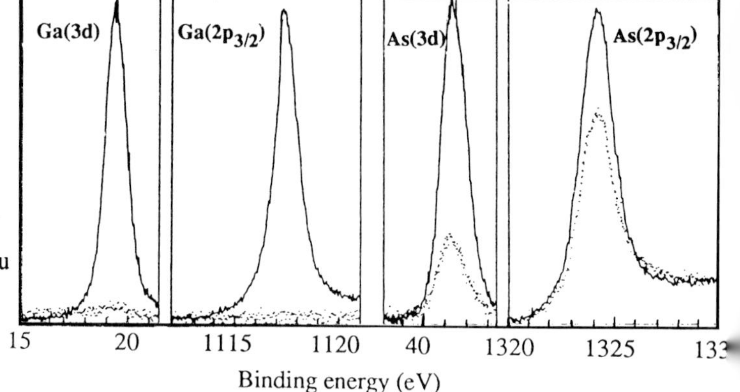

Figure 2. XPS spectra from GaAs(001) after deposition of 0.25Å Au (full lines) and 100 Å Au (dotted)

All spectra in Figure 2 were excited with MgK_α- radiation, except those from $As(2p_{3/2})$, in which case AlK_α- radiation was used. Considering the different kinetic energies of the As 2p and 3d electrons (~160 eV and ~1235 eV), one can directly conclude that the remaining As signal after 100 Å Au deposition originates from the surface region. Furthermore, the increased relative intensity at the high binding energy side of the As(2p) peak indicates that although the emission is excited near the surface, it does not come from the very first surface layer (Kanski et al., 1985). This conclusion is supported by the AES data, which show that the even more surface sensitive As(MVV) emission is attenuated to about 5% after 100 Å Au deposition. Assuming a mean free path of ~5Å for 50-150 eV electrons, and taking into consideration the relevant geo- metrical factors connected with different emission angles in our AES and XPS measurements (43° and 0° from the surface normal), we estimate that similar intensity relations as those ob- served here would be obtained from a monoatomic As layer embedded ~20 Å into the Au film.

In Figure 3 we show a set of UPS results, in which the Au overlayer formation can be followed via the development of the valence band. The Au induced features are dominated by emission from the 5d states. For the lowest coverages (up to ~0.5 Å) the energy separation between the 5d peaks is ~1.6 eV, a value characteristic for dispersed Au atoms. We note that the simple two-peak spectrum persists for higher coverages. Around 2 Å coverage the surface layer starts to become metallic (Kanski et al., 1986), as seen by the shift of the spectral onset to the Fermi level, but the inter- action between neighbour Au atoms becomes dominant first after ~4 Å deposition. The 5d emission is then modified by the crystal field and starts to resemble the spectrum of bulk Au. This is particularly clear in the difference spectra. However, the total spectrum differs significantly from that of bulk Au, even after 15 Å deposition.

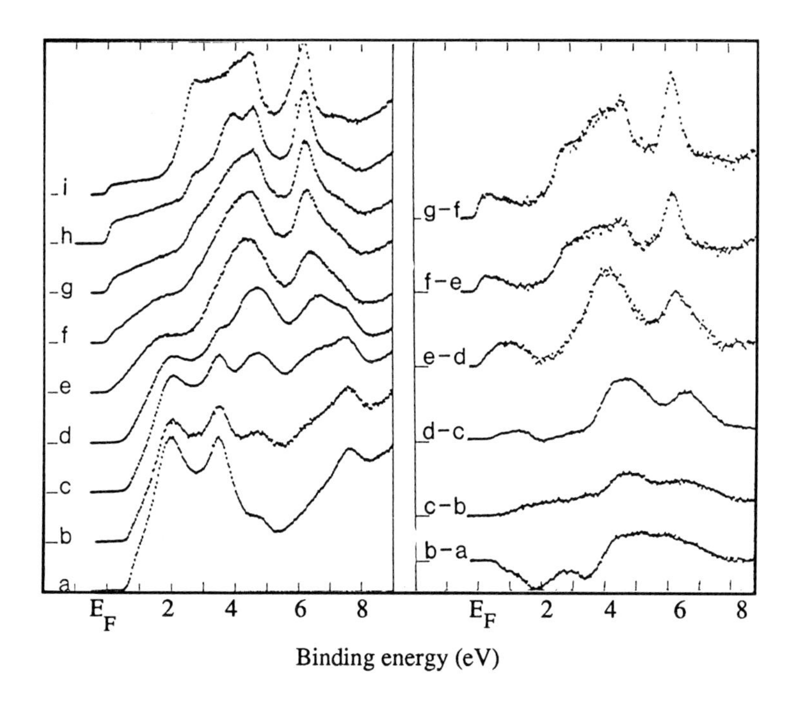

Figure 3. UV-photoemission from clean (a) and Au covered GaAs(001): b) 0.25 Å, c) 0.5 Å, d) 1.0 Å, e) 2.0 Å, f) 4.0 Å, g) 8.0 Å and h) 15 Å. For reference the spectrum from bulk Au is also included (i). To the right are shown the consecutive difference spectra.

4. CONCLUSION

Based on the above results we now present a model for the growth of an Au overlayer on the GaAs(001)-c(2x8) surface. The model contains two important ingredients: the As-induced disordering and the transport of deposited Au through an As-rich layer. The As found embedded in the overlayer must originate from the surface of the MBE-grown GaAs layer, as we have found no signs of Au or Ga interdiffusion. Thus, the interaction between Au and Ga is strong enough to liberate the surface As (which was also found by Narusawa et al.(1984)), with the underlying Ga and As layers remaining unaffected.

The liberated As atoms are incorporated in an initially disordered mixture with Au atoms, which are kept apart as witnessed by the UPS results. In this way the overlayer is stabilized against clustering. With increased coverage, the Au atoms coordinate with the first Ga layer, as was concluded by Waldrop and Grant (1979), and a basis is formed for epitaxial growth. The As layer is diluted and direct interaction between Au atoms becomes possible. As the dilution proceeds, crystalline order is obtained up to the surface (Andersson et al., 1986). This occurs above ~5 Å coverage, where the RHEED pattern is restored.Further deposition of Au results in growth of the thickness in such a way that the diluted As rich region remains near the overlayer surface, i.e. the deposited Au diffuses through this region.

5. ACKNOWLEDGEMENTS

The Swedish Natural Science Research Council and The Swedish National Board for Technical Development are acknowledged for financial support of this project.

6. REFERENCES

Andersson T G, Kanski J, Le Lay G and Svensson S 1986 Surface Sci. 168 301
Andersson T G, Le Lay G, Kanski J and Svensson S 1987 Phys. Rev.**B** (accepted for publication)
Hiraki A, Kim S, Kammura W and Iwami M 1979 Surface Sci. **86** 706
Kanski J, Andersson T G, Svensson S P and Le Lay G 1985 Solid State Commun. 54 339
Kanski J, Svensson S P, Andersson T G and Le Lay G 1986 Solid State Commun. 60 793
Narusawa T, Watanabe N, Kobayashi K L I and Nakushima H 1984 J. Vac. Sci. Technol. **A2** 538
Svensson S P, Kanski J, Andersson T G and Nilsson P O 1984 J. Vac. Sci. Technol. **B2** 235
Waldrop J R and Grant R W 1979 Appl. Phys. Letters **34** 630

Paper presented at Int. Symp. GaAs and Related Compounds, Heraklion, Greece, 1987

Mass production of 3-in diameter VPE GaAs wafers with excellent uniformity for microwave ICs

Y.MIURA, K.TAKEMOTO, T.TAKEBE*, T.SHIRAKAWA*, S.MURAI*, K.TADA*,S.IGUCHI
D.HARA, Y.NISHIDA** and S.AKAI*

Sumitomo Electric Industries Ltd. Semiconductor Division 1-1-1 Koya-kita, Itami 664, JAPAN TEL(0727)81-5151

*Basic High Technology Labolatories 1-1-3 Shimaya, Konohana-ku Osaka 554, JAPAN

**Sumitomo Electric U.S.A., Inc. 551, Madison Avenue(5th floor), New York, N.Y. 10022, U.S.A. TEL(212)308-6444

1.Introduction

Recently, various microwave devices based on GaAs have become indispensable in several fields including satellite communication. 2-in diameter epitaxial wafers have been used so far for such purposes, Nowadays, there are increasing demands for GaAs microwave ICs at low cost such as MMICs, so production of large size GaAs epitaxial wafers with good uniformity are in great demand to meet these requirements.
We have succeeded in the development and mass production of 3-in diameter GaAs wafers with thin S-doped active layers ranging from 0.05 to 0.5 μm with excellent uniformity by using a chloride VPE technique. We have developed large scale and multiple wafer charge VPE reactors, which combined our well-tested VPE technology, which has been accumulated over the years, with gas and thermal flow simulations using home-made software. In this report, we will explain our epitaxial conditions and demonstrate wafer quality and uniformity, including surface morphology, carrier concentration and thickness of the active layer, FET characteristics, reproducibility of uniformity and so on.

2.Epitaxial growth

Fig.1 shows a schematic diagram of the epitaxial growth system. This is the classical VPE growth method, but when the aperture of the reactor tube becomes larger, the nonuniform gas flow and thermal conventional effects cannot be ignored. We examined growth conditions by computer simulation (Y.MIURA et al 1987). On the basis of this simulation, we reconstructed the quartz parts, and created the best possible growth conditions. Table.1 shows an outline of typical epitaxial growth conditions. We used(100)2°off toward <101> Cr doped semi-insulating wafers(N_{cr}=0.05-0.35wt ppm) grown by the LEC method as the epitaxial substrate. The epitaxial wafers consist of a S-doped active layer of 0.05 to 0.5 μm thick, an undoped buffer layer of 0.7 to 2.0 μm thick, and LEC substrate.

3.Results and discussion

3-1 Surface morphology

Fig.2 shows an optical micrograph of the surface of a 3-in diameter wafer which was grown under the above conditions. We obtained a good mirror surface without haze and defects. Fig.3 shows the data on the surface defect density of 3-in diameter epitaxial wafers ,measured by surfscan 4500 (TENCOR INSTRUMENTS).

Table.1 Typical epitaxial conditions

F_{H_2}	(ℓ/min)	4.4~9.0
P_{AsCl_3}	(moℓ/min)	1.2~2.7 x 10^{-3}
$T_{SUB.}$	(°C)	723~738
ROTATION	(r. p. m)	6.5
DOPANT	(ppm)	1000 (H_2S/H_2)

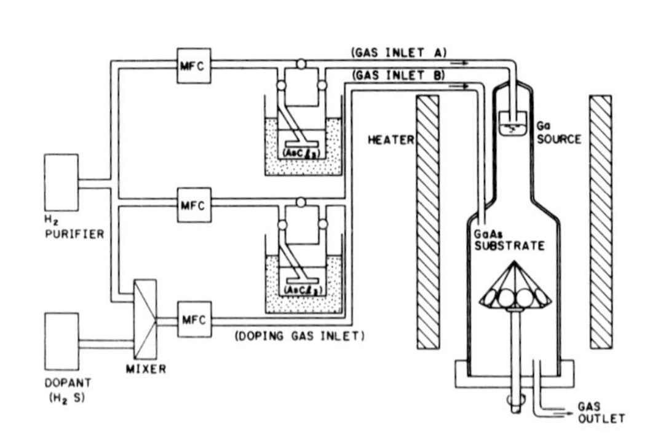

Fig.1 Schematic diagram of the apparatus for the epitaxial growth

Fig.2 Optical micrograph of the surface of a 3-in diameter wafer

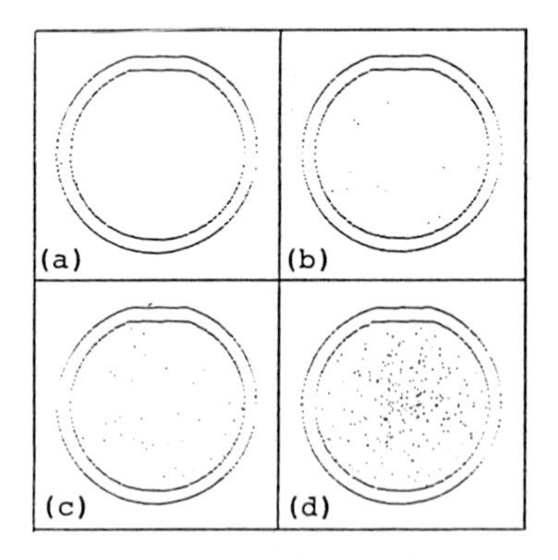

a) > 250 µm , 1 / wafers
b) > 12 μm^2 , 0.37 cm^{-2}
c) > 2.4 μm^2 , 1.14 cm^{-2}
d) > 0.06 μm^2 , 5.81 μm^{-2}

Fig.3 Surface defect density of the epitaxial wafers

3-2 Uniformity of active layer and FET characteristics

Fig.4 shows the distribution of the carrier concentration(na) and thickness(da) of the active layer, which were obtained at 17 points on the wafer from measurement by a schottky diode(diameter of schottky electrode is 500 µm). When we defind the distribution as follows.

$$\Delta \text{ na} = \frac{\text{na max} - \text{na min}}{\text{na max} + \text{na min}} \times 100 \quad \ldots (1)$$

$$\Delta \text{ da} = \frac{\text{da max} - \text{da min}}{\text{da max} + \text{da min}} \times 100 \quad \ldots (2)$$

The obtained epitaxial layer shows that both Δna and Δda are < 3 % over the whole wafer. Fig.5 examines the uniformity of na and da products by observing the interference color which results from anodic oxidation of the active layer. The interference color is nearly monochromatic over the whole wafer, indicating also good uniformity of the active layer. In order to demonstrate the uniformity of wafers on a microscopic scale, we fabricated an FET array in which identical 1 μm long 100 μm wide gate FETs are arranged at 1.1 mm intervals and we measured distribution of the threshold voltage Vth across the wafers. On a whole wafer with a non-recessed 0.15 μm thick active layer and 1.2 μm thick undoped buffer, the variation of Vth, defined by σVth/(0.7-Vth), where σVth and Vth are the standard deviation and the meam value of Vth, respectively, was only 6.0% with Vth = -1.95V for 3200 FETs (Fig.6). The Vth variation was still reduced to 4.0% for 1760 FETs in the 2-inch diameter central region of the wafer. Capacitance-voltage measurements on a 150 μm long and 200 μm wide gate FET adjacent to the above FET showed variations of na and da as low as 3.0% and 1.5% respectively, across the wafers. Fig.7 examines the results of the run-to-run reproducibility of na and da. The symbol ⌶ in Fig.7 shows the variation of na and da among the wafers grown in the same batch. With continued 30 growth runs, we could reproduce na and da within 3% variation over the whole wafer and among all the wafers.

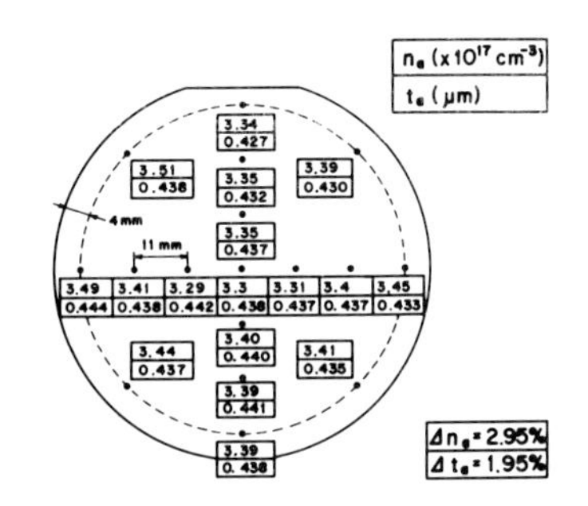

Fig.4 Results of C-V measurement

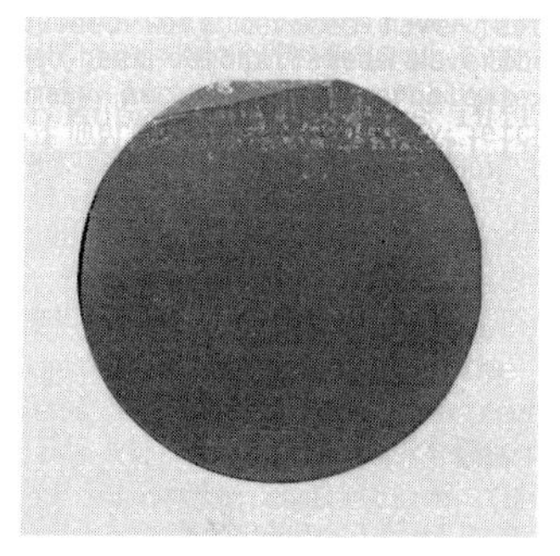

Fig.5 Check of active layer uniformity by anodic oxidation

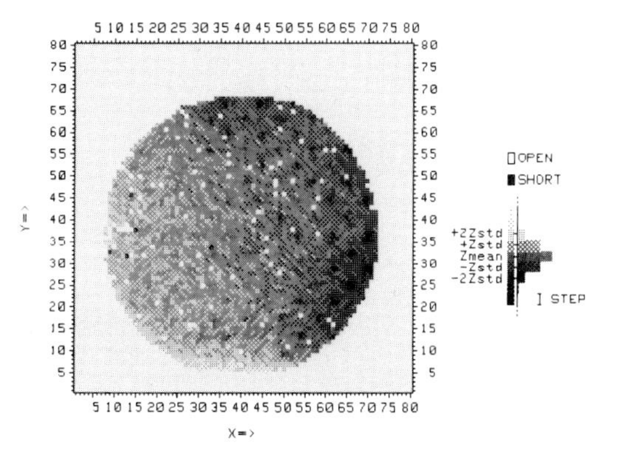

MEASUREMENT DATA

GATE LENGTH = 1 μm
GATE WIDTH = 100 μm
X INTERVAL = 1.1 mm
Y INTERVAL = 1.1 mm

RESULTS

MEAN VALUE = -1.947 V
STANDARD DEVIATION = 159 mV
UNIFORMITY = 6.01 %
SAMPLING NO. = 3200
SHADING STEP = 150 mV

Fig.6 Distribution of the threshold voltage Vth acroos the wafer

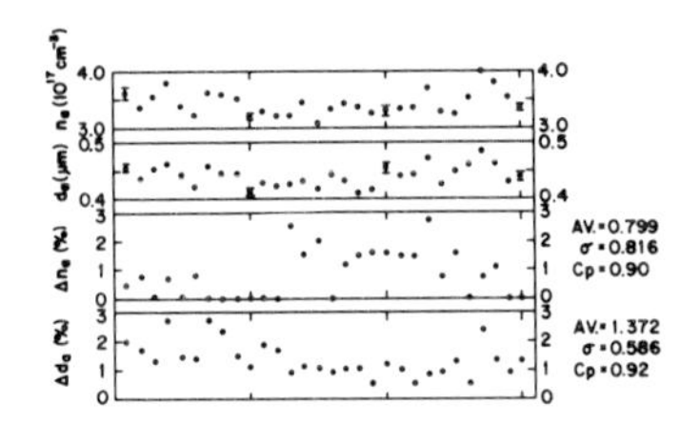

Fig.7 Results of run-to-run reproducibility of na, da, Δna and Δda

Table.2 Other properties of 3-in diameter epitaxial wafers

BUFFER LAYER	G. R. (μm/min)	0.15~0.30
	n_b (cm^{-3})	1~50 x 10^{14} (CONTROLLABLE)
	I_b (μA)	<5 at 1KV/cm
	μ_b ($cm^2 V^{-1} S^{-1}$)	>100,000 at 77K
STEEPNESS (Å)		500~530 at $n_a = 1 \times 10^{17} cm^{-3}$
ACTIVE LAYER	n_a (cm^{-3})	0.6 ~ 5 x 10^{17}
	t_a (μm)	0.1 ~1.0
	Δn_a (%)	0 ~ 3
	Δt_a (%)	0 ~ 3
	μ_a ($cm^2 V^{-1} S^{-1}$)	6,000 at $n_a = 1 \times 10^{17} cm^{-3}$

3-3 Other properties

Table.2 summarizes the characteristics of the 3-in diameter epitaxial wafers. As for the buffer layer properties, we see carrier concentration with high mobility at low temperatures, and low leakage current should be noted. Moreover, PL measurements at 4K revealed no emission peaks except from free-exciton, bound-exciton, and carbon acceptor possibily diffused from the LEC substrates, showing also high purity of the buffer layer. In addition, a growth rate about 10 times higher than that of OMVPE and MBE is very advantageous for mass production.

4.Conclusion

We have succeeded in development and mass production of 3-in diameter GaAs wafers with 0.05 to 0.5 μm thick S-doped active layers by using a chloride VPE technique. They are fully suited for MMIC applications. Measurement of the threshold voltage Vth of FETs fabricated on a 3-in diameter wafer showed variations of Vth as low as 6.0% at the mean value of -1.95 V for 3200 FETs on the whole wafer. The carrier concentration and the active layer thickness showed only 3.0% and 1.5% variations, respectively, proving the excellent uniformity of Vth distribution. The run-to-run and wafer-to-wafer reproducibility of the basic properties was also demonstrated. These results proved that our 3-in diameter VPE wafers have quality and uniformity.

Acknowledgements

The authors would like to thank Y.KAKINO, Y.SENDA for epitaxial growth equipement fabrication.

Reference

Y.MIURA, K.TAKEMOTO, M.KAZI, S.IGUCHI, Y.KAKINO, Y.SENDA
SUMITOMO ELECTRIC TECHNICAL REVIEW in Japanese 130 ,85

Paper presented at Int. Symp. GaAs and Related Compounds, Heraklion, Greece, 1987

A comparison of low pressure and atmospheric pressure MOCVD growth of InP

D J Nicholas, G J Clarke, B Hamilton, A R Peaker

Department of Electrical Engineering and Electronics and Centre for Electronic Materials, University of Manchester Institute of Science and Technology, P O Box 88, Manchester M60 1QD, England

E J Thrush

STC Technology Ltd., London Road, Harlow, Essex, England

M D Scott, J I Davies

Plessey (Caswell) Ltd., Caswell, Towcester, Northants, England

ABSTRACT: We have shown that zinc and silicon are the major impurities in epitaxial layers of InP when grown by the atmospheric pressure metal organic chemical vapour deposition (APMOCVD) technique. Carbon and an unidentified donor species are found to dominate when the layers are grown using a reduced pressure technique (LPMOCVD).

The dominant electron trap in APMOCVD layers is usually the E1 level associated with substitutional iron. Iron was not found in the LPMOCVD layers, and the dominant electron trap has not been reported in the literature.

1. INTRODUCTION

A wide range of binary and compound III–V semiconductors are now grown by the MOCVD technique. It offers a number of advantages over competing growth techniques such as sharp interfaces and abrupt doping profiles. Growth of device quality epitaxial layers has been demonstrated by Dapkus (1984) for APMOCVD growth and by Razeghi et al (1983) for LPMOCVD growth.

In this paper we present work on a comparison of the impurities introduced by the two techniques for MOCVD growth of the binary compound InP. We have chosen to study this materials system because of the problematic history of growing In based compounds (Bass et al 1983, Moss and Evans 1981) and because of its important role in the fabrication of a vast number of opto-electronic devices.

2. GROWTH DETAILS

Nominally undoped epitaxial layers were grown either on semi-insulating (Fe) for Hall measurements or n^+ substrates for DLTS measurements. PL was performed on S.I. and n^+ substrates for comparison. Atmospheric pressure growth was performed using TMI and PH_3 from various suppliers, in a horizontal reactor, details of which have been published by Thrush and Whiteaway (1980).

The horizontal low pressure MOCVD system (supplied by Thomas Swan and Co. Ltd.) has RF heating and is fitted with a linear vent/run manifold which is close-coupled to the reactor cell and also incorporates a comprehensive automatic pressure balancing system.

The highest quality layers were grown using TMI (supplied by Epichem Ltd.) and a 10% phosphine in H_2 mixture (supplied by BOC Ltd.) at Electra III–V grade. Growth conditions and typical electrical properties are summarised in Table 1.

TABLE 1 Growth Conditions and Typical Electrical Properties

Sample	Pressure (Torr)	Growth Temp (°C)	V/III	n(77K) cm^{-3}	μ(77K) cm^2/Vs
1	760	625	50	$8x10^{14}$	90,000
2	760	600	37	$2x10^{14}$	145,000
3	300	600	120	$5x10^{14}$	101,000
4	150	570	116	$3x10^{13}$	~300,000

3. DEEP LEVEL MEASUREMENTS

Deep level transient spectroscopy (DTLS) measurements were performed on evaporated titanium Schottky contacts in a Polaron S4700 DLTS unit. Barrier heights in the range 0.48 eV to 0.55 eV were achieved such that measurements could be performed up to 350 K. Ohmic contacts were fabricated using a 98% Au – 2% Ge eutectic alloy.

In earlier work (unpublished) we have monitored deep level concentrations as a function of growth conditions, and as a function of adduct formation (Nicholas et al 1984). In the next section we show that the growth conditions which yield lowest defect concentrations and highest mobilities are not necessarily those which give best optical quality.

Figure 1 shows a comparison of typical spectra obtained from extremely high quality epitaxial layers of InP grown by a) APMOCVD (sample 1) and b) LPMOCVD (Sample 3). The emission rate window for the plot is 200 s^{-1}. It is clear that the dominant trap in

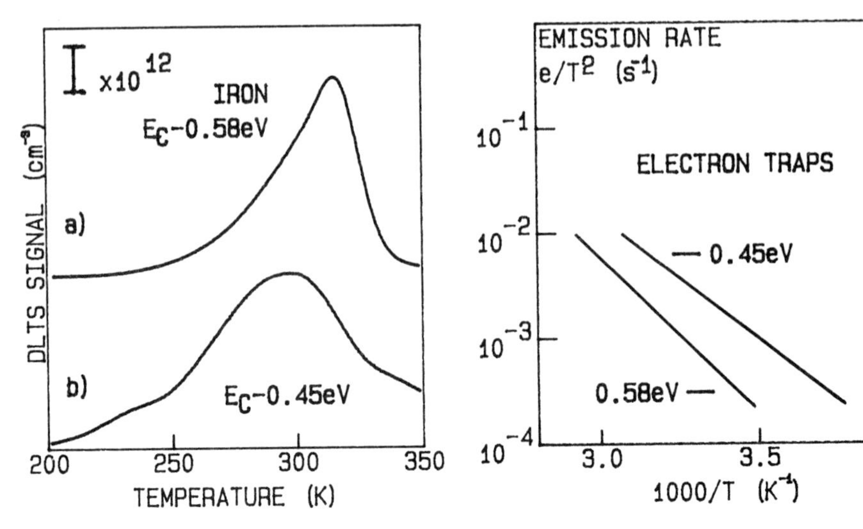

Figure 1 DLTS spectra obtained from a) APMOCVD and b) LPMOCVD

Figure 2 Arrhenius plot of dominant traps

APMOCVD layers is of a different origin to that in LPMOCVD layers. Figure 2 shows an Arhennius type plot of the temperature corrected thermal emission rates of the two levels. Level E1 dominant in the APMOCVD layers has been associated with the simple substitutional Fe centre (Tapster et al 1981) on the In sublattice. The concentration of this centre is seen to increase at high V/III ratios. From the slope of the Arhennius plot, a value of 0.58 eV is extracted for the activation energy of the iron level. The activation energy of the dominant trap in Figure 2b is 0.45eV. No previous reference to this trap could be found in the literature. Its identify is therefore unclear.

4 OPTICAL ASSESSMENT

Low temperature (4.2° K) photoluminescence (PL) measurements were performed as a means of identifying crystal quality, shallow impurities and deep level transition metal species. The 514.5 nm line of an argon ion laser was used at a power density of 160 mW/cm². A 1.0 m wavelength dispersive spectrometer and a cooled germanium detector, with conventional lock-in techniques, were used to obtain the spectra.

Figure 3 shows a comparison of typical spectra obtained from a) APMOCVD (sample 1) and b) LPMOCVD (sample 3) layers. The near band edge excitonic region of the spectra comprises a number of sharp, well resolved peaks confirming the high crystal quality of these layers. The startling difference between the two spectra is that the DAP transition in Figure 3a appears at 1.375 eV, whereas that in Figure 3b appears at 1.3803 eV. This is indicative of zinc as being the dominant background acceptor in APMOCVD layers and carbon the dominant background acceptor in LPMOCVD layers. In fact, over a large range of samples (~ 200) grown by APMOCVD, carbon was only detected in two layers.

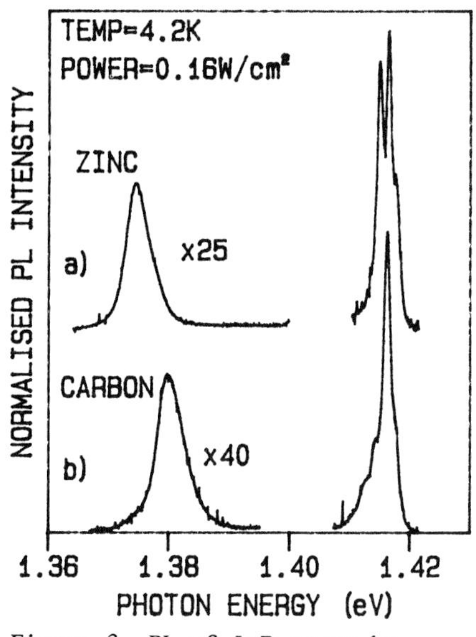

Figure 3 PL of InP grown by a) APMOCVD b) LPMOCVD

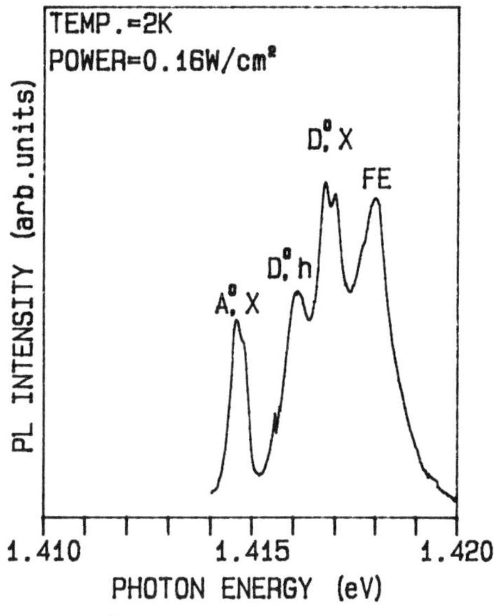

Figure 4 Expanded excitonic region of high quality InP

Figure 4 shows an expanded spectrum of the near band edge excitonic region of sample 2. This sample was grown by APMOCVD at a temperature of 600°C. This spectrum clearly demonstrates the high quality of these epitaxial layers. In the spectrum, the highest energy component is due to the recombination of free excitons (FE). The presence of this transition is a good indication of the quality. To lower energies we see sharp, well resolved transitions due to the recombination of free holes and electrons bound to donor levels (D,h), excitons bound to neutral donors (D°,X), the ionised versions of this (D^+,X) and excitons bound to neutral acceptors (A°,X).

5 DISCUSSION

Zinc, silicon and iron are invariably the dominant impurity species detected in our APMOCVD layers. However, zinc is rarely present in our LPMOCVD layers even when they are grown with TMI sources known to be contaminated with zinc. Silicon and iron have not been found in our LPMOCVD layers. A possible explanation for this behaviour is that the metal containing compounds responsible are not decomposed as effectively at the lower temperatures (570°C) and will also have a reduced surface coverage at the lower pressures employed in LPMOCVD growth. It is also possible that there is a reduction in the equilibrium In vacancy concentration during LPMOCVD growth since the three dominant impurities introduced during APMOCVD growth (absent in LPMOCVD) sit on the In sublattice. We are in the process of studying LPMOCVD growth as a function of temperature in an effort to confirm these suggestions.

It is temping to conclude that the presence of albeit small amounts of carbon in our LPMOCVD layers is a direct result of the reduction in hydrogen pressure. However, Kuech and Veuhoff (1984) have shown that the carrier gas does not play a role in impurity incorporation. If this is accepted, then we must conclude that carbon incorporation increases because of the high gas velocities employed in reduced pressure growth.

6. CONCLUSION

There is little difference in the electrical and optical quality of epitaxial layers of InP grown by APMOCVD and LPMOCVD. However, the chemical species and/or defect centres responsible for the electrical and optical properties are quite different. Zinc, silicon and iron are found to be the dominant impurity species in APMOCVD layers, whereas carbon, an unidentified donor species and an unidentfied defect dominate in LPMOCVD layers. Nevertheless, it is felt that the reproducible growth of high quality APMOCVD epitaxial layers relies much more heavily on the availability of high purity TMI sources, and, more significantly nowadays, on reliable supplies of high purity phosphine.

ACKNOWLEDGEMENTS

This work has been funded by the Science and Engineering Research Council under the Joint Opto Electronic Research Scheme.

REFERENCES

Bass S J, Pickering C, Young M L 1983 *J. Crystal Growth* **64** 68
Dapkus P D 1984 J. Crystal Growth **68** 345
Keuch T F and Veuhoff E 1984 J. Crystal Growth **68** 148
Moss R H and Evans R H 1981 *J. Crystal Growth* **55** 129
Nicholas D J, Allsop D, Hamilton B, Peaker A R and Bates S J 1984 *J. Crysta Growth* **68** 326
Razeghi M, Poisson M A, Larivain J P and Duchemin J P 1983 J. Electron. Mater. **12** 371
Tapster P R, Skolnick M S, Humphreys R G, Dean P J, Cockayne B and MacEwan W R 1981 *J. Phys. C,* **14** 5069
Thrush E J and Whiteaway J E A 1980 Inst. Phys. Conf. Ser. **56** Ch. 6 337

The effect of substrate orientation on deep levels in N-AlGaAs grown by molecular beam epitaxy

D.C. Radulescu, W.J. Schaff, G.W. Wicks*, A.R. Calawa,** and L.F. Eastman
School of Electrical Engineering, Cornell University, Ithaca, NY 14853

*Institute of Optics, University of Rochester, Rochester NY 14627
**MIT Lincoln Laboratory, Lexington MA 02173

ABSTRACT

Deep electron traps in Si-doped $Al_{0.25}Ga_{0.75}As$ grown by molecular beam epitaxy (MBE) on GaAs substrates deliberately misoriented (tilted 0 to 6 degrees) off the (001) plane towards either (111)A, (111)B or (011) have been investigated using deep-level transient capacitance spectroscopy (DLTS). Of the three dominant traps observed in AlGaAs, the concentrations of two of these are observed to be a direct function of the substrate tilt, while the concentration of the third dominant trap, which is related to the DX-center, is independent of the substrate misorientation.

INTRODUCTION

Recently, it has been shown that a deliberate substrate misorientation of a few degrees off (001) towards (111)A during MBE can improve the electron transport, optical and morphological properties of AlGaAs/GaAs heterostructures [Tsui et al 1986, 1986a, 1985b, Radulescu et al 1987, l987a]. However, deep electron traps in doped AlGaAs dominate the optical and electrical characteristics of heterojunction devices, while the impact of the misoriented substrate on these traps has not been investigated at all. This paper discusses the effect of substrate misorientation on deep electron traps in MBE grown AlGaAs. We have varied the substrate tilt angle and tilt direction in addition to the substrate temperature (610°C to 650°C) during growth. The dependence of deep-level concentrations on these MBE growth parameters will help in identifying which impurities and/or defects are affected by substrate misorientation during MBE growth.

EXPERIMENTAL

The structures used in this study were grown by MBE in a Varian GEN II on semi-insulating liquid encapsulated Czochralski (LEC) grown GaAs substrates cut either 0, 2, 3 or 6 ± 0.5 degrees off the (001) plane towards either (111)A, (111)B, or (011). The structures consist of a 7500Å GaAs buffer layer doped with silicon at 1×10^{18} cm^{-3} followed by 2000Å of 2×10^{17} cm^{-3} silicon-doped GaAs, 2500Å of 1.5×10^{17} cm^{-3} silicon-doped $Al_{0.25}Ga_{0.75}As$, and finally a 100Å undoped GaAs cap.

For each growth run (D1, D2, D3 and D4), three substrates with different orientations were mounted side by side, using indium on the molybdenum mounting block. The substrate temperature during growth of the AlGaAs was either 610, 625 or 650°C, while the GaAs was grown at 610°C for all growth runs.

The growth rate of GaAs and AlGaAs was 1.0 and 1.33 μm/h, respectively. Reflection high energy electron diffraction patterns taken along the <110> azimuth exhibited second and third order reconstructions for the GaAs and AlGaAs, respectively. Two arsenic sources (As_4) were used for the GaAs growth, while one was used for AlGaAs growth. Table 1 summarizes the substrate orientations and the AlGaAs growth temperatures used for the four growth runs presented in this study. Following epitaxial growth, conventional 600 μm diameter diodes were fabricated. DLTS measurements were performed with a system described previously [Kirchner et al 1981]. The diodes were reversed biased at 2.5V with a trap filling pulse of 3.0V for 2.5 mS. DLTS spectra were generated using the capacitance transient data at 4 and 20 mS, resulting in a rate window of 100 Hz.

Growth Run	Substrate Orientation Off (001)	AlGaAs Growth Temperature
D1	6° Toward (111)A 2° Toward (011) 6°Toward (111)B	625°C
D2	0° Nominally flat 3° Toward (111)A 6° Toward (111)A	610°C
D3	3°Toward (111)A 2°Toward (011) 3° Toward (111)B	625°C
D4	0° Nominally Flat 3°Toward (111)A 6°Toward (111)A	650°C

Table 1
Summary of the growth runs, substrate orientations, and AlGaAs growth temperatures used for this study.

RESULTS AND DISCUSSION

Fig. 1 shows typical DLTS spectra of electron traps in the silicon-doped AlGaAs structures from growth run D1 as a function of the substrate misorientation: 6° towards (111)A, 6° towards (111)B, and 2° towards (011). The spectra show the existence of seven electron trap levels labeled ME1 through ME7. The dependence of these commonly seen seven traps (ME1-ME7) on aluminum mole fraction, growth temperature, and V/III flux ratio have been previously reported [Yamanaka et al 1987, 1987a,Naritsuka et al I985, Mooney et al 1985] . In addition, their capture cross sections and thermal activation energies have been thoroughly studied. However, their fundamental origins not completely understood. In this paper we concentrate on the properties of the three dominant traps (ME3, ME5, ME6) observed in the spectra and their dependence on substrate tilt angle and tilt direction in addition to growth temperature. The trap labeled ME3 in Fig. 1 is attributed to the DX-center commonly observed in silicon doped AlGaAs [Yamanaka 1987, Lang 1979, Yamanaka 1984a] . As seen in the figure, its concentration is independent of the substrate misorientation. In contrast, the concentrations of the other two dominant traps (ME5 and ME6) are observed to be a function of the substrate misorientation. The concentrations of these traps are smallest for the substrate misoriented 6° towards (111)A and largest for the substrate misoriented 6° towards (111)B. A substrate misorientation of 2° towards (011) is seen to result in an intermediate concentration.

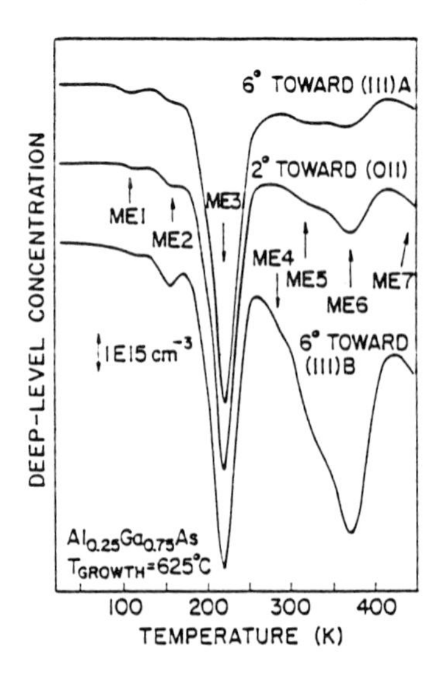

Fig. 1. Typical DLTS spectra showing electron traps for growth run D1 as a function of the intentional substrate misorientation off a nominal (001) surface.

Fig. 1 indicates that a substrate tilt toward (111)A results in a lower concentration of deep-levels than tilt toward (111)B. To investigate the significance of the angle of tilt toward (111)A growth run D2 was performed. In Fig. 2 the concentrations of levels ME3, ME5, and ME6 are plotted as a function of substrate tilt angle toward (111)A. As seen in the figure, the concentration of level ME3, which is related to the DX-center, is independent of substrate titlt angle consistent with growth run D1. However, the concentrations of levels ME5 and ME6 progressively decrease as the tilt angle increases. Fig. 2 indicates that a substrate titlt toward (111)A can improve the electrical quality of MBE grown AlGaAs in comparison to material grown on a nominally flat substrate at 610°C.

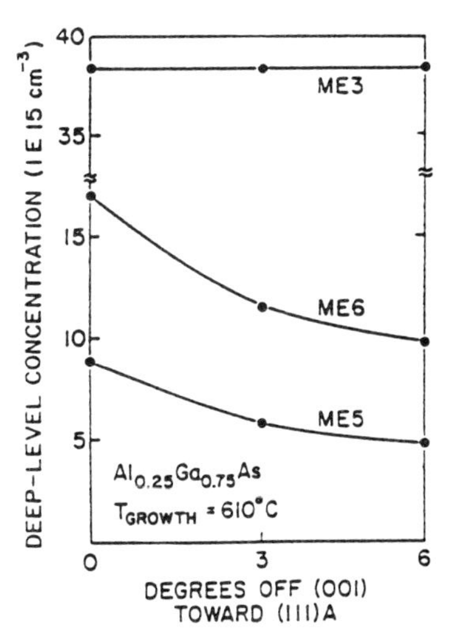

Fig. 2. Electron trap concentrations of ME3, ME5 and ME6 for growthrun D2 as a function of the angleof tilt off (001) toward (111)A.

A comparison between the effect of the tilt angle and tilt direction on the deep-level concentrations can also be made for a higher growth temperature of 625°C by plotting the deep -level concentrations for growth runs D1 and D3 versus substrate tilt. This is done in Fig. 3. As can be seen in the figure, trap level ME3 is again independent of substrate tilt, while trap levels ME5 and ME6 are dependent on the tilt angle and tilt direction. As the tilt angle toward (111)B progressively increases, the deep-level concentrations of ME5 and ME6 also progressively increase. This effect is in contrast to tilt toward (111)A where a larger tilt angle improves the electrical quality of MBE-grown AlGaAs.

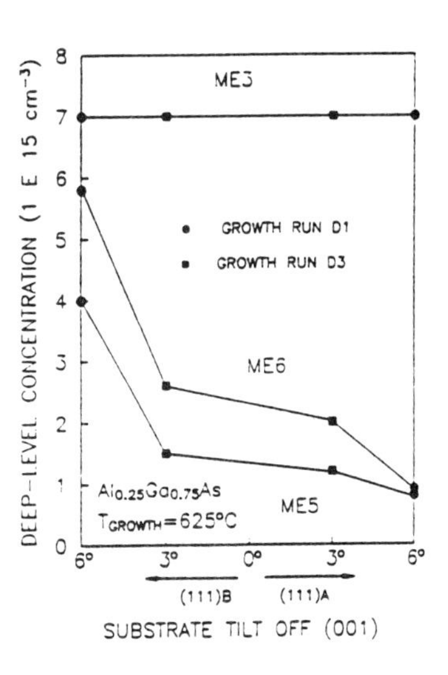

Fig. 3. Electron trap concentrations of ME3, ME5 and ME6 for growth runs D1 and D3 as a function of the substrate tilt angle and tilt direction off (001).

In Figs. 4a,b, and c the dependence of the deep-level concentrations of ME3, ME5 and ME6, respectively, on $Al_{0.25}Ga_{0.75}As$ growth temperature and tilt angle toward (111)A are plotted. As can be seen in the figures, all three trap concentrations are a minimum at 625°C in comparison to 610°C or 650°C. In addition, even at the highest growth temperature (650°C), the trap concentrations (ME5 and ME6) are still a function of the substrate tilt angle.

CONCLUSION

The influence of a substrate tilt off (001) during MBE on the incorporation of deep electron traps has been investigated. Tilt toward (111)A reduces the incorporation of ME5 and ME6, whereas tilt toward (111)B increases the incorporation of these deep levels, in comparison to nominally flat substrates. The DX-center (ME3) concentration is independent of substrate tilt.

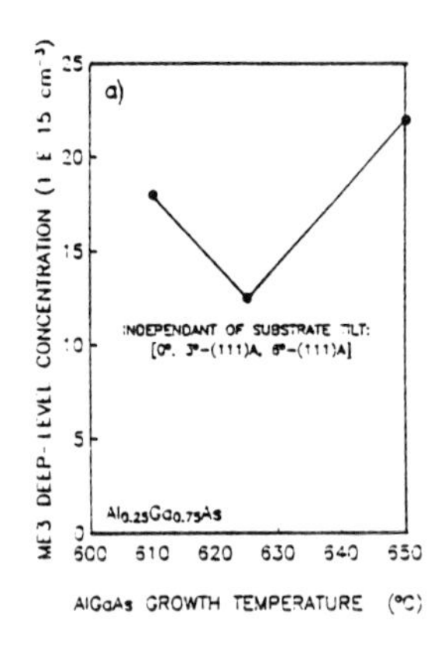

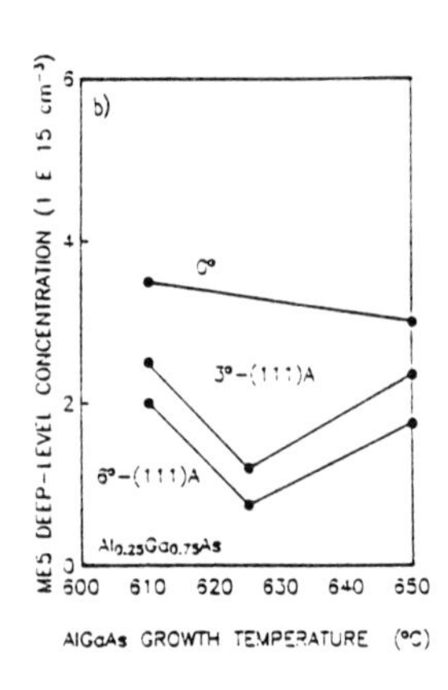

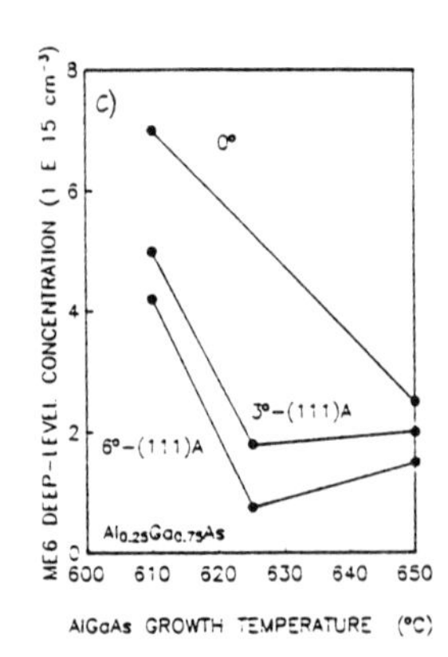

Figs. 4a, b, and c. Concentrations of ME3, ME5 and ME6, respectively, as a function of AlGaAs growth temperature and substrate tilt angle off (001) toward (111)A.

This work is partially supported by IBM, JSEP, the Office of Naval Research , Amoco and McDonnell-Douglas,.

References

Kirchner P D, Schaff W J, Maracas G N, Eastman L F, Chappell T I and Ransom C M, 1981 J. Appl. Phys. 52 6462.
Lang D V, Logan R A and Jaros M, 1979 Phys Rev B 19 1015.
Mooney P M, Fischer R and Morkoc H (1985) J. Appl. Phys. 57 1928.
Naritsuka S, Yamanaka K, Mannoh M, Mihara and Ishii M 1985 Jpn J. Appl. Phys. 24 1324.
Radulescu D C, Wicks G W, Schaff W J, Calawa A R and Eastman L F (1987) J. Appl. Phys. 62 954.
Radulescu D C , Wicks G W, Schaff W J, Calawa A R and Eastman L F (1987a) submitted to J. Appl. Phys.
Tsui R K, Curless J A, Karmer G D, Peffley M S and Wicks G W,(1986) J. Appl. Phys. 59 1508.
Tsui R K, Kramer G D, Curless J A and Peffley M S, (1986a) Appl. Phys. Lett. 48 940.
Tsui R K, Curless J A, Kramer G D, Peffley M S and Rode D L, (1985b) J. Appl. Phys. 58 2570.
Yamanaka K, Naritsuka S, Kanamoto K, Mihara M and Ishii M (1987) J. Appl. Phys. 61 5062.
Yamanaka K,Naritsuka S, Mannog M, Yuasa T, Nomura Y, Mihara M and Ishii M, (1984a) J. vac Sci Technol. B 2 229.

Paper presented at Int. Symp. GaAs and Related Compounds, Heraklion, Greece, 1987

Incorporation and excitation behaviour of rare earth ions in III–V semiconductors—InP:Yb

W Körber A Hangleiter K Thonke J Weber F Scholz K W Benz* and H Ennen**

Universität Stuttgart, 4. Physikalisches Institut, Pfaffenwaldring 57, D-7000 Stuttgart-80, West Germany
*Universität Paderborn, Fachbereich Physik, Warburger Str. 100, D-4790 Paderborn, West Germany
**Fraunhofer Institut für Angewandte Festkörperphysik, Eckerstr. 4, D-7800 Freiburg, West Germany

ABSTRACT: InP:Yb LPE layers have been grown by a supercooling process at high growth temperatures up to 800°C. Temperature dependent Hall measurements have revealed p-type conduction and a shallow acceptor ionization energy of about 45meV. The decay processes of internal transitions of $Yb^{3+}(4f^{13})$ incorporated in InP were investigated by means of time-resolved photoluminescence. From the temperature dependence of the excited state lifetime we find several decay mechanisms, including a bound-exciton-like Auger-process, energy transfer, and thermal depopulation.

1. INTRODUCTION

During the last years, the idea of incorporating rare earth (RE) - impurities in III-V semiconductors in order to make light emitting devices based on the sharp intra 4f-shell luminescence lines has attracted increasing interest. In contrast to other III-V optoelectronic devices, the luminescence (due to rare earth internal transitions) is practically independent of the III-V host crystal and of its band gap energy. Among others successful incorporation of Yb (e.g. Zakharenkov *et al* 1981, Ennen *et al* 1983a) and Er (e.g. Ushakov *et al* 1982, Ennen *et al* 1983b) in several III-V semiconductors and Si by ion implantation or by epitaxial growth has been reported. Up to now the work has been focused mainly on the investigation of the structure of RE related centers. However, only little information has been gained on the excitation and decay mechanisms of these luminescent centers (Kasatkin *et al* 1984, 1985). In this article we present the results of time-resolved photoluminescence investigations, which give us a quite detailed understanding of the decay processes of Yb centers in InP. Additionally we report on liquid phase epitaxial growth of InP:Yb layers and on their electrical properties.

Aszodi *et al* (1985) have proved that the Yb impurity in InP is incorporated as a cubic $Yb^{3+}(4f^{13})$ center on cation site (In). The sharp luminescence lines arising from crystal field split $^2F_{5/2} \rightarrow {}^2F_{7/2}$ inner transitions, which can be excited optically (e.g. Ennen *et al* 1985) as well as electrically (Dmitriev *et al* 1983, Haydl *et al* 1985) are shown in Fig.1. Because of the screening of the 4f states by outer closed $5s^2$ and $5p^6$ shells, the 4f states are little affected by the crystalline environment and therefore the crystal field splitting is small compared to the spin-orbit splitting.

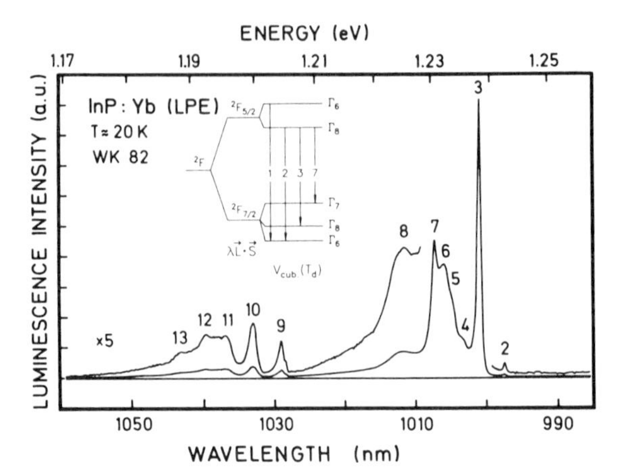

Fig.1. Yb^{3+} luminescence spectrum of a InP:Yb LPE layer at 20K, and energy level diagram of $Yb^{3+}(4f^{13})$ in InP, cubic symmetry (T_d).

2. LIQUID PHASE EPITAXY AND ELECTRICAL PROPERTIES OF InP:Yb LAYERS

The InP:Yb layers were grown in a graphite sliding-boat system by a supercooling process. The cooling rates were in the range 8°C h^{-1} to 30°C h^{-1} and were associated with supersaturations up to 6°C. Growth times varied from 10 min to 200 min. High growth temperatures between 650°C and 800°C were chosen, in order to increase the solubility of Yb in liquid In for efficient Yb incorporation in InP. The melt material In was prealloyed with Yb amounts between 0.001 and 0.005 mole fraction in a RF heater at 900°C for several hours under vacuum. The rapid evaporation of phosphorus from the InP substrates is another peculiarity in connection with high temperatures, but substrate covering by a countersubstrate and additional protection by a Sn-P melt is an adequate method to prevent or reduce this behaviour, even at temperatures up to 800°C.

All heavily doped InP:Yb layers exhibit p-type conduction, whereas undoped InP shows n-type behaviour. Net Hall carrier concentrations N_A-N_D of our epitaxial layers ranged from $1.5\cdot10^{16}$ to $6\cdot10^{17}$ cm^{-3} at 300K and from $1\cdot10^{16}$ to $8.5\cdot10^{16}$ cm^{-3} at 77K, respectively. The Hall mobilities varied from 80 to 110 cm^2/Vs at 300K and from 200 to 1400 cm^2/Vs at 77K, respectively. Temperature dependent Hall measurements reveal a shallow acceptor level at about 45meV above the valence band edge. Since a single acceptor behaviour of Yb in InP was predicted by Hemstreet (1986) on the basis of cluster calculations, we conclude that Yb indeed forms shallow acceptor states in InP, corresponding to a $Yb^{3+} \rightarrow Yb^{2+}$ transition.

3. TIME RESOLVED PHOTOLUMINESCENCE INVESTIGATIONS

The temporal decay of the Yb^{3+} emission is purely exponential with a lifetime of the excited state of 12.7±0.4 μs, which is the same for all our samples, independent of the Yb concentration. Our measurements indicate an identical decay time for all Yb lines (2-13, see Fig.1), confirming that all these lines originate from the same initial state. In contrast to this result the lifetimes of the excited $^2F_{5/2}$ state of Yb^{3+} in CaF_2 crystals (DeLuca ***et al*** 1977) and in garnet crystals (Bogomolova ***et al*** 1976) are ranging between 1 and 3 ms and are considered to be mainly radiative. Since the intra 4f shell transitions are little affected by the host crystal, there is no reason to expect a lower radiative lifetime in InP. Therefore we take these lifetimes as a lower limit for the radiative lifetime of excited Yb^{3+} ions in InP.

The large difference between our measured lifetime and the radiative one must be due to nonradiative processes. We easily can rule out any significant contribution of multiphonon relaxation to the quenching of the luminescence, since the luminescence spectrum reveals a very weak phonon coupling of the 4f states. Energy transfer processes to other centers can also be excluded since the decay time is independent of the Yb concentration. However, bearing in mind that Yb in InP acts as a shallow acceptor, the excited state of Yb^{3+} is somewhat akin to an exciton bound to a neutral acceptor. This similarity suggests that the excited state might decay nonradiatively, just like the Auger decay of acceptor-bound excitons (Nelson *et al* 1966, Schmid 1977). In such a process, the energy of the excited state is transferred to the bound hole, which is deeply excited into the valence band (Fig.2).

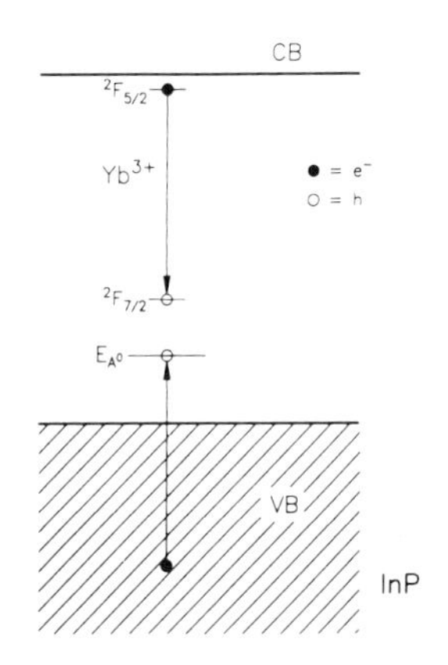

Fig.2. Localized Auger process at a Yb^{3+} center in InP with an energy transfer to the acceptor bound hole.

The efficiency of a bound-exciton-like Auger decay of the excited Yb^{3+} state can be estimated from calculations of free carrier Auger-processes involving localized transitions (Langer *et al* 1984). In order to do this, we represent the hole bound to the Yb^{2+} by a "local carrier density", which is approximately given by $n_{loc} = 4/3\pi a_0^3$, where a_0 is the Bohr radius of the acceptor. Applying the result of the theory of Langer and Le Van Hong to our case by using n_{loc} as the carrier density, we estimate the Auger lifetime of the excited Yb^{3+} state to be about 100ns, which is even lower than our experimental result of 12.7μs. Considering the fact that the wave function of the hole bound to Yb^{2+} is probably fairly different from an effective mass acceptor despite its shallow binding energy, it is not surprising that we are overestimating the Auger rate. Additional uncertainties arise from the nonparabolicity of the bands, and from the assumption that the radiative transition is a pure dipole transition.

Additional nonradiative decay processes are revealed by the temperature dependence of the decay time, which is shown in Fig.3 for 3 samples with different Yb concentrations. The decay time decreases sharply in the temperature range 50 - 110K. The decrease takes place at the highest temperatures (90 - 110K) for the sample of lowest Yb concentration (WK 129, n-type, Yb overcompensated by residual donors). For the more heavily doped samples (WK 82 and WK 107, p-type), the initial lifetime quenching occurs at lower temperatures (50 - 70K), but exhibits a step-like changeover to approximately the same high temperature behaviour as WK 129. It is also interesting to note that for WK 129 the decay remains exponential at all temperatures whereas the luminescence decay of WK 82 and WK 107 is nonexponential at intermediate temperatures (50 - 70K). The decrease of the decay time can be described by an activation energy in all cases, which is ≃ 160meV for the lightly doped sample, and ≃ 50meV for the more heavily doped samples.

First it might seem that the lifetime quenching observed for the heavily doped samples could be due to a free carrier Auger-process, since the decrease of the decay time coincides with the ionization energy of the shallow acceptor. However, this would not explain the nonexponentiality of the decay, since the density of free carriers involved in the Auger-process would be independent of time. It seems more likely that the lifetime quenching is due to an energy transfer process between excited Yb ions and ionized acceptors A^- (e.g. Yb^{2+}), the density of which is also increasing with an activation energy of about 50meV. Since the transfer rate for that kind of process will depend strongly on the distance between the $(Yb^{3+})^*$ and the A^-, it would lead to a nonexponential decay similar to the one well known from donor-acceptor pairs, where close pairs decay fast and distant pairs live longer.

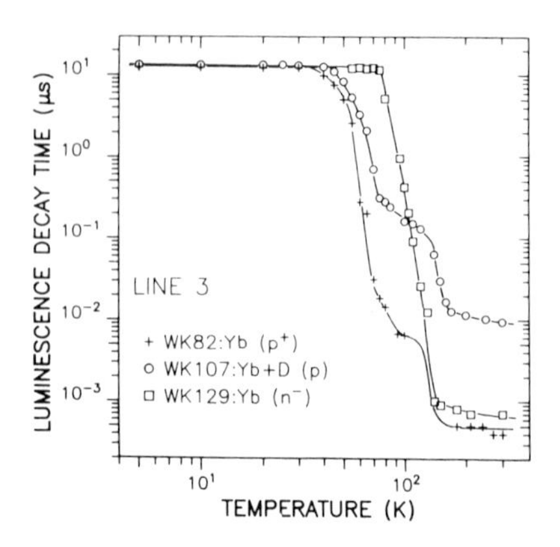

Fig.3. Lifetimes of the excited Yb^{3+} states in dependence of temperature for InP:Yb samples of various Yb doping levels.

Finally, there is a third decay process characterized by an activation energy of 160 meV, which is most clearly observed in the lightly doped sample, but also for the heavily doped ones. Since the energy difference between the InP band gap (1.41eV at 100K) and the excitation energy of $(Yb^{3+})^*$ (1.24eV) is about 170 meV, the final decrease of the decay time can be attributed to the thermal depopulation of excited $(Yb^{3+})^*$ ions by directly generating electron-hole pairs in the host bands or in shallow bound states.

4. REFERENCES

Aszodi G, Weber J, Uihlein Ch, Pu-lin L, Ennen H, Kaufmann U, Schneider J, and Windscheif J 1985 *Phys. Rev. B* **31** 7767
Bogomolova G A, Vylegzhanin D N, and Kaminskiĭ A A 1976 *Sov. Phys.–JETP* **42** 440
DeLuca J A and Ham F S 1977 *J. Electrochem. Soc.* **124** 1592
Dmitriev A G, Zakharenkov L F, Kasatkin V A, Masterov V F, and Samorukov B E 1983 *Sov. Phys. Semicond.* **17** 1201
Ennen H, Kaufmann U, Pomrenke G, Schneider J, Windscheif J, and Axmann A 1983a *J. Cryst. Growth* **64** 165
Ennen H, Schneider J, Pomrenke G, and Axmann A 1983b *Appl. Phys. Lett.* **43** 943
Ennen H, Pomrenke G, and Axmann A 1985 *J. Appl. Phys.* **57** 2182
Haydl W H, Müller H D, Ennen H, Körber W, and Benz K W 1985 *Appl. Phys. Lett.* **46** 870
Hemstreet L A 1986 in *Materials Science Forum*, ed H J von Bardeleben (Trans Tech Public. Ltd, Switzerland) Vol. **10-12**, pp 85-90.
Kasatkin V A and Savel'ev V P 1984 *Sov. Phys. Semicond.* **18** 1022
Kasatkin V A, Lavrent'ev A A, and Rodnyĭ P A 1985 *Sov. Phys. Semicond.* **19** 221
Langer J M and Le Van Hong 1984 *J. Phys. C* **17** L923
Nelson D F, Cuthbert J D, Dean P J, and Thomas D G 1966 *Phys. Rev. Lett.* **17** 1262
Schmid W 1977 *phys. stat. sol. (b)* **84** 529
Ushakov V V, Gippius A A, Dravin V A, and Spitsyn A V 1982 *Sov. Phys. Semicond.* **16** 723
Zakharenkov L F, Kasatkin V A, Kesamanly F B, Samorukov B E, and Sokolova M A 1981 *Sov. Phys. Semicond.* **15** 946

Reproducible, low temperature growth of high quality $Al_xGa_{1-x}As$ alloys by OMVPE

S. K. Shastry, S. Zemon, D. Dugger, and M. DeAngelis

GTE Laboratories Incorporated, 40 Sylvan Road, Waltham, MA 02254

Abstract: $Al_xGa_{1-x}As$ ($0 \le x < 1$) layers with high optical and electrical qualities and defect-free surface morphology have been reproducibly grown at low temperatures (≈ 650°C) by OMVPE. The as-grown layers are n-type ($n \le 4 \times 10^{15}$ cm^{-3}) and exhibit narrow (2 meV at x = 0.05; 5.4 meV at x = 0.35) 4.2 K photoluminescence spectral widths. Variable temperature (11 K - 301 K) mobility measurements indicate negligible space-charge scattering in these layers. Furthermore, carbon and oxygen contaminations are almost negligible and below the SIMS detection limit. High quality GaAs/AlGaAs heterostructures suitable for HEMT devices have also been synthesized.

Reproducible growth of high purity $Al_xGa_{1-x}As$ is essential for many high speed heterostructure device applications based on GaAs/$Al_xGa_{1-x}As$ materials. However, epitaxial growth of such $Al_xGa_{1-x}As$ by organometallic vapor phase epitaxy (OMVPE) has been a difficult task primarily due to the incorporation of oxygen and carbon into the epitaxial layers during the growth process (Stringfellow 1981). The oxygen incorporation decreases when use is made of high (750-800°C) growth temperature (Andre *et al.* 1981, Stringfellow 1981) or *in situ* gettering techniques, such as those utilizing quartz baffles (Stringfellow and Hom 1979), predeposition of $Al_xGa_{1-x}As$ (Hersee *et al.* 1981), or long entrance region (Kuech *et al.* 1985). Molecular sieves (Thrush and Whiteaway 1979) or bubblers of Ga:In:Al eutectic melt (Shealy *et al.* 1983) in the arsine line have also been used; this partially getters oxygen and water vapor originating from the arsine tank. In contrast, the behavior or origin of carbon incorporation into OMVPE $Al_xGa_{1-x}As$ is not well understood. Increasing concentration of carbon with increasing growth temperature (Kuech *et al.* 1986) has been reported. Use of triethylaluminum (TEA) instead of trimethylaluminum (TMA) has been shown to result in lower carbon contamination (Kobayashi and Makimoto 1985), suggesting that the aluminum precursor is the primary source of carbon impurity. However, uniformity and reproducibility of the $Al_xGa_{1-x}As$ layers grown using TEA have been very unsatisfactory (Kuech *et al.* 1986).

In this paper we report on the reproducible growth and properties of oxygen- and carbon-free $Al_xGa_{1-x}As$ ($0 \le x \le 1$) layers grown using the ethyl-based aluminum and gallium precursors. We have grown high quality $Al_xGa_{1-x}As$ layers at low temperatures (≈ 650°C) with growth conditions compatible with those used in the growth of high purity, high mobility GaAs (Shastry *et al.* 1987). Such matching of growth conditions is essential to the realization of optimized heterostructures.

Undoped semi-insulating GaAs substrates, 2 inches in diameter and oriented (100) with 2° off toward the nearest (110), were used in this work. The as-purchased substrate was dipped in dilute (20%) HC1 and loaded into a vertical reactor via a load-lock (Shastry *et al.* 1987). The epitaxial growth sequence consisted of the growth of a 0.5 μm-thick GaAs buffer layer at 600°C using triethlygallium (TEG) and arsine, and subsequent growths of a 0.2 μm-thick high aluminum content $Al_xGa_{1-x}As$ (x > 0.8) barrier layer, 1-3 μm-thick $Al_xGa_{1-x}As$ layer, and 20 nm-thick GaAs cap layer. TEA and, to a lesser degree, TMA were used as the Al precursors. Arsine flow rate of 56 sccm was maintained during the growth. A few $Al_xGa_{1-x}As$ layers were grown at different arsine flow rates (25-100 sccm) with fixed TEG and TEA fluxes. Quantum well and two-dimensional electron gas (2-DEG) structures were grown at 650°C without growth interruption. Silane (0.2%) was used for intentional n-doping. All the experiments were conducted at 50 Torr with hydrogen carrier gas flow rate of 5 SLM.

The $Al_xGa_{1-x}As$ epitaxial layers and heterostructures were characterized by mercury probe C-V, variable temperature Hall effect (van der Pauw), and 4.2 K photoluminescence (PL) measurements. Secondary ion mass spectroscopy (SIMS) was utilized to obtain depth profiles of aluminum and to detect trace impurities (Si, C, O, etc.).

The $Al_xGa_{1-x}As$ ($0 \leq x \leq 1$) epitaxial layers grown at and above 575°C had smooth, specular, and defect-free surface morphology over the 2"-dia. wafers. Layers grown at 550°C exhibited slight roughness when optically examined at a magnification of 1000. Layers grown at and above 600°C were n-type (net carrier concentration $n \leq 4 \times 10^{15}$ cm^{-3}) over a wide range of Al composition. Typical doping vs depth profiles obtained from C-V measurements are shown in Fig. 1 for x = 0.07, 0.28, and 0.4. Each set of the profiles in the figure consists of ten traces taken at ten different positions diagonally on the 2"-dia. wafers. These profiles indicate epitaxial layer thickness and doping uniformities within ±10% over a large part of the wafer surface. Furthermore, variations in the growth and material characteristics from run to run or due to change in the arsine tank were not observed, a significant improvement over earlier results (Kasemset *et al.* 1983).

Net carrier concentration and mobility values measured in several layers grown at 650°C are listed in Table 1. Both C-V and Hall measurements gave similar n values for all the layers, suggesting that deep level concentration is significantly less than the n values. Sheet carrier concentrations at 77 K (n_1) were nearly the same as the 300 K values (n_2), further suggesting that the total concentration of deep levels is less than 10^{15} cm^{-3}. Changes in the V:III ratio (obtained by changing the arsine flow rate only) did not significantly affect the net carrier concentration values (see first three rows in Table 1). Similarly, growth temperature (between 600°C and 700°C) did not drastically affect the carrier concentration values. Both 300 K and

Table 1. Electrical Properties of $Al_xGa_{1-x}As$ layers grown using TEA.

Sample #	V:III	x(%)	μ (cm²/V-s) 300 K	77 K	n (cm⁻³) C-V	Hall	n_1/n_2
022687A	8.6	8.7	4800	22,000	8.0×10^{14}	1.2×10^{15}	2.0
022687F	17.0	8.7	4700	28,000	2.8×10^{15}	2.2×10^{15}	1.43
022787A	35.0	8.7	5200	29,000	1.5×10^{15}	1.8×10^{15}	1.81
022487B	17.5	4.0	5200	31,000	2.5×10^{15}	4.0×10^{15}	1.11
022587B	15.0	14.0	4400	18,000	2.0×10^{15}	3.0×10^{15}	1.23
022887A	35.0	30.0	2400	9,100	5.0×10^{15}	3.8×10^{15}	0.97

77 K mobility values shown in Table 1 decrease with increasing Al mole fraction, which is common for $Al_xGa_{1-x}As$ alloys. However, these values are significantly higher than previously reported (Kaneko *et al.* 1977, Stringfellow 1979, Saxena 1981, Takagishi *et al.* 1986, Shiraki *et al.* 1987). Due to the presence of the barrier layer between the $Al_xGa_{1-x}As$ layer ($x \leq 0.3$) and the GaAs buffer layer, charge transfer effects (Matsumoto *et al.* 1982, Collins *et al.* 1983) on the C-V and mobility measurements were negligible; thus the n and mobility values listed in Table 1 reflect the high quality of the $Al_xGa_{1-x}As$ layers.

Figure 2 shows mobility vs T variations for three $Al_xGa_{1-x}As$ layers (x = 0.4, 0.15, and 0.25) grown at 650°C. The high-temperature slope decreases with increasing x, indicating an increase in space-charge scattering with increasing x. An analysis of these results and the various scattering processes indicates that space-charge scattering (Conwell and Vassell 1968) as well as polar optical scattering limit the mobility. The value of the scattering center density-cross section product N_sQ varies as $N_sQ \approx 5 \times 10^3 + 8 \times 10^4$ **x** cm^{-1} (**x** ≤ 0.4) for these layers.

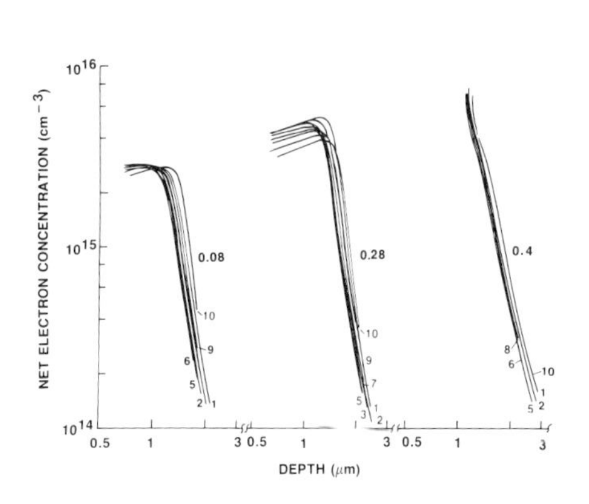

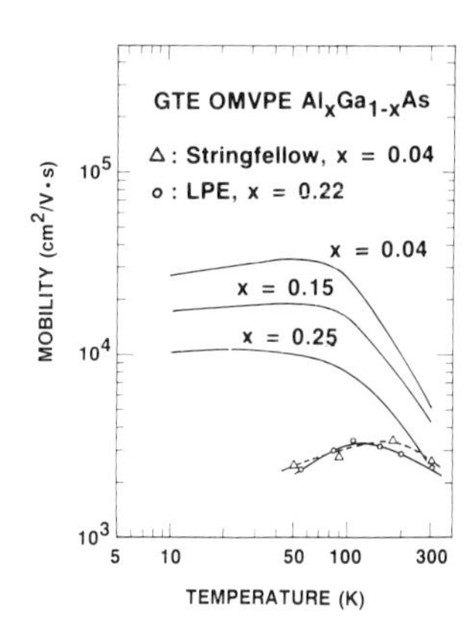

Figure 1. Doping vs depth profiles of OMVPE $Al_xGa_{1-x}As$ Layers (x = 0.08, 0.28, 0.4) grown at 650°C.

Figure 2. Mobility vs T variation of $Al_xGa_{1-x}As$ (x = 0.04, 0.15, 0.25) layers grown at 650°C. Published results (Stringfellow 1979) for OMVPE and LPE $Al_xGa_{1-x}As$ are also shown for comparison.

This is lower than previous reported values for LPE (Kaneko *et al.* 1977, Saxena 1981) and OMVPE (Stringfellow 1979) $Al_xGa_{1-x}As$. In particular, the rate of change of N_sQ with x is about an order of magnitude lower. Furthermore, judging from the low-temperature end of the curves in Fig. 2, negligible scattering due to ionized impurities (Falicov and Cuevas 1967) is present in our films.

Figure 3 shows 4.2 K PL spectra of $Al_xGa_{1-x}As$ (x ≈ 0.07) layers grown at different growth temperatures (T_G) at a V:III ratio of 17. With increasing T_G, the intensity of the lower energy line, identified as carbon acceptor luminescence, decreases and the intensity of the higher energy excitonic peak increases. The excitonic spectral widths also become narrower. These indicate that carbon incorporation *decreases* with increasing growth temperature ($T_G \leq$ 700°C). The intensity of the PL signal near 1560 nm, attributed to oxygen (Tsai *et al.* 1983), was also monitored. It comprised about 80% of the total luminescence power in the sample grown at 575°C and decreased to zero in the sample grown at 675°C. SIMS depth profiling of these samples indicated that the levels of silicon, carbon, and oxygen are below the detection limit (≈ 10^{16} cm^{-3}). We have further seen that the $Al_xGa_{1-x}As$ layers grown over a wide range of temperatures in the presence of trace oxygen contamination showed both the carbon and oxygen-related PL signals. In addition, when the oxygen-related signal was low or negligible, the carbon-related signal was low as well. Thus, we speculate that the mechanism of carbon incorporation into the $Al_xGa_{1-x}As$ films involves oxygen-related surface reactions.

Figure 4 shows the excitonic spectral widths (full width at half maximum) for various x values for $Al_xGa_{1-x}As$ layers grown at 650°C. Values reported in the literature for MBE and OMVPE $Al_xGa_{1-x}As$ are also shown for comparison. Our values are in accordance with the alloy broadening calculations of Schubert *et al.* (1984) but are above the calculated curve (dotted line in Fig. 4) as reported by Reynolds *et al.* (1986) based on expressions of Singh and Bajaj (1984). This reflects the state of the art of OMVPE $Al_xGa_{1-x}As$.

High quality GaAs/$Al_xGa_{1-x}As$ heterostructures have also been grown at ≈ 650°C (V:III = 17.0) using TEG and TEA. We have measured 2-DEG mobilities at 11 K as high as 130,000 cm^2/V-s with a high ($n_s \approx 1.2 \times 10^{12}$ cm^{-2}) sheet carrier concentration involving the *two* subbands in the 2-DEG structure. These values are significantly higher than the previous reports for OMVPE (Hersee *et al.* 1982, Maluenda and Frijlink 1983, Kobayashi and Fukui 1984).

In summary, we have demonstrated reproducible growth of high purity $Al_xGa_{1-x}As$ by

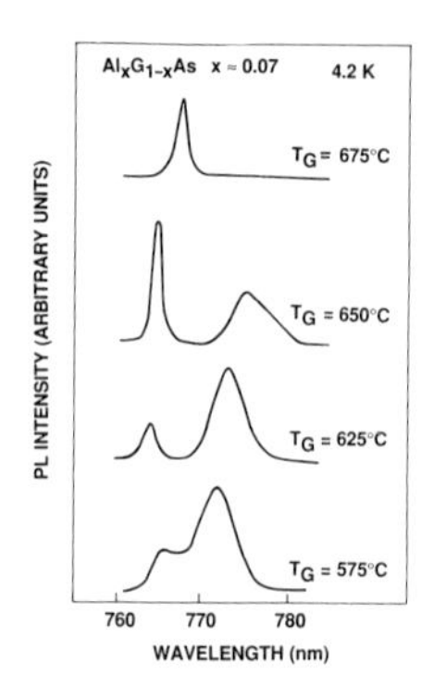

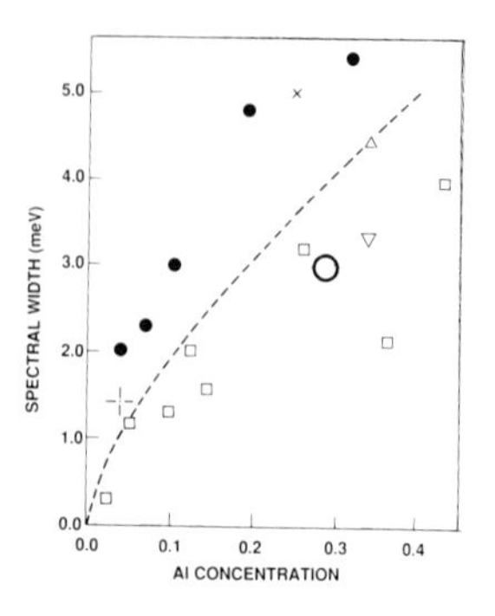

Figure 3. 4.2 K PL spectra of $Al_xGa_{1-x}As$ ($x \approx 0.07$) for four growth temperatures (T_G) using an excitation density ≈ 5 W/cm^2.

Figure 4. 4.2 K excitonic PL spectral widths of $Al_xGa_{1-x}As$ as a function of x: •, -¦- present work (-¦- denotes an extrapolated width of a feature in the excitonic line); ❐ Reynolds *et al.* (1986); O Ballingall and Collins (1983); Δ Heiblum *et al.* (1983); X Shealy *et al.* (1983); ∇ Cunningham *et al.* (1981).

OMVPE at growth conditions that are optimum for high purity GaAs. High quality heterostructures have also been grown.

We thank P. Tavilla and F. Lisherness for manuscript preparation, D. G. Kenneson for technical assistance, and Drs. P. Haugsjaa and H. Lockwood for encouragement and support.

References

Andre J P, Guittard P, Hallais J, and Piaget C 1981 J. Cryst. Growth **55** 235.
Ballingall J M and Collins D M 1983 J. Appl. Phys. **54** 34.
Collins D M, Mars D E, Fisher B, and Kocot C 1983 J. Appl. Phys. **54** 857.
Conwell E M and Vassell M O 1968 Phys. Rev. **166** 797.
Cunningham J E, Tsang W T, Chiu T H, and Schubert E F 1987 Appl. Phys. Lett. **50** 769.
Falicov L M and Cuevas M 1967 Phys. Rev. **164** 1025.
Heiblum M, Mendez E E, and Osterling L 1983 J. Appl. Phys. **54** 6982.
Hersee S D, Di Forte-Poisson M A, Baldy M, and Duchemin J P 1981 J. Cryst. Growth **55** 53.
Hersee S D, Hirtz J P, Baldy M, and Duchemin J P 1982 Electron. Letters **18** 1076.
Kaneko K, Ayabe M, and Watanabe N 1977 Inst. Phys. Conf. Ser. No. **33a** 216
Kasemset D, Hong C S, Patel N B, Dapkus P D, Mohammed K, and Merz J L 1983 Inst. Phys. Conf. Ser. No. **65** 79.
Kobayashi N and Fukui T 1984 Electron. Letters **10** 887
Kobayashi N and Makimoto T 1985 Jpn. J. Appl. Phys. **24** L824.
Kuech T F, Veuhoff E, Wolford D J, and Bradley J A 1985 Inst. Phys. Conf. Ser. No. **74** 181.
Kuech T F, Veuhoff E, Kuan T S, Deline V, and Potemski R 1986 J. Cryst. Growth **77** 257.
Maluenda J and Frijlink P M 1983 Jpn. J. Appl. Phys. **22** L127.
Matsumoto T, Bhattacharya P K, Darmawan J, and Ludowise M J 1982 Appl. Phys. Lett. **41** 1075
Reynolds D C, Bajaj K K, Litton C W, Yu P W, Klem J, Peng C K, Morkoç H, and Singh J 1986 Appl. Phys. Lett. **48** 727.
Saxena A K 1981 Phys. Rev. B **24** 3295.
Schubert E F, Gobel E O, Horiboshi Y, Ploong K, and Queisser H J 1984 Phys. Rev. B **30** 813.
Shastry S K, Zemon S, and Norris P 1987 Inst. Phys. Conf. Ser. No. **83** 81.
Shealy J R, Kreismanis V G, Wagner D K, and Woodall J M 1983 Appl. Phys. Lett. **42** 83
Shiraki Y, Mishima T, and Morioka M 1987 J. Cryst. Growth **81** 164.
Singh J and Bajaj K K 1984 Appl. Phys. Lett. **44** 1975
Stringfellow G B 1979 J. Appl. Phys. **50** 4178
Stringfellow G B and Hom G 1979, Appl. Phys. Lett. **34** 794.
Stringfellow G B 1981 J. Cryst. Growth **55** 42
Takagishi S, Mori H, Kimura K, Kamon K, and Ishii M 1986 J. Cryst. Growth **75** 545.
Thrush E J and Whiteaway J E A 1979 Electron. Lett. **15** 666.
Tsai M J, Tashima M M, Twu B L, and Moon R L 1983 Inst. Phys. Conf. Ser. No. **65** 85.

Plasma-assisted epitaxial growth of compound semiconductors on Si

T Hariu and Q Z Gao

Department of Electronic Engineering, Tohoku University, Sendai 980, Japan

ABSTRACT: Low temperature cleaning of Si surface in hydrogen plasma and the successive epitaxial growth of GaAs and InAs at temperatures as low as 500°C and 445°C, respectively, have been achieved. The simultaneous supply of Ga, In or As during the cleaning process has been found to be more effective than the treatment in pure hydrogen plasma, most likely due to the production of volatile compounds by reducible reaction of the native oxide of Si.

1. INTRODUCTION

Epitaxial growth of compound semiconductors grown directly on Si has been attracting wide attention in view of its application to the monolithic integration of Si and compound semiconductor devices, and also of possible replacement of bulk substrates of compound semiconductors with high quality and low cost Si wafers. For this purpose, it is essential to clean up the surface of Si wafers in order to make the successive layers grow epitaxially. The difficulty in this cleaning process of Si is the much higher temperature required to remove native oxide and carbon than such compound semiconductors as GaAs and InP, usually higher than the temperature of successive epitaxial growth of compound semiconductors on Si. For example, heat treatment around 850°C in high vacuum (Nishi et al 1985) in the case of MBE or in the flow of H_2+AsH_3 (Akiyama et al 1984) in the case of MOCVD has been employed for this cleaning purpose. It is then required to develop a cleaning process at lower temperatures, without raising above the temperature of epitaxial growth, for device application so that the designed structure processed before the successive epitaxial growth may not be destroyed.

The purpose of this paper is to describe the low temperature cleaning process of Si in hydrogen plasma and the successive epitaxial growth of InAs and GaAs by plasma-assisted epitaxy (PAE) in the same PAE apparatus. Particular attention will be paid on the effect of the simultaneous supply of Ga, In or As during the cleaning process in hydrogen plasma. This latter process is more useful in improving the reproducibility of cleaning and epitaxial growth, and also in further reducing the temperature of both processes.

2. EXPERIMENTAL

(100) n-type Si wafers were chemically etched by the RCA method (Kern and Puotinen 1970) prior to loading into the PAE chamber shown in Fig.1. The pressure of hydrogen gas during cleaning and the succesive epitaxial

growth was kept at 0.05Torr and variable rf power at 13.56MHz was applied to the reaction vessel with volume of $1000cm^3$ in order to hold the discharging plasma.

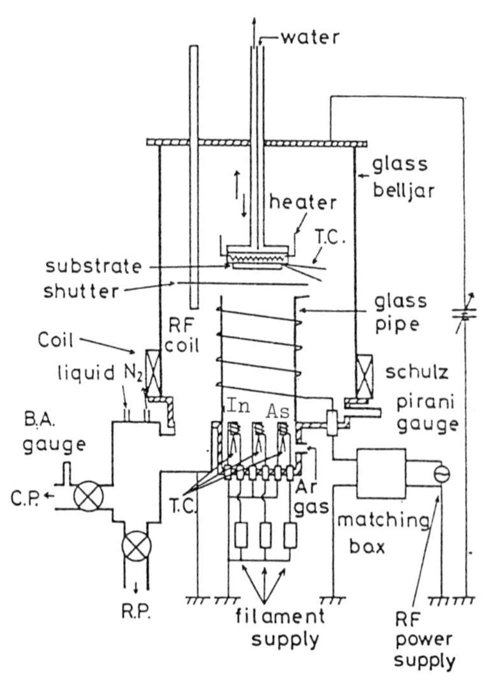

Fig.1. An experimental PAE apparatus for the growth of GaAs and InAs.

Two kinds of temperature programs shown in Fig.2 were employed. In the first program shown in Fig.2(a), the substrate wafers were treated in pure hydrogen plasma for surface cleaning, and then 6-nine purity In, Ga and As were evaporated from open sources by resistive heating and supplied through the plasma to deposit InAs or GaAs. In the second process shown in Fig.2(b), Ga, In or As was supplied during the cleaning process and the same deposition procedure as the first one was taken.

3. CLEANING IN PURE HYDROGEN PLASMA

It was possible to grow epitaxial single crystal InAs layers on Si at 530°C with applied rf power of 100W after the pure hydrogen plasma treatment, shown in Fig.2(a), at the same temperature, while polycrystalline layers grew on Si without this treatment. In-depth composition profile measurement by Auger Electron Spectroscopy did not detect oxygen at the interface between the Si substrate and the grown layer, while it was detected when the substrate was not chemically etched but treated in hydrogen plasma, probably due to a thicker oxide layer. Carbon was not detected in both cases. When the substrate was not treated in hydrogen plasma, it was not possible to reveal the clear interface

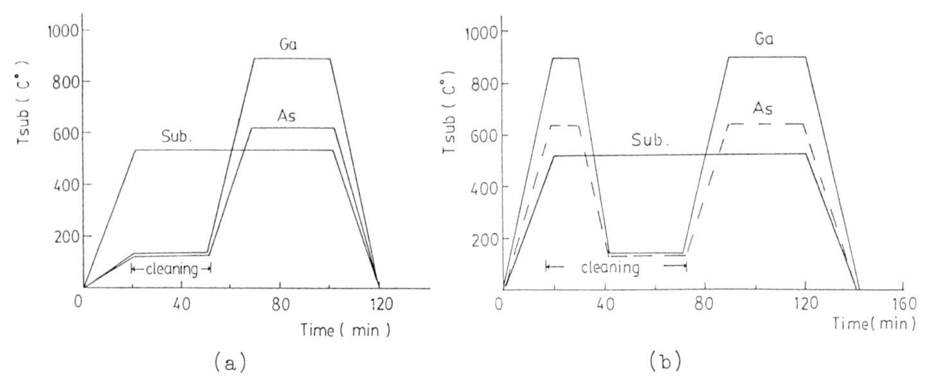

Fig.2. Temperature programs for cleaning process of Si and the successive epitaxial growth of GaAs and InAs.

because ion milling did not give flat surfaces enough to analyze the in-depth composition profile correctly.

It was, however, not always possible to grow epitaxial GaAs on Si treated in the same way as in the case of InAs growth, although epitaxail layers were sometimes obtained after pure hydrogen plasma treatment. The above fact seems to indicate that the epitaxial growth of GaAs is more critically dependent upon the surface condition of Si than InAs. It should be noted here that the PAE growth of GaAs on GaAs substrates is possible even at 350°C (Hariu et al 1982), because the surface cleaning of GaAs can be achieved at this low temperature.

4. SUPPLY OF Ga, In or As DURING CLEANING

The second process shown in Fig.2(b), where Ga or As is supplied during the cleaning process in hydrogen plasma, made it possible to grow epitaxial GaAs layers more reproducibly and also to further reduce both temperatures of cleaning and the successive epitaxial growth, compared with the treatment in pure hydrogen plasma, as indicated by the reflected electron diffraction patterns shown in Fig.3. It was also found that the supply of Ga is more effective than As in reducing these temperatures down to 500°C while in the case of As supply the lowest temperature so far achieved is 550°C.

The similar favorable effect of In supply during the cleaning process in hydrogen plasma was also observed in the case of PAE growth of InAs. and then the temperature for both

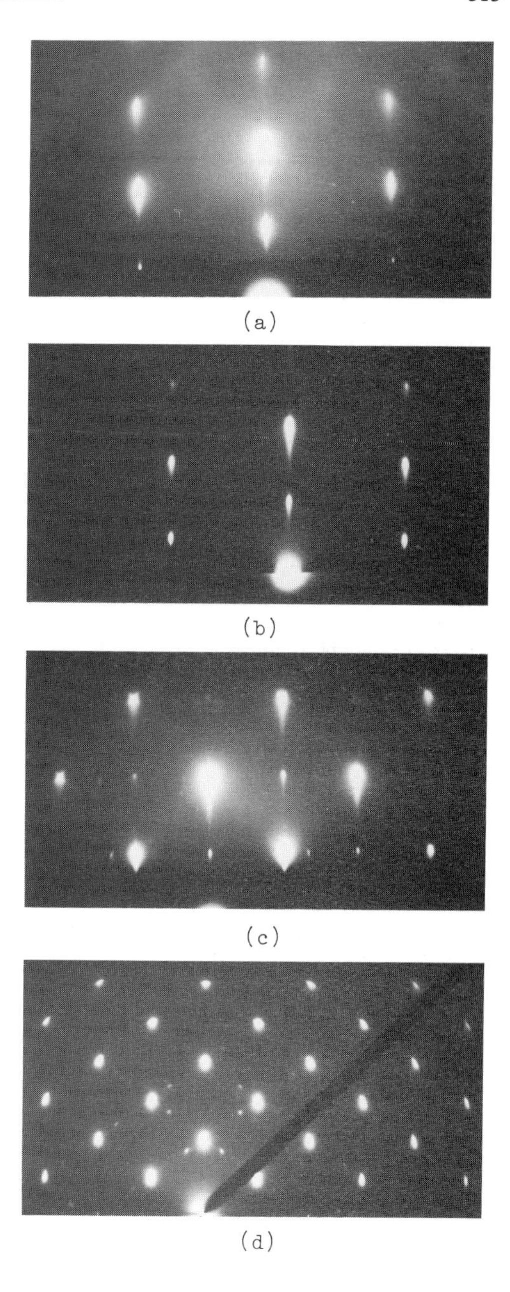

Fig.3. Reflected electron diffraction patterns of PAE-GaAs and -InAs. The temperatures for cleaning of Si surface and the successive growth are (a) GaAs, 600°C, with Ga supply, (b) GaAs, 550°C, with Ga supply, (c) GaAs, 500°C, with Ga supply and (d) InAs, 445°C, with In supply.

cleaning and epitaxial growth was reduced down to 445°C, compared to 530°C in the case of pure hydrogen plasma treatment.

The above observation on the effect of Ga, In or As supply during the cleaning process is likely to indicate the chemical reaction between the native oxide of Si and Ga, In or As, and the resultant production of volatile compounds like Ga_2O, In_2O or As_2O_3, in consistent with the observations by Wright and Kroemer (1980) and by Wang (1984). However, the fact that the present surface cleaning of Si has been achieved at much lower temperatures in hydrogen plasma with Ga, In or As supply suggests that the similar reducible reaction at Si surface can be remarkably enhanced by the coexistence of atomic hydrogen. In fact, similar treatment by the supply of Ga or As, but not in the plasma state, resulted in the growth of polycrystalline layers even at 600°C. Similar mechanism of oxygen removal from Si surface with simultaneous supply of Ge during heat treatment was suggested by Morar et al. (1987), but their cleaning in vacuum was down to 625°C.

5. CONCLUSIONS

The simultaneous supply of Ga or In during hydrogen plasma treatment has been found to be effective for low temperature cleaning of Si surface and the successive epitaxial growth of compound semiconductors. With this process, InAs and GaAs layers have been epitaxially grown directly on Si at 445 and 500°C, respectively. The optimization of plasma conditions will further reduce the temperature of these processes.

Acknowledgement

This work was supported in part by the Scientific Research Grant-in-Aid #62104004O for Special Project Research on 'Alloy Semiconductor Physics and Electronics' from the Ministry of Education, Science and Culture. The authors wish to thank Professor S. Ono, Dr. K. Matsushita, Mr. Y. Igarashi and Mr. S.F. Fang for their valuable discussions.

References

Akiyama M, Kawarada Y and Kaminishi K 1984 Jpn. J. Appl. Phys. 23 L843
Hariu T, Matsushita K, Komatsu Y, Shibuya S, Igarashi S and Shibata Y 1982 Inst. Phys. Conf. Ser. No.65 ed. Stillman G E (Bristol and London: The Institute of Physics)pp 141-8
Kern W and Puotinen D A 1970 RCA Rev. 31 187
Morar J F, Meyerson B S, Karlson U O, Himpsel F J, McFeely F R, Rieger D, Taleb-Ibrahimi A and Yarmoff J A 1987 Appl. Phys. Lett. 50, 463
Nishi S, Inomata H, Akiyama M and Kaminishi K 1985 Jpn. J. Appl. Phys. 24 L391
Wang W I 1984 Appl. Phys. Lett. 44 1149
Wright S and Kroemer H 1980 Appl. Phys. Lett. 36 210

Paper presented at Int. Symp. GaAs and Related Compounds, Heraklion, Greece, 1987

MO-Chloride VPE of AlAs and AlGaAs layers

F.Hasegawa, K.Katayama, H.Yamaguchi, T.Yamamoto and Y.Nannichi

Institute of Materials Science, University of Tsukuba,
Tsukuba Science City, 305 JAPAN.

ABSTRACT: AlAs and AlGaAs epitaxial layers were grown by MO-Chloride VPE for the first time. The growth method and temperature were exactly the same as those of GaAs Chloride VPE (750'C), except that the Al was supplied as MO (Metalorganic). $Al(CH_3)_3$ or $Al(CH_3)_2Cl$ was conveyed by He gas and was reacted with $AsCl_3$ to make $AlCl_3$. Metallic Ga was used as the Ga source. $AlCl_3$ and GaCl are reduced in Arsenic atmosphere at the deposition zone by introducing H_2, and $Al_xGa_{1-x}As$ is deposited. The growth rate was about 2-4um/hr. The n was $4\text{-}20\times10^{17}$ cm^{-3}. Bandedge emission was obtained, but emission at 970nm was also observed.

1. INTRODUCTION

There is a great demand for vapor phase epitaxy of AlGaAs, and MOCVD (Metal Organic Chemical Vapor Deposition) is being extensively developed for fabrication of LED, laser diodes etc.. For application in industries, however, MOCVD has some disadvantages such as high running cost, the use of highly toxic arsine(AsH_3), and therefore high utility cost.

On the other hand, chloride VPE (Vapor Phase Epitaxy) is commonly used for the production of GaAs FET's. The chloride VPE method, however, has a serious drawback, i.e.,there is no established way of growing Al containing alloy semiconductors. In the past, some attempts were made to grow AlGaAs by hydride VPE. Ettenberg et al.(1971) of RCA succeeded in growing AlAs, but their growth temperature was as high as 1000'C.

Bachem and Heyen (1981,1987) of Aachen Technical University were successful in growing AlGaAs layers by bubbling HCl/H_2 gas through a molten Al-Ga alloy which was placed in an alumina crucible. Their results were significant since they grew AlGaAs layers at around 750'C. Unfortunately, they had to rely on toxic AsH_3 for the arsenic supply and the Al composition did not seem to be controlled too well.

The aim of this paper is to investigate the possibility of chloride VPE of AlGaAs without using toxic AsH_3.

It was demonstrated that AlAs and AlGaAs can be grown by chloride VPE exactly same way as GaAs if we use the hydrogen reduction method (Koukitu and Seki et al., 1976).

2. GROWTH APPARATUS and METHOD

A schematic diagram of the reactor system is shown in Fig.1. It is essentially the same as the system used for the growth of GaAs (Hasegawa et al. 1985). The source region of the quartz reactor tube is separated into 3 chambers. The first one is the Ga source chamber where the Ga metal is placed on a quartz boat. The second chamber is for the Al source where metallorganic Al (TMA--$Al(CH_3)_3$ or DMAC--$Al(CH_3)_2Cl$) and $AsCl_3$ mix to form $AlCl_3$. The third chamber is used for introducing H_2 independently into the growth region. The optimum growth temperature was 750'C. The quartz reactor tube was coated with carbon by decomposing acetone.

The growth reaction is believed to occur in the following manner. In the source chamber, $AsCl_3$ carried by He+1%H_2 reacts with the Ga metal, and DMAC or TMA to form GaCl and $AlCl_3$ as follows:

$$3Ga + AsCl_3 \text{ -- } 3GaCl + (1/4)As_4 \qquad (1)$$

$$Al(CH_3)_3 + AsCl_3 + (3/2)H_2 \text{ -- } AlCl_3 + (1/4)As_4 + 3CH_4 \qquad (2)$$

GaCl and $AlCl_3$ are sent to the growth region where they are mixed with H_2. This causes GaCl and $AlCl_3$ to be reduced as in Eq.3,4, and thus AlGaAs deposition occurs.

$$4GaCl + As_4 + 2H_2 \text{ -- } 4GaAs + 4HCl \qquad (3)$$

$$4AlCl_3 + As_4 + 6H_2 \text{ -- } 4AlAs + 12HCl \qquad (4)$$

The optimum growth conditions for AlGaAs are as follows:

	bubbler temp.	flow rates	
TMA:	20'C	He	60 sccm
Al-side $AsCl_3$:	8'C	He+1%H_2	60 sccm
Ga-side $AsCl_3$:	-5'C 0'C	He+1%H_2	200 sccm
		by-path H_2	200 sccm

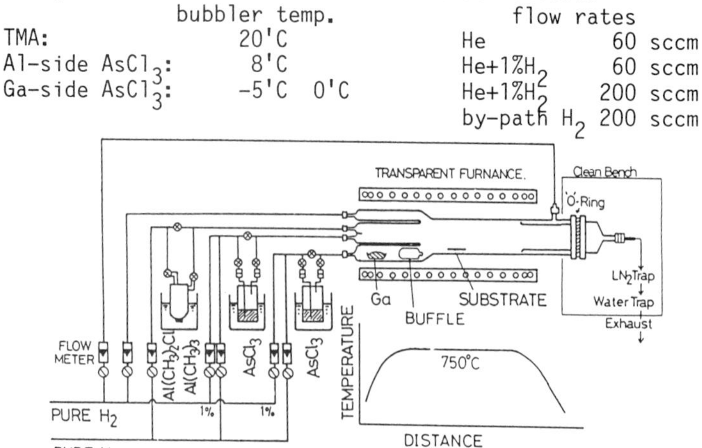

Fig.1, A schematic diagram of the chloride VPE system used for the growth.

3. EXPERIMENTAL RESULTS

A typical profile of the composition as measured by sputtering AES is shown in Fig.2. The chloride VPE AlGaAs used for this measurement was 1.9um thick with a mirror like surface. The figure shows that Al, with a composition of approximately 10%, is distributed uniformly throughout the entire epitaxial layer with exception of the surface and interface region. Oxygen peaks detected are thought to be due to surface oxidation during the AES measurements.

Carrier concentrations and mobilities of the grown layers are shown Fig.3 against the TMA or DMAC supply. When there is no Al supply, carrier concentration "n" of GaAs was about $1x10^{16}$ cm^{-3}, and mobility was more than 10^4 $cm^2/V.s$ at 77K. However, the "n" jumped up to more than or 10^{18} cm^{-3} by slight addition of TMA or DMAC, and the mobility came down to 1000 $cm^2/V.s$. Since the "n" for TMA is lower than the "n" for DMAC, these increase of the carriers might be thought to be due to impurities from MO materials.

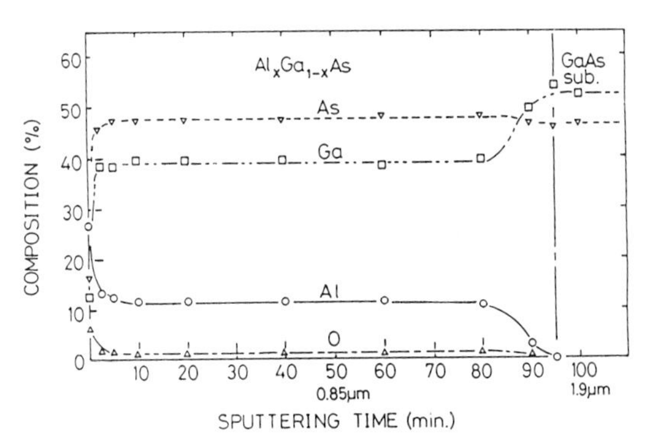

Fig.2, A composition profile of the AlGaAs film measured by a sputtering AES.

Contamination of Si due to reaction between quartz tube and $AlCl_3$ is still very suspicious, since the carbon coating of the quartz reactor might be incomplete.

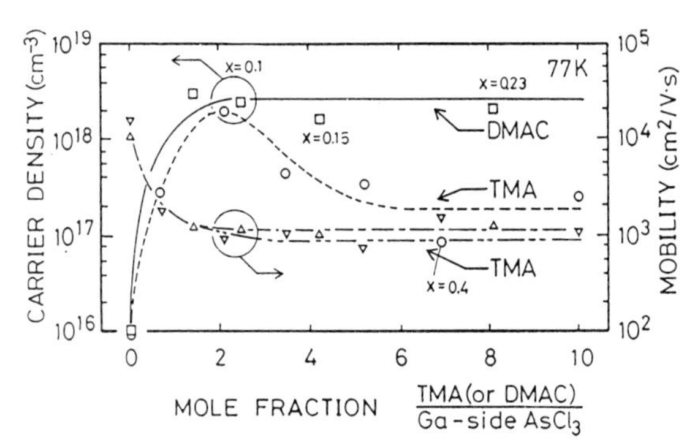

Fig.3, Dependence of the carrier density and mobility on the MO-Al supply.

When the quartz reactor is not coated with carbon, bandedge photoluminescence could not be observed even for the layer with very small Al content, and a broad luminescence peak was observed at around 950-1000nm.
If there is no Al supply, of course a strong bandedge emission of GaAs was observed for the layer grown in the same reactor. It suggests that the Al reacts with quartz tube and the produced oxygen is incorporated into the epitaxial layer which suppresses the bandedge emission.

In order to prevent the reaction between $AlCl_3$ and quartz, the reactor tube was coated with carbon by decomposition of acetone at 800'C. The bandedge emission could be observed for the grown layers after the carbon coating as shown in Fig.4. However, still a broad emission at around 970 nm is observed. The samples of the lower two spectra in Fig.4 were grown in the same conditions using TMA except that the connector of the rear hydrogen path was tightened to prevent leakage for the sample of the botom spectra. The broad emission is greatly reduced for the sample grown after the leakage was prevented. This longer wave length emission is quite similar to the one (shown in the insert) observed by Stringfellow and Hall (1979) for an AlGaAs layer grown by primitive MOCVD system. These facts seem to suggest that the broad emission at around 970nm is related to oxygen contamination.

On the other hand, there are several reports that a strong broad emission is observed at longer wave lengths than the bandedge for AlGaAs layers whose carrier concentration is more than 5×10^{17} cm^{-3} (Kajikawa et al. 1985, Ishikawa et al. 1986). This is thought to be due to so called Self Activated (SA) center, which is a complex of shallow donor impurity and vacancy. The SA center is widely observed for heavily doped bulk GaAs (Williams and Bebb, 1966). Photoluminecence due to SA center in AlGaAs is one order stronger than that in GaAs (Kajikawa et al. 1986). Since most of our samples are heavily doped, the broad emission at around 970nm might be due to SA center. Further investigations are necessary to judge whether this is due to oxygen contamination or due to SA center.

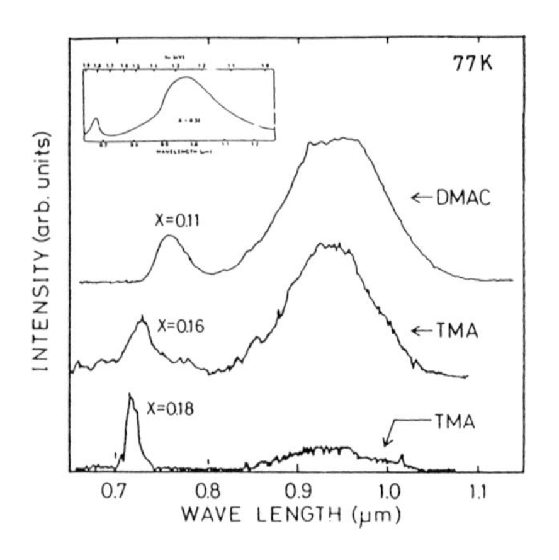

Fig.4, 77K photoluminecence spectra of AlGaAs layers grown by different way.

4. SUMMARY

It was demonstrated for the first time that AlGaAs layers can be grown by MO Chloride VPE without using toxic AsH_3. At the moment, carrier concentrations are too high probably either due to impure MO source or reaction between $AlCl_3$ and quartz reactor, but these results indicate that AlGaAs layer can be grown by the chloride transport method if one uses the hydrogen reduction method.

ACKNOWLEDGMENTS

The authors would like to express their sincere thanks to Profs. H.Seki and A.Koukitu of Tokyo University of Agriculture and Technology for their valuable suggestions. This work was supported by the Scientific Research Grant-in-Aid for Special Project Research on "Alloy Semiconductor Electronics," from the Ministry of Education, Science and Culture of Japan.

REFERENCES

Bachem K.H. and Heyen M. 1981, J. Crystal Growth, 55 pp.330-338; Deschler M. Cuppers M. Brauers A. Heyen M. and Balk P. 1987, J. Crystal Growth 82 pp.628-638.

Ettenberg M. Sigai A.G. Dreeben A. and Gilbert S.L. 1971, J. Electrochem. Soc. 118, pp.1355-1358

Hasegawa F. Yamate T. Yamamoto T. Nannichi Y. Koukitu A. and Seki H. 1985, Japan. J. Appl. Phys. 24 pp.1036-1042.

Ishikawa T. Inata T. Kondo K. Shibatomi A. 1986, Electronics Letts. 22 pp.198-190.

Koukitu A. Seki H. and Fujimoto M. 1976, Japan. J. Appl. Phys. 15 pp.1951-1952.

Koukitu A. Kouno S. Takashima K. and Seki H. 1984, Japan. J. Appl. Phys. 23 pp.951-955.

Kajikawa T. Hirano R. and Murotani T. 1985, Digest of Techical Meeting of Electron Device Group, IECE Japan, ED85-110. pp.67-74. (in Japanese).

Kajikawa T. Nakanishi M. Hirano R. and Murotani T. 1986, Digest of Techical Meeting of Semiconductor and Transistor Group, IECE Japan, SSD86-159, PP.15-21. (in Japanese).

Stringfellow G.B. and Hall H.Jr. 1979, J. Electrochem. Soc. 8 pp.201-226.

Williams E.W. Bebb H.B. 1966, Semiconductors and Semimetals (Academic Press), 8p.321.

Influence of reactor pressure and deposition temperature on the growth of InGaAs on InP by OMVPE

R. R. Saxena, V. M. Sardi, J. D. Oberstar and R. L. Moon

Hewlett-Packard Optical Communication Division
370 W. Trimble Road, San Jose, CA 95131

ABSTRACT: The influence of reactor pressure and growth temperature on InGaAs/InP structures grown by low pressure OMVPE using ethyl alkyl is described. A model for InGaAs growth is presented which takes into account the effect of TEIn alkyl and column V hydride interaction during epitaxial growth. Excellent InGaAs compositional uniformity over two substrates each 2" in diameter is demonstrated for PIN wafers grown at 60 Torr.

1. INTRODUCTION:

The epitaxial growth of InGaAsP alloys lattice matched to InP substrates by low pressure Organometallic Vapor Phase Epitaxy (OMVPE) using ethyl alkyls is now a mature technology, and excellent epitaxial layer quality and device performance have been reported. (Duchemin et al 1981, Razeghi and Duchemin 1984) We have established this technology for manufacturing of InGaAs/InP planar PIN detectors, as reported earlier (Saxena et al 1986). This paper describes the influence of reactor pressure (50-150 Torr) and growth temperature (550-650 C) on the properties of InGaAs/InP structures. The results are discussed in terms of a simple model for epitaxial growth which takes into account the effect of the reaction between TEIn and arsine or phosphine. This reaction was previously observed by Baliga and Gandhi (1975) in the OMVPE growth of InGaAs using TEIn at atmospheric pressure. It was described as an elimination reaction by Cooper et al (1980), plus Hess et al (1984) showed that there remains a finite interaction at low pressures. Our results confirm the presence of this reaction, however in spite of this reaction, sharper interfaces and improved uniformity of layer properties are obtained at lower pressures.

2. GROWTH MODEL:

The epitaxial growth model used by Aebi et al (1981) to describe the dependence of the growth rate and the composition of III-V ternary epitaxial layers on gas phase composition of constituent species, during OMVPE growth in a diffusion controlled mass transport limited growth regime, is extended to include the influence of TEIn - column V hydride elimination reaction. Assuming a first order elimination of In species, the effective gas phase TEIn concentration, [TEIn]eff, available for epitaxial growth can be written as:

$$[TEIn]_{eff} = [TEIn] \times (1 - K1as \times [AsH3] - K1p \times [PH3]) \quad (1)$$

where, [TEIn], [AsH3] and [PH3] represent the concentrations of In, As and P species input to the reactor. The constants, K1as and K1p, are determined by the overall interaction between TEIn and the respective hydride from the time mixing occurs inside the reactor to arrival at the growing surface. The growth rates for InP and ternary InGaAs can then be written

$$Gr(InP) = K2 \times [TEIn]eff \quad (2)$$
$$Gr(InGaAs) = K2 \times ([TEIn]eff + [TEGa]) \quad (3)$$

where, [TEGa] represents the concentration of TEGa input to the reactor, and it is assumed that the interaction between TEGa and V hydride is negligible. The constant K2 is determined by the diffusivity of species (assumed equal for TEGa and TEIn) and the effective thickness of the boundary layer. Substituting for [TEIn]eff from (1), and rearranging the growth rate expressions (2) and (3) gives,

$$Gr(InP) = (K2 - K1p \times [PH3]) \times [TEIn] \quad (4)$$
$$Gr(InGaAs)/C = K2([TEIn] + [TEGa])/C - K1as \times K2 \quad (5)$$

where, C denotes [TEIn]x[AsH3]. Finally the relative incorporation of In and Ga in the InGaAs layer can be written as,

$$[In]s/[Ga]s = (1 - K1as \times [AsH3]) \times [TEIn]/[TEGa] \quad (6)$$

3. RESULTS AND DISCUSSION:

The growth rate of InP and InGaAs layers was investigated as a function of the V/III ratio. In accordance with (4), the growth rate of InP layers reduces as the PH3 input is increased at a fixed TEIn flow, confirming a TEIn-PH3 interaction. The growth rate reduces by 12% as the V/III ratio is changed from 12.5 to 100, for InP layers grown at 60 Torr and 580 C. The growth rate data for InGaAs layers with varying V/III ratio is plotted in Figure 1, using the axes suggested by (5); growth rate (1E5 x microns/min.) and total column III input (1E3 x micromoles/min), both divided by [TEIn]x[AsH3] (micromoles/min.). A negative intercept on the y-axis confirms the presence of TEIn-AsH3 interaction.

The use of a quartz baffle at the reactor inlet reduced the growth rate significantly, which is shown in Figure 1 by a single data point. This is due to the increased interaction of the species resulting in a larger K1as value. A light brown deposit was clearly visible on the baffle, and the reactor inlet area after epitaxial growth runs with the baffle.

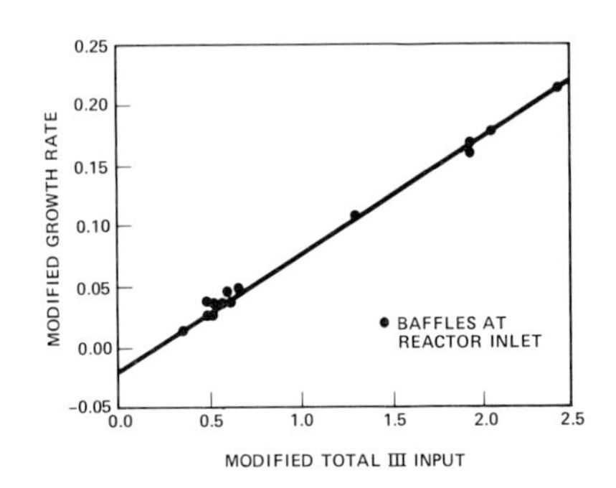

Figure 1. InGaAs growth rate data for layers grown at 100 Torr, 600 C

The incorporation of In and Ga in InGaAs epi layers grown on InP has been studied both as a function of growth temperature and pressure. Room temperature PL peak position is used for measuring the InGaAs layer composition. (Higher nm represents higher In content; Saxena et al 1986)

In incorporation in InGaAs epi layers increases substantially at higher pressures and by a smaller amount at higher temperatures (Figures 2a, 2b). The increased In incorporation is attributed to increased decomposition of arsine, which lowers Klas by reducing the amount of uncracked arsine that can participate in the TEIn-AsH3 elimination reaction. The background carrier concentration and carrier mobility of the InGaAs layers are not significantly affected by pressure and temperature in the range studied.

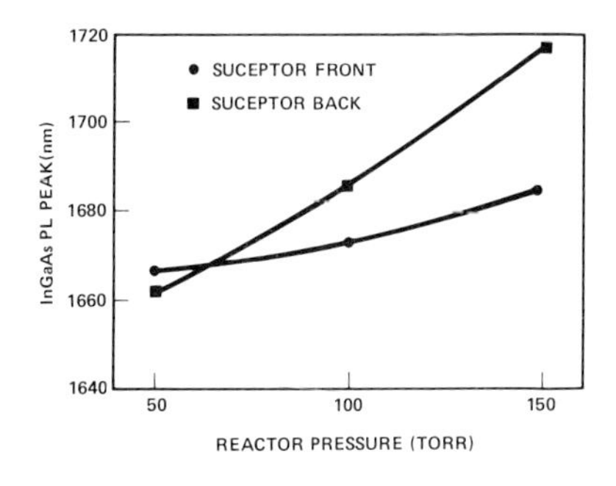

Figure 2a. InGaAs composition vs reactor pressure. (Tg = 580 C)

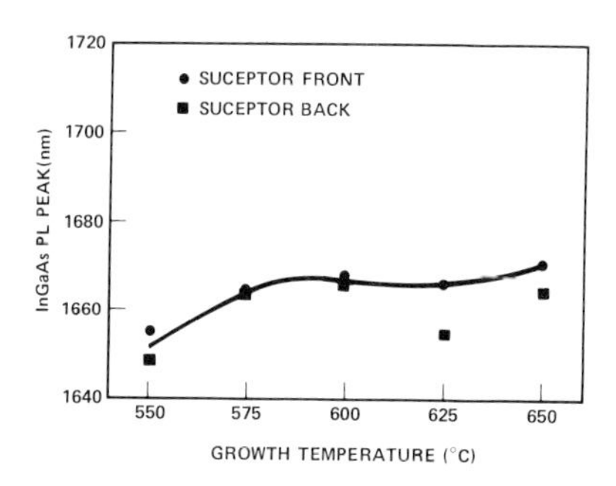

Figure 2b. InGaAs composition vs growth temperature. (P = 60 Torr)

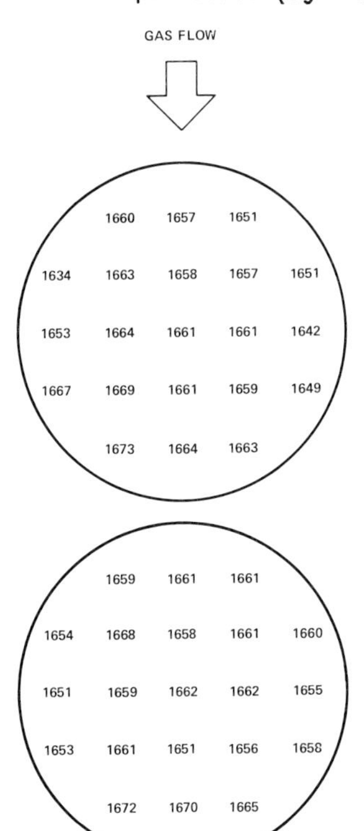

Figure 3. Composition uniformity. (Tg = 580 C)

The variation in InGaAs composition for PIN structures grown at 60 Torr on two substrates each 2" in diameter is shown in Figure 3. Improved compositional uniformity over the susceptor is obtained at 60 Torr compared to previously reported data for PIN detectors grown at 100 Torr (Saxena et al 1986). The primary improvement is near the edges due to a flatter velocity profile at lower pressure.

Multi quantum - well InGaAs/InP structures targeted for 1.3 micron wavelength were grown at pressures ranging from 50 to 150 Torr with a fixed well growth time. The peak wavelength shifts to longer wave length at higher pressure, and the full width at half maximum also increases corresponding to broader well thickness at higher pressure. The interfaces were examined by high resolution transmission electron microscopy. Both InGaAs/InP and InP/InGaAs interfaces were found to be many atomic layers broad in the sample grown at 150 Torr. The sample grown at 50 Torr showed one or two atomic layer InP to InGaAs transition, and a few monolayers broad InGaAs to InP transition. The results for quantum wells grown at atmospheric pressure using methyl alkyls are similar to our 50 Torr data. (Carey et al 1987) Secondary ion mass spectroscopy profiles of the interfaces in a PIN detector show that P turn-on and turn-off transitions are both equally sharp, but the As turn-on transition is much sharper compared to its turn-off transition. The grading of As in InP layer after the InGaAs/InP interface is shown in Figure 4. The difference between the two interfaces is attributed to the ease of As incorporation compared to that of P during the growth of InGaAsP alloys due to easier thermal decomposition of arsine and higher volatility of P species, (see Figure 1, Saxena et al 1986).

4. CONCLUSION

The influence of the reactor pressure and the growth temperature on properties of OMVPE grown InGaAs/InP structures has been studied. The electrical properties of InGaAs layers are not significantly affected. A simple growth model, assuming first order elimination of TEIn due to reaction between TEIn and column V hydride, has been used to understand the variation in growth rate and In incorporation. Excellent InGaAs compositional uniformity over two substrates each 2" in diameter is reported for InGaAs/InP PIN wafers.

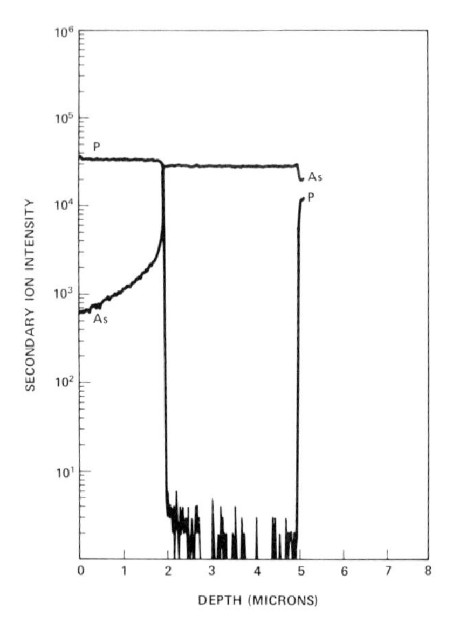

Figure 4. SIMS profile of a PIN detector structure grown at 60 Torr, 600 C.

5. ACKNOWLEDGEMENTS:

The authors wish to acknowledge expert technical assistance from Mark Tanner, Bill Perez and Jim Belden for OMVPE growth and characterization. Robert Hull for HRTEM studies, typing assistance from Carrie Patterakis, useful discussions with Mike Ludowise and Kent Carey and continued support and encouragement from Bob Weissman.

REFERENCES:

Aebi V., Cooper C. B., Moon R. L. and Saxena R. R., J. Elec. Matls, 55 517, 1981

Baliga B. J. and Ghandhi S. K., J. Electrochem. Soc., 122, 683, 1975

Carey K. W., Hull R., Fouquet J. E., Kellert F. G. and Trott G., Appl. Phys. Lett. to be published.

Cooper C.B., Ludowise M. J., Aebi V. and Moon R. L., J. Elec. Matls, 9, 299, 1980

Duchemin J. P., Hirtz J. P., Razeghi M., Bonnet M. and Hersee S.D., J. Crys. Growth, 55, 64, 1981

Hess K. L., Kasemset D. L. and Dapkus P.D., J. Elec. Matls., 13, 799, 1984

Razeghi M. and Duchemin J.P., J. Cryst. Growth, 70, 145, 1984

Saxena R., Sardi V., Oberstar J., Hodge L., Keever M., Trott G., Chen K.L. Moon R., J. Cryst. Growth, 77, 591, 1986

Highly uniform growth of modulation-doped N-AlInAs/GaInAs heterostructures by MBE

Hideo Toyoshima, Akihiko Okamoto, Yoshikazu Nakamura and Keiichi Ohata

Microelectronics Research Laboratories, NEC Corporation
4-1-1, Miyazaki, Miyamae-ku, Kawasaki 213 Japan

ABSTRACT: Highly uniform modulation-doped N-AlInAs/GaInAs heterostructures have been successfully grown on two inch InP substrates by a newly developed vertical growth chamber MBE. The deviation of 2DEG concentration was within ±1.5% over two inch substrates, in addition to the achieved high electron mobilities of 11,000 $cm^2/V{\cdot}s$ at room temperature and 51,700$cm^2/V{\cdot}s$ at 77K.

1. Introduction

$Ga_{0.47}In_{0.53}As/Al_{0.48}In_{0.52}As$ heterostructures, lattice-matched to InP, are very attractive for high speed and high frequency device applications. $Ga_{0.47}In_{0.53}As$ has high electron mobility, 13,800 $cm^2/V{\cdot}s$ at room temperature (Oliver et al 1980), high drift velocity more than 2.5×10^7cm/s (Bandy et al 1981) and large energy separation between Γ and L valley, 0.55eV (Cheng et al 1982). It is also advantageous that, compared with the GaAs/AlGaAs heterostructure, the $Ga_{0.47}In_{0.53}As/Al_{0.48}In_{0.52}As$ heterostructure has larger conduction band discontinuity, 0.50eV (People et al 1983), in addition to the higher doping capability (Cheng et al 1981) and lower persistent photoconductivity (Grien et al 1984) in $Al_{0.48}In_{0.52}As$. Up to the present, the potentialities of the GaInAs/AlInAs heterostructures have been demonstrated by their application to such devices as 2DEGFETs (Hirose et al 1985, Itoh et al 1985, Peng et al 1987), HBTs (Malik et al 1983, Furukawa et al 1986) and HETs (Imamura et al 1986). For further device development and their application to ICs, it is indispensable to grow high quality uniform layers on a large substrate. This paper presents the growth and properties of AlInAs/GaInAs system on InP substrates using a newly developed MBE. Highly uniform growth of modulation-doped N-AlInAs/GaInAs heterostructures has been achieved on two inch InP substrates with high electron mobility.

2. MBE growth

The thin GaInAs/AlInAs heterostructures growth by MBE requires more advanced technologies than GaAs/AlGaAs heterostructures growth, such as precise control of flux and substrate temperature, and adequate substrate pre-treatment before growth (Mizutani et al 1985, Hirose et al 1985). As shown in Fig.1, an MBE was newly developed whose growth chamber was vertical in design. Compared with the conventional horizontal growth chamber MBE, this growth chamber configuration increases the design feasibility and simplifies the chamber structure,

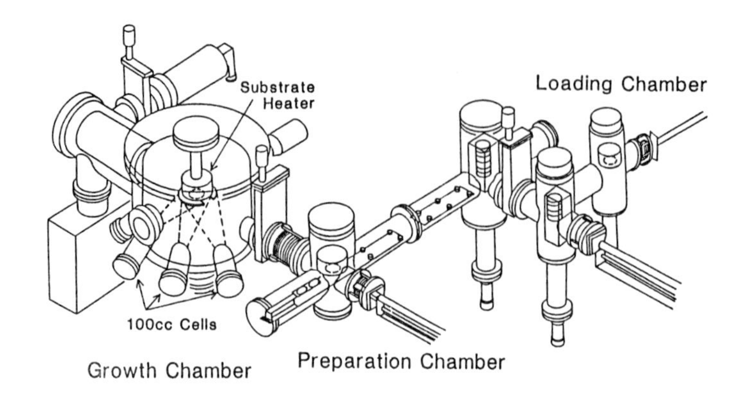

Fig.1. MBE system diagram with a vertical growth chamber.

attaining high growth performance. Large 100cc cells are used to obtain long time flux intensity stability and epilayer uniformity. Otherwise, beam flux distribution would soon be changed, due to source material consumption. These cells are attached diagonally to the bottom of the growth chamber to eliminate source material contamination caused by flakes which fall into the crucibles. In order to ensure a highly abrupt heterointerface, cell shutters are placed 5 cm from the crucibles, thus reducing flux intensity transient to a few percent, after the cell shutters are opened (Onabe et al 1986). A large four inch diameter graphite heater was adopted for uniform substrate heating. It is possible to handle a substrate up to three inches in diameter. The source materials employed were 7 nines purity gallium, 6 nines purity aluminum, 7 nines purity indium and 7 nines purity solid arsenic. GaInAs/AlInAs epilayers were grown on 2 inch diameter (100) oriented semi-insulating Fe-doped InP substrates at 540°C with an around 1.0µm/hr growth rate. The substrates were degreased then chemically etched by $H_2SO_4:H_2O_2:H_2O=3:1:1$ solution and Br(2%)-methanol before loading into the chamber. The substrate thermal cleaning, under As pressure before growth, were the same as that described in the Ref.(Mizutani et al 1985). The substrates were mounted with indium solder and rotated at 10 r.p.m. during the growth for attaining epilayer uniformity.

3. Results

Before GaInAs/AlInAs growth, GaAs/AlGaAs systems were grown on GaAs substrates to examine the GaAs and AlAs properties. The uniformity of the GaAs and AlAs thickness and Si doping were within ±2% over three inch substrate. The p-type background carrier concentration for GaAs epilayers was $\sim 1\times10^{14}cm^{-3}$. The highest obtained electron mobilities for the modulation-doped N-AlGaAs/GaAs heterostructures were $186,000cm^2/V{\cdot}s$ at 77K and $1,092,000cm^2/V{\cdot}s$ at 2K under light illumination conditions, demonstrating the high purity GaAs and AlAs. Then the 1µm thick GaInAs and AlInAs layers were directly grown on InP substrates, respectively, and (400) X-ray diffraction rocking curves were measured over the whole substrate area. Fig.2(a) and 2(b) show typical X-ray rocking curves for the nearly lattice matched GaInAs and AlInAs epilayers to InP substrates, respectively. The full width at half the maximum (FWHM) of

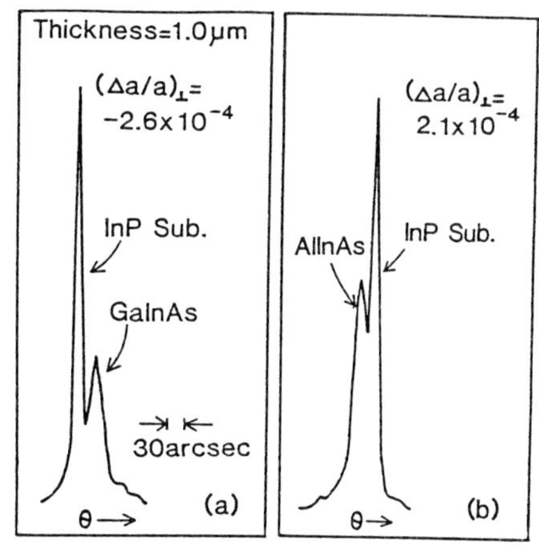

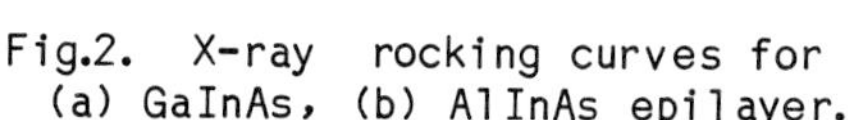
Fig.2. X-ray rocking curves for (a) GaInAs, (b) AlInAs epilayer.

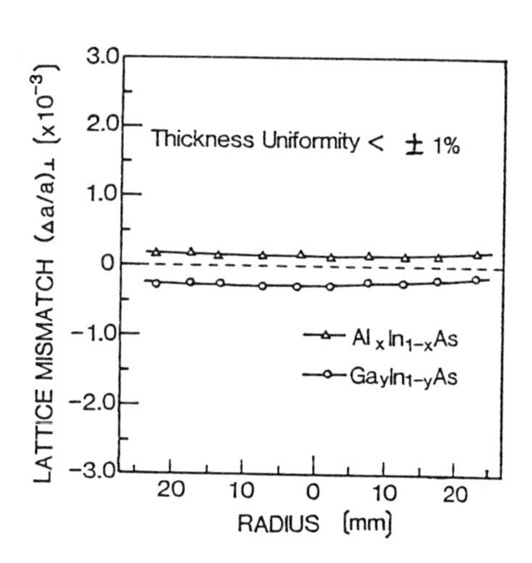

Fig.3. Lattice mismatch variation in GaInAs and AlInAs epilayers on two inch substrates.

GaInAs was as narrow as 30 arcsec and that of the AlInAs was also narrow to the same extent, confirming the high crystal quality of the epilayers. Fig.3 shows the variation in lattice mismatch for the GaInAs and AlInAs epilayers to the InP substrate in two inch diameter. For both epilayers, the lattice mismatch values ($\Delta a/a$) were within $\pm 3\times10^{-4}$, and the variations in ($\Delta a/a$) values were as small as 1×10^{-4}. This result indicates that the epilayer thickness uniformity is within ±1% over a two inch substrate. The undoped GaInAs epilayer had electron mobilities of as high as 11,700cm^2/V·s at room temperature and 51,100cm^2/V·s at 77K, with low n-type background carrier concentrations of 1×10^{15} cm^{-3} at room temperature and 5×10^{14} cm^{-3} at 77K. The obtained mobilities are comparable to the best for the thin GaInAs epilayers grown by MBE (Mizutani et al 1985). Then, uniform growth of modulation-doped N-AlInAs/GaInAs heterostructures was studied. As shown in Fig.4, the growth structure consisted of a 500Å thick N-AlInAs layer (Nd=2×10^{18} cm^{-3}), a 50 Å thick undoped-AlInAs spacer layer, a 1000 Å thick undoped-GaInAs channel layer and a 5000 Å thick undoped-AlInAs buffer layer. Hall measurement was carried out along the diameter of the two inch wafer. Fig.5 shows the variation in 2DEG carrier concentration and electron mobility at room temperature and at 77K versus wafer radius. Average 2DEG concentrations were 1.85×10^{12} cm^{-2} at room temperature and 1.62×10^{12}cm^{-2} at 77K, within the extremely small deviation, ±1.5%, over a two inch substrate. In addition, obtained average electron mobilities were as high as 11,000cm^2/V·s at room temperature and

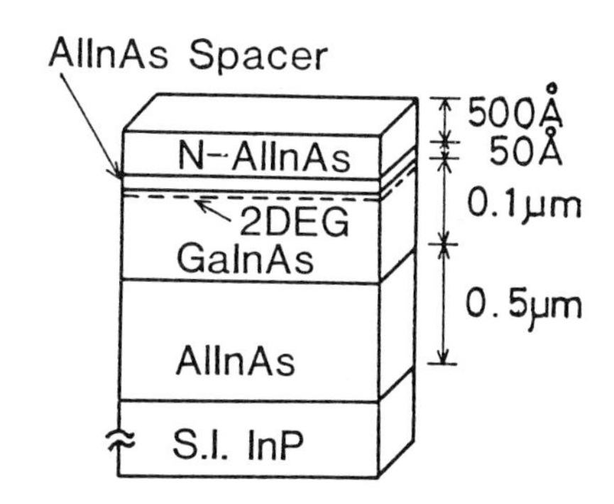

Fig.4. Diagram of a modulation -doped N-AlInAs/GaInAs heterostructure.

$51{,}700 cm^2/V{\cdot}s$ at 77K. Thus, markedly uniform growth of modulation doped N-AlInAs/GaInAs heterostructures with high electron mobility was successfully attained.

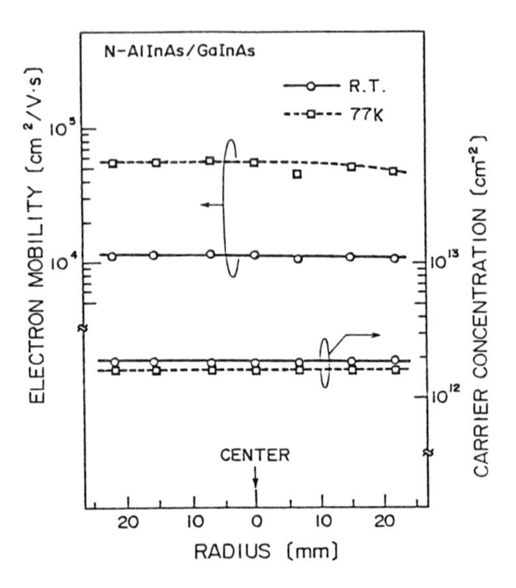

Fig.5. Variation in 2DEG carrier concentration and electron mobility at R.T. and 77K on a two inch substrate.

4. Conclusion

An original MBE with a vertical configuration growth-chamber has been developed and highly uniform growth of AlInAs/GaInAs heterostructures with high electron mobilities has been successfully achieved on two inch InP substrates. These results are very promising to apply AlInAs/GaInAs heterostructures for high speed electron devices and ICs.

Acknowledgment

The authors would like to thank T.Ishida, J.Sakai and K.Hirama of ANELVA for their help in MBE development. Thanks are due to K.Kasahara, T.Mizutani and K.Hirose for their meaningful discussion and help. The authors also thank Y.Takayama and H.Sakuma for their helpful suggestions and encouragement.

Reference

Bandy S, Nishimoto C, Hyder S and Hopper C 1981 Appl.Phys.Lett. 38 817
Cheng K Y, Cho A Y, Christman S B, Persall T P and Rowe J E 1982 Appl.Phys.Lett. 40 423
Cheng K Y, Cho A Y and Wagner W R 1981 J.Appl.Phys. 52 6328
Furukawa A and Baba T 1986 Jpn.J.Appl.Phys. 25 L862
Grim T, Nathan M, Wicks G W and Eastman L F 1984 Gallium Arsenide and Related Compounds Inst.Phys.Conf.Ser. 74 367
Hirose K, Ohata K, Mizutani T, Itoh T and Ogawa M 1985 Gallium Arsenide and Related Compounds Inst.Phys.Conf.Ser. 79 529
Imamura K, Muto S, Fuji T, Yokoyama N, Hiyamizu S, Shibatomi A 1986 Electron.Lett. 22 1148
Itoh T, Brown A S, Caminitz L H, Wicks G W, Berry J D and Eastman L F 1985 Gallium Arsenide and Related Compounds Inst.Phys.Conf.Ser.79 529
Malik R J, Hayes J R, Cappaso F, Alavi K and Cho A Y IEEE Electron Device Lett. EDL-4 383
Mizutani T and Hirose K 1985 Jpn.J.Appl.Phys. 24 L119
Oliver J D and Eastman L F 1980 J.Electron.Mater. 9 693
Onabe K, Tashiro Y, Ide Y 1986 Surface Science 174 401
Peng C K, Aksum M I, Ketterson A A, Morkoç H and Gleason K R 1987 IEEE Electron Device Lett. EDL-8 24
People R, Wecht K W, Alavi K and Cho A Y 1983 Appl.Phys.Lett. 43 118

Intermixing at III–V heterointerfaces

C Guille, F Houzay, JM Moison and F Barthe

Laboratoire de Bagneux, Centre National d'Etudes des Télécomunications, 196 Avenue Henri Ravéra, 92220 Bagneux, France

ABSTRACT: We report a quantitative study by in situ electron spectroscopies of segregation to the surface at the monolayer level in molecular beam epitaxy-grown heterojunctions combining AlAs, GaAs and InAs with each other. The InAs/AlAs and InAs/GaAs heterointerfaces are abrupt while the reverse ones are not; the topmost substrate ml is gradually distributed in the growing overlayer because of surface segregation of In atoms. No surface segregation is detected for AlAs-GaAs heterointerfaces. The tendency to surface segregation follows In > Ga ~ Al as in ternary alloys. The influence of growth parameters is reported and consequences of segregation on interface roughness along the growth axis are considered.

1. INTRODUCTION

The properties of heterostructures (HS) involving ultrathin layers of III-V compounds are mainly dictated by the abruptness of heterointerfaces. Since bulk diffusion is not possible at growth temperatures, the interface abruptness is determined by possible intermixing between substrate and impinging atoms at the growth surface, i.e. by surface segregation of substrate atoms. We present here a first systematic and quantitative study of the segregation of column III atoms at the most commonly used heterointerfaces (GaAs, AlAs and InAs with each other) by in situ surface-sensitive techniques.

2. EXPERIMENTAL

Three geometries of HS have been used: (I), 1 monolayer (1 ml=half the lattice parameter) of one material on a buffer layer of the other one; (II), layers of increasing thicknesses of one material on a buffer layer of the other one for the lattice-matched AlAs-GaAs systems; (III), HS (I) covered by layers of increasing thicknesses of the base material for the mismatched AlAs-InAs and GaAs-InAs systems in order to avoid the switch from 2D to 3D growth and to keep the systems homogeneous in the growth plane. Growth takes place in a MBE (molecular beam epitaxy) Riber 2300 chamber connected to a chamber with AES (Auger electron spectroscopy) and XPS (X-ray electron spectroscopy) facilities. Thicknesses are calibrated by RHEED (reflexion high-energy electron diffraction) oscillations during homoepitaxy (±5 %). Thin (~1 ml) layers are grown at a rate of 0.05 ml/s, thicker layers at about 0.3 ml/s and buffer layers at 1 ml/s. Growth temperature is 600 C, except for InAs layers where it is lowered to 470 C in order to avoid In desorption. The arsenic pressure is 1x10-5 torr in order to obtain "As stabilized" surfaces for AlAs and GaAs layers and "In- stabilized" surfaces for InAs layers. At each heterointerface, the growth is interrupted during one minute; it allows surface smoothening as demonstrated by an overall improvement of the RHEED pattern. Except during the initial buffer layer build-up, RHEED patterns display short streaks indicating layer-by-layer deposition; their integer-order-streak spacing corresponds to the lattice parameter of buffer layers (±3%) indicating pseudomorphism. After cooling down to room temperature, samples are transferred under ultra-high vacuum to the analysis chamber and their surface is analyzed by AES (primary beam=2KeV) and by XPS (AlKα line=1486.6eV) Some samples have been studied ex situ by Rutherford back scattering (RBS, 1 MeV He^+ ions).

3. RESULTS AND DISCUSSION

The AES/XPS peak energy positions for all HS are not significantly different and only the peak intensities are discussed here. In order to evaluate them on a quantitative basis, we assume a uniform primary excitation and an attenuation of the outgoing electron beams which follows an exponential law with an escape length depending on their kinetic energy (extrapolated from Chen et al 1982). Each column III atom signal is referred to a constant As signal. We use the available As signal corresponding to the largest escape length. The peak ratios are referred to corresponding ratios in bulk compounds. These reduced values then reflect only the repartition of column III atoms in HS. These classical assumptions are supported by results obtained on systems without segregation (Houzay et al 1987). The data are evaluated by a segregation coefficient S or number (in ml units) of atoms of the base material which segregate to the suface after 1 ml deposit of the other material.

We first consider the results for type-I HS. For InAs/AlAs and InAs/GaAs HS, the agreement between experimental and theoretical values obtained for a no-segregation model (S=0) is correct. On the other hand, the discrepancies in AlAs/InAs and GaAs/InAs HS exceed the experimental accuracy (30% for AES, 20% for XPS) and systematically indicate an excess of In atoms at the surface. The agreement is much better if a complete exchange (S=1) is assumed. A finer determination of S is obtained from the most surface-sensitive ratios; for InAs/AlAs and InAs/GaAs (S=-0.1±0.2), and for GaAs/InAs (S=1.1±0.2) and AlAs/InAs, we obtain respectively near- null and near-total segregation efficiencies. No significant segregation is detected in AlAs-GaAs couples (S<0.5).

This is confirmed by data on type-II and -III HS. For InAs/AlAs and InAs/GaAs HS, no atom intermixing is detected. For AlAs/InAs and GaAs/InAs, the decrease of In signal with increasing GaAs or AlAs overlayer thicknesses is clearly slower than the model prediction for S=0 (figure 1), so surface segregation does drive In atoms on top of the growing overlayer. These data are compared to expectations involving a gradual dissolution of the InAs ml in the overlayer, forming a $In_xGa_{1-x}As$ or $Al_xGa_{1-x}As$ ternary layer whose composition at the ith layer away from the interface is given by,

$$x_i = 1-S \text{ for } i=0,\ x_i=(1-S)S^i \text{ for } 0<i<n,\ x_i=S^i \text{ for } i=n$$

The interface thickness in ml units t is ~ 1/ln(S). Fitting of our data with such profiles gives S=0.95±0.15 (t~100Å) for AlAs/InAs and S=0.8±0.15 (t ~ 13Å) for GaAs/InAs. The corresponding t for GaAs-AlAs HS is < 5Å. We have confirmed these values for GaAs/InAs by RBS studies of type-III HS (Abel et al 1987). Under these conditions, the composition profile of an InAs-GaAs short-period superlattice is presented on figure 2; the two interfaces are not perfectly abrupt, GaAs/InAs by redistribution of In atoms in GaAs and InAs/GaAs by In atoms driven on top of GaAs.

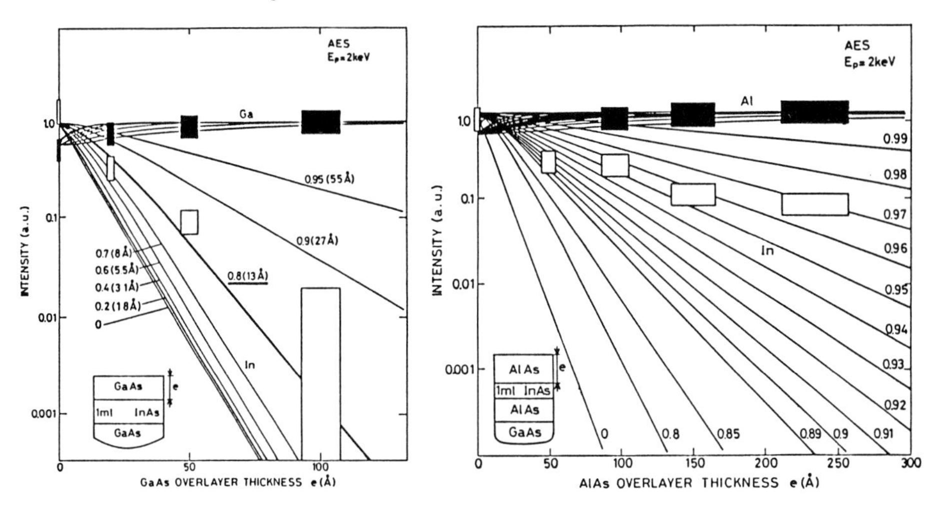

figure 1: III/As peak ratio intensities versus thickness of the top overlayer for type-III HS (see text)

We have examined the influence of various parameters on S. It does not depend significantly on the growth temperature from 420 to 560 C and on overlayer growth rate from 0.05 to 0.6 ml/s. We have also investigated the influence of compressive strain imposed by various base substrates (GaAs, InP, InAs) on In atoms and did not observe any significative variation of S with lattice mismatches ranging from 0 to 7%. Our data show that the defects likely responsible of interface roughness along the growth axis are intrinsic rather than extrinsic.

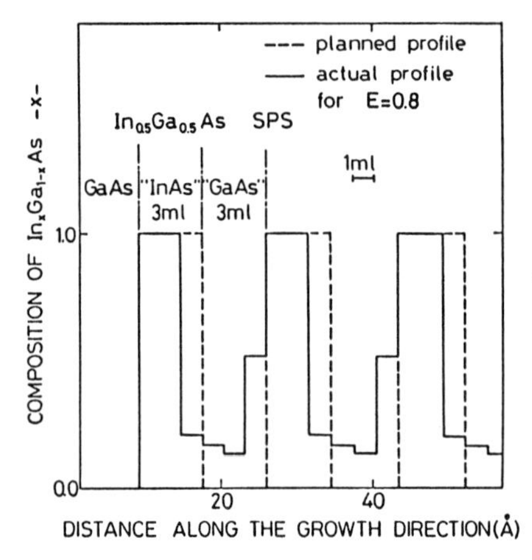

figure 2: In composition profile of a $(GaAs)_3(InAs)_3$ short-period superlattice

In summary, AlAs/GaAs, GaAs/AlAs, InAs/AlAs and InAs/GaAs heterointerfaces are abrupt in composition while AlAs/InAs and GaAs/InAs ones are not, because of the gradual distribution of the InAs top ml in the growing overlayer. The segregation efficiencies which are near-zero and near-unity respectively for the abrupt and diffuse heterointerfaces follow In>Ga ~ Al. We have observed a similar tendency in III-V ternary alloys (see also Massies et al 1987) and in metal-semiconductor contacts (Houzay et al 1986). The driving force involved is not clear at present. It has been suggested that it may originate from a bond-strength or a steric effect: the weakest-bonded or biggest atom goes out to the surface. These effects are probably biased by bond distortions due to pseudomorphic strain and to the surface reconstruction, both being variable during the growth of HS. The physical origin for the segregation process remains therefore an open question.

Abel F, Cohen C, Guille C, Houzay F, l'Hoir A, Moison JM and Schmaus D 1987 2nd Int. Conf.on the Structure of Surfaces (Amsterdam)

Chen P, Bolmont D and Sebenne CA 1982 J.Phys. **C15** 6101

Massies J, Turco F, Salettes A and Contour JP 1987 J.Cryst. Growth **80** 307

Houzay F, Bensoussan M and Moison JM 1986 Surf. Sci. **168** 347

Houzay F, Guille C, Moison JM, Henoc P and Barthe F 1987 J. Cryst.Growth **81** 67

Paper presented at Int. Symp. GaAs and Related Compounds, Heraklion, Greece, 1987

Lattice relaxation in InGaAs/GaAs strained layers

A. Aydinli[+], M. Berti, A.V. Drigo
Physic Departement, University of Padova, Italy
C. Ferrari, G. Salviati
MASPEC - CNR Institute, Parma, Italy
F. Genova, L. Moro
CSELT, Torino, Italy

ABSTRACT: $In_xGa_{1-x}As$ on GaAs strained single layers having composition $0.08 < x < 0.15$ and thicknesses ranging between 20 and 1000 nm have been grown by molecular beam epitaxy. The layer thickness the alloy composition and the lattice parameters have been measured. Imaging techniques have been employed for evidencing the presence of misfit dislocations. It has been found that there is not a definite layer thickness threshold for the lattice relaxation, but an extended transition region, where misfit dislocations are generated without a complete strain reduction.

1. INTRODUCTION

Recently heteroepitaxial structures of lattice mismatched materials have assumed increasing importance. It has been shown that epilayers without misfit dislocations (MDs), which have a detrimental effect on the electrical characteristics of the materials, can be grown if a critical thickness threshold is not overcome. In this case the lattice mismatch is accomodated by elastic strain and the epilayer unit cell is tetragonally distorted. In the case of larger thicknesses, the elastic strain is partially or totally reduced by the MD generation.
The aim of the present work is to investigate the MD efficiency in lessening the strain and to compare our experimental results with the predictions of the current models proposed in literature for critical thickness determination.

2. EXPERIMENTAL

$In_xGa_{1-x}As$/GaAs strained single heterostructures were grown, using a continuous substrate rotation, in a VG 80H twin chamber system. Undoped (001) oriented GaAs single crystals were used as substrates. For this study, three different nominal compositions (x = 0.08, 0.11 and 0.15) with thicknesses ranging from 20 to 1000 nm were grown. The layer composition was measured by Rutherford Backscattering Spectrometry (RBS) and in some cases by Auger Electron Spectrometry (AES) and Double Crystal Diffractometry (DCD) in order to have independent checks. A good agreement was always found among the techniques.

+ On leave from Hacettepe University, Ankara, Turkey

Strain measurements have been performed both by DCD and by channeling techniques. The DCD technique, by employing different asymmetric reflections, has been used for a direct measurement of the lattice parameter $a_{//}$ along the growth plane and $a_{\perp}$ normal to this plane, following Ishida et al. (1975). In addition to DCD, informations about the MDs presence has also been obtained by ion channeling spectra, X-ray topography (XRT) and integral cathodoluminescence (ICL) mode in the scanning electron microscope. Theoretical and experimental details about the application of the above mentioned techniques in InGaAs/GaAs strained layers, have already been reported by Aydinli et al. (1987).

3. RESULTS AND DISCUSSION

Fig. 1 shows a channeling investigation of a 68 nm thick specimen. The x value, determined by RBS, was 0.11.
The higher value of curves B and C with respect to curve A indicates the onset of MDs along the <110> directions. A higher MDs density along the [1$\bar{1}$0] direction evidences that the two <110> directions do not release the MDs equally. The analysis of the channeling data shows that MDs were located within the epilayer. This result has been confirmed for all the specimens investigated.

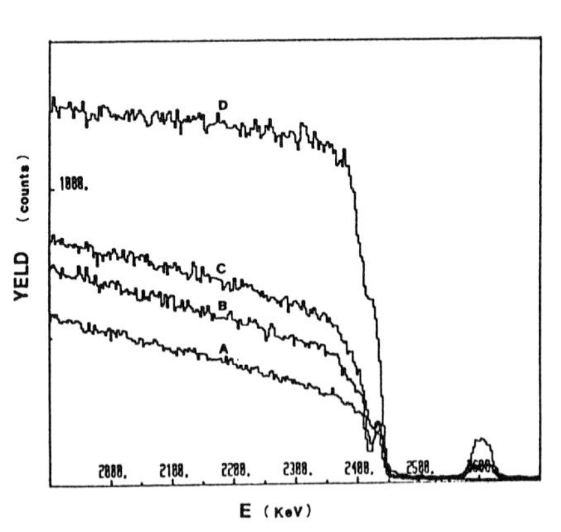

Fig. 1. Planar dechanneling spectra of a 68 nm thick sample, x = 0.11. Curve B and C (1$\bar{1}$0) and (110) plane respectively; curve D random direction. Curve A virgin GaAs (110) plane, for comparison.

XRT and ICL pictures of four specimens grown with different thickness and composition are shown in Fig. 2 as an example. All the pictures clearly show that, for a definite In molar fraction, the MDs density increases by increasing the epilayer thickness.
Fig. 3 shows $a_{//}$ and $a_{\perp}$ vs. the epilayer thickness obtained by employing channeling and DCD techniques on specimens grown with x = 0.08 and 0.11. Since the specimens with a thickness ranging between 50 and 1000 nm show a residual strain, it follows that there is not a definite layer thickness threshold for the complete lattice relaxation, but an extended transition region where MDs are generated without complete reduction in the strain. In addition to this, Fig 3 shows that in our case a layer thickness greater than 1000 nm is not sufficient for a complete strain reduction.
This result is not in agreement with the work of P.J. Orders and B.F. Usher (1987) where no variation of the lattice parameter $a_{\perp}$ is reported for t > 500 nm, in the same composition range as in our case. At the moment we cannot explain this difference, however, it should be pointed out that our epilayers have been grown without a buffer layer.

Fig. 4 shows a sketch of the residual strain vs. the layer thickness, obtained by employing channeling and DCD techniques. The curve of the critical strain obtained from the energy balance model is shown (curve B). For the three different compositions, the experimental points corresponding to the maximum strain can be fitted by curve A, where the slope is approximately $t^{-1/2}$. This result is in agreement with the People and Bean model, even though in our case the critical thicknesses are lower. The critical thicknesses for the beginning of strain reduction can be estimated by curve A. However, by employing XRT and ICL techniques, (see as an example Fig. 2a and 2b) MDs have been found also in the samples pointed out in Fig. 4.

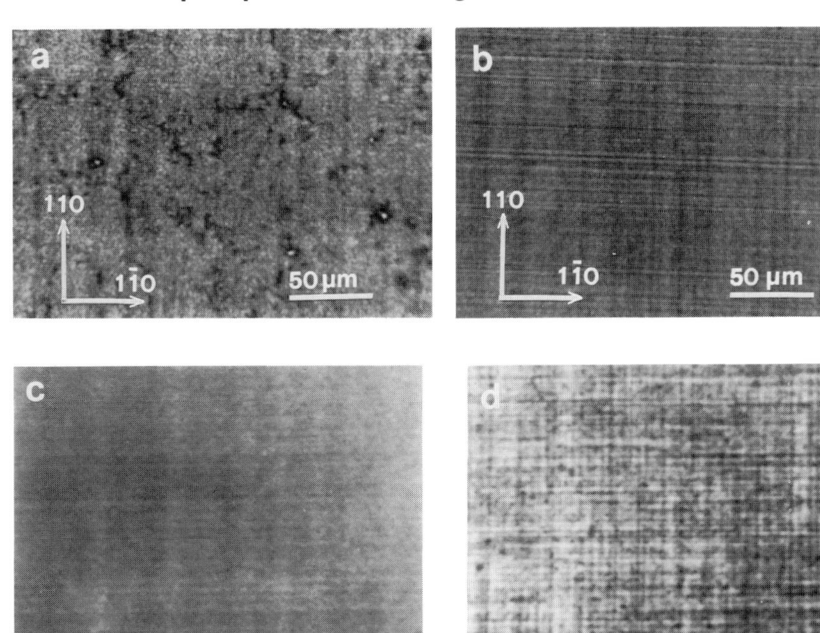

Fig. 2a,b. XRT pictures of 90 nm and 310 nm thick layers; x = 0.08.
Fig. 2c,d. ICL micrographs of 26 nm and 68 nm thick specimens; x = 0.11.

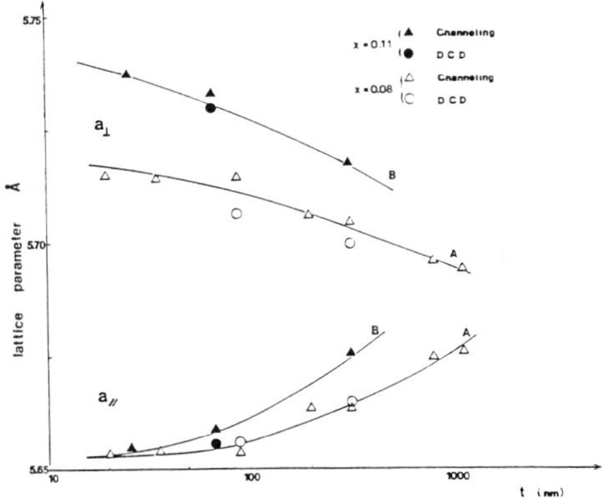

Fig. 3. Lattice parameters vs. the layer thickness. Curve A: $a_{//}$ and $a_\perp$ for x = 0.08. Curve B: the same for x = 0.11.

It follows that the critical thicknesses for the MD generation determined by strain measurements using RBS and DCD techniques, might be overestimated. In fact, RBS and DCD measurements cannot reveal MD linear densities less than 10^4 cm^{-1} and this leads to a strain reduction on the order of $2\ 10^{-4}$.

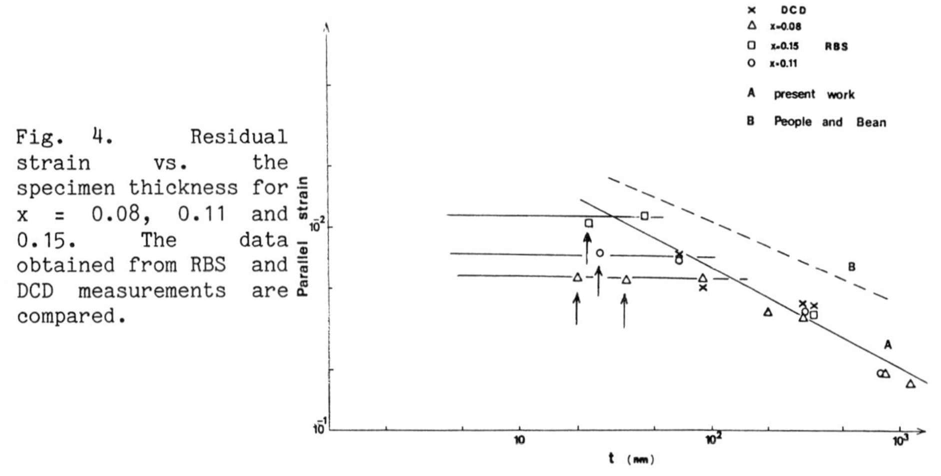

Fig. 4. Residual strain vs. the specimen thickness for x = 0.08, 0.11 and 0.15. The data obtained from RBS and DCD measurements are compared.

In addition to this, a careful analysis of the reported points shows that the residual strain is a function of the layer thickness and not of the composition. This result is in agreement with the predictions of the J.M. Mattews (1979) mechanical equilibrium model. This contraddiction (our results are partially in agreement both with the energy balance and mechanical equilibrium models) could be explained by the low growth temperature (520 C) which may prevent the thermodinamical equilibrium.

4. CONCLUSIONS

The role of the MDs in reducing the residual strain in InGaAs/GaAs single layers has been investigated. Work is underway for correlating the strain reduction with misfit dislocation density. Annealing experiments are planned in order to better discriminate between the energy balance and the mechanical equilibrium models.

REFERENCES

Aydinli A., Carnera A., Drigo A.V., Genova F., Moro L., Ferrari C., Franzosi P., Proceedings of the 1987 Conference of the European Materials Research Society, Strasburg 2-5June 1987.

Ishida K., Matsui J., Kamejima T., Sakuma I. 1975 a Phys. Stat. Sol. 1310

Mattews J. W. 1979 "Misfit dislocations" in "Dislocations in Solids" Nabarro (North-Holland) pp 461

Orders P.J. and Usher F. 1987 Appl. Phys. Lett. 50 980

People R. and Bean J.C. 1985 Appl. Phys. Lett. 47 322

This work has been partly supported by the Finalized Research Project"Materials and Devices for Solid State Electronics" of the Italian National Research Council under grant n. 8606652.

Inst. Phys. Conf. Ser. No. 91: Chapter 3
Paper presented at Int. Symp. GaAs and Related Compounds, Heraklion, Greece, 1987

MOCVD growth of GaAs on Si with strained layer superlattices

T. Soga, T. Imori*, M. Ogawa, T. Jimbo and M. Umeno

Department of Electrical and Computer Engineering, Nagoya Institute of Technology, Gokiso-cho, Showa-ku, Nagoya 466, Japan

ABSTRACT: GaAs is grown on Si with GaP, GaP/GaAsP superlattice and GaAsP/GaAs superlattice. The etch pit density of the top GaAs layer is less than $10^6 cm^{-2}$. The etch pit density, the curvature radius and the Si diffusion into the GaAs layer are investigated with changing the intermediate layer structure.

1. INTRODUCTION

The growth of GaAs on Si substrates has been actively studied recently aiming at the integration of GaAs optical and Si electronic devices on the same chip. Up to the present, several AlGaAs/GaAs devices such as lasers (Sakai et al 1986), solar cells (Fan 1984), light emitting diodes (Fletcher et al 1984) and FET's (Metze et al 1984) have been fabricated on Si. However, these device performances are inferior to those grown on GaAs substrates. These degradations are mainly due to the dislocations and the stress caused by the differences in the lattice constant and the thermal expansion coefficient.

We have proposed to use GaP and the GaP/GaAsP and GaAsP/GaAs strained layer superlattices (SLS's) to eliminate the dislocations in the top GaAs layer on Si and evaluated the electrical, optical and crystallographical properties of GaAs on Si grown by MOCVD (metalorganic chemical vapor deposition) by means of various method (Soga et el a,b,c). However, the evaluated GaAs/Si sample was concentrated to the GaAs/(GaAs/$GaAs_{0.5}P_{0.5}$)SLS/ ($GaAs_{0.5}P_{0.5}$/GaP)SLS/GaP/Si although the alternative structure which includes strained layer superlattices such as GaAs/(GaAs/$GaAs_{0.5}P_{0.5}$)SLS/($GaAs_{0.5}P_{0.5}$/GaP)SLS/GaAs/Si, etc., may be considered.

The purpose of this paper is to investigate which kind of intermediate layer structure is suitable for the growth of GaAs on Si. The intermediate layer structure is varied and the over-grown GaAs layers are characterized by various methods.

2. EXPERIMENTAL

The MOCVD growth was performed in an rf-heated horizontal reactor operating at atmospheric pressure. The Si substrate is 320μm-thick and n-type with the orientation of 2°off (100) toward [011]. The growth conditions are almost the same as previously reported (Soga et al 1986b). The annealing of the substrate is performed for 10 minutes in AsH_3 + H_2 or PH_3 + H_2 atmosphere when GaAs or GaP is grown on Si, respectively.

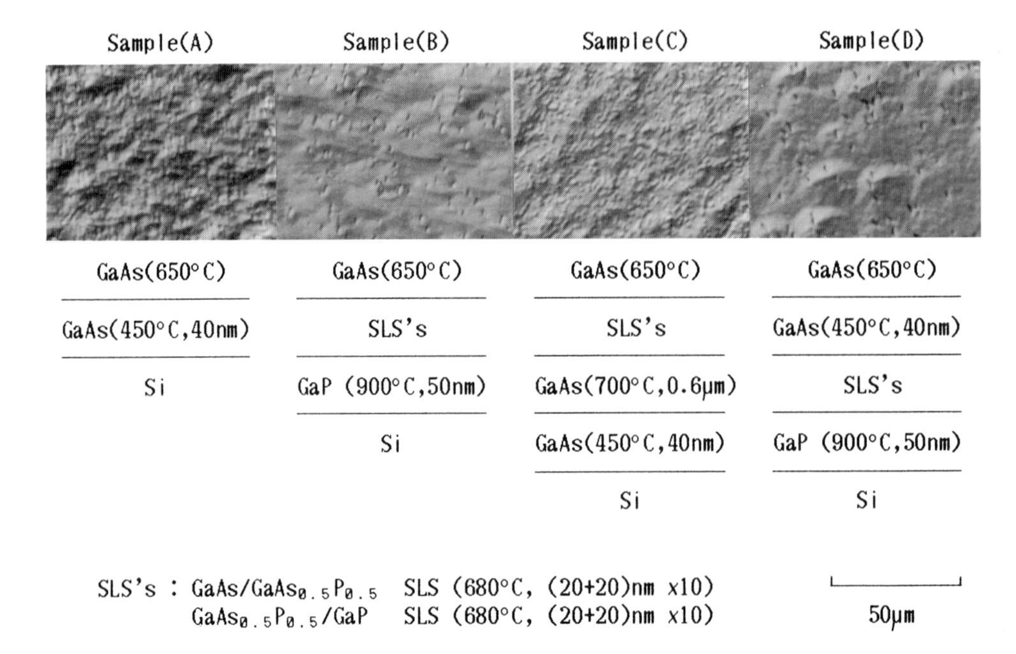

Fig. 1 Intermediate layer structures and etch pit patterns

The top layer is undoped and n-type GaAs.

Four kinds of intermediate layers were grown between GaAs and Si. The growth temperature, the thickness and the structure of the intermediate layers are shown in Fig. 1. The thickness of the top GaAs layer is about 3-4 μm. All the sample have a mirror-like surface and a single domain structure. When the GaAs thickness exceeds 4μm, cracks appear on the GaAs surface for sample (A) (two-step growth method) (Akiyama et al 1984) although no crack appears for other samples. The grown wafers are characterized by the etch pit density and the curvature radius. The SIMS (secondary ion mass spectroscopy) measurement is also performed to evaluate the Si diffusion into the epi-layer.

The etch pit density is revealed by the solution of H_2O (5ml) + $K_2Cr_2O_7$ (600mg) + HNO_3 (0.5ml) + HCl (0.5ml) + H_2SO_4 (1ml). The sample is dipped in this etchant for about 1 minutes at room temperature. The relationship between the dislocation and the etch pit will be published elsewhere (Nishikawa et al).

3. RESULTS AND DISCUSSION

3.1 Etch pit density and curvature radius

Figure 1 shows the Nomarski surface microphotographs of the etch pit patterns for the four samples. The etch pit density, the curvature radius and the thickness of the top GaAs layer of the four samples are summarized in Table 1. From this table, the following results are obtained.

(1) The etch pit density of the GaAs/Si grown by the two-step growth method is as high as 2 x 10^7 cm^{-2}.

(2) When the GaAs layer is directly grown on Si, the reduction of the etch pit density is small even if the strained layer superlattices are grown as the intermediate layer. The etch pit density is about 10^7 cm^{-2}. (sample (C))

(3) When the GaP layer is first grown on Si, the etch pit density is reduced to about $10^6 cm^{-2}$ together with the strained layer superlattices. (sample (B) and (D))

(4) When the strained layer superlattices are grown between GaAs and Si, the curvature radius is large compared with that of the GaAs/Si wafer grown by the two-step growth method. (sample(B), (C) and (D))

Table 1 GaAs thickness, curvature radius and etch pit density

	GaAs Thickness(μm)	Curvature Radius(m)	EPD(cm^{-2})
Sample(A)	4.0	4.2	$2x10^7$
Sample(B)	4.0	9.4	10^6
Sample(C)	3.1	14.4	10^7
Sample(D)	4.2	9.4	10^6

To reduce the etch pit density of GaAs on Si, the first layer on the Si substrate should be GaP which is lattice-matched to Si. The lattice mismatch between GaP and GaAs is relaxed by the strained layer superlattices. If the GaAs layer is grown on Si by the two-step growth method, a number of dislocations are generated. Once the misfit dislocations are generated, it is fairly difficult to eliminate the defects in the upper GaAs layer even if these superlattices are grown. The structual difference in sample (B) and (D) is still under consideration.

Some peole have confirmed that the annealing of the GaAs/Si wafer is

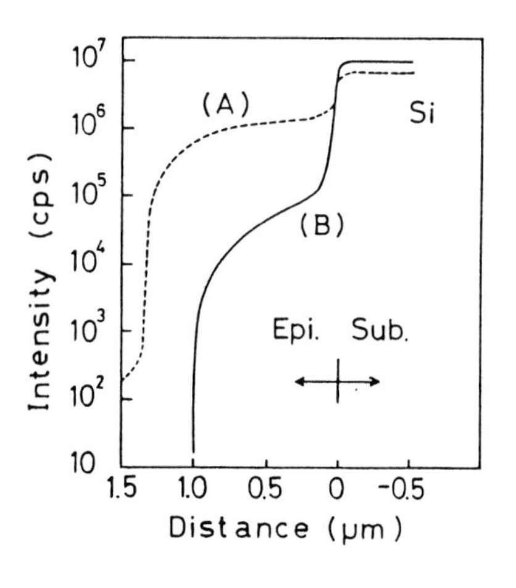

Fig.2 SIMS profile

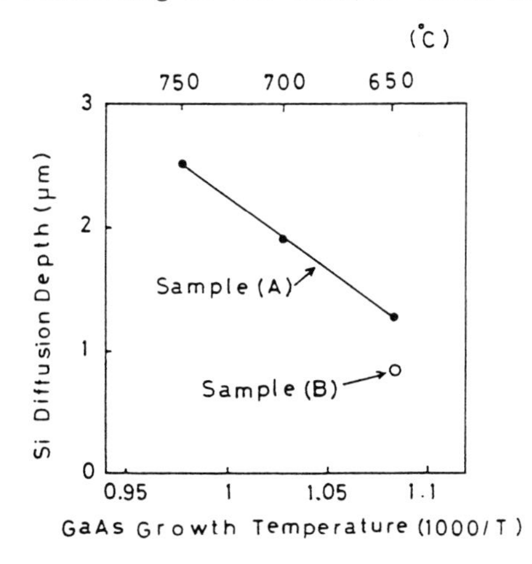

Fig. 3 Diffusion depth vs. growth temperature

effective to reduce the dislocations in GaAs on Si (Lee et al 1987). The GaAs/Si wafer with the extremely low etch pit density will be obtained by the annealing of this GaAs/Si sample grown with the strained layer superlattices.

3.2 SIMS

One problem in the GaAs/Si growth is the diffusion of Si into the epi-layer. To evaluate the Si diffusion, the SIMS measurement is performed for GaAs/Si samples. Figure 2 shows the in-depth SIMS profile of Si atom for sample (A) and (B). When GaP and strained layer superlattices are grown between GaAs and Si, the Si concentration in GaAs and the Si diffusion depth are low compared with those in GaAs/Si grown by the two-step growth method. Figure 3 shows the diffusion depth as a function of the growth temperature of the top GaAs layer. The diffusion depth increases with increasing the growth temperature. The low temperature growth and the use of GaP and the strained layer superlattices are effective to suppress the Si diffusion into the epi-layer.

4. Conclusion

GaAs was grown on Si substrate with GaP and the strained layer superlattices as the intermediate layer. The etch pit density of the top GaAs layer is less than $10^6 cm^{-2}$. Recently, we have succeeded in the growth of GaAs on Si with the etch pit density of the order of $10^5 cm^{-2}$ by improving the growth conditions. The further improvement is expected by the optimization of the superlattice structure and by the annealing of the GaAs/Si sample. GaP, GaP/GaAsP SLS and GaAsP/GaAs SLS are effective to reduce the etch pit density, to increase the curvature radius of the wafer and to lower the Si diffusion into the epi-layer.

Acknowledgement

The authors would like to thank Mr. H. Nishikawa for the measurement of the etch pit density and Analysis Research Center of Nippon Mining Company for the SIMS measurement. This work was partly supported by a Grant-in-Aid for Scientific Research (No.61114003) from the Ministry of Education, Science and Culture of Japan.

* Nippon Mining Company

References

Akiyama M, Kawarada Y and Kaminishi K 1984 J. Crystal Growth 18 21
Fan J C C 1984 Extended Abstracts of the 16th (1984 International) Conference on Solid State Devices and Materials, Kobe, p.115
Fletcher R M, Wagner D K and Ballantyne J M 1984 Appl. Phys. Lett. 44 967
Lee J W, Shichijo H, Tsai H L and Matyi R J 1987 Appl.Phys. Lett. 50 31
Metze G M, Choi H K and Tsaur B Y 1984 Appl. Phys. Lett. 45 1107
Nishikawa H, Soga T, Mikuriya N and Umeno M submitted
Sakai S, Soga T, Takeyasu M and Umeno M 1986 Appl. Phys. Lett. 48 413
Soga T, Sakai S, Takeyasu M, Umeno M and Hattori S 1986a Proc. of 12th Int. Symp. on GaAs and Related Comp. (Inst. Phys. Conf. Ser. No.79) p.133
Soga T, Hattori S, Sakai S and Umeno M 1986b J. Crystal Growth 77 498.
Soga T, Sakai S, Umeno M and Hattori S 1986c Jpn. J. Appl. Phys. 25 1510

Inst. Phys. Conf. Ser. No. 91: Chapter 3
Paper presented at Int. Symp. GaAs and Related Compounds, Heraklion, Greece, 1987

Photoluminescence characterisation of modulation-doped heterostructures grown by OM-VPE

Shlomo Ovadia

Joint Program for Advanced Electronic Materials: University of Maryland and Laboratory for Physical Sciences College Park, MD 20742

R. S. Sillmon and N. Bottka

U. S. Naval Research Laboratory Washington, DC 20375-5000

ABSTRACT: We report the observation of a low temperature photoluminescence peak from AlGaAs/GaAs modulation-doped heterostructures grown by OM-VPE, which is associated with the presence of a two-dimensional electron gas (2DEG) at the heterointerface. The observed emission has the characteristics of a bound exciton with a binding energy of 1.1 meV, and it is believed to be due to recombination of confined electrons in the 2DEG quantum well and free holes in the valence band.

1. INTRODUCTION

In recent years, modulation-doped heterostructures based on $Al_xGa_{1-x}As$/GaAs are being widely investigated because of carrier confinement in the quantum well at the heterointerface and significantly enhanced carrier mobility. The so-called modulation-doped field effect transistor (MODFET) or high electron mobility transistor (HEMT) is based on this heterostructure. In a typical structure, electrons from ionized donors in a doped AlGaAs layer transfer to an adjacent lower band gap energy GaAs layer in order to reduce their potential energy. These carriers are confined to a narrow triangularly-shaped quantum well at the heterointerface and form a two-dimensional electron gas (2DEG). These heterostructures are typically characterized by various transport measurements in a magnetic field such as Hall effect, magnetoresistance (Ando et al 1982) and cyclotron resonance (Chang et al 1977). However, these techniques are nontrivial and require elaborate sample preparation. Photoluminescence (PL) spectroscopy is a powerful, simple, and nondestructive optical technique that can be utilized to characterize such heterostructures. Recently, Aina et al (1987) reported the observation of a broad room temperature photoluminescence emission band above the GaAs fundamental band gap energy from high quality MODFET heterostructures grown by both organometallic vapor phase epitaxy (OM-VPE) and molecular beam epitaxy (MBE). Koteles et al (1986) have observed a 5 K PL peak in the excitonic region of GaAs which they associate with a 2DEG in a MODFET heterostructure. Other authors reported the observation of an additional feature in the photoreflectance spectra of MODFET heterostructures grown by MBE (Glembocki et al 1985 and Pearah et al 1986). In this letter, modulation-doped AlGaAs/GaAs heterostructures were grown and examined by photoluminescence spectroscopy in order to correlate the PL spectra with the presence of a 2DEG at the heterointerface.

Table I. MODFET HALL SHEET DENSITY (n_s) AND MOBILITY (μ)

	77K (Dark)		4.2K (Dark)	
OM#	$n_s(cm^{-2})$	$\mu(cm^2/v.s)$	$n_s(cm^{-2})$	$\mu(cm^2/v.s)$
372	$3.4x10^{11}$	79500	-	-
375	$4.9x10^{11}$	72000	$6.0x10^{11}$	106000
386	$4.1x10^{11}$	93200	$4.1x10^{11}$	144000
382[a]	insulating			

[a] No silicon doping

2. EXPERIMENTAL PROCEDURE

The heterostructures used in our experiments were grown on (100) semi-insulating GaAs substrates by OM-VPE. Trimethylaluminum was employed in a fast injection low-pressure OM-VPE system. These structures typically consisted of a 6000-Å thick nominally undoped p-type GaAs layer with an estimated carrier concentration of $N_A-N_D \simeq 8X10^{14}cm^{-3}$, followed by a 100-Å undoped $Al_{0.27}Ga_{0.73}As$ layer with an estimated carrier concentration of $N_A-N_D \simeq 1X10^{15}cm^{-3}$, and a 500-Å thick Si-doped $Al_{0.27}Ga_{0.73}As$ layer with a carrier concentration of $n \simeq 6X10^{17}cm^{-3}$. The structure was capped with a 100-Å thick Si-doped n^+ GaAs layer with a carrier concentration of $n \simeq 4X10^{18}cm^{-3}$. The undoped 100-Å AlGaAs layer was grown to separate the 2DEG from ionized donors in the n-type AlGaAs layer and thus reduce ionized impurity scattering. One sample had no Si doping in the heterostructure, and thus no 2DEG at the heterointerface.

The PL measurements were carried out at 4.3 K using a 50 mW air cooled argon laser at 5145 Å with a low excitation intensity of about 30 mW/cm^2. Then, the PL spectra were resolved by a 3/4 m monochromator and detected by a cooled 1024 elements intensified Si photodiode array with an optical

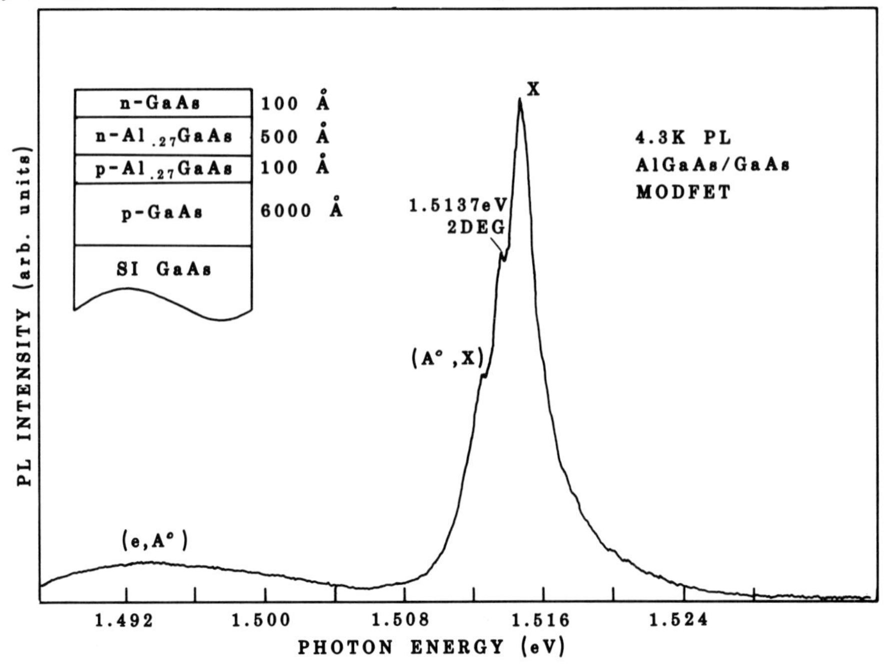

Fig. 1. The 4.3 K PL spectrum of an AlGaAs/GaAs MODFET heterostructure. The layer structure is depicted in the insert.

multichannel analyzer instead of the conventional lock-in system. This detector provides a spectral resolution 0.2 meV. The Hall effect measurements of carrier density and mobility were taken at 4.2 K, 77 K and room temperature in the Van der Pauw configuration.

3. EXPERIMENTAL RESULTS AND DISCUSSION

The Hall MODFET sheet carrier density (n_s) and mobility (μ) values for several heterostructures are summarized in Table I. We find that the increased Hall mobility at 4.2 K in comparison with the 77 K values implies the presence of a 2DEG in these samples. Additional confirmation of the presence of a 2DEG was obtained from magnetoresistance measurements in which oscillations and vanishing Hall resistance were observed.

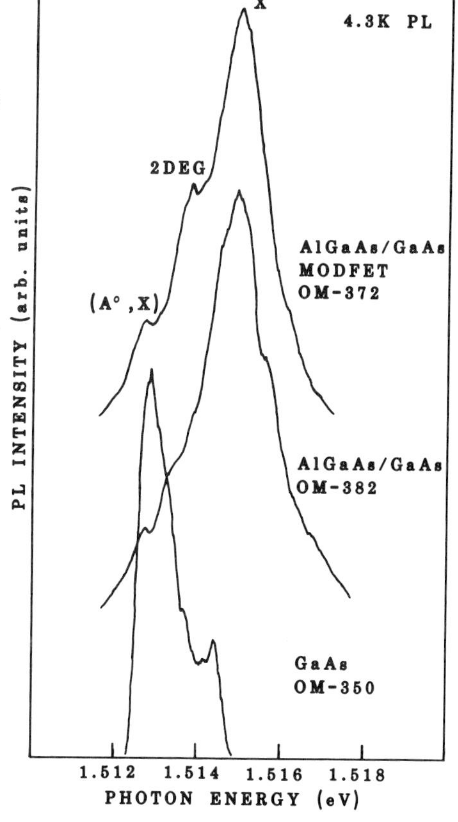

Fig. 2. The 4.3 K PL spectra of: OM-372, a MODFET structure with a 2DEG; OM-382, an undoped AlGaAs/GaAs structure without a 2DEG; and OM-350, an undoped p^--type GaAs single layer.

The 4.3 K PL spectra from an AlGaAs/GaAs MODFET heterostructure is shown in Fig. 1. It reveals a dominant free exciton peak (X) at 1.5148 eV with a full width at half maximum of 2.8 meV, a neutral acceptor bound exciton (A^o,X), and a free electron to acceptor (e, A^o) transition at 1.4932 eV probably due to carbon impurities. An additional peak at 1.5137 eV, labeled 2DEG, was observed from several MODFET heterostructures. The PL spectra shown in Fig. 1 was reproduced after etching off the 100-Å n^+ GaAs cap layer, but the 2DEG peak was not observed from the PL of the same sample after etching off the AlGaAs layers. Figure 2 compares the PL spectra of a MODFET heterostructure (OM-372), a similar AlGaAs/GaAs heterostructure but with no Si doping (OM-382), and a single undoped p-type GaAs layer (OM-350) which is grown under similar conditions as the MODFET heterostructures. The 2DEG peak is absent from the PL spectra of OM-382 and OM-350. As expected, a 2DEG is not present in these samples. It should also be noted that the proposed 2DEG peak is near the same energy as a neutral donor bound exciton emission (D^o,X). But this transition is absent or very weak in GaAs layers with low donor concentrations (Koteles <u>et al</u> 1985) and is expected to be weak in the GaAs grown during our study. The variation of the 2DEG PL intensity with laser excitation power, shown in Fig. 3, indicates that the transition has the characteristics of a bound exciton. As the laser exciton intensity is increased, we find that the 1.5137 eV 2DEG PL intensity increases superlinearly with an exponent of 1.54, and there is no discernible energy shift. It has been shown that quantum well PL intensity due to excitonic recombination has a superlinear variation with the laser power (Juang et al 1985). Also, it was observed in several samples with high mobility that the 2DEG peak intensity increases monotonically without significant broadening with an increase in 2DEG carrier density. Therefore, it is suggested that this emission peak has the characteristics of a bound exciton with a binding energy of 1.1 meV. This observation is consistent with Koteles et al (1986) data on modulation-doped AlGaAs/GaAs heterostructures grown by MBE. However, they do not report direct evidence of the presence of a 2DEG in their samples. We propose a mechanism of the 2DEG emission in which an electron confined in the triangular well recombines with a free hole in the valence band. The fact that the electron is confined or bound in two dimensions in a quantum well can explain the bound exciton characteristics of this emission.

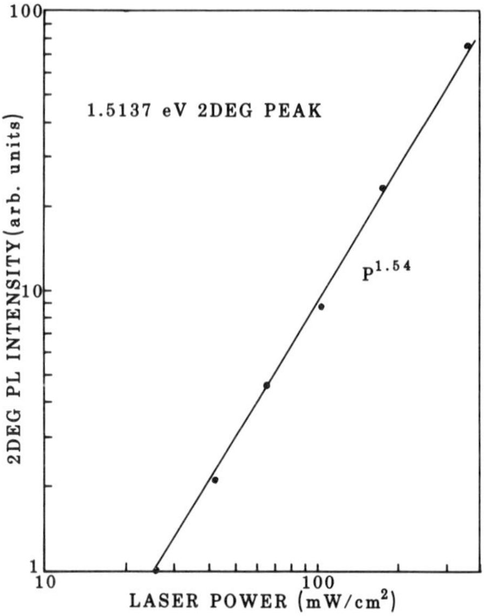

Fig. 3 The 2DEG PL peak intensity for a MODFET heterostructure as a function of laser excitation power.

4. CONCLUSIONS

We report the observation of an emission peak in the low temperature PL spectra of modulation-doped AlGaAs/GaAs heterostructures that is associated with the presence of a 2DEG at the heterointerface. Intensity measurements of this peak show that it behaves like a bound exciton with a binding energy of 1.1 meV. Although the exact physical mechanism that is responsible for this emission is unknown, we believe that it arises from transitions between confined electrons in the 2DEG quantum well and free holes in the valence band at the heterointerface. Potential applications exist for utilization of this emission in a simple, contactless nondestructive optical characterization of the 2DEG in modulation doped heterostructures.

5. ACKNOWLEDGEMENTS

The authors wish to thank Dr. D. K. Gaskill and Dr. O. J. Glembocki for the helpful discussions.

6. REFERENCES

Aina O, Mattingly M, Juan F Y, and Bhattachary P K, 1987, Appl. Phys. Lett. 50 43.
Ando T, Fowler A B, and Stern F, 1982, Rev. Mod. Phys. 54, 437.
Chang L L, Esaki H, Chang C A, and Esaki L, 1977, Phys. Rev. Lett. 38, 1489.
Glembock O J, Shanabrook B V, Bottka N, Beard W T, and Comas J, 1985, Appl. Phys. Lett. 46, 970.
Juang F Y, Nahimoto Y, and Bhattacharya P, 1985, J. Appl. Phys. 58, 1986.
Koteles E S, Lee J, Salerno J P, and Vassell M O, 1985, Phys. Rev. Lett. 55, 867.
Koteles E S, and Chi J Y, 1986, Superlattices and Microstructures, 2, 421.
Pearah P J, Klem J, Henderson T, Peng C K, and Morkoc H, 1986, J. Appl. Phys. 59, 3847.

Inst. Phys. Conf. Ser. No. 91: Chapter 3
Paper presented at Int. Symp. GaAs and Related Compounds, Heraklion, Greece, 1987

Reduction of dislocation density in GaAs/Si by strained-layer superlattice of $In_xGa_{1-x}As-GaAs_yP_{1-y}$ by MOCVD

T. Nishimura, N. Yoshida, K. Mizuguchi, N. Hayafuji and T. Murotani

LSI R&D Labolatory, Mitsubishi Electric Corp.,
4-1, Mizuhara, Itami, Hyogo 664, Japan

ABSTRACT: In-depth dislocation density profiles of the GaAs layers on Si (GaAs/Si) have been studied to clarify the dislocation reduction effect of $In_xGa_{1-x}As-GaAs_yP_{1-y}$ strained-layer superlattices (SLS's), which are interlaid in GaAs layers grown on annealed GaAs layers on Si substrates. Very thin SLS layer of 2000 Å in total thickness has strong effect for the reduction of dislocation density. The step-like reduction at the superlattice and continuous decrease after passing through the superlattice are observed. Further reduction has been achieved by annealing after the growth of SLS. Finally dislocation density is reduced to $1\text{-}2\times10^6 cm^{-2}$.

1. INTRODUCTION

In recent years, GaAs/AlGaAs epitaxial layers grown on Si substrates have been successfully applied mainly for majority carrier devices(Fischer et al. 1984, Metze et al. 1984, Nonaka et al. 1984). However, the performances of minority carrier devices on GaAs/Si are inferior to those grown on GaAs substrates due to the large density of dislocations. Lee et al. (1987) reported the defect is reduced to $10^7 cm^{-2}$ by the post annealing, however, this density is still too high for the application of minority carrier devices. Therefore, making GaAs epitaxial layers on Si substrates without unacceptably high dislocation density has been a matter of great concern. Matthews and Blakeslee (1974,1977, Matthews et al. 1976) first proposed to use GaAsP-GaAs strained-layer superlattice (SLS) for the GaAs/GaAs. Fischer et al.(1986) have reported that by a transmission electron microscope (TEM), the dislocation density of GaAs/Si is reduced by InGaAs-GaAs SLS's. However, TEM observation is apt to lead to an error in quantitative estimation due to the limited region of view when the density is less than about $10^7 cm^{-2}$. In order to achieve the reproducible optimization of growth conditions for GaAs/Si, more quantitative evaluation of dislocation density is indispensable. In-depth profile of etch pit density (EPD) by molten KOH etching is well-known technique for quantitative estimation of dislocation density on GaAs substrate. In this work, we have measured the In-depth EPD profiles of many samples with and without SLS. By comparing these profiles, we show that the dislocation density is greatly reduced in the GaAs layer with SLS and that the further reduction occurs through annealing after SLS growth.

2. EXPERIMENTAL

The epitaxial layers were grown by metalorganic chemical vapor deposition

(MOCVD). Figure 1 and Table I show the layer structures and their parameters, respectively. The growth sequence is as follows. The Si (100) 4° off toward <011> substrates were heated up to about 900°C in the flow of H_2 and AsH_3 for 20 min. Then, the substrates were cooled down to 400°C and a 100Å thick GaAs layer was grown as the first buffer layer. Subsequently, the substrates were heated to the conventional GaAs growth temperature of 700°C and then 1.5µm thick GaAs layer were grown. Then these layers were annealed at 900°C for 30 min in the flow of H_2 and AsH_3. On these layers, 3000Å of GaAs was grown followed by a ten period 100Å $In_xGa_{1-x}As$-100Å $GaAs_yP_{1-y}$ (x=0.03, 0.06, 0.12 y=0.76, 0.88, 1.00) strained-layer superlattice buffer layer. Finally, the 2.5um thick top GaAs layer was grown. In addition, two of the samples with SLS were annealed at 900°C for 30 min after the growth of the top GaAs layer. In-depth profiles of EPD were measured to study the dislocation reduction effect by SLS's. A slow mirror etchant was used for step etching of every sample and molten KOH etching was employed to reveal the dislocation on the epitaxially grown layer.

The etching rate of GaAs layers on Si at 400°C by molten KOH was about 400Å/s. A few seconds of etching revealed the small etchpits clearly on the GaAs layer(Ishida et al. 1987). The density of the etch pits agreed with the dislocation density estimated by the TEM for the sample whose dislocation density was over $10^8 cm^{-2}$.

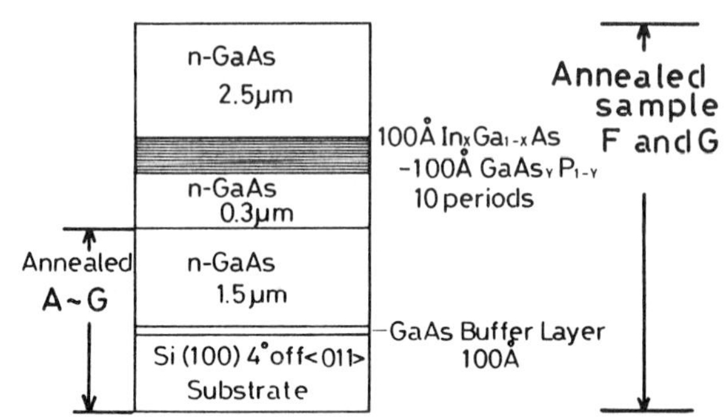

Fig. 1. Schematic of the layer structure.

Table I. Layer parameters.

Sample	Si cleaning condition		Composition of SLS		Annealing after growth of SLS
	Temp.	AsH_3 flow	$In_xGa_{1-x}As$	$GaAs_yP_{1-y}$	
A	900°C	40cc/min	x=0.12	y=1	
B	900°C	40cc/min	x=0.06	y=1	
C	900°C	26cc/min	x=0.03	y=1	
D	900°C	40cc/min	x=0.12	y=0.76	
E	900°C	26cc/min	x=0.06	y=0.88	
F	900°C	40cc/min	x=0.12	y=1	900°C 30min
G	900°C	26cc/min	x=0.06	y=0.88	900°C 30min

3. RESULTS AND DISCUSSION

The In-depth profiles of etch pit density are shown in Figs. 2 and 3 for the samples with InGaAs-GaAs and InGaAs-GaAsP SLS's, respectively. Profiles of as-grown and annealed samples are also shown in Fig. 2.

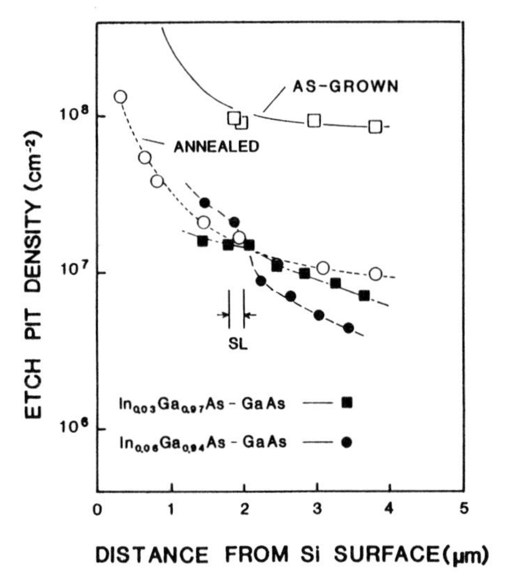

Fig.2. In-depth EPD profiles of the GaAs/Si with and without InGaAs-GaAs SLS.

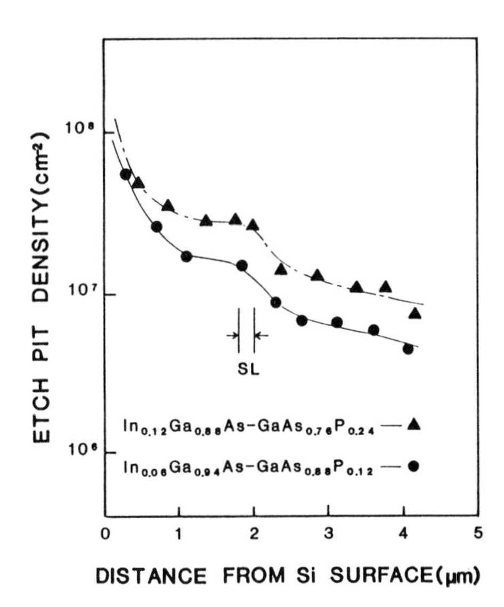

Fig.3. In-depth EPD profiles of the GaAs/Si with InGaAs-GaAsP SLS.

The dislocation density of as-grown and annealed samples drastically decreases with increasing layer thickness and seem to saturate at the thickness of about 2μm. By the post annealing, the dislocation density is reduced from $10^8 cm^{-2}$ to $10^7 cm^{-2}$ at the saturation region. These results agree with the report of Lee et al. (1987).

The dislocation reduction effect of SLS's grown on annealed GaAs/Si can be clearly seen by EPD profiles. The position of SLS's is shown in Figs. 2 and 3 by "SL". In sample B with InGaAs-GaAs SLS (In:x=0.06), step-like reduction of the dislocation density is observed at the superlattice. This is due to the termination of dislocation by the formation of looping at SLS's. The EPD profile of sample A with InGaAs-GaAs SLS (In:x=0.12) was the same as that of sample B. On the contrary, in the case of x=0.03, only a slight reduction of dislocation is observed. It seems that the force exerted by the misfit strain for x=0.06 is sufficiently large for the dislocations to bend. After passing through the superlattice position, the dislocation density reduces continuously with the thickness without saturation. The superlattice probably bends the threading dislocation and drive away toward the periphery. Consequently, the reduction of dislocation density by a factor of more than 3 at the thickness of 4μm is observed, compared with the annealed sample. InGaAs-GaAsP superlattice buffer layers also reduce the dislocation density by a factor of 2, and again the dislocation density continuously decreases with GaAs layer thickness (Fig.3).

The EPD profiles of samples F and G with InGaAs-GaAs SLS (In:x=0.12) and InGaAs-GaAsP SLS (In:x=0.06 As:y=0.88) respectively are shown in Fig. 4. These samples were annealed again after the growth of the SLS and the top GaAs layer. Further reduction of the dislocation has been achieved and the dislocation density reduced to $1\text{-}2\times10^6$ cm^{-2} at the thickness of 4μm. The strain is caused by the difference of the thermal expansion coefficients between Si and GaAs. The strain induced during the annealing at higher temperature than the growth seems to enhance the dislocation reduction effect of SLS's.

The comparative study of the EPD profiles shows that InGaAs-GaAs SLS's are more effective than InGaAsP SLS's. However, InGaAs-GaAs SLS's are not lattice-matched to the GaAs as a whole and total thickness of 1000Å is so great that cross-hatched morphology appears in the GaAs layer on the monitor GaAs substrate. And the undoped GaAs layers grown on the InGaAs-GaAs SLS's showed high resistivity, due to the deep traps related to the defects. In contrast, InGaAs-GaAsP SLS's are free from these problems, because the compressive strain and the tensile strain exactly cancel each other within each period of the superlattice. It seems that the InGaAs-GaAsP SLS is one of the promising candidates and further reduction of the dislocation can be expected by optimizing the superlattice structure.

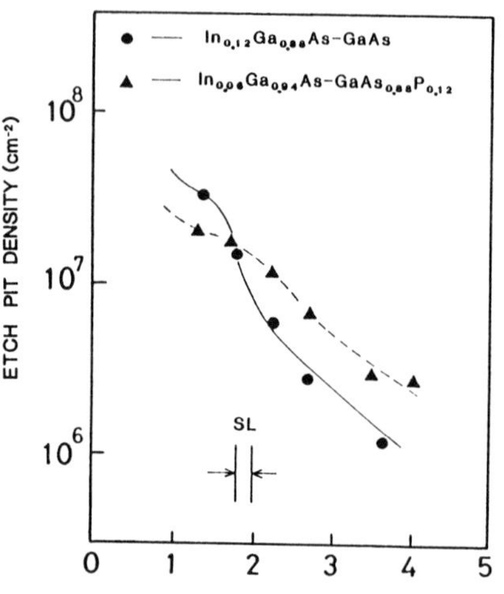

Fig.4. In depth EPD profiles of the GaAs/Si annealed after SLS growth.

The reason why the dislocation density at the 1.5μm of GaAs thickness varies from sample to sample is probably due to the different cleaning conditions of Si surface prior to growth. Low AsH_3 flow rate leads to low EPD background(see Table I), suggesting that the oxide reduction by Ga (Wright et al. 1980) is the main cleaning process and evaporation of Ga coming from susceptor is suppressed under excess arsenic pressure.

In conclusion, we have observed the dislocation reduction effect of the $In_x Ga_{1-x}As$-$GaAs_y P_{1-y}$ strained-layer superlattice grown on the annealed GaAs/Si substrate by means of depth profile measurement of the dislocation density using molten KOH etching. In spite of very thin SLS layer of 2000Å in total thickness these superlattices have strong effect on the reduction of dislocation density. By annealing after growth of SLS, the dislocation density is reduced to $1\text{-}2\times10^6 cm^{-2}$ as a result of step-like reduction at the superlattice and continuous decrease after passing through the superlattice.

REFERENCES

Fischer R, Henderson T, Klem J, Masselink W T, Kopp W, Morkoc H and Litton C W 1984 Electron. Lett. **20** 945.
Fischer R, Neuman D, Zabel H, Morkoc H, Choi C and Otsuka N 1986 Appl. Phys. Lett. **48** 1223.
Ishida K, Akiyama M and Nishi S 1987 Jpn. J. Appl. Phys. **26** L163.
Lee J W, Tsai H L and Matyi R J 1987 Appl. Phys. Lett. **50** 31.
Matthews J W and Blakeslee A E 1974 J. Cryst. Growth **27** 118.
Matthews J W, Blakeslee A E and Mader J 1976 Thin Solid Films **33** 253.
Matthews J W and Blakeslee 1977 J. Vac. Sci. Technol. **14** 989.
Metze G M, Choi H K and Tsaur B Y 1984 Appl. Phys. Lett. **45** 1107.
Nonaka T, Akiyama M, Kawarada Y and Kaminishi K 1984 Jpn. J. Appl. Phys. **23** L919.
Wright S and Kroemer H 1980 Appl. Phys. Lett. **36** 210.

Inst. Phys. Conf. Ser. No. 91: Chapter 3
Paper presented at Int. Symp. GaAs and Related Compounds, Heraklion, Greece, 1987

Photoluminescence studies of thin strained Ga(As,P) layers grown by metal organic vapor phase epitaxy

L Samuelson, M-E Pistol and M R Leys

Department of Solid State Physics, University of Lund, Box 118, S-221 00 LUND, Sweden

ABSTRACT: Single layers of strained Ga(As,P) have have been grown in between GaP by using Metal Organic Vapor Phase Epitaxy (MOVPE). From a comparison between the luminescence of thick and relaxed layers of Ga(As,P) and intra-well luminescence of thin and strained Ga(As,P) layers we can determine the effect of strain on the band edges of the ternary layer, which is found to be in good agreement with deformation potential theory. A novel effect of quantum confinement is observed in layers thinner than ~100 Å where quantum effects on higher-lying conduction bands reduce the zero-phonon transition probability in the indirect Ga(As,P) material.

1. INTRODUCTION

The possibility of growing structures of III-V compounds where the various layers have different compositions offers the opportunity of creating material structures with controlled potential steps at the interfaces between different layers and with special electronic properties (Esaki 1986). Greater ability to tailor the desired material properties is obtained when, due to lattice mismatch, strain or tension is deliberately introduced in one or several layers of a multi-layer structure, by which the various portions of the band structure are shifted corresponding to the strain response (strain tensor) of these states (Osbourn 1986). Using advanced epitaxial growth techniques, with which one can grow atomically sharp interfaces and layers with thicknesses corresponding to only a few atomic layers, a third means of controlling the electronic structures in multi-layer structures is offered, namely the quantum confinement effect (Bastard and Brum 1986). In this case spatial confinement of charge carriers replaces the three-dimensional density of states (DOS) of the bulk material by a step-wise DOS, the energy steps of which shift to higher energies with decreasing layer thickness.

In this paper we describe the optical properties of MOVPE-grown thin layers of Ga(As,P), laterally strained to the lattice constant of GaP. We mention how the strain affects the band edges of the ternary layer and that the observed shifts are in good agreement with a calculation based on deformation potential theory. Finally, we describe effects of quantum confinement in the Ga(As,P) layer where, besides the expected shift of the effective band gap with decreasing layer thickness, we also observe a previously not reported phenomenon where the effective phonon coupling, seen in the luminescence spectra, changes as a function of layer thickness. The reduction in the zero-phonon transition probability is interpreted as being due to a lower inmixing of Γ character in the wavefunction of band edge electrons when quantum confinement shifts the Γ_6 conduction band minimum.

2. EXPERIMENTAL

The GaP/Ga(As,P)/GaP double heterostructures studied here have been grown by atmospheric pressure MOVPE using trimethylgallium (TMGa), arsine (AsH_3) and phosphine (PH_3) as sources in a hydrogen ambient atmosphere. The structures were grown on (100)-oriented n-type (sulfur-doped, $\approx 5\cdot 10^{17}$ cm^{-3}) GaP substrates. The heterostructure region was grown at 715°C with a growth rate of 1 monolayer per second. Temperature and V/III ratio were chosen based on a separate study by Leys *et al.* (1987) on the conditions for growth of GaP, optimized in terms of morphological and doping ($n \leq 5\cdot 10^{15}$ cm^{-3}) properties. The heterostructures were grown without interruption of growth at the interfaces. Strained single Ga(As,P) layers were grown with the As fraction varying between 10% and 50%, and with layer thickness ranging from 11Å (corresponding to 4 atomic layers) to an upper limit of 50Å to 1300Å. The maximum thickness of the strained ternary layer is limited by the strain-dependent critical thickness for dislocation generation.

The As/P ratio was measured by SIMS and by EDAX. Transmission electron micrographs show (see Pistol *et al.* 1987) that the transitions between adjacent layers are achieved within one or two monolayers and that the interfaces are extremely flat. The luminescence properties reported here were obtained at 1.5-2 K with the luminescence excited by ~100 W/cm^2 of the 364 nm line from an Ar-ion laser and with the light detected by a GaAs photomultiplier.

3. RESULTS AND DISCUSSION

3.1 Effects of strain on the luminescence of Ga(As,P)

In the composition ranges reported on here, Ga(As,P) has an indirect band gap. Pure bulk Ga(As,P) is known to have a characteristic low- temperature PL feature, labeled M_0^x, which is dominated by the indirect nature of the conduction band and by the alloy composition dependence of band edges, as discussed by Samuelson *et al.* (1986). An example of the M_0^x luminescence is shown in Figure 1a for thick and relaxed $GaAs_{0.3}P_{0.7}$. The zero-phonon line is due to a quasi-direct recombination between holes at the top of the valence band and electrons at the X-point of the conduction band. This inmixing is believed to occur as the result of the perturbation by the disorder potential of the random alloy, by which Γ character is borrowed from the higher Γ_6 conduction band minimum.

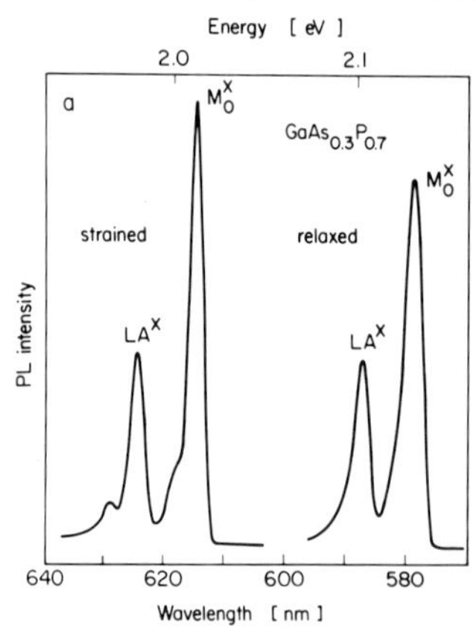

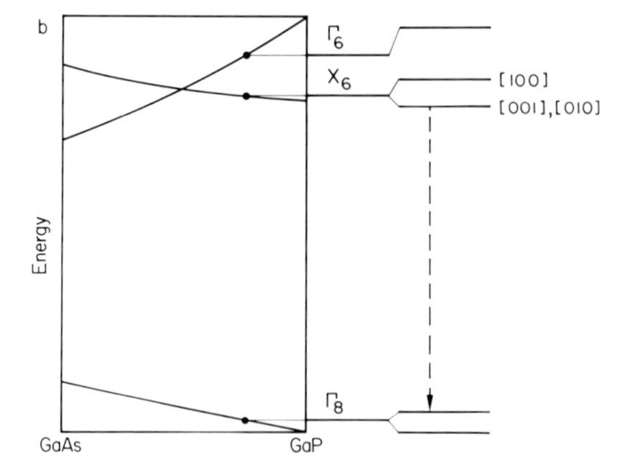

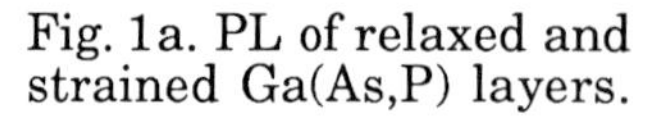

Fig. 1a. PL of relaxed and strained Ga(As,P) layers.

Fig. 1b. Qualitative effect of strain on points of high symmetry in the Ga(As,P) band structure.

Figure 1a also shows that the PL of a 120 Å thick strained layer of $GaAs_{0.3}P_{0.7}$ strongly resembles the $M_o{}^x$ line but is shifted to a lower energy by more than 150 meV. The origin of this shift is explained in Figure 1b, where we show how strain affects the different band extrema, forming wells for both holes and electrons. In a separate publication by Pistol *et al.* (1987) it is shown that deformation potential theory, with stress tensor components obtained from uniaxial stress experiments, can also account quantitatively for the strain-induced shifts. A detailed discussion about effects of alloying (Samuelson *et al.* 1986) and hydrostatic (Samuelson and Nilsson 1987) as well as tetragonal strain on band edges is, however, beyond the scope of the present paper.

3.2 Effects of quantum confinement on the luminescence of Ga(As,P)

The luminescence spectrum of the strained Ga(As,P) layer in Figure 1a is only weakly affected by quantum shifts, since the heavy mass of the electrons in the indirect minimum requires layers to be thinner than 100 Å for such effects to be of significance. In Figure 2a we show three PL spectra obtained from samples with identical alloy composition, and hence with the same amount of strain, but with the thickness varying from 11 Å to 120 Å. It is seen that quantum well confinement shifts the recombination energy by about 0.13 eV, almost half the energy difference between the alloy and the GaP in which the alloy is contained. As discussed by Pistol *et al.* (1987), this is in good agreement with estimates from the well depths and carrier masses involved. Figure 2b illustrates the electronic band structures of GaP and of strained Ga(As,P), with the resulting potential wells created for electron states having X_6 and Γ_6 characters and for holes at the Γ_8-valence band maximum. These wells are schematically drawn for a thick well and for a thinner well, which could correspond to the 50 Å and 11 Å wells studied experimentally.

With decreasing thickness of Ga(As,P) one observes a change in the spectral shape. The intensity ratio of the phonon-assisted to the zero-phonon peaks increases from approximately 0.4 for the 120 Å layer, to ≈0.5 for the 50 Å layer and to ≈0.7 for the 11 Å well case. Since both TEM images and the energy shifts of the PL peaks suggest that even the thinnest wells have close to ideal compositional profiles, we suggest that also the variation of the phonon coupling is due to quantum confinement, but of a type which has not been discussed previously.

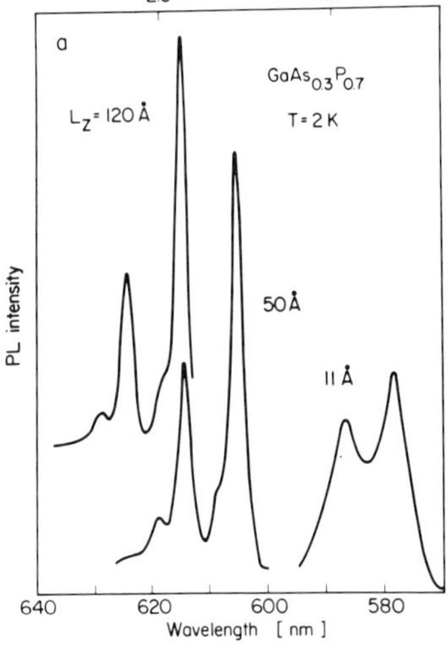

Fig. 2a. PL of Ga(As,P) wells of constant composition but varying thickness.

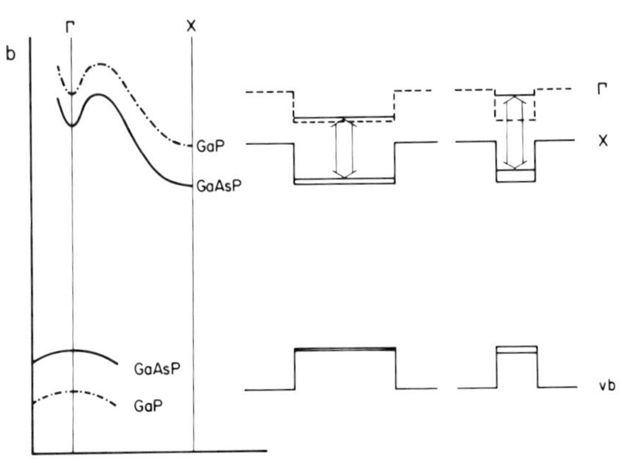

Fig. 2b. Schematic band structure of GaP and strained Ga(As,P) with resulting potential wells for electrons and holes.

In the case of recombination of donor-acceptor pairs and of excitons bound to isolelectronic traps in indirect gap materials such as GaP, Morgan (1968) and Dean (1970) have suggested that the intensity of the zero-phonon line is proportional to the amount of Γ character mixed into the electron wavefunction by the defect potential, V. This admixture of Γ character can formally be written as: $\psi = \alpha_X\varphi_X + \alpha_\Gamma\varphi_\Gamma$, with the different φ characters mixed in by the different α factors. Obviously, $\alpha_X >> \alpha_\Gamma$ since the electron is located close to the X-point of the Brillouin zone. However, the zero-phonon transition is allowed only for the φ_Γ portion, which can be estimated using perturbation theory as: $\alpha_\Gamma \propto V/(E_\Gamma - E_X)$.

In Figure 2b the energy denominator can be identified as the minimum energy difference between the Γ and X conduction band minima. For bulk conditions this amounts to the energy difference between the Γ_6 and X_6 points in the band structure. In the case of a quantum well structure (as for the thin well drawn in Figure 2b) one has to add the difference in confinement energies of the wells for Γ and for X electrons. Since the effective masses of electrons in these minima differ by a factor of at least 10, and since the carrier confinement energy, to a first approximation, scales inversely proportionally to the masses, we expect the energy denominator (which enters squared in transition probability expressions) to increase rapidly with decreasing well width. Since the α_X portion should not change appreciably with well width, the model describes qualitatively the well-width dependence of the intensity ratio between zero-phonon and phonon-assisted PL transitions.

4. CONCLUSIONS

High-quality single quantum well structures of strained Ga(As,P)/GaP have been grown by MOVPE. Effects of strain are in good agreement with deformation potential theory. Quantum confinement is found to give rise to both spectral shifts and to variations in the effective phonon coupling in luminescence. This latter effect is interpreted as being due to confinement in higher conduction band edges, reducing the amount of Γ-character coupled into the wavefunction of electrons in the indirect gap alloy. We believe this to be the first manifestation of such an indirect effect of quantum confinement.

5. ACKNOWLEDGEMENTS

This work was supported by grants from the National Swedish Board for Technical Development and by the Swedish Natural Science Research Council. Special thanks to H Titze for assistance with crystal growth.

REFERENCES

Bastard G and Brum J A 1986 *IEEE* ***QE-22*** * *1625* (and references therein)
Dean P J 1970 *J. of Luminescence* ***1, 2*** *398*
Esaki L 1986 *IEEE* ***QE-22*** * *1611* (and references therein)
Leys M R, Pistol M-E, Titze H and Samuelson L 1987 to be published
Morgan T N 1968 *Phys. Rev. Lett.* ***21*** *819*
Osbourn G C 1986 *IEEE* ***QE-22*** * *1677* (and references therein)
Pistol M-E, Leys M R and Samuelson L 1987 submitted to *Phys. Rev.* ***B***
Samuelson L, Pistol M-E and Nilsson S 1986 *Phys. Rev.* ***B33*** *8776* (and references therein)
Samuelson L and Nilsson S 1987 *Proc 1987 Int Conf. on Luminescence, Beijing, to be published in J. of Luminescence*

*) Special issue on Physics and Applications of Semiconductor Quantum-Well Structures.

Investigation of DX-center-free selectively doped GaAs/AlGaAs heterostructures

T. Ishikawa and K. Kondo

FUJITSU LIMITED
10-1 Wakamiya, Morinosato, Atsugi 243-01, Japan

Abstract. Modified selectively doped GaAs/N-AlGaAs heterostructures with undoped AlGaAs barrier layers at the heterointerface have been investigated to obtain DX-center-free HEMTs with high performance. The AlAs mole fraction, x, in the N-AlGaAs layers was fixed to 0.15 to eliminate the influence of DX centers. In this case, two-dimensional electron gas characteristics were improved by introducing barrier layers. By optimizing the barrier layer structure, HEMTs with good performance and no drain current collapse were successfully fabricated.

1. Introduction

The DX centers in N-$Al_xGa_{1-x}As$ layers degrade the performance of high electron mobility transistors (HEMTs). The most serious problem is the collapse in drain current at low temperatures. This prevents the stable operation of HEMTs and must be eliminated to achieve a HEMT-IC which can operate reliably at 77 K. To avoid the influence of DX centers, some novel structures, including an n-GaAs/AlAs superdoped structure (Baba et al. 1984) and an $Al_{0.15}Ga_{0.85}As/In_{0.15}Ga_{0.85}As$ pseudomorphic structure (Ketterson et al. 1985), have been proposed. However, these structures are rather complicated for IC fabrication. It is thus very important to obtain high-performance HEMTs fabricated on simple DX-center-free selectively doped GaAs/AlGaAs heterostructures. In this work, we examined the two-dimensional electron gas (2DEG) characteristics of heterostructures with N-$Al_{0.15}Ga_{0.85}As$ layers and thin undoped AlGaAs barrier layers, and also the characteristics of HEMTs fabricated on such heterostructures.

2. DX centers and drain current collapse

Reports on DX centers indicate that n-type dopants in $Al_xGa_{1-x}As$ layers form two types of donors; shallow donors mainly in the composition region below x=0.2, and DX centers above x=0.3 (Watanabe et al. 1984, Schubert et al. 1984). These conclusions are based on experiments using lightly doped (about 1×10^{17} cm^{-3}) N-$Al_xGa_{1-x}As$ layers. We recently investigated the dependence of shallow donors and DX centers on Si doping concentrations and showed that a considerable number of DX centers were formed even in the low Al content region around x=0.2 by heavy doping above 1×10^{18} cm^{-3}, which is common in HEMTs (Ishikawa et al. 1986). Consistent with this result, the collapse of drain current at 77 K was observed in an HEMT having this composition. Figure 1 shows the operating-time dependence of the drain current in the HEMTs at 77 K. The drain current at V_{DS} = 0.1 V is normalized by its initial value. Even in this HEMT with an N-$Al_xGa_{1-x}As$ of x=0.19, as shown by open circles, the drain current collapsed after several seconds of operation. The detailed mechanism of the collapse is not yet clear, but is apparently caused by DX centers (Kastalsky et al. 1986). The AlAs mole fraction x in the Si-doped layers of HEMTs must thus be reduced to less than 0.15, where DX center concentration becomes negligibe (Ishikawa et al. 1986).

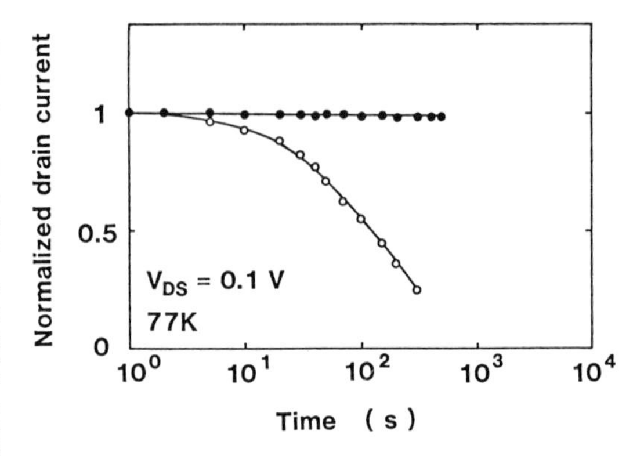

Fig. 1 Normalized drain current in HEMTs at 77 K as a function of operating time. The open circles represent a conventional HEMT with x=0.19, and the solid circles a modified one with x=0.15.

3. 2DEG characteristics

In this section, we discuss the 2DEG characteristics of selectively doped heterostructures with various AlAs mole fractions of the Si doped layer, and their improvement by introducing barrier layers. In Figure 2, the mobility and sheet concentration of 2DEG as a function of the AlAs mole fraction are shown by open circles. The samples were grown by conventional MBE. Undoped spacer layers are 60 Å thick. With a reduced AlAs mole fraction of less than 0.2, mobility was greatly reduced and electron concentration was decreased slightly. This reduction of mobility is due to the increased penetration of the 2DEG wave function into the low mobility AlGaAs layers, because of a low barrier height at the hetero-interface. For a DX-center-free composition of x=0.15, the mobility was as low as 20,000 cm^2/Vs, even with a 60-Å spacer.

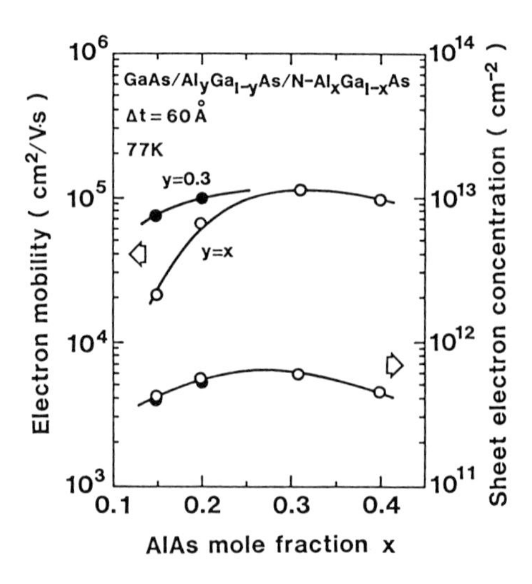

Fig. 2 2DEG characteristics at 77 K in a conventional selectively doped heterostructure (open circles) and in the modified structure (solid circles) as a function of the AlAs mole fraction in the Si-doped layers.

To prevent the penetration of 2DEG into AlGaAs layers having a low Al content, thin undoped AlGaAs barrier layers were introduced at the hetero-interface. Figure 3 shows the modified structure used in this experiment. The mole fraction of Si doped layers, x, was decreased to reduce DX center concentration. The mole fraction of the undoped barrier layers, y, was larger than x to prevent the penetration of the 2DEG wave function. The solid circles in Figure 2 show the mobility and sheet concentration of 2DEG in such modified structures, in which the mole fraction of the barrier layers was fixed to 0.3 and the thickness to 60 Å. By introducing an AlGaAs barrier layer, 2DEG mobility was improved. This indicated that barrier layers are very effective in preventing penetration of 2DEG into the AlGaAs layers. Mobility was especially improved for x=0.15, which showed a value as high as 70,000 cm^2/Vs.

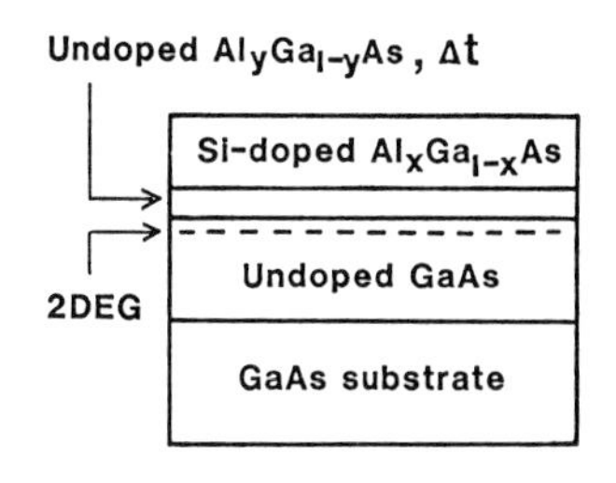

Fig. 3 Structure of this experiment.

Since a high carrier concentration is important for a practical application of HEMTs, we examined the 2DEG characteristics in this modified structure of x=0.15 having a thinner barrier thickness. Figure 4 shows the mobility and sheet concentration of 2DEG at 77 K in the modified structures as a function of barrier layer thickness. For comparison, the data of conventional structures with an AlAs mole fraction of x=y=0.30 is shown using broken lines. Although the electron concentration was less in the modified structure compared to conventional structures, relatively high mobility was obtained. For a thickness of 20 or 30 Å, the barrier layers seem to be effective, and electron mobility of more than 40,000 cm^2/Vs with a sheet concentration of more than $6x10^{11}$ cm^{-2} was obtained.

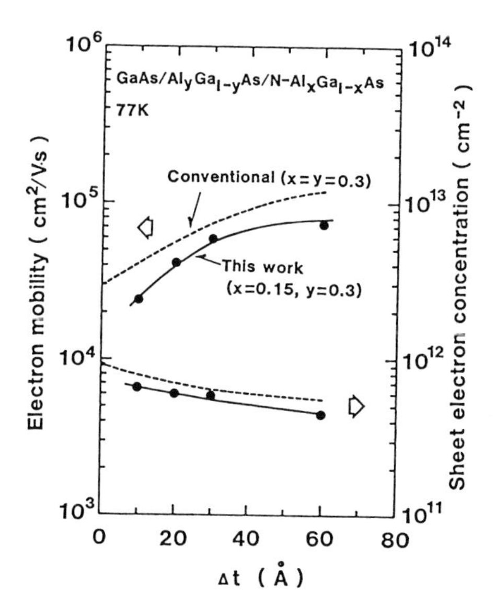

Fig. 4 2DEG characteristics at 77 K in the modified structure as a function of barrier layer thickness. The broken lines represent the data for a conventional structure.

4. HEMT characteristics

The HEMTs fabricated using the modified structure with barrier layers had good performance. Table 1 lists the performance values at room temperature and at 77 K without light for conventional and modified HEMTs. Their structural parameters are also given. To increase carrier concentration, a barrier layer thickness of 30 Å was used.

	Structural parameters			300K		77K	
	Δt (Å)	x	y	g_m (mS/mm)	K (mA/V^2·mm)	g_m (mS/mm)	K (mA/V^2·mm)
This work	30	0.15	0.30	170	170	230	540
Conventional	30	0.30	0.30	180	170	260	570

Table 1 HEMT characteristics of the modified and conventional structures.

The gate was 1.2 μm long and 50 μm wide. Good performance comparable to that of the conventional structure was obtained for the modified structure. With decreased temperature, the K-value showed excellent enhancement in the both structures; this may be due to enhanced mobility at low temperatures. This modified structure also prevents drain current collapse because of a low AlAs mole fracton in the Si-doped layer. The time dependence of the normalized drain current at 77 K for the same HEMT as in Table 1 is shown by the solid circles in Figure 1. The drain current was stable even after 500 seconds of operation. This structure is thus promising for HEMT ICs with good reliability at low temperatures.

5. Summary

In summary, the 2DEG mobility in selectively doped heterostructures with an Al content of less than x=0.2 was improved. This was accomplised by introducing AlGaAs barrier layers at the hetero-interface. Good performance without drain current collapse was obtained in the modified HEMTs with a Si-doped layer of x=0.15.

Acknowledgments

The authors thank T. Yokoyama, M. Suzuki and T. Mimura for their valuable discussions and T. Maeda and T. Yamamoto for their work on MBE growth. They also thank M. Abe and M. Kobayashi for their support. The present research effort is the National Research and Development Program on the "Scientific Computing System," conducted under a program set by the Agency of Industrial Science and Technology, Ministry of International Trade and Industry, Japan.

References

Baba T, Mizutani T, Ogawa M and Ohta K 1984 Jpn. J. Appl. Phys. 23 L654.
Ishikawa T, Yamamoto T, Kondo K, Komeno J and Shibatomi A 1986 Inst. Phys. Conf. Ser. 83 99.
Kastalsky A and Kiehl R A 1986 IEEE Trans. Electron Devices ED-33 414.
Ketterson A, Moloney M, Masselink W T, Peng C K, Klem J, Fischer R, Kopp W and Morkoc H 1985 IEEE Electron Device lett. EDL-6 628.
Schubert E F and Ploog K 1984 Phys. Rev. B. 30 7021.
Watanabe M O, Morizuka K, Mashita M, Ashizawa Y and Zohta Y 1984 Jpn. J. Appl. Phys. 23 L103.

Evidence for the simple substitutional nature of the DX centre in GaAs

L Eaves+, T J Foster+, D K Maude+, G A Toombs+, R Murray°, R C Newman°, J C Portal', L. Dmowski', R B Beall#, J J Harris#, M I Nathan* & M Heiblum*

+Department of Physics, University of Nottingham, NG7 2RD, UK.
°J J Thomson Physical Laboratory, University of Reading, RG6 2AF, UK.
'Dept. de Genie Physique, INSA, 31077 Toulouse and SNCI-CNRS, Avenue des Martyrs, 38042 Grenoble, France.
#Philips Research Laboratories, Redhill, Surrey RH1 5HA, UK.
*IBM, T J Watson Research Center, Yorktown Heights, NY 10598, USA.

ABSTRACT: Previous measurements have shown that, under hydrostatic pressure, DX-like centres trap out free electrons in Si-doped GaAs. Their concentration is comparable to the Si doping level. Two models have been proposed for this centre, a Si-defect complex or a simple substitutional donor, Si_{Ga}. Local vibrational mode measurements on such a sample show that 90% of the Si is incorporated as Si_{Ga}, with no evidence of Si-defect complexes.

The DX centre in n-(AlGa)As has been extensively studied for many years [1], but the exact nature of this level remains controversial. It has the character of a deep donor and is present in all n-type (AlGa)As at a concentration close to that of the dopant impurity, which can be from group IV or VI. A microscopic potential barrier to electron capture at DX levels (e.g. arising from lattice relaxation) has been proposed to explain the well-known persistent photoconductivity (PPC) observed in n-(AlGa)As at low temperatures, though it should be noted that PPC observed in some GaAs and (AlGa)As layers can be quantitatively explained by macroscopic separation of carriers [2]. Although DX has not been observed previously in GaAs at atmospheric pressure, Mizuta et al [3] have used DLTS measurements to show that such levels become occupied in GaAs and low Al content (AlGa)As under hydrostatic pressure. The critical pressure at which these levels become occupied decreases with increasing Al fraction, x; for $x > 0.22$ the level is occupied at atmospheric pressure, and appears to follow the L-conduction band edge in the direct gap alloy composition range [4,5,6]. Several alternative models have been proposed for the chemical nature of the level giving rise to the PPC in (AlGa)As [1,7-9]. In particular, it has been suggested recently that PPC is associated with states arising from simple substitutional donor impurities [3,8], rather than from defect-impurity complexes [1]. We present here a study of heavily-doped GaAs which strongly supports the substitutional donor model.

We have previously reported [10] the results of Shubnikov-de Haas (SdH) measurements made under hydrostatic pressure on a range of heavily Si- and Sn-doped GaAs samples, and observed the onset of carrier freeze-out at critical pressures which decreased with increasing doping level.

This behaviour is ascribed to the presence of DX-like centres whose energy is lowered relative to the Γ-minimum with increasing pressure. Freeze-out occurs when this level falls below the Fermi level. So less pressure is required to start depopulating the more heavily doped material.

Of particular interest is the most heavily Si-doped layer, prepared by MBE at the Philips Research Laboratories, using a low growth temperature of 400°C to increase the solubility limit of Si on the donor (Ga) sites [11]. The carrier density in this sample (thickness d = 11.8 μm) is 1.1×10^{19} cm^{-3}, compared with a value of 1.3×10^{19} cm^{-3} expectedfrom the calibrated Si flux used for growth, and 1.4×10^{19} cm^{-3} as measured by SIMS. The carrier density and mobility deduced from SdH measurements on this layer as a function of hydrostatic pressure are shown in Figure 1, together with the calculated carrier density variation (solid line) based on standard semiconductor statistics and a fitted pressure dependence for the DX centre of -9.4 meV/kbar relative to the Γ-minimum [12]. Also shown is the effect of illuminating the sample after cooling to 4.2 K in the dark under a pressure of 9 kbar. A persistent increase in carrier density, up to precisely the same level measured at zero pressure is observed, accompanied by a significant drop in mobility. These results are completely consistent with the presence in this sample of a donor-like DX centre at a concentration close to the total Si content. Note that the sheet carrier concentration (10^{16} cm^{-2}) is about 10^4 times larger than that employed in recent studies of the growth and decay kinetics of PPC in n-GaAs and (AlGa)As which have been interpreted in terms of macroscopic separation of photogenerated carriers [2]. We believe that both models for the PPC have physical validity under appropriate experimental conditions.

In order to investigate the microscopic nature of this centre, we have performed Fourier Transform Infrared (FTIR) absorption measurements on this sample to determine the local vibrational mode (LVM) energies of the Si impurities. The samples are irradiated with 2 MeV electrons to eliminate free carrier absorption before taking the spectra. These measurements give detailed information about the atomic location of the impurities, since the vibrational frequencies depend on the local force constant and the masses of neighbouring host lattice atoms or other impurities, as well as the mass of the impurity itself. This is illustrated in Figure 2a for heavily Si-doped bulk GaAs, where seven Si-related absorption lines are present. These are interpreted in the following way: in lightly-doped material ([Si] < 5×10^{17} cm^{-3}), most Si atoms are present as Si_{Ga}, giving LVM lines at 384, 379 and 373 cm^{-1} for ^{28}Si (92.3% abundant), ^{29}Si (4.7%) and ^{30}Si (3%) respectively. At higher levels, absorption is also observed at 399 cm^{-1} from Si_{As}. Yet higher doping levels produce LVM lines at 464 and 393 cm^{-1} from adjacent Si_{Ga}-Si_{As}. (The former line, due to the longitudinal mode of the pair, is not shown in Figure 2a). As the doping level increases, the carrier concentration reaches a limiting value around 4×10^{18} cm^{-3}; this is the situation in the sample shown here. At this stage, other LVM lines are produced at around 370 cm^{-1} from two distinct complexes, Si-X and Si-Y, of unknown structure [13]. Si_{Ga}-Si_{As} pairs should be electrically neutral, but there is evidence that Si-X is an acceptor, while Si-Y is a donor [14]. The defects X and Y may be intrinsic, but it is also necessary to consider the possibility of clusters such as Si_{As}-Si_{Ga}-Si_{As}, which may also absorb around 370 cm^{-1}.

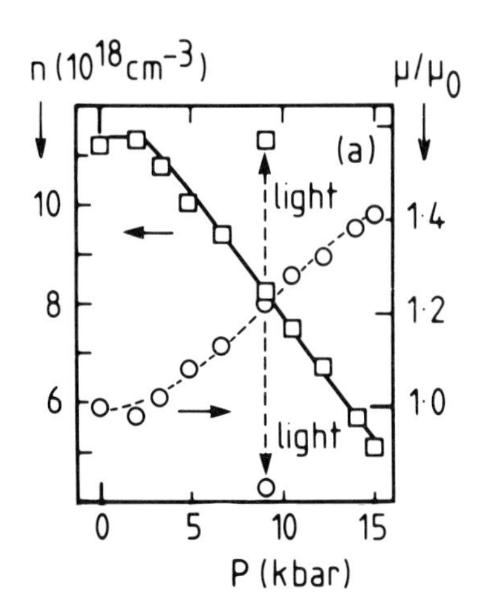

Figure 1. Carrier concentration n (open squares) and mobility μ (open circles) vs. pressure P for the MBE layer heavily doped with Si. The mobility is normalised to its atmospheric value. The solid lines show the expected variation of n(P) - see text. The persistent increase of n and decrease of μ induced by light at 9 kbar are indicated by arrows.

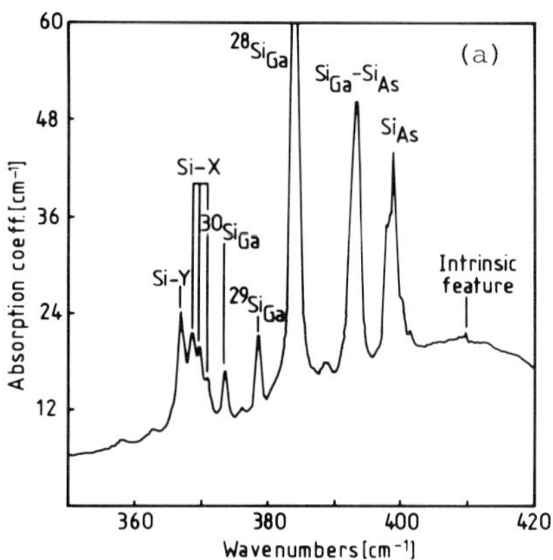

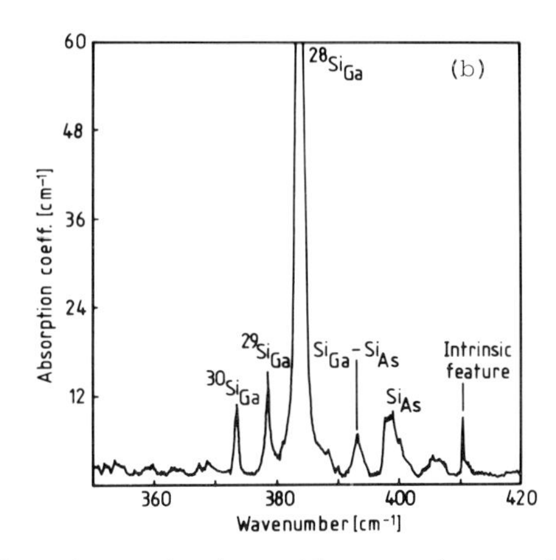

Figure 2. (a) Far infrared local vibration mode absorption spectrum of Si doped bulk GaAs - see text. (b) Corresponding spectrum for the Si-doped MBE layer.

Figure 2b shows the same spectral region for the MBE-grown sample described above. Despite the high doping level there is no detectable absorption from Si-X or Si-Y defects, although Si-X has been observed in layers grown at higher temperatures [15]. The concentrations of impurities giving rise to the remaining LVM lines are:

Line	Position (cm^{-1})	Concentration ($\times 10^{19}$ cm^{-3})
Si_{Ga}	384	1.25 ± 0.21
Si_{Ga}-Si_{As}	393	0.06 ± 0.01
Si_{As}	399	0.13 ± 0.02

where the calibration of Woodhead et al [16] for Si_{Ga} has been used, and it has been assumed that the same calibration holds for the other lines. These LVM results indicate a total Si content of 1.44×10^{19} cm^{-3}, in good agreement with SIMS and the intended Si doping level. The carrier concentration, assuming the Si impurities to be the only electrically active species, is $n = [Si_{Ga}]-[Si_{As}] = 1.12 \times 10^{19}$ cm^{-3}, in excellent agreement with the SdH results (1.13×10^{19} cm^{-3} at P = 0).

This is very strong evidence that the DX centres observed in this sample under pressure are not due to a Si complex, since we have shown that ~90% of Si is present as the substitutional donor Si_{Ga}; the remaining 10% is much smaller than the reduction in carrier density observed under hydrostatic pressure.

Finally, we note the recent work of Nabity et al [17], who demonstrated that exposure to a hydrogen plasma passivates both shallow donors and DX centres in $Al_{0.3}Ga_{0.7}As$ grown by MBE. Subsequent heat treatments result in a recovery of both defects at around 400°C with the same activation energy (2.0 eV). These observations are fully consistent with the above conclusions, namely that the DX centres are simply isolated Si impurities, perhaps with a lattice distortion from T_d symmetry. Complexing with hydrogen to form Si-H pairs reduces the activity of all Si_{Ga} while annealing around 400°C breaks the Si-H bond and both shallow donors and DX centres are regenerated [18].

This work is supported by SERC and CNRS. L Dmowski was on leave from the Polish Academy of Sciences, Warsaw. We wish to thank J B Clegg for the SIMS measurements and P E Simmonds for useful discussions.

REFERENCES

[1] Lang D V and Logan R A 1977, Phys Rev Lett **39** 653; see also Nelson R J 1977, Appl Phys Lett **31** 351.
[2] Queisser H J and Theodolou D E 1979, Phys Rev Lett **43** 401; 1986 Phys Rev B**33** 4027, and references therein.
[3] Mizuta M, Tachikawa M, Kukimoto H and Minomura S 1985, Jap J Appl Phys **24** L143.
[4] Theis T N 1986 "Defects in Semiconductors", ed H J von Bardeleben, Material Science Forum **10-12** p393.
[5] Tachikawa M, Mizuta M, Kukimoto H and Minomura S 1985, Jap J Appl Phys **24** L821.
[6] Theis T N and Wright S L 1986, Appl Phys Lett **48** 1374.
[7] Saxena A K 1982, Sol St Electr **25** 127.
[8] Hjalmarson H P and Drummond T J 1986, Appl Phys Lett **48** 656.
[9] Henning J C M and Ansems J P M 1987, Semicon Sci Technol **2** 1.
[10] Maude D K, Portal J C, Dmowski L, Foster T, Eaves L, Nathan M, Heiblum M, Harris J J and Beall R B 1987, Phys Rev Lett **59** 815.
[11] Neave J H, Dobson P J, Harris J J, Dawson P and Joyce B A 1983, Appl Phys A**32** 195.
[12] Portal J C, Maude D K, Foster T J, Eaves L, Dmowski L, Nathan M, Heiblum M, Harris J J, Beall R B and Simmonds P E, Proc 3rd Int Conf on Superlattices, Microstructures and Microdevices, Chicago 1987, to be published in Superlattices & Microstructures.
[13] Brozel M R, Newman R C and Ozbay B 1979, J Phys C**12** L785.
[14] Chen R T, Rana V and Spitzer W G 1980, J Appl Phys **51** 1532.
[15] Maguire J, Murray R, Newman R C, Beall R B and Harris J J 1987, Appl Phys Lett **50** 516.
[16] Woodhead J, Newman R C, Tipping A K, Clegg J B, Roberts J A and Gale I 1985, J Phys D**18** 1575.
[17] Nabity J C, Stavola M, Lopata J, Dautremont-Smith W C, Tu C W and Pearton S J 1987, Appl Phys Lett **50** 921.
[18] Pajot B, Newman R C, Murray R, Jalil A, Chevallier J and Azoulay R 1987, submitted to Phys Rev B.

Inst. Phys. Conf. Ser. No. 91: Chapter 4
Paper presented at Int. Symp. GaAs and Related Compounds, Heraklion, Greece, 1987

The role of DX centres in limiting the free carrier density in GaAs

P. M. Mooney, T. N. Theis, and S. L. Wright

IBM T.J. Watson Research Center, PO Box 218, Yorktown Heights, NY 10598 USA

ABSTRACT: The DX center can trap electrons, even when it lies above the bottom of the conduction band. At high free carrier densities, when the Fermi level approaches the trap level, electrons enter the trap state. Thus trapping by DX centers acts to limit the carrier concentration independent of other mechanisms. Using DLTS we have observed this effect in Si-doped GaAs with carrier concentrations $\gtrsim 6\times10^{18}$ cm^{-3}. The results suggest a maximum attainable equilibrium free carrier concentration $\sim 2\times10^{19}$ cm^{-3} in n-GaAs.

1. INTRODUCTION

The electrical properties of n-type $Al_xGa_{1-x}As$ (x>0.22), including the persistent photoconductivity (PPC) observed at low temperatures, are controlled by deep donors known as DX centers (Lang et al 1979). DX centers have been observed for all n-type substitutional dopants in concentrations approximately equal to the donor concentration, independent of the method or conditions of epitaxial growth. Hall measurements show that the deep level follows the L minimum of the conduction band (Chand et al 1984) as shown in Figure 1a. Extrapolation of the data to alloy compositions with x<0.22 suggests that in GaAs it lies about 170 meV above the Γ minimum of the conduction band. Both emission and capture of an electron at this deep level are thermally activated processes. The emission energy, E_e, is independent of the alloy composition and doping concentration, but the capture energy, E_c, increases rapidly with decreasing AlAs mole fraction in the direct gap alloy (Mooney et al 1986). Thus the phenomenological barrier for $Al_{0.4}Ga_{0.6}As$ (Figure 1b) extrapolates smoothly to the barrier of a metastable state in GaAs (Figure 1c) (Theis 1986).

The DX center has been observed in GaAs with applied hydrostatic pressure using capacitance transient spectroscopy (DLTS) techniques (Mizuta et al 1985). The application of hydrostatic pressure modifies the conduction band structure in much the same way as increasing the AlAs mole fraction does. Thus above a critical pressure the DX level, which moves with the L minimum of the conduction band, is pushed below the Γ minimum. The activation energy for thermal emission was found to be smaller in GaAs than in the alloy, but the thermally activated capture behavior which is characteristic of the DX center was observed. The concentration of the DX center was equal to the donor concentration. PPC, the other signature of the DX center, was also observed in GaAs under pressure (Tachikawa et al 1985).

It has been demonstrated that the DX center can trap hot electrons even in $Al_{0.14}Ga_{0.86}As$ where the level lies above the bottom of the conduction band (Theis et al 1986). The emission kinetics of the resonant level were observed to be unchanged from those at higher Al mole fractions where the level lies in the direct gap. We have now studied the thermal emission and capture kinetics of the DX center in both $Al_xGa_{1-x}As$ (x<0.22) and GaAs using capacitance spectroscopy (DLTS). We demonstrate not only that the DX center maintains its ability to trap charge even when it is energetically resonant with the conduction band, but also that by trapping electrons it significantly reduces the free carrier density in heavily doped GaAs.

2. EXPERIMENT

Large area modulation doped FETs fabricated next to the small devices used for the hot electron studies (Theis et al 1986) were measured by DLTS. The net donor concentration was 1×10^{18} cm^{-3}. The emission energy and the shape of the DLTS peak is the same as was observed in samples grown under similar conditions but with x>0.22 (Mooney et al 1986). The thermally activated capture characteristic of the DX center is also observed. Filling times of about 1 second at T~200 K are needed to saturate the trap due to the very large capture barrier at x=0.14. The small amplitude of the DLTS peak indicates a very low equilibrium concentration of *occupied* DX centers, ~ 1×10^{15} cm^{-3}. However, increasing the free carrier concentration and thus pushing the Fermi level higher into the conduction band would increase the occupation of the DX level. The result would be a larger DLTS signal as found by Ishikawa et al (1986). We show here that at sufficiently high free carrier concentrations the DX center can be observed in GaAs without using hydrostatic pressure.

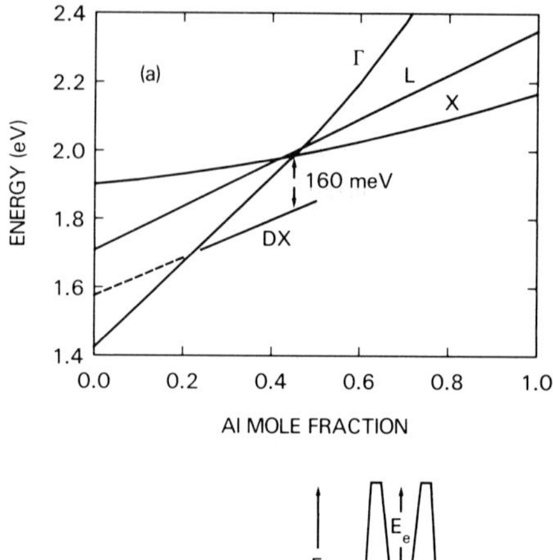

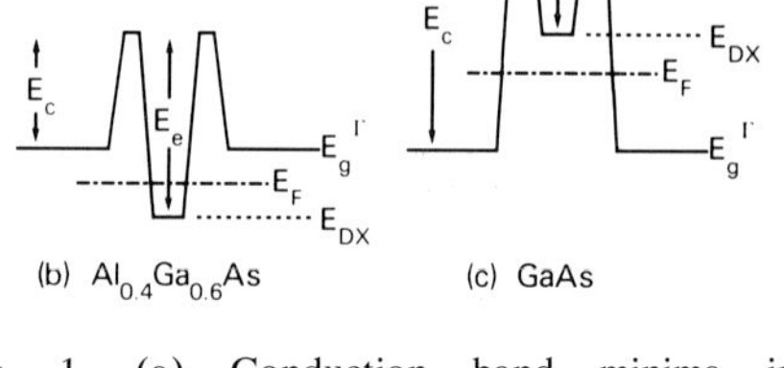

Fig. 1. (a) Conduction band minima in $Al_xGa_{1-x}As$ as a function of x. Effective potential barriers for capture to and emission from the DX center (b) when x=0.4 and (c) in GaAs.

A series of GaAs samples with increasing free carrier concentrations was grown by molecular beam epitaxy (MBE). The depletion region of a heavily doped Schottky contact would normally be so thin that excessive tunneling currents would prevent capacitance measurements from being made. To avoid this problem a 25 nm layer of undoped GaAs was grown upon each heavily doped GaAs layer, and the Schottky contact was made to the undoped layer, thereby reducing the tunneling current. Each layer was grown on a conducting GaAs substrate so that a AuGe/Ni/Au alloy contact could be made to the back side of the substrate. Sample characteristics are listed in Table I. The Si concentration, N_{Si}, was measured by secondary ion mass spectroscopy (SIMS), the net donor concentration, N_D-N_A, was determined from capacitance vs. voltage measurements (C-V), and the free carrier concentration, n_o (300 K), was determined from the Raman shift of the bulk plasmon frequency as discussed by Theis et al (1987). $\Delta C/C$, the relative amplitude of the DLTS capacitance transients is also listed for each sample. The transients were obtained with forward bias (filling condition) and reverse bias (emptying condition) voltages of +0.5 V and -1.0 V, respectively.

TABLE I: Characteristics of the heavily doped GaAs samples.

Sample	T_g (°C)	N_{Si}(cm^{-3})	$N_D - N_A$(cm^{-3})	n_o(cm^{-3})	$\Delta C/C$
1	570	4.3x10^{19}	6.0x10^{18}	6.0x10^{18}	1.4x10^{-4}
2	570	2.3x10^{19}	1.0x10^{19}	8.4x10^{18}	2.4x10^{-4}
3	530	2.2x10^{19}	1.1x10^{19}	1.0x10^{19}	5.2x10^{-4}
4	530	1.9x10^{19}	1.1x10^{19}	1.1x10^{19}	5.8x10^{-4}

DLTS spectra for the GaAs samples are shown in Figure 2. As was found for GaAs under pressure (Mizuta et al 1984) the peak is shifted to lower temperature and is narrower than in the alloy. The emission transient is exponential, consistent with the absence of alloy broadening (Calleja et al 1986) and demonstrating strong spatial localization of the state which is not measurably perturbed by neighboring Si atoms (donors or acceptors) at an average distance ~ 35 Å. The thermal emission energy, 0.33 eV, agrees exactly with that found in lightly doped GaAs under pressure. This is expected since in the alloy the emission energy is independent of both the alloy composition (and hence the band structure) and the DX concentration. A thermally activated capture rate was observed, leaving no doubt that this is indeed the DX center.

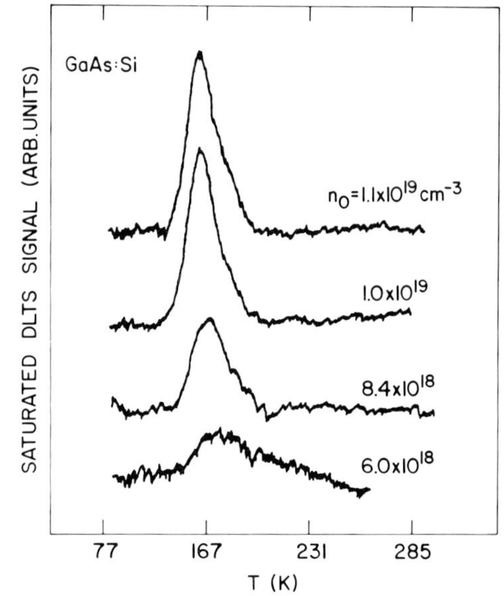

Fig. 2. DLTS spectra for the DX center in GaAs. The rate window is 91.7 Hz.

3. DISCUSSION

Since care was taken to use filling times sufficiently long to saturate the DLTS signals, $\Delta C/C$ is a measure of the equilibrium occupation of the DX level in each sample. However, because of the limited bias voltages applicable to these samples, the depletion depth in each heavily doped layer was very small, making the abrupt depletion approximation invalid. Thus a numerical calculation was used to relate $\Delta C/C$ to the equilibrium occupation of the DX level in the charge neutral interior of each heavily doped layer (Theis et al 1987). Poisson's equation was solved to calculate the sample capacitance at reverse bias with the traps filled and with the traps empty. From the experimental capacitance difference the energy difference between the Fermi level and the DX level was determined. The energy difference between the Fermi level and the Γ band edge was determined with the aid of our nonparabolicity model.

The results are shown in Figure 3. The dotted line indicates the Fermi level, E_F, at T=160 K obtained from our nonparabolicity model, and the data points indicate the energy of the DX level, E_{DX}, required to explain the measured values of $\Delta C/C$. The position of the Fermi level and the DX level in each sample are also shown. The energy of the DX level with respect to the Γ band edge is found to be considerably larger than the value of 170 meV predicted from Figure 1, and to increase with the free carrier density. This upward movement of the level with respect to the band edges is

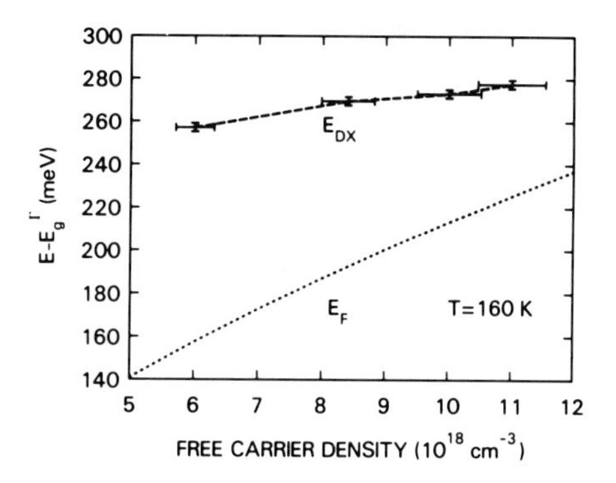

Fig. 3. Energy of the DX level, E_{DX}, and the Fermi level, E_F, with respect to the conduction band edge in GaAs at T=160 K as a function of the free carrier density. Error bars indicate the reproducibility of the measurements. Possible systematic errors are not included.

also found in hydrostatic pressure experiments (Maude et al 1987). The fractional occupation of the DX level is small in these samples, reaching a maximum of 6 percent of the net donor concentration. However, as the free carrier density increases, the occupation of the DX center increases rapidly so that higher free carrier densities are achieved only at the cost of ever increasing occupation of the DX center. Extrapolation of the data in Figure 3 suggests a crossing of the DX level and the Fermi level at a free carrier density ~ 1.5 - 2×10^{19} cm^{-3}. If the concentration of DX centers is greater than the net uncompensated donor concentration, this will be the maximum equilibrium free carrier concentration attainable in n-type GaAs. A similar limit is inferred from the hydrostatic pressure experiments of Maude et al (1987).

The maximum carrier concentration typically obtained in n-GaAs grown near thermal equilibrium is ~ 8×10^{18} cm^{-3}, limited by self-compensation mechanisms such as site switching and formation of acceptor complexes. Using low temperature growth techniques such as MBE, self-compensation can be suppressed and higher free carrier densities are achieved. Nevertheless, for all donors except Ge (Metze et al 1980, Li et al 1985) we are unaware of any published reports which significantly exceed ~ 2×10^{19} cm^{-3}. (We exclude C-V measurements which measure the fixed charge, including ionized DX centers, in a depletion region.) The exceptional case of Ge doped GaAs (or *any* n-GaAs with carrier density >> 2×10^{19} cm^{-3}) is of great interest, since it suggests that donors may, under proper conditions, be incorporated in GaAs in such a way as to avoid formation of the localized DX level. However, this does not alter our principle conclusion: Charge trapping by the DX center is a previously unknown mechanism, independent of and acting in addition to self-compensation, which limits the equilibrium free carrier density in heavily doped n-GaAs.

4. ACKNOWLEDGMENTS

We thank R. F. Marks for assistance with MBE growth, G. J. Scilla for SIMS analysis, J. C. Tsang for Raman measurements. We are grateful to F. Koch of the Technical University of Munich for some very helpful comments on this manuscript.

5. REFERENCES

Calleja E, Mooney P M, Wright S L, and Heiblum M 1986 *Appl. Phys. Lett.* **49** 657
Chand N, Hendersen T, Klem J, Masselink T, Fischer R, Chang Y C, and Morkoc 1984 *Phys. Rev. B* **30** 4481
Ishikawa T, Yamamoto T, and Kondo K 1986 *Jpn. J. Appl. Phys.* **25** L484
Lang D V, Logan R A and Jaros M 1979 *Phys. Rev. B* **19** 1015
Li A, Milnes A G, Chen Z Y, Shao Y F, Wang S B, 1985 *J. Vac. Sci. Technol. B* **3** 629
Maude D K, Portal J C, Dmowski L, Foster T, Eaves L, Nathan M, Heiblum M, Harris J J, and Beall R B 1987 *Phys. Rev. Lett.* **59** 815
Metze G M, Stall R A, Wood C E C and Eastman L F 1980 *Appl. Phys. Lett.* **37** 15
Mizuta M, Tachikawa M, Kukimoto H, and Minomura S 1985 *Jpn. J. Appl. Phys.* **24** L143
Mooney P M, Calleja H, Wright S L, and Heiblum M 1986 *Proc. of the 14th Int. Conf. on Defects in Semiconductors,* Ed. H J von Bardeleben (Trans Tech Publications, Switzerland) Vol. 10-12 pp. 417-22
Tachikawa M, Fujisawa T, Kukimoto H, Shibata A, Oomi G and Minimura S 1985 *Jpn. J. Appl. Phys.* **24** L893
Theis T N 1986 *Proc. of the 14th Int. Conf. on Defects in Semiconductors,* Ed. H J von Bardeleben (Trans Tech Publications, Switzerland) Vol. 10-12 pp. 393-8
Theis T N, Parker B D, Solomon P M, and Wright S L 1986 *Appl. Phys. Lett.* **49** 1542
Theis T N, Mooney P M, and Wright S L 1987 unpublished.

Acoustic deep level spectroscopy of semi-insulating GaAs and InP

Yutaka Abe

Laboratory of Quantum Instrumentation, Hokkaido University, Sapporo, 060, Japan

ABSTRACT: The acoustic internal friction of undoped, In-doped GaAs, and Fe-doped InP has been measured along the various crystal directions. It is found that the observed relaxation peaks in the specimen are well explained taking account of the screening of piezoelectric field by free-carriers emitted from deep-lying impurity level.

1. INTRODUCTION

Zinc-blende type semiconductors are piezoelectric active due to the lack of center of symmetry in their crystal structures, though their mechanical coupling constants are much smaller compared with those of II-VI semiconductors such as CdS and ZnO. The semi-insulating(SI) GaAs and Fe-doped InP have high resistivities of $10^6 - 10^8$ ohm-cm at room temperature, and this is very favourable situation for observing the relaxation phenomena due to the piezoelectric interaction with free-carriers excited externally or with lattice defects. This situation is in marked contrast with one in deep-level transient spectroscopy, which is very difficult to apply for the investigation of SI materials.

The acoustic internal frictions in transition-metal doped SI GaAs and GaP have been measured by Mitrokhn et al (1984, 1985) using flexural mode of vibrations. In our experiments, we have used a standard resonance method for measuring the internal friction of SI GaAs and InP. The temperature dependence of the internal frictions were analyzed taking account of the screening of piezoelectric field by free-carriers emitted from the deep-lying impurity level.

2. EXPERIMENT

The rod- and plate-shaped specimens used in this experiment were cut from the LEC grown ingot along (100), (110) and (111) directions. In the case of InP, only the specimens along the (110) direction were available at this time.

The internal friction, Q^{-1}, were measured using the composite resonator which consists of a quartz resonant exciter and the specimen. The internal friction is given by

$$Q^{-1} = \Delta f / f_r \qquad (1)$$

where f_r is the eigen resonant frequency of the specimen and Δf is the half-width of the resonance. From the internal friction of the composite resonator and the one due to the exciter, Q^{-1} is determined experimentally.

The bonding between the exciter and the specimen is very crucial in this experiment. We have used several bonders of fine ceramic powder, and of thermal compounds. The specimens were set into small dewar and the internal friction was measured as function of temperature between 200 to 550K. We have used several quartz exciters of which the resonant frequencies were ranged between 100Khz to 450Khz.

3. RESULTS AND DISCUSSION

The temperature dependence of the internal friction of SI GaAs, In-doped GaAs, and Fe-doped InP are shown in Fig. 1 and Fig. 2. Each specimen has its own relaxation peak in the measured temperature region. The position of the peak depends on the resistivity at room temperature and the peak is shifed to higher temperature as the resistivity is increased. This is a strong contrast against the results reported by Mitrokhn et al (1985). The magnitude of the peak has marked dependence on the crystal orientation, and this fact indicates that the observed relaxation is due to the piezoelectric interaction with free-carriers.

Along the (100) direction, GaAs and InP have very weak coupling of free-carriers with sound waves through deformation potential, where as along (110) and (111) directions, they have much stronger coupling due to piezoelectric polarization of lattice. In piezoelectric semiconductors, stress-strain relation is expressed in terms of complex elastic stiffness constants and is given by (Hutson and White, 1962)

$$\bar{T} = (C' - i C'') \bar{S} \tag{2}$$

where $\bar{T}$ is the stress vector, $\bar{S}$ is the strain vector and C' and C" is the real and imaginary part of the elastic stiffness constants, respectively.

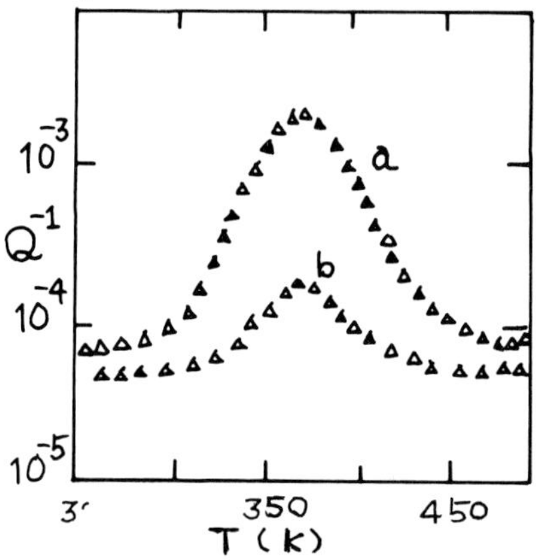

FIG. !. Temperature dependence of the internal friction in SI GaAs. a; (110) b; (111) direction.

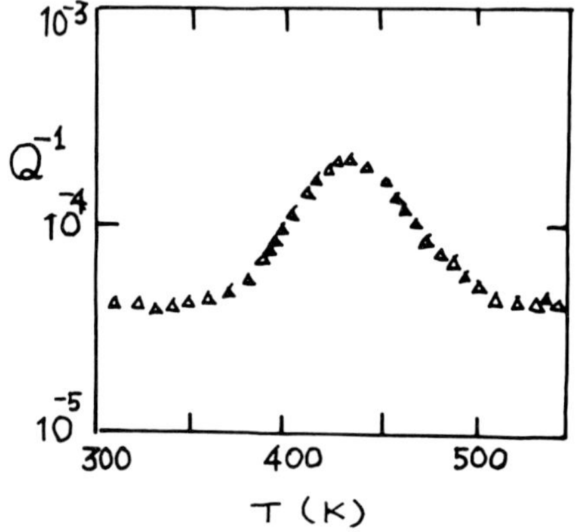

Fig. 2. Temperature dependence of the internal frictio in Fe-doped InP. (110) direction.

The internal friction Q_c-1 under a resonant condition along the piezoactive axis is given by

$$Q_c^{-1} = K_{ij}^2 \; P \qquad (3),$$

where $P = \dfrac{(\omega_c/\omega)}{1+(\omega_c/\omega)^2+(\omega/\omega_d)^2}$

$$\omega_c = \sigma/\varepsilon \;,\; \omega_d = v_s^2/D_n \;.$$

Here, K_{ij}^2 is the mechanical coupling constant, σ is the conductivity, v_s is the sound velocity, and ε is the dielectric constant. There exist the additional internal friction due to elastic damping of crystal lattice. In the temperature region of our experiments, we assume the elastic damping is simply given by elastic viscous damping term, which is proportional to the velocity of the elastic displacement. The resultant Q-1 is expressed as

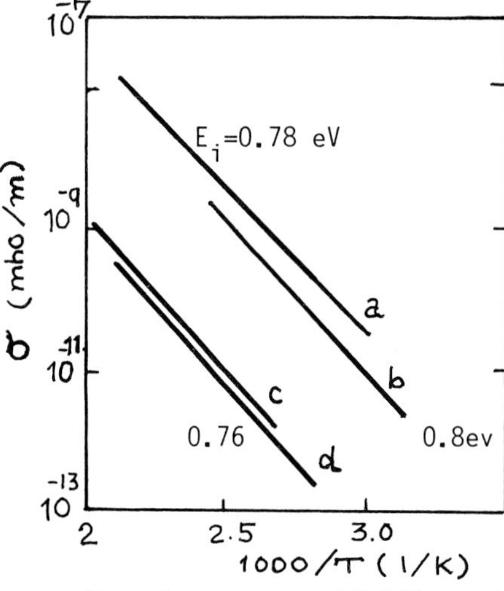

Fig. 3. versus 1000/T. a, b and c: GaAs. d: Fe-doped InP.

$$Q^{-1} = Q_c^{-1} + Q_L^{-1} \qquad (4),$$

where $Q_L^{-1} = \omega\eta_{ij} / C_{ij}$, and η_{ij} is the viscous damping constant.

The resonant frequency of composite resonator is low enough so that we can safely neglect the diffusion term in Eq.(3). The value of $1/Q_L$ is obtained from the value of 1/Q in the low temperature region where 1/Q is temperature-independent.

The temperature dependence of the conductivities of various specimens derived from $1/Q_c$ is shown in Figure 3. The activation energies of the conductivities have the excellent agreements with ones from 4 points probes dc measurements.

Mitrokhn et al (1985) have proposed another mechanism for the relaxation of internal friction in SI GaAs and SI GaP. They have suggested the above mentioned relaxation is due to the process of generation of non-equilibrium carriers from a deep level by modulation of the emission rate under unscreened piezoelectric field. The electric field effect on the thermal emission rate of deep trap has been investigated by several authors (Martin et al, 1981; Markram-Ebeid, 1980; Vincent et al, 1979). These investigations show that the electric field of order 10^5 v/cm is necessary to observe a substantial variation in the emission rate. In the case of GaAs and the related compound semiconductors, the unscreened piezoelectric field $\bar{E}$ is

$$\bar{E} = e_{ij}\,\bar{S}/\varepsilon_{ij} \qquad (5)$$

where e_{ij} is the piezoelectric stiffness constant, this is the order of 10^3 v/cm with strain of 10^{-3}, which is far from the limiting value of the field for the observation of the proposed process.

The obtained activation energy in Fe- doped InP has very close d value to the one of Fe^{++} centers reported by Juhl and Bimberg (1986), and we conjecture that the relaxation in Fe-doped InP is due to the free holes from the Fe ++ centers.

The frequency dependence of $\omega\eta_{ij}/C$ in Si GaAs and In-doped GaAs is shown in Figure 4, It is found that the viscous damping of In-doped GaAs is larger than the one of undoped GaAs , and the reason is still unclear. However, it has been reported that In-doped GaAs involved large amount of microdefects which acts as nonradiative centers (Nakamura et al, 1986). Further investigations are necessary to clarify the above difference.

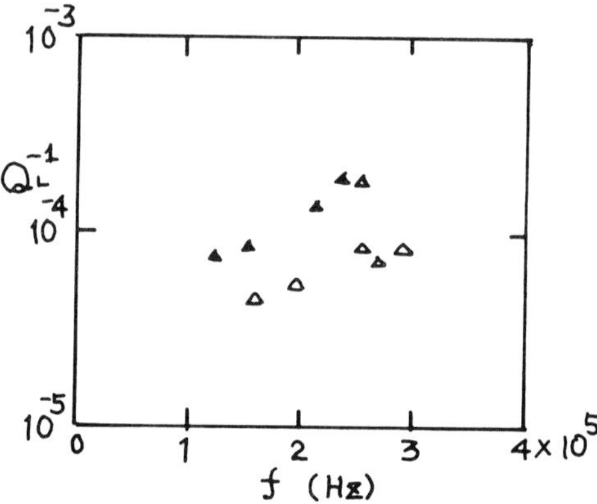

Fig. 4. Frequency dependence of in SI GaAs. : undoped. : In-doped.

We have investigated the relaxation process in the internal friction of SI GaAs and SI InP, It was also demonstrated that the present contactless method offer a new tool for the investigation of deep levels in SI III-V compounds.

REFERENCES

Hutson A R, and White D L 1962 J. Appl. Physics. 33 40

Juhl A and Bimberg D 1986 Proc. Int. Conf. Semi-Insulating III-V matrials,(Amsterdam, North-Holland) 477-438

Makram-Ebeid S 1981 Appl. Phys.Lett. 37 464

Martin P A , Streetman B G, and Hess K 1981 J. Appl. Phys. 52 7409

Mitrokhn V I, Rembeza S I, Sviridov V V and Yaroslavtev, Soviet Phys. Solid State 27 1247

Mitrokhn V I, Rembeza S I and Yaroslovtev N P 1984 Soviet Phys. Solid State 26 1354.

Nakamura H, Tubouchi K, Mikoshiba N, and Fukuda T 1986 IEEE Ultrasonic Symposium, Proc. 481-485

Inst. Phys. Conf. Ser. No. 91: Chapter 4
Paper presented at Int. Symp. GaAs and Related Compounds, Heraklion, Greece, 1987

Interdiffusion coefficient and its relation to thermal stability of superlattice

Naoki Hara and Takashi Katoda

Institute of Interdisciplinary Research, Faculty of Engineering,
The University of Tokyo
4-6-1 Komaba Meguro-ku, Tokyo 153, Japan

Abstract: Effects of the number of layers in superlattice and stress on interdiffusion of atoms constituting superlattices were studied theoretically. Effective interdiffusion coefficient of atoms in superlattice with a larger stress has a larger dependence on the number of layers. $(GaAs)_n(InAs)_n$ superlattice with $n \geq 7$ is thermally stable from the standpoint of the interdiffusion. However misfit dislocation will be introduced if $n>15$. Therefore $(GaAs)_n(InAs)_n$ superlattice with $7 \leq n \leq 15$ is thermally stable. A comparison between theoretical calculations and experimental results is also discussed.

1. Introduction

The thermal stability of superlattice is very important because it relates to the stability of devices during fabrication and operation. We showed (Katoda et al. 1986) experimentally that superlattice with large stress is thermally unstable. Theoretical estimation of thermal stability of superlattice, however, has not been reported. We calculated interdiffusion coefficient in superlattice theoretically. In this paper we report calculations of interdiffusion coefficient and compare them with the experimental results based on Raman spectra a part of which we reported last year at this conference (Katoda et al. 1986).

2. Theory

At first interdiffusion of atoms between two layers with a large concentration gradient is analyzed. We consider here a solid solution consisting of M and N atoms. Suppose that composition varies in one direction (x direction). Taking into account the strain effects, the effective interdiffusion coefficient along the (100) direction in fcc crystals, D_λ, is approximately expressed as follows (Cahn et al. 1958, Cahn 1962, Cook et al. 1969, and Yamaguchi et al. 1972).

$$D_\lambda = \left(1+\frac{2\kappa B^2}{f''_0}+\frac{2\eta^2 Y}{f''_0}\right) D \qquad (1)$$

$$\kappa = \frac{4a^2}{z}\, h^m_{0.5} \qquad (2)$$

$$B^2 = \frac{2}{d^2}\left[1-\cos\left(\frac{2\pi d}{\lambda}\right)\right] \qquad (3)$$

$$\eta=\frac{1}{a}\frac{da}{dc} \tag{4}$$

$$Y=C_{11}+C_{12}-\frac{2C_{12}^2}{C_{11}} \tag{5}$$

where f_0'' is second derivative of the free energy per unit volume with respect to c which is the fraction of M atom, D is the macroscopic interdiffusion coefficient, a is the lattice parameter, z is the coordination number, $h_{0.5}^m$ is the mixing enthalpy per unit volume at c=0.5, d is the space between the atomic planes, C_{11} and C_{12} are the elastic constants, and λ is the wavelength of the compositional wave.

We made assumptions as follows to obtain Eqs.(1)-(5).
[1] Higher order terms in Taylor series form of the free energy are negligible.
[2] The parameters f_0'' and κ are independent of the composition.
[3] Y defined by Eq.(5) is independent of the wavelength of the compositional wave.
Therefore it is possible that behavior predicted by Eqs.(1)-(5) contains some quantitative error. Qualitative behavior, however, can be made sufficiently clear.

The expression derived above can be used to calculate the effective interdiffusion coefficient in semiconductor superlattices. We use the expressions of mixing enthalpy per mole at c=0.5, $H_{0.5}^m$, and second derivative of the free energy per mole with respect to c, F_0'', given as

$$H_{0.5}^m=\frac{\alpha}{4} \tag{6}$$

$$F_0''=4RT-2\alpha \tag{7}$$

where α and R are the interaction parameter and the gas constant, respectively. Since the values of the elastic constants in superlattices are unknown the average values of constitutional binary semiconductors are used. The value of η is calculated by using the Vegard's law.

3.Results and Discussion

The ratio D_{4d}/D_{8d} is calculated at first in order to compare the thermal stability between superlattices whose number of layers n are 2 and 4. D_{4d} and D_{8d} are effective interdiffusion coefficients of atoms in superlattices with n=2 and 4, respectively. The results obtained for some superlattices are shown in Fig.1. From Fig.1 the ratio D_{4d}/D_{8d} is found to be large for superlattices $(GaP)_n(InP)_n$ or $(GaAs)_n(InAs)_n$. On the contrary for superlattices $(GaAs)_n(AlAs)_n$ the difference between D_{4d} and D_{8d} is negligibly small. Therefore the more the stress is large, the more D_λ

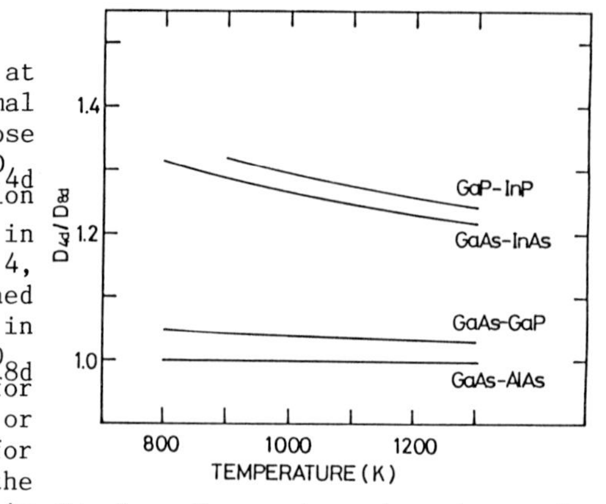

Fig.1 Temperature dependence of the ratio D_{4d}/D_{8d} for various superlattices.

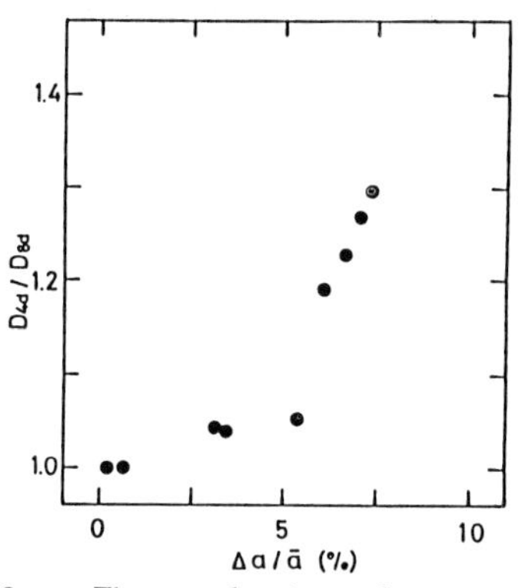

Fig.2 The relation between the difference of lattice parameters and the ratio D_{4d}/D_{8d} at 1000K.

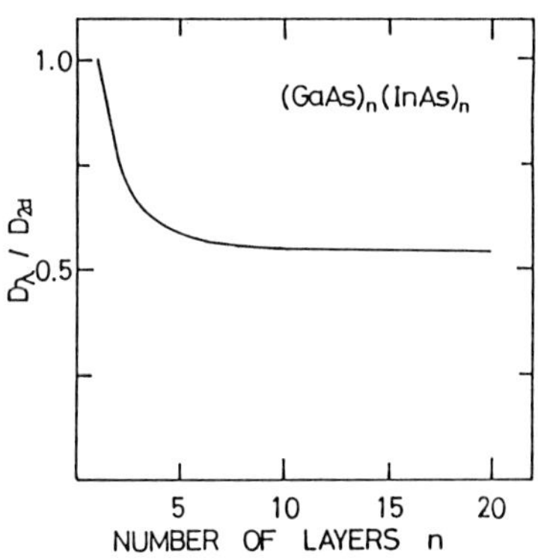

Fig.3 The relation between the effective interdiffusion coefficient and the number of layers.

changes with the number of layers largely. Figure 2 shows the relation between the difference of lattice parameters and the ratio D_{4d}/D_{8d} in many systems. This figure confirms the above consideration.

We calculate also the effective interdiffusion coefficient, D_λ, for superlattices $(GaAs)_n(InAs)_n$. Figure 3 shows D_λ normalized by the value at n=1. It can be seen that change in D_λ is very large when n is small while D_λ is almost constant when n is larger than about 7. In highly mismatched systems, however, superlattices cannot be grown without misfit dislocations when the layer thickness is larger than a critical thickness. A critical thickness for $(GaAs)_n(InAs)_n$ superlattices calculated in the manner Matthews et al. proposed (1974) is about 45Å and it corresponds to n=15. Therefore, when $7 \leq n \leq 15$, $(GaAs)_n(InAs)_n$ superlattices are most stable.

Finally we compare the calculations described above with experimental results based on Raman spectra a part of which were reported at this conference last year (Katoda et al. 1986). Critical temperatures for disordering were examined for superlattices $(GaAs)_n(InAs)_n$ (n=2,4) and $(GaAs)_n(AlAs)_n$ (n=4,9). Annealing of superlattices was done in an atmosphere of nitrogen with 6% hydrogen. A SiO_2 film was formed on the surface of samples before annealing in order to protect the surface. A temperature at which a new peak corresponding to phonon from alloy semiconductor appears in Raman spectrum is assigned to be a critical temperature. Raman spectra from $(GaAs)_2(InAs)_2$ superlattice before and after annealing are shown in Fig.4 as examples. A new peak at about 235cm^{-1} which corresponds to InAs-mode LO phonon from $Ga_{1-X}In_XAs$ ($X \simeq 0.5$) is observed in the spectra obtained after annealing at 580°C and 750°C for 30min. Therefore the critical temperature for disordering for $(GaAs)_2(InAs)_2$ superlattice is defined to be 580°C. Figure 5 shows the relation between the critical temperature for disordering and the number of layers, n, for $(GaAs)_n(InAs)_n$ and $(GaAs)_n(AlAs)_n$. From the figure we can see that dependence of the critical temperature for disordering on n is large for $(GaAs)_n(InAs)_n$ superlattices which have large stress. On the other hand the dependence is small for $(GaAs)_n(AlAs)_n$ superlattices which have negligibly small stress. These results can be explained by calculations described above.

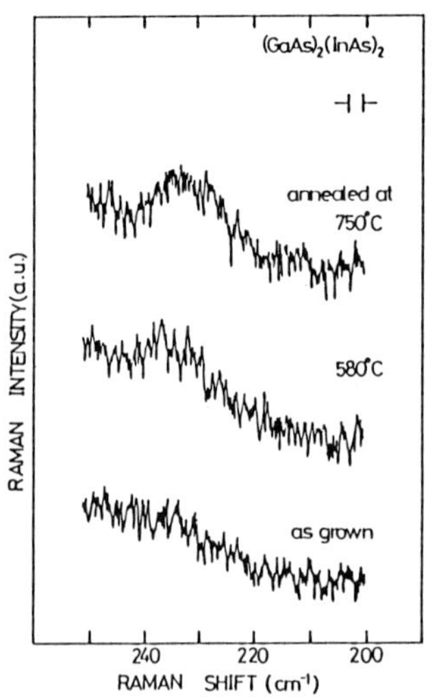

Fig.4 Raman spectra from $(GaAs)_2(InAs)_2$ superlattice before and after annealing.

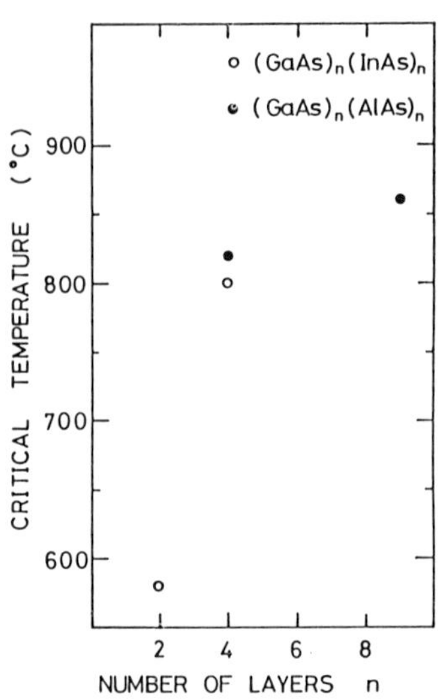

Fig.5 The relation between critical temperature for disordering and the number of layers.

4.Conclusions

Interdiffusion coefficient in superlattices one constituting atom of which is common was calculated in order to estimate thermal stability of superlattice. Effects of stress and compositional variation were taken into account in the calculation. Important results are as follows. (i) Interdiffusion coefficient in $(GaAs)_n(InAs)_n$ superlattice has a large dependence on the number of layers, n, when n is smaller than about 7. (ii) Dependence of interdiffusion coefficient on n changes with temperature more largely for superlattice with larger stress such as $(GaAs)_n(InAs)_n$ and $(GaP)_n(InP)_n$ than that with smaller stress such as $(GaAs)_n(AlAs)_n$. (iii) The results described in (i) and (ii) explain the experimental results based on Raman spectra. (iv) From another standpoint that misfit dislocation is introduced when both the thickness of layers and lattice mismatch between heteroepitaxial layers are large. Therefore $(GaAs)_n(InAs)_n$ superlattices with $7 \leq n \leq 15$ are thermally stable. However it is a different problem whether a superlattice with a large n can be grown in practice.

This work was supported in part by the Scientific Research Grant-in-Aid for Special Project Research on "Alloy Semiconductor Physics and Electronics", from the Ministry of Education, Science and Culture.

References

Cahn J W 1962 Acta Metall. 10 179.
Cahn J W and Hilliard J E 1958 J.Chem.Phys. 28 258.
Cook H E, deFontaine D, and Hilliard J E 1969 Acta Metall. 17 765.
Katoda T and Hara N 1986 Int.Symp.GaAs and Related Compounds, Las Vegas, Nevada (1987 Inst.Phys.Conf.Ser. 83 263).
Matthews J W and Blakeslee A E 1974 J.Cryst.Growth 27 118.
Yamaguchi H and Hilliard J E 1972 Scr.Metall. 6 909.

Inst. Phys. Conf. Ser. No. 91: Chapter 4
Paper presented at Int. Symp. GaAs and Related Compounds, Heraklion, Greece, 1987

Chemical composition and crystallographic orientation of oval defects in $Ga_{1-x}Al_xAs$. A microprobe analysis of MBE layers

J. Sapriel, J. Chavignon, F. Alexandre, P. De Souza and A.C. Papadopoulo

Centre National d'Etudes des Télécommunications
196 Avenue H. Ravéra 92220 BAGNEUX - FRANCE

ABSTRACT : The chemical composition, crystallographic orientation and crystalline quality of oval defects are deduced from spatially resolved Raman and Photoluminescence studies.

1. INTRODUCTION

The existence of macroscopic surface defects such as the "oval defects" is considered as a troublesome problem for practical applications such as integrated circuits. The factors which provoke their creation depend on both the system employed and the growth conditions of the epilayers. Since many optoelectronic and microelectronic devices consist of $GaAs/Ga_{1-x}Al_xAs$ heterostructures, it is crucial to perform thorough investigation of the oval defects in $Ga_{1-x}Al_xAs$ $(0 < x < 1)$. Surface and cross sectional microscopic observations on such layers have mainly focused on their morphology[1]. The orientation of the long-axis is parallel to $[1\bar{1}0]$ directions. Defects with macroscopic core particulates are named β-type (the others are called α-type).
This is a combined study of photoluminescence (PL) and Raman Scattering (RS) performed with a spatial resolution of 1 μm^3 using a Raman microprobe at room temperature. RS gives information on the crystallographic orientation and both techniques (RS and PL) can characterize the composition and the crystalline quality of the probed volume.

2. RESULTS AND DISCUSSION

We used a Ramanor U1000 doubled monochromator from Jobin Yvon equiped with a Microanalysis attachment. The incident light was focused with a microscope objective on the sample surface on a 1 μm-diam spot. The RS and PL were performed with the same objective in the bakscattering configuration. Though the RS and PL signals corresponded to the same point on the surface, they probed different depths. In the case of RS the distance from the surface which was probed corresponded to the light penetration depth d, though the PL signal corresponded to the diffusion length ℓ of the free carriers created by the incident photons. In our samples one can estimate $d \sim 0.1$ μm and $\ell \sim 1$ μm.
The composition, doping and thickness of the different samples are reported on Table I. In every sample the RS and PL investigations were performed in different oval defects and the resulting spectra were compared to

those obtained on a defect free region of the layer (good region). The RS and PL signals are actually very different in the oval defects. Besides, our systematical study led to the conclusion that the orientation and the composition of oval defects vary from one defect to the other and also from point to point in the same defect (which cannot be thus considered as homogeneous in the scale of one micron cube). However certain characteristics are common to all defects:

Sample	x	doping	thickness (μm)
6328	0	$p=6.10^{14}$	3
61256	0.31	$n=2.10^{17}_{(Si)}$	2
61112	0.31	$n=2.10^{17}_{(Si)}$	2
6247	0.55	$n=1.1\ 10^{17}_{(Si)}$	2

Table I-Characteristics of the of the investigated $Ga_{1-x}Al_xAs$ layers which were grown on (100) oriented semi-insulating GaAs substrates, in a second generation MBE system (Riber 2300). The growth conditions were chosen in order to improve the optical properties of $Ga_{1-x}Al_xAs$ ($T_{substrate}=650°C$, low flux ratio and growth rate of 1.8μm per hour) For GaAs the growth temperature was 600°C.

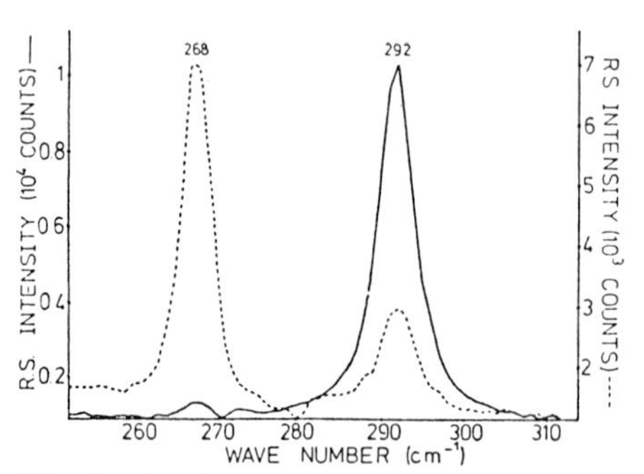

Fig 1- Good region —— and oval defect ---- in GaAs (R.S.).

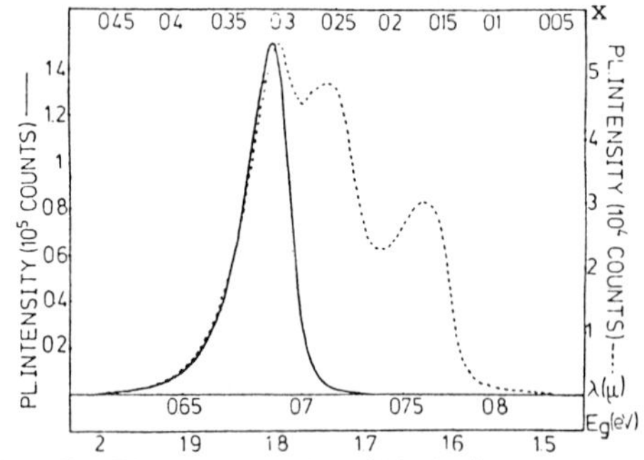

Fig 2- PL. in sample 61112 (good region —— and oval defect ----). The upper abscissa x is the Al conc. corresp. to a direct gap luminescence

- the crystallographic orientation changes with respect to the umperturbed crystal. These changes are evidenced by the activation of the Raman TO which is normally forbidden for (100) oriented layers. In all Raman spectra (Fig. 1,3,4c) the TO intensity is much larger than the LO in the oval defect. One can take advantage of this feature to improve the focusing and positionment of the light spot inside the oval defect. The width of the TO is generally small, thus indicating a rather good crystalline state in the oval defects.
- several aluminium concentrations x can coexist in every probed zone in the oval defects, i.e., x takes several values as shown particularly by the observation of several resolved luminescence peaks (Fig. 2,4a-b). Actually the luminescence issued from Ga rich parts of the defects is exalted because they attract the carriers (wells for electrons and holes).

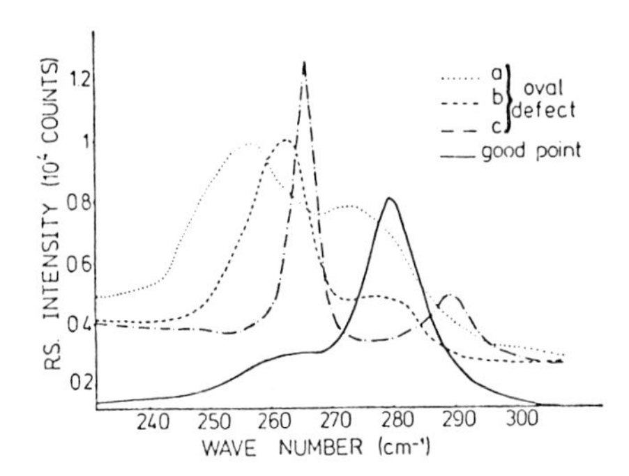

Fig 3- R.S. in different points (a, b, c) of a large oval defect showing a clear change in the concentration x, $x_c < x_b < x_a$ (sample 61256).

Fig4 - Results on an indirect gap epilayer (sample 6247).

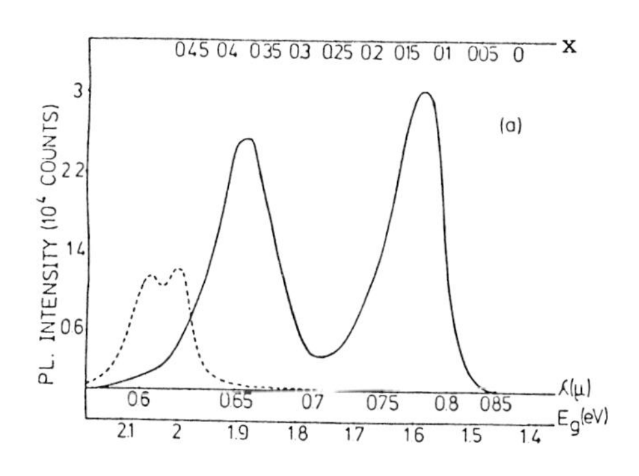

a-PL for incident photons at 2.41eV
---- good region
——— oval defect

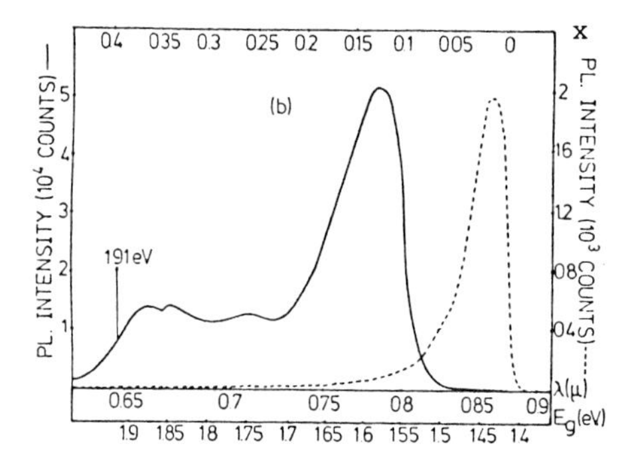

b-PL for incident photons at 1.91eV One can see the luminescence of the substrate in the good region (----). Same point as (a).

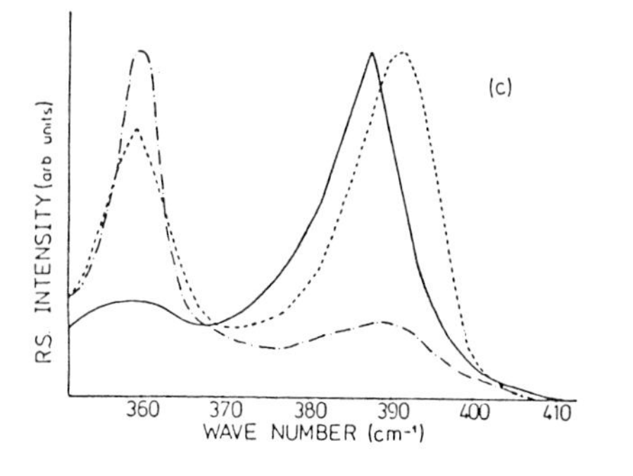

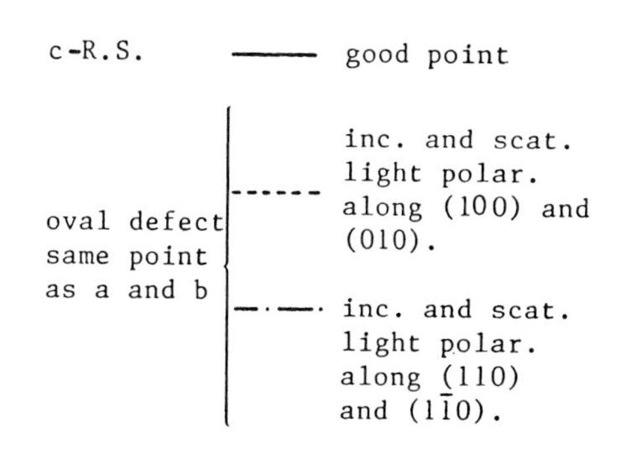

c-R.S. ——— good point

oval defect same point as a and b
----- inc. and scat. light polar. along (100) and (010).
—·—· inc. and scat. light polar. along (110) and (1$\bar{1}$0).

- most of x values correspond to an increase of Ga concentration in the oval defect with respect to the layer. This results from the RS and PL measurements and is in good agreement with the observations made on a $Ga_{1-x}Al_xAs$ MBE epilayer by spatially resolved cathodoluminescence and Auger spectroscopy[2]. As the Al concentration x decreases, the LO and TO frequencies of the GaAs type increase and the LO and TO of the AlAs-type frequencies decrease[3]. Besides the PL peaks shift to lower energies for decreasing x values (Fig. 2,4a-b).
Particularly striking are the PL results obtained on an indirect gap alloy (x=0.55) for incident photons corresponding to λ=0.54μm and λ=0.6471μm. The luminescence in the good region is either very low (for λ=0.514μm it corresponds to the Γ and X extrema points) or inexistant (for λ=0.641μm), though the oval defect zone displays an important luminescence at λ=0.66μm and λ=0.78μm. The Raman spectra in this case don't indicate a Gallium enrichment in the oval defect. The whole set of results (RS and PL) are consistent with an increase of Ga concentration far from the surface (near the interface $Ga/Ga_{1-x}Al_xAs$). By photoluminescence topography the $Ga_{0.45}A_{0.55}As$ layer appear as a dark background though the oval defects are visualized by bright regions. This topography experiment consists of using an Ar laser to excite a 0.5 mm^2 area and collecting with an infrared camera the total luminescence emitted by the sample.

3. CONCLUSION

In conclusion, important variations of the crystallographic orientation and of the Al composition have been evidenced inside oval defects of $Ga_{1-x}Al_xAs$ MBE layers. Thus it is not surprising that the properties of devices constituted of $GaAs/Ga_{1-x}Al_xAs$ heterostructures should be affected by such kind of defects.

REFERENCES

1) K. FUJIWARA, K. KANAMATO, Y.N. OHTA, Y. TAKUDA, AND T. NAKAYAMA, J. Crystal Growth 80, 104, 1987
2) A.C. PAPADOPOULO, F. ALEXANDRE and J.F. BRESSE to be published in Appl. Phys. Lett.
3) B. JUSSERAND and J. SAPRIEL, Phys. Rev. B, 24, 7194 (1981)

Inst. Phys. Conf. Ser. No. 91: Chapter 4
Paper presented at Int. Symp. GaAs and Related Compounds, Heraklion, Greece, 1987

Kinetics of microprecipitates formation in GaAs obtained from high resolution infrared tomography (HRIT) and A–B etching

P. Suchet*, M. Duseaux**

* LEP (a), 3 avenue Descartes, 94451 Limeil-Brévannes Cedex (FRANCE)
** Present address : P.D.O. 26 avenue W. Churchill 27400 Louviers (FRANCE)

ABSTRACT : The presence of microprecipitates in the matrix of GaAs has been observed by Transmission Electron Microscopy (TEM) and by chemical A-B etching. Besides these two characterization methods, we have developed a high resolution infrared tomography (HRIT) system enabling to conclude that infrared scattering in GaAs is due to the presence of microprecipitates. We have thus been able to study the influence of crystal growth and different annealing conditions on the microprecipitates concentration.

1. INTRODUCTION

Crystallographic defects can induce non-uniformities in the electrical properties of the substrate by their own presence or by impurity segregation around them. Therefore, they have to be eliminated or controlled. Among these defects, microprecipitates have been observed by TEM (Cornier et al, 1985 ; Duseaux et al, 1986) and A-B etching (Stirland, 1986 ; Stirland et al, 1984). They have been identified as arsenic hexagonal precipitates (Cullis et al, 1980, Stirland et al, 1984). Recently, infrared light scattering has been assumed to be caused by these precipitates (Ogawa 1985, Moriya 1986).
In this paper we will first summarize the main results obtained with the high resolution infrared tomography system (Suchet et al, 1987) we have built in our laboratories in order to conclude on the origin of infrared scattering. Comparison between tomography and A-B etching observations have enabled us to estimate the concentration and "nature" of the precipitates, according to their behaviour during crystal growth and annealing. These results, together with kinetics of the microprecipitates formation, will be presented in the second part of the paper.

2. HIGH RESOLUTION INFRARED TOMOGRAPHY(HRIT)

Infrared light scattering tomography has been proposed to characterize GaAs substrates in order to reveal inhomogeneities in bulk materials. Many observations and theoretical calculations on infrared scattering (Moriya et al, 1978) have suggested that it is caused by microprecipitates.
In order to reach a clear statement, we have built an experimental set-up for HRIT. The optical resolution of the system is of the order of one micron. Comparison between tomography and A-B etching observations has definitely proven that microprecipitates are responsible for infrared scattering (Suchet et al, 1987).

a) LEP : **Laboratoires d'Electronique et de Physique appliquée - A member of the Philips Research Organization**

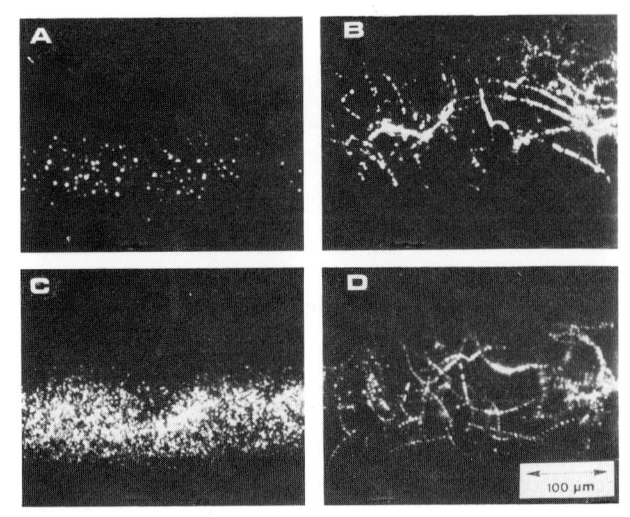

FIGURE 1 : Different types of HRIT observations

Furthermore, tomography observations have shown three different types of microprecipitates distributions as illustrated on picture 1 : i) isolated dots distribution corresponding to a low density of precipitates (1a), ii) network distribution similar to dislocation network revealed by A-B etching (1b), iii) "milky way" distribution composed of a network of bright dots superimposed on a "background" of lower intensity points (1c). Furthermore, one notices different dot intensities as observed on one crystal (1a) or between different crystals (1b and 1d). Since the dot intensity is related to the size and magnitude of the electron distribution heterogeneity responsible for the scattering (Moriya, 1978), this suggests that the different intensity dots relate to precipitates of different size or chemical nature.

3. MICROPRECIPITATES OBSERVATIONS IN AS-GROWN AND ANNEALED BULK GaAs

As-grown - Using A-B etching and HRIT, we have studied the influence of crystal growth conditions on the presence of microprecipitates. Three different parameters have been tested :

i) dislocation density (undoped dislocated crystal and nearly dislocation-free In-doped crystal)

ii) Stoechiometry (As-rich, Ga-rich, stoechiometric crystals)

iii) Cooling conditions (standard LEC cooling , i.e. seed cooling slower than tail cooling, or fast cooling (from 1100°C to 700°C) after encapsulated growth).

All the samples have been cut along the (001) plane. In each case, (A-B

	Thermal History		As-grown				Annealed at 900°C for 60 hours				Annealed at 1000°C for 60 hours			
	Crystals (LEC)		Tomography			A-B	Tomography			A-B	Tomography			A-B
			Density	Type	Intensity	Density	D	T	I	D	D	T	I	D
1	undoped	Seed	2	#	+	1	3	☆	+	2	1	#	-	0
2	undoped	Tail	2	#	+	0	3	☆	+	2	2	#	-	0
3	As-rich	Middle	2	#	++	2	3	☆	++	2	2	#	++	1
4	Ga-rich	Middle	0	/	/	0	1	#	-	0	1	#	-	0
5	fast cooling	Middle	1	•	-	0	3	☆	+	1	2	#	-	0
6	In-doped	Dislocat. free areas	0	/	/	0	0	/	+	0	0	/	/	0
		Slip bands	2	#	++	2	2	#	++	2	2	#	++	2

TABLE I

Density (cm^{-3}) :
0 $\leq 10^6$
1 $10^6 < \leq 5.10^7$
2 $5.10^7 < \leq 10^9$
3 $10^9 <$

Type :
• Isolated dots
Network
☆ Milky way pattern

Intensity :
+(+) high
- weak

etching or HRIT), the density of the microprecipitates has been scaled from 0 (Dm $\leqslant 10^6 cm^{-3}$) up to 3 (Dm $> 10^9 cm^{-3}$). The tomography observations of microprecipitates have been related to their distribution types (isolated dots, network or "milky way" pattern) and their scattering intensity (weak or high).
Column 1 of Table I summarizes the observations and suggests the following comments :
i) Excess of arsenic enhances the microprecipitate density (see samples 1, 3 and 4)
ii) A-B etching and tomography reveal that most (more than 95 %) of microprecipitates are pinned on dislocations which seem necessary for the formation of precipitates. As a matter of fact, the only material which does not present any precipitate, neither by **tomography** nor by A-B etching, is the dislocation-free one (ingot 6).
iii) the number of particles observed by tomography is at least equal to and very often larger than the number of particles revealed by A-B etching. In the case of samples 2 and 5, tomography revealed precipitates which are not detected in A-B etching. In those cases, the density is definitely lower. This can be related to the fact that these materials were cooled down more quickly, i.e. they did not remained at 900-1000°C more than 5 minutes (Farges et al, 1987). This suggest that this is the temperature range for formation of precipitates. This hypothesis has been confirmed by annealing experiments.

Annealed material - Results of 900 and 1000°C annealing (followed by fast cooling) are reported in Table I which leads to new comments :
i) 900°C annealing for 60 hours (column 2) strongly increases the density of particles revealed by A-B etching, but only weakly increases the density of particles observed by tomography. The extra dots revealed by HRIT have a weaker scattering intensity which changes the network aspect of the tomographic observation into a "milky way" pattern.
ii) 1000°C annealing for 60 hours (column 3) reduces the density of particles revealed by A-B etching and has nearly no influence on the density of particles observed by tomography. However, the scattering intensity of these dots is lower.

4. DISCUSSION

Since tomography evidences a larger number of precipitates than A-B etching, it is clear that there are two classes of precipitates.

The first class of precipitates, only detected by tomography (weak scattering intensity) has been identified by high resolution TEM X ray probe, to correspond to pure As precipitates. These results confirm previous studies made by Cullis et al (1980), Stirland et al (1984) and Cornier (1986) who have particularly shown that dislocations allow precipitates formation by the presence around them of arsenic interstitials and vacancies (necessary to the formation of cavities). Furthermore, they have suggested that dislocations may act as "pipes" for the arsenic atoms.

The second class of precipitates very likely corresponds to As precipitates with impurities trapped either inside the As precipitates (as checked by X ray probe (Suchet et al, 1987)) or around it. They will be refered to as As + I precipitates. As a matter of fact, the A-B observation of that second class is strongly dependent on the impurity migration. In order to clarify the corresponding kinetics, A-B etching relief around dislocations and

precipitates has been recorded using a talistep (model Tencor Alphastep 200) on undoped LEC substrate after several annealings (t = 0.5 h - 60 h/T = 700°C - 1000°C) (figure 2). This relief is related to the impurity concentration around the dislocations (Brown et al, 1986). During the crystal annealing, the impurities drift thermally into the substrate. For 700°C $\leqslant$ T $\leqslant$ 800°C we see that impurities get attracted by dislocations (relief increases). On the other hand, for T = 1000°C, impurities have enough energy to "jump over" dislocations which results in an homogeneous impurity distribution across the substrate (relief decrease).
In the intermediate temperature range (850-950°C), a dip occurs right in the middle of the bump indicating a new phenomenon which can be interpreted as trapping of impurities by already existing As precipitates. This is only in that case that we have been able to visualize precipitates by A-B etching, because they become bigger and introduce large electrical inhomogeneity in the matrix (tomography high scattering intensity).

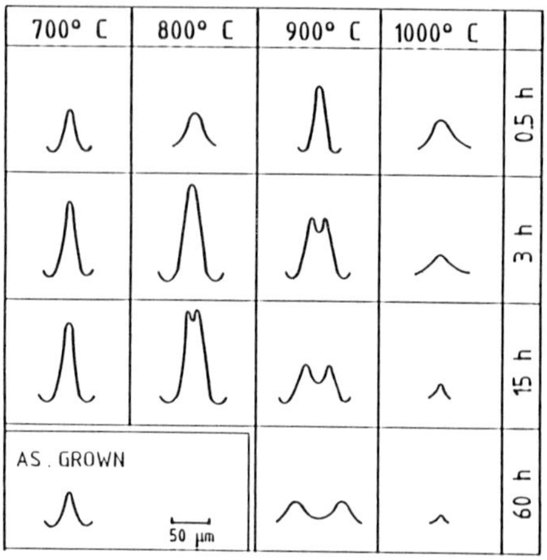

Figure 2 : A-B etching relief around a dislocation after several annealing conditions

We can also conclude from all our reported observations that the pure As precipitates form at temperature between 950 and 1000°C. This seems to be a fast, dislocation assisted process, since only a few minutes are sufficient for the formation (according to fast cooling experiments). It seems to be an irreversible process.
On the other hand, the process [As precipitates] $\underset{2}{\overset{1}{\rightleftharpoons}}$ [As + I precipitates] is reversible. Equilibrium moves in direction 1 when T $\sim$ 900°C, while it moves in direction 2 when T $>$ 950°C. In that later case, impurities can move away from precipitates which are left as pure As precipitates.

REFERENCES

Brown GT and Warwick CA 1986 J. Electrochem. Soc. 133 n°12, 2576
Cornier JP 1986 Thesis University of Paris VI
Cornier JP, Duseaux M and Chevalier JP 1985 Inst. Phys. Conf. Ser. 74(2) 95
Cullis AG, Augustus PD and Stirland DJ 1980 J. Appl. Phys. 51 2556
Duseaux M, Martin S, Chevalier JP 1986 Semi-insulating III-V Materials, Hakone (Japan) 221
Farges JP, Suchet P 1987 private communication
Moriya K and Ogawa T 1978 J. of Cryst. Growth 44 53
Moriya K 1986 Semi-insulating III-V Materials, Hakone (Japan) 151
Ogawa Y 1985 D.R.I.P. in III-V Compounds, Montpellier (France) 1
Stirland DJ 1986 Semi-insulating III-V Materials, Hakone (Japan) 81
Stirland DJ, Augustus PD, Brozel DR, Foulkes EJ, Semi-insulating III-V Materials, Kah-nee-ta (USA) 91
Suchet P, Duseaux M, Gillardin G, Le Bris J and Martin GM 1987 to be published in J. Appl. Phys.
Suchet P, Chevalier JP 1987, private communication

This work has been partly supported by an EEC ESPRIT contract

Shallow and deep donors in ^{119}Sn doped $Ga_{1-x}Al_xAs$ studied by Mössbauer spectroscopy

P. Gibart*, D.L. Williamson**, B. El Jani* and P. Basmaji*

*Laboratoire de Physique du Solide et Energie Solaire - CNRS, Sophia Antipolis 06560 Valbonne, France
** Colorado School of Mines, Golden, Colorado 80401, USA

ABSTRACT : Epitaxial layers of $Ga_{1-x}Al_xAs$ ($0 \leqslant x \leqslant 1$) doped with ^{119}Sn - enriched tin have been grown by metal organic vapor phase epitaxy (MOVPE) and characterized by ^{119}Sn - Mössbauer spectroscopy. The GaAs and AlAs samples yield Mössbauer spectra that are interpreted in terms of two Sn sites corresponding mainly to substitutional shallow donors and clustered species. The x = 0.3 - 0.4 samples yield significantly altered Mössbauer resonance with a new type of dominating site. This is interpreted as due to the DX-center with an electronic structure altered due to electron localization and a local distortion of cubic symmetry.

1. INTRODUCTION

The nature of substitutional donors in $Ga_{1-x}Al_xAs$ is fairly well established (Lang 1985) : independant of the nature of the dopant and the epitaxial growth method, donor incorporation leads to similar features : for x < 0.2 the n impurity introduces a shallow level a few meV from the Γ valley ; for x > 0.2 , in addition to the effective mass level, a deep state appears with an activation energy of the order of 10^2 meV. These deep states ("DX" centers) can be ionized to produce electrons in the Γ valley. From there they must surmount a large potential barrier such that thermal recapture to the deep level is impossible at low temperature. This is known as the persistent photoconductivity effect (PPC). These features have been explained by a model of large lattice relaxation. It is concluded from previous studies that the metastability of DX centers is connected with the particular band structure of $Ga_{1-x}Al_xAs$.

In a previous paper, the behaviour of Sn in $Ga_{1-x}Al_xAs$ grown by metal-organic vapor-phase epitaxy (MOVPE) was deduced from Hall and photoluminescence (PL) measurements (El Jani 1987).

The present investigation uses Mössbauer spectroscopy to obtain further information about the electronic structure and the local symmetry of both types of donors in $Ga_{1-x}Al_xAs$. The pertinent models proposed to explain the behaviour of DX centers have been reviewed (Gibart 1987).

2. EXPERIMENTAL

The preparation of ^{119}Sn- $Ga_{1-x}Al_xAs$ by MOVPE and the details of the Mössbauer experiments are given by Williamson (1986, 1987). The electrical activity of Sn in $Ga_{1-x}Al_xAs$ was deduced from Hall measurements on the same samples studied by Mössbauer spectroscopy and by PL on samples

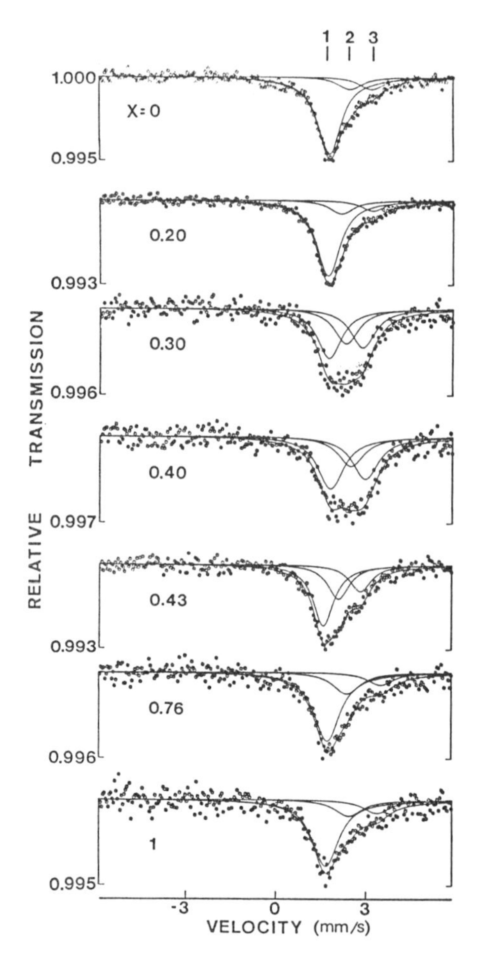

Fig. 1. Mössbauer spectra from ^{119}Sn-doped $Ga_{1-x}Al_xAs$. The solid line passing through the data is a least square fit of the three Lorentzian lines indicated the Mössbauer spectral parameters are given by Gibart (1987).

prepared under similar conditions but with a total Sn concentration in the epilayers of 5.10^{18} cm^{-3}. It was shown - as in other reports - that for x > 0.2 Sn is incorporated simultaneously as shallow donors and as deep localized states.

3. RESULTS

Figure 1 shows the 76 K Mössbauer spectra of 7 $Ga_{1-x}Al_xAs$ samples ranging from x = 0 (Ga As) to x = 1 (Al As). To obtain a good fit by least squares refinement, three Lorentzian lines were required. The fits shown on Fig. 1 were obtained with the single restriction that the three lines have the same linewidth. The salient feature is the enhanced resonance from lines 2 and 3 for x = 0.3, 0.4 and 0.43 ; this variation can be compared to the variation of DX centers concentration deduced either from PPC measurements (Chaud 1984) or by theoretically evaluated (Shubert 1984). Since the evaluation of DX concentration is not a direct measurement and since the concentration of Sn_{Ga} (N_{SD} shallow donor) from Mössbauer experiments (ME) could also be overestimated (the shallow acceptors, Sn_{Ga} - V_{Ga} have approximately the same hyperfine parameters as Sn_{Ga}), the agreement is good. Hence, the lines 2 and 3 are assumed to correspond to DX centers. Furthermore, assignment of lines 2 and 3 to DX centers is supported by ME under pressure (Williamson 1986,1987, Moser 1987).

4. DISCUSSION

4.1 Isomer shift (I.S, δ)

The Mössbauer IS is directly related to the nature of the chemical bond via its proportionality to the electron contact density, $\rho(0)$.

For ^{119}Sn, changes in $\rho(0)$ are produced primarily by changes in the 5s-like valence electron occupation number, n_s, although substantial

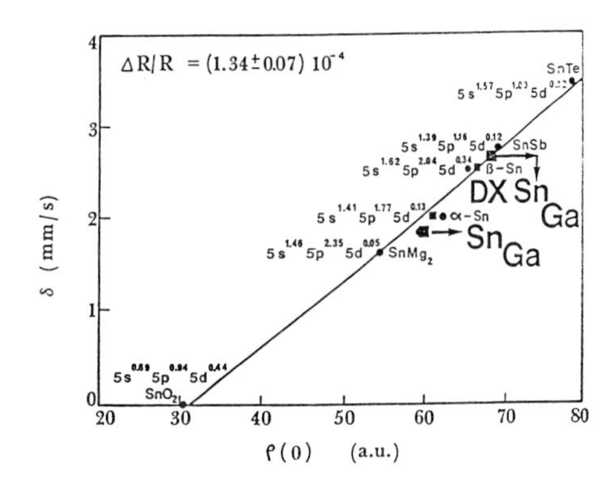

Fig. 2. From ref. 10 : measured isomer shifts in tin compounds versus the calculated electron densities. Occupation numbers are given for the different tin compounds (●),■according to Svane (1985) are calculated here using Eq. (1) near the δ values of interest in the present study.

changes in 5p-like and 5d-like occupations, n_p and n_d, can also cause changes in ρ(0) because of their shielding effects on the 5s-like electrons. In solids, the occupation numbers of valence electrons are non-integers and realistic numbers should be deduced from band structure calculations. Recently Svane (1985, 1986) and Antoncik (1977, 1982) calculated ρ(0) in several compounds containing tin in an effort to establish the proportionality constant between δ and ρ(0). Their calibration will be used as a starting point for discussing the electronic structure of Sn in $Ga_{1-x}Al_xAs$. Figure 2 shows the δ <u>vs</u> ρ(0) correlation for different tin compounds together with the calculated electronic configurations. Since a detailed theoretical investigation of the electronic structure of Sn in (Ga Al)As is beyond the scope of the present paper, we use an approximation discussed by Antoncik (1977).

$$\rho(Sn)(n_s,n_p)=61.0n_s-8.25n_s^2-3.3n_sn_p \qquad (1)$$

which is found to give a reasonable agreement between δ and the theoretical fit of Svane and Antoncik (1986) (squares on figure 2). Equation (1) will be used in the following discussion as an approximate means of estimating the Sn electronic structure.

4.2 Electronic structure of Sn_{Ga} and Sn_{Al} in $Ga_{1-x}Al_xAs$

The electronic configurations of Ga and As in GaAs deduced from a tight binding approximation are $5s^{1.48}$ $5p^{1.38}$ and $5s^{1.54}$ $5p^{3.60}$, respectively (Gu 1982). When an atom acts as an electrically active impurity in a semiconductor the correct evaluation of its electronic configuration should involve compression effects. Furthermore, the electronic structure of the host semiconductor and its influence on the impurity are to be taken into account also. These features have been discussed by Gu (1982) and his main conclusions can be summarized as follows : In the case of Sn_{Ga} in GaAs, it is of very little importance whether this level is occupied or not ; the wave function associated with a shallow donor is highly delocalized and does not contribute to ρ(0). The best agreement with δ corresponds to Sn donors having three valence electrons and transferring no charge to the neighbouring As atoms. Then, in this evaluation the occupancy numbers of Sn_{Ga} in GaAs are $5s^{1.28}$ $5p^{1.72}$ and the corresponding δ calculated according to Eq.(1) is plotted in Fig. 2. The electronic structure of Sn_{Al} in AlAs was estimated to be $5s^{1.21}$ $5p^{1.26}$ $sd^{0.12}$ (Basmaji 1987). Antoncik and Gu (1982) discussed the limits of validity of their evaluation ; they argue that the lack of relevant information on the compression effect does not allow one to give a unique value for U(Sn) the extra potential produced in the Hamiltonian by the Sn impurity. A redistribution of the electrons of th Sn atom could also be restricted to $\delta\, n_s(Sn) + \delta\, n_p(Sn) = 0$: in this case the deduced occupation numbers for Sn_{Ga} would be $5s^{1.19}5p^{1.67}$ which gives the same value for $\rho(0)$ (Fig. 2).

4.3 Electronic structure of Sn DX-center in $Ga_{1-x}Al_xAs$

The information brought by ME is that the DX centers attributed to lines 2 and 3 in Fig. 1 for x = 0.3 - 0.4 could be either (3) : **(i)** a single substitutional atom with a quadrupole splitting ; **(ii)** two or more deep Sn-related centers in undistorted symmetry ; or **(iii)** even a distribution of similar centers. In all cases the electronic structure is quite different from that of the Sn_{Ga} shallow donors (SD). The occupancy numbers of 5s electrons in DX centers - deduced from Eq. (1) and δ values of lines 2 and 3 - exceed by 0.4 to 0.6 electron those of the Sn_{Ga} (SD). As suggested by various atomic-scale DX models, the hybridization of s and p orbital in DX centers is completely different from the SD, and even different from donors tied to the X valley (like Sn_{Al} in AlAs). Thus photoionization of a DX center involves not only the transfer of an electron to the Γ valley, but also a bond reconstruction. If we assume a unique Sn - DX center, then lines 2 and 3 correspond to a quadrupole splitting. In this case, δ = 2.67 mm/s, Δ = 0.6 mm/s. According to Eq. 1 the electronic structure would be $5s^{1.74}$ $5p^{2.25}$, and the observed quadrupole splitting means that a significant local distortion of the As surrounding occurs. This is in agreement with DX models involving local distortion (Oshiyama 1986, Morgan 1986, Hasegawa 1986).

5. CONCLUSION

Mössbauer spectroscopy yields evidence that substitutional tin in $Ga_{1-x}Al_xAs$ binds electrons for x > 0.2 in deep states associated with lattice relaxation. More detailed insight will be provided by ME under pressure and under light illumination. In addition EXAFS studies would also bring crucial new information on the local structure of the DX-center. These experiments are all currently in progress.

REFERENCES

Antoncik E, 1977, Phys. Stat. Sol., b 179, p 605, and refs therein.
Antoncik E and Gu BL, 1982, Physica Scripta 25, p 836.
Basmaji P, Raymond F, Gibart P, Williamson DL (submitted).
Chaud N., Henderson T, Klem J, Ted Masselink W, Fisher R, Chang YC and Morkoç H, 1984, Phys. Rev., B 30, p 4481.
El Jani B, Kölher K, N'Guessan K, Bel Hadj A and Gibart P, 1987, J. Appl. Phys. (submitted).
Gibart P, Williamson DL, El Jani B and Basmaji P, 1987, Phys. Rev. B (submitted).
Gu BL, 1982, thesis, Aarhus University.
Hasegawa H and Ohno H, 1986, Jpn. J. Appl. Phys., 25, L 319.
Lang DV, 1985, in Deep Levels in Semiconductors, S.T. Pantelides ed., Gordon and Breach, New-York, p. 489 and refs therein.
Morgan TN, 1986, Phys. Rev., B 34, p 2664.
Moser J, 1987, private communication.
Oshiyama A and Ohnishi S, 1986, Phys. Rev., B 33, p 4320.
Shubert E and Ploog K, 1984, Phys. Rev., B 30, p 7021.
Svane A and Antoncik E, 1986, Phys. Rev., B 34, p 1944 ; Svane A, 1985 thesis Aarhus University.
Williamson DL, 1986, J. Appl. Phys., 60, p 3466.
Williamson DL, Gibart P, El Jani B and N'Guessan K, 1987, J. Appl. Phys., 62, p 1739

Paper presented at Int. Symp. GaAs and Related Compounds, Heraklion, Greece, 1987

Correlation of oscillations in the far-infrared photoresponse with conductance oscillations in GaAs/AlGaAs single barrier tunnel structures

P.M. Campbell, J. Comas, R.J. Wagner, and J.E. Furneaux
Naval Research Laboratory, Washington, DC 20375

ABSTRACT: We report the observation of oscillations of period $E_{LO\ phonon}/e$, coincident with conductance oscillations, in the far-infrared photoresponse of GaAs/AlGaAs single-barrier heterostructures. In addition, the photoresponse shows a strong resonant peak at the value of the magnetic field at which the $1s \rightarrow 2p(m = = +1)$ transition energy of the shallow donors equals the energy of the incident radiation. These results suggest that the oscillations in both the photoresponse and the conductance are caused by the periodic ionization and deionization of shallow donor impurities.

1. INTRODUCTION

The observation by Hickmott *et al.* (1984) of anomalous conductance oscillations (of period E_{LO}/e = 36 mV, where E_{LO} = LO phonon energy) in the tunnel current of appropriately biased GaAs/AlGaAs/GaAs single-barrier heterostructures has generated considerable interest in the transport properties of these structures. Figure 1 illustrates the conduction band of such a device under voltage bias. Electrons from the emitter tunnel through the AlGaAs barrier into the n^-GaAs drift region, where they lose their energy by the emission of LO phonons. It is therefore not surprising to find LO phonon-related structure in the transport characteristics of these structures. Figure 2 shows typical I *vs.* V and dI/dV *vs.* V traces of such a device. The remarkable aspects of this effect include not only the size of the oscillations ($\Delta I/I$ as large as 10%) but also the large number of oscillations, with up to thirty periods evident in some samples. In some cases the application of magnetic field is necessary to observe the oscillations, as in Hickmott *et al.* (1984), while others (Guimaraes *et al.* 1985) report the effect in zero magnetic field. The structures used in this report all show clear oscillations with no magnetic field, but the application of a magnetic field parallel to the current transport increases the size of the effect by up to an order of magnitude.

Considerable controversy has developed over the cause of these oscillations, and numerous theoretical models have been proposed to explain their origin, including: sequential single-phonon emission by ballistic electrons (Hickmott *et al.* 1984); oscillations in the substrate collector density of states (Ihm, 1985); space-charge accumulation due to polaron formation (Hellman *et al.* 1986); and the periodic change in the charge state of shallow donors, either by impact ionization (Eaves *et al.* 1985) or by phonon ionization (Leburton 1985), which alters the voltage profile across the device and hence modulates the tunneling current through the barrier. Most of these theories agree that the tunneling electrons lose energy in the n^-GaAs drift region by LO phonon emission, but disagree over the mechanism coupling this emission to the conductance oscillations. All of the proposed models can predict qualitatively at least some of the features of the data. However, detailed analysis has raised serious doubts about many of the proposed mechanisms, with only the shallow donor ionization model remaining relatively unchallenged in principle. In addition, Eaves *et al.* (1985) have reported the observation of a sharpening of the conductance peaks in a magnetic field of 5.5T, where the separation between Landau levels equals the energy difference between the first Landau level and the ground state of the shallow donors. Although this observation is suggestive of a shallow donor mechanism, a direct experimental probe of the charge state of the shallow donors is needed to establish the shallow donor link conclusively.

2. EXPERIMENTAL APPROACH

In order to test the shallow donor hypothesis, we have fabricated a set of specially-designed single-barrier tunnel structures which allow direct probing of the shallow donors in the n^-GaAs drift region by far-infrared radiation. The cross-section of this structure is illustrated in Figure 3. These structures were grown by molecular beam epitaxy at a temperature of 630°C. A 1μm buffer layer of n^+ GaAs ($n = 2 \times 10^{18}$ cm^{-3}) was followed by an undoped AlGaAs barrier of thickness 112Å and aluminum mole fraction equal to 0.35. Next, an n^-GaAs drift region ($n = 1 \times 10^{15}$ cm^{-3}) of thickness 1.0μm was grown, followed by a thin (2000Å) n^+ GaAs contact layer ($n = 2 \times 10^{18}$ cm^{-3}). Annular Au-Ge-Ni contacts were patterned on the top layer, and planar contacts were made to the bottom of the substrate. Mesas were then etched to form isolated devices. Finally, approximately 1000Å of the top n^+GaAs contact layer was etched away from the central portion of the annulus in order to permit greater light transmission through the top contact into the GaAs drift region. This design allows probing of the interior of the device with far-infrared radiation while still maintaining an ohmic contact for applying voltage bias across the device. This structure showed strong conductance oscillations similar to those of Hickmott *et al.* (1984) and Guimaraes *et al.* (1985). However, the oscillations in our structure did tend to become weaker and somewhat distorted at large bias, probably due to the thinness of the top ohmic contact layer.

Shallow donors in GaAs are, to a very close approximation, hydrogenic in nature. As such, they will exhibit all the spectroscopic features of the hydrogen atom, but scaled by the effective Rydberg $R_y^* = (m^*/m_0)\,(\epsilon_0/\epsilon)^2\, R_y = 5.83$ meV where, respectively, m_o and m^* are the free and effective electron masses, ϵ_o and ϵ are the permittivity of free space and GaAs, and $R_y = 13.6$eV is the binding energy of the hydrogen atom. As in the free hydrogen atom, the application of a magnetic field will lift the degeneracy of the four $n = 2$ states, and this splitting will increase

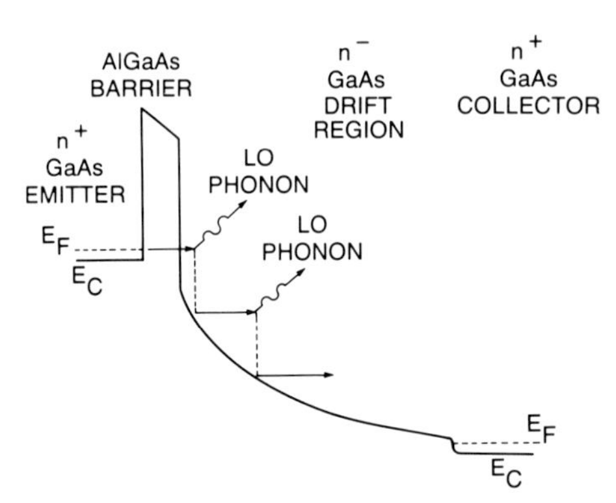

Fig. 1. Conduction band diagram of single barrier tunnel structure under voltage bias (above).

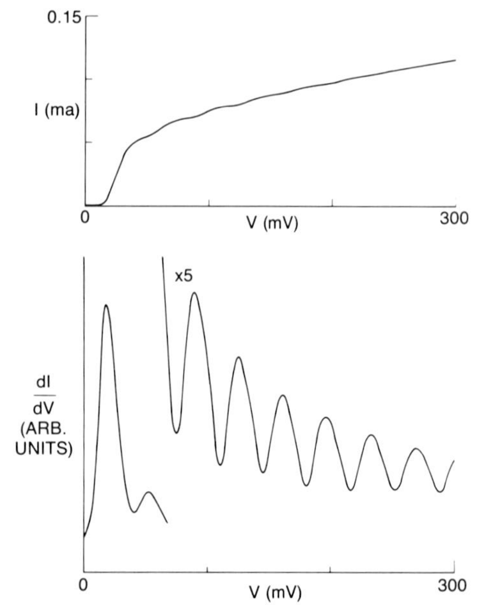

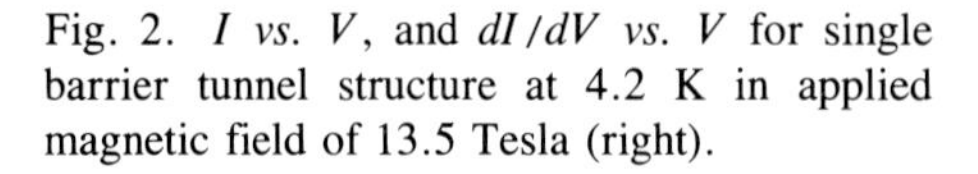

Fig. 2. *I vs. V*, and *dI/dV vs. V* for single barrier tunnel structure at 4.2 K in applied magnetic field of 13.5 Tesla (right).

with field. However, because the effective Rydberg is reduced by a factor of several thousand, transitions between different states of the shallow donors will be in the far-infrared rather than the visible or near-ultraviolet region of the spectrum. Therefore, far-infrared radiation of energy equal to the energy difference between states can induce resonant absorption between these states. If the excited state is within kT of the conduction band, the excited electron can transfer thermally to the conduction band and change the conductance of the sample. This combined process of photothermal ionization has been reported in GaAs by Stillman *et al.* (1969).

3. RESULTS AND DISCUSSION

We have measured the photoconductive response of the structure shown in Figure 3 to far infrared radiation of wavelength $\lambda = 70.5\mu m$. At a temperature of 4.2K, the photoconductive response showed a large resonant maximum at an applied magnetic field value of 8 Tesla. At this value of magnetic field, the energy of the infrared radiation is equal to the energy separation between the $1s$ and $2p(m = +1)$ hydrogenic states of the shallow donors. We therefore conclude that the photoconductive response at this value of magnetic field is dominated by photothermal ionization involving the $1s \rightarrow 2p(m = +1)$ transition of the shallow donors. The $1s \rightarrow 2p(m = +1)$ transition rate is proportional to the number of occupied $1s$ states. Therefore, under these experimental conditions, the resonant photoconductive response is a sensitive probe of the charge state of the shallow donors, so that an increase in the number of shallow donors in the ground state will increase the photoconductive response.

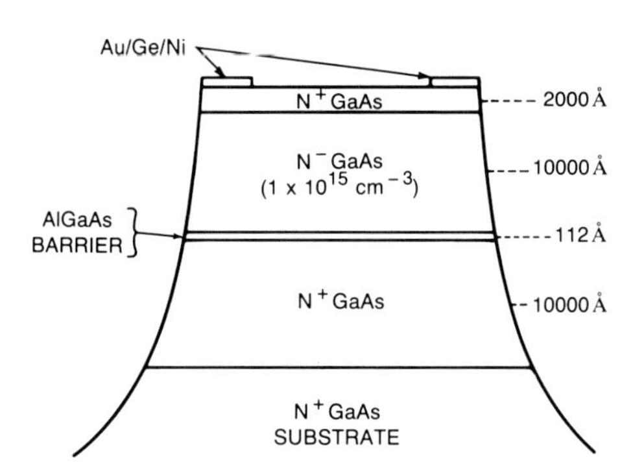

Fig. 3. Cross sectional diagram of device used to measure far-infrared photoconductive response of single barrier tunnel structures (above).

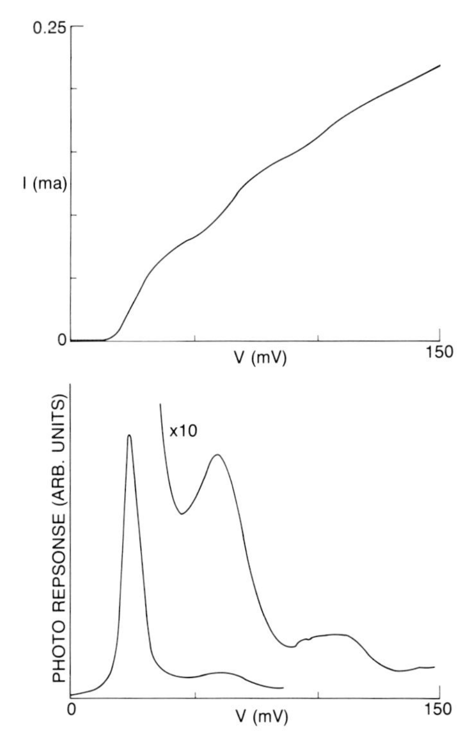

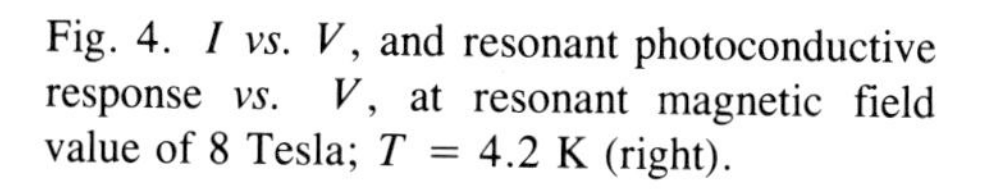

Fig. 4. I *vs.* V, and resonant photoconductive response *vs.* V, at resonant magnetic field value of 8 Tesla; $T = 4.2$ K (right).

In the next experiment, we fixed the magnetic field at the resonant value of 8 Tesla and measured the photoconductive response as a function of voltage bias across the device. At the resonant magnetic field, the photoconductive response showed oscillations, of period 36 mV, similar to the conductance oscillations observed in this structure. Furthermore, the maxima in the photoresponse oscillations occur at the voltage biases corresponding to the current minima in the conductance oscillations. This is shown in Figure 4. Because of this correlation, and because at the resonant magnetic field the photoconductive response is sensitive to the number of shallow donors in the $1s$ state, we conclude that the conductance oscillations are related to the periodic ionization and deionization of shallow donor impurities, as suggested by Eaves *et al.* (1985), who assumed an impact ionization model, and Leburton (1985), who invoked a phonon assisted ionization mechanism. Both models agree that the electrons tunnel through the AlGaAs barrier and gain energy as they traverse the depleted portion of the GaAs drift region. The electrons lose energy by emitting an integral number of LO phonons, each of energy 36 meV, and thus cool to within one LO phonon energy of the bottom of the conduction band. It is this mechanism which imposes the 36 meV periodicity in voltage bias. This same periodicity is likewise imposed upon the conditions leading to the ionization of the shallow donors in the undepleted region of the n^- GaAs layer. The details of the ionization mechanism vary depending upon whether one uses Eaves' or Leburton's model, but the general features of the process producing the conductance oscillations can be explained in the following way. For a given voltage bias across the device, an increase in the percentage of neutral donors in the undepleted portion of the GaAs drift region will decrease the conductivity of this layer, and hence a larger portion of the total voltage bias across the device will be dropped across this region. This implies that the voltage drop across the rest of the structure must decrease, including the drop across the AlGaAs barrier. The tunneling current is a strong function of the bias across the barrier. Hence, even a small drop in the bias across the barrier will result in a relatively large decrease in the tunnel current. Calculations by both Eaves *et al.* (1985) and Leburton (1985), each using their own ionization model, have produced estimates for the change in the current by this feedback mechanism which are in reasonably good agreement with the measured values.

REFERENCES

Eaves L, Guimaraes P S S, Sheard F W, Snell B R, Taylor D C, Toombs G A, and Singer K E 1985 J. Phys. C. 18 L885.

Guimaraes P S S, Taylor D C, Snell B R, Eaves L, Singer K E, Hill G, Pate M A, Toombs G A, and Sheard F W 1985 J. Phys. C. 18 L605.

Hellman E S and Harris J S 1986 Phys. Rev. *B*33 8284.

Hickmott T W, Soloman P M, Fang F F, Stern F, Fischer R, and Morkoc H 1984 Phys. Rev. Lett. 52 2053.

Ihm J 1985 Phys. Rev. Lett. 55 999.

Leburton J P 1985 Phys. Rev. *B*31 4080.

Stillman G E, Wolte C M, and Dimmock J O 1969 Solid State Commun. 1 921.

Magnetotunnelling in single-barrier III–V semiconductor heterostructures

F W Sheard, K S Chan, G A Toombs, L Eaves and J C Portal*

Department of Physics, University of Nottingham, Nottingham NG7 2RD, U.K.
*LPS, INSA, 31077 Toulouse and SNCI-CNRS, 38042 Grenoble, France.

ABSTRACT: The tunnelling of electrons between a 2DEG and n^+ electrode is studied theoretically. In the presence of a quantising magnetic field parallel to the plane of the barrier, it is shown that the electrons tunnel into interfacial Landau states, which correspond to skipping orbits along the barrier interface. Experimental data on InP/(InGa)As and (AlGa)As/GaAs single-barrier structures are discussed.

We have studied the tunnelling of electrons between a degenerate 2DEG and n^+ electrode in a semiconductor heterostructure, in the presence of a magnetic field parallel to the plane of the tunnel barrier. The band structure is shown in Figure 1 for a n^-(InGa)As/InP/n^+(InGa)As device under a forward bias voltage V. The effect of the magnetic field on the bound state of the 2DEG can be shown to be small (Stern and Howard, 1968). From Figure 1, an electron at the Fermi level of the 2DEG tunnels through the barrier and emerges into the RHS n^+ region with kinetic energy $E = eV + E_{FR}$, where E_{FR} is the Fermi energy on the RHS. Owing to magnetic quantisation of the electron motion in this region, a resonance in the tunnel current occurs whenever E coincides with a Landau level E_n. This results in oscillatory structure in the voltage or field dependence of the current. However we shall show that in typical heterostructures the electrons can only tunnel into interfacial Landau states. These are states which correspond to skipping orbits of the electron along the barrier interface. Coupling to bulk Landau states, which correspond to closed orbits in which the electron does not strike the interface, does not occur.

We show this by considering the effect of the Lorentz force on the electron motion. Taking the x axis perpendicular to the tunnel barrier (Figure 1) and $\underline{B}\|Oz$, the equation of motion of the momentum component m^*v_y of an electron (effective mass m^*, charge -e) is

$$m^* dv_y/dt = ev_xB = eBdx/dt \ .$$

Hence the y component of canonical momentum $p_y = m^*v_y - eBx$ is a constant of the motion (as is $p_z = m^*v_z$). For electrons in the degenerate 2DEG, m^*v_y lies in the range $\pm p_{FL}$, where $p_{FL} = (2m^*E_{FL})^{\frac{1}{2}}$ is the Fermi momentum on the LHS. The linear variation of m^*v_y with distance x is shown in Figure 2(a) for electrons with extremal momenta $\pm p_{FL}$ in the 2DEG. Taking $x = 0$ to be the RHS of the barrier, the 2DEG is situated at $x = -(b + a)$, where b is the barrier width and a is the mean distance of the 2DEG from the LHS of the barrier. Thus for the case when $m^*v_y = -p_{FL}$, the constant $p_y = -p_{FL} + eB(b + a)$. When an electron emerges from the barrier into the RHS n^+ region, the orbit (in the xy plane) is a circular arc, as in Figure 2(b), and the position of the orbit centre $x = X_-$ corresponds to $v_y = 0$. This gives $X_- = -p_y/eB = (p_{FL}/eB) - (b + a)$. If the orbit intersects the barrier interface x = 0, the electron will be repeatedly reflected (since the tunnelling probability is small) and we have a skipping orbit as illustrated. In order for the electron to tunnel into a closed (bulk) orbit it is necessary that $X_- > R$, where R is the orbital radius. Setting the electron kinetic energy $E = \frac{1}{2}m^*v^2$, we have $R = v/\omega_c = (2m^*E)^{\frac{1}{2}}/eB$, where $\omega_c = eB/m^*$ is the cyclotron frequency. Thus the condition for coupling to a bulk orbit is

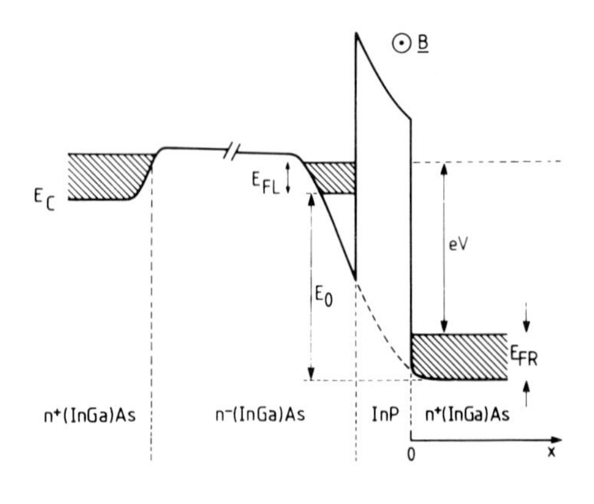

Fig. 1. Spatial variation of electron energy at conduction-band edge E_c for InP/(InGa)As device under forward-bias voltage V.

$$p_{FL} - eB(b + a) > (2m^*E)^{\frac{1}{2}}.$$

But in heterostructures with barrier widths ~ 200 Å the charge in the 2DEG, and hence E_{FL}, is limited by the small capacitance and, as in Figure 1, $E_{FL} << E$ at all biases. Hence $p_{FL} << (2m^*E)^{\frac{1}{2}}$ and the above condition cannot be satisfied. For the $+p_{FL}$ case, the electrons also tunnel into a skipping state and the orbit centre $X_+ = -(p_{FL}/eB) - (b + a)$ lies on the LHS of the 2DEG as shown in Figure 2.

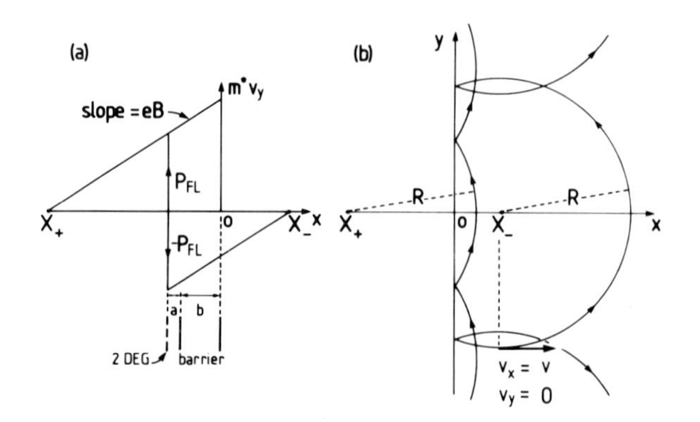

Fig. 2. (a) Variation of transverse electron momentum m^*v_y with distance x for extremal $\pm p_{FL}$ electrons. (b) Skipping orbits corresponding to centre coordinates X_-, X_+.

Thus the electrons tunnel from the degenerate 2DEG into skipping states whose orbit centres lie in the range $X_+ < X < X_-$. But the electron motion in the n^+ region is quantised and the kinetic energy takes discrete values $E_n(X)$ ($n = 0,1,2...$), which depend on X. This gives a discrete set of orbit centres X_n into which the electrons can tunnel. A resonant change in current occurs whenever, as V or B is varied, a discrete value X_n enters or leaves the allowed range. This gives rise to two series of oscillatory structure associated with the two extremal values of momentum $\pm p_{FL}$.

Since the x component of the Lorentz force

$$F_x = -eBv_y = -\omega_c(eBx + p_y) = -m^*\omega_c^2(x - X) = -\partial U/\partial x,$$

can be written as the gradient of a simple harmonic potential $U(x) = \frac{1}{2}m^*\omega_c^2(x - X)^2$, the energy levels are those of a quantum oscillator, centred at $X = -p_y/eB$, but restricted by the barrier at $x = 0$. For sufficiently large X, we have bulk Landau levels with $E_n = (n + \frac{1}{2})\hbar\omega_c$ independent of X. As X decreases the wave function is squeezed closer to the barrier, $E_n(X)$ increases and we have interfacial Landau states (skipping orbits). In the WKB approximation the energies of the interfacial states are given by

$$E_n(X)\left\{\frac{1}{2} + \frac{1}{\pi} f(X/R_n)\right\} = (n + \frac{3}{4})\hbar\omega_c ,$$

$$f(u) = \sin^{-1}u + u(1 - u^2)^{\frac{1}{2}},$$

where $R_n = \{2m^*E_n(X)\}^{\frac{1}{2}}/eB$ is the classical orbit radius.

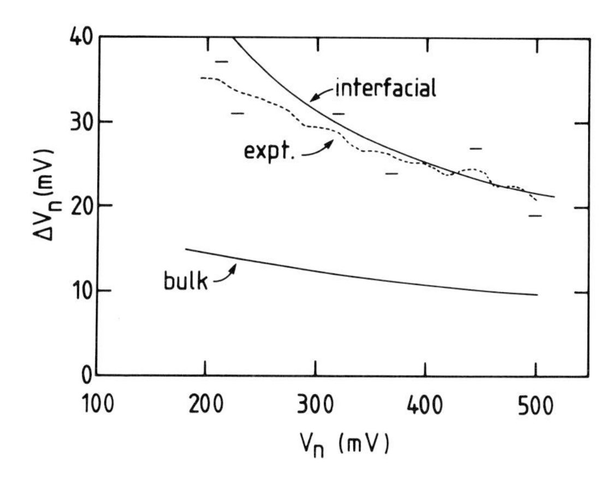

Fig. 3. Variation of voltage period ΔV_n with applied bias V_n. Dashed line: average of experimental results taken at 1.6 - 3.2 K, horizontal bars show range of data. Solid lines: theory for interfacial levels and bulk Landau levels for same value of nonparabolicity.

To find the voltages (or fields) at which the resonant structure occurs, we equate the tunnelling energy $E = eV + E_{FR}$ to the Landau-level energy $E_n(X_\pm)$, for the extremal $\pm p_{FL}$ electrons:

$$eV + E_{FR} = E_n(X_\pm), \qquad X_\pm = \mp(p_{FL}/eB) - (b + a).$$

These equations must be solved numerically for each value of n.

Experiments on InP/(InGa)As structures revealed two series of oscillatory structure in the current-field dependence at a fixed bias V (Chan et al., 1987, Snell et al., 1987). The $+p_{FL}$ series was appreciably weaker than the $-p_{FL}$ series. For each series the structure was periodic in 1/B occurring at fields B_n given by $1/B_n = (n + \phi)/B_f$, where ϕ and B_f are voltage dependent. This is consistent with the above theory. The fundamental periodicity field B_f varies approximately linearly with voltage for both $\pm p_{FL}$ series. Deviations from linearity arise principally from nonparabolicity of the conduction band in (InGa)As. Using $m^*(E) = m(0)(1 + \alpha E)$, comparison with experiment gave $m(0) = 0.04m_e$ and $\alpha = 1.1 \times 10^{-3}$ meV^{-1}.

Similar experiments on (AlGa)As/GaAs structures (Hickmott 1987) revealed only the $-p_{FL}$ series of oscillatory structure in the current-voltage dependence at a fixed field B = 13 T. We have calculated the voltages V_n corresponding to tunnelling of the $-p_{FL}$ electrons. The voltage period $\Delta V_n = V_n - V_{n-1}$ decreases with increasing V_n owing to nonparabolicity and unequal spacing of the interfacial Landau levels. Using the measured values of capacitance C = 2930 μF m^{-2}, barrier width b = 23 nm (Hickmott 1987) and a mean value $a \cong 8$ nm, good agreement with experiment is obtained (Figure 3) for tunnelling into interfacial levels with $m(0) = 0.067m_e$ and $\alpha = 1.2 \times 10^{-3}$ meV^{-1}. The nonparabolicity parameter α is comparable with that obtained by Heiblum et al (1987). It is clear from Figure 3 that tunnelling into bulk levels cannot account for the experimental data. Deviations between theory and experiment at low voltages may be due to neglect of band bending in the n^+ region in calculating the energies of the interfacial levels. The absence of the $+p_{FL}$ series in the experimental data is attributed to the effect of the magnetic potential U(x) which at 13 T significantly raises the tunnel barrier and reduces the tunnelling probability for $+p_{FL}$ electrons as compared with the effect for $-p_{FL}$ electrons.

ACKNOWLEDGEMENT: This work is supported by the SERC.

REFERENCES

Chan K S, Eaves L, Maude D K, Sheard F W, Snell B R, Toombs G A, Alves E S, Portal J C and Bass S 1987 Conference on Hot Electrons (Boston) to be published

Heiblum M, Fischetti M V, Dumke W P, Frank D J, Anderson I M, Knoedler C M and Osterling L 1987 Phys. Rev. Lett. 58 816

Hickmott T W 1987 Solid St. Commun. 63 371 and private communication

Snell B R, Chan K S, Sheard F W, Eaves L, Toombs G A, Maude D K, Portal J C, Bass S J, Claxton P, Hill G and Pate M A 1987 submitted to Phys. Rev. Lett.

Stern F and Howard W E 1967 Phys. Rev. 163 816

The microscopic structure of EL2 defects, their thermal stability and distribution across s.i. GaAs wafers

J.-M. Spaeth, D.M. Hofmann, M. Heinemann and B.K. Meyer

University of Paderborn, FB 6 - Physik, Warburger Str. 100A,
D-4790 Paderborn, FRG

ABSTRACT: Previous ODENDOR results showed that the EL2 defect in GaAs is a (As_{Ga}-As_i) pair. From a theoretical estimate of the quadrupole interaction constants of the As_i and the 4 nearest and 12 next nearest As-neighbours of As_{Ga} is inferred that the As_i is on the interstitial T_{d2} site $\sim$ 4.9 Å from As_{Ga} and positively charged. A mapping investigation of both the IR absorption of $EL2^o$ and the magnetic circular dichroism of $EL2^+$ showed that the EL2 defects are distributed rather homogeneously and that shallow acceptors must exist of yet unknown nature in the concentration range of about $2x10^{16}$ cm^{-3} with a M-shaped distribution. Annealing at 1200 ^oC under As pressure and rapid quenching destroys the EL2 defects. New paramagnetic centers appear. Annealing at 850 ^oC restore the EL2 defects, the new centers disappear.

1. INTRODUCTION

In order to obtain good s.i. GaAs substrate material it is important to understand the structure of the deep level defects rendering the material semi-insulating and the compensation mechanism. In s.i. LEC GaAs the dominant mid gap donor is the EL2 defect. It is also necessary to analyse and understand the defect distribution across a wafer (inhomogeneities) and their thermal stability.

2. STRUCTURE OF THE EL2 DEFECTS - SITE AND CHARGE STATE OF THE INTERSTITIAL ARSENIC

In our previous work (Meyer et al 1987) we identified the EL2 defect in s.i. GaAs as being an arsenic antisite-arsenic interstitial pair (As_{Ga} - As_i) with C_{3v} symmetry (Fig.1). This was accomplished by measuring and analysing the optically detected ESR and ENDOR spectra of $EL2^+$ and correlating the ODESR/ODENDOR spectra with the energy levels of $EL2^+$ ($As_{Ga}{}^+$ - $As_i{}^+$) and $EL2^o$ ($As_{Ga}{}^o$ - $As_i{}^+$). The presence of the As_i as an integral part of the EL2 defect is not only seen by its ENDOR lines between 80 and 140 MHz, but also in the 3rd shell of ^{75}As neighbours. Without the interstitial all 12 3rd shell As atoms would be equivalent with respect to the As_{Ga}, they would have the same superhyperfine (shf) and quadrupole interactions. They are, however, not equivalent. There is one shell (III_a) with 3 equivalent As atoms having interactions (a = 35.2 MHz, b = -1.3 MHz, q = 2.8 MHz), which differ considerably from those of shells III_b with 3 equivalent atoms and of shell III_c (6 equivalent atoms, a = 19.5 MHz, b = 3.2 MHz, q = 0.9 MHz), which are similar.

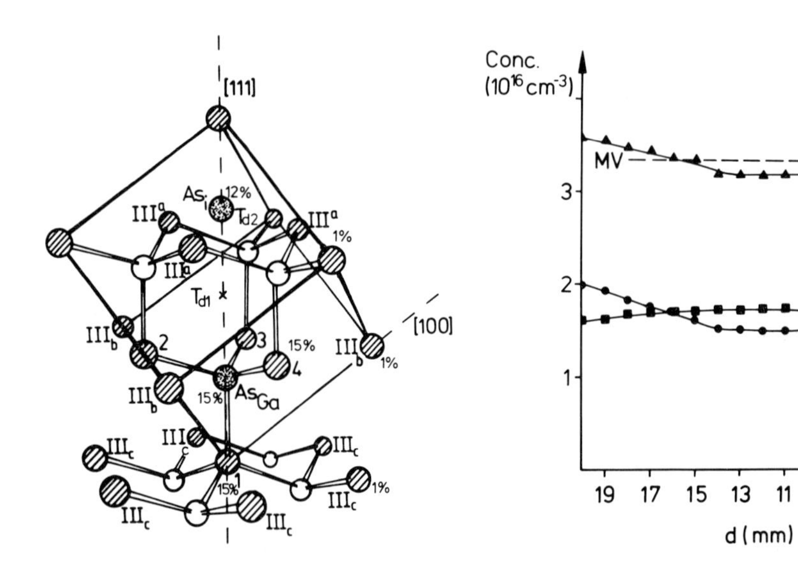

Fig. 1. Detailed model of the As_{Ga}-As_i pair. The nearest As ligands are labeled 1 to 4, the next nearest neighbours are labeled by III.

Fig. 2. Spatially resolved MCD ($EL2^+$, sqares) and IR absorption ($EL2^o$, circles) measurement across half a s.i. as-grown GaAs wafer. (Triangles: total EL2 concentration, MV: mean value).

The charge state and the site of the As_i can be inferred from a theoretical analysis of the quadrupole interaction constants. Only the charge states As_i^+ and As_i^- are possible, since the As_i is diamagnetic. The quadrupole interaction constant is approximately determined by two major terms (van Engelen 1980)

$$q = q(p) + q(b) \tag{1}$$

q(p) is due to the electrical field gradient caused by the (point)charges outside the atomic radius of the considered ligand, q(b) is due to the charge density moving in the As outer p-orbitals and can therefore be determined from the experimental anisotropic shf constant b. The charge density distribution can be estimated from a LCAO analysis of the data assuming that both charge and spin distribution are identical (see Fig.1). The nuclei of shell III_a see the charged As_i in the nearest neighbour distance of 2.44 Å. The As_i charge gives the dominant contribution to q of this shell, q(b) is small. In shells III_b (d = 4.68 Å) and III_c (d = 7.32 Å) the point charge contribution is rather small as q(p) decreases with $1/d^3$. A calculation of q taking into account the full charge distribution gives q ≈ 0.8-1 MHz (for As_i^+ or As_i^-) for the shells III_b and III_c in fair agreement with the measured value of 0.9 MHz. A decision As_i^+ or As_i^- can be inferred from the q-value of shell III_a. Assuming As_i^+ one calculates a q of 3.2 MHz compared to the exerimental value of 2.8 MHz, whereas for As_i^- q = -4.3 MHz. Although the sign of q cannot be determined from the experiment, the resuls show, that the explanation of the quadrupole interactions of shell III_a is much better for As_i^+.

The As_i must therefore be positively charged and occupies a site which is rather far away from the As_{Ga} site. Whether or not this is exactly the T_{d2} site cannot be concluded since the estimate of the quadrupole interactions is too approximate for that. A displacement from the T_{d2} site of 10-15 % towards As_{Ga} is well conceivable.

3. DISTRIBUTION OF EL2 DEFECTS ACROSS AN AS-GROWN WAFER

Imaging (Windscheif et al 1985, Dobrilla et al 1985) is widely used to measure the two-dimensional nonuniform EL2-distribution. In s.i. LEC-grown GaAs the IR absorption ascribed to EL2 is W-shaped across a wafer and is believed to reflect the EL2 defect distribution. This technique, however, monitors only one charge state of EL2, and neglects the occupation of the $EL2^+$ charge state if the Fermi level is such, that both charge states of EL2, $EL2^+$ and $EL2^o$, are partially occupied, which is the case in s.i. material. Therefore, the IR measurement is not sufficient to investigate the distribution of the EL2 defect, which requires the knowledge of the sum of the occupation of $EL2^+$ and $EL2^o$.

$EL2^+$ can be monitored by its magnetic circular dichroism (MCD). The paramagnetic MCD signal of $EL2^+$ originates from the level at E_v + 0.52 eV. The occupation of the mid gap level at E_v + 0.74 eV, the diamagnetic two electron state is determined by measuring the IR absorption at 1.24 eV (the charge of the As_i is not included in the notation). The IR absorption constant at 1.24 eV is proportional to the $EL2^o$ concentration (1 cm^{-1} ≙ $1x10^{16}$ cm^{-3}) (Martin 1981) under the assumption that the hole ionisation transition $\sigma_p{}^o$ of $EL2^o$ ($EL2^+$--> $EL2^o$ + $h^+{}_{vb}$) at 1.24 eV is small. The MCD was calibrated using ESR (the MCD value in % at 0.96 eV measured at B = 1 Tesla and T = 1.8 K, 1 % ≙ $3x10^{16}$ cm^{-3}). For the validity of this calibration it was confirmed that ESR and ODESR detect the same species, which was done by an independent investigation of the saturation behaviour of the ESR signal at low temperatures. (Wilkening et al 1987).

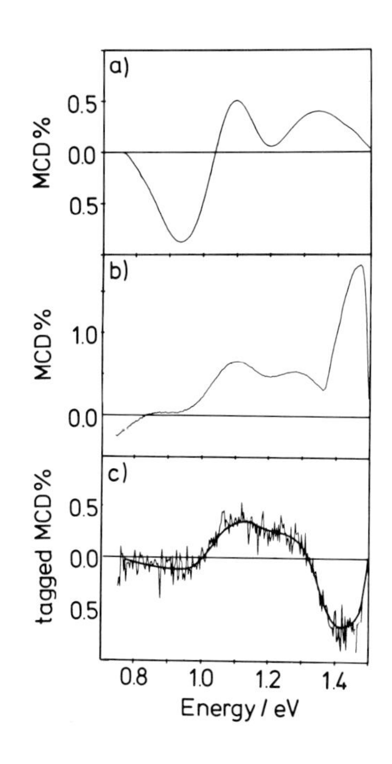

Fig. 3. MCD of $EL2^+$ in s.i. as-grown GaAs (a), MCD in ITC (1200 °C) heat treated GaAs (b, c).

The IR absorption across half a wafer clearly shows the known W-shaped distribution (Fig.2, circles), while the MCD is anticorrelated and W-shaped (squares). The total EL2 concentration (triangles) is rather homogeneous. The deviations from the mean value (MV) are of the order of 8 %, whereas e.g. the $EL2^o$ fluctuations are of the order of 30-40 % ($2.4x10^{16}$ cm^{-3} at position 4, $1.5x10^{16}$ cm^{-3} at position 11). The fluctuations in the IR absorption and the MCD must therefore be caused by an inhomogeneous distribution of the acceptors compensated by EL2. The acceptors must have an M-shaped profile and their total concentration must be of the order of $2\text{-}3x10^{16}$ cm^{-3}. Carbon, which is believed to be a dominant acceptor can be ruled out since it is homogeneous (Walukiewicz et al 1987). Transition metals as Zn and Fe are both too low in concentration ($\sim 10^{15}$ cm^{-3} range) and not M-shaped (Wagner et al 1986). It is not clear at present, which acceptors are compensated by EL2. Little is known about B_{As} and the role of intrinsic defects such as Ga vacancies and cation antisite defects or complexes involving them.

4. THERMAL TREATMENT OF BULK GaAs

S.i. GaAs crystals subjected to 1200 ^{o}C annealing (12 hours) under equilibrium As-pressure followed by rapid quenching to room temperature (inverted thermal conversion: "ITC" treatment) converts to p-type (Lagowski et al 1986). Fig.3b shows the MCD after ITC treatment compared to the normal MCD of $EL2^{+}$ (Fig.3a). The ITC material is EL2-free ($< 10^{15}$ cm^{-3}), the IR absorption shows no measurable contribution due to EL2. The MCD spectrum is due to a superposition of three new paramagnetic defects. For the high energy absorption band with an approximate onset at 1.35 eV no ODESR signal was obtained indicating a very short spin lattice relaxation time at 1.5 K (< 1 μsec) probably from an acceptor, the ODESR spectrum of which can only be obtained under uniaxial stress. The prominent double hump structure (Fig.3b) is caused by a defect with the following ODESR signature: a single ESR line with halfwidth 66 mT and a g-value of 2.07. The lineshape indicates unresolved hyperfine structure. The structure determination of the respective defect with ODENDOR is under way. Fig.3c shows the MCD spectrum of the third paramagnetic defect. Its ODESR spectrum is within experimental error identical to the well known quadruplet ODESR signature of $EL2^{+}$ (hf and g-value) in s.i. GaAs with, however, remarkably different optical properties (compare Fig.3a with 3c). Upon subsequent annealing to 800 ^{o}C all 3 new defects disappear and the $EL2^{+}$ MCD reappears.

References:

Dobrilla P, Blakemore J S, McCamont A J, Gleason K R, and Koyama R Y 1985 Appl. Phys. Lett. 47, 602
von Engelen P 1980 Phys. Rev. B 22, 3144
Lagowski J, Gatos H C, Kang C H, Skowronski M, Ko K Y, and Lin D G 1986 Appl. Phys. Lett. 49, 892
Martin G M 1981, Appl. Phys. Lett. 39, 747
Meyer B K, Hofmann D M, Niklas J R, and Spaeth J M 1987 Phys. Rev. B 36, 1332
Wagner J, Seelewind H, and Kaufmann U 1986, Appl. Phys. Lett. 48, 1054
Walukiewicz W, Bourret E D, Yan W F, Murray R E, Haller E E, and Bliss D F 1986 Proc. Int. Symp. Defect Recognition and Image Processing (Shiva Publ.) in press
Wilkening W,Kaufmann U, and Baeumler M 1987, private communication
Windscheif U, Baeumler M, and Kaufmann U 1985 Appl. Phys. Lett. 46, 661

Quantitative correlation between the EL2 midgap donor, the 1.039 eV zero phonon line, and the EPR arsenic antisite signal

J. Lagowski, M. Matsui, M. Bugajski, C.H. Kang, M. Skowronski and H.C. Gatos and M. Hoinkis,* E.R. Weber,* and W. Walukiewicz*

Massachusetts Institute of Technology, Cambridge, MA 02139, USA

*Center for Advanced Materials Research, Lawrence Berkeley Laboratory University of California, Berkeley, CA 94720, USA

ABSTRACT: We provide the first quantitative evidence that in GaAs the EL2 midgap donor, the 1.039 eV zero phonon absorption line, and the EPR quadruplet are manifestations of one and the same defect. The EL2 donor corresponds to the (o/+) state while the zero phonon line and the EPR quadruplet correspond to the neutral (o) and singly ionized (+) defect, respectively. The defect also exhibits the (+/++) state 0.54 eV above the valence band and no other states in the energy gap.

1. INTRODUCTION

The native midgap donor EL2 in GaAs has been the subject of intensive studies motivated by a fundamental interest in the identity of this metastable defect and by the practical need for stringent control of EL2 during fabrication of integrated circuits. In a continuing search for the atomic structure of the EL2 defect three characteristic features have been the center of attention: energy levels of the defect, (Gatos and Lagowski 1985), the 1.039 eV zero phonon optical absorption line (ZPL) (Kaminska et al 1985), and the electron paramagnetic resonance (EPR) quadruplet signal (Weber et al 1982). All these features play an important role in linking the EL2 and the arsenic antisite defect As_{Ga}. A defect symmetry deduced from the behavior of ZPL in the uniaxial stress implies an isolated antisite As_{Ga}. The interpretation of EPR and ODENDOR is experiencing a continuous evolution of models from an isolated antisite through a distorted antisite to antisite aggregates and to As_{Ga}-As_I (Meyer et al 1986, von Bardeleben et al 1986). Recent theoretical calculations (Baraff and Schluter 1987) indicate that the As_{Ga}-As_I pair may be in conflict with the energy level structure of the EL2 defect. The above discrepancies created uncertainty and confusion which require careful re-examination of EL2-related properties.

2. EXPERIMENTAL APPROACH

Determination of quantitative relationships among the EL2, ZPL, and EPR quadruplet have been thus far impeded by (1) material incompatibility between EPR and DLTS techniques (which require high and low resistivity samples, respectively); (2) difficulties in controlling and measuring the Fermi energy in SI GaAs; (3) ambiguities associated with the "EL2 family", i.e., midgap levels other than EL2; and (4) very low intensity of the EPR quadruplet in SI GaAs. In the present study special attention was devoted to overcoming the above difficulties. Thus, the material incompatibility was resolved using conducting and high resistivity crystals subjected to identical thermal treatment. Optical characterization (applicable to both crystals) was then employed to verify EL2 behavior in SI GaAs. With the most recent version of the optical method (Lagowski et al 1987) we achieved the determination of the total EL2 concentration as well as the EL2 occupancy and the Fermi energy. The inverted thermal conversion (ITC) treatment (Lagowski et al 1986) (i.e., 1200°C annealing-rapid quenching followed by 800-850°C annealing) was used to produce GaAs containing only EL2 and no other midgap levels. The DLTS spectrum of electron traps in such material shown in Fig. 1 indeed reveals only one high temperature peak. The corresponding capacitance transient is exponential (with the emission rate equal to the signature of EL2) (Gatos, Lagowski 1985).

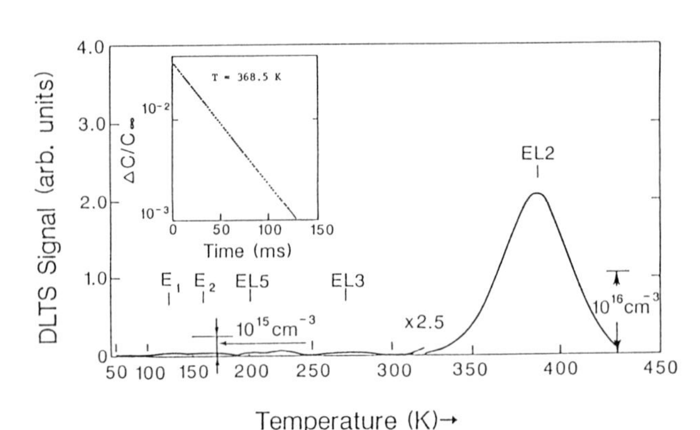

Fig. 1. DLTS spectrum of electron traps in GaAs after ITC+800°C, 1 h treatment, t_1/t_2=5ms/10ms.

In ITC GaAs EL2 concentration, N_{EL2} is typically $10^{15}cm^{-3}$. 800-850°C annealing increases N_{EL2} to as much as $3x10^{16}cm^{-3}$ depending on the annealing time and temperature. For undoped GaAs we used this treatment in order to achieve predetermined concentration and occupancy of EL2. As shown in Fig. 2, a very strong EPR quadruplet signal is obtained in SI LEC GaAs after ITC+800°C treatment. The hyperfine splitting constant of the quadruplet is the same as the value originally reported by Wagner et al 1980 for the As^+_{Ga} signal in as-grown GaAs. The strong quadruplet and readily resolvable ZPL illustrates advantages of our approach for the quantitative correlation of the EPR signal and ZPL.

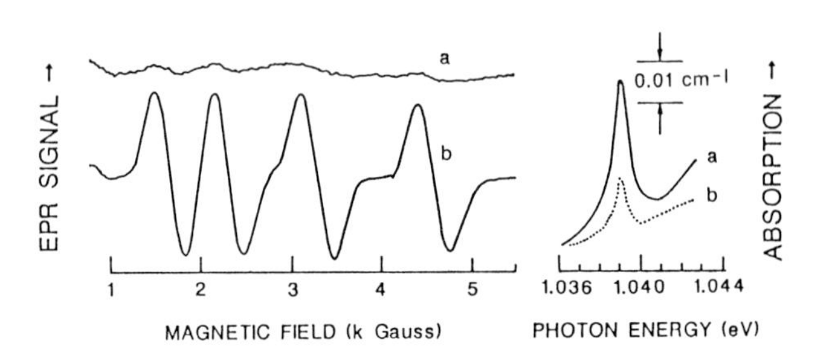

Fig. 2. EPR and ZPL in as-grown GaAs (a), and after ITC+800°C, 1 h treatment (b).

EPR measurements (4-300 K) were carried out using an X-band spectrometer interfaced to a computer. The apparatus had provisions for illumination of the sample in the cavity with either a monochromatic or white light. The absolute spin density corresponding to doubly integrated EPR quadruplet was determined using P-doped silicon samples as standard and applying a very careful calibration procedure. The differences in microwave power characteristics and in temperature dependence of the EPR quadruplet (maximum of EPR absorption signal of the quadruplet occurred near 10 K) in GaAs and P-signal in silicon were taken into account. The results of EPR and optical absorption discussed below correspond to measurements done at 10 K.

The exact sequence of high resolution optical absorption and EPR measurements on undoped LEC GaAs subjected to ITC+800°C, 1 h, treatment is listed in Table I. The EPR measurements were made at Berkeley, while the optical absorption measurements on the sample were made at MIT. Special attention was placed on making the experimental conditions of the two experiments as similar as possible. The results given in Table I constitute a fragment of a larger experimental program involving measurements on a series of 24 samples: eight representing the as-grown material, eight after ITC treatment, and eight after ITC+800°C annealing. The values of N^o_{EL2} (i.e., the concentration of occupied EL2) corresponding to a given zero phonon line absorption α_{ZPL}, were calculated using calibration provided by Skowronski et al 1986 for GaAs containing only one midgap level EL2. Estimated uncertainties are about $2 \times 10^{15} cm^{-3}$ for N^o_{EL2} and about 20% for the spin density.

Table I. Sequence of measurements employed for correlation of EPR quadruplet (QT) and ZPL

	ZPL		EPR		
EXPERIMENTAL STEPS	α_{ZPL} ($10^{-2} cm^{-1}$)	N^o_{EL2} ($10^{16} cm^{-3}$)	QT (arb. units)	N_s ($10^{16} cm^{-3}$)	$N^o_{EL2}+N_s$ ($10^{16} cm^{-3}$)
1. Cooling from 300 K in dark	0.75	0.6	27	1.8	2.4
2. White light illumination[a]	0	0	0	0	--
3. 150 K; 10 min[b]	1.9	1.7	11	0.7	2.4
4. Illumination; 0.9 μm, 30 min[c]	1.3	1.2	20	1.3	2.5
5. 150 K, 10 min[b]	1.1	1.0	23	1.5	2.5
6. White light illumination[a]	0	0	0	0	--
7. 150 K, 10 min[b]	2.6	2.3	4.6	0.3	2.6
8. Illumination, 2.5 μm, 30 min[c]	2.4	2.2	6	0.4	2.6
9. Illumination, 1.3 μm, 30 min[c]	2.5	2.25	0.2	0.1	2.4

[a]Transfer to metastable state; [b]Recovery of normal state; [c]Photoionization.

3. RESULTS AND DISCUSSION

Quantitative results given in Table I were also used to construct a plot of the intensity of the EPR QT signal vs. the intensity of the zero phonon line. This plot shown in Fig. 3 demonstrates the one-to-one correlation between a decrease in spin density and increase in the concentration of the neutral EL2. The data in Table I prove that this correlation remains valid during the photoionization of EL2 (0.9 μm illumination) or photo-population of EL2 by photoexcitation of holes to the valence band (1.3 μm illumination) as well as upon recovery of EL2 from the metastable state. It is also seen that the transfer of EL2 to a metastable state (by white light illumination) eliminates both the ZPL and the EPR quadruplet. With the exception of this specific case (steps 2 and 6) the concentration of the neutral EL2 plus the spin density remains constant $N^o_{EL2}+N_s=(2.5\pm0.1) \times 10^{16} cm^{-3}$. This value coincides with the total EL2 concentration, N_{EL2}. We thus conclude that the changes in α_{ZPL} and QT signal shown in Fig. 3 are brought about by the changes in the occupation of EL2. We further conclude that the midgap EL2 donor, the ZPL and EPR quadruplet are the manifestations of different charge states of one and the same defect. The EL2 donor corresponds to the (o/+) state of the defect. The zero phonon line and the EPR quadruplet correspond to the neutral (o) and singly ionized (+) defect, respectively. An additional (+/++) level of the defect is located 0.54 eV above the valence band. This level acts as a hole trap and is visible in photo EPR (Weber et al 1982), DLTS, and photo-DLTS (Gatos and Lagowski 1985). No other EL2 level exists in the upper half of the energy gap. Such level, if present, would have to be seen in the DLTS spectrum of Fig. 1.

The results of the present study should be viewed as a quantitative empirical unification of EL2 properties. This unification, however, leads to the interpretational dilemmas and the following sequence of conflicts: ZPL interpretation that the isolated As_{Ga} is the EL2 defect (Kaminska et al 1985) is in conflict with the pair As_{Ga}-As_i postulated from ODENDOR (Meyer et al 1986) and EPR (von Bardeleben et al 1986). The theory (Baraff and Schluter 1987) states, however, that the pair cannot be the EL2 defect unless additional EL2 levels (other than (o/+) and the (+/++) at E_c-0.75 eV and E_v+ 0.54 respectively) exist in the upper half of the energy gap; experiments show that such levels do not exist. The above sequence must contain elements which are false. Their disclosure and the search for alternative interpretations of ZPL or ODENDOR are critical for resolving the EL2 dilemma.

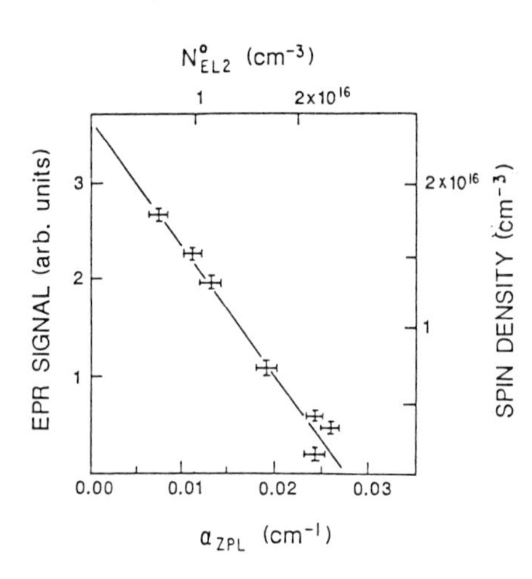

Fig. 3. See text.

The MIT contribution was sponsored by NASA and the USAFOSR. The contribution from Berkeley was supported by the US Dept. of Energy under contract No. DE-AC03-76SF00098.

REFERENCES

1. G.A. Baraff and M. Schluter, Phys. Rev. B35, 6154 (1987).
2. H.J. von Bardeleben, D. Stievenard, D. Deresmes, A. Huber and J.C. Bourgoin, Phys. Rev. B34, 7192 (1986).
3. H.C. Gatos and J. Lagowski, Mat. Res. Soc. Symp. Proc. 46, 153 (1985).
4. M. Kaminska, M. Skowronski, and W. Kuszko, Phys. Rev. Lett. 55, 2204 (1985).
5. J. Lagowski, M. Bugajski, M. Matsui, and H.C. Gatos, Appl. Phys. Lett. 51, 511 (1987).
6. J. Lagowski, H.C. Gatos, C.H. Kang, M. Skowronski, K.Y. Ko, and D.G. Lin, Appl. PHys. Lett. 49 892 (1986).
7. B.K. Meyer, D.M. Hofmann, and J.-M. Spaeth, Materials Science Forum, vol. lo, 311 (1986).
8. M. Skowronski, J. Lagowski, and H.C. Gatos, J. Appl. Phys. 59, 2451 (1986).
9. R.J. Wagner, J.J. Krebs, G.M. Stauss, and A.M. White, Solid State Commun. 36, 15 (1980).
10. E.R. Weber, H. Ennen, U. Kaufmann, J. Windscheif, J. Scheider, and T. Wosinski, J. Appl. Phys. 53, 6140 (1982).

Inst. Phys. Conf. Ser. No. 91: Chapter 4
Paper presented at Int. Symp. GaAs and Related Compounds, Heraklion, Greece, 1987

On the metastable state of EL2 in GaAs

H.J. von Bardeleben, J.C. Bourgoin
Groupe de Physique des Solides de l'Ecole Normale Supérieure, C.N.R.S., Tour 23, 2 place Jussieu, 75251 Paris Cedex 05, France

D. Stievenard, M. Lannoo
Département de Physique des Solides, Institut Supérieur d'Électronique du Nord, 41 boulevard Vauban, 59046 Lille Cedex, France

ABSTRACT: We present an atomic model for the metastable configuration of the EL2 defect in GaAs : it consists of a As split interstitial, the $(As_i)_2$ defect pair at a Gallium vacancy site. Its electronic structure, characterized by the absence of a gap level, explains its non observation by technique such as DLTS, EPR or optical absorption.

1. INTRODUCTION

Due to its technological importance for the production of semi-insulating GaAs substrates the native point defect EL2 has been the subject of numerous studies (1). Nevertheless, it is not until recently that the atomic configuration of this defect had been identified : a weakly interacting arsenic antisite - arsenic interstitial defect pair (2,3). A main complication in the study of this defect is the fact that no simple post-growth formation mechanism is known and in particular this defect is not introduced by electron irradiation. Due to the existence of multiple charge states of this center, corresponding to the 2+, +, 0 charge states of the arsenic antisite defect, it gives rise to different optical absorption bands in the near infrared corresponding to photoionization, internal transitions to excited states of the two constituants as well as a charge transfer transition within the center. An essential property of the EL2 defect is the existence of a low temperature metastable configuration, which can be optically induced by excitation in the charge transfer band centered at 1.2 eV (4). The transformation of EL2 into the metastable configuration, EL2*, has two consequences : first, the disappearance of all optical, electrical and paramagnetic properties associated with the stable configuration; surprisingly, however, these properties are not replaced by spectra attributable to EL2 in the metastable state : no optical absorption band, no electrically active level and no paramagnetic spectrum have been reported for EL2*; it is as if EL2* gave not rise to any level in the gap. Second, the free carrier concentrations are also metastably modified by the transformation. In consequence no direct experimental observation of the metastable configuration has been reported and its microscopic structure is completely unknown. However, the indirect information available from the characteristics of the transitions between the two configurations as well as the knowledge of the properties of the isolated As_{Ga} defect allow us to propose an atomic model : it consists of a split interstitial configuration of the original As_{Ga}-As_i defect pair. This model explains simply from the electronic structure of the split interstitial (a pair of As interstitial on each side of a Ga vacancy : $As_iV_{Ga}As_i$) the absence of any direct observation of EL2* as well as the free carrier concentration variations. The paper is structured in three parts : a short recall of the basic properties of the stable configuration of EL2; a description of the transformation of the stable to the metastable state and vice versa and, finally, the interpretation of the observations in light of the new model.

1.1 Stable Configuration of EL2

The stable configuration of EL2 is the high temperature configuration (T > 140 K) or the low temperature configuration when the crystal is cooled to low temperature in the dark, irrespective of the Fermi level position. It has been attributed to an arsenic antisite defect As_{Ga} interacting with an arsenic interstitial at a second nearest neighbour position in the [111] direction (2,3). The charge state of the As_i ion has been assigned to 1+ (3). For the As_{Ga} the charge states 0, 1+ and 2+ have been observed corresponding to the energy levels at E_c - 0.75 eV and E_V + 0.5 eV (5). The total charge states of the EL2 defect are then 1+, 2+, 3+. Nevertheless, the atomic configuration has only been directly determined for the 2+ charge state of EL2 (As_{Ga}^{+}-As_i^{+}) and it is tacitely assumed that it does not change for the other charge states. From DLTS measurements under uniaxial stress it has been concluded that the As_i ion is mobile and can reorient between the 4 equivalent interstitial lattice sites for T ≥ 150° C (6). From the information on the charge states of the As_{Ga} and As_i defects it is clear that the pair formation is not due to a Coulombic interaction. In fact, recent calculations by M. Lannoo et al. (7) show a binding energy of ~ 0.8 eV due to an enhanced relaxation of the As_i interstitial.

1.2 Transition from the stable to the metastable state

The only reported means to transform EL2 into the metastable state is by low temperature absorption in the band centered at 1.2 eV. It leads at 16 K for example to a 100 % transformation of EL2 defects with a time constant of minutes. This absorption band is superposed on the photoionization spectrum σ_n. Additionally, it presents on the low energy side a zero phonon line, which is generally believed to be associated with this band (8). However, a simple intensity comparison based on the Huang Rhys factor S ~ 12 shows that they must belong to different transitions. An important information is the fact that this photo-induced transformation can equally been performed in the space charge region of a reverse biased diode, in which the capture of free carriers is negligible. This demonstrates that it is due to an intracenter transition and not to photoionization. Further, there is experimental evidence that the transition is charge state conserving and proceeds only from the EL2$^+$ (As_{Ga}^{0}-As_i^{+}) charge state. A consequence of these two features is that, when done in material with the Fermi level pinned at E_c - 0.7 eV where only a part of the EL2 is in the + charge state (As_{Ga}^{0}-As_i^{+}), the photo-quenching is accompanied by a carrier transfer from the ionized acceptors leading to a metastable p-type conductivity (9).

The transition has the characteristics of a charge transfer, which we have tentatively modelized as follows :

$As_{Ga}^{0}-As_{i}^{+} \xrightarrow{h\nu} As_{Ga}^{-} - As_{i}^{2+} \rightarrow (As_{i}V_{Ga}As_{i})^{+}$

Thus the formation of the metastable configuration of the pair is believed to be due to a Coulomb attraction.

1.3 Transition from the metastable to the stable state

The stable configuration can be regenerated thermally (4) or, as it has been reported recently, optically (10). In the first case the regeneration rate is given by two terms, the first one corresponding to a pure thermal barrier of ~ 300 meV and the second being dependent on the free electron concentration n (11) with a reduced barrier of ~ 100 meV.

From the polarization dependence of the quenching transient under uniaxial stress the C_{3V} symmetry of the EL2 defect has been confirmed (6). The optically induced regeneration proceeds via an absorption band at ~ 0.8 eV (10). It is equally thermally activated, a feature, which can be understood in a configuration coordinate model. This optical transition is once again an intracenter transition within the metastable state this time.

1.4 Model of the metastable configuration

The model we propose is that of a split arsenic interstitial pair, composed by the initial As_{Ga} and the As_i defect (fig. 1). In this configuration the two central As ions are threefold coordinated, while the four nearest neighbour ions (As) retain their tetravalent coordination. This picture was developped from some very simple arguments : as has been shown for example by the recent uniaxial stress measurements (6) the interstitial As ion of the pair is a highly mobile defect (T = 150° C). On an other side the isolated As_{Ga} defect is of high thermal stability (T ≥ 450° C). Thus it seemed logical to attribute the strong changes in the atomic configuration to the changed lattice position of the As_i ion. As the electronic properties of the isolated As_{Ga} defect are known the absence of observation of the metastable state implies a displacement of the As_i towards the As_{Ga} defect (2). The split interstitial appears to be an energetically favorable configuration (12). Indeed, it allows for the

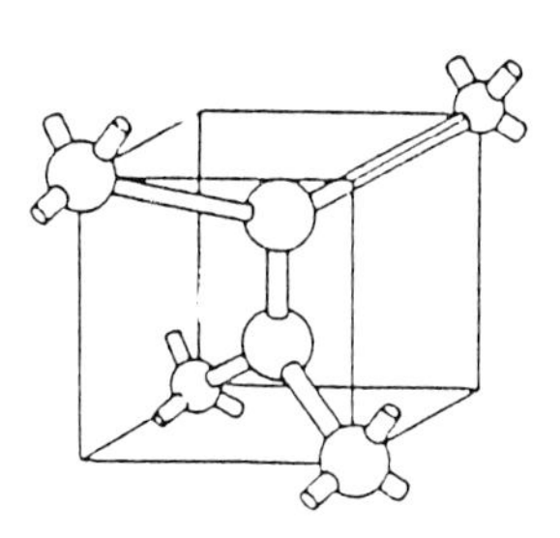

Fig. 1: Atomic structure of EL2*.

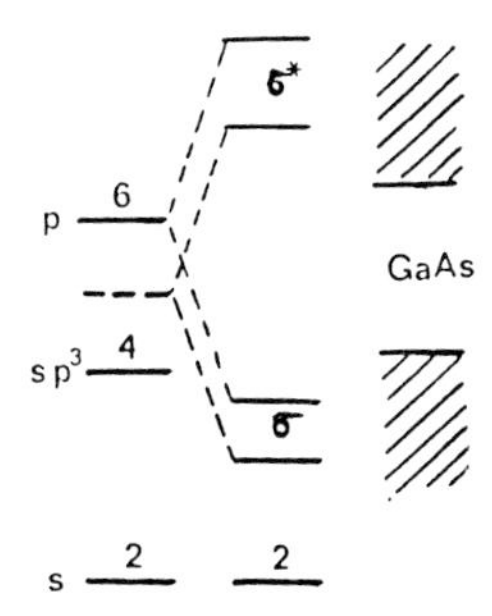

Fig. 2: Electronic structure of EL2*.

two central As ions to refind the normal threefold coordination which arsenic ions have both in the cristalline and the amorphous phases. Additionally, bond lengths and bond angles of As in elemental As and in the split interstitial correspond very closely. The high stability of the split interstitial as compared to a nearest neighbor As_{Ga}, As_i pair can also be obtained from thermodynamic arguments (12).

Let us now derive the electronic structures of this complex in a tight binding molecular description. Similar to the case of elemental As, bond formation will take place between 4 of the 6 p orbitals of the two As_i ions and 4 of the sp^3 orbitals of the 4 As neighbours. This gives rise to 4 p-sp^3 bonds and one p-p bond. Thus there will be 5 σ bonding and 5 σ^* antibonding states, the two remaining s-like orbitals giving rise to deep atomic states (fig. 2). As the σ - σ^* splitting is larger than the gap of GaAs and the bonding states being resonant with the valence band, no gap states are expected for this defect. Additionally its charge state is determined : it contains 2 electrons per bonding state and 2 electron per s state, that is a total of 14 electrons; as a neutral complex would contain 15 electrons (5 on the sp^3 orbitals and 5 per interstitial As atom) the charge state of the metastable complex is 1+, just as has been found experimentally.

In conclusion, we have proposed an atomic model for the metastable configuration of the EL2 defect in GaAs. This model, based on the general tendancy of As to form threefold coordinated bonds, explains the puzzling absence of any direct experimental observation of EL2* up to now, as well as the metastable variations of the free carrier concentrations. Nevertheless, the validity of this model has to be confirmed by more quantitative theoretical studies and hopefully direct experimental results.

2. REFERENCES

1. Martin G M and Makram-Ebeid S 1986 *Deep Centers in Semiconductors* ed Pantelides S (New York: Gordon & Breach Sc. Publ.) p 399
2. von Bardeleben H J, Stievenard D, Deresmes D, Huber A and Bourgoin J C 1986 *Phys. Rev.* **B34** 7192
3. Meyer B K, Hofmann D M, Niklas J R and Spaeth J M 1987 *Phys. Rev.* **B30** 1332
4. Vincent G and Bois D 1978 *Solid State Com.* **27** 431
5. Weber E R, Ennen H, Kaufmann U, Windscheif J, Schneider J and Wosinski J 1982 *J. Appl. Phys.* **53** 6140
6. Levinson M and Kafalas J A 1987 *Phys. Rev.* **B35** 9383
7. Baraff G A, Lannoo M and Schluter M 1987 (to be published)
8. Kaminska M, Skowronski M, Lagowski J, Parsey J M and Gatos H C 1983 *Appl. Phys. Lett.* **43** 302
9. von Bardeleben H J, Bourgoin J C, Zelsmann H R and Bonnet M *European MRS Meeting*, Strasbourg 1987 (to be published).
10. von Bardeleben H J, Bagraev N T and Bourgoin J C 1987 *Appl. Phys. Lett.* 1987 (to be published)
11. Mitonneau A and Mircea A 1979 *Sol. State Com.* **30** 157
12. von Bardeleben H J, Bourgoin J C, Stievenard D and Lannoo M *unpublished*

X-ray analysis of superlattices and quantum wells of GaAs/$Ga_{1-x}Al_xAs$ on Si

M Fatemi and A Christou

Electronics Technology Division, US Naval Research Laboratory, Washington DC 20375-5000

ABSTRACT: Quantum wells and superlattice structures of GaAs/$Ga_{1-x}Al_xAs$ grown on (100) Si substrates by MBE have been analyzed by X-ray diffraction techniques. Line broadening, dislocation density, epilayer-substrate misorientations, and strain have been measured, showing the effect of substrate orientation and defect disordering in these systems.

I. INTRODUCTION

In the past decade several researchers have examined the growth of GaAs/$Ga_{1-x}Al_xAs$ superlattices on GaAs substrates by x-ray diffraction techniques (1-6). It is well-known that parameters such as layer thickness, average period, and aluminum concentration can be obtained by these methods in a straight-forward manner.

The growth of the same type of superlattices on Si substrates, although as yet somewhat difficult, is technologically desirable, since the electronic properties of both GaAs superlattices and Si would be combined into one package. Molecular beam epitaxy (MBE) as one of the two commonly used deposition techniques, would permit accurate control of deposition parameters. However, the experimental conditions for fabrication of functioning superlattices on Si substrates have not yet been completely formulated. This work describes some of our findings by x-ray diffraction which suggest that the problem should be approached both from the stand point of operating conditions and substrate orientation and preparation.

Because of major differences between GaAs and Si in lattice parameter and crystal structure (GaAs: 5.6538 Å, Zincblende; Si: 5.4307 Å, diamond) neither the congruity, i.e., the similarity of growth habit, nor a reduced dislocation density at the interface is *a priori* guaranteed. For this reason, it is customary to start the growth process with a thin buffer layer at the interface, which would serve both as a transition region and as a barrier to the propagation of dislocations. Even with this buffer layer, however, defects associated with GaAs/Si mismatch are still abundant. Although the familiar superlattice diffraction pattern has not been observed in the GaAs/Si system, several important results were obtained by x-ray diffraction, which are briefly described here, and will be published in detail elsewhere.

Specimens were prepared at the MBE facility of the Research Center of Crete, Greece. The typical procedure is shown in Table I. Different barrier and well thicknesses of steps (4) and (5) were repeated in various samples to achieve a thickness of $\sim$ 1 μm. X-ray powder diffraction measurements were made on an automated Philips horizontal goniometer equipped with an independent theta motion. This feature is necessary to provide adjustments in the substrate orientation with respect to the beam. A Blake double crystal instrument with a nearly-perfect (100) Si monochromator was used for rocking curve and strain measurements.

Table I

Step No.	Layer Type	Temperature (°C)	Thickness	Time
1	GaAs	—	1000 Å	6 min
2	n GaAs	944	2000 Å	12 min
3	n+ GaAs	1073	0.5 μm	30 min
4	GaAlAs		50 Å	18 sec
5	n- GaAs	820	50 Å	18 sec
6	GaAlAs		50 Å	18 sec
7	n+ GaAs	1073	0.5 μm	30 min

Typical sample preparation procedure for quantum wells and superlattices of GaAs/$Ga_{1-x}Al_xAs$

II. EXPERIMENTAL RESULTS

A. GaAs/Si Epitaxies

Two Si substrate orientations of (100), 4° → (100) and (112) were used for the growth of GaAs layers. Powder diffraction particle size analysis was performed on both (400) and (422) lines. An example of this analysis is shown in Table II, comparing particle size effects for the two substrate orientations as well as the effect of rapid thermal anneal (RTA) on the rate of coarsening. To perform the calculations, the $K_{\alpha 1}$, $K_{\alpha 2}$ components were first deconvoluted. Instrumental broadening was then removed and the resulting "excess" broadening was used in the calculation (7).

Table II

Sample	Broadening B in degrees	Particle Size (Å)
A-PRE-ANNEAL	.47	230
A-POST-ANNEAL	.33	330
B-PRE-ANNEAL	.36	300
B-POST-ANNEAL	.21	520

Comparison of particle sizes in epitaxially grown GaAs/Si (112).
A=(100),4°→(100);B=(112).

Double-crystal measurement on sample #125 B (112-orientation) after annealing (Fig. 1) shows a broad diffraction profile (β = 2200 sec) resulting from large defect densities, estimated to be about 6.5×10^{10} cm^{-2}. From this density an average inter-dislocation distance of about 400 Å is obtained, which is comparable to the measured particle size of about 520 Å. The similarity between these numbers is an indication of the limits of the powder diffractometer in determining dislocation densities (8). For the same sample, a misorientation $\Delta\psi$ of ~1000 arc sec was noted between the epilayer and the substrate, and a fractional decrease in the (224) GaAs lattice spacing of $\sim 1.6 \times 10^{-3}$.

B. Quantum Wells and Superlattices

These structures were fabricated only on Si (100), 4° → (100) substrates. Relevant measurements at various diffraction angles are shown in Table III. Relatively high intensities at (111) GaAs, (200) Si and (200) GaAs were sometimes observed. Although, as expected, the intensity of polycrystalline GaAs at (111) diffraction angle is high, double crystal measurements at that

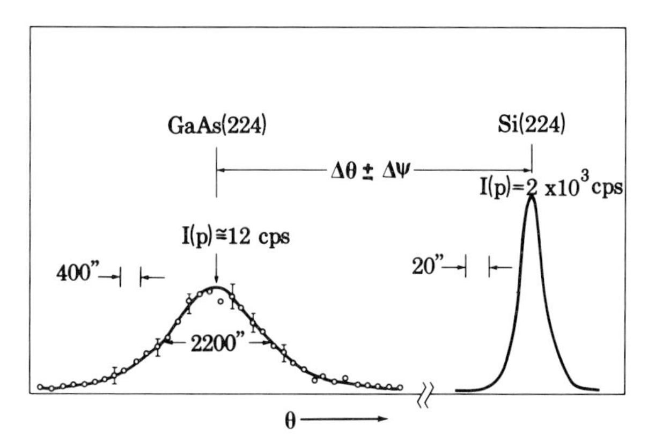

Fig. 1 — Double crystal diffraction pattern from epitaxially-grown GaAs/Si (112), rapid-thermal annealed at 1050°C for 10 s.

Table III
Diffraction Angle 2θ, degrees

Sample	27.28[1]	31.66[2]	32.95[3]	66.05[4]
135	0.2	8.5	2	240
137	12.0	9.7	1.7	260
143	7.7	0.2	1.0	7.5
144	3	21	2.5	417
110	160	0.4	—	0.2
111	210			2.2
112	21	1.5		35
156	160			0.5

Relative intensities in 10^3 counts per second for typical superlattice structures. 1: (111) GaAs; 2: (200) GaAs, AlAs, 3: (200) Si, 4: (400) GaAs

angle were nevertheless attempted. The resulting low count rates together with the extremely broad "rocking curve" verified the absence of single crystal growth along (111).

An interesting feature of the intensities measured at GaAs (400), 2θ ~66.05° and (111), 2θ = 27.28°, is their competing, "see-saw" behavior. In most cases a high value obtained for one reflection e.g., (400), is correlated with a low value for the other reflection, i.e. (111); in a few cases, however, both intensities have "moderate" values. In the group shown, samples 143 and 112 show more or less equal intensities while the remaining samples exhibit the "competing pattern."

III. DISCUSSION

We observed the effects of substrate orientation on powder diffraction line widths, described in the previous section, Table II, in a number of other specimens prepared under similar conditions. It can be seen that both the original particle size and the rate of coarsening appear to be larger on the (112) Si substrates than on (100). Similarly, the large strain of 1.6×10^{-3} in the (224) direction of GaAs indicates the tendency of the GaAs to form a distorted lattice atop the Si substrate. The growths on (100) Si, unlike (112) Si, could not always be measured easily due to their very low (400) intensities, suggesting the presence of randomly oriented polycrystalline material.

In the case of superlattice structures, the ''high-low'' intensity pattern associated with the GaAs (400) and GaAs (111) suggests the degree of epilayer-substrate congruity: a high (400) powder diffraction peak is correlated with a measurable, albeit broad double-crystal diffraction peak, as well as a weak (111) reflection. Conversely, a low (400) GaAs intensity on the powder diffractometer corresponds to a high (111) reflection which again yields an extremely weak ''double crystal pattern.'' The large dislocation densities of the order of 10^{11} cm^{-2} are thus consistent with the absence of superlattice reflection, which are frequently observed when GaAs substrates are used.

IV. CONCLUSIONS

In this paper, the first x-ray diffraction results on the formation of GaAs/$Ga_{1-x}Al_xAs$ superlattices on Si were presented. The improved growth of GaAs on (112) Si compared to (100) Si and the appearance of competing intensities of GaAs in superlattices grown on (100) Si were discussed. It is concluded that the growth of GaAs superlattice structures on Si is predominantly influenced by the disordering caused by rather large dislocation densities. A detailed paper based on these preliminary results is under preparation.

V. ACKNOWLEDGMENTS

The authors thank Dr. P. Thompson for rapid thermal annealing of the samples and Mr. M. Geiger for assistance in X-ray measurements.

VI. REFERENCES

1. A Segmüller, P Krishna, and L Esaki, J. Appl. Cryst. **10**, 1 (1977).

2. V S Speriocu and T Vreeland, Jr., J. Appl. Phys. **56**, 1591 (1984).

3. H Terauchi, S Sekimoto, K Kamigaki, H Sakashita, N Sano, H Kato, and M Nakayama, J. Phys. Soc. Japan, **54**, 4576 (1985).

4. M Quillec, ''Springer Proceedings in Physics,'' **13**, 121 (1986).

5. J F Petroff, M Sauvage-Simkin, S Benoussan, and B Capelle, J. Appl. Crys. **20**, 111 (1987).

6. A Regreny, P Auvray, A Chomette, B Deveaud, G Dupas, J Y Emery, and A Poudoulec, Revue Phys. Appl. (France) **22**, 273 (1987).

7. H P Klug and L E Alexander, *X-ray Diffraction Procedures* (New York: John Wiley and Sons, 1954).

8. M Fatemi, Proc. 36th Annual Denver Conference on Applications of X-Ray Analysis, 3-7 Aug. 1987, to be published.

Inst. Phys. Conf. Ser. No. 91: Chapter 4
Paper presented at Int. Symp. GaAs and Related Compounds, Heraklion, Greece, 1987

High-field characteristics of two-dimensional electron gas in InGaAs/N-InAlAs heterostructures

Shigehiko Sasa, Yoshiaki Nakata, Toshio Fujii, and Satoshi Hiyamizu

Fujitsu Laboratories Ltd., 10-1 Morinosato-Wakamiya, Atsugi 243-01, Japan

ABSTRACT: We studied the electric-field dependence of the two-dimensional-electron-gas mobility in selectively doped, InGaAs/N-InAlAs heterostructures, grown by MBE. The measurements were made by a pulsed Hall method with electric fields up to 2-4 kV/cm at 77K and 300K. It was found that the saturation velocity at 300K in the InGaAs/N-InAlAs heterostructures was comparable with that at 77K for GaAs/N-AlGaAs heterostructures. This implies that, even at room temperature, the performance of InGaAs/N-InAlAs HEMTs is comparable with that shown by GaAs/N-AlGaAs HEMTs at 77K.

1. INTRODUCTION

Selectively-doped InGaAs/N-InAlAs heterostructures lattice matched to an InP substrate, have great potential because their properties made them well-suited to high-speed device applications such as high-electron-mobility transistors (HEMT) (Hirose et al 1985, Kamada et al 1987): (a) a large conduction-band discontinuity between InGaAs and InAlAs (Sugiyama et al 1986), (b) a high saturation velocity of electrons in an InGaAs (Bandy et al 1981), and (c) the heavy doping potential of InAlAs, resulting in a very high two-dimensional electron gas (2DEG) concentration at the heterointerface (Nakata et al 1987). An understanding of the high-field characteristics is important for device design. However, only a few reports on this system have been published to date (Drummond et al 1982). Very recently, Hong et al. reported the high-field characteristics of InGaAs/InAlAs heterostructures. Their data, however, showed some sample dependence, and a discrepancy between the I-V (velocity saturation) and Hall measurements (no velocity saturation) (Hong et al 1987).

In this paper, we report on our examination of the high-field characteristics of the 2DEG in InGaAs/N-InAlAs heterostructures and our observation of the velocity saturation by high-field Hall measurement. Measurements were made using the pulsed Hall method at both 77K and 300K for the electric field, E, up to 2-4 kV/cm, which is comparable to the electric field in actual devices. The saturation velocity, v_s, was found to be $2.0\text{-}2.2 \times 10^7$ cm/s at 77K, which is 40-50% greater than that obtained for GaAs/N-AlGaAs heterostructures (Inayama et al 1983, Tsubaki et al 1983). Calculation also predicted similar values for GaAs/N-AlGaAs (Lei et al 1986). Besides, the saturation velocity at 300K for InGaAs/N-InAlAs ($v_s = 1.4 \times 10^7$ cm^{-3}) is comparable to the v_s at 77K of GaAs/N-AlGaAs. This implies that, even at room temperature, InGaAs/N-InAlAs heterostructure devices, such as HEMTs, have a performance comparable to that of GaAs/N-AlGaAs devices at 77K.

2. EXPERIMENTAL

The cross-section of the samples studied is shown in Fig. 1. An undoped InAlAs buffer layer (0.35 µm), an InGaAs channel layer (0.1 µm), a Si-doped InAlAs layer (d, N_d), and a Si-doped InGaAs cap layer (10 nm, N_d) were successively grown on a semi-insulating InP substrate by MBE. Two types of samples were grown with different Si-doping concentrations to explore the electron concentration dependence of the transport properties, e.g. those due to inter-subband scattering or real-space transfer. The respective values of the thickness of the N-InAlAs layer, d, and Si doping concentration, N_d, for the two samples were 90 nm and 45 nm, and 3×10^{17} cm^{-3} and 1×10^{18} cm^{-3}. The thickness d was adjusted for each sample to minimize the effect of parallel conduction through the N-InAlAs layer. The growth rate was 1 µm/h and the substrate temperature was 470°C. No spacer layer was used because no significant reduction in the mobility was observed in the InGaAs/N-InAlAs heterostructures (Nakata et al 1987), while it was very pronounced in GaAs/N-AlGaAs heterostructures (Störmer et al 1981). Hall bar geometry was fabricated for the high-field measurements. Ohmic contacts were formed from AuGe/Au eutectic alloy. The low-field mobility and the sheet electron concentration for each sample were measured by the van der Pauw method. The mobility-field characteristics were measured by the pulsed Hall method with a pulse duration of about 20 us, a repetition frequency of 100 Hz, and the electric field range of up to 2-4 kV/cm comparable to that in actual devices.

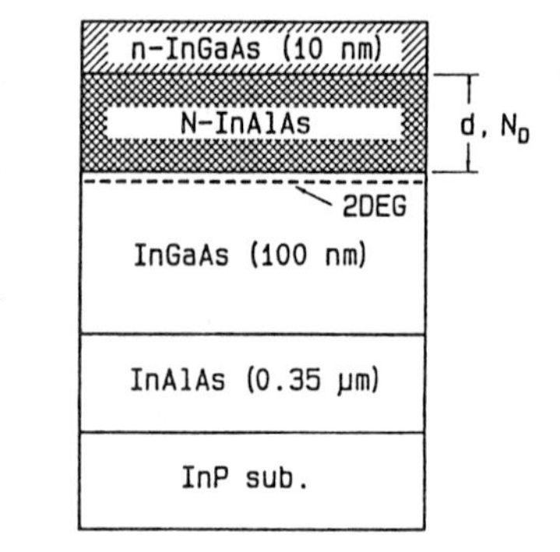

Fig. 1. Schematic cross-section of the selectively-doped InGaAs/N-InAlAs heterostructure.

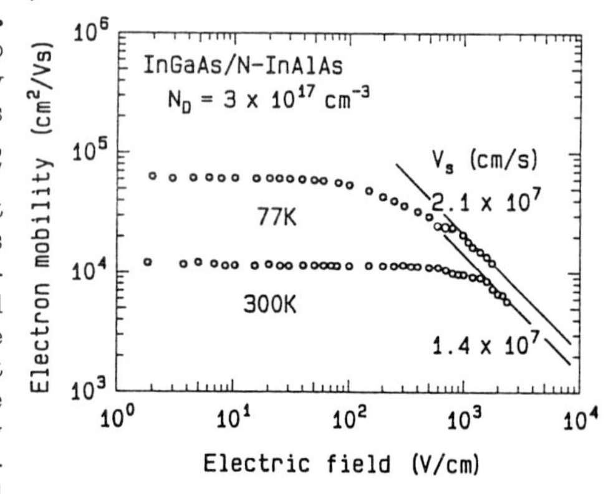

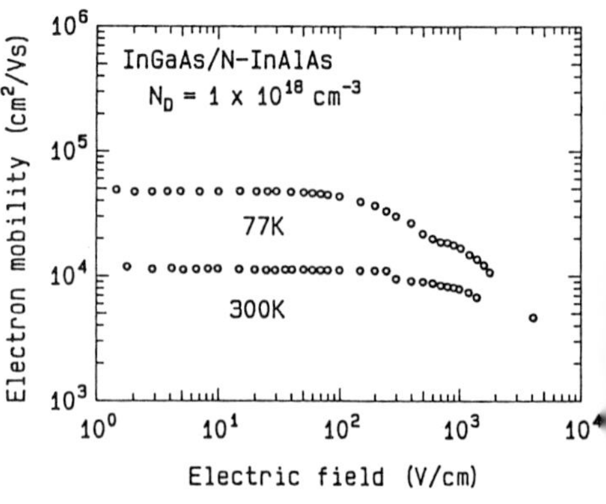

Fig. 2. The mobility-field characteristics of the InGaAs/N-InAlAs heterostructure at both 300K and 77K for $N_d = 3 \times 10^{17}$ cm^{-3} and $N_d = 1 \times 10^{18}$ cm^{-3}.

3. RESULTS AND DISCUSSION

The low-field mobility and the 2DEG concentration for both samples are summarized in Table I. Figure 2(a) and 2(b) show the mobility-field characteristics for the two samples at 300K and 77K. At 300K, the mobility is almost constant for $E < 0.5$ kV/cm. Therefore, the velocity increases linearly as the electric field increases. A slight mobility reduction occured for $E > 0.5$ kV/cm and velocity saturation was observed for $E > 1.5$ kV/cm. However, at 77K, a reduction in the

Table I. Low-field characteristics of the InGaAs/N-InAlAs heterostructures.

Sample	N_d (cm^{-3})	μ (cm^2/Vs) 300K	μ (cm^2/Vs) 77K	N_s (10^{12} cm^{-2}) 300K	N_s (10^{12} cm^{-2}) 77K
#1	3×10^{17}	10600	52800	1.42	1.34
#2	1×10^{18}	10100	44600	2.36	2.12

mobility with $E^{-0.5}$ was observed for the intermediate electric-field range ($0.1 < E < 1$ kV/cm), and velocity saturation occured for about $E > 1$ kV/cm. We also observed a dip around $E = 0.6$ kV/cm for each sample at 77K. The reason for this dip is not clear at present. The saturation velocity was 1.4×10^7 cm/s at 300K and 2.0-2.2×10^7 cm/s 77K, for $N_d = 3 \times 10^{17}$ cm^{-3}. The saturation velocity and the electric-field dependence of the mobility was almost the same for the two samples, regardless of the different, low-field mobilities and 2DEG concentrations.

The mobility at various temperatures and electric fields is ascribed to the different scattering mechanisms. It is plausible that the electron scattering processes in InGaAs/N-InAlAs are similar to those in GaAs/N-AlGaAs heterostructures, except for additional alloy scattering. At 300K, electrons are predominantly scattered by polar-optical phonons in both low electric fields and high electric fields range used in the present study. At 77K, instead, acoustic phonon and alloy scattering are the dominant scattering mechanisms in the low electric-field region. The polar-optical phonon scattering becomes strong as the electric-field strength increases, leading to a noticeable reduction in the mobility. Inter-subband scattering and real-space transfer do not have a significant effect on the high-field characteristics for this materials system but do for GaAs/N-AlGaAs (Tsubaki et al 1983).

Next, we made a detailed comparison between the high-field characteristics of the InGaAs/N-InAlAs heterostructure mentioned above and GaAs/N-AlGaAs heterostructures with a low-field mobility of 9.0×10^4 and 1.0×10^5 cm^2/Vs at 77K. Figure 3 shows the mobility-field characteristics at 300K and at 77K for both systems. At 300K, the mobility of the InGaAs/N-InAlAs is almost 50% greater than that of the GaAs/N-AlGaAs over the entire electric-field range. Hence, the superiority of the InGaAs/N-InAlAs in high-speed performance over GaAs/N-AlGaAs at 300K is clear. However, at 77K, the low-field mobility of the InGaAs/N-InAlAs is a half that of GaAs/N-AlGaAs. As can be seen from the figure, the reduction in the mobility as the electric field is increased is pronounced in the GaAs/N-AlGaAs for $E < 0.2$ kV/cm, and one can see a mobility intersection at about $E = 0.2$ kV/cm. Thereafter, the InGaAs/N-InAlAs shows a higher mobility than the GaAs/N-AlGaAs. From the point of view of device applications, such low-field characteristics ($E < 0.2$ kV/cm) are not so significant, as explained below.

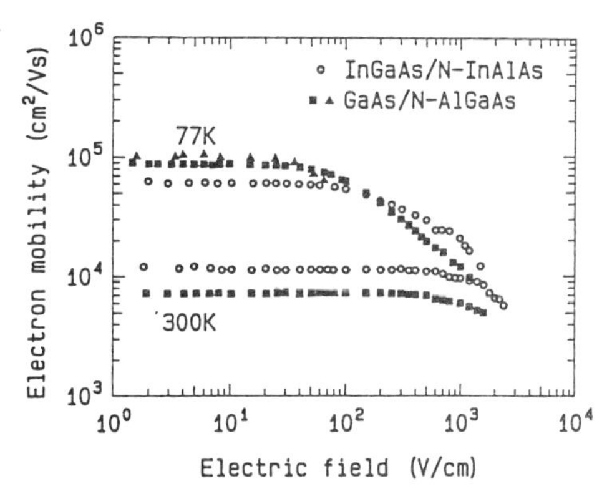

Fig. 3. The mobility-field characteristics for both InGaAs/N-InAlAs (O) and GaAs/N-AlGaAs (■,▲) heterostructures at 300K and 77K.

Figure 4 shows the electron drift velocity as a function of the electric field characteristics. This clearly shows the superiority in high-speed performance of the InGaAs/N-InAlAs over the GaAs/N-AlGaAs. A great improvement in the drift velocity at both temperatures was achieved with the InGaAs/N-InAlAs heterostructures. Moreover, the saturation velocity of the InGaAs/N-InAlAs system at 300K was comparable with that of the GaAs/N-AlGaAs system at 77K. This indicates that even at room temperature, the high speed performance of InGaAs/N-InAlAs is comparable with that of GaAs/N-AlGaAs at 77K.

4. CONCLUSION

The high-field characteristics of InGaAs/N-InAlAs heterostructures were studied and were compared with those of the GaAs/N-AlGaAs heterostructures. The saturation velocity of the InGaAs/N-InAlAs is 40-50% greater than that of the GaAs/N-AlGaAs at both 300K and 77K. It was found that the saturation velocity of the InGaAs/N-InAlAs heterostructures at 300K is comparable with that at 77K for the GaAs/N-AlGaAs. These results indicate that the high-speed performance of a InGaAs/N-InAlAs system is greatly superior to that of a GaAs/N-AlGaAs system. Even at room temperature, the potential of InGaAs/N-InAlAs is comparable to that exhibited by GaAs/N-AlGaAs at 77K.

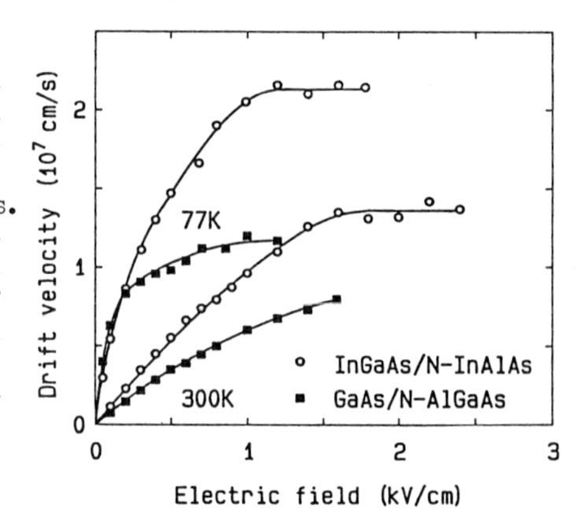

Fig. 4. The drift velocity-field characteristics for the two materials systems.

5. ACKNOWLEDGEMENT

The authors thank O. Ryuzan, K. Dazai, and T. Yamaoka for their support in this work.

References

Das M B, Koop W and Morkoc H 1984 IEEE Electron Device Lett. **EDL-5** 446.
Drummond T J, Morkoc H, Cheng K Y and Cho A Y 1982 J. Appl. Phys. **53** 3654.
Hirose K, Ohata K, Mizutani T, Itoh T and Ogawa M 1985 GaAs and Related Compounds (Inst. of Phys., 1986) p. 529.
Hong W-P and Bhattacharya P. K. 1987 IEEE Electron Devices **ED-34** 1491.
Inayama M, Inoue M, Inuishi Y and Hiyamizu S 1983 Technol. Repts. Osaka Univ. **33** 271.
Inoue K, Sakaki H and Yoshino J 1985 Appl. Phys. Lett. **47** 614.
Kamada M, Kobayashi T, Ishikawa H, Mori Y, Kaneko K and Kojima C 1987 Electron. Lett. **23** 298.
Lei X L, Zhang J Q and Birman J L 1986 Phys. Rev. B **33** 4382.
Nakata Y, Sasa S, Sugiyama Y, Fujii T and Hiyamizu S 1987 Jpn. J. Appl. Phys. **26** L59.
Stormer H L, Pinczuk A, Gossard A C and Wiegmann W 1981 Appl. Phys. Lett. **38** 691.
Sugiyama Y, Inata T, Fujii T, Nakata Y, Muto S and Hiyamizu S 1986 Jpn. J. Appl. Phys. **25** L648.
Tsubaki K, Livingstone A, Kawashima M and Kumabe K 1983 Solid State Commun. **46** 517.

Inst. Phys. Conf. Ser. No. 91: Chapter 4
Paper presented at Int. Symp. GaAs and Related Compounds, Heraklion, Greece, 1987

Effect of superlattices band structure on spontaneous emission lineshapes in GaAs multiple quantum wells

M. Krahl, J. Christen, D. Bimberg
Institut für Festkörperphysik I der Technischen Universität Berlin,
Hardenbergstr. 36, 1000 Berlin 12, Germany

G. Weimann and W. Schlapp,
Forschungsinstitut der Deutschen Bundespost
beim Fernmeldetechnischen Zentralamt,
6100 Darmstadt, Germany

A comparison of the low temperature luminescence of MBE grown GaAs coupled and uncoupled quantum wells is presented. The emission narrows dramatically with increasing coupling. A realistic lineshape model of superlattices (SL) accounts for this observation: The SL density of states calculated in the framework of envelope function approximation is broadened by Gaussians, thus accounting for the structural disorder at the interfaces. An extended averaging mechanism of SL excitons as compared to quantum well excitons explains the obseved narrowing.

1. INTRODUCTION

Quantum well lasers and light emitting diodes are in many respects superior to classical light emitting devices (Bimberg et al. 1984). The drastically altered electronic density of states (Dingle et al. 1981), the existence of a monomolecular radiative recombination channel at room temperature and the large enhancement of the bimolecular recombination coefficient as shown by Böttcher et al. (1987) are the main physical causes for the superiority of two-dimensional structures for light emitters.

Presently two different types of such structures are used. In most of the initial work stimulated by Tsangs (1981a) pioneering experiments the active area of the device consists of a multiple quantum well (e.g. Dingle et al. 1981). Closer examination, however, shows that this term is misleading, since quite often (e.g. Böttcher et al. 1981 or Tsang 1981b) the barriers between the wells are relatively narrow. Consequently the wells are coupled, like in a superlattice, and their electronic properties are not truly two-dimensional. In particular, the density of states is not steplike.
It is the purpose of the present letter to demonstrate the impact of a coupling of quantum wells on their spontaneous emission lineshape and to establish a theory of emission lineshape of superlattices. GaAs is investigated as a model system.

The emission lineshape of narrow uncoupled QW's with thickness $L_Z \leqslant 10$ nm shows strong statistical broadening effects (Weisbuch et al. 1981) which can be well approximated by simple Gaussians (Bimberg et al. 1986). This broadening is a consequence of the roughness of the heterointerfaces. We will show below that dramatic linewidth narrowing is observed upon coupling at low temperatures, in contrast to what one expects from a simple model. A new excitonic interface averaging mechanism is proposed to account for this narrowing.

2. EXPERIMENTAL RESULTS

All samples are grown by MBE in a Varian GEN2 system at a constant substrate temperature of 680°C and a growth rate of 3.4 Å/s. The Al concentration in the barriers is 0.4-0.45. Each sample contains 40-100 QW's. The 2nd and 3rd column of table I list the well and barrier thicknesses L_Z and L_B, respectively, of some of the samples investigated.

Photoluminescence experiments are performed at various temperatures T $\geq$ 1.5 K and at excitation densities 0.05 W/cm^2 - 500 W/cm^2. We concentrate here on a few typical low temperature results.

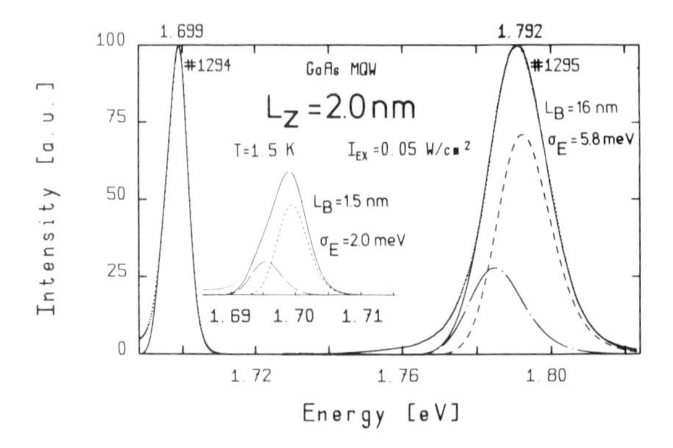

Fig. 1: Luminescence spectra for two samples with the same well width (2.0 nm) but different barrier widths (1.5 nm and 16.0 nm) shown by dotted lines. Full line: line shape fits using two components (dashed, dashed dotted lines). In the inset the SL-spectrum is shown on an extended energy scale.

In fig. 1 spectra of samples #1294 and #1295 having the same L_Z=2 nm but different L_B's of 1.5 nm and 16 nm are compared (dotted lines). The peak of the luminescence shifts from 1.792 eV to 1.699 eV by 93meV towards lower energy upon the reduction of the barrier width. More important, the halfwidth decreases from 19 meV to 6.8 meV.

3. LINESHAPE THEORY

The spectra of the uncoupled (fig. 1, sample 1295) wells are fitted using a product of a thermal distribution function with a Gaussian broadened steplike density of states (Bimberg et al. 1987). All lines consist of doublets due to the localisation induced enhancement of the exchange interaction (Bauer et al. 1987). The variance of the Gaussian is determined from the fit as σ_E=5.8 meV. A variance σ_L=0.2 monolayers of the L_Z-distribution function accounts for this energy broadening. Details on how to calculate σ_L from σ_E were reported elsewhere (Bimberg et al. 1987). This lineshape theory is now being extented to consider additionally coupling of wells. Using the common tight-binding-ansatz in envelope function approximation (Voisin et al. 1984) the eigenenergies E^{SL} can be calculated analytically. Taking only next neighbours into account the resulting density of states D^{SL} (E) is:

$$D^{SL}(\xi) \sim \begin{cases} 0 & , \xi < 0 \\ \arccos(1-2\xi), & 0 \leq \xi < 1 \\ \pi & , 1 \leq \xi \end{cases} \quad (1)$$

where ξ is the reduced energy (Krahl et al. 1987).

A realistic lineshape model has again to take into account the structural disorder at the interfaces which induces a statistical distribution of L_Z and L_B. We assume σ_L to be the same for L_Z and L_B and broadening of L_Z and L_B to be uncorrelated. The distribution of layer widths induces a Gaussian distribution function of the quantisation energies, whose variance σ_E is given by the equation,

$$\sigma_E \approx \left|\frac{\partial E^{SL}}{\partial L_Z}\right| \cdot \sigma_L + \left|\frac{\partial E^{SL}}{\partial L_B}\right| \cdot \sigma_L \quad (2)$$

Based on eq. 2 one should expect a larger energy broadening for coupled wells because in uncoupled QW's the second term in eq. 2 vanishes. The theoretical luminescence lineshape is eventually obtained by multiplying the broadened density of states $Z^{SL}(E)$ with a carrier distribution function (here a Maxwellian).

4. EXCITONIC AVERAGING MECHANISM IN SUPERLATTICES

Column 4 of Table I gives the variances σ_E obtained from lineshape fits of the different samples investigated. Column 5 gives the expected σ_E -values assuming that the interface roughness is the same for all samples (σ_L=0.2 monolayers). This assumption was tested to be valid by investigating samples with varying L_Z but large L_B. Obviously the experimentally observed lineshape narrowing (Fig. 1) upon a coupling of wells is opposite to what one expects theoretically (column 5).

sample	L_Z [nm]	L_B [nm]	experimental σ_E (lineshapefit) [meV]	simple theory σ_E (σ_L=0.2 ML) [meV]	extended theory σ_E (σ_L=0.2ML*(d/29nm)$^{1/2}$) [meV]
1294	2.0	1.5	2.0	6.2	2.1
1295	2.0	∞	5.8	5.8	SQW
1301	4.2	1.5	0.9	2.2	1.0
1297	3.4	1.5	1.1	2.4	1.0
1303	2.8	1.5	1.5	6.3	1.5
1292	1.7	2.2	2.2	6.3	2.3
1296	1.7	4.5	2.8	4.8	2.2

Table I: Comparison of variances σ_E for different samples: obtained from lineshapefits (column 4), predicted for 0.2 monolayers (ML) interface roughness (column 5), estimated from exciton extension (column 6).

We conclude that the excitonic averaging process in a SL is strongly modified as compared to a SQW. In order to understand this effect Fig. 2 shows in a qualitative way the extension of an exciton wavefunction in growth direction for several limiting cases. In a SQW a (2D-) exciton is essentially confined to the well. It's extension is much less than the diameter of the exciton in bulk GaAs of 29 nm. In a SL the SL-exciton is no longer localized in one well but extends over several wells. Due to the

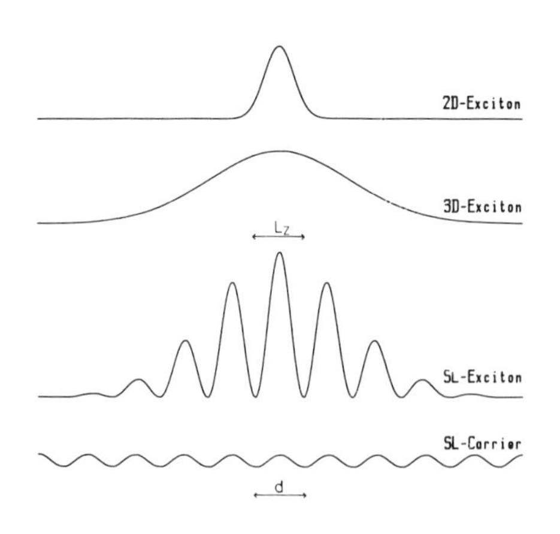

Fig. 2: Schematic representation of excitonic wavefunction in a SQW, bulk-material and in a SL.

periodically varying potential it's wavefunction should be given qualitatively by a periodically modulated (3D-) envelope function: A SL-exciton "sees" the uncorrelated interface roughness of several wells. Thus coupling of wells has the same effect as an increase of the Bohr-radius. The number of contributing wells N can be estimated from the 3D-exciton diameter and the superlattice constant d: $N \approx 29nm/d$. Since the averaging volume is extended by a factor N, the width of the distribution functions σ_L and σ_E should narrow by a factor $(N)^{1/2}$. The last column of table I shows the variances σ_E predicted in this way. The agreement between the predicted variance and the experimental one is surprisingly good.

5. CONCLUSION

In conclusion we presented results of a comparative experimental and theoretical study of interface roughness induced luminescene broadening of coupled and uncoupled QW's (true superlattice) made of GaAs/GaAlAs heterostructures. Coupling results in pronounced suppression of Gaussian line broadening. An excitonic averaging mechanism accounts quantitatively for this observation. A previously established theory of QW luminescence lineshapes is extended to the case of coupled wells and used to derive the statistical contribution of the lineshape.

References

Bauer R, Bimberg D, Christen J, Oertel D, Mars D, Miller J N, Fukunaga T and Nakashima H 1987 Proc. 18th Intern. Conf. Phys. Semic. O.Engström ed., World Scientific Singapore 1987, p.525

Bimberg D, Christen J, Fukunaga T, Nakashima H, Mars D and Miller J N 1987, J. Vac. Sci. Technol. B6

Bimberg D, Christen J and Steckenborn A 1984 in Springer Series in Solid State Science 53, edited by G.Bauer, F.Kuchar, and H.Heinrich (Springer, Berlin, 1984) p.136

Bimberg D, Mars D, Miller J N, Bauer R and Oertel D 1986 J. Vac. Sci. Technol. B4, 1014

Böttcher E H, Ketterer K, Bimberg D, Weimann G and Schlapp W 1987 Appl. Phys. Lett. 50, 1074

Dingle R and Henry C H 1976 U.S. Patent No.3982207, 21.September 1976

Krahl M, Christen J, Bimberg D, Weimann G, Schlapp W 1987 Appl. Phys. Lett., to be published

Tsang W T 1981a Appl. Phys. Lett. 38, 204

Tsang W T 1981b Appl. Phys. Lett. 39, 786

Voisin P, Bastard G and Voos M 1984 Phys. Rev. B 29, 935

Weisbuch C, Dingle R, Petroff P M, Gossard A C and Wiegmann W 1981 Appl. Phys. Lett. 38, 840

Quantitative optical analysis of residual shallow acceptors in semi-insulating GaAs

J. Wagner, M. Ramsteiner, W. Jantz, and K. Löhnert*

Fraunhofer-Institut für Angewandte Festkörperphysik, Eckerstr. 4, 7800 Freiburg, West-Germany
* Wacker-Chemitronic, Postfach 11 40, 8263 Burghausen, West-Germany

ABSTRACT: State of the art LEC grown undoped semi-insulating GaAs often exhibits a low concentration (< 10^{15} cm^{-3}) of residual shallow acceptors. The characterization of this material requires analytic techniques that detect all shallow acceptors, e.g. C and Zn, with sufficiently high sensitivity. The present paper discusses electronic Raman scattering as a quantitative tool for the assessment of residual acceptors with a detection limit below 5 x 10^{14} cm^{-3}. Calibration factors for both C and Zn acceptors are given.

The quality of semi-insulating (s.i.) GaAs is influenced by the incorporation of residual shallow acceptors. Carbon on the arsenic lattice site (C_{As}) has generally been considered to be the most significant contaminant. It can sensitively be identified by local vibrational mode (LVM) spectroscopy (Homma et al. 1985, Brozel et al. 1986), a convenient analytic tool to measure impurities that are much lighter than the matrix elements.

Meanwhile, state of the art liquid encapsulated Czochralski (LEC) grown s.i. GaAs often exhibits a low concentration of C_{As} (< 5 x 10^{14} cm^{-3}), such that other elements may become the dominant shallow acceptors. Indeed sometimes Zn concentrations up to 3 x 10^{15} cm^{-3} have been found, exceeding the very low C_{As} content by almost an order of magnitude. The occurence of shallow acceptors other than C_{As} requires analytic techniques that are not limited to a certain class of elements. Electronic Raman scattering (ERS), which is a nondestructive optical technique like LVM spectroscopy, is sensitive to all shallow acceptors in undoped s.i. LEC GaAs including C and Zn in particular (Wan and Bray 1985, Wagner et al. 1986a, Wagner and Ramsteiner 1986b). In this material the residual acceptors are predominantly ionized by electrons of the compensating EL2 donor level. As ERS by acceptor bound holes requires neutral impurities, the ionized acceptors have to be photoneutralized by optical excitation into the EL2 absorption band prior to the Raman scattering process (Wan and Bray 1985). Using e.g. the 1.16 eV (1064.4 nm) line of a Nd-YAG laser both the photoneutralization and the excitation of the Raman spectra can be done by the same light source (Wan and Bray 1985). In the present study ERS has been calibrated as a quantitative tool for the characterization of residual acceptors.

The ERS spectra were excited with the 1064.4 nm line of a Nd-YAG laser and recorded in backscattering from a (100) surface. The scattered light was not analyzed for its polarization and the transmission of the spectrometer was about the same for vertical and horizontal polarization. The spectral resolution was set to 7 cm^{-1}. The samples were cooled to 6 K. Their C_{As} content was measured by LVM spectroscopy using the calibration factor given by Homma et al. (1985). The concentration of residual Zn was determined by spark source mass spectroscopy (SSMS).

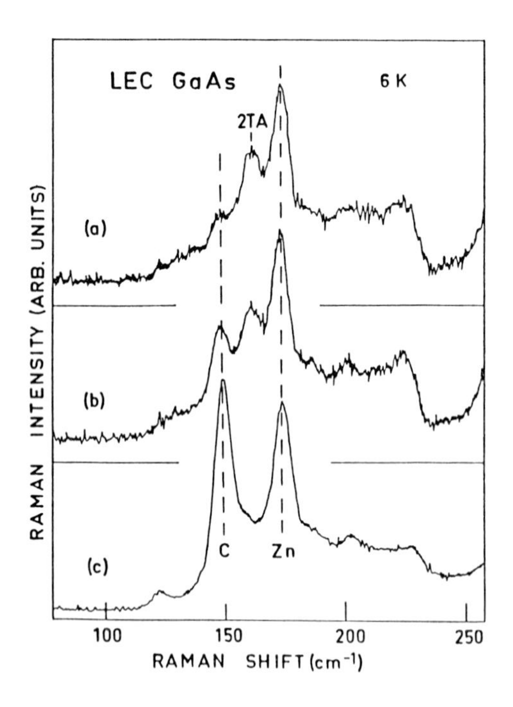

Fig. 1. Low-temperature Raman spectra of three different s.i. LEC GaAs samples.

Fig. 1 displays a series of electronic Raman spectra of undoped s.i. LEC GaAs. The spectra show, superimposed on a background of intrinsic second-order (2TA) phonon scattering (Trommer and Cardona 1978), electronic 1s-2s excitations of C_{As} and Zn acceptors at 148 and 173 cm^{-1}, respectively (Wan and Bray 1985, Wagner et al. 1986a). Going from Fig. 1a to Fig. 1c the C_{As} concentration increases from 4 x 10^{14} cm^{-3} (a) to 1.2 x 10^{15} cm^{-3} (b) and to 6 x 10^{15} cm^{-3} (c). The Zn content in the range 1-3 x 10^{15} cm^{-3} is comparable in all samples. Fig. 1a clearly shows, that the Zn concentration may exceed the C_{As} content considerably. It also demonstrates the sensitivity of ERS, which is $< 5 \times 10^{14}$ cm^{-3} for samples of standard wafer thickness.

In Fig. 2 the scattering intensity $I(C)/I_{2TA}(225\ cm^{-1})$ of the C_{As} 1s-2s Raman line, normalized to the intrinsic second-order phonon scattering intensity at 225 cm^{-1} (Wagner et al. 1986a), is plötted versus the carbon content. A perfect linear correlation is found over the whole concentration range, demonstrating that quantitative photoneutralization of the acceptors is achieved. A normalized ERS intensity of 1 corresponds to a C_{As} concentration of 2 x 10^{15} cm^{-3}. Note that this calibration factor is only valid for the present experimental conditions, i.e. backscattering from a

(100) surface and a spectral resolution of 7 cm^{-1}. The accuracy of this calibration factor is essentially given by the uncertainty in the calibration of the LVM absorption strength (Homma et al. 1985, Brozel et al. 1986). Using the known Raman scattering efficiency of the intrinsic transverse optical phonon (Flytzanis 1972) as a reference, the absolute scattering cross section per carbon acceptor is found to be 4 x 10^{-24} $Sr^{-1}cm^{2}$ for an incident photon energy of 1.16 eV.

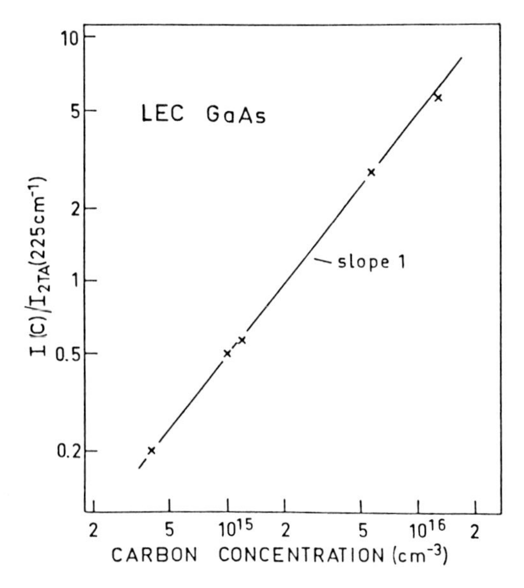

Fig. 2. Normalized electronic Raman scattering intensity for the C_{As} acceptor plotted versus the C_{As} concentration measured by local vibrational mode absorption.

Fig. 3. Normalized electronic Raman scattering intensity for the Zn acceptor plotted versus the Zn concentration measured by spark source mass spectroscopy.

Fig. 3 shows the normalized ERS intensity for residual Zn acceptors plotted versus the Zn concentration. The larger scatter of these data is caused by the uncertainty in the measurement of the Zn content. This uncertainty is partly due to the fact that the present Zn concentrations are close to the detection limit (≈ 4 x 10^{14} Zn/cm^{3}) of SSMS. Discrepancies may also be due to the large spatial variations sometimes found in the Zn concentration (Wagner et al. 1986a). If the ERS and SSMS are not performed on the same spot of the sample, a precaution not observed in the present study, this spatial variation may contribute to the uncertainty of the data. However, from the C_{As} Raman data a linear relation between Zn concentration and ERS intensity is expected. Based on this evidence a calibration factor of (0.9 ± 0.4) x 10^{15} cm^{-3} can be deduced, as illustrated in Fig. 3.

This factor is smaller than the one for C_{As} acceptors indicating a larger scattering cross section per defect for Zn as compared to C_{As} acceptors.

However, more accurate data for Zn are necessary to draw a final conclusion on this point.

It has to be noted that ERS in s.i. LEC GaAs measures the concentration of only those residual acceptors which are compensated by EL2 (Wan and Bray 1985). In other terms, ERS measures $N_A - N_D$ where N_A and N_D are the concentrations of shallow acceptors and shallow donors, respectively (Wagner and Ramsteiner 1987). In the present set of samples N_D was much smaller than N_A, as can be seen from the good correlation found between the total C_{As} concentration and the C_{As} ERS signal intensity (see Fig. 2). For the same reason the present ERS technique can neither be applied to n-type semi-conducting material, nor to Cr compensated s.i. GaAs (Wan and Bray 1985).

In summary we have shown that ERS provides a sensitive and quantitative tool for residual acceptor assessment in s.i. LEC GaAs with a detection limit better than 5×10^{14} cm^{-3}. Calibration factors for C_{As} and Zn acceptors have been deduced.

ACKNOWLEDGMENTS: We would like to thank H. Seelewind for performing the LVM spectroscopy, P. Koidl and H.S. Rupprecht for supporting this work.

REFERENCES:

Brozel M R, Foulkes E J, Series R W, and Hurle D T J 1986 Appl. Phys. Lett. 49, 337
Flytzanis C 1972 Phys. Rev. B 6 1278
Homma Y, Ishii Y, Kobayashi T, and Osaka J 1985 J. Appl. Phys. 57 2931
Trommer R and Cardona M 1978 Phys. Rev. B 17 1865
Wagner J, Seelewind H, and Kaufmann U 1986a Appl. Phys. Lett. 48 1054
Wagner J and Ramsteiner M 1986b Appl. Phys. Lett. 49 1369
Wagner J and Ramsteiner M 1987 (unpublished)
Wan K and Bray R 1985 Phys. Rev. B 32 5265

Inst. Phys. Conf. Ser. No. 91: Chapter 4
Paper presented at Int. Symp. GaAs and Related Compounds, Heraklion, Greece, 1987

Impurity behaviour in hydrogenated high purity GaAs

N Pan, S S Bose, M S Feng, M A Plano, M H Kim, B Lee, and G E Stillman

Center for Compound Semiconductor Microelectronics
Coordinated Science Laboratory and Materials Research Laboratory
University of Illinois at Urbana-Champaign, Urbana, Illinois 61801

The effects of hydrogenation in both high purity n and p-type GaAs grown by molecular beam epitaxy (MBE) have been investigated by photothermal ionization spectroscopy (PTIS), low temperature photoluminescence (PL), deep level transient spectroscopy (DLTS), capacitance-voltage (C-V), and Hall-effect measurements. After hydrogenation, a significant decrease in the concentration of electrically active Si donors in n-type GaAs is observed. Similarly, the concentrations of electrically active C acceptors in both n and p-type GaAs show a large decrease. An increase in mobility is observed for the n-type samples, while the p-type samples are highly resistive. The defects and traps that are present in these samples are also passivated.

1. INTRODUCTION

A significant change in the electrical properties of high purity GaAs is observed after exposure to a hydrogen plasma (Pan et al. 1987a, Pan et al. 1987b). There have been numerous reports on the neutralization of shallow donors in GaAs (Chevallier et al 1985, Pearton et al 1986, Chung et al. 1985). However, there have been few reports on the neutralization of shallow acceptors in GaAs and the results have not been consistent (Johnson et al. 1986, Pearton et al. 1986). Since many of the conclusions regarding the effects of hydrogenation were made from samples that were heavily doped, it was not possible to provide detailed spectroscopic measurements in addition to electrical measurements in order to study the behavior of each of the shallow impurities after exposure to a hydrogen plasma.

In this study, the concentration of both donors and acceptors and traps in the high purity n and p-type (MBE) GaAs before and after hydrogenation has been investigated by PTIS, PL, C-V, DLTS and Hall-effect measurements.

2. EXPERIMENT

The growth conditions of the high purity MBE GaAs samples and the plasma reactor configuration including the details of the hydrogenation experiments and the characterization techniques have been described previously (Pan et al. 1987a, Pan et al. 1987b).

The samples were first characterized by the described techniques before the hydrogenation experiments. During hydrogenation, the temperature of the sample was maintained at 250°C, the pressure at 750 mTorr and the power density at 0.40W/cm^2. The exposure time was varied from 30 min to 3 hours. The samples were again characterized to determine the effects of hydrogen plasma on the deep and shallow impurities.

3. RESULTS AND DISCUSSION

The results indicating the neutralization of Si donors in the high purity n-type MBE GaAs have been reported (Pan et al. 1987a). A large reduction of the electrically active Si donor concentration along with an increase in mobility was observed in the samples after hydrogenation. The results of the Hall-effect measurements on several samples before and after hydrogenation, including the separate values for N_d and N_a determined from fits to the empirical curve of Wolfe et al. (1970) are summarized in Table 1. Since the carrier concentration profile after hydrogenation was not homogeneous, the absolute value of N_d and N_a after hydrogenation may not be accurate. However, the results indicate that in addition to the neutralization of Si donors, some of the acceptors may also be neutralized.

Table 1. Donor and Acceptor Concentration (n-type MBE GaAs) Before and After (estimates) Hydrogenation

Sample	μ_{77} Before (cm^2/V-s)	μ_{77} After (cm^2/V-s)	N_d Before (cm^{-3})	N_d After (cm^{-3})	N_a Before (cm^{-3})	N_a After (cm^{-3})
#1	63,500	103,000	1.23×10^{15}	5.66×10^{14}	5.41×10^{14}	1.72×10^{14}
#2	83,400	95,500	7.44×10^{14}	6.35×10^{14}	3.14×10^{14}	2.13×10^{14}
#3	124,600	142,300	3.03×10^{14}	2.42×10^{14}	1.37×10^{14}	7.53×10^{13}

PL and Hall-effect measurements on the hydrogenated p-type MBE GaAs samples have revealed that C acceptors can be passivated by hydrogenation (Pan et al. 1987b). The hydrogenated samples are highly resistive and reliable Hall-effect measurements could not be obtained. The most interesting feature in the PL results of the hydrogenated sample is the appearance of the Be (D°-A°) transition. This result can be explained by the large reduction of electrically active C by hydrogenation (Pan et al. 1987b).

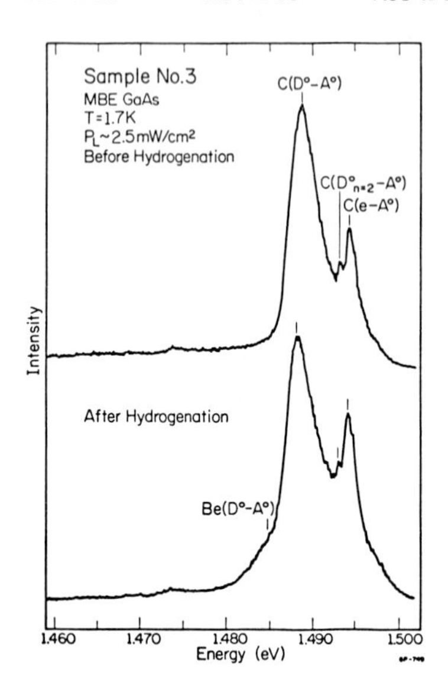

Fig. 1 PL spectra of (D°-A°) and (e-A°) transitions for an n-type MBE GaAs sample before and after hydrogenation. The luminescence intensity is reduced and the Be (D°-A°) peak appears.

Figure 1 shows the PL spectra of the conduction band and/or donor to acceptor transitions for the high purity n-type GaAs sample. The spectrum of the sample before hydrogenation shows that only C acceptors are present. After hydrogenation, the Be (D°-A°) peak appears in the spectrum and the luminescence intensity is reduced by a factor of 2. The reduction of the luminescence intensity of the (D°-A°) or (e-A°) peaks has been previously shown to be related to the reduction of the total impurity concentration (Skromme et al. 1985). The appearance of the Be (D°-A°) peak in this sample can be explained using the same argument described above for the p-type MBE GaAs sample. The combined results of PL and Hall-effect measurements show that C acceptors in high purity n-type samples are also passivated by hydrogenation.

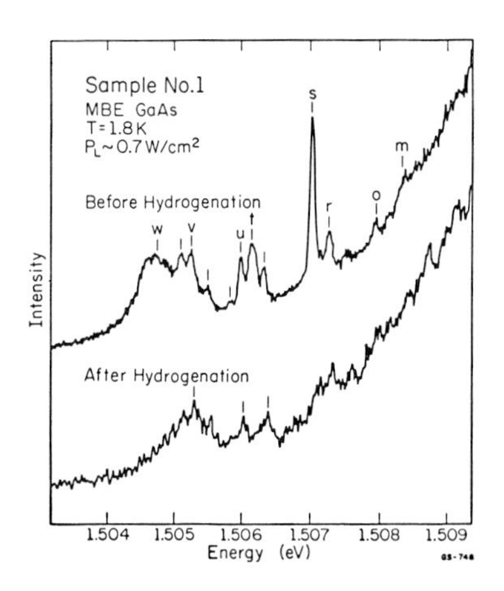

Fig. 2 PL spectra of the sharp defect bound exciton transitions for an n-type MBE sample before and after hydrogenation. The defect related peaks are almost completely passivated and the luminescence intensity is reduced by a factor of 7.

Figures 2 and 3 show the defect related spectra of the high purity n-type and p-type GaAs samples before and after hydrogenation. The peaks are labeled according to the convention of Kunzel and Ploog (1980). There are numerous defect bound exciton peaks in the original samples. After hydrogenation the luminescence intensity of the peaks is significantly reduced by a factor of 7 and almost all of the defect bound exciton peaks are passivated. Since the defect related peaks have been reported to be associated with the C acceptor (Skromme et al. 1985), the passivation of these peaks in both n and p-type GaAs may be related to the passivation of C acceptors by hydrogenation.

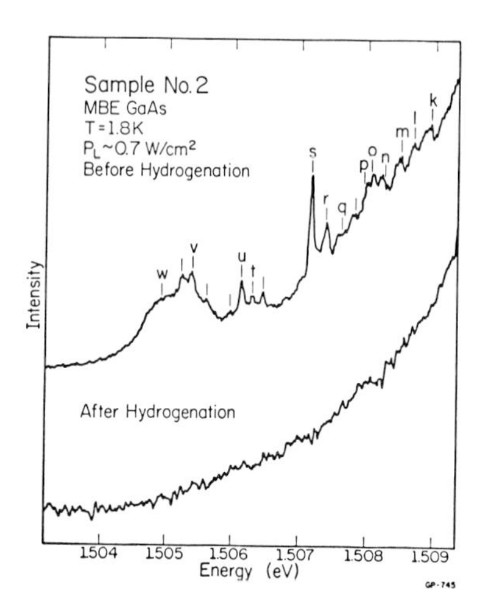

Fig. 3 PL spectra of the sharp defect bound exciton transitions for a p-type MBE sample before and after hydrogenation. The luminescence intensity is significantly reduced and the defect related peaks are almost completely passivated.

The DLTS spectra for the high purity n-type GaAs sample before and after hydrogenation is shown in Fig. 4. The peaks (M1, M3, M4) are labeled according to the convention of Lang et al. (1976). The shallowest trap M00 has been reported by DeJule et al. (1985) in various high purity MBE GaAs samples. The distinct feature of the hydrogenated spectrum is the almost complete passivation of the traps. The initial total trap concentration was determined to be $2\times10^{13}/cm^{-3}$, while after hydrogenation the total trap concentration is reduced by an order of magnitude. These results on the passivation of deep levels are in agreement with Dautremont-Smith et al. (1987) who reported that M1, M3, and M4 can be passivated by hydrogenation.

4. CONCLUSION

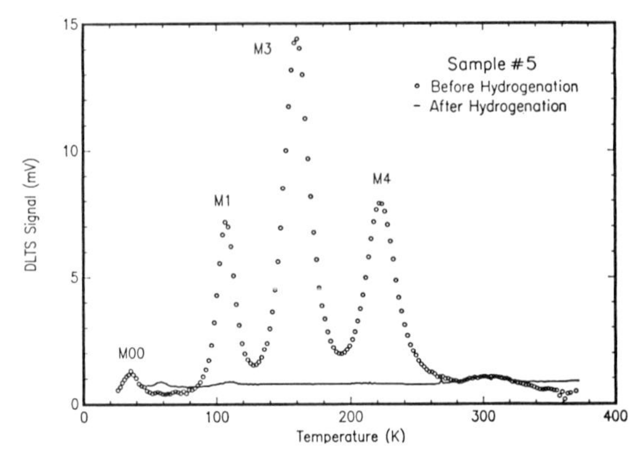

Fig. 4 DLTS spectra of the traps of a high purity n-type MBE GaAs sample before and after hydrogenation. A significant reduction of the traps is observed.

The combined results of spectroscopic and electrical measurements show that the effects of hydrogenation in high purity GaAs is significant. The concentration of Si donors in n-type GaAs and the concentration of C acceptors in both n and p-type GaAs are significantly decreased after hydrogenation. An increase in the mobility is observed for the n-type samples, while the p-type samples are highly resistive. The defects for both n and p-type GaAs are almost completely passivated. A large reduction in the trap concentration is observed in the n-type samples. The hydrogenation mechanism for the electrical deactivation of the shallow impurities has been previously described (Pearton et al. 1986, Johnson et al. 1986). However, the mechanism of the passivation of the defects and traps is not well understood. Because hydrogenation of p-type GaAs results in highly resistive material without introducing additional defects or impurities, this process may be quite useful in providing isolation regions in various devices.

The authors would like to thank B L Payne and B MacFarlane for assistance. This work is supported by the National Science Foundation under Contract NSF DMR 83-16981 and NSF CDR 85-22666 and SDIO/IST managed by Army Research Office under Contract DAAL 03-87-K-0013.

REFERENCES

Chevallier J, Dautremont-Smith W C, Tu C W and Pearton S J 1986 Appl. Phys. Lett. **47** 108

Chung Y, Langer D W, Becker P and Look D 1985 IEEE Trans. Electron Devices **ED-32** 40

Dautremont-Smith W C, Nabity J C, Mostefaoui R, Pajot B, Murawala P and Azoulag R 1987 Appl. Phys. Lett. **49** 1098

DeJule R Y, Haase M A, Stillman G E, Palmateer S C and Hwang J C M 1985 J. Appl. Phys. **57** 5287

Johnson N M, Burnham R D, Street R A and Thornton R L 1986 Phys. Rev. **B33** 1102

Kunzel H and Ploog K 1987 Appl. Phys. Lett. **37** 416

Lang D V, Cho A Y, Gossard A C, Ilegems M and Wiegman W 1976 J. Appl. Phys. **47** 2558

Pan N, Lee B, Bose S S, Kim M H, Hughes J S, Stillman G E, Arai K and Nashimoto Y 1987a Appl. Phys. Lett. **50** 1832.

Pan N, Bose S S, Kim M H, Stillman G E, Chambers F, Devane G, Ito C R and Feng M 1987b Appl. Phys. Lett. **51** 596.

Pearton S J, Dautremont-Smith W C, Chevallier J, Tu C W and Cummings D 1986 J. Appl. Phys. **59** 2821

Skromme B J, Bose S S, Lee B, Low T S, Lepkowski T R, DeJule R Y, Stillman G E and Hwang J C M 1985 J. Appl. Phys. **58**, 4685

Wolfe C M, Stillman G E and Lindley W T 1970 J. Appl. Phys. **41** 3088.

Inst. Phys. Conf. Ser. No. 91: Chapter 4
Paper presented at Int. Symp. GaAs and Related Compounds, Heraklion, Greece, 1987

Characterization of a neutral As_{Ga}–Cu_{Ga} pair defect in GaAs

B Monemar, Q X Zhao, H P Gislason*, W M Chen, P O Holtz and M Ahlström

Department of Physics and Measurement Technology, Linköping University, S-581 83 Linköping, Sweden.

*Permanent address: Science Institute, University of Iceland, Dunhaga 3, 107 Reykjavik, Iceland.

ABSTRACT: The creation and probable identity of a Ga-site substitutional neutral pair defect Cu_{Ga}-As_{Ga} in GaAs is discussed. The defect is associated with a bound exciton (BE) at 1.429 eV, which has been studied in detail by optical spectroscopy. The electronic structure of the BE can be understood in detail, in terms of contributions of a similar magnitude from the electron-hole exchange interaction and the local strain field from the pair of defect atoms.

1. INTRODUCTION

Cu-doped or Cu-contaminated p-type GaAs, which has been annealed at temperatures below 650°C, often shows a broad structured band in low temperature photoluminescence (PL), peaking at about 1.37eV and with a no-phonon line at 1.429 eV at 2K (Hwang 1968, Monemar et al 1986, Holtz et al 1987). A very similar spectrum was observed in an early study of electron-irradiated bulk GaAs:Zn, after annealing at 200°C (Arnold 1966, Arnold and Brice 1969). In early work it was concluded that the corresponding defect was a complex including V_{As} or another As-site donor paired with Cu_{Ga} (Hwang 1968). In more recent work it has been argued that the defect corresponding to the 1.429 eV bound exciton (BE) is instead a Ga-site pair, probably Cu_{Ga}-As_{Ga} (Monemar et al 1986, Holtz et al 1987). In this paper we report on a more detailed study on the conditions for creation of this defect, and also a consistent analysis of the electronic structure of the corresponding 1.429 eV BE.

2. CONDITIONS FOR CREATION OF THE 1.429 eV DEFECT AND ITS POSSIBLE IDENTITY

From the present work (Monemar et al 1986, Holtz et al 1987) as well as from previous reports (Hwang 1968) it seems clear that the defect is only created when the material is p-type during the diffusion process, indicating a Fermi-level dependence of the defect reaction. The early suggestion of a V_{As}-Cu_{Ga} pair is clearly ruled out by the <110> orientation of the defect, established from Zeeman data for the 1.429 eV BE (Monemar et al 1986). The identity must be either a Ga-site pair or an As-site pair. The probable involvement of Cu_{Ga} seems to rule out the latter possibility. A deep double donor is necessary to compensate the Cu_{Ga} double acceptor, thus creating a neutral ground state for the 1.429 eV BE, as observed (Holtz et al 1987). As_{Ga} is the only obvious candidate for this donor, and occurs in sufficient concentration ($\sim10^{16}cm^{-3}$) in bulk GaAs to account for the observed 1.429 eV defect creation. The concentration of the Cu_{Ga}-As_{Ga} defect has not been determined, but is concluded to be well below $10^{16}cm^{-3}$ (Monemar et al 1986). The details of the defect reactions during the low temperature Cu-diffusion (or annealing) process are not yet clear.

3. SPECTROSCOPIC CHARACTERIZATION OF THE 1.429 eV BE

The PL spectra of the 1.429 eV BE at different temperatures are shown in Figure 1. It is obvious from these spectra that the electronic (no-phonon) line structure is dominated by three lines, L3 at 1.4285 eV (2K), L2 at 1.4299 eV (4K) and L1 at 1.4308 eV (10K). The phonon coupling is rather strong, dominated by a single low energy quasilocalized mode of about 11 meV, similar to the situation documented for nearest neighbour

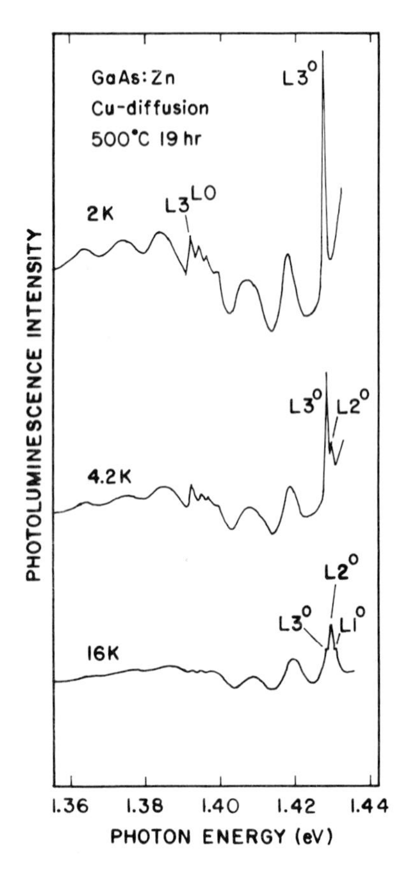

Figure 1. Photoluminescence spectra of the 1.429 eV BE at three different temperatures.

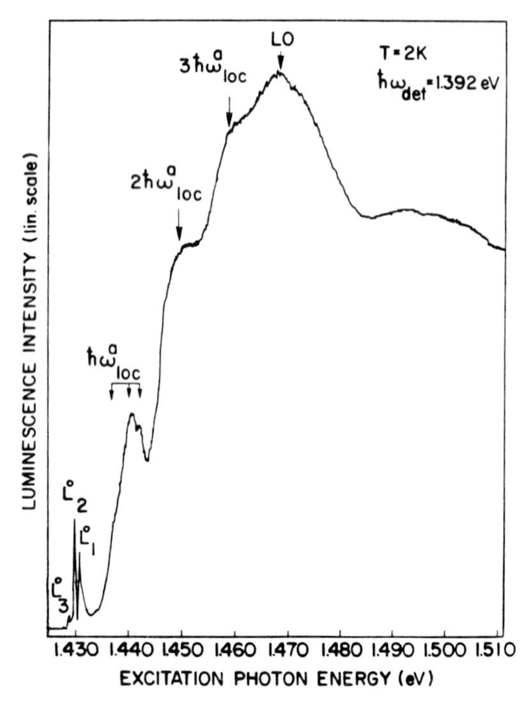

Figure 2. Photoluminescence excitation (PLE) spectra for the 1.429 eV BE, obtained by scanning a tunable dye laser(Styryl 9 dye) with detection via a monocromator set at 1.392 eV. The spectrum has not been corrected for the slow variation in the dye laser intensity over the photon energy region shown.

substitutional pairs in GaP(Vandevyver et al 1978), notably $Cd_{Ga}-O_P$ and $Zn_{Ga}-O_P$ (Henry et al 1968, Morgan et al 1968, Dean 1973).

The oscillator strength of this BE is too weak to allow direct transmission studies, but photoluminescence excitation (PLE) spectra with the aid of a tunable dye laser could be obtained with a good signal to noise ratio, as shown in Figure 2. The same three main electronic lines L1-L3 are seen at 2K in the PLE spectrum, together with strong phonon replicas towards higher photon energies. The phonon energies are similar to the 11 meV replica seen in the PL emission spectra, these phonon energies vary slightly for the three different electronic lines, though (Holtz et al 1987).

In Figure 3 is shown in more detail the PLE spectrum at 2K, in the region of the electronic lines. A simulation of this spectrum is done with the aid of the zero field Hamiltonian for the BE:

$$H=-a\mathbf{J}_h\cdot\mathbf{S}_e-D(J_{hz}^2-\tfrac{1}{3}J_h(J_h+1))-E(J_{hx}^2-J_{hy}^2), \qquad (1)$$

where $\mathbf{J}_h$ is the angular momentum of the bound hole and $\mathbf{S}_e$ is the same quantity (spin) for the electron(x,y,z refer to the local crystal field axes). The first term denotes the electron-hole exchange interaction, and the two additional terms describe the local crystal field. Diagonalization of Eq.1 was performed with a suitable set of basis functions (in this case spin-like for the electron and p-like (J_h=3/2) for the hole, i.e. 2×4=8 BE basis functions) (Chen 1987, Zhao et al 1987). The eigenenergies obtained correspond to the lines in the observed PLE-spectrum (Figure 3). To obtain a reliable

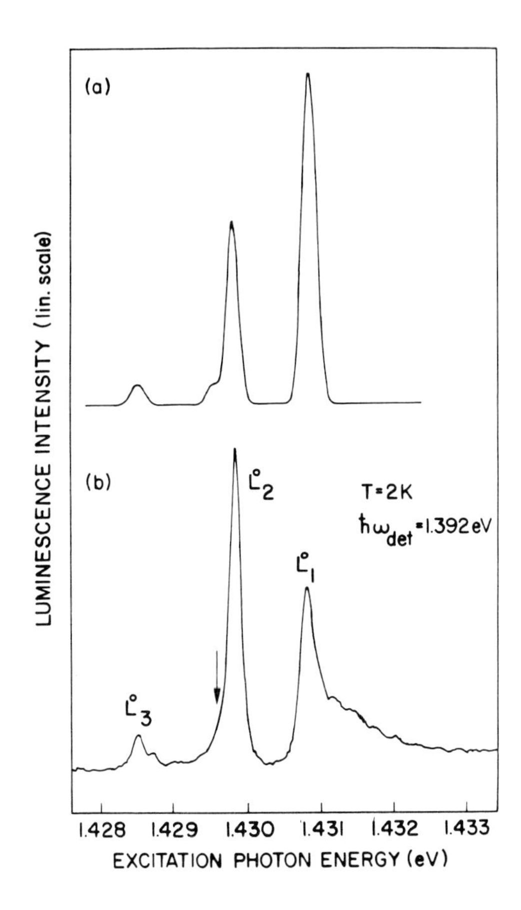

Figure 3. Comparison between the theoretically simulated spectrum (a) for the electronic lines of the 1.429 eV BE and the experimental PLE spectrum(b) (part of Figure 2). The high energy line L_1^o seems to experience a coupling to low energy phonons, which is not accounted for in (a).

comparison with the experimental spectrum, the oscillator strengths of the different electronic lines also have to be computed, with the aid of dipole matrix elements for the BE eigenfunctions. Such a computed spectrum for the electronic lines of the 1.429 eV BE is shown in Figure 3(a), obtained with the parameter values a=0.71 meV, D=0.37 meV and E=0.28 meV. The positive value of D indicates that the local strain field is tensional. These values show the importance of the local strain field parameters in determining the zero field electronic structure of this BE. The reminiscence of an exchange dominated BE , such as for NN pairs in GaP(Zhao et al 1987), is lost when the local field parameters (D and E) are comparable to the strength of the exchange interaction(represented with the a parameter in Eq.1). It seems like the strong background in the PLE spectrum observed just above the L1 line (Fig.3(b)) is due to phonon coupling to very low energy phonons, not unusual for such BE spectra (Monemar et at 1985).

4. CONCLUSIONS

The spectroscopic studies presented here and in previous work (Monemar et at 1986,Holtz et al 1987) give a consistent physical picture of the electronic structure of a substitutional deep nearest neighbour donor-acceptor pair in GaAs, dominated by the deep As_{Ga} donor potential (as shown by the Zeeman data reported separately(Monemar et al 1986)). The local strain field strongly influences the electronic structure of the defect, as simulated in detail in this work. Studies of similar complex defects are of interest to understand the properties of the As_{Ga}-related EL2 defect in GaAs (Meyer et al 1987).

REFERENCES

Arnold G W 1966 Phys. Rev. **149** 679
Arnold G W and Brice D K 1969 Phys. Rev. **178** 1399
Chen W M Thesis Linköping University 1987
Dean P J 1973 J.Luminescence **7** 51
Henry C H,Dean P J and Cuthbert J D 1968 Phys. Rev. **166** 754
Holtz P O,Zhao Q X and Monemar B 1987 Phys. Rev. **B36** to be published
Hwang C J 1968 J. Appl. Phys. **39** 4307 and **39** 4313
Meyer B K,Hofmann D M,Niklas J R and Spaeth J M 1987 Phys. Rev. **B36** 1332
Monemar B,Gislason H P,Chen W M and Wang Z G 1986 Phys. Rev. **B33** 4424
Monemar B,Molva E and Dang L S 1985 Phys. Rev. **B33** 1134
Morgan T N,Welber B and Bhargava R N 1968 Phys. Rev. **166** 751
Vandevyver M and Plumelle P 1978 Phys. Rev. **B17** 675
Zhao Q X and Monemar B unpublished

Paper presented at Int. Symp. GaAs and Related Compounds, Heraklion, Greece, 1987

Observation of the *n* = 2 excited states of the light and heavy hole excitons in GaAs grown directly on Si by OMVPE

S. Zemon, C. Jagannath, S. K. Shastry, and G. Lambert

GTE Laboratories Incorporated, 40 Sylvan Road, Waltham, MA 02254

Abstract: The first evidence for n=2 excited states of both the light hole and the heavy hole excitons is presented for GaAs grown directly on Si by OMVPE. The excited states are about 3 meV above the corresponding n=1 ground states, similar to results for homoepitaxial GaAs. Using selective excitation a spectral width of ≈ 2 meV is observed for the n=1 transition of the light hole exciton, the narrowest yet reported. In addition, exciton-mediated excitation of shallow donors is described.

There is intense interest in GaAs grown directly on Si (GaAs/Si) because of the device possibilities inherent in combining GaAs and Si technologies. Furthermore, this configuration provides an opportunity to study the properties of a III-V compound under uniform biaxial stress (Zemon et al 1986, Zemon et al 1987, Gourly et al 1986). The epitaxial growth of GaAs/Si, however, has severe problems stemming from the thermal and lattice mismatch between the two materials. Despite these problems, light hole excitons (comprised of an electron bound to a $m_J = \pm 3/2$ hole) and heavy hole excitons (associated with the $m_J = \pm 1/2$ hole) have been observed and acceptor transitions identified (Zemon et al 1986, Zemon et al 1987). However, the narrowest photoluminescence (PL) excitonic spectral widths have been found to be about 4 meV for both the heavy and light hole excitons (Zemon et al 1987, Duncan et al 1986) as compared to less than ~ 0.5-meV for excitons in high purity, homoepitaxial GaAs at 4.2 K (Heim and Hiesinger 1974, Koteles et al 1987). In addition, excited state excitonic transitions and donor transitions, commonly observed for the latter material (Heim and Hiesinger 1974), have not yet been reported for GaAs/Si.

We present new studies of a GaAs/Si layer grown by a previously described organometallic vapor phase epitaxy (OMVPE) process (Shastry and Zemon 1986). The epitaxial quality is sufficiently high so that, for the first time, the n=2 excited state of the light hole exciton ($E^{n=2}_{\pm 3/2}$) and the heavy hole exciton ($E^{n=2}_{\pm 1/2}$) are observed. Exciton binding energies are estimated to be about 4 meV, similar to the value for homoepitaxial GaAs. Employing selective PL excitation the $E^{n=1}_{\pm 3/2}$ spectral width (full width at half maxima) is found to be ≈ 2 meV, the narrowest yet reported for GaAs/Si. Furthermore, exciton-mediated excitation of shallow donors from the n=1 to the n=2 states is observed. The monitoring of these new features can be used as a sensitive indicator of the epitaxial quality and should lead to an increased understanding of the GaAs/Si layers.

The GaAs/Si layers were epitaxially grown by OMVPE on 2" diameter (100)Si by a two-step, low temperature process (Shastry and Zemon 1986). The layer thicknesses ranged from 3.5 μm to 7.5 μm. The selective PL and the PL excitation (PLE) spectroscopy procedures have been described previously (Zemon et al 1986). The 4.2 K measurements were made with the sample freely suspended in liquid helium using excitation power densities ≤ 1 mW/cm^2 and an unfocused excitation beam at normal incidence. Temperature dependent studies were performed with the sample mounted in a variable tempera-

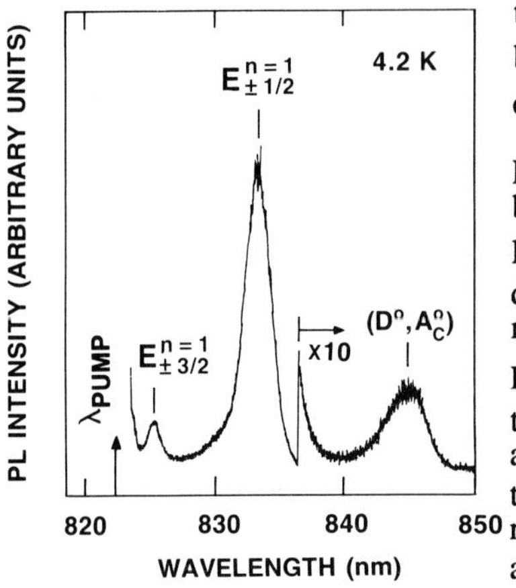

Fig. 1. Photoluminescence spectrum of undoped GaAs/Si at 4.2 K using λ pump = 822.7 nm.

ture, cold finger cryostat using a focused excitation beam incident at ~ 60° to the sample normal and an excitation power density of about 200 mW/cm^2.

Figure 1 shows a 4.2 K PL trace for wavelengths between 823.2 nm and 850 nm with the pump wavelength (λ_{pump}) set at 822.7 nm. Three lines are evident with peaks at 825.3 nm, 833.3 nm, and 844.9 nm, which have previously been identified as $E^{n=1}_{\pm 3/2}$, $E^{n=1}_{\pm 1/2}$, and donor-to-carbon-acceptor [(D°, A°$_C$)] transitions, respectively (Zemon et al 1986, Zemon et al 1987). The spectral widths of the $E^{n=1}_{\pm 1/2}$ and $E^{n=1}_{\pm 3/2}$ transitions are 4.3 meV and ≈ 2 meV respectively. We note that the 2-meV value for the $E^{n=1}_{\pm 3/2}$ transition is about a factor of two narrower than the value found using non-resonant excitation ($\lambda_{pump} < 820$ nm). The peak energies of the exciton transitions can be utilized to determine a uniform biaxial tensile stress of about 2.5 kbar in the plane of the layer at 4.2 K (Zemon et al 1987).

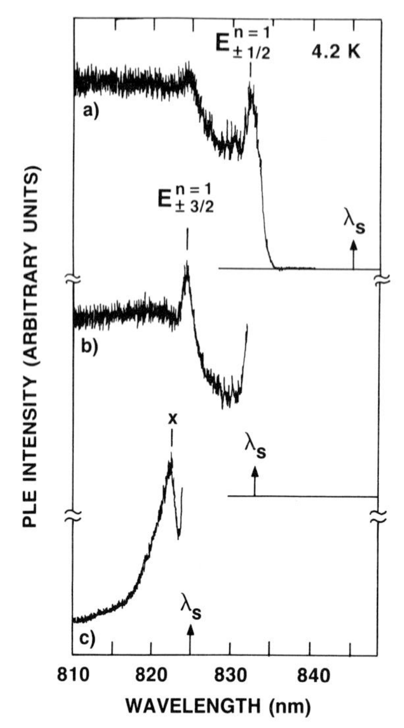

Fig. 2. 4.2 K photoluminescence excitation (PLE) spectra. The spectrometer settings are indicated by the arrows.

In Fig. 2 we present three 4.2 K PLE spectra where the spectrometer wavelength (λ_S) is set at the values indicated by the arrows in the figure and the pump wavelength is scanned. Such data provide a measure of the strength of the absorption. In the PLE spectrum of Fig. 2a [λ_S set on the (D°, A°$_C$) transition wavelength] two peaks are observed representing the $E^{n=1}_{\pm 3/2}$ and $E^{n=1}_{\pm 1/2}$ exciton transitions as expected. In Fig. 2b where the spectrometer is set at the peak of the $E^{n=1}_{\pm 1/2}$ transition, the $E^{n=1}_{\pm 3/2}$ transition can be seen more prominently and the peak wavelength is determined to be 824.3 nm. In Fig. 2c [λ_S set on the $E^{n=1}_{\pm 3/2}$ transition] a new peak (labeled x) is observed at 822.6 nm, strongly suggesting the presence of a higher excited excitonic state coupled to $E^{n=1}_{\pm 3/2}$. Confirming evidence for this PLE peak was obtained using selective PL. The energy difference between the $E^{n=1}_{\pm 3/2}$ peak of Fig. 2b and the peak x of Fig. 2c is ≈ 3.0 meV. The intensity of feature x monotonically decreases with increasing temperature, becoming unobservable above 16.5 K. This indicates a weakly bound state.

We identify peak x with the n=2 excited state of the light hole exciton, $E^{n=2}_{\pm 3/2}$. It is well known that n=1 excitonic features can be resonantly excited by pumping the n=2 excited state transition (Heim and Hiesinger 1974), in agreement with the spectrum of Fig. 2c. The 3.0-meV energy difference between the $E^{n=2}_{\pm 3/2}$ and $E^{n=1}_{\pm 3/2}$ transitions is close to the value found for free excitons in homoepitaxial GaAs at zero stress (Sell 1972). Interestingly enough, it has been shown that this energy difference is not a function of uniaxial stress, at least for values up to 1 kbar (Jagannath and Koteles 1986). Using a hydrogenic model an estimate of the binding energies of the $E^{n=1}_{\pm 3/2}$ and the $E^{n=2}_{\pm 3/2}$ states are found to be 4.0 meV and 1.0 meV, respectively. In terms of an equivalent temperature, the latter value corresponds to about 12 K. Thus, it is reasonable to expect that the $E^{n=2}_{\pm 3/2}$ state would be ionized in this temperature region, resulting in a quenching of the resonance transfer in agreement with the results shown in Fig. 3.

Further evidence for the strong coupling between $E^{n=2}_{\pm 3/2}$ and $E^{n=1}_{\pm 3/2}$ is seen in the polarization dependence of the 4.2 K PLE spectra with the spectrometer set at the peak of the $E^{n=1}_{\pm 3/2}$ transition. Utilizing a right circularly polarized (RCP) pump beam, the PLE was analyzed for both left circularly polarized (LCP) and RCP signals. The PLE signal detected for LCP light was observed to be identical to the unpolarized spectrum of Fig. 2c. In marked contrast, the spectrum analyzed for RCP light exhibited a low level signal indicating the PLE was virtually 100% polarized. Optical selection rules indicate that a state excited with RCP light will, assuming no depolarization, lose its energy by emitting LCP light. Since a strong polarization memory is apparent in the data, we conclude that there is a fast transfer of energy between feature x and $E^{n=1}_{\pm 3/2}$. These striking polarization results are consistent with the strong coupling one would expect to find between the ground and first excited state of an exciton.

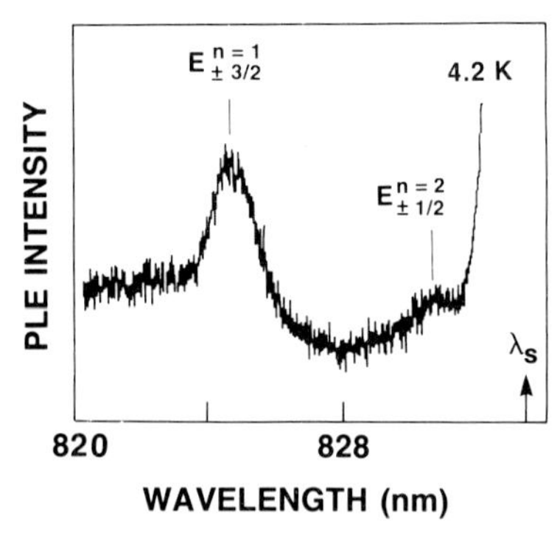

Fig. 3. The 4.2 K PLE spectrum of the heavy hole exciton $E^{n=1}_{\pm 1/2}$. The excitation source is focused to ~ 200 mW/cm^2. The arrow indicates the spectrometer setting λ_S.

$E^{n=2}_{\pm 1/2}$ transitions are not observed in the PLE spectrum of $E^{n=1}_{\pm 1/2}$ in Fig. 2b. Possibly the $E^{n=1}_{\pm 1/2}$ transition is sufficiently broad so that its high energy tail would obscure a weak $E^{n=2}_{\pm 1/2}$ peak ≈ 3 meV higher in energy. However, upon focusing the beam (≈ 100 μm in diameter) a weak $E^{n=2}_{\pm 1/2}$ transition was observed in some regions of the sample. This positional dependence of the intensity is a consequence of growth related inhomogeneities. Figure 4 shows a PLE spectrum of the $E^{n=1}_{\pm 1/2}$ transition where a small peak is evident at 830.5 nm, about 3 meV higher in energy than the peak of the $E^{n=1}_{\pm 1/2}$ transition. Data taken at higher temperatures indicate that the line becomes undetectable above ~ 11 K. Using arguments similar to those presented for peak x, we identify the 830.5-nm line with a weakly bound $E^{n=2}_{\pm 1/2}$ state. Consistent with the results found for the light hole excitons, we note that the 3-meV energy difference between the $E^{n=2}_{\pm 1/2}$ and $E^{n=1}_{\pm 1/2}$ transitions is close to the value for free excitons in homoepitaxial GaAs (Jagannath and Koteles 1986).

When pumping the wavelength region in the vicinity of the $E^{n=1}_{\pm 3/2}$ PLE transition (see Fig. 3), a "Raman-like" spectral peak (at a wavelength λ_{PEAK}), which tracks the laser pump, is observed when 823.8 nm ≤ λ_{PUMP} ≤ 825.8 nm. It is found that $\lambda_{PEAK} - \lambda_{PUMP}$ = 2.5 nm, i.e. 4.5 meV, corresponding to the energy difference between the n=1 and n=2 donor states in GaAs/GaAs. Since donor energy level differences in GaAs are not expected to be a function of stress, we identify the new feature with a process in which the donor in GaAs/Si is excited from the n=1 ground state to an n=2 final state. This represents the first observation of an impurity excited state transition in GaAs/Si. Work is in progress to understand the origin of this effect.

In summary, these data indicate the good epitaxial quality that can be attained with OMVPE GaAs/Si layers. However, issues of uniformity remain.

Acknowledgement—We thank Dr. W. Miniscalco for helpful conversations, Drs. P. Haugsjaa and H. Lockwood for their support of this work, and M. DeAngelis for technical assistance.

REFERENCES

Duncan W M, Lee J W, Matyi R J and Liu H-Y 1986 J. Appl. Phys. **59** 2161. We identify the 1.499-eV PL line of Duncan et al with a $E^{n=1}_{\pm 3/2}$ transition instead of the free-to-carbon-acceptor identification employed by these authors.

Gourley P L, Wiczer J J, Zipperian E E and Dawson L R 1986 Appl. Phys. Lett. **49** 100 and references therein

Heim U and Hiesinger P 1974 Phys. Stat. Sol. (b) **66** 461

Jagannath C and Koteles E 1986 Solid State Commun. **58** 417

Koteles E S, Elman B S and Zemon S 1987 Solid State Commun. **62** 703

Sell D D 1972 Phys. Rev. B **6** 3750

Shastry S K and Zemon S 1986 Appl. Phys. Lett. **49** 467

Zemon S, Shastry S K, Norris P, Jagannath C and Lambert G 1986 Solid State Commun. **58** 457

Zemon S, Jagannath C, Koteles E, Shastry S K, Norris P, Lambert G, Choudhury A N M M and Armiento C A 1987 in Gallium Aresnide and Related Compounds 1986 (Institute of Physics, Bristol), Inst. Phys. Conf. Ser. **83**, p. 141

Quasi-one-dimensional planar GaAs wires fabricated by focused ion beam implantation

Toshiro Hiramoto, Kazuhiko Hirakawa, and Toshiaki Ikoma

Institute of Industrial Science, University of Tokyo, Roppongi, Minatoku, Tokyo 106, Japan

Abstract: The focused-ion-beam (FIB) implantation was applied, for the first time, to fabricating one-dimensional GaAs wires. The FIB implantation forms high-resistivity regions which define a very narrow conductive wire. The minimum width of the fabricated GaAs wire was 20 nm. Magnetoconductance of the wires showed one-dimensional electron localization and conductance fluctuations due to a quantum interference effect. The temperature dependence of phase coherent length was determined and compared with the existing theory. It is shown that measured amplitudes of the conductance fluctuations clearly depend on the width of quantum wires.

1. INTRODUCTION

The carrier confinement into one-dimensional systems is now of great interest in the fields of semiconductor physics and technology. The recent advances in microfabrication techniques enable us to fabricate very thin semiconductor wires with submicron dimensions, which are comparable to phase coherent length L_ϕ of electron waves. The transport properties of electrons in such wires exhibit weak localization and random quantum interference effects which appear as negative magnetoresistances and conductance fluctuations, respectively. However, quantum wire structures are mainly fabricated by using the electron-beam (EB) lithography, a rather complicated process.

The focused-ion-beam (FIB) implantation is one of the most promising techniques to fabricate semiconductor microstructures (Hiramoto *et al.* 1986). In this technique, an ion beam focused into less than 100 nm is directly implanted into semiconductors under a precise computer control. This is a simple process and allows us to make arbitrary planar microstructures with a submicron resolution. In this work, we demonstrate a feasibility of the FIB technique to fabricate one-dimensional GaAs wires.

2. FABRICATION PROCESS

Conductive GaAs layers become high-resistive when high-density of defects are introduced by the FIB implantation. We made use of this phenomenon to define very narrow GaAs wires. First, focused Si ions were implanted into semi-insulating GaAs at 40 keV (projected range: 34 nm) with $4\times10^{13}cm^{-2}$ dose in an area of 100 $\mu m \times 20$ μm (the 1st-stage implantation). Rapid thermal annealing (RTA) was performed to activate the implanted species and to form a conductive thin layer (Hiramoto *et al.* 1985). Then, a focused-Si^{2+} beam with 0.1 μm diameter was implanted (the 2nd-stage implantation) to form high-resistivity regions which confine the conducting layer into a very narrow channel as shown in the inset of Fig.1. The designed gap spacing d between the 2nd-stage implanted patterns was varied from 0.2 μm to 1.0 μm and the designed length of the channel was 2.0 μm. This is one of the simplest methods ever reported to fabricate one-dimensional structures.

3. CHARACTERIZATION OF WIRES

Figure 1 shows the conductance G of the fabricated GaAs wires as a function of d at three temperatures. The conductance decreases linearly with decreasing d. The extrapolated value of d which corresponds to $G = 0$ is $d = d_0 = 0.48$ µm at 4.2 K. $d_0/2$ is equal to the sum of the lateral spread of thc implanted ions and the depletion layer thickness formed between the conducting wire and the high-resistivity regions. The effective width of the wire is, then, derived by $d_{eff} = d - d_0$ and the effective length L_{eff} of the wire is 2.48 µm (= 2.0 µm + d_0). The conductivity of the fabricated wires deduced from the slope of the line in Fig. 1 is found to be almost the same as that of the original conducting layer. This indicates that the wires are free from the implantation damage. The minimum width of the fabricated wires was 20 nm.

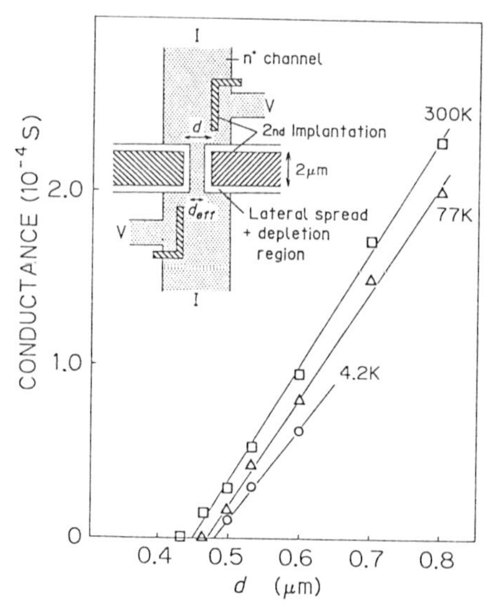

Fig. 1. Measured conductance G of the fabricated GaAs wires as a function of d. The inset shows the top view of the sample structure.

Figure 2 shows the magnetoconductance of three wires with different d_{eff} (120 nm, 53 nm and 20 nm) measured at 1.3 K at low magnetic fields. The conductance is normalized to e^2/h, where e is the electron charge and h the Planck constant. The current level I in the wires was set to 3.0 nA in this case. Each wire exhibits a positive magnetoconductance (*i.e.* a negative magnetoresistance), indicating the delocalization of electronic states by magnetic field. Based on the theory of the one-dimensional localization (Al'tshuler and Aronov 1981, Hiramoto *et al.* 1987), the deviation of the magnetoconductance from the value at $B = 0$ is given as,

$$dG(B) = \frac{e^2}{h}\frac{2}{L}\left(L_\phi - \left(\frac{1}{L_\phi^2} + \frac{1}{L_H^2}\right)^{-1/2}\right), \qquad (1)$$

where $L_H = \sqrt{3}h/(2\pi eBW)$ is the "magnetic length" in a one-dimensional system, L_ϕ is the phase coherent length, and L and W are length and width of the wire, respectively.

Fig. 2. Magnetoconductances of the GaAs wires measured at 1.3 K at low magnetic fields. The conductance is normalized to $G_{univ} = e^2/h$. The solid curves are the theoretical fittings by Eq. (1).

The measured data (dots) in Fig. 2 are compared with the calculated results (solid lines) of Eq. (1), with L_ϕ as a fitting parameter. The effective wire width d_{eff} obtained from Fig. 1 is used for W. As shown in Fig. 2, the theoretical results with $L_\phi = 0.12$ µm ~ 0.13 µm are best fitted to the experimental data. It is noted that the deduced value of L_ϕ is larger than W, indicating that the fabricated channels behave as one-dimensional quantum wires in the weak localization regime.

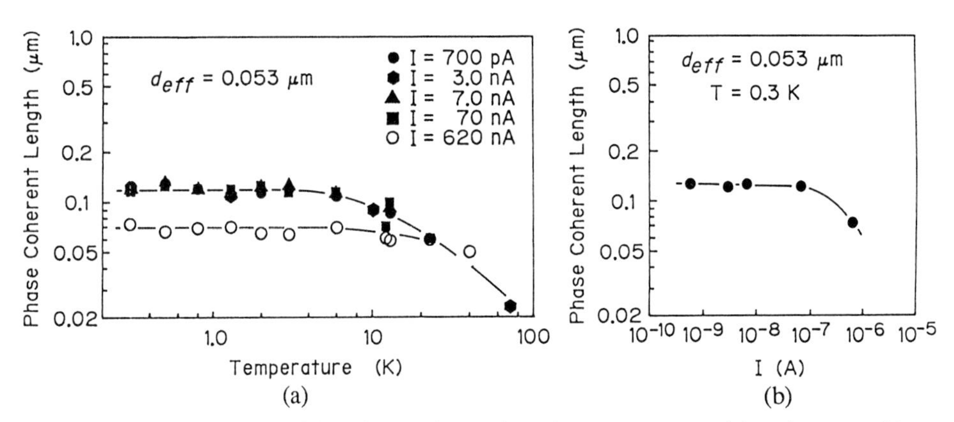

Fig. 3 The dependences of the phase coherent length on temperature (a) and current (b).

Figure 3 (a) shows the temperature dependence of L_ϕ for the sample with d_{eff} = 53 nm. L_ϕ was estimated at five different current levels. When I < 70 nA, L_ϕ increases with decreasing temperature down to 3 K, and becomes constant (~0.12μm) at lower temperatures. This small dependence of L_ϕ on T in the lower temperature range is not caused by electron heating, because no change of L_ϕ was observed even when the current was reduced down to 700 pA, as shown in Fig. 3 (b). This is in contrast with the theories of electron-electron scattering and electron scattering by electromagnetic fluctuations, which predict temperature dependence of L_ϕ as $T^{-1/4}$ and $T^{-1/3}$, respectively (Al'tshuler *et al.* 1981). The origin of this discrepancy is not clear at present. A possible explanation is the Fermi surface smearing effect due to frequent elastic (impurity) scattering in the present sample, whose electron mobility is only 2000 cm^2/Vs. Such smearing of the Fermi surface may well hide the lattice temperature effect on L_ϕ and give a temperature independent L_ϕ. A shorter L_ϕ obtained when I = 620 nA comes from the electron heating effect (Hirakawa and Sakaki 1986).

Figure 4 shows the magnetoconductance measured in the three wires at higher magnetic fields at 1.3 K. The magnetoconductances randomly fluctuate and their patterns are reproducible. However, they change after an external stimulus such as light or heat is applied. This phenomenon is known as the "universal conductance fluctuation" due to a quantum interference effect (Umbach *et al.* 1984). The recent theory (Al'tshuler 1985) predicts that the amplitude of conductance fluctuation is approximately equal to the universal value $G_{univ} = e^2/h \sim 3.9 \times 10^{-5}$ S for a completely phase coherent system. In one-dimensional wires longer than L_ϕ, the amplitude is reduced to (Skocpol *et al.* 1986)

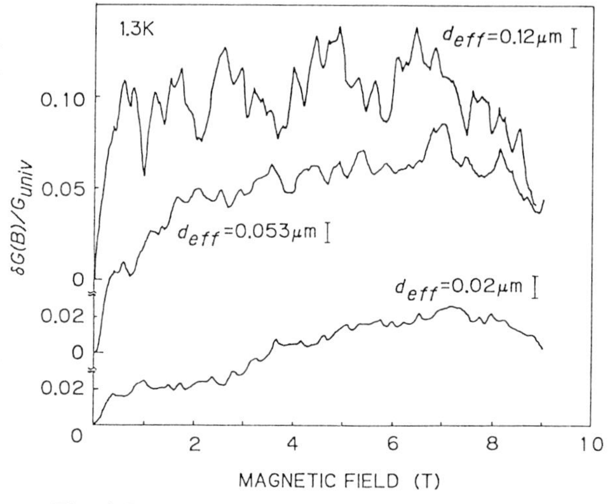

Fig. 4. Magnetoconductances of the three wires at higher magnetic fields at 1.3 K. The bars indicate the amplitudes of the conductance fluctuations predicted by Eq. (2).

$$\Delta G = (L_\phi/L)^{3/2}\, G_{univ}, \qquad (2)$$

which predicts that in one-dimensional wires narrower than L_ϕ the amplitude of the fluctuation depends only on L_ϕ and L, and is not dependent on W. The amplitudes of the fluctuation calculated from Eq. (2) using L_ϕ determined in the weak localization regime are indicated by bars in Fig. 4. The order of magnitude of the fluctuations is in good agreement with the experimental results. However, the measured amplitude is larger than the predicted ΔG in the 120 nm-wide wire. Furthermore, the observed fluctuation becomes smaller as the GaAs wire becomes narrower, clearly suggesting a W-dependence of ΔG.

It was reported by Skocpol *et al.* (1986) that the measured magnitude of conductance fluctuations in Si inversion-layer systems is in excellent agreement with the theoretical prediction of Eq. (2). However, several researchers (Ishibashi *et al.* 1987, Thornton *et al.* 1986, Timp *et al.* 1987) observed conductance fluctuations larger than the prediction of Eq. (2). Our data clearly show that the amplitude of the fluctuation depends on the width of the wires. This unresolved problem calls for further study.

4. CONCLUSIONS

We have developed a novel, simple technique to fabricate one-dimensional GaAs wires by using the FIB technology. It is confirmed by measuring the localization effect at very low temperatures that the wires behave as one-dimensional systems. The phase coherent length of electron waves is found to increase with decreasing temperature and become temperature independent below 3 K. A random quantum interference effect is also observed as a conductance fluctuation in magnetoconductance. The measured amplitudes of conductance fluctuations cannot be fully understood by the existing theory, which calls for further refinement of the theory.

ACKNOWLEDGEMENTS

The authors would like to thank Prof. Y. Iye and Dr. T. Tamegai for useful discussions and magnetoconductance measurements. This work was in part supported by the Grant-in-Aid for Special Project Research from the Ministry of Education, Science and Culture, Japan, and also in part supported by the Joint Research Project of the Institute for Solid State Physics, University of Tokyo.

REFERENCES

Al'tshuler B L and Aronov A G1981 *JETP Lett.* **33** 499
Al'tshuler B L, Aronov A G and Khmelnitsky D E 1981 *Solid State Commun.* **39** 619
Al'tshuler B L 1985 *JETP Lett.* **41** 648
Hirakawa K and Sakaki H 1986 *Appl. Phys. Lett.* **49** 889
Hiramoto T, Saito T and Ikoma T 1985 *Jpn. J. Appl. Phys.* **24** L193
Hiramoto T, Odagiri T, Oldiges P, Saito T and Ikoma T 1986 *Gallium Arsenide and Related Compounds Inst. Phys. Conf. Ser.* No.83 pp 295-300
Hiramoto T, Hirakawa K, Iye Y and Ikoma T 1987 *Appl. Phys. Lett.* (in press)
Ishibashi K, Nagata K, Gamo K, Namba S, Ishida S, Murase K, Kawabe M and Aoyagi Y 1987 *Solid State Commun.* **61** 385
Skocpol W J, Mankiewich P M, Howard R E, Jackel L D, Tennant D M and Stone A D 1986 *Phys. Rev. Lett.* **56** 2865
Thornton T J, Pepper M, Davies G J and Andrews D 1986 *Proc. 18th Int. Conf. on the Phys. of Semicond.* ed O Engstrom (Singapore: World Scientific Publishing) pp 1503-1506
Timp G, Chang A M, Mankiewich P, Behringer R, Cunningham J E, Chang T Y and Howard R E 1987 *Phys. Rev. Lett.* **59** 732
Umbach C P, Washburn S, Laibowitz R B and Webb R A 1984 *Phys. Rev.* **B30** 4048

Inst. Phys. Conf. Ser. No. 91: Chapter 5
Paper presented at Int. Symp. GaAs and Related Compounds, Heraklion, Greece, 1987

Growth of low dislocation density GaAs by As pressure controlled Czochralski method

K Tomizawa, K Sassa, Y Shimanuki and J Nishizawa*

Mitsubishi Metal Corporation,
1-297 Kitabukuro, Omiya, Saitama, 330 Japan
*Research Institute of Electrical Communications, Tohoku University, 2-1-1 Katahira, Sendai, 980 Japan

ABSTRACT: A new pulling apparatus has been developed in order to grow GaAs crystals under controlled As pressures. The effect of As pressure on the EPD was studied on crystals grown by this apparatus. A low EPD of 500 cm^{-2} was realised on 10 mm in diameter crystals grown under the As pressure controlling temperatures from 610 to 624°C.

1. INTRODUCTION

With the progress of the development of GaAs devices , there is increasing demand to the perfect crystal that is defect-free, pure and homogeneous. Among many ways to choose to reach such a goal, stoichiometry control during growth seems to be most fascinating. Nishizawa et al (1981) studied the effect of the As pressure on the quality of GaAs crystals during LPE process or heat treatment process. They found that there existed As pressures to give the minimum values of several characteristics of crystals such as carrier concentration and lattice dilation. On the basis of their experimental data ,the optimum pressure is given as follows.

$$P_{opt}(\text{torr})=2.6\times10^{6}\times\exp(-1.05/kT_g) \qquad (1)$$

From this relation, a value of 813 torr is obtained for the optimum As pressure for the bulk growth where T_g is 1511K. In this paper we report the results of growth of GaAs crystals by using a new pulling apparatus in which As pressure can be precisely controlled during the growth. This method, Arsenic Pressure Controlled Czochralski method (PCZ)is endowed with both advantages of HB and LEC methods ,i.e., (1)precise stoichiometry control is possible , (2)thermal stress can be lowered by decreasing the temperature gradient across the growth interface and a round <100 > crystal can begrown without Si contamination from a crucible.

2. THE PULLING APPARATUS

Fig.1 shows a schematic diagram of the PCZ puller we developed. The apparatus consists of a double chamber. Crystals are grown

in the inner chamber. Before the growth procedure, GaAs is directly synthesized. Following requirements should be fulfilled to grow stoichiometric crystals, and improvements of the puller were performed to meet these requirements.
1. The temperature distribution on the chamber wall should be such that the vapor pressure within the chamber can be controlled by the temperature of a specified portion and the temperature of the other portions must be maintained above this temperature to avoid As deposition onto the chamber wall.
2. A view port of quartz rod through which growing crystal is observed is disposed on the upper flange. To avoid As and/or GaAs deposition onto the front end of the rod, the temperature of this part should be kept in an optimum range.
3. The chamber material must bear repeated uses. It must also not become sources of contamination. The cylindrical structure was made of PBN-coated graphite and the upper flange of the inner chamber was made of Molybdenum.
4. An effective seal is required at divided portions of the chamber. We adopted gaskets to attain the sufficient sealing and repeated use of the chamber.
5. The rotation seals are necessary on the flanges at which the pull rod and the crucible supporting rod are introduced into the chamber. Molten B_2O_3 was used as a sealing material.

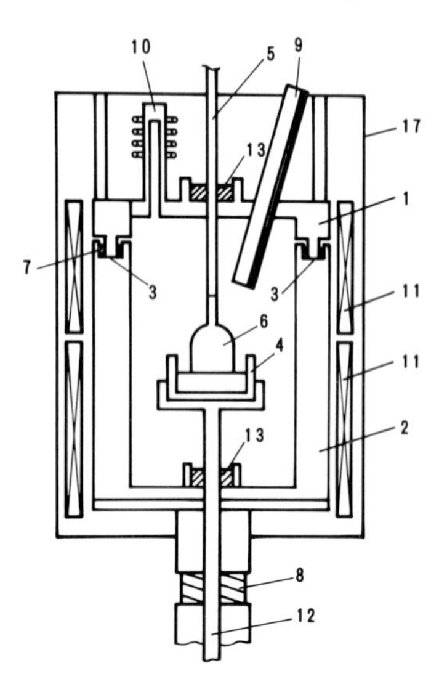

Fig.1 Schematic diagram of the crystal puller

1. Upper flange
2. Inner chamber
3. Seal
4. Crucible
5. Upper shaft
6. GaAs
7. Groove
8. Spring
9. Optical window
10. As pressure controlling furnace
11. Heater
12. Lower shaft
13. B2O3
17. Outer chamber

3. RESULT AND DISCUSSION

3-1 The effect of As vapor pressure on the EPD

To clarify how the dislocation density is affected by the As pressure over the melt during the growth, a <100> crystal of 10mm in diameter was pulled with the temperature gradient at the growth interface of about 10°C/cm. After pulling of each interval 10 mm under a fixed As pressure, T_{As} was changed by 2°C for the next 10 mm pulling. Thus T_{As} was changed from 604 to 632 °C. An average EPD value corresponding to each T_{As} was measured in the central part of each segment of the crystal. Fig.2 shows the result of such measurements. The result obtained by Parsey et al 1982 on HB crystals of a similar diameter to this case is also shown in this figure. In the PCZ method, the EPD was as low as 500 cm^{-2} over the range of T_{As} from 610 to 624 C and it increased gradually in the outside of the range. The minimum EPD values with both methods are nearly the same, However it is noticeable that the PCZ method realized a low EPD over such a wide range of T_{As}, while in the HB method, the EPD increased in a drastic manner from 500 to $2x10^4$ cm^{-2} when T_{As} was slightly changed from the optimum value. In the HB method, since a crystal grows in contact with a quartz boat, a slight deviation from stoichiometry may cause a critical change in adhesiveness, bringing about generation of dislocations.

Undoped GaAs crystals of 75 mm in diameter were grown; the temperature gradient was 10 °C/cm and T_{As} was 616° C. Fig.3 shows the X-ray topograph of the wafer sliced from the shoulder part of this crystal. The average EPD measured along <110> axis was $1.6x10^4$ cm^{-2}; this EPD level is as low as that of HB crystals of the similar diameter.

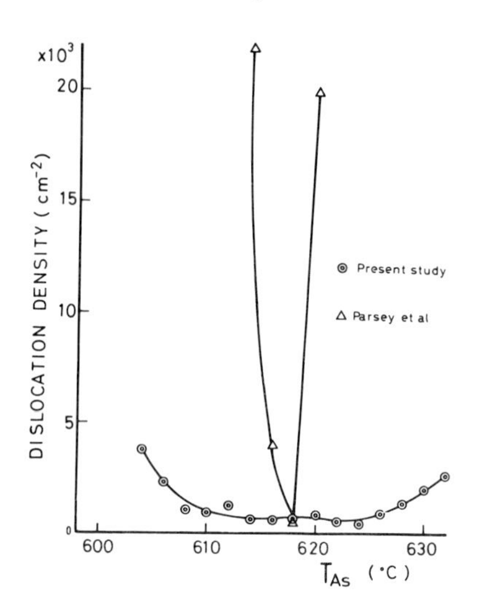

Fig.2 Effect of As pressure on the EPD

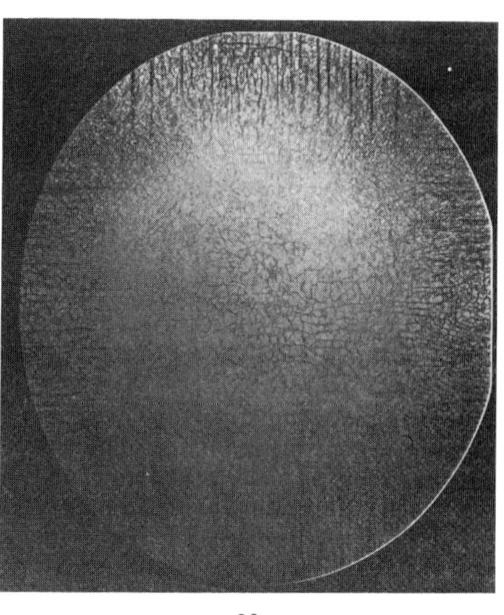

Fig.3 X-ray topograph of the crystal grown by PCZ method

3-2 Purity of the PCZ crystal

Table 1 shows the impurity concentrations in various crystals. Impurity concentrations in PCZ crystals are as low as those in typical LEC crystals except for S concentration. We suppose that the source of S contamination are materials used for the inner chamber and parts within it. Purer crystals can be grown by selecting these materials.

Table 1 Impurity concentrations of GaAs crystals grown by various methods (cm-3)

	Si	S	C	B	
PCZ	3E14	3E15	3E15	1E15	
LEC (1)	2E14	1E14	3E15	8E16	(1) Osaka J 1984
HB (2)	1E16	?	?	?	(2) Akai S 1981

Se, Te, Mg, Cr, Mn, Fe, Mo $<$1E15 in PCZ crystals

3-3 The improvement of electrical homogeneity

In the present experiment, crystals were slowly cooled down in the puller to 600 °C at the rate of 60 °C/hour in order to avoid additional formation of dislocations. But this process made the crystal inhomogeneous. In order to improve the homogeneity, ingots were annealed at 950°C for 7 hours in N2. Fig. 4 shows the two dimensional resistivity distribution of the wafer after annealing. The resistivity was measured by the three-electrode guard method (Obokata et al 1984). The interval of electrodes was 250 µm. The average resistivity is $8.5x10^7$ Ω-cm and the standard deviation is 10.2 %. This homogeneity of resistivity distribution in PCZ crystals is much superior to that of previous report.(Obokata etal 1984)

Fig.4 Resistivity distribution measured by three electrode guard method at 250µm intervals
ρ (Ave.)= 8.5E7Ω-cm
S.D. = 10.2 %

4. SUMMARY

A new pulling apparatus was developed to grow GaAs crystals under a controlled As pressure. The effect of As pressure on the EPD was studied by changing the As pressure controlling temperature during the growth. Minimum dislocation density as low as $500cm^{-2}$ was realized over a wide range of T_{As} from 610 to 624° C. It was shown the new pulling apparatus had a potential to grow large diameter crystals of low dislocation density, high purity and homogeneity.

Acknowledgment

This work was carried out as part of Nishizawa Perfect Crystal Project under the support of the Research Development Corporation of Japan.

References

Akai S ,Fujita K, Sasaki M and Tada K 1981 9th Int.symp.GaAs and Related Compounds Oiso p13

Nishizawa J, Toyama N, Oyama Y and Inokuchi K 1981 Proc. 3rd Int. School on Semiconductor Optoelectronics, Centniewo

Obokata T,Matsumura T,and Fukuda T 1984 Jpn.J.Appl. Phys. 23,L602

Osaka J and Hoshikawa K 1984 in Semi-Insulating III - V materials,kah-nee-ta p126

Parsey J M, Nanishi Y, Lagowski J and Gatos H C 1982 J Electrochem. Soc. 129,388

Growth of low-dislocation density InP by the modified CZ method in the atmosphere of phosphorus vapour pressure

K. Tada, M. Tatsumi, M. Nakagawa, T. Kawase and S. Akai
Sumitomo Electric Industries, Ltd.
1-3, Shimaya 1-chome, Konohana-ku, Osaka 554, Japan

Abstract. Undoped and Fe-doped semi-insulating InP single crystals of 50 mm diam. with low EPD of ca. 2,000 cm^{-2} were successfully grown under the low temperature gradient dT/dZ = 30 °C/cm by applying phosphorus vapor pressure in order to suppress the dissociation of InP at the crystal surface. The reduction of the EPD was explained by use of computer simulations on the basis of Jordan's criteria.

1. Introduction

Indium phosphide single crystals have been commonly used exclusively as the conductive substrates for optical devices and have been under development as the semi-insulating substrates applicable to MISFETs, higher frequency microwave devices and optoelectronic integrated circuits (OEICs). These devices require low dislocation density of the InP substrates for high emission efficiency in LDs and for low dark current in APDs (Chin 1984). On the electronic devices, unlike GaAs crystal, the effects of dislocations in the InP crystal have not yet been made clear (Emeis 1987). However, there is a great demand to develop semi-insulating InP crystal with low dislocation density. The reduction of dislocations is thus assumed to be one of the basically important problems to eliminate the dislocation-induced defects such as micro-precipitations or spatial redistributions of the residual impurities which will affect the device performance. The reduction of the dislocation density is usually realized by keeping the temperature gradient low around the crystal during growth. However, that causes the surface temperature of B_2O_3 to rise to an extent that the dissociation of InP occurs at the surface of the crystal pulled out from the B_2O_3 layer. This drawback can be overcome by using the FEC method or by using the modified LEC method of growth under phosphorus pressure. By employing the modified LEC method, we have been able to realize a very low temperature gradient without the dissociation of InP crystal occurring at the surface and also keep the stoichiometry of the melt by depressing the release of phosphorus through the B_2O_3 layer. Then we have successfully obtained undoped and Fe-doped semi-insulating InP crystals with extremely low EPDs.

2. Experimental

The growth of InP single crystals and the temperature measurements were performed in a high pressure Czochralski puller with multi-zone heaters. A 100 mmϕ crucible containing InP polycrystals (ca. 1 Kg) and B_2O_3 was loaded in a closed vessel sealed with liquid boric oxide. Solid phosphorus or indium phosphide was also placed in the vessel to supply phosphorus vapor pressure. The minimum temperature of the vessel at the inner surface was controlled to be kept above the deposition temperature of the phosphorus vapor. The pressure of the dry nitrogen (35 ∿ 40 atm) in the puller was balanced against the total pressure of nitrogen and phosphorus in the vessel. The crystals were pulled along a <001> direction at constant rates of 5 ∿ 6 mm/h. The crystal and the crucible were counter-rotated at rates of 3 rpm and 5 rpm, respectively. In the case of semi-insulating crystals the concentration of Fe was 1 ∿ 2 × 10^{16} cm^{-3} in a crystal. Measurements of temperature profiles along the pulling direction were performed both with and without dummy crystal made of graphite. The conditions during the temperature profile measurements were the same as those during the crystal growth. Dislocation densities and distributions were observed by etch pit densities (EPD) revealed by H_3PO_4 and HBr mixture etchant and X-ray transmission topography on the (001) wafers. Experiments on infrared light scattering tomography were tried in order to obtain information about defects such as inclusions or aggregates of dislocations. The observation region of 2 × 2 mm^2 was exposed from the edge by a He-Ne laser beam (λ= 1.15 μm) with 15 mW power.

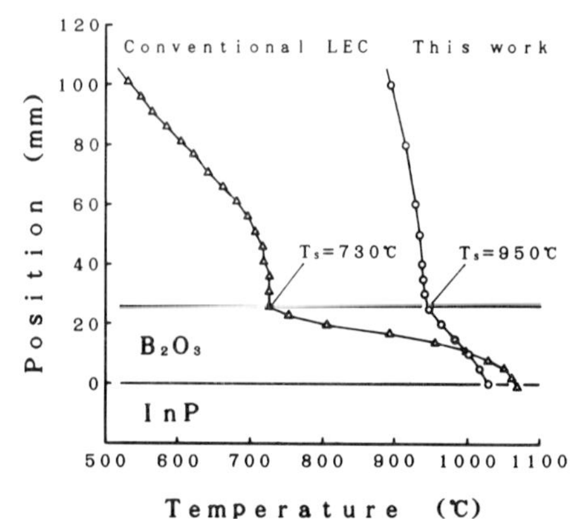

Fig.1 Temperature profiles above the InP melt.

3. Results and Discussions

The axial temperature profile in the modified LEC system is shown in Fig.1 together with that in the conventional LEC. The temperature gradient along the axial direction in the B_2O_3 layer is about 30 °C/cm, which is much lower than 100 ∿ 150 °C/cm in the conventional LEC. The temperature of B_2O_3 at the surface rises with the reduction of the axial temperature gradient, that is, dissociation of the

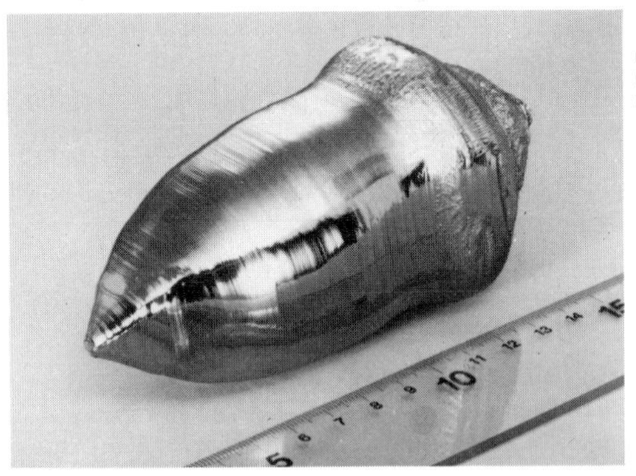

Fig.2 Fe-doped InP single crystal grown by modified LEC system.

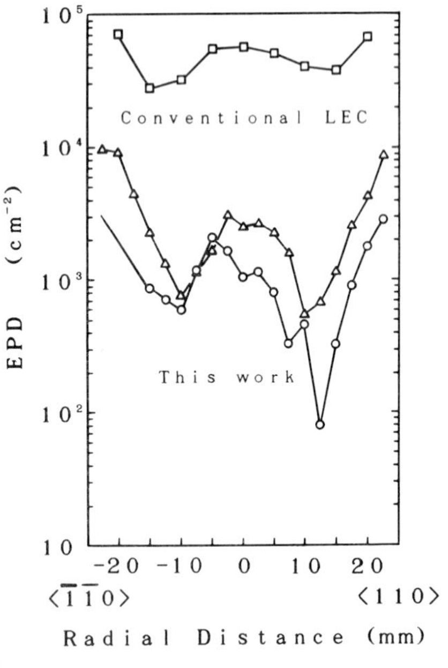

Fig.3 EPD profiles on the InP (001) wafers.

InP crystal into $P_2(g)$, $P_4(g)$ and In(l) exhibits maximum pressure near the B_2O_3 surface where the InP crystal is not encapsulated by B_2O_3. The pressure (P_1) of phosphorus in equilibrium with InP crystal exponentially increases with the rise of the temperature (Ts) of the B_2O_3 surface, namely with the reduction of the corresponding axial temperature gradient in the B_2O_3 layer. The dissociation pressure is given by following equation (Panish 1974, Backmann 1974a)

$$\log(P_1/\text{atm}) = -17.8 \times 10^3/Ts + 13.72 \qquad \text{where } P_1 = P_{P_2} + P_{P_4}.$$

The suppression of the dissociation is attained by applying the equilibrium phosphorus vapor pressure. The minimum pressure within the vessel is determined by the lowest temperature (Tmin), where the phosphorus vapor is in equilibrium with solid phosphorus at the sublimation pressure (P_0) described by the following equation (Backmann 1974b)

$$\log(P_0/\text{atm}) = -6.99 \times 10^3/T_{min} + 9.86$$

It is expected that the modified LEC method can produce a much lower temperature gradient of less than 10 °C/cm, in principle, which can satisfy Shinoyama's (1980) criteria for growing a dislocation free crystal i.e.

$$(dT/dZ)\, Dc^2 = 124 \text{ (Kcm)} \qquad \text{where Dc is critical diameter for DF.}$$

In this work we adopted the condition of Ts = 950 °C and dT/dZ = 30 °C/cm by taking into account of the controllability of the crystal diameter. The phosphorus vapor pressure (P_0) in the vessel was more than 5 atm which has much higher than dissociation pressure (P_1). The Fe-doped semi-insulating InP crystal grown under these conditions was 50 mm diam. and 120 mm long with surface of metallic luster as shown in Fig.2.

The average EPD on the wafer with 40 mm diam. cut from the middle part in the crystal is about 2 × 10 cm^{-2} with a W-shape profile which is one order of magnitude less than those of crystals grown by the conventional LEC method. Fig.3 shows the EPD profiles on the (001) wafers grown both by the modified LEC method and by the conventional one. The great difference of EPD on both wafers is expected to be due to the reduction of the temperature gradient from 100 °C/cm in a conventional LEC to 30 °C/cm in this work. To clarify this, we estimated the thermal stress profiles in both crystals on the basis of Jordan's (1984) criteria by using the result of temperature measurement in the dummy crystal. The temperature dependence of the critical resolved shear stress (σ_{CRSS}) is taken into account, because of its exponential change with temperature. Fig.4 shows the excess shear stress profiles obtained by computer simulations. The excess shear stress (σ_{ex}) is the sum of the differences between the σ_{CRSS} and each shear stresses which exceed σ_{CRSS}. The σ_{ex} in the crystal grown by the conventional LEC method are ten times as large as those grown by the modified LEC method. According to Jordan's assumption that the dislocation density is proportional to the σ_{ex}, most of the great reduction of about one order of magnitude in EPD is explained by the temperature profiles surrounding the InP crystal.

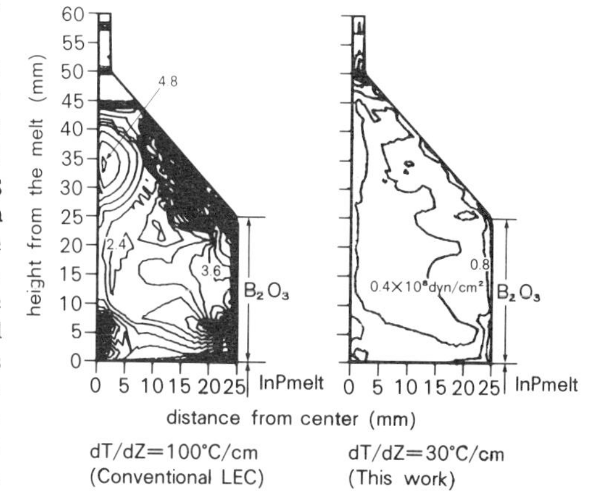

Fig.4 Estimated contour lines of iso-excess shear stress in the crystal.

Fillard et al. (1987) reported that many scatterers were found in a Fe- doped crystal grown by an usual LEC in the IR light scattering images, as shown in Fig.4(a). On the other hand, very few scatterers are shown in the crystal grown by the modified LEC (Fig.4(b)). These scatterers are presumed to be aggregates induced by dislocations or inclusions of indium. Above results can be attributed to the effect of the reduction of dislocation densities.

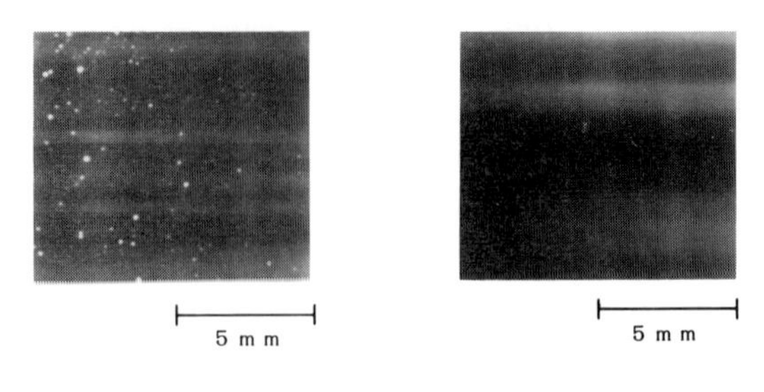

Fig.5 Infrared-light scattering images of Fe-doped InP: (a) usual LEC and (b) modified LEC.

The inclusions in the InP crystal are said to be related to the incorporation of the phosphorus from the molten InP into the B_2O_3 encapsulant including moisture (Chin 1984). The incorporation may be limited by establishing the equilibrium with phosphorus vapor over the B_2O_3 layer. This effect may be additional merit in the modified LEC method and is expected to become more effective by applying higher phosphorus vapor pressure.

4. Conclusions

By newly developing the modified LEC system, we have realized the low temperature gradients of $dT/dZ \approx 30$ °C/cm without the dissociation of the InP crystal at the surface under the phosphorus pressure of more than 5 atm. As a result, we have successfully obtained undoped and semi-insulating Fe-doped InP single crystals of 50 mm diam. with an extremely low EPD of 1,000 ∿ 3,000 cm^{-2}. The reduction of EPD was explained by a computer simulation of the thermal stress on the basis of Jordan's criteria. The defects such as precipitation or dislocation-induced infrared scattering centers in the crystals are not found in the improved InP single crystals grown by our new method.

Acknowledgements

The author would like to thank the colleagues of the Semiconductor Division and Characterization groups of Sumitomo Electric Industries, Ltd. for precious experimental cooperations and discussions.

References

Backmann K J and Buehler E 1974a J. Electrochem. Soc. 121 835
Backmann K J and Buehler E 1974b J. Electrochem. Mater. 3 279
Chin AK 1984 J. Crystal Growth 70 582
Emeis N and Beneking H 1987 J. Crystal Growth 83 286
Fillard J P, Gall P, Baroudi A, George A and Bonafe J 1987 Jpn. J. Appl. Phys. 26 L1255
Jordan A S, Von Neida A R and Caruso R 1984 J. Crystal Growth 70 555
Shimizu A, Nishine S, Morioka M, Fujita K and Akai S 1986 Semi-Insulating III-V Materials (Hakone) 41
Shinoyama S, Uemura C, Yamamoto A and Tohno S 1980 Jpn. J. Appl. Phys. 19 L331

Paper presented at Int. Symp. GaAs and Related Compounds, Heraklion, Greece, 1987

H_2 plasma induced effects on GaAs examined by photoluminescence, ellipsometry and X-ray photoemission

P.Boher, M.Renaud, J.Schneider, J.P.Landesman, R.Mabon, J.N.Patillon, Y.Hily, A. Barrois

Laboratoires d'Electronique et de Physique Appliquée (LEP[1]),
3, avenue Descartes, 94451 Limeil Brévannes Cédex (France)

ABSTRACT

Using photoluminescence ellipsometry and X-ray photoemission, we have separated the effects of band-bending, surface doping and surface defects on both n-type and p-type GaAs, during the native oxide removal by a H_2 multipolar plasma. We noticed a direct dependence between the Fermi level position deduced from XPS measurements, the enhancement of the excitonic structure E_1 , $E_1 + \Delta_1$ of GaAs around 3eV observed by spectroscopic ellipsometry, and the intensity of the integrated photoluminescence signal. Both on n-type and on p-type GaAs, the Fermi level pinned near mid-gap on the oxidized surface, is shifted towards the valence band of $\simeq$200meV after short H_2 plasma exposures. In addition, a dopant neutralization appears for longer plasma exposures.

Many of the III-V compounds are of considerable technical interest both because of their narrow band gaps (suitable for long wavelength optical applications), as well as because of their large values of electron mobility, which makes them appropriate candidates for high-speed devices. Unfortunately, the use of these materials is limited by their poor surface electrical properties. A passivation scheme using a multipolar plasma associated with a UHV system has been developed at LEP and described elsewhere ([1]). The native oxide removal is performed in a H_2 plasma, the passivation in a N_2 plasma , and the protection by a Si_3N_4 layer. Interesting results have been obtained both on GaAs I.C. postencapsulation ([2]), and on GaInAs / Si_3N_4 interface with low density of states ([3]). From these experiments, it is clear that in order to obtain high quality interfaces, the key point is to remove the native oxide without creation of additionnal surface defects.

Photoluminescence (PL) has long been recognized as a method well adapted to such studies. Non radiative recombination has been phenomenologically taken into account through the concept of surface recombination velocity V_s ([4]). Nevertheless a correlation between PL intensity and surface electronic properties is generally difficult due to the overlap of bulk and surface contributions to the PL signal. The PL signal depends on three parameters: the surface Fermi level position $E_{f,s}$ and the doping level at the surface of the sample N_s via the width of the space charge layer, and the surface recombination velocity V_s. We also use two other techniques: X-ray photoemission (XPS) which provides $E_{f,s}$, and spectroscopic ellipsometry (SE) which is sensitive both on surface electric field and doping level via the excitonic structure (E_1, $E_1 + \Delta_1$) of GaAs around 3eV. The results provided by these three techniques allow us to decorrelate the part due to Fermi level variations, doping level changes, and surface defects in the PL signal variations measured during H_2 multipolar plasma on both n-type and p-type GaAs substrates.

I EXPERIMENTAL DETAILS

The samples are highly doped CZ (100) GaAs wafers with mechano-chemical polishing. The n-type samples are Se-doped ($7.10^{17}cm^{-3}$) and p-type samples are Zn-doped ($6.10^{18}cm^{-3}$). Prior to the introduction in the UHV system, they are chemically etched by HCl/Ethanol (1:1). This scheme has been shown to minimize the carbon contamination and to preserve the surface stoichiometry of GaAs is preserved ([5]).

[1] LEP, Laboratoire d'Electronique et de Physique Appliquée: a member of the international Philips Organization.

The multipolar plasma system consists in an electron emitter (a hot filament) negatively biased (typically 20 to 100 V) with respect to the walls of a magnetic container. The primary electrons emitted by the filament are confined by permanent magnets mounted around the walls of the chamber, and ionize the low pressure gases. A plasma without high energy ions is produced and can be controlled by three parameters: the filament bias, the discharge current, and the pressure of the gases. Our experimental system, described elsewhere ([6]), is equipped with an ellipsometer and an in-situ photoluminescence system. XPS measurements are performed in another system with the same plasma chamber. The sample is heated with infrared lamps at 230°C during the hydrogen plasma. This heating is necessary to remove completely the native oxide ([6]). The integrated PL yield is measured from the front polished and illuminated surface by a Si p-I-n diode detector with appropriated filters. Excitation radiation is supplied by an Ar laser at 514.5nm (penetration depth $\simeq$1100Å), with a power density from 100 to 300mW/cm^2. The measurement is made at room temperature and also during the cooling of the sample after all treatments. PL variation versus temperature is reported on Figure 1 for two different samples (n-type and p-type) after a first heating under UHV, and a typical H_2 plasma. PL maxima occurs at about 160°C for n-type and 110°C for p-type GaAs. After a H_2 plasma the shape of the two curves is maintained but with a general decrease. The PL variations reported in the following are evaluated at PL maxima.

The spectroscopic ellipsometer consists in a Xe lamp, a rotating polarizer, a fixed analyser, a double monochromator and a photomultiplier ([3]). The effect of the plasma treatment is monitored in "real time" in the kinetic ellipsometry (KE) mode ([2,9,3]). We are especially interested here in the excitonic structure (E_1 , $E_1 + \Delta_1$) observed around 3eV for GaAs ([7]). This structure is sensitive to high electric fields ($>10^5$volts/cm) induced by the depletion layer, and to the doping level (screening of the exciton by free carriers). Important effects can be measured on samples in which the depletion depth is of the same order of the penetration depth of the light (140Å at 3eV). This structure has been analysed for GaAs as a function of the doping of the material using a parameter labelled S ([7, 8]). The graphical meaning of this parameter is in the insert on Figure 3.

XPS experiments have been carried out using a Leybold EA 11 spectrometer, with a VSW unmonochromated $Al - K_\alpha$ source. Ga-3d and As-3d core levels where measured at normal emission to the surface of the sample (escape depth $\simeq$25Å). After substraction of the background, the decomposition of each peak into different components is made using a least square fit procedure with experimental "calibration spectra" measured on a very clean GaAs surface to avoid profile problems due to the non-monochromaticity of the source.

II EXPERIMENTAL RESULTS

We have reported elsewhere ([2]) that a standard H_2 multipolar plasma at 230°C applied after a heating under UHV at 280°C, allows to obtain a complete removal of native oxides and amorphous As layers. This step of the treatment does not induce any additional surface roughness (confirmed by Transmission Electron microscopy), and the surface stoichiometry (As/Ga ratio of the bulk contributions on the As-2p, Ga-2p, As-3d, Ga-3d XPS spectra) seems to be preserved. Together with the chemical effects, peak position shifts of all bulk peaks have been observed.

On Figure 2, we have reported the kinetic energy variations of the bulk contributions to the experimental As-3d and Ga-3d XPS spectra obtained on three samples (two n-type and one p-type GaAs), after chemical cleaning, UHV heating at 280°C, and different plasma exposures. The standard state is taken after heating because the chemical treatment has shown to be no reproducible in terms of native oxide thickness. Both on p-type and n-type samples, the kinetic energy variations correspond to Fermi level movements of about 200 ± 50 meV towards valence band maximum. Band bending is then increased on n-type GaAs and decreased on p-type GaAs. The As-3d and Ga-3d oxide contributions are eliminated after the shorter plasma exposures. Longer plasma exposures are not able to induce other important Fermi level variations.

Figure 3 shows the shape of $\tan \Psi$ (one of the ellipsometric parameters) near the transitions (E_1, $E_1 + \Delta_1$) for a n-type sample just after a UHV heating at 280°C, and after different plasma exposures (10s and 6mn). The S parameter decreases after the first plasma, indicating an increase of the electric field (Indeed a increase of N_c^s due to H_2 seems improbable), and then increase slowly at higher exposures. The electric field increase after ten seconds exposures can be directly related to the Fermi level movement measured by XPS. The important values reached by S (1790 after 7mn) compared with values for semi-insulating materials (2100) seem to indicate a dopant neutralization effect of the H_2 plasma.

The different S parameter behaviors are reported on Figure 4 both for n-type and p-type materials. Two different H_2 plasma conditions have been used, the same conditions as for the XPS experiments (Primary electron energy at 75eV), and also treatments without additional bias applied to the electron source (Primary electron energy $<$ 9eV). In this second condition only atomic H can be formed. All the other plasma conditions (pressure, discharge current...) are kept constant. For the two cases and for short exposures, the S parameter behavior can be directly related with the Fermi level movements observed by XPS. The n-type GaAs case has been examined above. For p-type GaAs, The electric field decrease after short exposures is certainly related with the Fermi level decrease of 200meV towards valence band. For longer exposures at 75eV primary electron energy plasma, the S parameter increases both for n-type an p-type GaAs. On the other hand, for low primary electron energy plasma it is constant.

The PL variations measured during the same experiments as on Figure 4 are reported on Figure 5 versus the H_2 plasma exposure time. 75eV primary electron energy plasma induces a reduction of the PL signal both on n-type and p-type samples. This reduction presents two different regimes: a rapid reduction at the beginning of the exposure, and then a slower behavior certainly related to the decrease of electric field deduced from SE (Cf. §III.2). On the contrary, for low primary electron energy plasma (< 9eV) no reduction is observed. On p-type samples a little PL increase is some times observed at the beginning of the exposure. The great importance of the electron source applied bias on the integrated PL behavior is confirmed by the experiment performed on p-type GaAs with 40eV primary electron energy. The PL behavior is in that case intermediate between 0eV and 75eV.

III DISCUSSION AND CONCLUSION

The systematic reductions of electric field and PL signals reported after long plasma exposures both on n-type an p-type GaAs are certainly due to an electrical compensation of the material. This property has been reported using RF plasma or hydrogen implantation on n-type ([10], [11]) and also on p-type GaAs ([12]). To demonstrate this effect without ambiguity we have measured the emission spectra of different samples (issued from the same wafer to ensure a same doping level), which have been submitted to different plasma exposure times. The surface have been protected by a Si_3N_4 layer and the intensities have been corrected from the dielectric thickness. The emission spectra of three p-type samples (two exposure times and sample without treatment) are represented on Figure 6. Besides the intensity values which are in agreement with integrated ones (measured in-situ), ***the emission linewidth decreases when the plasma exposure time increases, which is a direct proof of the acceptor neutralization.*** In spite of the inhomogeneity of the hydrogen neutralization we can estimate that the doping level is reduced by two order magnitude for a 7min exposure ([13]). On n-type GaAs, the initial doping level is too low to permit the same type of demonstration, but according to the litterature results, the neutralization certainly exists.

At the first stages of the plasma exposure, the doping neutralization has certainly no consequence on the PL signal because of the occurence of a "dead layer" formed by the depletion layer. The PL variation can be controlled by the Fermi level variations and also by the increase of V_s due to the creation of additional surface defects. In our system, we can change the ion and electron energies varying the bias applied to the "hot source" (Cf. Figure 5). Without bias applied (only atomic H can be formed) the PL is kept constant, is to say no de-doping effect and no creation of additional surface defects. The de-doping effect is then entirely due to H ionized species. Applying a bias just sufficient to ionize H_2 ($>$ 20eV), increases V^s immediately and de-doping effect appears. The variation is related to the applied bias. The creation of additional surface defects is due to the ionized H species or to the primary electrons. ***Nevertheless, applying a bias not sufficient to ionize H_2 we have been able to remove completely the native oxide from the surface without creation of additional surface defects.***

Using three complementary in-situ techniques (photoluminescence, ellipsometry, and X-ray photoemission), we have separated the effects of band-bending, surface doping and surface defects on both n-type and p-type GaAs, during the native oxide removal by a hydrogen multipolar plasma. A direct dependence between the Fermi level position deduced from XPS measurements, the enhancement of the excitonic structure E_1 , $E_1 + \Delta_1$ of GaAs around 3eV observed by spectroscopic ellipsometry, and the intensity of the integrated photoluminescence signal has been noticed. Both on p-type and on n-type GaAs, the Fermi level pinned near mid-gap on the oxidized surface, is shifted towards the valence band of $\simeq$200meV after short H_2 plasma exposures. ***An additional doping neutralization demonstrated by an energy analysis of the photoluminescence spectra on p-type GaAs, appears after longer plasma exposures. This doping neutralization effect is induced by the hydrogen ionized species. The creation of additional surface defects is related to the electronic and ionic bombardment. Optimized plasma conditions have been achieved to eliminate these two parasitics effects.***

1. J.B. Theeten, S. Gourrier, P. Friedel, M. Taillepied, D. Arnoult, D. Bennaroche Mat. Res. Soc. Symp. Proc., vol 38 p. 499 (1985)
2. P. Boher, J.F. Pasqualini, J. Schneider, Y. Hily, CIPG 87, Antibes, June (1987)
3. P. Boher, M. Renaud, J.M. Lopez-Villegas, J. Schneider, J.P. Chane, INFOS 87, Leuven, April 13th-15th (1987)
4. J. Vilms, W.E. Spicer, J. Appl. Phys., 36, 2815 (1965)
5. R.P. Vasquez, B.F. Lewis, F.J. Grunthaner, J. Vac. Sci. Technol., B1 (2),328 (1983)

6. D. Arnoult, Thèse de docteur Ingénieur, INSA de Lyon, France (1986)
7. M. Erman, J.B. Theeten, N. Vodjdani, Y. Demay, J. Vac. Sci. Technol., B4 (4), 931 (1983)
8. M. Erman, These de doctorat, Paris, France (1986)
9. P. Boher, M. Renaud, L.J. Van Ijzendoorn, J. Barrier, Y. Hily, J. Appl. Phys., to be published, (1987)
10. K. Steeples, G. Dearnaley, A.M. Stoneham, Appl. Phys. Lett., 36, 981 (1980)
11. J. Chevallier, W.C. Dautremont-Smith, C.W. Tu, S.J. Pearton, Appl. Phys. Lett., 47, 108 (1985)
12. N.M. Johnson, R.D. Burnham, R.A. Street, R.L. Thornton, Phys. Rev. B, 33, 1102 (1986)
13. J.I. Pankove, J. Phys. Soc. (Japan) 21, 298 (1966)

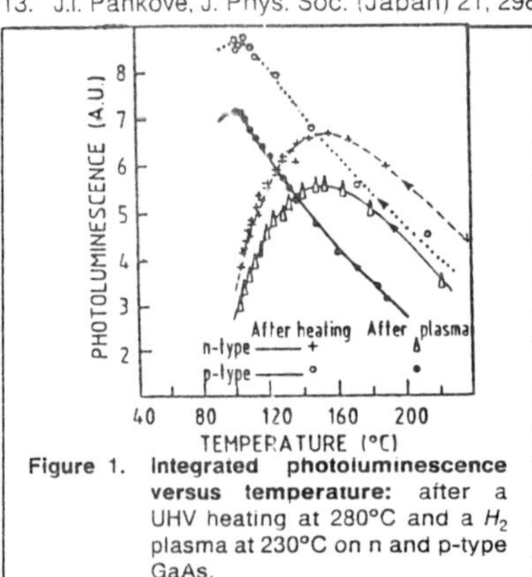

Figure 1. Integrated photoluminescence versus temperature: after a UHV heating at 280°C and a H_2 plasma at 230°C on n and p-type GaAs.

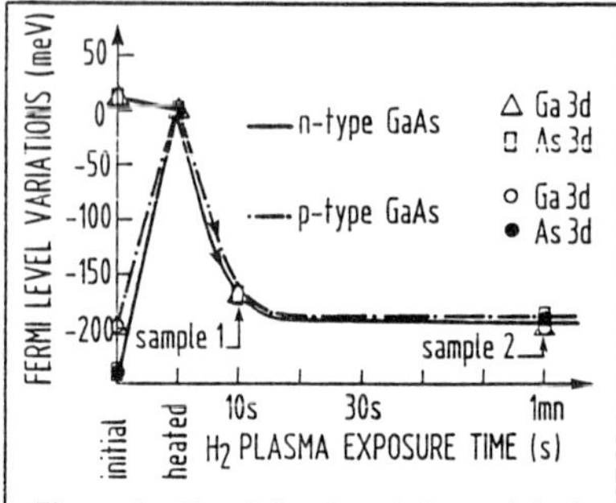

Figure 2. Fermi level variations of bulk contributions to As-3d and Ga-3d XPS peaks.: We have been reported three different samples after chemical treatment, heating under UHV and different H_2 plasma exposures.

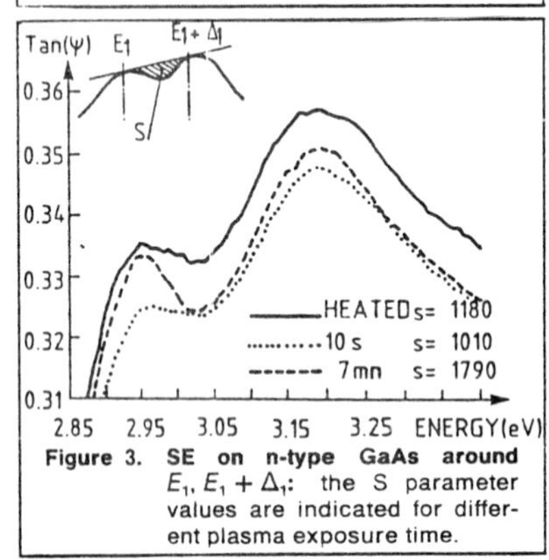

Figure 3. SE on n-type GaAs around E_1, $E_1 + \Delta_1$: the S parameter values are indicated for different plasma exposure time.

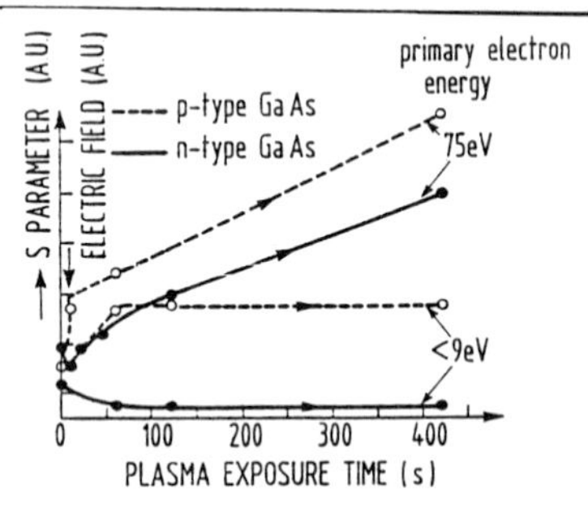

Figure 4. S parameter versus plasma exposure time on n-type and p-type GaAs: For p-type and n-type GaAs, two different primary electron energies have been used (9eV and 75 eV).

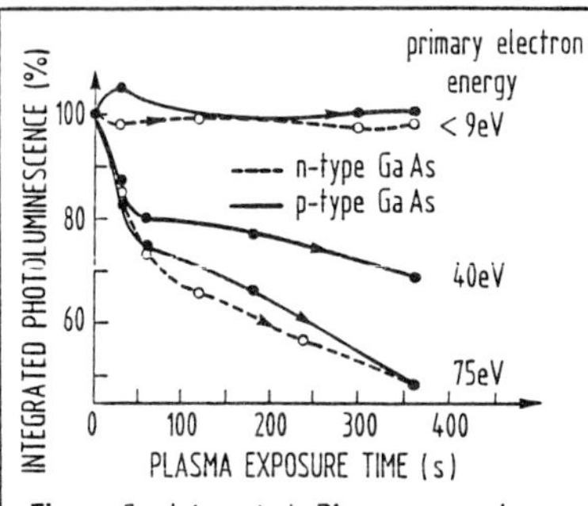

Figure 5. Integrated PL versus plasma exposure for n-type and p-type GaAs: A 40eV primary electron energy experiment on p-type GaAs has been added to the experiements also reported on figure 4.

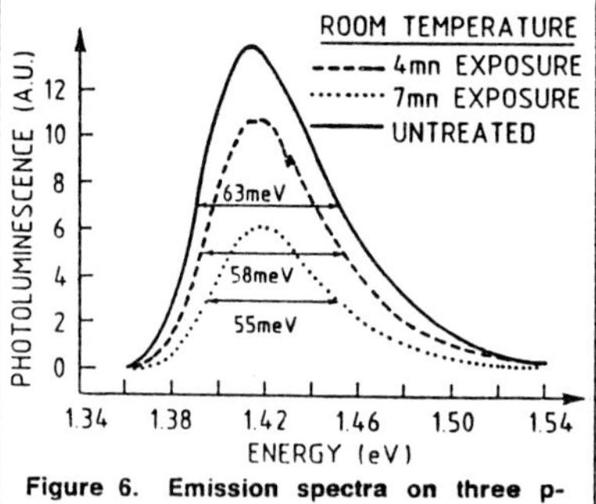

Figure 6. Emission spectra on three p-type GaAs samples: (H_2 plasma temperature 230°C, primary electron energy 75eV). Linewidths are indicated.

Inst. Phys. Conf. Ser. No. 91: Chapter 5
Paper presented at Int. Symp. GaAs and Related Compounds, Heraklion, Greece, 1987

RIE induced carbon-related shallow acceptor in GaAs

Z. Shingu, K. Uetake, A. Higashisaka and G. Mitsuhashi

Compound Semiconductor Device Div., NEC Corporation
1753 Shimonumabe, Nakaharaku, Kawasaki 211, Japan

ABSTRACT: The influence of the RIE-induced damage and contamination on the electrical properties of n-type GaAs active layer is described. Low temperature photoluminescence measurements demonstrated that the CF_4-RIE produces a shallow acceptor level (Ea 20mV), compensating for the carriers of the active layer. The newly observed level, which increases with increasing the RIE power and duration, was estimated to be a carbon at As-site.

1. INTRODUCTION

Reactive ion etching (RIE) is widely applied in GaAs devices fabrication process, for example, to micro-fabricate the refractory metal gate electrode and the dielectric film on the wafer. Because a thin active layer is usually used in GaAs devices, the surface irregulality by RIE brings about a serious problem on the device characteristics. The purposes of this paper are to estimate quantitatively the influences of the RIE-induced damage and contamination on the electrical properties of n-type GaAs layer, and to investigate the physical mechanism of the surface reconstruction by RIE, as well as its recovery treatments.

2. EXPERIMENTAL

In the experiments, Sulfur-doped vapor-phase epitaxial GaAs layer ($n \sim 1 \times 10^{17} cm^{-3}$, $t \sim 0.6$ um) on a Cr-doped semi-insulating substrate with a 2 um-thick undoped GaAs as an intermediate layer was used.
GaAs surface was exposed to CF_4, O_2 or $CF_4 + O_2$ mixed gas plasma in the parallel plate reactive ion etcher. RF power of RIE was varied from 50 Watt (0.05 Watt/cm^2) to 200 Watt to evaluate the RIE power dependence of the surface irregulality. The sheet carrier concentration of the epitaxial layer was measured by means of Hall measurement. Low temperature (4.2°K) photoluminescence measurement was also carried out to investigate the impurity levels introduced by RIE. SIMS analysis was employed to evaluate the depth profile of the impurities in GaAs. In order to seek for the recovery treatments, all measurements were performed not only as RIE-condition but also after heat treatments up to 500°C in H_2 atmosphare with SiO_2 encapusulation at 500°C.

3. DEGRADATION OF THE GaAs SURFACE CARRIER CONCENTRATION AND ITS RECOVERY BY A HEAT TREATMENT

Degradation of the sheet carrier concectration of the epitaxial layer due to RIE with CF_4 and its recovery by the heat treatments are shown in Fig.1. Hall measurement was not performed as RIE condition because of its extraordinary high resistivity. Even after 300°C (20 min.) heat treatment, degradation of the sheet carrier concentration was measured to be as large as 5.5 x 10^{11} cm^{-2}, corresponding to the carrier concentration degradation of about 10^{16} cm^{-3} under the assumption of the uniform carrier decrease along 0.6 um-depth. After 500°C (20 min.) heat treatment, the sheet carrier concentration is almost recovered to the untreated level. Another important experimental result was that the sheet carrier concentration decrease for O_2-RIE was easily recovered by 300°C annealing, suggesting that the carbon contamination plays a crucial role to kill the carriers of n-type GaAs.

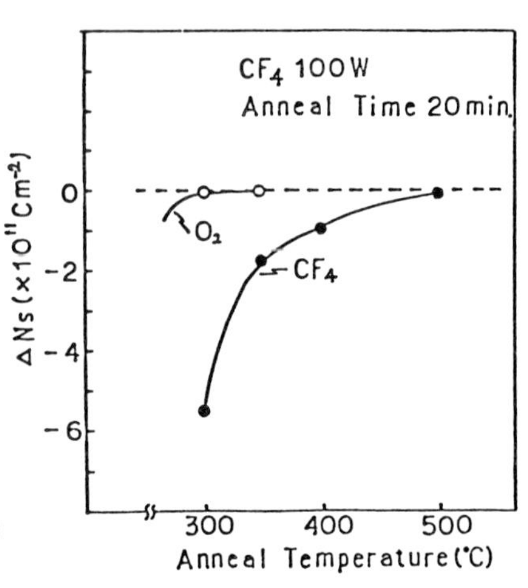

Fig.1 Carrier concentration degradation in GaAs by RIE and its recovery by heat treatment.

4. PHOTOLUMINESCENCE MEASUREMENTS ON RIE-DAMAGED GaAs SURFACE

Figure 2 shows the low temperature (4.2°K) photoluminescence spectra for samples with and without CF_4-RIE, and after 500°C heat treatment. RIE power and duration, in this case, are 100 Watt and 10 minutes, respectively. In addition to the (D^0,X) luminescence at 1.515 eV, a new (D,A) pair luminescence is observed at 1.493 eV (830 nm) in RIE sample. The (D,A) pair luminescence intensity increases with increase of RIE power as shown in Fig.3. The intensity was increased by increasing the duration of RIE. In the case of O_2-RIE or by increasing O_2 content in CF_4+O_2 mixed gas, the (D,A) pair luminescence was almost undetectable even for the same RIE power and duration. This fact implies that the shallow acceptor, an origin of the (D,A) pair luminescence,

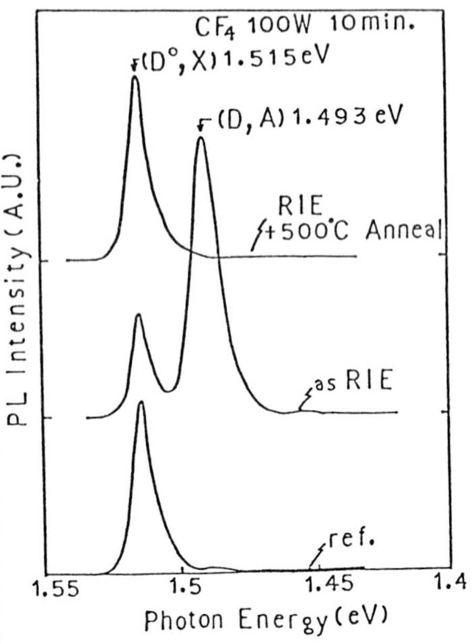

Fig.2 PL spectra from GaAs surface with and without RIE, and after 500°C heat treatment.

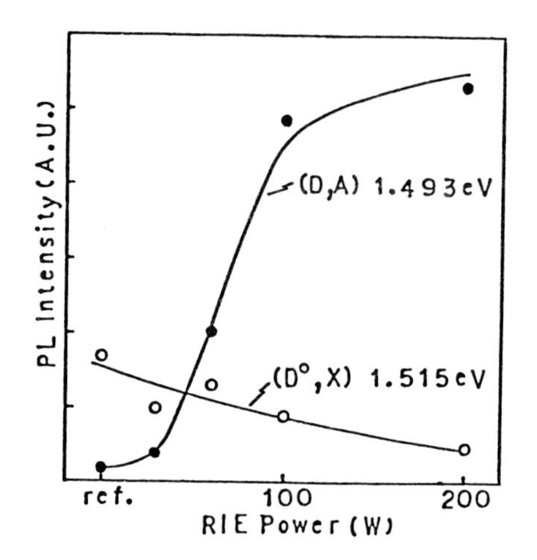

Fig.3 RIE power dependences of (D^0,X) and (D,A) PL intensities.

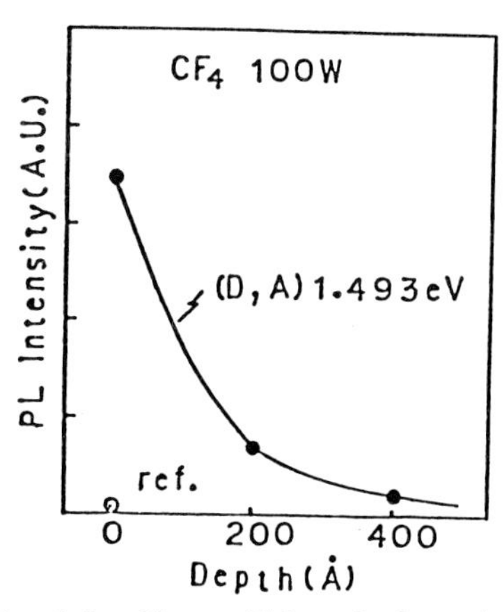

Fig.4 Depth profile of the (D, A) pair luminescence intensity.

is related to carbon in CF_4 gas. The decrease of (D^0,X) pair luminescence in Fig's 2 and 3, which is due to the physical damage by the incident ions, becoms remarkable as RIE power increases. In order to obtain the in-depth profile of the shallow acceptor, the photoluminescence measurements were performed after step-etching of GaAs surface. It is found from Fig. 4 that the (D,A) pair luminescence intensity declined with the depth, but it was detected to as far as 400 A-depth from the surface. This means that the newly introduced acceptor penetrates deep in GaAs. The depth profile of the impurities in GaAs was also obtained by SIMS (Fig. 5). As RIE condition, carbon was detected until about 500 A beneath the GaAs surface, giving a circumstantial evidence for the aforementioned assumption that the origin for the shallow acceptor is carbon.

5. THE RECOVERY TREATMENT FOR CF_4-RIE INDUCED DAMAGE AND CONTAMINATION

The experimental results in Section 2 showed that the carrier concenration was recovered to the non-RIE level after 500°C, 20 minutes heat treatment. The same investigations were carried out for the low temperature photoluminescence (Fig.2) and for the carbon depth profile (Fig. 5). The (D,A) pair luminescence disappered after 500°C, 20 minutes heat treatment. On the contrary, the (D^0,X) luminescence at 1.515 eV is almost recovered for the same treatment. These results are shown in detail in Fig. 6, which predicts that the (D,A) pair luminescence intensity decreases with increasing the annealing temperature from 300°C to 500°C, and that 500°C (20 min.) anneal is needed for the sufficient recovery. According to SIMS analysis, the carbon concentration at 500 A depth declines to the background level after the same treatment (Fig. 5). These experimental results on the photo-

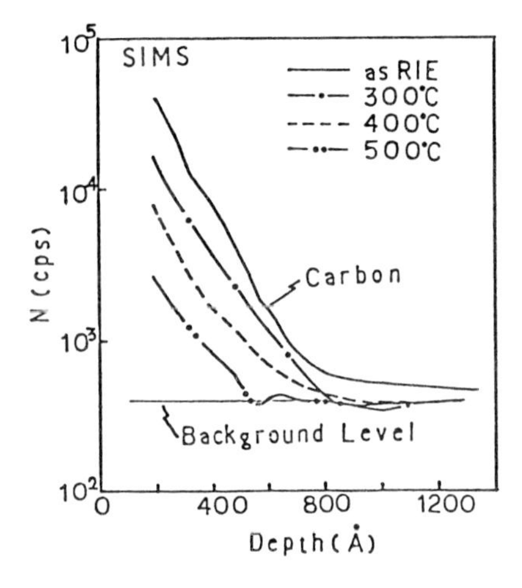

Fig.5 Carbon concentration in RIE GaAs as a function of annealing temperature.

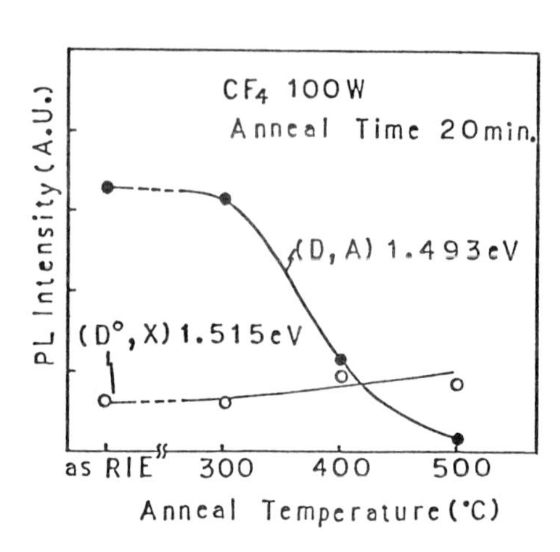

Fig.6 Annealing temperature dependence of (D^0,X) and (D,A) luminescence intensities.

luminescence and on SIMS analysis are in a good correspondence to the change in sheet resistivity in Fig. 1. From these results, it can be concluded that the degradation in carrier concentration, which is recovered by 500°C, 20 minutes annealing, is originated from the carbon acceptor. (D,A) luminescence energy level (1.493 eV) is coincident with a luminescence from carbon at As-site, which is commonly observed in GaAs substrates (Ashen 1975).

6. CONCLUSION

Effects of RIE-induced damage and contamination on the electrical properties on n-type GaAs active layer were studied by means of Hall, photoluminescence and SIMS measurements. Shallow acceptor level (Ea= 1.493 eV) related to the carbon contamination was produced by CF_4-RIE, compensating for the carriers of the GaAs surface region. The degradation in the carrier concentration due to RIE is recovered by 500°C, 20 minutes annealing.

ACKNOWLEDGEMENT

The authors would like to thank Y.Takayama and N.Kitagawa for their continuous encouragements throughout this work.

REFERENCE

1) Ashen, D.J., Dean, P.J., Hurle, D.T.J., Mullin, J.B., and White, A.M. 1975 J. Phys. Chem. Solids. 36 1041.

Paper presented at Int. Symp. GaAs and Related Compounds, Heraklion, Greece, 1987

Characterizing-process-induced microstructural damage in III–V materials

T. S. Ananthanarayanan, J. I. Soos, R. G. Rosemeier, D. C. Leepa and A. L. Wiltrout

Brimrose Corporation of America, 7720 Belair Road, Baltimore, MD 21236

ABSTRACT: X-ray rocking curve analysis is emerging as a powerful technique for quantifying micro-lattice strain state. Ananthanarayanan, etal. (1986-1987) in the past have clearly demonstrated the efficacy of this technique for characterizing surface microstructural quality. The present study utilizes x-ray rocking curve topography to estimate process induced microstructural damage in materials such as GaAs and GaP. These materials are currently being used for sophisticated acousto-optic devices. Their surface/subsurface strain state is critical to device fabrication (transducer bonding) and performance. Several surface conditions obtained by varying the grinding/polishing grit size and Brinell hardness indentation have been evaluated by the digital automated rocking curve (DARC) topography technique. The 2-D topographic maps of Bragg peak shift, Bragg peak broadening and Bragg peak integrated intensity have been used to discern grown-in and process induced microstructural inhomogenieties. The DARC technique is amenable to advanced computing and artificial intelligence (AI) environments. Both research and production oriented application will significantly benefit from such a tool. Currently the entire analysis for a 1" wafer requires about 5 minutes at 100μm spatial resolution with a personal computer based system.

1. INTRODUCTION

Berg-Barrett (1945) x-ray diffraction topography is a fairly popular tool for evaluating single crystal microstructure. This technique is highly sensitive to microstructure and fairly rapid in execution. Several modifications to this technique have been developed to improve spatial and defect resolution such as ACT (Asymmetric Crystal Topography) by Green, etal. (1976) and Double Crystal topography by Bonze (1958). Typically with higher spatial resolution the diffracted beam intensity diminishes and so requires long data acquisition times. This study utilizes the state-of-the-art modification to the Berg-Barrett topography called rocking curve topography. This technique has been used to monitor process induced microstructural damage. Surface grinding/polishing and single point Brinell hardness indentation have been used to simulate process damage.

2. ROCKING CURVE TOPOGRAPHY

Rocking curve topography involves the combination of rocking curve analysis and diffraction topography. The diffracting domain of a given crystal is measured for a set of geometric conditions and x-ray beam optics. The diffracting domain of the crystal is a measure of the reciprocal volume of the crystal. The reciprocal volume is in-turn related to the micro-lattice structure of the crystal. In general (kinematic theory) the reciprocal volume is inversely related to crystal perfection i.e., as the amount of defects (imperfections) increase the reciprocal volume (diffracting domain) increases. However, for special cases such as thin epitaxial films the above generalization will not be valid. It will be necessary to invoke the dynamic theory of x-ray diffraction in such cases. The measurement will still be appropriate but the analysis will be different. The current study uses the kinematic approach for the analysis. These substrates were prepared to minimize the effect of surface geometry on the rocking curve measurements. They were ground, polished and lapped with minimal surface curvature.

Ananthanarayanan and Trivedi have presented a detailed description of the hardware and software involved in digital rocking curve topography.

3. RESULTS & DISCUSSIONS

Figure 1 depicts the rocking curve topograph obtained from a GaP crystal with (111) surface orientation and two Brinell hardness indentations at 60kg preload. The series of Berg-Barrett topographs (over 0.5^{O} rocking angle range at 0.1^{O} interval) show the image contrast obtained with the conventional technique. Immediately below this series of topographs is the x-ray rocking curve half-width map obtained by analyzing the entire rocking range of the GaP crystal. The enhanced contrast due the the local disturbance in the crystal lattice state (microstructure) is evident. The 3-D perspective view of the rocking curve topograph also shows the crystal dimensions. The entire rocking curve analysis for this specimen took no longer than 40 secs.

Figure 2a depicts the rocking curve profiles obtained from a GaAs ((100) surface orientation) sample subjected to varying surface damage (by grinding) showing the K alpha 1 and K alpha 2 peaks for individual pixels. The cleaved surface was used as the starting surface for measuring grown-in defect density. This surface was then ground with several grinding grit sizes (40μm, 15μm, 9μm, 3μm, 0.3μm). The rocking curve measurements were made over the entire specimen at 100μm spatial resolution and the mean value over 10 X 10 pixels was computed to obtain the error bar at each data point in Figure 2b which shows the sensitivity of the rocking curve half width (unmonochromated) to surface grind.

In conclusion, the combination of the conventional 2-D Berg-Barrett topography with rocking curve analysis through digital data acquisition and image processing yields significantly enhanced defect contrast both grown-in as well as process induced. The non-destructive, non-contacting nature of this technique is the paramount advantage both in the production and research environments.

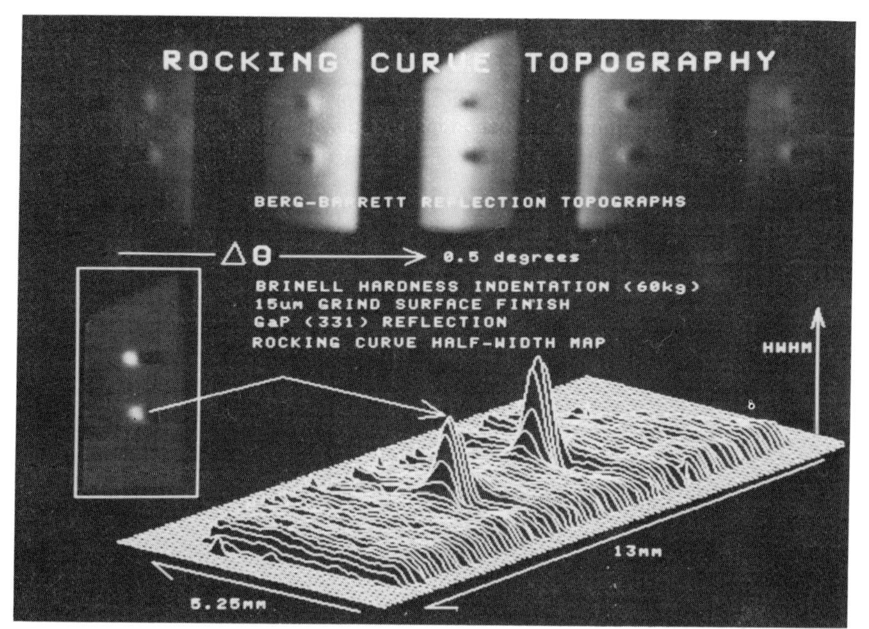

Figure 1: GaP (331) Berg-Barrett Topographs, Rocking Curve Half-width Map of Brinell Hardness Indentations Cu Radiation.

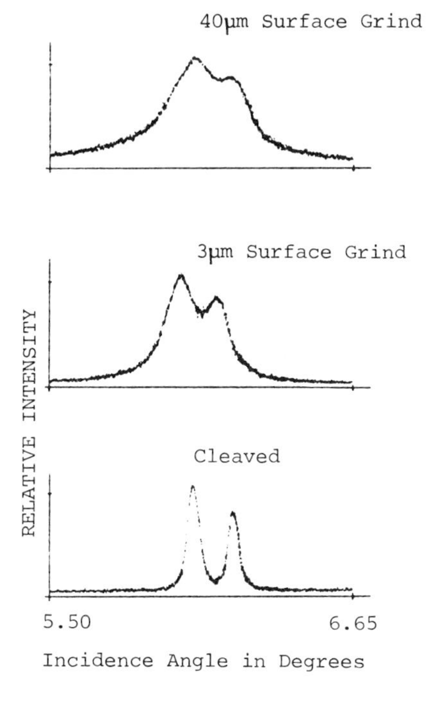

Figure 2a: Rocking Curve Pixel Profile GaAs (331) Cu Radiation.

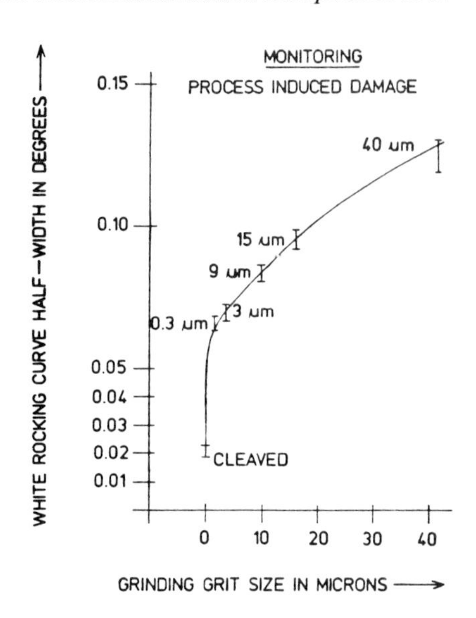

Figure 2b: GaAs (331) Rocking Curve, Cu Radiation.

4. REFERENCES

Ananthanarayanan, T. S., **Renaissance in X-Ray Diffraction Topography,** International Advances in Non-Destructive Testing, Gordon & Preech, NY, Vol. 13, 1987.

Ananthanarayanan, T. S, Rosemeier, R. G., Mayo, W. E. and Dinan, J. H., **Digital X-ray Rocking Curve Topography,** published in the 1986 MRS Proceedings, 1986.

Ananthanarayanan, T. S., Rosemeier, R. G., Mayo, W. E. and Becla, P., **Subsurface Micro-Lattice Strain Mapping,** published in the 1986 MRS Proceedings, 1986.

Ananthanarayanan, T. S. and Trivedi, S. B., **DARC, a Novel Topographic Technique for Rapid Non-destructive Characterization of III-V compounds,** Proceedings of DRIP II International Symposium on Defect Recognition and Image Processing in III-V Compounds, Monterey, Ca., 1987.

Barrett, C. S., **Trans. AIME,** Vol. 161, p15, 1945.

Green, R. E., Boettinger, W. J., Burdette, H. E. and Kuriyama, M., **Asymmetric Crystal Topography Camera,** Rev. Sci. Instr. Vol. 47, No. 8, 1976.

Bonze, U., **Z. Physik,** Vol. 153, p278, 1958.

RESEARCH SPONSOR: DEFENSE ADVANCED RESEARCH PROJECTS AGENCY (DARPA)

Paper presented at Int. Symp. GaAs and Related Compounds, Heraklion, Greece, 1987

Passivation of GaAs IC MESFETs by photo-CVD SiN_x

N Arnold, L Schleicher, T Grave

Siemens AG, Corporate Research and Development
D-8000 Munich 83, P.O.Box 830952, F.R.G.

ABSTRACT: Photo-enhanced CVD has been used to deposit SiN_x for inter-metal isolation in GaAs ICs. The damage induced by plasma-processing is thereby avoided, but the device characteristics can still be affected by the stress in the silicon nitride layer. This paper demonstrates the capability of the photo-enhanced deposition to vary the intrinsic film stress by the change of the deposition parameters. The parameters pressure, $NH_3:SiH_4$ ratio and deposition temperature have been investigated. Out of these, the temperature is the most important one. Raising it from 70 °C to 200 °C changes the film stress from $-2 \cdot 10^9$ dyn/cm² (compressive) to $1.8 \cdot 10^9$ dyn/cm² (tensile). The sign of the stress changes at around 90 °C. Due to this stress reduction the threshold voltage shift in MESFETs could be reduced from 300 mV to 25 mV.

1. INTRODUCTION

It is well known from literature that passivating dielectrics on GaAs MESFETs influence the device characteristics. One reason is the plasma-related surface damage observed in the commonly used PECVD-processes (e.g. Yamane et al 1983). Additionally, the intrinsic stress in the film changes the device characteristics. The stress results in orientation dependent threshold voltage shifts (e. g. Lee et al 1980). Asbeck et al (1980) explained these effects by calculating the piezoelectric charge induced in GaAs by the dielectric overlayer. This model has been confirmed by various articles and reviewed recently by Chen et al (1987). The additional effect of the stress in the gate metal has been measured by Schnell and Schink (1987) and could be explained also by the piezoelectric charge model. Therefore, good control of the intrinsic stress in dielectric overlayers has been proven to be a very important factor for good threshold voltage reproducibility. In this paper we demonstrate the possibility of changing the internal stress in the photo-CVD SiN_x just by varying the deposition temperature.

2. EXPERIMENTAL

For the deposition of the SiN_x-layers we used the Hg sensitized decomposition of silane and ammonia as described by Peters et al. (1980). The layer characteristics obtained so far have been published recently (Arnold and Schleicher 1987). In general, they agree with the results of Meliga et al (1986): The step coverage is excellent, the inter-metal isolation resistance is high and uniformity as well as reproducibility are very suitable for this application.

Here we report on the variation of the intrinsic stress in the nitride layers due to different deposition processes. The process parameters investigated were:

- SiH_4 to NH_3 ratio; range 1...10 %
- Chamber pressure; range 0.5...2 mbar
- Deposition temperature; range 70...200 °C.

The experiments showed that the deposition temperature is the most relevant among these parameters. Compared to the changes due to the temperature, the influences of silane to ammonia ratio and pressure are negligible. We therefore discuss only the temperature dependence.

To determine the stress in the nitride layer, the bow of the GaAs wafers due to the nitride deposition was measured by an interferometric measurement system (Autoselect by GCA/Tropel). The stress was calculated from the bow according to Chen and Fatemi (1986). Youngs modulus E and Poisson-ratio v were chosen to be $E = 1.2 \cdot 10^{12}$ dyn/cm² and $v = 0.23$ after Landolt-Börnstein (1982). The accuracy of the stress measurement was limited to about $3 \cdot 10^8$ dyn/cm² by the resolution of the bow-measurement at the given wafer diameter (2 inch), wafer thickness (400 µm) and film thickness (around 300 nm).

3. RESULTS

Fig. 1 shows the dependence of the film stress on the deposition temperature. By lowering the temperature from 200 °C to 70 °C the stress changes from $1.8 \cdot 10^9$ dyn/cm² (tensile) to $-2.1 \cdot 10^9$ dyn/cm² (compressive). The correlation between stress and temperature is not linear. Whereas in the region of tensile stress (above about 90 °C) the gradient is relatively small

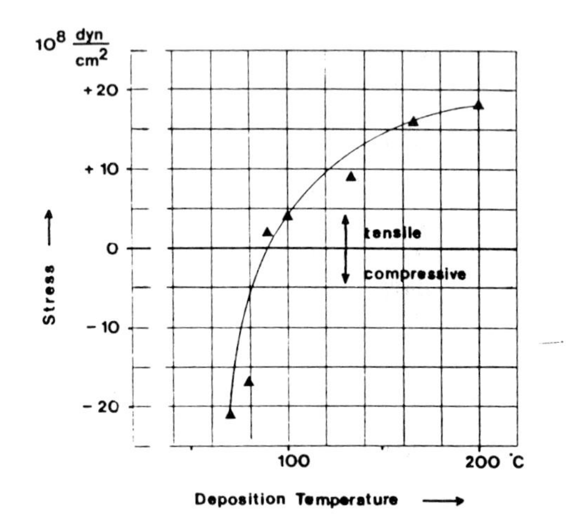

Fig. 1. Film stress vs deposition temperature

(about 2•10⁹ dyn/cm² within 100 °C), this gradient is about 5 times higher in the compressive region. Therefore, a very good control of the deposition temperature is essential for reproducible results in the region of low stress.

The dielectric behaviour (characterized by the refractive index n) did not change during these depositions. In contrast to the results of Meliga et al (1986) n was always 1.86, independent of other parameters. As stated above, these parameters showed only negligible influence on stress.

4. EFFECT OF THE DIELECTRICS ON THE THRESHOLD VOLTAGE OF MESFETs

Fig. 2 shows the effect of dielectric deposition at 200 °C with a tensile stress of $2 \cdot 10^9$ dyn/cm². The threshold voltage is substantially shifted from -1.06 V before to -0.76 V after the deposition of the SiN_x layer. The gate orientation is perpendicular to the major flat. For gates parallel to the flat the threshold voltage shifts in opposite direction (Schnell and Schink 1987). In contrast, the deposition at 100 °C, resulting in a stress value about one order of magnitude lower, influences the threshold voltage only marginally. Here, the shift in threshold voltage is only 25 mV (fig. 3). In both cases the sheet resistance of the implanted layers (measured by a transmission line structure) is 5 to 10 % lower after the nitride deposition. This highly desired behaviour is assumed to be caused by a reduction of the surface state concentration.

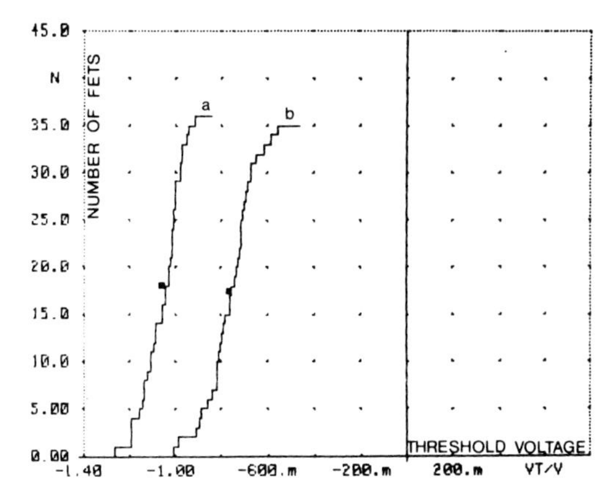

Fig. 2. Cumulative histograms of the threshold voltage variation over a wafer before (a) and after (b) the deposition of SiN_x with high stress.
Mean values: a) V_t = -1.06 ± 0.08 V
b) V_t = -0.76 ± 0.1 V

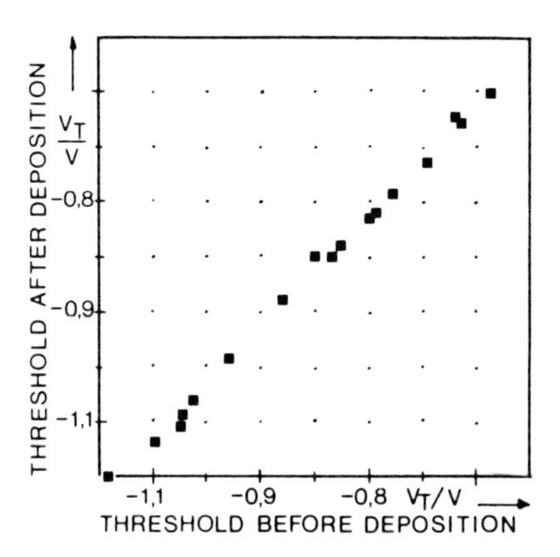

Fig. 3. Correlation between the threshold voltages before and after the deposition of low stress silicon nitride

Thus, the reduced sheet resistance demonstrates that the photoenhanced deposition process does not add damage to the shallow implanted layers used. Therefore and due to the orientation dependence of the threshold voltage shift, this shift must be caused by the intrinsic stress in the film.

5. CONCLUSIONS

Photo-enhanced deposition of silicon nitride layers has been found to be very suitable for the fabrication of inter-metal dielectrics on GaAs ICs. The step coverage is excellent and the isolation resistance is high due to a low pin hole density as reported by Meliga et al (1986). This deposition process does not introduce damage to the shallow implanted layers necessary for high speed ICs. In contrary, the sheet resistance is lowered by the deposition due to a reduction of the surface state density. MESFET characteristics have been shown to be substantially affected by the dielectric only if the intrinsic stress in the film is high. However, the photo-CVD process allows the optimization of the film stress by varying the deposition temperature. The intrinsic film stress can be varied from $-2 \cdot 10^9$ dyn/cm^2 (compressive) at 70 °C to $1.8 \cdot 10^9$ dyn/cm^2 (tensile) at 200 °C. The accurate control of the temperature is essential for a reproducible deposition process with low stress. The deposition parameters pressure and silane to ammonia ratio have been found to have less influence on stress and can be used to optimize other film characteristics like uniformity.

ACKNOWLEDGEMENTS

The authors would like to thank T. Ohle for the preparation of the nitride layers and for the stress measurements. This work has been supported by the Bundesministerium für Forschung und Technologie, FRG, Contract No. NT 2717C8.

LITERATURE

Arnold N and Schleicher L 1987
  to be published in Applied Surface Science
Asbeck P M, Lee C P and Chang M C F 1984
  IEEE Trans. El. Dev. **ED 31** 1377
Chen C H, Peczalski A, Shur M S and Chung H K 1987
  IEEE Trans. El. Dev. **ED 34** 1470
Chen Y S and Fatemi H 1986 J. Vac. Sci. Technol. EA4 645
Landolt-Börnstein 1982 **III/17a** pp 236, 529
Lee C P, Zucca R and Welch B M 1980 Appl. Phys. Lett. **37** 311
Meliga M, Stano A and Tamagno S 1986
  SPIE Thin Film Technologies II **652** 243
Peters J W, Gebhart F L and Hall T C 1980
  Solid State Technol. **23** 121
Schnell R D and Schink H 1987
  to be published in Jap. Jour. Appl. Phys.
Yamane Y, Ishii Y and Mizutani T 1983 J. Appl. Phys. **22** L350

Characterization of SiF_x and SF_x molecular ion implanted layers in semi-insulating GaAs

Akiyoshi Tamura, Kaoru Inoue and Takeshi Onuma

Semiconductor Research Center
Matsushita Electric Industrial Co., Ltd.
3-15, Yagumonakamachi, Moriguchi, Osaka 570, Japan

ABSTRACT: We have investigated fundamental electrical and optical characteristics of GaAs implanted with SiF_x and SF_x (x=1, 2 and 3) molecular ions using Hall effect, capacitance-voltage (C-V), secondary ion mass spectrometry (SIMS) and 77K-photoluminescence (PL) measurements. A new annealing method which can enhance activation efficiencies of these molecular ion implanted layers is proposed.

1. INTRODUCTION

In order to improve the performance of a GaAs MESFET, it is important to form a thin active layer with high carrier concentration by means of low-energy ion implantation. The minimum implantation energy is, however, limited by the apparatus. By implanting molecular ions that include a dopant atom, the implantation energy can be reduced by the mass ratio between molecule and dopant atom. Recently, ^{47}SiF (Jaeckel et al. 1987) and $^{66}SiF_2$ (Kuzuhara et al. 1986) implantations into GaAs to form thin active layers in GaAs MESFETs have been reported. However, fundamental characteristics of GaAs implanted with molecular ions have not been reported. In this work, we report the results of a systematic investigation of fundamental electrical and optical characteristics of GaAs implanted with SiF_x and SF_x (x=1, 2 and 3) molecular ions and a new annealing method which can enhance activation of molecular ion implanted layers.

2. EXPERIMENTS

The substrates used in this work were undoped semi-insulating (SI) LEC (100) GaAs crystals. ^{28}Si, ^{47}SiF, $^{66}SiF_2$ and $^{85}SiF_3$, and ^{32}S, ^{51}SF, $^{70}SF_2$ and $^{89}SF_3$ were implanted into GaAs. The energies of molecular ions were chosen to give the same dopant atom (^{28}Si or ^{32}S) ion ranges from theoretical considerations. Table I shows the implant conditions used. After implantation, these samples were annealed in an Ar atmosphere. For annealing, SiO_2-capped conventional furnace annealing (FA) at 820°C for 15min, capless rapid thermal annealing (RTA) using the GaAs proximity technique, and SiO_2-capped RTA at 900°C for 10s were used. For the sample implanted with S-dopant, only RTA methods were used in order to reduce the diffusion of S atoms. After annealing, these layers were characterized using the Hall effect, capacitance-voltage (C-V) and photoluminecence (PL) and secondary ion mass spectrometry (SIMS) measurements.

Table I. The implant conditions.

Implanted ion	Mass	Implantation energy (keV)	Dopant atom implantation energy (keV)	Dose (cm^{-2})
Si	28	40		
SiF	47	67		
SiF_2	66	94	40	$8x10^{12}$
SiF_3	85	121		
Si	28	80		
SiF	47	134		
SiF_2	66	189	80	$7x10^{12}$
SiF_3	85	243		
S	32	40		
SF	51	64		
SF_2	70	87	40	$8x10^{12}$
SF_3	89	111		

3. RESULTS AND DISCUSSIONS

3.1 SIMS MEASUREMENTS

Figures 1(a) and (b) show the depth profiles of ^{28}Si, ^{32}S and ^{19}F atoms measured by SIMS using a CAMECA IMS-3f ion microprobe for unannealed and annealed samples. From these figures, it is found that the dopant atom depth profiles for as-implanted samples using molecular ion were almost the same as that of the samples implanted with only the dopant ion, and F atoms were not observed for all the annealed samples because of outdiffusion. After annealing, a slight diffusion of S atom was observed in spite of the short RTA annealing time.

3.2 ELECTRICAL CHARACTERIZATION

Figure 2 shows the dependence of sheet carrier concentration (Ns) obtained from Hall effect measurement on the mass of implanted ion. For the samples implanted with Si-dopant, The Ns for Si- and SiF-implanted layers were almost the same and decreases almost linearly with a implanted ion mass. For both SiF_2- and SiF_3-implanted layers, the SiO_2-capped RTA method gave the highest activation efficiency. For the samples implanted with S-dopant, the decrease of Ns with the increase of implanted ion mass was not so large with the exception of SF_2-implanted layer, which showed the lowest Ns, although the values of Ns were lower than those of the sample implanted with Si-dopant. The reason for this result is not clear at present. Figures 3(a) and (b) show typical carrier concentration profiles obtained from C-V measurements for the samples implanted with Si- and

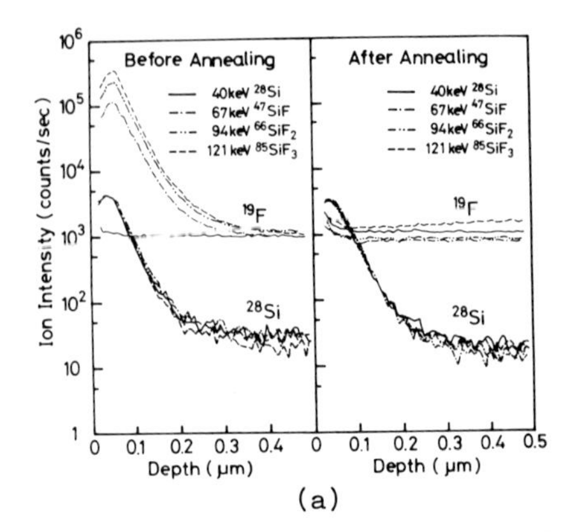

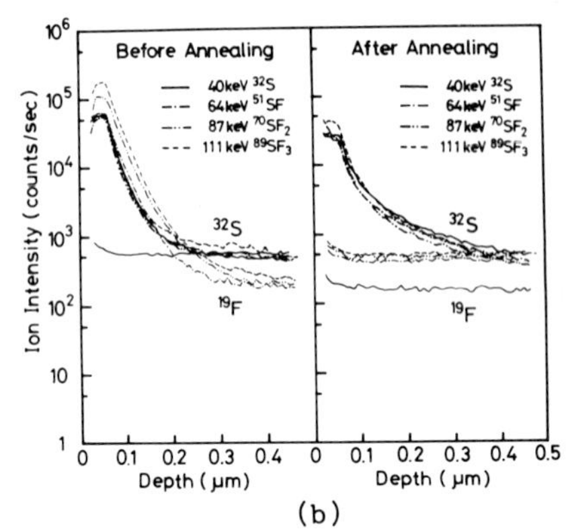

Fig. 1. SIMS depth profiles of Si, S and F atoms for GaAs implanted with (a) Si-dopant and (b) S-dopant.

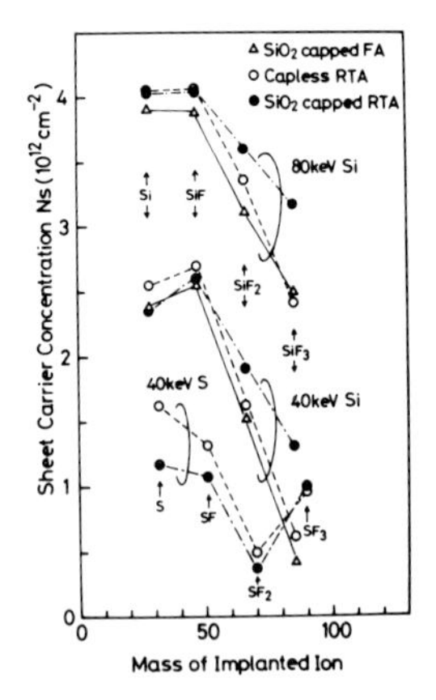

Fig. 2. The dependence of Ns on the mass of implanted ion.

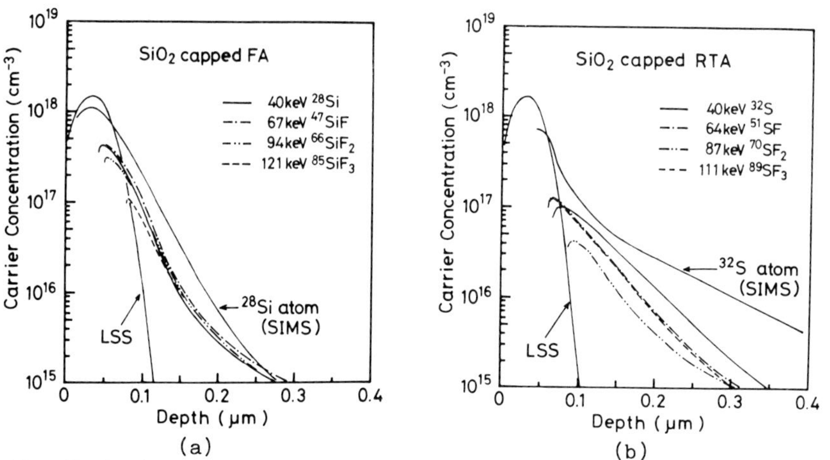

Fig. 3. Typical carrier concentration profiles for GaAs implanted with (a) Si-dopant and (b) S-dopant.

S-dopants, respectively. SiO_2-capped FA and RTA methods were used, respectively. The curve of LSS theory and Si and S atomic concentration profiles measured by SIMS are also shown. From Fig. 3(a), it is found that the peak carrier concentrations for Si- and SiF-implanted layers were almost the same and decrease for SiF_2- and SiF_3-implanted layers corresponding to the results of Ns in Fig. 2. This result suggests that residual implanted damage caused by high-dose F implantation was detrimental to activation. SiF implantation is promising for forming thiner active layers of MESFET without degrading activation efficiency compared with Si implantation. For SiF_3-implanted layers, The SiO_2-capped RTA method gave the highest peak carrier concentration. For the samples implanted with S-dopant, the peak carrier concentrations were lower and all profiles were broader than those for Si-dopant because of the diffusion of S atom during annealing, as shown in Fig. 3(b). This result indicates that S-dopant implanted layers are unsuitable for the active layers of MESFETs.

3.3 OPTICAL CHARACTERIZATION

Photoluminescence (PL) measurements were carried out at 77K using the 5145A Ar laser as an excitation source and photomultiplier tube for detection. PL specta for the sample implanted with Si- and S-dopants by different annealing methods are shown in Figs. 4 and 5, respectively. In these figures, the emission at 1.50-1.51eV was due to band-to-band transition, and the 1.44-eV band was attributed to Ga antisite defect, Ga_{As} (Elliott 1983, Hiramoto et al. 1986). In Fig. 4(a), three PL peaks are present and the intensity of the 1.32-eV broad emission, whose origin is not clear at present but it may relate to residual damage, is seen to increase with the increase of implanted ion mass. The similar PL spectra was obtained for SiO_2-capped FA methods. In Fig. 4(b), on the other hand, the intensity of broad emission at 1.2-1.3eV with LO-phonon replica is small and almost constant, and a new emission band at 1.42eV appears for SiF_2- and SiF_3-implanted layers. For the sample implanted with S-dopant,the broad emission at 1.2-1.3eV is similar, and the 1.42-eV band is again seen in Fig. 5(b).

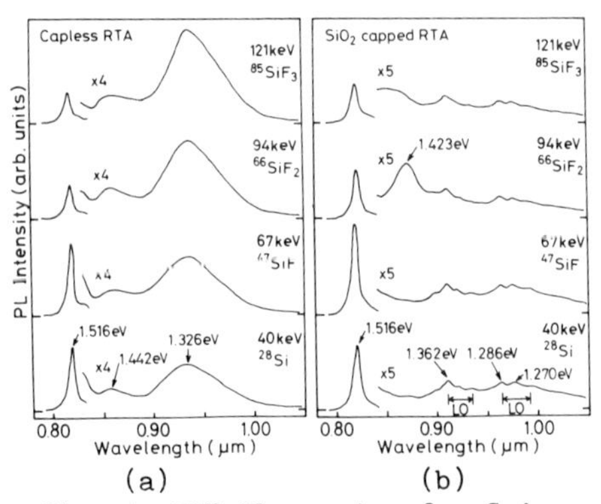

(a) (b)

Fig. 4. 77K-PL spectra for GaAs implanted with Si-dopant using different annealing methods.

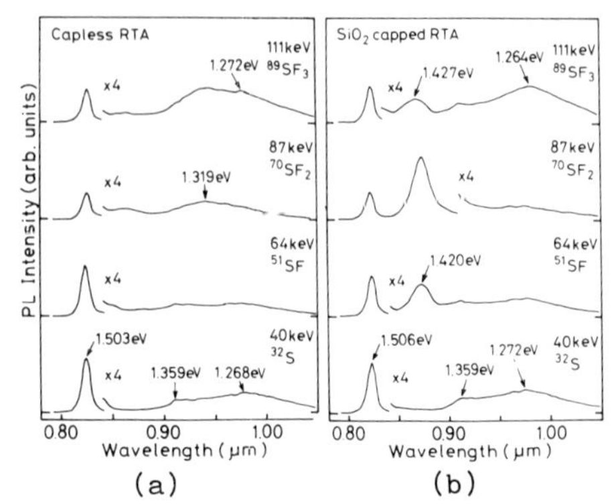

(a) (b)

Fig. 5. 77K-PL spectra for GaAs implanted with S-dopant using diffent annealing methods.

3.4 A NEW ANNEALING METHOD FOR MOLECULAR ION IMPLANTED LAYERS

We have found that the thermal treatment of SI GaAs with SiO_2 encapsulation before implantation can enhance the activation efficiency of SiF_3-implanted GaAs layers. In this work, thermal treatments before implantation were carried out by using SiO_2-capped RTA and FA methods, and post-implantation annealing was carried out by using SiO_2-capped RTA methods at 900°C for 10s. The dependence of peak carrier concentration on the thermal-treatment temperature (850∼1000°C) was investigated. It is found that the peak carrier concentration has the maximum value at 950°C-thermal treatment using SiO_2-capped RTA for 15s, which is about 1.5 times higher than that for the sample without this thermal treatment. The reason why this treatment enhances the activation is not clear at present, but it may relate to Ga vacancy generation during this treatment with SiO_2 encapsulation. On the other hand, these thermal treatments did not change the activation for Si-implanted layers.

4. SUMMARY

In summary, we have investigated fundamental electrical and optical characteristics of SiF_x- and SF_x- (x=1, 2 and 3) implanted GaAs layers. SiF implantation into GaAs is promising for forming thin active layers in MESFETs. The SiO_2-capped RTA method is effective for SiF_2- and SiF_3-implanted layers and we propose a new annealing method which can enhance activation. S- and SF_x-implanted layers are unsuitable for MESFETs due to broad carrier concentration profiles and low activation.

REFERENCES

Elliot K R, 1983 Appl. Phys. Lett. **42**, 274.

Hiramoto T, Mochizuki Y and Ikoma T, 1986 Jpn. J. Appl. Phys. **25**, L830.

Jaeckel H, Graf V, Van Zeghbroeck B J, Vettiger P, and Wolf P, 1987 Inst. Phys. Conf. Ser. **83**, 471.

Kuzuhara M, Ogawa Y, Asai S, Furutsuka T and Nozaki T, 1986 IEDM. Tech. Dig. Papers, pp. 763.

Inst. Phys. Conf. Ser. No. 91: Chapter 5
Paper presented at Int. Symp. GaAs and Related Compounds, Heraklion, Greece, 1987

Modification of CV-profiles in GaAs due to substrate influences

R D Schnell, H Schink and R Treichler

Siemens AG, Corporate Research and Development,
D-8000 München 83, P.O.Box 830952, FRG

ABSTRACT: A comparison of CV- and SIMS-profiles has been performed for Si-implants into different s.i. LEC GaAs substrates. It can result in distinct and reproducible differences in CV-profile steepness without a change of the atomic dopant distribution. This effect is explained by a modification of the channel substrate depletion layer due to variations of residual impurity concentrations. This substrate influence is significantly reduced by buried p-layer implantation.

1. INTRODUCTION

Channel layers of GaAs MESFETs are commonly formed by selective ion implantation into the semi-insulating substrate. A high degree of threshold voltage (V_T) reproducibility and homogeneity is necessary for LSI circuit applications (see e.g. Packeiser 1987). Therefore a tight control of the activation of the dopants and the shape of the carrier profiles is indispensible. In this paper we demonstrate how the use of different s.i. LEC grown substrates can cause significant differences in the shape of the carrier profile of Si-implanted active layers and give a qualitative interpretation of the results. Furthermore we will discuss to what extent the material influence can be reduced by an intentional doping of the substrate.

2. EXPERIMENTAL

LEC grown GaAs wafers of two different suppliers have been used. The wafers were either unintentionally doped (undoped) or doped with a low Cr-concentration ($\approx 5 \cdot 10^{15} cm^{-3}$). A comparison with highly Cr-doped ($\approx 5 \cdot 10^{16} cm^{-3}$) wafers has also been performed. ^{28}Si has been implanted with an energy of 60keV in the dose range $3\text{-}7 \cdot 10^{12} cm^{-2}$ into the blanket wafer. During implantation appropriate tilt and rotation angles are used to avoid axial and planar channeling. Capless annealing has been performed in AsH_3 overpressure. SIMS measurements have been done with a CAMECA IMS 3F using a Cs^+ ion beam of 14.5keV. A depth resolution of better than 10nm/decade is obtained (Treichler et al. 1987). A high resolution

(100µm*100µm) CV-pattern is delineated by lift off technique using Ti/Pt/Au metallisation. Details of this CV-pattern and the measuring technique have been reported elsewhere (Schink 1987).

3. RESULTS AND DISCUSSION

By comparing the SIMS profiles taken before and after annealing (Fig.1) it can be seen that during the annealing step no significant redistribution of the implanted dopants takes place. Fig.2 shows a comparison of CV- and SIMS-profiles after Si implantation (60keV, $3 \cdot 10^{12} cm^{-2}$) into substrates of two different suppliers (#1,#2). The SIMS-profiles are identical. The CV-profiles, however, are steeper compared to the SIMS measurements and differ for the substrates of the two suppliers used. In Fig.3 the CV-measurements are shown on a linear scale. The comparison between the undoped and low Cr-doped substrates of one supplier exhibits no influence of the Cr-doping in this range of concentrations.

The difference between the two kinds of substrates is seen only in the tail of the profiles whereas they are convergent towards the peak. This is more clearly seen in Fig.5 where a higher dose implant (60keV, $5 \cdot 10^{12} cm^{-2}$) is used and CV analysis reveals the profile peak. The observed differences can be explained neither by an activation nor a simple compensation effect. Activation changes are most pronounced at the profile peak (see e.g. Schnell and Schink 1987). The effect of a pure compensation is demonstrated in Fig.4. The carrier profile in highly Cr-doped substrates is shifted compared to the profile in undoped substrates by a constant value. We explain the

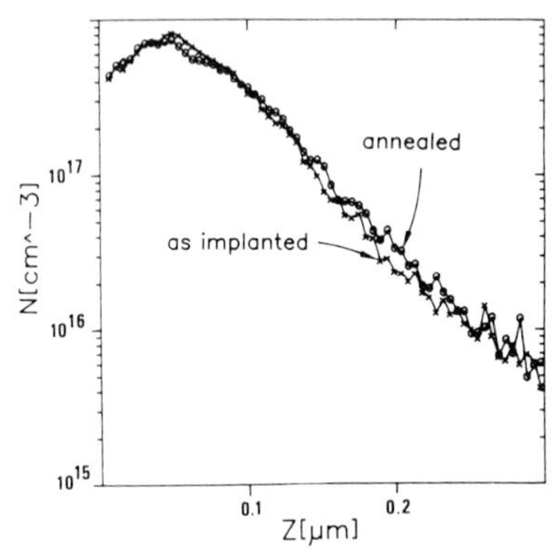

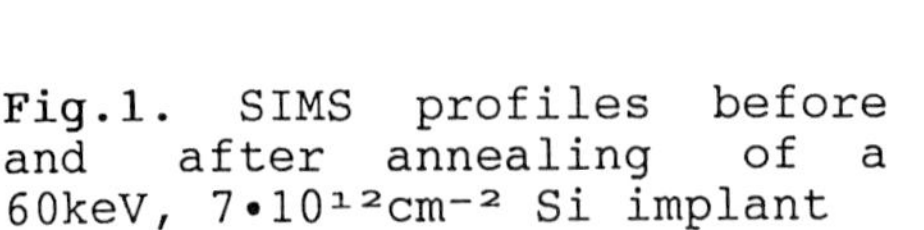
Fig.1. SIMS profiles before and after annealing of a 60keV, $7 \cdot 10^{12} cm^{-2}$ Si implant

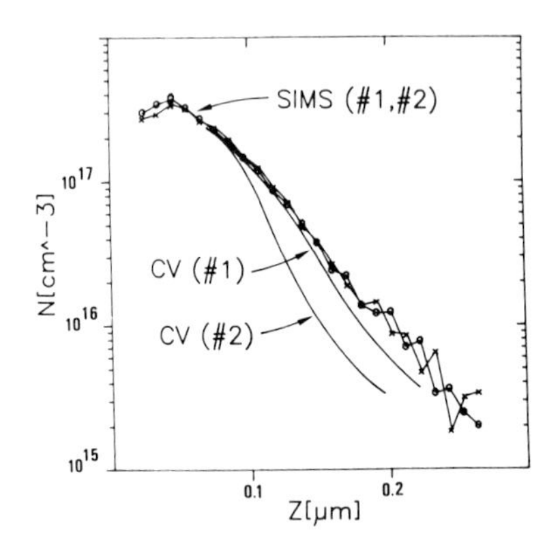

Fig.2. CV- and SIMS-profiles after Si implantation (60keV, $3 \cdot 10^{12} cm^{-2}$) into different undoped substrates (#1,#2)

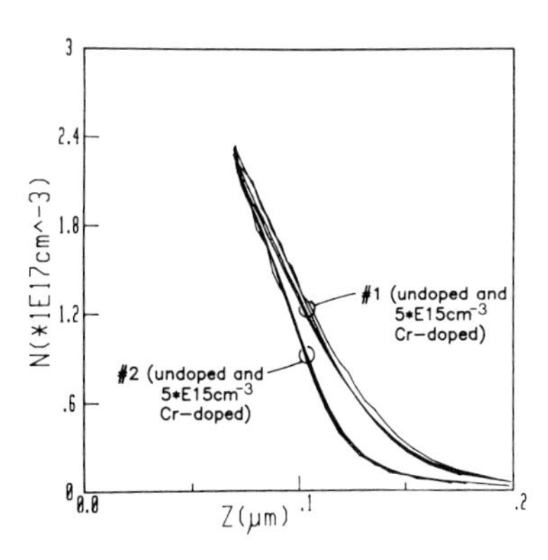

Fig.3. CV-profiles after a Si implant into substrates of two different suppliers (linear scale)

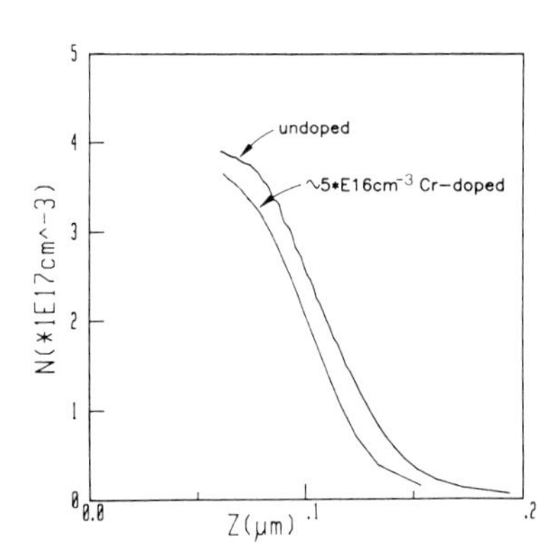

Fig.4. Comparison of CV-profiles in an undoped and a highly Cr-doped substrate

above mentioned substrate influence by a modification of the depletion layer between the channel region and the s.i. GaAs. This effect is also responsible for the increase of steepness of the carrier profile compared to the doping profile at the channel substrate interface (Lehovec and Zulegg 1974). A variation of the concentration of impurities or electrically active defects can cause the differences between the two materials. First measurements of the carbon concentration with FTIR spectroscopy (Alt 1987) indeed give values of $[C]=2.6\cdot10^{15}cm^{-3}$ for substrate #2 whereas for substrate #1 the carbon concentration is below the detection limit of $8\cdot10^{14}$ cm^{-3}.For both materials $[EL2]=1.6\cdot10^{16}cm^{-3}$ has been obtained by NIR measurements (Alt and Packeiser 1986). Further SIMS analysis is necessary to obtain concentrations of other impurities.

The observed influence of the s.i. GaAs on the carrier profile can be a severe limitation of threshold voltage reproducibility when changing from one ingot to another. With buried p-layers the channel substrate interface region can intentionally be modified. We have investigated how Be-coimplantation can reduce the above described substrate effect. The results are shown in Fig.5 and Fig.6. The profiles in Fig.5 qualitatively demonstrate the reproducibility improvement by Be-coimplantation. For a quantitative analysis V_T-mapping over the wafer has been performed using the CV-pattern. V_T is defined as the voltage necessary to deplete the profile to a depth where a carrier concentration of $1\cdot10^{16}cm^{-3}$ is present. The results are summarised in Fig.6 for different Be doses. Be-coimplantation reduces the differences between the two substrates from ΔV_T= 150 mV to 80 mV.

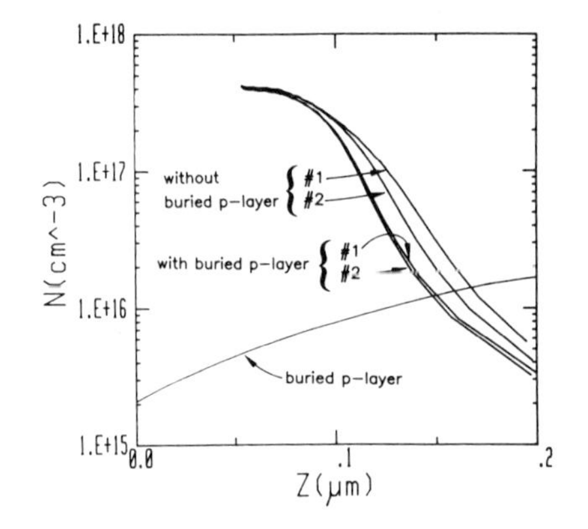

Fig.5. Influence of a buried p-layer (Be,90keV,$7 \cdot 10^{11}cm^{-2}$) on CV-profiles in the two undoped substrates

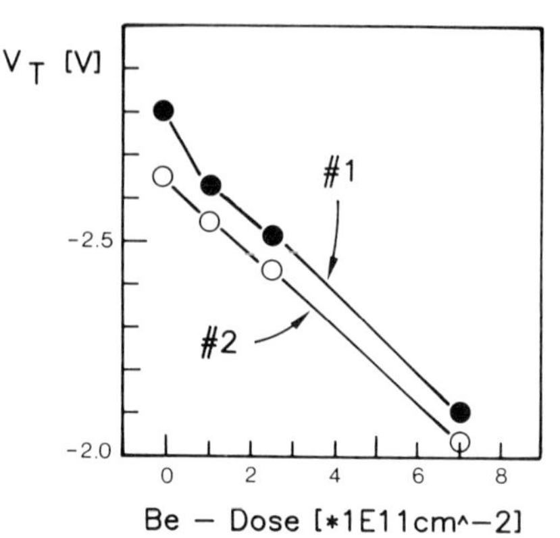

Fig.6. Threshold voltage as a function of Be-dose for a constant Si-implant (60keV, $5 \cdot 10^{12}cm^{-2}$) in the two undoped substrates

4. CONCLUSIONS

GaAs substrate material of different manufacturers results in remarkable and reproducible differences in the threshold voltage of ion-implanted MESFET layers. We have proven that these differences are caused by a modification of the channel substrate space charge layer and not by activation or simple compensation effects. This is attributed to differences in the individual concentration of background acceptors. The intentional p-doping by Be-implantation reduces the material effect significantly.

Acknowledgements

We like to thank our colleague H C Alt for providing results of NIR- and FTIR-measurements. This work has been supported by the Bundesministerium für Forschung und Technologie, FRG, contract No. 2716C.

References

Alt H C and Packeiser G 1986, J.Appl.Phys. **60** 2954
Alt H C 1987, to be published
Lehovec K and Zulegg R 1974, Proc. Int. Conf. on GaAs and Related Compounds (Deauville,France)
Packeiser G 1987, Proc. NATO-Workshop on Microscopic Inhomogeneities of Bulk GaAs (Oxford,England)
Schink H 1987, Proc. European-MRS Meeting (Straßbourg,France)
Schnell R D and Schink H 1987, Proc. European Solid State Device Research Conf. (Bologna,Italy)
Treichler R, Korte L, Cerva H and von Criegern R 1987,Proc. SIMS VI (Versailles,France)

Correlation between carrier and atomic distributions in Si-implanted semi-insulating InP

G. BAHIR and J. L. MERZ
Dept. Electrical and Computer Engineering, Univ. of California, Santa Barbara, CA 93106

ABSTRACT: We report the SIMS profiles of implanted Si, and the Fe redistribution in Si-implanted SI InP as a function of implant temperature. These results are correlated with the profile of electrically-active impurities measured by electrochemical profiling. The range and standard deviation of the implanted Si is 30% and 50% larger than the theoretical prediction, respectively. No effect of the annealing temperature on the atomic profile has been observed. The profile of the electrically-active carriers for a room-temperature Si implant into SI InP is asymmetric, and differs from either a hot implant profile or the Si profile obtained in undoped InP. These differences are interpreted in terms of three compensating mechanisms.

1. INTRODUCTION

InP is becoming increasingly important with regard to optoelectronic and microwave devices. It is envisioned that ion implantation will be used in the manufacture of such devices, due to its control and reproducibility. As with any doping technique, it is desirable to know the impurity distribution as a function of depth. However, ion implantation requires high temperature annealing after implantation to provide both regrowth of the damaged substrate and electrical activation of the implanted impurity, and other impurities are invariably present, resulting in n-type material with net electron concentration ranging from 10^{15} to low 10^{16} cm^{-3}. Iron, a deep acceptor in InP, is often added to the melt to compensate these background impurities. Device fabrication requires semi-insulating (SI) substrates, and Fe is presently the most common compensating impurity for InP. It is therefore important to understand the behavior of this compensating dopant in InP during thermal processing.

The purpose of this work is to study the effect of different implant and annealing conditions on the redistribution and electrical activation of a common n-type dopant, Si, and the usual compensating deep center, Fe. We report on the SIMS study of Fe redistribution in Si-implanted SI InP as a function of the implant temperature of the substrate, and the post-implant thermal processes, furnace annealing (FA) or rapid thermal annealing (RTA). These results are correlated with electrochemical profiling of the electrically-active Si impurities, and with Rutherford backscattering (RBS) measurements of the damage, reported previously [Bahir *et al.*, 1986]. In order to gain a better understanding of the behavior of the profile of implanted and electrically-active Si in SI InP, we have also implanted Si into undoped substrates, with the aim of studying the parameters which govern the carrier distribution and the effect of Fe redistribution as a result of annealing.

2. RESULTS AND DISCUSSION

The Si atomic profiles as measured by SIMS for room temperature (RT) and 200°C implants are shown in Fig. 1. The most probable depth (maximum atomic density) is nearly equal for the two implants, about 2200 Å, but the hot implant profile is wider, as expected. Also plotted is the LSS profile calculated for GaAs and scaled to InP. The LSS range (R_p) and standard deviation (ΔR_p) are 1750Å and 680Å, respectively; the measured values for the RT implant are 2200Å and 920Å. The solid line is a fit of a PEARSON IV distribution to the experimental results for the RT implant. Annealing (750°C FA for 15 minutes or 850°C RTA for 5 seconds) has no effect on the Si profiles for either RT or hot implant. Fig. 2 shows typical carrier distributions obtained on SI InP implanted with $3x10^{14}$ Si ions/cm^2 at RT or 200°C, followed by 750°C FA for 15 minutes. The RT implant profile is asymmetric and there is an anomalous decrease in carrier concentration in the region between 1000-3000Å. The RT electrical profile is in reasonable agreement with the theoretical LSS profile, and does *not* overlap the atomic profile on the deep side. The carrier profile of the hot implant is fairly flat and considerably wider than the RT implant; it is in reasonable agreement with the atomic (SIMS) profile for doses lower then 3-4 x 10^{18} cm^{-3}. The difference between RT and hot implant carrier profiles is larger than the difference in atomic profiles between the two implants (Fig. 1). For the two carrier profiles, considerable compensation exists for concentrations higher than $4x10^{18}$ cm^{-3}.

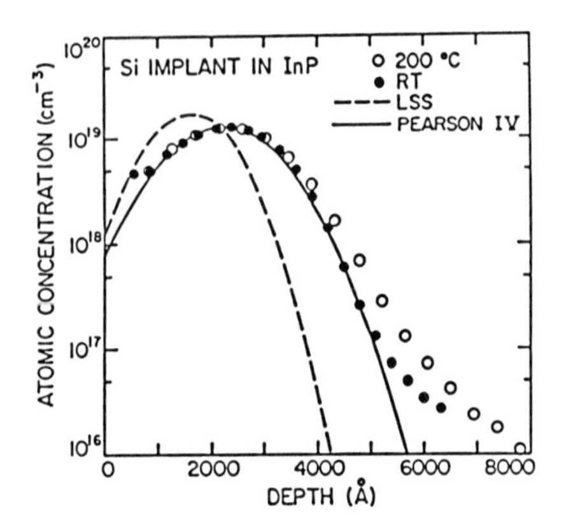

Fig.1. Si SIMS profile for RT and 200°C implants of 180 keV, $3x10^{14}$ Si/cm^2. Solid line is the best fit of the SIMS data to a Pearson type IV distribution; dashed line is the calculated LSS profile.

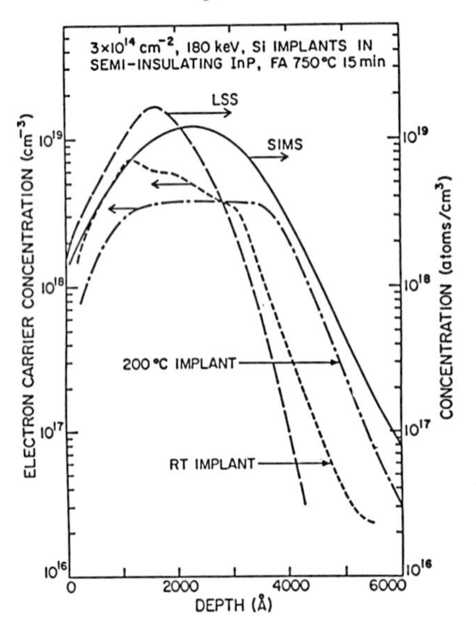

Fig.2. Comparison of carrier distribution for RT and 200°C Si implants into SI InP. Solid line is RT implant atomic distribution SIMS data, dashed line is theoretical LSS profile.

Results obtained for implants into *undoped* samples show similarities but significant differences. The RT implant is nearly symmetric and we do not observe an analogous decrease in the region 1000-3000Å. However, the electrical profile is also narrower then the atomic profile, as in the case of the SI samples.The hot-implant electrical profiles are similar in the two samples.

The atomic distribution of Fe, measured by SIMS, is shown in Fig. 3. A weak peak at $0.8R_p$ (2000Å) is generally attributed to gettering by residual damage resulting from the implant. This can be verified by using hot implantation to prevent the residual damage; in this case the 2000Å peak observed for the RT-implanted sample is not seen. The dominant feature in Fig. 3 is the peak at about 3200Å, which is attributed either to the defects created by the disturbed stoichiometry along with Fe pile-up in the region where phosphorus is in excess, or to residual defect clusters remaining at the original amorphous/crystal boundary [Lecrosnier 1983]. The nonstoichiometry resulting from implantation into compound semiconductors has been described by Christel *et al.* [1980, 1981], who simulated this process by solving the Boltzmann transport equation (BTE). To test this model we have compared SIMS profiles with BTE calculations. The results of this comparison indicate that the Fe is gettered into a range which is close to the amorphous-crystalline interface (3100Å), and somewhat deeper (3200Å) than the calculated P interstitial profile peak (2600Å). This is consistent with the absence of any Fe-accumulation peak in hot-implant samples. The proximity of the Fe peak to the amorphous/crystalline interface suggests that Fe diffuses rapidly in the amorphized region and nucleates at damage sites in the interfacial region [Lecrosnier 1983]. The comparison between the atomic and electrical profiles described above reflects various compensation mechanisms which result from different conditions, e.g. SI vs. undoped substrate, or RT vs. hot implants. These compensation effects may arise from (a) formation of compensating acceptors by the amphoteric dopant, (b) compensation due to the redistribution of the deep acceptor Fe, or (c) implant-induced damage.

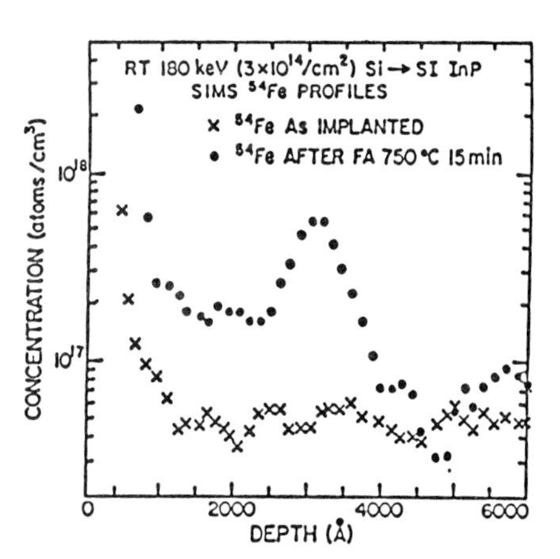

Fig.3. ^{54}Fe depth distribution in InP measured by SIMS for RT implant (180 keV 3 x 10^{14} Si/cm^2) into SI InP, showing the distributions obtained as implanted, and following FA at 750°C.

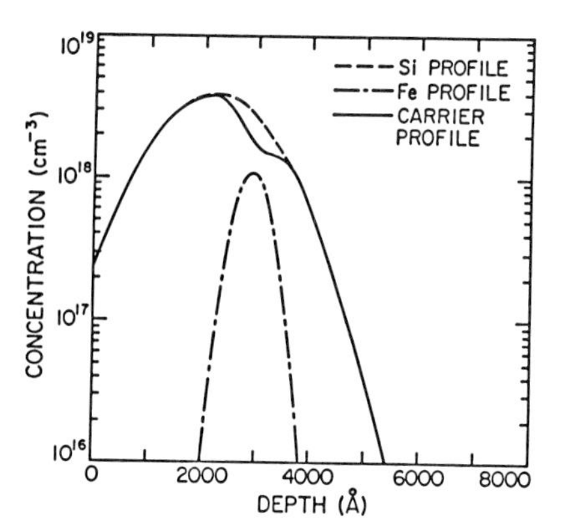

Fig. 4. (a) Simulated Si atomic profile, (b) Simulated Fe redistribution profile, (c) Simulated carrier profile.

The effect of the excess Fe in the SI substrate on the threshold dose for the onset of electrical activation of Si-implanted SI InP has been reported. Asymmetric or multiply-peaked electrical profiles in implanted SI InP have been observed by Woodhouse *et al* . [1984],Lorenzo *et al.* [1983], and Duhamel *et al.* [1987]. The asymmetric peak could be the result of some (as yet unidentified) compensation mechanism, or it could be due to Si redistribution. Because there is no change in the Si profile following annealing, as observed both in this work and by others [Oberstar, *et al.* 1982, Duhamel, *et al.* 1987], we rule out the latter possibility. Comparison of the differences between the redistributed Fe profiles following RT or hot Si implants with the active electrical profiles of SI and undoped samples (Fig.2) can be explained on the basis of either of two compensating mechanisms: (1) the amphoteric nature of Si (discussed above) which shows itself in the case of hot implants into undoped or SI InP by the fact that there is no deep Fe peak, and (2) a new compensation mechanism which results from Fe redistribution and subsequent compensation of the shallow Si donors by the redistributed Fe. Fig.4 shows the calculated atomic profile of Si-implanted InP with the parameters extracted from the SIMS experimental data. The redistributed Fe peak is described by a Gaussian peaking at the center of the Fe experimental peak (≈3000Å). The difference between the two curves, assuming 100% activation of the implant and 100% compensation due to available Fe, yields the measured electrical profile. This simulated electrical profile is very similar to the published results mentioned above. In this simulation the atomic concentration is reduced to a peak concentration less then $4x10^{18}$ cm^{-3} to avoid the first compensation mechanism. The actual profile that one obtains in this case (high dose RT implant into InP) results from a combination of both mechanisms, taking into account various parameters of the material, such as the initial Fe concentration.

A third mechanism, implantation-induced damage centers which can also cause carrier compensation, is believed to cause the narrow profile (in comparison with the atomic profile) of the RT implant into either SI or undoped material (e.g., Fig.2).

3. CONCLUSIONS

It is evident from the data presented here that significant Fe redistribution occurs in InP under commonly-used annealing conditions. The location of the redistributed Fe depends on the implant conditions as well as the exact nature of the subsequent thermal processing. Several different compensation mechanisms control the resulting profile of electrically-active carriers in the case of Si-implanted InP; among them are the amphoteric nature of Si, the Fe redistribution profile, and implantation-induced damage mechanisms.

ACKNOWLEDGEMENT

This work was supported by the Rome Air Development Center, Hanscom AFB, Mass.

REFERENCES

Bahir G, Merz J L, Abelson J R and Sigmon T W 1986 *Proc. SPIEE* **623** 149
Christel L A and Gibbons J F 1981 *J. Appl. Phys.* **52** 5050
Christel L A, Gibbons J F and Mylroie S 1980 *J. Appl. Phys.* **51** 6176
Duhamel N, Descouts B, Krauz P, Rao K, Dangla J and Henoc P April 1987 *Proc. MRS Conference*, Anaheim, CA
Lecrosnier D 1983 *Nucl. Instr. and Methods* **209** 325
Lorenzo J P, Davies D E, Soda K J, Ryan T G and McNally P J 1983 *Laser Solid Interaction and Transcient Processing of Materials* (New York: R.A. Lemons) p 683
Oberstar J D, Streetman B G, Baker J E and Williams P 1982 *J. Electrochem. Soc.* **129** 1312
Woodhouse J D, Donnelly J P, Nitishin P M, Owens E B, and Ryan J L 1984 *Solid State Electron* **27** 677

Radiation induced displacement damage in GaAs devices

W.T. Anderson,[a] A. Meulenberg,[b] J.M. Beall,[c] A.H. Kazi[d], R.C. Harrison[d], J. Gerdes[d], and S.D. Mittleman[e]

ABSTRACT: Radiation induced lattice damage has been studied in GaAs FETs, MMICs, and PIN diodes. Neutron lattice damage is significant in FETs and MMICs at 0.2 to 4 $\times$ 10^{15} n/cm^2. It appears that transient neutron damage was observed. GaAs PIN diodes were found to degrade at 3 $\times$ 10^{13} n/cm^2. A study was also made of the energy dependence of electron induced lattice damage in GaAs.

1. INTRODUCTION

The type of lattice damage in a semiconductor depends on the kind of incident particle radiation and its energy. High energy neutrons of (above 1 MeV) have small capture cross sections but interact with the lattice nuclei by the nuclear (strong) force and defects result from inelastic scattering. Such scattering results in clusters of defects which trap conduction electrons and thus carrier removal predominates. Incident electrons interact with lattice atoms by the electromagnetic force and must have energy above a certain threshold value to result in displacement of an atom to an interstitial position. Point defects result which trap few conduction electrons compared to neutron radiation but the mobility decreases from the scattering of conduction electrons at the large number of point defects. Gamma rays from a Co^{60} source result in no displacement damage by themselves but produce Compton electrons which can cause lattice damage providing they have energy above the threshold value.

2.0 Neutron radiation effects

Degradation resulting from neutron irradiation has been reported in low noise GaAs Field Effect Transistors (FET) (Borrego et al 1978, Anderson et al 1985), GaAs power FETs (Moghe et al 1981), and GaAs junction FETs (JFET) (Zuleeg et al 1978). It was found that degradation in the D.C. characteristics and high frequency (X-band) operation was the result of carrier removal induced by high energy neutron irradiation. In the present paper a report is made of neutron degradation effects in two types of GaAs monolithic microwave integrated circuits (MMIC) and long gate GaAs FETs (FATFET).

2.1 Total Fluence Effects

2.1.1 MMICs

The Texas Instruments MMICs investigated were feedback amplifiers designed for use as a broadband, low to medium power gain stage (Anderson et al. 1985). To study total fluence effects the pulsed reactor at the Army Pulse Radiation Facility (APRF) at Aberdeen, Maryland was used in the continuous mode. Degradation in TI MMICs resulting from neutron irradiation is shown in Figure 1. It was found that the gain falls off faster than with individual FETs (Anderson 1985)

*This work was supported by DNA, DARPA, and the Communications Satellite Corporation
[a] Naval Research Laboratory, Code 6815, Washington, D.C. 20375-5000
[b] COMSAT Laboratories, Clarksburg, MD
[c] Texas Instruments, Dallas; TX
[d] Army Pulse Radiation Facility, Aberdeen Proving Ground, MD
[e] Rome Air Development Center, Hanscom AFB, MA

with increasing neutron fluence. To investigate this in more detail the individual passive components that comprise the bias and tuning elements of the MMIC were also irradiated. The MIM capacitors exhibited little measurable change but the implanted GaAs resistors showed large increases with increasing neutron fluence. Gain degradation occurs in the MMICs partly as a result of a change in bias of the FETs and tuning as a consequence of changes in the on-chip resistors. Changes in the carrier concentration and mobility in the FETs are also responsible for gain degradation. FATFETs fabricated in the same GaAs material with a similar (but not identical) implant profile as used for the TI MMICs were irradiated at the same time as the passive elements and received the same neutron fluence. The carrier concentration and mobility were calculated using the FATFET model of Williams (1984). It is evident that carrier removal is not the only mechanism responsible for the performance degradation of the FETs and MMICs above 1×10^{15} n cm^{-2} but that mobility decrease is also important. The results for carrier concentration and mobility are similar to that reported by Lehovec et al (1975).

2.1.2 GaAs PIN Diodes

Figure 2 shows the affect of neutron radiation on the forward resistance of GaAs PIN diodes. The PINs were fabricated (Barratt 1983) by M/A-COM and were of the mesa type with a 16 μm undoped intrinsic (I) region with a background doping of 7×10^{12} cm^{-3}. Based on a 20% increase in forward resistance, the hardness level is 3×10^{13} n cm^{-2}. At this low fluence, carrier removal and changes in mobility due to neutron damage are insignificant. Instead, the main contribution to the forward resistance increase is attributed to the decrease in carrier lifetime in the I region resulting from neutron induced increase in recombination centers.

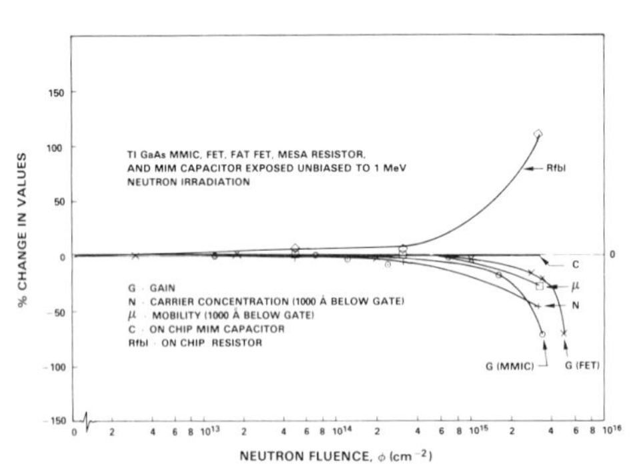

Fig. 1 — Percent change in TI GaAs MMICs, FETs, FAT FETs, mesa resistors, and MIM capacitors as a function of neutron fluence.

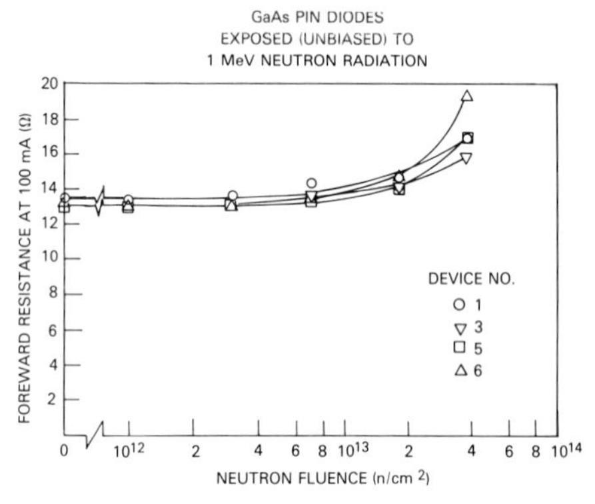

Fig. 2 — D.C. forward resistance of GaAs PIN diodes vs fluence following exposure to 1 MeV equivalent neutrons.

2.2 Pulsed Neutron Effects

A study was made at APRF of combined radiation effects to investigate synergistic effects and to separate the transient effects that occur during pulsed irradiation by neutron and gamma particles. The long term transient response observed when an MMIC (TI EG8300) was exposed to combined flash X-ray (FXR) and pulsed neutron radiation is shown in Figure 3. Positive gate and drain current photoresponses were observed, but the drain current exhibited a large negative long term transient of about 3 ms. Similar gate and drain photocurrent and drain long term transients were observed when an MMIC was irradiated by an order of magnitude larger neutron pulse alone. Because a FXR pulse resulted in a positive long term drain current transient, it is unlikely that the accompanying gamma radiation during the neutron pulse is responsible for the long term negative drain current transient. Therefore, the negative long term drain current transient induced by pulsed neutron irradiation is attributed to carrier removal resulting from lattice displacement damage that anneals out in 3 ms at room temperature. Similar transient effects have been observed in GaAs (Zuleeg et al. 1980) and Si (Sander et al. 1966, McMurray et al. 1981) devices following pulsed neutron irradiation.

3.0 Electron and Gamma Radiation Effects

Gamma rays from a CO^{60} source generate electron-hole pairs, photo electrons, and Compton electrons in GaAs, but cause no displacement damage by themselves. It is the Compton electrons that produce displacement damage in GaAs. There is an energy below which electrons cannot displace a lattice atom. Above this onset energy, the lattice atom may be knocked further from its site or more than one atom may be displaced. The onset energy for silicon is approximately 150 keV and for GaAs it is approximately 250 keV for damage detectable by DLTS (Pons 1980). The effective GaAs onset level detectable by carrier removal and mobility degradation data (reported by Kalma et al 1975 and the present results) appears to be closer to 600 keV. The importance of a higher onset level to radiation testing in GaAs is seen when it is compared to the maximum energy of electrons generated by the gamma rays of a Co^{60} source (Garth et al. 1985). The energy dependence of displacement damage in GaAs, as revealed by carrier removal and mobility degradation, was carried out at the Naval Research Laboratory Electron Van de Graaff (at 0.6, 0.8, 1, 1.5, and 2 MeV) and LINAC (10 MeV) facilities using FATFETs. Gamma irradiation was carried out at the Rome Air Development Center Radiation Facility, Hanscom AFB. In contract to the neutron irradiation results discussed above, mobility changes are comparable to changes in pinchoff voltage (proportional carrier centration) for electron irradiation. Figure 4 shows the change in carrier concentration for a typical epitaxial FATFET as a function of electrons energy. Carrier removal was observed only above 600 keV and only small changes were measured at 2×10^7 rad. Figures 5 and 6 compare the degradation of mobility and pinch-off voltage for 1-MeV-electron-irradiated devices with gamma irradiated devices. It is seen that the total dose of gamma radiation to produce the same change in mobility or V_p as for 1 MeV electron irradiation is larger by at least a factor of 15, and is 30 times larger in the case of V_p for implanted devices.

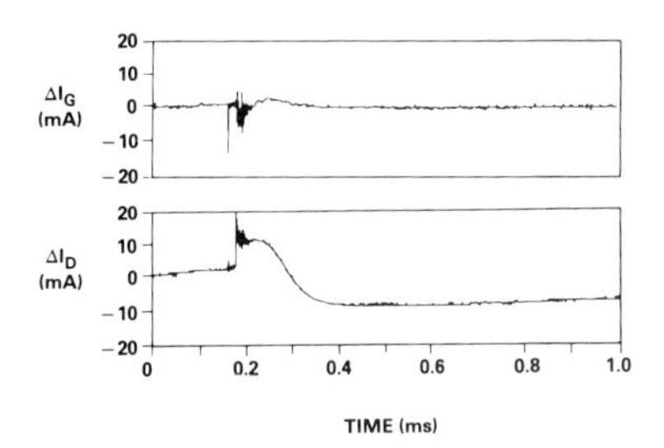

Fig. 3 — Transient change in gate current, ΔI_G, and drain current, ΔI_D, of GaAs MMIC EG8300, #36 following combined FXR (85 ns, 1×10^{11} rad/s) and pulsed neutron (83 μs, 10^{17} n/cm^{2}s) exposure. FXR alone resulted in a positive long term transient in I_D of 50 μs.

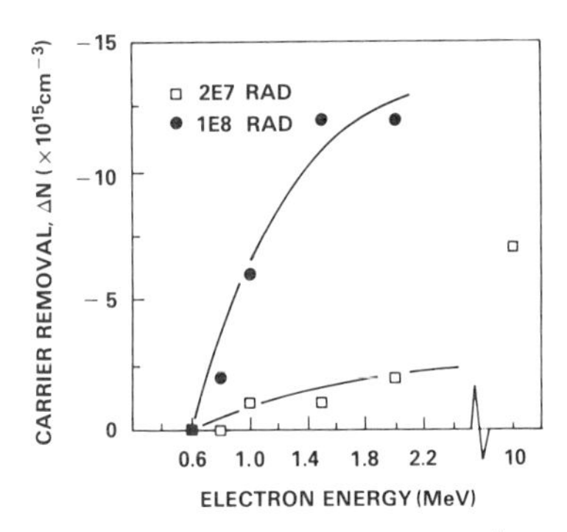

Fig. 4 — Carrier Removal at 2×10^7 and 10^8 rad (Si) vs electron energy in GaAs FATFETs fabricated on epitaxial layers.

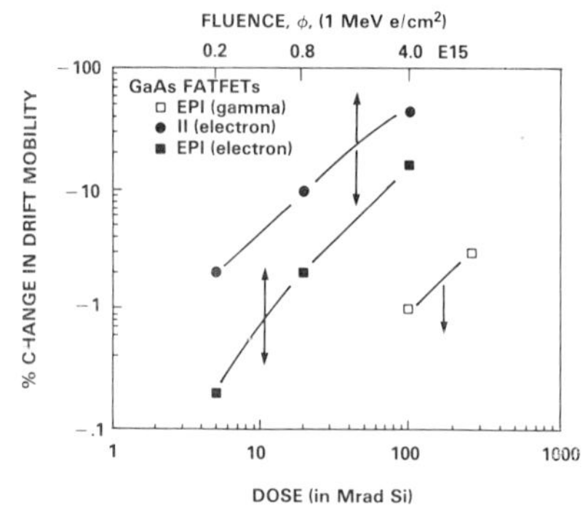

Fig. 5 — Percent degradation in drift mobility vs. fluence, ϕ, and rad dose for 1 MeV electron and gamma irradiations.

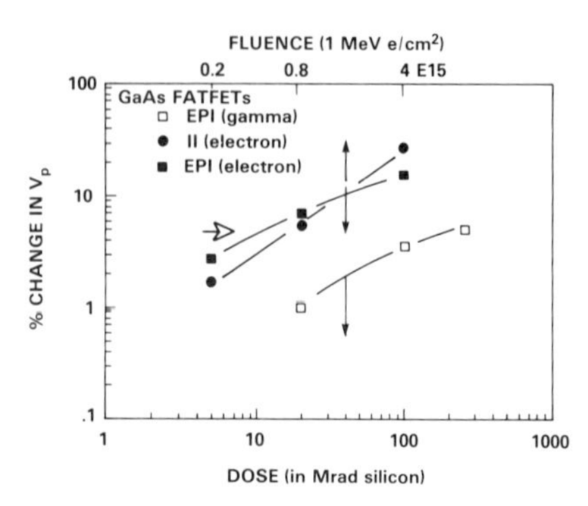

Fig. 6 — Percent degradation in pinch-off voltage, V_P, vs rad dose and fluence for gamma and 1 MeV electron irradiations.

4.0 Conclusions

Neutron induced lattice damage is significant in FETs and MMICs at 0.2 to 4 $\times$ 10^{15} n/cm^2. PIN diodes degrade at a neutron fluence of 3 $\times$ 10^{13} n/cm^2 attributed to neutron induced increase in recombination centers in the I region. Based on pulsed neutron experiments it appears that transient neutron damage occurs in GaAs with a room temperature recovery time of 3 ms. An energy dependence study of electron induced lattice damage in FATFETs revealed an onset energy of 0.6 MeV, below which little decrease in mobility of carrier concentration was observed.

References

Borrego, J.M., et al., 1978, IEEE Trans. on Nucl. Sci. *NS-25* pp. 1436-1443.
Anderson, W.T., et al., 1985, IEEE Trans. on Nucl. Sci. NS-32, pp. 4040-4045.
Moghe, S.B., et al., 1981, IEEE Trans. on Nucl. Sci. NS-25, pp. 1010-1013.
Zuleeg, R., and Lehovec, K., 1978, IEEE Trans. on Nucl. Sci. NS-25, pp. 1444-1449.
Williams, R.E., 1984, *GaAs Processing Techniques,* (Artech, Dedham, MA), p. 371.
Lehovec, K., et al., 1975, Inst. Phys. Conf. Ser. (24), (Inst. Phys., London) p. 292.
Barratt, C., 1983, 1983 IEEE MTT-S Digest (IEEE, NY, 1983), pp. 507-509.
Garth, J.C., 1985, IEEE Trans. on Nucl. Sci. Vol. NS-32, No. 6, pp. 4382-4387.
Pons, D., et al., 1980, J. Appl. Phys., 51(4), April 1980, pp. 2038-2042.
Kalma, A.H., et al., 1975 IEEE Trans. on Nucl. Sci. Vol. NS-22, No. 6, pp. 2038-2042.
Zuleeg, R. et al., 1980, IEEE Trans. on Nucl. Sci. Vol NS-27, No. 5, pp. 1343-1354.
Sander, H.H. et al., 1966, IEEE Trans. on Nucl. Sci, Vol. NS-13, No. 6, pp. 53-62.
McMurray, L.R. et al., 1981, IEEE Trans. on Nucl. Sci. Vol. NS-28, No. 6, pp. 4392-4396.

Redistribution of implanted hydrogen and substrate dopants in annealed (100 to 600°C) substrates of GaP(S), InP(S), InP(Sn), GaAs(Si), and GaAs(Zn)

R.G. Wilson, S.W. Novak,* and J.M. Zavada**

Hughes Research Laboratories, Malibu, CA 90265
*Charles Evans and Associates, Redwood City, CA 94063
**USARDSG (UK), London NW1 5TH, UK

ABSTRACT: We implanted H (300 keV and $5x10^{15}$ and $1x10^{16}$ cm^{-2}) into samples of GaP(S), InP(S), InP(Sn), GaAs(Si), and GaAs(Zn), and annealed them at temperatures from 100 to 700°C. We then profiled the resulting hydrogen and substrate dopant depth distributions using SIMS. Two regions and temperature regimes of H redistribution are observed in general, one toward the surface, through the implant damage, in a higher temperature regime, and one deeper into the undamaged substrate, in a lower temperature regime. The temperature regimes are different in the different materials. The associated redistribution (depletion or carrier removal) of the substrate dopants (S, Sn, Si, Zn) are also described.

1. INTRODUCTION

The purpose of this work was to measure any redistribution of n- or p-type dopants in III-V materials implanted with H and subsequently thermally processed for waveguide fabrication, or implanted with dopant ions and subsequently annealed for device fabrication. Another aspect of this work was to measure the redistribution of the implanted H associated with the annealing. The materials studied were GaP(S), GaAs(Si), GaAs(Zn), InP(S), and InP(Sn). All of these materials were implanted with H, Be, Mg, Si, and Se, and thermally processed at temperatures from 100 to 600 or 700°C. The depth profiles of the implanted atoms and the substrate dopant atoms were measured using the appropriate secondary ion mass spectrometry (SIMS) technologies. H is a light ion that penetrates more deeply and creates little damage at commonly used implantation energies; the dopant ions have shallower depth profiles that are characterized by greater damage.

2. EXPERIMENTAL TECHNIQUES

H, as ^1H or ^2H, was implanted into all substrates as ions of 300, 333, or 350 keV energy and fluences of $5x10^{15}$ or $1x10^{16}$ cm^{-2}, and pieces of these substrates were thermally processed at 100, 200, 300, 400, 500, and 600°C, and in some cases, also at 275, 325, 350, 375, 450, 550, or 700°C to better define aspects of H redistribution at intermediate temperatures. Be, Mg, and Si were implanted at energies from 100 to 700 keV, and Se, from 200 to 1050 keV. Implantation fluences were $5x10^{12}$, 2 or $3x10^{13}$, and/or $1x10^{14}$ cm^{-2}. A few samples were selected for each doped substrate for study, to be representative of the cases for substrate dopant atom redistribution. All thermal processing was performed using proximity capping. H, Si, S, Se, and Sn were profiled using Cs negative SIMS, and Be, Mg, Zn, and Sn were profiled using O positive SIMS.

3. EXPERIMENTAL RESULTS AND OBSERVATIONS

Redistribution of H with thermal processing was observed for all substrates. Two regions and temperature regimes of redistribution are observed in general, one toward the surface, through the implant damage, in a higher temperature regime, and one deeper into the undamaged substrate, in a lower temperature regime. The threshold temperatures for these redistributions, the details of the shape and temperature dependence of the depth (rate or diffusion coefficient) of redistribution, and the densities of H in the redistributions vary slightly among substrates, but the general natures of the two regimes of redistribution are consistent, and a detailed example is illustrated in Figure 1 for 333 keV H in (111) GaP.

For the H and dopant implants in GaAs(Si) and InP(Sn), no redistribution of Si or Sn was observed for thermal processing at 600°C, so no other temperatures were studied.

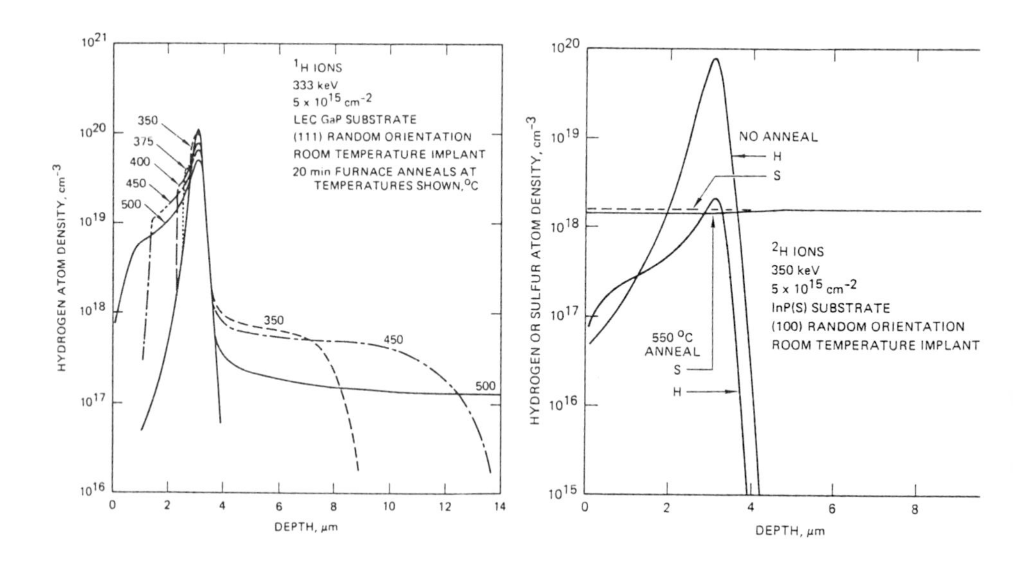

Fig. 1. H depth distributions in annealed GaP.

Fig. 2. H and S depth distributions in annealed InP(S)

For the H and dopant implants in GaP(S) and In(S), S redistribution was observed in both materials, as illustrated in Figure 2 for a 350 keV H implant into (100) InP(S) annealed at 550°C, and in Figure 3 for a 400 keV Se implant into (111) GaP annealed at 600°C. A maximum depletion (carrier removal) of about 11% in the S density was observed for H and dopant implants, in the region of the implant damage, which is about 4 μm for the H implant, and about 0.5 μm for the Se implant. An accumulation of S within the first ~0.1 μm is seen for the annealed Se implant. Higher processing temperatures could be studied to determine whether this effect increases with temperature above 600°C.

A different result was observed for Zn redistribution in H-implanted, p-type Zn-doped GaAs, namely: 1) Essentially the same Zn depletion (redistribution) profile was measured for 100, 200, 300, 400, 500, and 600°C, as illustrated in Figure 4. 2) This Zn depletion extends only to about 1μm, while the as-implanted H depth and damage distributions approach 3μm in all these materials (for 350 keV). 3) The Zn is depleted to (40±10)% of its original density, for all these processing temperatures. These results for Zn, which seem not to correlate with implant or thermal processing parameters, are not understood or explained as yet.

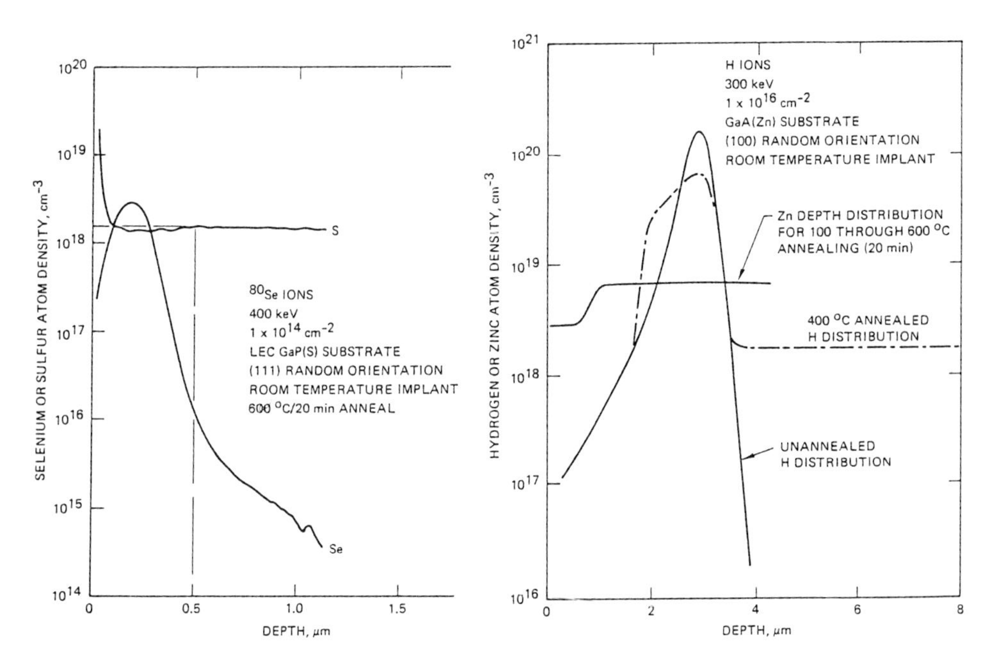

Fig. 3. Se and S depth distributions in annealed GaP(S)

Fig. 4. H and Zn depth distributions in annealed GaAs(Zn)

4. SUMMARY AND CONCLUSIONS

Implanted H depth distributions redistribute in doped GaP, GaAs, and InP in two regions and temperature regimes, one toward the surface and through the implant damage, in a higher temperature regime (typically 350 to 700°C), and one deeper into the undamaged substrate, in a lower temperature regime (typically 200 to 500°C). The substrate dopants, Si and Sn, were not observed to redistribute for annealing temperatures up to 600 or 700°C, in either GaAs(Si) or InP(Sn). S was depleted about 11% in the implant damaged region at 550 or 600°C, for both GaP(S) and InP(S). Zn depleted to about 40% of its initial density to a constant depth of about 1μm in H-implanted GaAs(Zn) for all temperatures from 100 to 600°C, even though the implanted H atom and damage distributions extended 3 or more μm in depth.

Depth distributions of Be and Si implanted into GaP, InP and InSb, after implantation and after furnace or flash lamp annealing, compared with GaAs

R.G. Wilson, S.W. Novak,* and J.M. Zavada**

Hughes Research Laboratories, Malibu, CA 90265
*Charles Evans and Associates, Redwood City, CA 94063
**USARDSG (UK), London NW1 5TH, UK

ABSTRACT: We have implanted 100 to 700 keV Be and Si ions into GaP, InP, and InSb in random orientations and at a representative channel fluence of $5x10^{12}$ cm^{-2}, a representative source/drain fluence of $1x10^{14}$ cm^{-2}, and an intermediate fluence of $1\text{-}3x10^{13}$ cm^{-3}, and measured their depth distributions using secondary ion mass spectrometry (SIMS), for both unannealed and for furnace and flash lamp annealed samples (except for InSb). We have also implanted Be and Si ions in the <110> or <111> channeling directions of GaP, InP, and InSb. One result is that Be does not redistribute in GaP during annealing the same way that it does in GaAs or InP (at higher atom densities). We do not observe Si redistribution in any material at these fluences.

1. INTRODUCTION

The two dopants of greatest interest today for GaAs and other III-V device and circuit technologies are Be and Si. Implantation profiles for Be and Si in GaAs have been studied and reported, both unannealed and annealed. Similar information is needed for other III-V compounds so that devices and circuits can be designed in those materials. Channeled implants are useful to achieve significantly deeper implanted depth distributions using 200 and 400 keV implanters.

2. EXPERIMENTAL TECHNIQUES

Implants were performed at room temperature with Be or Si ions of 100, 200, 350 or 400, and 700 keV energy, using 200 or 350 kV implanters, and fluences of $5x10^{12}$, 1, 2, or $3x10^{13}$, and $1x10^{14}$ cm^{-2}. Samples of <100>, <111>, or <110> GaP, GaAs, InP, and InSb were implanted with the ion beam oriented 8° from the crystal axis for random implants, and aligned to within 0.1° of the crystal direction using Rutherford backscattering, for the channeled implants. Pieces of these implanted samples were subsequently annealed in a furnace under flowing nitrogen gas, or in a Heat Pulse flash lamp annealer at temperatures between 600 and 800°C. Be depth distributions were measured using O positive secondary ion mass spectrometry (SIMS), and Si depth distributions, using Cs negative SIMS. Depth scales were determined by measuring the SIMS crater depths using a surface profilometer, with an error of about ±7%.

3. EXPERIMENTAL RESULTS

The results of our study of unannealed and annealed 100 keV Be channel and source/drain implants into GaP and InP are shown in Figures 1 through 4.

For $5x10^{12}$ cm^{-2} fluence, no Be redistribution occurs for GaP or InP. For $1x10^{14}$ cm^{-2} fluence, Be redistribution occurs both toward the surface and deeper into the bulk for InP, but for GaP, it occurs only toward the surface and not into the bulk. The behavior for InP is similar to that for GaAs; data for $4x10^{14}$ cm^{-2} Be implanted into GaAs are shown in Figure 5 for reference. This redistribution is density driven and not temperature dependent (thermal diffusion); it saturates at a fixed density and the depth of redistribution depends on the initial implanted density (in the temperature range from 550 to 900°C). Essentially the same depth distribution is measured for all annealing temperatures (to 900°C) once the Be has redistrbuted to its density limit. which is a function of the original Be density (implant fluence). The data of Figure 3 are the interesting ones, where no Be redistribution deeper into the GaP is observed. No Be redistribution occurs in GaAs for the $5x10^{12}$ cm^{-2} fluence.

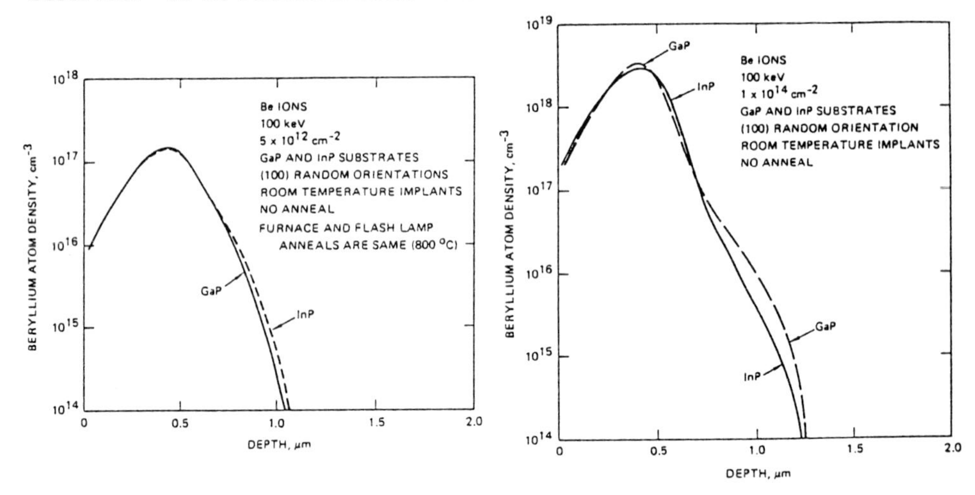

Fig. 1. Low fluence profiles of Be in GaP and InP

Fig. 2. Moderate fluence profiles of Be in GaP and InP

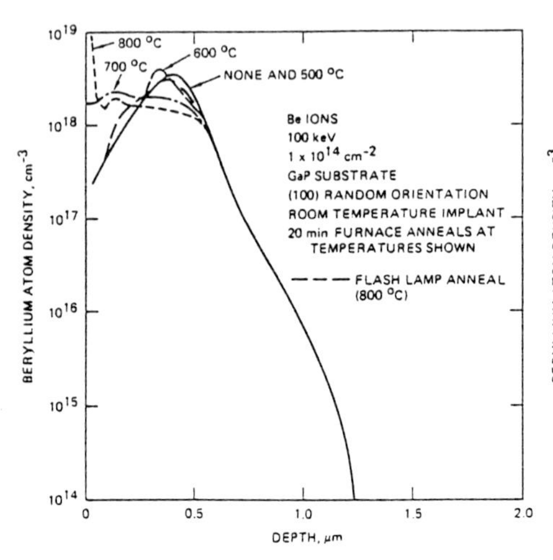

Fig. 3. Ann. Profiles of Be in GaP

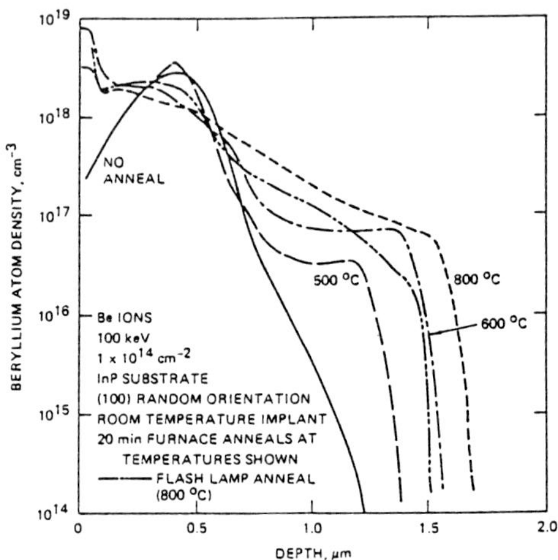

Fig. 4. Ann. Profiles of Be in InP

Examples of other depth distributions for Be and Si in GaP, InP, and InSb are shown in Figures 6 through 12. Various range and maximum channeling density data for the profiles for various energies and crystal orientations are listed in Table I. One additional representative annealed redistribution for a <110> channeled implant into InP is shown in Figure 13, where Be has redistributed toward the surface and deeper into the crystal, as well as being retained in the region of maximum implant damage (0.8 μm).

Table I. Range and density parameters for random and channeled implants of Be and Si in III-V materials

Matrix	Ion	Energy keV	Fluence cm^{-2}	Orient. <hkl>	R_m μm	R_p μm	ΔR_p μm	R_{max}<hkl> μm	ρ<hkl> cm^{-3}
GaP	Be	100	5[12]	100	0.44	0.41	0.18		
		100	1[14]	100	0.40	0.36	0.15		
		100	2[13]	111				1.3	3[17]
		200	2[13]	111	0.64	0.55	0.23	1.7	3[17]
		400	2[13]	111	1.05	0.98	0.25	2.25	2[17]
	Si	100	1[13]	111	0.105	0.065	0.11		
		200	1[13]	111	0.215	0.16	0.19		
		350	1[13]	111	0.39	0.306	0.25	2.6	[15]-[16]
		700	1[13]	111	0.77	0.75	0.255	3.7	[15]-[16]
InP	Be	100	5[12]	100	0.44	0.41	0.19		
		100	1[14]	100	0.42	0.36	0.16		
		100	2[13]	100	0.38	0.34	0.155	1.3	[16]-[17]
		200	2[13]	100	0.65	0.56	0.225	1.9	[16]-[17]
		400	2[13]	100	1.06	0.93	0.273	2.7	[16]-[17]
		100	2[13]	110				2.8	3[16]
		200	2[13]	110				3.7	3[16]
		400	2[13]	110				4.5	3[16]
	Si	200	1[13]	100	0.25	0.16	0.22	1.85	1[15]
		350	1[13]	100	0.45	0.35	0.30	2.7	1[15]
		700	1[13]	100	0.81	0.72	0.32		
		200	1[13]	110				3.0	3[15]
		350	1[13]	110				4.0	2-3[15]
		700	1[13]	110				5.5	2-3[15]
InSb	Be	200	3[13]	111	0.61	0.48	0.29		
		200	3[13]	110				4.0	1[17]
	Si	200	1[13]	111	0.23	0.20	0.19		
		350	1[13]	111	0.38	0.37	0.24		
		700	1[13]	111	0.72	0.69	0.33		
		350	1[13]	110				4.3	6[15]
		700	1[13]	110				6.0	6[15]

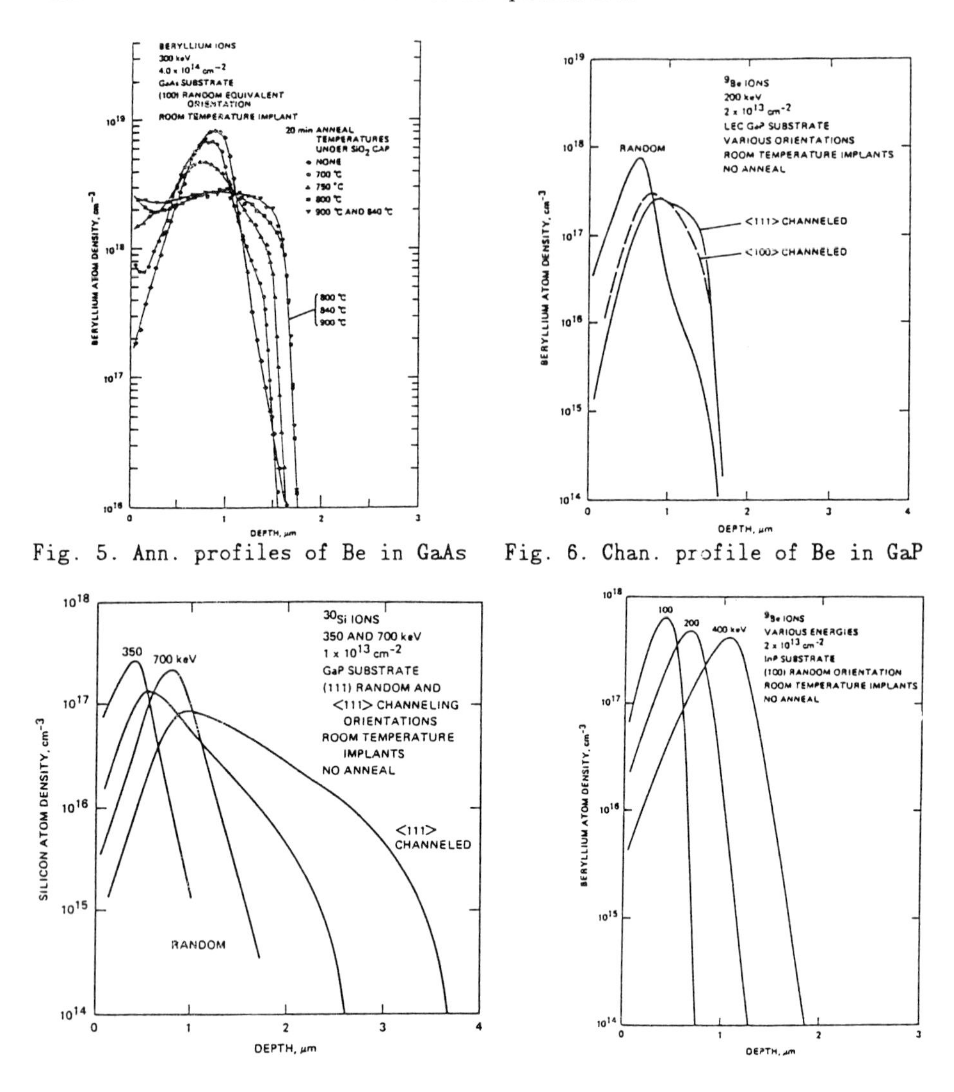

Fig. 5. Ann. profiles of Be in GaAs

Fig. 6. Chan. profile of Be in GaP

Fig. 7. Chan. profiles of Si in GaP

Fig. 8. Rand. profiles of Be in InP

CONCLUSIONS AND SUMMARY

Be does not redistribute with annealing for implant fluences of 10^{13} cm^{-2} or lower (Be densities of 10^{18} cm^{-3}). For higher fluences (at least 10^{14} cm^{-2}), Be redistributes toward the surface for GaP, GaAs, and InP, and deeper into the bulk for GaAs and InP, but not for GaP. Implanted Si does not redistribute with annealing (to 800°C) in any of these materials for fluences up to at least 10^{15} cm^{-2}. We have measured random and channeled range parameters for Be and Si implanted into these materials for energies from 100 to 700 keV. Channeled ranges are approximately the same for all four materials (GaP, GaAs, InP, and InSb), generally varying between 1 and 5 µm for Be, and between 1.5 and 6 µm for Si, in this energy range.

5. ACKNOWLEDGMENTS

This work was largely supported by the U.S. Army Research Office

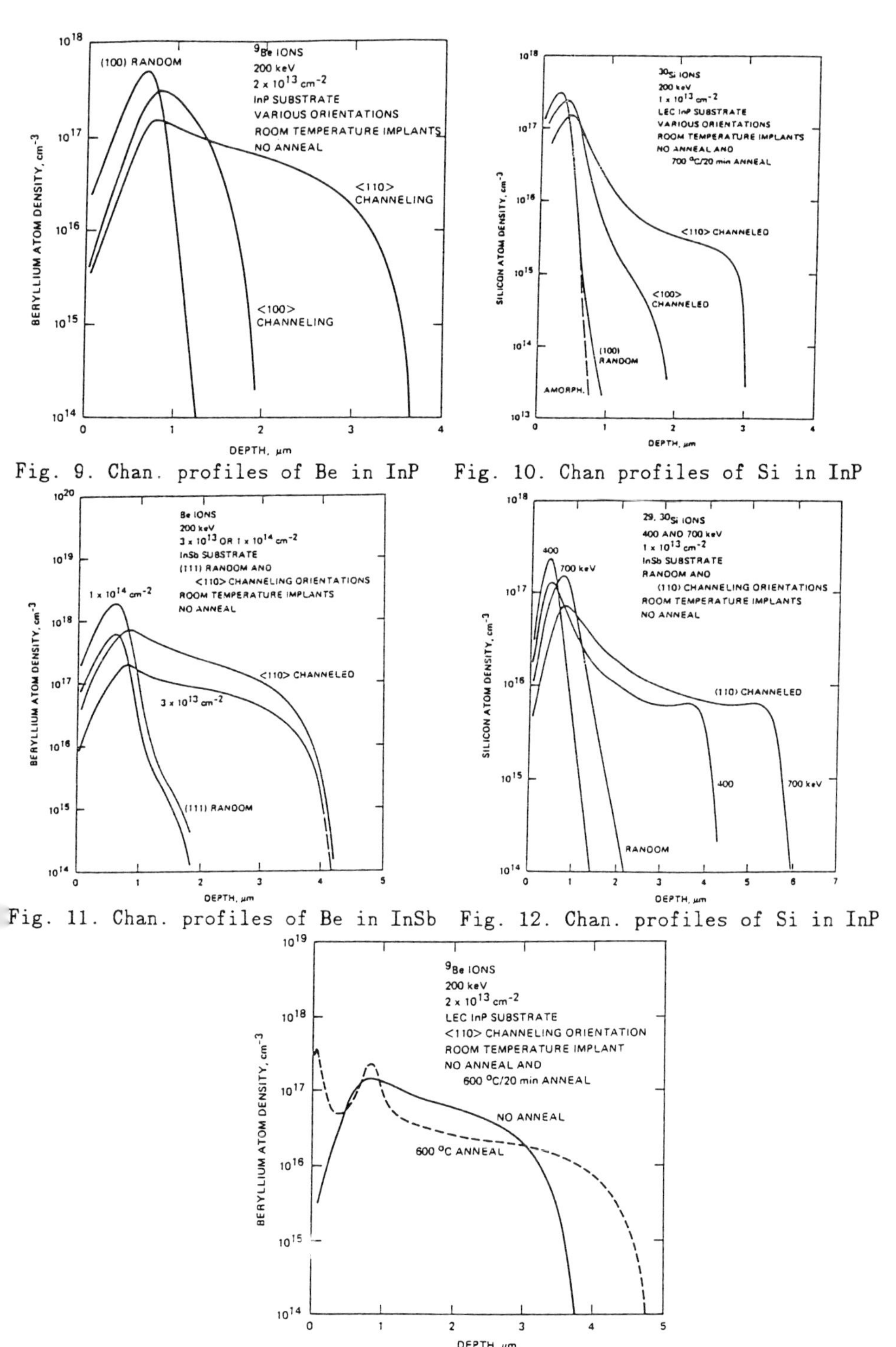

Fig. 9. Chan. profiles of Be in InP

Fig. 10. Chan profiles of Si in InP

Fig. 11. Chan. profiles of Be in InSb

Fig. 12. Chan. profiles of Si in InP

Fig. 13. Annealed channeled profile of Be in InP

Depth profiles and redistribution during annealing of 300-keV hydrogen (protons) implanted into an AlAs/GaAs superlattice

R.G. Wilson, J.M. Zavada,* S.W. Novak,** and S.P. Smith**

Hughes Research Laboratories, Malibu, CA 90265
*USARDSG (UK), London NW1 5TH, UK
**Charles Evans and Associates, Redwood City, CA 94063

ABSTRACT: AlAs/GaAs superlattices are important for optoelectronics and electronics applications. We implanted 300-keV hydrogen into an AlAs/GaAs superlattice, and annealed pieces of the superlattice at temperatures of 300, 500, 600, and 700°C. We then depth profiled the H, together with the Al, in these samples, using SIMS, and found the H to have redistributed and accumulated at the superlattice/substrate interface, and not to have penetrated into the GaAs substrate. Evidence of superlattice layer mixing is seen in high resolution Al profiles for 500°C annealing.

1. INTRODUCTION

The synthesis of quality III-V semiconductor superlattices has initiated the development of a new class of materials with important applications in optoelectronics. Many of these applications require additional material processing, including ion implantation and annealing, to produce necessary electrical isolation, optical index definition/modification, or layer mixing. Hydrogen implantation of GaAs multilayered structures has been a valuable processing technique for such purposes and has yielded optical waveguides, coupled laser arrays, millimeter wave mixers, and photodiodes, as described by Zavada et al. (1986). Hydrogen atoms have been shown to be active entities in semiconductors and able to produce optical effects, by Liou et al. (1986), and electronic effects, by Cheviallier et al. (1985). Because these effects may interfere with device performance, it is important to understand the behavior of H in potential device structures. Here we describe the distribution of ^{1}H atoms implanted into an AlAs/GaAs superlattice. The depth distributions of both implanted ^{1}H and Al substrate atoms were measured using secondary ion mass spectrometry (SIMS). The changes in these distributions caused by furnace annealing were also investigated and evidence of superlattice layer mixing is presented.

2. EXPERIMENTAL TECHNIQUES

The superlattice used in this study consisted of alternating layers of AlAs and GaAs grown on an undoped, semi-insulating GaAs substrate as described by Laidig et al. (1984). The total thickness of the superlattice was ~ 6 μm with the individual layers being 4 nm of AlAs and 7.3 nm of GaAs, for an average Al concentration in the superlattice of 0.35. The superlattice was implanted at room temperature with $1x10^{16}$ cm^{-2} 300 keV ^{1}H ions at an angle of 8° from the <100> direction. The sample was subsequently cleaved into sections that were annealed in flowing nitrogen for 20 min at temperatures from 300 to 700°C. Both ^{1}H and Al were profiled in the annealed and as-implanted pieces, using Cs SIMS for the H, and O SIMS for the Al. The background-subtracted detection sensitivity for ^{1}H in GaAs was approximately 1×10^{17} cm^{-3}, and the depths were measured using surface profilometry of the SIMS craters, with an error of about ±7%.

3. EXPERIMENTAL RESULTS

The ^{1}H depth distributions measured using SIMS are shown in Figure 1, together with an Al profile for reference. On this depth scale the Al layers cannot be resolved and the density appears constant to a depth of 6 μm where it drops sharply at the superlattice/substrate interface. The as-implanted ^{1}H depth profile is similar to that obtained for ^{1}H implanted into bulk GaAs described by Wilson et al. (1985), as shown in Figure 2. The peak of the distribution at R_m is deeper than that measured in the bulk material, as expected, because the presence of Al in the superlattice lowers the average atomic number of the target. The SIMS profile for the sample annealed at 300°C shows that the H has begun to redistribute, with a slight movement toward the surface, but mostly deeper into the superlattice. However, a significant fraction of the H has accumulated at the superlattice/substrate interface. At 300°C, deep diffusion of H into the bulk at densities below 10^{17} cm^{-3} is observed, as was measured for annealing of H implants into bulk GaAs [Wilson, et al. (1985)]. The density in this diffused profile drops below the detection limit for ^{1}H of 3×10^{16} cm^{-3} for annealing temperatures of 500°C and higher. This development is continued in the profile for the sample annealed at 500°C. Movement of the H toward the surface is more pronounced, and the H has nearly reached the surface, as it did for bulk GaAs. The ^{1}H density at R_m has decreased to less than 10% of its as-implanted value, and has increased to more than 10^{19} cm^{-3} at the interface, and there exceeds the H density at R_m. For annealing at 600°C, some H has reached the surface and escaped from the crystal. The H density at R_m is about 1% of the as-implanted amount, in general agreement with the behavior in bulk material. The H density at the interface is still ~10^{19} cm^{-3}, showing little change from the 500°C anneal. At this temperature, most of the residual H lies in the interface region. This trend continues for annealing at 700°C, with the remaining H concentrated primarily at the surface. There is also a near-surface component in this distribution, indicating some H accumulation at the surface. The H density at R_m is now ~0.1% of the as-implanted density. At the interface, the ^{1}H density shows some reduction, but is still ~10^{19} cm^{-3}.

Figure 3 shows the results for near-surface SIMS profiling of Al done with greater depth resolution (~0.6 nm/s using 2.75-keV O). Measurements for the unannealed superlattice reveal an oscillatory pattern in the Al density with a period of 11.5 nm, which was also measured for the layers using x-ray diffraction. As shown in Figure 3, these results are unchanged by the H implantation or by subsequent annealing at 300°C. Only after annealing at 500°C, is there a significant change in the Al depth distribution. The oscillations in the Al density begin to deteriorate at about 25 nm and are gone by about 60 nm, indicating layer mixing for greater depth. Referring to Figure 1, this layer mixing occurs at the same annealing tempertaure at which the H redistribution approaches the surface, having decorated the implant damage associated with the original H implant. The redistributed H profile for the 500°C anneal corresponds to the calculated depth

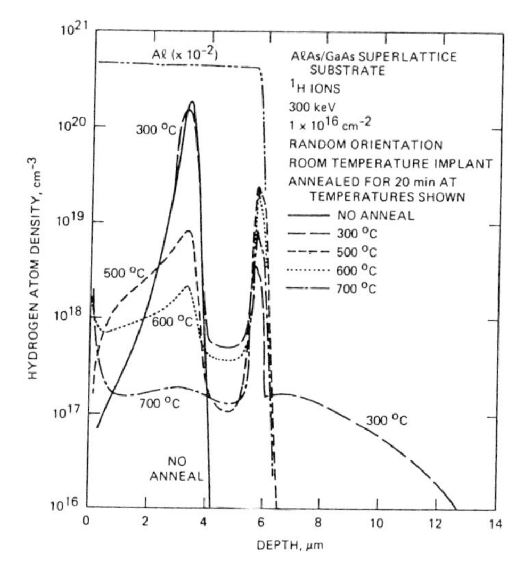

Fig. 1. Depth profiles of 300 keV ^{1}H implanted into an AlAs/GaAs superlattice, then annealed

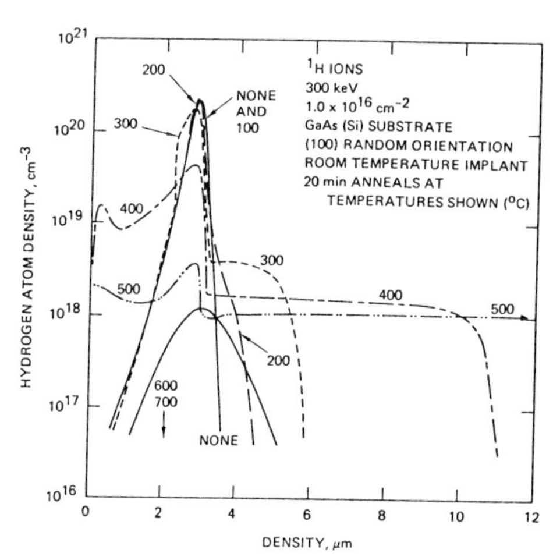

Fig. 2. Depth profiles of 300 keV H implanted into bulk GaAs, then annealed as indicated

distribution of energy deposited into damage for the 300-keV, the diffusing H atoms having become attached in a one-to-one relationship to the dangling bonds associated with the damage. For higher anneal temperatures, 600 and 700°C, this damage is observed to anneal, shown by the decreases in the H atom density. In the 500°C annealed H profile shown in Figure 1, the H density is low in the first few tens of nm, as is the calculated depth distribution. This is the region in which mixing is not observed in Figure 3, the region of slight damage. Layer mixing caused by ion implantation is known to occur, but usually involves more massive ions and higher temperatures, as shown by Schwarz, et al. (1987).

4. DISCUSSION

While the range and general movement of H atoms are nearly the same in bulk GaAs and in this AlAs/GaAs superlattice, the redistribution seems to occur less rapidly in the superlattice. However, the presence of the superlattice/substrate interface had a major effect on the redistribution of the H deeper into the structure. H atoms accumulate at the interface and remain there at 10^{19} cm^{-3} density even after annealing at 700°C. This accumulation is probably associated with dislocations (strain) at the superlattice/substrate boundary. as described by TEM measurements of Rajan et al. (1987) for an InGaAs/GaAs superlattice structure. Rajan et al. also demonstrate that "the strained-layer interface can be an effective barrier to dislocation propagation." These parallel observations support the concept that H atoms redistribute to regions of defects caused by implantation or in early MBE layer growth, and that H atom propagation is promoted by or impeded together with the propagation of dislocations or other defects.

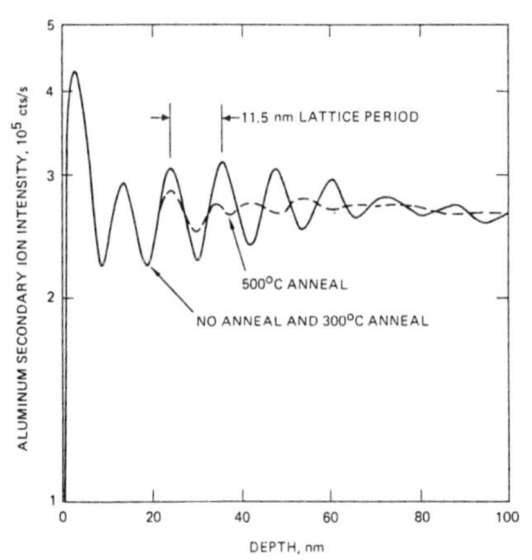

Fig. 3. Depth profiles of Al in a H-implanted and annealed AlAs/GaAs superlattice

In an optical study of a related superlattice by Zavada et al. (1987), it was necessary to introduce a graded optical index in the interfacial region to obtain a best-fit solution to an infrared reflectance spectrum. These observations can also be attributed to dislocations (strain) at the superlattice/substrate interface. Furthermore, because the H density can be 10^{19} cm^{-3} at the interface, it is possible that unwanted optoelectronic effects may occur in devices that involve such processing. Indications that H redistribution may be associated with layer mixing of the superlattice is independently important. H implantation causes relatively minor damage to the AlAs/GaAs superlattice crystal. If the observed layer mixing occurs over a wide range of superlattice depth, this method might form an attractive processing technique to alter material composition in selected regions of a superlattice while maintaining crystal perfection throughout the remainder of the wafer.

5. ACKNOWLEDGMENTS

The authors thank W.D. Laidig for growing the superlattice used in this study. Portions of this work were supported by the U.S Army Research Office.

6. REFERENCES

Chevallier J, Dautremont-Smith W C, Tu C W, and Pearton S J 1985 Appl Phys Lett 47 108

Laidig W D, Blanks D K, and Schetzina J F 1984 J Appl Phys 56 1791

Liou L L, Spitzer W G, Zavada J M, and Jenkinson, H A 1986 J Appl Phys 59 1936

Rajan K, Devine R, Moore W T, and Maigue P 1987 J Appl Phys 62 1713

Schwarz S A, Venkatesan T, Hwang D M, Yoon H W, Bhat R, and Arakawa Y 1987 Appl Phys Lett 50 281 and 1823

Wilson R G, Betts D A, Sadana D K, Zavada J M, and Hunsperger R G 1985 J Appl Phys 57 5006

Zavada J M, Jenkinson H A, and Larson D C 1986 Proc Soc Photo-Opt Instrum Eng 623 144

Zavada J M, Hubler G K, Jenkinson H A, and Laidig W D 1987 Mat Res Soc Proc 90 257

Effects of deposition thickness and substrate type on the properties of GaAs-on-Si

A. S. Jordan, S. J. Pearton*, C. R. Abernathy*, R. Caruso* and S. M. Vernon+*
*AT&T Bell Laboratories, Murray Hill, N.J. 07974, USA
+Spire Corporation, Patriots Park, Bedford, MA 01730, USA

ABSTRACT The evolution of the electrical and structural properties of GaAs layers deposited directly on Si or SOI substrates as a function of GaAs thickness (0.01-4 μm) is presented. At a GaAs thickness of 4 μm the near surface region exhibits a donor activation efficiency for low dose ^{29}Si implants (3×10^{12} cm^{-2} dose at an energy of 60 keV) of ~70%, identical to that for similar implants into bulk GaAs. The crystalline quality of this region is also comparable to bulk material. The thermal stability of the heterointerface to rapid annealing at 900°C provides evidence for defect-modulated Si diffusion, and is correlated with the microscopic structure of the interface.

INTRODUCTION

The heteroepitaxial growth of GaAs on Si substrates is potentially useful for the eventual integration of optical devices with electrical circuits, and in the more immediate future for the replacement of brittle GaAs substrates with larger diameter, mechanically stronger Si wafers (Windhorn et. al. 1984, Kroemer 1986). The growth of a polar semiconductor (GaAs) on a non-polar substrate (Si) has two major impediments-the difference in lattice constant between the two systems, and the dissimilar thermal expansion coefficients (Duncan et. al. 1986, Shastry and Zemon, 1986). The former manifests itself in the presence of ~10^{12} cm^{-2} dislocations at the heterointerface. Although only a small fraction of these thread to the surface of the GaAs, the resultant dislocation density in layers deposited directly on Si (~10^8 cm^{-2}) is four orders of magnitude larger than in typical homoepitaxial or bulk GaAs. The difference in thermal expansion coefficients results in significant warpage of the wafer which must be alleviated if large scale circuits are to be fabricated.

In this paper we report the electrical, and structural properties of GaAs layers grown by Metal Organic Chemical Vapor Deposition (MOCVD) on misoriented Si or Si-on-insulator (SOI) substrates. In particular, the evolution of these properties as a function of GaAs layer thickness (0.01-4 μm) will be given.

EXPERIMENTAL

We investigated the growth of GaAs on both Si and SOI wafers. The advantage to growth on SOI is that the buried oxide layer provides electrical isolation of the GaAs from the conducting Si substrate and offers the potential for three dimensional integration. Silicon substrates with (100) (n-type, 0.3 Ω cm) or 2° off (100) towards (110) (n-type 0.01 Ω cm and p-type 0.02 Ω cm) orientations were employed. This enabled comparison of the GaAs quality for layers grown on misoriented Si with that deposited on exact orientation substrates. The SOI structures were formed by a buried, high dose (~10^{18} cm^{-2}) 0^+ implantation into 1 Ω cm n-type Si (100). After implantation the top ~0.2 μm of Si was recrystallized by a high temperature annealing step that leads to a defect density of ~10^8 cm^{-2} and leaves a buried 0.4 μm thick SiO_2 film. The whole structure was then overgrown with ~0.5μm of Si by conventional chemical vapor deposition. All of the GaAs layers were deposited in a Spire barrel reactor. A standard two stage deposition method was used with a nucleation stage at ~425°C and growth of the GaAs layers at 675°C at a rate of ~4 μm hr^{-1}. (Vernon et. al. 1986, Pearton et. al. 1987). Layer thicknesses between ~0.01 and 4 μm were deposited on both the Si and SOI substrates.

RESULTS AND DISCUSSION

Scanning electron micrographs of the surface morphologies of representative areas of the 4 μm thick GaAs films deposited on the different types of substrate are shown in Figure 1. They are clearly inferior to the specular surfaces generally obtained with homoepitaxial GaAs, and the films grown on SOI exhibit large craters apparently due to particulates that adhered to the wafers during the 0 implantation and which are avoidable by a suitable cleaning process. We do not observe any signficant difference in the surface quality between layers grown on exactly oriented or misoriented Si substrates, apart from some indication of antiphase disorder on exact (100) Si. There is an improvement in surface morphology with increasing layer thickness, as the GaAs islands where growth initiated merge to form a continuous layer.

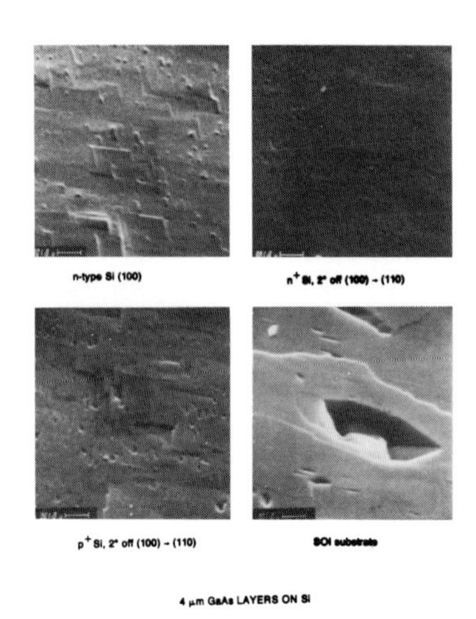

Figure 1. Scanning electron micrographs of the surface of 4 μm thick GaAs layers deposited on Si or SOI substrates.

It is now well established that GaAs layers grown directly on Si contain $10^7 - 10^9$ cm^{-2} threading dislocations. In all of our samples regardless of substrate type, we observe similar defect densities and structures, with all having densities at about 10^8 cm^{-2}. We also find that annealing near 900°C can eliminate most of the stacking faults, and cause the other defects to align themselves along crystallographic directions, but that the total defect density remains invariant. The high defect density is correlated with substantial generation-recombination leakage currents in devices fabricated on the GaAs layers on Si. Thus, the reduction of threading dislocation density in the GaAs is the major problem that must be solved before the use of heteroepitaxial wafers in place of bulk material becomes more widespread.

All of the samples we have measured show a tetragonal distortion of the GaAs lattice, indicating a high level of misfit with the Si substrate. The x-ray peak widths measured on a single crystal diffractometer show a near monotonic decrease with increasing GaAs layer thickness. For example the (400) $CuK\beta_1$ reflection peak width was ~800 arc-sec for 0.5 μm thick films on Si or SOI, and was typically in the 400 arc-sec range for 4 μm thick GaAs layers. The comparable value for a bulk GaAs sample is 125 arc-sec. The average lattice constant measured perpendicular to the layer surface as a function of GaAs film thickness is shown in Table 1, along with the corresponding tensile stress in the film necessary to create the observed lattice distortion. The measured radii of curvature for 2"ϕ heteroepitaxial wafers were of the order of 4m for 4 μ m thick GaAs layers, and 20 m for 0.5 μm thick films. In thick films (> 2 μm) we also observed significant cracking in the GaAs. Problems in device processing as a result of bowing will necessitate some novel steps such as deposition of GaAs or another substance onto the backside of the heteroepitaxial wafer, or the use of low temperature growth methods.

Table 1. Thickness dependence of perpendicular lattice constant and tensile stress in GaAs/Si (100) films.

GaAs Thickness(μm)	$a_\perp$ (Å)	Tensile Stress (dyne.cm^{-2})
0.2	5.647	1.73×10^9
0.5	5.647	1.73×10^9
1.5	5.648	1.50×10^9
4	5.643	2.70×10^9

The crystalline quality of the GaAs films as measured by Rutherford backscattering (RBS) also improved significantly with increasing layer thickness, consistent with the x-ray results. The backscattering yield fell from 100% for the highly defective 0.01 μm thick films to ~4% in the near surface of the 4 μm layers. Values measured for GaAs grown on Si or SOI were virtually the same. As previously reported (Vernon et. al. 1987, Chand et. al. 1986, Lee et. al. 1987), annealing at 900°C caused significant reductions in the backscattering yield near the heterointerface, but little improvement near the surface of GaAs. As shown in Fig. 2, Si exhibits enhanced diffusion near the heterointerface due to the high degree of crystalline disorder. Clearly, diffusion is modulated by the presence of line and point defects. On the other hand close to the surface Si redistribution is minuscule because the interface progressively improves and the thickness of the film itself (2 μm) provides a diffusion barrier. Another consequence of Si outdiffusion is that autodoping in GaAs during growth produces typical background donor concentrations of 1-5$\times10^{16}$ cm^{-3} compared to 10^{14} – 10^{15} cm^{-3} in homoepitxial GaAs grown under identical conditions. The total deep level concentration is similar in both types of material (~10^{13} cm^{-3} in 4 μm thick films), but there is a substantially higher density in thin GaAs layers on Si. For example the average total deep level concentration observed in 0.5 μm thick GaAs layers was ~5×10^{14} cm^{-2}.

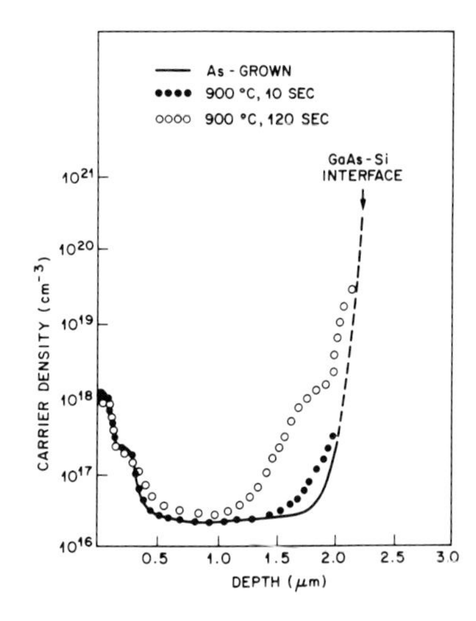

Figure 2. Carrier profiles obtained by electrochemical capacitance-voltage measurements in $n^+/n/$ undoped GaAs layers on Si, as a function of annealing time at 900°C.

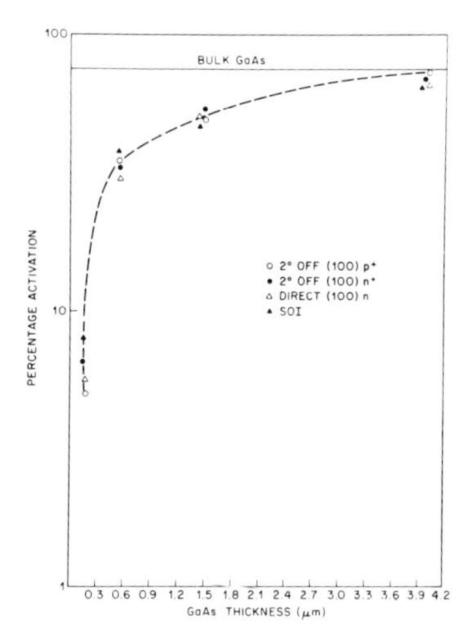

Figure 3. Net activation percentage of 60 keV ^{29}Si ions implanted into GaAs-on-Si at a dose of 3×10^{12} cm^{-2}, after annealing at 900°C for 10 sec.

The production of large scale GaAs-on-Si based circuits utilizing a planar technology requires the formation of selectively doped regions on the wafer for use as channel and contact layers. For this application we have studied the GaAs film thickness dependence of low dose Si implant activation. Figure 3 shows the net donor activation percentage in GaAs layers of different thicknesses implanted with ^{29}Si ions (dose of 3×10^{12} cm^{-2} at 60 keV) energy after rapid capless annealing at 900°C for 10 sec. For the thinner layers, the poor activation is most likely the result of the presence of a relatively high level of crystalline disorder which compensates the donor activity. As the film thickness increases the crystalline quality of the near-surface region improves significantly, as does the implant activation efficiency until, for 4 μm films, the activation ~70% is similar to that observed in bulk GaAs.

In conclusion, we have demonstrated a significant improvement in the crystalline quality of GaAs films on Si or SOI substrates with increasing film thickness. The major remaining problem with currently available GaAs-on-Si is the high defect density in the GaAs. This will require the development of effective defect reduction and filtering techniques.

ACKNOWLEDGEMENTS The authors acknowledge the support of their colleagues K. T. Short, M. Stavola, J. M. Brown, J. M. Gibson, S. N. G. Chu, W. S. Hobson, D. C. Jacobson and J. M. Poate at AT&T Bell Labs, and V. E. Haven, S. N. Bunker and R. G. Wolfson at Spire Corporation.

References

Chand N, People R, Baiocchi F. A., Wecht K. W., and Cho A. Y. 1986 Appl. Phys. Lett. *49* 815.

Duncan W. M., Lee J. W., Matyi R. J. and Lui H. Y. 1986 J. Appl. Phys. *59* 2161.

Kroemer H. B., 1986, Mat. Res. Soc. Symp. *67* 3.

Lee J. W., Shichijo H., Tsai H. L. and Matyi R. J., 1987, Appl. Phys. Lett. *50* 31.

Pearton S. J., Vernon S. M., Abernathy C. R., Short K. T., Caruso R., Stavola M., Gibson J. M., Haven V. E., White A. E. and Jacobson D. C. 1987a J. Appl. Phys. *62* 862.

Shastry S. K. and Zemon S. 1986 Appl. Phys. Lett. *49* 467.

Vernon S. M., Haven V. E., Tobin S. P. and Wolfson R. G. 1986 J. Cryst. Growth *77* 530.

Vernon S. M., Pearton S. J., Gibson J. M., Short K. T. and Haven V. E. 1987 Appl. Phys. Lett. *50* 1161.

Windhorn T. H., Metze G. M., Tsaur B. Y. and Fan J. C. C. 1984, Appl. Phys. Lett. *45* 309.

Paper presented at Int. Symp. GaAs and Related Compounds, Heraklion, Greece, 1987

Anneal behaviour of MeV Si implants in undoped InP

W Häussler, J Müller, R Trommer

Siemens Research Laboratories, Otto-Hahn-Ring 6, 8000 München 83, F.R.G.

ABSTRACT: The annealing behavior of low-dose Si implants (E= 2 MeV) into undoped InP was studied using secondary ion mass spectrometry (SIMS) and capacitance-voltage (CV) carrier profiling. Rapid thermal annealing was employed with temperatures up to 850 °C. It is shown that the Si profile is rather insensitive to a variation in annealing parameters. However, for short-time anneals, there is a pronounced n-type surface peak. The surface peak disappears for longer anneals and/or higher implantation doses and is tentatively attributed to implantation-induced defects.

1. INTRODUCTION

Ion implantation at MeV energies is increasingly gaining interest as a means to introduce a precisely defined, locally confined doping level deep into the substrate. Thompson and Dietrich (1985) have extensively investigated MeV Si implants into GaAs and have recently applied this technology to the fabrication of a mixer diode (1987). High-energy implantation into InP is important for devices needed for optoelectronic integration, such as heterojunction bipolar or field-effect transistors.

In this paper results of low-dose Si implants at 2 MeV into undoped InP are reported. Several implantation runs were performed using different implanters with doses below or slightly above 1E+13 cm^{-2}. The parameters investigated include implantation dose, annealing temperature and duration of the anneal.

2. EXPERIMENTAL DETAILS

Si is implanted into InP using a Van-de-Graaff- or a tandem- accelerator. Atomic implantation profiles were measured using SIMS. Implanted layers were non-intentionally doped LEC wafers having a background carrier concentration of about 5E+15 cm^{-3}. It has been shown by Gauneau et al (1986) that the two dominant species in undoped LEC-InP are S and Si, being responsible for n-type conductivity. In order to be able to measure the implanted Si profile without interference from the residual Si impurity in the substrate the mass-30 Si isotope (representing only 3.1 % of naturally occuring Si) is implanted. By monitoring the P-Si-molecule, a SIMS detection limit of about 1E+15 cm^{-3} is obtained. SIMS data were calibrated with a low-energy Si implant of known dose.
Annealing is performed in a rapid thermal annealing system employing a graphite strip. No protecting cap is used during annealing, and evaporation of P is avoided in a PH_3/H_2 atmosphere. This procedure yields mirror-like surfaces even for a 6-minute anneal at 850 °C.

Carrier concentration profiles are obtained using a commercial electrochemical CV-profiler. Etched craters are essentially flat, but always exhibit striations. The measured profiles are very reproducible: measured depths for the peak of the implanted profile scatter less than for the SIMS measurements.

3. RESULTS

Fig.1 shows the result of a Si implant at a dose of 2E+12 cm^{-2}, which is annealed at 850 °C for 6 minutes. Comparison of the atomic and the carrier profiles indicates an electrical activation of about 70 %. Such activation values are obtained for all dose investigated.

SIMS profiles were measured before and after a 6-minute anneal at 850 °C. No redistribution of the implanted Si atoms is observed.

The effect of varying the duration of the anneal on the carrier profiles is shown in Fig. 2a for an anneal temperature of 800 °C and in Fig. 2b for 850 °C. It is apparent that short-time anneals result in an n-type surface peak which decreases in size for longer anneals. However, this peak completely disappears only for long anneals at higher temperatures. With a 20 second anneal at 800 °C, the surface peak dominates the carrier profile completely.

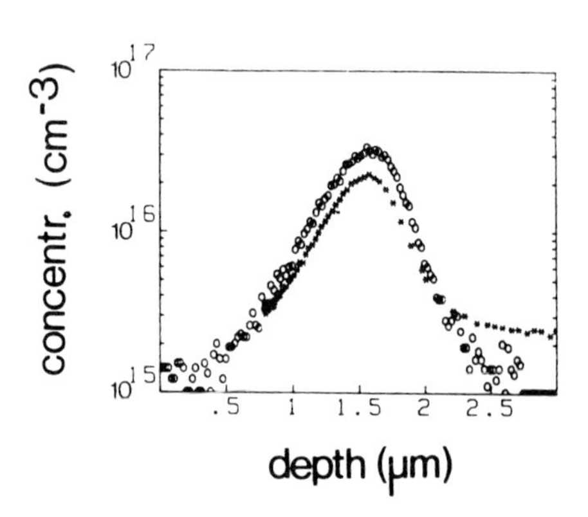

Fig. 1. SIMS-profile (O) and electrical profile (*) of a 2 MeV Si implant; implantation dose is 2E+12 cm^{-2}; annealed at 850 °C for 6 minutes.

It is evident that the Si doping peak is activated very early in the annealing process and that most of the supplied thermal energy is required to remove the surface peak.

An identical set of annealing experiments was carried out for an implantation dose of 4E+12 cm^{-2}. Again the carrier profiles exhibit a surface-peak. However, longer anneal durations lead to a faster reduction of the surface peak, it being hardly detectable for a 6-minute anneal at 800 °C.

For an implantation dose of 2E+13 cm^{-2} a surface peak could only be detected when the annealing temperature was reduced to 700 °C.

The same effects were seen for implantations into undoped substrates from a different supplier.
In addition, a Si implant using a conventional ion implanter (E= 680 keV) showed the same behavior during annealing with the surface peak being compressed in shape and hardly separable from the doping peak. For low-dose Si implants of lower energy, both peaks can thus be expected not to be resolvable. High electrical activation may then erroneously be measured.

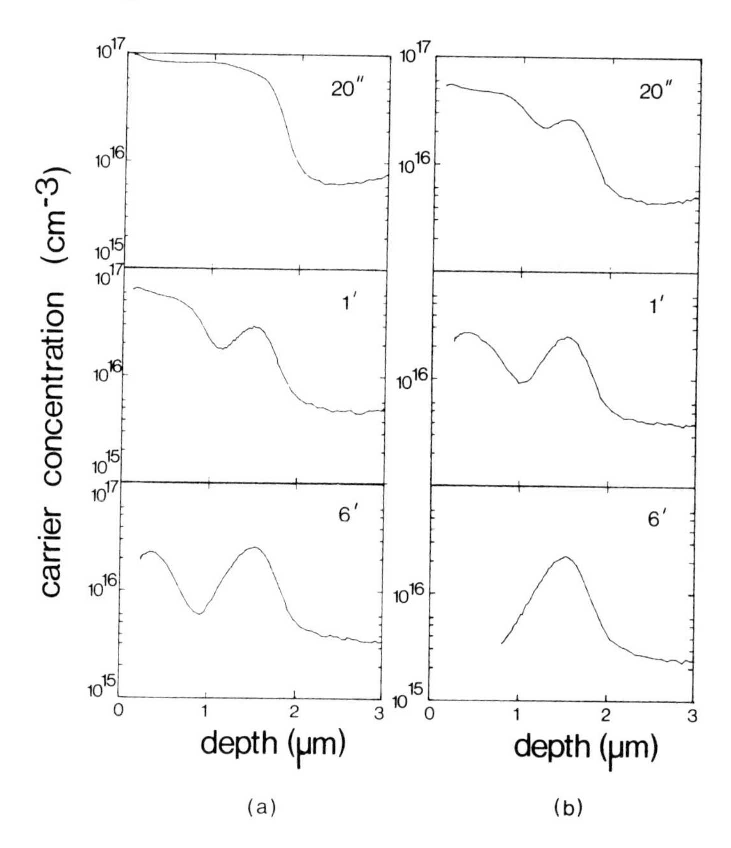

Fig. 2. Carrier profiles of a 2 MeV Si implant into undoped InP. Implantation dose is 2E+12 cm^{-2}. Profiles for different durations of the anneal are shown for an anneal temperature of 800 °C (a) and 850 °C (b).

4. DISCUSSION

The salient feature of the results reported above is the appearance of an n-type surface peak and its dependence on implantation dose and annealing parameters. For higher implantation doses the surface peak becomes smaller for the same annealing conditions. Implants of high dose can thus be annealed at lower temperatures and/or shorter intervals. The experiments clearly show that rapid thermal annealing is not suitable for low-dose Si implants into undoped InP.

The data presented in Fig. 2 raise the following questions: What causes the surface peak, and how is it reduced during annealing ?
The surface peak might have one of two origins: (a) electrically active defects produced by the penetrating ion, (b) a surface pile-up of donors diffusing out of the substrate.
When a high-energy ion penetrates a target, it displaces hundreds or even thousands of target atoms. Defects recombine during annealing. A primitive approach to modeling annealing would be to set the defect recombination rate proportional to the square of the number of implanted ions and the defect number directly proportional to the implantation dose. Then, annealing of

defects should become more rapid when the implantation dose is increased, even though the initial implantation damage has been larger. This is in accordance with our observations and supports the hypothesis that the surface peak is directly caused by the implantation process.
Redistribution of residual donor impurities and accumulation at the surface may also be invoked to explain the surface peak. However, no such peak was found when annealing unimplanted samples. On the contrary, we observe a reduction in donor concentration at the surface for unimplanted InP similar to that reported by Guha et al (1977). Implantation-induced redistribution of residual **acceptor** impurities has been reported for Fe-doped InP by several authors. Duhamel et al (1983) reported a dose-dependent surface pile-up of residual Zn in Si-implanted InP:Fe. Gauneau et al (1985) reported a surface pile-up of Fe, Cr and Zn during implantation-annealing. It is not known whether similar redistribution effects can be expected in undoped InP and it is doubtful whether **donors** redistribute at all. In addition, the dose of the surface peak is much higher than the measured decrease in background donor concentration. This leads to the conclusion that the surface peak is not caused by impurity redistribution effects, thus favoring explanation (a).

5. CONCLUSION

Annealing results of low-dose, high-energy Si implants into undoped InP were reported. Carrier profiling revealed an implantation-dose-dependent n-type surface peak. This peak can be removed by suitably chosen annealing parameters. Care should therefore be exercised when annealing Si implants of different dose. The results indicate that short-time anneals are not appropriate for Si implants of low dose.

ACKNOWLEDGEMENTS

We would like to thank Mrs. Frenzel of Gemetec, Munich, for SIMS measurements.

This work was supported by the German BMFT under contract TK 0271/1.

REFERENCES

Duhamel N, Rao E V K and Gauneau M 1983 *J. Cryst. Growth* **64** 186
Gauneau M, Chaplain R and Rupert A 1985 *J. Appl. Phys.* **57** 1029
Gauneau M, LeCorre A and Chaplain R 1986 *Proc. E-MRS* 203
Guha S and Hasegawa F 1977 *Solid-St. Electron.* **20** 27
Thompson P E, Dietrich H B and Ingram D C 1985 *Nucl. Instr. & Meth.* **B6** 287
Thompson P E and Dietrich H B 1987 *El. Lett.* **23** 725

Doping of $Ga_{0.47}In_{0.53}As$/InP during OMVPE-growth for heterobipolar devices

E Woelk and H Beneking

Institut für Halbleitertechnik, RWTH Aachen, Templergraben 55, 5100 Aachen, Germany

ABSTRACT: InP and GaInAs have been doped using Si_2H_6 and DMCd as precursors in an atmospheric pressure OMVPE-system. Si-doping of InP and Cd-doping of GaInAs have been studied in order to fabricate heterojunctions for bipolar devices. Results of these studies, such as the carrier concentration vs dopant precursor flow, temperature dependence of doping and diffusion constant estimates are given. To show the usefulness of both dopants in bipolar devices a p^+N GaInAs/InP heterodiode is presented.

1. INTRODUCTION

In the quest for dopants and dopant sources in the organometallic vapor phase epitaxy (OMVPE) of InP, Si was soon discovered to be a promising candidate. Monosilane as precursor was conveniently available in suitable purity. Si proved to be an n-type dopant allowing doping levels up to 10^{19} cm^{-3} in InP without saturation effects (Clawson et al. 1987) nor severe compensation through amphoteric incorporation. The observed temperature dependence of InP-doping by means of monosilane is usually smaller (Hsu et al. 1986) than that observed with GaAs-doping (Kuech et al. 1984). In the latter case there were attempts to eliminate this dependence by using disilane (Si_2H_6) which is less thermally stable than monosilane (Bowery and Purnell 1971) but most studies only revealed a slight reduction (e.g. Shimazu et al. 1987). This indicates that the limiting factors of dopant incorporation are probably surface reactions (Field and Ghandhi 1984) rather than gas phase reactions and bulk properties (Giling and de Moor 1983) or gas transport (Stringfellow 1985). On the doping behavior of Si in GaInAs only ion implantation data are available which report this element to have a very small diffusion coefficient and to allow doping levels up to 10^{19} cm^{-3} (Splettstößer et al. 1987).

Cd is well known as an acceptor in GaAs (Manasevit and Thorsen 1972) and InP (Nelson and Westbrook 1984) and as a precursor dimethylcadmium (DMCd) which is widely used in II-VI OMVPE is available. Nelson and Westbrook (1984) report Cd to have a somewhat smaller diffusion coefficient in InP than alternate dopants such as magnesium or zinc recommending it for use in bipolar devices where abrupt doping profiles are desired.

It is possible to describe the OMVPE doping behavior of a semiconductor / dopant couple by a useful empirical equation

$$N \propto \Phi_{dop}^{\alpha} \cdot \Phi_{III}^{\beta} \cdot \Phi_{V}^{\gamma} \cdot \exp(-E_a/kT) \qquad (1)$$

where Φ_{dop}, Φ_{III} and Φ_V are the precursor flows of the dopant, group-III-element and group-V-element respectively. α, β, γ are some empirical exponents, and E_a is an activation energy. For the discussion it is sometimes useful to define the doping efficiency k of a certain precursor by relating gas phase composition and solid-state composition

$$k = (N/x_{SLS}) \;/\; (x_{dop}/x_{CE}) \qquad (2)$$

where N is the carrier concentration, x_{SLS} is the concentration of sublattice sites on which the doping atom is incorporated, x_{dop} is the concentration of dopant precursor molecules in the gas phase and x_{CE} is the concentration of the precursor molecules of the competing element for sites on the particular sublattice.

2. EXPERIMENTAL

All films reported here were grown under atmospheric pressure at a set of standard parameters indicated in Figure 1. TMGa, TMIn and DMCd were held at -13°C, 19.5°C and 19.5°C, respectively. To calculate the flows of pure gases the vapor pressure of TMGa was assumed to be 39.7 hPa, 1.46 hPa that of TMIn and 28.2 hPa that of DMCd. The used disilane source was a mixture of 100 ppm Si_2H_6 in hydrogen. The reactor used was the one of Roehle and Beneking (1981). Arsine and phosphine have been introduced into the reactor through the susceptor, while TMGa, TMIn, Si_2H_6 and DMCd have been introduced through the alkyls line.

The epitaxial films were electrically characterized by van der Pauw- Hall measurements and C-V profiling. In the case of the ternary material occasional checks of lattice constant were carried out by means of double X-ray diffraction spectroscopy. No dependence of lattice constant on doping was observed.

3. RESULTS AND DISCUSSION: InP:Si

Figure 1 shows the dependence of electron concentration on Si_2H_6-flow in the log-log representation. The slope of the line is the exponent α in eq. (1). In our case of InP-doping using disilane it is α=0.75 ± 0.07. This means that the doping efficiency k decreases if the concentration of disilane is increased. If surface reactions are assumed to be dominant it could be that the sticking coefficients of growth relevant In and Si spezies are influenced by silicon coverage of the surface.

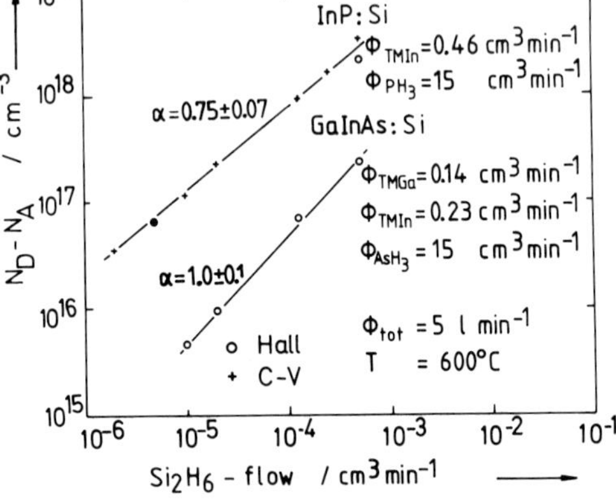

FIGURE 1: Doping of InP and GaInAs using disilane

The exponent β was determined to be -0.73±0.07 so that within the error $\beta = -\alpha$. This means that doubling both, TMIn and Si_2H_6 flow, does not change the doping level. Only the growth rate is doubled than, and $\beta = -\alpha$ implies that the doping efficiency k of disilane is not growth rate dependent in the investigated range from 0.5 to 2 μm/h.

Activation energy was found to be 1.2±0.4 eV (Figure 2). As to be seen in Table I this is a high value compared to GaAs-doping using disilane. It is remarkable that all E_a's for the low vapor pressure element silicon lie in the range of 0.6 to 1.8 eV and no rule can be given how E_a is related to the growth conditions. Interestingly enough, the E_a's are different for InP- and GaAs-doping using monosilane which contradicts clearly to the common assumption that the activation energy is related only to the thermal decomposition of monosilane. Much more this observation favors the model of Field and Ghandhi (1986) who assume the E_a's to be related with surface reactions occuring during growth. They point out that activation energies between 0.6 and 1.5 eV can be explained by adsorption of atomic hydrogen which saturates surface sites needed for doping elements.

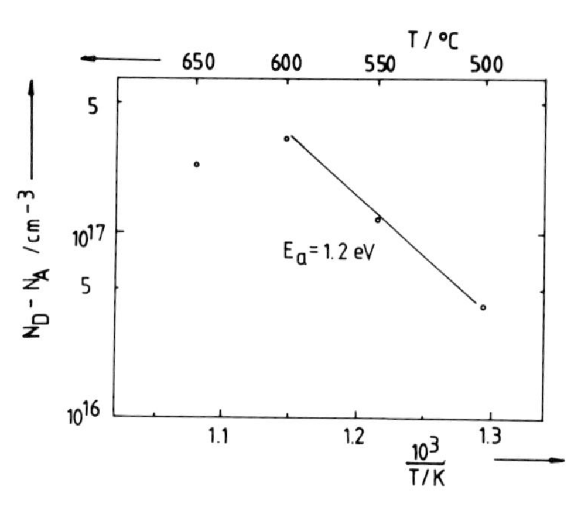

FIGURE 2: Temperature dependence of InP-doping using Si_2H_6

Between 600°C and 650°C the temperature dependence of Si-doping changes. This could either be due to a different incorporation mechanism becoming dominant or due to the fact that above this temperature the incorporation of Si into P-sites starts, creating compensation and a decrease of free carrier concentration.

Diffusion behavior of Si in InP was investigated by a test structure consisting of three ∂-doped layers grown at 600°C. The time used to grow the succeeding layers was taken as the annealing time for the preceding layers. This structure was C-V profiled to yield three Gaussian curves. From their widths the diffusion constant $D = 1.4 \times 10^{-14} cm^2 s^{-1}$ was calculated.

4. RESULTS AND DISCUSSION GaInAs:Si AND GaInAs:Cd

The empirical exponent is α=1±0.1 for GaInAs-doping with disilane (Figure 1). Therefore we can calculate a doping efficiency of $k=9.3 \times 10^{-4}$ which is

	α	β	γ	E_a /eV	reactor-pressure	T_{sub} /°C	precursor compound
InP:Si							
Clawson (1987)	1	-1	/	0.5	76 torr	650	SiH_4
this work	0.75	-0.73	0	1.2	760 torr	600	Si_2H_6
GaAs:Si							
Bass (1979)	1	-0.7	-0.6	1.8	760 torr	710	SiH_4
Shimazu (1987)	1	/	/	0.8	76 torr	630	Si_2H_6
GaInAs:Si							
this work	1	/	/	0.6	760 torr	600	Si_2H_6
GaInAs:Cd							
this work	0.5	/	/	-2.8	760 torr	600	DMCd

TABLE I : Doping behavior (see equation 1)

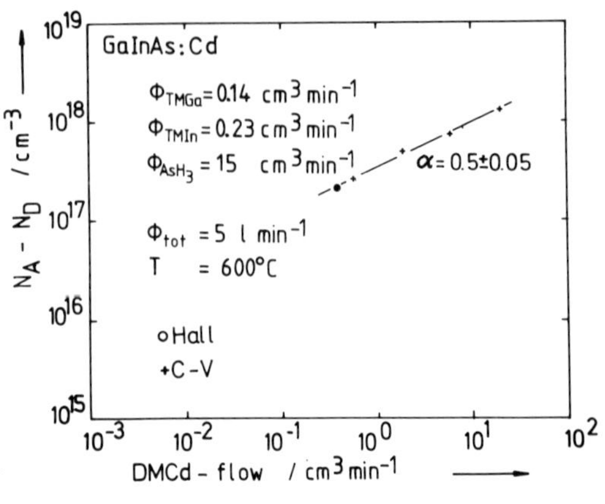

FIGURE 3: Cadmium-doping of GaInAs

in contrast to InP constant over the whole doping range from background of 1-3 x 10^{15} cm^{-3} to 5 x 10^{17} cm^{-3} limited by the gas mixing system. 300 K Hall measurements revealed that the carrier mobilities of the grown layers conicide with the mobility vs carrier concentration curve of Kuphal (1983) indicating low compensation.

Figure 3 shows the dependence of acceptor concentration on DMCd flow. In this case is α=0.5±0.05, a value which is frequently observed. A rough estimation shows that only one of 10^5 to 10^6 offered Cd atoms incorporates and forms an acceptor. In contrast to Si the p-type dopant Cd is a high-vapor-pressure element. This could be the reason for the different temperature dependence. In the case of Cd doping an activation energy of -2.8 eV was observed. This means that an elevation of growth temperature by 100°C lowers the doping level to 1/75.

5. HETEROJUNCTION DIODE

To show the usefulness of the Si/Cd dopant combination in the GaInAs/InP system a p^+N heterodiode was made. The I-V characteristic is shown in Figure 4. The ideality factor is 1.2 indicating a good quality of the junction. C-V measurements of this diode give a built-in voltage of 0.7 V proving that only little Cd has diffused into the InP.

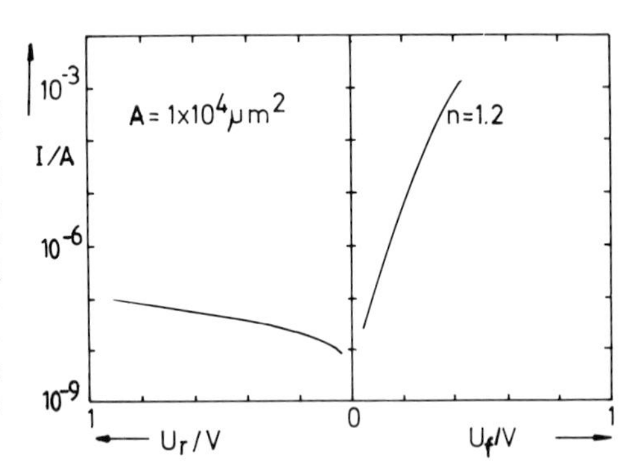

FIGURE 4: Diode characteristic

Bass S J 1979 J. Cryst. Growth 47 613
Bowery M and Purnell J H 1971 Proc.Roy. Soc. Lond. A.321 341
Clawson A R, Vu T T and Elder D I 1987 J. Cryst. Growth 83 211
Field R J and Ghandhi S K 1986 J. Cryst. Growth 74 543
Giling L J and de Moor H H 1983 Proc. 4th European Conf. on CVD, Eindhoven NL 184
Hsu CC, Huan JS, Cohen RM and Stringfellow GB 1986 J. Cryst. Growth 74 535
Kuech T F, Veuhoff E, Meyerson B S 1984 J. Cryst. Growth 68 48
Kuphal E and Fitzsche D 1983 J. Electron. Mat. 12 743
Manasevit H M and Thorsen A C 1972 J. Electrochem. Soc. 119 99
Nelson A W and Westbrook L D 1984 J.Cryst. Growth 68 102
Roehle H and Beneking H 1981 J. Cryst. Growth 55 79
Shimazu M, Kamon K, Kimura K, Mashita M, Mihara M and Ishii M 1987 J. Cryst. Growth 83 327
Splettstößer J, Heesel H, Breuer U, Albrecht W, Schmitz D, Selders J and Beneking H 1987 17th ESSDERC Bologna 1987 47
Stringfellow G B 1986 J. Cryst. Growth 75 91

Simultaneous fabrication of very low resistance ohmic contacts to n-InP and p-InGaAs

R Kaumanns, N Grote, H-G Bach, F Fidorra*

Heinrich-Hertz-Institut für Nachrichtentechnik Berlin GmbH
Einsteinufer 37, D-1000 Berlin 10 (FRG)
*on leave from Philips Kommunikations Industrie AG,
Thurn-und-Taxis-Str. 10, D-8500 Nürnberg (FRG)

ABSTRACT: A process to simultaneously form high-quality ohmic contacts to n-InP and p-InGaAs has been developed. Ti-Au metallization was applied to n-InP following an Ar^+-sputter etching treatment and to p-InGaAs after Zn diffusion from doped spin-on glass. Utilizing a selective-wet etching step for the p type material permits simultaneous fabrication of these contacts with specific contact resistances of well below $1*10^{-6}$ Ωcm^2.

1. INTRODUCTION

Very low resistance ohmic contacts are of essential importance for the electrical performance (high-frequency, power consumption etc.) of optoelectronic devices such as transistors and lasers. From a technological point of view a common contacting system for n- and p-materials is desirable to simplify the design and fabrication of devices. This is specially true for integrated optoelectronic devices on semi-insulating substrates requiring all contacts to be made on the top side. This paper reports on a procedure to simultaneously fabricate nonalloyed ohmic contacts to n-InP and p-InGaAs of very low specific resistance using Ti-Au.

2. Ti-Au CONTACTS TO n-InP

Ti-Au contact films on n-type InP were found to yield ohmic characteristics equivalent or even superior to those of conventionally used Au/Sn or Au/Ge metallisations without the necessity of post-deposition alloying (Dautremont-Smith et al 1984, Bach et al 1986). The crucial step in the contacting process is the bombardment of the InP surface with Ar-ions performed immediately prior to metal deposition. In this work this was accomplished by rf sputter-etching through the contact windows in the photoresist layer used for "lift-off". The metal films were applied using electron-beam evaporation at a residual pressure of $\leq 5*10^{-7}$mbar. Thicknesses were 20nm and 300nm for the Ti- and Au-layer, respectively. In fig. 1 the I/V characteristics of as fabricated Ti-Au contacts to InP ($n=2*10^{18}cm^{-3}$) and of contacts involving a treatment with diluted HF instead of Ar^+bombardment are shown for comparison proving the great impact of the latter process on the ohmic contact behaviour. Excellent I/V-linearity up to current densities in excess of $50kA/cm^2$ was attained with the sputter-etched samples. Specific contact resistances of $R_c \leq 1*10^{-6}\Omega cm^2$ were measured using the "Transmission Line Model (TLM)" (Woelk et al 1986) and were found to be virtually independent of doping concentration in the investigated range of $n=1\text{-}10*10^{18}cm^{-3}$. The TLM-test structures used comprise 4 contact stripes (width: 10μm) separated by 15μm, 65μm and 90μm which traverse 10μm to 80μm wide isolated InP mesa stripes on s.i. substrate.

3. Ti-Au CONTACTS TO p-InGaAs

The electrical characteristics of Ti-Au contacts to p-type InGaAs appear to be very sensitive to the acceptor concentration at the semiconductor surface. To achieve perfect ohmic contacts very high hole densities of around $p=1*10^{20}cm^{-3}$ were found necessary. This is illustrated in fig. 2 which shows the I/V curves for Ti-Au contacts to InGaAs layers with surface concentrations of $p_{surf}=1*10^{19}cm^{-3}$ and $p_{surf}=1.5*10^{20}cm^{-3}$ the former being close to the maximum value attainable with Zn-doping in LPE-grown InGaAs (Kuphal 1982).

To enhance the surface concentration well above this level a shallow Zn-diffusion was performed using spin-on glass as diffusion source (Arnold et al 1984, Amann and Franz 1987) which reproducibly produced the above value of $p_{surf}=1.5*10^{20}cm^{-3}$. Fig.3 represents the well-behaved hole profile measured on a diffused InGaAs-layer with an epitaxial "background" doping level of $p=1*10^{19}cm^{-3}$.

In the experiments Zn-doped silica films (DEMESOL, DEMETRON GmbH) were used. Although these films do not match the thermal expansion coefficient of InGaAs they could be successfully employed by properly adjusting the temperature cycle of the diffusion procedure. This process was carried out in an N_2 atmosphere in a specially designed furnace equipped with IR-lamps. This set-up allows for uniform heating of 1"-wafers and exhibits fast thermal response.

Diffusion was carried out typically at 700 °C for 20 s, after drying the films at 300 °C. Following the removal of the silica film with HF-A solution and an additional cleaning step in an oxygen plasma, TLM test structures were processed as described above. However, in contrast to n-type InP layers Ar^+-bombardment of p-doped InGaAs was found to adversely effect the quality of ohmic Ti-Au contacts resulting in nonlinear I/V characteristics even for the highest doping concentrations attained (Fig. 4). Hence, the InGaAs surface was treated with diluted HF only.

For as deposited contacts TLM evaluation gave specific resistances of $R_c=2\text{-}5*10^{-6}\Omega cm^2$.

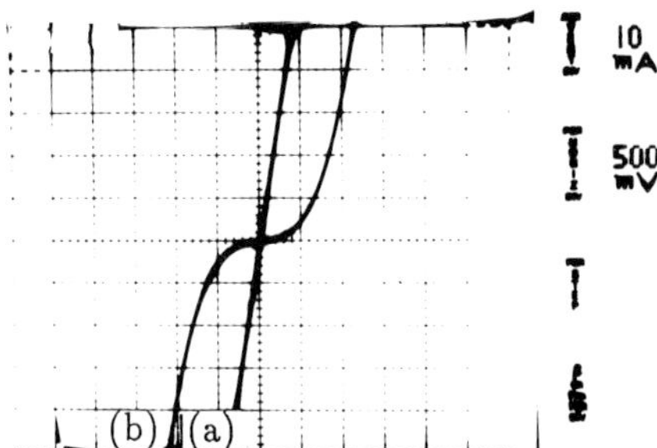

Fig. 1
I/V characteristics of Ti-Au contacts to n-InP ($n=2*10^{18}cm^{-3}$) treated with (a) Ar^+-sputter etching (b) diluted HF ($A_c=10\mu m*80\mu m$)

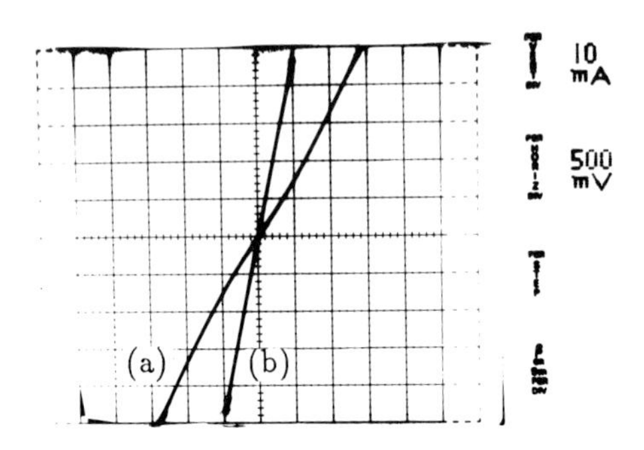

Fig. 2
I/V characteristics of Ti-Au contacts to p-InGaAs
(a) $p=1*10^{19}cm^{-3}$
(b) $p_{surf}=1.5*10^{20}cm^{-3}$
($A_c=10\mu m*40\mu m$)

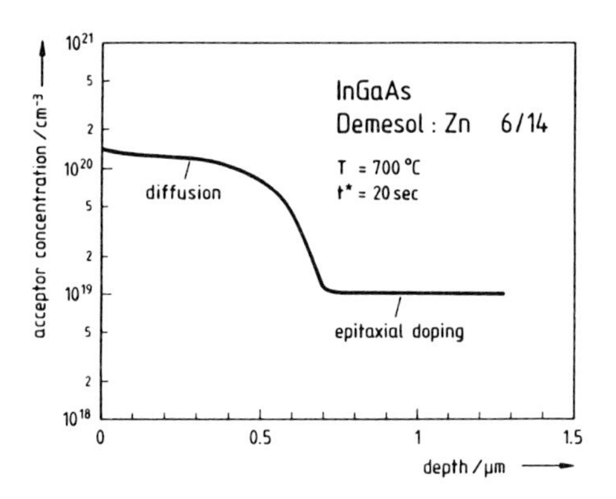

Fig. 3
Carrier profile of Zn diffused InGaAs layer as measured by a C/V etch profiler

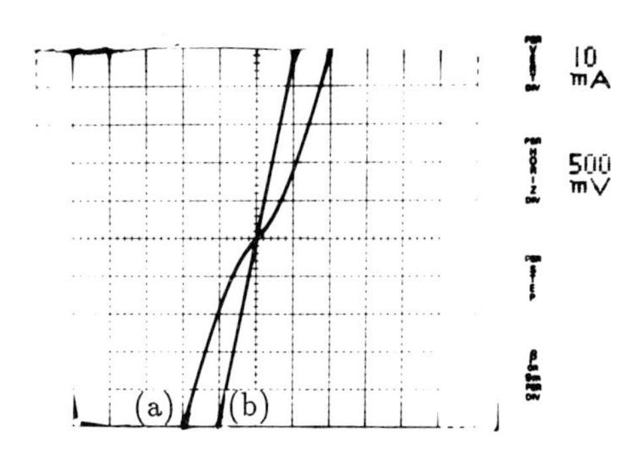

Fig. 4
I/V characteristics of Ti-Au contacts to p^+-InGaAs ($p_{surf}=1.5*10^{20}cm^{-3}$), treated with (a) Ar^+ sputter etching (b) diluted HF ($A_c=10\mu m*80\mu m$)

4. SIMULTANEOUS FABRICATION OF n- and p-TYPE CONTACTS

In order to enable the simultaneous fabrication of ohmic Ti-Au contacts to n-InP and diffused p-InGaAs the complication originating from the opposite effect of the sputter-etching treatment was overcome by introducing an additional wet-etching step to selectively remove a thin surface layer (≈25nm) from the InGaAs material. For this citric acid:H_20:H_20_2 (1:1:8) and subsequently diluted HF were utilized which on the other hand did not affect the bombarded InP surface. The supplementary HF-treatment was found to yield a reduction of R_c for both materials by a factor of about 2 as compared to using the citric acid-based etchant solely. As a result of the combined procedure the R_c-values achieved were essentially equivalent to those attained under the separate processing conditions outlined above.

Fig. 5 shows the temperature dependence of R_c of non-annealed contacts measured in the range of -30 ° C...+80 ° C. No significant change was observed proving the suitability of the fabricated Ti-Au contacts for high-temperature and high-power operation.

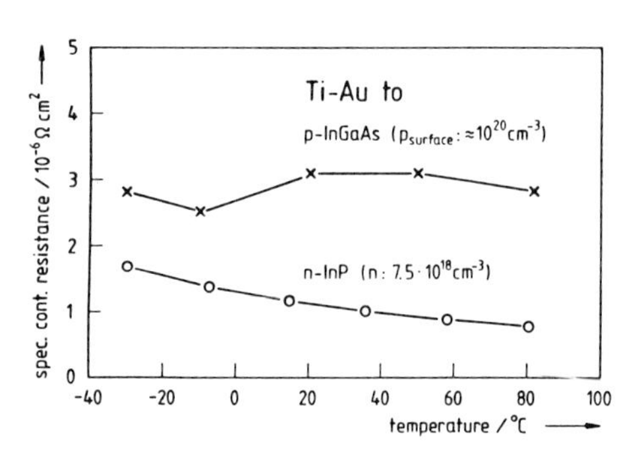

Fig. 5
Temperature dependence of contact resistance of as deposited Ti-Au films on p^+-InGaAs and n-InP

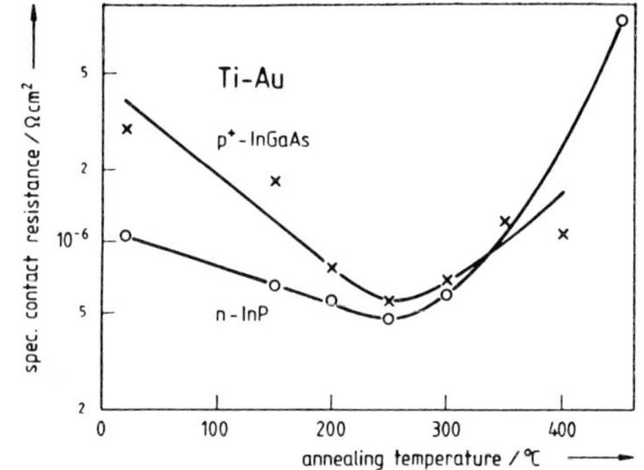

Fig. 6
Specific contact resistances of Ti-Au contacts to n-InP and p^+-InGaAs after heat treatment (N_2, 15 min)

To examine the thermal stability against high-temperature processes required for device fabrication (e.g. CVD of dielectrics, curing of polyimide) the TLM-samples were subjected to heat treatments at temperatures up to 400 °C in N_2 for 15 min each. The heating time was not varied. The results summarized in fig. 6 indicate the ability of the contacts to withstand such temperatures. Yet more, they show that a further noticeable reduction of R_c is attainable upon annealing at 200 °C-300 °C leading to very low R_c-values of $5*10^{-7}\Omega cm^2$ and $6*10^{-7}\Omega cm^2$ in the case of n-InP and p^{+}-InGaAs, respectively.

5. CONCLUSION

Ti-Au contacts with excellent ohmic characteristics were made on n-InP and p-InGaAs employing Ar^{+}-bombardment and Zn-diffusion from spin-on glass, respectively. Introducing an additional selective etching step allowed simultaneous fabrication of these contacts. Although the contacts were ohmic without any further heat treatment, annealing at 200 °C-300 °C was found to give lower specific contact resistances which were $5*10^{-7}\Omega cm^2$ and $6*10^{-7}\Omega cm^2$ for n-InP and p^{+}-InGaAs, respectively. In particular, the latter value is among the lowest figures reported for p-type contacts to InGaAs, if not the best at all.

Acknowledgements

The authors wish to appreciate the assistance of I. Tiedke and I. Gross in the processing of the TLM-samples and of B. Lehmann in the assessment work.

This work was partly conducted under the ESPRIT programme (project 263). The authors alone are responsible for the content.

6. REFERENCES

Amann M-C, Franz G, 1987 J. Appl. Phys. **62** (4) 1541
Arnold N, Schmitt R, Heime K, 1984 J.Phys. D **17** 443
Bach H-G, Grote N, Niggebrügge U, 1986 E-MRS (Strasbourg), in: *Advanced Materials for Telecommunications 1986,* XIII, Les Editions des Physique 461
Dautremont-Smith W G, Barnes P A, Stayt Jr. J W, 1984 Vac. Sci. Technol. **B2** 620
Kuphal E, 1982 *Techn. Ber. 65 TBr 20, Teil II*, Forschungsinstitut der DBP beim FTZ (in German)
Woelk E G, Kräutle H, Beneking H, 1986 IEEE Trans. Electron. Dev. **ED-33** 19

Heat tolerance of the In–GaAs ohmic contact up to 900 °C

T. Otsuki, H. Aoki, H. Takagi, G. Kano and I. Teramoto

Electronics Research Laboratory, Matsushita Electronics Corporation
Takatsuki Osaka, 569 Japan

Abstract: The In-GaAs ohmic contact resistance exhibits practically insignificant increase after a heat treatment at temperatures up to 900°C. The studies of the diffusion behavior of In in GaAs and of the dissolution of GaAs into In revealed that the high stability of the contact resistance we have found is attributable to the low diffusion constant of In in GaAs and to the growth of an InGaAs interfacial layer.

1. INTRODUCTION

In this paper, we report our new finding that the contact resistance exhibits practically insignificant increase after a heat treatment at temperatures up to 900 °C. In order to clarify the cause of the high stability of the In ohmic contact resistance, secondary ion mass spectroscopy (SIMS) measurements and thermodynamic analyses were made. The results of the analyses showed that the high stability of the contact resistance of In to GaAs under the heat treatment is due to the low diffusion constant of In into GaAs and to the growth of an InGaAs interfacial layer.

2. DEPENDENCE OF OHMIC CONTACT RESISTANCE ON HEAT TREATMENT TEMPERATURE

The dependence of the In-GaAs ohmic contact resistance on the heat treatment temperature was measured using the samples of which the cross section is shown in Figure 1. The In film thickness and the carrier concentration of the n type GaAs substrate are 50 nm and 3×10^{17} cm^{-3}, respectively. The samples were annealed at various annealing temperatures with a face to face method in N_2 ambient gas for 10 min. After the annealing process, little number of hillocks caused by irregular alloyings are observed on the surface of the In contacts.

The results of the measurements are shown in Figure 2. The data shows that the contact resistance Rc's are of the order of 10^{-5} ohm cm^2 which are the same level as those of the AuGe ohmic contact to n type GaAs (Braslau 1981). An important point to be noted is that the In contact resistance keeps the low values even after the heat treatment at such a high temperature as 900°C.

3. SIMS ANALYSES OF THE In OHMIC CONTACT TO GaAs

It seems that such high stability of the In contact resistance is closely related to the diffusion characteristics of In into GaAs. In order to clarify the above, SIMS analyses were made on the samples annealed at above 800 °C. The In concentration was calibrated by using the standard In-doped GaAs substrate of which the In concentration was measured by a chemical analysis.

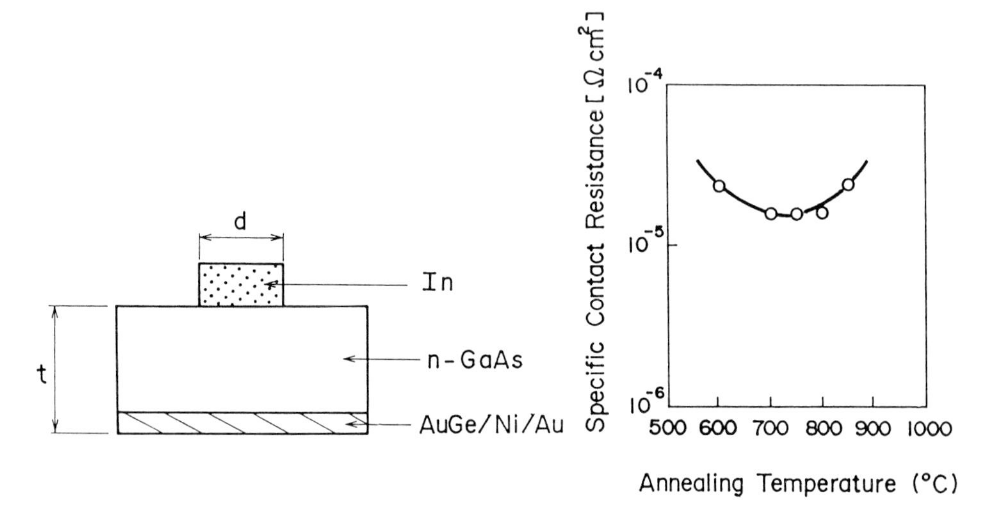

Fig. 1. Cross section of sample

Fig. 2. Ohmic contact resistance of In

In Figure 3, is shown the diffusion constant of In obtained by using the result of SIMS measurements as a function of the annealing temperature, together with the diffusion constant of Au (Casey 1973) for comparison. It is seen from this figure that the diffusion constant D_{In} of In is given as

$$D_{In}=3.3 \times 10^{-5}\exp(-1.9/kT)\ cm^2/sec. \quad (1)$$

This value is about 1/100 of that of Au.
If the diffusion constant of In is very large, all of the In deposited on the GaAs substrate diffuses into the GaAs during the annealing process so that the concentration of the In at the surface of the GaAs substrate can no longer maintain the high value after annealing. In other words, the low diffusion constant of In leads to the fact that the eutectic layer at the surface exhibits no significant change after high temperature annealing. As a result, the ohmic contact resistance of In keeps a low value.

4. THEORETICAL STUDIES ON GaAs DISSOLUTION INTO In

Dissolution of GaAs into the In film possibly occurs during the heat treatmtne because the melting point of In is as low as 156.6°C. If the dissolution occurs, an InGaAs layer is formed between In and GaAs from the In-Ga-As melt in the same manner as the liquid phase epitaxy during the course of cooling down to the room temperature. In a view of a possibility that the InGaAs layer plays an important role on the contact resistance, we analyzed the dissolution by calculating the phase diagram of the ternary In-Ga-As at a given annealing temperature (Otsuki et al.). The mole fraction x_i^ℓ of each element in the ternary liquid phase is shown in Figure 4 as a function of annealing temperature. By use of these results, the dissolution thickness y of GaAs normalized by the In film thickness t is given by

$$\frac{y}{t} = \frac{N_A\, p\, a^3}{16\, w} \left(\frac{1}{x_{In}^\ell} - 1 \right), \qquad (2)$$

where N_A is Avogadro's constant, p and w are the density and the atomic weight of In, respectively. Substituting the literature data and the mole fraction of In into Equation (2), the dissolved thickness of GaAs is calculated to be 30% of that of the In film thickness at an annealing temperature of 900 °C. During the course of cooling down to the room temperature, an InGaAs interfacial layer is formed from the In-Ga-As melt in the same manner as liquid phase epitaxy. According to the previous works on the metal-InGaAs-GaAs contact (Woodall et al. 1981a & 1981b, Kajima et al. 1973, Rao et al. 1987), a thin InGaAs layer between metal and GaAs is effective for decreasing the ohmic contact resistance since the InGaAs layer lowers the Schottky barrier height. The above analyses conclude that the growth of the InGaAs interfacial layer is one of the reasons why In ohmic contact resistance exhibits the high stability after the heat treatments at such a high temperature as 900°C.

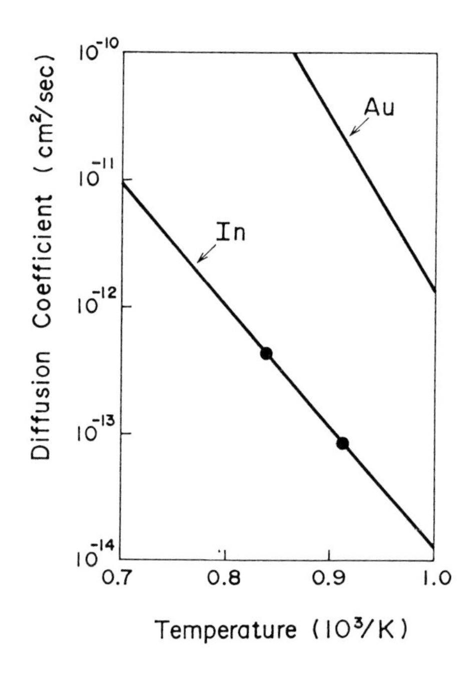

Fig. 3. Diffusion coefficient of In in GaAs

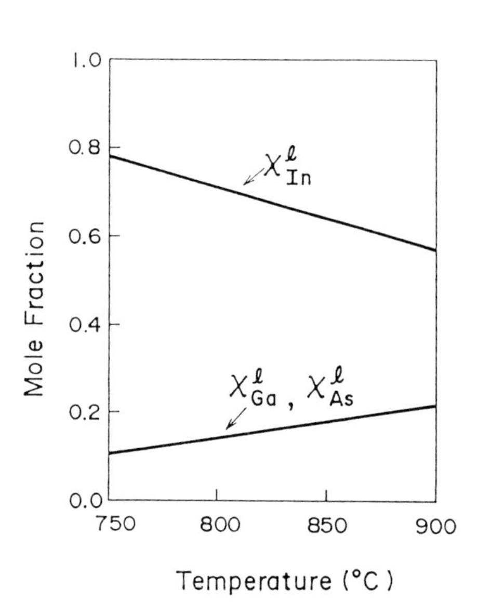

Fig. 4. Mole fraction of each element in In-Ga-As liquid phase

5. SUMMARY

Heat treatment effects on the In-GaAs ohmic contact have been studied. The specific contact resistance of In exhibits high stability under heat treatments up to 900°C. As a result of the studies on the high stability, it is concluded that the high stability of the In ohmic contact is due to the low diffusion constant and to the growth of the InGaAs interfacial layer.

ACKNOWLEDGEMENT

Authors would like to thank to Dr. Takeshima and Mr. Kazumura for their helpful discussions.

REFERENCES

Braslau N 1981 J.Vac.Sci.Technol, 19, 803

Casey H C 1973 Atomic Diffusion in Semiconductors, Diffusion in the III-V Semiconductors (New York: Plenum Press) Chap.6

Kajima K, Mizushima Y and Sakata S 1973 Appl.Phys.Lett. 23 458

Otsuki T, Aoki H, Takagi H and Kano G to be submitted

Rao M A, Caine E J, Long S I and Kroemer H 1987 IEEE Electron Device Lett. EDL-8 30

Woodall J M, Freeouf J L, Pettit G D, Jackson T and Kirchner P 1981a J.Vac.Sci.Technol. 19 626

Woodall J M, Freeouf J L 1981b J.Vac.Sci.Technol. 19 794

Paper presented at Int. Symp. GaAs and Related Compounds, Heraklion, Greece, 1987

Zinc diffusion in InGaAsP

G.J. van Gurp, P.R. Boudewijn, G.M. Fontijn and D.L.A. Tjaden

Philips Research Laboratories
PO Box 80.000, 5600 JA Eindhoven, The Netherlands

ABSTRACT: Ampoule diffusion of Zn in undoped LPE InGaAsP layers between 425 and 525°C shows the Zn solubility, as measured with SIMS, to be much larger than that in InP and to be slightly less than that in GaAs. The acceptor concentration, as determined by C-V measurements, is 60-90% of the Zn concentration. The diffusion can be described with the interstitial-substitutional model. The diffusion depth is slightly smaller than that in InP and much greater than that in GaAs. In N-type InGaAsP profiles are found with a cut-off, similar to the behaviour in InP and GaAs.

1. INTRODUCTION

In(x)Ga(1-x)As(y)P(1-y) epitaxially grown on InP is often used as a contacting layer, as the contact resistance to Zn diffused P-type quaternary material is much smaller than that to InP, where the acceptor concentration is much smaller than that of Zn: only 10 to 25% of the Zn is electrically active.

In this paper the Zn and acceptor solubility and the diffusion of Zn in InGaAsP (emitting at 1.3 μm) are reported.

2. EXPERIMENTAL

InGaAsP layers (x=0.72, y=0.60) were grown by liquid phase epitaxy at 650 °C on undoped <100> InP. The quaternary layers, which emitted at 1.3 μm, were not intentionally doped and were N-type with a donor concentration of about 10^{17} cm^{-3}. Also Te-doped layers were grown with a donor concentration of 8 x 10^{18} cm^{-3}. Layer thickness was up to about 2 μm.
Closed ampoule diffusion of Zn was carried out using Zn_3P_2 and $ZnAs_2$ powder together with the InGaAsP/InP sample. Diffusion temperatures were between 425 and 525 °C.

Zinc profile measurements were made with Secondary Ion Mass Spectrometry (SIMS) using a Cs^+ beam and detecting the $CsZn^+$ cluster in a Cameca IMF-3F instrument. Acceptor profiles were determined by capacitance-voltage measurements using a Polaron Profiler PN 4200.

3. MODEL

Zinc incorporation in III-V semiconductors is usually described in terms of diffusion of Zn interstitials and their trapping on the site of a metal vacancy V.

$$Zn_i^{m+} + V \rightleftarrows Zn_s^- + (m+1)h \tag{1}$$

where Zn_i is the interstitial with charge m+, Zn_s is substitutional Zn and h is a hole.

To account for the difference in Zn and acceptor concentration, two models can be used.
Firstly, it has been suggested that Zn can also be built in as an electrically inactive complex of Zn on a metal site and two phosphorous vacancies (Tuck and Hooper 1975).

$$Zn_i^{m+} + 2V_P + V \rightleftarrows V_PZnV_P + mh \tag{2}$$

We assume this equation also to hold for InGaAsP, where V stands for an In or Ga vacancy. It is further assumed that the diffusion coefficient of substitutional Zn and that of the complex can be neglected with respect to that of interstitial Zn and that the concentration of interstitial Zn is much smaller than the substitutional Zn and complex concentrations.

A second model is that in which there are no complexes, but the interstitial Zn donor concentration is not negligible and compensates part of the substitutional Zn acceptors. The substitutional diffusion coefficient is still neglected.

The diffusion equation in one dimension can then be written

$$\frac{\partial C'}{\partial t} + \frac{\partial C_s}{\partial t} = \frac{\partial}{\partial x}\left[D_i \frac{\partial C_i}{\partial x}\right] \tag{3}$$

where C' stands for C_n (model 1) or C_i (model 2). C_n, C_s and C_i are the concentrations of the complex and substitutional and interstitial Zn, respectively, and D_i is the diffusion coefficient of interstitial Zn.

Using expressions for the various concentrations, derived from the laws of mass action for the reactions (1) and (2), the diffusion equations can be solved, resulting in calculated concentration profiles with m as a parameter (Van Gurp et al 1987 and Tjaden and Van Gurp unpublished).

In heavily doped N-type material the interstitial Zn concentration becomes zero when the acceptor concentration falls below the donor concentration N and interstitial diffusion is suppressed. As a result the acceptor profile exhibits a jump from the value N to zero.

4. RESULTS

Figure 1 shows Zn profiles for two diffusion treatments, compared with calculated profiles, using model 1 or model 2. The difference between the results for the two models is small. Profiles calculated with m=1 (singly ionized Zn interstitials) provide a good description of the measurements.

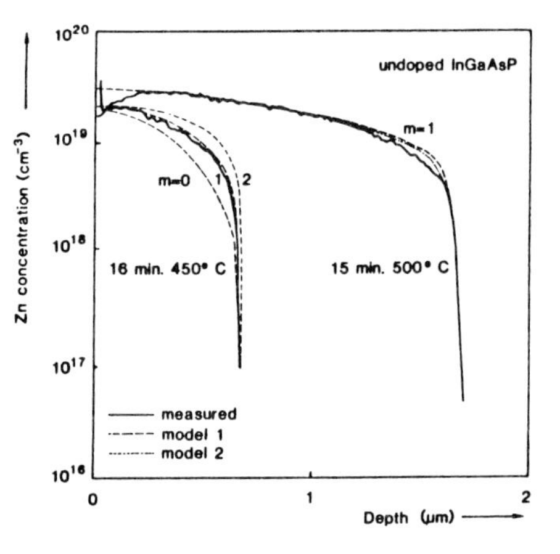

Figure 1. Measured and calculated Zn profiles

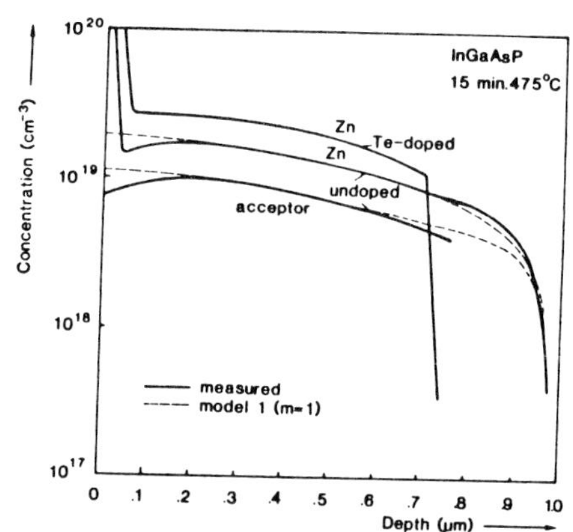

Figure 2. Measured and calculated Zn and acceptor profiles in undoped and Te-doped InGaAsP.

Figure 2 shows Zn and acceptor profiles in undoped InGaAsP, compared with calculated profiles for m=1. The figure also shows a profile in N-type Te-doped InGaAsP ($N = 8 \times 10^{18}$ cm^{-3}), which exhibits a jump and is less deep than the profile in undoped InGaAsP, in agreement with theoretical prediction.
Measurements show that the electrical activity of Zn in the InGaAsP layers increases from about 60% at 425 °C to about 90% at 525 °C.

The solubility of Zn in undoped InGaAsP, for which we take the maximum concentration in the profile, has an activation energy of 0.55 eV. The solubility in InGaAsP is close to that in GaAs and much larger than that in InP. In N-type InGaAsP the Zn and acceptor solubility is higher than in undoped material, as in InP and GaAs (Van Gurp et al 1987).

The diffusion depth x, defined as the depth where the Zn concentration has dropped to 10^{17} cm^{-3}, is proportional to $t^{1/2}$, where t is the diffusion time, as expected for a diffusion process. The activation energy of x^2/t is 1.6 eV. The diffusion depth is close to that in InP and much larger than that in GaAs.
The table gives values of the maximum Zn concentration and the diffusion depth both at 500 °C, as well as the acceptor concentration as related to the Zn concentration, and values for the activation energies for undoped InGaAsP, InP and GaAs.

5. CONCLUSIONS

1. Zn incorporation in InGaAsP can be described by the interstitial-substitutional model with singly ionized Zn interstitials.
2. Acceptor concentration is 60 - 90% of the Zn concentration.
3. In N-type material the diffusion of Zn is suppressed when the acceptor concentration falls below the donor concentration.
4. Zn solubility in InGaAsP is comparable to that in GaAs.
5. Zn diffusivity in InGaAsP is comparable to that in InP.

6. REFERENCES

Tuck B and Hooper A 1975 J. Phys. D 8 1806

Van Gurp G J, Boudewijn P R, Kempeners M N C and Tjaden D L A 1987 J. Appl. Phys. 61 1846

Tjaden D L A and Van Gurp G J unpublished

TABLE

	InP	InGaAsP	GaAs	
Max. Zn conc. at 500 °C	10^{19}	3×10^{19}	4×10^{19}	cm^{-3}
act. energy	1.0	0.55	0.6	eV
Acceptor conc.	10 - 25	60 - 90	100	%
x^2/t at 500 °C	4×10^{-11}	2.5×10^{-11}	10^{-14}	cm^2/s
act. energy	1.4	1.6	2.3	eV

Unpinned Schottky barrier in GaP(110)

P Chiaradia, M Fanfoni, P De Padova, P Nataletti

Istituto di Struttura della Materia, CNR, Frascati, Italy

L J Brillson, R E Viturro, M L Slade

Xerox Webster Research Center, Webster, N Y 14580, USA

G Margaritondo, M K Kelly, D Kilday, N Tache

Dept of Physics, University of Wisconsin, Madison, Wisconsin 53706, USA

A photoemission study of the Schottky barriers obtained by deposition of several metals onto GaP(110) cleaved in UHV has been performed. The results indicate a wide range of barrier heights, from less than 0.5 eV to about 1.5 eV. Surprisingly, the value of the Schottky barrier is nearly proportional to the metal work function, as in the original Schottky model of metal-semiconductor interfaces. This is the first case in which such a nearly ideal behavior is observed. The results are clearly at variance with all other models implying pinning of the Fermi level in a narrow range, such as the Defect Model and the Metal-Induced-Gap-States Model. On the other hand, data are consistent with the Effective-Work-Function Model proposed by Freeouf and Woodall.

1. INTRODUCTION

A physical description of the mechanisms controlling the Schottky barrier height (SBH) in metal-semiconductor (M-S) interfaces is still lacking (Spicer 1986)). Numerous models have been elaborated and are presently under discussion. It is not even clear whether or not a single model can account for the great variety of experimental behaviors observed.
Generally speaking, models can be divided into two groups. Those stating that SBH is related to intrinsic properties of the semiconductor alone (and therefore is not dependent on the metal work function) belong to one category. Besides the old Bardeen model, Spicer's Defect model (Spicer 1986) and the Metal-Induced-Gap-States (MIGS) model proposed by Heine and developed by Flores and Tersoff (1984 and 1985) are examples of the first

category. Recently both these models (Defects and MIGS) have been used in a complementary way in order to explain the SBH in metal-silicon and silicide-silicon interfaces (Monch 1987). On the opposite side are models derived from Schottky's old theory of M-S interfaces. For instance the Effective Work Function model of Freeouf and Woodall (1981) belongs to this second group.
From a phenomenological point of view, a well established result is that in cleaved GaAs(110) practically no variation of SBH with different metals occurs. GaAs is then a typical material for which a model of SB formation of the first category is required.
The present results indicate that GaP(110) shows an opposite behavior with respect to GaAs, in spite of the close similarity between these two compounds. Therefore GaP can be considered a prototypical case of SB formation belonging to the second group.

2. EXPERIMENT AND METHOD

Schottky barrier formation upon room temperature deposition of Au, Ag, Cu, Al and In onto n-type GaP(110) cleaved in UHV has been studied by photoemission with synchrotron radiation (Brillson 1987 and Chiaradia 1987). Metals were evaporated from filaments. The pressure was always in the low 10^{-10} mbar range. The experiments have been carried out at the SRC of Stoughton, Wisconsin and the Adone national facility of Frascati, Italy, using Grasshopper monochromators.
The initial position of the top of the valence band (Ev) with respect to the Fermi level (Ef) has been determined by linearly extrapolating the leading edge of the valence band spectrum. This method, although affected by a small systematic error (Kraut 1980), is currently used when relative rather than absolute values are required. The Fermi level position was determined by evaporating a thick layer of gold "in situ." The band bending change at each metal coverage was determined by following the position of the substrate core levels. In some cases chemically shifted peaks due to chemical reactions (for instance with Al and Cu) were observed in the core level lineshape. This effect was particularly evident in surface sensitive spectra and has been taken into account in the data analysis (Brillson 1987 and Chiaradia 1987).

3. RESULTS AND DISCUSSION

The evolution of the Fermi level position with respect to the top of the valence band upon metal coverage is shown in Figure 1, for In, Al, Ag, Cu and Au. Also datapoints for Ge and Si are reported. The initial position of Ef varies over a fairly broad range and is presumably correlated to the quality of the cleave. As the case of Ge demonstrates, however, different initial positions of Ef evolve to the same saturation value, for a given material. Figure 1 shows that in GaP(110) Schottky barriers differing by more than 1 eV can occur, depending on the metal. This behavior is strikingly different with respect to other 3-5 compounds, as GaAs(110), in which a strong Ef pinning has been observed (Spicer 1986). Also the rate of the process is much different in the two cases. The Schottky barrier formation in GaAs(110) is already accomplished at submonolayer coverages (Spicer 1986), while in GaP(110) an overlayer 5 to 10 Å thick is necessary.

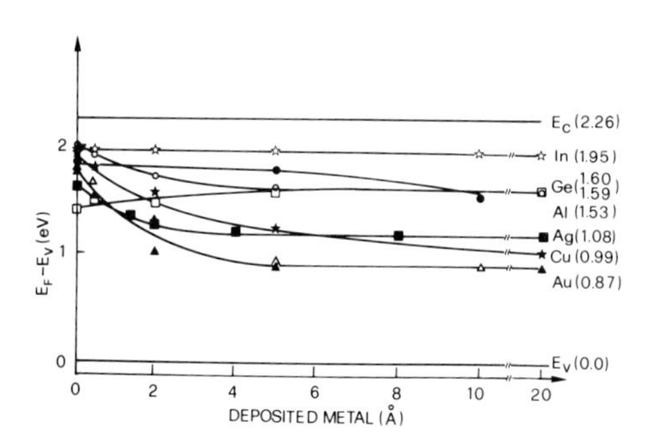

Fig.1. Fermi level movement for cleaved GaP(110) as a function of coverage with various metals (In, Al, Ag,Cu and Au) and also Ge and Si.

The most important result is that the experimental data show a surprisingly good correlation between Schottky barrier height and metal work function, like in a textbook description of M-S interfaces (Sze 1969). This is clearly visible in Figure 2, in which the experimental values of SBH's are plotted versus the corresponding metal work function. Although one might argue that electronegativity is a better parameter than work function in order to characterize the metal properties, we have found that either parameter can be used for heuristic purposes. The straight line at 45 degrees in Figure 2 represents the Schottky limit. It is remarkable how well this line (which is not a linear best fit) is compatible with the datapoints. A marked deviation occurs only in the case of Cu. Significantly this metal strongly reacts with the substrate, as does Al. Datapoints for Ge and Si are also included.

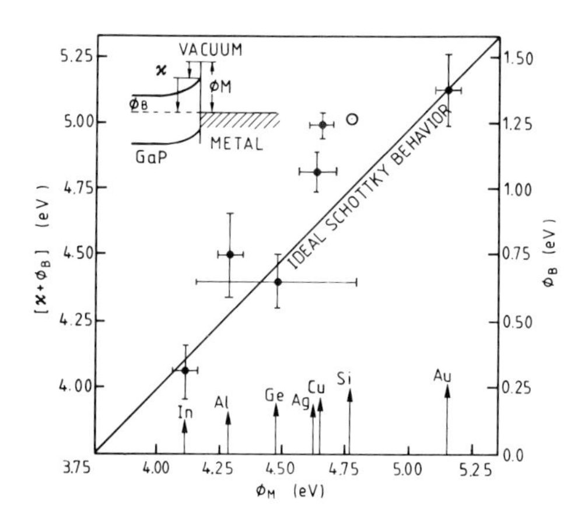

Fig.2. Schottky barrier values (right vertical scale) versus metal work function for cleaved GaP(110). Datapoints for Ge and Si are also shown.

The observed broad range of SBH's cannot be explained by models, such as the Defect and MIGS models, that are instead appropriate for M-S interfaces exhibiting strong Ef pinning. The case of the MIGS model, however, deserves further discussion. In fact, on one hand this model may not apply to highly ionic compounds as GaP because of "poor interfacial bonding" (Tersoff 1985), as recently proposed by Lince and coworkers for

SB on MoS_2 (1987). On the other hand, we have evidence of a strong metal-substrate reaction at least in two cases out of five. Moreover it is not clear why metal interfaces involving semiconductors with dielectric properties so similar as GaAs and GaP should behave so differently.
The only model that seems to account for the experimental results reported in Figure 2 is the Effective Work Function model. The basic idea underlying this model is that in the case of reacted interfaces the anion work function should be used instead of the metal work function, in plots such as Figure 2. The Phosphorus work function being of the order of 5 eV, it is clear that a correction of the Al and Cu datapoints (reacted interfaces) according to this model would improve the fit.

In conclusions, we have studied Schottky barrier formation in a number of metal-GaP(110) interfaces and observed pinning of Ef in a range of more than 1 eV. This finding, together with analogous results on In-based semiconductors and MBE-grown GaAs (Brillson 1987 and Chiaradia 1987), demonstrates that in 3-5 compounds the cases of Ef pinning in a broad range are at least as numerous as those of pinning in a narrow range.

4. ACKNOWLEDGEMENTS

The authors wish to thank S.Priori, of the Frascati staff, for technical assistance.

5. REFERENCES

Brillson L J, Viturro R E, Slade M L, Chiaradia P, Kilday D, Kelly M K and Margaritondo G 1987 Appl.Phys.Lett. 50 1379
Chiaradia P, Brillson L J, Slade M L, Viturro R E, Kilday D, Tache N, Kelly M K and Margaritondo G 1987 J.Vac.Sci.Technol., in press
Freeouf J L and Woodall J M 1981 Appl.Phys.Lett. 39 727
Kraut E A, Grant R W, Waldrop J R and Kowalczyk S P 1980 Phys.Rev.Lett. 44 1620
Monch W 1987 Phys.Rev.Lett 58 1260
Spicer W E, Kendelewicz T, Newman N, Chin K K and Lindau I 1986 Surf.Sci 168 240
Sze S M 1969 "Physics of Semiconductors Devices" (New York: Wiley)
Tersoff J 1984 Phys.Rev.Lett. 52 465
Tersoff J 1985 Phys.Rev B32 6968

Schottky and FET fabrication on InP and GaInAs

S.LOUALICHE, H. L'HARIDON, A. LE CORRE, D. LECROSNIER, M. SALVI, P.N. FAVENNEC

Centre National d'Etudes des Télécommunications
BP 40 22301 LANNION CEDEX FRANCE

Abstract
Schottky contact with barrier heights of 0.76 eV on n type InP and 0.65 eV on n type GaInAs are realized by a new surface treatment. These contacts are used as a gate for the fabrication of field effect transistors in these materials. A transconductance of 100 mS/mm has been obtained on a device realized on GaInAs.

The optoelectronic devices for telecommunications in the wavelength range of 1.3 to 1.5 μm are based on InP and ternary and quaternary related compounds. The carrier saturation velocity in these materials exceed that of GaAs and can theoretically reach 4×10^7 cm/s (Nag et al) in GaInAs where a value of 3.6×10^7 cm/s has been measured (Raulin et al). In addition the thermal conductivity of InP substrate exceed that of GaAs. For these reasons InP and related compounds present potential properties for high speed and high power devices. The fields effect transistor (FET) fabrication on these materials is difficult due to the absence of good quality Schottky contacts. The Schottky barrier height is lower than 0.3 eV in GaInAs (Kajiyama et al). The metal insulator FET (MISFET) devices are not usually used because the poor long term stability of their electrical characteristics. In the present work we have been able to improve considerably the Schottky barrier height in GaInAs and InP.

The Schottky diodes on InP material are fabricated on undoped residual n type (~ 10^{16} cm^{-3}) Czochralski grown substrates. The InP FETs are realized on Silicon implanted semi-insulating (SI), Fe doped substrates. The GaInAs Schottky diodes and FETs are realized on layers grown by molecular beam epitaxy (MBE). The substrates used for MBE layers are 100 InP (SI). The substrate preparation has been improved to obtain oxide free surface. After a classical cleaning the substrate is bonded to a molybdenum block using In as solder. It is introduced in a glove box connected to the introduction chamber of a commercial RIBER 2300 molecular beam epitaxy system. Then under nitrogen atmosphere, the sample is etched with HF : ethanol solution (10 %), rinsed in ethanol and dried with boiling ethanol. The observation of RHEED pattern lines after heating at 200°C indicates that the surface is oxide free. The growth temperature is 500°C. The undoped GaInAs MBE layers are of good quality with low residual doping (~ 10^{15} cm^{-3}) and Hall mobilities as high as 55000 cm^2/vs for thick samples (~5μm).

The barrier height of the Schottky contact on GaInAs and InP has been increased by a soft oxidation of the sample surface before the Schottky metal deposition. The sample is cleaned using organic sovents,the unwanted oxide is removed in HF (10 %) solution. The surface oxidation begins in deionised water (18 MΩ) at 100°C for one hour under oxygen flow. This operation produces a clean native oxide (Massies et al). Then a solution is added to the boiling deionized water to obtain a liquid with a final composition (100:5:1 ; $H_2O:NH_4OH:H_2O_2$) in which the sample stands another 20 minutes. The ohmic contact of the device has been fabricated before the oxidation and consists on a vacuum (10^{-6}Torr) evaporation of an AuGe compound followed by an annealing at 320°C under N_2 + 10 % H_2 atmosphere. The gold (Au) metal is used as Schottky contact and is evaporated in vacuum (10^{-6} Torr). The Schottky I(V) characteristic is then measured and the device parameters are deduced from this measure. The I(V) characteristic obtained at room temperature on a non doped (n type residual (~ $10^{16}cm^{-3}$) GaInAs layer is given in figure 1. The diode diameter is 0.4 mm. The barrier height and ideality factor obtained from the I(V) curve are 0.65 eV and 1.10 respectively. At our knowledge this is the highest Schottky barrier obtained on n type GaInAs layer. The leakage current is 0.25 μA at 1 V reverse voltage. The breakdown voltage is about 2.5 V. A measure of the saturation current of the device between 90 K and 300 K shows that the diode acts like a MIS tunnel device (Sze) with an oxide layer of 20 A. This oxide thickness is in close agreement with the measure of Massies et al in GaAs.

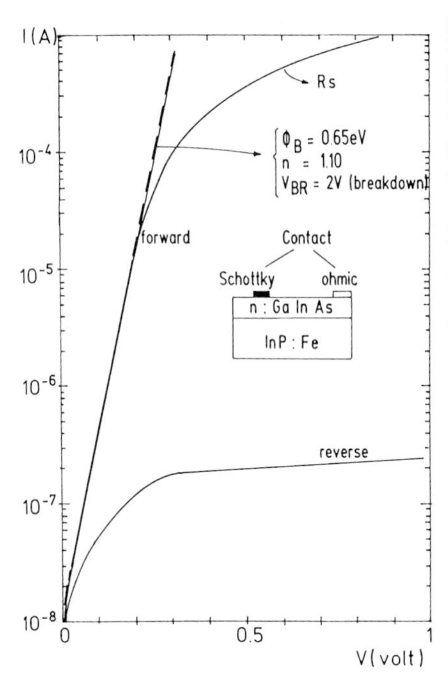

Fig. 1 : Schottky/n GaInAs

The Schottky device has been used as a gate of a FET realized on MBE GaInAs layers. In our MBE chamber we cannot grow InP buffer layers and a p type GaInAs layer is used for this purpose. The epitaxy for the FET begins by a 0.3 μm thick p type ($10^{18}cm^{-3}$) Berylium doped layer. It is followed by a 0.2 μm thick n type ($10^{17}cm^{-3}$) silicon doped active layer. The carrier concentration has been measured by a commercial capacitance voltage (C(V)) electrochemical profiler (Polaron). The Hall effect method confirms the C(V) measure and gives 7000 cm^2/vs for the mobility of the active layer at 300 K. The FET area on GaInAs has been isolated by a selective etching to the substrate. After the ohmic contact evaporation and annealing, the surface preparation to increase the semiconductor barrier height, begins for the Schottky fabrication. Then the gold Schottky metal is deposited to perform the FET gate. The gate dimensions are 140 μm x 1μm (Fig.2a). The Schottky gate has been tested and shows a reverse leakage current below 1μA at 1 V. The Ids-Vds network of the FET is in Figure 2b. The FET of Figure 2 presents an extrinsic tranconductance of 100 ms/mm. This value is obtained without a recess below the gate and without an optimisation of the source and drain ohmic contact (there is, no n^+ over doping below the source and drain). The highest transconductance that can be reached by our structure is 400 mS/mm ($\frac{\varepsilon V_s}{W}$: ε = 1.03 pF/cm : W = 140 μm : gate width ; V_s = $4 10^7$cm/s). A value of $1.3 x 10^7$cm/s for the carrier velocity is ob tained from the slope of the Ids-$\sqrt{Vg - \emptyset}$ plot for the FET of the figure 2.

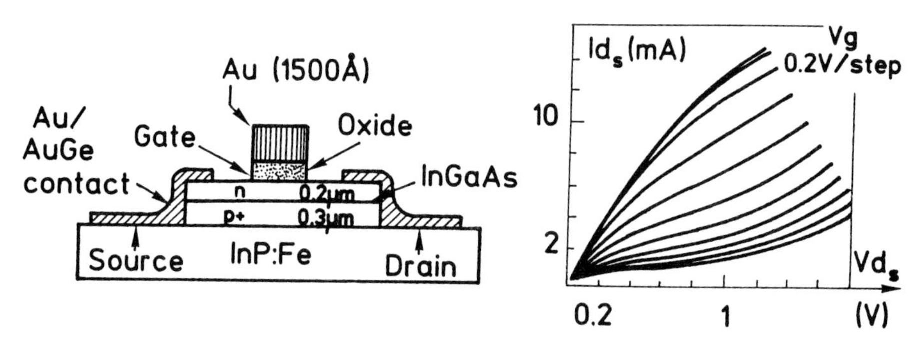

Fig. 2a) : Schematic diagram of the FET on n type MBE GaInAs

Fig. 2b) : Ids-Vds network of the GaInAs FET

The Schottky device obtained on non doped n type (~ 10^{16}cm^{-3}) InP substrate by the present method presents a barrier height of 0.76 eV and an ideality factor of 1.48 at room temperature (Figure 3). The reverse leakage current is below 100 nA at 1 V for a device with a diameter of 1 mm. The breakdown voltage reaches 5 V and is higher than in GaInAs device (~ 2.5 V). An FET has been performed on InP using this kind of Schottky device as a gate. The active part of the device is obtained by an ion implantation of Silicon on a semi-insulating InP substrate. The ion implantation consists on two doses and two energies (10^{12}cm^{-2} - 40 keV and $2 x 10^{12}$cm^2 - 100 keV). After a rapid thermal annealing (750°C - 30 s) the carrier profile measured by C(V) method presents a constant concentration of $2.3 x 10^{17}$cm^{-3} from the surface to 0.1 μm and falls to 10^{17}cm^{-3} at 0.2 μm and to 10^{16}cm^{-3} at 0,27 μm. The transconductance of the FET on InP is 7,5 mS/mm (Figure 4).

In conclusion, a way is found to perform Schottky contact on n type InP and GaInAs. The method is very easy to realize and leads to barrier height of 0.65 eV on GaInAs and 0.76 eV on InP. The oxidation technique does not introduce any additional surface trap (Loualiche et al). The Schottky device has been used as a gate on a FET fabricated on these material and a transconductance of 100 mS/mm has been obtained on GaInAs FET s.

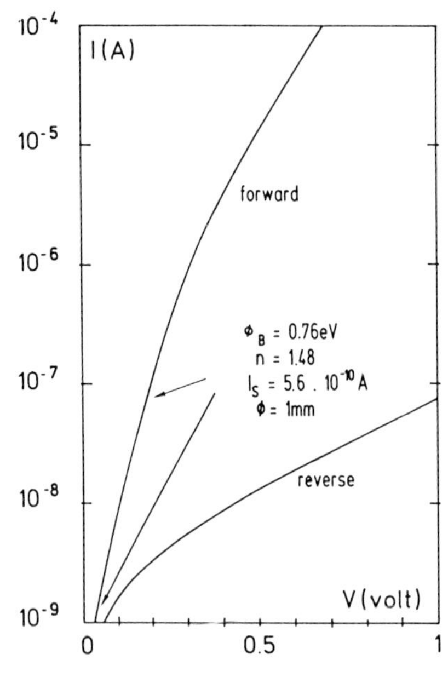

Fig. 3 : Schottky on n-InP

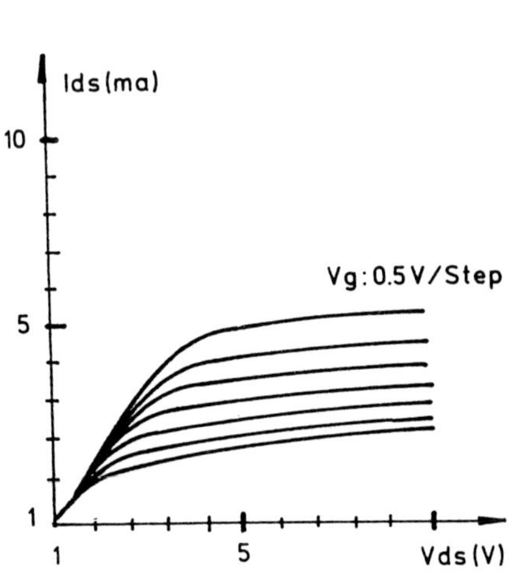

Fig. 4 : InP implanted FET

REFERENCES

- Kajiyama K., Misoshima M., Skakata S.
  Appl. Phys. Lett. 23, p. 758, 1973

- Loualiche S., Gauneau M., Le Corre A., Lecrosnier D., L'haridon H.,
  Appl. Phys. Letters, to be published (~ october 1987)

- Massies J., Contour J.P.
  Appl. Phys. Lett. 46, p. 1150, 1985

- Nag B.R., Ahmed S.R., Deb Roy M.
  IEEE Trans. El. Dev. ED33, p. 188, 1986

- Raulin J.Y., Thorngren E., Di Forte Poisson M.A., Colomer G.
  Appl. Phys. Lett. 50, p. 535, 1987

- Sze S.M., Wiley J. New York 1981

Paper presented at Int. Symp. GaAs and Related Compounds, Heraklion, Greece, 1987

Room temperature exciton transitions in partially intermixed GaAs/AlGaAs superlattice

J.D. Ralston, S. O'Brien, G.W. Wicks and L.F. Eastman
School of Electrical Engineering and National Nanofabrication Facility
Phillips Hall, Cornell University, Ithaca, NY 14853 USA

Abstract

Substantial increases are observed in the energies of room temperature exciton transitions in GaAs/AlGaAs superlattices which have been partially intermixed via the impurity-free vacancy diffusion process. In the samples studied, the above process allows for continuously variable energy shifts of at least 61 meV while still maintaining clearly resolved excitonic behavior. The degree of intermixing is characterized using Raman spectroscopy; shifting and broadening of the exciton transistions are studied using room temperature photoluminescence and photocurrent spectroscopy.

1. Introduction

An important consequence of two-dimensional carrier confinement in GaAs/AlGaAs MQW structures is an enhancement of exciton binding energies to the degree that clear excitonic absorption features can be observed at room temperature [Miller et al. 1982]. The associated optical nonlinearity (Chemla et al. 1984] and electric field dependence [Miller et al. 1985a] have been exploited for a variety of novel applications [Miller et al. 1985b].

The ability to selectively intermix GaAs/AlGaAs and other semiconductor superlattices has also found application in the design of novel optical device structures. Several processes have been utilized to achieve selective intermixing, including localized diffusion or implantation of various impurities [Laidig et al. 1981, Coleman et al. 1982, Camras et al. 1983, Meehan et al. 1984], laser irradiation [Epler et al. 1986, Ralston et al. 1987], and impurity-free vacancy diffusion [Deppe et al. 1986].

In the present work the behavior of room temperature exciton resonances is studied in GaAs/$Al_{0.3}Ga_{0.7}As$ MQW samples which have been partially intermixed by the impurity-free vacancy diffusion process. As the degree of intermixing increases, exciton absorption peaks broaden and shift to higher energies.

2. Sample Preparation

The samples used in these experiments consisted of a 4000 Å superlattice (25 periods of 80 Å GaAs layers alternating with 80 Å $Al_{0.3}Ga_{0.7}As$ layers) grown by molecular beam epitaxy. A 3000 Å layer of electron beam evaporated SiO_2 was deposited on one half of each sample. Localized intermixing of the layered structure was accomplished by rapid thermal annealing (RTA) at temperatures between 850° and 950°C for 15 seconds. As previously reported [Deppe et al. 1986], intermixing is attributed to diffusion of Ga into the SiO_2, accompanied by Ga vacancy diffusion into the superlattice.

3. Raman Scattering

In order to obtain a non-destructive assessment of the degree of intermixing following RTA, Raman scattering measurements were performed on both the SiO_2-capped and uncapped regions of each sample using a CW argon ion laser operating at 514.5 nm. The Raman signal from a superlattice is simply the superposition of spectra from the individual layers [Merz et al. 1977]. Furthermore, the optical phonon spectra from $Al_xGa_{1-x}As$ mixed crystals show two distinct bands, with frequencies near those of pure GaAs and pure AlAs [Ilegems and Pearson 1970]. For each band, Raman scattered light shifts to lower frequencies as the concentration of the corresponding compound is decreased in the ternary alloy, allowing the composition x of individual superlattice layers to be determined.

Fig. 1 shows Raman scattering profiles of the GaAs-like phonon band from samples which underwent RTA at 850°, 900°, and 950°C. The spectra from uncapped regions all exhibit longitudinal optical (LO) phonon peaks at 292 cm^{-1}, corresponding to the pure GaAs and the $Al_{0.3}Ga_{0.7}As$ layers, respectively. The above spectra are virtually identical to those of as-grown samples without RTA, i.e., regions not capped with SiO_2 demonstrate negligible thermal intermixing due to RTA alone. In contrast, the spectra from the SiO_2 capped regions show a shifting and merging of the two LO phonon peaks as the anneal temperature is increased, indicative of some slight degree of intermixing following the 900°C RTA and extensive, but not complete, intermixing following the 950°C RTA.

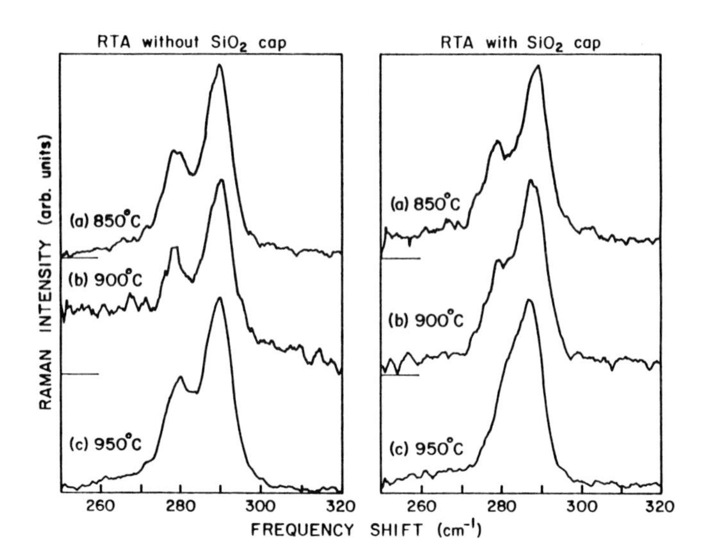

Fig. 1. Room temperature Raman spectra of uncapped and SiO_2-capped MQW samples following 15 seconds rapid thermal anneal at (a) 850°C; (b) 900°C; (c) 950° (λ_o = 5145 Å).

4. Photoluminescence and photocurrent

Figs. 2 and 3 present room temperature photoluminescence (PL) and photocurrent (PC) spectra, respectively, from the various samples. PL was measured using the 632.8 nm line of a HeNe laser. PC measurements were performed using a lateral p-i-n structure, with p-type (InZn) and n-type (Sn) contacts alloyed simultaneously into the undoped superlattice samples following RTA. Light from a1/2-meter scanning spectrometer with a tungsten lamp source was focussed between the contacts, and with a dc bias applied, sample current was recorded as a function of incident wavelength. The resulting photocurrent spectra strongly resemble optical absorption profiles for MQW samples [Collins et al. 1986, Tsang et al. 1987].

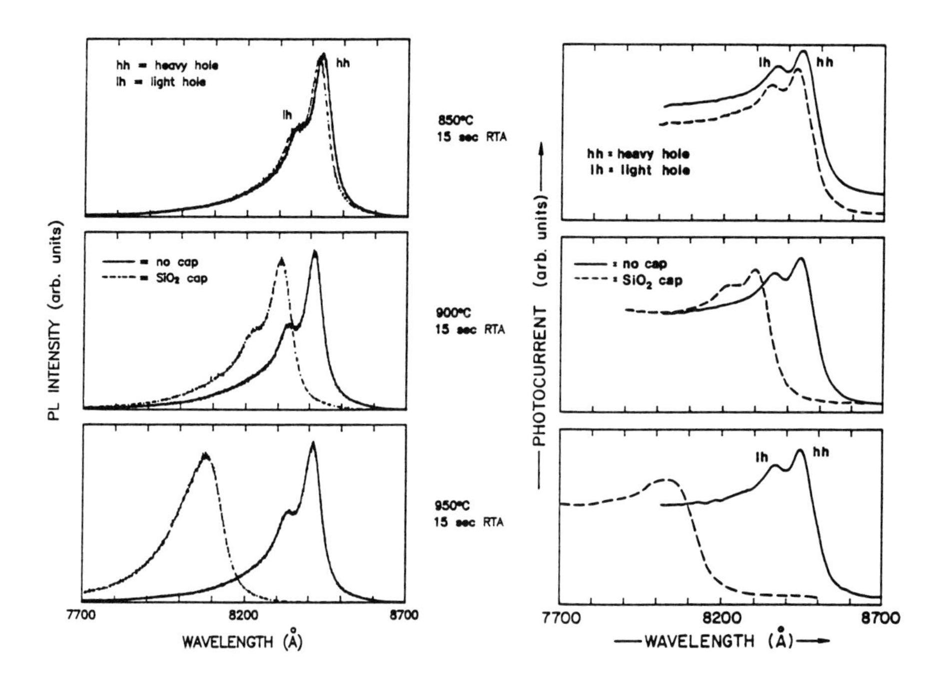

Fig. 2. Room temperature photoluminescence spectra of uncapped (solid lines) and SiO_2-capped (dashed lines) MQW samples following RTA at 850°,900°, and 950°C.

Fig. 3. Room temperature photocurrent spectra of uncapped (solid lines) and SiO_2-capped (dashed lines) MQW samples following RTA of 850°C, 900°, and 950°C. The two curves corresponding to 850°C have been shifted vertically for clarity.

Both PL and PC spectra exhibit clearly resolved n = 1 heavy hole (hh) and light hole (lh) exciton transitions which, for those regions not capped with SiO_2, remain essentially unaltered over the entire RTA temperature range. In contrast, exciton transitions in the SiO_2 capped regions are shifted to higher energies following RTA, which can be attributed both to a decrease in the effective width of the quantum wells and an increase in the Al composition within the wells. Measured shifts in exciton peak energies are 2 meV (850° RTA), 19 meV (900° RTA), and 61 meV (950° RTA). Due to compositional grading of the quantum well/barrier interfaces, the exciton peaks are also broadened in the partially intermixed samples. Both hh and lh peaks remain resolved following the 850°and 900°C anneals, whereas only a single broad excitonic peak is observed in the photocurrent spectra following the 950° anneal.

5. Summary

Room temperature exciton behavior has been studied in partially intermixed GaAs/AlGaAs MQW samples, with energy shifts as great as 61 meV demonstrated while still maintaining excitonic features. The ability to continuously vary the effective bandgap and still retain the exciton-related absorptive and refractive properties of MQW structure has a variety of potential optical device applications.

Acknowledgements

The authors wish to thank W.J. Schaff for technical assistance. This work was funded by the Joint Services Electronics Program.

References

Camras M D, Coleman J J, Holonyak N Jr., Hess K, Dapkus, P D and Kirkpatrick C G Proc. 10th Intnat. Symp. on GaAs and Related Compounds, Albuquerque, 1982, ed. G.E. Stillman (Inst. Phys. Conf. Ser. No. 65, London, 1983), p. 233.
Chemla D S, Miller D A B, Smith P W, Gossard A C, and Wiegmann W, 1984 IEEE J. Quantum Electr. QE-20 265.
Coleman J J, Daplus P D, Kirkpatrick C G, Camras M D, and Holonyak N Jr. 1982 Appl. Phys. Lett. 40 905.
Collins R T, v. Klitzing K, and Ploog K, 1986 J. Vac. Sci. Technol. B4 986.
Deppe D G, Guido L J, Holonyak N Jr., Hsieh K C, Burnham R D, Thornton R L, and Paoli T L, 1986 Appl. Physics Lett. 49 510."
Epler J E, Burnham R D , Thornton R L, Paoli T L, and Bashaw M C, 1986 Appl. Phys. Lett. 49 1447.
Ilegems M and Pearson G L, 1970 Phys. Rev. B1 1576.
Laidig W D, Holonyak N Jr., Camras M D, Hess K, Coleman J J, Dapkus P D and Bardeen J, 1981 Appl. Phys. Lett. 38 776.
Meehan K, Holonyak N Jr., Brown J M, Nixon M A, Gavrilovic P and Burnham R D, 1984 Appl. Phys. Lett. 45 549.
Merz J L, Barker A S Jr., and Gossard A C, 1977 Appl. Phys. Lett. 31 117.
Miller D A B, Chemla D C, Eilenberg D J, Smith P W, Gossard A C, and Tsang W T, 1982 Appl. Phys. Lett. 41 679.
Miller D A B, Chemla D S, Damen T C, Gossard A C, Wiegmann W, Wood T H, and Burrus C A, 1985a Phys Rev. B32 1043.
Miller D A B, Chemla D S, Damen T C, Wood T H, Burrus C A, Gossard A C and Wiegmann W, 1985b IEEE J. Quantum Electr. QE-21 1462.
Ralston J D, Moretti A L, Jain R K, and Chambers F A, 1987 Appl. Phys. Lett. 50 1817.
Tsang W T, Schubert E F, Chu S N G, Tai K, and Sauer R, 1987 Appl. Phys. Lett. 50 540.

MBE growth of InAsSb strained-layer superlattices

L. R. Dawson

Sandia National Laboratories, Albuquerque, New Mexico, 87185, USA

ABSTRACT: Tensile strain parallel to [100] interfaces in $InAs_xSb_{1-x}$ strained-layer superlattices (SLSs) has been used to reduce the energy gap below that of unstrained alloys. Dislocation-free SLSs with $\bar{x}$ = .10 have been prepared. Cracking in structures with $\bar{x}$ > .10 has been overcome by the initial growth of a severely mismatched $InAs_{.35}Sb_{.65}$ buffer layer, followed by compressively strained $InAs_{.27}Sb_{.73}$ and $InAs_{.20}Sb_{.80}$ buffer layers. SLSs with $\bar{x}$ ~ .20 show substantial dislocation filtering and significant optical absorption at wavelengths beyond those for equivalent bulk alloys.

1. INTRODUCTION

Conventional III-V semiconductor materials are not suitable for the fabrication of intrinsic detectors with cutoff wavelength, λ_c, as large as 12μm, since no bulk III-V alloys have small enough energy gap, E_g, at typical operating temperatures (77K). The minimum Eg for III-V materials occurs in the $InAs_xSb_{1-x}$ system for x ~ .35 (Woolley 1964), with 77K values corresponding to λ_c ~ 9μm (Osbourn 1984). One means of obtaining III-V materials structures with sufficiently small E_g is the use of tensile strain parallel to [100] interfaces in $InAs_xSb_{1-x}$ strained-layer superlattices (SLSs) (Osbourn, 1984). Layers of $InAs_{.35}Sb_{.65}$ (minimum unstrained E_g), when alternated with layers of $InAs_{.35-y}Sb_{.35+y}$ (larger bulk lattice constant), experience tensile strain parallel to the interfaces and a reduction in E_g. Figure 1 shows an SLS structure suitable for λ_c = 12μm. The well layers are relatively thick (≥ 175 Å) to minimize quantum size effects, while the barrier layers are even thicker to increase the fraction of lattice mismatch accommodated by the well layers. The structure shown in Figure 1 has a lattice constant equal to that of bulk material with x = .22, suggesting the use of InSb substrates to minimize the net mismatch between the epitaxial structure and its parent substrate.

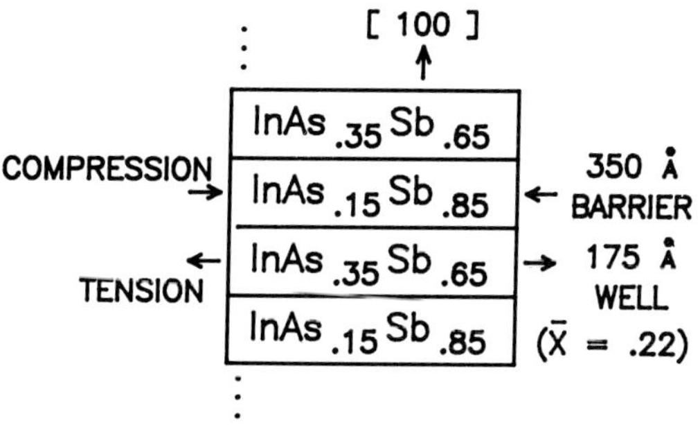

Figure 1. SLS structure for $\lambda_c \geq 12\mu m$.

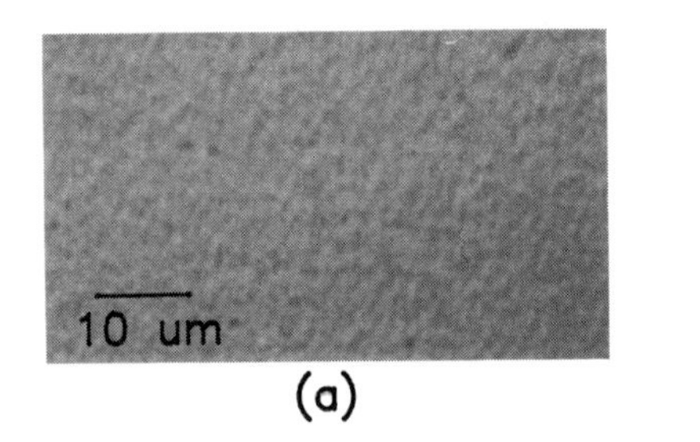

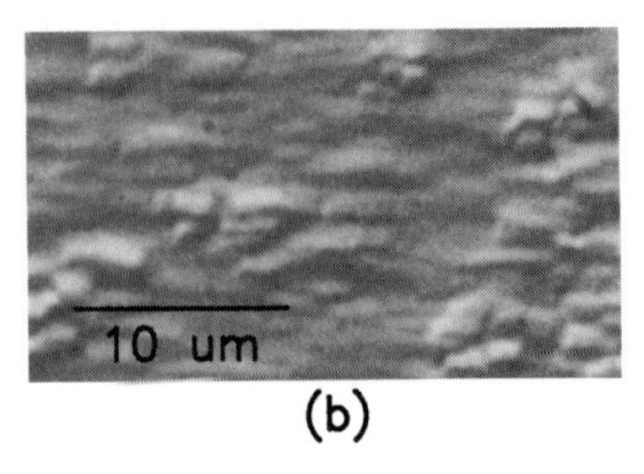

Figure 2. InSb surface for a) optimized Sb/In flux ratio; b) Sb-rich growth conditions.

2. MBE GROWTH

The low melting temperatures of both the InSb substrate (525°C) and $InAs_xSb_{1-x}$ alloys (<540° for $x \leq .35$) (Woolley 1958, Stringfellow 1971) require low growth temperature, typically 425 - 450°C, to prevent surface deterioration and interdiffusion effects. At these temperatures Sb has low vapor pressure [$P_{Sb\ (bulk)} \sim 10^{-5}$ torr] and does not readily evaporate from the surface. Consequently little excess Sb can be tolerated without severe degradation of the surface morphology by macroscopic Sb accumulation. Figure 2a shows the surface of an InSb layer grown under an optimum Sb/In flux ratio. Figure 2b shows the surface when the Sb flux is increased 25% beyond that at which the surface is clearly In-rich, i.e., there is macroscopic accumulation of In droplets. Since the Sb/In ratio must be held within narrow limits, very stable Sb and In sources are required for reproducible growth of high quality InSb and InAsSb alloys.

In the growth of InAsSb alloys, arsenic is not incorporated efficiently using an As_4 source, since the vapor pressure of As at these temperatures [$P_{As\ (bulk)} \sim 1$ torr] is much higher than that of Sb and the ensuing short surface residence time is not sufficient for decomposition of the arsenic tetramer. The growth of alloys with $x > .10$ requires large As_4 overpressures, which seriously degrade the surface quality. Efficient arsenic incorporation is obtained by using an As_2 source, produced by thermal cracking at 900°C. When used in conjunction with fixed In and Sb_2 sources, arsenic incorporation in the alloy follows the relationship shown in Figure 3. The dependence of x on As_2 flux is consistent with an Sb_2/As_2 sticking coefficient ratio of 3.

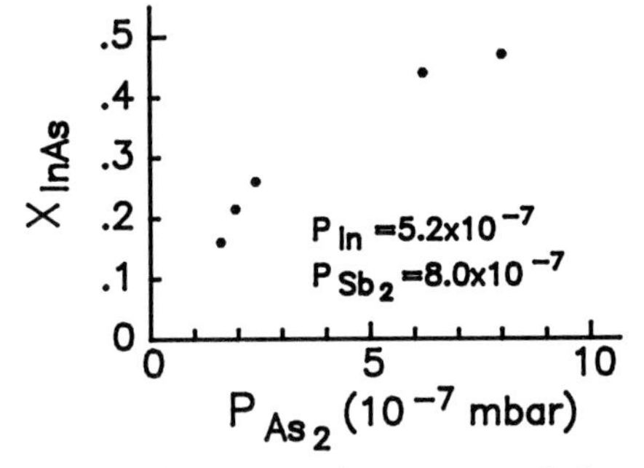

Figure 3. $InAs_xSb_{1-x}$ composition as a function of As_2 flux for fixed In and Sb_2 flux.

3. SLS RESULTS

SLS structures of the form shown in Figure 1 are grown on [100] InSb substrates with intervening InAsSb buffer layers chosen to be lattice matched to the overlying

SLS. Such epitaxial structures are in tension with respect to the parent substrate, and for x $\geq$ .07, thick $InAs_xSb_{1-x}$ buffer layers show severe cracking. This behavior is in agreement with the theory of Matthews (Matthews 1972) if it is assumed that equilibrium densities of dislocations are not formed to relieve the strain in thick layers, allowing strain energy to accumulate as growth proceeds. In low net mismatch structures ($\bar{x} \leq .10$) the use of step-graded buffers is effective in eliminating cracks, and provides an excellent medium on which to grow high quality, nearly dislocation-free $InSb/InAs_{.2}Sb_{.8}$ SLSs, as shown in Figure 4. Attempts to approach the $\bar{x}$ = .22 structure of Figure 1 by incremental increases in x beyond 0.10 result in severe cracking of the epitaxial layers, even with the use of a wide variety of buffer layer structures, including step-grading, linear-grading, variable-period superlattices, and the total omission of a buffer layer.

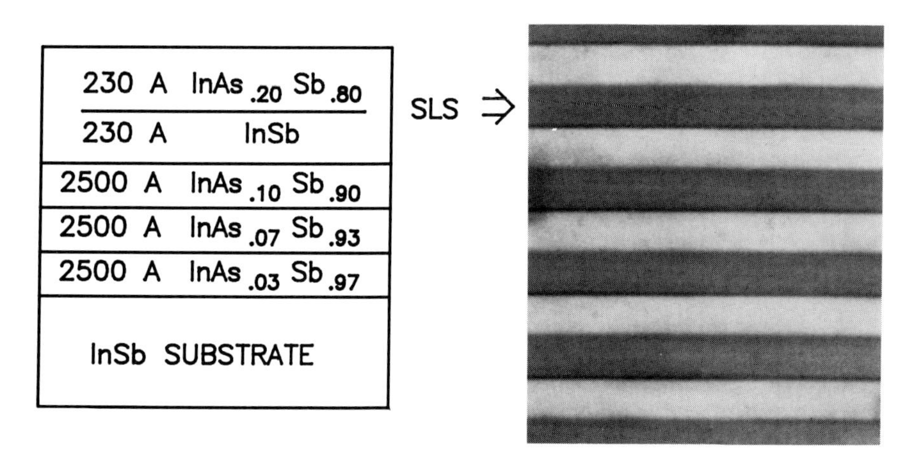

Figure 4. Dislocation-free SLS (x = .10) grown on a step-graded buffer layer.

The cracking problem has been eliminated by use of a more complex buffer structure. Initially a severely mismatched $InAs_{.35}Sb_{.65}$ buffer layer is grown directly onto the InSb substrate. The high level of strain at this interface induces the nucleation of a large enough density of dislocations to relieve much of the strain in the ensuing epitaxial layer, avoiding the accumulation of enough strain energy for the formation of cracks. Subsequent buffer layers with x = .27 and x = .20 are grown to establish lattice matching for the desired SLS. These layers are in compression and do not exhibit cracks. They also provide beneficial strain fields for filtering many of the dislocations created during the growth of the initial severely strained layer. $InAs_{.3}Sb_{.7}/InAs_{.1}Sb_{.9}$ SLSs grown on such buffers are crack free and show substantial reduction in dislocation density relative to that in the buffer layers, as shown in Figure 5.

For optical characterization SLS structures similar to that of Figure 5 have been prepared with the total SLS thickness exceeding 5μm. Preliminary FTIR absorption measurements indicate significant absorption for λ beyond 12μm in materials with unstrained $\lambda_c \sim 9$ μm (Dawson).

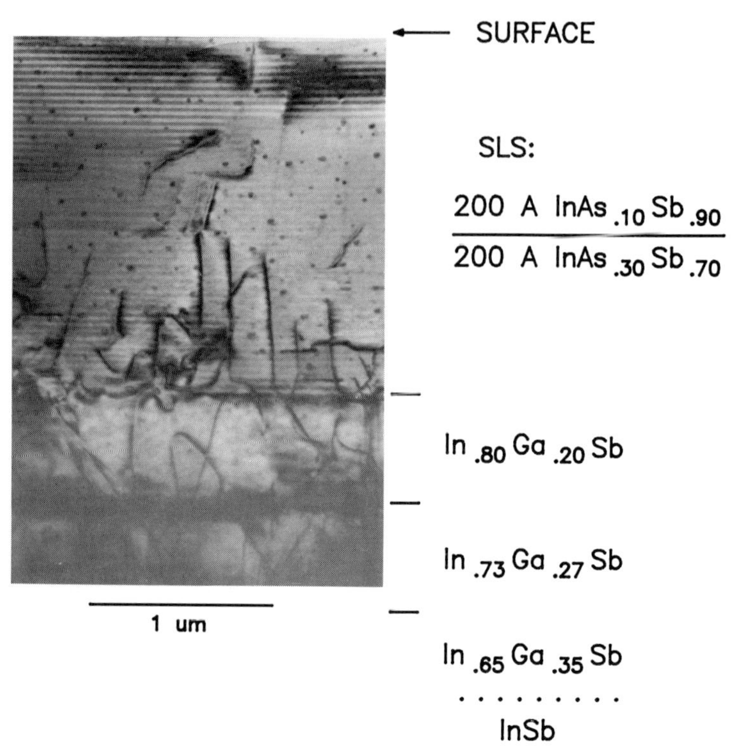

Figure 5. Dislocation filtering in an SLS (x = .20) grown on a crack-free buffer layer.

4. CONCLUSIONS

MBE Growth conditions have been established for a wide range of $InAs_xSb_{1-x}$ alloys using In, Sb_2, and As_2 sources. $InSb/InAs_{.2}Sb_{.8}$ SLSs are crack-free and essentially dislocation-free. Cracking is avoided in more As-rich SLSs by using buffer layers which combine severe tensile strain with modest compressive strain to provide a lattice-matched environment for the SLS. Initial FTIR measurements on thick $InAs_{.3}Sb_{.7}/InAs_{.1}Sb_{.9}$ SLSs show substantial absorption for $\lambda \geq 12$ μm, suggesting significant shift in Eg due to the strain field in the SLS.

5. REFERENCES

Dawson, L. R., et al., to be published.
Matthews, J. W. and Klokholm, E., 1972, Mat. Res. Bull. 7, 213
Osbourn, G. C., 1984 J. Vac. Sci. Technol. B 2, 176
Stringfellow, G. B. and Greene P. E., 1971, J. Electrochem. Soc. 118, 805
Woolley, J. C., and Warner J., 1964, Can. J. of Physics 42, 1879
Woolley, J. C., and Smith, B. A., 1958, Proc. Phys. Soc. 72, 214

Acknowledgements

The author wishes to thank R. E. Hibray and C. R. Hills for excellent technical assistance, G. C. Osbourn for many helpful discussions, and the late R. J. Chaffin for his persistent support of the work.

Electronic states in the ultrathin superlattices $(GaAs)_n(AlAs)_n$ and $(GaAs)_n(ZnSe)_n$

G.P. Srivastava
Physics Department, University of Ulster, Coleraine BT52 1SA, UK.

ABSTRACT: Using the first-principles pseudopotential method we have studied the electronic states in the ultrathin superlattices $(GaAs)_n(AlAs)_n(110)$ and $(GaAs)_n(ZnSe)_n(110)$, $1 \leq n \leq 4$. The energetics of growth of these superlattices is studied with the help of total energy calculations.

1. INTRODUCTION

Molecular-beam epitaxy and metalorganic vapour phase epitaxy techniques can be used to grow superlattices of desired layer thicknesses. It is important, therefore, to investigate the electronic properties of ultrathin superlattices. Using the first-principles pseudopotential method we present in this paper the electronic band structure of $(GaAs)_n(AlAs)_n(110)$ and $(GaAs)_n(ZnSe)_n(110)$, $1 \leq n \leq 4$. Also, we use the total energy method to study the energetics of growth of these superlattices.

2. DETAILS OF CALCULATION

We considered the superlattices based on the (110) face. The crystal system is simple tetragonal for n=1, and orthorhombic for $n \geq 2$. We used the norm-conserving non-local pseudopotentials of Bachelet et al (1982), a plane-wave basis set with 7 Ryd kinetic energy cut-off, and Ceperley-Alder's correlation scheme (Perdew and Zunger, 1981).
Self-consistency in the electronic charge density was achieved by using four special $\underline{k}$- points in the Brillouin zone (see e.g. Evarestov and Smirnov, 1983).

3. RESULTS AND DISCUSSION

The calculated equilibrium lattice constants of $(GaAs)_n(AlAs)_n$ and $(GaAs)_n(ZnSe)_n$, 5.52 Å and 5.22 Å respectively, are within 0.5% of the average of the theoretical lattice constants of their constituents. The much smaller lattice constant for $(GaAs)_n(ZnSe)_n$ may be due to inclusion of the 3d electrons in the core of Zn.

3.1 $(GaAs)_n(AlAs)_n$

The decrease in the theoretical band gap of these superlattices with increasing layer thickness n, shown in Figure 1(a), is in agreement with the photoluminescence measurements of Ishibashi et al (1985). The much lower theoretical band gaps are due to the local density approximation used in this work. When internal bond lengths are relaxed, the band gap of the monolayer thin

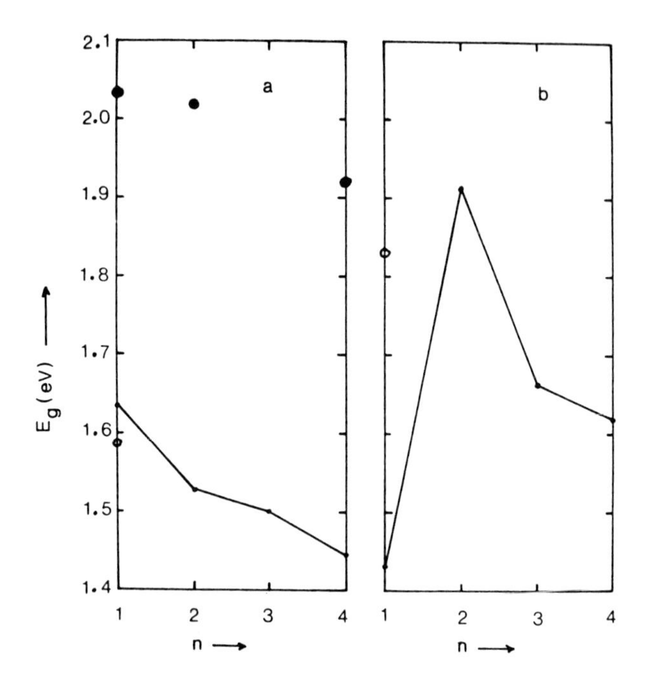

Fig. 1 Variation of the band gap with the number of layers n in (a) $(GaAs)_n(AlAs)_n$ and (b) $(GaAs)_n(ZnSe)_n$. The experimental results of Ishibashi et al are shown by heavy dots. The open circle indicates the band gap when internal bond lengths are fully relaxed.

superlattice decreases by 0.05 eV. The band gap in $(GaAs)_1(AlAs)_1$ is very nearly equal to the average of the band gaps in the constituent compounds GaAs and AlAs. There are superlattice gaps at the zone boundary $Y=\pi/a$ in the valence band structure of $(GaAs)_1(AlAs)_1$ as shown in Figure 2. The band structure does not significantly change as the layer thickness n increases from 1 to 4, except for more folding effects and vanishing of gaps at the zone boundary. The electronic states at the band edges are mixed states of folded bands of GaAs and AlAs. The mixed nature of the highest valence band at Γ for n=2 is shown in Figure 3. The enthalpy of formation of $(GaAs)_1(AlAs)_1(110)$, $\Delta H = E[(GaAs)_1(AlAs)_1] - E(GaAs) - E(AlAs)$, is ~+6 meV/atom. Thus we suggest that the monolayer thin superlattice is metastable, in agreement with the recent work of Wood et al (1987). The total energy of the superlattice increases with the layer thickness n. We calculate $E[(GaAs)_4(AlAs)_4] - E[(GaAs)_1(AlAs)_1]$ ~ 7.5 mRyd/atom. Thus thicker superlattices are inherently more unstable than the monolayer thin superlattice.

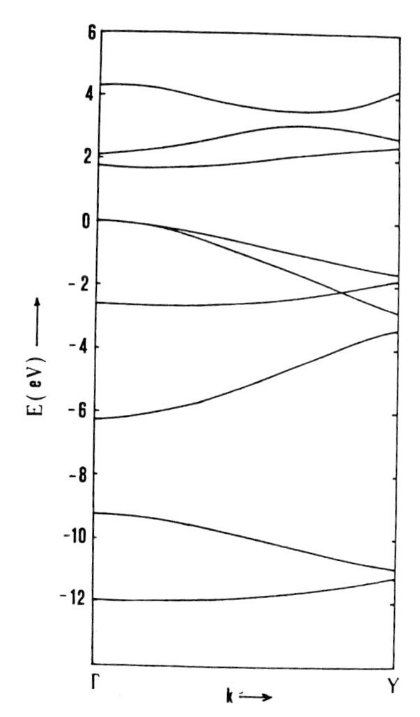

Fig. 2 Band structure of $(GaAs)_1(AlAs)_1$.

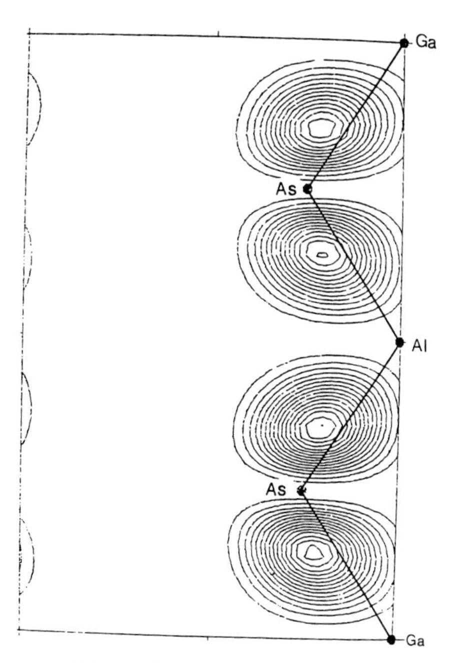

Fig. 3 Electronic charge density of the highest valence band in $(GaAs)_2(AlAs)_2$ at Γ.

3.2 $(GaAs)_n(ZnSe)_n$

As shown in Figure 1(b), for unrelaxed bonds the variation of the band gap in $(GaAs)_n(ZnSe)_n$ with n is very different from that in $(GaAs)_n(AlAs)_n$. Internal relaxation of bonds changes the band gap of $(GaAs)_1(ZnSe)_1$ dramatically. The band gap in the monolayer thin superlattice with relaxed bond lengths is about 0.44 eV lower than the average of the band gaps in GaAs and ZnSe. Thus the monolayer thin superlattice shows a large positive optical bowing. The band structure of $(GaAs)_1(ZnSe)_1$ is very different from the folded band structures of GaAs and ZnSe. Due to coexistence of four different bondings the valence band structure shows various splittings. The band structure of $(GaAs)_n(ZnSe)_n$ gradually changes as n increases. This is due primarily to less pronounced effects of interfacial bonds Ga-Se and Zn-As for $n \geq 3$ than for $n=1$. Figures 4(a) and (b) show a comparison of the band structures of $(GaAs)_1(ZnSe)_1$ and $(GaAs)_4(ZnSe)_4$. The most noticeable change is the vanishing of the band gap around -13 eV in $(GaAs)_4(ZnSe)_4$. These superlattices are direct band gap materials.

The enthalpy of formation of $(GaAs)_1(ZnSe)_1$ is ~+3 mRyd/atom, suggesting that the monolayer thin superlattice is metastable.

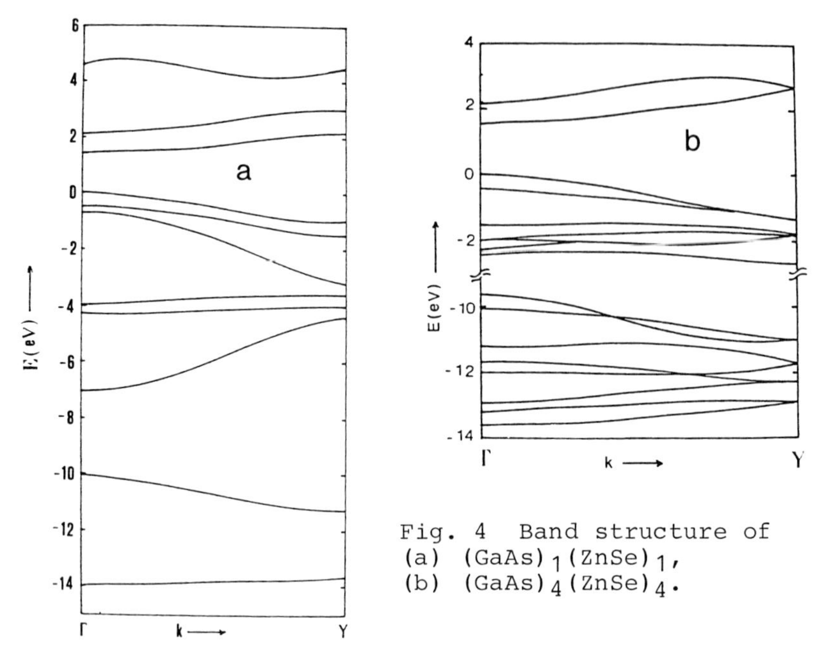

Fig. 4 Band structure of (a) $(GaAs)_1(ZnSe)_1$, (b) $(GaAs)_4(ZnSe)_4$.

The total energy increases with n, reaching $E[(GaAs)_4(ZnSe)_4] - E[(GaAs)_1(ZnSe)_1] \sim +8$ mRyd/atom. Thus thicker superlattices are more unstable than is the monolayer thin superlattice. This finding lends support to the observation of Tu and Kahn (1985) that GaAs-ZnSe interfaces are reactive and thermodynamics may play a big role in their growth.

4. CONCLUSIONS

Most of the electronic states in the ultrathin superlattices $(GaAs)_n(AlAs)_n$ can be fairly understood in terms of folded bands of GaAs and AlAs. The electronic states in $(GaAs)_1(ZnSe)_1$ are very different from those in thicker superlattices. The band edges of the ultrathin superlattices studied in this work are mixed states of their constituents. Monolayer thin superlattices are inherently unstable towards disproportionation into compounds. Thicker superlattices are even more unstable.

Bachelet GB, Hamann Dr. and Schlüter M 1982 Phys. Rev. B26 4199.
Evarestov RA and Smirnov VP 1983 Phys. Stat. Sol. (b) 119 9.
Ishibashi A, Mori Y, Itabashi, M. and Watanabe, N. 1985 J. Appl. Phys. 58 2691.
Perdew JP and Zunger A 1981 Phys. Rev. B23 5048.
Tu D-W and Kahn A 1985 J. Vac. Sci. Technol. A3 922.
Wood DM, Wei S-H and Zunger A 1987 Phys. Rev. Lett. 58 1123.

Inst. Phys. Conf. Ser. No. 91: Chapter 6
Paper presented at Int. Symp. GaAs and Related Compounds, Heraklion, Greece, 1987

X-ray Bragg-diffraction analysis of one-dimensional quasiperiodic AlAs/GaAs superlattices

L. Tapfer* and Y. Horikoshi

NTT ECL, Musashino-shi, Tokyo 180, Japan
* present address: Max-Planck-Inst. f. Festkörperforschung, Stuttgart, F.R.G.

ABSTRACT: The structural properties of one-dimensional quasiperiodic AlAs/GaAs superlattices (Fibonacci superlattices) as determined by X-ray diffractometry are discussed. We show simple algebraic formulae which allow an easy interpretation of the X-ray diffraction pattern and yield detailed knowledge on the structural parameter of the quasiperiodic superlattice. The effect of interface roughness and defects (mistakes) in the quasiperiodic chain are also investigated.

1. INTRODUCTION

There is presently a large interest in the electrical, optical and magnetic properties of one-dimensional quasiperiodic structures (Merlin et al. 1985, Fujita and Machida 1986, Das Sarma et al. 1986, Karkut et. al. 1986). The unique properties are strongly correlated to the structure of the system. In this paper we report on a detailed structural study of 1-dimensional quasiperiodic AlAs/GaAs superlattices by using X-ray diffractometry.

2. THEORETICAL

It was first shown by Merlin et al. (1985) that it is possible to realize quasiperiodic (or incommensurate) superlattices by using the Fibonacci sequence to line up two different materials. The Fibonacci sequence F_1, F_2, F_3,... is defined recursively by $F_{n+1}=F_n+F_{n-1}$ where F_n is called the n-th Fibonacci number. The arithmetic properties of the Fibonacci sequence is correlated to the gold section number $\tau = (1+\sqrt{5})/2 = 1.618...$ (Dodd 1983).A very useful algebraic formula to determine the type of the n-th layer in a Fibonacci chain is given by $\Omega=INT(n\tau^{-1}+1)-INT((n-1)\tau^{-1}+1)$; if $\Omega=0$ then layer A, otherwise layer B. A characteristic property of the Fibonacci superlattice (FSL) is that there is no superlattice unit cell which may forbid us to apply a recursive formalism to calculate the X-ray diffraction pattern. Therefore, a layer by layer calculation is performed to determine the X-ray reflectivity as a function of the scattering angle by using the dynamical diffraction model by Bartels (1986).

3. EXPERIMENTAL

The X-ray diffraction pattern (XRDP) from the FSL are recorded with a computer- controlled high-resolution double-crystal X-ray diffractometer. The Fibonacci superlattices were grown in a molecular beam epitaxy (MBE) system. The epitaxial layers were deposited onto <001>-oriented GaAs substrates at a growth temperature of 580 ^{0}C.

4. RESULTS AND DISCUSSION

In Fig.1 we show the theoretical XRDP a perfect AlAs/GaAs FSL around the symmetrical (004) and the quasiforbidden (002) and (006) GaAs peaks. This FSL is composed of 233 layers in total (13 generations of the Fibonacci sequence), i.e., 144 GaAs and 89 AlAs layers. The GaAs layers are 2.6 nm (d_A = 9 monolayers) and the AlAs layers are 4.2 nm (d_B = 15 monolayers) thick. The characteristic dense set (Cantor set) of satellite peaks are shown in the vicinity of the main SL-peak denoted by "0". The self-similiar property of the satellite peak distribution is also clearly observed, i.e., the FSL peaks are "equidistant" if the x-axis (angle) is plotted in a logarithmic scale. In accordance with the projection method (V. Elser 1985, Zia and Dallas 1985) the angular distance $\Delta\Theta_{q,p}$ between the strongest satellite peaks (q anp p are integers), and the main SL-peak "0" is given by (Merlin et al. 1985)

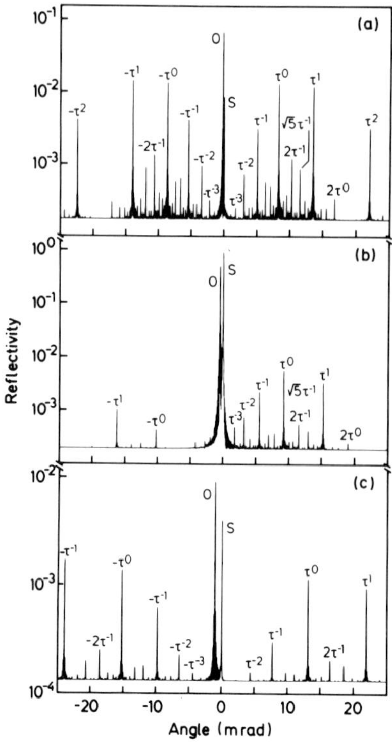

Fig.1 Calculated XRDP of a perfect AlAs/GaAs FSL for the 002 (a), 004 (b), and 006 (c) CuKα_1 - reflections

$$(1) \quad \Delta\theta_{q,p} = \lambda|\gamma_h| q\tau^p \, (\Lambda_F \sin 2\theta_B)^{-1}$$

where Θ_B is the kinematic Bragg angle, $\gamma_h = \sin(\Theta_B + \alpha)$ with α the angle between crystal surface and diffraction plane, λ is the X-ray wavelength and $\Lambda_F = d_B + \tau d_A$ is the average FSL period length. The average mole fraction $\tilde{x}$ is related to the angular distance $\Delta\Theta_e$ between the main epilayer peak "0" and the GaAs substrate peak by the equation

$$(2) \quad \tilde{x} = \Delta\theta_e \, (a_{GaAs} \, \Delta a \, (1 + c_{12}/c_{11}) \tan \theta_B)^{-1}$$

Here, c_{11} and c_{12} are the elastic stiffness constants and $\Delta a = a_{AlAs} - a_{GaAs}$ is the difference in the lattice constant. The average composition $\tilde{x}$ is also related to the thickness of the individual AlAs and GaAs layers which is

$$(3) \quad \tilde{x} = d_B/\Lambda_F = d_B \, (d_B + \tau d_A)^{-1}$$

The individual layer thicknesses of the AlAs and GaAs layers are then easily obtained by using equations (1)-(3). The theoretical intrinsic linewidth of the FSL satellite peaks $q\tau^p$ is given in the kinematical limit by

$$(4) \quad \Delta\omega_{q,p} = 2 \arcsin \, (\lambda (4\cos\theta_B \, T_F)^{-1})$$

where the total thickness of the FSL is $T_F = F_{n-2} {}^* \Lambda_F$. Especially the (002) reflection exhibit a dense set of peaks very close to the main peaks "0" and "S" and even the intensity of higher order peaks are well pronounced. This finding can be explained by the large difference of the structure factors for the AlAs and GaAs materials at the (002) reflection (Tapfer and Ploog 1986). However, a better angular resolution of the peak distributions is obtained by reflections of higher diffraction order.

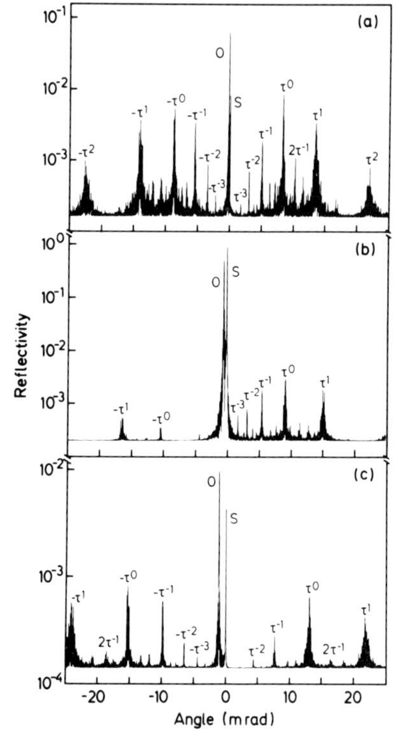

Fig.2 Calculated XRDP of a FSL assuming a roughness at the AlAs/GaAs heterointerface for the 002 (a), 004 (b), and 006 (c) $CuK\alpha_1$ - reflections

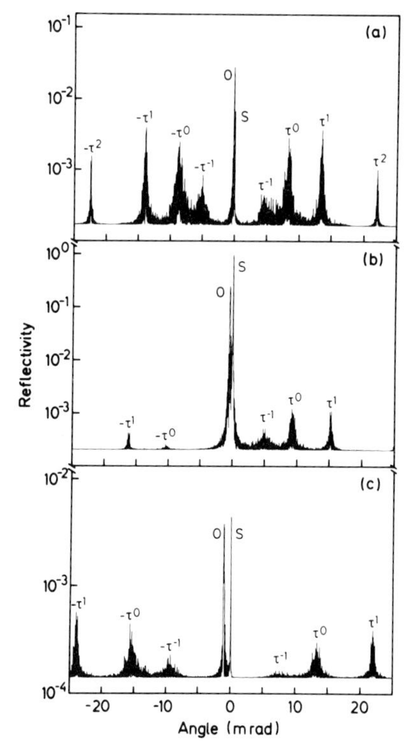

Fig.3 Calculated XRDP of a FSL if the AAB units are placed randomly in the Fibonacci chain for different $CuK\alpha_1$ - reflections: 002 (a), 004 (b), and 006 (c)

In Fig.2 the calculated XRDP of a FSL with the same structural parameters as for the perfect FSL shown in Fig.1 are presented assuming a roughness at the AlAs/GaAs heterointerface. We considered in our calculation that the thickness of the AlAs and GaAs layers fluctuates by ±1 monolayer with a probability of 0.4 which corresponds to an average interface roughness of 0.11 nm. The results show a broadening of the FSL satellite peaks and a decrease of their intensities while the weak satellite peaks are smeared out. The peak broadening $\omega_{q,p}$ can be given analytically in the kinematical limit by the approximation

$$(5) \quad \omega_{q,p} = \lambda q \tau^p \delta_i \, (4\cos\theta_B \Lambda_F{}^2)^{-1}$$

where δ_i is the average heterointerface roughness. From Fig.2 as well as from equation (5) it may be deduced that the line broadening (FWHM) increases with the angular distance between the FSL satellite peak and the main "0" peak.

The Fibonacci sequence can be considered to be a 1-dimensional periodic lattice of the unit ABAAB and the substitutional defect unit AAB. If we introduce now the unit AAB randomly in the lattice chain (intentional mistake), but keeping the number of ABAAB units equal to the number of AAB units muliplied by τ, we obtain the calculated X-ray diffraction patterns as shown in Fig.3. Similar mistakes (phasons) were found experimentally by Lubensky et al. (1986) in 3-dimensional quasicrystals. Like the defects at the heterointerface these mistakes in the Fibonacci chain produce also a broadening of the satellite peaks but the broadening decreases with the distance between the satellite peak and the main peak "O". It is remarkable that the strongest satellite peaks $p\tau^p$ are still at their angular position and well pronounced while only the weaker peaks have disappeared. This finding suggests that this quasi-Fibonacci structure with mistakes represent still a FSL but with a large random component. It will be interesting to investigate the electrical and optical properties of this structure in comparison with the properties of a perfect FSL.

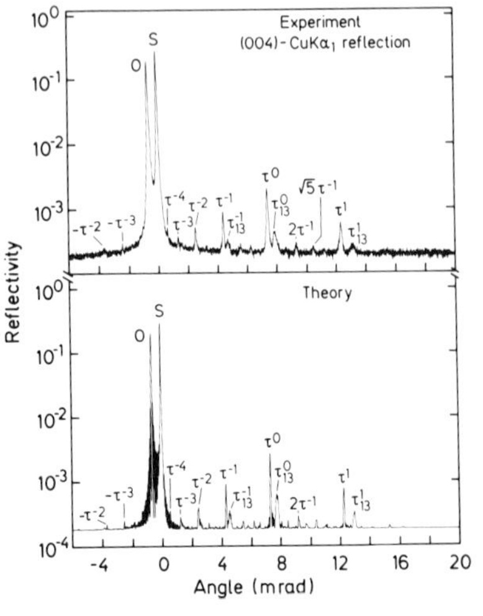

Fig.4 Experimental XRDP of a FSL grown by MBE for 233 AlAs and GaAs layers and the corresponding calculated XRDP

Fig. 4 shows the experimental and theoretical XRDP of an AlAs/GaAs FSL. An excellent agreement between theory and experiment is observed, demonstrating the high quality of the grown FSL. The thickness of the individual GaAs and AlAs layers are d_A=3.92 nm and d_B = 4.69 nm, respectively. The 13-th Fibonacci generation was however grown with a different AlAs layer thickness (d_B = 3.33 nm), which is also observed by the splitting of the FSL satellite peaks (the peaks arising from the 13-th generation are noted by the index 13). The angular splitting follows equation (5) too. It should be noted that even very weak satellite peaks, for example τ^{-3}, τ^{-4} and $-\tau^{-3}$, can be still observed as predicted by the calculated diffraction pattern.

5. CONCLUSION

We reported a theoretical and experimental study of the structural properties of one-dimensional quasiperiodic superlattices (Fibonacci superlattices) by using the X-ray diffraction method. We demonstrated that the structural parameters can be obtained directly from the experimental diffraction data through simple algebraic formulae.

ACKNOLEDGEMENT

The authors would like to thank M. Kawashima and H. Yamaguchi for the sample preparation. They also wish to thank Drs. T. Kimura and T. Izawa for their encouragement throughout this work.

Bartels W J, Hornstra J and Lobeek D J W 1986 Acta Cryst. A42 539
Das Sarma S, Kobayashi A, Prange R E 1986 Phys. Rev. B34 5309
Dodd F W 1983 Number Theory in the Quadratic Field with Golden Section Unit (Passaic: Polygonal Publishing House)
Elser V 1985 Phys. Rev. B32 4892
Fujita M and Machida K 1986 Solid State Comm. 59 61
Karkut M, Triscone J, Ariosa D and Fischer O 1986 Phys. Rev. B34 4390
Lubensky T C, Socolar J E, Steinhardt P J, Bancel P A and Heiney P A 1986 Phys. Rev. Lett. 57 1440
Merlin R, Bajema K, Clark R, Juang F-Y and Bhattacharya P K 1985 Phys. Rev. Lett. 55 1768
Tapfer L and Ploog K 1986 Phys. Rev. B33 5565
Zia R K P and Dallas W J 1985 J. Phys. A: Math. Gen. 18 L341

Electronic properties of $(GaAs)_n(AlAs)_n(001)$ superlattices and of $Ga_{1-x}Al_xAs$ random alloys

M. Posternak and A. Baldereschi*

Institut de physique appliquée, Ecole polytechnique fédérale
CH-1015 Lausanne, Switzerland

S. Massidda and A.J. Freeman

Material Research Center and Physics Department, Northwestern University,
Evanston, Il 60201, USA

ABSTRACT : Results of full-potential linearized augmented plane-wave calculations are reported for $(GaAs)_n(AlAs)_n(001)$ superlattices with n = 1,2 and for the ideal simple-cubic compounds Ga_3AlAs_4 and $GaAl_3As_4$. The most relevant energy gaps are calculated self-consistently using the local-density ground-state approximation (LDA) and are then corrected using experimental data on the pure compounds GaAs and AlAs. The final values of the energy gaps are compared with existing experimental data both as a function of composition and of superlattice period (= n). In particular, our results indicate that the lowest gap of $(GaAs)_1(AlAs)_1$ is smaller than that of $(GaAs)_2(AlAs)_2$. Furthermore, corrections to LDA are essential for explaining the order of the lowest conduction states at Γ in the superlattices, and the direct-indirect gap cross-over in $Ga_{1-x}Al_xAs$.

The electronic structure of $(GaAs)_n(AlAs)_m$ superlattices has been studied widely in recent years, mainly for m = n (Nakayama and Kamimura 1985, Christensen *et al* 1985, Gell *et al* 1986, Nelson *et al* 1987, Oshiyama and Saito 1987, Cardona *et al* 1987), while the more complex problem of $Ga_{1-x}Al_xAs$ alloys has received considerably less attention (Baldereschi *et al* 1977).

In this paper, we study the electronic states of $(GaAs)_n(AlAs)_n$ (001) superlattices with n = 1,2 corresponding to the composition x = 0.5, and of the simple-cubic compounds Ga_3AlAs_4 and $GaAl_3As_4$ for the compositions x = 0.25 and x = 0.75, respectively. The superlattices have a tetragonal structure (space group D^1_{2d}) with 4n atoms per unit cell, while the simple-cubic compounds have the germanite structure (de Jong 1930) with space group T^1_d and 8 atoms per unit cell. Due to the almost perfect lattice matching between GaAs and AlAs, lattice parameter values have been chosen in order to correspond to the same lattice constant a = 10.66 a.u. of the related zinc-blende structure for all these systems. Electronic energies and wave functions are calculated with the self-consistent full-potential linearized augmented plane wave (FLAPW) method (Jansen and Freeman 1984). The local-density approximation (LDA) to density-functional theory with the Hedin-Lundqvist exchange-correlation potential is used. Valence energies are computed semi-relativistically, while the core states are treated fully relativistically and updated at each iteration. Great care has been devoted to the determination of the muffin-tin sphere radii, in order to minimize the amount of the semi-core Ga 3d and As 3d charge in the interstitial region (R_{MT} = 2.5 a.u. for Ga and Al, and 2.1 a.u. for As). The residual core charge spill-out is described by using the overlapping charge method.

The LDA band structures of the n = 1 and n = 2 superlattices are given in Figure 1 along the k_z axis (where single and double folding occur for n =1 and n =2, respectively). The 2π/a (0,0,0) and 2π/a (0,0,1) zinc-blende states are found at Γ for n = 1, while the additional 2π/a (0,0,1/2) state is present for n = 2. According to LDA, the n = 1 is a direct-gap material, but conduction states at R are only 34 meV higher in energy, so that the exact nature (direct/indirect) of the n = 1 system is in doubt. In contrast, the n = 2 superlattice has a well defined direct character since higher conduction states are located 0.4 eV higher in energy.

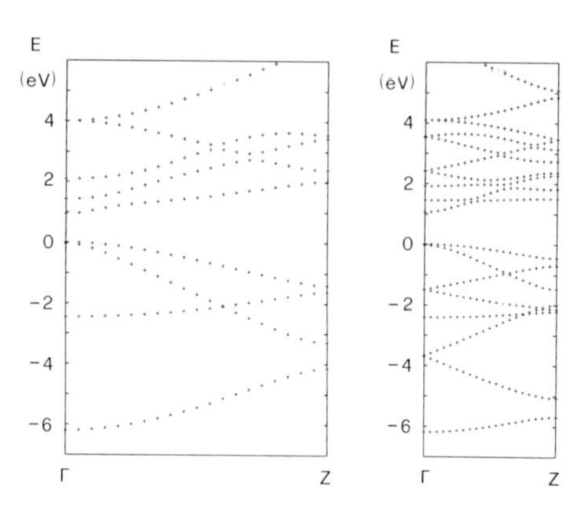

Fig. 1 Energy-band structure of the n = 1 (left panel) and n = 2 (right panel) superlattices along the Γ-Z direction. The lowest valence bands are not displayed.

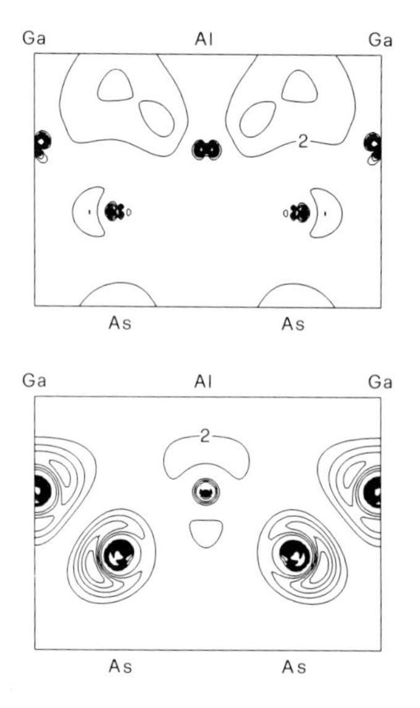

Fig. 2 Charge-density contour plots of the lowest (bottom panel) and second (top panel) conduction state at Γ for the n = 1 superlattice. Contour values are given in units of 0.001 $e/(a.u.)^3$, with subsequent contours separated by 0.002 $e/(a.u.)^3$.

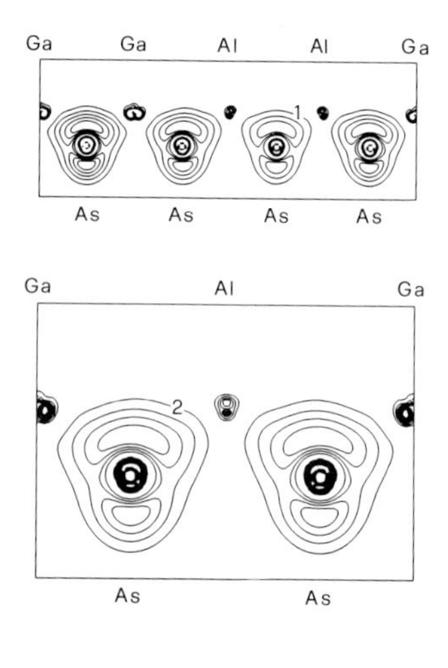

Fig. 3 Charge-density coutour plots of the highest valence state at Γ for the n = 1 (bottom panel) and n = 2 (top panel) superlattices. Contour values are given in units of 0.001 $e/(a.u.)^3$ with subsequent contours separated by 0.002 and 0.001 $e/(a.u.)^3$ for n = 1 and n = 2, respectively.

The LDA value of the lowest direct gap E_g for n =1 is 0.96 eV, which is much less than the experimental value of 2.2 eV. The underestimation of the gap by more than 1 eV is a typical consequence of LDA. A similar underestimate of E_g occurs for n = 2. Our LDA calculation predicts a value of E_g~ 0.1 eV larger than in the n =1 case, in agreement with experimental findings (Garriga *et al* 1987) and with existing ab-initio reports. For both superlattices, the lowest conduction state at Γ originates from a Γ state before folding. On the contrary, the second conduction state at Γ originates from the k = 2π/a(0,0,1) state of the zinc-blende structure. This state is located 0.48 eV (n =1) and 0.39 eV (n =2) higher in energy. This order of the two conduction states is well inside numerical accuracy, and agrees with LDA-LMTO results of Christensen *et al* (1985), while it is at variance with LDA-pseudopotential calculations of Nelson *et al* (1987). The reason of this discrepancy is not clear at present. The inversion of order is important since the two Γ conduction states have different density distributions (see Figure 2 for n = 1). The state with Γ origin is strongly localized in the GaAs region, while the state with 2π/a(0,0,1) origin is nearly equally shared between the GaAs and AlAs regions of the superlattice (with only slightly more charge density in the AlAs region).

The topmost valence state at Γ in both superlattices originates, of course, from Γ before folding. The density distribution of this two-fold degenerate state is given in Figure 3 for both superlattices (the average of the 2 degenerate states is given). Comparing the distributions for n = 1 and n = 2, it is evident that with increasing n, the state tends to become localized in the GaAs region. For these low-period superlattices, however, the state still extends over the full crystal.

The energy bands have also been calculated for the simple-cubic compounds Ga_3AlAs_4 and $GaAl_3As_4$. In this case, the four points (distinct in the zinc-blende structure) 2π/a(0,0,0), 2π/a(0,0,1), 2π/a(0,1,0) and 2π/a(1,0,0) fold into Γ. According to our LDA calculation, Ga_3AlAs_4 is a direct semiconductor at Γ. The lowest conduction state has Γ origin before folding. The conduction states originating in folding 2π/a(0,0,1) and equivalent points are 0.76 eV higher in energy . $GaAl_3As_4$ is instead an indirect-gap semiconductor with the minimum of the conduction band at the R point of the simple-cubic Brillouin zone. The conduction band minimum at Γ is 0.17 eV higher in energy.

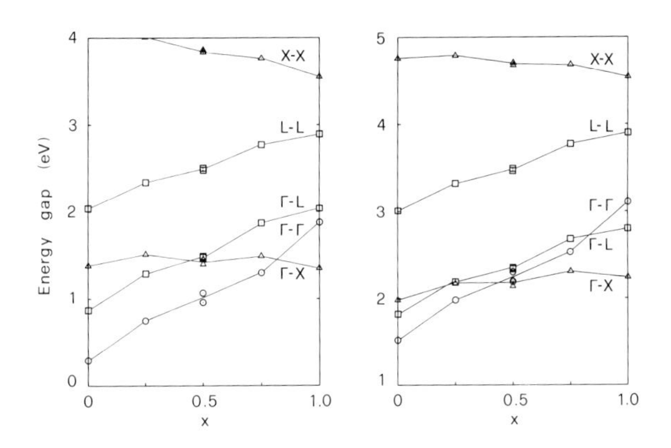

Fig. 4 Most relevant direct and indirect energy gaps in $Ga_{1-x}Al_xAs$ random alloys as a function of composition x. Left panel : LDA-FLAPW values. Right panel : results corrected as explained in text.

The n = 1, 2 superlattices and the two simple-cubic compounds may be used to obtain information on relevant energy gaps in $Ga_{1-x}Al_xAs$ random alloys. In order to do so, we must remove the splittings introduced by the periodicity of the supercrystals. This is done by averaging over states which are split in the supercrystal and are instead degenerate with each other in a random zinc-blende alloy. Identification of the states whose energies have to be averaged, and their classification according to the zinc-blende structure is done by considering state degeneracies and inspection of the angular-momentum decomposition inside APW spheres of the corresponding wave functions. The results obtained from the FLAPW energy levels are displayed on the left panel of Figure 4 for the most relevant direct and indirect gaps. The LDA results are in considerable disagreement with experimental data since (i) all gaps are ~1 eV smaller than the experimental values, and (ii) the alloy is predicted to be direct for values of x up to 0.8 whereas the accepted experimental value is x = 0.405. The origin of this disagreement is the use of the LDA ground-state density functional theory, as explained by Godby *et al* (1987) for pure GaAs and AlAs. These authors have shown that the corrections to be added to a given energy gap are about the same for the two end-point compounds. We can therefore correct the LDA results in Figure 4 by adding to a given energy gap, as a function of composition, the composition-weighted average of the corrections valid for the two end-compounds. The resulting energy gaps are given in the right panel of Figure 4, and have been obtained by using the experimental gap values in GaAs and AlAs in order to compute the x = 0 and x = 1 corrections. Comparison of the two panels of Figure 4 leads to the following observations : (i) the fact that corrections to LDA depend on the particular gap reduces considerably the composition where the direct-indirect cross-over occurs. This composition (x ~ 0.43) is now in much better agreement with experiment. (ii) a further effect of these corrections is the change of order between the two lowest conduction states at Γ for the $(GaAs)_n(AlAs)_n$ superlattices.While within LDA the lowest conduction state has Γ origin as discussed above, it becomes, after the corrections done in this work, the one with X origin for both n = 1 and n =2. Considering however the uncertainties introduced by these corrections, we can only conclude that in both superlattices the two lowest conduction states are nearly degenerate.

This work was supported by the Swiss National Science Foundation and partly by the U.S. National Science Foundation (through the NU Material Research Center) and by the U.S.-Swiss Cooperative Program (Grant No NSF-INT83-04346).

*Also at University Trieste, Italy

REFERENCES

Baldereschi A, Hess E, Maschke K, Neumann H, Schulze K R and Unger K 1977 *J. Phys. C : Solid St. Phys.* **10** 4709
Cardona M, Suemoto T, Christensen N E, Isu T and Ploog K 1987 *Phys. Rev. B* to appear
Christensen N E, Molinari E and Bachelet G B 1985 *Solid St. Commun.* **56** 125
Garriga M, Cardona M, Christensen N E, Lautenschlager P, Isu T and Ploog K 1987 *Phys. Rev. B* to appear
Gell M A, Ninno D, Jaros M and Herbert D C 1986 *Phys. Rev. B* **34** 2416
Goodby R W, Schlüter M and Sham L J 1987 *Phys. Rev. B* **35** 4170
Jansen H J F and Freeman A J 1984 *Phys. Rev. B* **30** 561
de Jong W F 1930 *Z. Kristallogr.* **73** 176
Nakayama T and Kamimura H 1985 *J. Phys. Soc. Japan* **54** 4726
Nelson J S, Fong C Y and Batra I P 1987 *Appl. Phys. Lett.* **50** 1595
Oshiyama A and Saito H 1987 *J. Phys. Soc. Japan* **56** 2104, and preprint

Deep levels in GaAs–AlGaAs superlattice structures

Y.J. Huang*, D.E. Ioannou*, Z. Hatzopoulos**, G. Kyriakidis**, A. Christou**,+ and N. A. Papanicolaou***

* Joint Program for Advanced Electronic Materials, Electrical Engineering Department, University of Maryland, College Park, MD 20742, and Laboratory for Physical Sciences, College Park, MD 20742.
** Research Center of Crete, Heraklion, Greece, and Physics Department, University of Crete, Heraklion, Greece.
*** Naval Research Laboratory, Washington, DC 20375.

ABSTRACT: DLTS was applied to investigate the deep levels in various GaAs-AlGaAs structures incorporating superlattices with varying period magnitudes and numbers. The structures were grown by MBE on n^+ GaAs substrates, and included a GaAs buffer layer and a top GaAs cap. A trap (T_{SL}) was observed in the (50Å/50Å) GaAs-AlGaAs superlattice with thermal activation energy 0.34±0.02 eV, and it is argued that this trap emits to the first miniband in the GaAs well. Traps (M1, M2, and M4) were also observed in the GaAs caps of the samples containing a small number of superlattice periods. The trap density decreased (below the detection limit of our spectrometer) as the number of periods increased, which may be explained entirely by strain accommodation.

1. INTRODUCTION

There is currently a great interest in superlattices either as active layers in device structures or as buffer layers. For successful operation of both minority and majority carrier devices with superlattices, it is important to detect and control the presence of electrically active defects in the superlattice structures. Previous studies have shown that by replacing the undoped AlGaAs spacer layer in an inverted modulation-doped heterostructure with a superlattice structure, it is possible to eliminate the persistent photoconductivity effect related to AlGaAs traps and enhance the electron mobility (Drummond et al 1983), and that low current threshold densities can be achieved in double heterojunction lasers by the incorporation of a superlattice interface (Fischer et al 1984). Moreover, it has been reported recently that with regard to deep traps within the superlattice, both the period magnitude and the number of periods involved may play an important role on the concentration of deep levels and the interpretation of their characteristics (Martin et al 1986).

In this paper we report the results of an investigation of deep traps in a number of GaAs(cap) layers grown by MBE on various AlGaAs-GaAs superlattice structures, as a function of superlattice period size and total thickness. We also report on the

---

+ Also with Naval Research Laboratory, Washington, DC 20375

observation of a trap in the superlattice, and its thermal emission properties. The conclusions resulting from this investigation may provide a method of reducing the deep trap concentrations in superlattice structures.

2. EXPERIMENTAL

The samples were grown by MBE on n^+ GaAs substrates, by first growing a n-GaAs buffer layer, followed by a $Al_xGa_{1-x}As$-GaAs (x=0.31) superlattice, and a top n-GaAs cap layer. The details of the growth conditions are shown in Table 1. All the superlattices (expect for sample 147) were Si doped.

Table 1: Sample growth conditions and material information

sample number	As_4/Ga	T_G (°C)	superlattice barrier/well	superlattice thickness (μm)	cap thickness (μm)	C-V data N_D (cm^{-3})
123	11.6	635	50Å/50Å	0.5	-	1.0×10^{16}
141	9.2	600	80Å/80Å	2.0	2.00	8.1×10^{15}
143	9.2	600	120Å/120Å	2.0	2.00	3.2×10^{15}
146	22	630	200Å/200Å	0.1	0.35	4.1×10^{17}
147	22	630	200Å/200Å	0.1	0.35	2.9×10^{17}
155	20	620	-	-	0.50	1.8×10^{17}

Schottky diodes for DLTS examination were prepared by first forming ohmic contacts on the backside by evaporation of Au-Ge-Ni and alloying. Al Schottky contacts were then formed by depositing the metal on carefully prepared surfaces and using standard photolithography to define an array of 500 μm dots with guard rings for surface leakage reduction. The deep traps and their concentrations were evaluated by DLTS experiments in the temperature range 80-400 K, and carrier concentrations were obtained from C-V measurements.

3. RESULTS AND DISCUSSION

DLTS measurements were performed under bias conditions such that for present cap thicknesses and doping levels the observed traps were in the GaAs cap layer, and only for sample 123 they were in the superlattice. Fig.1 shows typical DLTS spectra, and Fig.2 the corresponding Arrhenius plots. It is seen that sample 155 contains a substantial amount of the well known traps M1, M2, and M4 (Lang et al 1976), with energy levels 0.17 0.22 and 0.49 eV below the conduction band edge, and corresponding capture cross sections 3.8×10^{-15}, 5.3×10^{-15}, and 2.9×10^{-13} cm^2, respectively, as obtained from the Arrhenius plots. At the present sensitivity level no traps are revealed for samples 141 and 143. Further measurements showed that one out of four diodes in sample 143 and two out of four diodes in sample 141 revealed traps M2 and M4, just above the detection limit of our spectrometer (here $\sim 10^{11}$ cm^{-3}). Traps M1, M2 and M4 are also observed in substantial amounts in samples 146 and 147.

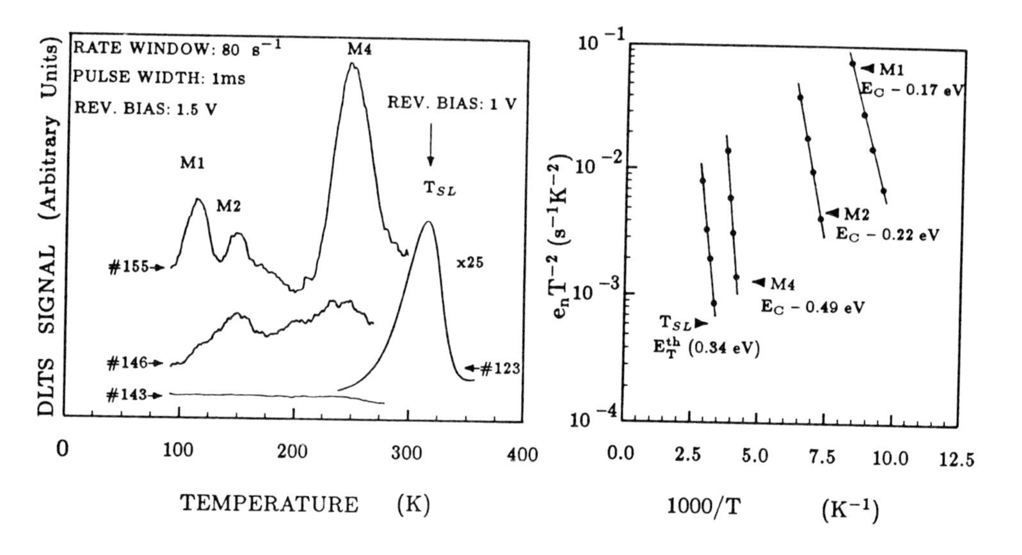

Fig. 1. DLTS spectra of electron traps in GaAs layers obtained from samples 155, 146, 143, and 123.

Fig. 2. Arrhenius plots for the peaks labeled M1, M2, M4 and T_{SL} in Fig. 2.

Finally, in sample 123 a trap(T_{SL}) is observed with thermal activation energy of 0.34 ± 0.02 eV. A summary of the concentrations of these traps in the various samples is given in Table 2.

Table 2: Measured concentrations of the various traps

(cm^{-3})	155	147	146	143	141	123
M4	8.83×10^{14}	4.43×10^{14}	2.41×10^{14}	1.03×10^{12}	-	-
M2	2.41×10^{14}	3.13×10^{14}	1.86×10^{14}	5.89×10^{11}	5.33×10^{12}	-
M1	3.84×10^{14}	2.60×10^{14}	6.42×10^{13}	-	-	-
T_{SL}	-	-	-	-	-	5.54×10^{14}

With regard to the GaAs traps M1, M2, and M4 in the cap, from the data in Tables 1 and 2 it is seen that a dramatic reduction (more than 10^2) in the trap concentration is achieved by the incorporation of a large number of spuerlattice periods grown at a As_4/Ga ratio of 9.2 (sample 141, 143). The incorporation of only a small number (three) of superlattice periods grown at a As_4/Ga ratio of 22, on the other hand, results in substantial trap concentrations (samples 146, 147), albeit less than when the superlattice is completely absent (sample 155). It is known that trap densities in MBE GaAs increase strongly with decreasing growth temperature (Blood et al 1984). However within the temperature range in Table 1, this behavior was not observed in our samples. The predominant behavior observed was the effect of strain accommodation of the superlattice and the effect of the As_4/Ga ratio. With regard to trap T_{SL} it should be noted that for this superlattice (period: 50Å/50Å) the width of the lowest miniband, as calculated by a Kronig-Penney model, is 0.096 eV - 0.081 eV = 0.015 eV. This seems to be wide enough for miniband conduction to be plausible (Martin et al 1986). In this case, the observed thermal activation energy is measured with respect to the lowest edge of the miniband E_{50} (=0.081 eV). Assuming the relationship $\Delta E_C=0.7\Delta E_G$ (Martin et al 1986) we obtain in this case ΔE_C=0.27 eV. We can now calculate the position of the trap with reference to the conduction band edges of the GaAs and AlGaAs. The values obtained are $E_{TSL}^{GaAs}=0.26\pm0.02$ eV, and $E_{TSL}^{AlGaAs}=0.53\pm0.02$ eV respectively. It is not clear in which of the two layers the trap lies, but from a review of published data on bulk traps in MBE grown GaAs and AlGaAs layers it is concluded that most likely this trap is the DX2 center often observed in AlGaAs (Dhar et al 1986).

4. CONCLUSIONS

In conclusion, we have presented the results of a study of deep traps in GaAs layers in various superlattice structures. Traps M1, M2, and M4 were observed in the GaAs cap layers of the samples containing a small number of superlattice periods. The concentration of these traps decreased dramatically with increasing number of superlattice periods. On the other hand, increasing the As_4/Ga ratio from 9 to 22 and reducing the total superlattice thickness, resulted in an increase of the trap concentration. A trap (T_{SL}) was observed in the 50Å/50Å superlattice with energy 0.34±0.02 eV below the lower edge of the first miniband. This may be related to the DX2 center in AlGaAs.

REFERENCES

Blood P and Harris J J 1984 J. Appl. Phys. 56 993

Dhar S, Hong W P, Bhattacharya P K, Nashimoto Y and Jung F Y 1986 IEEE Trans. on Electron Devices ED-33 698

Drummond T J, Klem J, Arnold D, Fisher R, Thorne R E, Lyons W G and Morkoc H 1983 J. Appl. Phys. Lett. 42 615

Fischer F, Klem J, Drummod T J, Kopp W and Morkoc H 1984 Appl. Phys. Lett. 44 1

Lang D V, Cho A Y, Grossard A C, Ilegems M and Wiegmann W 1976 J. Appl. Phys. 47 2558

Martin P A, Hess K, Emanuel M and Coleman J J 1986 J. Appl. Phys. 60 2882

Paper presented at Int. Symp. GaAs and Related Compounds, Heraklion, Greece, 1987

Energy band structure of $(GaAs)_1(InAs)_1$ (001) superlattice and $Ga_{4-n}In_nAs_4$ ($n=1, 3$) crystals

P.Bogusławski[1,2] and A.Baldereschi[2,3]

1. Institute of Physics, Polish Academy of Sciences, PL-02668 Warsaw
2. Institute of Applied Physics, EPFL, CH-1015 Lausanne
3. Department of Theoretical Physics, University of Trieste

ABSTRACT: Reduction of the optical band gap induced by long-range order in GaInAs and GaInP alloys has recently been reported. Analysis of ab-initio pseudopotential calculations for $Ga_{4-n}In_nAs_4$ (n=1,2,3) supercrystals shows that this effect is caused by a repulsion between the lowest conduction band and higher states folded from the edge of the Brillouin Zone. Distortion of the anion sublattice due to the lattice mismatch between GaAs and InAs closes the gap in Ga_3InAs_4 and $(GaAs)_1(InAs)_1$, but opens it in $GaIn_3As_4$.

1. INTRODUCTION

Since the observation of an ordered phase in AlGaAs alloys (Kuan et al. 1985), long-range order has been reported for GaAsSb (Jen et al. 1986), GaInAs (Nakayama and Fujita 1986, Shahid et al. 1987), AlInAs (Hull et al. 1986), and GaInP (Gomyo et al. 1987), which were intentionally grown as random alloys. Intentional growth of ultrathin superlattices has been achieved for AlGaAs (Gossard et al. 1976) and GaInAs (Fukui and Saito 1986, Katsumi et al. 1986). Bogusławski and Baldereschi (1987) have shown that the ordered phases of lattice mismatched alloys are unstable against phase segregation at T=0 due to the excess elastic energy, and that the observed ordering may be due to an even larger instability of the random alloys. In this work, we study the effects of the structural ordering on the energy band structure in lattice mismatched GaInAs alloy.

The sublinear dependence (the bowing) of the optical band gap on the composition is a general feature shared by all random semiconductor alloys. Within the virtual crystal approximation (VCA), the bowing is due to the second order corrections in the random scattering potential (Baldereschi and Maschke 1975). Since the random potential is obviously absent in ordered phases, one would expect that the band gap of a supercrystal is larger than that of the corresponding random alloy. Surprisingly, the opposite relation was recently observed in GaInAs (Fukui and Saito 1986) and GaInP (Gomyo et al. 1987). This enhancement of bowing accompanying the disorder-to-order transition shows that modifications of the VCA band structure by the coherent supercrystal potential are larger than those due to the disorder. The present study supports this conclusion and indicates that the effect is in fact quite general.

2. EQUILIBRIUM CONFIGURATIONS OF $Ga_{4-n}In_nAs_4$ SUPERCRYSTALS

We consider two ordered phases of GaInAs. The first is the $(GaAs)_1(InAs)_1$ (001) superlattice, whose tetragonal unit cell is characterized by the lattice parameters a, η=c/a, and u, where (u-1/4) measures the anion internal displacement along the z axis (i.e., for u=1/4, the Ga-As and In-As bond lengths are the same). The second structure, suited for Ga_3InAs_4 and $GaIn_3As_4$, is a simple cubic structure with lattice constant a. Cations occupy ideal fcc sites with the $AuCu_3$ arrangement, while anions are at a(u, u, u) and tetrahedrally equivalent sites, where again (u-1/4) measures the internal distortion.

Crystal total energies and band structures are calculated using local density approximation and ab-initio atomic pseudopotentials of Bachelet et al.(1982). Details and discussion of structural properties are presented elsewhere (Bogusławski and Baldereschi 1987). The calculated lattice constants of pure GaAs (5.57 Å) and InAs (5.95 Å) compare well with experimental values (5.65 Å and 6.05 Å, respectively). For intermediate compositions n, the lattice constants a_n follow Vegard law (i. e. linear dependence on n) to within 0.1%. The equilibrium value of the tetragonal distortion of the superlattice differs from $\eta_{ideal}=\sqrt{2}$ by less than 1%. The equilibrium values of the internal distortion parameter u_{eq} are 0.242, 0.267, and 0.259 for n=1, 2, and 3, respectively.

3. BOWING OF THE BAND GAP IN GaInAs

All supercrystals are direct-gap semiconductors at Γ. The computed values $E_g(n)$ of the optical gap in $Ga_{4-n}In_nAs_4$ (n=0,..,4) at theoretical equilibrium are shown in Fig. 1 by full dots. A considerable bowing, given by the reduction $\Delta E_g(n)$ of the supercrystal band gap with respect to the interpolated value $\overline{E}_g$ (full line) between the GaAs and InAs band gaps, is

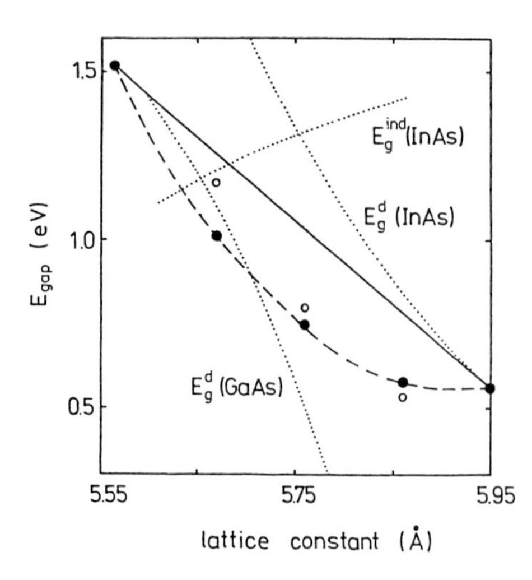

Fig.1. Band gap energies of $Ga_{4-n}In_nAs_4$ (n=0,...,4) for the ideal (open circles) and distorted (full dots) configurations. For GaAs and InAs, the lattice constant dependence of direct Γ-Γ and indirect Γ-X band gaps is given by dotted lines. The dashed line is to guide the eye.

well approximated by

$$\Delta E_g(n) = E_g(n) - \overline{E}_g(a_n) = -b\,\frac{n(n-4)}{16} ,$$

or $\Delta E_g=-bx(1-x)$, with $x=n/4$. The bowing parameter obtained by us $b\simeq 1$ eV is larger than the experimental value $b\simeq 0.45$ eV for random GaInAs alloy (Goetz et al. 1983). Using these values, one may estimate that the disorder-to-order transition reduces E_g by $\simeq 120$ meV for $x=0.5$, while the observed reduction is 35 meV for GaInAs (Fukui and Saito 1986) and 50 meV for GaInP (Gomyo et al. 1987).

To explain the enhancement of bowing we note first that the VCA renders the same band structure for both supercrystal and a corresponding random alloy with $x=n/4$. In the random alloy, this structure is modified by the second order corrections in the random scattering potential. In a supercrystal, modifications of the VCA band structure are due to a coherent potential ΔV_{coh} (given by the difference of the supercrystal and the VCA potentials), which has the supercrystal periodicity. The increase of the periodicity leads to the folding of the Brillouin Zone, and to the coupling between states which differ by a reciprocal lattice vector of the supercrystal.

We analyse first the effects due to folding. For the cubic phases, the three X^* points (the zinc-blende representations are denoted here by asterisks) fold back to Γ and one has: $3\,X_1^* \rightarrow \Gamma_{15}$, $3\,X_3^* \rightarrow \Gamma_1+\Gamma_{12}$, and $3\,X_5^* \rightarrow \Gamma_{15}+\Gamma_{25'}$. The coherent potential ΔV_{coh} affects the folded VCA band structure in the first order, and splits degenerate X^* levels by $\simeq 0.5$ eV. More importantly, the conduction level X_3^{c*} of the virtual crystal after folding to Γ is energetically just above the bottom of the conduction band Γ_1^{c*}. The repulsion between Γ_1^{c*} and X_3^{c*}-derived Γ_1 states reduces the gap. Similar considerations apply to the (001) superlattice, where only $Z^*=(2\pi/c)(001)$ folds back to Γ. The energy bands are shown in Fig. 2, and one has $\Gamma_1^*\rightarrow\Gamma_1$, $\Gamma_{15}^*\rightarrow\Gamma_4+\Gamma_5$; $Z_1^*\rightarrow\Gamma_4$, and $Z_3^*\rightarrow\Gamma_1$. In the superlattice, the

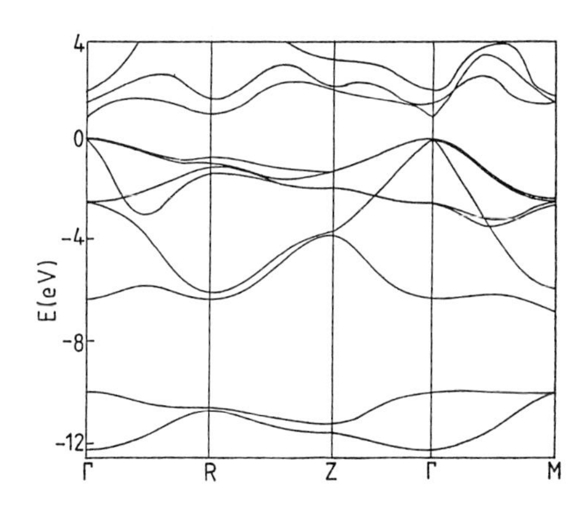

Fig. 2. Energy bands of $(GaAs)_1(InAs)_1$ (001) at theoretical equilibrium.

gap reduction is due to the interaction of Γ_1^{c*} with the Z_3^{c*}-derived Γ_1 state, and, to a lesser extent, to the tetragonal splitting of the Γ_{15}^* top of the valence band.

Considering the conduction band at L^*=R, one has for the cubic crystals 4 $L_1^* \rightarrow R_1 + R_{15}$, and for the superlattice 2 $L_1^* \rightarrow R_1 + R_4$. The calculated splittings are $\simeq$0.7 eV, similarly to what reported by Bylander and Kleinman (1986) for AlGaAs.

4. EFFECT OF INTERNAL DISTORTION ON OPTICAL ENERGY GAP

Zunger and Jaffe (1983) and Bernard and Zunger (1986) have shown that the energy gap in lattice-mismatched superlattices is reduced by internal lattice distortion. The present computations for GaInAs demonstrate that the effect of distortion is strongly composition-dependent. The gaps of the undistorted (u=1/4) supercrystals are shown in Fig. 1 by open circles. We see that for n=1 and 2 the distortion closes the gap, contributing ~70% and ~20% of the calculated bowing, respectively, while for n=3 the ~20% distortion-induced contribution opens the gap. In order to explain this behaviour, we have projected conduction band wave functions at Γ on s-, p-, and d-symmetry atomic states. For n=1, the contribution of In orbitals decreases with internal distortion, and the wave function becomes more GaAs-like. This fact corresponds to the closure of the gap since, as we see in Fig. 1, the direct band gap of GaAs is smaller than that of InAs for all values of the lattice constant. On the contrary, for n=3, the contribution of In orbitals increases with distortion, thus opening the gap.

5. SUMMARY

Self-consistent computations of the energy band structure of $Ga_{4-n}In_nAs_4$ (n=1,2,3) ordered crystals show that the large bowing of the optical gap results from the repulsion between the lowest conduction bands and higher levels folded from the X point of the Brillouin Zone. The calculated bowing of the ordered phases is larger than that measured in random alloys of equal composition, in agreement with experimental data. Finally, the effect on the optical gap of the internal distortion depends strongly on composition, being negative for the Ga-rich system, and positive for the In-rich one.

REFERENCES

Bachelet G B, D.R. Hamann, and M. Schluter 1982, Phys. Rev. B 26 4199
Baldereschi A and Maschke K 1975, Sol. St. Commun. 16 99
Bernard J E and Zunger A 1986, Phys. Rev. B 34 5992
Bogusławski P and Baldereschi A 1987, to be published
Bylander D M and Kleinman L 1986, Phys. Rev. B 34 5280
Fukui T and Saito H 1986, Inst. Phys. Conf. Ser. 79 p. 397
Goetz K H et al. 1983, J. Appl. Phys. 54 4543
Gomyo A et al. 1987, Appl. Phys. Lett. 50 673
Gossard A C et al. 1976, Appl. Phys. Lett. 29 323
Hull R et al. 1986, Proc. Int. Symp. GaAs and Related Compounds, in press
Jen H R, M.J. Cherng, and G.B.Stringfellow 1986, Appl. Phys Lett. 48 1603
Katsumi R et al.1986, Inst. Phys. Conf. Ser. 79 p. 391
Kuan T et al.1985, Phys. Rev. Lett. 54 201
Nakayama H and Fujita H 1986, Inst. Phys. Conf. Ser. 79 p. 289
Shahid M A et al. 1987, Phys. Rev. Lett. 58 2567
Zunger A and Jaffe J E 1983, Phys. Rev. Lett. 51 662

Inst. Phys. Conf. Ser. No. 91: Chapter 6
Paper presented at Int. Symp. GaAs and Related Compounds, Heraklion, Greece, 1987

Sequential resonant tunnelling characteristics of AlAs/GaAs multiple-quantum-well structures

S. Tarucha* and K. Ploog

Max-Planck-Institut für Festkörperforschung, D-7000 Stuttgart-80, FRG
*On leave from NTT Electrical Communications Laboratories, Tokyo/JAPAN

ABSTRACT: The resonant tunneling (RT) characteristics of an AlAs/GaAs multiple-quantum-well structure are studied using time-resolved photo-current (PC) as well as static PC and photoluminescence measurements. The high potential barrier formed by AlAs allows to observe RT from the ground state (1e) in one well to the first three excited states (2e, 3e, and 4e) in the neighbouring well. Pronounced features due to RT are observed up to temperatures as high as 260 K. The time-resolved PC exhibits an initial decay with a delayed upper-convex rather than an exponential profile, which reflects the multiple repetition of RT from 1e to 3e (4e) followed by relaxation back to 1e.

1. INTRODUCTION

Resonant tunneling (RT) is one of the most essential characteristics of vertical transport in quantum wells. RT between the excited states and the ground state in the respective adjacent wells was originally proposed by Kazarinov and Suris (1971), and later exemplified by the observation of peaks in the static photocurrent (PC)/voltage characteristics of an $Al_xIn_{1-x}As/Ga_xIn_{1-x}As$ multiple-quantum-well structure (MQWS) (Capasso et al. 1986) and an $Al_xGa_{1-x}As$/GaAs MQWS (Furuta et al. 1986). The PC peak indicates that PC is enhanced by the RT of electrons from the ground state (1e) in one well to the excited state (2e, 3e, etc.) in the neighbouring well. Recently we have studied time-resolved PC as well as static PC and photoluminescence (PL) characteristics of an $Al_xGa_{1-x}As$/GaAs MQWS. We observed the reduction of tunneling time and PL quenching, both of which are induced by RT from 1e to 2e and 3e (Tarucha et al. 1987). This paper reports a detailed study on RT characteristics of an AlAs/GaAs MQWS grown by molecular beam epitaxy. The AlAs/GaAs heterostructure enables RT associated with the higher excited states because of the high potential barriers. In addition, it completes a sequence in which an electron after RT from 1e to 2e (3e, 4e, etc.) relaxes back to 1e. These features are investigated using time-resolved PC and static PC and PL measurements. The exciting energy is chosen to create carriers homogeneously in the entire MQWS along the axis of the layer sequence.

2. EXPERIMENTAL RESULTS AND DISCUSSION

The sample is a p-i-n heterostructure diode whose intrinsic region consists of a 50-period AlAs(3.4 nm)/GaAs(14 nm) MQW, sandwiched by undoped 3o-nm GaAs layers. The p and n layers are $Al_{0.5}Ga_{0.5}As$/GaAs (0.8 μm, n = p = 5 x 10^{17} cm^{-3}). The diode is processed into a high-mesa cylindrical geometry having a diameter of 200 μm. A built-in-voltage of the diode is 1.6V at 10 K. A cut-off frequency evaluated from the RC time constant is about 5 GHz. Details of the experimental setup is the same as described previously (Tarucha et al. 1987).

Figure 1 shows the static PC and the spectrally integrated PL vs bias voltage (V_b) measured under the same excitation condition. The 753-nm line of a Kr^+ laser with 0.3 W/cm^2 is used as exciting light source. The negative sign of V_b denotes the backward of the diode. The observed PL is associated with excitons of electrons and heavy holes in the ground states. The PC shows four peaks labeled a, b, c, and d. At the same voltages where the peaks b and c appear, the PL intensity shows pronounced dips, labeled B and C, and also a complete quenching as V_b is decreased down to the value where the peak d appears. These PC peaks b, c, and d are assigned to RT between 1e and 2e, 3e, and 4e in the adjacent wells, repsectively, according to the analysis of the corresponding electric fields (Tarucha et al. 1987). The real electric fields free from the field-screening of the photogenerated carriers are determined from the PC/V_b characteristics at sufficiently low excitation intensity. The values are -2.4, -7.5, and -13.8 V for the peaks b, c, and d, respectively, which are invarient within an error of 0.2 V below 20 mW/cm^2. These voltages give the electric field of 43, 98, and 165 kV /cm, respectively. On the other hand, the effective-mass calculation gives the subband energy differences of 63 meV between 1e and 2e, 163 meV between 1e and 3e, and 296 meV between 1e and 4e, respectively. These values predict RT of electrons at 36, 94, and 170 kV/cm, which agree well with those for the peaks b, c, and d, respectively. The PC peak a and the PL dip A_1 are probably due to the field-induced change of the absorption coefficient at the exciting energy because the peak a is not observed under white light excitation (see Fig. 2). In addition, the dip A_1 accompanies a PC dip at the same voltage. The small PL intensity dip A_2 accompanies a small excess PC indicated by a_2. Their origin is not yet well understood.

Figure 2 shows the temperature dependence of the PC/V_b characteristics. White light excitation is used to eliminate the field-induced change of the absorption coefficient. The PC peaks b and c due to RT between 1e and 2e, and between 1e and 3e, respectively, are observed up to 260 K. Most of the previous observations of RT in MQWS's were achieved only at low temperatures: below 50 K for the $Al_xIn_{1-x}As/Ga_xIn_{1-x}As$ MQWS (Capasso et al. 1986) and below 77 K for the $Al_xGa_{1-x}As/GaAs$ MQWS's (Furuta et al. 1986), Tarucha et al. 1987). The present result implies the existence of well defined subband levels formed by the high potential barriers and also a high crystal quality. As the temperature is increased, the PC becomes smaller up to 120K

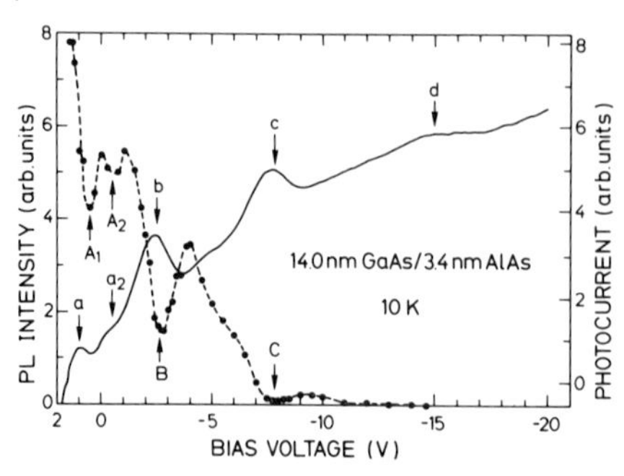

Fig.1 Static PC and integrated PL intensity vs bias voltage. The broken line is a guide for the eyes.

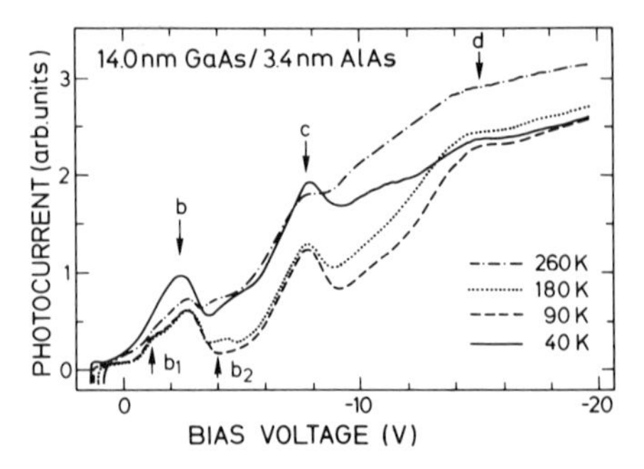

Fig.2. Temperature dependence of static PC vs bias voltage characteristics.

and then larger again. In addition, shoulder-like features labeled b_1 and b_2 become more pronounced at higher temperatures. The increase of the PC above 120 K is attributed to phonon-assisted tunneling. Particularly, the shoulder-like features indicate a field-induced enhancement of LO-phonon assisted tunneling because the voltage difference between the peak b and features b_1 and b_2 gives a potential drop across the MQW period in the range of 30 to 40 meV, which is comparable to the LO-phonon energy in GaAs. This assignment is more likely for the feature b_2 because of the larger probability for phonon emission than for phonon absorption. The decrease of PC below 120 K is attributed to thermally activated nonradiative centers because the PL intensity is also reduced in this temperature range. Defects or impurities at the hetero-interfaces or in the AlAs barriers are a possible origin for these nonradiative centers.

In Fig. 3 we show the time-resolved PC around V_b providing the resonances. A 785-nm $Al_xGa_{1-x}As$/GaAs heterostructure laser diode with 350-ps optical pulses is used as exciting light source. The photogenerated carrier density is (1 to 2)x10^{14} cm^{-3} per pulse. As V_b approaches the values giving the resonance peaks, fast initial decay components appear and become larger. The peak values of the initial PC component reach maxima at -2.5, -8, and -14.85 V, which agree well with the values corresponding to the PC peaks b, c, and d, respectively, in Fig. 1. This V_b dependence confirms RT from 1e in one well to 2e (3e, 4e) in the neighbouring well, which are observed in the static PC/V_b measurement (Tarucha et al. 1987). The observation of RT even from 1e to 4e confirms the confinement by the Γ band of the AlAs layer. The subsequent slow decay of the time-resolved PC can be more related to bipolar transport properties, which are not yet well understood. Therefore, we discuss only the initial PC component observed under resonance conditions by assuming electron transport. An interesting feature in Fig. 3 is that the initial PC component under resonance condition has a more upper-convex decay profile rather than an exponential one and the pronounced decay starts several nsec after the excitation. A typical example is shown in Fig. 4 with linear scale plots of the time-resolved PC at -8V. The rise profile of the initial component is probably affected by the profile of the exciting pulse, which is indicated by the dotted line in Fig. 4, as well as by the initial distribution of the photogenerated carriers introduced by the absorption coefficient. The delayed upper-convex profile of the initial decay component reflects the multiple repetition of RT followed by nonradiative relaxation. An electron after tunneling from 1e to 2e (3e, 4e) through one barrier either relaxes back to 1e or further tunnels through the

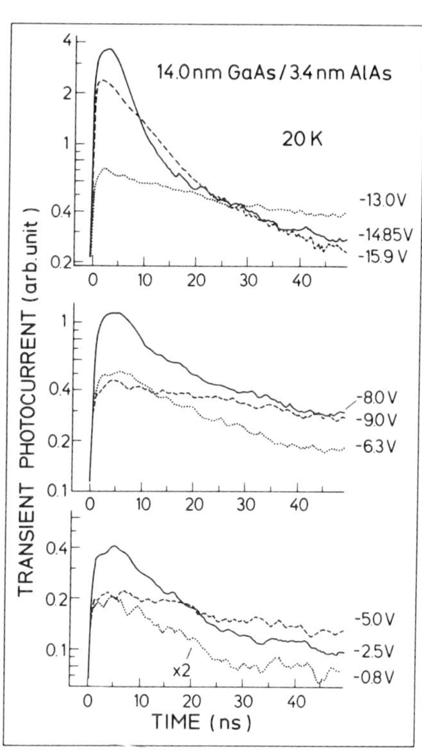

Fig.3. Time-resolved PC around the bias voltages providing RT from 1e to 2e (-2.5V), from 1e to 3e (-8V), and from 1e to 4e (-14.85V).

next barrier as shown in the inset of Fig. 4. In the present AlAs/GaAs MQWS, electrons cross the entire MQWS by repeating the former sequence (RT followed by back-relaxation) because of the high potential barriers. During this repetition the electron PC flows with a duration given by $N_{eff}\ \tau_t$, where τ_t and N_{eff} are RT time between 1e and 2e (3e, 4e) in the adjacent respective wells and an effective number of RT repeated by an electron, respectively. N_{eff} is significantly affected by the homogeneity of the electric field across the entire MQWS, which is temporally disturbed by the space charge, as well as by recombination. When we assume pure electron transport across the homogeneous field for simplicity, $N_{eff} \sim 25$ under the present homogeneous excitation condition. This assumption gives τ_t = 150 - 200 ps at -8 V and 100 - 150 ps at -14.85 V, where excitonic recombination can be neglected. In contrast, in an $Al_xGa_{1-x}As$/GaAs MQWS, we observe an initial decay of the time-resolved PC which has a more exponential profile with the much shorter time constants of 430 and less than 60 ps for RT from 1e to 2e and 3e, respectively (Tarucha et al. 1987). This is explained by the different RT sequence involving tunneling out of the well, which will be described in a separate paper.

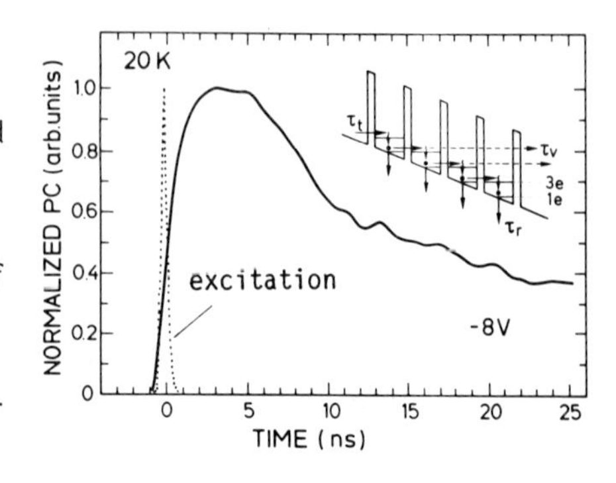

Fig.4. Time-resolved PC on the linear scale for V_b=-8V.

3. CONCLUSION

RT characteristics of an AlAs/GaAs MQWS are studied using time-resolved PC as well as static PC and PL measurements. The high potential barrier of AlAs allows the observation of RT between 1e and 2e (3e, 4e) in the respective adjacent wells. The pronounced features due to RT are observed in the static PC/V_b characteristics even at temperatures as high as 260 K. The time-resolved PC exhibits an initial decay with a delayed upper-convex profile rather than an exponential one under the resonance conditions, which reflects the multiple repetition of RT of electrons from 1e to 3e (4e) followed by back-ralaxation to 1e.

ACKNOWLEDGEMENT

The authors are grateful to M. Hauser for sample growth, and K. von Klitzing and H. Schneider for valuable discussions. Part of this work is sponsored by the Bundesministerium für Forschung und Technologie of the Federal Republic of Germany.

REFERENCES

Capasso F, Mohammed K and Cho A Y 1986 Appl. Phys. Lett. 48 478.
Furuta T, Hirakawa K, Yoshino I and Sakaki H 1986 Jpn. J. Appl. Phys. 25 L151.
Kazarinov R F and Suris R A 1971 Sov. Phys. Semicond. 5 707.
Tarucha S, Ploog K, and von Klitzing K 1987 Phys. Rev. B 36.

Evidence of defect induced disordering in $Al_{0.3}Ga_{0.7}As$–GaAs undoped superlattices

E.V.K. Rao, F. Brillouet, P. Ossart, Y. Gao, J. Sapriel and P. Krauz
Centre National d'Etudes des Télécommunications
Laboratoire de Bagneux
196 avenue Henri Ravera - 92220 BAGNEUX - FRANCE

ABSTRACT : We have investigated the influence of implant damage on disordering of $Al_{0.3}Ga_{0.7}As$-GaAs superlattices (SLs) by separating it from impurity charge associated effects. The implants of electrically inactive isoelectronic phosphorus (P) have been performed in MBE grown SLs at 25°C and 250°C to generate different damages prior to anneals (850°C up to 6 h). Using several characterization techniques to monitor intermixing, an unambiguous evidence to implant damage-induced-disordering is presented and discussed.

1. INTRODUCTION

Compositional disordering of $Al_xGa_{1-x}As$-GaAs SLs is a potentially useful process in the technology of optoelectronic devices. This can be obtained by introducing dopant impurities either by thermal diffusion (Lee et al 1984, Meehan et al 1984, Rao et al 1985a), or by doping during growth (Kawabe et al 1984, Rao et al 1987b) or ion implantation followed by annealings (Coleman et al 1982). Recently there have been numerous reports (Gavrilovic et al 1985, Hirayama et al 1985, Matsui et al 1986, Dobisz et al 1986, Schwarz et al 1987) on the use of ion implantation to investigate the impurity-induced-disordering (IID). However, as regards to the nature of implant damage influence namely, enhancement (Gavrilovic et al 1985, Ralston et al 1986) or supression (Matsui et al 1986, Venkatesan et al 1986, Schwarz et al 1987) of Al/Ga interdiffusion, the situation is not totally clarified. The work described here, is therefore aimed at investigating the implant damage influence on Al/Ga interdiffusion by separating it from impurity charge associated effects.

2. EXPERIMENTAL

Implants of electrically inactive isoelectronic P at 100keV to a dose of $10^{15}P^+/cm^2$ were realized in 30 periods SLs containing alternating layers of ∿ 90Å $Al_{0.3}Ga_{0.7}As$ and 80Å GaAs grown by MBE. During implantation, the samples were held at 25°C or ∿ 250°C to generate different damage densities. All implanted and control samples were annealed at 850°C for durations up to 6 h using close contact configuration. The evolution in the properties of SLs was monitored using non destructive optical techniques like, Raman and photoluminescence (PL) measurements, or depth profiling techniques like Auger electron spectroscopy (AES coupled to Ar^+ ion sputtering) and Secondary ion mass spectroscopy (SIMS with Cs^+ primary beam).

3. RESULTS AND DISCUSSION

Fig. 1 shows Raman spectra (backscattering mode using 5145 Å line of Ar^+ laser) taken to assess different damage densities in 25°C and ∿ 250°C implanted SLs. Also shown here for comparison is the spectrum of a bulk GaAs

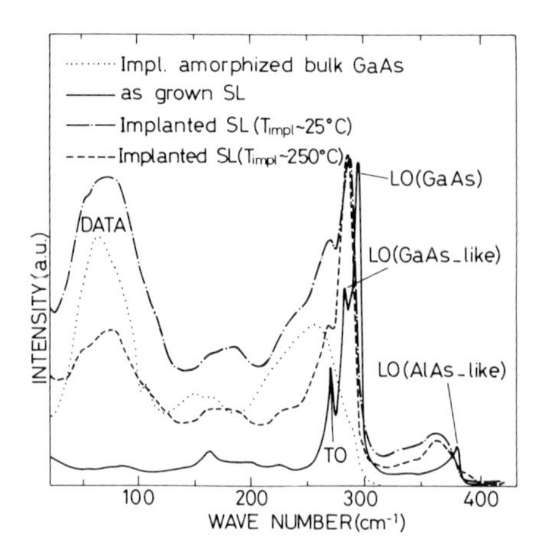

Fig. 1: Room temperature Raman spectra of SLs recorded before (as-grown) and after (25°C and 250°C) P^+ implants. Data taken on a bulk GaAs, sample amorphized after a similar implant is also shown.

Fig. 2 compares the evolution in room-temperature PL spectra of 25°C and 250°C P^+ implanted SLs with anneal duration at 850°C. Fig. 3 summarizes this evolution where we have plotted the high energy shift of PL peak (ΔE) against anneal time. It is now clear that a progressive intermixing is indeed occuring with annealing time and that the implant with higher initial damage density (25°C implant) leads to a greater degree of intermixing.

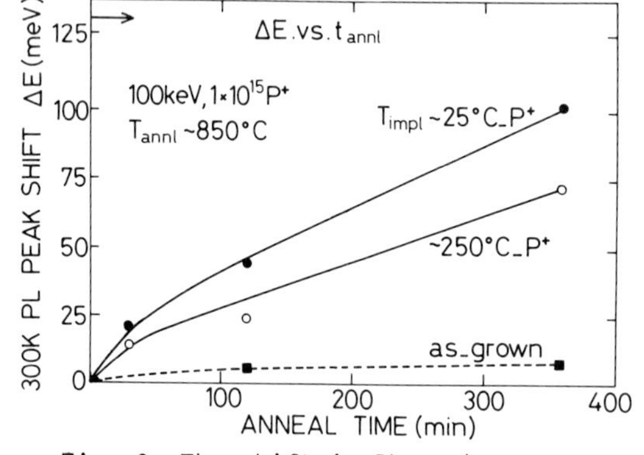

Fig. 3: The shift in PL peak energy as a function of annealing time at 850°C. The data taken on as-grown SL is shown to illustrate the high thermal stability of unimplanted structure.

sample that turned amorphous after an identical implant. In addition to characteristic LO phonon features (Sapriel et al 1987), the emergence of the broad phonon band DATA is indicative of damage presence in the implanted layers. Comparing the intensity ratios of DATA to LO bands we deduce the following. The 25°C implant in SL leads to heavier damage than the 250°C one. Secondly the density of damage generated in the SLs is much below amorphization level in comparison to GaAs bulk sample (fig. 1).

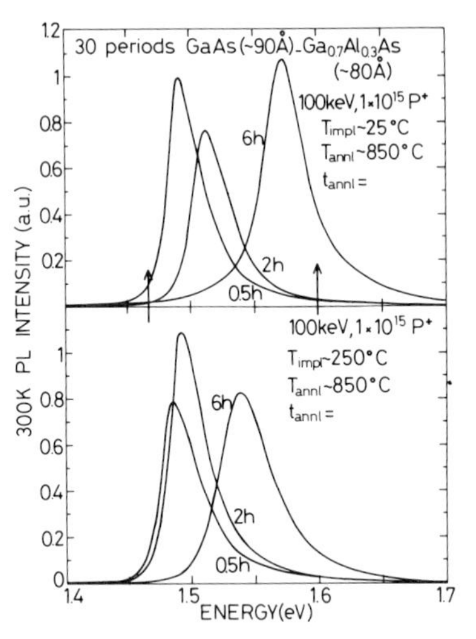

Fig. 2: Evolution in room temperature PL spectra of P^+ implanted (25°C and 250°C) SLs with annealing time at 850°C. Left arrow : peak position in the as-grown SL ; right arrow : expected peak position in a totally intermixed case.

3.1 Damage-induced-disordering (DID)

A further insight into the nature of damage induced disordering is obtained by recording the depth distributions of Al and P using respectively AES coupled to Ar^+ ion sputtering and SIMS with Cs^+ primary beam. These results are shown in Fig. 4 for each implant after 2 h and 6 h anneals.

The decrease in Al oscillations confined to the region <Rp of the as-implanted SL (shown in the bottom

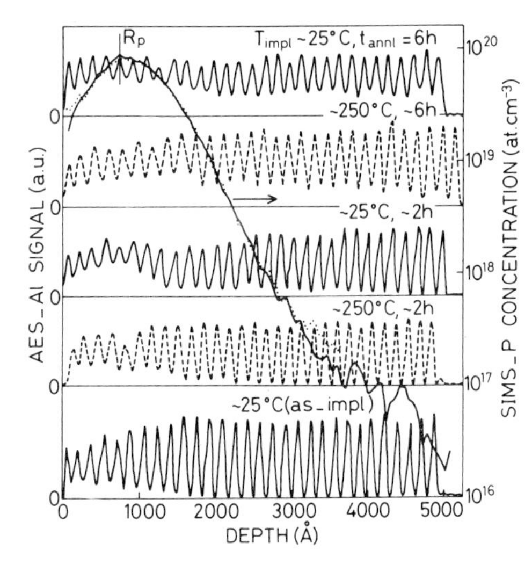

Fig. 4: Depth distributions of Al (AES-Al : left ordinate) and P (SIMS : right ordinate) in P^+ implanted (25°C and 250°C) and annealed SLs (850°C, 2 h and 6 h). SIMS distributions are shown as dotted (as-implanted at 25°C) and continuous curves (annealed at 850°C for 6 h).

of fig. 4) is essentially a consequence of collisional mixing during implantations. On the other hand, with increasing anneal time (see spectra of 2 h and 6 h for each implant) the intermixing is seen to proceed progressively into the depth of SLs far beyond Rp. Also, confirming the data of PL measurements (figs 2 and 3), the 25°C implant with high damage density leads to a more efficient intermixing.

The above data on AES-Al depth distributions strongly suggest a diffusive nature for the species enhancing Al/Ga interdiffusion in the depth of SLs. Since the distribution of implanted P (SIMS data of Fig. 4) has shifted scarcely during anneals this species cannot be phosphorus impurity, but rather a defect comming from the damaged region. As a first explanation, we suggest the migration of this defect to higher depths can be due to misfit strain induced by P substitution on As site (Rao et al c, 1987).

4. CONCLUSIONS

In conclusion, by separating damage effects from impurity charge associated effects, we have examined the influence of P implant damage on disordering in $Al_{0.3}Ga_{0.7}As$-GaAs SLs. We have shown here that higher damage densities lead to more efficient intermixing. Also, with increasing anneal duration, intermixing is shown to proceed into the depth of SLs far beyond Rp. Even though the basic understanding of DID mechanism necessitates further studies, its application in device technology is quite promising since intermixed regions with no modifications in electrical properties can be obtained.

REFERENCES

Coleman JJ, Dapkus PD, Kirkpartrick CG, Camras MD and Holonyak Jr N, 1982, Appl. Phys. Lett. 40, 904.

Dobisz EA, Tell B, Craighed HG and Tamargo MC, 1986, J. Appl. Phys. 60, 4150.

Gavrilovic P., Deppe DG, Meehan K, Holonyak Jr N, Coleman JJ and Burnham RD, 1985, Appl. Phys. Lett. 47, 130.

Hirayama Y, Suzuki Y and Okamoto H, 1985, Jpn. J. Appl. Phys. 24, 1498.

Kawabe M, Matsuura, Shimizu N, Hasegawa F and Nannichi Y, 1984, Jpn. J. Appl. Phys. 23, L 623.

Lee JW and Laidig WD, 1984, J. Electron. Mater., 13, 147.

Matsui K, Kobayashi J, Fukunaga T, Ishida K and Nakashima H, 1986, Jpn. J. Appl. Phys. 25, L 651.

Meehan K, Holonyak Jr N, Brown JM, Nixon NA, Gavrilovic P and Burnham RD, 1984. Appl. Phys. Lett. 45, 549.
Rao EVK, Thibierge H, Brillouet F, Alexandre F and Azoulay R, 1985a, Appl. Phys. Lett. 46, 867.
Rao EVK, Ossart P, Alexandre F and Thibierge H, 1987b, Appl. Phys. Lett. 50, 588.
Rao EVK, Brillouet F, Ossart P, Gao Y, Sapriel and Krauz P, Proceedings of the 3rd International Conference on Modulated Semiconductor Structures, Montpellier 1987c.
Sapriel J, Rao EVK, Brillouet F, Chavignon J, Ossart P, Gao Y and Krauz P, 1987, Proceedings of the 3rd International Conference on Superlattices, Microstructures and Microdevices, Chicago.
Schwarz SA, Venkatesan T, Hwang DM, Yoon HW, Bhat R and Arakawa Y, 1987, Appl. Phys. Lett. 50, 281.
Venkatesan T, Schwarz SA, Hwang DM, Bhat R, Koza M, Yoon HW, Mei P, Arakawa Y and Yariv A, 1986, Appl. Phys. Lett. 49, 701.

Electric field behaviour of excitons in GaAs/AlGaAs coupled quantum wells

B.S. Elman, Emil S. Koteles, Y.J. Chen, C. Jagannath,
S. Brown and C.A. Armiento

GTE Laboratories Incorporated
40 Sylvan Road
Waltham, MA 02254

ABSTRACT: Optical spectroscopic techniques have been employed in a detailed experimental study of the influence of electric fields on exciton states in a GaAs/AlGaAs coupled double quantum well structure. Both intra- and inter-well exciton transitions were observed and energy splittings of the coupled quantum confined levels were derived without recourse to a theoretical model. The coupling between wells leads to an enhancement of the quantum confined Stark effect so that it is as much as five times larger than in the case of a single quantum well.

INTRODUCTION

The quantum confined Stark effect (QCSE) in semiconductor quantum wells (QW) has attracted a great deal of interest, for both technological and fundamental reasons.[1] The energies and oscillator strengths of confined excitons are modified over a wide range of values by application of large electric fields[1-3] and there has been considerable effort to further enhance the effect, in particular, to increase the size of the energy shift for a given applied field. To achieve this goal a coupled double quantum well structure (CDQW) has been proposed.[4-7] A study of a symmetric CDQW system is presented in this work.

This structure consists of a pair of identical QWs separated by a barrier narrow enough that the wavefunctions of the electronic states in the conduction/valence bands in the adjacent wells overlap and therefore split. In the absence of external perturbations (i.e., flat-band conditions) state symmetries are well defined and only symmetry-allowed transitions can occur (Fig. 1, $E=0$). In the presence of an electric field (Fig. 1, $E>0$) wavefunction symmetries are distorted, selection rules are relaxed and all transitions become possible. As the field is increased, intra-well-like transitions (2, 4, 5 and 7 in Fig. 1) experience the normal QCSE. However, the energies of inter-well-like transitions undergo much more rapid decreases (1 and 3) or increases (6 and 8) with increasing applied field.[8] We have utilized room temperature photoreflectance (PR) and low temperature photoluminescence (PL), photoluminescence excitation (PLE) and photocurrent (PC) spectroscopies to study all the transitions in a CDQW PIN structure as a function of applied electric field.

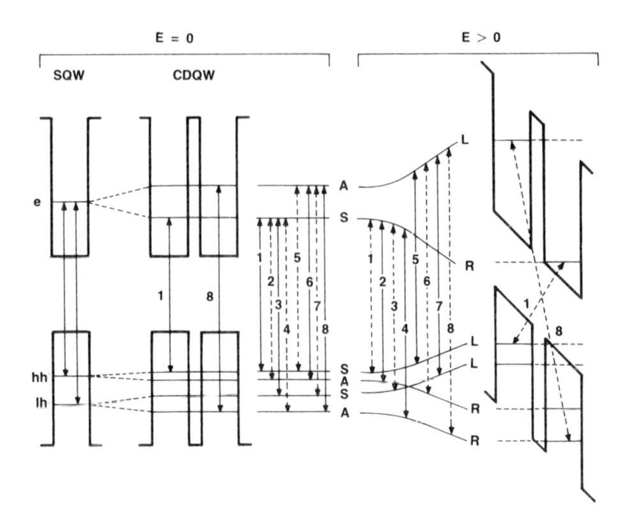

Fig.1 Schematic diagram of the energy levels of a SQW and a symmetric CDQW under flatband conditions (middle) and in the presence of an external electric field. S and A correspond to symmetric and anti-symmetric coupled states.

EXPERIMENTAL, RESULTS AND DISCUSSION

The sample was grown by molecular beam epitaxy on an n^+ GaAs substrate and consisted of a single pair of $L_z = 7.5$ nm GaAs quantum wells separated by a $L_B = 1.8$ nm $Al_{.27}Ga_{.73}As$ barrier, two 85 nm outer undoped $Al_{.27}Ga_{.73}As$ barriers, and a 20 nm p^+ GaAs cladding layer, all grown on an n^+ GaAs buffer. The electric field was applied perpendicular to the layers.

Low temperature PL and PLE spectra from a CDQW sample at several bias voltages are shown in Figures 2 and 3, respectively. A shift to lower energies of the excitonic transition 1 (Fig. 1) with concomitant quenching of the PL intensity (Fig. 2) and dramatic changes in the PLE spectra (Fig. 3) are observed as the bias voltage is changed. At the flat-band condition (1.6V), when the external and build-in internal electric fields exactly cancel each other, only allowed transitions are observed. Additional

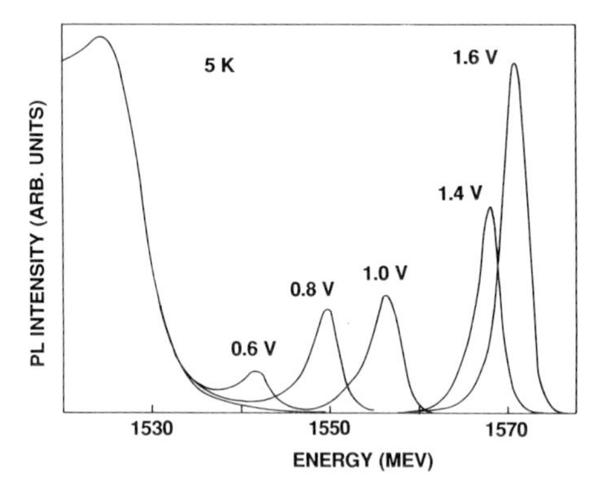

Fig.2 5K PL spectra of a CDQW under various bias voltages.

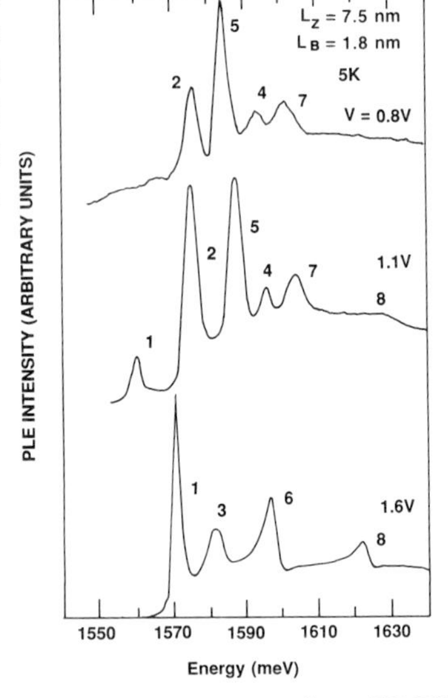

Fig.3 PLE spectra of a CDQW under three bias voltages. The peaks are labelled according to the notation in Fig.1.

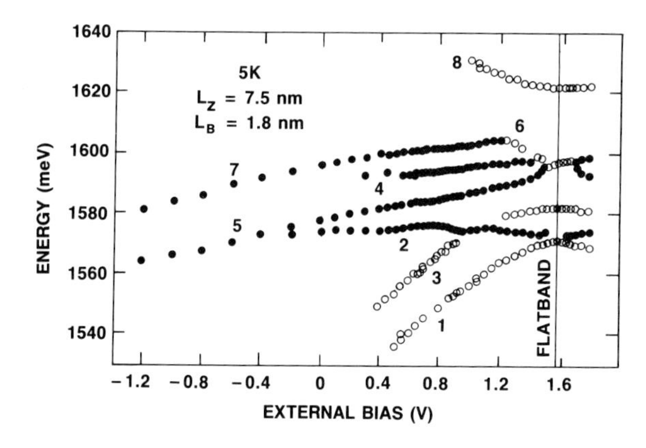

Fig.4 Energies of exciton peaks in a CDQW as a function of applied voltage.

peaks corresponding to symmetry-forbidden transitions appear in the PLE spectrum as the electric field is increased to ~$2x10^4$ V/cm (Fig. 3, V=1.1V). With a net applied field small enough that the Stark effect was negligible but large enough that most of the transitions were observable, we were able to derive all of the splittings due to inter-well coupling using measured exciton energies.[9] Splittings of the ground state electron, heavy-hole (hh) and light-hole (lh) states in our CDQW structure were determined to be 22.1, 3.9 and 15.5 meV, respectively. With these values and assuming symmetrical splittings, we were able to derive the hh exciton energy (1.582 eV) and hh-lh splitting, (18.8 meV) in the QWs in the absence of coupling by employing simple arithmetic manipulation. Agreement with values measured in a reference single quantum well sample grown under identical conditions (1.584 eV and 19 meV, respectively) was very good. As the electric field is increased to ~$3.5x10^4$ V/cm (V=0.8V, Fig. 3) intra-well-like transitions, normally forbidden under flat-band conditions dominate the PLEspectra. The photocurrent spectra (not shown) exhibited much stronger peaks than the PLE spectra at bias voltages less than 0.7V (E>$4x10^4$ V/cm). PC spectroscopy was used to continue our studies down to V=-1.4V. A summary of the observed exciton transitions as a function of applied voltage is presented in Fig. 4 where intra- and inter-well-like transition energies are shown inclosed and open circles respectively. It is clear that the field-induced shifts of interwell transitions (1, 3 and 8) are much larger (~5X) than those analogous to single QWs (e.g., intra-well transitions 5 and 7 at large fields). Room temperature electric field effects were studied using PR spectroscopy. Room temperature PR and low temperature PLE spectra taken under flat-band conditions (1.3V and 1.6V respectively) are compared in

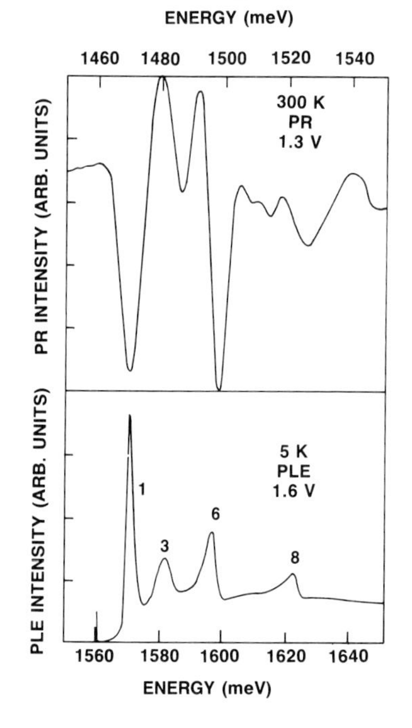

Fig.5 Room temperature PR and low temperature PLE spectra of a CDQW under flat-band conditions.

Fig. 5 (the energy scales were shifted by ~ 100 meV to compensate for the energy shift due to the temperature dependence of the bandgap). The PR data was fit using the third-derivative functional form for a 2-D critical point.[10,11] The three peaks corresponding to transitions 1, 3 and 6 (Fig. 1) are deduced to be at 1.469, 1.480 and 1.496 eV respectively. These are in excellent agreement with the adjusted low temperature PLE data. The electric field dependence of the room temperature PR spectra produced results similar to those in Fig. 4. Analysis of the spectra are presently in progress.

SUMMARY

In summary, we have studied a coupled double quantum well structure subject to perpendicular electric fields using a variety of optical spectroscopic techniques. Substantial enhancement of the Stark effect for inter-well-like transitions was demonstrated. These large field induced shifts of exciton resonance energies have potential utility in electro-optic devices.

We would like to thank Joseph Powers and Doug Owens for expert technical assistance.

References

1. D.A.B. Miller, D.S. Chemla, T.C. Damen, A.C. Gossard, W. Wiegmann, T.H. Wood, and C.A. Burrus, *Phys. Rev. Lett.* **53**, 2173 (1984).

2. E.J. Austin and M. Jaros, *Appl. Phys. Lett.* **47**, 274 (1985).

3. G.D. Sander and K.K. Bajaj, *Phys. Rev*. B35, 2308 (1987) and references therein.

4. E.J. Austin and M. Jaros, *J. Phys. C.* 19, 533 (1986).

5. H. Kawai, J. Kaneko, and N. Watanabe, *J. Appl. Phys.* **58**, 1263 (1985).

6. A. Yariv, C. Lindsey and U. Sivan, *J. Appl. Phys.* **58**, 3669 (1985).

7. H.Q. Le, J.J. Zayhowski, W.D. Goodhue and J. Bales, 1987 **Proceedings of 13th Int. Symposium on GaAs and Related Compounds** (Inst.Phys.Conf.Ser.) pp319-24

8. Y.J. Chen, Emil S. Koteles, B.S. Elman, and C.A. Armiento, *Phys. Rev.* **B** September 15, 1987 issue.

9. Emil S. Koteles, Y.J. Chen, and B.S. Elman, to appear in the **Proceedings of the International Meeting on Excitons in Confined Systems, Roma, 1987**.

10. D.E. Aspnes in **Handbook on Semiconductors**, edited by T.S. Moss (North-Holland, New-York, 1980), Vol.2, p.109 and references therein.

11. B.V. Shanabrook, O.J. Glembocki and W.T. Beard, *Phys. Rev.* **B35**, 2540 (1987).

DLTS measurements on MBE-grown narrow GaAs/n-AlGaAs single quantum wells

D J As* and P W Epperlein

IBM Research, Zurich Research Laboratory, 8803 Rüschlikon, Switzerland

P M Mooney

IBM Research, T J Watson Research Center, Yorktown Heights, NY 10598, USA

* Postdoctoral Associate from University of Linz, Linz, Austria

ABSTRACT: Molecular-Beam Epitaxy (MBE) grown GaAs/n $-$ $Al_xGa_{1-x}As$ (x=0.24-0.39) Single Quantum Wells have been studied by Deep-Level-Transient-Spectroscopy techniques (DLTS, DDLTS). A series of electron traps with energies at 0.12, 0.22, 0.29, 0.52 and 0.63 eV and with concentrations of about 5×10^{15} cm^{-3} has been detected in the upper AlGaAs layer near the Quantum Well (QW). The widths of the spatial trap distribution are typically 15 nm and are independent of the QW width. The capture process of the spatially localized states is discussed.

1. INTRODUCTION

Nonradiative carrier recombination in the active region, at the interfaces and in the wide-gap regions of Double Heterostructure (DH) GaAs/$Al_xGa_{1-x}As$ lasers is important for laser threshold current densities (Tsang 1978). To obtain information on the responsible defect centers and laser-degradation mechanisms, Deep-Level-Transient Spectroscopy (DLTS) (Lang 1974) has been performed primarily on DH lasers (Lang *et al.* 1976, Uji *et al.* 1980, McAfee *et al.* 1982); fewer experiments are available for Single-Quantum-Well (SQW) structures (Hamilton *et al.* 1985).

This paper reports several deep levels measured with capacitance DLTS in GaAs/n $-$ $Al_xGa_{1-x}As$ SQWs with different well widths and x-values of the $Al_xGa_{1-x}As$ cladding layers. Depth profiling of the different traps has been studied by DDLTS (Lefevre and Schulz 1977) and shows a spatial accumulation near the upper QW interface. The capture process has been investigated by varying the majority-carrier pulse length, and a non-exponential filling of all traps has been observed.

2. EXPERIMENT

The samples under study were grown by MBE on semi-insulating LEC-GaAs (100) substrates. The SQWs with GaAs well widths L_z 3 – 40 nm and 100 nm thick n-type $Al_xGa_{1-x}As$ cladding layers (Si:4×10^{17} cm^{-3}, $x = 0.24 - 0.39$) were deposited on a 0.3 μm $GaAs/Al_{0.3}Ga_{0.7}As$ superlattice buffer layer. The parameters of the samples are listed in Table I. The x-values were determined by low-temperature photoluminescence. Capacitance-DLTS and C-V measurements were carried out using Schottky diodes, formed by evaporating both the (Ti, Pt, Au) barrier and (Au, Ge, Ni) ohmic-contact spots on top of the multi-heterostructure. Since the zero-voltage depletion width of the Schottky diodes is smaller than the 100 nm, measured for the QW depth by C-V probing, the depletion edge could be controllably swept through the QW. The spatial resolution determined by the Debye length is 8 nm. DLTS and DDLTS measurements were performed using a high-sensitivity lock-in spectrometer (Ferenczi *et al.* 1980).

Table I: Parameters of the different SQW samples

Sample	x-value	L_z (nm)	Spacer (nm)	QW-doping N_d (cm^{-3})
Su 1082	0.24	8	-	-
Su 1029	0.24	40	3	-
Su 1030	0.25	3	3	-
Su 1446	0.25	8	3	-
Su 1447	0.24	8	-	5×10^{17}
Su 1031	0.39	8	3	-

3. RESULTS AND DISCUSSION

The C-V doping profile of a 3 nm $GaAs/Al_{0.24}Ga_{0.76}As$ SQW is shown in Figure 1. The characteristic peak-and-valley shape of the profile is the same as that calculated by Blood (1986). From the peak position, the QW can be located at about 120 nm from the surface. As already discussed by Kroemer *et al.* (1980) the peak and valley do not indicate actual variations in the true doping profile but rather accumulation and depletion effects, respectively, owing to the discontinuities in the conduction band at the interfaces.

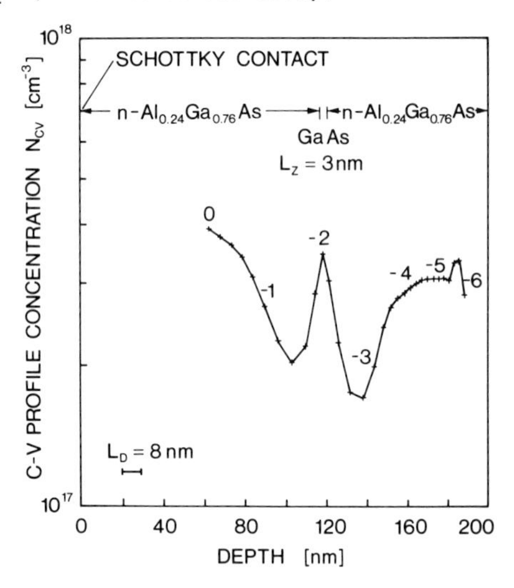

Figure 1. C-V doping profile of a 3 nm $GaAs/Al_{0.24}Ga_{0.76}As$ SQW. The numbers on the profile indicate the values of the reverse-bias voltage in the C-V measurement (Debye length $L_D = 8$ nm)

DLTS spectra recorded for a fixed majority-carrier pulse $V_p = 0.5$ V superimposed on various reverse-bias voltages V_r are shown in Figure 2. At $V_r = 0$ V and at $V_r = -5.5$ V, the QW is outside the region probed by DLTS. Here, the spectra are dominated by one single peak which can be attributed to the well-known DX center (Lang *et al.* 1979). With the QW in the DLTS-probed zone, at least five additional peaks appear. These electron traps E1, E2, E3, E4 and E5 (trap numbering is arbitrary) have T^2-corrected activation energies of 0.12, 0.22, 0.29, 0.52 and 0.63 eV, respectively. Maximum trap concentrations are of the order of 5×10^{15} cm^{-3}. For further investigations, the DX peak could be suppressed by using a sufficiently short filling pulse (1 μs).

The depth distributions of traps E1 to E5 were determined by DDLTS. Figure 3 shows the DDLTS signal $(\delta(\Delta C/C)/\delta V_p)$ for trap E5 as a function of depth which clearly exhibits a maximum at 88 nm. The same distribution was obtained for the other traps. To obtain the real trap concentration $N_t(x)$, the DDLTS(x) has to be convoluted with the measured C-V profile concentration $N_{cv}(W_p)$ (Lang 1979) according to

$$N_t\big(W_p - L(t_p)\big) = \frac{q \cdot W_b^2 \cdot N_{cv}(W_b)}{\varepsilon} \cdot N_{cv}(W_p) \cdot \frac{\delta\left(\frac{\Delta C}{C}\right)}{\delta V_p}$$

where W is the depletion-layer width either at the steady-state bias W_b or during the majority-carrier pulse W_p. $N_{cv}(W_p)$ is the measured carrier concentration at W_p by C-V measurements. $L(t_p)$ is the distance into the edge region of the depletion layer (from W_p) over which traps are filled by a majority-carrier pulse V_p and duration t_p. The spatial scale of N_t for the E5 trap is also shown in Figure 3. Decreasing t_p shifts the DDLTS maximum to higher V_p. This is typical for spatially localized levels and reflects the

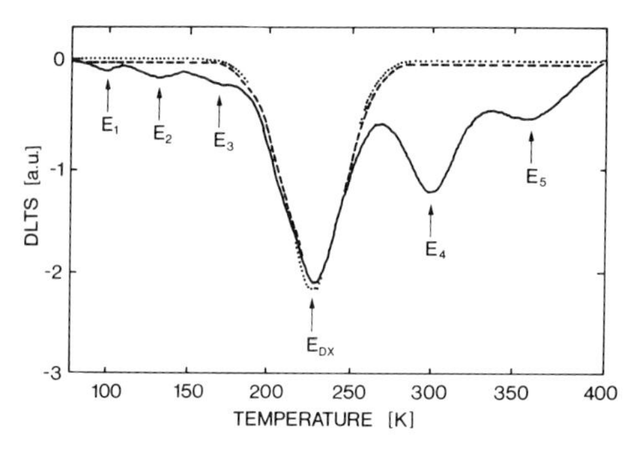

Figure 2. DLTS spectra at different reverse biases V_r and constant pulse-voltage amplitude V_p ($V_p = 0.5$ V, ... $V_r = 0$ V, — $V_r = -3.5$ V, --- $V_r = -5.5$ V, rate window $e_n = 278$ s^{-1})

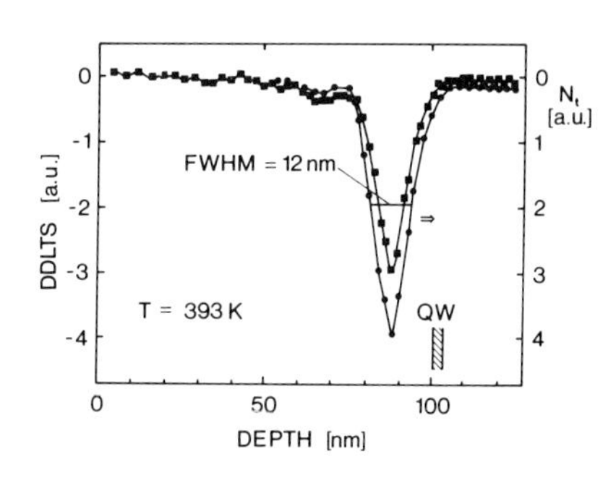

Figure 3. Depth profile of E5 = 0.63 eV of sample Su 1030 ($L_z = 3$ nm). (● ... N_t, ■ ... DDLTS signal, pulse duration $t_p = 100$ μs, rate window $e_n = 2787$ s^{-1})

change in $L(t_p)$. The results indicate that all five traps have their true maximum concentration in the upper AlGaAs layer just in front of the QW. The widths of the distributions are independent of the QW width, and vary randomly between 12 and 26 nm for the samples measured.

The filling kinetics of the different levels were studied by varying the filling pulse duration t_p. For all traps, we found a nonexponential capture process which is logarithmic in time over three decades in t_p. This effect can be well-explained by the so-called edge effect. Because of the spatial spread of the defects at the interface, the capture time varies by at least three orders of magnitude. From the shortest applied pulse duration (40 ns) the capture cross section of all traps could be estimated to be larger than 10^{-17} cm^2.

4. CONCLUSION

In summary, using DLTS, we have detected several deep electron traps spatially localized in MBE-grown GaAs/n-AlGaAs single-quantum-well structures. The traps are localized in the top AlGaAs layer next to the QW interface within a region typically 15 nm wide. The location is independent of the GaAs-well width. In addition, it is supported by the trap-filling kinetics, which showed a logarithmic dependence of the DLTS signal. The origin of these defects may be due to the non-optimum growth condition for the first 10-20 nm of the top AlGaAs layer (low substrate temperature). For SQW structures, such as GRINSCH lasers, these states may increase the laser threshold current densities or may be responsible for enhanced laser degradation. A study of this subject is underway.

ACKNOWLEDGEMENT

The authors would like to acknowledge H. Meier for growing the MBE samples, A. Moser for the C-V measurements, and W. Bucher, U. Deutsch, H.P. Dietrich, L. Perriard, H. Richard and W. Walter for technical assistance.

REFERENCES

Blood P 1986 *Semicond. Sci. Technol.* **1** 7
Ferenczi G, Horvath P, Toth F, Kiss J and Boda J 1980 Hungarian Patent Appl., manufacturer MTA MFKI (Research for Technical Physics) Budapest, Hungary
Hamilton B, Singer K E and Peaker A R 1985 *Inst. Phys. Conf. Ser.* No **79** Chap 4 241
Kroemer H, Chien W Y, Harris J S, Jr and Edwall D D 1980 *Appl. Phys. Lett.* **36** 295
Lang D V 1974 *J. Appl. Phys.* **45** 3014 and 3023
Lang D V 1979 in *Thermally Stimulated Relaxation in Solids, Topics in Applied Physics,* ed P Braeunlich (Berlin: Springer) **37** 93
Lang D V, Hartman R L and Schuhmaker N E 1976 *J. Appl. Phys.* **47** 4986
Lang D V, Logan R A and Jaros M 1979 *Phys. Rev. B* **19** 1015
Lefevre H and Schulz M 1977 *Appl. Phys.* **12** 45
McAfee S R, Lang D V and Tsang W T 1982 *Appl. Phys. Lett.* **40** 520
Tsang W T 1978 *Appl. Phys. Lett.* **33** 245
Uji T, Suzuki T and Kamejima T 1980 *Appl. Phys. Lett.* **36** 655

Linear and nonlinear electrical conduction in quasi-two-dimensional quantum wells

P. Vasilopoulos, M. Charbonneau, and C. M. van Vliet

Université de Montréal, CRM, Montréal, Canada H3C 3A7

The dc electrical transport parallel to the walls of a quasi-two-dimensional quantum well, with a magnetic field normal to its barriers, is considered. For scattering by optical phonons, strong electric fields convert the usual magnetophonon maxima into minima and vice versa. For scattering by impurities and weak electric fields the dc conductivity oscillates with period $(\varepsilon_F - \varepsilon_0 n^2)/\hbar\omega_0$, where ε_F is the Fermi level, ε_0 is the lowest subband energy, and ω_0 is the cyclotron frequency.

1. INTRODUCTION

In the past years the electrical transport in quasi-two-dimensional systems has received considerable attention both for weak and strong electric fields. Unusual effects, such as the breakdown of the integral quantum Hall effect (Ebert et al. 1983) and a new type of conduction in $n^+n^-n^+$ GaAs structures (Eaves et al. 1984) have been observed at strong electric fields.

In this paper we consider dc electrical transport parallel to the walls of a thin quantum well in the presence of a magnetic field normal to its barriers, both for weak and strong electric fields. We evaluate the dc conductivity σ_{xx} for scattering by polar optical phonons (magnetophonon effect, Sec. III) and by impurities (Sec. V) including screening. We also evaluate the Hall conductivity σ_{yx}. The main results are mentioned in the abstract. Both σ_{xx} and σ_{yx} depend on the thickness of the well.

2. FORMALISM

We consider an infinitely deep square well with a magnetic field B along the z direction (normal to the barriers) and an electric field E along the x direction. The thickness of the well is L_z and the other dimensions L_y, L_x. In the Landau gauge the one-electron Hamiltonian, eigenfunctions and eigenvalues read

$$h^\circ = (\vec{P} + q\vec{A})/2m^* + eEx, \quad \vec{A} = (0, Bx, 0) \tag{1}$$

$$(\vec{r}\,|\zeta) = (2/L_y L_z)^{1/2}\, \phi_N(x - x_0)\, e^{ik_y y} \sin(n\pi z/L_z), \quad n = 1,2,3,\ldots \tag{2}$$

$$\epsilon_\zeta \equiv \epsilon_{N,n,k_y} = (N + 1/2)\hbar\omega_0 + n^2\,\epsilon_0 - \hbar V_d k_y + m^*\,V_d^2/2, \quad N = 0,1,2,\ldots \tag{3}$$

where V_d, ω_0, and ϵ_0 are the drift velocity, the cyclotron frequency, and the energy of the lowest well subband, respectively ($V_d = E/B, \omega_0 = eB/m^*$,

$\epsilon_0=\hbar^2\pi^2/2m^*L_z^2$). We have assumed a spherical effective mass m^* but the results hold for $m_x^*\neq m_z^*$ as well. N is the Landau level index and the harmonic oscillator functions $\phi_N(x-x_0)$ are centered at $-x_0=\ell^2(k_y+eE/\hbar\omega_0)$, where ℓ is the radius of the ground state orbit. The last two terms in Eq. (3) are the potential and kinetic energy due to the electric field. In the absence of this field these terms as well as the second parts of h° and x_0 are zero. Thus, the main effect of including E in h° is to lift the k_y degeneracy of the energy levels (3) and to shift the center of the orbits by $eE\ell^2/\hbar\omega_0$.

For <u>linear</u> transport, i.e., for weak electric fields the terms which depend on E, in Eqs. (1)-(3), are neglected. In this case, the dc conductivity σ_{xx} is given by (Charbonneau et al. 1982)

$$\sigma_{xx} = (\beta\ e^2/\Omega) \sum_{\zeta,\zeta'} f_\zeta(1 - f_{\zeta'})\ W_{\zeta\zeta'}\ R^2_{\zeta\zeta'}; \qquad (4)$$

here f_ζ is the Fermi-Dirac distribution function, Ω is the volume, $\beta=1/k_BT$, $W_{\zeta\zeta'}$ is the binary transition rate given by the golden rule, and $R_{\zeta\zeta'}$ is the mean distance involved in the transition $\zeta \rightarrow \zeta'$. For scattering by phonons formula (4) has been evaluated by Vasilopoulos (1986a); here it will be evaluated for scattering by impurities only. As for the Hall conductivity σ_{yx} it is given (Charbonneau et al. 1982) by

$$\sigma_{yx}=(i\hbar e^2/\Omega) \sum_{\zeta,\zeta'} f_\zeta(1-f_{\zeta'})(\zeta\ |V_x|\zeta')(\zeta'|V_y|\zeta\ )(1-e^{\beta\Delta})/\Delta^2, \Delta=\epsilon_{\zeta'}-\epsilon_\zeta\ , \zeta\neq\zeta', \qquad (5)$$

where V is the velocity operator. The main feature of (5) is that it is independent of the interaction, assumed nondiagonal in the representation of h°; if it has a diagonal part, it is included in ϵ_ζ and formula (5) remains valid.

For <u>nonlinear</u> transport, i.e., for strong electric fields, the following formula has been derived for the current density J_x, by Calecki et al. (1984), provided that the scattering system (e.g. phonons, impurities) remains at equilibrium (the same assumption is used in (4)):

$$J_x = (e/\Omega) \sum_{\zeta,\zeta'} R_{\zeta\zeta'} \left[f_\zeta(1-f_{\zeta'})\ W_{\zeta\zeta'} - f_{\zeta'}\ (1 - f_\zeta)\ W_{\zeta'\zeta}\right] \qquad (6)$$

3. SCATTERING BY POLAR OPTICAL PHONONS: NONLINEAR TRANSPORT

We consider bulk three-dimensional and dispersionless ($E_{\vec{q}} = \hbar\omega_L$) phonons in the deformation potential model. That is, we neglect interface or slab modes. When screening is taken into account the constant $|F(\vec{q})|^2=A/\Omega q^2$, entering $W_{\zeta\zeta'}$, is modified to $Aq^2/\Omega(q^2+q_s^2)^2$, where q_s is the inverse screening length. We assume $q_s\ell > 1$ and consider high temperatures ($1-f_\zeta\approx1-f_{\zeta'}\approx1$). To simplify the calculations we consider only narrow wells for which transitions between the levels n are not possible. Now $R_{\zeta\zeta'} \propto q_y$, the y component of the phonon wave vector, and the argument of the delta function, in $W_{\zeta\zeta'}$, contains the term $\hbar V_d q_y$. For transitions between Landau levels $q_y\approx1/\ell$ and this allows the main approximation <u>$\hbar V_d q_y\approx eE\Delta x\approx eE\ell$.</u> Corresponding to Eq. (3.8) of Vasilopoulos et al. (1987), which neglects screening, we now obtain

$$J_x \approx L_z^{-2} \sum_{N,n} C_{N,n} f_{N,n} \left[1 + 2 \sum_{s=1}^{\infty} e^{-2\pi s(\Gamma_N/\hbar\omega_o)} \cos(2\pi s \bar{\omega}_L/\omega_o) \right], \quad (7)$$

where $C_{N,n} = (3e^2BA/2\pi\hbar^3\omega_o q_z^2 \ell^4)\, n(1+2N_o)(1+2N-\bar{\omega}_L/\omega_o)$, Γ_N is the level width, $\bar{\omega}_L = \omega_L + eE\Delta x/\hbar$, and where, for a uniform system, we took $f_{N,nk_y} \equiv f_{N,n} \approx \exp[\beta_e(\epsilon_F - \epsilon_{N,n})]$ with $\beta_e = 1/k_B T_e$, T_e being an electron temperature, and $\epsilon_{N,n} = (N+1/2)\hbar\omega_o + n^2\epsilon_o$. The width Γ_N is estimated from $\hbar/\tau$, where τ is the inverse scattering rate, see Vasilopoulos (1986a). At resonance, $\bar{\omega}_L = P'\omega_o$, P' integer, and the quantity [...] is equal to $\coth(\pi\Gamma_N/\hbar\omega_o)$. The resonances are shifted: in the linear case $\omega_L = P\omega_o$ whereas now $\bar{\omega}_L = \omega_L + eE\Delta x/\hbar = P'\omega_o$. Thus, by varying the electric field the resonance maxima convert into minima and vice versa, as observed, in $n^+n^-n^+$ structures, by Eaves et al. (1984), and in GaAs/AℓGaAs heterostructures by Leadley et al. (1987). Assuming that only the term s=1 contributes to the oscillatory part of J_x, J_x^{osc}, we find that J_x^{osc} vanishes for values of the electric field E_c such that

$$eE_c\Delta x \approx (2m+1)\,\hbar\omega_o/4\,. \quad (8)$$

E_c, upon approximating Δx by the arithmetic mean of the mean square deviations of neighboring Landau levels, becomes .7 times the value E_c, of Eaves et al. (1984), which is in good agreement with the experiment. The physical interpretation of (8) is that for $E = E_c$ the energy gained by the electric field is of the order of $\hbar\omega_o$ and the electrons make transitions to neighboring Landau levels.

4. SCATTERING BY IMPURITIES

We assume that the electrons are scattered elastically by randomly distributed impurities via short-range or long-range potentials. In the latter case the screening is treated as in inversion layers. We consider only very low temperatures for which $\beta f_\zeta(1-f_\zeta) \approx \delta(\epsilon_\zeta - \epsilon_F)$.

In the <u>linear</u> case, we evaluate σ_{xx} from Eq.(4). For $\ell \sim L_z$, $q_z\ell > 1$, and neglecting transitions between the levels n, we obtain:

$$\sigma_{xx} \approx (C/\hbar\omega_o L_z^2) \sum_n \bar{\epsilon}_F \left[1 + 2 \sum_{s=1}^{\infty} (-1)^s e^{-2\pi s(\Gamma_N/\hbar\omega_o)} \cos(2\pi s \bar{\epsilon}_F) \right], \quad (9)$$

where $C = (3N_I/4\hbar)(eV_o/\ell)^2$, $V_o = e^2\sqrt{2}/\epsilon q_z$, and $\bar{\epsilon}_F = (\epsilon_F - \epsilon_o n^2)/\hbar\omega_o$. N_I is the impurity density and ϵ the dielectric constant. The conductivity oscillates with period $\bar{\epsilon}_F$ which changes as the Fermi level passes through the subbands n, as observed in heterostructures (Störmer et al. (1982)). The zero temperature limit of (9), for n = 1, is easily obtained ($\epsilon_F = (N + 1/2)\hbar\,\omega_o + \epsilon_o$);

$$\lim_{T\to 0} \sigma_{xx} \approx (C/\pi\,\Gamma_N L_z^2)\,(N + 1/2), \quad \pi\,\Gamma_N \ll \hbar\,\omega_o\,, \quad (10)$$

In the <u>nonlinear</u> case, with $\hbar V_d k_y \approx eE\Delta x$, we obtain from Eq. (6)

$$J_x \approx (C'/L_z^2) \sum_n \Delta x \left[1 + 2 \sum_{s=1}^{\infty} (-1)^s e^{-2\pi s(\Gamma_N/\hbar\omega_o)} \cos(2\pi s\bar{\epsilon}_F)\right]$$

$$\times \left[1 + 2 \sum_{m=1}^{\infty} e^{-2\pi m(\Gamma_N/\hbar\omega_o)} \cos(2\pi m\delta)\right] \qquad (11)$$

where $C' = (3BN_I/\beta\hbar^3\omega_o)(eV_o/\ell)^2$, $V_o = e^2\sqrt{2}/\epsilon q_\infty$, and $\delta = eE\Delta x/\hbar\omega_o$. The second line of Eq. (11) shows an oscillatory structure induced by the strong electric field.

5. THE HALL CONDUCTIVITY

Using Eq. (5) we obtain in the linear case (Vasilopoulos et al. (1987))

$$\sigma_{yx} = (e^2/hL_z) \sum_{N,n} f_{N,n}. \qquad (12)$$

Near zero temperature, $f_{N,n} \to 1$ and (12) becomes $\sigma_{yx} = (e^2/hL_z)(N+1)n$, whereas at high temperatures, with $f_{N,n} \approx \exp[\beta(\epsilon_F - \epsilon_{N,n})]$, $\sigma_{yx} \approx (e^2/hL_z)\ [2\sinh(\beta\hbar\omega_o/2)]^{-1} \sum_n \exp[\beta(\epsilon_F - \epsilon_o n^2)]$. Eq. (12) gives also the quantum Hall effect in heterostructures and superlattices (Vasilopoulos 1985, 1986b) for n = 1.

6. SUMMARY

We have considered magnetotransport paraller to the walls of a thin quantum well for weak and strong electric fields. For scattering by polar optical phonons the low field magnetophonon resonances $\omega_L = P\omega_o$ change, at strong fields, to $\omega_L + eE\Delta x/\hbar = P'\omega_o$ and, upon increasing E, maxima convert into minima and vice versa. A similar behavior, at strong E, is obtained for impurity scattering. In both cases we have included the effect of screening. For weak fields and impurity scattering the conductivity σ_{xx} oscillates, at very low temperatures, with period $\bar{\epsilon}_F = (\epsilon_F - \epsilon_o n^2)/\hbar\omega_o$. Both σ_{xx} (or J_x) and σ_{yx} depend on the thickness of the well. The results for J_x are tied to the approximation $\hbar V_d q_y \approx eE\Delta x$ but they do not change qualitatively, as shown by Warmenbol et al. (1987), if this approximation is not made.

REFERENCES

1. Calecki D. et al., J. Phys. C17, 5017 (1984).
2. Charbonneau et al., J. Math. Phys. 23, 318 (1982).
3. Eaves L., et al., Phys. Rev. Lett. 53, 608 (1984).
4. Ebert G. et al., J. Phys. C16, 5441 (1983).
5. Leadley R. D. et al. Preprint.
6. Störmer L.H. et al., Solid State Commun. 41, 707 (1982).
7. Vasilopoulos P., Phys. Rev. B32, 771 (1985).
8. Vasilopoulos P.,Phys.Rev.B33, 8587 (1986a);Phys.Rev.B34,3019 (1986b).
9. Vasilopoulos P. et al., Phys. Rev. B35, 1334 (1987).
10. Warmenbol P. et al., Phys. Rev. B (1987), in press.

Photoluminescence of GaInAs/AlInAs quantum wells grown by OMVPE

T. Kato, H. Kamei, M. Murata, G. Sasaki, K. Ono and K. Yoshida

Advanced Semiconductor Devices R & D Department,
Sumitomo Electric Industries, Ltd.
1, Taya-cho, Sakae-ku, Yokohama 244, JAPAN

ABSTRACT: We have investigated optical properties of GaInAs/AlInAs quantum wells by photoluminescence (PL) measurement at 4.2K. GaInAs/AlInAs quantum wells were grown by OMVPE and the thickness of GaInAs quantum wells is from 6 Å to 100 Å. The emission peak energy of a 6 Å quantum well was obtained at 1.432 eV. Further, it was found that emission linewidth of the 100 Å quantum well increased with increasing sheet carrier densities. This suggests that the PL linewidth of the 100 Å quantum well was mainly broadened by the band-filling of the quantum well.

1. INTRODUCTION

GaInAs/AlInAs quantum well structures are very attractive for their application to optical devices used for fiber optic communication. The emission wavelength can be shifted from 1.3 μm to 1.55 μm by varing the well thickness.

GaInAs/AlInAs quantum wells have been grown by MBE (Temkin et al 1983, Welch et al 1985, Scott et al 1987) and by OMVPE (Kamada et al 1986).

Welch et al (1985) studied PL linewidth broadening mechanisms of GaInAs/AlInAs quantum wells grown by MBE. It was reported that the PL linewidths of quantum wells thicker than 50 Å were broadened by carriers within the quantum well (band-filling effect) and that these carriers could originate from deep donors in AlInAs layers. In this study, we investigated optical properties of GaInAs/AlInAs quantum wells grown by OMVPE. The thickness of AlInAs barriers was varied in order to examine the band-filling effect.

2. EXPERIMENTAL

GaInAs/AlInAs quantum wells lattice-matched to InP substrates have been grown by reduced pressure OMVPE. The substrates were Fe-doped and oriented 2° off (100). Triethylgallium(TEG), trimethylindium(TMI), trimethylaluminium(TMA) and arsine(AsH_3) were used as the source materials. Growth conditions are summarized in Table I.

A 0.2 μm-thick GaInAs buffer layer (reference layer), 100 Å, 40 Å and 20 Å quantum wells with AlInAs barriers were subsequently grown and a 20 Å GaInAs cap layer was grown on the surface. Three samples with different barrier thickness of 50 Å, 100 Å and 300 Å were grown (Figure 1 (a)). Another sample with 500 Å barriers, in which a 6 Å GaInAs quantum well was added to the three GaInAs quantum wells described above, was grown (Figure 1 (b)). The thickness of the quantum wells and the barriers was estimated from the steady-state growth rate.

PL spectra of the samples immerced in liquid helium were measured. These samples were illuminated by a 5145 Å line argon ion laser with the intensity of ∿ 1 W/cm^2. A Ge detector was used and the correction for the photo-response of the detector and the grating was performed.

Table I. Growth conditions

Carrier Gas	H_2
Growth Temperature	650°C
Growth Pressure	90 Torr
Total Flow Rate	8 slm
Substrate Rotation Rate	12 rpm
GaInAs Layers	
TMI Concentration	9.8×10^{-6} m.f.
TEG Concentration	7.3×10^{-6} m.f.
AsH_3 Concentration	1.5×10^{-3} m.f.
[V/III]	88
Growth Rate	0.65 μm/hr
AlInAs Layers	
TMI Concentration	9.8×10^{-6} m.f.
TMA Concentration	5.4×10^{-6} m.f.
AsH_3 Concentration	1.5×10^{-3} m.f.
[V/III]	99
Growth Rate	0.65 μm/hr

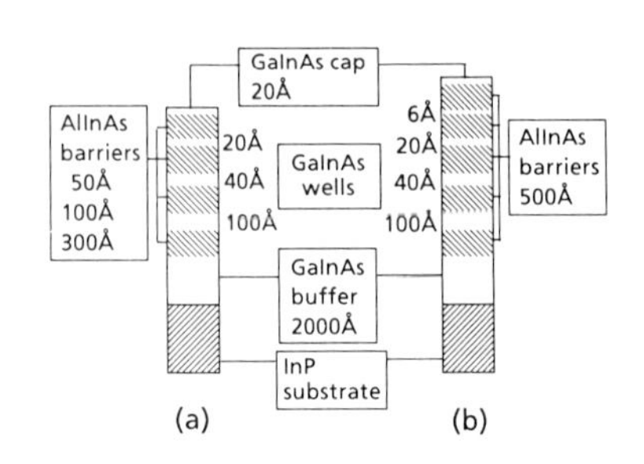

FIG. 1. The sample structures: (a) the three samples with 50 Å, 100 Å, 300 Å - thick AlInAs barriers, respectively, (b) the sample with 500 Å AlInAs barrier, in which a 6 Å quantum well was contained.

3. RESULTS AND DISCUSSION

Figure 2 shows a PL spectrum of the sample shown in Figure 1 (b). The emission peak energy of a 6 Å quantum well was obtained at 1.432 eV. This energy shift in a GaInAs quantum well is presumably larger than any other previous works. The emission intensity of the quantum wells is almost independent of the well thickness. Figure 3 shows the dependence of their peak energy shifts on the well thickness. The solid curve in Figure 3 was calculated by Alavi et al (1983). The band parameters used in this calculation are listed in Table II. The band discontinuity ΔE_C was used as a fitting parameter. The best fit is obtained at ΔE_C=502 meV. This value agrees well with the value of ΔE_C=500 meV reported by People et al (1983).

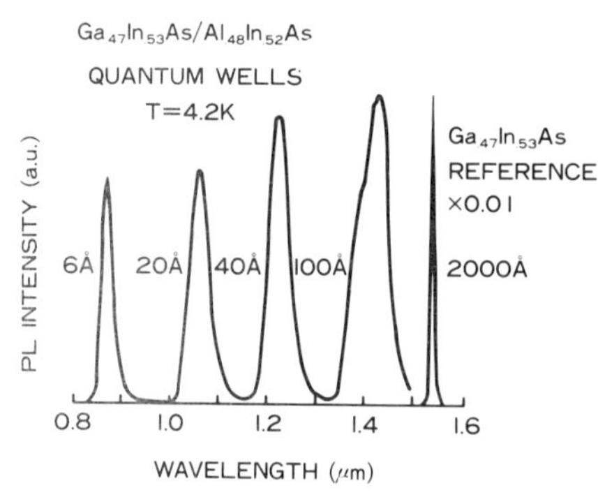

FIG. 2. PL spectrum from a stack of quantum wells at 4.2K.

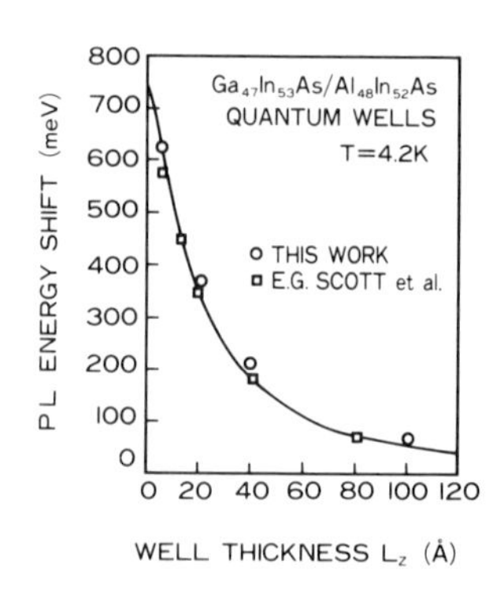

FIG. 3. Measured PL peak energy shift vs the well thickness.

Figure 4 shows the dependence of emission linewidths (FWHM) on the quantum well thickness L_Z. Welch et al (1985) proposed that PL (at 4 K) linewidths were mainly broadened by the band-filling for $L_Z > 50$ Å and the fluctuation of the well thickness for $L_Z < 50$ Å. The broadening due to the fluctuation of the well thickness is given by

$$\Delta E=(dE/dL_Z)\Delta L_Z \tag{1}$$

where ΔL_Z is the fluctuation of the well thickness.
The broadening due to the band-filling is given by

$$\Delta E=N_S\pi^2\hbar^3/L_Z m_e(2m_eE_e)^{1/2} \tag{2}$$

where N_S is the sheet carrier density of the well, me is the effective mass of electron and E_e is the energy shift of the 1st quantized electron state (Welch et al 1985). The broadenings calculated from eq.(1) and eq.(2) are also plotted in Figure 4. The thickness fluctuation is presumably the main broadening mechanism in the case of narrow wells. According to this PL linewidth data, the fluctuation of the narrow wells is estimated about one monolayer.

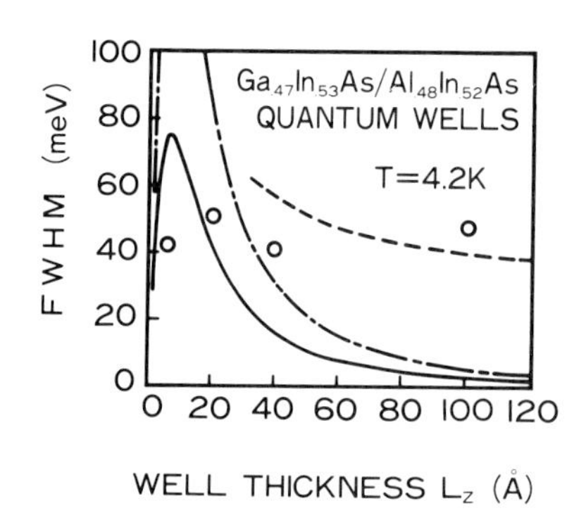

FIG. 4. The PL linewidth (FWHM) vs the well thickness. Solid and dot-dash curves show broadenings due to the well thickness fluctuation ΔL_Z of one monolayer ($a_0/2=$ 2.93 Å) and two monolayers(a_0), respectively. Dash curve shows broadening due to the band-filling. The sheet carrier density of 5×10^{11} cm^{-2} was assumed.

Table II. Band parameters of GaInAs and AlInAs. m_e and m_{hh} are electron and heavy -hole effective masses, respectively (Alavi K et al 1980, Olego D et al 1982, Wagner et al 1985). E_g is the bandgap energy. (Welch D F et al 1984, Goetz K H et al 1983)

	m_e/m_o	m_{hh}/m_o	E_g (eV)
$Ga_{47}In_{53}As$	0.041	0.47	0.811
$Al_{48}In_{52}As$	0.075	0.8	1.56

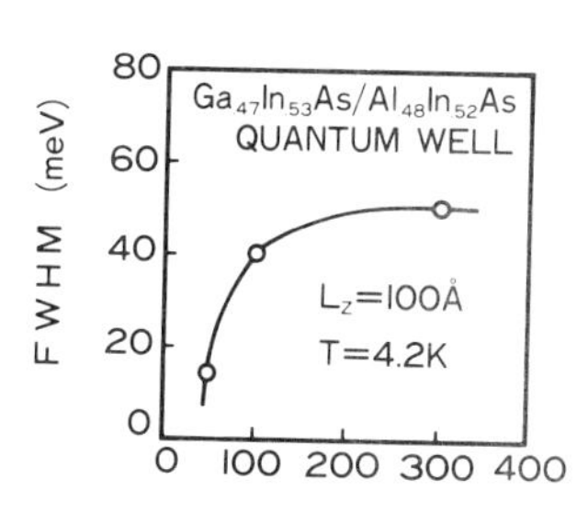

FIG. 5. The PL linewidth (FWHM) of the 100 Å quantum well vs the barrier thickness.

Figure 5 shows the dependence of the PL linewidth (FWHM) of the 100 Å quantum well on the barrier thickness. The PL linewidths increased from 14 meV to 50 meV with increasing the barrier thickness. This effect can be explained by the band-filling effect. We have measured the sheet carrier density N_S of multilayer with a GaInAs buffer, in which the quantum well thickness is 20 Å, 40 Å and 100 Å, using the Hall effect. Figure 6 shows the dependence of sheet carrier density N_S on the barrier thickness. The sheet carrier densities of the samples increase approximately proportional to the barrier thickness. This suggests that carriers of each GaInAs layer, i.e., 20 Å, 40 Å and 100 Å-thick quantum wells and a 0.2 µm buffer layer, are mainly supplied from AlInAs layers on both sides. The sheet carrier density of each well is estimated to be approximately $N_S/4$. Figure 7 shows the dependence of the PL linewidth of the 100 Å quantum well on $N_S/4$. In the region of low sheet carrier densities, PL linewidths agree well with the theoretical curve given by the eq.(2). Thus, the band-filling of the GaInAs quantum well, caused by the donor of AlInAs barrier layers, could dominantly broaden the PL linewidth. However, in the region of high sheet carrier densities, PL linewidth is much narrower than the theoretical results. In this region, the energy difference between the 1st and the 2nd subband can be smaller than that in the region of low sheet carrier densities due to the bending of GaInAs and AlInAs conduction band. Therefore, PL linewidth may possibly be narrower than the theoretical results due to the effect of the 2nd subband-filling in a GaInAs quantum well with high sheet carrier density.

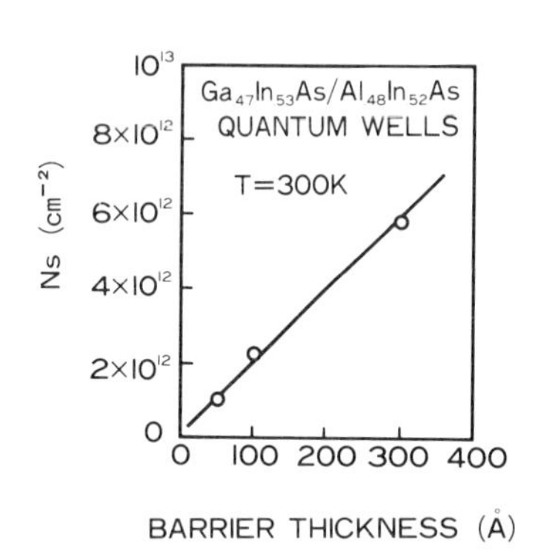

FIG. 6. The sheet carrier density N_S vs the barrier thickness.

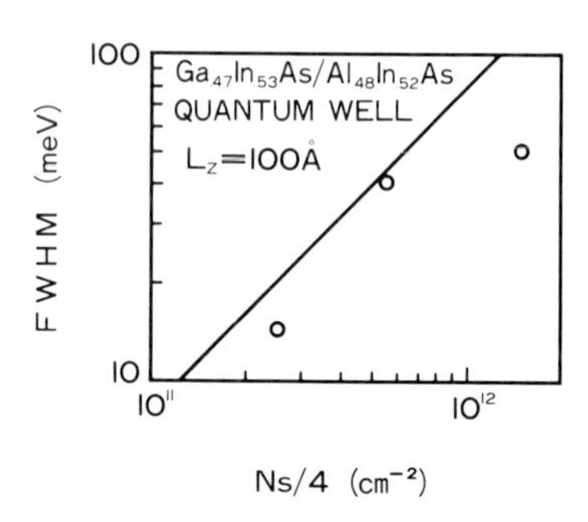

FIG. 7. The PL linewidth (FWHM) of the 100 Å quantum well vs the sheet carrier density $N_S/4$ in a quantum well. The theoretical curve shows the bloadening of the 100 Å well due to the band-filling.

REFERENCES

Alavi K and Aggarwal R L 1980 PHYSICAL REVIEW B **21** 1311
Alavi K, Pearsall T P, Forrest S R and Cho A Y 1983 ELECTRONICS LETTERS **19** No.6 229
Goetz K H, Bimberg D, Jürgensen H, Selders J, Solomonov A V, Glinskii G F and Razeghi M 1983 J.Appl.Phys. **54** 4543
Kamada M, Ishikawa H, Ikeda M, Mori Y and Kojima C 1986 GaAs and Related Compounds 1986 No.83 575
Olego D, Chang T Y, Silberg E, Caridi E A and Pinczuk A 1982 Appl.Phys.Lett. **41** 476
Scott E G, Davey S T and Davies G J 1987 ELECTRONICS LETTERS 23 No.14 762
Temkin H, Alavi K, Wagner W R, Pearsall T P and Cho A Y 1983 Appl.Phys.Lett. **42** 845
Welch D F, Wicks G W and Eastman L F 1984 J.Appl.Phys. **55** 3176
Welch D F, Wicks G W and Eastman L F 1985 Appl.Phys.Lett. **46** 991
Wagner J, Stolz W and Ploog K 1985 The American Physical Society **32**. 4214

Inst. Phys. Conf. Ser. No. 91: Chapter 6
Paper presented at Int. Symp. GaAs and Related Compounds, Heraklion, Greece, 1987

Fourier analysis of universal conductance fluctuations in the magnetoresistance of submicron-size n$^+$ GaAs wires

M L Leadbeater, R P Taylor, P C Main, L Eaves, S P Beaumont*, I McIntyre*, S Thoms* and C D W Wilkinson*

Department of Physics, University of Nottingham, Nottingham NG7 2RD, U.K.
*Department of Electronics and Electrical Engineering, University of Glasgow, Glasgow, GL2 8QQ, U.K.

ABSTRACT: We have measured the oscillatory magnetoresistance of small n$^+$GaAs stripes of thickness 50 nm, length 10 µm and widths between 90 nm and 260 nm. By using a Fourier Transform technique we find evidence that the negative magnetoresistance associated with weak localistion and universal conductance fluctuations (UCF) are manifestations of the same phenomenon. The microscopic potential distribution which determines the scattering and hence the UCF is shown to be stable at low temperatures. At temperatures ≥ 100 K this distribution is altered in what appears to be an activated manner.

Quantum inteference between scattered electron waves has been studied in considerable detail recently. Advances in lithographic techniques have led to the observation of universal conductance fluctuations (UCF) in single metallic and semiconducting wires (see for example, Skocpol et al., 1986) and of the Bohm-Aharanov effect in multiply connected ring structures (see for example, Webb et al., 1985). Also, weak localisation and an associated negative magnetoresistance (NMR) has been described as a quantum interference effect between back-scattered electron waves (Bergmann 1985), although this process is thought to be qualitatively different from the other two effects. We present here measurements on narrow n$^+$ GaAs stripes fabricated by electron beam lithography from MBE grown material(Thoms et al., 1986) in which the conducting layer was 50 nm thick and doped with Si to give a free carrier concentration of 5 x 10^{18} cm^{-3}. The low temperature magnetoresistance R(B) of these stripes shows, in addition to the UCF, the NMR usually associated with weak localisation. We have Fourier analysed the magnetoresistance curves and conclude that the UCF and NMR are associated with the <u>same</u> physical phenomenon.

Typical results for the transverse magnetoresistance of wires of dimensions 260 nm x 50 nm x 10 µm and 90 nm x 50 nm x 10 µm are shown in Figure 1 and Figure 2(a) respectively. Note that the amplitude of the UCF is comparable with that of the NMR and that the width of the NMR is comparable with a typical period of the UCF. Note also that as the temperature increases, the positions of the peaks and troughs do not change. The UCF persist with undiminished amplitude up to our highest field, (11.5 T) and are still visible at temperatures as high as 100 K. Our previous work has shown that the UCF in these n$^+$ GaAs wires are associated with electron trajectories of three-dimensional character (Whittington et al., 1986).

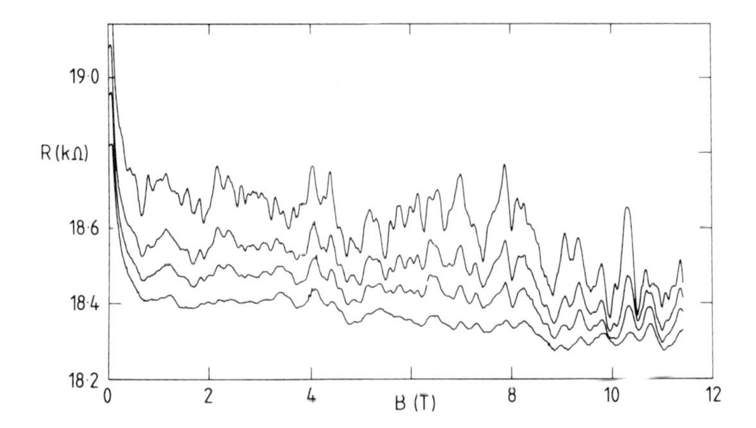

Fig. 1. Transverse magnetoresistance of 260 nm wide n^+GaAs wire at temperatures of 4.2 K (top), 11.8 K, 17.7 K, 30.0 K (bottom)

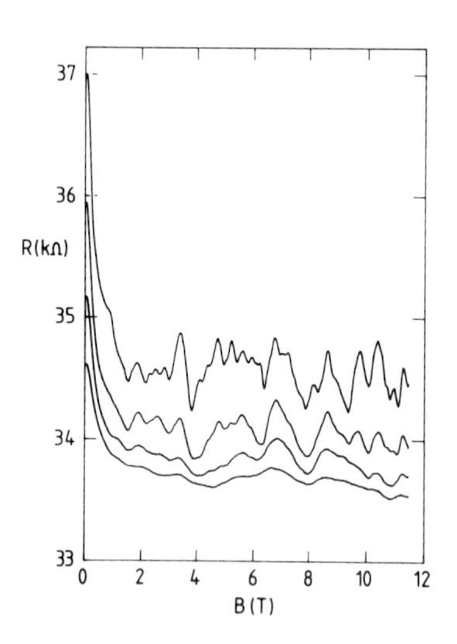

Fig. 2(a). Transverse magnetoresistance of 90 nm wide n^+GaAs wire at temperatures of 4.2 K (top), 13.4 K, 26.0 K and 51.5 K (bottom)

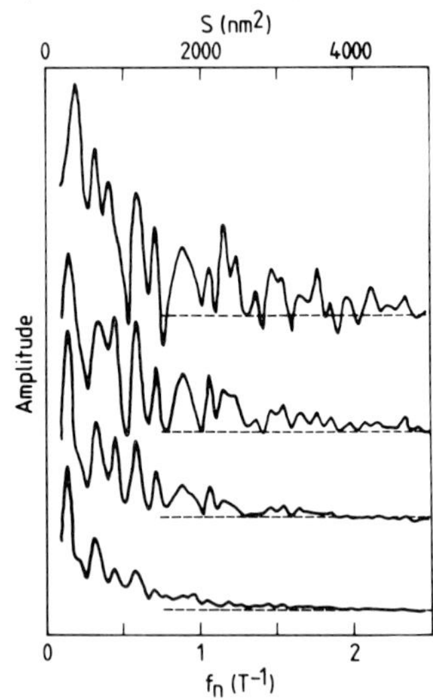

Fig. 2(b). Fourier Transforms of the curves shown in Fig. 2(a). The curves are displaced for clarity and the dashed lines indicate the respective zeros.

Figure 2(b) shows the corresponding Fourier Transforms (FT) of the results shown in Figure 2(a). The positions of the peaks, f_n, shown in Figure 2(b) are independent of the magnetic field range over which the FT is performed, although naturally the highest resolution is obtained when we use the full range of data. Note that the positions of the features in the FT do not depend on temperature. However, as the temperature increases, the amplitudes of the higher frequency peaks decrease more rapidly than those at lower frequency. We stress here that the negative magnetoresistance is included in the Fourier analysis. Originally we treated the NMR and the UCF as essentially different effects and we attempted to Fourier Transform only the oscillatory part of the magnetoresistance. Using this approach we were unable to obtain any sort of consistency between traces at different temperatures. We are therefore led to the empirical observation that, in this system at least, the UCF and weak localisation are manifestations of the same phenomenon.

Weak localisation has been described in terms of coherent back scattering of electrons, the constructive interference between an electronic partial wave scattered around a closed loop and its time-reversed equivalent, which corresponds to a wave traversing the same loop but in the opposite direction. Following this, we assume that the UCF have a similar origin i.e. each fluctuation "period" corresponds to a loop of area $\underline{S}$. Application of a magnetic field $\underline{B}$ then introduces a phase difference between the partial waves of $\Delta\phi = 2e\underline{B}.\underline{S}/\hbar$. The areas corresponding to the Fourier Transform periods are shown on the upper axis of Figure 2(b). The probability of an electron completing such a loop depends on the electron phase-breaking length ℓ_ϕ. Since the elastic scattering length in the material is known from conductivity and Shubnikov-de Haas measurements to be 32 nm, the loops comprise fewer than ten or so elastic scattering events. Thus a ballistic model for the electronic motion is more appropriate here than a diffusive one. We assume that the amplitude, a_n, of a particular peak in the FT is controlled by the perimeter ℓ of the loop so that

$$a_n \propto \exp\left(\frac{-2\ell}{\ell_\phi}\right) \quad ,$$

where the factor 2 arises because the electron waves have to traverse the loop in both directions. Detailed quantitative analysis of this simple model is described elsewhere (Taylor et al., 1987). We obtain values of the phase-breaking scattering rate, τ_ϕ^{-1}, as a function of temperature. These are shown in Figure 3 for two wires. The lower curve, for a wire of dimensions 10 μm x 260 nm x 50 nm, shows $\tau_\phi^{-1} = AT + C$ where $A \sim 4 \times 10^{10}$ s^{-1} K^{-1} and the constant C corresponds to a length of 270 ± 20 nm which suggests a size-dependent phase-breaking process, possibly due to surface scattering. The upper curve in Figure 3 is for a wire 10 μm x 90 nm x 50 nm. For T < 30 K many loops have $S^{\frac{1}{2}}$ greater than the width of the wire and our analysis is not applicable. However, above that temperature the data are consistent with those of the other wire, with the same value of A but with a value of C which is equivalent to ℓ_ϕ = 100 ± 10 nm.

The magnetoresistance traces are reproducible within the noise level so long as the temperature is held constant at close to liquid helium temperatures. If a trace is taken at, say, 4.2 K and the temperature is raised to 300 K then cooled again to 4.2 K, a qualitatively similar trace of R(B) is obtained. However it differs in the detailed positions of the peaks and troughs. We have quantified this observation by the following procedure. A magnetoresistance trace was taken at 4.2 K (Trace r). The temperature was then increased to some value T, then reduced back to 4.2 K and another trace taken (Trace r + 1). Each trace R(B) comprises data points at 1000 different values of B. These two traces are then correlated using the function

$$F = \left\langle \frac{\langle \Delta R(B) \rangle - \mathrm{ABS}\left(R_{r+1}(B) - R_r(B)\right)}{\langle \Delta R(B) \rangle} \right\rangle_B$$

The normalisation parameter <ΔR(B)> is the value of $(R_n(B) - R_{n-1}(B))$ averaged over all field values and over six totally uncorrelated traces. The final averaging $\langle \ \rangle_B$ is performed over the 1000 field values for the two traces. Thus F = 1 for two identical traces and F = 0 for two uncorrelated ones. A graph of F plotted against T is shown in Figure 4. There is a sharp drop in the correlation at around 100 K. This can be understood by realising that the magnetoresistance oscillations are determined by the microscopic positions of the charged impurities and defects giving rise to scattering. Increasing the temperature will have no effect on the microscopic configuration until kT is large enough to activate the charge bound to defects. We are currently extending the data to provide a quantitative test for the model.

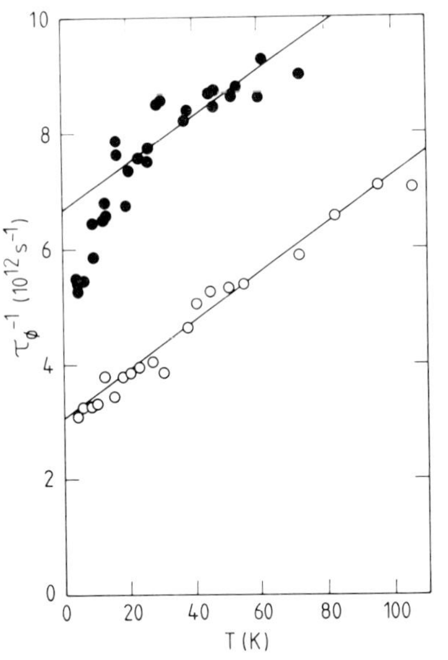

Fig. 3. Phase breaking scattering rate, τ_ϕ^{-1} plotted against T for 90 nm (upper) and 0.26 μm (lower) wide wires

Fig. 4. Correlation function F plotted against T. The calculation of F and the definition of T are described in the text

REFERENCES

Bergmann G 1985 Phys. Rep. 107 1

Skocpol W J, Mankiewicz P M, Howard R E, Jackel L D and Tennant D M 1986 Phys. Rev. Lett. 56 2365

Taylor R P, Leadbeater M L, Whittington G P, Main P C, Eaves L, Beaumont S P, McIntyre I, Thoms S and Wilkinson C D W 1987 to be published in the Proceedings of the Seventh Int. Conf. on the Electronic Properties of Two-dimensional Systems, Santa Fe

Thoms S, Beaumont S P, Wilkinson C D W, Frost J and Stanley C R 1986 Proc. Int. Conf. on Microcircuit Engineering, Interlaken

Webb R A, Washburn S, Umbach C P and Laibowitz R B 1985 Phys. Rev. Lett. 54 2696

Whittington G P, Main P C, Eaves L, Taylor R P, Thoms S, Beaumont S P, Wilkinson C D W, Stanley C R and Frost J 1986 Superlattices and Microstructures 2 381

Electronic properties of the two-dimensional electron gas in the GaInAs/InP system studied using hydrostatic pressure

M A Fisher, D Lancefield, A R Adams, J M Boud, M J Kane*, L L Taylor* and S J Bass*

Department of Physics, University of Surrey, Guildford, Surrey GU2 5XH, UK
*Royal Signals and Radar Establishment, St Andrews Road, Great Malvern, Worcestershire WR14 3PS, UK

ABSTRACT: We report results of Hall measurements made on different heterostructures in the GaInAs/InP system as a function of hydrostatic pressure both at room temperature and down to 4K. We deduce the effective mass and carrier density dependence of polar phonon scattering and the effective mass dependence of the low temperature mobility. These results indicate deficiencies in current theories of scattering in two dimensions.

When hydrostatic pressure is applied to either InP or $Ga_{0.47}In_{0.53}As$ the direct band gap increases leading to an increase in the effective mass, m*, of the carriers and hence to a change in the electron scattering rate. This has provided important information for comparison with theories describing the electron mobility in bulk material. In this paper we report measurements of the electron mobility as a function of pressure and temperature in modulation doped $Ga_{0.47}In_{0.53}As$/InP single and multiple quantum well systems and in a two-dimensional electron gas (2DEG) at a single interface. The structures were all grown by atmospheric pressure metalorganic chemical vapour deposition (MOCVD). Controlled doping was achieved using sulphur introduced into the reactor as hydrogen sulphide from a diffusion cell. Details of the growth techniques have been published elsewhere (Bass et al, 1986). Sample 1 is a single heterojunction grown as an inverted structure (i.e. the doped layer between the junction and the substrate). Sample 2 is a single 100Å quantum well and 3 is a multiple 100Å quantum well with 16 periods. Details of these samples have been published elsewhere (Kane et al, 1986, Skolnick et al, 1987). Their room temperature carrier concentrations and electron mobilities were $n_s = 2\times10^{11}cm^{-2}$, $\mu = 9700cm^2/Vs$ and $n_s = 1.8\times10^{11}cm^{-2}$, $\mu = 6200cm^2/Vs$ for 1 and 2 respectively and they had been designed to have no parallel conduction. The multiple well structure has a room temperature mobility of $7200cm^2/Vs$ and a Hall carrier density n_H of $1.68\times10^{13}cm^{-2}$ corresponding to $1.05\times10^{12}cm^{-2}$ carriers per well, in good agreement with Shubnikov-de Haas measurements. This confirms that there is no parallel conduction in the InP.

Experiments were performed on van der Pauw structures to measure the resistivity and Hall coefficient. Pressure to 15 kbar at room temperature was generated in a piston and cylinder apparatus and to 8 kbar down to 4.2K in a capillary fed high pressure cell described by Lambkin et al (1987).

Room temperature experimental results already briefly presented by Fisher et al (1987) are shown in Figures 1 and 2. Figure 1 shows the pressure dependence of the Hall carrier density plotted normalized to its atmospheric pressure value

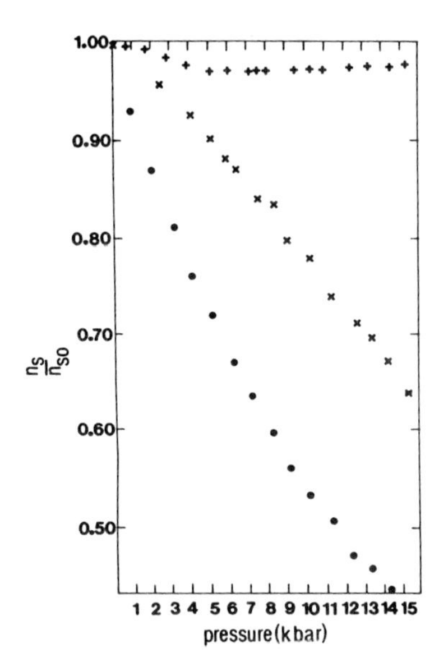

Fig.1 Pressure dependence of electron concentration at 300K.
x-Sample 1, •-Sample 2, +Sample 3.

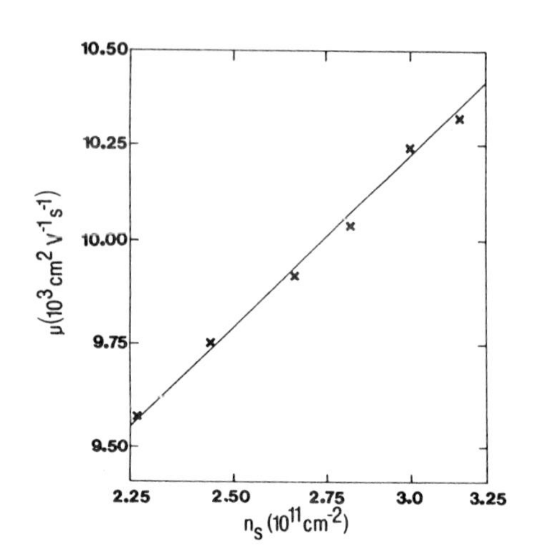

Fig.3 Carrier density dependence of mobility for heterojunction measured under low illumination.

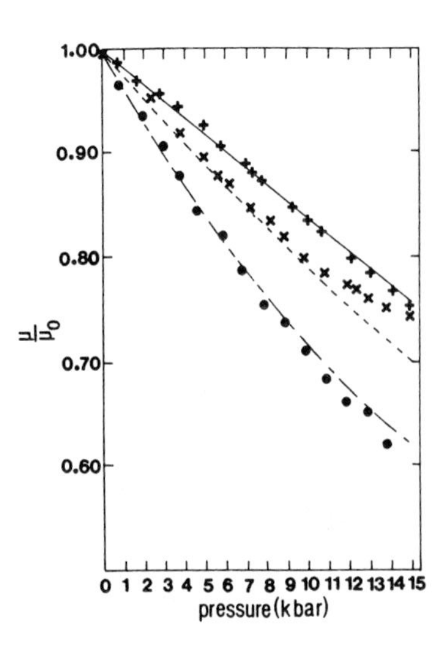

Fig.2 Pressure dependence of electron mobility at 300K.
x-Sample 1, •-Sample 2, +-Sample 3.
Solid line - fit to data.
Dashed and dot-dashed lines - theoretical curves corrected for n_s variation.

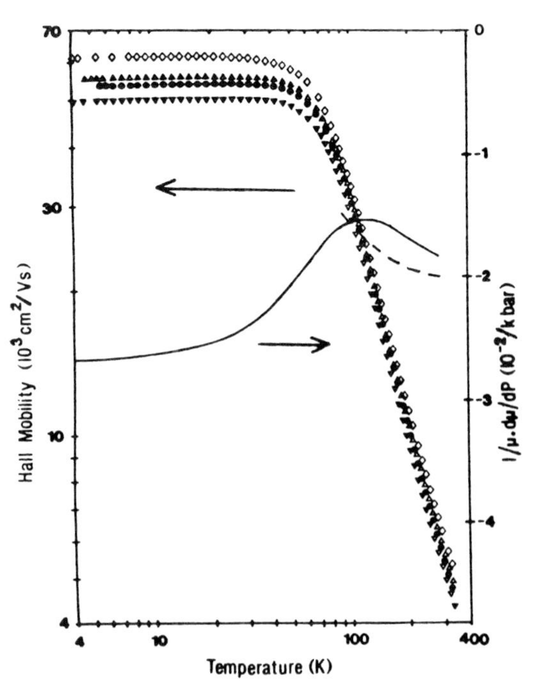

Fig.4 Temperature dependence of mobility at different pressures.
◇ -0 kbar, ▲-4 kbar, •-5 kbar, ▼-8kbar.
Solid line - pressure coefficient of mobility obtained from these data.
Dashed line - variation of pressure coefficient expected including increase of polar phonon frequency with pressure.

for each of the three samples. There is considerable difference in their behaviour. In the multiple well sample there is negligible trapout whereas in the single well and the heterojunction samples a substantial fraction of the carriers present at atmospheric pressure have trapped out by 15 kbar. The reasons for the trapout are not completely clear but it seems to be associated with interaction between the 2DEG and the substrate and to be closely related to persistent photoconductivity which is observed only in samples showing carrier trapout with pressure. Figure 2 shows the pressure dependence of the mobility at room temperature where polar phonon scattering dominates for each of the three systems studied. In each case there is a fairly strong decrease in mobility with pressure and we will show that the difference between the samples is due only to their different carrier concentration behaviour. When carrier loss is occurring with pressure, mobility analysis is complicated since μ is a function of both n_s and m^*. Fortunately in sample 3 there is no change in n_s and the mobility variation may be principally ascribed to an effective mass change alone allowing analysis of all three systems.

Let us assume $\mu \propto n_s^{\alpha} m^{*\beta}$. From **k·p** theory we can take $1/m^* \cdot dm^*/dP$ as constant to 15 kbar as is observed in bulk semiconductors. We give it a value γ so we may write $\mu \propto n_s^{\alpha} \cdot e^{\beta\gamma P}$. For the multiple well sample n_s is a constant and taking the full line in Fig.2 we obtain $\beta\gamma$ = -0.018/kbar. To determine the n_s dependence of μ independent of any m^* change we have measured the mobility and carrier concentration in the heterojunction sample at room temperature under low illumination. The results are shown in Fig.3. From this we determine the exponent α to be 0.22. The predicted pressure dependence of mobility in the heterojunction and single quantum well samples are shown by the dashed and dot-dashed curves in Fig.2 respectively. The agreement with the experimental results is very good indicating that the mobility dependence on effective mass is the same in all three structures. It is not clear why there is some deviation from the general trend in the heterojunction sample at high pressures but this may be due to the 2DEG becoming non-degenerate and the effect of pressure on the scattering processes changing. The measured dependence of the mobility on carrier concentration (α = 0.22) is quite close to that measured by Hirakawa and Sakaki (1986) at 300K (α = 0.3) in gated GaAs/AlGaAs heterojunctions and cannot be described by polar optical phonon scattering theories which predict a negative value for α.

For direct comparison with theory the effective mass dependence of the mobility is more useful than the pressure dependence. The pressure dependence of the effective mass is thus required. In bulk GaInAs, Shantharama et al (1985) obtain a value of 1.9% /kbar. In two dimensions Gauthier et al (1986) have reported a value of 1% /kbar, but with some reservations as to the general applicability of this figure. This leads to a variation of $\mu \propto m^{*(-1.8\pm0.3)}$. For electrons in the lowest subband, as here, Ridley (1982) predicts a dependence $\mu \propto m^{*-2}$, which agrees with our results.

We now consider measurements of the multiple quantum well mobility at high pressures and low temperatures. This sample shows negligible trapout with pressure which simplifies the interpretation of the results. Figure 4 shows the temperature variation of the mobility at different pressures. At all temperatures the mobility decreases with pressure. It is convenient to consider two temperature regions. Firstly, below about 60K, where scattering is thought to be dominated by alloy and ionised impurity scattering the mobility is essentially independent of temperature. Secondly in the temperature range 100-300K, where polar phonon scattering becomes increasingly important with increasing temperature and the mobility shows a strong temperature dependence.

In the plateau region at low temperatures pressure may be particularly useful to study the dominant scattering mechanisms as little experimental data on the functional dependence of the mobility can be obtained by temperature dependent measurements. The solid curve plotted in Fig.4 is the temperature dependence of the pressure coefficient of the electron mobility determined from the experimental data of that figure. Between 4 and 20K the pressure coefficient is nearly constant and assuming the same pressure dependence of the effective mass used above we deduce that $\mu \propto m^{*-(2.7\pm0.3)}$. Lee et al (1983) derive expressions for the mobility limited by remote and background ionised impurities in a number of models. The strongest negative effective mass dependence of the mobility they obtain is m^{*-1}. Alloy scattering is poorly understood in 2D but Bastard (1983) has calculated that $\mu \propto m^{*-2}$. Thus it is clear that current theory does not predict as strong an effective mass dependence as our results indicate.

In the higher temperature region, above 100K, where polar phonon scattering dominates, we obtain a pressure coefficient of the electron mobility which decreases with decreasing temperature. This we believe to be associated with the pressure dependence of the polar phonon energy, $\hbar\omega_l$. Using a value of $1/\omega_l\ d\omega_l/dP = 1.9\times10^{-3}\mathrm{kbar}^{-1}$ obtained for bulk material by interpolation between measured values for the binary constituents (Pearsall, 1981), the dashed curve in Fig.4 was calculated and is in good agreement with experiment. Such a dependence is due to the reduction in the density of phonons $n(\omega_0)$ with pressure which partially cancels the mobility reduction caused by the increase in m^* with pressure. As the phonon energy increases with pressure, $n(\omega_0)$ becomes more sensitive to temperature through the Bose-Einstein factor and so the phonon density is reduced proportionately more at low temperatures than at higher temperatures. This produces the observed reduction in the pressure coefficient of the mobility. In the intermediate temperature range 60-100K, the pressure coefficient goes through a transition, reflecting the changing importance of the various scattering mechanisms.

We have presented results of measurement of the electron mobility in 2DEGs in GaInAs/InP made as a function of pressure both at room temperature and low temperature. While accurate determination of the pressure dependence of the effective mass in 2D systems is still required, our results indicate some shortcomings of current theories of electron scattering. Pressure should prove to be as valuable a technique in studying scattering in 2D as it has been in 3D, particularly in the low temperature regime.

REFERENCES

Bass SJ, Barnett SJ, Brown GT, Chew NG, Cullis AG, Pitt AD and Skolnick MS 1986 *J. Cryst. Growth* **79** 378

Bastard G 1983 *Surf. Sci.* **142** 284

Fisher MA, Adams AR, Lancefield D, Kane MJ, Taylor LL and Bass SJ 1987 *Appl. Phys. Lett.* to be published

Gauthier D, Dmowski L, BenAmor S, Blondel R, Portal JC, Razeghi M, Maurel P, Omnes F and Laviron M 1986 *Semicond. Sci. Tech.* **1** 105

Hirakawa K, Sakaki H 1986 *Phys. Rev.* B**33** 8291

Kane MJ, Anderson DA, Taylor LL and Bass SJ 1986 *J. Appl. Phys.* **60**(2) 657

Lambkin J D, Lancefield D, Gunney B J, Dunstan D J and Bristow F G 1987 *J. Phys. E* submitted

Lee J, Spector HN and Arora VK 1983 *J. Appl. Phys.* **54**(12) 6995

Pearsall TP 1981 *GaInAsP Alloy Semiconductors* (New York; Wiley)

Ridley BK 1982 *J. Phys.* C**15** 5889

Shantharama LG, Nicholas RJ, Adams AR and Sarkar CJ 1985 *J. Phys.* C**18** L443

Skolnick MS, Taylor LL, Bass SJ, Pitt AD, Mowbray DJ, Cullis AG and Chew NG 1987 *Appl. Phys. Lett.* to be published

Magnetic field studies of resonant tunnelling double barrier structures

G A Toombs, E S Alves, L Eaves, T J Foster, M Henini, O H Hughes, M L Leadbeater, C A Payling, F W Sheard, P A Claxton*, G Hill*, M A Pate* and J C Portal+

Department of Physics, University of Nottingham, Nottingham NG7 2RD, U.K.
*Department of Electronic and Electrical Engineering, University of Sheffield, Sheffield S1 3JD, U.K.
+SNCI-CNRS, Grenoble 38042 and LPS, INSA, Toulouse 31077, France.

ABSTRACT: Resonant tunnelling double barrier structures based on n-GaAs/(AlGa)As and n-(InGa)As/InP are investigated by examining the oscillatory structure in the magnetotunnelling current for $\underline{J} \parallel \underline{B}$. Evidence for charge build-up in the well controlling the resonant tunnelling current is obtained. The current bistability of the devices is discussed and a simulation of the bistability is presented.

In this paper, we present magnetotunnelling studies of two different types of resonant double barrier structures (DBS). The first is a symmetric structure based on n-GaAs/(AlGa)As. It requires a bias voltage ~ 100 mV for the first subband of the quantum well formed between the two barriers to be brought into resonance with the emitter states. The other is based on n-(InGa)As/InP and, in contrast, is resonant at zero bias.

The current-voltage characteristics of a typical GaAs/(AlGa)As structure are shown in figure 1a. The growth parameters have been given by Payling et al (1987a). The negative differential conductivity (NDC) and associated bistability for voltages below -0.4 V are very similar to those reported by Goldman et al (1987a) who suggested that the bistability is intrinsic. Intrinsic bistability is theoretically possible due to electrostatic feedback produced by the build-up of charge in the well on the tunnelling current. Sollner (1987) has proposed, as an alternative explanation of the data, that the bistability is extrinsic owing to the oscillations inherent in this type of device. Figure 1b shows the average current $\bar{I}$ in a circuit as a function of applied voltage V for a simulation of a device with NDC in parallel with a capacitor and connected by resistive and inductive leads to a voltage source. It can be seen that the characteristics are remarkably similar to those obtained experimentally. Both regions of bistability are present and the current profile resembles closely that of figure 1a. The intermediate values of $\bar{I}$ occur for voltages when the circuit oscillates. The simulated oscillatory current waveforms are also similar to those observed. The bistability in the simulation arises because the turn-on and turn-off voltages for oscillation as V is increased are not the same as the corresponding turn-off and turn-on voltages as V is reduced.

Payling et al (1987a) have shown that the tunnelling current oscillates as a function of magnetic field $\underline{B} \parallel \underline{J}$. Each series of oscillations is periodic in 1/B. Therefore a fundamental field B_f can be defined by the periodicity $\Delta(1/B) = 1/B_f$. Results for voltages in the range where current bistability is observed have been given by Payling et al (1987a). In this range, there are two distinct values of B_f corresponding to high and low current. This confirms that there are two possible states of accumulation, as expected from the current bistability, but they could be of either extrinsic or intrinsic origin.

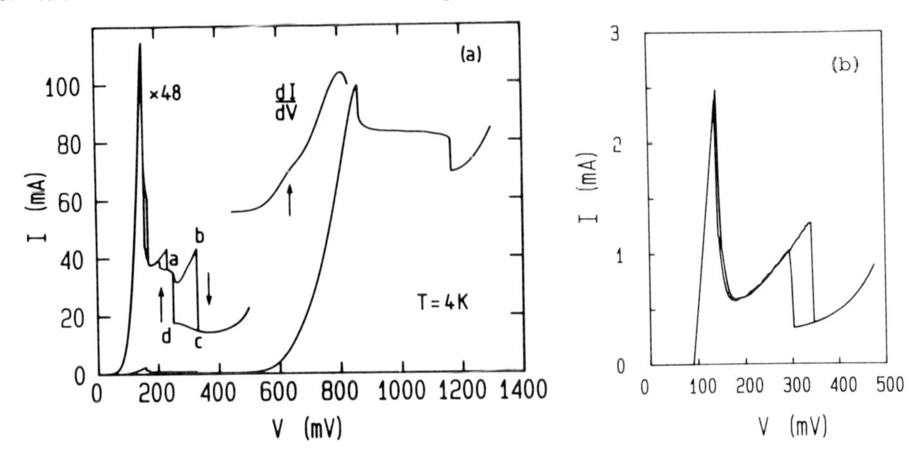

Fig. 1. (a) I(V) characteristics of n-GaAs/(AlGa)As device at 4 K with negative substrate bias. The low voltage part is magnified x48. The conductance near 0.6 V is also shown. (b) Simulation of I(V) characteristics for a device with NDC in parallel with a capacitor and connected by resistive and inductive leads to a voltage source.

We next discuss the magnetotunnelling for bias less than 170 mV, above which NDC commences for the GaAs/(AlGa)As structures. Figure 2 shows the fundamental fields B_f as a function of voltage in this range. They correspond to data points which fall into two distinct groups. At low currents, the oscillatory structure observed in the tunnel current has a fundamental field $B_f \equiv 8$ T, almost independent of applied voltage. At these low voltages ($\lesssim$ 130 mV) the tunnel current is predominantly non-resonant. The oscillations are associated with Landau levels passing through the Fermi energy E_F of the emitter contact layer as B is increased. The fundamental field is therefore given by $B_f = m^*E_F/\hbar e$. This expression gives reasonable agreement with the observed value (8 T) for the nominal doping level in the emitter (2×10^{17} cm^{-3}).

The second group of oscillations appears at the onset of resonant tunnelling when the current increases rapidly with voltage (see figure 1a). The corresponding value of B_f increases steadily with voltage up to the onset of NDC. It can be shown (Leadbeater et al 1987, Sheard and Toombs 1987) that magnetotunnelling oscillations associated with resonant tunnelling are given by an equation of the form

$$\frac{\hbar eB}{m^*}(n + \tfrac{1}{2}) = E_F - E_k \, , \qquad (1)$$

where E_k is the energy of the bound state in the well relative to the

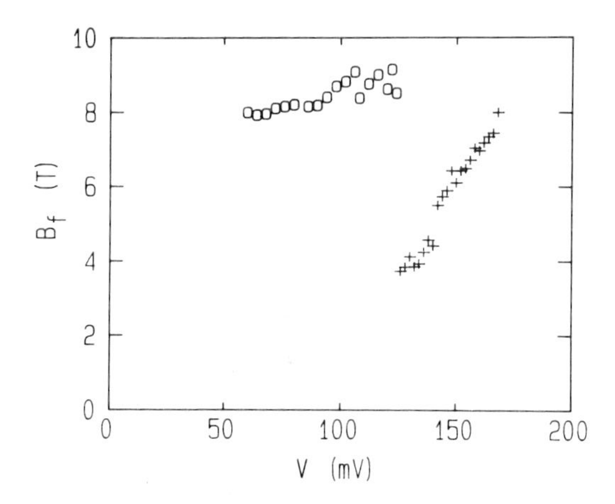

Fig. 2. Fundamental fields, B_f, versus bias up to 170 mV for the magneto-oscillations with $\underline{B} \parallel \underline{J}$ for the GaAs/(AlGa)As device

conduction-band edge of the emitter. E_k is related to the charge Q in the well through the electrostatic equation for the total potential drop across the device. The steady-state Q obtained by considering the dynamic equilibrium of the sequential tunnelling process also depends on E_k. Elimination of Q from these two relations gives an expression for E_k as a function of voltage (Payling et al 1987b). If we use this E_k in equation (1), the fundamental field for resonant tunnelling is

$$B_{fr} = \frac{m^*}{2e\hbar} \frac{eV - 2(E_1 - E_F)}{1 + \frac{L}{a_0} \frac{T_e}{T_e + T_c}} , \qquad (2)$$

where $L = 2b+w + \lambda_e + \lambda_c$, b and w are respectively the widths of the barriers and well, λ_e and λ_c are screening lengths in the emitter and collector contacts respectively, T_e and T_c are respectively transmission coefficients for the emitter and collector barriers and a_0 = 10 nm is the Bohr radius in GaAs. E_1 is the energy of the lowest well state relative to the bottom of the well. Thus $B_{fr} = 0$ when $V = 2(E_1 - E_F)/e$, which is the condition for the onset of resonance. At the peak of the resonant current, $E_k = 0$, and, from equation (1), $B_{fr} = m^*E_F/\hbar e$. Therefore the fundamental field rises to the same value as that for non-resonant tunnelling. This is exactly what is observed experimentally as can be seen from figure 2. Typically λ_e = 6.5 nm for this doping level, $\lambda_c \sim 10$ nm and $T_e \lesssim T_c$. Therefore the rate of increase of B_{fr} with voltage is ~ 0.11 T mV^{-1}, in excellent agreement with the experimental value of 0.1 T mV^{-1}. Equation (2) is also consistent with the results of Goldman et al (1987b) who have recently deduced charge build-up from an analysis of their data but they estimated the charge from the experimental current instead of calculating it theoretically. Our results confirm that charge build-up in the well occurs during resonant tunnelling.

Finally, we discuss briefly a very interesting feature of the magneto-tunnelling current for our (InGa)As/InP structures. Figure 3 shows the

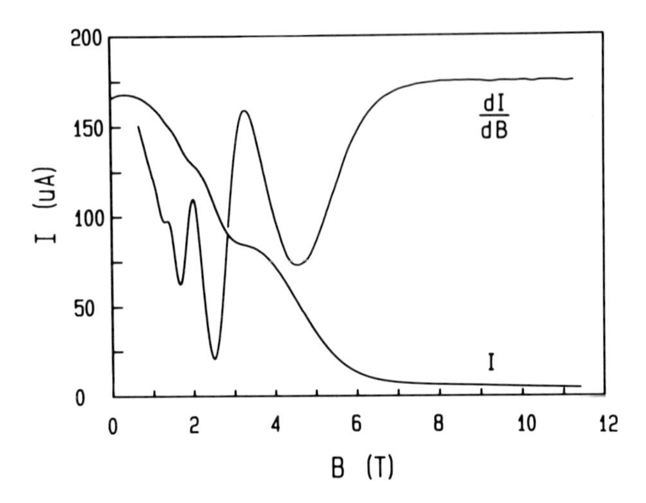

Fig. 3. The magnetotunnelling current I(B) and dI/dB of the (InGa)As/InP device for $\underline{B} \parallel \underline{J}$ at 4 K and bias 1 mV.

observed oscillations as a function of $\underline{B} \parallel \underline{J}$ for V = 1 mV. It can be seen that the low-bias current is strongly suppressed by the magnetic field. The suppression of the current can be understood in terms of the basic physics of resonant tunnelling through a DBS in the presence of a magnetic field (Sheard and Toombs 1987). The Landau levels in the well have energies $\varepsilon_n = E_K + (n + \frac{1}{2})\hbar eB/m^*$. Electrons tunnel into these levels from the collector and the emitter contact layers. The nett current is given by the difference between the rates of tunnelling from the two contacts. Therefore, if ε_n is greater than the Fermi energy of the emitter for all n, electrons cannot tunnel into the well and the current is suppressed. This result is in marked contrast to that for non-resonant tunnelling where there are always states available for electron tunnelling. The B_f for these oscillations is 5.5 T which shows that the well state is some 13 meV below the emitter Fermi enegy at zero magnetic field and zero bias for this (InGa)As/InP structure.

ACKNOWLEDGEMENTS: This work is supported by SERC and CNRS. ESA is supported by CNPq, Brazil. We acknowledge helpful discussions with P E Simmonds and T C L G Sollner.

REFERENCES

Goldman V J, Tsui D C and Cunningham J E 1987a Phys. Rev. Lett. 58 1256

Goldman V J, Tsui D C and Cunningham J E 1987b Phys. Rev. B 35 9387

Leadbeater M L, Eaves L, Simmonds P E, Toombs G A, Sheard F W, Claxton P A, Hill G and Pate M A 1987 Proc. Hot Electrons Conf., Boston, USA

Payling C A, Alves E S, Eaves L, Foster T J, Henini M, Hughes O H, Simmonds P E, Portal J C, Hill G and Pate M A 1987a Proc. 3rd Int. Conf. on Modulated Semiconductor Structures, Montpellier, France

Payling C A, Alves E S, Eaves L, Foster T J, Henini M, Hughes O H, Simmonds P E, Sheard F W and Toombs G A 1987b Proc. Conf. on Electronic Properties of Two Dimensional Systems - VII, Santa Fe, USA

Sheard F W and Toombs G A 1987 to be published

Sollner T C L G 1987 Comment at Hot Electrons Conference, Boston, USA

Inst. Phys. Conf. Ser. No. 91: Chapter 6
Paper presented at Int. Symp. GaAs and Related Compounds, Heraklion, Greece, 1987

Evidence of nearest neighbor hopping in GaAlAs/GaAs heterojunctions in the zero magnetoresistance state. Study of the conduction at high electric fields

J.L. Robert, A. Raymond, J.Y. Mulot, C. Bousquet

G.E.S. Université des Sciences et Techniques du Languedoc
34060 - Montpellier-Cédex, France

J.P. André

Laboratoire d'Electronique et de Physique Appliquée
94450 - Limeil-Brevannes, France

ABSTRACT : The temperature dependence of the conductivity σ_{xx} of GaAs/GaAlAs heterojunctions is investigated in the zero magnetoresistance state for several values of the filling factor υ, using the pressure as an external parameter to change the density of the 2D electron gas.
We report on the first evidence of a conduction involving nearest neighbor hopping, and on the possibility to achieve ultra-low resistance in the quantum Hall regime.
This conduction process involves localized states distributed over a band of energies.

1. INTRODUCTION

In a quantizing magnetic field B, the energy spectrum of a 2D system consists of a series of discrete Landau levels separated by the cyclotron energy $h\omega_c$. Each Landau level has a degeneracy N = eB/h. If n_s is the density of the 2D electron gas, the filling factor is given by $\upsilon = n_s/N$. In real systems, each level is broadened into a band of states (Landau sub-band). In the commomly accepted model, two kinds of states are assumed : the central part of each Landau sub-band contains the extended states, while the localized states lie on the flanks. When the Fermi level lies in the localized states, the longitudinal conductivity σ_{xx} is equal to zero and the off-diagonal Hall conductivity $\sigma_{xy} = n_s/eB$ is equal to ie^2/h, where i is an integer equal to the number of filled Landau levels ν.
The conductivity $\sigma_{xx} \sim 0$, $\sigma_{xy} = ie^2/h$, implie that the Hall current is free of dissipation and that :

$$\rho_{xx} = \frac{\sigma_{xx}}{\sigma_{xx}^2 + \sigma_{xy}^2} = 0.$$

In this study, we investigated the temperature and the electric field dependences of σ_{xx} in the zero magnetoresistance state, to get information on localized states. For the first time in GaAlAs/GaAs heterojunctions, we report on direct evidence of a conduction process involving nearest neighbor-hopping. This process involves localized states distributed over a band of energies.Moreover, we show that, for $\upsilon = 2$ and at T = 1.4K,value of ρ_{xx} lower than 7.10 $^{-14}$ Ω/□ can be achievable, on one sample (782), which is more than three order of magnitude lower than the previously determined values. (Tsui et al 1982, Syphers et al 1986).

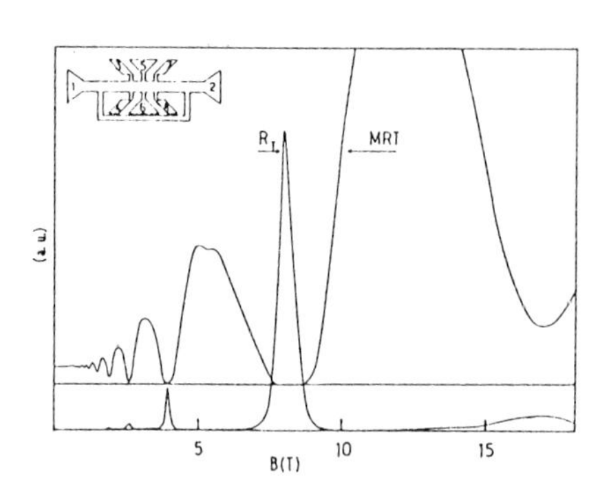

Fig. 1. Transverse resistance ($R_T = V_{56}/I_{34}$) and magnetoresistance ($MRT = V_{35}/I_{12}$) as a function of B. (Sample 518, T = 4.2K).

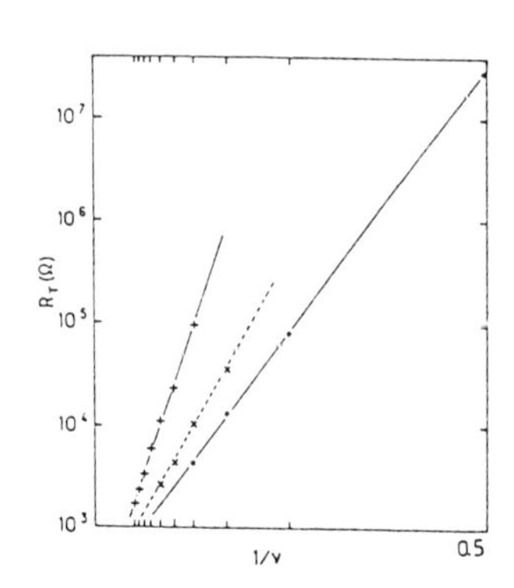

Fig. 2. Logarithm of R_T (maximum of R_t) as a function of $1/\upsilon$. + sample 784, x sample 518 pressure p = 0, • sample 518 p ≠ 0 (pressure is applied to change n_s) T = 4.2K.

Such low values of the longitudinal resistivity are correlated with the observation of persisting Hall current, circulating in samples with shunted bridge configuration.

2. EXPERIMENTS

Our experiments in the quantum Hall regime were performed on samples with a particular shape. The geometry, corresponding to the so-called shunted bridge configuration (S.B.C.) (Fig.1), was defined in order to observe on the same sample the classical quantized magnetoresistance and to use the σ_{xx} =0 nature of the Corbino geometry (ring geometry) in the zero magnetoresistance state.
To measure σ_{xx}, we use the ring geometry : a current I is injected between probes 3-4 and the transverse voltage V_H is measured on two other probes 5-6. This non-conventional method is very convenient since the transverse resistance R_t is directly proportional to V_H. In this experiment, the transverse resistance can be so high that the transverse voltage reaches its maximum value long after the application of the current. A striking feature in this experiment is that the Hall current circulating along the ring is persistent and decays slowly after switching the d.c current off. In the absence of external circuit, the time constant of the test-sample itself is about 1500 s when ρ_{xx} is equal to 7.10 $^{-14}$ Ω/□.

3. RESULTS

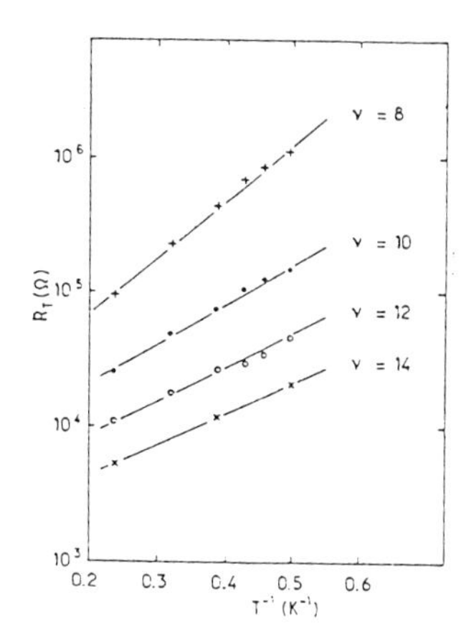

Fig. 3. Logarithm of R_T versus 1/T (sample 784).

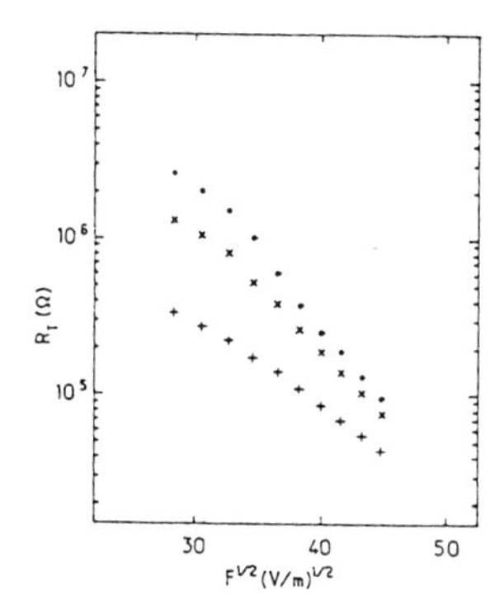

Fig. 4. Variation of R_T as a function of the electric field F. (sample 784, υ = 6 ; + T = 4.2K, X T = 2.9K, . T = 2.2K).

The characteristics of the MOCVD samples are given in Table I. Fig.1 shows the variations of the resistance R_t as a function of the magnetic field B (sample 518). R_t passes through a maximum value R_T for each integer value υ_i of ν . In the same figure, the classical magnetoresistance is measured conventionally (V_{35}/I_{12}). From these experiments, a quasi-linear dependence of Ln R_T versus $1/\nu_i$ is obtained (Fig.2). The conduction is activated with an activation energy ϵ_3 which increases when ν decreases (Fig.3).
The resistance R_T obeys the following law.

$$R_T = R_o \exp \alpha/\nu_i \exp \frac{\beta B_M - \Gamma}{kT}$$

(B_M is the value of B at the maximum value of R_t)
The values of R_o, α, β, Γ are given in Table 1.
It is then possible to deduce the value of R_T for each υ as a function of T. This has been done to obtain the value of ρ_{xx} given above for sample 782.
The existence of a preexponential factor $R_o \exp \alpha/\nu_i$ is characteristic of a nearest neighbor hopping process. The results can be compared to those of Fowler et al (1980) and Hayden et al (1979). By comparing samples 518 and 784, it can be seen that a two-fold increase in R_o corresponds to an equivalent decrease in α. This result is in good agreement with the expected change of R_o and α which vary inversely when the number of hopping sites varies.
The existence of such a conduction process is confirmed by the study of R_T as a function of the applied electric field F. When F is high enough, the current voltage characteristic becomes non linear (Fig. 4-5). The activation energy ϵ_3 decreases with increasing F. This effect is comparable to that observed in the reduction of a barrier by the electric field in the Poole-Frenkel effect. The results show that the change in ϵ_3 is proportional to $F^{1/2}$.
When F is high enough to reduce the "barrier" to zero the current-voltage characteristic is again linear (Fig.5)

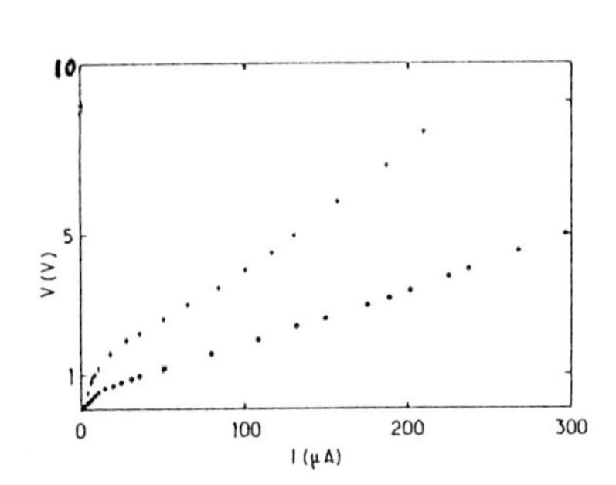

Fig. 5. I-V characteristics for two values of ν (sample 784, T = 4.2K, + ν = 2, • ν = 4).

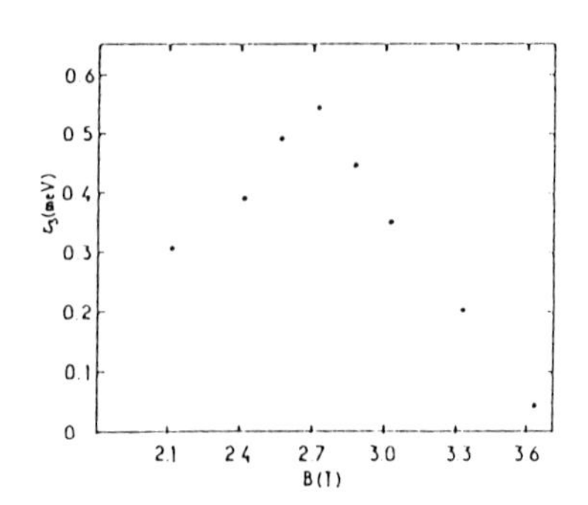

Fig. 6. Activation energy ϵ_3 versus B. (sample 784, ν = 10, T = 4.2K).

It must be noticed that ϵ_3 passes through a maximum when B is varied around B_M (Fig.6). Assuming that ϵ_3 is proportional to the reverse of the density of states, this result would indicate the presence of a Coulomb gap at the Fermi level.

Table I

	n_s (m^{-2})	μ(m^2/vs)	R_o (Ω)	α	βk	Γ (meV)
518	6.1 10^{15}	18.3	337	17	3.11	0.23
784	6.7 10^{15}	24.9	168	34	3.25	0.16

REFERENCES

Tsui D G, Stormer H L, Gosard A C, 1982 Phys. Rev. B.25-2, 1405.
Syphers D A, Martin K P, Higgens R J, 1986 Applied Physics Letters 48-4 193.
Fowler A B, Hartstein A, 1980 Phil. Mag. B, Vol. 42, nb 6-949.
Hayden K J, Butcher P N, 1979, Phil. Mag. B, Vol. 38, 603.

Study on generation mechanism of antiphase disorder in heteroepitaxial GaAs on $Ca_xSr_{1-x}F_2$(100) films

K. Tsutsui, T. Asano, H. Ishiwara and S. Furukawa

Dept. of Appl. Electronics, Tokyo Institute of Technology
4259 Nagatsuda, Midoriku, Yokohama 227 Japan

ABSTRACT : Generation mechanism of antiphase disorder in GaAs grown on epitaxial $Ca_xSr_{1-x}F_2$(100) film is discussed by evaluating original antiphase domain size at the initial stage of GaAs growth and observing micro {111} facet pyramids on the $Ca_xSr_{1-x}F_2$(100) films. A new concept about the micro facet enhancement in generation of the antiphase disorder in this system is proposed.

1. INTRODUCTION

In the recent progress in researches for semiconductor-on-insulator (SOI) structure using GaAs aiming future high-speed and/or three-dimensional integrated circuit applications, heteroepitaxy of GaAs on group-IIa fluoride, $Ca_xSr_{1-x}F_2$, has become promising (Siskos et al. 1984, Sullivan et al. 1985, Tsutsui et al. 1986a and 1987). It has been shown that GaAs/$Ca_xSr_{1-x}F_2$/Substrate(GaAs or Si) structures can be grown on both (100) and (111) orientations. Comparing these two orientations, growth on (100) orientation would be more preferable than that on (111) because high quality n-type GaAs layer can be usually obtained on (100) in the conventional GaAs MBE. We have found, however, that GaAs layers grown on (100) orientation often has the antiphase disorder (Tsutsui et al. 1986a and 1986b). This antiphase disorder tends to be understood similar to that observed on the GaAs/Si(100) structure. But, actually, the antiphase disorder in the GaAs/$Ca_xSr_{1-x}F_2$(100) originates from the growth of twofold symmetrical crystal(GaAs) on fourfold symmetrical crystal ($Ca_xSr_{1-x}F_2$), which is definitely different from that in the GaAs/Si case. In this work, we have investigated initial stage of generation of antiphase disorder to propose an advanced model of generation mechanism by introducing a new concept of micro facet enhancement.

2. GROWTH OF GaAs-SOI STRUCTURES

GaAs-SOI structures as shown in Fig.1 were grown by MBE. Chemically cleaned GaAs(100) or Si(100) wafers were loaded in a vacuum chamber and 100-300nm-thick fluoride films were epitaxially grown on these substrate at a growth temperature of 450°C for GaAs or 600°C for Si, after in-situ cleaning. Mixing ratios of the films were $Ca_{0.5}Sr_{0.5}F_2$ on the GaAs substrates and CaF_2 on the Si substrates, so that the lattice matching conditions between the films and substrates were satisfied. Then, they were transfered to another MBE chamber to grow SOI-GaAs films. They were grown by two-step growth method (450°C - 580°C) (Tsutsui et al. 1986c). Thicknesses of the films were 0.3-1.5μm, growth rate and As_4/Ga ratio were 1μm/h and 3-5 respectively.

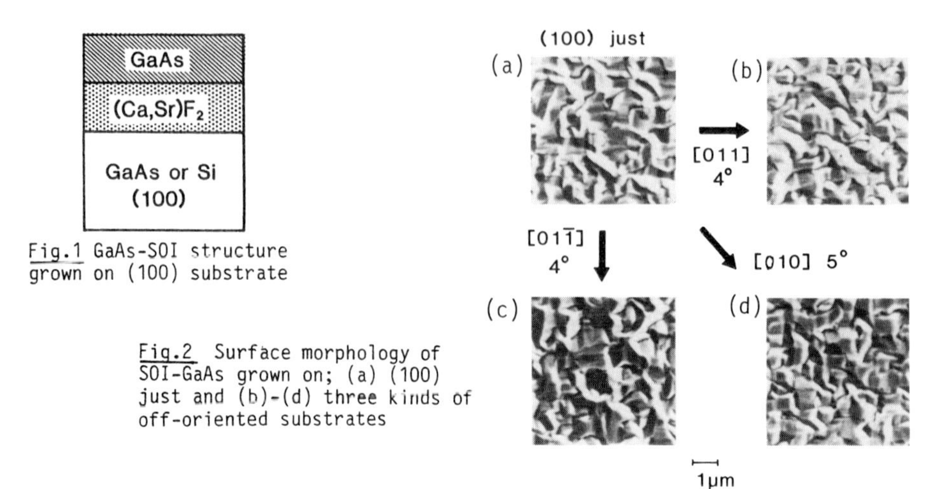

Fig.1 GaAs-SOI structure grown on (100) substrate

Fig.2 Surface morphology of SOI-GaAs grown on; (a) (100) just and (b)-(d) three kinds of off-oriented substrates

3. GENERATION OF ANTIPHASE DISORDER IN SOI-GaAs

Figure 2(a) shows surface morphology of a GaAs(1.0μm)/$Ca_{0.5}Sr_{0.5}F_2$(0.1μm) /GaAs(100) structure observed by SEM. It is found that textural structures where there are domains whose ridges are 90° rotated with respect to each other around the surface normal [100]. Indeed, an unisotropical etching revealed the relation of 90° rotation of crystalographic orientation between neighbouring domains. To be interesting, such a domain structure observed in GaAs(100) is similar to that in the case of antiphase disorder generated in heteroepitaxy of GaAs on Si(100).

Use of off-oriented substrate was tried to suppress the antiphase disorder of SOI-GaAs film, considering analogy of GaAs on Si. Three kinds of GaAs wafers tilted to [011], [010], or [01̄1̄] by 4-5° were set with a (100) just one on the same holder, and SOI structures were grown on them at the same time. But, as shown in Fig.2(b)-(d), it was found that the antiphase disorder could scarcely be controled by only introducing off-orientation of the substrates.

4. GENERATION MECHANISM OF ANTIPHASE DISORDER

4.1 Uncertainty of Epitaxial Orientation Resulting From Fourfold Symmetry

Basically, origin of generation of antiphase disorder in GaAs/$Ca_xSr_{1-x}F_2$(100) is considered due to the higher order symmetry in the crystalline structure of the fluoride, which is quite different in the case of antiphase disorder in GaAs/Si(100) (Tsutsui et al. 1986b). Figure 3 shows atomic arrangements on (100) surfaces of GaAs and CaF_2. It is noted that the GaAs(100)face has a twofold symmetry,

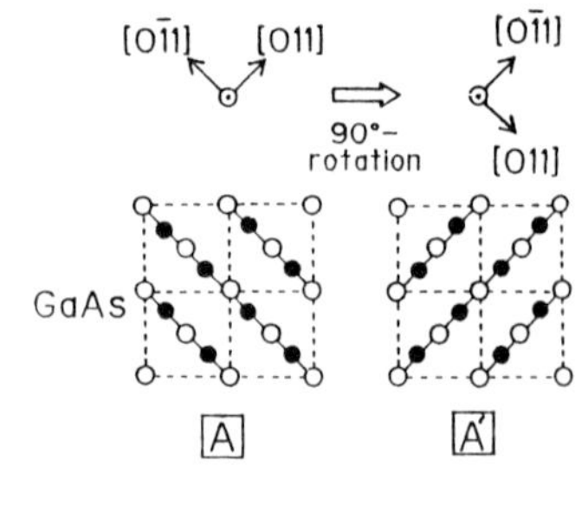

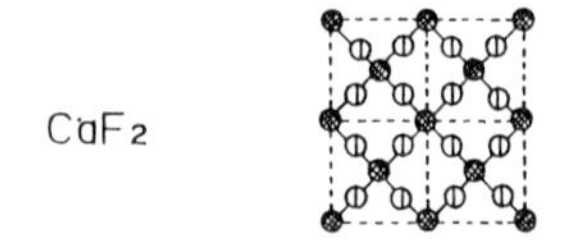

Fig.3 Atomic arrangements of (100) surface of GaAs and CaF_2 structures

whereas the $CaF_2(100)$ face has a fourfold symmetry. So, GaAs with both of the two different directions indicated by A and A' in Fig.3 can grow on the $CaF_2(100)$ face. If the A and A' regions coexist in a grown film, this will result in the existence of the antiphase disorder.

4.2 Enhancement of Antiphase Nucleations by {111} Micro Facets

Figure 4 shows relation between average size of antiphase domain at the surface evaluated from SEM micrographs and thickness of SOI-GaAs films. Three regions which were different in thickness of a grown GaAs films were formed on the same sample by sliding a shutter in parallel with growth surfaces. Though the values of observed domain sizes were different in each samples, it can be said commonly that domain sizes are decreased as the films becom thinner, and that extrapolation to the thickness of 0nm shows that original domain sizes at the initial stage of GaAs growth must be very small, maybe, less than about 50nm.

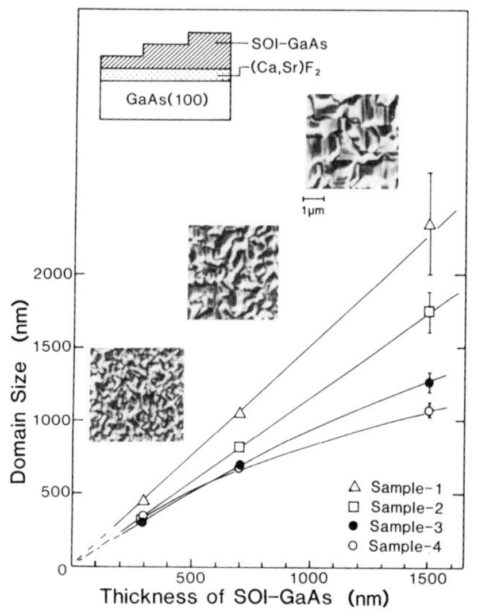

Fig.4 Relation between sizes of antiphase domain and thicknesses of SOI-GaAs

In such a small scale, it must be considered such that the surfaces of the fluoride films grown on (100) substrates are not flat. It was reported that there were many {111} facets on the surface of CaF_2 and $(Ca,Sr)F_2$ grown on (100) substrates (Schowalter and Fathauer 1986, Heral et al. 1987), and Schowalter(1986) explained this phenomenon by the larger surface energy of (100) face of the fluoride than that of (111) face. Figure 5 shows TEM micrograph of a replica made by shadowing evaporation of Pt-C on the surface of CaF_2/Si(100). It is found that the surface of CaF_2 film is fully covered with pyramids formed by four facets, and their average size is about 20nm which is consisitent with the extrapolated value in Fig.4. Off-angle streaks appearing in the RMEED pattern confirmed for the facets to be {111} facets. That is, GaAs nucleate not on the flat (100) face but on the {111} facets of the fluoride.

From the above observations, generation mechanism of antiphase disorder in GaAs on the fluoride (100) faces are considered as follows: Atomic arrangements on the surfaces are completely the same in all four {111} facets of a pyramid because of the fourfold symmetrical structure of the fluoride. So, it is natural that GaAs molecules nucleate on each pyramid side in the same direction in relation to the surface normal of each facet face. However, such nucleations result in that crystallographic orientations of GaAs on neighbouring sides of a pyramid are rotated by 90° around [100] axis with respect to each other, as shown in Fig.6. In other words, the micro facets on the top of the fluoride films enhance the generation of antiphase disorder. This disorder would be generated only by the uncertainty of orientation for the case when the surface of fluoride film was flat one having no facets.

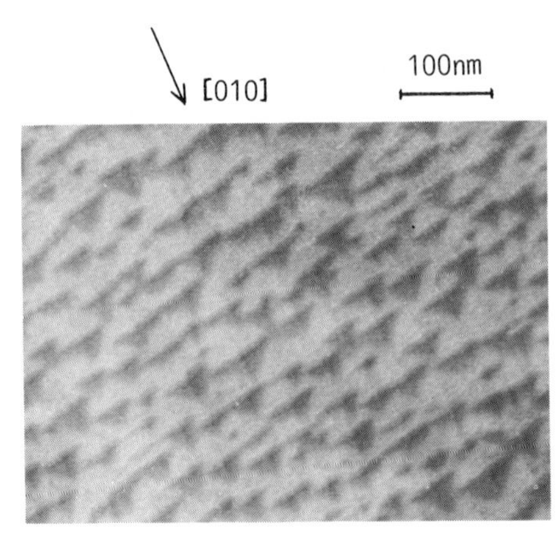

Fig.5 Faceted surface morphology of CaF_2 films grown on Si(100). (TEM micrograph of a replica)

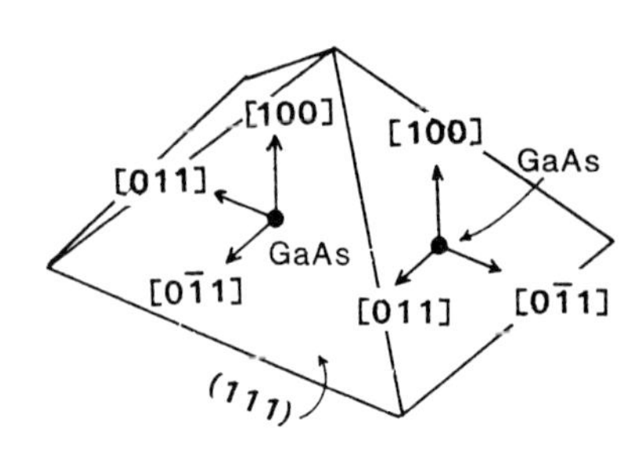

Fig.6 Ilustration of GaAs nucleations on micro {111} facet pyramid of the fluoride.

This concept of micro facet enhancement can explain the experimental result why the use of off-oriented substrates were not so effective (Fig.2) as for the case of GaAs/Si. The off-orientation would introduce many atomic steps onto the surfaces if they were completely flat as in the case of Si(100) surface. On the other hand, it can scarcely introduce drastic changes onto the morphology of system, of which surfaces are originally faceted as usually observed in the case of $Ca_xSr_{1-x}F_2$(100) films.

5. CONCLUSION

Antiphase disorder in SOI-GaAs grown on epitaxial $Ca_xSr_{1-x}F_2$(100) films originates from the uncertainty in the growth of twofold symmetrical GaAs on the fourfold symmetrical $Ca_xSr_{1-x}F_2$. Moreover, micro {111} facets existing on actual surfaces of the $Ca_xSr_{1-x}F_2$(100) films enhance the nucleations of GaAs in the antiphased relation.

This work was supported by the grant-in-aid for special distinguished research (No.59060002) from the Ministry of Education, Science and Culture of Japan.

References

Heral H, Bernard L, Rocher A, Fontaine C, and Yague A M 1987, J. Appl. Phys. 61, 2410
Schowalter L J, and Fathauer R W 1986, J. Vac. Sci. Technol. A4, 1026
Siskos S, Fontaine C, and Yague A M 1984, Appl. Phys. Lett. 44, 1146
Sullivan P W, Bower J E, Metze G M 1985, J. Vac. Sci. Technol. B3, 500
Tsutsui K, Asano T, Ishiwara H, and Furukawa S 1986a, Tech. Dig. of Electron Devices Meeting (IEDM) 1986, 775
Tsutsui K, Ishiwara H, and Furukawa S 1986b, Appl. Phys. Lett., 49, 1705
Tsutsui K, Ishiwara H, and Furukawa S 1986c, Appl. Phys. Lett., 48, 587
Tsutsui K, Nakazawa T, Asano T, Ishiwara H, and Furukawa S 1987, Electron Device Letters, EDL-8, 277

Paper presented at Int. Symp. GaAs and Related Compounds, Heraklion, Greece, 1987

Valence band discontinuity modification induced by hydrogen and cesium intralayers in GaP–Si heterojunctions

B. Russo, C. Quaresima, M. Capozi, E. Paparazzo, and P. Perfetti
Istituto di Struttura della Materia (CNR), Via E. Fermi 38, 00044 Frascati, Italy
C. Coluzza
Dipartiamento di Fisica, Universita' La Sapienza, P.le A. Moro 2, 00185 Roma, Italy
G. Margaritondo
Department of Physics and Synchrotron Radiation Center, University of Wisconsin, Madison, Wisconsin 53706, USA

ABSTRACT: We present photoemission measurements of the valence band discontinuity in the GaP-Si heterojunction. Large ΔE_v *modifications were found using both hydrogen and cesium intralayers. Hydrogen seems to be quite more effective in reducing the discontinuity; we found 0.4 reduction of* ΔE_v*, while Cs increased the valence band discontinuity by 0.1-0.2 eV. The observed intralayer-induced changes in* ΔE_v *are due to modifications of the interface dipoles. A simple model calculation, based on electronegativity arguments, gives a reduction of 0.3 eV (hydrogen effect) and an increase of 0.2 eV (cesium effect).*

The valence and conduction band discontinuities, ΔE_v and ΔE_c, result from the different energy gaps of the two semiconductors forming a heterojunction. The problem of how the gap difference is shared between ΔE_v and ΔE_c is perhaps the most important problem in heterojunction research.

The quantities play a leading role in determining the transport and optical properties of heterojunction devices (Milnes 1972). Microscopic factors, as interfacial electrostatic dipoles could play an important role in determining the heterojunction discontinuities. A check of the above hypothesis can be performed using artifical dipoles, created by ultrathin intralayers of materials inserted between the two semiconductors.

The first evidence that the band lineup can be changed by modifying the microscopic interface dipoles was obtained by photoemission experiments with an aluminium ultrathin (0.5-2 Å intralayer on the interfaces between CdS and Si or Ge (Niles 1985). It was consistently found a ΔE_v increase by 0.1-0.2 eV depending on the intralayer thickness. The above results generated some controversy, since the observed changes were quite close to the experimental uncertainty in the measurements of the band lineup. Changes, beyond any conceivable experimental uncertainty, were found later in the SiO_2-Si interface using two different materials with very different electronegativity, cesium and hydrogen (Perfetti 1986). ΔE_v increased by 0.25 eV with Cs and dcreased by 0.5 eV with hydrogen.

We present here an extension of the above analysis to GaP-Si interfaces. Again large ΔE_v modifications were found in both cases. Hydrogen seems to be quite more effective in reducing the discontinuity; we found 0.4 eV reduction of ΔE_v in GaP-Si (see Figures 1 and 2), while cesium increased the valence band discontinuity by 0.1-0.2 eV (see Figure 3). The observed intralayer-induced changes in ΔE_v are due to modifications of the interface dipoles, and a simple model calculation based on electronegativity arguments [3] have been used. The theory predicts a reduction of 0.3 eV in ΔE_v, due to the hydrogen intralayer effect and an increase of 0.2 eV due to the cesium intralayer effect (Russo 1987).

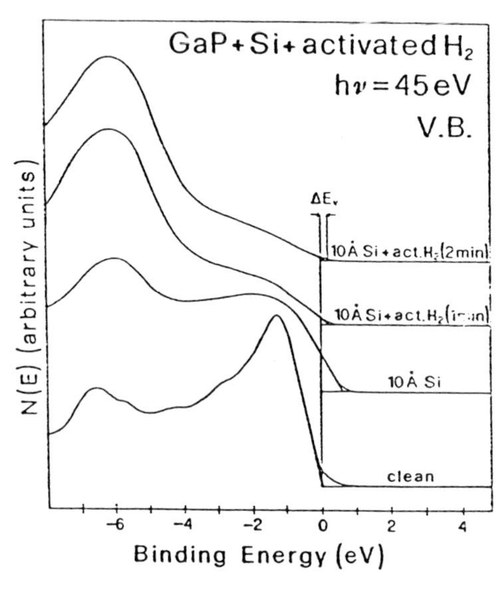

Fig.1 Photoemission spectra of GaP-Si interface. The spectra are aligned to the Ga3d core level. From the bottom: the first spectrum is the valence band of the cleaved in situ GaP; the second one is the interface GaP-Si (10 Å). The next two spectra are the same heterojunction after hydrogen ion bombardment (E_{beam} = 100 eV, Dose = $2 \cdot 10^{13}$ ions/cm^2 and Dose = $4 \cdot 10^{13}$ ions/cm^2 respectively)

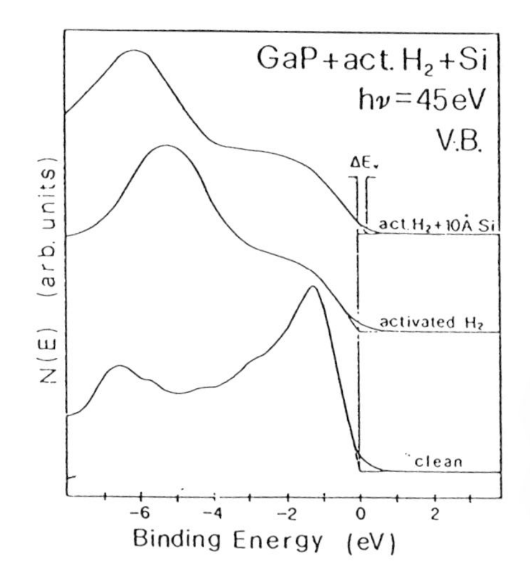

Fig.2 Valence band spectra of the GaP-Si (10 Å) structure with hydrogen at the interface. The hydrogen intralayer was obtained bombarding the surface of the GaP cleaved in situ by hydrogen ions (E_{beam} = 100 meV, Dose = $1 \cdot 10^{14}$ ions/cm^2).

The general agreement between the experimental results and the calculations gives to the model a general validity which is beyond its simplicity and the approximations that the models contains. The general validity resides in the prediction that interfacial dipoles play a substantial role in determining the discontinuities at the heterojunction interface.

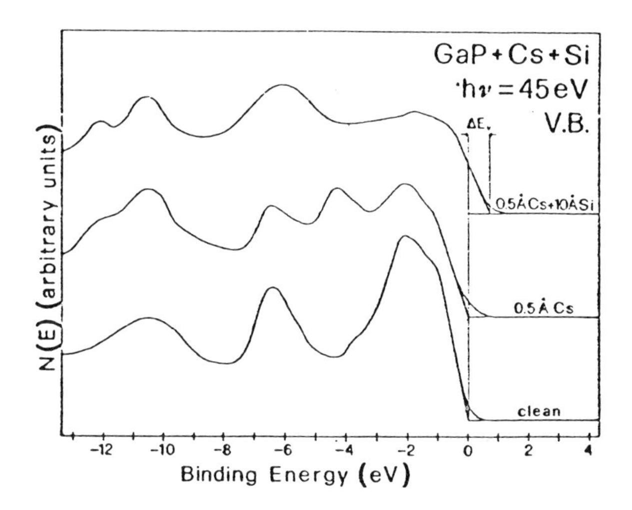

Fig.3 Photoemission spectra of the GaP-Si heterojunction with 0.5 Å of cesium at the interface. The cesium was evaporated in situ on the cleaved GaP.

References

Niles DW, Margaritondo G, Perfetti P, Quaresima C, and Caozi M. 1985 Appl. Phys. Lett. **47** 1092

Milnes AG and Feucht DL 1972 "Heterojunctions and Metal-Semiconductor Junctions" (Academic Press, New York)

Perfetti P, Quaresima C, Coluzza C, Fortunato G. and Margaritondo G 1986 Phys. Rev. Lett. **57** 2065

Russo B 1987 Thesis, Department of Physics, University La Sapienza, Rome

Paper presented at Int. Symp. GaAs and Related Compounds, Heraklion, Greece, 1987

Two- and three-dimensional characterization of AlGaAs heterostructures using SIMS

S W Novak[1] and R G Wilson[2]

[1]Charles Evans & Associates, 301 Chesapeake Dr., Redwood City, CA 94063 USA
[2]Hughes Research Laboratories, 3011 Malibu Canyon Rd., Malibu, CA 90265 USA

ABSTRACT: We have calibrated SIMS ion yield and sputtering rate variations in $Al_xGa_{1-x}As$ using O_2^+ and Cs^+ primary ion beams by analyzing ion implant profiles of fourteen elements in thick MBE-grown films having x from 0 to 0.9. These data allow us to produce quantitative two-dimensional composition versus depth profiles of AlGaAs/GaAs heterostructures. Modern SIMS instruments also allow ion images, which reflect the lateral distribution of dopants and impurities, to be acquired while sputtering the sample. By defining a plane through these images, a vertical cross section of a layered sample may be examined.

1. INTRODUCTION

The most common semiconductor material used for optoelectronic devices is $Al_xGa_{1-x}As$. Band-gap engineering has opened up the possibility of using many Al compositions to tailor the properties of the resulting device. Secondary ion mass spectrometry (SIMS) is among the most sensitive of the depth profiling techniques available for characterizing such devices, however the variable matrix compositions can significantly affect ion yields and sputtering rates during profiling resulting in distortion of the true depth profile (Benninghoven, et al., 1987). We have examined the effects of variable Al compositions for profiling AlGaAs by analyzing ion implants of fourteen ions into $Al_xGa_{1-x}As$ having $x = 0$ to 0.9 using both O_2^+ and Cs^+ primary ion beams. Bombardment by O_2^+ is useful for determining p-type dopants such as Be and Mg and transition metal impurity contents, whereas Cs^+ bombardment is typically used for analysis of n-type dopants Si and Se, and the impurities C and O.

2. EXPERIMENT

The MBE-grown AlGaAs films used in this study are 1 to 2.3 μm thick. Stoichiometry of the films was checked by Rutherford backscattering spectrometry and electron microprobe analyses and found to be within 7% of the desired compositions x = 0.17, 0.30, 0.50, 0.70 and 0.90. The films, together with GaAs, were implanted with ^{9}Be, ^{13}C, ^{18}O, ^{24}Mg, ^{30}Si, ^{34}S, ^{52}Cr, ^{55}Mn, ^{56}Fe, ^{63}Cu, ^{64}Zn, ^{80}Se, ^{120}Sn and ^{130}Te with fluences of 1 to $5x10^{14}$ cm^{-2} and 80 to 600 keV. In most cases, a given piece of material was implanted with only one ion, however Sn and Te were co-implanted.

Depth profile measurements of the implanted films were carried out using CAMECA IMS-3f SIMS instruments. All samples implanted with a given ion were measured in the same run to minimize variations in instrument conditions. Measurement of Be, Mg, Cr, Mn, Cu, Fe, and Zn were performed using an O_2+ primary ion beam having 8 keV net impact energy. For samples having x = 0.7 and 0.9, automated control of the sample bias (nominally +4500 V) was necessary in order to compensate for charging during ion bombardment. A mass resolution factor of approximately 1500 was used during measurement of ^{9}Be in order to eliminate interference from $^{27}Al^{+++}$. All ions measured using O_2^+ bombardment were detected as positive ions. Profiles of the remaining implants were measured using a Cs^+ primary ion beam having a net impact energy of 14.5 keV. Negative secondary ions were detected and automated control of the sample bias was again used for profiling samples having x = 0.7 and 0.9. Crater depths were measured using an Alphastep-200 stylus profilometer to establish sputtering rates.

3. RESULTS

Sputtering rates were corrected for variations in primary ion current and normalized to the sputtering rate of GaAs determined under identical conditions. Normalized sputtering rates are plotted against x_{Al} in Figure 1.

For both O_2^+ and Cs^+ bombardment, sputtering rates are nonlinear functions of composition up to x = 0.5 above which sputtering rates are approximately constant. This result contrasts with the data of Meyer et al. (1983) who show an approximately linear relationship between sputtering rate and x_{Al} (Fig. 1). At present, we cannot explain this discrepancy, although the data of Taga (1985) and several studies cited in Benninghoven et al. (1987) show nonlinear dependancies of sputter yield on alloy composition.

Relative sensitivity factors (RSF) are commonly used to convert measured ion count rates to to concentrations in SIMS

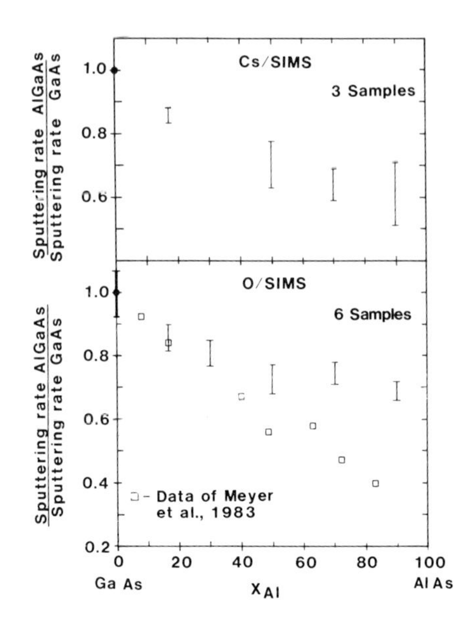

Fig. 1. Sputtering rates in AlGaAs normalized to GaAs

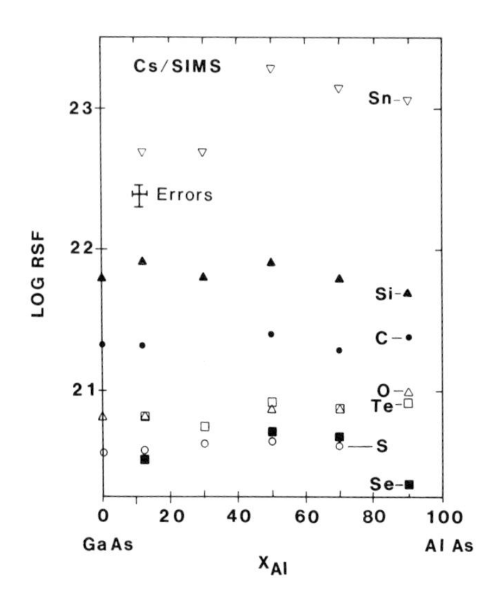

Fig. 2. Logarithm of relative sensitivity factors for Sn, Si, C, O, Te, S and Se plotted versus x_{Al} in AlGaAs.

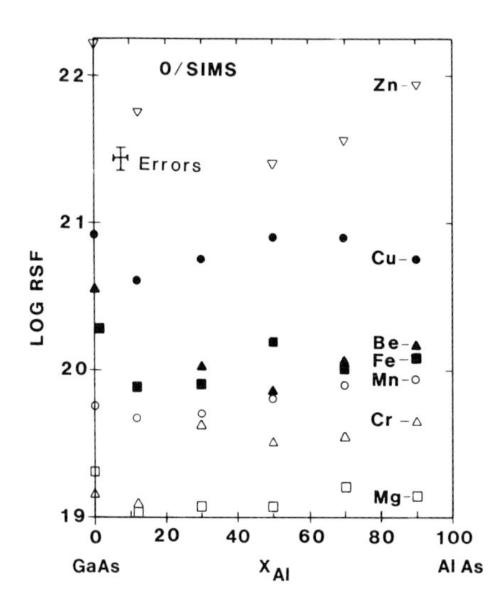

Fig. 3. Logarithm of relative sensitivity factors for Zn, Cu, Be, Fe, Mn, Cr and Mg versus x_{Al} in AlGaAs.

measurements. We have calculated RSF's from the ion implant profiles measured in this study using standard procedures (Benninghoven, et al., 1987) and plotted them against x_{Al} in Figures 2 and 3. These figures show that RSF's are constant (within 20%) for all AlGaAs compositions, and that the factors for AlGaAs are within 20% of those for GaAs. These factors, together with the known sputtering rates, allow SIMS depth profile data from AlGaAs-GaAs heterostructures to be readily converted to quantitative concentration versus depth profiles. In addition, constancy of the relative sensitivity factors in the AlGaAs system suggests that known sensitivity factors for GaAs (Wilson and Novak, 1987) should be applicable to other III-V ternary alloy systems (e.g., InGaAs) in which the As content remains fixed.

Modern SIMS instruments (specifically the CAMECA IMS series instruments in this work) have the capability of collecting ion images of a sample having lateral resolution of approximately 1 μm (Benninghoven, et al., 1987). By recording ion images sequentially, i.e., as the sample is sputtered, a three-dimensional compositional representation of the sample can be recorded. Magnetic peak switching of the mass spectrometer allows images of several ions to be recorded as the sample is sputtered. Once the images are stored, digital image processing can be used to examine any cross section through the images. The depth resolution of such cross-sections will be a function of the sputtering rate, the image acquisition time, and the primary ion impact energy, but will typically be greater than 5 nm. We have utilized a position senstive detector (Odom, et al., 1983) to acquire ion images from an ion implanted AlGaAs multi-quantum-well structure. This sample consists of 50 periods of 6.5 nm AlAs/6.5 nm GaAs with a 20 nm GaAs cap. The sample has been ion implanted through a photoresist mask with ^{28}Si ($2x10^{15}$ cm^{-2} 300 keV) to disorder the layer structure in selected areas. An ion cross-sectional image of this sample is shown in Figure 4. The lighter vertical stripes show implanted areas (20 μm wide) in which the compositional contrast between layers is less than in unimplanted regions because of interdiffusion of the AlAs and GaAs layers. Because such images are representations of digitally recorded ion count rates, the relative sensitivity factors determined in this study can be used to extract

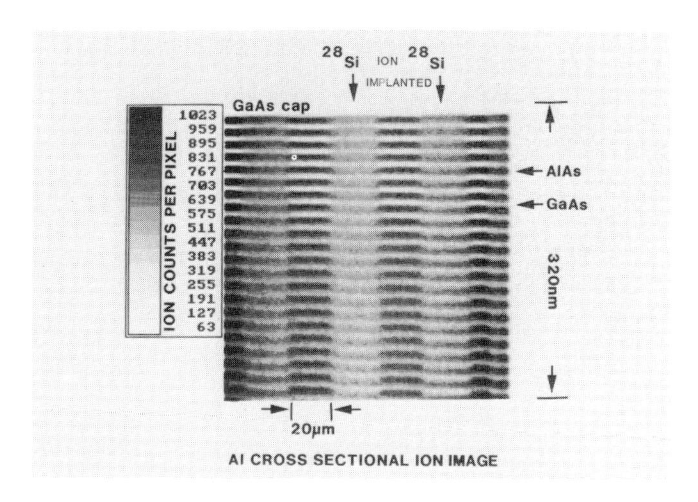

Fig. 4. Cross-sectional Al^+ ion image of a ^{28}Si ion implanted AlAs-GaAs heterostructure.

quantitative compositional information from the image data. The advantages of such an experiment are that both implanted and unimplanted areas of the sample are profiled simultaneously, thus minimizing instrumental variations, and that lateral diffusion of either dopants or major elements can be examined. In addition, this technique is useful for examining small structures within semiconductor devices that would otherwise be impossible to analyze using standard SIMS techniques.

The multi-quantum well sample was grown by Kanji Yoh of Stanford University. Partial support was provided by the Army Research Office Contract No. DAAL03-86-C-0010 and the Office of Naval Research Contract No. N00014-83-C-0532.

Benninghoven A, Rudenauer F G and Werner A H 1987 *Secondary Ion Mass Spectrometry* (New York: Wiley)
Meyer C, Maier M and Bimberg D 1983 *J Appl Phys* **54**, 2672
Odom R G, Furman B K, Evans C A, Bryson C E, Petersen W A, Kelley M A and Wayne D H 1983 *Anal. Chem.* **55**, 574
Taga Y 1985 *in Secondary Ion Mass Spectrometry SIMS-V* A Benninghoven et al. (eds.) (Berlin: Spring-Verlag) 38-41
Wilson R G and Novak S W 1987 *in Secondary Ion Mass Spectrometry SIMS-VI*

Infrared PL emission spectroscopy of deep levels in AlGaAs/GaAs, InP/InGaAs and InAlAs/InGaAs heterostructures

L. AINA, H. HIER, M. MATTINGLY, J. O'CONNOR, AGIS ILIADIS*
BENDIX AEROSPACE TECHNOLOGY CENTER, ALLIED CORPORATION, 9140 OLD ANNAPOLIS ROAD, COLUMBIA, MARYLAND 21045, *UNIVERSITY OF MARYLAND, DEPARTMENT OF ELECTRICAL ENGINEERING, COLLEGE PARK, MARYLAND 20742

ABSTRACT

Numerous deep radiative emissions have been observed in AlGaAs/GaAs, InP/InGaAs and InAlAs/InGaAs heterostructures. The most prominent of these is a 0.755eV emission observed in AlGaAs/GaAs heterostructures. This emission is sharp (FWHM ~2meV), originates from the AlGaAs/GaAs interface and correlates quite well with both the electrical and optical properties of the materials. Other deep radiative centers at 1.35, 1.24, 1.114 and 0.93eV have been identified as arising from either the doped AlGaAs or from the doped GaAs cap layer. Correlation of the PL energies of some of these centers with DLTS measurements is presented.

1. INTRODUCTION

Heterostructures based on combinations of III-V materials are excellent candidates for high speed digital and microwave circuits. Since their electrical and optical properties are adversely affected by the presence of impurities and defects in the material, it is important to properly characterize these defects.

Infrared photoluminescence spectroscopy is a versatile technique which probes recombination centers with energy levels well below the midgap. It therefore complements the DLTS technique which can most conveniently measure shallower deep levels. Although DLTS measurements and studies of AlGaAs/GaAs heterostructures have been widely reported (S. Dhar et al, 1985), no clear identification of these results with the heterojunction and its properties has been made. In view of the detrimental effects of interface defects on heterojunction devices, photoluminescence and DLTS investigations of these deep radiative levels is important.

We report the observation of deep radiative emissions using infrared photoluminescence in AlGaAs/GaAs, InP/InGaAs and InAlAs/InGaAs selectively doped heterostructures. The AlGaAs/GaAs heterostructures show the most intense and the most numerous deep level emissions. The most prominent of these is a 0.755eV emission which is sharp (FWHM ~2meV) and originates from the AlGaAs/GaAs interface. Other deep radiative centers at 1.35, 1.24, 1.114 and 0.93eV have been identified as arising from either the doped AlGaAs or GaAs layers. These emissions correlate quite well with DLTS levels measured on our samples. Previously reported DLTS studies are also in agreement with some of these emissions.

2. EXPERIMENTAL

The MBE layers were grown using a Physical Electronics Model 425B system at 600°C. The layers consisted of a 5000A GaAs layer grown on LEC GaAs substrates, followed by a 40A undoped AlGaAs spacer layer, a 350A n^+ si-doped AlGaAs layer and a 200A n^+ GaAs cap layer. Layers grown in this manner have electron mobilities at 300K and 77K in excess of 8000 cm^2/volt-sec and 110,000 cm^2/volt-sec respectively, with sheet electron concentrations in the range of 0.8 - 1.2 x 10^{12} cm^{-2}. The growth and properties of the OMVPE layers have been described elsewhere (Aina et al, 1987). MODFETs from both types of materials had transconductances of 220 mS/mm and 170 mS/mm at 300K for the MBE and the OMVPE grown layers respectively. The layer structures of the InGaAs/InAlAs and InGaAs/InP have been reported in other publications (Fathimulla et al, 1987 & Aina et al, 1987). The electron mobilities at 300K are in excess of 10,000 cm^2/volt-sec, while the electron mobility at 10^oK is as high as 100,000 cm^2/volt-sec. The PL apparatus is a standard set-up using a PIN Ge photodetector (Aina et al, 1987). The DLTS apparatus is a standard Polaron boxcar system.

3. PHOTOLUMINESCENCE SPECTRA

Figure 1 shows the photoluminescence spectra for AlGaAs/GaAs grown by MBE, InGaAs/InP grown by OMVPE and InAlAs/InGaAs grown by MBE. The AlGaAs/GaAs heterostructure has the most intense spectrum with the most numerous PL peaks and therefore we concentrate on this in this work. This kind of spectrum may be due to better optical qualities of the AlGaAs/GaAs layers or to a greater concentration of deep radiative levels in the heterostructures. That the former is the case is shown in Figure 2 where the PL spectra is compared for OMVPE and MBE grown layers. While the OMVPE (Figure 2a) and MBE (Figure 2b) layers with similar optical qualities show almost identical spectra, the MBE layer (Figure 2c) with the poorest optical properties shows a less intense spectrum with fewer deep levels. In particular, the latter spectrum does not show the 0.755eV peak.

Table 1 details the peaks observed in the AlGaAs/GaAs structures that we investigated. It can be seen here that most of the peaks in the AlGaAs/GaAs heterostructures are also found in doped AlGaAs and GaAs layers (i.e., 1.345 - 1.369eV, 1.01 - 1.17eV, 0.906 - 0.922eV). The slight energy variation may be due to differences in doping concentrations in the n^+ layers. The 0.755eV emission originates only from samples with a AlGaAs/GaAs interface; it is not observed in either bulk AlGaAs or GaAs.

4. 0.755eV EMISSION

The emission peak at 0.755eV is unusually sharp for a deep radiative center as Figure 3a shows. The width at half-maximum is as low as 2meV. The second peak is 11meV below the main emission and is possibly a transverse acoustic phonon replica, since the TA phonon energy is ~10meV (Xin et al. 1982). No features corresponding to longitudinal optical phonon replicas were observed. The energy of this peak at 300K is 0.705eV. As Figure 3b and c show, the intensity of this peak varies with the intensity of the GaAs band edge PL emission and varies inversely as the electron mobility of the AlGaAs/GaAs heterostructure. The former is indicative of the existence of deep non-radiative centers in the materials with low GaAs bandedge PL emision. This is why the 0.755 eV PL emission is not prominent in the MBE grown material with poor optical quality (Figure 2).

5. DLTS MEASUREMENTS

The activation energies of the DLTS peaks shown in Figure 4 are 0.43eV for peak A and 0.36eV for peak B. These are similar to activation energies obtained for D-X centers in AlGaAs/GaAs heterostuctures (S. Dhar et al, 1986) and AlGaAs layers (Mooney et al, 1985 and Wagner et al, 1980). Because of the large lattice relaxation of the D-X center, the optical emission energy is about 1.00eV (Lang et al, (1979), Therefore the emissions around 1.0eV in Table II found in the AlGaAs/GaAs heterostructures and n^+ AlGaAs are probably due to the D-X centers.

6. DISCUSSION

The origin of the 0.755eV emission cannot be easily determined. However, it is safe to conclude from the foregoing results that it is probably due to an acceptor at the midgap of the GaAs side at the AlGaAs/GaAs interface. Such an acceptor would be an efficient recombination center and could therefore give rise to luminescence. McAfee et al (1982) have identified by DLTS an interface trap in MBE grown AlGaAs/GaAs. This trap was deduced to be a band of states on the GaAs side of the interface and had an activation energy of 0.66eV with an energy spread of 0.13eV and was concluded to be due to interface roughness possibly caused by extrinsic impurities. The inverse variation of the intensity of the 0.755eV PL peak with electron mobility confirms the existence of these kinds of midgap states.

The optical process leading to this 0.755eV PL may involve optically generated electrons in the 2DEG making transitions into the midgap level where they can then recombine with similarly generated holes. Alternately, it may involve the recombination of excitons bound to the midgap level. Only the latter recombination process could account for the sharpness of the emission, and we feel that it is the most likely explanation.

7. SUMMARY

A study of deep radiative emissions in III-V heterostructures has uncovered emission peaks especially in AlGaAs/GaAs which has more intense and numerous peaks than in InGaAs/InP and InGaAs/InAlAs heterostructures. The most important of these emissions is a sharp peak at 0.755eV the intensity of which correlates with the optical and electrical quality of the heterostructure and is believed to be due to a midgap level at the AlGaAs/GaAs interface. The other emissions are due to n^+AlGaAs and GaAs and are related to deep levels measured by DLTS.

ACKNOWLEDGMENT

The authors wish to acknowledge Dawn Smith for manuscript preparation, Lisa Stecker for PL measurements, S. C. Laih for DLTS measurements, and Eric Martin and John McKitterick for useful comments and discussions.

REFERENCES

Aina, O., Mattingly, M., Juan, F. Y., Bhattacharya, P. K., 1987 Appl. Phys. Lett. 50, 43
Aina, O., Mattingly, M., Potter, R., 1987 Appl. Phys. Lett., To Be Published
Dhar, S., Hong, W. P., Bhattacharya, P. K., Nashimoto, Y., Juang, F. Y., 1986 IEEE Trans. on Electr. Dev.
Fathimulla, A., et al, 1987 IEEE IEDM Conference Proceedings
Lang, D. V., Logan, R. A., 1979 Physical Review B19, p. 1015
McAfee, S. R., Lang, D. V., Tsang, W. T., 1982 Appl. Phys. Lett. 40, 520
Mooney, P. M., Fischer, R., Morkoc, H., 1985 J. Appl. Phys. 57, 1928
Wagner, E. E., 1980 J. Appl. Phys. 51, 5434
Xin, S. H., Wood, C. E. C., Desimone, O., Palmateer, S., Eastman, L. F., 1982 Electron. Letts. 18, 3

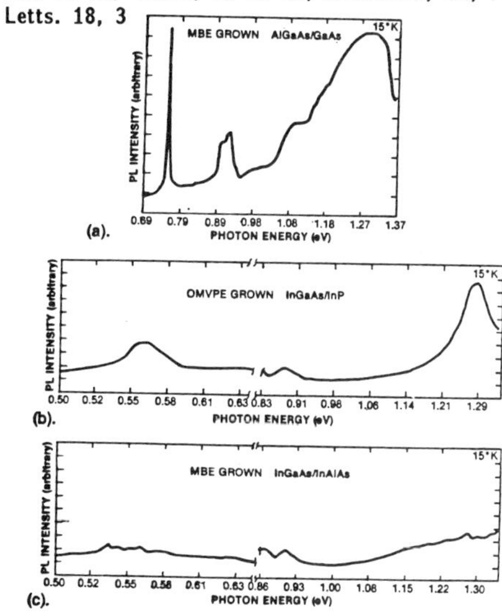

Figure 1. Photoluminescence spectra of
(a) AlGaAs/GaAs heterostructure
(b) InGaAs/InP heterostructure
(c) InGaAs/InAlAs heterostructure at 15°K.

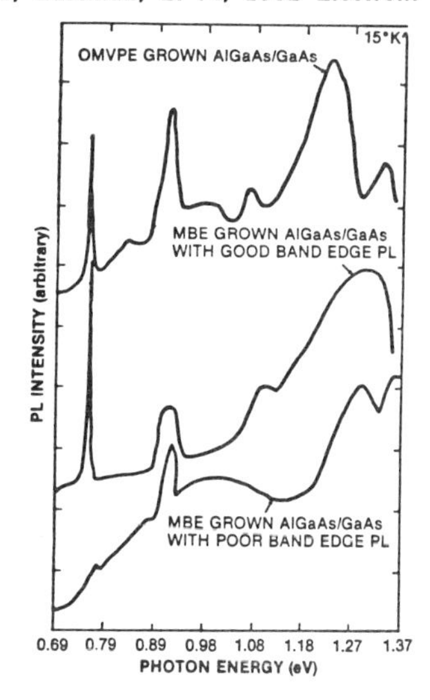

Figure 2. Photoluminescence spectra of AlGaAs/GaAs heterostructures grown by OMVPE and MBE.

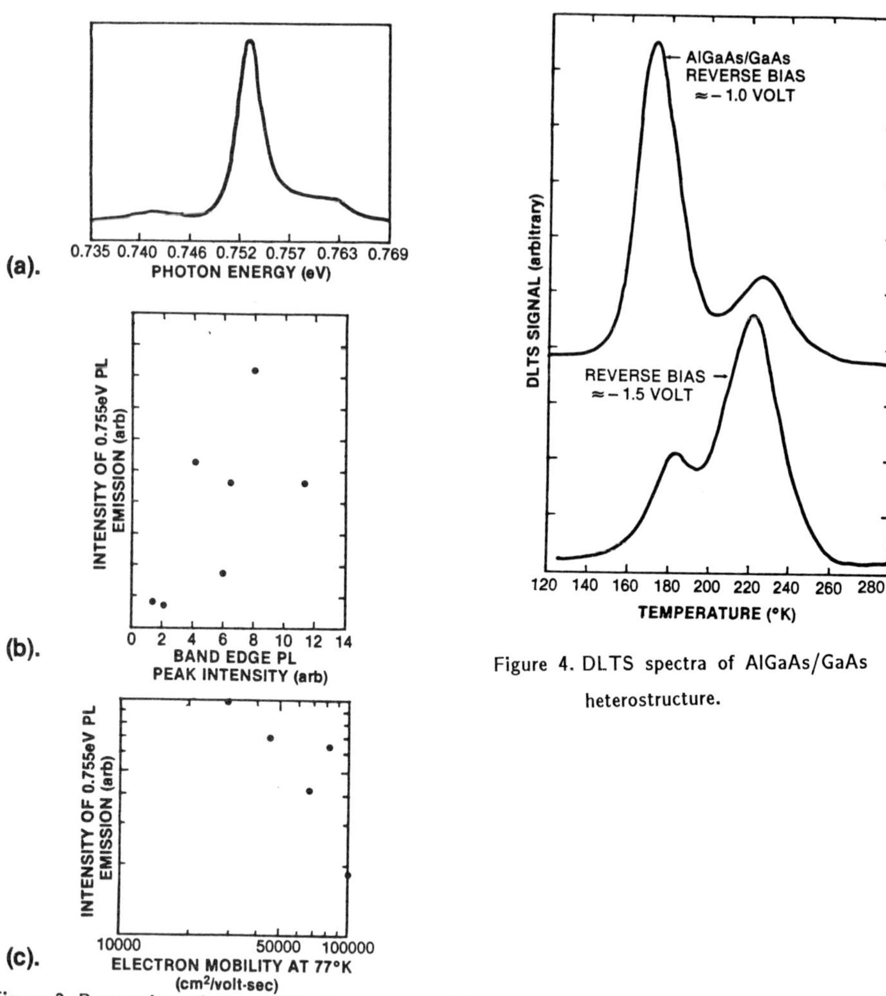

Figure 4. DLTS spectra of AlGaAs/GaAs heterostructure.

Figure 3. Properties of the 0.755eV emission, (a) photoluminescence spectra, (b) variation of the intensity of 0.755eV emission with the intensity of the GaAs bandedge peak, (c) variation of the intensity of 0.755eV emission with 2DEG electron mobility at 77°K.

SAMPLE	SAMPLE STRUCTURE	DEEP PL EMISSION ENERGY (eV)									
MBE 277	AlGaAs/GaAs HEMT - MBE		1.310		1.106			0.906			0.755
MBE 156	AlGaAs/GaAs HEMT - MBE	1.369	1.301		1.010			0.918			
OMA 167	AlGaAs/GaAs HEMT -OMVPE	1.349		1.240	1.073	0.987		0.915			0.755
OMA 166	DOPED GaAs - OMVPE	1.345		1.235		0.978		0.922			
OMA 165	DOPED AlGaAs - OMVPE	1.333			1.165			0.916		0.841	
OMA 158	UNDOPED AlGaAs - OMVPE						0.945	0.912	0.898	0.831	

TABLE 1 - PL EMISSIONS FROM VARIOUS SAMPLE STRUCTURES

Paper presented at Int. Symp. GaAs and Related Compounds, Heraklion, Greece, 1987

Interface recombination in GaAs–GaAlAs double heterostructures and quantum wells

B Sermage, M F Pereira, F Alexandre, J Beerens, R Azoulay and N Kobayashi

Centre National d'Etudes des Télécommunications, Laboratoire de Bagneux, 196 Avenue Henri Ravera, 92220 Bagneux, France

ABSTRACT : Interface recombination has been studied by luminescence decay measurement in different MOCVD and MBE grown undoped GaAlAs-GaAs-GaAlAs double heterostructures. We show that for usual GaAs thicknesses (d < 0.2 μm), non radiative recombination is dominated by interface recombination. In the case of MBE growth, interface recombination velocity can vary by nearly two orders of magnitude. In thin GaAs layers(< 200 Å), interface recombination increases due to quantum effects.

1. INTRODUCTION

It is well known that non radiative recombination is very prejudicial to semiconductor lasers, not only because it increases the threshold current but also because it contributes to the degradation of the device. It has been shown by R.J. Nelson and R.G. Sobers (1978) that GaAs-GaAlAs interface recombination was small (~ 500 cm/s) in LPE grown structures. In MBE and MOCVD grown layers, the situation is not so simple : B. Sermage et al. (1985).
In a GaAs-GaAlAs double heterostructure, the effective non radiative lifetime τ_{nr} in the GaAs layer is given by the following equation

$$\frac{1}{\tau_{nr}} = \frac{1}{\tau_{nrb}} + \frac{S}{d} \qquad (1)$$

where $S = S_1 + S_2$ is the sum of the recombination velocities at the two interfaces, d is the thickness of the GaAs layer and τ_{nrb} is the non radiative lifetime in a thick GaAs layer. This equation is valid when the carriers diffusion length L is large compared to d.

2. EXPERIMENT

We have studied three series of samples : two were MBE grown at 690°C in the same growth chamber and as we will see the qualities of the two series were very different. The main difference between the growth of the two series was the change of the Aluminium source shutter which was probably contaminated after too long openings of the growth chamber. The third series was grown by atmospheric pressure MOCVD. The samples consisted of a GaAs substrate on which was grown a GaAs buffer layer 1 μm thick, a first GaAlAs confinement layer with 30 % Aluminium and a thickness d_1 between

0.2 μm and 1 μm, a GaAs layer with a thickness d and a second confinement layer with the same composition as the first one and a thickness d_2 between 0.2 μm and 1 μm. The substrate was n doped and the epitaxial layers were undoped to avoid radiative recombination at low excitation. For each series, the samples were grown in the same conditions within one week to avoid changes in the growth conditions.

Carriers lifetime was obtained from luminescence decay measurements. The samples were excited by 3 ps pulses from a mode locked dye laser and the luminescence selected by a small monochromator was detected with a streak camera (time resolution 10 ps and good sensitivity). Luminescence decay curves at low excitation are shown in Fig. 1.

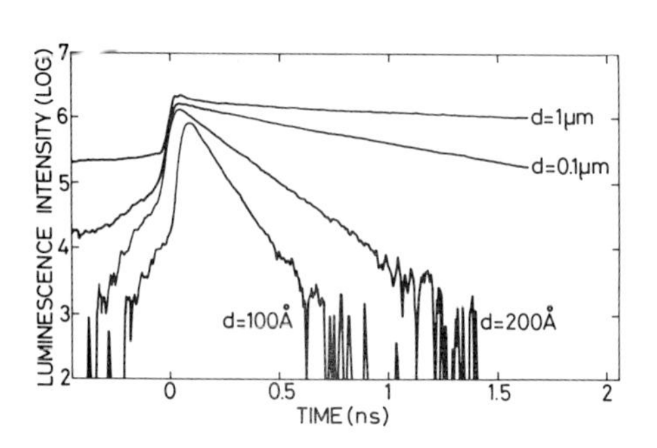

Fig. 1. Luminescence decay curves at room temperature and low excitation in first series MBE grown GaAs-GaAlAs double heterostructures.

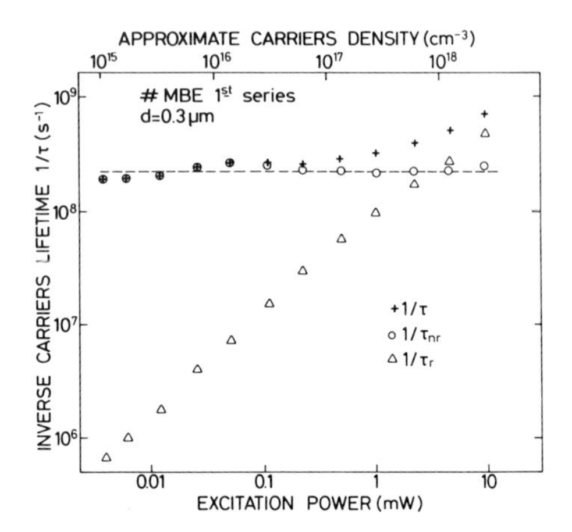

Fig. 2. Variation with excitation of the total (+), the radiative (△) and the non radiative (o) inverse carrier lifetime

The excitation pulse being short compared to the carriers lifetime, the carriers do not recombine during the excitation and the carriers density n at the end of the excitation pulse is proportionnal to the excitation power P_{ex}. Thus the luminescence intensity I_L (o) just after the excitation pulse is :

$$I_L(o) \sim n.\frac{1}{\tau_r} \sim P_{ex} \cdot \frac{1}{\tau_r} \qquad (2)$$

and the ratio between the luminescence intensity at time o and the excitation power is proportionnal to $1/\tau_r$. In fig. 2 we show the variation with excitation power of the inverse lifetime $1/\tau$ and of the inverse radiative lifetime $1/\tau_r$. At high excitation $1/\tau$ increases like $1/\tau_r$. At low excitation $1/\tau_r$ decreases nearly as a bimolecular reaction and $1/\tau$ is constant which means that at low excitation carriers lifetime is the non radiative ones τ_{nr} assumed to be independent on excitation power. $1/\tau_{nr}$ is proportional to the density of recombination centers N_I, to their capture cross section 6_I, and to the velocity of the carriers v_e

$$\frac{1}{\tau_{nr}} \sim \sum N_I \, \sigma_I \, v_e \quad (3)$$

In fig. 3 the inverse non radiative lifetime $1/\tau_{nr}$ is plotted as a function of the inverse GaAs thickness 1/d in the case of large thicknesses (d > 200 Å). The data follow equation 1 and we can see that

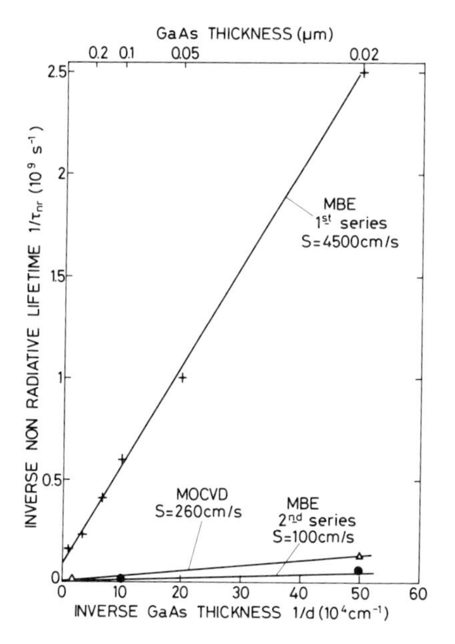

Fig. 3. Inverse non radiative lifetime versus inverse GaAs thickness for the first MBE series (+), the second MBE series (•) and the MOCVD series (Δ)

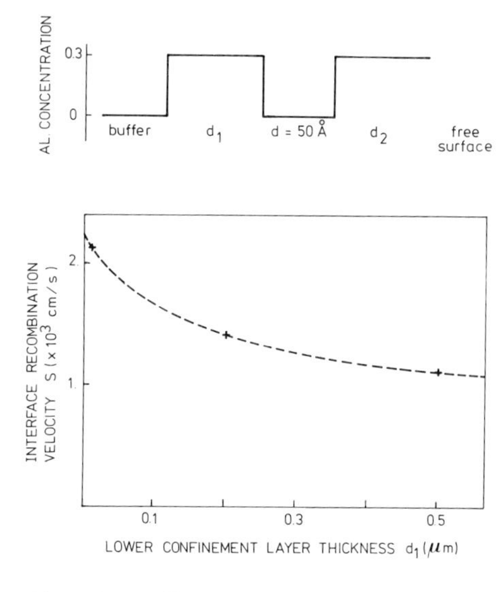

Fig. 4. Influence of the lower confinement layer thickness on the interface recombination velocity

for d = 0.1 μm most of the inverse non radiative lifetime corresponds to interface recombination ($S/d > 1/\tau_{nrb}$). It shows also that GaAs-GaAlAs interface recombination can be rather high in MBE growth. (The first MBE series was grown in 84 but we have not observed any change in non radiative recombination lifetime on these samples between 84 and 87). Lasers structures grown at the same period as the first MBE series did not work, but at the same period as the second MBE series or the MOCVD series, the threshold current density was small (~ 1 KA/cm^2). Concerning MBE growth, we have observed that carriers lifetime in the buffer layer was very short (150 ps) which means probably that the first interface (between the buffer layer and the lower confinement layer) is the worst. This is consistent with the fact that the non radiative carriers lifetime in a 50 Å well increases when the thickness of the lower confinement layer increases as shown on fig. 4.

3. NON RADIATIVE RECOMBINATION IN GaAs-GaAlAs QUANTUM WELLS

Within the three previous series, we have also grown samples with small GaAs layer thicknesses (d < 200 Å). Non radiative carriers lifetimes were measured as previously on these samples. The monochromator was utilized to separate the luminescence of the quantum well from that of the buffer layer. This was particularly useful when d_1 and d_2 were small since the laser beam was not entirely absorbed in the structure. The interface recombination velocity is still defined by eq. 1 and we observe for the three series an increase of the interface recombination velocity for the thin layers compared to the thick ones (S_0) as shown in fig. 5.

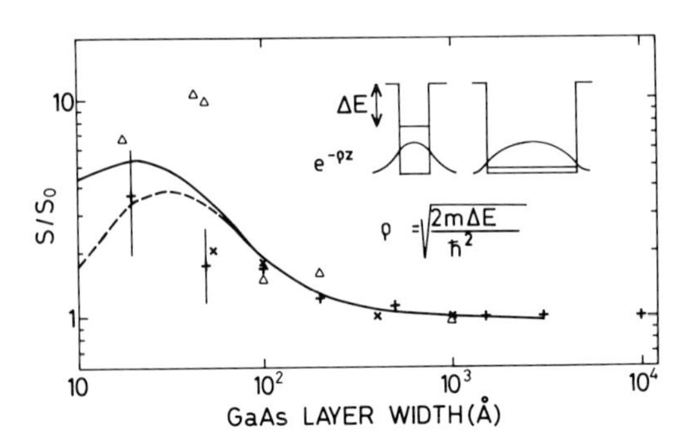

Fig. 5. Increase of the interface recombination velocity in quantum wells for first MBE series (+), second MBE series (Δ) and MOCVD(×). Models with recombination centers at the interfaces (---) and in the barriers (—) and a band offset of 68 % as measured by Danan et al. (1987).

This effect was theoretically foreseen by Duggan et al. (1986) as due to the increase of the leakage of carriers wave functions into the barriers when the thickness of the GaAs layer decreases. The model supposes either that the recombination centers are located at the interfaces or that there is a constant density of recombination centers in the barriers.
This increase of the interface recombination velocity in thin GaAs layers is prejudicial to quantum well lasers. For example, we have calculated that even in the case of good quality interfaces such as the second MBE series (S_0 = 100 cm/s), and for an optimised separate confinement heterostructure 400 μm long in which the optical guide has an internal $Ga_{0.7}Al_{0.3}As$ layer 0.16 μm thick and external $Ga_{0.2}Al_{0.8}As$ layers, interface recombination increases the threshold current density from 130 A/cm² to 220 A/cm² and that it changes the optimum well thickness from 60 Å to 100 Å.
We gratefully aknowledge J.P. Noblanc for useful suggestions and J. Worlock for fruitful discussions.

REFERENCES

Danan G, Etienne B, Mollot F, Planel R, Jean Louis AM, Alexandre F, Jusserand B, Le Roux G, Marzin JY, Savary H and Sermage B (1987) Phys. Rev. B, 35, 6207

Duggan G, Ralph HJ and Elliott RJ (1986) Solid State Commun 56, 17

Nelson RJ, Sobers RG (1978), Appl. Phys. Lett. 32, 761

Sermage B, Alexandre F, Liévin JL, Azoulay R, El Kaim M, Le Person H and Marzin JY 1985, Inst. Conf. Ser. 74, 345.

Channelled substrate (100) GaAs MBE growth and lateral p–n junction formation of lasers

H P Meier, E van Gieson*, R F Broom, W Walter, D J Webb, C Harder and H Jäckel

IBM Research, Zurich Research Laboratory, 8803 Rüschlikon, Switzerland

* Permanent address: 427 Phillips Hall, Cornell University, Ithaca, NY 14853

ABSTRACT: Results of investigations into MBE growth on textured substrates are presented. The relative growth rates on facets and ridges, faceting, plane-dependent doping and morphology of the films grown are described. Conditions, suitable for the fabrication of light-emitting devices are developed, and a prototype transverse junction stripe device described.

1. INTRODUCTION

Epitaxial growth, on substrates with etched ridges, is a means of achieving lateral optical confinement and non-absorbing mirrors in semiconductor lasers (Gavrilovic *et al.* 1985, Wu *et al.* 1984). It is advantageous to develop processes for MBE instead of LPE hitherto used. Lateral p-n junctions may be made in MBE on textured substrates by utilizing the plane selective doping behavior of Si (Ballingall and Wood 1982). Figure 1 shows an example where the junctions are made visible by stain etching. Two of the problems observed in MBE growth on patterned substrates are the degradation of the crystal morphology on the higher-order planes (Wu *et al.* 1984, Miller and Asbeck 1986), and the growth of extra undesired facets. This paper investigates the growth rates on facets and ridges. Growth conditions which eliminate morphology and faceting problems are developed.

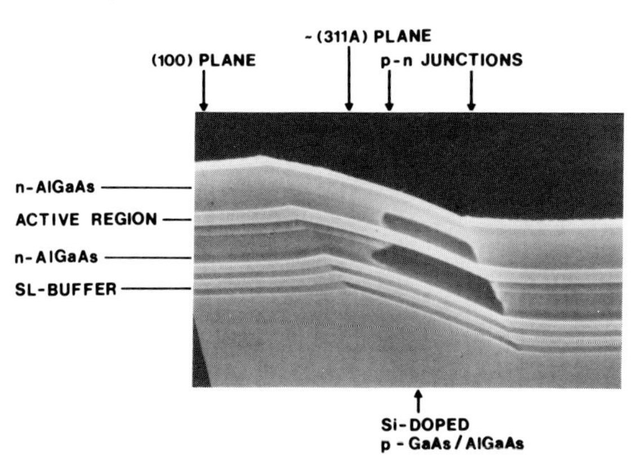

Figure 1. SEM Cross-section of a TJS-device structure. The p-type region (dark) is stain etched

2. MBE GROWTH ON TEXTURED SUBSTRATES

The (100)-oriented GaAs substrates were cleaned and patterned photolithographically. The desired (A or B) facets were prepared by etching in a 2:1:20 mixture of buffered HF: H_2O_2:H_2O (Meier *et al.* 1987). The H_2O_2 concentration determines the angles of the facets, which are the same in both A and B directions. The photoresist was then stripped and the wafers cleaned. The layers were grown in a Varian GEN II MBE with wafer rotation at 12 rpm.

The incident angles of the group-III and group-V fluxes are not normal to the wafer surface (100) plane. Consequently, V:III ratio of the evaporant flux, incident on the sloping facets, varies as the wafer is rotated (around the 100 axis) during growth. This affects the morphology of the films grown on these facets. The substrate temperature and V:III ratio [i.e., As_4:(Ga + Al)] should be about 720°C and 1, respectively, in order to grow good quality lasers on the (100) facets. Under these conditions, (Meier *et al.* 1987) good growth morphology can only be achieved when the facet makes an angle of 35° or less with the (100) plane.

The sticking coefficient of the group-III elements depends on the crystal orientation (Kapon *et al.* 1987, Smith *et al.* 1985). The growth rates on the A-planes are lower than on the (100) surface. Growth on the B-planes proceeds slightly faster up to a facet angle of 25° as shown in Figure 2. The diffusion length and desorption rate of Al are much smaller than for Ga since films of AlAs grown at 720°C on (211A) and (100) planes have no additional facets and a uniform layer thickness. Therefore, A-planes should have a higher Al concentration than the (100) surface. This has been confirmed by Auger spectroscopy. B-facets do not develop additional facets whereas A-facets, with a tilt of 30° or more, develop additional (111A) and (411A) facets as shown in Figure 3. The dependence of the relative growth rates on the substrate temperature for all facets appearing on a (211A) facet is shown in Figure 4.

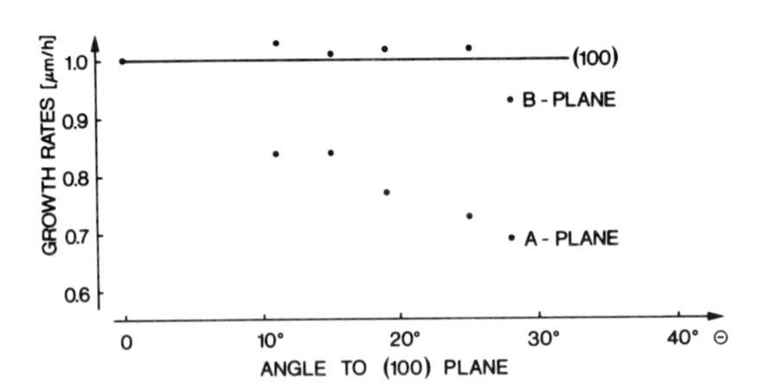

Figure 2. Growth rates for $Al_{0.3}Ga_{0.7}As$ at 720°C and a V:III ratio of 1 on facets at an angle θ with respect to the (100) surface. The growth rate on the (100) surface is 1 μm/h

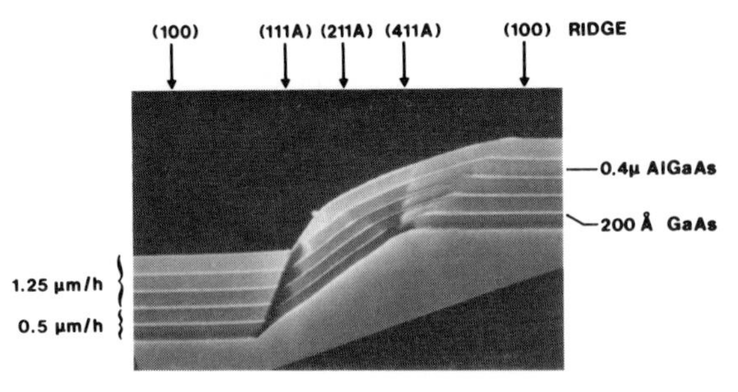

Figure 3. SEM cross-section of (211A) facet. Additional (111A) and (411A) planes appear. The substrate temperature was 720°C and the V:III ratio 1

The Ga incorporation rate on the top of ridge (100) consists of three components: the incident vapor, the diffusing Ga from any adjacent A-facets, and desorption. For our growth

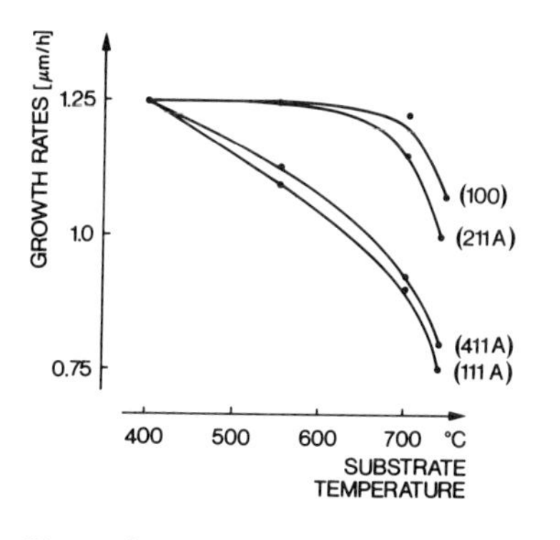

Figure 4. Dependence of growth rate on substrate temperature for different facets

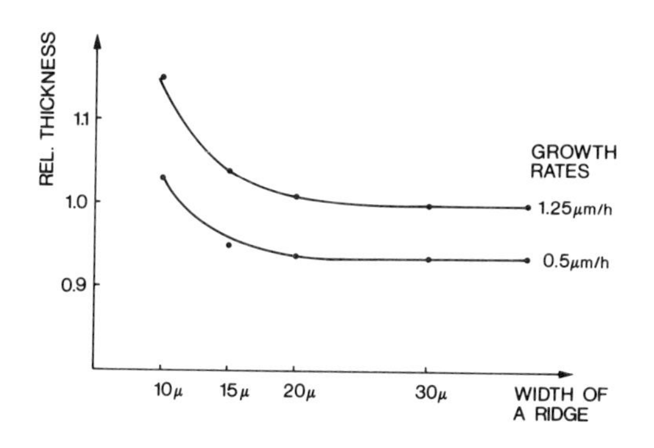

Figure 5. Dependence of the relative layer thickness on ridge width

conditions the top of the ridge remains almost flat, showing that diffusion of the Ga is fast enough to distribute the Ga evenly over the (100) face. The growth rate of the (100) face will then be higher (and Al concentration lower) when the (100) face area is smaller. This is shown in Figure 5, where the growth rate on the (100)-ridge top is shown as a function of ridge width. The growth temperature is 720°C. Each curve refers to a different incident flux. Ga desorption reduces the relative film thickness when the incident flux is reduced. These variations of thickness and Al concentration with ridge width must be considered when device structures are designed.

At a growth temperature of about 720°C, at a V:III ratio of 1, Si dopes GaAs and AlGaAs n-type on (100) planes but p-type on facets which make an angle of more than about 21° with the (100) plane. If non-selective Si n-doping is also required, then the growth temperature may be reduced to 450°C and the V:III ratio increased to 4 (Meier *et al.* 1987).

3. DEVICE STRUCTURES

Prototype structures have been grown which are suitable both for lateral optical confinement and non-absorbing mirrors. We have also made a transverse junction stripe light-emitting device where growth over etched facets is used for lateral optical confinement, and where the plane-selective doping is used to form transverse p-n junctions. The device structure is shown in Figure 6(a). The lateral p-n junctions are displayed in the EBIC SEM of such a structure in Figure 6(b). The near-field light emission from the structure is shown in Figure 6(c). The light emission is concentrated at the etched facet, as expected, but some light is emitted from the layers on the (100) face. We have not yet achieved lasing in this structure at room temperature. This may be due to non-radiative recombination in the active region on the facets. In addition the optical confinement may need further optimization.

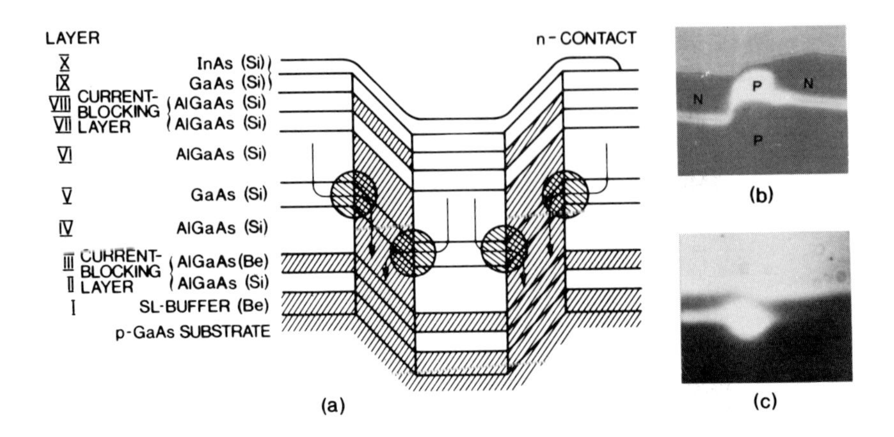

Figure 6. Transverse junction stripe device: (a) structure; (b) EBIC visualization of p-n junctions, and (c) near-field radiation pattern

4. CONCLUSIONS

MBE growth of AlGaAs over ridges has been investigated. At growth temperatures in the region of 720°C and a V:III ratio of 1, Ga diffuses from A facets to the (100) facets. Therefore the A-facets grow slower (and have a higher Al concentration) than the (100) facets. Also, growth rates and Al concentration of the (100) facets depend on ridge width. Extra facets grow on A-facets angled more than 30°; Si dopes AlGaAs p-type on facets angled more than about 21°. A transverse junction stripe light-emitting device was grown using these techniques.

ACKNOWLEDGEMENT

The authors are greatly indebted to E. Latta for the Auger measurements.

REFERENCES

Ballingall J M and Wood C E C 1982 *Appl. Phys. Lett.* **41** 947
Gavrilovic P, Meehan K, Epler J E, Holonyak H, Burnham R D and Thornton R L 1985 *Appl. Phys. Lett.* **46** 857
Kapon E, Tamargo M C and Hwang D M 1987 *Appl. Phys. Lett.* **50** 347
Meier H P, Broom R F, Epperlein P W, Van Gieson E, Harder C H, Jaeckel H, Walter W and Webb D J 1987 *J. Vac. Sci. Technol.* to be published
Miller D L and Asbeck P M 1986 *Proc. 4th Int. Conf. on Molecular Beam Epitaxy* eds C T Foxon and J J Harris (Amsterdam: North-Holland) pp 368-373
Smith J S, Derry P L, Margalit S and Yariv A 1985, *Appl. Phys. Lett.* **47** 712
Wu Y H, Werner M, Chan K L and Wang S 1984 *Appl. Phys. Lett.* **44** 834

GaInAs/InP multi-quantum well structures for lasers by LP–MOVPE

D. Grützmacher, K. Wolter, M. Zachau*, H. Jürgensen, H. Kurz and P. Balk

Institute of Semiconductor Electronics
Aachen Technical University D-5100 Aachen, FRG
* Physics Department Technical University München
D-8000 München, FRG

Abstract

GaInAs/InP MQW structures have been grown at conditions optimized for LASER fabrication. Across 2" wafers a homogeneity in thickness, composition and doping was achieved (to better than 2 %) at a total pressure of 20 mbar and high flow rates. The abruptness of the transition for quantum wells ranging from 0.5 to 50 nm in width is on the monolayer level. Photoluminescence line shifts (2K) are among the highest observed so far (max. 528 meV); the line widths are very small (i.e. 2.2 meV for 20 nm wells).

Introduction

GaInAs/InP multi quantum well (MQW) lasers are of increasing interest because of low threshold current (Tsang et al, 1981), narrow emission (Dingle et al, 1976), reduced temperature sensitivity (Holonyak et al, 1980) and wavelength tunability in the range of 0.8 to 1.35 eV (Dingle et al, 1974). Although MOVPE and MBE present themselves for production of InP and GaInAs films, abrupt heterojunctions and precise adjustment of the GaInAs wells down to 1 nm have been difficult to achieve uniformly over large substrates. Low pressure metal organic vapor phase epitaxy (LP-MOVPE) appears to be an attractive approach to tackle this problem; at high linear flow velocities a distinctly improved homogeneity of film thickness, composition and doping is achieved (Grützmacher et al, 1987). For these conditions a very rapid exchange of the gas phase in the reactor may be achieved and extremely abrupt interfaces produced.

In the present paper we discuss optimized conditions for uniform growth of GaInAs/InP heterostructures. These allow MQW structures with abrupt interfaces and uniform well widths down to 0.5 nm. Experimental data indicative of the quality of these structures will be presented.

Experimental

The growth studies were performed in a LP-MOVPE apparatus which will be described in detail elsewhere (Grützmacher et al, to be published). TMG (trimethyl Ga), TMI (trimethyl In), AsH_3 (arsine) and PH_3 (phosphine) were used as source materials; H_2S was chosen to introduce donors. Short gas switching times were realized by the use of special 5-way vent-run valves with minimized dead space (Jürgensen et al, 1985). IR strip heaters in combination with a SiC coated graphite susceptor were employed to reach a very uniform substrate temperature at reduced pressures.

To prepare quantum well structures with controlled well width strongly reduced growth rates (between 0.1 and 0.65 nm/s) were obtained by decreasing the reactant pressures. Evaluation of the structures was performed by Secondary ion mass spectrometry (SIMS), 2K photoluminescence (PL) and electron microscopy studies. Electrical properties were evaluated by means of Shubnikov-de Haas and quantum Hall effect measurements.

Results and Discussion

Fig. 1 demonstrates the uniformity of thickness and composition of a GaInAs layer grown on a 2"InP-substrate at 20 mbar overall pressure and a flow velocity of 140 cm/s. The variation of thickness in the flow direction is approx. 1 %. The variation in thickness of the layers in the direction perpendicular to the flow is similar except atthe substrate edges, where a variation of 5% is observed. The change in composition leads to a fractional variation in lattice parameter (da/a) of $7x10^{-5}$ accross the central section (3 cm * 2 cm) of the wafer; towards the front and rear edges the Ga concentration increases. The increase can be suppressed by placing dummywafers before and after the substrate. This behavior suggests that the small divergent behavior at the periphery is caused by a phenomenon which is typical for the edges of the wafer, like slight fluctuations in the gas phase. The background doping gives rise to an electron concentration of $1.7x10^{15}cm^{-3}$ at 300K, which varied by less than 2 % across the entire wafer. A 300K mobility around 10.000 cm^2/Vs was found.

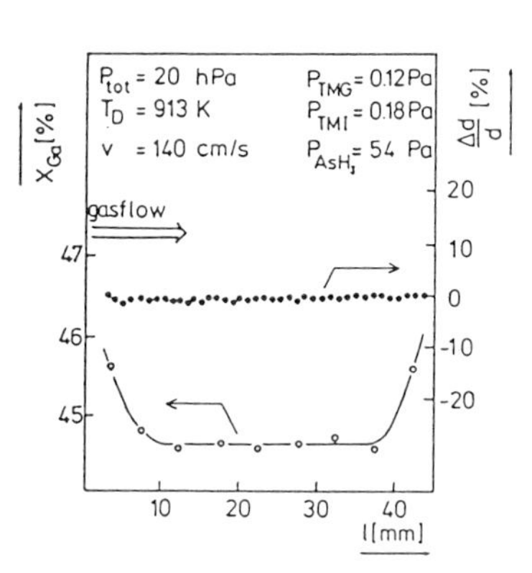

Fig.1: Homogeneity of film thickness and composition of GaInAs

A complete listing of the optimized growth parameters (temperature, total and partial pressures) is presented in table 1. An important feature of this set of parameters is that they permit the uniform deposition of GaInAs and InP at identical conditions.

	P_{tot} [Pa]	T_D [K]	v [cm/s]	P_{PH_3} [Pa]	P_{AsH_3} [Pa]	P_{TMI} [Pa]	P_{TMG} [Pa]
InP	2×10^3	913	140	94	–	0.18	–
GaInAs	2×10^3	913	140	–	54	0.18	0.12

Tab.1: Optimized growth parameters for GaInAs/InP heterostructures

The SIMS data for a MQW structure with 20 nm GaInAs wells and 20 nm InP barriers in fig. 2 give an impression of the reproducibility of compositon and thickness of the individual layers. They are identical within the accuracy of the measurement. The higher intensity of the phosphorus signal at the surface is due to preferential sputtering of this element.

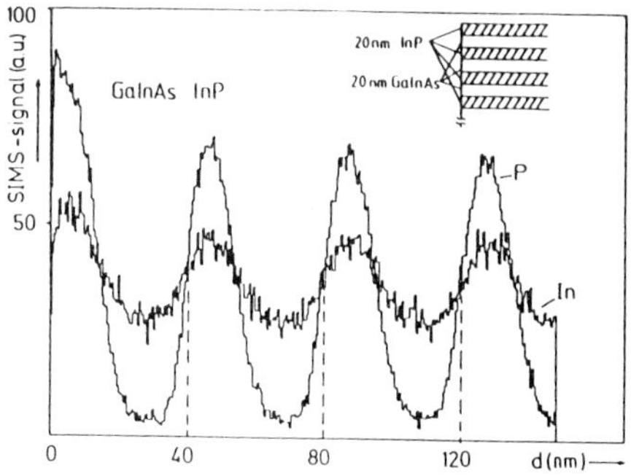

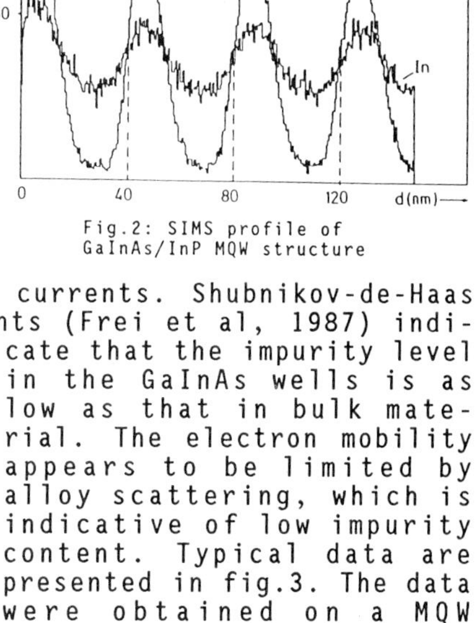

Fig.2: SIMS profile of GaInAs/InP MQW structure

High material purity is an important condition for obtaining a high output of recombination radiation and low laser threshold currents. Shubnikov-de-Haas and quantum Hall effect measurements (Frei et al, 1987) indicate that the impurity level in the GaInAs wells is as low as that in bulk material. The electron mobility appears to be limited by alloy scattering, which is indicative of low impurity content. Typical data are presented in fig.3. The data were obtained on a MQW structure with 15 GaInAs wells of 10 nm width, separated by 30 nm InP-barriers, each containing a 10nm sulfur doped ($n=10^{18}cm^{-3}$) central region. From the measured data an electron mobility in the wells of 103.800 cm^2/Vs may be calculated, at a carrier concentration per unit area of $8.5x10^{11}cm^{-2}$. These measurements also show the high reproducibility of the technology used for fabricating the MQW structures.

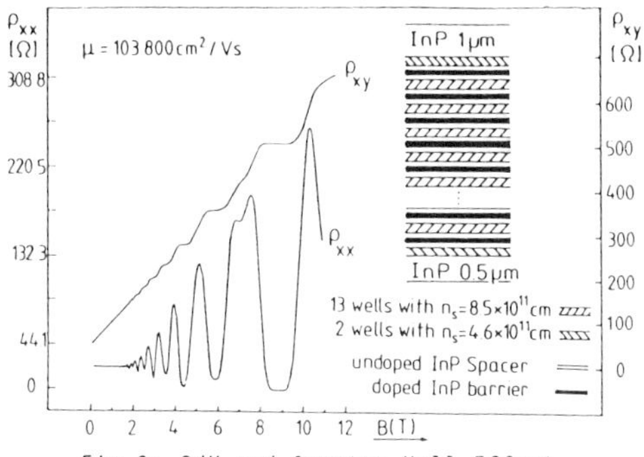

Fig.3: SdH and Quantum Hall Effect of 15 well MQW structure

The energy shift of the exciton line in the PL spectrum due to the quantum well effect is demonstrated in fig. 4 for wells of different widths. The shifts observed on single and multi quantum wells in our samples are comparable to those published for samples prepared by AP-MOVPE (Miller et al, 1986), LP-MOVPE (Razeghi et al, 1983) and MBE (Tsang et al, 1986, Panish et al, 1986). For a 0.5 nm well we observed a shift of 528 meV, which corresponds to approx. 90 % of the discontinuity in the bandgaps. Using the optimized growth paramters from table 1 the energy of emission at 2 K can be tuned via the wellwidth from 800 to 1330 meV.

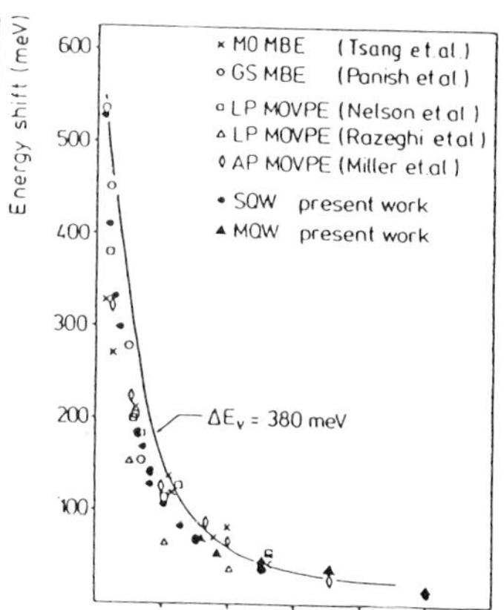

Fig.4: PL (2K) Energy shift due to quantum size effect vs. well width

PL data at 2 K obtained on a quantum well structure containing 6 GaInAs wells of different widths and a 100 nm GaInAs reference layer are given in fig. 5. The line width for very narrow wells is quite sensitive to interface roughness. The small line widths (2.2 meV for the 20 nm well to 37 meV for the 1 nm well on top of the stack) indicate that the uniformity of the well widths is approximately one monolayer. Our PL and electrical data do not show any evidence of the presence of extended defects. The same tolds true for the SEM pictures, which show only very few small defects.

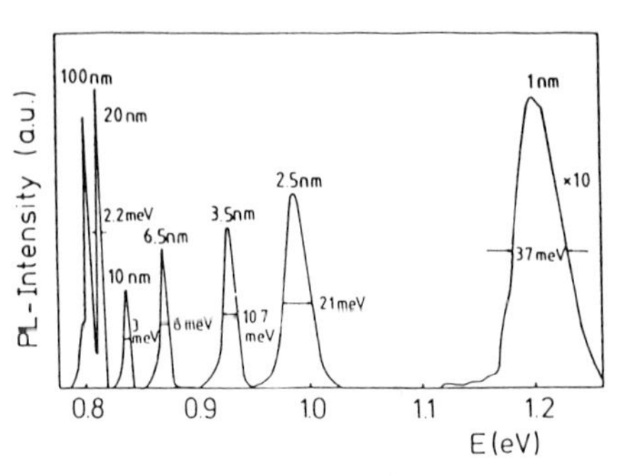

Fig.5: Typical PL(2K) measurement of SQW structure

Conclusions

Using low pressure MOVPE and high linear flow velocities high purity GaInAs/InP heterostructures can be prepared. Excellent homogeneity in thickness, composition and doping on a 2" InP substrate can be realized by this approach for optimized conditions. The low growth rates required for the deposition of very narrow well structures are achieved by selecting reduced pressures of the group III and group V compounds used for deposition. The method yields structures with high electron mobilities of the two dimensional electron gas in the well and narrow PL line widths, which is indicative of low impurity incorporation and abrupt heterojunctions. The observed energy shifts (up to 528 meV) demonstrate of the large range of band gap variation attainable by this method.

Literature

Dingle R. and Henry C.H. US Patent 3982207

Dingle R., Wiegmann W. and Henry C.H. Phys.Rev.Lett. 33,827 (1974)

Frei M., Tsui D.C. and Tsang W.T. Appl.Phys.Lett. 50, 606 (1987)

Grützmacher D., Schmitz D., Jürgensen H., Heyen M. and Balk P. to be puplished

Grützmacher D., Schmitz D., Jürgensen H., Heyen M. and Balk P. First European workshop on MOVPE, Aachen (1987)

Holonyak N.Jun., Kolbas R.M. and Dapkus P.D. IEEE J. Quant. Electr. QE 16, 170 (1980)

Jürgensen H., Aixtron, German Patent DE 3537544C1 (1985)

Miller B.I., Schubert E.F., Koren U., Ourmazd A., Dayem A.H. and Capik R.J. Appl.Phys.Lett. 49, 1384 (1986)

Panish M.B., Temkin H., Hamm R.A. and Chu S.N.G. Appl.Phys.Lett. 49, 164 (1986)

Razeghi M. and Hirtz P. Appl.Phys.Lett. 43, 585 (1983)

Tsang W.T. Appl.Phys.Lett. 39, 768 (1981)

Tsang W.T. and Schubert E.F. Appl.Phys.Lett. 49, 220 (1986)

Optimization of GaInAs/GaInAsP/InP and GaInAs/AlInAs/InP quantum well lasers

J.Nagle* and C.Weisbuch
Laboratoire Central de Recherches, Thomson-CSF,
B.P. 10, 91401 Orsay Cedex, France

ABSTRACT: We present an optimization of the structure design of GaInAs/GaInAsP/InP and GaInAs/AlInAs/InP quantum well lasers. We emphasize the decisive influence of the optical cavity on the current-gain relation for separate confinement structures. The optimum structures are fairly different from those actually grown up to now and show clear improvement over conventional lasers. Nevertheless this improvement is smaller than in the GaAs case and the threshold properties are much more sensitive to any nonideal material property.

1. INTRODUCTION

The GaAs/AlGaAs Quantum Well Laser (QWL) has now several well established advantages as compared to the usual Double-Heterostructure (DH) laser (Tsang 1984, Arakawa and Yariv 1986). The observation of lower threshold currents (Hersee et al. 1984), reduced temperature sensitivity of the threshold current (Tsang and Hartman 1981) as well as increase of the modulation bandwidth (Uomi et al. 1985) has been reported. The use of QW structures also leads to clear improvements in the case of GaSb/AlGaSb(GaSb) (Ohmori et al. 1985), GaInP/AlGaInP(GaAs) (Ikeda et al. 1987) and PbTe/PbEuSeTe(PbTe) (Partin 1985) lasers. QWL's of other material systems are not so well developed. In particular, no improvement has been obtained yet in the GaInAs/GaInAsP/InP or GaInAs/Al(Ga)InAs/InP system despite intense efforts towards high quality materials and interfaces. Our purpose is to establish by using a realistic model the limits imposed by specific properties of these systems (Auger recombination, small optical confinement factor, small electron effective mass leading to high band filling, conduction band offset,...). We consider the various structure parameters (well width, number of wells, width and composition of the optical cavity) for Multi Quantum Well (MQW), Separate Confinement Heterostructure (SCH), and Modified Multi Quantum Well (MMQW) lasers in order to determine the best structures for low threshold current and weak temperature dependance (high T_0).
The basic calculation technique has been presented in detail elsewhere (Nagle et al. 1986a). The main new feature of our model is the inclusion of the cavity effect on the gain-current relation. The confinement factor is calculated using the wavelength-dependent refractive index values indicated by Henry et al.(1985) for GaInAsP alloys. Auger recombination is included by means of a constant temperature-independent coefficient taken to be C_A = 4×10^{-29} cm^6s^{-1} for GaInAs (Sermage et al. 1986). We first do not consider any additional non-radiative process since we are investigating intrinsic

* Part of this work has been done at the Max-Planck-Institut für Festkörperforschung, D-7000 Stuttgart 80, Fed. Rep. Germany

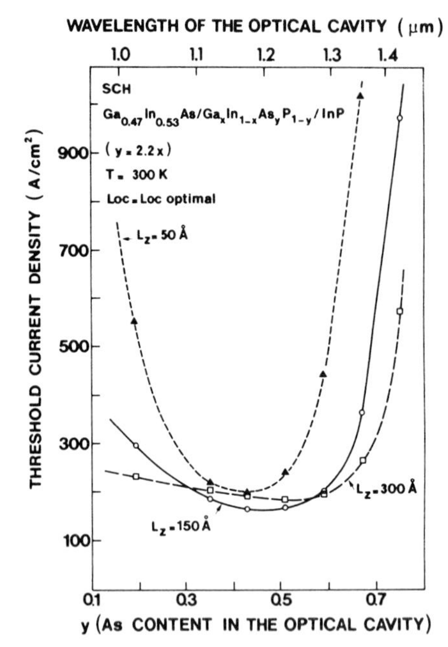

Fig.1 Threshold current density vs composition of the optical cavity for SCH lasers (α = 45 cm^{-1}). The width of the optical cavity maximizes the optical confinement factor.

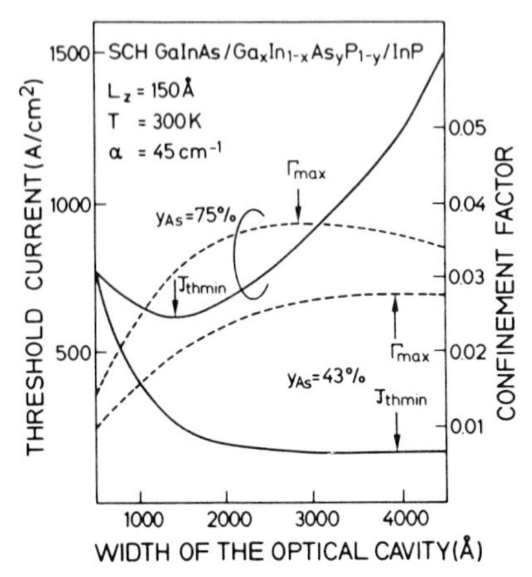

Fig.2 Threshold current density and optical confinement factor of a 150 A-SCH laser vs width of the optical cavity.

fundamental limitations of perfectly grown materials. We have supposed that the carrier temperature equals the lattice temperature. Similar calculations yielding a separate optimization of threshold current and its temperature dependence have been presented earlier by Sugimura (1983) and Asada et al.(1984) using different approximations but none of them included the tremendous effect of the optical cavity.

2. OPTIMIZATION OF THRESHOLD CURRENT

In separate confinement lasers the optical confinement factor is mainly determined by the step in refractive index at the interface between the optical cavity and the outer confinement layers. For a given well width, a decrease of the band-gap of the optical cavity corresponds to an increase of the confinement factor. This leads to a smaller threshold carrier density and hence a strong reduction of the Auger component. On the other hand, the contribution of the cavity recombination current increases with the reduction of the barrier height for carrier confinement. The best tradeoff determines the optimum composition of the optical cavity and is exemplified in Fig.1 for several well widths. It can be shown that an As percentage around 43% (λ_g = 1.17 μm) in the optical cavity gives good performance independently of the width and number of QW's. This holds also for enhanced optical losses. The corresponding value in the GaInAs/AlGaInAs/AlInAs system lies around 20% Al in the optical cavity.
Obviously, in the range where the cavity recombination current dominates, it would be advantageous to reduce the size of the optical cavity (Loc) below the usually adopted value maximizing the confinement factor (Γ). We present the results for two compositions of the optical cavity (Fig.2). When y_{As} = 43% (λ_g = 1.17 μm), the threshold current is entirely determined by the variation of Γ. This is not true for a low-gap cavity (y_{As} = 75%, λ_g = 1.42 μm) where the optimum width (Loc = 1400 Å) is two times smaller than the usual Γ-optimized one (Loc = 2800 Å). It should be noted that this two-fold reduction in Loc corresponds only to a 20% decrease of Γ.
Results for SCH and MMQW GaInAs/GaInAs$_{0.43}$P/InP lasers are compared in

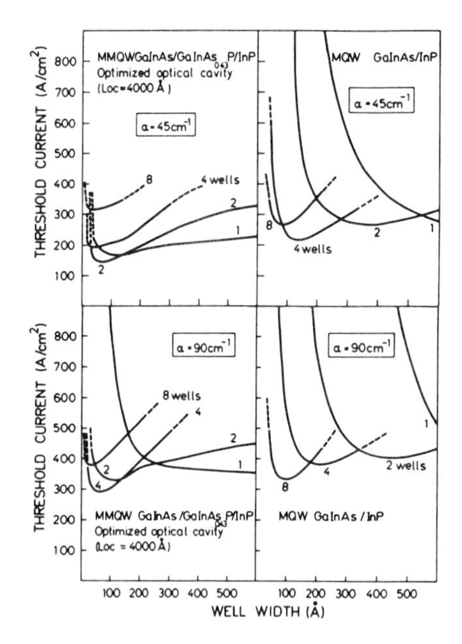

Fig.3 Comparison of threshold current density for MQW and optimized MMQW lasers. The width of the optical cavity is adjusted to maximize the optical confinement factor.

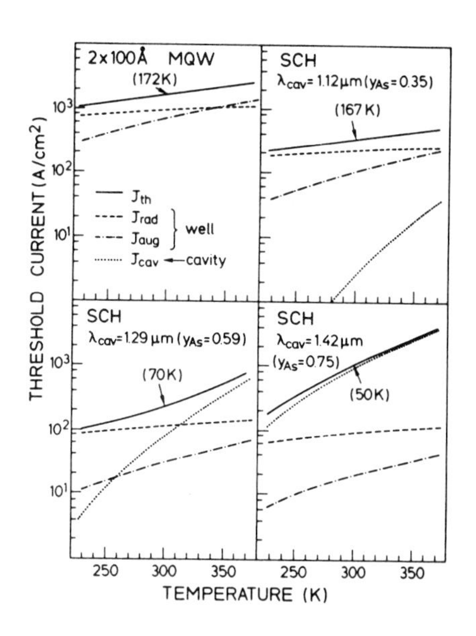

Fig.4 Decomposition of the temperature variation of the threshold current for one MQW and three GaInAs/GaInAsP/InP SCH lasers (L_z = 100 Å, Loc = 2000 Å). The value of T_0 at 300 K is indicated.

Fig.3 with the case of MQWL's without cavity. Threshold currents for MQWL's are slightly underestimated because we assume unity injection efficiency, uniform carrier injection between wells, and neglect possible leakage current in the outer confinement layer. The MMQWL's are better than the MQWL's due to the improved Γ and reduced threshold carrier density. The best structure for low-loss (long) lasers (α = 45 cm^{-1}) is a 2-well MMQWL with well thickness between 50 and 100 Å having a threshold current density as low as 150 A/cm^2. For high optical losses (α = 90 cm^{-1}), a small improvement subsists for a 4-well MMQW laser with well thickness ranging between 50 and 100 Å. The extreme sensitivity of the performance of QWL's is illustrated by the behaviour of 1-well SCH lasers which do not exhibit any improvement with reduction of the quantum well thickness if the optical losses are high. Band filling being very high in these QWL's with narrow wells, a bad quality of the optical cavity should be very detrimental to their operation. On the other hand the intervalence band absorption efect remains small because of the very small Γ (as long as the cavity is not significantly populated). Carrier heating has recently been demonstrated in narrow-well SCH lasers (Nagle et al. 1986b). The small advantage exhibited by perfect QW structures might also be suppressed in the presence of a too high carrier temperature.

3. DISCUSSION OF THE TEMPERATURE DEPENDENCE

The different terms of the threshold current have different temperature variation as shown in Fig.4. The T_0 associated to the radiative recombination in the well alone is ranging from around 250 K (low inversion) to 400 K (high inversion). The Auger current has a typical T_0 around 80 K and the cavity current between 30 and 50 K. The T_0 is reduced when the strongly temperature dependent Auger current or cavity current becomes dominant. For low-gap optical cavities, the T_0 is limited by the cavity recombination, for higher-gap cavities by the Auger recombination. When compared to the optimum for threshold current, the conditions for high T_0 are always corresponding to a reduction of the available higher-lying density of states which means a higher-gap and narrower optical cavity. It is interesting to note that because of the importance of Auger recombination the optimum for T_0 is reached for MMQWL's ($y_{As} \simeq 0.2$ in the cavity) and not for MQWL's, contrary to the case of GaAs/AlGaAs lasers. The threshold-current-optimized structures of the preceding paragraph have a T_0 ranging between 110 and 150 K, limited by the optical cavity. The 600 Å well SCH laser of Fig.3 should have a high T_0 of 230 K with only J_{th} = 230 A/cm^2 for α = 45 cm^{-1}. Except in the domain where it is dominated by the effect of the cavity, the T_0 is relatively insensitive to any decrease of the non-radiative lifetime or increase of the optical losses. This is consistent with the observation of slightly improved T_0 values for MQWL's that exhibit at the same time relatively high threshold currents (Dutta et al. 1985).

4. CONCLUSIONS

The intrinsic difficulties with the GaInAs active layer QWL's are mainly due to the nonlinear components of the threshold current, namely Auger recombination and cavity recombination current. We have performed an optimization of the structure design of QWL's in this material system with particular attention to these nonlinear components. We emphasize the extreme importance of a careful design of the optical cavity for SCH and MMQW lasers. The optimum structures are fairly different from those actually grown up to now and show clear improvement over conventional lasers. Nevertheless this improvement is not as large as in the GaAs case and the threshold properties are much more sensitive to any nonideal material property, like non-radiative recombination, surface recombination, or enhanced optical losses. This explains probably why no improvement has been obtained up to now for QWL's in these material systems.

REFERENCES

Arakawa Y. and Yariv A. 1986, IEEE J. Quantum Electron. QE-20 1119

Asada M., Kameyama A., and Suematsu Y. 1984, ibid. QE-20 745

Dutta N.K., Wessel T., Olsson N.A., Logan R.A., Koszi L.A., and Yen R. 1985, Appl. Phys. Lett. 46, 525

Henry C.H., Johnson L.F., Logan R.A., and Clarke D.P. 1985, IEEE J. Quantum Electron. QE-21 1887

Hersee S.D., de Crémoux B., and Duchemin J.P. 1984, Appl. Phys. Lett. 44 476

Ikeda M., Toda A., Nakano K., Mori Y., and Watanabe N. 1987, ibid. 47 1184

Nagle J., Hersee S.D., Krakowski M., Weil T., and Weisbuch C. 1986a, ibid. 49 1325

Nagle J., Hersee S.D., Razeghi M., Krakowski M., de Crémoux B., and Weisbuch C. 1986b, Surf. Sci. 174 148

Ohmori Y., Suzuki Y., and Okamoto H. 1985, Jpn. J. Appl. Phys. 24 L657

Partin D.L. 1985, J. Vac. Sci. Technol. B3(2) 576

Sermage B., Chemla D.S., Sivco D., and Cho A.Y. 1986, IEEE J. Quantum Electron. QE-22 774

Sugimura A. 1983, ibid. QE-19 932

Tsang W.T. 1984, ibid. QE-20 1119

Tsang W.T. and Hartman R.L. 1981, Appl.Phys. Lett. 38 502

Uomi K., Chinone N., Ohtoshi T., and Kajimura T. 1985, Jpn. J. Appl. Phys. 24 L539

Paper presented at Int. Symp. GaAs and Related Compounds, Heraklion, Greece, 1987

Analysis and modelling of 1.3 μm laser diodes

S Mottet, A Changenet*, J E Viallet, E Dudda**, A Accard**
R Blondeau***, M Krakowski***

C.N.E.T. Lannion, Route de Trégastel 22301 Lannion France
* C.N.E.T., 38 Rue du Général Leclerc 92131 Issy les Moulineaux France
** CGE-LdM, Route de Nozay 91140 Marcoussies France
*** THOMSON CSF-LCR, Domainde Corbeville 91401 Orsay France

ABSTRACT: Numerical simulations and analytical models derived on lasers and test structures show which are the parameters needed to fit experimental measurements and their evolution under aging.

1. INTRODUCTION

The main difficulties encountered today in the 1.3 μm lasers are to obtain reproducible III-V epitaxial layers, to obtain small leakage currents and to control the evolution of the device under long term operation. To achieve experimental curves interpretation, numerical simulations are used to identify the signatures of the different current contributions in the different layers.

The two types of 1.3 μm lasers studied are BH lasers from THOMSON CSF realized by MOCVD and DCPBH lasers from CGE obtained by LPE growth. Moreover, test structures, issued from the same fabrication lines as the lasers, have been studied in order to characterize each of the layer interfaces.

2. NUMERICAL SIMULATION

The simulation of the electrical behaviour of the laser is performed by solving the general equation set describing heterojunctions under Fermi-Dirac statistics (parabolic band assumption) (Viallet, Mottet 1985).

$$\left.\begin{aligned} \mathrm{div}\,(\epsilon\cdot\overrightarrow{\mathbf{grad}}\,\varphi) &= q\cdot(n-p-dop) \\ -\frac{1}{q}\cdot\mathbf{div}\,\overrightarrow{J_n} &= -U \\ \frac{1}{q}\cdot\mathbf{div}\,\overrightarrow{J_p} &= -U \end{aligned}\right\}\ \text{with}\ \left\{\begin{aligned} \overrightarrow{J_n} &= -q\cdot n\cdot\mu_n\cdot\overrightarrow{\mathbf{grad}}\,\varphi_n \\ \overrightarrow{J_p} &= -q\cdot p\cdot\mu_p\cdot\overrightarrow{\mathbf{grad}}\,\varphi_p \\ n &= N_c\cdot\mathcal{F}_{1/2}[(q\cdot\varphi+\chi-q\cdot\varphi_n)/k\cdot T] \\ p &= N_v\cdot\mathcal{F}_{1/2}[(-q\cdot\varphi-\chi-E_g+q\cdot\varphi_p)/k\cdot T] \end{aligned}\right.$$

φ , φ_n, φ_p are the electrical and electrochemical potentials, χ the electronic affinity and E_g the forbiden band gap energy. The generation recombination term $U = U_T + U_{sp} + U_A$ takes into account the three main phenomena involved up to the threshold current:

- *Deep center thermal recombination*: the whole thermal recombination can be described considering, in each layer, one dominant deep level center with energy level E_T (usually located at midgap) and carriers extrinsic lifetimes τ_n, τ_p. In Fermi-Dirac statistics it is written

$$U_T = \frac{n\cdot p - n_1\cdot p_1}{\tau_n\cdot(p + p_1) + \tau_p\cdot(n + n_1)}; \quad n_1 = n\cdot\exp\left[\frac{E_T + q\cdot\varphi_n}{k\cdot T}\right]; \quad p_1 = p\cdot\exp\left[-\frac{E_T + q\cdot\varphi_p}{k\cdot T}\right]$$

- *band to band spontaneous optical emission* is expressed to be valid at thermal equilibrium under Fermi-Dirac statistics.

$$U_{sp} = B_0\cdot(n\cdot p - n_0\cdot p_0) \text{ with } n_0\cdot p_0 = n\cdot p\cdot\exp\left[\frac{q\cdot\varphi_n - q\cdot\varphi_p}{k\cdot T}\right]$$

- *Auger recombination* term describes a mean value for hole and electron processes since n = p when Auger recombination is significant.

$$U_A = C_A\cdot(n + p)\cdot(n\cdot p - n_0\cdot p_0)$$

Most of the physical parameter values, temperature dependant, have been found without ambiguity in the litterature, so that the simulation can be performed for different device temperatures. But some discrepancy on the values of the emission constant B_0 and Auger coefficienr C_A appears in the litterature. The choosen values are :

□ $B_0 = 10^{-10}\cdot(T/300)^{1.5}$ $cm^3\cdot s^{-1}$ for 1.3 µm emission in GaInAsP.

□ $C_A = 3.6\ 10^{-27}\cdot\exp[-1280/T]$ $cm^6\cdot s^{-1}$ for Auger coefficient in the quaternary layer ($C_A = 5\ 10^{-29}$ $cm^6\cdot s^{-1}$ at room temperature).

Agreement of these laws have been obtain by comparison between simulations and measured characteristics of electroluminescent test structures as a function of the temperature. These physical laws are **now fixed once for all throughout the computations further described.**

3. SIMPLIFIED MODEL

From the numerical simulation it has been found that the current due to recombination processes, outside the quaternary active layer, could be neglected in these type of structures. The carrier densities in the quaternary increase rapidly when increasing the bias voltage and the assumption n = p can be made when $n \geqslant 10^{15} cm^{-3}$.

On these hypothesis an $I(\nu)$,$V(\nu)$ parametric formulation can be derived, where parameter ν is the carrier density in the active layer:

$$I(\nu) = \nu^2\cdot\left(\frac{1}{\tau_0(2\nu + n_1)} + B_0 + 2C_A\cdot\nu\right)\cdot v_{quater}$$

where τ_0 is the carrier lifetime in the quaternary and n_1 is related to the main recombination center energy level. Let $\overline{\mathcal{F}_{1/2}}$ be the inverse function of $\mathcal{F}_{1/2}$ which can be numerically computed:

$$V(\nu) = \frac{k\cdot T}{q}\cdot\left(E_g + \overline{\mathcal{F}_{1/2}}\left[\frac{\nu}{N_c}\right] + \overline{\mathcal{F}_{1/2}}\left[\frac{\nu}{N_v}\right]\right) + R_s\cdot I(\nu)$$

All the physical parameters used in this formulation are those used in the numerical simulation and are temperature dependant. The main difference between the two models lies in the description of the serial resistance R_s. The resistivity of the p InP semiconductor material is implicitly taken into account within the numerical simulation, whereas it has to be added to the simplified model. In both cases the contact resistance has to be included.

In this model the **only adjustable parameters are** τ_0 **and** $n_1(E_T)$, which concern the main recombination center, **and the serial resistance** R_s.

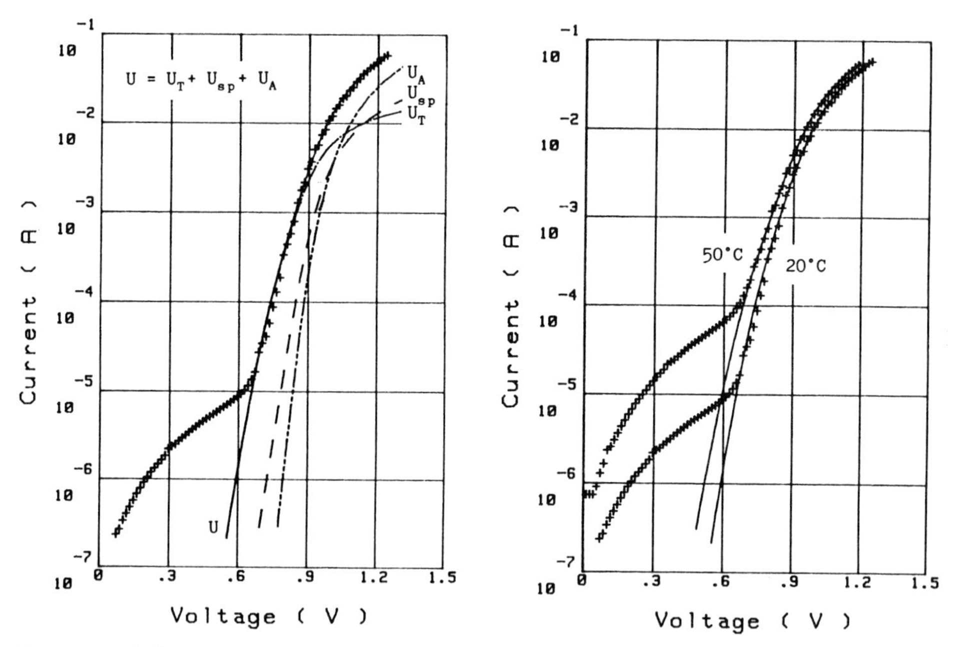

Fig. 1. I(V) curves from experiment and model: contribution of the three recombination terms.

Fig. 2. Comparison between experiment and model for BH laser at 20 and 50°C

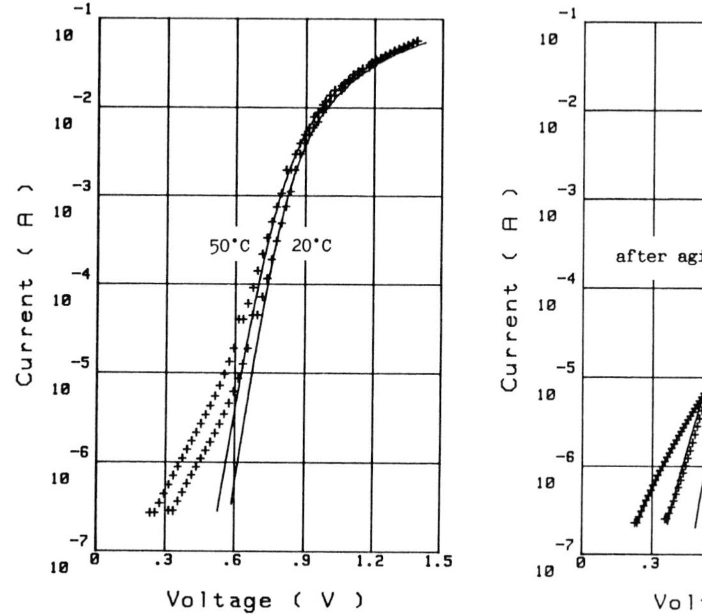

Fig. 3. Comparison between experiment and model for DCPBH laser (20, 50°C).

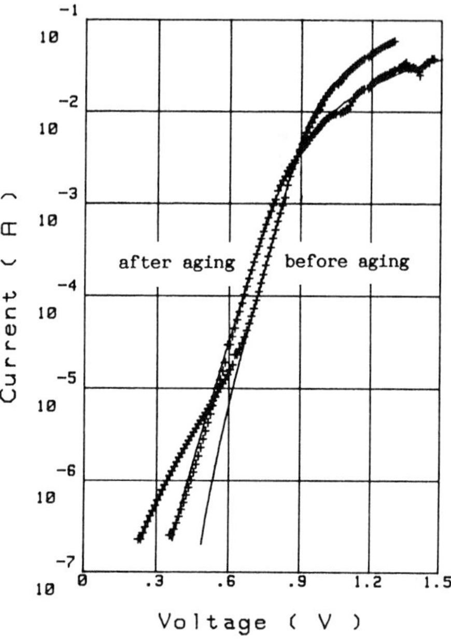

Fig. 4. I(V) characteristics of a laser before and after aging.

4. RESULTS AND DISCUSSION

The experimental I(V) characteristic of a laser and the curves given by the model are plotted in figure 1. This figure shows the contribution of each of the three recombination terms.

The low current range principally depends on the thermal recombination term U_T . The parameters τ_0 and n_1 (E_T), are independant and act respectively on the amplitude and the slope of the curve thus allowing their identification.

The high current range depends on the spontaneous emission, the Auger recombination and serial resistance. In fact Auger recombination dominates the upper range region so that the stimulated current can be neglected when only describing the current characteristic. Both Auger recombination and spontaneous emission laws are fixed. The only fitting parameter is the serial resistance R_s.

Figure 2 and 3 show comparison between experiment and model for both laser types at two temperatures. The model parameters have to be temperature independant. The identification of the parameters is performed at one temperature. The curve for the other temperature is obtained by direct application of the model at the new temperature with the same parameter values. The analysis of the two types of lasers gives close results concerning the quality of the quaternary layer: the lifetime of the carriers, τ_0, ranges between some 10^{-8}s and 10^{-9}s.

For both type of lasers the model does not describe the very low current range. First it was thought that those excess currents (with regard to the model) had their origins in the lateral blocking layers. In fact, measuments and simulations, performed on test structures, suggest that the excess current takes its origin at the interface region between quaternary layer and lateral blocking layers.

Figure 4 shows the critical evolution of the characteristic of a laser when aging and the capability of the model to fit this evolution. The aging conditions are: current 150 mA, temperature 70 °C, duration 120 hours. The analysis of the experiments leads to an evolution of the life time τ_0 from 1.4 10^{-8} to 4 10^{-9}s and of the serial resistance R_s from 5.5 to 15 Ω. The increase of the thermal recombination current, due to the degradation of the carriers lifetime, induces the increase of the threshold current which can be experimentally confirmed. It must be noticed that this laser has been choosen for its large degradations of both lifetime and serial resistance and is not representative of the mean evolution of these lasers.

5. CONCLUSION

For laser reliability studies the proposed model allows to identify and separate the main contributions to the current characteristic of the lasers and to describe their evolutions when aging.

6. REFERENCE

Viallet J E, Mottet S 1985 *Proc. NASECODE IV Conf.* ed J Miller (Dublin: Boole) pp 530-535

Monolithic integration of a Schottky photodiode and a FET using a $GA_{0.49}In_{0.51}P/Ga_{0.47}In_{0.53}As$ strained material

*M.Razeghi,**A.Hosseini Therani,**J.P.Vilcot,**D.Decoster

*L.C.R.Thomson-C.S.F., Domaine de Corbeville 91401 ORSAY- ** C.H.S. Université des Sciences et Techniques de Lille F.A. 59655 Villeneuve d'Ascq CEDEX-FRANCE.

ABSTRACT: We present an optoelectronic integrated circuit (O.E.I.C.) associating a Schottky photodiode and a F.E.T. which has been fabricated, for the first time, on a $Ga_{0.49}In_{0.51}P/Ga_{0.47}In_{0.53}As$ strained heteroepitaxy deposited by L.P.M.O.C.V.D. on a S.I. InP substrate. Static, dynamic and noise properties of the O.E.I.C., the Schottky photodiode and the F.E.T. have been investigated and are reported. As an example, responsivity and F.E.T. transconductance close to 0.1 A/W and 30mS/mm respectively have been achieved. All our experimental results are discussed taking into account the particular aspects of the material and the integrated circuit structure.

1. INTRODUCTION

For the fabrication of optoelectronic integrated circuits (O.E.I.C's) suitable for long wavelengths optical communications systems (1,3-1,55µm), the $Ga_{0.47}In_{0.53}As$, lattice matched to InP, is a potentially important material. Metal-Schottky barrier heights on GaInAs are too low to be used as MESFET gates, for example. Various solutions have already been given ($Al_{0.48}In_{0.52}As/Ga_{0.47}In_{0.53}As$/InP : Eastman 1980; GaAs/GaInAs : Chen 1985; Selders 1986; Rogers 1987). In this communication, we report, for the first time, the fabrication of an O.E.I.C., constituted of a Schottky photodiode, with a transparent Schottky contact, associated with a F.E.T. using a $Ga_{0.49}In_{0.51}P/Ga_{0.47}In_{0.53}As$ strained heteroepitaxy (figure 1). The energy bandgap of $Ga_{0.49}$

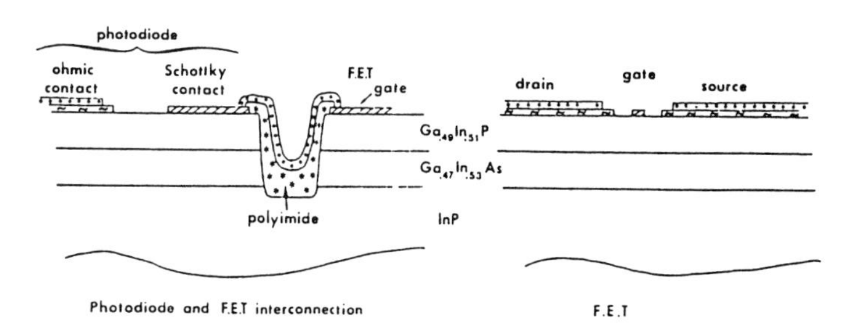

Fig. 1. Schematic cross section of the O.E.I.C.

$In_{0.51}P$ is 1.9eV and this superficial epilayer would improve the quality of Schottky barriers (Razeghi 1986).

2. DEVICE FABRICATION

The various epilayers were grown by L.P.M.O.C.V.D. on a S.I. InP substrate (figure 1). The $Ga_{0.47}In_{0.53}As$ epilayer ($N_D = 2.10^{17}$ At/cm^3; 2000A° thick) is suitable for long wavelengths (1,3 - 1,55µm) photodetection and F.E.T. channel fabrication. The undoped (residual n type) $Ga_{0.49}In_{0.51}P$ epilayer is 1000 A Thick. A Schottky photodiode, a F.E.T. and an O.E.I.C. have been fabricated on the same chip. A photograph of the O.E.I.C. and its electrical circuit is shown in figure 2. The classical common source amplification was chosen. The Schottky contact (Ti : 100 A, Pt 150 A) of the photodiode is deposited on the $Ga_{0.49}In_{0.51}P$ layer; its size is 40µm x 40µm. The F.E.T. has a 3µm x

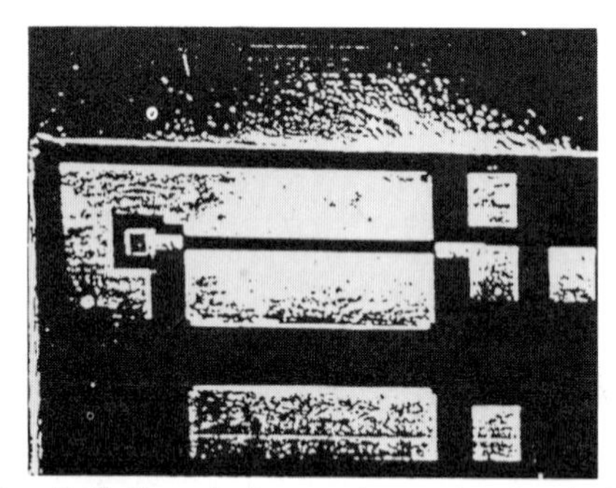

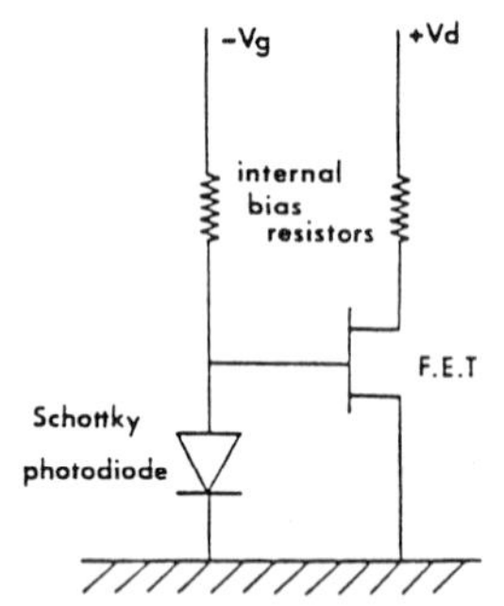

Fig. 2. S.E.M. photograph and electrical equivalent circuit of the photoreceiver

600µm gate (Ti : 1000 A, Pt : 1000 A, Ti : 1000 A, Au : 1500 A) and a 12µm source-drain spacing. The source of the F.E.T. is also the ohmic contact of the Schottky diode (Au-Ge-Ni 2000 A). The interconnection between the F.E.T. gate and the Schottky contact of the photodiode is made using a polyimide bridge.

3. EXPERIMENTAL RESULTS

The dark current of the photodiode under reverse bias voltage is plotted figure 3. Even with the large lattice mismatch the leakage current was found to be close to 10µ A for 1V, which is lower than for Schottky contacts made on GaAs/ GaInAs strained material (Selders 1986). The barrier height of the Schottky contact, determined using both current-voltage and capacitance-voltage measurements ($0.3pF < C < 0.7pF$), is approximatively equal to 0.4V. The ideality factor, deduced from I-V characteristic is close to 2. The static responsivity of the photodiode versus the reverse bias voltage for a 1.3µm wavelength optical signal is reported figure 4. The maximum value (0.1A/W) corresponds to electron-hole pairs created in the GaInAs layer whose thickness (0.2µm) is lower than the penetration length of the light. The dynamic responsivity of the Schottky photodiode has been measured up to several gigahertz using a sinusoïdaly modulated 1.3µm laser diode. The experimental results show, for a 50 Ω load resistor, a cutoff frequency

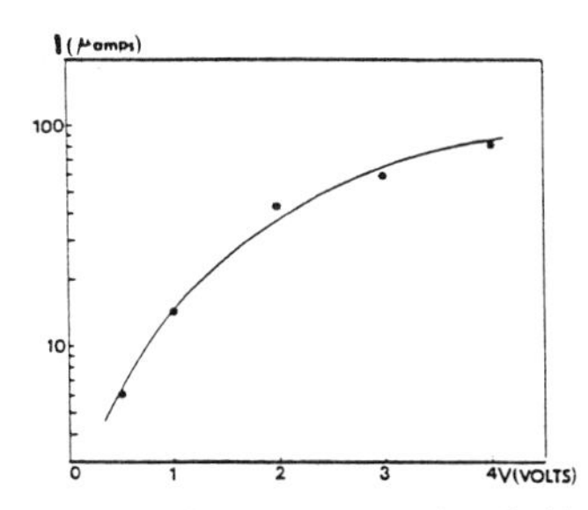

Fig. 3. Dark current of the Schottky photodiode under reverse bias voltage

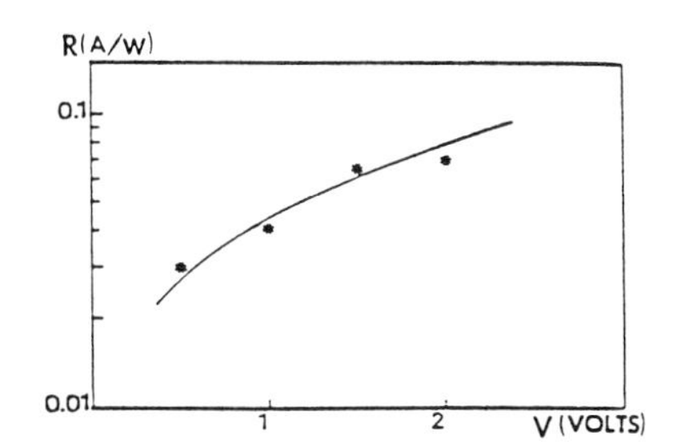

Fig. 4. Static responsivity of the Schottky photodiode under reverse bias voltage. P_L = 12μW, λ= 1.3μm

close to 1 GHz, corresponding to a response time lower than 1ns(figure 5). The long tail could be explained either by a diffusion phenomenon of the photocreated carriers before being collected at the ohmic contact, as for GaAs planar Schottky photodiode (Verriele 1985), or by trapping effects at the GaInP/GaInAs interface. Test experiments have been performed with a F.E.T. which gate dimensions are 3μmx300μm. The static I.V characteristic is given in figure 6; the transconductance is close to 30mS/mm. This result can be related to the thickness and the doping

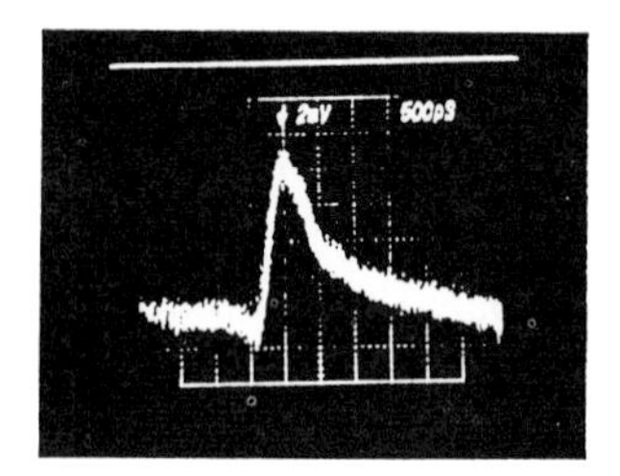

Fig. 5. Pulse response of the photodiode λ=1.3μm; P_L=9μW. V= -3V

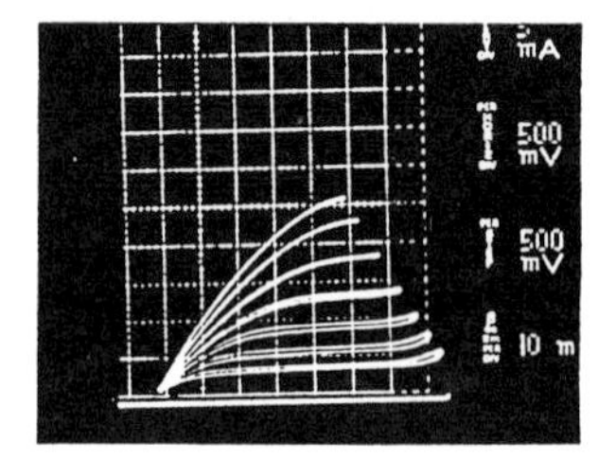

Fig. 6. Static I.V. characteristics of the 300μm F.E.T

level of the GaInAs epilayer and to the drain and source contact resistance (E.Constant 1987). The dynamic properties of this F.E.T. have also been studied. A maximum available gain close to 10 dB is obtained in the 100MHz frequency range and the cutoff frequency F_t is equal to 1.5GHz. Our simulations show that the experimental cutoff frequency is mainly due to the length of the gate. The static and dynamic properties of the O.E.I.C. have been investigated. For example, a dynamic responsivity of 50 A/W has been achieved at 1.3μm for a 11.5 KΩ bias resistor (figure 7). Noise measurements for the Schottky photodiode, the F.E.T. and the O.E.I.C. have been performed in the 10MHz-1.5GHz frequency range using a HP 8970A noise figure meter. The noise due to the illumination of the photodiode agrees with the well known equation i^2 = 2q I_{ph} B (Sze) where I_{ph} is the photocurrent and B the bandwidth of the noise figure meter. As would be expected for frequencies lower than 100MHz, a 1/f noise appears for the F.E.T. and the O.E.I.C. For higher frequencies our experimental results show that the noise of the O.E.I.C.

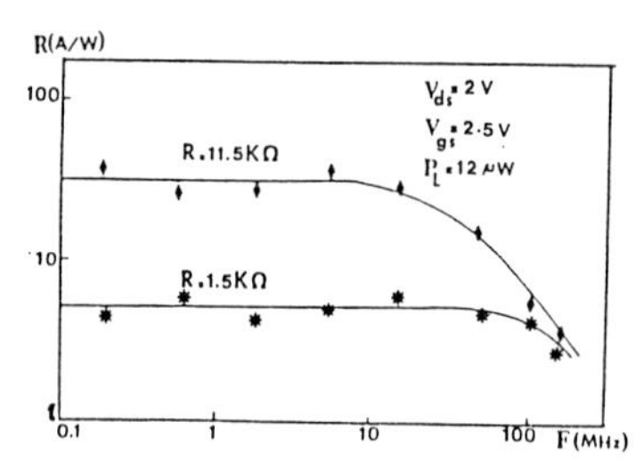

Fig. 7. Dynamic responsivity of the O.E.I.C. For various bias resistors

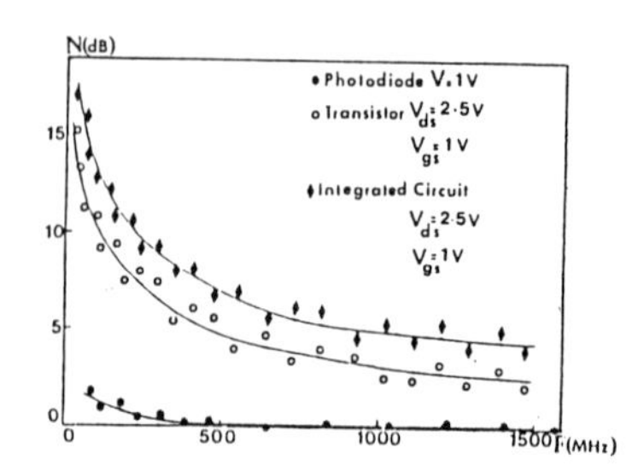

Fig. 8. Noise power versus frequency

is mainly due to the F.E.T. From our noise and dynamic responsivity measurements, an optical sensivity of -23dBm at a data rate of 250 Mbit/s with a 10^{-9} bit error rate has been deduced.

4. CONCLUSION

We have shown the feasibility of such O.E.I.C. with Schottky contact, prepared on single heterostructure (with lattice mismatch) GaInP/ GaInAs grown by L.P.M.O.C.V.D. Several improvements can obviously be introduced to increase the performance of the O.E.I.C.: a strong enhancement of the transconductance of the FET can be obtained by reducing the thickness and increasing the doping level of the GaInAs layer and by reducing the drain and source ohmic contacts using convenient GaInP layer etching or n-type supplementary epilayer (E.Constant 1987); moreover the cutoff frequency can be shifted to higher frequencies by decreasing the gate length; a better responsivity will be achieved by growing a thicker GaInAs layer for the fabrication of the photodiode or by coupling the light with an optical waveguide.

5. REFERENCES

Barnard J., Ohno H., Wood C.E.C., and Eastman C.F.,1980 IEEE Electron Device letters EDL-1 n°9, 174

CHEN C.Y., CHO A.Y., Garbinski P.A., 1985 IEEE Electron Device Lett. EDL-6 n°1,20

Selders J., Roentgen P., Beneking H., 1986 Elect.Lett., 22, pp 14-16

Rogers D.L., Woodall J.M., Pettit G.D., Inturff D.MC., 22-24 June 1987,paper VI-A8, University of California at Santa Barbara

Razeghi M., Maurel P., Omnes F., Thörngren E., 29 July 1986 Nato Invited paper

Verriele H., Maricot S., Constant M., Ramdani J. and Decoster D., 1985 Elect.Lett. 21, pp 878-879

Constant E., Depreeuw D., Godts P. and Zimmermann J., 1987 This FET structure has been simulated in our laboratory using a novel method for the modelling of FET presented at this conference (paper DE(2).1)

Sze S.M., 1981 Physics of semiconductor devices (Wiley.J and Sons, 2nd Ed)

Monolithic photoreceiver integrating InGaAs PIN/JFET with diffused junctions

JC Renaud, L Nguyen, M Allovon, P Blanconnier, F Lugiez, A Scavennec

Centre National d'Etudes des Télécommunications, Laboratoire de Bagneux, 196 avenue Henri Ravera, 92220 Bagneux, FRANCE

ABSTRACT: The process and the characteristics of a new integrated PIN/JFET photoreceiver are reported. The structure employs an original three layer InGaAs system grown by MBE in order to optimize separately the two devices which call for conflicting requirements. Such a choice has led to optimized doping levels of active layers allowing a same diffusion step to be used for the formation of both the photodiode and the JFET junctions. Thanks to the high performances of individual components, the monolithic receivers exhibit sensitivities as low as -33.7 dBm for a 10^{-9} B.E.R. at 140 Mbit/s.

1. INTRODUCTION

The high performances now achieved by optical fiber communications in the 1.3 - 1.55 µm spectral range have generated an intense research activity concerning the associated optoelectronic devices. For use in low-noise, large-bandwidth photoreceivers, avalanche photodiodes (Fujita and al 1986), photoconductors with internal gain (Rao and al 1986), or InGaAs PIN photodiodes (Renaud and al 1986) are presently considered. The latter can offer additional attractiveness provided they can be monolithically integrated with low noise InGaAs FETs which have great high frequency potential. In fact, such a choice can allow the realization of high sensitivity devices through the elimination of parasitic capacitances and thus noise degradation, in parallel with the improvement of compactness and reliability.

However, besides the fact that a technology for InP-based transistors is not yet well established, both photodiode and FET require different characteristics in their epilayer structure : the FET channel must be highly doped (about 5.10^{16} - 10^{17} cm^{-3}) on a few 1000 Å thickness to give high transconductance while the PIN photodiode intrinsic region must be undoped (about $2.10^{15}.cm^{-3}$) with a several microns thickness to allow high speed and high sensitivity. Thus, an optimized monolithic photoreceiver will involve separate layers for the FET and the photodetector. These requirements have led Tell and al (1985) and Kasahara and al (1984) to combine InGaAs PIN photodiodes with InP MISFETs using either selective implantation or two epigrowths and one Zn diffusion. However, to date InP MISFETs are prone to current drift phenomena and so the long-term functionality of such photoreceivers is not yet ensured.

This paper reports on the process and characteristics of a new PIN-JFET structure based on a simple three-layer InGaAs structure ($n^-/n^+/n$/S.I. InP) grown by Molecular Beam Epitaxy, for the PIN absorption and contact regions and JFET channel respectively. With this epitaxial structure, an independent optimization of the two optical and electrical components is possible.

2. DEVICE FABRICATION

In order to use the same Zn diffusion step for the formation of both PIN and JFET junctions, the lower (JFET) layer is first exposed by locally etching the upper n^- and n^+ layers. After deposition and patterning of a P.E.C.V.D. SiN_x film, acting as a diffusion mask, Zn diffusion is performed in a semi-closed ampoule (500°C - 0.7 µm) ; following a second etching step for isolation of the JFET active region, AuGeNi is evaporated for n contacts. Then, polyimide is spun on and patterned to insulate the mesa edges before evaporating TiAu for the gate and diode p^+ side contacts. The last step consists in depositing the biasing thin film $CrSi_2$ resistor (100 kΩ) by sputtering, resulting in the structure illustrated in Figure 1.

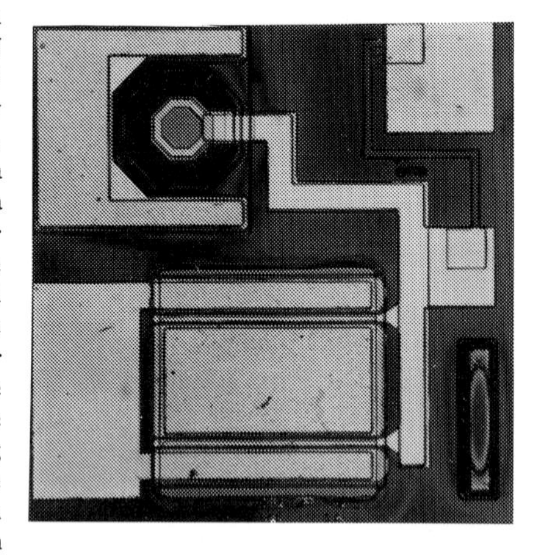

Fig.1 : PIN/JFET/R Structure

3. DEVICE CHARACTERISTICS

Figure 2 illustrates the voltage dependence of the leakage current (I_D) and the capacitance of a PIN diode with a diameter of the active area of 40 µm. At 10 V reverse bias, I_D is lower than 15 nA showing thus the good electrical quality of MBE grown InGaAs and the efficient passivation provided by the P.E.C.V.D. SiN_x film. The variation of the dark current with temperature shows an activation energy of 0.4 eV, close to the midgap energy. A minimum capacitance of 0.17 pF is obtained with these integrated photodiodes. This value is in good agreement with the doping level (~ 3.10^{15} cm^{-3}), active layer thickness and connection parasitics.

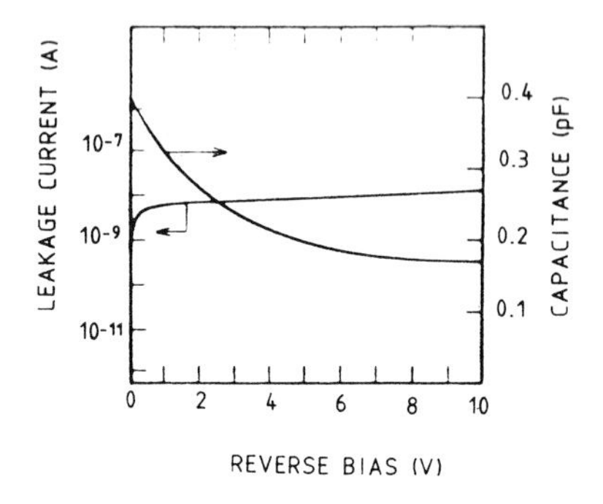

Fig.2 : Dark current and capacitance versus reverse bias for photodiodes with a 40 µm active diameter.

The photodiode responsivity is determined by comparison with a calibrated Ge detector. Without anti-reflection coating, the front-illuminated devices exhibit good responsivity (0.5 A/W) at 1.48 μm and high speed (~ 60 ps F.W.H.M. response to an optical pulse).

At 5 V reverse bias, the leakage current of the gate-source junction is as low as 24 nA for a 1.5 μm gate length (length of the window opened in the SiN_x film) and 300 μm gate width. This value is very similar to the dark current of the PIN diode with an equivalent junction area. The gate to source capacitance decreases from 1.6 pF/mm to 0.8 pF/mm as the gate voltage is reduced from 0 to -2 V.

The best results in terms of transconductance are obtained with devices operating in enhancement regime : on devices with a 5.5 μm spacing between drain and source contacts, transconductances of 140 mS/mm have been measured for 1.5 μm gate lengths at a gate voltage of + 0.6 V. But, for a PIN-FET receiver, operation of the JFET with a negative gate bias is mandatory (to keep the gate current at a low value). Thus, at -0.25 V, the transconductance decreases to 30 mS/mm (Figure 3), significant of an active channel which is too thin for its doping level. High frequency measurements have been performed and a Maximum Available Gain cutoff frequency of about 12 GHz has been determined.

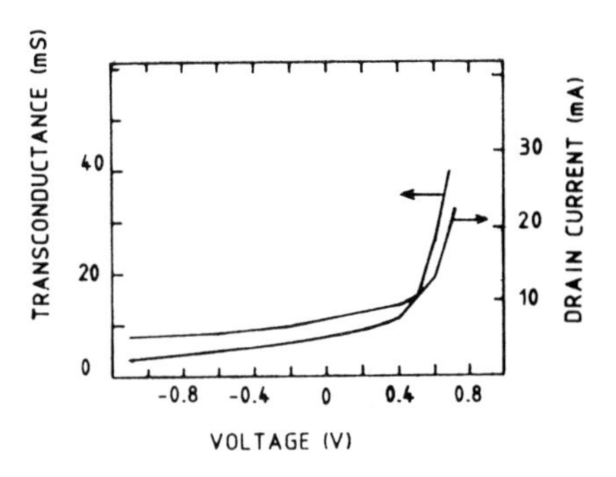

Fig.3 : Transconductance and drain current as a function of reverse gate voltage for 1.5 x 150 μm^2 gate JFETs (V_{DS} = 3 V).

Monolithic PIN-FETs have been mounted as input stages of hybrid three-stage photoreceivers. Following bandwidth measurement and equalization to give a total bandwidth of more than 100 MHz, signal and noise power measurements were performed for sensitivity evaluation of these photoreceivers. The 3 MHz signal (~ 0.6 μW peak power) was provided by a Light Emitting Diode (1.48 μm). The noise power was measured with a 110 MHz low-pass filter. From the measured signal to noise ratio, minimum detectable powers of - 33 dBm and - 33,7 dBm at 140 Mbit/s (for a Bit Error Rate of 10^{-9}) were infered for several InGaAs monolithic PIN-FETs with 40 μm photodiodes associated with 1.5 x 300 μm^2 and 1.5 x 150 μm^2 gate JFETs respectively. The received eye pattern for a 140 MHz NRZ sequence is shown in Figure 4.

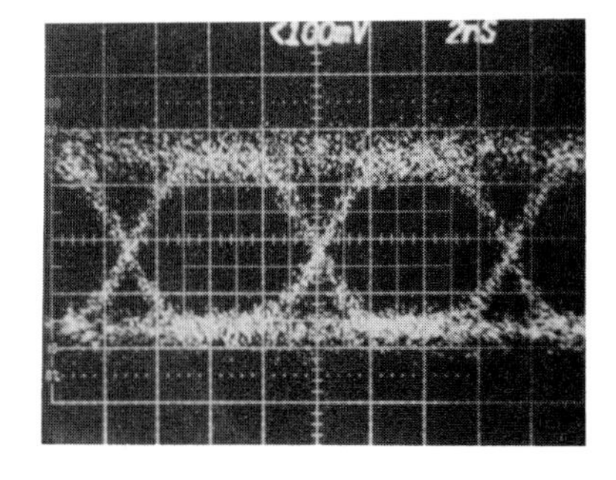

Fig.4 : Received eye pattern for a 140 MHz NRZ sequence.

These values of sensitivity are somewhat higher than expected from predictions based on devices characteristics. Part of the degradation seems to be attributable to the low JFET transconductance which gives a small voltage gain to the input stage of the receiver. Nevertheless, thanks to the high performances of the individual components in terms of capacitance and leakage currents, the sensitivity of such photoreceivers is one of the best reported up to now, using monolithic PIN-FETs. Moreover, the natural passivation provided by the SiN_x for both PIN and JFET junctions ensures a high reliability of the circuits (which exhibited no drift in characteristics over several weeks).

4. CONCLUSION

A new integrated PIN-JFET has been developed, using an original three-layer InGaAs structure which allows a separate optimization of the optical and electronic devices. The same diffusion step is used for the formation of both photodiode and JFET junctions. Encouraging characteristics have been measured on these integrated circuits which exhibit a remarkable stability with time. The sensitivity of these monolithic photoreceivers is close to - 33.7 dBm for 10^{-9} B.E.R. at 140 Mbit/s ; further optimization, including the incorporation of a thin buffer layer between the active JFET channel and the substrate, is likely to bring a large improvement of the sensitivity of these first circuits.

This work is partly supported by the EEC through the ESPRIT Programme n° 263 B.

The authors wish to thank F. Héliot, D. Arquey, L. Bricard, S. Vuye and B. Bourdon from CGE (Marcoussis) for technical assistance and fruitful discussions.

REFERENCES :

Fujita S, Takano I, Henmi N, Shikada S, Mito I, Taguchi K and Minemura K 1986, 12th European Conference on Optical Communication, pp.507-4

Kasahara K, Nayashi J, Makita K, Taguchi K, Suzuki A, Nomura H and Matushita S 1984, Electronics Letters, vol.20, pp.314-2

Rao M, Bhattacharya P and Chen C 1986, IEEE Trans. on Electron Devices, vol. ED33, n°1, pp.67-5

Renaud JC, Allovon M, Dimitriou P, Lugiez F and Scavennec A 1986, 12th European Conference on Optical Communication, pp.285-4

Tell B, Liao A, Brown-Goebeler K, Bridges T, Burkhardt G, Chang TY and Bergano N 1985, IEEE Transactions on Electron Devices, vol. ED32, n°11, pp.2319-3

Inst. Phys. Conf. Ser. No. 91: Chapter 7
Paper presented at Int. Symp. GaAs and Related Compounds, Heraklion, Greece, 1987

High efficiency blue LED utilizing GaN film with AlN buffer layer grown by MOVPE

I.Akasaki, H.Amano, K.Hiramatsu and N.Sawaki

Department of Electronics, Nagoya University, Nagoya 464, JAPAN

ABSTRACT: The GaN film with a smooth surface free from cracks has been grown by MOVPE using a thin AlN buffer layer. High efficiency blue LED's utilizing this GaN film have been successfully developed.

1. INTRODUCTION

The GaN blue LED with an external efficiency of about 0.03%, which is applicable for practical uses, was developed by using m-i-n structure grown on a sapphire substrate by hydride VPE.(Ohki et al. 1981) However, it still has serious problems to be overcome: it is fairly difficult (1) to grow the GaN film with a smooth surface free from cracks, because of the large lattice and thermal mismatches between GaN and sapphire, and (2) to control the diode operating voltage, which depends mostly on the thickness of Zn-doped insulating(i)-layer.

By MOVPE using a thin AlN buffer layer, we have succeeded to solve these problems and to grow the GaN film with specular surface free from cracks. High efficiency blue LED's have been developed by using this GaN film with good reproducibility, for the first time.

2. GROWTH AND CHARACTERIZATION OF GaN FILMS

Before the GaN growth, the thin AlN layer was deposited on a (0001) sapphire substrate kept at 900-950°C for about 1min by feeding trimethyl-aluminum (7μmol/min) and NH_3(2l/min) diluted with H_2(3l/min). Characterization by means of SEM(Fig.1(a)) and RHEED (Fig.1(b)) showed that this AlN layer does not a single crystal but an amorphous-like structure with about 50nm thick. GaN single crystal films of about 2um thick was grown at about 1000° C by feeding trimethylgallium(TMG)(17μmol/min) and NH_3(1.5l/min) diluted with H_2(2l/min) for about 20min on the substrate covered with the AlN layer. Fig.2 is SEM photographs of GaN films thus grown (with AlN layer), together with an ordinary one (without the AlN layer) for comparison. The former is quite smooth and have no cracks. Thus we call hereafter the former the "newly-developed GaN film" and the latter a "conventional one". Fig.3 shows the x-ray rocking curves of both GaN films. The FWHM of the curve for the new film is 1.9min, which is much narrower than values ever reported indicating that the crystal quality is fairly improved.(Amano et al. 1986) Fig.4 is photoluminescence spectra of these GaN films. In the new film, the I_2 line, which is the emission line due to excitons bound to neutral donors, is very strong and sharp (its FWHM is 1.1meV). And the broad band in long wavelength region due to deep-level impurities and/or defects are hardly observed. These facts show that the new film is of high

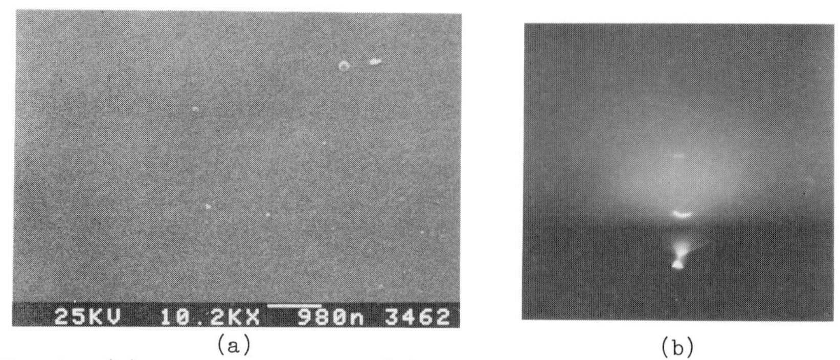

Fig.1. (a)SEM photograph and (b)RHEED pattern of a thin AlN layer deposited on (0001) sapphire at 900°C for 1min.

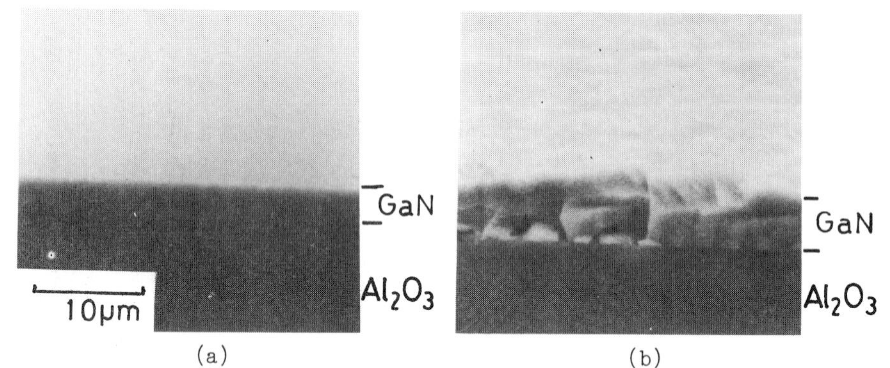

Fig.2. SEM photographs of GaN films grown at 1000°C on (0001) sapphire (a) with and (b) without the AlN buffer layer.

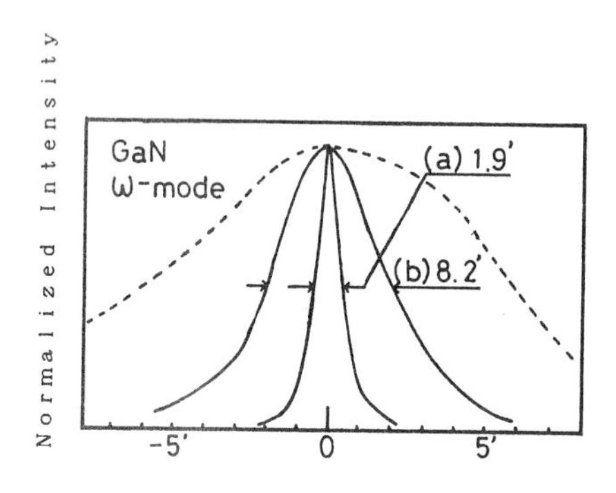

Fig.3. X-ray rocking curves for (0002) diffraction from GaN films same as shown in Fig.2. Dotted line shows one for (0006) diffraction obtained from GaN film grown by HVPE (Shintani et al. 1978).

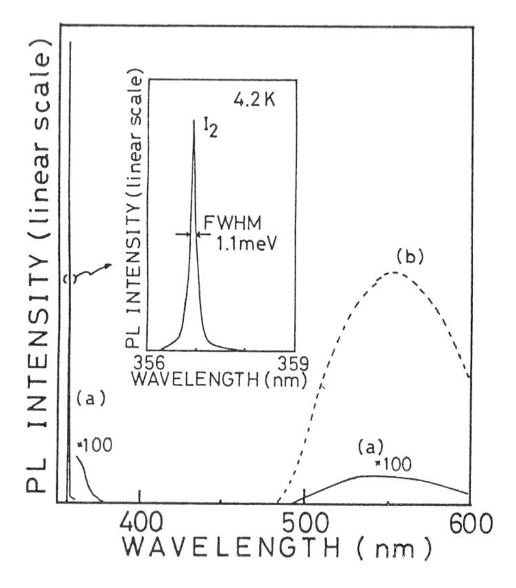

Fig.4. PL spectra of GaN films same as shown in Fig.2.

purity. The electron mobility of the new film was of about 300 $cm^2/v \cdot s$, which is about three times higher than those of conventional ones. These results show that the electrical and optical properties as well as the surface quality of GaN single crystal films grown in this way are of high quality.

If deposition time of the AlN exceeds 3min (layer thickness becomes thicker than 150nm), however, and/or the deposition temperature is higher than about 1100°C, both the AlN layer and the successively grown GaN film become polycrystalline as shown in Fig.5.

Thus, both the crystallographic structure and the thickness of the predeposited AlN layer were found to be optimal as a buffer layer to relax the strain due to the mismatches in this heteroepitaxial growth.

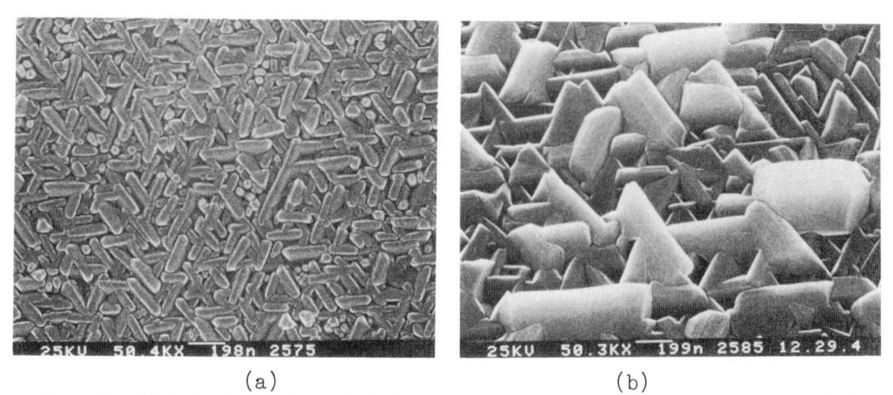

(a) (b)

Fig.5. SEM photographs of (a) AlN layer deposited for 3min and (b) GaN film grown successively on the AlN layer.

3. CHARACTERISTICS OF BULE LED'S

In order to fabricate a m-i-n structure, at the end of the growth of undoped n-type GaN film of about 10μm thick in this way, diethylzinc(DEZ) was introduced for 1min, by which Zn-doped i-GaN layer of about 100nm thick was formed. Zn forms blue luminescent centers as well as behaves as an acceptor in GaN, and its concentration was proportional to the flow rate of DEZ in the range of Zn concentration above about $1x10^{19}$ cm^{-3}. The growth rate of the GaN does not depend on the growth temperature, but is proportional to the flow rate of TMG. The surface quality had not been affected by the Zn doping. These enable us to control the diode operating voltage which depends mostly on the i-layer thickness. Both an ohmic contact on n-GaN film and a Schottky barrier electrode on i-GaN layer, were formed by evaporation of matallic Al. Fig.6 is a typical I-V characteristics of the m-i-n LED, indicating that the main conduction mechanism of the diode current is the "tunnel-induced impact ionization". An example of EL spectrum at room temperature is shown in Fig.7, where the violet-blue band peaking at about 430 nm originates in Zn-doping. No long wavelength emission bands due to deep-level impurities and/or defects are observed.

The external quantum efficiency of about 0.3% typically was obtained with good reproducibility, and the operating voltage ranged from 5.0 to 7.0 V.

In conclusion, by MOVPE utilizing AlN buffer layer, blue LED's with high performance have been developed.

ACKNOWLEDEMENTS

The authors would like to thank Messrs. N.Okazaki, K.Manabe and M.Hashimoto of Toyoda Gosei Co., Ltd. for help with PL measurement and for fabricating diodes. This work was supported in part by the "Joint Research with Industry" from the Ministry of Education, Science and Culture of Japan.

Amano H, Sawaki N and Akasaki I 1986 Appl.Phys.Lett. 48 353.
Ohki Y, Toyoda Y, Kobayashi H and Akasaki I 1981 Inst.Phys.Conf.Ser. 63 479.
Shintani A, Takano Y, Minagawa S and Mari M 1978 J.Electrochem.Soc. 125 2076.

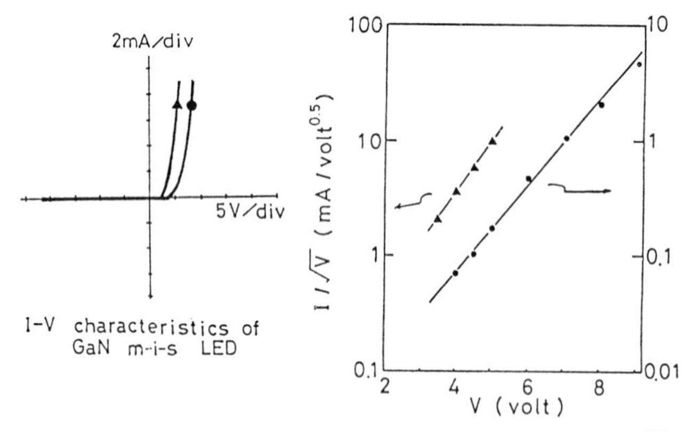

Fig.6. I-V characteristics of m-i-n GaN LED. $(\mathrm{Log}(I/\sqrt{V}) \propto V)$

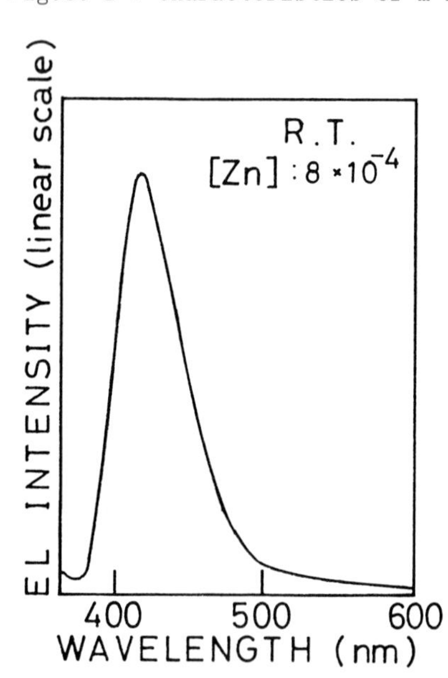

Fig.7. EL spectrum of m-i-n GaN LED at room temperature. Zn concentratiion in i-layer is about 8×10^{-4}.

Inst. Phys. Conf. Ser. No. 91: Chapter 7
Paper presented at Int. Symp. GaAs and Related Compounds, Heráklion, Greece, 1987

Technology for submicron recessed gate GaAs MESFETs on thin MBE layers using electron-beam lithography

* W Patrick, K Dätwyler, B J Van Zeghbroeck and P Vettiger.

IBM Research Division, Zürich Research Laboratory, Switzerland.
* Now with Fraunhofer I.A.F, Freiburg, W. Germany.

ABSTRACT: Processing techniques which have been developed for fabricating MESFET devices on thin, highly doped MBE grown layers, are described. Emphasis will be given to resist technology, contact formation (ohmic and Schottky) and gate recessing.

1. INTRODUCTION

Reduction of the gate-length of MESFET devices is normally associated with improved device performance. For full optimisation however, there are other factors to consider. Thin, highly doped channel layers are essential to maintain a reasonable gate/channel aspect ratio (Daembkes et al, 1984) and to reduce parasitic resistances. High quality, epitaxially grown layers with abrupt interfaces reduce substrate injection, hence improving pinch-off characteristics (Patrick et al 1985, Lee et al 1987). A recessed-gate technology reduces the device access resistance and provides a means for threshold voltage control. Short gate to n^+ GaAs and gate to source spacing (R and S, Fig. 1) further improve device performance. A near to ideal structure of a high performance device is shown in Fig. 1; further details are given in a recent letter by Van Zeghbroeck et al (1987a) where MESFETs with excellent characteristics were reported. These devices had a gate-length of 0.5μm while R and S were 0.1 and 0.2-0.3μm respectively. Device fabrication requires high resolution lithography with accurate overlay and reproducability of small dimensions. Electron beam lithography was therefore chosen for pattern definition. We have found that, because of the thin highly doped channel layers, standard ohmic, Schottky and recessing technologies can not be utilised. It is the purpose of this paper to describe the optimisation of these technologies for MESFET fabrication.

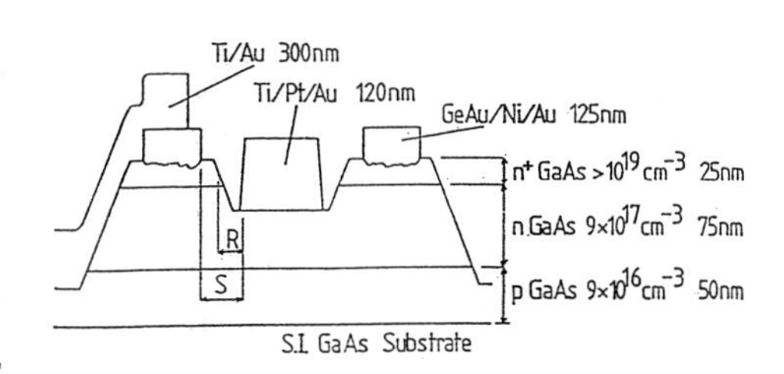

Fig. 1. MESFET Structure

2. RESIST SYSTEMS

For definition of thick metallic layers (300nm) a Terpolymer/PMMA resist system was adopted (Fig. 2a). This system is particularly useful for large area exposures (high sensitivity) and for defining closely spaced (1μm) contacts. For reliable lift-off, the resist stripe defining such a gap should be relatively thick and have a T shaped cross section (Fig. 2a). This can be achieved through optimisation of the development of the two resist layers. 1:3 MEK:MIBK is used to develop the Terpolymer layer. Toluene, which does not dissolve the Terpolymer significantly, is used to develop the PMMA. Suitable choice of development times produces the desired undercut resist profile. Ti/Au pads, 75μm wide, 300nm thick, 0.8μm apart, have been defined with this resist.

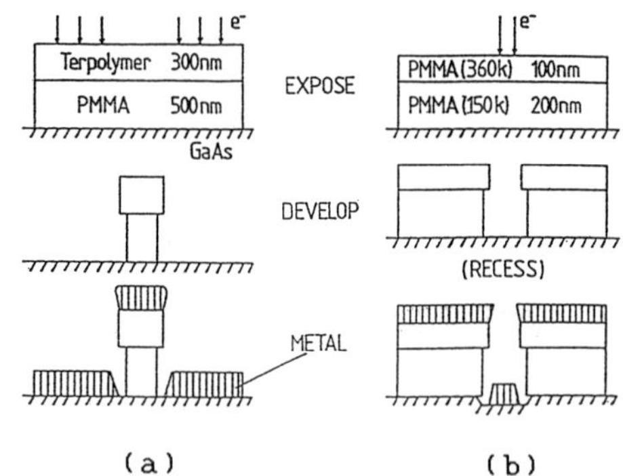

Fig. 2a. Terpolymer/PMMA
Fig. 2b. Bi-layer PMMA

For gate definition, a bi-layer resist consisting of 100nm PMMA 2041 (Mol. Wt 360k) on 200nm PMMA 2010 (150k) is used (Fig.2b). Exposure dose varies from 100 to 700μC/cm² depending on gate-length. The developer is 1:2 MIBK:IPA. Because of the difference in sensitivities of the two PMMA layers, the lower layer develops faster than the upper layer leading to a desirable undercut profile (Mackie et al, 1985). The undercut determines (in part) the gate to n^+ GaAs spacing, R in Fig. 1, and provides an ideal profile for reliable lift-off. Linewidths down to 0.15μm have been resolved with this resist system. This dimension is machine and not resist limited since the minimum e-beam spot size is in the order of 0.1μm (FWHM).

3. OHMIC CONTACT TECHNOLOGY

The use of standard ohmic contact technologies may not produce optimum contacts to thin (100nm) layers. It has been surmised that to achieve high quality contacts to thin layers, the diffusion of the contact metals should be reduced; conveniently achieved by low temperature annealing. Optimisation of the GeAu/Ni/Au contact composition can produce low resistivity contacts at 300°C annealing temperatures (Patrick et al, 1986). For this MESFET work a composition of 17nm Ge / 48nm Au / 10nm Ni / 50nm Au was found to produce reliable contacts with $R_{\square}$ below 0.2 Ωmm, when annealed at 360°C. This value is sufficiently low not to limit MESFET device performance.

4. GATE RECESSING

The gate recess determines the device threshold voltage and should therefore be controllable to within about 20Å. Wet chemical etches cannot be relied upon to give reproducible etch rates in this range. A recess monitoring technique was therefore developed. Fat FETs (100μm gate-length), situated at various sites over a wafer, can be probed, after gate development, through windows on the source and drain contacts. The drain current can then be measured in the device at intermediate steps during the recess etching. A correlation, determined from an accumulation of data, exists between the measured Fat FET drain current and the ultimate threshold voltage of these and other devices. Solutions of 1:8:1000 and 5:1:200 H_2SO_4:H_2O_2:H_2O are used for recess etching, the latter being more reliable.

Reasonable control of the threshold voltage is achieved by current monitoring. Fig. 3 compares threshold voltage with Fat FET control current from several wafer runs. The seemingly wide spread of results can be attributed to varying diode quality between runs; the lower the ideality factor, the more positive the threshold voltage. Fortunately, good diodes can now be consistently fabricated (following section), enabling this current monitoring technique to be satisfactorily used in threshold voltage control.

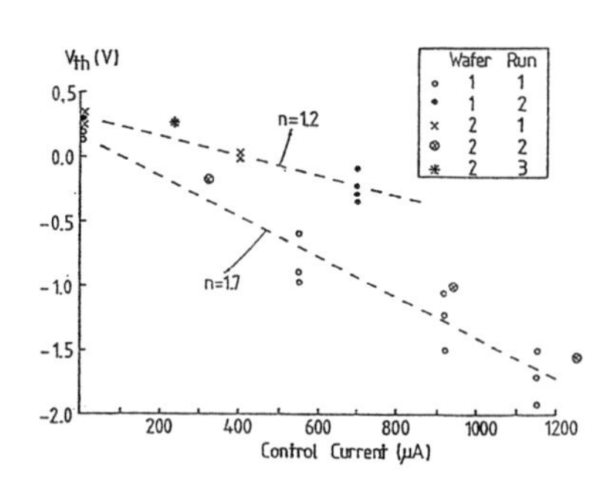

Fig. 3. Threshold Voltage versus Control Current.

5. SCHOTTKY GATE TECHNOLOGY

The highly doped channel layers of our MESFET devices ($10^{18}/cm^3$) make it difficult to produce good Schottky gate diodes. Stringent surface preparation prior to metal deposition is essential. A series of pre-evaporation cleaning processes was studied to find the most suitable for fabricating high quality diodes. The results are tabulated below. The best diodes were obtained using the following pre-evaporation sequence; a) IPA Resist wetting, b) 2min. HCl:H_2O Solution, c) Blow dry in N_2, d) Evaporate gate metal less than 3hrs after loading vacuum system. A slightly improved process utilises a gentle ultra-sonic agitation during the 2 minute HCl etch. The best diodes had n = 1.08 (1.17 Ave.); a good value for such

TREATMENT	AVERAGE IDEALITY	BEST IDEALITY	BARRIER HEIGHT
BF	1.36	1.15	0.66V
ABF	1.20	1.10	0.65V
ABDF	1.17	1.08	0.65V
ACF	1.29	1.17	0.72V
BEG	1.70	N/A	>0.80V
ABDG	>1.50	N/A	>0.80V

A. IPA Wetting, B. HCl:H_2O 1:1, 2 min, C. 20% NH_3OH, 2 min, D. Ultrasonic Agitation, E. Rinse DI Water, F. Pump Down <3h G. Pump Down Overnight, N/A. Not Available.

Table 1 Diode quality dependence on pre-evaporation treatment.

highly doped layers. When ammonia is used instead of HCl or if the IPA wetting step is omitted, the diode quality decreases significantly. If the wafers are left in the vacuum system overnight the diodes are dramatically degraded. This is probably due to incomplete removal of the surface oxide or, in the latter case, to regrowth of an oxide layer in the vacuum system. This may be confirmed by the noted increase of barrier height of the poorer diodes; ie up to 0.8V on poor diodes compared with 0.65V on good diodes.

6. DISCUSSION

Technologies have been developed to be applied in the fabrication of GaAs MESFET devices on thin, highly doped epitaxial layers. In our laboratory, good discrete devices have been fabricated on successive wafer runs, using these technologies. 0.5μm devices were fabricated with high transconductance (440mS/mm) and the highest ever reported K-value for a MESFET device (580mS/Vmm) (Van Zeghbroeck et al, 1987a). Recently a wafer run was completed with operational circuits utilising enhancement/depletion direct coupled MESFET logic. The gate-length of the devices used in these circuits was 0.35μm. Even though discrete devices on this wafer had significantly inferior characteristics, compared with the near optimum devices reported earlier (due to lower channel doping), ring oscillator circuits with 16ps stage delay were measured using a spectrum analyser. On the same wafer, high speed opto-electronic receivers were fabricated and were found to be operational up to 5.2 GHz. Details of this circuit will be published by Van Zeghbroeck et al (1987b). These examples illustrate the advantage (in terms of device performance) to be gained combining high resolution e-beam lithography and high quality epitaxially grown GaAs in device manufacture. The submicron recessed-gate MESFET will have an important role in certain high speed digital or optoelectronic applications.

7. ACKNOWLEDGMENTS

We thank H. Meier for wafer growth and H.P. Dietrich, G. Sasso, W. Walter and many others for technical support within the Dept. Thanks are due to S. Rishton (IBM Yorktown) for information on PMMA resits. Some preliminary work was carried out by W. Patrick at the Department of Electronics, University of Glasgow.

8. REFERENCES

Daembkes H, et al, 1984 IEEE Trans. Elect. Dev, ED-31, p1032.
Lee K Y, et al, 1987 Elect. Lett. 23, p11.
Mackie W S, et al, 1985 Solid State Tech, p117 (Aug).
Patrick W, et al, 1985 IEEE Elect. Dev. Lett, EDL-6, p471.
Patrick W, et al, 1986 Appl. Phys. Lett, 48, p986.
Van Zeghbroeck B J, et al, 1987a IEEE Elect. Dev. Lett, EDL-8, p118. See also Proc. IEDM, Los Angeles 1986, p832.
Van Zeghbroeck B J, et al, 1987b to be presented IEDM 87, Washington, (Dec 1987).

GaAs-gate field effect transistor utilizing self-aligned diffusion

A.T. Yuen, E.P. Zucker, E.L. Hu, S.I. Long, D. Hirschnitz *

Department of Electrical and Computer Engineering, University of California,
Santa Barbara, CA. 93106
*Hewlett Packard Company, High Speed Device Laboratory,
Palo Alto, CA. 94301

Abstract: A semiconductor-"insulator"-semiconductor field effect transistor (SISFET) is fabricated for the first time using a self-aligned diffusion technique. No ion implantation was used to process the SISFETs, which demonstrates the feasibility of rapid Si diffusion and resulting interface grading in fabricating MBE grown heterostructure devices. Fabricated SISFETs showed very little light sensitivity as well as the absence of hysteresis in the I-V characteristics. Backgating effects were minimal for backgate biases up to -10 V.

1. INTRODUCTION

Currently, several laboratories are working on SISFET devices of similar design [Matsumoto et al. 1984, Solomon et al. 1984]. Almost all device processes employ a self-aligned implantation. We report the first fabrication of a SISFET using a self-aligned Si-diffusion, with no ion implantation. The technique incorporates a novel process which utilizes a sputtered silicon film to simultaneously dope the source and drain regions, and to grade the GaAs/AlGaAs interface. This is accomplished with little damage to the critical GaAs/AlGaAs interface in the vicinity of the two-dimensional electron gas (2DEG).

Impurity-enhanced disordering (IED) or interface grading, has been previously described, and has been widely utilized in the fabrication of optoelectronic devices [Ishida et al. 1986, Deppe et al. 1987a]. However, there has been no previous report of its use in high speed electrical devices of small dimensions. IED can be accomplished using a variety of implanted species [Hirayama et al. 1985]. Silicon is one such species; in this case, IED is a natural consequence of the rapid or *silicon pair* diffusion [Greiner et al. 1984, Laidig et al. 1981, Kobayashi et al. 1986a]. Loosely coupled Si atoms residing at Ga or As sites, respectively, rapidly diffuse through the GaAs substrate. Pair diffusion is expected to predominate over single Si diffusion for Si concentrations exceeding ~ $3x10^{18}$ cm^{-3}. The IED is therefore selective, occurring in regions that are heavily doped with silicon, and absent for conventional implant doses, such as ~ $2x10^{13}$ cm^{-2}. Studies on the IED of AlGaAs/GaAs superlattices, have shown a diminished homogenization of the material as the Si implantation dose was lowered from $1x10^{16}$ cm^{-2} to $3x10^{14}$ cm^{-2} [Matsui et al. 1986]. This selectivity allows us to devise a process that will permit adequate doping of the source and drain regions of a SISFET, while at the same time grading the AlGaAs/GaAs interface. Feuer has suggested that such a grading of the heterostructure interface, or elimination of the AlGaAs/GaAs conduction band discontinuity, would lead to reduced parasitic source to gate resistance [Feuer 1985].

2. DEVICE STRUCTURE AND FABRICATION

The initial MBE grown structure starts with a semi-insulating LEC GaAs substrate. Subsequent layers are: 1 μm unintentionally doped (u.i.d.) GaAs buffer layer, 700 Å u.i.d. $Al_{0.5}Ga_{0.5}As$ barrier layer, 3000 Å n^+ GaAs gate layer, followed finally by a unconventional n^- GaAs *diode* layer. The top MBE layer, n^- GaAs diode layer, was used to decrease the gate conduction [Yang et al. 1986].

The self-aligned diffusion process flow is provided in Figure 1. Mesa isolation is done using a H_3PO_4:H_2O_2:H_2O etch down to the u.i.d. GaAs buffer layer. Next, a Ti (2500 Å)/ Ni (100Å) gate contact is deposited. The Ti/Ni acts as the mask for the following wet etch, which undercuts the Ti/Ni as it etches down to the u.i.d. $Al_{0.5}Ga_{0.5}As$ barrier layer. Before removing the photoresist, a self-aligned Si e-beam evaporation is done [Omura et al. 1987]. A 3000 Å thick layer of conformal PECVD SiO_2 is then deposited on the entire sample [14]. Next, the sample is loaded into a rapid thermal annealing system and annealed at 1000 °C for ~3 sec. [Greiner et al. 1984, Kobayashi et al. 1987b]. In addition to serving as a cap for the RTA of the device, the SiO_2 further enhances the diffusion of the underlying silicon in the source/drain regions into the GaAs/AlGaAs material [Deppe et al. 1986b]. The PECVD and the residual Si layer on the sample surface are then removed in a CF_4 (90%) and O_2 (10%) plasma etch. The final step is a conventional Au/Ge/Ni ohmic contact deposition and alloy (450 °C).

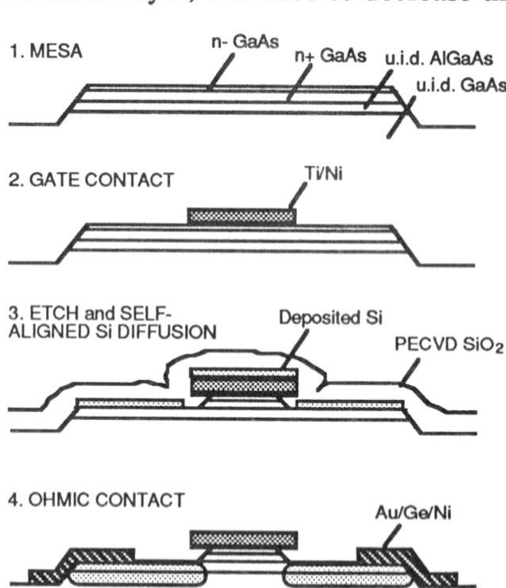

Fig. 1 Self-aligned diffusion process.

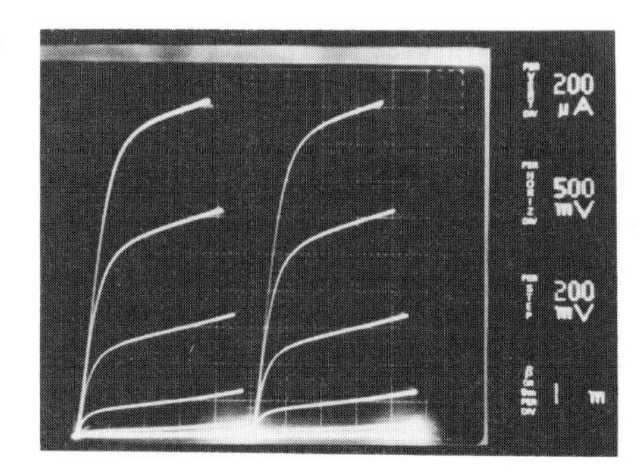

Fig. 2 I-V characteristic of a 1x50 μm^2 SISFET at 300 K (V_{Gmax} = +1.7 V).

3. DEVICE RESULTS

Figure 2 shows the obtained current-voltage characteristic of a 1x50 μm^2 SISFET fabricated using the self-aligned diffusion process (both with and without illumination, left and right traces, respectively, are provided). Note the small amount of gate conduction (near $V_{DS} = 0$ V) of the SISFET for gate-biases as high as +1.7 V. Also, even after the high temperature cycle, the SISFET displays little light sensitivity as well as no hysteresis in the I-V characteristic. This would suggest that minimal diffusion occurred for the Si in the n^+ GaAs-gate into the u.i.d. $Al_{0.5}Ga_{0.5}As$. If substantial Si diffusion had occurred, there would be a large increase in deep level traps, due to unintentional Si doping of the high Al content $Al_{0.5}Ga_{0.5}As$ barrier layer. This would show up as persistent photoconductivity or increased hysteresis in the current-voltage traces, due to the formation of DX centers.

All the fabricated devices exhibited threshold voltages that were positive (~ 0.7 V). This is due to the deterioration of the coupling diode on top of the SISFET. After the RTA, test patterns of Ti/GaAs Schottky diodes showed very large turn on voltages. Therefore, most of the forward voltage on the complete SISFET is initially dropped across the Schottky diode until the diode turns on. After turning-on the Schottky diode, further increases in the gate voltage are shared between the diode and the intrinsic SISFET. This positive threshold voltage also implies that little diffusion of the $Al_{0.5}Ga_{0.5}As$ barrier layer took place, because Si doping of the barrier layer would tend to result in depletion-mode SISFET devices [Barratte et al. 1986].

At room temperature, the on-resistance (R_{ON}) was 11.8 Ω-mm, the output-resistance (R_{DS}) was 411 Ω-mm and the source resistance (R_S) was ~ 4 Ω-mm. The transconductance increases from a room temperature value of 78 mS/mm to 108 mS/mm at 77 K. The low values for Gm is due to the high source resistance as well as the sharing of the gate voltage with the coupling diode. The source resistance can be substantially reduced by increasing the temperature-time cycle of the RTA in order to get deeper Si diffusion which would decrease the sheet resistance. Higher temperature cycles were not tried with the Ti/Ni system, because the current conditions were as high as the Ti/Ni could withstand. The use of a silicide (i.e. WSi_x) system would solve this problem.

Finally, backgating characteristics were measured with a sidegate contact spaced 20 μm from the SISFET. Figure 3 shows the drain current (I_D) and backgate current (I_{BG}) versus gate bias (V_{GS}) for backgate voltages from 0 to -10 V (with illumination). As can be seen, there is a negligible shift in I_D as well as the threshold voltage (linearly extra-polated from I_D vs V_{GS}) with increasing backgate biases.

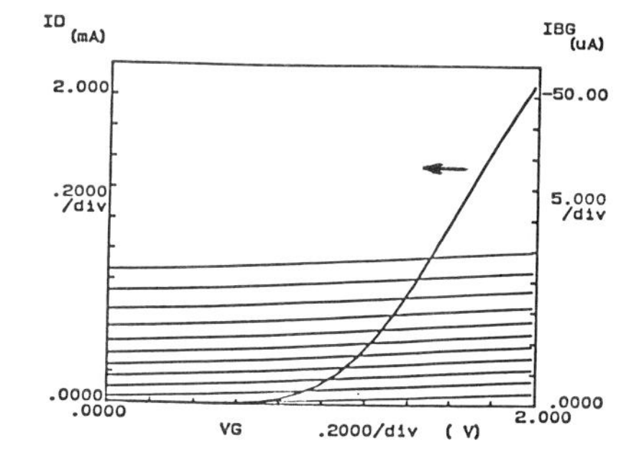

Fig. 3 I-V characteristic showing the effect of the backgate bias from 0 to -10 V.

4. CONCLUSIONS

We have demonstrated the first use of a self-aligned Si diffusion process in the fabrication of SISFETs. The high temperature-time diffusion cycle did not lead to detectable mixing of the

n^+ GaAs/u.i.d. AlGaAs interface which is critical for 2DEG conduction. The SISFETs exhibited very little light sensitivity as well as the absence of hysteresis in the I-V characteristics. The backgating effect was found to be minimal for backgate voltages from 0 to -10 V. Further optimization of the process parameters will enable the SISFET device to be appropriate for LSI or eventual VLSI applications.

5. ACKNOWLEDGEMENTS

The authors are grateful to G.L. Snider and L. Yang for fruitful discussions, G. A. Patterson, M. Lightner and L. Strouse for the MBE growth, S. Corzine and A. Vawter for useful information on Si diffusion, J. Bangs for helpful suggestions on Si deposition and A. Stromberg and P. Shor for continual encouragement.

Work is being supported by a grant from the Semiconductor Research Corporation under Contract SRC-84-01-044.

6. REFERENCES

Baratte, H., LaTulipe, D.C., Knoedler, C.M., Jackson, T.N., Frank, D.J., Solomon, P.M., and Wright, S.L., 1986, IEEE International Electron Devices Meeting Digest, Los Angeles, pp. 444-447.

Deppe, D.G., Jackson, G.S., Holonyak, N., Jr., Burnham, R.D., and Thornton, R.L.,1987a Appl. Phys. Lett., 50, pp. 632-634.

Deppe, D.G., Guido, L.J., Holonyak, N., Hsieh, K.C., Burnham, R.D., Thornton, R.L., and Paoli, T.L., 1986b, Appl. Phys. Lett., 49 (9), pp. 510-512.

Feuer, M.D., 1985, IEEE Trans. Electron Devices, ED-32, pp. 7-11.

Greiner, M.E., and Gibbons, J.F., 1984, Appl. Phys. Lett., Vol. 44, pp. 750.

Hirayama, Y., Suzuki, Y., and Okamoto, H., 1985, Jpn. J. Appl. Phys., Vol. 24, pp. 1498-1502.

Ishida, K., Matsui, K., Fukunaga, T., Takamori, T., and Nakashima, H., 1986, Jpn. J. Appl. Phys., Vol. 25, pp. L690-L692.

Kobayashi, J., Nakajima, M., Bamba, Y., Fukunaga, T., Matsui, K., Ishida, K., and Nakashima, H., 1986a, Jpn. J. Appl. Phys., Vol. 25, No. 5, pp. L183-L185.

Kobayashi, J., Fukunaga, T., Ishida, K., and Nakashima, H., Flood, J.D., Bahir, G., Merz, J.L., 1987b, Appl. Phys. Lett. 50 (9), pp. 519-521.

Laidig, W.D., Holonyak, N., Jr., Camras, M.D., Hess, K., Coleman, J.J., Dapkus, P.D., and Bardeen, J., 1981, Appl. Phys. Lett., Vol. 38, No. 10, pp. 777-778.

Matsui, K., Kobayashi, J., Fukunaga, T., Ishida, K., and Nakashima, H., 1986, Jpn. J. Appl. Phys., Vol. 25, pp. L651-L653.

Matsumoto, K., Ogura, M., Wada, T., Hashizume, N., and Hayashi, Y., 1984, Electron. Lett., 20, pp. 462-463.

Omura, E., Wu, X.S., Vawter, G.A., Hu, E.L., Coldren, L.A., and Merz, J.L., 1987, Appl. Phys. Lett., 50, pp. 265-266.

Solomon, P.M., Knoedler, C.M., and Wright, S.L., 1984, IEEE Electron Device Lett., EDL-5, pp. 379-381.

Yang, L., Yuen, A.T., and Long, S.I., 1986, IEEE Electron Device Lett., EDL-7, pp. 145-148.

Fully self-aligned shallow implanted GaAs MESFET

V Graf, R F Broom, P Buchmann, Th Forster, W Heuberger, G Sasso, P Vettiger and P Wolf

IBM Research, Zurich Research Laboratory, 8803 Rüschlikon, Switzerland

ABSTRACT: We have used successfully a new technique for the fabrication of fully self-aligned MESFETs using a high-temperature-stable dummy T-gate as a mask for implantation as well as for ohmic and gate metallization. The T-gate consists of a trilevel SiO_x or SiN_x PE-CVD sandwich and withstands annealing temperatures as high as 975°C. This new process reduces parasitics and allows a reduction in cell size by a factor of 2 in comparison to the well-known SAINT process. The best MESFETs, having gate lengths of 0.7 μm, exhibit excellent g_m and K-factors of 400 mS/mm and 430 mS/Vmm, respectively. The high values are due to the reduction of the source-gate series resistance in the fully self-aligned structure.

1. INTRODUCTION

Improvements in high-speed GaAs integrated circuits require the reduction of lateral as well as vertical dimensions, by advanced lithography techniques in combination with self-aligned structures and shallow implantation. Of the many implanted MESFET processes invented and developed for LSI/VLSI GaAs IC's, the "SAINT" process (Yamasaki *et al.* 1982) is the most advanced example of a dummy-gate approach. To achieve high performance, self-alignment between different process steps is important for the reduction of parasitics and MESFET cell size. The SAINT process allows self-alignment between the contact implantation and gate deposition. The separation between gate

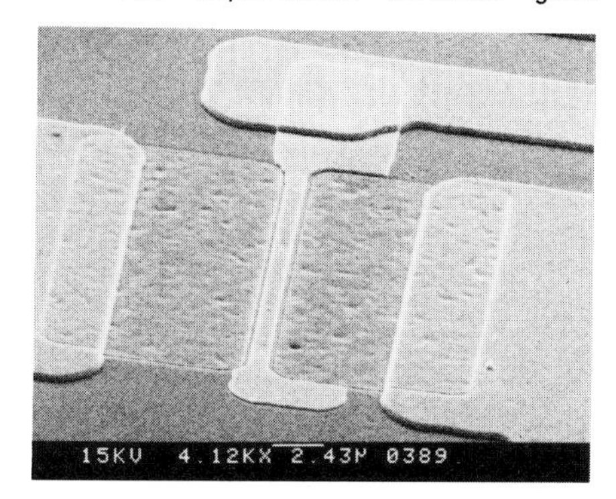

Figure 1. SEM photographs of dummy T-gate and the fully self-aligned MESFET structure

and n^+ region is achieved by a photoresist dummy T-structure with an undercut of 0.2 μm. Gate capacitance can be further reduced by self-alignment between gate metallization edge and n^+ implant region (Terada *et al.* 1983, Enoki *et al.* 1986). However this is not possible with the "standard" SAINT process, because the photoresist dummy gate must be removed before annealing. Self-alignment between gate edge and ohmic contacts would help to reduce parasitic series resistance.

The new technique described here allows full self-alignment between gate edge, n^+ region and ohmic contacts by use of a high-temperature-stable and chemically inert T-structure, which permits post-implantation annealing with the dummy gate in place.

2. MESFET PROCESS

Figure 1 shows SEM photographs of the new MESFET structure. The process is as follows (see Figure 2): First, the n channels, as thin as 40 nm, are formed by direct implantation of 20 keV SiF(47) ions into the undoped semi-insulating LEC GaAs wafers masked by standard optical lithography. In addition, buried p-layers are implanted (Mg at 80 keV, 5×10^{11} cm^{-2}) under the channel areas. Next, a trilevel $SiO/SiO_2/SiO$ sandwich is deposited by PE-CVD and structured by photoresist and RIE. Then the dummy T-gate is formed by taking advantage of the etch-rate selectivity between SiO and SiO_2 in BHF which yields a controlled undercut of 0.2 μm. Also the through-cap n^+ contact implants are scaled to lower implant energies (Si at 80 keV, 1×10^{14} cm^{-2}) and with the top SiO layer defining the implantation [Figure 2(a)]. After rapid thermal annealing for 10 sec at 950°C, vias are etched into the SiO capping layer by a CF_4 RIE step. Then the T-structure is used for a self-aligned ohmic (AuGeNi) contact metallization by evaporation and alloying at 430°C for 40 sec in N_2/H_2 [Figure 2(b)]. As a next step, the bottom part of the T-structure has to be transferred into a lift-off mask for the gate metallization for full self-alignment between contact implant, ohmic contacts and gate. This requires embedding and planarization of the T-structure with Polyimide/resist and O_2-RIE [Figure 2(c)]. The dummy gate is then removed by sputter etching of the ohmic metals, by wet etching the top SiO and the SiO_2 layers and by CF_4-RIE of the bottom SiO layer for gate definition. Finally,

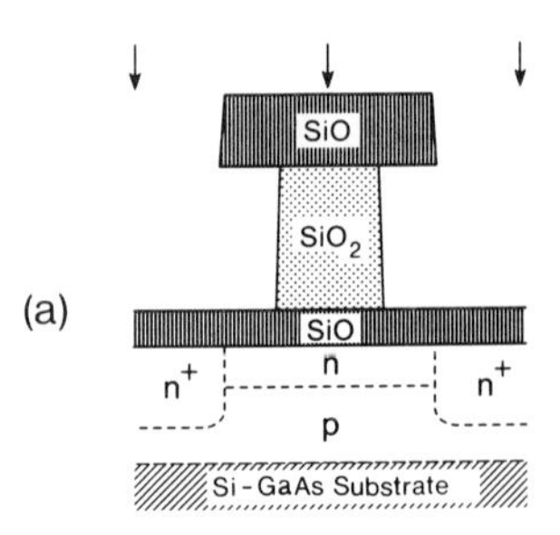

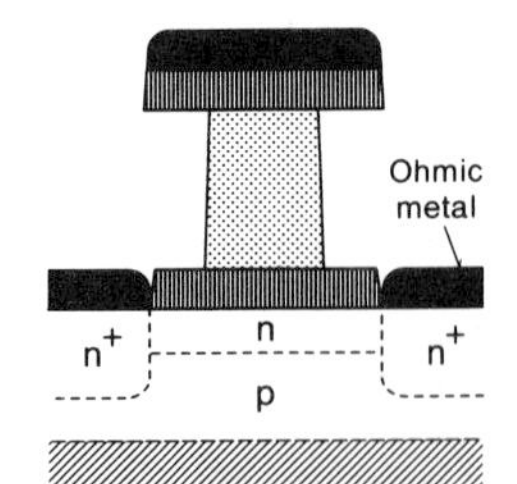

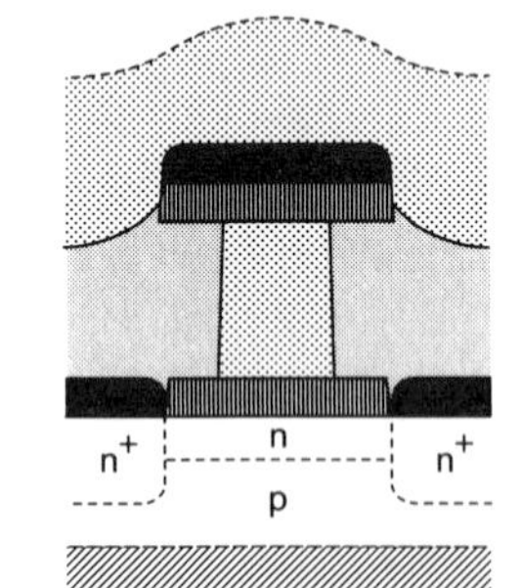

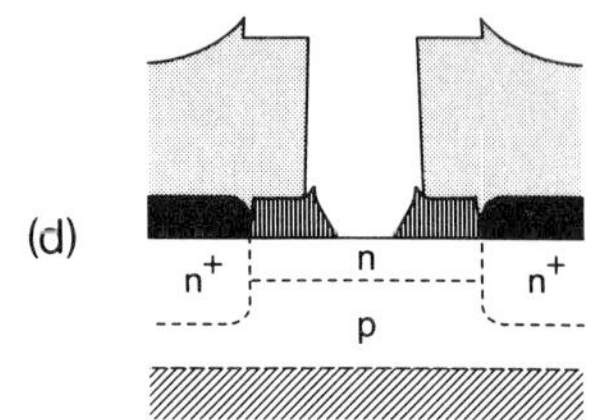

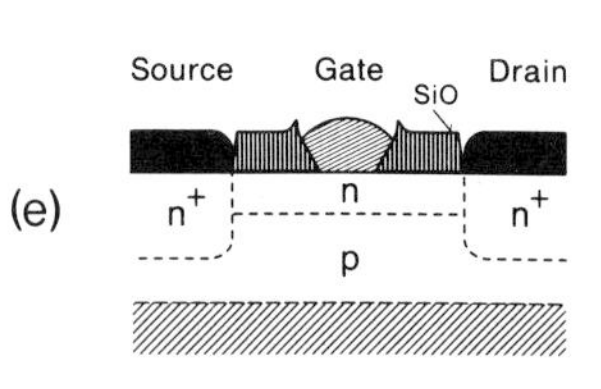

Figure 2. Process flow (see text)

a perfect lift-off mask with slight overhang is achieved owing to the sloping sidewalls of the undercut part of the original dummy T-gate. Submicron gate length (0.6 μm) is obtained with a one-micron lithography process because of the 0.2 μm undercut on each side of the T-structure.

3. RESULTS

The doping-concentration and drift-mobility profiles, obtained by capacitance-voltage measurements of a large area MESFET of length and width 135 and 230 μm, respectively, are shown in Figure 3.

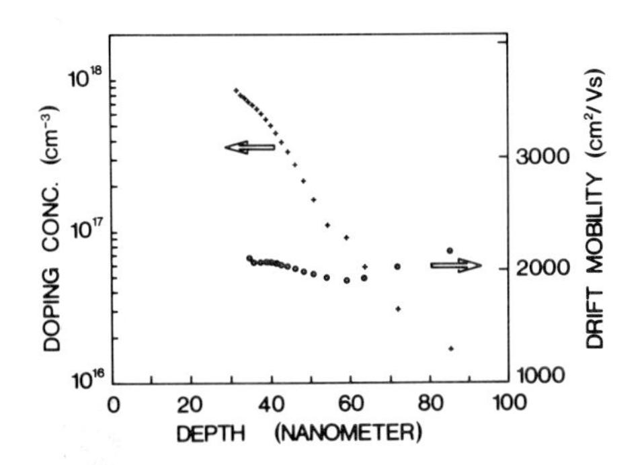

Figure 3. Doping-concentration and drift-mobility profiles for an implantation dosage of 3.17×10^{13} cm^{-2} with SiF at 20 keV

Curve fitting of the doping profile in the region above 10^{17} cm^{-3} leads to approximate values of the implantation range and straggle of 12 and 20 nm, respectively. This corresponds to a free-electron concentration of 4.8×10^{12} cm^{-2} and hence to an activation efficiency of 15%. The maximum free-electron concentration in the channel is estimated to be 1.2-1.4×10^{18} cm^{-3}. The mean drift mobility of 2000 cm^2/Vs is about 2/3 of that expected for bulk material, the reason being the fairly high concentration of neutral Si atoms. The drain-current versus drain-voltage characteristics of a MESFET having a gate length of 0.7 μm and width 15 μm is shown in Figure 4. The gate voltage V_g is varied between -1 and 0.6 V in steps of 0.2 V. At V_g = 0, g_m and g_d are 330 and 16.7 mS/mm, respectively. The ratio g_m/g_d = 20 is quite high for the 0.7 μm gate length owing to the shallow implantation depth of the channel and the buried p-layer.

Figure 5 shows the K-factor and g_m of one of the best 0.7 μm gate-length MESFETs as a function of V_g. The maximum g_m of 417 mS/mm is comparable to that obtained by Yamasaki *et al.* (1985) for MESFETs having a gate length of 0.6

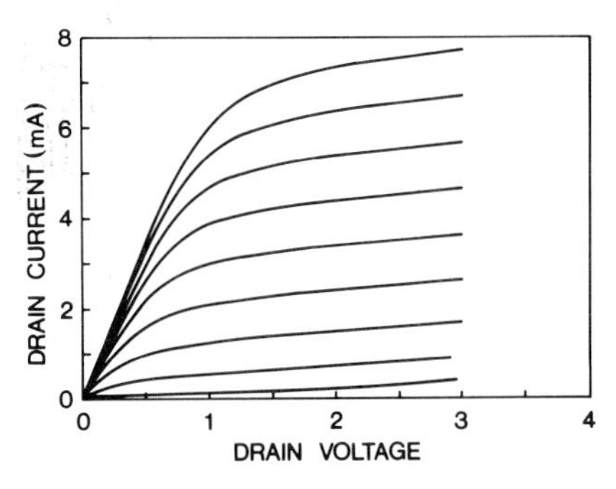

Figure 4. Drain current versus drain voltage for a MESFET of gate length 0.7 μm and width 15 μm. V_g is varied between -1 and 0.6 V in steps of 0.2 V. The channel-implantation conditions are the same as in Figure 3

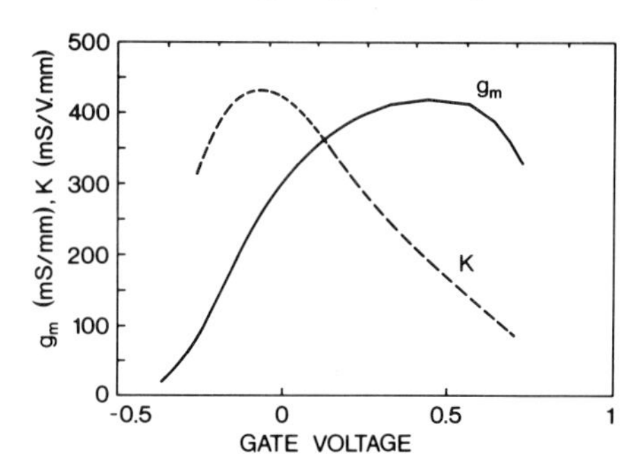

Figure 5. K-factor and g_m versus V_g. The gate length is 0.7 μm and the width 10 μm. The channel-implantation dose is 2.15×10^{13} cm^{-2} at 20 keV

μm, fabricated by the buried p-layer "SAINT" process. The maximum K of 430 mS/Vmm is very high and exceeds the best value obtained from non-fully self-aligned MESFETs (4 μm separation between gate and contact metals) by a factor of 1.7.

As to be expected from a process with no gate overlap, the measured drain-gate capacitance with $C_{dg} = 0.3...0.4$ pF/mm is smaller by about 0.25 pF/mm than in our previous FET's made with a SAINT-like process and a gate overlap of 2 μm. However, this is still much larger than the present theoretical value of about 0.1 pF/mm, for unknown reasons. The measured value of the transit frequency $f_t = 18$ GHz for $L = 0.9$ μm is high, despite the high value of C_{dg}. However, owing to the highly doped, shallow channels the gate capacitance is high, too, so that parasitic effects have less influence. For $L = 2$ μm, $V_g = 0$ V and $V_d = 2$ V we measured $C_g = 4.3$ pF/mm, and for $L = 0.9$ μm we found $C_g = 2.2$ pF/mm.

4. CONCLUSION

We developed a fully self-aligned single-mask process for defining contact implant, ohmic contacts and gate of a GaAs MESFET structure (Buchmann *et al.* 1986). The fully self-alignment reduces parasitic series resistance, gate capacitance and MESFET cell size considerably. Depending on the overlay capability of the optical lithography system, cell-size reduction between a factor of 2 to 3 can be achieved with the process described compared to the "standard" SAINT stable T-structure. Submicron gate length (0.6 μm) is obtained with a one-micron lithography process owing to the 0.2 μm undercut on each side of the T-structure.

ACKNOWLEDGEMENT

The authors would like to acknowledge the cooperation and stimulating discussions with many members of the device department at the IBM Zurich Research Laboratory. In particular, the contributions of H. P. Dietrich and W. Walter for metallization, and W. Bucher for measurements are greatly appreciated.

REFERENCES

Buchmann P, Graf V, Mohr Th O and Vettiger P 1986 *Microcircuit Engineering* **5** 395
Enoki T, Yamasaki K, Osafune K and Ohwada K 1986 *Electron. Lett.* **22** 68
Terada T, Kitaura Y, Mizoguchi T, Mochizuki M, Toyoda N and Hojo A 1983 *Proc. GaAs IC Symposium* (IEEE: Technical Digest) pp 138-140
Yamasaki K, Asai K, Mizutani T and Kurumada K 1982 *Electron. Lett.* **18** 119
Yamasaki K, Kato N and Hirayama M 1985 *IEEE Trans. Electron Devices* **ED-32** 2420

Step doped HEMT structure for e–d logic and MMIC applications

R.K. Surridge, T. Lester, A.J. SpringThorpe, P. Mandeville, C. Miner, and D.J. Day

Bell-Northern Research, P.O. Box 3511, Station C,
Ottawa, Ontario, Canada K1Y 4H7

ABSTRACT: We have investigated a step doped HEMT structure which facilitates the simple fabrication of devices with two pinch-off voltages, using a single gate recess. DC, S-parameter low frequency noise and ring oscillator measurements performed on HEMTs made with this structure show that devices with excellent characteristics may be readily fabricated for use in both logic and microwave integrated circuits.

I INTRODUCTION

High electron mobility transistors (HEMTs) have been shown to offer performance advantages over MESFETs in both high speed logic and high frequency microwave applications. For example, Sheng et al have fabricated high speed 1kb SRAMs, and Smith et al have demonstrated HEMT amplifiers operating at 94GHz. To optimise circuit performance, it is often necessary to fabricate devices with two pinch-off voltages on the same chip; for example in enhancement-depletion (e-d) logic circuits, and circuits with integrated logic and analog components. Pinch-off voltages of +0.2 and -1.0 volts are typical requirements for e-d logic circuits, whereas -1.5 and -3.5 volts would be realistic for a mixed d-mode logic/microwave amplifier circuit. In the latter case a higher drain breakdown voltage than is commonly observed with HEMTs would also be advantageous.

We have investigated a step doped HEMT structure which facilitates the simple fabrication of devices with two pinch-off voltages on the same wafer, variation of pinch-off voltage being achieved by recessing one of the gates. The structure, which is similar to that used by Lin et al, allows the use of a selective etch to accurately determine the depth of the gate recess. An additional advantage of the step doped structure is that both the recessed and non-recessed gates are situated on relatively low doped material, and the devices consequently display high breakdown voltages.

II MATERIAL STRUCTURE

The step doped structure is compared with that of a 'standard' HEMT in Figure 1. The most significant difference is the use of n^- GaAs cap and AℓGaAs etch stop layer in the step doped case. The thicknesses of the cap, etch stop and doping layer are adjusted to achieve the required combination of pinch-off voltages for devices with gates located at the surface of the cap layer or recessed to the cap/etch stop layer interface. Also, by reducing the doping level in the cap and etch stop layers, shunt conductance effects are minimised.

Material was grown by MBE using an As_4 flux, which in our case gives a p^- background. The aluminum mole fraction in the AℓGaAs was limited to 21%, to minimise the occupancy of DX centers. Prior to the growth of the active layers, a GaAs/AℓAs super lattice was grown to improve material quality.

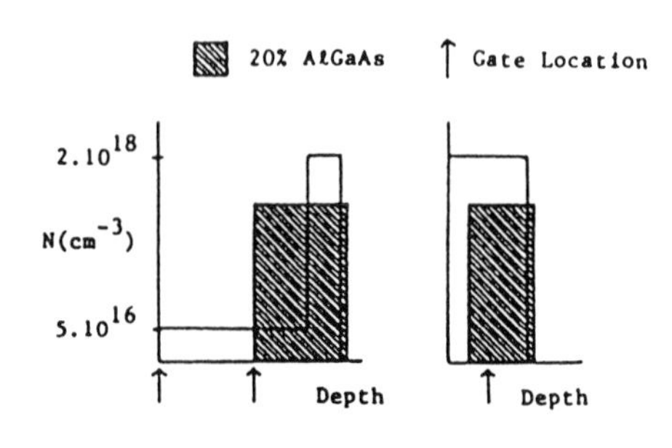

Fig. 1 Comparison of Depth Profiles of Step Doped and Standard HEMTs

III DEVICE FABRICATION

All devices were fabricated using a mask set which allows a comprehensive range of I-V, C-V, S-parameter and ring-oscillator measurements to be performed. The fabrication process used is entirely compatible with our standard integrated circuit process, and includes selective n+ implantation of the source and drain regions which is activated by rapid thermal annealing. Ni/Ge/Au ohmic contacts, and Ti/Pt/Au Schottky contacts were deposited using a dielectric assisted lift-off technique. A bi-layer resist was used to pattern the Si_3N_4 (deposited in a downstream microwave plasma system) which was then removed locally with a CF_4 plasma. A thin sputtered AℓN film deposited below the Si_3N_4 was used to isolate the GaAs from the plasma, to avoid carrier removal by H^+ ions, and was subsequently removed by wet chemical etching.

IV DISCUSSION OF RESULTS

Hall measurements on Van der Pauw structures at room temperature and 77K show that the 2D electron concentration is approximately $1.10^{12}cm^{-2}$, with Hall mobilities of 7000 and 50,000 at 300 and 77K respectively. Because of this relatively low concentration, which results from the small conduction band discontinuity in the $Al_{0.2}Ga_{0.8}As$/GaAs system, it is necessary to locate the 2D gas relatively deep in order to achieve high pinch-off voltages. This results in comparitively low transconductances; as shown in Figure 2 for a non-recessed and a recessed device fabricated on a structure with a 1000Å cap, 500Å etch stop and 100Å, $2.10^{18}cm^{-3}$ doping layer. The I-V curves of devices with 1μm gate lengths and 200μm total gate widths (Figure 2a, 2b) display low output conductances, high breakdown voltages, no hysteresis and almost no sensitivity to light. By reducing layer thicknesses, devices with lower pinch-off voltages (to +0.4V) and higher transconductances have also been fabricated.

The two-port S-parameters measured for the above devices at V_{DS} = +3V and V_{gs} = -0.6 and -2V for the recessed and non-recessed devices respectively, are shown in Figure 3. These measurements were performed using an HP8510 network analyser in conjunction with coplanar waveguide probes fabricated at BNR. The device parameters derived for a simplified

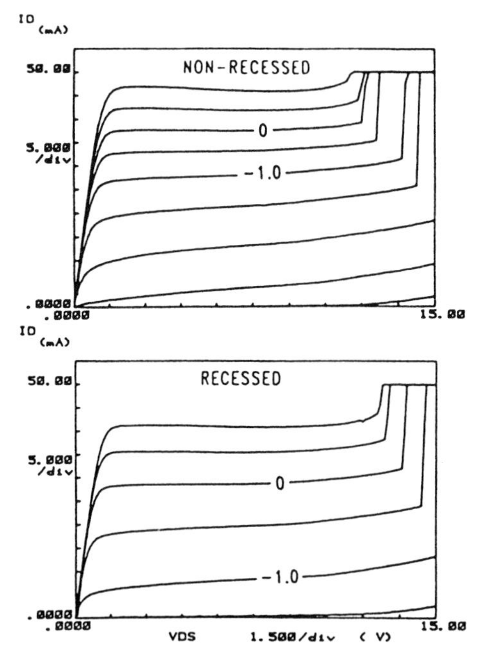

Fig. 2a, b. I-V Curves of Non-recessed and Recessed devices on same wafer. ΔV_{gs} = 0.5

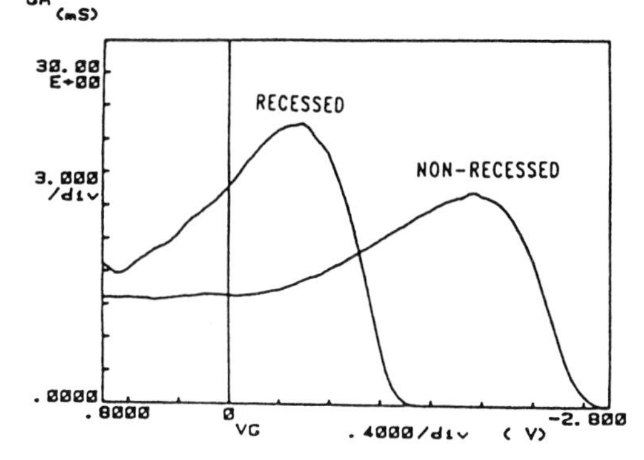

Fig. 2c. Variation of Transconductance with Gate Voltage

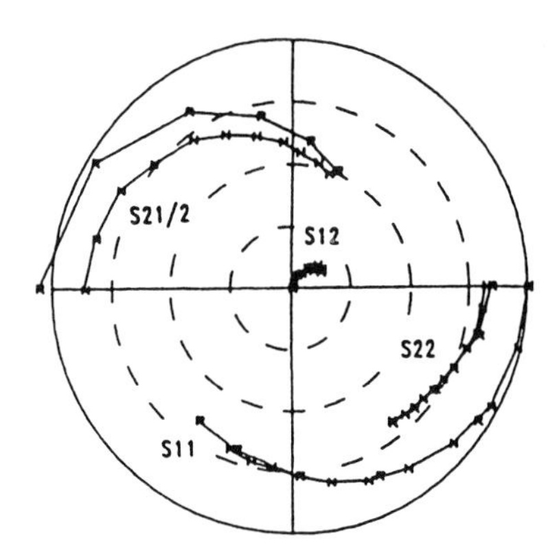

Fig. 3a. Measured S-Parameters of a Recessed (R) and Non-Recessed (N) HEMT on the Same Slice

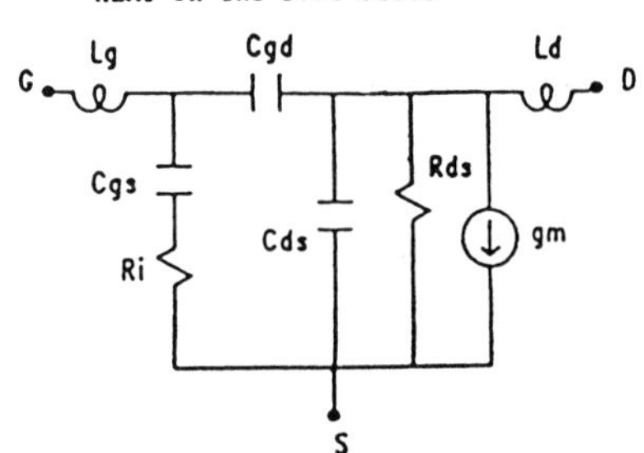

Fig. 3b. Simplified Equivalent Circuit of HEMT

	Non-Recessed	Recessed
C_{gs}	172 fF	208 fF
C_{gd}	25 fF	25 fF
C_{ds}	44 fF	44 fF
R_i	16.2	16.5
R_{ds}	488	586
g_m	19.2 mS	22.6 mS
τ	2.5 pS	2 pS
L_g	.03 nH	.03 nH
L_d	.06 nH	.06 nH

Table 1 Comparison of Equivalent Circuit Parameters of Non-Recessed and Recessed HEMTs on Same Wafer, Determined at Optimum Bias Level

equivalent circuit (Figure 3b) are compared in Table 1. The f_T of these devices, determined from the frequency at which h_{21} = 1, shows a significant peak, with a maximum of ~16GHz at the gate voltage where g_m is also a maximum. This arises because the depletion region is effectively pinned at the 2D gas, giving a relatively constant gate-source capacitance, whereas the transconductance varies significantly with gate voltage.

Whilst the insensitivity of C_{gs} to gate voltage ensures that a good circuit match may be maintained over a range of bias conditions, the variation of f_T with gate bias has serious implications regarding the performance of HEMT microwave circuits operating at non-optimum bias conditions.

Two types of 1μm gate length ring oscillator were included in the mask set, a direct coupled (DCFL) design using resistive loads, and a capacitor coupled (CCFL) design using active loads. Because of the dependance of f_T upon gate bias voltage, ring oscillator results only give an accurate prediction of minimum gate delay if the peak transconductance occurs at a slightly positive bias (DCFL design) or zero bias (CCFL design). To date the lowest gate delay we have measured using the step doped structure is 50 ps for enhancement devices using the DCFL ring oscillators.

Capacitance and conductance DLTS measurements on 1μm gate length devices and FATFETs show the presence of two deep levels at ~800meV and ~200meV, the 800meV level predominating with the channel undepleted, and the 200meV level predominating close to pinch-off.

The low frequency current noise measured in step doped devices is significantly less than we have observed in standard HEMTs. We propose that this is due to two effects. Firstly, in the step doped devices there is a reduction in g-r noise (α $1/f^2$) which we associate with decreased trapping in the AlGaAs due to lower shunt conductance in the step doped case. Secondly we have observed a reduction in diffusion related g-r noise (α $1/f^{3/2}$). This noise predominates in all the HEMTs and MESFETs we have measured, when biased close to pinch-off. We have also shown that the magnitude of this noise scales with transconductance, and so the reduced levels present in step doped devices is consistent with the lower transconductance already described in this paper.

V CONCLUSION

We have investigated a step doped structure which is suitable for the co-fabrication of HEMTs with two breakdown voltages. These devices display high breakdown voltages, high f_Ts, low propagation delays and low noise, making them suitable for both microwave circuit and logic applications. The ability to tailor the pinch-off voltages makes them particularly suitable for e-d logic circuits.

VI ACKNOWLEDGEMENTS

The authors wish to thank P. Jay and R. Streater for their support and encouragement, and members of the Advanced Technology Department of BNR, particularly N. Bonneau, D. Kelly, R. Bruce and B. MacLaurin for their assistance in device fabrication and measurement.

This work was financed in part by the National Research Council of Canada through the Program for Industy/Laboratory Projects.

VII REFERENCES

Lin B J, Kofol S, Kocot C, Luechinger H, Miller J N, Mars D E, White B and Littau E 1986 GaAs Ic Symposium pp 51-53

Sheng N H, Wang H T, Lee C P, Sullivan G J and Miller D L 1987 IEEE Trans. Electron Devices ED-32 pp 1670-1675

Smith P M, Chao P C, Duh K H, Lester L F and Lee B R 1986 Electronic Letts. 22 no 15 pp 780-781

Quartermicron gate inverted HEMT for high speed ICs

H.Inomata Fujishiro, T.Saito, S.Nishi, S.Seki, Y.Sano and K.Kaminishi

Research Laboratory, Oki Electric Industry Co., Ltd.
550-5, Higashiasakawa, Hachioji, Tokyo 193, Japan

ABSTRACT: I-HEMTs with gate lengths ranging from 0.15μm to 1.6μm were fabricated. A new gate fabrication technique (angle evaporation technique) was demonstrated to be effective to realize a quartermicron gate I-HEMT for high speed digital ICs. These I-HEMTs showed the extremely small drain conductances. The confinement of 2DEG by the AlGaAs layer was experimentally confirmed to be sufficient. As a result of good saturation characteristics, an anomalous increase of the drain current was observed at high drain voltages, which was analized to come from the impact ionization in the channel. Extremely uniform threshold voltages were measured for 0.25μm gate I-HEMTs (σV_{th}=9mV, 798FETs). The propagation delays of 11.8 and 9.0ps/gate were obtained even for 0.4μm gate I-HEMTs at R.T. and 110K, respectively, using the 21 stage DCFL ring oscillator.

1. INTRODUCTION

The selectively doped GaAs/AlGaAs heterostructure FETs have been developed for high speed logic applications (Nishi et al 1986, Watanabe et al 1986). At this stage, many attempts are concentrated on scaling down the dimension of FET for further improvements in the characteristics (Matsuoka et al 1986). However, the reduction of the gate length causes short channel effects and these effects make it difficult to control the electrical characteristics of FETs, precisely. We have been studying an inverted HEMT (I-HEMT) in which GaAs exists on top of AlGaAs. In this structure, short channel effects are effectively suppressed, because the 2DEG is confined by the heterobarrier. Moreover, the mobility of 2DEG is high. The source resistance can be easily lowered by the top n^+ layer, because there is no heterobarrier between the surface and the 2DEG. Then the short channel I-HEMT is considered to be suitable for high speed ICs at R.T. We fabricated quartermicron gate I-HEMTs and studied short channel effects in I-HEMTs. In this paper, a new gate fabrication procedure is described and electrical characteristics of very short gate I-HEMTs are shown.

2. EPITAXIAL GROWTH AND FABRICATION PROCEDURE

The inverted heterostructure was grown by a conventional III-V MBE system on a 2 inch Cr-O doped semi-insulating GaAs substrate. Figure 1 shows the schematic cross section of an I-HEMT structure. In this structure, the mobility at 77K and the maximum sheet carrier concentration were about 2×10^4 $cm^2/V\cdot sec$ and $9\times10^{11} cm^{-2}$, respectively. In the previous report, the growth condition and the optimization of the layer structure were described (Nishi et al 1986).

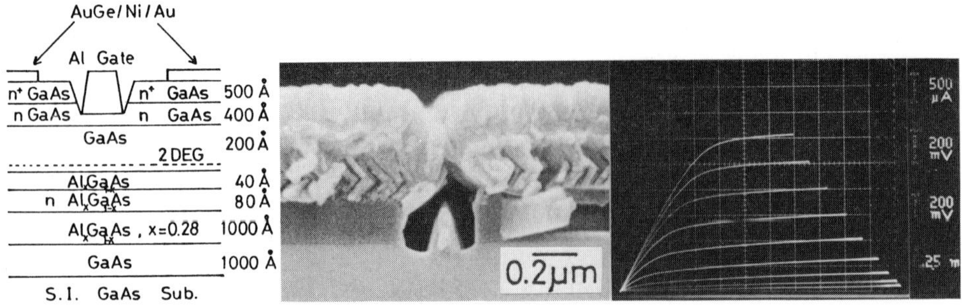

Fig. 1. Schematic cross section of an I-HEMT.

Fig. 2. A SEM photograph of 0.15μm gate I-HEMT.

Fig. 3. I-V characteristics of 0.15μm I-HEMT

I-HEMTs were fabricated using the following processing technique. After O^+ implantation for the device isolation, ohmic electrodes were made by alloying the evaporated AuGe/Ni/Au. The gate electrode was formed by a conventional lift off technique. First, an LMR photoresist was developed by a deep UV contact lithography. A recess etching was performed by a 50eV Ar ion beam. Then the Al metal was evaporated and consequently a self aligned gate electrode was patterned by the lift off. We also developed a new gate fabrication technique, using an angle evaporation technique (Saito et al 1987). In the new process, Ti was evaporated at a small incident angle onto both sides of the mask walls before the gate metal evaporation. By employing the angle evaporation technique, not only a gate length was reduced but also a gate metal was effectively separated from an n+ layer about 0.1μm. Figure 2 shows the cross sectional SEM photograph after the gate metal evaporation. In the figure, 1000Å thick Ti layers could be seen on both mask walls and even the 0.15μm length gate was easily formed with a photoresist opening of about 0.35μm.

3. DEVICE CHARACTERISTICS

I-V characteristics of the 0.15μm gate I-HEMT are shown in fig. 3. In spite of the very short gate length, the drain current saturation characteristics are good enough. The drain conductance, g_d, is only 7.8mS/mm, which is extremely low compared with those ever reported (Matsuoka et al 1986). This result suggests the good suppression of short channel effects in the I-HEMT structure. To confirm the contribution of the AlGaAs layer on this suppression, we fabricated I-HEMTs with various Al contents of AlGaAs layers. Figure 4 shows the dependence of g_d on the gate length at various Al mole fractions, x. These I-HEMTs show the extremely small drain conductances even at x=0.18. The rather high g_d at x=0.18 may be caused by the elec-

Fig. 4. The dependence of g_d on the Al content.

Fig. 5. V_{th}, g_d and g_m versus Lg.

tron injection into the AlGaAs layer because of the rather low hetero-barrier potential. This results suggest that in the I-HEMT structure, the confinement of 2DEG by the AlGaAs layer is sufficient.

In fig. 5, we present the electrical characterisics of I-HEMTs with gate lengths ranging from 0.25μm to 1.6μm. These I-HEMTs were made by a conventional gate fabrication technique without the angle evaporation. As shown in the figure, the K-value and the transconductance, g_m, increase monotonically with reducing the gate length. No reduction of K-value and g_m are observed even at 0.25μm gate length. Even though g_d increases at the gate length of less than about 0.5μm, these values are extremely low. The shift of the threshold voltage due to the reduction of the gate length from 1.6μm to 0.25μm is about -200mV. This value includes the influence of the dependence of the etching depth on the photoresist mask opening, because the Ar ion beam has an angular spread of about 15°. The threshold voltage shift is reduced by employing the angle evaporation technique and/or a thinner photoresist mask. When the angle evaporation technique was employed, the shift of the threshold voltage at the gate length of 0.25μm was only -80mV.

In fig. 6, we present the I-V characteristics of an I-HEMT at high drain voltages. The gate length and the gate width are 0.4μm and 10μm, respectively. As shown in the figure, a kink on the drain current is clealy observed around the drain voltage of 4V. To analyze the mechanism, we examined the temperature dependence of this phenomenon. Figure 7 shows the dependence of g_d on temperature. The kinks are more clearly observed on g_d curves. The kink voltage shifts to lower drain voltages and g_d increases at the high drain voltage region as decreasing the temperature. No abrupt change of the gate leak current was observed within the applied drain voltages. The temperature dependence of the kink voltage suggests the impact ionization mechanism. At the high drain voltage, the electric field at the drain side of the channel reaches to the value high enough to cause the impact ionization which generates electron-hole pairs. The resultant holes are injected into the interface region between the substrate and the epitaxial layer to form the p domain at the source side. Then the substrate is positively biased and the drain current increases. In this mechanism, the substrate potential is governed by the diode characteristics between the source and the substrate. These phenomena were well known in the short channel MOS FET as the parasitic bipolar action (Sun et al 1978). More clear observations were reported in ESFI MOS transistors where the substrates were insulating (Tihanyi et al 1974). In the GaAs FET, the similar experimental result was reported (Zeghbroeck et al 1987). In the I-HEMT, the anomalous increase of the drain current was clearly observed because of the good saturation characteristics at low drain voltages.

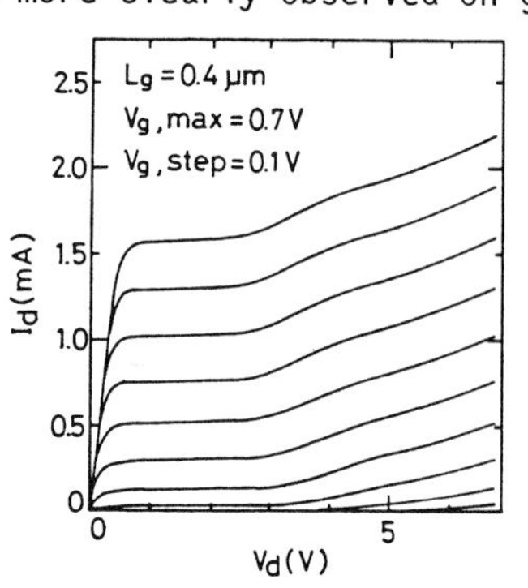

Fig. 6. High drain voltage characteristics at R.T.

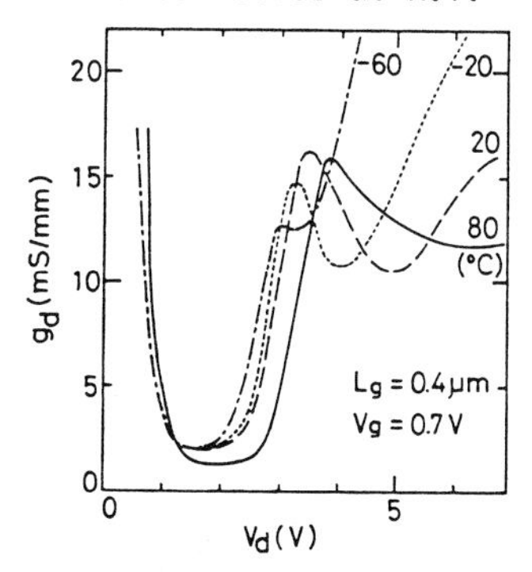

Fig. 7. Temperature dependence of g_d.

4. UNIFORMITY AND DYNAMIC CHARACTERISTICS

For application to the digital IC, the threshold voltage uniformity and the gate propagation delay were evaluated. I-HEMTs were fabricated using the angle evaporation technique. For the uniformity measurements, 0.25μm gate I-HEMTs were used. On a whole area of 2 inch wafer, the standard deviation, σV_{th}, of 33mV was obtained at the mean V_{th} of -159mV. The microscopic uniformity was measured using the 50μmx50μm pitch FET array. The extremely small σV_{th} of 9mV was obtained at the mean V_{th} of -141mV in the area of 0.7mm x 2.85mm (798FETs). The propagation delay of an I-HEMT E/D DCFL circuit was measured at R.T. and 110K using the 21 stage ring oscillator (L_g=0.4μm). The minimum propagation delay of 11.8ps/gate with a power dissipation of 0.83mW/gate was obtained at R.T. At 110K, the minimum propagation delay was reduced to 9.0ps/gate with a power dissipation of 1.06mW/gate. This improvement may be attributed to the increase of g_m. These excellent values were obtained even for 0.4μm gate I-HEMTs.

5. CONCLUSION

Reduction of the gate length and improved FET characteristics are key parameters for high speed digital ICs. The I-HEMT is a strong candidate for this application at R.T. because small short channel effects are expected from the confinement of 2DEG by the AlGaAs layer. In this report, small short channel effects in the I-HEMT were experimentally demonstrated. The drain conductance as low as 7.8mS/mm even for the gate length of 0.15μm was obtained. As the saturation characteristics are good, the anomalous increase of the drain current was clearly observed at the high electric field region, which comes from the impact ionization in the channel. Using a new gate fabrication technique, quartermicron I-HEMTs with extremely uniform threshold voltages were fabricated. Propagation delays of 11.8ps/gate at R.T. and 9.0ps/gate at 110K were obtained even for 0.4μm gate I-HEMTs. From these results, it is confirmed that the I-HEMT with the quartermicron gate length is promising for high speed digital ICs.

ACKNOWLEDGEMENT

The present research effort is part of the National Research and Development program on "Scientific Computing System", conducted under a program set by the Agency of Industrial Science and Technology, Ministry of International Trade and Industry.

REFERENCES

Nishi S, Saito T, Seki S, Sano Y, Inomata H, Itoh T, Akiyama M and Kaminishi K 1986 Proc. 13th Int. Symp. on GaAs and Related compounds P515.
Watanabe Y, Kajii K, Nishiuchi K, Suzuki M, Hanyu I, Kosugi M, Odani K, Shibatomi A, Mimura T, Abe M and Kobayashi M 1986 IEEE ISSCC Tech. Dig. P80.
Matsuoka Y, Sugitani S, Kato N and Yamazaki H 1986 Proc. 13th Int. Symp. on GaAs and Related Compounds P459.
Saito T, Fujishiro H, Nishi S, Sano Y and Kaminishi K 1987 Extended Abstracts 19th Conf. Solid State Devices and Materials Tokyo P267.
Sun E, Moll J, Berger J and Alders B 1978 IEDM Tech. Dig. P478.
Tihanyi J and Schlotterer H 1975 Solid State Electron. 18 309.
Zeghbroeck B J V, Patrick W, Meier H and Vettiger P 1987 IEEE Electron Device Lett. EDL-8 188.

High performance AlGaAs/GaAs/AlGaAs selectively-doped double-heterojunction FET and its application to digital ICs

Katsunori Nishii, Kaoru Inoue, Toshinobu Matsuno, Akitoshi Tezuka
and Takeshi Onuma

Semiconductor Research Center
Matsushita Electric Industrial Co., Ltd.
3-15 Yagumonakamachi, Moriguchi, Osaka 570, Japan

Abstract: A high-performance AlGaAs/GaAs/AlGaAs selectively-doped double-heterojunction FET (SD-DH FET) has been fabricated by optimizing the MBE growth condition and layer structures. The structure showed a sheet electron concentration as high as $3x10^{12}cm^{-2}$ and 77K electron mobility of $40000cm^2/Vs$. The room temperature sheet resistance was reduced to 350 Ω/□ which is less than a half of that for conventional HEMT. The maximum extrinsic transconductance of 500mS/mm was achieved at room temperature for 1μm gate SD-DH FET. Using the SD-DH FETs, 4-bit multiplexer IC operating up to above 2GHz was successfully fabricated for the first time.

1. Introduction

GaAs/AlGaAs selectively-doped single-heterojunction FETs such as HEMT have been widely used for high speed digital ICs and low noise microwave devices in recent years. In order to further improve the device performance, several novel structures of heterojunction FETs have been proposed (Baba et al. 1983). Among them, an approach that uses a selectively-doped double-heterojunction (SD-DH) structure (Inoue et al. 1984) is very promising because this structure offers sheet electron concentration two times higher than conventional single heterojunction structure, which leads to lower source resistance and higher transconductance of an FET. Moreover, the reduction of short channel effect and drainconductance is expected for SD-DH FETs because electrons are confined in a GaAs quantum well.
In order to fully extract the high performance features of SD-DH FETs, it is important to optimize the MBE growth condition and the epitaxial layer structure. We report here successful fabrication of 1μm gate length SD-DH FETs with high transconductance of 500mS/mm at room temperature by using a highly-conductive SD-DH structure. Using the high-performance SD-DH FETs, a 4-bit multiplexer IC composed of source coupled FET logic operating up to above 2GHz was successfully fabricated for the first time.

2. MBE growth of SD-DH structure

The SD-DH structure used in this work is shown in Fig. 1. The structure was grown by MBE on a Cr-doped semi-insulating (100) GaAs substrate.

An epitaxial layer consists of 1000Å undoped GaAs and 2000Å undoped AlGaAs buffer layer, 150Å bottom N-AlGaAs, 50Å undoped AlGaAs bottom spacer layer, 200Å undoped GaAs quantum well, 20Å undoped AlGaAs top spacer layer, 500Å N-AlGaAs and 100Å N-GaAs cap layer. The AlAs mole fraction in AlGaAs layer was 0.3 and the doping level of Si in N-AlGaAs layer was 1.0×10^{18} cm^{-3}. The growth rate of GaAs was 1.0µm/h. The dependence of sheet electron concentration (Ns) and Hall electron mobility (µ) for SD-DH structures on the substrate temperature (Tsub) were studied, the result of which is shown in Fig. 2. One can see in Fig. 2 that high electron mobilities at 77K exceeding 30000cm^2/Vs can be obtained in the limited Tsub region from 530°C to 550°C, which is very similar to the previously reported results (Inoue et al. 1984). In this Tsub region, sheet electron concentration as high as $1.8-2.0\times10^{12}$ cm^{-2} has been achieved with room temperature mobility of 6600cm^2/Vs. However, at low temperature below 530°C, reduction in Ns is observed, which is probably due to the low activation rate of Si in N-AlGaAs layer. The mobility reduction at high temperature above 570°C is considered to be caused by the Si surface segregation effects (Inoue et al. 1985).

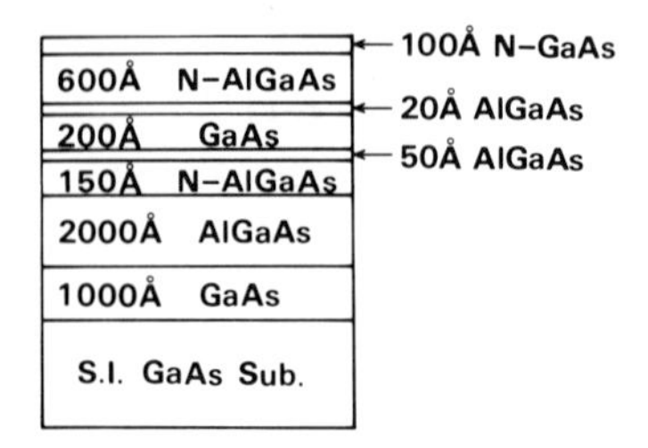

Fig. 1. Sample structure used for MBE growth experiments.

To achieve high-performance SD-DH FETs, the layer structure must be optimized. Thinning of AlGaAs spacer layer is desirable because the transconductance of FETs tends to be reduced with the increase of AlGaAs spacer layer thickness. Figure 3 shows the dependence of Ns and µ on the spacer layer thickness (Wsp). In terms of conductivity, Wsp of 20-50Å was found to be optimum at both room temperature and 77K. Thus, the minimum sheet resistance at room temperature of 350 Ω/□ has been achieved for the structure shown in Fig. 1, with high electron concentration of 3×10^{12} cm^{-2} and 77K electron mobility of 40000cm^2/Vs.

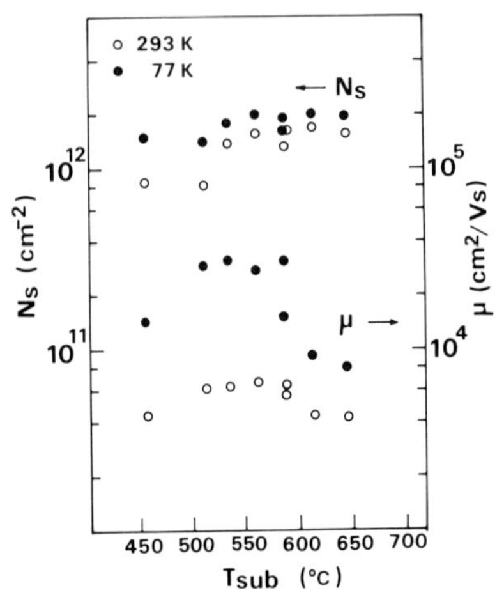

Fig. 2. Dependences of electron concentration (Ns) and Hall mobility (µ) on substrate temperature (Tsub).

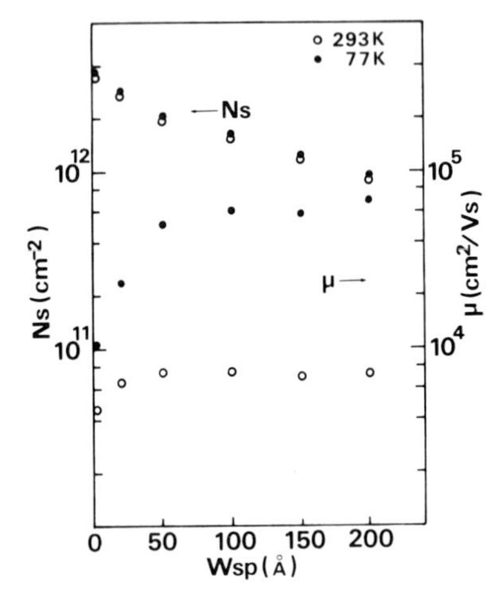

Fig. 3. Dependences of electron concentration (Ns) and Hall mobility (µ) on AlGaAs spacer layer thickness (Wsp).

S G D

100 A N-GaAs
500 A N-AlGaAs
20 A AlGaAs
200 A GaAs
50 A AlGaAs
150 A N-AlGaAs
2000 A AlGaAs
1000 A GaAs
S.I. GaAs Sub.

Fig. 4. Cross section of SD-DH FET

Table I. Sheet electron concentration (Ns) and Hall mobility (μ) for SD-DH structure.

T (K)	Ns (cm^{-2})	μ_H (cm^2/V·sec)
293	2.25×10^{12}	6100
77	2.50×10^{12}	37000

3. Fabrication and characteristics of SD-DH FET

Figure 4 shows the structure of SD-DH FET used for the FET fabrication. The structure is similar to that explained in the section of the MBE growth. The electrical characteristics are listed in Table I. The SD-DH structure had a very high sheet electron concentration of $2.5\times10^{12}cm^{-2}$ with high electron mobility of $37000cm^2$/Vs at 77K. The sheet resistance for SD-DH structure 450 Ω/□ at room temperature, which is about one half of that for conventional HEMT.

The device fabrication process begins with the mesa-etching for the definition of the channel region. Source and drain ohmic contacts were formed by AuGe/Ni/Au. Pd/Ti/Au was used as the gate metal because the threshold voltage control can be easily done by heat treatment after gate metallization. Heat treatment for the threshold voltage control was done at 300°C in Ar atmosphere. Source to drain spacing, gate length and gate width were 2.5μm, 1.0μm and 10μm, respectively.

Figure 5 shows the dependence of maximum transconductance (gm_{Max}) and the gate voltage (Vgsm) that gives gm_{Max} on the threshold voltage (V_T) for SD-DH FET. Initial V_T was -1.5V. With the increase of V_T, monotonical increase of gm_{Max} was observed for V_T up to 0.1V, at which a very high transconductance of 500mS/mm was obtained. The sudden decrease of transconductance for V_T above 0.1V is probably due to the increase of gate leakage current. In fact, the value of Vgsm becomes almost constant in the corresponding V_T region. The dotted lines show the calculated extrinsic transconductance which are in good agreement with the experiment. The drain

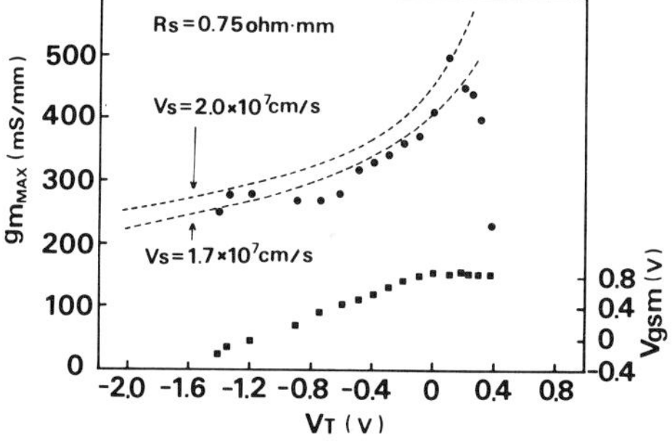

Fig. 5. Dependeces of maximum transconductance (gm_{Max}) and gate voltage (Vgsm) on threshold voltage (V_T).

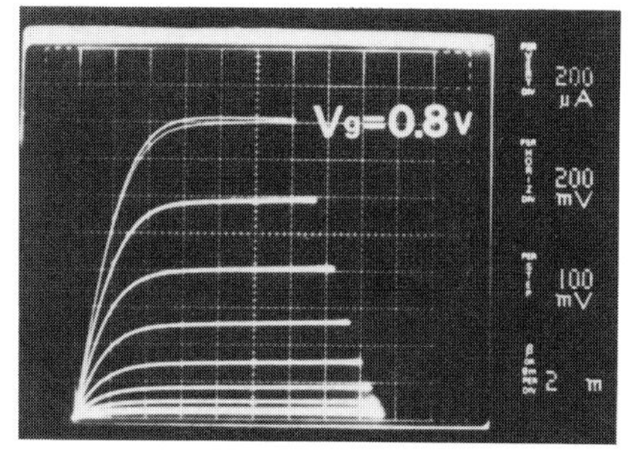

Fig. 6. Drain I-V characteristics for SD-DH FET. The gate size is 10 μm x 1 μm.

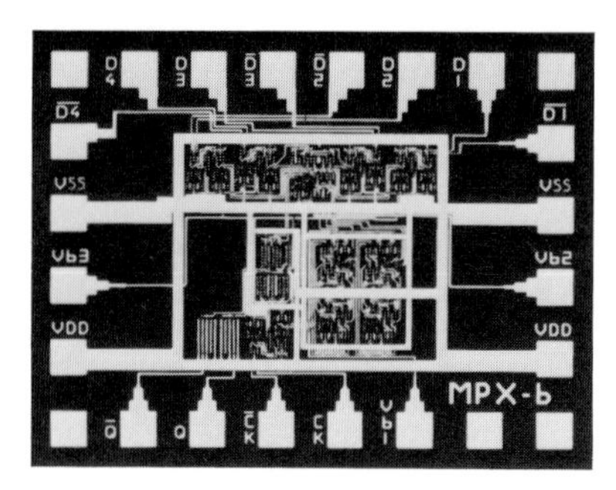

Fig. 7. Photograph of the SCFL multiplexer IC

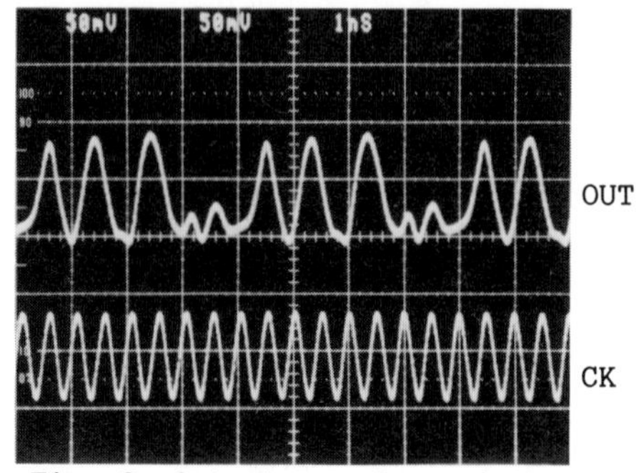

Fig. 8. Output waveform of the multiplexer IC

characteristics for SD-DH FET with V_T of 0.2V are shown in Fig. 6. The K-values obtained from the square law characteristics were $327mA/V^2mm$. These results show that SD-DH FET is very promising for realizing high speed circuits.

4. Application to digital IC

Using the high-performance SD-DH FETs, we have fabricated a 4-bit multiplexer IC composed of source-coupled FET logic (SCFL). This IC is consisted of 1μm gate length enhancement-mode SD-DH FETs and resistors. Figure 7 shows the photograph of the IC chip. Threshold voltage of enhancement-mode SD-DH FETs was adjusted to 0.1V by heat treatment of Pd gate. Resistors were formed with channel region by the mesa-etching. The first and second interconnection lines were formed with about 1μm-thick Ti/Pt/Au by a plating technique. 2μm-thick polyimide was formed as the cross-over insulator to reduce the interconnection capacitance. Figure 8 shows the output waveform of the multiplexer IC at input frequency of 2GHz. This result has proved that SD-DH FET was very promising for digital ICs.

5. Conclusion

A high performance AlGaAs/GaAs/AlGaAs SD-DH FET with 1μm gate length has been successfully fabricated by optimizing the MBE growth condition and layer structure. The structure showed sheet electron concentration as high as $3x10^{12}cm^{-2}$ with 77K electron mobility of $40000cm^2/Vs$. The maximum extrinsic transconductance of 500mS/mm was achieved at room temperature. Using the SD-DH FETs, 4-bit multiplexer IC operating up to above 2GHz was successfully fabricated.

Acknowledgement

The authors would like to thank Dr. H. Mizuno and T. Kajiwara for their continuous encouragements throughout this work.

References

Baba T, Mizutani T and Ogawa M, 1983 Jpn. J. Appl. Phys. **22**, L627.
Inoue K and Sakaki H, 1984 Jpn. J. Appl. Phys. **23**, L61.
Inoue K, Sakaki H and Yoshino J, 1984 Jpn. J. Appl. Phys. **23**, L767.
Inoue K, Sakaki H and Yoshino J, 1985 Appl. Phys. Lett. **46**, 973.

Inst. Phys. Conf. Ser. No. 91: Chapter 7
Paper presented at Int. Symp. GaAs and Related Compounds, Heraklion, Greece, 1987

A novel method for the modelling and the design of field effect transistors: application to MESFET, MODFET, SISFET, DMT and multiple HEMT simulation

E.Constant, D.Depreeuw, P.Godts and J.Zimmermann

Centre Hyperfréquences et Semiconducteurs, UA CNRS N°287 Bâtiment P3
Université des Sciences et Techniques de Lille Flandres Artois
59655 Villeneuve d'Ascq CEDEX-FRANCE.

ABSTRACT: A novel method for modelling field effect transistors is described. It is applied to the study of a number of devices like FETs, MODFETs, SISFETs, DMTs and multiple HEMTs. This method includes the most important physical effects occuring in these devices, but it is very simple and easy to implement on a microcomputer. It is used for the control of technological device processes realized in the laboratory as well as for designing more efficient structures.

1. INTRODUCTION

The development of field effect transistors for microwave and fast logic applications requires more and more efficient structures to be designed in which active layers are thinner and doping levels are higher. They also most often entail the use of heterojunctions. Physical phenomena governing the creation and transport of electrons in these structures become much more complex and as a result modelling and fabrication of these new kinds of transistors become much more difficult and hazardous. In the present paper we describe a model which is simple enough in order to be used on a microcomputer and also accurate enough in order to conceive and control technological processing of any kind of field effect transistor.

2. PHYSICAL EFFECTS

Our computation method is based on the experience acquired in our laboratory in device modelling (Fauquembergue et Al, Heliodore et Al, Shawki et Al 1987), these device modellings are simplified and adapted to the structures that we want to study. Our method applies to all the type of field effect transistors GaAs/AlGaAs and this whatever the Al mole fraction x and the impurity concentrations are. The goal is to treat the classical field effect transistor as well as the MODFET, the SISFET and so on.

Unlike many other models requiring more computation time and bigger computers, the main advantage of our model is that it is able to properly take into account a number of important physical effects which are peculiar to heterojunction and/or short channel devices :
- quantization of the electron motion when heterojunctions are used;

- lateral diffusion and carrier drift through heterojunctions and/or metal-semiconductor barriers which result in a gate current Ig which can be no longer overlooked;
- trapping of electrons into and from deep levels (DX centers in doped AlGaAs, for example);
- non-stationnary transport (such as overshoot effect) characterizing charge carriers moving along the conducting channel;
- carrier injection below the conducting channel when the gate is short enough, giving rise to the so-called "short channel effect".

3. NOVEL METHOD OF MODELLING

The input data are the temperature, the doping, alloy composition and thickness of the various layers, the depth of the recess, the position of the recessed zone, the source and drain contact resistances. In addition, Joule heating due to DC power dissipation is taken into account by introducing the channel-ambient thermal resistance, as a first approximation.

In a first step one computes the two-dimensional electron gas concentration n_s as a function of the gate potential, the electron concentration in the other active layers and finally the gate current density using the drift diffusion approximation (Ponse et Al 1986). The starting point of this computation requires, in the case of an heterotransistor, the knowledge of the energies of the first subbands as a function of n_s (in practice we take five subbands). They are determined by a self-consistent solution of Schrödinger and Poisson equations. If there are no heterojunctions in the structures we impose initial condition in the buffer. The electron concentration is computed using the Fermi-Dirac statistics and a two parabolic valley model whose characteristics depend on x. Since the layers may be highly n-doped we have to consider, first, that the shallow donor levels form a continuous band connected to the lower Γ -valley and not a single level; second, there exists a deep donor level connected to the upper L-valley (Chand et Al 1984). Then one proceeds with a one-dimensional computation of carrier and ionized donor concentrations along the direction perpendicular to the gate plane which is used to compute Ig(n_s), Vg(n_s). These results are used to compute the transistor characteristics once the geometry of the device is introduced. For this purpose we use a one-dimensional computation of carrier concentration and velocity along the direction of the conducting channel. This is an extension of the model developed in the laboratory successively by Salmer, Constant, Carnez, Cappy to study FETs and MODFETs (Cappy et Al 1985 and Carnez et Al 1980). The simulation gives the main characteristics of the device namely the DC Id(Vds, Vgs) and Ig(Vds, Vgs), the usual dynamical parameters gm, gd, Cgd, Cgs, Ig, fT. It usually takes a few tens of minutes on a PC AT3 for all these data to appear in plotted form.

Our programs were already employed in the laboratory for controlling processes in the course of fabrication. For example, figure 1 shows a comparison between the theoretical and experimental characteristics of a MODFET device realized in the laboratory.

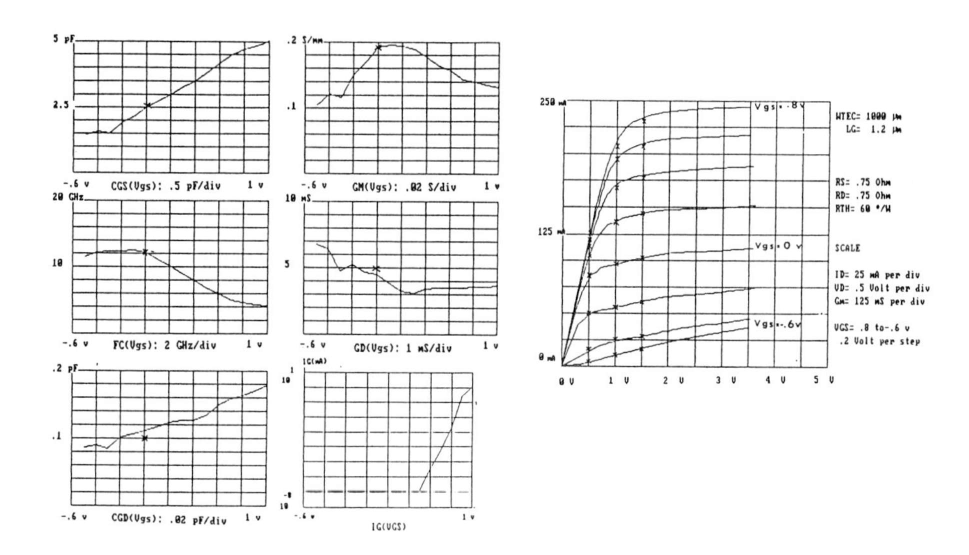

Fig. 1. Full lines show the results from MODFET simulations. (X) : Experimental data obtained with the same MODFETs realized in the laboratory. These data are average values obtained over a total of twelve transistors with three different gate widths (75, 150 and 300μm)

From a more prospective point of view our method of modelling can serve to design more efficient structures. For example high power (or logic) devices must have high cut-off frequency at high drain current. We were led to study a number of devices which we summarize in figure 3. Then, in figure 2, we show the variation of the cut-off frequency fT as a function of Id (at Vds= 3V), which we take from Id(Vgs) and fT(Vgs). For a MESFET, no noticeable variation of cut-off frequency is observed. For a TEGFET, fT is reduced when Id increase because of the variations of free and trapped carriers in the AlGaAs layer. If DX centers are ignored in the simulation noticeable improvement of fT is noticed. For a SISFET, the limiting factor is the gate current. However from the results shown in figure 2, it can be noted, that multiple HEMTs and DMTs seem to be the best structures for power and logic application.

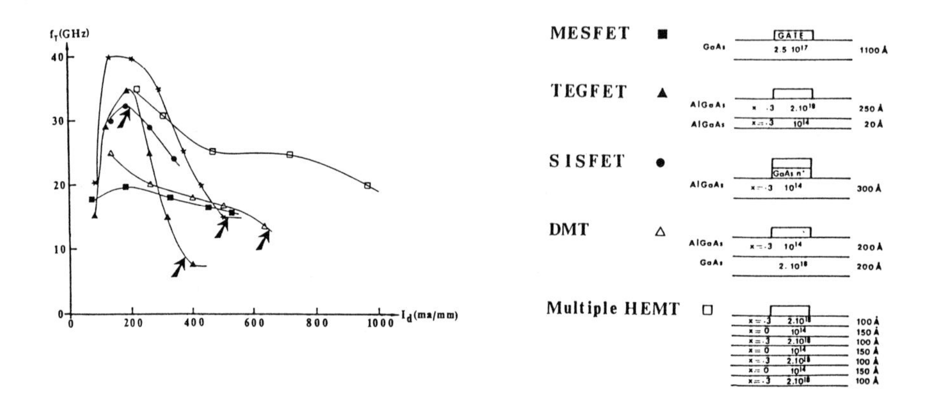

Fig. 2. Fig. 3.

Fig. 2 . Variation of cut-off frequency versus drain current of the various structures represented in Fig. 3. This is obtained by changing the gate voltage keeping the drain voltage at the same value of 3 volts. The arrows indicate where the gate current is 1mA.

4. CONCLUSION

A very efficient novel model which applies to any kind of heterojunction field effect transistors has been carried out. Our programs give results in good agreement with experiment. They are well suited for the control and practical realization of transistors and will be very convenient for conception of improved structures for the future.

5. REFERENCES

Cappy A., Vanoverschelde A., Schortgen M., Versnaeyen C. and Salmer G. 1985 IEEE Trans. Electron Devices ED-32 2787.

Carnez B., Cappy A., Kaszynski A., Constant E. and Salmer G. 1980 J. Appl. Phys. 51 784.

Fauquembergue R., Pernisek M., Thobel J.L. and Bourel P. 1987 Gallium Arsenide and Related compounds.

Heliodore F., Lefebvre M. and Salmer G. 1987 Gallium Arsenide and Related compounds.

Ponse F., Masselink W.T. and Morkoc H. 1986 IEEE Trans. Electron Devices ED-32 1017.

Shawki T.A. and Salmer G. 1987 Gallium Arsenide and Related compounds.

Paper presented at Int. Symp. GaAs and Related Compounds, Heraklion, Greece, 1987

Velocity-field characteristics of carriers in AlGaAs/GaAs modulation-doped heterostructures

W.T. Masselink, N. Braslau, D. LaTulipe, W.I. Wang†, and S.L. Wright
IBM T.J. Watson Research Center, P.O. Box 218, Yorktown Heights, NY 10598 USA

ABSTRACT: We report first measurements of the velocities of quasi-two-dimensional electrons and holes at high electric fields in AlGaAs/GaAs modulation-doped heterostructures and negative differential mobility (NDM) for the electrons. These data, measured from 0 to 8000 V/cm, reveal that peak velocities for the 77 K two-dimensional electron gas occur at much lower fields than observed for bulk GaAs. This difference is explained in terms of phonon scattering and intervalley transfer. At 300 K, the peak velocity for the 2DEG is somewhat lower than observed for lightly doped bulk GaAs because of modified intervalley transfer. Electrons in heavily doped GaAs ($n = 10^{18}$ cm^{-3}), however, have peak velocities which are lower than electrons in modulation-doped structures at both 300 and 77 K.

I. Introduction

This study is motivated by the need to understand the mid and high field transport of electrons and holes in modulation-doped GaAs. It is well known that carriers in AlGaAs/GaAs modulation-doped heterostructures may have extremely high mobilities at low lattice temperatures and low electric fields — that is, so long as the carrier temperature is very low the mobility may be very high. Recent studies indicate that the electron mobility may be as high as 5×10^6 cm^2/Vs [English *et al.*]. Previous studies demonstrate that at cryogenic lattice temperatures, as the carrier temperature increases, the mobility is seriously degraded [Masselink *et al.* and references contained therein]. The details of this dependence of mobility (and therefore of velocity) on electric field are crucial for understanding the operation of most electronic devices making use of transport parallel to heterostructures because such devices operate with large electric fields.

These previous studies, however, have been made with dc electric fields which will induce charge and field domain formation if the differential mobility is negative at the electric field being applied. Thus, measurements of the velocity of electrons in GaAs made with dc electric fields greater than the critical field (the electric field at the onset of NDM), will not be reliable. Furthermore, it may not even be apparent when the critical field has been reached. In the present experiment, we apply an electric field oscillating with a frequency of 35 GHz and avoid the charge redistribution which will occur during a dc measurement. In this way, we reliably measure the electron and hole velocities as functions of electric field up to 8000 V/cm in modulation-doped heterostructures at both 300 and 77 K. The electron results are compared to similarly measured velocities of electrons in lightly and heavily doped bulk GaAs at the same temperatures. The results indicate that in the modulation-doped structures, there exist uniquely two-dimensional scattering mechanisms which, at 77 K, lead to negative differential mobility at fields significantly lower than the critical field in bulk GaAs.

II. Experimental Technique

The experimental technique used in these experiments has been described in detail by Braslau and Hauge. Briefly, a small dc electric field is superimposed upon a large ac field in the sample to be measured. This is realized by inserting a sample with ohmic contacts at each end through a microwave waveguide; the microwave field provides the 35 GHz ac electric field and the dc field is supplied via the ohmic contacts. The experiment consists of measuring the dc conductivity between the two ohmic contacts as a function of microwave power or peak ac electric field. The 35 GHz microwave radiation is generated by a pulsed 10 kW magnetron with a pulse width of 0.2 μs at a repetition rate of 50 to 100 Hz. The microwave pulse is timed to occur in the middle of a dc pulse of 20 V/cm lasting about 1.5 μs. So long as the dc electric field $E_0 \ll E_1$, the dc conductivity for a peak ac field E_1 may be written

$$\sigma(E_1) = \frac{2qn_{2d}}{\pi E_1} \int_0^{E_1} \mu(E) \frac{dE}{\sqrt{1-(\frac{E}{E_1})^2}},$$

where q is the electronic charge, n_{2d} is the sheet electron concentration, and μ is the differential mobility of the electrons. Thus, by knowing the relation between the dc conductivity and the ac electric field, one can extract the mobility and therefore the velocity as functions of electric field, so long as sheet carrier concentration remains constant.

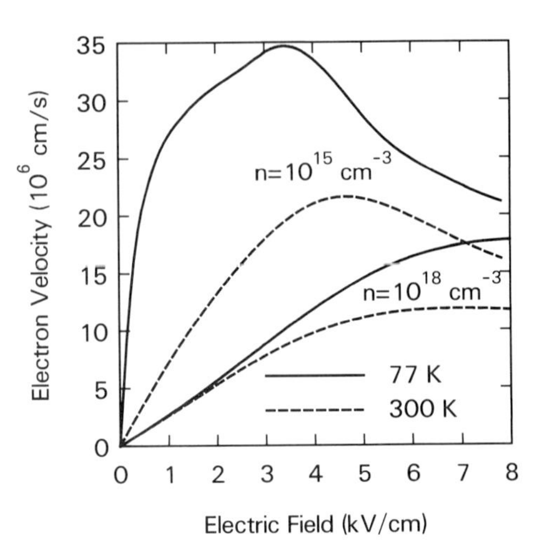

Fig. 1. Measured velocity versus electric field relations for bulk GaAs with $n=10^{15}$ and $n=10^{18}$ cm^{-3} at 77 and 300 K.

III. Results and Discussion

Bulk GaAs samples with two doping levels were studied in this experiment: $n=1 \times 10^{15}$ and 1×10^{18} cm^{-3}. Both samples were grown by MBE on semi-insulating substrates. The sample with $n=10^{15}$ cm^{-3} is 11 μm thick and Hall measurements indicate that its low-field mobility is 63000 (7600) cm^2/Vs at 77 (300) K. The $n=10^{18}$ cm^{-3} sample is 2000 Å thick and has a mobility of 2590 cm^2/Vs at both 77 and 300 K. Figure 1 depicts the velocity versus electric field relations for these two samples at 300 and 77 K determined using the technique described above. The 300 K result for the lightly doped sample is in excellent agreement with earlier measurements [Braslau and Hauge]. The authors are not aware of previous measurements of the velocity of electrons in bulk GaAs at 77 K in this electric field range, but note that the measurements of the lightly doped samples agree quite well with Monte Carlo calculations [Ruch and Fawcett, Wang]. The authors are also not aware of previous measurements of the electron velocity of GaAs with n = 10^{18} cm^{-3}.

Several n-type AlGaAs/GaAs modulation-doped samples were also similarly studied. One such structure has an Al mole fraction of 0.3 and a sheet carrier concentration of 4.1 (4.9) $\times 10^{11}$ cm^{-2} and a low field mobility of 127000 (7600) cm^2/Vs at 77 (300) K. Another structure has an Al mole fraction of about 0.5 and a sheet carrier concentration of 3.2 (2.7) $\times 10^{11}$ cm^{-2} and a low field mobility of 189000 (8130) cm^2/Vs at 77 (300) K. Figure 2 depicts the velocity versus electric field characteristics of these two modulation-doped samples at 300 K, along with the 10^{15} cm^{-3} bulk GaAs sample also at 300 K. These data indicate that although all three of these samples have similar low-field mobilities, the peak velocity of the two-dimensional electron gas (2DEG) is lower than that of electrons in lightly doped bulk GaAs and occurs at a somewhat lower electric field. This behavior is explained

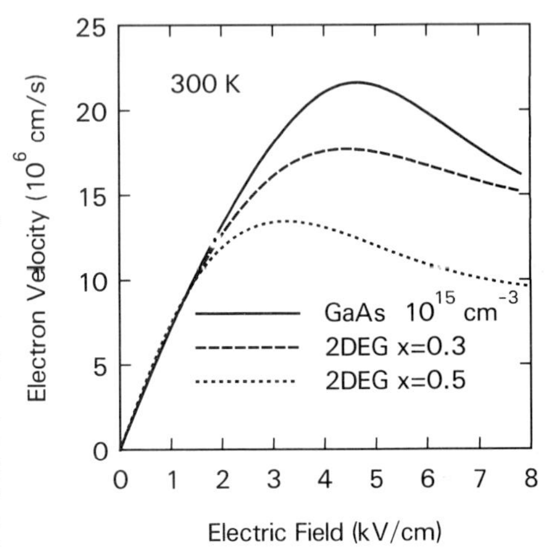

Fig. 2. The 300 K electron velocity as a function of electric field. The bulk GaAs has $n=10^{15}$ cm^{-3}. The two 2DEG samples are $Al_xGa_{1-x}As$/GaAs modulation-doped heterostructures with x=0.3 and x=0.5.

through two mechanisms. First, because the electrons are spatially confined, the density of states is step-like with zero states at the bottom of the Γ valley conduction band. The lowest energy available for electron states is at the lowest Γ valley subband with an energy of about 40 meV above the bottom of the band [Stern and Das Sarma]. This results in the Γ—L energy separation being reduced from about 0.31 eV to about 0.27 eV. Since intervalley transfer and, therefore, peak velocity followed by NDM depends on this energy difference in an exponential manner, this reduction will be important. A simple calculation based on the electrons being thermal equilibrium with each other indicates that this 40 meV reduction in intervalley energy will result in about a 10% decrease in the peak velocity. Real space transfer is probably also partially responsible for the lower peak velocity in the modulation-doped samples. Although most theoretical considerations of real space transfer have only included the transfer from one Γ valley to the other, the very high density of states in the L and X valleys suggest that transfer from the Γ valley in GaAs into the L or, for higher Al mole fractions, the X valleys in the AlGaAs may be more important. A simple model which treats the coupling between the different valleys only approximately indicates that in the range of Al mole fractions between 0 and 0.5, higher mole fractions result in lower peak velocities because of the transfer into the X valleys. This agrees with the present data and could explain the peak velocity in the x=0.5 sample appears so much lower than that of the x=0.3 sample.

Figure 3 shows the 77 K velocity-field relations for the same samples whose 300 K behavior was discussed above. Similar to that measurement, at 77 K the modulation-doped samples have lower peak velocities than does the lightly doped bulk GaAs sample. The two reasons given above for this behavior are still valid. In addition, when the lattice is 77 K, at about 1 kV/cm the differential mobility of the electrons in the 2DEG samples abruptly decreases, whereas this effect is not so pronounced in the bulk GaAs. Such behavior is consistent with calculations by Ridley *et al.* [Ridley, Riddoch and Ridley, Al-Mudares and Ridley] who predict a sharp decrease in the differential mobility because of an enhanced scattering of electrons with polar optical phonons. Their calculations indicate that this decrease should occur at about 1 kV/cm for the AlGaAs/GaAs system which is consistent with our measurements. The NDM occurring at higher fields is probably due to intervalley and real space transfer. The steady decrease in velocity above 2.2 kV/cm in the x=0.5 sample may be in part due to electrons cooling in the AlGaAs and not transferring back to the GaAs for relatively long times. Thus, the true velocity for the x=0.5 sample may be somewhat higher than indicated in Figure 3 at fields above 2.2 kV/cm.

At both 300 and 77 K, the 2DEG samples exhibited lower peak electron velocity than the lightly doped bulk GaAs. The comparison of 2DEG to electron in very lightly doped GaAs demonstrates the additional scattering mechanism resulting from the heterostructure and the two-dimensional confinement. For analysis appropriate for FET performance, however, it is more useful to compare the modulation-doped samples to bulk GaAs doped $n=10^{18}$ cm^{-3}. Only with such heavy doping can one obtain the gate capacitance (i.e., charge control) and shallow channel (necessary for short gate FETs) as one can obtain with the modulation-doped structures. Comparing the heavily doped GaAs results depicted in Figure 2 with the 2DEG results of Figures 2 and 3, one can easily see that the modulation-doped heterostructures have both superior low field mobility and higher field velocity than the heavily doped bulk GaAs. This result is consistent with the

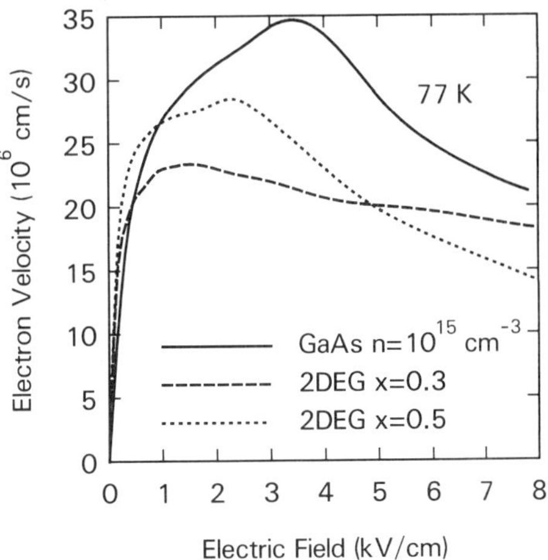

Fig. 3. The 77 K electron velocity as a function of electric field. The bulk GaAs has $n=10^{15}$ cm^{-3}. The two 2DEG samples are $Al_xGa_{1-x}As/GaAs$ modulation-doped heterostructures with x=0.3 and x=0.5. The continued decrease of velocity in the x=0.5 sample may be partially due to cold electron trapping in the indirect valleys of the $Al_{0.5}Ga_{0.5}As$.

generally superior performance of MODFETs compared with MESFETs.

Two single heterointerface p-type AlGaAs/GaAs modulation-doped samples were also investigated using this technique. Samples 1 and 2 have two-dimensional hole gas (2DHG) concentrations of 3.3 and 4.2 $\times 10^{11}$ cm^{-2} and Hall mobilities of 3300 and 4000 cm^2/Vs at 77 K. Both samples have Al mole fractions of 0.5. The measured velocity versus electric field relations for these samples are depicted in Figure 4. This measurement accurately demonstrates the mobility degradation in 2DHG samples at 77 K. In addition, we have analyzed the warm electron regime of the velocity versus field in order to determine the energy relaxation time, τ_e, in a similar fashion as has been done for the 2DEG [Tsubaki *et al.*] We calculate this time to be 2.0 ps. From Figure 4, one can see that at 10 kV/cm, the velocity of the 2DHG is approaching 0.8 × 10^7 cm/s. Such information should be useful for the analysis and modelling for p channel heterojunction FETs such as are currently being investigated [Hirano *et al.*]. Preliminary results from a ten period modulation-doped quantum well sample indicate that this structure may have significantly superior transport properties up to 10 kV/cm.

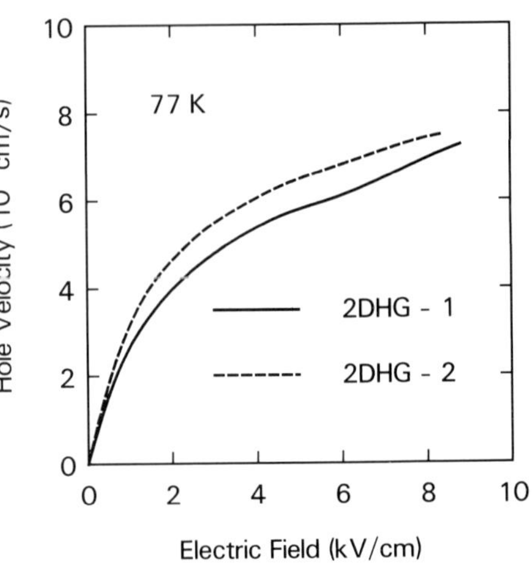

Fig. 4. The 77 K hole velocity as a function of electric field for the two single heterointerface 2DHG samples described in the text.

IV. Conclusion

In conclusion, we have measured the velocity of electrons and holes in AlGaAs/GaAs modulation-doped heterostructures at electric fields up to 8000 V/cm at both 300 and 77 K. We have also measured electron velocity in bulk GaAs with both heavy and light doping. The measurement uses 35 GHz radiation to supply the electric field. This technique allows us to avoid field and charge domain formation which prevents the measurement of NDM regimes when a dc measurement is used. These measurements indicate that the electrons in the heterostructures have higher peak velocities than those in the heavily doped bulk GaAs, but lower peak velocities than those in the low-doped bulk GaAs. This latter behavior is explained in terms of modified intervalley transfer, real space transfer, and, at 77 K, an enhanced scattering with polar optical phonons.

Acknowledgement — One of the authors (W.T.M.) acknowledges useful conversations with many colleagues, in particular with M. Artaki and P. Solomon.

REFERENCES

† Present address: Electrical Engineering Department, Columbia University, New York, NY 10027 USA

Al-Mudares M.A.R. and Ridley B.K., *J. Phys. C: Solid State Phys.* **19,** 3179 (1986).
Braslau N. and Hauge P.S., *IEEE Trans. Electron Devices* **ED-17,** 616 (1970).
English J.H., Gossard A.C., Störmer H.L., and Baldwin K.W., *Appl. Phys. Lett.* **50,** 1826 (1987).
Masselink W.T., Henderson T., Klem J., Kopp W.F., and Morkoç H., *IEEE Trans. Electron Devices* **ED-33** 639 (1986).
Riddoch F.A. and Ridley B.K., *J. Phys. C: Solid State Phys.* **16,** 6971 (1983).
Ridley B.K., *J. Phys. C: Solid State Phys.* **15,** 5899 (1982).
Ruch J.G. and Fawcett W., *J. Appl. Phys.* **41,** 3843 (1970).
Stern F. and Das Sarma S., *Phys. Rev. B* **30,** 840 (1984).
Tsubaki K., Livingstone A., Kawashima M., Okamoto H., and Kumabe K., *Solid State Commun.* **46,** 517 (1983).
Wang T., Ph.D. Thesis, University of Illinois at Urbana-Champaign, 1986.

Inst. Phys. Conf. Ser. No. 91: Chapter 7
Paper presented at Int. Symp. GaAs and Related Compounds, Heraklion, Greece, 1987

Low-temperature current drift and its origin in heterostructure and quantum-well MODFETs

V. Iyengar, S. T. Fu, S. M. Liu, and M. B. Das*

Center for Electronic Materials and Devices
Department of Electrical Engineering
The Pennsylvania State University
University Park, PA 16802

C. K. Peng, J. Klem, T. Henderson, and H. Morkoç

Coordinated Science Laboratory
Department of Electrical Engineering
University of Illinois
Urbana, IL 61801

ABSTRACT: Drift and hysteresis of drain I-V characteristics in conventional AlGaAs and pseudomorphic InGaAs quantum-well FET's have been studied at low temperatures. A method of measurement has been developed that utilizes isothermal emission of high field trapped electrons at several energy levels. The devices tested do not show significant threshold voltage shift nor do their I-V characteristics collapse. It is suggested that trapping of high energy electrons from the 2-D electron gas in the gate-drain separation region is responsible for the observed current drift.

1. INTRODUCTION

It is well known that due to the presence of high concentration DX centers in Si doped AlGaAs with Al-molefraction greater than 0.2, the conventional heterostructure MODFET's show large threshold voltage shift under appropriate gate bias stress, and drain I-V collapse at low temperatures [1,2,3]. These effects have been overcome by reducing the DX centers by improving material growth-technique and use of low Al and In molefractions in the pseudomorphic InGaAs quantum-well MODFET's [3]. However, a detailed examination of these devices has revealed that their drain I-V characteristics display hysterisis at low temperatures. It is observed that as the drain current saturates with increasing V_{DS}, the current drifts to a lower value, and when V_{DS} is reduced a different I-V curve below the original one is obtained. If sufficient time is allowed, or the temperature is raised, or the sample is illuminated while V_{DS} is held near zero, then by applying V_{DS} the original I-V curve can be obtained. This phenomenon

* Chalmers University of Technology, Dept. of Radio- & Space Science, Sweden. (Fall 1987)

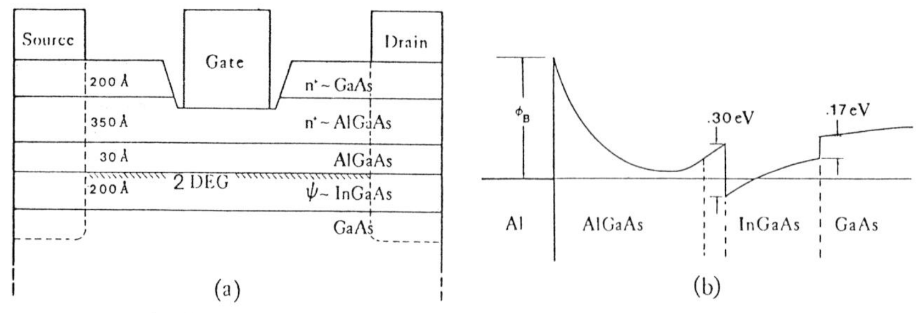

Figure 1 (a) Schematic cross-sectional details of the InGaAs quantum-well MODFET structure. (b) The conduction band diagram.

is studied for an understanding of carrier emission process and results are reported in this paper.

2. EXPERIMENTAL

Pseudomorphic MODFET's with low In and Al molefractions (~0.15) were fabricated at the University of Illinois. A schematic cross-sectional diagram of this structure and its associated conduction band diagram are shown in Figure 1. The details of material growth and fabrication can be found elsewhere[1,4]. The conventional heterostructure devices with improved material were fabricated at the GE Electronics Laboratory, with n-AlGaAs thickness comparable to that given in Figure 1(a). The test devices were mounted on carriers suitable for mounting on the cryostat stage. Before proceeding with experiment trial runs were made to ensure that no instability or oscillations occurred while the drain I-V characteristics were displayed at all temperatures.

First the drain I-V curves were obtained at selected temperatures. Then a suitable gate bias was selected for the carrier trapping and emission experiment. The drain voltage was raised to display the saturated I-V curve and held at a higher V_{DS} until I_{DS} drifted to a steady (lower) value. Finally, a low constant I_{DS} supply was switched on so that a low V_{DS} was attained in the 'knee' region. This V_{DS} was monitored as a function of time. A suitable range of constant temperatures was used to repeat this experiment, and before obtaining the initial I-V curve light was used to ensure total emission of carriers.

3. RESULTS AND DISCUSSIONS

The typical results showing the isothermal I-V curves at different gate bias voltages are presented in Figure 2. There is a considerable difference between the I-V curve when increasing V_{DS} and that when decreasing V_{DS}. It is also apparent that the 'knee' or the linear parts of all I-V curves when V_{DS} is decreasing tend to trace a nearly constant slope. This clearly suggests the existence of a high series resistance associated with the channel after carrier trapping has occurred. It is most likely that this carrier

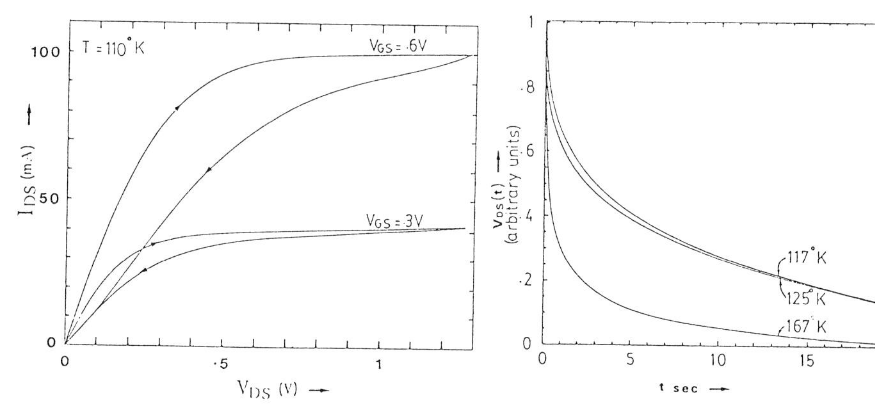

Figure 2. Typical Drain I-V characteristics with increasing and decreasing drain voltage for a pseudomorphic MODFET.

Figure 3. Time decay of "Below-the knee" drain voltage for a constant I_{DS} due to carrier emission from traps.

trapping occurs in the gate-drain separation region where high energy electrons are transferred from the underlying 2-D electron gas; and this in turn reduces the 2-D electron gas **sheet** concentration, thus limiting any further reduction of I_{DS}. **Carrier trapping could also occur in InGaAs layer.**

The emission characteristics of trapped carriers have been studied by examining the time decay of the 'knee' voltage V_{DS} when low constant I_{DS} is maintained after carrier trapping has occurred (See Figure 3). These decay curves and their time derivatives when plotted in log/linear manner, clearly demonstrate that there are multiple time constants representing different traps involved in this process.

Using a numerical curve fitting approach we have been able to model and determine these multiple time constants (τ's) and the relative magnitudes (a's) of the trapped charge as given in Table I. A comparison of a typical experimental curve and a corresponding model curve is presented in Figure 4.

4. CONCLUSION

Conventional and pseudomorphic MODFET's based on high quality n-AlGaAs active gate material, which normally show low threshold voltage shift and collapse-free drain I-V characteristics at low temperatures, are not necessarily free from degradation and hysteresis of saturation drain current caused by high energy electron trapping at multiple centers identified in this work. Elimination and reduction of these traps would be essential by further improving the material quality for the realization of low current drift MODFET's.

Temp °K	$V_{DS}(0)-V_{DS}(\infty)$ mV	τ_1, a_1 sec - Trap #1	τ_2, a_2 sec - Trap #2	τ_3, a_3 sec - Trap #3	τ_4, a_4 sec - Trap #4
83	12.3	.10 .35	.40 .081	1.2 .187	12 .407
100	16.5	.07 .115	.35 .121	1.4 .218	12 .545
111	27.0	.04 .148	.20 .111	1.1 .222	12 .518
117	29.1	.04 .034	.50 .137	1.8 .199	13 .629
125	40.9	.04 .132	.24 .122	1.4 .196	14 .550
133	41.23	.06 .088	.40 .131	2.0 .172	18 .609
167	6.8	.04 .243	.20 .294	1.2 .250	7 .213
200	1.59	.04 .635	--	0.7 .245	5 .119

TABLE 1: Experimental and calculated results characterizing different traps causing high field current drifts.

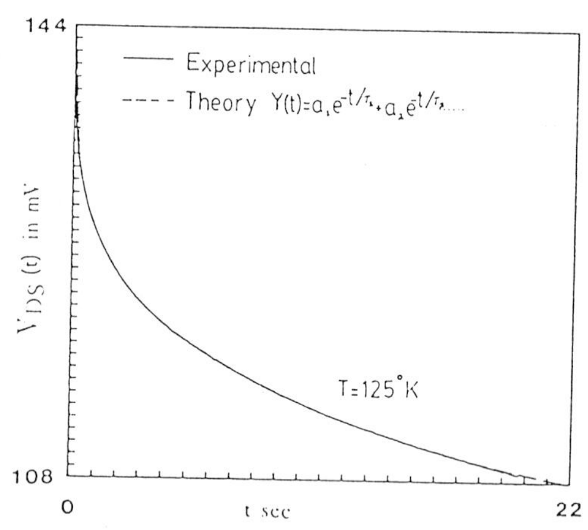

Figure 4. Typical experimental and theoretical time decay curves with identified multiple time constants and relative amplitudes for the various traps.

5. ACKNOWLEDGEMENTS

The work done at Penn State in cooperation with GE Electronics Laboratory was supported by the National Science Foundation under Grant No. EC85-03894. The work done at the University of Illinois was supported by the Air Force Office of Scientific Research and NASA.

6. REFERENCES

1. R. Fischer et al, IEEE Trans. on Electron Dev., ED-31, P. 1028, 1985.

2. A. Kastalasky and R. A. Kiehl, IEEE Trans. on Electron Dev., ED-33, p. 414, 1986.

3. S. M. Liu et al, J. Cryst. Growth, 81, p. 359, 1987.

4. A. Ketterson et al, IEEE Trans. on Electron Dev., ED-33, p. 564, 1986.

Inst. Phys. Conf. Ser. No. 91: Chapter 7
Paper presented at Int. Symp. GaAs and Related Compounds, Heraklion, Greece, 1987

High performance HEMT structure with GaAs Schottky gate

E. Kohn, C.J. Wu, H. Lee, M. Schneider, T. Bambridge*, and H. M. Levy

SIEMENS Research and Technology Laboratories, 105 College Road East, Princeton, N.J.,08540
*Microwave Semiconductor Corporation, 100 School House Road, Somerset, N.J. 08873

ABSTRACT: HEMT structures have been fabricated with buried (Al,Ga)As , where all contacts, Schottky and ohmic , are made to GaAs as in a MESFET. Data achieved are: $R_S < 0.05\ \Omega$–mm, g_m (300 K) = 590 mS/mm and g_m (77K) = 1.0 S/mm for 0.75 μm gate length. First results for short gate devices (0.2 μm) indicate an intrinsic $f_t > 65$GHz.

INTRODUCTION

High performance FETs for microwave and mm wave applications are mainly fabricated by recessed gate technologies. Thereby, very low parasitic capacitances and series resistances, can be achieved. However, the incorporation of (Al, Ga)As into the material structure can create a number of processing problems. Reliable technological processes are well established for MESFETs, and it thus seems attractive to fabricate HEMTs with all contacts made to GaAs. One can then combine conventional GaAs processing with heterojunction FET performance.

The following aspects are important for conventional HEMTs, where the gate recess reaches into the (Al,Ga)As:

- Differences in the recess etching behavior of (Al,Ga)As with respect to GaAs have to be taken into account even for relatively nonselective etches
- The stability of the (Al,Ga)As surface during the various processing steps is an important factor.
- Surface preparation for Schottky barrier metal deposition has to be examined in light of additional Al-oxide formation during etching.
- The influence of interfacial oxide films, metal reactions and interdiffusion on the thermal stability of the (Al,Ga)As Schottky barrier characteristics have to be identified.
- The alloying of AuGe based ohmic contacts has to be reexamined in the presence of Al.

Figure.1 compares the stability of Ti/Pt/Au contacts to (Al,Ga)As relative to those on GaAs for rapid thermal treatments in a forming gas atmosphere. As can be seen, the I-V barrier height on GaAs improves with annealing due to rearrangement of the interfacial oxide layers before deteriorating due to interdiffusion at around 600 °C. The same metal system on (Al,Ga)As shows a gradual decrease above 300 °C, most likely due to metallic interdiffusion of Ti and Al as deduced from a similar behavior of Ti/Al diodes on GaAs (Wada et al., 1983). Thus it seems questionable if any advantage can be drawn from the initially larger Schottky barrier height due to the (Al,Ga)As in the HEMT.

The above aspects of device processing have to be considered in conjunction with the need for structures providing a high aspect ratio of channel length to vertical gate / channel separation. To realize advanced HEMT structures with all-GaAs contacts makes it therefore necessary to compress the entire (Al,Ga)As layer system into 250 Å or less when short gates (<0.25 μm) are used.

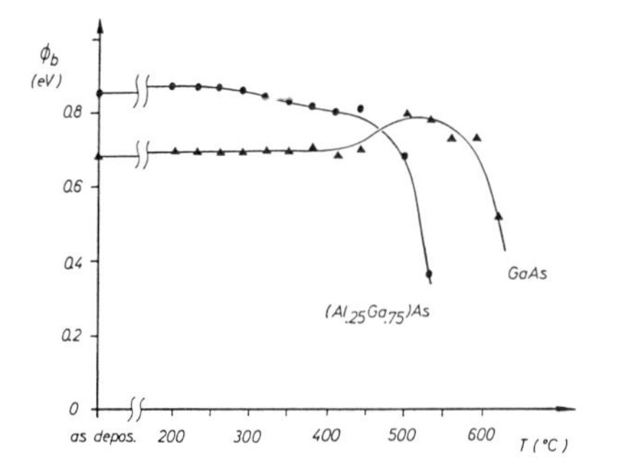

Fig.1

IV-barrier height ($\varnothing_b$) vs annealing temperature (T) for 20 sec rapid thermal treatment of Ti/Pt/Au Schottky diodes on GaAs and (Al,Ga)As.

DEVICE STRUCTURE

The device structure is sketched in Figure.2. The top GaAs cap layer has a thickness of 800 Å which is larger than the alloying depth of the ohmic contact metallization as illustrated in the following.

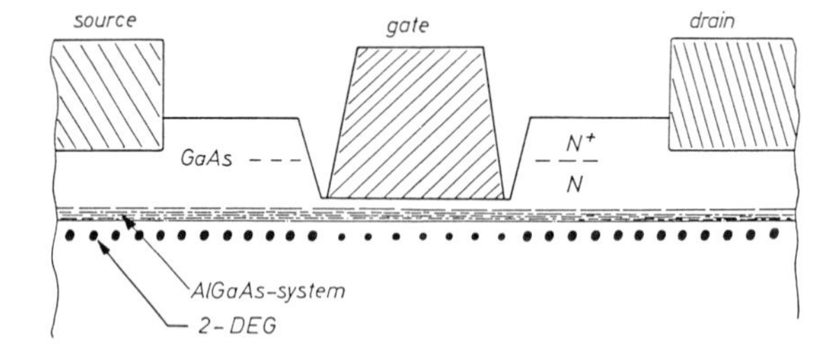

Fig.2

Schematic cross-section of HEMT structure with all contacts to GaAs

The ohmic contact metallization applied is a standard metallization as used in MESFET fabrication and consists of 1000 Å of AuGe (88%:12%), followed by 300 Å of Ni and 500 Å of Au. This is alloyed by rapid thermal treatment for 20 s at 380 °C. Fig.3 shows a TEM cross-section of such a contact in a HEMT structure with thick cap layer (1800 Å). One can clearly identify the ohmic metal penetration depth of approx. 500 Å and see the (Al,Ga)As layer system approx. 1400 Å deeper in the material. Planar contact resistances of such HEMT structures with thick cap layers, doped above 1×10^{18} cm^{-3}, have been studied systematically by transmission line measurements. Very low values of <0.05 Ω-mm have been found even after elimination of parallel conduction through the cap and (Al,Ga)As by recess etching. This finding is consistent with results obtained from non-alloyed InGaAs contacts in HEMTs, where the device series resistance is mainly attributed to the tunnel resistance across the (Al,Ga)As layer at the recess edge (Koruda et al., 1987).

The "buried (Al,Ga)As" (Al content of 27%) HEMT structure consists of a thin (15 Å) undoped spacer layer, a spike doped layer consisting of either an 80 Å doping spike (wafer A) or a 40 Å doping spike (wafer B) of 3×10^{18} cm^{-3} Si doping followed by a graded layer of reduced doping (1×10^{18} cm^{-3}). In this layer the composition is graded down exponentially by setting the Al source to a low temperature. The Al concentration drops below the resolution limit for TEM and

Auger within approx. 170 Å. This grading profile has been used to avoid a second conduction band discontinuity above the 2-DEG interface and facilitate tunneling and charge transfer into the 2-DEG.

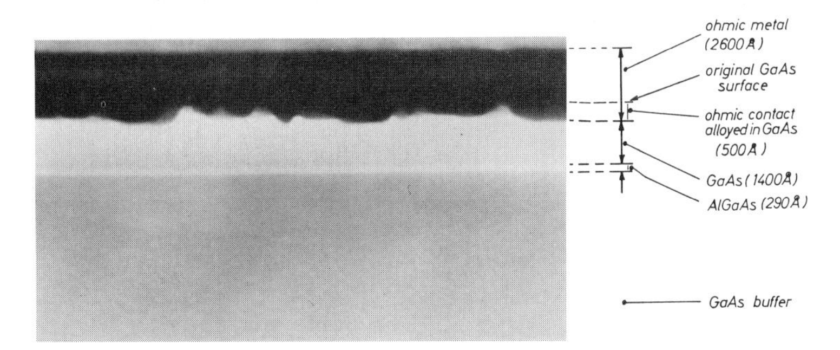

Fig.3 TEM micrograph of material structure and alloyed ohmic contact (TEM by SRI, Princeton)

The doping spike configuration has been chosen to obtain a high 2-DEG density above 10^{12} cm^{-2}. Simulation by a numerical charge control model shows a high charge transfer efficiency compared to a thick layer of 98% for the 80 Å spike and 95%for the 40 Å spike. Indeed, a saturated 2-DEG channel current of approx. 320 mA/mm has been measured for both spikes as well as for a thick layer. This confirms high active doping in the (Al,Ga)As, high charge transfer efficiency and a high electron velocity.

DEVICE CHARACTERISTICS

HEMTs with 0.75 μm gate length were fabricated on the 80 Å spike doped wafer (A). An optical gate definition process was used, which results in a narrow recess profile, where parasitic series resistances are virtually eliminated (for details see Wu et al. 1987).

Fig.4 shows the peak transconductances (g_m) for various recess depths and thus threshold voltages. A maximum g_m of 590 mS/mm is obtained for V_{th} = +0.2 V (see fig.5). The dependence of g_m on V_{th} closely follows the prediction of the charge control model.(Hofmann, 1986) The peak g_m is obtained at approx. 200 mA/mm. Cooled operation at 77 °K is illustrated in Figure 6. The threshold voltage is shifted by 250 mV, due to the freeze out of DX-centers in the (Al,Ga)As. A peak g_m = 1.0 S/mm is observed. In dark, no IV-collapse is seen due to the absence of an ungated part within the narrow recess area.

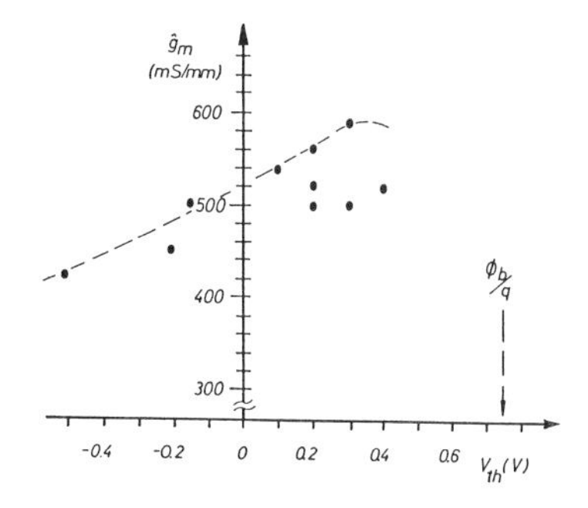

Fig.4 HEMT peak transconductance (g_m) vs. threshold voltage (V_{th})

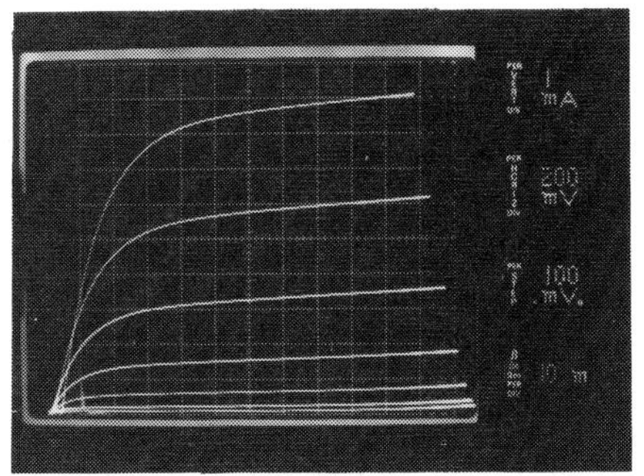

Fig.5 0.75 μm HEMT output characteristic W=50 μm, gate offset +0.7V, negative steps

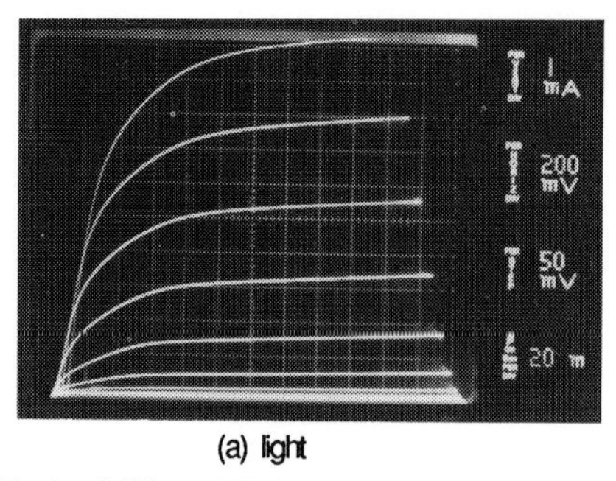

(a) light

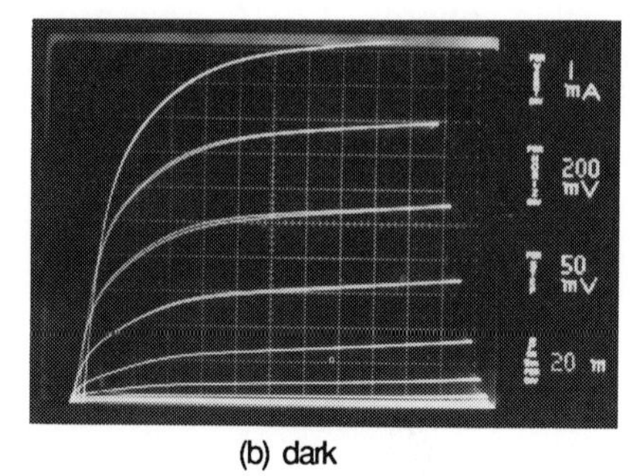

(b) dark

Fig.6 0.75 μm HEMT (W=50 um) output characteristics at 77°K; gate offset +.9V, neg.steps.

The source resistance has been determined to 0.16 Ω-mm by gate diode forward bias measurement. Modeling of microwave S-parameters leads to an RF source resistance of 0.038 Ω-mm. This seems to verify that a very low contact resistance is present. The low contact resistance however, has been achieved at the expense of a gate metal overlap capacitance at the side walls of the recess trench. Thus, an RF input capacitance of 4.9 pF/ mm is measured.

Wafer (B) with 40 Å doping spike has been used to fabricate FETs with 0.2 μm gate length using e-beam lithography. In a first attempt depletion mode devices were fabricated, because here the control of parasitic resistances is easier to achieve. The saturated FET current is 310 mA/mm. The source resistance as obtained by the forward gate diode measurement is 0.18 Ω-mm, which is as low as that determined for wafer (A). For a threshold voltage of V_{th} = -0.7 V a peak DC transconductance of 410 mS/mm is obtained at 190 mA/mm; the input capacitance is 0.76 pF/mm and the voltage gain g_m/g_o = 14.4. An f_t of 50 GHz is extrapolated although the performance is limited by gate resistance and pad capacitance. An intrinsic gain bandwidth product above 65 GHz is estimated.

CONCLUSION

It has been shown, that high performance HEMTs can be fabricated with a buried heterostructure. In such a structure all contacts are made to GaAs. This does not only facilitate processing, but shows also that very thin (Al,Ga)As layers can be used to obtain a high aspect ratio FET. It has been shown that charge transfer efficiency can be near unity from $3x10^{18}$ cm^{-3} Si doped spikes as narrow as 40 Å. Further improvements are expected for 0.1 um gate devices with quantum well channel as in the case of the pseudomorphic devices.

ACKNOWLEDGEMENT

Sincere thanks are to T. Jones and L. Harner for help in device processing, for R. Cheponis on the diode annealing experiments, K. Hofmann for simulations with the charge control model, A. Lapore and P. Tasker (Cornell University) for microwave evaluation and R.Tiberio (NNF at Cornell University) for help with e-beam lithography. We thank K. Zaininger for the steady support of this project.

REFERENCES

Hofmann K et al. 1986, Electron. Lett. **22**, 335
Koruda S et al. 1987, IEEE Electron. Dev. Lett. **EDL-8**, 389-391
Wada Y et al. 1983, Solid State Electron. **26**, 559-564
Wu C J et al. 1987, Journ. Electrochem. Society **134**, 2613-2616

Paper presented at Int. Symp. GaAs and Related Compounds, Heraklion, Greece, 1987

A high effective barrier height GaAs M – p + – n SAGFET

S. P. Kwok and S. K. Cheung

Ford Microelectronics, Inc., Colorado Springs, Colorado, 80921, USA

ABSTRACT: A LDD M-p+-n SAGFET with a thin p+ layer has been fabricated using ion implantations of Mg and Se into semi-insulating GaAs. Enhancement mode Vth of 1.3 V and gate Von of 1.6 V were obtained. FET 'K'-factor of 8.4 mS/mm and 'transconductance' equal to 800 mS/mm were obtained with 1.4 X 20 micron gate, under a typical drain/gate current ratio of 4 to 5. Comparison with fully depleted M-p+-n model suggests a strong p+ layer driving barrier enhancement of the fabricated device into saturation.

1. INTRODUCTION

A large effective gate barrier height is desirable for increased noise margin of digital circuits; concomitantly, a strong current drive capability is needed for high speed LSIs. Therefore, FET with a high effective barrier height and a high transconductance are currently pursued. Introduction of an ultra-thin insulator between Schottky metal and GaAs FET channel was reported by Kwok (1986) and limited interface reaction of refractory metal nitride with GaAs to enhance the Schottky barrier height was also reported by Zhang (1987). Increase of effective Schottky barrier height by oppositely doping surface layer between the metal and semiconductor has been known since early 70s. Schottky diode barrier height enhancement using various fabrication technologies were reported by: Lubberts (1974), Shannon (1974), Li (1978), Chen (1982) and Zhang (1987). In this paper an implanted Lightly-Doped-Drain (LDD), Self-Aligned Gate FET (SAGFET) using a M-p+-n gate structure having a thin p+ layer will be described.

2. DEVICE DESIGN AND FABRICATION

Barrier height enhancement equations (Lubberts) for M-p+-n structure as a function of device parameters using a fully depleted surface-doped zone are given in Figure 1. For large barrier heights, minority carriers collect at the vicinity of the potential hump, eventually causing saturation of barrier height. Schwartz (1986) calculated numerically the barrier height saturates to an asymtotic value of: $\Delta\Phi = 0.87 Eg - \Phi_{bo}$, where Eg is band gap energy and Φ_{bo} barrier height described in Figure 1. It is well known that a shallow and highly doped channel increases FET's

BAND DIAGRAM OF M–P–N STRUCTURE

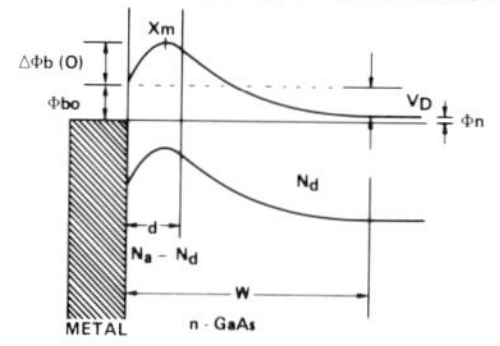

$$\Delta\Phi b(V) = \frac{q}{\epsilon}(N_A - N_D)\, Xm^2 = \frac{q}{\epsilon} N_D \frac{(Kd - W)^2}{K-1}$$

$$Xm = \frac{Kd - W}{K-1} \geq 0$$

$$W = \left[\frac{2\epsilon}{qN_D}(V_D - V) + Kd^2\right]^{1/2}$$

$$K = \frac{N_A}{N_D} > 1$$

V_D = DIFFUSION POTENTIAL BARRIER

V = APPLIED BIAS VOLTAGE

Figure 1

transconductance and immunity to short channel effects. A similar approach is adopted in the design of the enhancement M-p+-n gate FET here. The n-channel is formed by implantation of of Se with 100-200 KeV and a typical dose of 1.9E12 (A) and 3.8E12 (B) cm-2. The corresponding projected range is typically 44 nm with a standard deviation of 21 nm. The corresponding LSS peak channel concentrations range from 3. to 6.E17 cm-3, based on activation efficiency of 80 percent. The p+ layer is formed by a shallow implantation of Mg at 5-15 KeV and a typical dose of 1E13 cm-2. First order estimate of typical projected range and standard deviation are 15 nm and 10 nm respectively. By assuming implant-activation efficiency of 50 percent for such a high dosage of Mg, the p+ typical peak concentration is 2.E18 cm-3. Figure 2 presents the barrier enhancement for uniformly doped p+ layer with a density of (NA-ND) and n layer with ND. The thickness for the p+ layer is d and the n-channel is assumed to be sufficiently long to accommodate the extent of space-charge depletion. Depletion model given by the equations in Figure 1 is used. The carrier saturation derived by Schwartz is also included in the design plot. To ensure an enhancement mode FET, the total depletion width, W, of the M-p+-n structure is made to be at least equal to or greater than the n-channel depth.

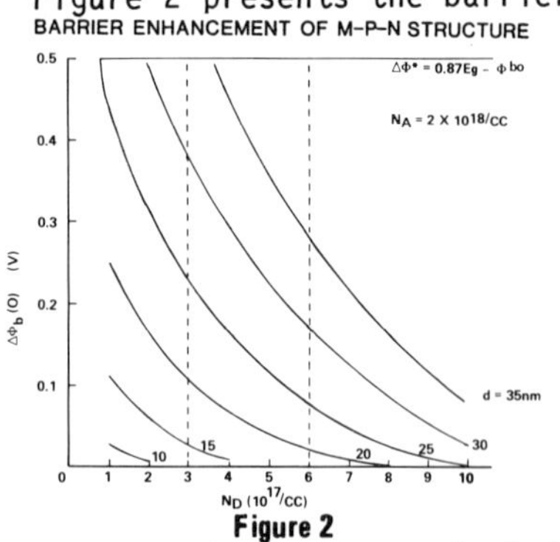

Figure 2

The ranges of parameters have been chosen, for this paper to include those of the implanted M-p+-n structure fabricated for this paper. In order to minimize the thickness of the p+ layer, the channel is recess etched typically 15 nm after post implantation annealing. The p+ layer depth is estimated to be 3 standard deviations of the typical projected range which is 30 nm. The two vertical broken lines represent the peak n-channel doping densities used. The maximum barrier height increase can be estimated to be 0.38 eV with the p+ layer fully depleted for ND=3E17 cm-3 and d=30nm.

Selective ion implantation technology into semi-insulating GaAs was used for the fabrication of the device. Two M-p+-n SAGFETs were implanted according to the design mentioned earlier. The wafer was then furnace annealed under arsine overpressure. A dielectric film was deposited; and then the substitutional gate with its sidewall formed by RIE. Heavy dose of Si species was then implanted to form the N+ drain and source regions. The sidewall was then stripped and the wafer furnace annealed under arsine overpressure. Au-Ge-Ni ohmic contact was then formed. Planarization resist was then applied and etched back to expose the dielectric substitutional gate, which was subsequently removed. After the channel had been recess etched for typically 15 nm, a Ti/Pt/Au gate was deposited and delineated by resist lift-off to form the LDD self-aligned M-p+-n gate FET (Kwok, 1986).

3. DEVICE CHARACTERISTICS

The I-V characteristics of M-p+-n SAGFET A, with typically n-channel implantation of 1.9E12 cm-2., having a 1.4 X 20 micron gate are given in Figure 3a. The I-V characteristics exhibit a square law relation between the drain current and the gate voltage with a threshold voltage of +1.3 V and 'K'-factor of 8.4 mS/V. The corresponding gate-source diode characteristics are shown in Figure 3b, which indicates a turn-on voltage, Von, of 1.6 V (arbitrarily defined at 2E-4 A current). The high Von suggests a large barrier increase of the M-p+-n structure. According to the full depletion model given in Figure 2, a barrier enhancement of approximately 0.38 eV can be expected from the Device A. The Schottky barrier height of Ti/Pt/Au on n-type GaAs is typically 0.74 eV; so the estimated total effective barrier height for the device A is 1.12 eV. Because of the large positive Vth of 1.3 V, large gate voltages are required to drive the FET on. For voltages larger than the diode's Von, significant gate current is injected.

I-V CHARACTERISTICS OF M-P-N GATE FET

(1.4 X 20 μm)

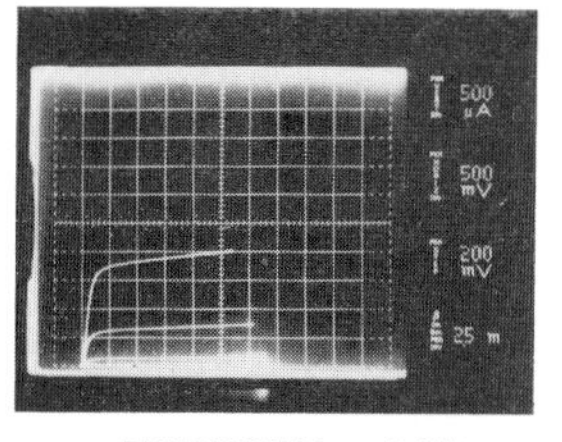

a. FET MODE Vgs = 1.8V
gm = 325mS/mm

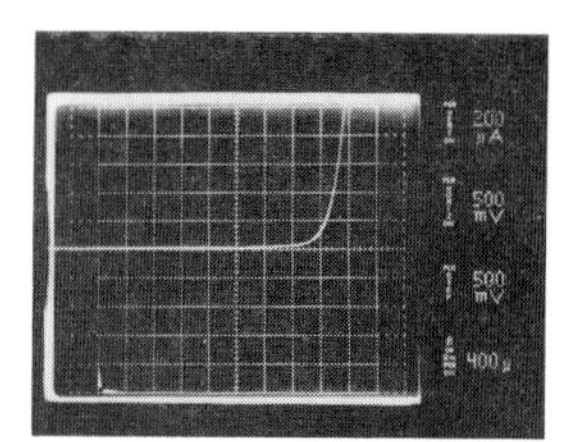

b. GATE - SOURCE $V_{on} \simeq 1.6$ V

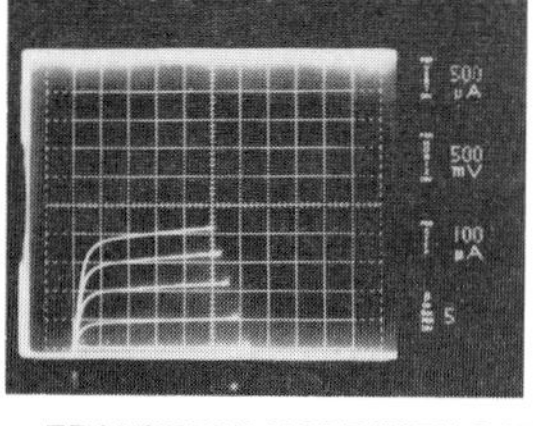

c. TRANSISTOR MODE WITH GATE CURRENT $\beta = 4 - 5$

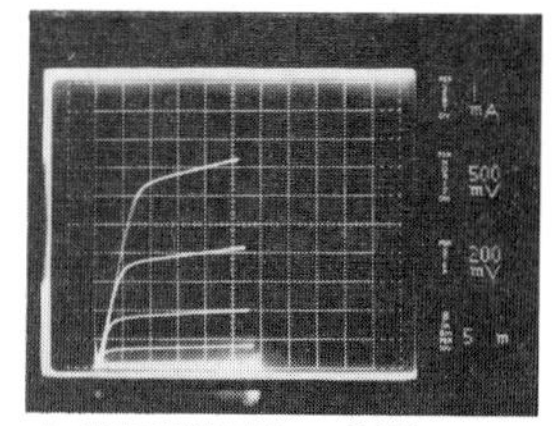

d. FET MODE Vgs = 2.2V
gm = 800mS/mm
k = 8.4 mS/V

Figure 3

Figure 3c shows the drain-source I-V characteristics using gate current drive. The drain-to-gate current ratio is typically 4 to 5. Therefore, the device is not operated purely in the FET mode. For even higher gate voltage of 2.2 V, large drain current of 7ma and 'transconductance' of 800 mS/mm are demonstrated with a 1.4 X 20 micron gate geometry (Figure 3d). The log(I)-V characteristics of gate-source diode (2.5 X 200 micron) test structure was also measured. Because of large leakage current, partly due to the contribution of surface leakage of large diode perimeter/area ratio, the forward biased log(I)-V exhibits multiple slopes. No meaningful barrier height parameter can be extracted from it thus far. The C-V characteristics of the same diode (2.5 X 200 micron) and dopant profiles were also measured using complex impedance and exclusion of resistive component to extract the capacitance. Because of strongly depleted M-p+-n, proper interpretation of the data remains under investigation at the time of this writing. However, it is clear that the structure is depleted even at a positive gate voltage of 1 V. The depleted capacitance value was 0.35 pF.

The I-V characteristics of two neighboring M-p+-n SAGFETs with type A and B doping densities are given in Figure 4. Both devices are biased identically with a maximum gate voltage of +1.8 V. Device A having lower n-concentration exhibits drain current of 1.7 mA, while B

M-P-N SAGFET I-V CHARACTERISTICS

(1.4 x 20 MICRON GATE)

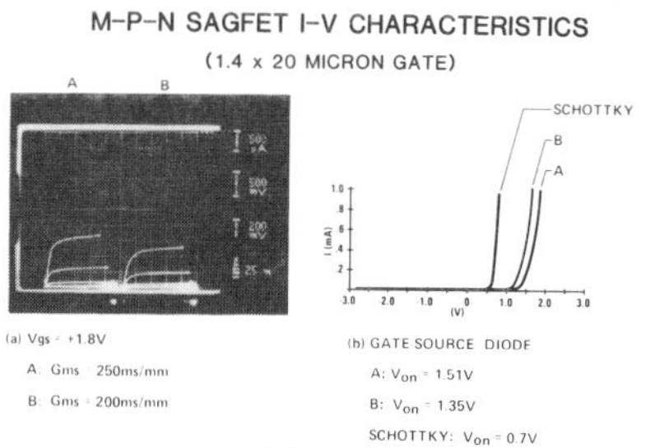

(a) Vgs = +1.8V
A: Gms = 250ms/mm
B: Gms = 200ms/mm

(b) GATE SOURCE DIODE
A: V_{on} = 1.51V
B: V_{on} = 1.35V
SCHOTTKY: V_{on} = 0.7V

Figure 4

having twice the concentration exhibits 1.3 mA. The Vth of both devices are approximately +1.25 V, which are insensitive to the n-channel doping variation by 2 to 1. However, the Von of the gate-source diode with a lower n-channel doping density is 1.51 V compared to 1.35 V of higher doping, consistent with prediction of Figure 2.

4. DISCUSSION

A strong enhancement mode M-p+-n SAGFET has been fabricated, which has a high gate barrier height. The gate turn-on voltage has been increased from 0.7 V of conventional Schottky diode to as large as 1.6 V. Also, the barrier height enhancement was found to decrease with increasing n-channel doping density as predicted by the M-p+-n depletion model. It is suggested that partly because of significant diffusion of Mg of the p+ layer during post implantation annealing, the width of the p+ layer increases significantly causing large barrier height enhancement to saturation. Not only holes accumulate in the vicinity of the barrier hump, the p+n junction causes a strong depletion of the short n-channel, which was designed to enhance the SAGFET transconductance. As a result, larger than expected Vth of +1.2 to 1.3 V was obtained. A 'K'-factor equal to 8.4 mS/V and 'transconductance' up to 800 mS/mm were obtained at gate voltages significantly higher than the gate turn-on voltage and drain/gate current ratio of 4 to 5. The SAGFET Vth was also found to be relatively insensitive to n-channel doping density. It is suggested that with a shallower p+ layer or deeper n-channel, an enhancement mode M-p+-n SAGFET with a high gate barrier height and small Vth (0.3 to 0.4 V) can be realized, thus high transconductance without significant gate current.

5. ACKNOWLEDGEMENT

Prof. N. W. Cheung for discussion and encouragement and D. Bullock for process assistance.

6. REFERENCES

Chen C Y, Cho A Y, Cheng K Y and Garbinski P A Appl Phys Lett 40 1982.
Kwok S P, Feng M and Kim H B Proc of 13th Int Sym on GaAs & Rel Comp 1986.
Kwok S P, Chang Y et al GaAs IC Sym Tech Digest Grenelefe Fl 1986.
Kwok S P J Vac Sci Technol B 4 1986.
Li S S Solid State Electrons 21 1978.
Lubberts G, Burkey B C, Bucher H K, Wolf E L J Appl Phys 45 1974.
Shannon J M Appl Phys Lett 24 & 25 1974.
Zhang L C, Cheung S K, Liang C L and Cheung N W Appl Phys lett 50 1987.
Zhang L C et al J Vac Sci Technol B 1987 (Nov)

Paper presented at Int. Symp. GaAs and Related Compounds, Heraklion, Greece, 1987

Device performance and transport properties of HFETs at low temperatures

W.Prost, W.Brockerhoff, K.Heime
Universität Duisburg Halbleitertechnik/Halbleitertechnologie Sonderforschungsbereich 254, D-4100 Duisburg

W.Schlapp, G.Weimann
Forschungsinstitut der DBP, D-6100 Darmstadt

Abstract

Transport and device performance of HFETs at cryogenic temperatures in absolute darkness are optimized by use of a 20 period superlattice (SL) with 1,75nm center doped GaAs wells sandwiched between 1,0nm AlAs barriers instead of a n-AlGaAs carrier supplying layer. At 30K in the dark the following data are achieved: μ_H=230.000cm^2/Vs, n_s=6,5$\cdot 10^{11}$cm^{-2}, R_{2DEG}=42 Ohm/square. SL HFETs with a gate length of 1,3μm exhibit a transconductance g_m=240mS/mm below 70K, which is two times the room temperature value. From cryogenic DC performance and room temperature RF measurements we have calculated the maximum oscillation frequency at 77K to 110GHz.

1. Introduction

Conventional AlGaAs/GaAs heterostructure field effect transistors (HFETs) provide strongly enhanced transport properties at low temperature. But excellent device performance at cryogenic temperature are reported (e.g. Pospialski 1986) under illumination, only. We have correlated this effect mainly with a donor like level of 0.44eV and 0.47eV below the conduction band edge (Heuken 1986). This level is called DX-center and is related to the heavily Si doped AlGaAs layer (Ishikawa 1986). Based on the assumption that the DX-center is due to the neighbourhood of Al, Ga and a dopant atom, Baba (1983) replaced the AlGaAs donor layer by an n-GaAs/AlAs superlattice (SL) which provides a spatial separation of Al from the dopant. In addition Baba (1986) applied the center doping technique (Schubert 1985) to the GaAs quantum wells. Using this doping scheme the highest spatial separation and best results are obtained (Baba 1986, Tu 1986, Prost 1987). Center doped SL HFETs were fabricated. Transport, deep level and device (DC and RF) measurements were carried out giving rise to the contention that this structure will overcome the problems of the n-AlGaAs/GaAs HFET at cryogenic temperatures.

2. SL HFET structure and device fabrication

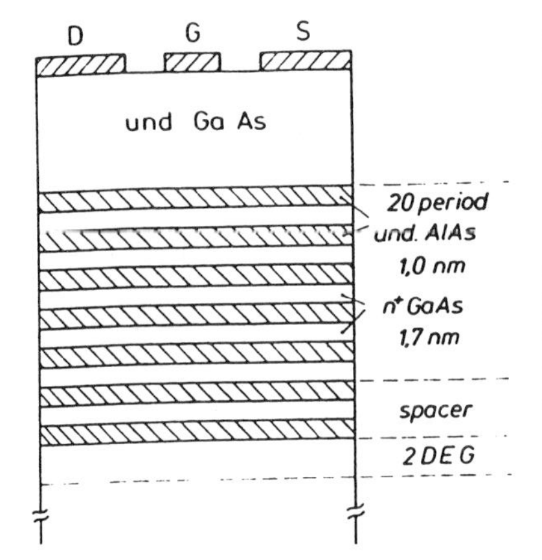

Fig. 1: Schematic layer sequence of a SL HFET

Fig. 1 illustrates the layer sequence. A 1μm thick undoped GaAs buffer is grown on (100) oriented semiinsulating substrate by MBE, followed by an undoped SL spacer. (three 1,0nm AlAs barriers, two 1,75nm GaAs wells). The carrier supplying layer consists of a 20 period SL with same layer dimensions as the spacer. At each center of the GaAs wells the growth was interrupted an $1 \cdot 10^{12} cm^{-2}$ Si dopants were in corporated at a temperature of 915K. A 20nm undoped GaAs toplayer serves as surface protection and facilitates ohmic contact formation. AuGe/Ni ohmic contacts were evaporated and annealed at 840K for 6 minutes in a radiation heater in vacuum. This high temperature annealing process was necessery in order to contact the 2DEG through the AlAs barriers. The gate was recessed by use of a sulphuric etchant and a Cr/Au gate with a gate length of 1.3μm was evaporated directly into the gate recess.

3. Experimental results and discussion

The samples were cooled down to the minimum temperature of about 30K and held there for about 12 hours without bias applied, but under the illumination conditions of the measurement (dark or illuminated by a red LED).

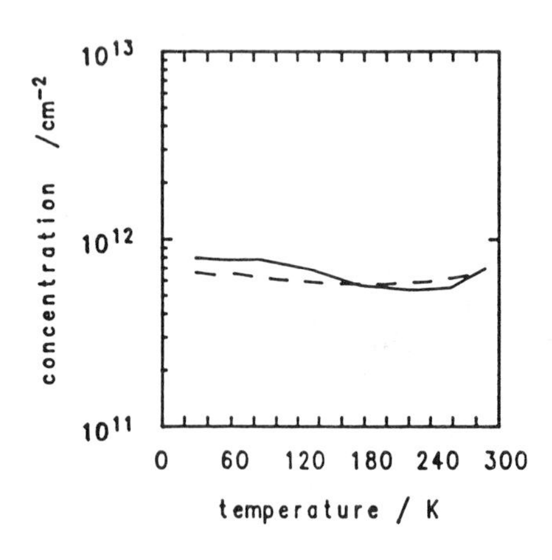

Fig. 2: 2DEG sheet carrier concentration in the dark and under red LED illumination of a SL HFET

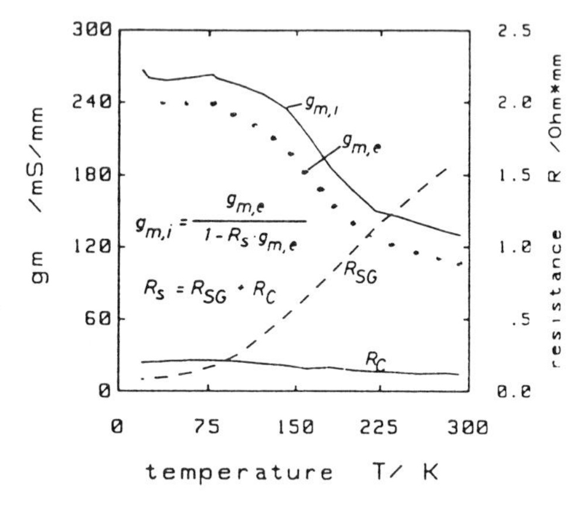

Fig. 3: Influence of parasitic contact (R_C) and conduction region (R_{SG}) resistance on transconductance g_m (i:intrinsic, e:extrinsic) of a SL HFET

The HALL mobility of the two dimensional electron gas (2DEG) in the sample increases from $8300cm^2/Vs$ (300K) to $230.000cm^2/Vs$ (30K). Illumination influence is very small. The sheet carrier concentration (cf. fig.2) remains constant in the dark. Under illumination there is small nonpersistant enhancement of less than 20% below 140K. Transmission-line measurements were performed in order to discriminate which part of the feed-back source resistance in the device is related to the contact resistance R_C and which to the conduction region between source and gate (R_{SG}). In fig. 3 the influence of R_C and R_{SG} on extrinsic transconductance is plotted versus temperature. We have used an undoped caplayer resulting in a high R_{SG} at room temperature. Cooling down the device R_{SG} decreases according to the higher mobility. The thermionic emission through the AlAs barriers decreases which results in a higher contact resistance. Below 100K R_C is dominant in SL HFETs.

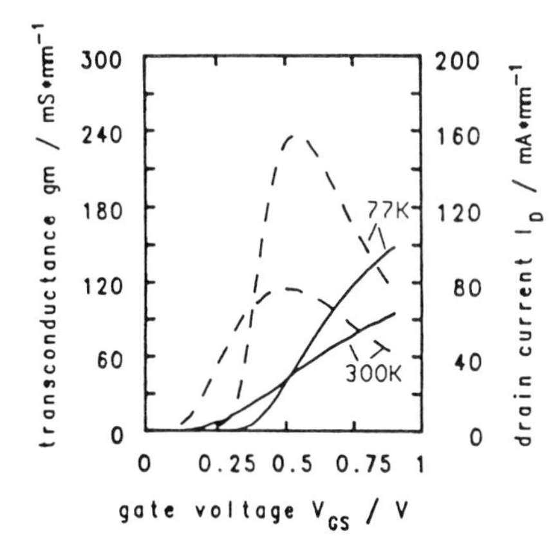

Fig. 4: Drain current and extrinsic transconductance (dashed lines) in the dark versus gate bias of a SL HFET at 77K and 300K.

In fig.4 drain current and transconductance are plotted versus gate bias at 77K and 300K in the dark. The threshold voltage shift is less than 200mV. The drain saturation current increases up to 40% at high forward bias. As the number of carriers is held constant (cf. fig.2) the average velocity under the gate is 40% higher at 77K in comparison to 300K. The transconductance $g_m=\Delta I_D/\Delta V_{GS}$ increases from 120mS/mm to 240mS/mm, indicating that the higher average velocity is not able to explain the main part of enhancement. The HFET is mainly a one subband device as most carrier are confined in the lowest subband (Yoshida 1986). The peak point of transconductance corresponds to that gate bias where the Fermi level crosses the lowest subband. At low temperaturethe Fermi distribution is much smaller resulting in lower gate bias swing in order to (de)populate the subband. This results in a higher transconductance and a small threshold voltage shift at low temperatures. The variation of 2DEG carrier concentration from 300K to 77K is as low as $3\cdot10^{10}cm^{-2}$ leading to a threshold shift of about +80mV, only. Tiwari (1984) estimated a thermal shift of +85mV. The sum of these shifts is in good agreement with our result. RF measurements were performed at 300K: (MAG=1) at f_c=18GHz. An equivalent circuit was evaluated. The undoped caplayer results in a high source resistance of 2 Ohm·mm. The maximum oscillation frequency is f_{max}=29GHz. Taking the improvement of transport properties and transconductance at low temperature into account the maximum oscillation frequency at 77K is calculated to $f_{max,77K}$=110GHz.

Conclusion

DX-center related device degradation in HFETs has been eliminated by use of a center-doped n-GaAs/AlAs carrier supplying layer. Although PhotoCAP measurements have proven that a total spatial separation of Al from dopant Si has not been achieved (Prost 1987) the remaining trap concentration does not affect device performance. Cooling the device down from 300K to 30K the carrier concentration in the dark is unchanged, whereas the mobility increases by a factor of 27. The average saturation velocity is about 40% higher, only. The main part of transconductance enhancement at low temperature is a result of the smaller Fermi distribution function resulting in a lower gate voltage swing and a small threshold voltage shift. RF performance at low temperature was carefully predicted ($f_{max,77K}$=110GHz) from both room-temperature RF measurement and from low temperature transport and device performance. These results proofe that this device is an excellent candidate for microwave amplification at cryogenic temperatures in the dark.

Acknowledgement:

The authors are indepted to G.Howahl and U.Doerk for device fabrication and measurements. This work was in part supported by Deutsche Forschungsgemeinschaft und Stiftung Volkswagenwerk.

References

Baba T.,Mizutani T.,Ogawa M., Japan J.Appl.Phys.22(1983) L627-629
Baba T., Microelectronic Engineering 4 (1986) 195-206
Heuken M.,Prost W.,Kugler S.,Heime K.,Schlapp W.,Weimann G., Inst. Phys. Conf. Ser. No.83, 563, 1986
Heuken M.,Loreck L.,Heime K.,Ploog K.,Schlapp W.,Weimann G., IEEE Trans. Electron Devices, ED-33, 693, May 1986
Ishikawa T.,Yamamoto T.,Kondo K.,Komeno J., Inst. Phys. Conf. Ser. No.83, 99, 1986
Pospialski M.W.,Weinreb S.,Chao P.C.,Mishra C.K., Palmateer S.C.,Smith P.M.,Hwang J.C.M., IEEE Trans. Electron Devices, Vol.ED-33, 218, 1986
Prost W.,Brockerhoff W.,Heuken M.,Kugler S.,Heime K., Schlapp W.,Weimann G., in "Properties of Impurity States in Semiconductor Superlattices", Plenum Press.1987 (to be published)
Schubert E.F.,Ploog K., Jap.J. of Appl.Phys.,Vol.24, No.8, 1985
Tiwari,S., IEEE Trans. Electron Devices, Vol.ED-31, 879, 1984
Tu C.W.,Jones W.L.,Kopf R.F.,Urbanek L.D.,Pei,S.S., IEEE Electron Device Letters, Vol.EDL-7, No.9,1986
Yoshida J., IEEE Trans. Electron Devices, Vol.ED-33,No1, 1986

Influence of extended defects on the electrical behaviour of GaAs field-effect transistors

W De Raedt, M Van Hove, M de Potter, M Van Rossum
IMEC v.z.w., Kapeldreef 75, 3030 Leuven, Belgium

JL Weyher
Catholic University, Faculty of Science, Solid State III
Toernooiveld, 6525 ED Nijmegen, The Netherlands

ABSTRACT: MESFET arrays have been processed on LEC GaAs wafers after Si implantation and rapid thermal annealing. A systematic mapping of the threshold voltage (V_{th}) has been performed. The influence of grown-in and process-induced extended defects on the V_{th} shift of nearby transistors has been studied by photoetching the wafers with diluted Sirtl-like solutions. Transistors close to the dislocation cell boundary and the surrounding impurity atmosphere show a diffuse decrease in V_{th}, while stress-induced (glide) dislocations have no observable influence on the device parameters.

1. INTRODUCTION

It is well known that the quality of GaAs substrates has a direct influence on the device performance. This quality can be affected by the crystal growth procedure as well as by several device or circuit processing steps, among which ion implantation and subsequent thermal annealing are of foremost importance. Various annealing procedures have been developed to remove ion implantation damage and activate the implanted dopants. In recent years rapid thermal annealing (RTA) by incoherent light heating (Pearton et al 1986) has attracted much attention because of the excellent quality of the active layers obtained by this technique. Although evaluation of the active layers is usually performed by electrical measurements (sheet resistivity and V_{th} determination), it is not possible to obtain a detailed picture of the defect structures from electrical data alone.

Recently, a new etching system based on CrO_3-HF-H_2O solutions has been introduced under the "DS(L)" denomination (Diluted Sirtl-like solutions with the use of Light) (Weyher and van de Ven 1983, van de Ven et al 1986). These solutions exhibit a very sensitive defect revealing action on different III-V compounds. Moreover, their etching properties are strongly influenced by light absorption at the etched surfaces, allowing to detect chemical inhomogeneities due to dopant concentration, precipitates, defect aggregation, etc. The DSL technique is thus well suited for correlating crystal defects and their influence on the electrical characteristics of doped substrates.

In the present paper, we report the first results of an extended study on the correlation between various defects revealed by DSL and the V_{th} of GaAs MESFET's.

2. EXPERIMENTS

Standard commercially available undoped semi-insulating LEC-grown 2" GaAs substrates were used for this study. After degreasing and chemical etching (5:1:1 H_2SO_4-H_2O_2-H_2O) the wafers were implanted with 70 keV ^{29}Si ions at a dose of 3 10^{12} at/cm^2. To minimize inhomogeneities due to channeling a rotated and tilted implantation angle was used and part of the wafers were implanted through a 90nm SiN cap layer (120 keV, 5 10^{12} at/cm^2). After removal of the nitride, RTA of the samples was carried out under forming gas flow, in a HEATPULSE© 410 system at 900°C during 10sec in proximity mode face-to-face with GaAs. To minimize additional process-induced defects a simple MESFET fabrication process was used. The devices were isolated by mesa etching, AuGeNi was used as ohmic metallisation and TiW/Au as the gate metallisation.

The 1.5x30μm^2 FET's were patterned in 6x8 arrays with common gates and source-drain contacts (fig.1). Electrical measurements were carried out on an automatic system. The V_{th} of each FET was determined by extrapolating the linear region of the gate voltage-transconductance curve at a source-drain voltage of 100mV. The metallisation was removed by a short wet etch prior to DSL photo-etching.

Fig.1. MESFET array (350x500 μm^2) for V_{th} mapping

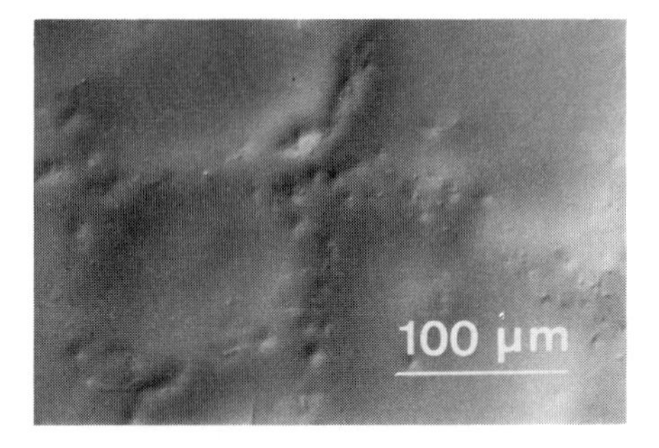

Fig.2. Dislocation cell walls revealed by DSL photoetching of the GaAs wafer after Si implantation and RTA

3. RESULTS AND DISCUSSION

In a previous paper we already listed some defect structures revealed by DSL on implanted and annealed substrates (de Potter et al 1987). In the present study, we focus our attention on dislocation cell walls and glide dislocations.

Dislocation cell walls only appear after deep photoetching. They form periodic patterns of 100μm wide ridges with a 100 to 200 μm average separation (fig.2). Numerous examples of cell walls crossing the FET arrays were found on the wafers (fig.3a). The electrical signature of these defects seems to relate with the occurence of diffuse periodic fluctuations of V_{th} on a 100 μm scale with an average 50 mV amplitude. We observe that the cell wall corresponds with a more negative V_{th} (fig.3b). Glide dislocations appear preferentially at the wafer edges. It is generally believed that they are due to high thermal stresses during growth and RTA. In some occasions, we could clearly identify slip lines running parallel to the FET rows along the <110> direction (fig.4). Their influence on V_{th} seems hardly noticeable. Finally, a fairly low density of isolated (possibly process-induced) hillocks was detected over the wafers. Some of them were located in the immediate vicinity of a partic-

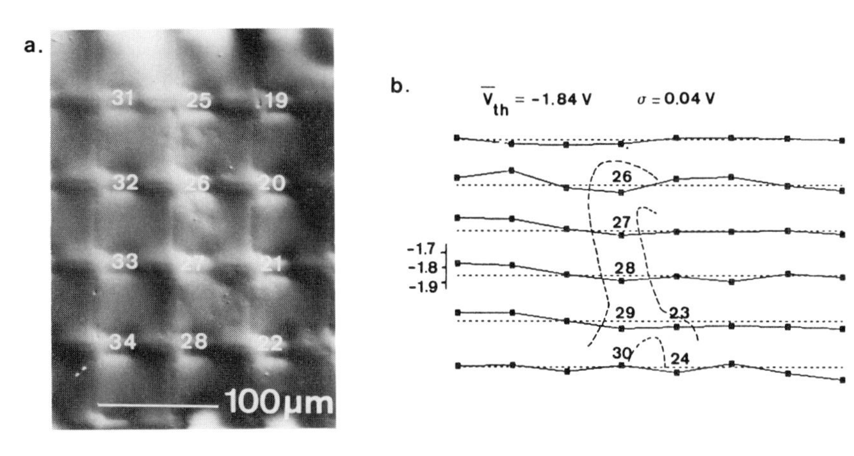

Fig.3. Fet array with dislocation cell wall (a), and V_{th} mapping of the same array (b)

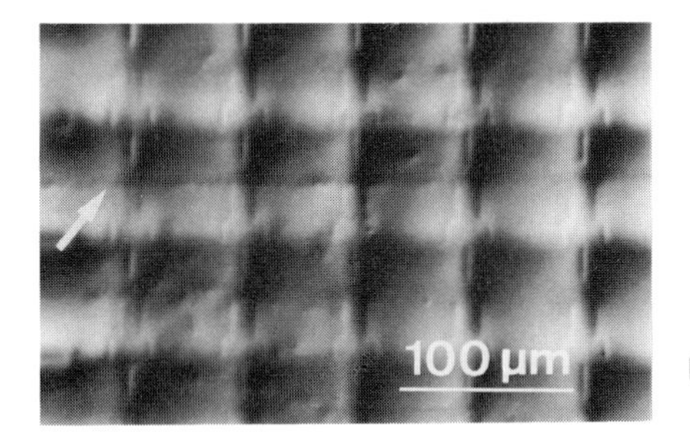

Fig.4. FET array with a glide dislocation

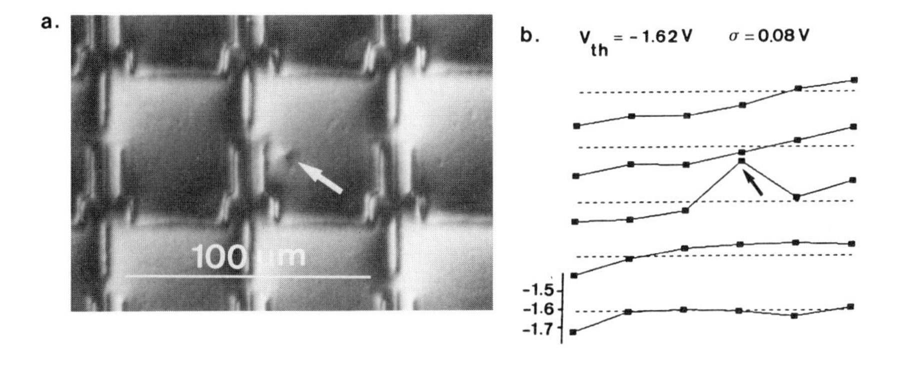

Fig.5. FET array with isolated hillock (a) and corresponding V_{th} mapping (b)

ular transistor (fig.5). In such cases, a dramatic increase of the corresponding V_{th} (up to 200 mV) was observed. It should be emphasized that this shift indicates a decrease in the local electron concentration, contrary to what happens on the dislocation cell walls. The exact nature of these hillocks is still uncertain.

The proximity effects observed around cell walls follow the general trend reported in earlier studies in which dislocations were identified using molten KOH (Miyazawa and Hyuga 1986a, Egawa et al 1986, Suchet et al 1987 and Kitching 1987). Miyazawa and Wada (1986b) have proposed a model in which a more negative V_{th} is attributable to an increase in the Si_{Ga}-Si_{As} net concentration. This model does not take into account the presence of other electrically active centers in the impurity atmosphere surrounding the cell wall. However, the DSL etch patterns, combined with high resolution photoluminescence spectroscopy, clearly indicate that high concentrations of impurities are present indeed (de Potter et al 1987). At the same time no impurity atmosphere is detected around stress-induced dislocations (Weyher et al 1987). This might indicate that dislocations themselve do not contribute to V_{th} shifts. Since the cell walls are known to contain dislocations, precipitates and impurities, it remains to be established what is the relative influence of the latter two on the V_{th} shift.

In conclusion, after application of photoetching on implanted wafers processed by RTA, we found evidence that the DSL-revealed impurity atmosphere around the dislocation cell walls is the dominant factor affecting V_{th} variations.

ACKNOWLEDGMENTS

The authors are pleased to thank P.Richardson for wafer processing, and S.Naten and D.Verhaeghe for their help with electrical measurements and data analysis. Financial support was received from the Instituut voor Wetenschappelijk Onderzoek in Nijverheid en Landbouw (IWONL). One of the authors (JLW) wishes to thank the Stichting voor Fundamenteel Onderzoek der Materie (FOM) and the Nederlandse Organisatie voor Zuiver-Wetenschappelijk Onderzoek (ZWO).

REFERENCES

de Potter M, De Raedt W, Van Hove M, Van Rossum M and Weyher JL, 1987 E-MRS Strasbourg (in press)
Egawa T, Sano Y, Nakamura H and Kaminishi K 1986 Jpn. J. of Appl. Phys. **25** L973
Kitching SA 1987 E-MRS Strasbourg (in press)
Miyazawa S and Hyuga F 1986a IEEE Trans on Electron Devices**33** 227
Miyazawa S and Wada K 1986b Appl. Phys. Lett. **48** 905
Pearton SJ, Gibson JM, Jacobson DC, Poate JM, Williams JS and Boerma DO 1986 Mat. Res. Soc. Symp. Proc. **52** 351
Suchet P, Duseaux M, Maluenda J and Martin GM 1987 J. Appl. Phys. **62** 1097
van de Ven J, Weyher JL, van den Meerakker JEAM and Kelly JJ 1986 J. Electrochem. Soc. **133** 799
Weyher JL and van de Ven J 1983 J. Crystal Growth **63** 285
Weyher JL, Le Si Dang and Visser EP 1987 (this conference)

Paper presented at Int. Symp. GaAs and Related Compounds, Heraklion, Greece, 1987

GaAs microwave MESFET fabricated by ion implantation on a three-inch diameter MOCVD-grown GaAs on Si substrate

Y. Chang, M. Feng, C. Ito, D. McIntyre, R.W. Kaliski, V. Eu, R. Laird, T.R. Lepkowski, D. Williams, Z. Lemnios, and H.B. Kim

Ford Microelectronics, Inc., 10340 State Highway 83 North, Colorado Springs, Colorado 80908, USA

ABSTRACT: GaAs grown on a silicon substrate by MOCVD is used to fabricate ion-implanted, microwave MESFETs. A 0.7 x 300 micron gate FET exhibits transconductance of 110 mS/mm. The unilateral gain is 24dB at 1GHz and the cutoff frequency is 11 GHz for a device with a 150 micron gate. The parasitic capacitance caused by the conductive interface between the GaAs layer and silicon substrate is found to degrade RF performance.

1. INTRODUCTION

Since the first demonstration of single crystal GaAs epitaxial growth directly on silicon substrates by MBE (Wang, 1984) and MOCVD (Akiyama, 1984), the GaAs on silicon material system has gained great interest for both discrete device and integrated circuit applications. Discrete devices include room temperature CW lasers (Kaliski, 1987; Deppe, 1987) and MBE grown MESFETS (Fischer, 1986). Integrated circuits include a digital 1K SRAM (Shichijo, 1987) and an X-band MMIC amplifier (Eron, 1987).

Previous demonstrations of microwave performance for FETs on GaAs/silicon material have used an epitaxially-grown active layer fabricated by either MBE (Fischer, 1987) or MOCVD (Inomata, 1986). In order, however, to achieve a selectively-doped, planarized process suitable for production of monolithic microwave and digital ICs, development of ion implantation and annealing techniques for GaAs/silicon material are critical. Therefore, this work investigates the microwave performance of ion-implanted GaAs MESFETs fabricated on GaAs/silicon material.

2. MATERIAL GROWTH AND CHARACTERIZATION

A MOCVD growth technique is used to grow GaAs directly on the silicon substrates (Ito, 1986). The reactor is a customized system with a maximum capacity of 39, 3-inch wafers and can accomodate wafers up to 8 inches in diameter. Typical reactor loads are 8 to 10 wafers per run.

Hall measurements for an implant dose of 5×10^{13} cm^{-3} at 150 kev for both GaAs/silicon and GaAs buffer/GaAs substrate have been compared. The average Hall mobility and concentration for the N^+ implant is 2300 cm^2/V-sec and 6.4×10^{17} cm^{-3} for GaAs/silicon and 3100 cm^2/V-sec and 8.4×10^{17} cm^{-3} for GaAs/GaAs. This data indicates that the implant activation of GaAs/silicon material is about 75% compared activation in GaAs/GaAs.

A C-V plot of the background impurity concentration for GaAs/silicon shown in Figure 1, indicates an n-type concentration of about 5×10^{15} cm^{-3}. The free carrier concentration increases to about an order to magnitude to 4×10^{16} cm^{-3} at the GaAs-silicon interface (about 3.5 microns from the surface). This high concentration results in a conduction layer at the interface, and hence a parasitic capacitance which can degrade high frequency operation.

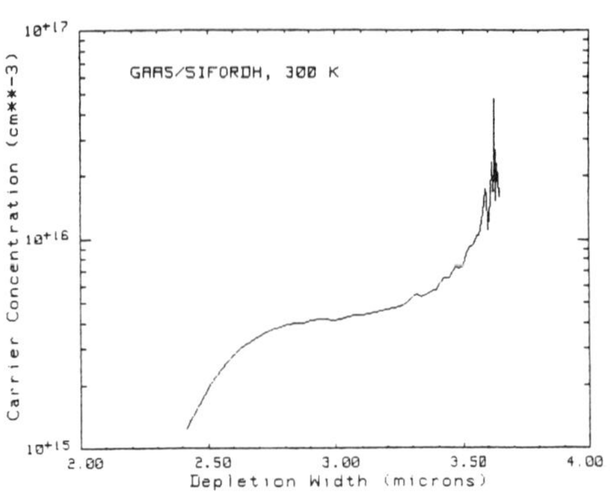

Figure 1. C-V plot of background and interface concentration

3. MESFET DEVICE FABRICATION

The active channel layer is formed by selective ion implantation of Si^{28} at an energy of 100keV. To activate the implant, rapid thermal annealing is done at 925C for 6 seconds. The peak concentration is about 2.7×10^{17} cm^{-3}. The ohmic contacts to the N^+ region is formed by evaporation of Au-Ge/Ni/Au and alloyed at 400C for 2 minutes. The channel current is then adjusted by recessing the implanted layer prior to Ti/Pt/Au gate deposition. A typical gate length is 0.8 microns.

4. DC DEVICE CHARACTERIZATION

The local uniformity of I_{sat} measured prior to the channel etch is shown in Figure 2. These uniformity measurements are made on four 300 micron wide devices, each 200 microns apart. The average I_{sat} is 180 mA ± 5 mA. This uniformity data is comparable to GaAs buffer/GaAs substrate material, however, the average Isat for GaAs/GaAs material is about 210 mA.

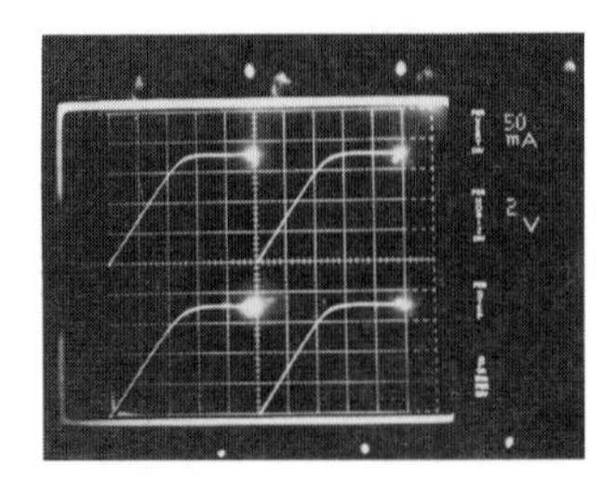

Figure 2. Local I_{sat} Uniformity

The FET I-V curves are shown in Figure 3 for the same devices described previously. For local uniformity, the average I_{dss} is 44 mA ± 5 mA at V_{ds} = 4 V, the average pinch-off voltage is 2.6 V ± 0.2 V at 5% I_{dss} and the average transconductance is 70 mS/mm ± 3 mS/mm. This local uniformity is also comparable to GaAs/GaAs material. The drain-to-source low field resistance, however, is about 1.7 higher for GaAs/silicon when compared to GaAs/GaAs. The highest transconductance

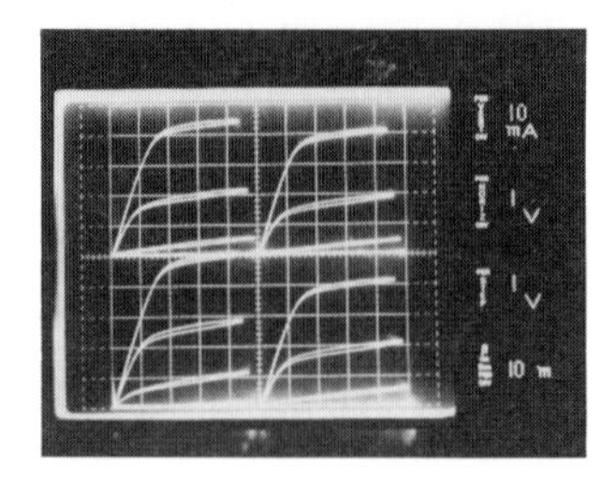

Figure 3. Local I-V Uniformity

achieved is 110 mS/mm for a MESFET with gate dimensions of 0.7 x 300 microns.

5. RF DEVICE CHARACTERIZATION

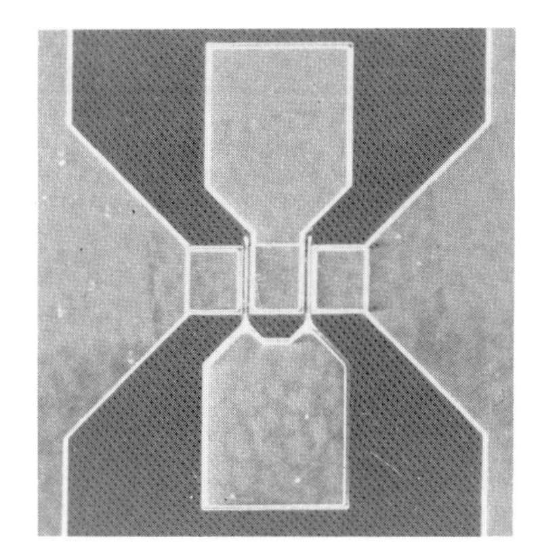

Figure 4. GaAs MESFETs

The RF characteristics are measured at wafer-level on Cascade-probeable, GaAs MESFETS (shown in Figure 4) with gate widths of 60, 100, and 150 microns and at frequencies from 0.1 to 10 GHz. For low frequency operation between 0.1 to 1.0 GHz, the unilateral gain for devices fabricated on GaAs/silicon material is the same as for comparable devices on GaAs/GaAs material. Also, for a 150 micron device at 1 GHz, the maximum stable gain (Figure 5) and unilaterial gain are 20 dB and 24 dB, respectively. The typical performance of a 0.6 x 300 micron, ion-implanted GaAs/GaAs FET is also shown in Figure 5 for reference, with f_{max} at approximately 70 GHz.

When the frequency increased from 1 to 10 GHz, the capacitance between the source pad and the conductive GaAs-silicon interface of GaAs/Si (as well as the low resisitivity of silicon substrate) degrades the RF gain. This parasitic capacitance effect is demonstrated by varying the gate width while holding the source area constant. Since the intrinsic input capacitance decreases with smaller gate widths while the source capacitance remains constant, the effect of this parasitic capacitance should be more pronounced for the smaller gate geometries. This effect is clearly demonstrated in Figure 5.

The S-parameters of ion-implanted MESFETs for both GaAs/GaAs and GaAs/silicon material are presented in Figure 6. S21 and S12 are the key parameters for determining amplifier gain since gain = [S21]/[S12]. The forward gain, [S21], for GaAs/GaAs is about 10 dB better than GaAs/silicon over the frequency range of 1.5 to 26.5 GHz while the reverse gain, [S12], for GaAs/GaAs is -6 dB better than GaAs/silicon over the same frequency range.

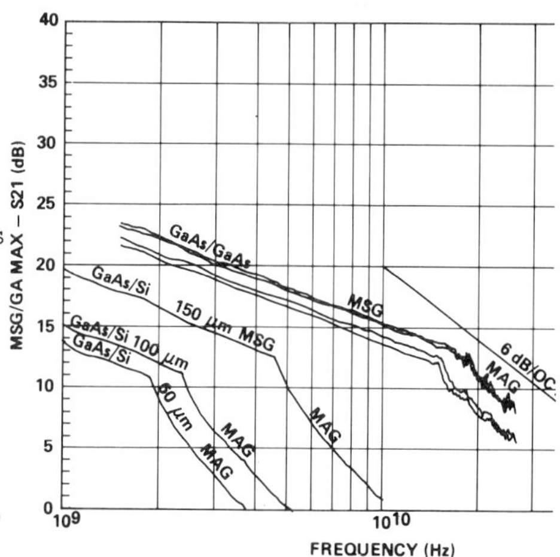

Figure 5. Gain vs Frequency

The Smith charts for GaAs/silicon and GaAs/GaAs are shown in Figure 7. For GaAs/GaAs, the S11 parameter is typically represented by a series RC equivalent circuit. For GaAs/silicon, however, the addition of the parasitic capacitance and resistance cause S11 to move inward on the Smith chart. S22, a parallel RC equivalent circuit for GaAs/GaAs, experiences the same complication due to the parasitic capacitance in GaAs/silicon material and moves inward.

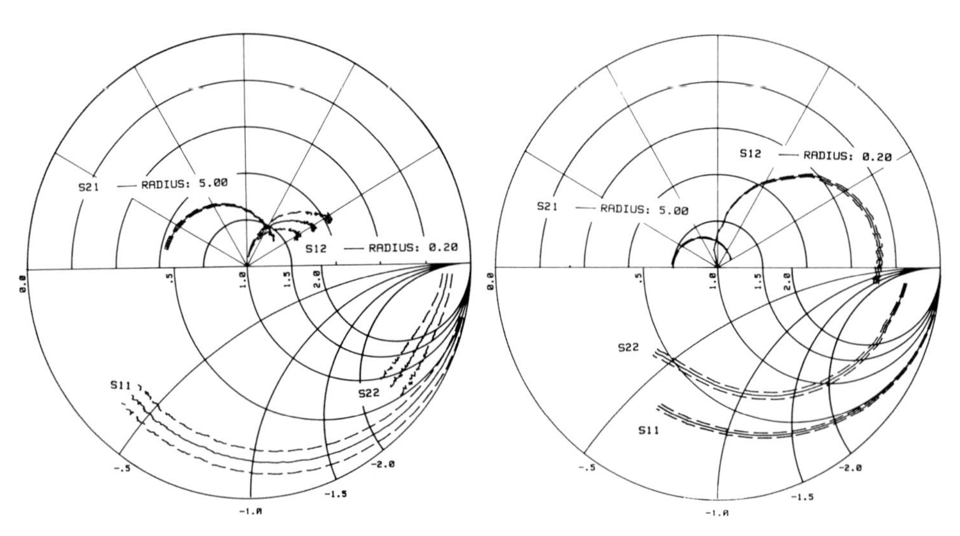

Figure 6. GaAs/GaAs S-Parameter Figure 7. GaAs/Si S-Parameter

REFERENCE

Akiyama M, Kawarada Y, Kaminishi, 1984 Jpn.J. Appl. Phys. 23, 843.

Deppe D G, Holonyak Jr. N, Nam D W, Hsieh K C, Jackson G S, Matyi R J, Shichijo H, Epler J E, Chung H F, 1987 Appl. Phys. Lett. 51, 637 (1987)

Eron M, Taylor G, Menna R, Narayan S Y, Klaskin J, 1987 IEEE EDL-8 350.

Fisher R, Chand N, Kopp W, Peng C K, Morkoc H, Gleason K R, Scheitlin D, 1986 IEEE-ED, ED-33, 206.

Inomata H, Nishi S, Akiyama M, Itoh M, Takahashi S, Kaminishi, 1985 Gallium Arsenide and Related Compounds (Inst. Phys. Conf. Ser. No. 79) 481.

Kaliski R W, Holonyak Jr. N, Hsieh K C, Nam D W, Lee J W, Shichijo H, Burnham R D, Epler J E, Chung H F, 1987 Appl. Phys. Lett. 50, 836.

Shichijo H, Lee J W, Mclevige, Taddiken A H, 1987 IEEE-ED Lett. EDL-8, 121.

Wang W I, 1984 Appl. Phys. Lett. 44, 1149.

Ito C, 1986 MRS Symposia Proceedings, 67, 197.

Inst. Phys. Conf. Ser. No. 91: Chapter 7
Paper presented at Int. Symp. GaAs and Related Compounds, Heraklion, Greece, 1987

Ultra-high-speed emitter coupled logic circuits using AlGaAs/GaAs HBTs

Y.YAMAUCHI, K.NAGATA, O.NAKAJIMA, H.ITO, T.ISHIBASHI and K.HIRATA

NTT LSI Laboratories
3-1, Morinosato Wakamiya, Atsugi-shi, Kanagawa 243-01 Japan

Abstract Emitter coupled logic gates and static frequency dividers using AlGaAs/GaAs HBTs were designed and the performance of fabricated circuits was characterized. A gate delay time of 4.5 ps was obtained with an internal logic swing of 250 mV. The divide-by-four circuits have successfully been operated with a maximum toggle frequency of 22.15 GHz.

1. Introduction

For future digital systems operating at extremely high clock rates, the HBT is a promising device because of its high-speed switching performance, high threshold uniformity, and large current driving capability. Recently, AlGaAs/GaAs HBTs with a cutoff frequency, f_T, of 105 GHz (Ishibashi et al. 1987) and a maximum oscillation frequency, f_{max}, of 105 GHz (Chang et al. 1987a) have been reported. These devices were fabricated to have very low parasitic resistances and capacitances using a self-alignment technique, which makes it possible to realize high-speed logic circuits (Nagata et al. 1987b, Chang et al. 1987b).

In practical logic gates using HBTs, emitter coupled logic (ECL) circuits are quite suitable for high-speed operation. The switching speed of ECL circuits is closely dependent on the internal logic swing level, V_L. High-speed switching requires a low V_L, which in turn requires HBTs with low emitter resistance.

This paper investigates the effect of V_L on the switching speed of ECL circuits using AlGaAs/GaAs HBTs with low emitter resistance. For the purpose of estimating circuit switching speed, 21-stage ECL ring oscillators and divide-by-four frequency dividers were designed and fabricated. A gate delay time, t_{pd}, of 4.5 ps with a logic swing of 250 mV and a maximum toggle frequency, f_{tog}, of 22.15 GHz in a divide-by-four circuit with a logic swing of over 400 mV have been achieved.

2. HBT characteristics

HBTs were fabricated on MBE-grown wafers with an InGaAs emitter cap layer to obtain low emitter contact resistivity (Woodall et al. 1981, Nagata et. al 1987a). As a base layer, a graded P^+-$Al_yGa_{1-y}As$ (y = 0.12 - 0) layer was utilized to reduce base transit time. This layer was heavily doped with Be to 4 x 10^{19} cm^{-3} to reduce base resistance. The collector layer consists of two n-GaAs layers, a 2000 A thick layer doped to 5 x 10^{16} cm^{-3}

and a 3000 A thick layer doped to 2 x $10^{17}cm^{-3}$. In this collector structure, the 2000 A thick layer determines the depletion layer thickness. Thus, this structure realizes a short collector transit time even under conditions of a high collector bias voltage and high collector current density. The relatively short depletion layer thickness also effectively takes advantage of velocity overshoot in the collector depletion layer (Yamauchi et.al 1986).

The cross sectional structure of the HBT is shown in Fig. 1. The HBTs were fabricated using a self-alignment technique. Proton ion implantation into the external collector layers (2000+3000 A, as described above) was also applied to reduce external collector capacitance. The HBT emitter area was 2 x 5 μm^2. For the final HBTs to be applied in output buffer circuits, the emitter area was 5 x 10 μm^2 to drive a 50 Ω line.

The fabricated HBTs with an emitter area of 2 x 5 μm^2 operated stably up to a very high collector current density of 3 x 10^5 A/cm^2 with a current gain of about 25. Cutoff frequencies, f_T, and maximum oscillation frequencies, f_{max}, were derived from S-parameter measurements. In the circuit operating bias condition of J_C = 1.5 - 3 x 10^4 A/cm^2, the f_T and f_{max} were 30 - 45 GHz and about 40 GHz, respectively. Maximum values of f_T = 80 GHz and f_{max} = 60 GHz were obtained at J_C = 1.2 x 10^5 A/cm^2. HBT equivalent circuit parameters were estimated from the static and microwave characteristics. In spite of the small emitter area of 2 x 5 μm^2, a relatively small emitter resistance of 1.7 Ω was obtained. Base resistance, R_B, collector resistance, R_C, base capacitance, C_{BE}, and collector capacitance, C_{BC}, were 50 Ω, 42 Ω, 72 fF and 15 fF, respectively.

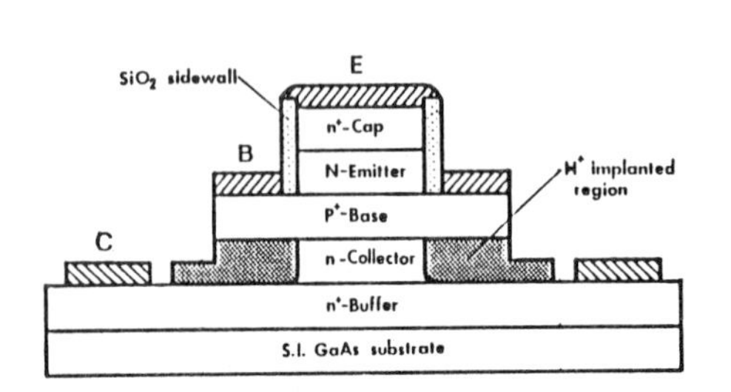

Fig.1 Cross-sectional structure of self-aligned HBT with InGaAs emitter cap layer.

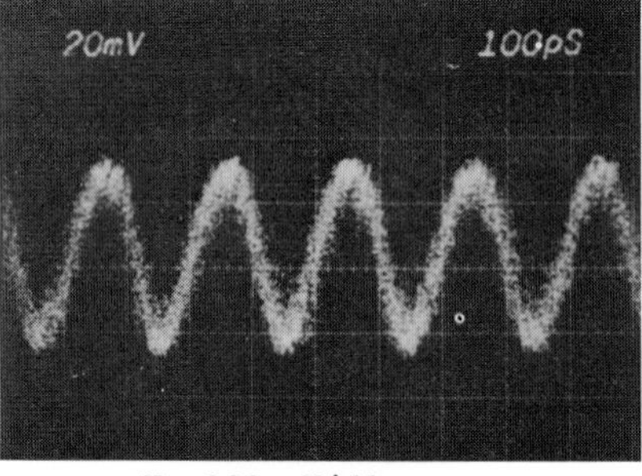

V: 100 mV/div.
H: 100 ps/div.
(t_{pd} = 4.5 ps/gate)

Fig.2 Output wave form of 21-stage ECL ring oscillator with logic swings of about 250 mV.

3. <u>Circuit Performance</u>

ECL 21-stage ring oscillators with the fan-in and fan-out of unity were designed. Load resistances were designed to provide a V_L of 250 mV - 800 mV at a power supply voltage of 7 V.

A minimum t_{pd} of 9.0 ps/gate and 5.5 ps/gate was obtained for V_L = 800 mV and 400 mV, respectively. Thus, the compression of V_L shortened the t_{pd}. Further compression to 250 mV resulted in an even shorter t_{pd} of 4.5

ps/gate. The ring oscillator output wave form of 4.5 ps/gate operation is shown in Fig. 2.

The simulation results for t_{pd} and power dissipation, P_d, for a V_L of 800, 400, and 250 mV are shown in Fig. 3. In this figure, experimental results are plotted by crosses for V_L = 800 mV, dots for 400 mV, and open circles for 250 mV. Good agreement was obtained between the measured and simulated results. Through these trials, it was confirmed that small emitter resistance results in stable operation with small logic swings.

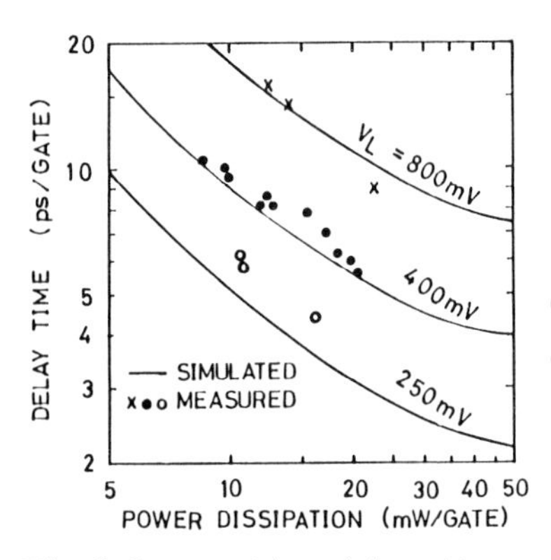

Fig.3 Propagation delay times versus power dissipation for ECL gate with logic swings of 800, 400 and 250 mV.

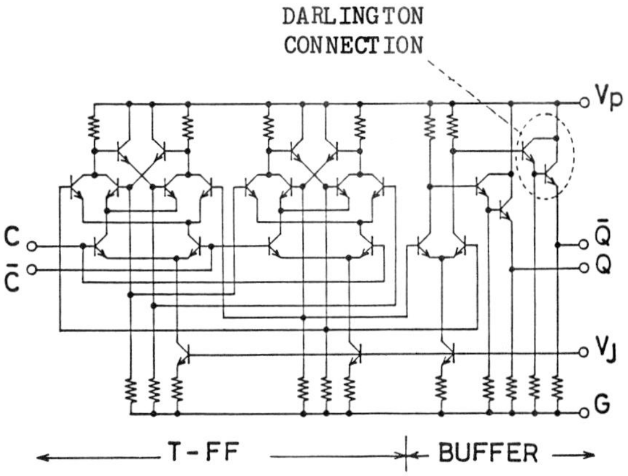

Fig.4 Schematic of half the 1/4 frequency divider circuit.

The 1/4 frequency divider circuit consisting of series gate T-type flip-flops and buffer circuits operating as level shifters was designed. A schematic of half the 1/4 frequency divider circuits is shown in Fig. 4. The divider operates with a single phase input signal, and double-phase (true and complementary) internal and output signals. For the level shifters, Darlington connections are employed to provide stable voltage shift operations, as also shown in Fig. 4.

The divider circuit operated stably changing the V_L from 800 mV to 400 mV at a power supply voltage of 9 V. The minimum input power was measured as a function of frequency. The results are plotted in Fig. 5. From this figure, it can be seen that the free-running frequency is about 11 GHz and the maximum toggle frequency is 15.6 GHz at a V_L of 800 mV. At a V_L of 400 mV, the free-running frequency is as high as 20 GHz. Moreover, a maximum toggle frequency of 22.15 GHz was obtained at a total power dissipation of 712 mW. A low minimum input power of 0 dBm was also achieved.

The frequency response of the buffer circuit was measured. The measured gain versus frequency characteristics were shown in Fig. 6. As shown in this figure, the cutoff frequencies of $|S_{21}|$ and G_u (unilateral gain) are 11 GHz and 12 GHz, respectively. In large input voltage swings, the buffer circuit operated as a voltage limiter and output the power over 0 dBm in the frequency range of up to 12 GHz. Therefore, the buffer circuit with Darlington connection has a frequency range wide enough for divider circuits.

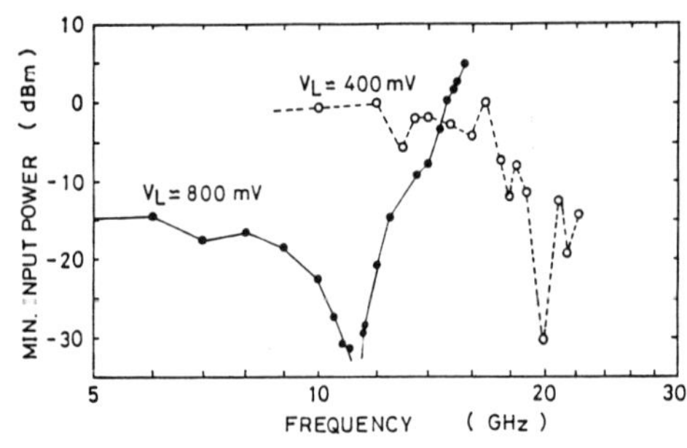

Fig.5 Frequency dependency of the minimum input power for divide-by-four operation.

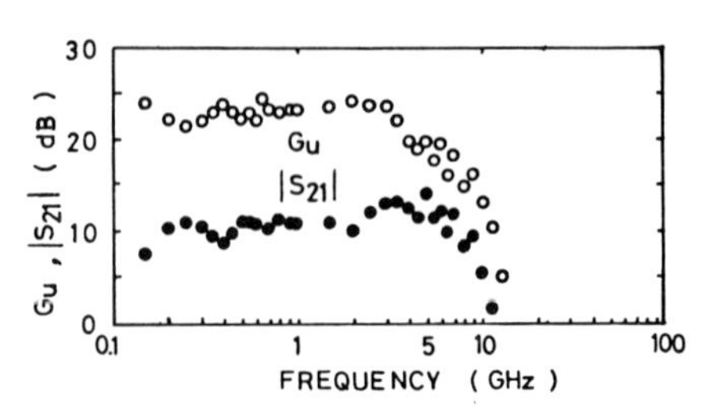

Fig.6 Measured frequency response $|S_{21}|$ and unilateral gain Gu of a buffer circuit with Darlington connection.

4. Summary

Emitter coupled logic gates and static frequency dividers using AlGaAs/GaAs HBTs were designed and the performance of fabricated circuits was characterized. A gate delay time of 5.5 ps/gate was obtained with a logic swing of 400 mV. Moreover, low emitter resistance made it possible to compress the logic swings to 250 mV, and a gate delay time of 4.5 ps/gate was obtained. The divide-by-four circuits have successfully been operated with a maximum toggle frequency of 22.15 GHz with a logic swing of 400 mV. The buffer circuits with Dalington connection were confirmed to have a wide frequency range for divider operation. To our knowledge, the switching speed obtained in this work are the fastest ever reported for transistor circuits.

Acknowledgement

The authors wish to thank T. Nittono for valuable discussions.

References

Chang M F, Asbeck P M, Wang K C, Sullivan N S, Higgins J A and Miller D L 1987a IEEE Electoron Dev. Lett. EDL-8 303

Chang M F, Asbeck P M, Wang K C, Sullivan G J and Miller D L 1987b IEEE 45th Annual Device Reserch Conf., IVA-1

Ishibashi T and Yamauchi Y 1987 45th Annual Device Research Conf., IVA-6

Nagata K, Nakajima O, Ito H, Nittono T and Ishibashi T 1987a Electron. Lett. 23, 566

Nagata K, Nakajima O, Yamauchi Y, Ito H, Nittono T and Ishibashi T 1987b 45th Annual Device Research Conf.,IVA-2

Woodall J M, Freeouf J L, Pettit G D, Jackson T and Kircher P 1981 J. Vac. Sci. Technol. 19 626

Yamauchi Y and Ishibashi T 1986 IEEE Electron Dev. Lett. EDL-7 655

Paper presented at Int. Symp. GaAs and Related Compounds, Heraklion, Greece, 1987

GaAs inversion-base bipolar transistor (GaAs IBT) with various types of emitter barrier

K. Matsumoto, T. Kinosada, Y. Hayashi, N. Hashizume, T. Nagata, T. Yoshimoto

Electrotechnical Laboratory MITI JAPAN
1-1-4, Umezono, Sakura-mura, Niihari-gun, Ibaraki, 305, JAPAN

ABSTRACT: The effects of the emitter barrier thickness and of the graded emitter barrier on the characteristics of the GaAs Inversion-Base Bipolar Transistor (GaAs IBT) were examined. By reducing the AlAs emitter barrier layer from 400Å to 100Å, the current gain, on-resistance, transconductance, etc. were greatly improved. The current transport through the 100Å AlAs emitter barrier turned out to be a tunnel current. The graded emitter barrier enhances the emitter injection efficiency by more than two order of magnitude.

1. Introduction

Recently, a GaAs Inversion-Base Bipolar Transistor (GaAs IBT), which uses a two dimensional inversion hole layer as a base, was proposed and realized. (Matsumoto 1986) The GaAs IBT has a structure of n^{+}GaAs emitter/undoped emitter barrier/inversion hole layer base/n^{-}GaAs collector. Thus the base of the GaAs IBT is more than ten times thinner than that of the conventional GaAs HBT. The base resistance remains low owing to the high density and the high mobility of the holes in the inversion layer. In the previous study, however, an obtained current gain of the IBT was very low. The current gain of the GaAs IBT was mainly determined by the emitter injection efficiency which depends on the emitter barrier structure. In the present study, four types of emitter barriers are examined in order to improve the characteristics of the GaAs IBT. Furthermore, the electron transport through the emitter barrier is also examined.

2. Structure of GaAs IBT

The structure of the GaAs IBT is shown in Fig. 1. The transistor has an n^{+}GaAs as an emitter, an undoped emitter barrier layer, an n^{-}GaAs as a collector, and a self-aligned external base p^{+}GaAs region. The principal feature of the GaAs IBT is that there is no metallurgical base layer present, and that an undoped emitter barrier layer is present between the emitter

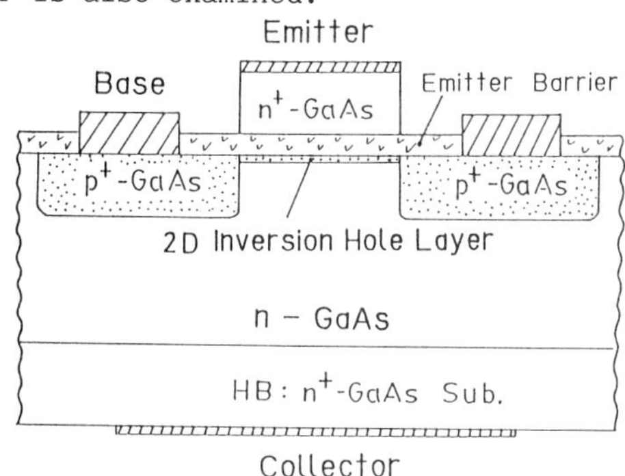

Fig. 1, Structure of GaAs IBT

and the collector layers. In order to improve the characteristics of the IBT, effects of the emitter-barrier thickness and the emitter-barrier grading were examined. Four types of IBT's with the following emitter barrier layers were fabricated, i.e., 1) an undoped 100Å AlAs layer (We designate this device as "100Å AlAs IBT".), 2) an undoped 400Å AlAs layer ("400Å AlAs IBT"), 3) an undoped 300Å $Ga_{0.6}Al_{0.4}As$ layer ("300Å GaAlAs IBT"), 4)an undoped 300Å $Ga_{0.6}Al_{0.4}As$ layer plus 150Å $Ga_{1-x}Al_xAs$ ($x=0.4 \rightarrow 0$) graded layer ("graded IBT"). The substrate is HB n^+GaAs (Si doped). The collector is MBE grown Si-doped n^-GaAs ($1x10^{16}/cm^3$, 1.5µm), the emitter Si-doped n^+GaAs ($1x10^{18}/cm^3$, 0.5µm). The emitter size is 50µmx50µm.

3. Comparison of Characteristics (I) : 100Å & 400Å AlAs Emitter Barrier

Figure 2(a),(b) shows the transistor characteristics of the 100Å AlAs IBT and 400Å AlAs IBT, respsctively, measured at 77K for the common emitter configuration. For the 100Å AlAs IBT, the collector current reaches such a high value of 180mA ($7.2x10^3 A/cm^2$) at the small base current of 2mA, and a high current gain of B=150 is obtained. The on-resistance in the saturation region is R_{on}=6.2 ohm. While for the 400Å AlAs IBT, the collector current is only about 10mA at the same base current of 2mA. Furthermore, the current gain is a small value of B=20 at the colector current of I_c=80mA, and begins to saturate above that collector current. The on-resistance is R_{on}=19 ohm, which is about three times higher than the 100Å AlAs IBT. Thus, by reducing the emitter barrier thickness, the great improvement in the IBT characteristics was obtained.

Figure 3 shows the dependence of the collector current I_c on the emitter-base bias V_{EB} for the common-base configuration at 77K. For the 100Å AlAs IBT (closed circle), the on-bias at which the collector current begins to flow and the GaAs IBT begins to work as a transistor is V_{EB}=1.6V. The collector current increases rapidly from 1mA to 22mA with the increase of emitter-base bias from V_{EB}=1.7V to 2.0V. The transconductance g_m obtained from the slope of the I_c-V_{EB} curve is as high as 90mS. For the 400Å AlAs IBT, the collector current begins to flow at the higher on-bias of V_{EB}=1.8V and increases gradually than that of the former case. The transconductance is g_m=21mS. The calculated emitter-base bias which is necessary for the formation of the inversion hole layer is V=1.49V and V=1.76V for the 100Å and 400Å IBT, respectively, which almost correspond to the on-biases of the GaAs IBT's in Fig. 3. For the thinner emitter

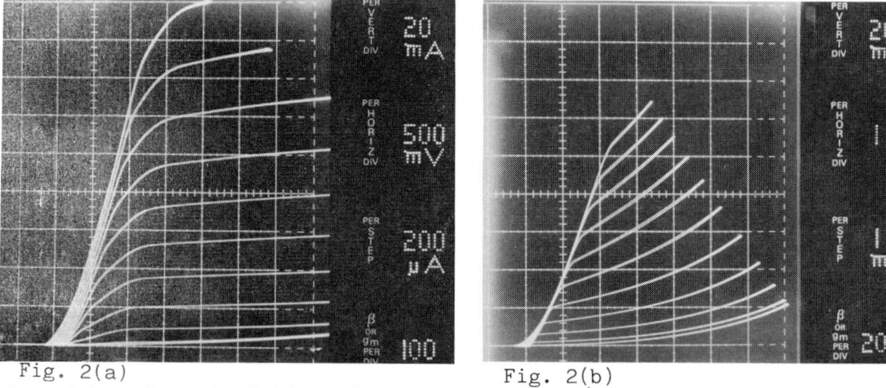

Fig. 2(a) Fig. 2(b)

Transistor characteristics of GaAs IBT with 100Å AlAs emitter barrier (a) and 400Å AlAs emitter barrier (b) at 77K for common emitter configuration.

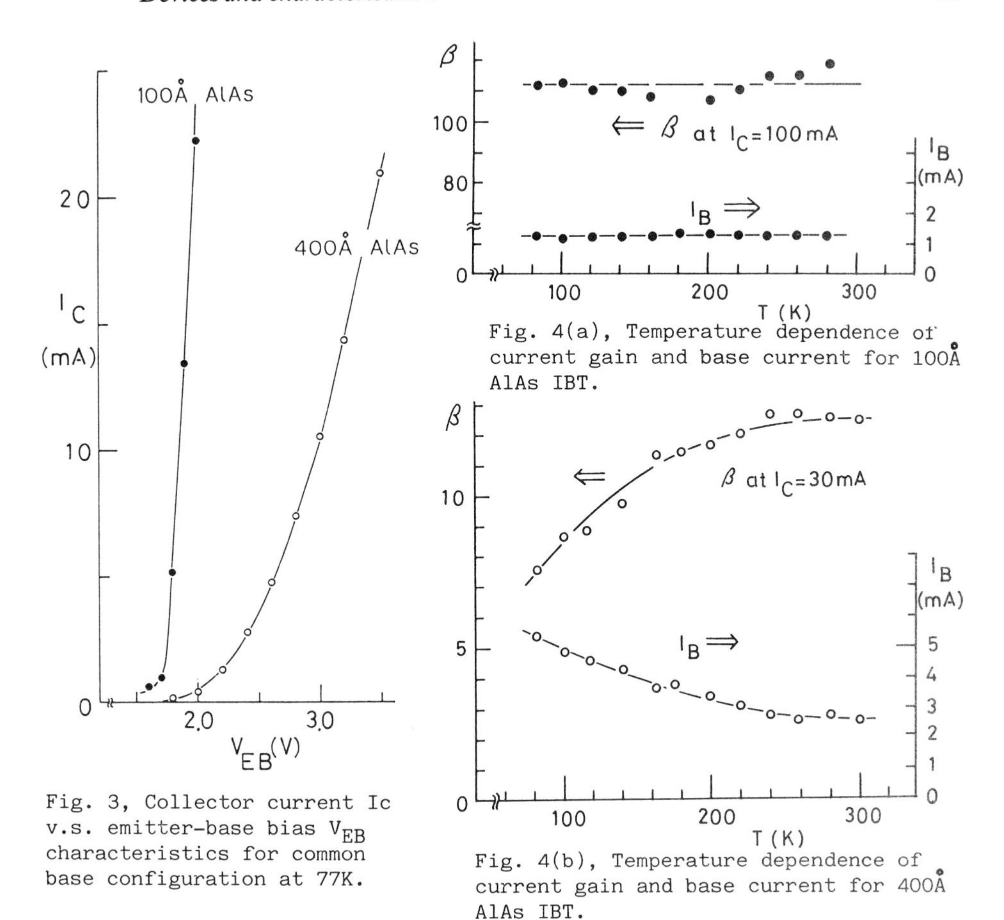

Fig. 3, Collector current Ic v.s. emitter-base bias V_{EB} characteristics for common base configuration at 77K.

Fig. 4(a), Temperature dependence of current gain and base current for 100Å AlAs IBT.

Fig. 4(b), Temperature dependence of current gain and base current for 400Å AlAs IBT.

barrier layer, the electric field along the barrier becomes higher. Therefore, the transconductance becomes higher and the on-bias becomes lower.

The temperature dependences of the current gain for the 100Å and 400Å AlAs IBT's are shown in Fig. 4(a) and (b), respectively. The base current I_B, also shown in the figure, was controlled to keep the collector current constant. For the 100Å AlAs IBT, the current gain keeps the constant value of B=110 at the constant collector current of I_C=100mA for the wide range of the temperature. (80K - 280K) The base current also keeps the constant value of I_B=1.2mA for the same temperature range. While for the 400Å AlAs IBT, both of the current gain and the base current show the temperature dependence for the temperature range of 80K to 300K.
If the recombination of the hole and electron in the base region is assumed to be negligible, the collector current is the electron flow from the emitter, and the base current is the hole flow from the base to the emitter. Then the above results show that both electron and hole transport through the 100Å AlAs emitter barrier is dominated by the tunnel transport. While in the 400Å AlAs IBT, the some current transport other than the tunnel current, which may be the thermal stimulated current, also contributes to the current transport. This difference of the current transport by the difference of the barrier thickness in the IBT may be one of the reasons of the current gain improvement.

4. Comparison of Characteristics (II) : Graded Emitter Barrier Layer

Another method to improve the transistor characteristics is to introduce a undoped $Ga_{1-x}Al_xAs$ graded layer to an emitter barrier. In this section, the characteristics of 300Å GaAlAs IBT and the graded IBT are compared. The features of the graded layer is effective to improve the injection efficiency of the transistor, i.e., the higher electron flow from the emitter to collector, while the lower hole flow from the base to the emitter. This prediction is confirmed by the measurement of the current gain of the two kinds of GaAs IBT's. The graded IBT shows the current gain of B=36 at 77K for the common emitter configuration. While the current gain of the 300Å GaAlAs is only B=0.28 at 77K, which is more than two order of magunitude smaller than that of the graded IBT.

Figure 5 shows the current-voltage characteristics between the emitter and the collector with the external base terminal open for the GaAs IBT at 77K, i.e., the diode characteristics of the n^+GaAs/i GaAlAs/n^-GaAs structure. The conduction energy band diagrams for the 300Å GaAlAs IBT and the graded IBT at the zero bias condition are also shown in Fig. 5. When the emitter is negatively biased, the current of the graded IBT (closed circle) increases steeply and is more than two orders of magnitude higher value than that of the 300Å GaAlAs IBT (open circle) at the bias of -1V, though the graded IBT has the thicker emitter barrier. This is attributed to the graded layer which becomes flat when the emitter is negatively biased, lowers the effective conduction band discontinuity between the n^+GaAs emitter and the undoped GaAlAs barrier layer, and make it easy for electrons in the n^+GaAs emitter to go over the barrier layer. The lowering of the effective conduction band discontinuity by applying the bias was confirmed by the measurement of the temperature dependence of the current for the various bias condition. Thus the graded layer is effective to greatly improve the current gain of the GaAs IBT.

5. Conclusion

In conclusion, we have shown two methods to improve the characteristics of the GaAs IBT. One method is to reduce the thickness of the emitter barrier, and the other method is to introduce the graded layer to the emitter barrier. And we succeeded in obtaining the improved characteristics of the GaAs IBT. By combining these methods, further improvement of the characterristics of the GaAs IBT will be expected.

Matsumoto K, Hayashi Y, Hashizume N, Yao T, Kato M, Miyashita T, Fukuhara N, Hirashima H, and Kinosada T 1986 IEEE Electron Device Lett. EDL-7, pp627-8.

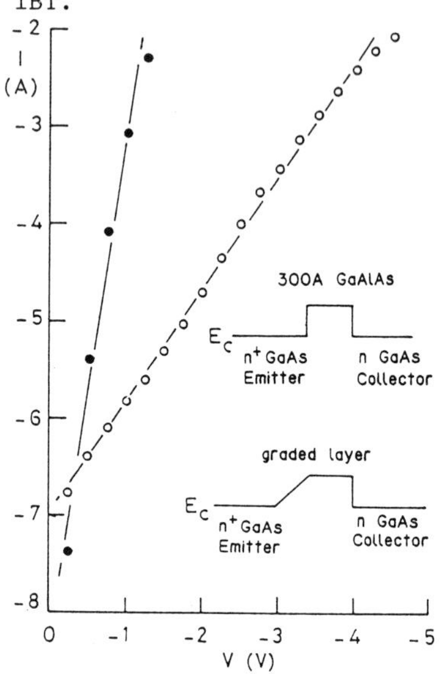

Fig. 5, Current-voltage characteristics between emitter and collector of GaAs IBT at 77K. Closed circle is for graded IBT, open circle for 300Å GaAlAs IBT. Emitter is negatively biased.

Inst. Phys. Conf. Ser. No. 91: Chapter 7
Paper presented at Int. Symp. GaAs and Related Compounds, Heraklion, Greece, 1987

A new process for the fabrication of field effect transistor using the neutralization of shallow donors by atomic hydrogen in n-GaAs (Si)

E. CONSTANT*, J. CHEVALLIER**, J.C. PESANT**, N. CAGLIO*

* Centre Hyperfréquences et Semiconducteurs, U.A CNRS N° 287
Université des Sciences et Techniques de Lille Flandres-Artois
59655 VILLENEUVE D'ASCQ CEDEX - France
** Laboratoire de Physique des Solides, CNRS
1 Place Aristide Briand 92195 MEUDON - France

ABSTRACT :

A new process for the fabrication of GaAs field effect transistors is described. This process is based upon the neutralization of shallow donors by atomic hydrogen diffused into a highly silicon doped epilayer. This original process should be able to produce field effect transistors with low access resistances. The first 1.2 μm and 0.4 μm gate length HFET's (hydrogenated FET) show very encouraging characteristics. For a 1.2 μm gate length, the typical transconductance is 330 mS/mm and the cut-off frequency is larger than 15 GHz.

1. INTRODUCTION

When atomic hydrogen is diffused into n-type silicon doped GaAs, formation of silicon related hydrogen complexes occurs (JALIL and all, 1987]. As a result, the concentration of active silicon donors drops down as seen through the decrease of the free carrier concentration [CHEVALLIER and all, 1985]. The existence of these complexes has been readily demonstrated from infrared absorption spectroscopy. It is thought that the extra free electron of the silicon donor enters into the silicon related hydrogen complex leaving a neutral entity after hydrogenation [CHEVALLIER and all, 1985]. This effect explains that the decrease of the free electron concentration is accompanied by a simultaneous increase of the electron mobility [JALIL and all, 1986]. In this paper, we, firstly, give and explain typical results concerning the passivation, of shallow impurities in GaAs epilayers. Then, we describe how to realize HFET's by using such a phenomenon and what are the potential advantages of this new method of fabrication. Then, we present the performances of this new type of GaAs FET.

2. PASSIVATION OF THE SILICON DONORS :

Before developing the different steps for the realization of the transistor and the presentation of its characteristics, let us study first the effect of the hydrogen plasma exposure on the electrical properties of a highly silicon doped GaAs epilayer. For this purpose, we consider a typical example obtained on a 1 µm n^+ GaAs Si doped layer grown on a semi-insulating substrate by molecular beam epitaxy. Hall and conductivity measurements performed by the Van Der Pauw method indicated a free carrier concentration of 2×10^{18} cm^{-3} and a mobility value of 2000 cm^2/v.s. at 300 K. Three parts of this wafer have been exposed to a R.F. hydrogen plasma at three different plasma powers 10W, 30W and 50W. Hall and conductivity measurements were then performed on each sample. After hydrogen plasma exposure, the free concentration is found to decrease and the electron mobility increases ; this is shown on fig 1 where we have reported the corresponding values of µ and n. This confirms our previous results obtained on a series of MOCVD epilayers [JALIL and all, 1986] however, in this case, the exposure duration was significantly longer (6 hours)). We note that the free carrier concentration gets lower and the electron mobility gets higher as the R.F. power increases. It has been shown previously that the amont of hydrogen introduced in the layer increases with the R.F. power [CHEVALLIER and all, 1986]. As a result, the probability for the silicon related hydrogen complexes to be formed increases and the remaining active donor concentration decreases. On figure 1, we have also reported the µ(n) dependence currently obtained for a series of non hydrogenated silicon doped n-GaAs epilayers with various doping levels. We note that the electron mobility in the hydrogenated samples is slightly above the mobility versus concentration curve deduced from the series of non hydrogenated samples. These results, clearly show the good electronic quality of the hydrogenated silicon doped GaAs layers. For this reason, such layers may be used as active regions in field effect transistors.

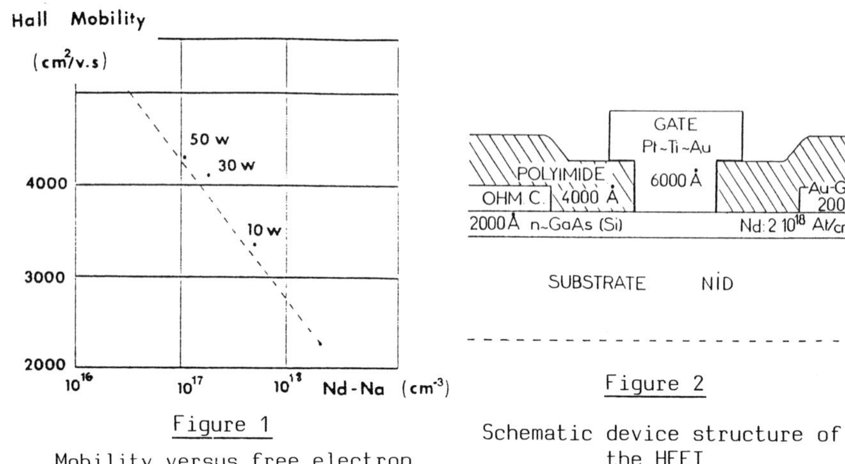

Figure 1

Mobility versus free electron concentration for various RF plasma power : substrate temperature : 230°C and exposure time 15 min

Figure 2

Schematic device structure of the HFET

3. FABRICATION PROCESS

The process can be described as follows. We start from a single highly doped GaAs epilayer. After the mesa and the source and drain ohmic contacts fabrication, a protecting dielectric layer is deposited and we proceed to the gate lithography (fig. 2). The processed sample is then exposed to a hydrogen plasma for hydrogen diffusion in the non protected region. The hydrogen diffusion results in the neutralization of a more or less large fraction of the silicon donors depending on the plasma exposure parameters. As the neutralization becomes more effective, the electron mobility increases. After the gate deposition, the devices are expected to have low access resistances because these resistances will occur in highly doped regions. Other potential advantages are worthy to be mentionned :

- the possibility of having a completely planar process. Recess is not necessary to achieve low access resistances
- the concentration of active donors below the gate is expected to be lower than deeper from the surface because of the hydrogen diffusion profile. Then we expect this active donor gradient to improve the transistor linearity and to increase the breakdown voltage

Practically, we have used MBE grown silicon doped GaAs epilayers. Their characteristics are presented on Fig. 2. The gate length was determined by the opening width made in the polyimide protection layer. It was 1.2 µm after exposure. The processed sample was then left in a R.F. capacitively coupled hydrogen plasma for 10 minutes at 220°C. In such conditions, only a few per cent of the total donor concentration remains active after exposure. We then proceeded to a small recess of 0.1 µm in the gate region before the gate deposition. Three different geometries have been used with gate width of 75, 150 and 300 µm. We finally annealed the transistor at 300°C for 15 minutes in order to adjust the active donor concentraton in the active region to the desired value. Actually, it has been shown that the neutralization effect is stable for several years or more at 120°C and below and that, above 280°C annealing leads to a deneutralization of the donors. A very simular process has been used to realize 0.4 µm gate HFET but electronic lighography has been employed.

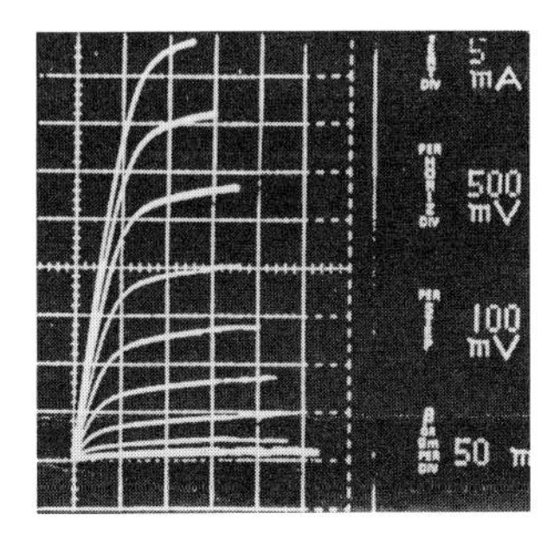

DC voltage characteristics of 300 µm gate HFET.
The gate voltage varies from + 200 mV to - 600 mV

Figure 3

Figure 3 shows the static I-V characteristics of the HFET with a 300 µm gate. The corresponding gm value measured for a constant source-drain voltage of 2 volts is as high as 290 mS/mm. A more complete characterization of these devices has been performed by magnetoresistance experiments [Sites, 1980]. Applying this technique to our HFET, we find an electron mobility of 3500 $cm^2/V.s.$ in the active layer. Taking into account a doping level of 6.10^{17} cm^{-3} deduced from gate-source capacitance, the electron mobility value is satisfactorily good. The equivalent circuit of these transistors in the high frequency regime has been deduced [DAMBRINE and all, 1987]. The cut-off frequency is higher than 15 GHz and 35 GHz for a gate length of 1.2 µm and 0.4 µm respectively. The high frequency transconductance in both cases is higher than 300 mS/mm. The frequency dependences of the

Current Gain, Maximum Available Gain, and Unilateral Gain are given on figure 4. For a 1.2 μm gate length transistor, the performances appear somewhat remarkable.

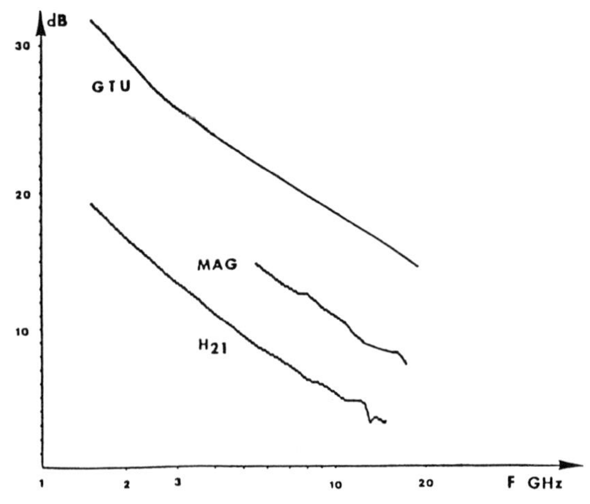

Variation of the current gain, Maximum Available Gain and Unilateral gain versus the frequency for 1.2 μm gate length HFET

Figure 4

4. CONCLUSION

We have presented a new type of field effect transistor based on the neutralization of shallow donors by atomic hydrogen in the active region, the static and the high frequency characteristics can be considered as very encouraging. A more extended study is necessary to optimize the process and to evaluate its potential use in micro-electronics.

REFERENCES

CHEVALLIER J., DAUTREMONT-SMITH W.C., TU C.W., PEARTON S.J., 1985, Appl. Phys. Lett., 47, 108
CHEVALLIER J., JALIL A., AZOULAY R., MIRCEA A., 1986, Material Science Forum, 10-12, 591
JALIL A., CHEVALLIER J., AZOULAY R., MIRCEA A., 1986, J. Appl. Phys., 59, 3774
JALIL A., CHEVALLIER J., PESANT J.C., MOSTEFAOUI, PAJOT B., MURAWALA P. AZOULAY R., 1987, Appl. Phys. Lett., 50, 439
SITES J.R., WIEDER, H.H., 1980, IEEE Trans., 27, 2277
DAMBRINE G., CAPPY A., PLAYEZ E., 22-24 June 1987, JNM Nice, Communication 0905

ACKNOWLEDGEMENTS

The authors are very grateful to the technological staff for device processing and characterization in particular to J. VANBREMEERSCH, B. GRIMBERT and E. PLAYEZ.

The financial support of the DRET is acknowledged.

Carrier injection and base transport in heterojunction bipolar transistors with InGaAs base

Y.Ashizawa *, S.Akbar, G.W.Wicks ** and L.F.Eastman
School of Electrical Engineering, Phillips Hall,
Cornell University, Ithaca, NY 14853 USA

Abstract
Carrier transport in the base of MBE grown AlGaAs/GaAs heterojunction bipolar transistors with compositionally graded InGaAs bases has been investigated. Room temperature electroluminescence exhibited luminescence peaks due to radiative recombinations in the GaAs and the InGaAs bases. The accumulation of electrons in the undoped GaAs spacer layer in the base at the emitter-base junction has been confirmed. The superiority of the abrupt emitter to the graded emitter in terms of launching higher energy electrons has been substantiated.

Introduction

Among the extensive efforts to improve the high speed performance of heterojunction bipolar transistors (HBTs), use of highly doped bases has become one of the key issues [Chang 1987]. Minority carrier transport, however, is influenced by increased scattering due to hole plasmons and optical phonons. Temperature of hot minority carriers, T_c, has been deduced by measuring luminescence, assuming Maxwell-Boltzman distribution [Shah 1982, Ishibashi 1984). In the HBTs,however, carriers are not in equilibrium [Levi 1985] but go through the base with a velocity 10^7 cm/s and lose their energy in the process. Since the location of radiative recombinaton has not been clarified, the dynamics of the electron transport is not described only by T_c.

In this paper, HBTs with compositionally graded InGaAs bases were considered. The compositionally graded bases offer a unique opportunity to investigate the base transport spectroscopically. Grown devices were characterized by DC current gain measurements and electroluminescence measurements, focussing on the comparison between graded emitter and abrupt emitter.

Experimental

Five HBTs with compositionally varying structures of the emitter (abrupt or graded) and of the $In_yGa_{1-y}As$ base (Uniform $y = 0.5$, slightly graded from $y = 0$ to 0.05 or highly graded from $y = 0$ to 0.1) were grown by MBE as shown in table 1. Devices with various areas were fabricated using a two level mesa process. Electroluminescence from large area devices with emitter-base (E-B) junction areas of 90x2000 μm² was collected through the unmetallized edge of the emitter mesas. A pulsed voltage with a duty cycle ratio 10% was biased between the emitter and the base, while a DC voltage, V_{CE}, was applied between the collector and the emitter. Photoluminescence from the base was collected through the n-AlGaAs emitter with n + GaAs contact layer being removed by wet etching. An Ar ion laser (5145A) was used to excite the samples and the luminescence was detected by a S1 photomultiplier.

* On leave from Toshiba Research and Development Center, 1Komukai Toshiba-cho, Saiwai-ku, Kawasaki 210 Japan
** Now at Institute of Optics, University of Rochester, Rochester, NY 14627

Table 1 STRUCTURES OF MBE GROWN HBTs

		HBT1	HBT2	HBT3	HBT4	HBT5
n^+ GaAs	2000Å	2×10^{18}	2×10^{18}	2×10^{18}	2×10^{18}	2×10^{18} (cm^{-3})
n-$Al_xGa_{1-x}As$ ($x = 0.25 \to 0$)	250Å	2×10^{18}	2×10^{18}	2×10^{18}	2×10^{18}	2×10^{18} (cm^{-3})
n-$Al_{0.25}Ga_{0.75}As$	2500Å	3×10^{17}	3×10^{17}	3×10^{17}	3×10^{17}	3×10^{18} (cm^{-3})
n-$Al_xGa_{1-x}As$ ($x = 0 \to 0.25$)	250Å	---------	----------	3×10^{17}	3×10^{17}	--------(cm^{-3})
GaAs spacer	200Å	undoped	undoped	undoped	undoped	undoped
p^+-$In_yGa_{1-y}As$	800Å	1×10^{19}	1×10^{19}	5×10^{18}	1×10^{19}	1×10^{19} (cm^{-3})
		$y = 0.05$	$y = 0.05 \to 0$	$y = 0.05 \to 0$	$y = 0.1 \to 0$	$y = 0.1 \to 0$
n-$In_yGa_{1-y}As$	500 Å	2×10^{16}	2×10^{16}	2×10^{16}	2×10^{16}	2×10^{16} (cm^{-3})
		$y = 0 \to 0.05$	$y = 0 \to 0.05$	$y = 0 \to 0.05$	$y = 0 \to 0.1$	$y = 0 \to 0.1$
n-GaAs	4500Å	2×10^{16}	2×10^{16}	2×10^{16}	2×10^{16}	2×10^{16} (cm^{-3})
n^+ GaAs	5000Å	2×10^{18}	2×10^{18}	2×10^{18}	2×10^{18}	2×10^{18} (cm^{-3})

Results and Discussions

The DC current gains, β, of the HBTs were measured using devices with the E-B junction areas of 3x20μm² and 5x36 μm². Smaller devices showed β between 50 and 120, while larger devices showed β higher than 100. No significant trends were seen between HBTs with the abrupt emitter and with the graded emitter. HBTs with steeped grading gave lower β. This is considered to be due to the higher dislocation density of the steeped graded bases which cause increased base recombintion. The existence of the misfit dislocations was confirmed by cathodo luminescence measurements. Detailed analysis of defect formation and the electrical properties will be given elsewhere [Ashizawa 1987]. Although there exist defects associated with this lattice mismatch, these high current gains reconfirm the potential of this material system as an alternative to high speed GaAs base HBTs [Sullivan 1986, Enquist 1986]. Photoluminescence spectra from HBT2 and HBT3 are shown in Fig. 1. Two distinct peaks were observed: PL1 at 1.43eV and PL2 at 1.38eV. PL1 arises from

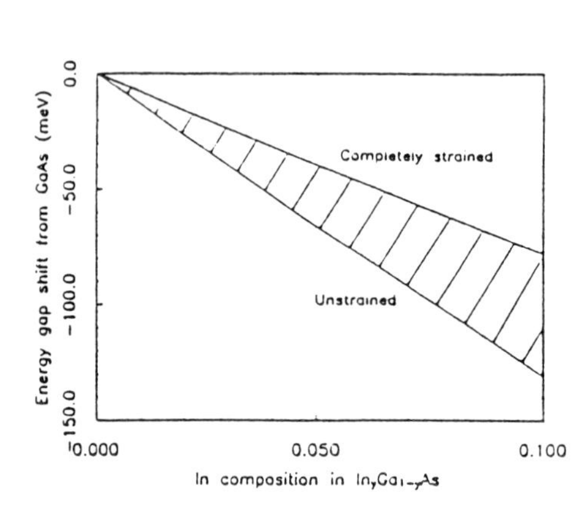

Fig. 1. Energy gap shift of $In_yGa_{1-y}As$ from GaAs.

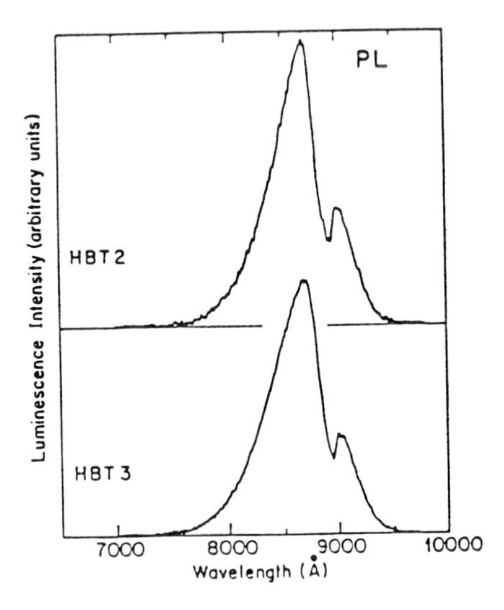

Fig. 2. Room temperature photoluminescence from HBTs.

the GaAs spacer layer in the base and/or GaAs collector and PL2 from the p+ InGaAs base. From the peak energy in the base the location of the radiative recombina-tions can be deduced. The effects on the bandgap caused by the high Be doping and by the strain have to be taken into account. Although Be doping causes band shrinkage [Casey 1978], Be atoms are assumed to have diffused into the undoped GaAs spacer layer in the HBTs under study. Therefore it is expected that the difference in the peak energy of luminescence from the GaAs spacer layer and the InGaAs in the base is only slightly affected by high Be doping. The strain caused by high Be doping was confirmed to be negligibly small by the double crystal X-ray diffraction measurements done on a p+ GaAs layer grown on a GaAs substrate. Due to the lattice mismatch between GaAs and InGaAs, the InGaAs bases suffer biaxial stress, resulting in a compressive splitting of the valence band. The effective bandgap is then formed between the conduction band and the light hole band. The energy gap shift from the unstrained condition, E(y), can be estimated using the material parameters [Asai 1983, Kuo 1985].The lattice mismatch is in fact accommodated by both the elastic strain and the misfit dislocations. Therefore the energy gap difference between highly Be doped InGaAs and GaAs as a function of In composition is expected to lie between the two extreme cases (unstrained and completely strained) shown in Fig. 2 by the shaded area.

Based on the discussion above, the In composition of the InGaAs base where PL2 in Fig. 1 arises is estimated to be 0.05 ± .01.This means that photoexcited electrons are effectively swept during their energy relaxation by the built-in electric field produced by compositional grading and recombine with holes at or very close to the base-collector (B-C) junction. Also it was confirmed that HBT2 and HBT3 have almost identical compositional grading structures.

Fig. 3 shows the elctroluminescence spectra, taken under the conditions of the pulse base current, I_B = 100mA/pulse (55 A/cm2 pulse) and V_{CE} = 0V. Two kinds of luminescence peaks were observed: P1with peak energy 1.42-1.43eV and P2 with various lower peak energies. P1 is considered to come from GaAs spacer layer in the base at the E-B junction and not from the collector. To confirm this the luminescence from HBT2 was measured by

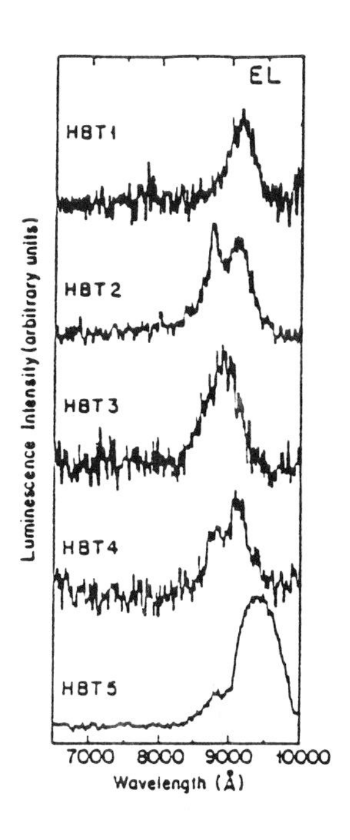

Fig. 3. Electroluminescence from HBTs.

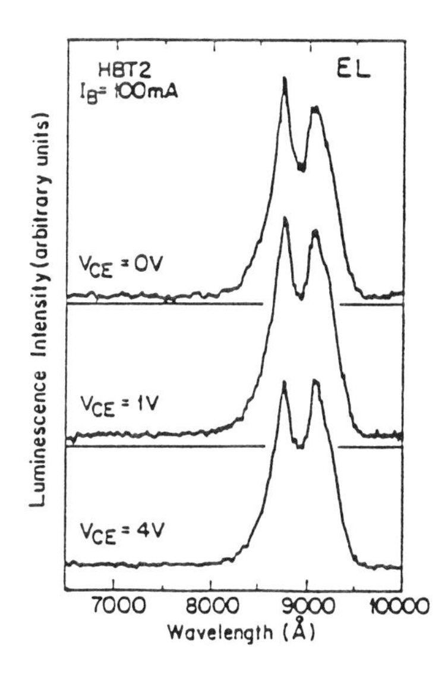

Fig. 4. Electroluminescence from HBT2 with various V_{CE}.

varying V_{CE}, while I_B was kept constant (Fig. 4). As V_{CE} increases, the peak intensity ratio of P1 to P2 decreases. If P1 arose in the collector, P1 should become larger relative to P2. The electrons accumulate in the potential dip at the depletion region in the base at the E-B junction. It is suggested that the thickness of the undoped GaAs layer could be made thinner to minimize the depletion region and hence to reduce the population of electrons in the potential dip.

As seen in Fig.3, the HBTs with abrupt emitters (HBT2,5) gave lower P2 peak energies than the HBTs with graded emitters (HBT3,4). Since almost identical compositional grading structures were achieved as has been confirmed by the photoluminescence measurement, this difference in the location of radiative recombination reflects the difference of the average velocity of electrons as well as the energy distribution of injected electrons depending on the emitter structure. Using the relation in Fig. 2, it is concluded that in HBTs with the abrupt emitter, the electrons go through the base without recombining radiatively and reach the B-C junction. On the other hand in HBTs with the graded emitter, the electrons start to recombine radiatively in the middle of the base, because of reduced injection energy and hence reduced velocity, which caused electrons to spend more time in the base than in the HBTs with abrupt emitter. Above data are the clear evidence of the superiority of abrupt emitter to graded emitter in terms of launching high energy electrons. Therefore it is concluded that use of abrupt emitter is essential in the HBTs with highly doped base, if a short base transit time has to be achieved.

Conclusion

HBTs with compositionally uniform or graded bases have been grown by MBE and characterized by DC curent gain measurements and electroluminescence measurements. Direct evidence of the superiority of abrupt emitters over graded emitters in terms of launching higher energy electrons has been exhibited by the use of a compositionally graded InGaAs base. The observed superiority will not be unique in the HBTs with InGaAs but reflects the common nature of HBTs with highly doped base.

Acknowledgement

The authors would like to thank P.Tasker, L.Knoch, E.A.Fitzgerald, W.J. Schaff and J.Berry for their assistance and useful discussions. Ken Miller and Associates is acknowledged for X-ray diffraction measurements. They are also grateful to E.Weaver for typing this manuscript. This work was partially supported by the Joint Services Electronics Program, IBM and Brush Wellman

References

Asai H, and Oe K 1983 J. Appl. Phys. 54 2052.
Ashizawa Y, Fitzgerald E A, et al 1987 Unpublished.
Casey H C, and Panish M B 1978 *Heterostructure Lasers*, Academic Press p157.
Chang M F, Asbeck P M, Wang K C, Sullivan G J, Sheng N, Higgins J A, and Miller D L, 1987 IEEE Electron Device Lett. EDL-8, No7.
Enquist P M, Ramberg L R, Najjar F E, Schaff W J, and Eastman L F, 1986 Appl. Phys. Lett. 49 179.
Ishibashi T, Ito H, and Sugeta T, 1984 Inst. Conf. Ser. No.74, Ch.7. 593.
Kuo C P, Vong S K, Cohen R M, and Stringfellow G B, 1985 J. Appl. Phys. 57 5428.
Levy A F J, Hayes J R, Platzman P M and Wiegman W 1985 Phys. Rev. Lett. 55 2071.
Shah J, Nahory R E, Leheny R F, Degani J , and A.E.DiGiovanni 1982 Appl. Phys. Lett. 40 505.
Sullivan G J, Asbeck P M, Chang M F, Miller D L, and Wang K C 1986 Proc. 44th Annual Device Research Conf..

Inst. Phys. Conf. Ser. No. 91: Chapter 7
Paper presented at Int. Symp. GaAs and Related Compounds, Heraklion, Greece, 1987

Superlattice bipolar transistors

J.F.Palmier, A. Sibille, J.C.Harmand, C.Dubon-Chevallier,
C.Minot, H.Le Person and J.Dangla.
C.N.E.T. 196 Avenue H.Ravera 92220 BAGNEUX (FRANCE).

Two superlattice bipolar transistor structures are reported. The first one, involving a superlattice base allows to directly measure the perpendicular minority carrier diffusion length in the superlattice, and gives evidence of a Bloch to hopping conduction transition. In the second one, the emitter is a modulation doped superlattice which avoids the need of graded layers with a nevertheless low V_{ce} offset.

1. INTRODUCTION

Semiconductor superlattices have been studied since the 70's. Most of the published data concern optical and electrical transport parallel to the layers. Although at the origin of the superlattice [1], the perpendicular transport properties have still many questions to be answered: role of the finite wavefunction coherence length, excitation to states involving X or L bands, reality of the formation of Stark ladders at high electric fields, etc... The present communication is devoted to the study of specially designed heterojunction bipolar transistors (HBT) incorporating superlattices (SL) in the base or in the emitter. The aims are the study of perpendicular transport (SL in the base), or of the SL specificity in novel device structures.

2. THE SL BASE BIPOLAR TRANSISTOR : A TEST STRUCTURE FOR PERPENDICULAR TRANSPORT.

In superlattices, the low field mobility and the diffusion coefficient perpendicular to layers are interesting parameters, not so easy to measure. Time-dependent photoconduction experiments can be performed, in order to separate the electron and the hole contributions [2]. However, in this case, due to the band bending, the measured data are not zero-field mobilities, but some average value. In the neutral base of a transistor the minority carrier transport is limited by a pure diffusion regime (no drift). Therefore, the superlattice bipolar transistor (SBT) can provide a good test structure for the measurement of the perpendicular minority carrier diffusion length L_d:the base transport factor α is directly related to it through the relation $\alpha = 1-W_b^2/2L_d^2$. In figure 1 are plotted the room temperature variations of the common emitter gain h_{FE} versus the superlattice barrier width b. The other parameters have been kept almost constant for the three samples, namely : well width a = 40 Å, Al % in the barrier x =0.3, base width W_b =0.3 um, respective doping of emitter, base and collector : $N_d=2.10^{17}cm^{-3}$, $N_a=2$ to 5.10^{16} cm^{-3}, $N_d=5\ 10^{16}cm^{-3}$. In the case of an unity injection efficiency, the saturated value is $h_{FE} = 2(L_d/W_b)^2$. The experimental points (with large error bars due to the data dispersion) can be compared with the exponential variation of the mean electron velocity in a miniband which approximately follows the square of the envelope wavefunction in the barrier, $\exp(-\chi b)$. As proposed in earlier papers [3][4], there is a possibility that in the lowest barriers SL (b<30Å), the conduction mechanism would be coherent Bloch motion

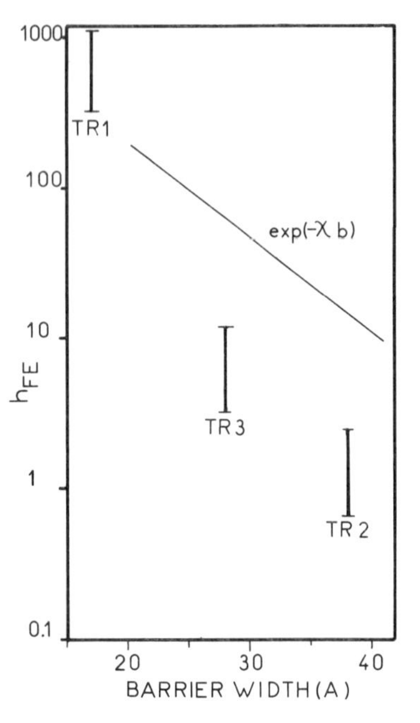

Fig. 1. Variation of the common emitter gain with the barrier width.

with random collisions whereas in the larger barrier SLs the layer to layer hopping or phonon-assisted tunneling prevails [5][6]. This picture appears nicely confirmed by the temperature dependence of an effective mobility μ_{eff} for the two extreme structures (b=17Å and b=38Å)[7]. μ_{eff} is defined by $L_d = \sqrt{kT\mu_{eff}\tau/e}$, where we assume a lifetime τ = 7 ns. The variations of μ_{eff} with T (figure 2) are exactly opposite for the two samples above 200K. As expected, μ_{eff} decreases with T for TR1 (b=17Å) and increases with Tfor TR2 (b=38Å). This is respectively consistent with Bloch and hopping conduction in this temperature range, in which the dominant collision mechanism is the electron-polar optical mode scattering. Below 200K the temperature dependence of μ_{eff} for TR2 is reversed. This may be explained by a mixed Bloch-hopping regime [7], assuming a mono-

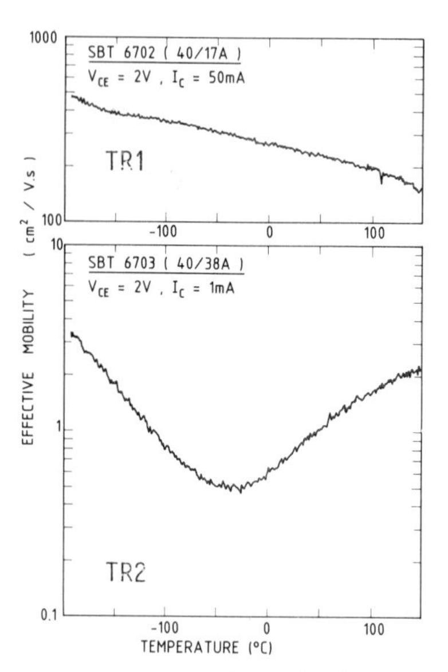

Fig. 2. Variation of the effective mobility versus T for TR1 and TR2.

-tonous variation of with T.

Further SBT have been grown by MBE recently and processed. Their relevant data are those of superlattices with 10Å < b < 22Å , and the h_{FE} values within the error bar of TR1. Beside its physical interest the SBT structure may allow to achieve integrated structures including a multi-quantum well laser and SBT drivers.

3. BIPOLAR TRANSISTORS WITH A MODULATION-DOPED SUPERLATTICE EMITTER(MBT).

One of the basic principles of HBTs is the injection efficiency improvement due to the full band-gap difference at the emitter-base (EB) junction. This implies a precise design and fabrication of a graded layer at the EB interface. Otherwise, abrupt heterojunctions lead to a poor injection efficiency due to the conduction band spike; also, the assymmetry of their EB and BC junctions may result in large V_{ce} offsets [8] in static characteristics. A novel concept of the EB junction is demonstrated here . Its principle is based upon the perpendicular transport properties of a modulation doped AlGaAs/GaAs superlattice. It is designed in such a way as the electron current is maximised whereas the hole leakage current is minimized. The effect tends to the effective mass filter [9] at the low barrier thickness limit. Using modulation doping increases the valence-band barrier and transforms the conduction-band barrier in two spikes in which the tunneling is higher. The band gap engineering of the structure is described in figure 3 which reproduces a numerical simulation of the real device under study. The practical achievement of that structure has been already described [10].

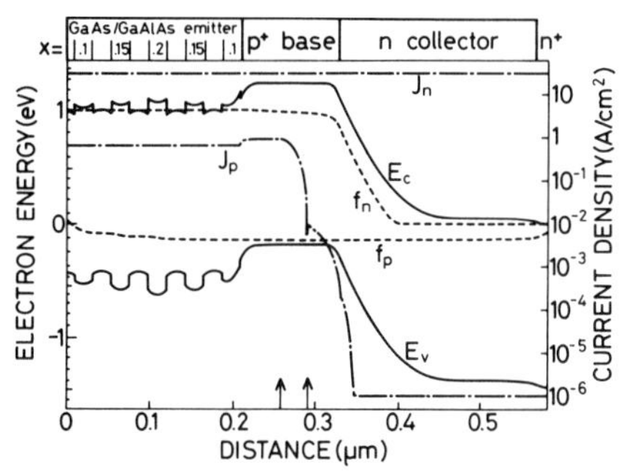

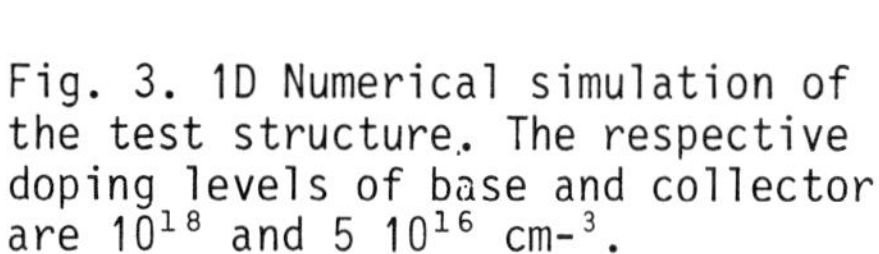
Fig. 3. 1D Numerical simulation of the test structure. The respective doping levels of base and collector are 10^{18} and 5 10^{16} cm^{-3}.

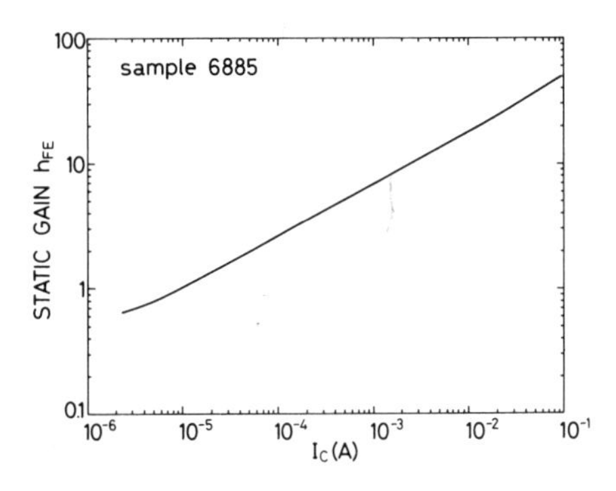

Fig. 4. Variation of h_{FE} with collector current.

In figure (4) the room temperature variations of h_{FE} versus the base current I_b is drawn. The measured maximum gain of 40 is of the order of the theoretical value of 50 provided by the numerical simulation of figure 3 . Both values are pessimistic. Another important feature of these results are the very low V_{ce} (collector-emitter voltage offset) = 80 mV. This is confirmed by very similar behavior of the EB and BC junctions at 300K. At 77K, however, the junctions are less symmetric, revealing perhaps their basic physical differences. We point that in spite of their interesting properties, these test structures are not optimized with respect to the highest attainable gain values. Single barrier structures without modulation doping have been reported [11-12]. Further studies will also cover the possibility of MBTs equivalent to the double heterojunction HBTs.

4. CONCLUSIONS

Perpendicular transport in superlattices has been studied in SBT. For AlGaAs/GaAs superlattices with a=40Å and x=0.3 two transport mechanisms are clearly distinguished, according to the barrier width : the miniband transport or the hopping transport (incoherent phonon-assisted tunneling). A modulation doped emitter structure has also been proposed, which may present an interesting alternative to the conventional heterojunction structure, in terms of low V_{ce}, lower Al content,fabrication simplicity as shown by experimental results in relation with device simulation.

AKNOWLEDGEMENTS . The authors whish to thank Mrs Vuye and C.Besombes for their help in sample processing, G.Le Roux for X-ray data, Y.M.Gao for the SIMS data, D.Ankri and F.Alexandre for fruitful discussions.

REFERENCES

[1] L.Esaki and R.Tsu, IBM J. Res. Dev. 14, 61 (1970)

[2] C.Minot, H. Le Person, F. Alexandre and J.F. Palmier to appear in Applied Physics Letters.

[3] J.F.Palmier, C.Minot, J.L.Lievin, F.Alexandre, J.C.Harmand, J.Dangla, C.Dubon-Chevallier and D.Ankri, Applied Physics Letters 49,19(1986) 1260.

[4] F.Alexandre, J.C.Harmand, J.L.Lievin, C.Dubon-Chevallier, D.Ankri, C.Minot and J.F.Palmier, Journal of Crystal Growth 81(1987) 391.

[5] G.H.Dohler, R.Tsu and L.Esaki, Solid State Commun., 38(1975) 709.

[6] D.Calecki, J.F.Palmier and A.Chomette, Journal of Physics C 17(1984),5017.

[7] A.Sibille, J.F.Palmier, C.Minot, J.C.Harmand and C.Dubon-Chevallier, third Int. Conf. Superlattices, Microstructures and Microdevices(1987), to appear in Superlattices and Microstructures.

[8] J.Dangla, E.Caquot, M.Campana, R.Azoulay, F.Alexandre, J.L.Lievin, J.F.Palmier, and D.Ankri, Physica B, 129 (1985), 366.

[9] F. Capasso, K.Mohammed, A.Y.Cho, R.Hull and A.L.Hutchinson Phys. Rev. Letters 55(1985),1152.

[10] J.F.Palmier, A.Sibille, J.C.Harmand and J.Dangla, Electronics Letters 23,18 (1987), 938.

[11] J.Xu and M.Shur IEEE Electron Device Letters 7,7(1986), 416

[12] F.E. Najjar, D.C.Radulescu, Y.K.Chen, G.W.Wicks, P.J.Tasker and L.F.Eastman , Appl. Phys. Lett. 50,26(1987), 1915

Paper presented at Int. Symp. GaAs and Related Compounds, Heraklion, Greece, 1987

InP and $In_{0.53}Ga_{0.47}As$ metal–insulator–semiconductor field effect transistors with PAs_xN_y as the gate insulator

Y. Takahashi*, T. Takahashi, T. Shitara, Y. Iwase*, Y-H. Jeong**, S. Takagi***, F. Arai and T. Sugano

Department of Electronic Engineering, The University of Tokyo, Tokyo 113, Japan

Abstract. PAs_xN_y insulating films were deposited on InP and $In_{0.53}Ga_{0.47}As$ substrates by decomposing PCl_3 and $AsCl_3$ in NH_3 gas. Phosphorous-arsenic nitride film showed good interface characteristics and the density of trap states on InP substrate surface was reduced to less than $10^{11}cm^{-2}eV^{-1}$, but that on $In_{0.53}Ga_{0.47}As$ was $10^{12}cm^{-2}eV^{-1}$. Effective mobility of electron and the drain current drift for 10^3s were about $1200cm^2V^{-1}s^{-1}$ and about 10% for accumulation-mode PN_y/InP metal-insulator-semiconductor field effect transistors (MISFETs), and about $1300cm^2V^{-1}s^{-1}$ and as small as about 1% for $PAs_xN_y/In_{0.53}Ga_{0.47}As$ MISFETs, respectively.

1. Introduction

A key issue for realizing metal-insulator-semiconductor field effect transistors (MISFETs) using III-V compound semiconductor substrates is how to reduce the density of trap states on the substrate surface. The trap states are created during the growth of natural oxide film on the surface and also due to decomposition of the surface in high temperature processing.

Vapor phase etching of the substrate surface and in-situ deposition of phosphorous nitride film on the surface of InP substrates under the excess pressure of phosphorus has been found useful to reduce the density of trap states at the interface between the InP substrate and the phosphorous nitride film (1).

Here the results of application of this technique to InP MISFETs' and $In_{0.53}Ga_{0.47}As$ MISFETs' fabrication, especially using PAs_xN_y film as the gate insulators, will be presented. Chemical vapor deposition (CVD) of PAs_xN_y, in-situ etching of the substrates by reagent gases, interface properties between PAs_xN_y films and InP or $In_{0.53}Ga_{0.47}As$ surfaces and electrical characteristics of MISFETs built by PAs_xN_y/InP and $PAs_xN_y/In_{0.53}Ga_{0.47}As$ will be described.

* On leave from Electronic Materials & Components Laboratories, Nippon Mining Co., Ltd., Saitama 335, Japan.
** Present address : Department of Electrical & Electronic Engineering, Pohang Institute of Science & Technology, Pohang, Korea.
*** Present address : VLSI Research Center, Toshiba Corporation, Kawasaki 210, Japan

2. Chemical vapor deposition of PAs_xN_y film

The CVD system used in this experiment is schematically shown in Fig.1. PCl_3 (11 Torr. in H_2), $AsCl_3$ (2.3 Torr. in H_2), and NH_3 were introduced into the reaction zone, and the PAs_xN_y films were deposited on the substrates. The flow rate of PCl_3, $AsCl_3$, and NH_3 was ranged from 0 to 30cc/min, 0 to 150cc/min, and 20 to 200cc/min, respectively. However, the total flow rate was maintained to 450cc/min by diluting with H_2. The deposition temperature was ranged from 300 to 600 °C. Insulating PAs_xN_y films with good surface morphology were deposited when NH_3 flow rate was higher than 80cc/min, molar ratio As/P+As was smaller than 0.25, and the deposition temperature was in the range from 400 to 500°C. The deposition rate is shown in Fig.2, where the film thickness was measured by ellipsometry. The deposition rate of the PAs_xN_y film decreased with increase of deposition temperature from 400 to 500°C and with increase of the molar ratio As/P+As. The etching rates by one of the reagent gases for both InP and $In_{0.53}Ga_{0.47}As$ were measured. InP was etched either by PCl_3 or by $AsCl_3$ when the etching temperature was higher than 400°C at almost the same rate, that increased with temperature. The results for $In_{0.53}Ga_{0.47}As$ were shown in Fig.3. $In_{0.53}Ga_{0.47}As$ was etched by $AsCl_3$ when the etching temperature was higher than 300°C and the etching rate was much higher than that for InP. To confirm the effect of vapor etching by each gas on the properties of interfaces, one of the reagent gases was introduced into the reaction zone a few minutes before the beginning of the deposition.

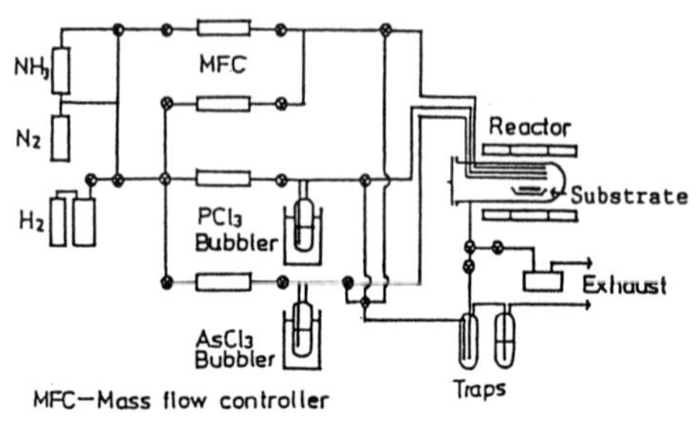

Fig.1. Apparatus for PAs_xN_y CVD.

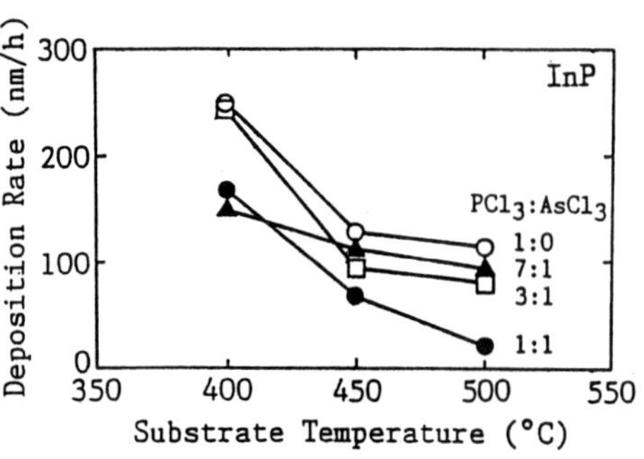

Fig.2. PAs_xN_y deposition rate vs. substrate temperature.

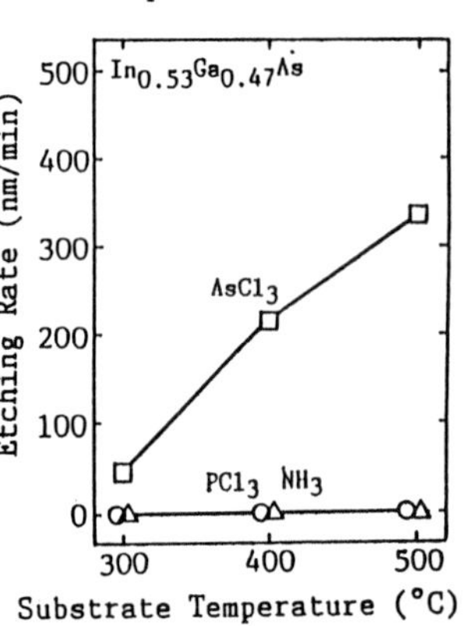

Fig.3. Etching rate of $In_{0.53}Ga_{0.47}As$ vs. substrate temperature.

3. Electrical characteristics of PAs_xN_y/InP or $PAs_xN_y/In_{0.53}Ga_{0.47}As$ interfaces

PAs_xN_y/InP and $PAs_xN_y/In_{0.53}Ga_{0.47}As$ interface properties were measured by the capacitance-voltage (C-V) characteristics of MIS diodes. The substrates were undoped InP wafers whose electron density was $5x10^{15}cm^{-3}$ or $In_{0.53}Ga_{0.47}As$ films grown by liquid phase epitaxy (LPE) on undoped InP whose electron density was $1.5x10^{16}cm^{-3}$, with (100) orientations. The measuring frequency and the sweep rate of the bias voltage were 1 MHz and 0.1 V/s, respectively. The density of trap states at the interface, N_{ss}, was obtained by using the Terman method.

3.1 PAs_xN_y/InP interfaces

The density of interface trap states in the energy gap of InP for PAs_xN_y/InP system is shown in Fig.4. The minimum density of states for PN_y/InP system was about $4x10^{11}cm^{-2}eV^{-1}$ with PCl_3 etching. Phosphorous-arsenic nitride reduced the density of the trap states to $1x10^{11}cm^{-2}eV^{-1}$ or less, particularly with $AsCl_3$ etching. These results may be caused by the annihilation of phosphorous vacancies on InP surface by As atoms, and are consistent with results previously reported by Yamaguchi et al. (2), Blanchet et al. (3), and Chave et al. (4).

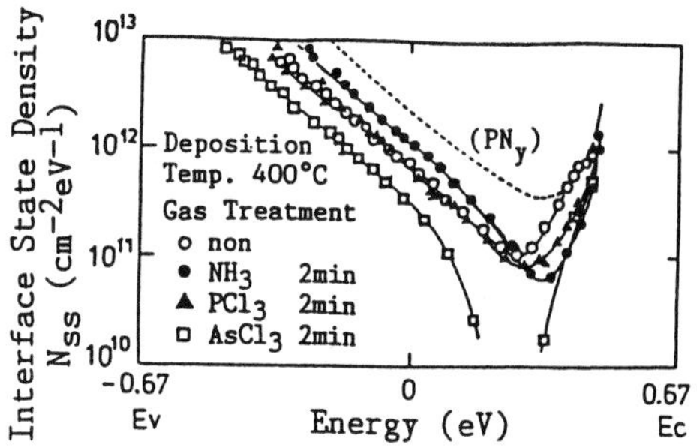

Fig.4. Distribution of the density of interface trap states in the energy gap of InP for PAs_xN_y/InP system.

3.2 $PAs_xN_y/In_{0.53}Ga_{0.47}As$ interfaces

The density of interface trap states for $PAs_xN_y/In_{0.53}Ga_{0.47}As$ system is shown in Fig.5. The dependence of the density of trap states on the film composition is weak in comparison with that of InP system, even with $AsCl_3$ etching. The minimum density of trap states for $PAs_xN_y/In_{0.53}Ga_{0.47}As$ system was $1x10^{12}cm^{-2}eV^{-1}$ or more.

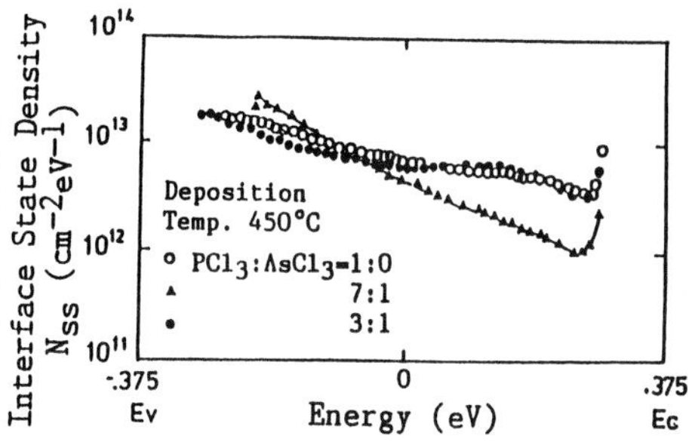

Fig.5. Distribution of the density of interface trap states in the energy gap of $In_{0.53}Ga_{0.47}As$ for $PAs_xN_y/In_{0.53}Ga_{0.47}As$ system.

4. Electrical characteristics of PAs_xN_y/InP or $PAs_xN_y/In_{0.53}Ga_{0.47}As$ MISFETs

4.1 Device structure

Accumulation-mode PAs_xN_y/InP MISFETs were fabricated on Fe-doped semi-insulating (S.I.) substrates whose resistivity was greater than 10^6 ohm-cm and Zn-doped p-type substrates whose hole density was $2x10^{16}cm^{-3}$, and $PAs_xN_y/In_{0.53}Ga_{0.47}As$ MISFETs on Cd-doped p-type LPE layers whose hole density was $2\sim7x10^{16}cm^{-3}$ with (100) orientations. The MISFETs prepared in the present work have ring-gates and the inner and outer radii are 100 and 200 μm, respectively.

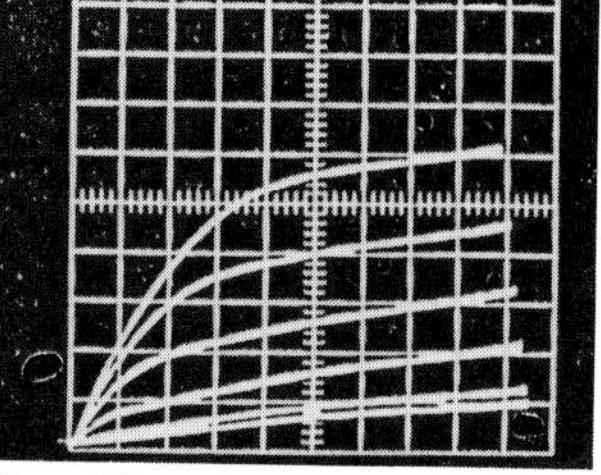

Repetition frequency = 100 Hz
Vertical : 2 mA / div.
Horizontal : 1 V / div.
V_g = 2 V / step from V_g = -6.5 V

Fig.6. Drain-current/drain-voltage characteristics for a $PAs_xN_y/In_{0.53}Ga_{0.47}As$ MISFET.

4.2 PAs_xN_y/InP MISFETs

Effective electron mobility for PN_y/InP MISFETs was about $1200cm^2V^{-1}s^{-1}$ for S.I substrates and $400cm^2V^{-1}s^{-1}$ for p-type substrates. It was found that the effective electron mobility decreased with increase of PAs_xN_y deposition temperature, probably due to the thermal degradation of InP surface. The drain current drift after applying 6V to the gate for 10^3s was about 10% for the MISFETs fabricated on the S.I. substrates. Phosphorous-arsenic nitride film did not provided any improvement in the effective electron mobility or the drain current stability in comparison with phosphorous nitride film although the density of interface trap states was reduced.

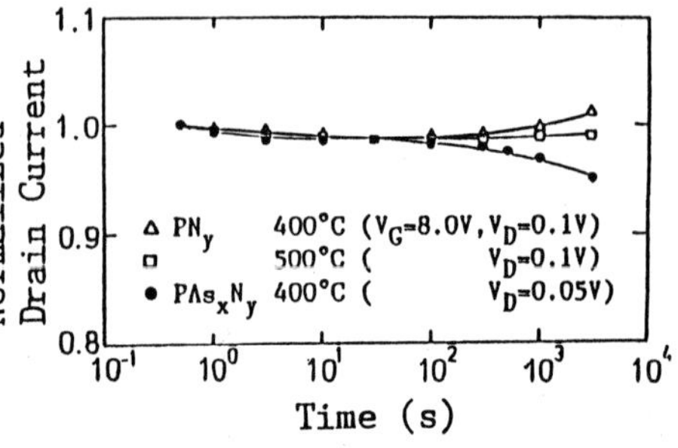

Fig.7. Drain current drift in $PAs_xN_y/In_{0.53}Ga_{0.47}As$ MISFETs.

4.3 $PAs_xN_y/In_{0.53}Ga_{0.47}As$ MISFETs

The observed drain current-voltage characteristics are shown in Fig.6. The effective electron mobility was about $1300cm^2V^{-1}s^{-1}$, higher than InP MISFETs. The drift of drain current is shown in Fig.7. The drain current drift after applying 8V to the gate for 10^3s was as small as about 1%. The effect of deposition temperature and the film composition was small.

5.Conclusions

PAs_xN_y insulating films were deposited on InP and $In_{0.53}Ga_{0.47}As$ surfaces by a chemical vapor deposition technique. Insulating PAs_xN_y films with good surface morphology were deposited when NH_3 flow rate was higher than 80cc/min, molar ratio As/P+As was smaller than 0.25, and the deposition temperature was in the range from 400 to 500°C. It was confirmed that InP was etched either by PCl_3 or by $AsCl_3$, and $In_{0.53}Ga_{0.47}As$ by $AsCl_3$. Phosphorous-arsenic nitride reduced the density of interface trap states on InP substrates to less than $10^{11}cm^{-2}eV^{-1}$, but had little effect on $In_{0.53}Ga_{0.47}As$ system. The effective electron mobility for accumulation-mode PN_y/InP MISFETs was about $1200cm^2V^{-1}s^{-1}$, and the drain current drift for 10^3s was about 10%. The effective electron mobility for $PAs_xN_y/In_{0.53}Ga_{0.47}As$ MISFETs was about $1300cm^2V^{-1}s^{-1}$ and the drain current drift for 10^3s was as small as about 1%, respectively.

Acknowledgment

This work was supported by the Scientific Research Grant-in-aid #62104007 for Special Project Research on "Alloy Semiconductor Physics and Electronics", from the Ministry of Education, Science and Culture of Japan.

References

(1) Y-H. Jeong, S. Takagi, F. Arai and T. Sugano, J. Appl. Phys., to be published.
(2) E. Yamaguchi, Y. Hirota and M. Minakata, Thin Solid Films, 103 (1983), 201.
(3) R. Blanchet, P. Viktorovitch, J. Chave and C. Santinelli, Appl. Phys. Lett., 46 (1985) 761.
(4) J. Chave, A. Choujaa, C. Santinelli, R. Blanchet and P. Viktorovitch, J. Appl. Phys. 61 (1987) 257.

Inst. Phys. Conf. Ser. No. 91: Chapter 7
Paper presented at Int. Symp. GaAs and Related Compounds, Heraklion, Greece, 1987

Microwave characteristics of an InGaAs junction field effect transistor grown by MOCVD

J.Y. Raulin, E. Vassilakis, M.A. di Forte-Poisson, C. Brylinski, M. Razeghi

Laboratoire Central de Recherches, Thomson-CSF, B.P. N° 10, 91401 Orsay

ABSTRACT: GaInAs/InP junction field effect transistors were made by metalorganic low pressure chemical vapor deposition. A gate as short as 0.5 m was achieved by chemical etching. The transistor exhibits a transconductance of 260 mS/mm and a microwave gain of 10 dB at 18 GHz.

1. INTRODUCTION

Because of its compatibility with optical device material, Ga In As is an attractive material choice for a field effect transistor. Integration of such a transistor on the same chip as optoelectronic devices fabricated on InP substrate would result in an enhancement of the circuit abilities.

Furthermore, the high electron velocity (Windhorm 1982) achievable in InGaAs even at the doping levels usually used in field effect transistor channels lets one expect high performances in the millimeter wave range.

The metal-semiconductor barrier is too low, and interface state density too difficult to control, to allow easy fabrication of InGaAs MESFETs and MISFETs (Kaumanns 1982, Mullin 1983). On an other hand, JFET structures have been demonstrated with promising dc results (Wake 1985, Raulin 1987).

In spite of a large amount of attempts, only poor microwave performances have so far been reported for an InGaAs FET (Schmitt 1985, Cheng 1985), while simulation studies let one expect high gains and high cutoff frequency (Brockerhoff 1986). Going on with our investigation of the InGaAs JFET, we have performed microwave measurements on our structure.

In this paper, we report experimental evidence of the high capabilities of InGaAs FETs in the millimeter wave range. A maximum available gain (MAG) of 10 dB at 18 GHz has been brought to the fore. We will first review in section I the layer growth process, then in section II the device preparation. Section III deals with dc and microwave results.

2. MATERIAL GROWTH

InGaAs and InP layers were grown by low pressure metalorganic vapor deposition (LPMOCVD) (di Forte-Poisson 1985). General presentation of the low pressure method used in our laboratory and schematic diagram of the reactor has already been published (Duchemin 1979). In order to fulfill the specific material requirements of a junction field effect transistor, we studied the following points : (i) influence of the buffer layer thickness ; (ii) background concentration level in the buffer layer ; (iii) N type doping of InGaAs and P type doping of InGaAs and InP.

A high quality buffer layer is an imperative demand. Propagation of defects or diffusion of impurities from the substrate to the active layer would result in a degradation of electron velocity in the channel, bad pinchoff, and poor device performances. We found that a buffer thickness of 1 μm is required to eliminate the defect propagation from the substrate. We made a 0.5 m thick buffer for the transistor, which was considered as being a good compromise between crystallographic quality and low leakage current. Background impurity concentration is $1 \times 10^{14} cm^{-3}$ in such a layer, while $8 \times 10^{13} cm^{-3}$ can be achieved in a 10μ m layer.

Silicon from silane was used as the donor and zinc as the acceptor for doping. A very precise control of the doping level in the range 10^{16} $10^{18} cm^{-3}$ has been obtained by varying the molar fraction of silane in the reactor. The active layer of the FET, a 0.2 μm thick InGaAs layer, was doped to $1.3 \times 10^{17} cm^{-3}$. On the other hand, accurate control of zinc doping is hard because of the high diffusion coefficient of this element. Nevertheless, we achieved to avoid zinc diffusion into the active layer while depositing the 0.5 μm thick $5 \times 10^{18} cm^{-3}$ P doped InP layer. Barrier height of the so formed junction was between 0.3 and 0.5 volts. A top InGaAs layer, heavily P doped, was finally deposited in order to facilitate a good ohmic contact of the gate metallization.

A cross section of the layers grown for the FET is shown in figure 1a.

3. DEVICE PREPARATION

Gate metallization (AuZn) is used as a mask for the chemical etching of the P-type layers. The later is made in two steps : the InGaAs top layer first, using a citric acid solution, Figure 1b, then the InP layer using an HCl-H_3PO_4, Figure 1c. Each of those solutions is material selective, and no etching of the adjacent layers occurs.

Gate length is found to be determined by the etching time of the InGaAs top layer. No further undercut happens during the etching of the InP layer. A citric acid solution allows an etching rate slow enough to enable us to control accurately the gate length.

We made submicrometer gates using this process. Gates as short as half a micron have been made, and it should soon be possible to reach 0.3 microns.

Ohmic contacts (AuGe Ti Pt Au) are deposited self aligned with respect to gate metallization, Figure 1d. By this means, the gate is centered in the Source Drain spacing which is about 1.5 microns long. The self alignement technique combined with the low contact resistance achievable on N-type InGaAs leads to very low access resistance (Rs) from source to channel. A typical value of Rs is 0.5 Ohms.mm.

Mesa isolation is finally etched, Figure 1e, using a non selective acid solution (Br_2CH_3OH).

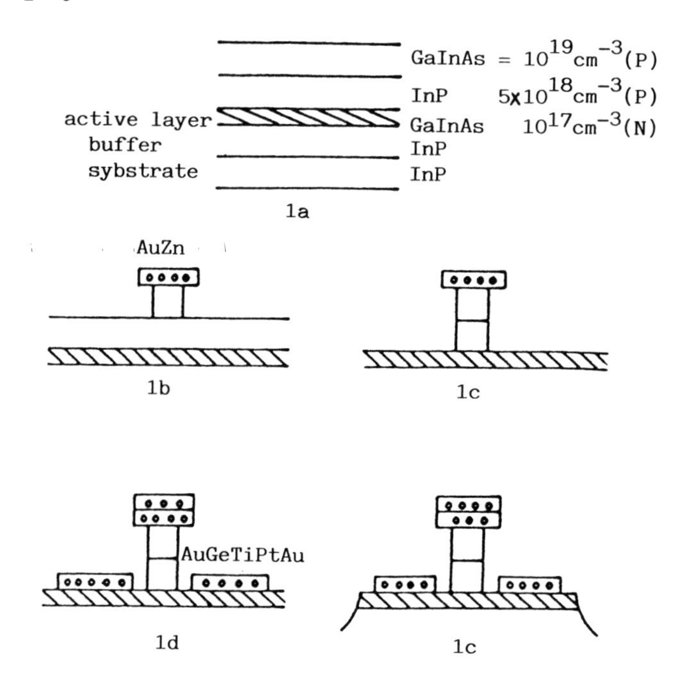

Fig. 1. JFET fabrication process : (a) layers, (b) and (c) the gate was obtained from wet etching using the gate metal as a mask, (d) ohmic contacts deposition, (e) mesa isolation

4. DEVICE CHARACTERISTICS

The dc characteristics of the JFET are shown in figure 2. The transistor presents a good saturation behaviour even at high currents (the JFET is 150 µm wide). Pinchoff is effective at a gate voltage of -3.5 volts. A very high transconductance is achieved (235 mS/mm) ; taking into account the effect of the access resistance Rs, we derive an intrinsic transconductance of 260 mS/mm.

In order to perform microwave measurements, a wire bonding of the chip is done. After this step, a degradation of dc characteristics is noticed. The transconductance is reduced to about 200 mS/mm. Further improvement on the bonding conditions should allow us to eliminate this phenomenon.

S scattering parameters have been measured from 2 to 18 GHz. Very high gains are achieved : 12.2 dB at 12 GHz, and 10 dB at 18 GHz. These values stand comparison with the ones of a GaAs FET of same gate length. They are the best ones to our knowledge ever reported on a GaInAs FET. Elimination of the FET characteristics degradation during bonding and further optimization of the device should lead to gains and cutoff frequency higher than what can be obtained at present using GaAs.

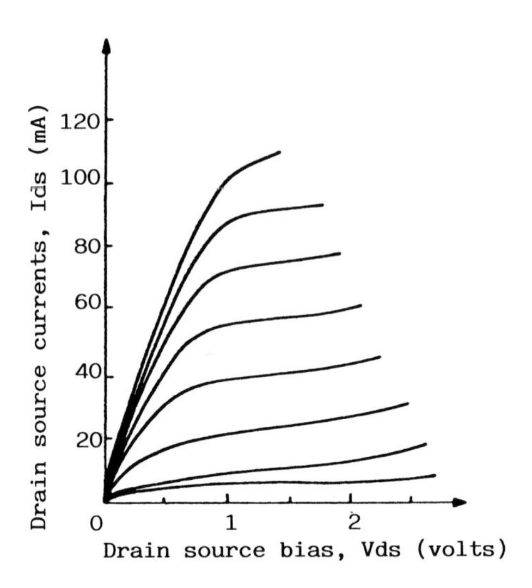

Fig. 2. dc characteristics of the junction field-effect transistor. The gate is 0.5μm long by 150 μm wide. The transconductance is 235 mS/mm. (Gate bias step is - 0.5V starting at 0 V).

CONCLUSION

A first experimental evidence of the abilities of InGaAs as a material for high performance microwave field effect transistors has been established. A gain as high as 10 dB at 18 GHz has been obtained with an InGaAs junction field effect transistor whose gate length was 0.5 μ m. Further improvements of the gate etching process should allow shorter gate lengths and therefore still better microwave performances.

ACKNOLEDGEMENT

The authors would like to thank J.P. Duchemin, J.P. Hirtz, H. Derewonko and D. Delagebeaudeuf for constant help and beneficial discussions. This work was supported by the DRET.

REFERENCES

Brockerhoff 1986 Comp. Semic. Integ. Circ. Visby Gotland Sweden
Cheng 1985 Appl. Phys. Lett. 46 (9) p 885.
Di Forte Poisson 1985 Appl. Phys. Lett. 46 (5) p 476.
Duchemin 1979 Inst. Phys. Conf. Series n° 45 Chap. 1
Kaumanns 1982 Inst. Phys. Conf. Series n° 63 p 329.
Mullin 1983 Vac. Sci. Tech. B1(3) p 782.
Raulin 1987 Appl. Phys. Lett. 50(9) p 535.
Schmitt 1985 Elect. Lett. Vol 21 p 449.
Wake 1985 IEEE Elec. Dev. Lett. vol EDL-6 p 626.
Windhorn 1982 IEEE Elec. Dev. Lett. vol EDL-3 p 18.

Paper presented at Int. Symp. GaAs and Related Compounds, Heraklion, Greece, 1987

High transconductance submicron self-aligned InGaAs JFETs

K. Steiner, U. Seiler, W. Brockerhoff, K. Heime
Universität Duisburg, Halbleitertechnik/ Halbleitertechnologie,
Sonderforschungsbereich 254, D-4100 Duisburg 1, FRG
E. Kuphal
Forschungsinstitut der DBP, D-6100 Darmstadt, FRG

Abstract

A novel self-aligned gate process for high-transconductance InGaAs-JFETs using diffused gates is described. DC-characteristics of normaly-on transistors exhibit more than 330mS/mm and 220mS/mm with gate lengths of 0.6 and 1.6μm. Short channel effects are supressed significantly by introducing a p-type InP buffer layer between the LPE grown channel and the substrate. Mobility degradation during gate formation is strongly reduced, the drift mobility being 6500 cm^2/Vs at a channel carrier concentration of $1.4 \cdot 10^{17}$cm^{-3} with an appertaining average Hall mobility of 7330cm^2/Vs.

Introduction

Self-aligned FETs offer advantage over conventional structures since the series resistances are reduced considerable while the increase in gate-source and gate-drain capacitances is less pronounced /Brockerhoff et.al.1985/. Thus these devices should exhibit higher transconductances /Schmitt et.al.,1985/. Several self-aligned gate processes have been developed to fabricate InGaAs-FETs. There are silicon-oxide enhanced Schottky-gate FETs /Cheng et.al.,1984a/, InGaAs MISFETs /Cheng et.al.,1984b, Gardner et.al., 1986/, diffused p$^+$-InGaAs gates /Albrecht et.al.,1985/ MBE grown p$^+$-InGaAs gates /Wake et.al.,1984/ and MOCVD grown p$^+$-InP gates /Wake et.al.,1985, Raulin et.al.,1987/. Only three normaly-on devices /Cheng et.al.,1984b, Wake et.al.,1985, Raulin et.al.,1987/ in the submicron and micron regime achieved high transconductances above 200mS/mm as expected from simulations /Schmitt et.al.,1985/. However, these devices exhibit high output conductances and/or threshold voltages, which depend on both drain-source voltage and gate length. In addition the measured transconductances are lower than those expected from carrier concentrations and mobility as determined before the gate formation. This indicates a mobility degradation by the process of gate formation, e.g. by ion-implantation and subsequent annealing or by diffusion /Chai et.al.,1985, Schmitt et.al.,1985/. Normally-off devices should have higher transconductances and indeed the first normally-off InGaAs-JFET with transconductances above 50mS/mm (at 1μm gate length) was achieved recently /Albrecht and Lauterbach,1987/.

In this work several efforts were combined in order to reduce short-channel effects and series resistances: a novel self-aligned gate process for InGaAs-JFETs allows the reduction of series resistances, a low diffusion temperature reduces the mobility degradation, and the insertion of a p-InP buffer layer eliminates short-channel effects even for submicron (0.6μm) gate lengths.

Device fabrication

LPE growth started with a Zn-doped p-InP buffer layer ($p=1 \cdot 10^{17}$cm^{-3}, $\approx$0.8μm) grown on i. InP:Fe substrate (Sumitomo). It is finished by the n-InGaAs channel layer ($1.4 \cdot 10^{17}$cm^{-3}, .35μm), which is lattice matched at growth temperature to the InP buffer layer. The p$^+$-InGaAs layer is diffused from Zn doped spin-on film sources into the n-InGaAs /Schmitt et.al.,1985/. The sample is diffused at 535°C within 10 min. After the diffusion step a CF_4-plasma is applied to the surface to remove organic remains of diffusion source.

For the T-shaped gate Al is evaporated onto the whole sample (200nm). It is followed by a photolithography step and a second metallization (10nm Cr, 200nm Au). After lift-off the CrAu-patterns define the gate. The Al is selectively etched with a phosphoric acid (H_3PO_4:CH_3COOH) /Ruge,1984/ using the CrAu cap layer as an etch mask. The T-shaped gate is completed by etching the p$^+$-mesa. Source and drain metal is directly evaporated onto the undercut structure to ensure the self-alignement. The process is finished by etching the n-mesa for device isolation.

Experimental Results and Discussion

Fig. 1 displays two output characteristics of such InGaAs-JFETs with different gate lengths locate in the same chip of the wafer. Due to the T-shaped gate the gate lengths are 0.6μm and 1.6μm The channel thickness is about 0.2μm and the gate width 100μm. The DC-characteristics show n short channel effects even in the submicron regime. Especially there are no threshold voltage shift or increasing output conductances. The pinch-off and saturation behaviour is excellent.

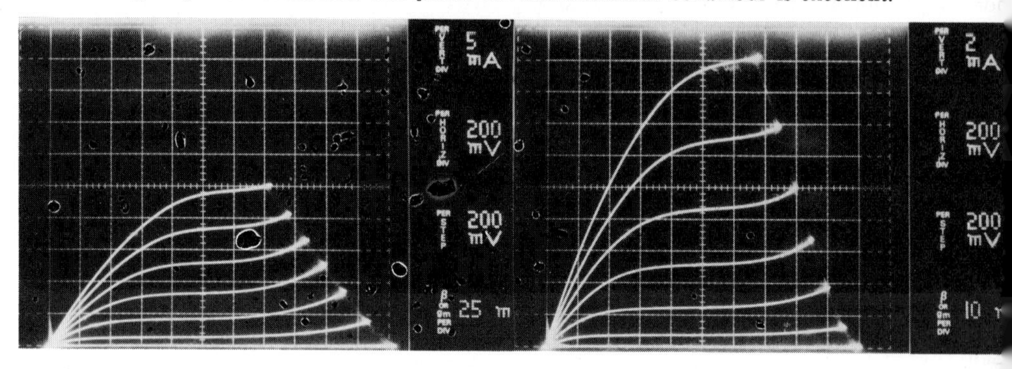

Fig. 1: DC-characteristics of self-aligned InGaAs-JFETs; L_G=0.6μm (left), L_G=1.6μm (right)

Electron transfer from the channel into the buffer or the substrate results in short channel effect /Matsumoto et.al.,1984/. This was experimentally observed in /Schmitt et.al.,1985/. Without buffer layer DC-characteristics of InGaAs-JFETs exhibit increasing short-channel effects wit decreasing gate length below 6μm . In this work the absence of short channel effects underline the carrier confinement in the channel. This was achieved by the high overall potential barriers c the p^+n-InGaAs gate-diode and the anisotype heterojunction at the channel substrate heterointer face, which confine the electrons to the active region of the FET. Thus especially at smaller gat lengths carrier injection into the buffer or the substrate is suppressed and short channel effects d not appear.

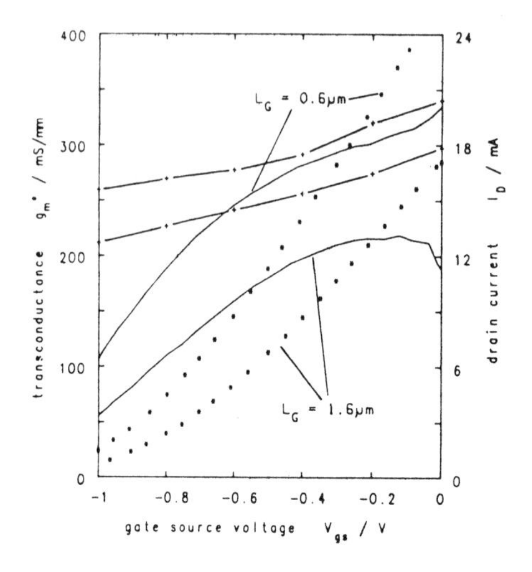

Fig. 2: Measured transfer characteristics of InGaAs-JFETs shown in fig. 1; U_{DS}=1V; line: transconductance; dots: drain current; broken line: simulated transconductance results, upper curve L_G=0.5μm, lower curve L_G=1μm

Fig. 2 shows transfer characteristics of both o the two JFETs of fig. 1. The drain sourc voltage is V_{DS}=1V. At V_{GS}>-0.4V the drai current I_D varies linearly with the gate voltag V_{GS} as expected in the case of consta carrier velocity in the channel. At lowe voltages (V_{GS}<-0.4V) deviations are observe which will be explained later by mobilit degradation. Transconductance values ar normalized to 1mm gate width. Maximu extrinsic transconductances achieve more tha 330mS/mm with a gate length of L_G=0.6μ and 220mS/mm with a gate length c L_G=1.6μm. To the authors knowledge, thes values are among the highest values eve reported for normally-on InGaAs-FETs in th submicron and micron regim Transconductance values obtained fro numerical simulations are also shown in fig. and compared with the best extrinsi transconductance curve of this work. Detai of the device simulation are describe elswhere /Carnez et.al.,1980, Dämbke et.al.,1984, Brockerhoff et.al.,1985/. Th calculations were made using the following vice dimensions and properties: drain-source voltage V_{DS}=1V; threshold voltage $_T$=-1.6V, channel doping concentration $N_D=1.5\cdot10^{17}cm^{-3}$; drain gate and source gate distance $_{DG}=L_{SG}$=0.5μm; effective channel thickness a=0.5μm, low field mobility μ_o=6250cm^2/Vs; gate ngth L_G=0.5μm for the upper and L_G=1μm for the lower simulated curve. For higher gate urce voltages (V_{GS}>-0.4V) the simulation results are a good fit to the extrinsic transconductance rve of the 0.6μm JFET. At lower gate source voltages discrepancies appear and may be explained a decrease of the channel carrier concentration towards the buffer layer and a drop of the drift obility, which is discussed later. In contrast to experimental results both parameters are kept nstant in the channel during the numerical simulations. For the 1.6μm gate JFET the fit is less od, mainly because of the smaller gate length (1μm) in the simulations.

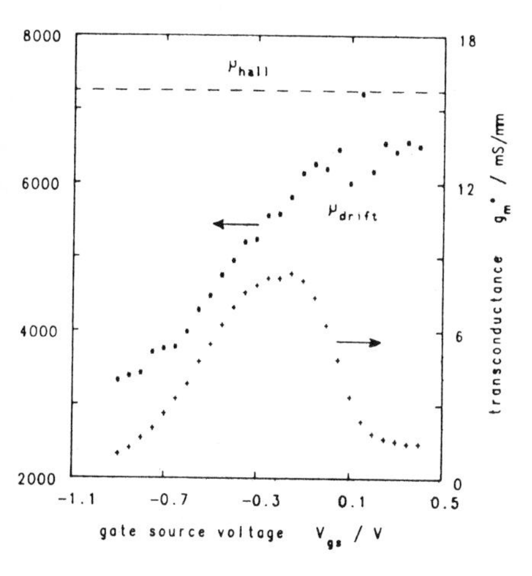

g. 3: Mobility profile of an InGaAs-JFET asured at 300K under daylight illumination $_{DS}$=100mV)

Channel mobilities on FET-structures can be deduced from magnetotransconductance measurements /Jay and Wallis, 1981/. In fig. 3 a drift mobility profile and the related transconductance values are illustrated. The curves have to be taken in the ohmic regime of the FET, in this case at V_{DS}=100mV. Therefore transconductances are low. The mobility displays a plateau at 6500cm^2/Vs between 0 and 0.4V gate source voltage. In contrast to the mobility profile shown earlier by Schmitt et.al.,1985, the profile does not show a degradation towards the gate. This is due to lower diffusion temperatures here which avoid long diffusion tails. Such tails are known from Zn-diffusion in InP, which occur under certain diffusion temperature conditions in addition to the steep diffusion profile /Arnold et.al.,1984/. The tails are also known from implanted and annealed profiles /Chai et.al.,1985/. There the mobility degradation is tremendous.

e measured Hall mobility μ_H=7330cm^2/Vs is also indicated in fig. 3. This value was measured er gate formation on an ungated Hall-bar. The maximum drift and Hall mobility give a Hall tor of $r_H=\mu_H/\mu_d$=1,13. This value agrees well with calculated Hall factors /Takeda et.al.,1981, keda and Littlejohn, 1982/. The result indicates that mobility degradation during the gate fusion is eliminated. Fig. 3 also exhibits a strong drift mobility degradation towards the p-InP fer layer. Near pinch-off the mobility decreases to 3300cm^2/Vs. At this point the mobility de-dation may be in part caused by Zn-outdiffusion from the buffer. An influence on the whole ive layer of the FET seems less probable due to the Zn gettering at InGaAs/InP heterointerfaces eva and Seidel, 1986/. The degradation at higher gate source voltages might be correlated to the ce charge region at the anisotype channel buffer heterointerface, which dips several Debye-gths into the active region of the FET. The screening of Coulomb potential by free carriers omes less effective. Since the mobility is a function of free carrier density /Störmer,1983/ it reases towards the p-InP buffer layer. To create a wider drift-mobility plateau the space charge er in the channel has to be decreased. This can be done by lowering the Zn concentration in the ype buffer layer. As a result linearity between I_D and V_{GS} of the InGaAs-JFET will be proved.

nclusion

novel self-aligned gate process for high-transconductance InGaAs-JFETs is discussed. The -characteristics exhibit transconductances above 330mS/mm in the submicron regime $_G$=0.6μm). At 1.6μm gate length the transconductances achieve 220mS/mm. This improvement r previously published results is in part due to the care that was taken during the gate nation, especially through the use of diffusion at low temperatures (535°C). This process serves high mobilities in the channel near the gate (drift mobility μ_d=6500cm^2/Vs, average Hall mobility μ_H=7330cm^2/Vs at n=1.4·10^{17}cm^{-3}). Additional improvement in both transconductance and the elimination of short channel effects is due to the use of the p-InP bu layer. Remaining mobility degradation towards the buffer should be reduceable by lower acce concentrations in the buffer layer.

Acknowledgement

The authors are indepted to Dr. G.Schlamp and G.Ptaschek (Demetron, Hanau, FRG) for prep tion of the spin-on films, M.Böhm and U.Doerk for assistance in device preparation and meas ments. The work was supported by Deutsche Forschungsgemeinschaft.

Albrecht H.,Huber H.,Lauterbach Ch.,Plihal M.,
Siemens Forsch.- u. Entwickl.-Ber.,14,1985,pp.295-298
Albrecht H.,Lauterbach Ch.,IEEE El.Dev.Lett.,8,1987,pp.353-354
Arnold N.,Schmitt R.,Heime K.,J.Phys.D.:Appl.Phys.,17,1984,pp.443-479
Brockerhoff W.,Dämbkes H.,Heime K.,Proc. 3.Int. Workshop Phys.
Sem.Dev.,S.C.Jain,S.Radhaknishna(Eds.),Madras,India,1985,pp.23-30
Carnez B.,Cappy A.,Kaszynski A.,Constant E.,Salmer G.,Appl.-Phys.51,1980,pp.784-790
Chai Y.G.,Yuen C.,Zdasiuk G.A.,IEEE Trans.El.Dev.,32,1985,pp.972-977
Cheng C.L.,Liao A.S.H.,Chang T.Y.,Cari di E.A.,Coldrin L.A., Lalevic B.,
IEEE El.Dev.Lett.,5,1984a,pp.511-514
Cheng C.L.,Liao A.S.H.,Chang T.Y.,Leheny R.F.,Coldren L.A.,Lalevic B.,
IEEE El.Dev.Lett.,5,1984b,pp.169-171
Dämbkes H.,Brockerhoff W.,Heime K.,Cappy A.,IEEE Trans.El.Dev.,31,1984,pp.1032-1037
Gardner P.D.,Liu S.G.,Narayan S.Y.,Colvin S.D.,Paczkowski J.P., Capewell D.R.,
IEEE El.Dev.Lett.,7,1986,pp.363-364
Geva M., Seidel T.E.,J.Appl.Phys.,59,1986,pp.2408-2415
Jay P.R., Wallis R.H.,IEEE El.Dev.Lett.,2,1981,pp.265-267
Matsumoto K.,Hashizume N.,Atoda N.,Awano Y.,Inst.Phys.Conf.Ser. 74(7),
Int.Symp.GaAs and Rel.Comp.,Biarritz,1984,pp.515-520
Raulin J.Y.,Thorngren E.,di Forte-Poisson M.A.,Razeghi M.,Colomer G.,
Appl.Phys.Lett.50,1987,pp.535-536
Ruge I.,1984,Halbleiter-Technologie,(Berlin:Springer Verlag),p.296
Schmitt R.,Steiner K.,Kaufmann L.M.F.,Brockerhoff W.,Heime K., Kuphal E.,
Inst.Phys.Conf.Ser.79(11),Int.Symp.GaAs and Rel.Comp., Karuizawa,Japan,1985,pp.619-624
Störmer H.L.,Surface Science,132,1983,pp.519-52.
Takeda Y.,Littlejohn M.A.,Hutchby J.A.,Trew R.J.,El.Lett.,17,1981,pp.686-688
Takeda Y.,Littlejohn M.A.,Appl.Phys.Lett.,40,1982,pp.251-253
Wake D.,Livingstone A.W.,Andrews D.A.,Davis G.J.,IEEE El.Dev.Lett.,5,1984,pp.285-287
Wake D.,Nelson A.W.,Cole S.,Wong S.,Henning I.C.,Scott E.G.,IEEE El.Dev.Lett.,6,1985,pp.62

DC-characterization of normally-off InGaAs/InP:Fe junction field-effect transistor inverters

H. Albrecht

Siemens Research Laboratories, Otto-Hahn-Ring 6, D-8000 München 83, FRG

ABSTRACT: A detailed DC-characterization of normally-off InGaAs junction field-effect transistor (JFET) inverters with 1 μm gate length is presented. The JFET´s show a threshold voltage of 0.25 V and at operation conditions (U_{GS} = o.7 V, U_{DS} = 3 V, I_D = 19 mA) a transconductance of 82 mS or 430 mS/mm. With time resolved pulse measurements neither the gate-lag nor the drain-lag effect is observed. With a monolithically integrated load resistor of 120 Ω in the drain region of the JFET the inverters show a voltage transfer characteristic with a gain of -6 , an inverter threshold voltage of 0.45 V, a logic voltage swing of 0.6 V and a noise margin of 0.3 V .

1. INTRODUCTION

$In_{0.53}Ga_{0.47}As$ shows properties such as high electron mobility, high peak electron velocity and a large Γ-L separation (Pearsall 1982), which makes this material very attractive for FET´s in high-speed device applications (Chai 1985) and in OEIC´s (Wake 1986). In this paper for the first time InGaAs inverters are presented which consist of a normally-off InGaAs JFET with 1 μm gate length and a drain load resistor monolithically integrated in the drain region of the JFET.

2. FABRICATION AND TECHNOLOGY

Fig. 1 shows a schematic representation of the chip with the InGaAs JFET and the drain load resisitor. The epitaxial layers were grown lattice matched onto a semi-insulating (100) InP:Fe substrate by liquid

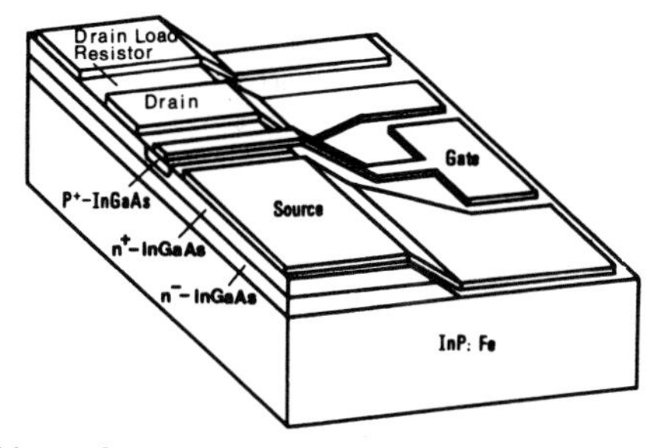

Fig. 1. Schematic representation of an InGaAs JFET chip with a drain load resistor.

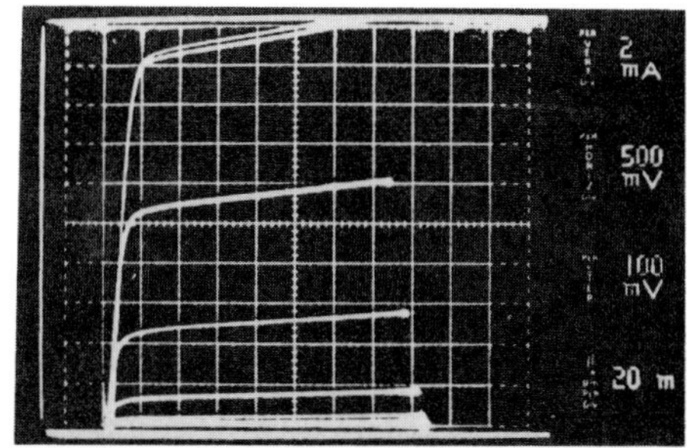

Fig. 2. I-U characteristic of a normally-off InGaAs JFET with 1 μm gate length (U_{GS}=0 to 0.7V).

phase epitaxy. First an unintentionally doped n-InGaAs buffer layer of 0.5 µm thickness was grown ($1x10^{15}cm^{-3}$). This was followed by a Sn-doped 0.4 µm thick InGaAs active layer ($8\cdot10^{16}cm^{-3}$). The junction gate was formed by Zn diffusion at 650° C for 15 min through a plasma-assisted CVD SiN_x mask. The resulting p^+-n junction is 0.3 µm beneath the semiconductor surface. Details for further technological steps for the fabrication of the InGaAs JFET have recently been described (Albrecht 1987). A 1 µm thick Ti/Au metallization has been used for both contacts and bond pads. All bond pads are on the semi-insulating InP:Fe substrate.The appropriate value of the drain load resistor can be adjusted by the InGaAs layer thickness using a chemical etching technique.

3. InGaAs JFET CHARACTERISTIC

A typical drain current versus drain-source voltage characteristic for a normally-off InGaAs JFET with 1 µm gate length and 190 µm gate width is shown in Fig. 2. The drain current I_D saturates at U_{DS} = 0.4 V and is 19 mA or 100 mA/mm at U_{GS} = 0.7 V. The JFET has an excellent pinch-off behaviour with a saturated drain current I_{Dss} of 1 µA at U_{GS} = 0 V. The threshold voltage U_{TH} is 0.25 V. The transconductance increases rapidly above U_{GS} > 0.3 V to a maximum value of 82 mS (430 mS/mm) at U_{GS} = 0.7 V. This extremely high transconductance at room temperature is one of the highest values for an InGaAs FET with 1 µm gate length reported so far.

4. TIME RESOLVED PULSE MEASUREMENTS

Several investigations have been published about parasitic effects limiting the performance of FETs. In the case of logic circuits, e. g. normally-off direct coupled field-effect logic (DCFL), the gate-lag effect and the drain-lag effect influence markedly the properties of the devices. The measurement principles and some subsequent analysis of these effects have been reported by Rocchi (1985). The gate-lag effect reduces the high frequency drain current and the transconductance. This corresponds in time resolved pulse measurements with low repetition rates to a drain current transient in response to a pulse on the gate. The drain-lag effect, which reduces the high frequency FET output resistance, corresponds to a drain current transient associated with a voltage pulse on the drain. These measurement techniques have been used to investigate the normally-off InGaA JFETs. As shown in Fig. 3 there is no gate-lag effect observed in the entire time domain of the gate pulse widths from 40 µs to 400 ms. Similarly, no drain-lag effect could be observed.

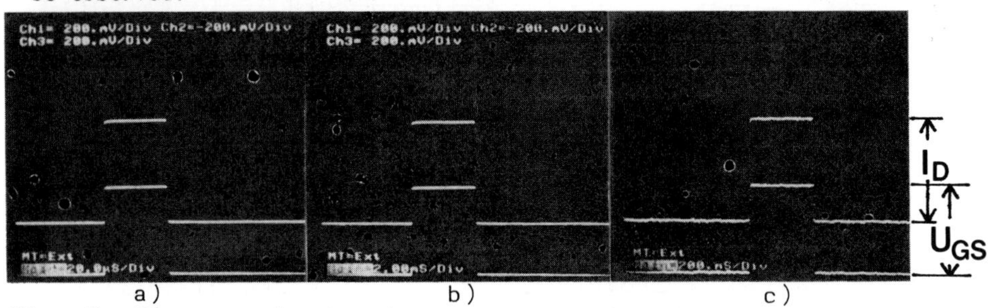

Fig. 3. Time resolved pulse measurements of the gate-lag effect pulse width: a) 40 µs b) 4 ms c) 400 ms

5. DC-CHARACTERIZATION OF InGaAs INVERTERS

Normally-off DCFL inverters have been realized with these InGaAs JFETs by monolithically integrating an InGaAs drain load resistor. Fig. 4 shows the load line diagram for DCFL inverter, which contains the measured drain current I_D versus drain-source voltage U_{DS} characteristic at U_{GS} = 0.6 V, the drain load resistor characteristic and the input characteristic of the next stage (Lehovec 1980). For a supply voltage of U_{DD} = 0.8 V and a drain current of I = 5 mA (which should be nearly half of the maximal drain current of 10 mA at U_{GS} = 0.6 V), the drain load resistor must be adjusted to 120 Ω. Under these conditions the static inverter transfer characteristic has been measured as shown in Fig. 5 (full line). Also superimposed on the curve is the same curve (broken line) with the axis interchanged, which represents the input-output characteristic of the next stage (Hyun 1986, Ketterson 1986). The output levels U_{OH} = 0.8 V and U_{OL} = 0.2 V correspond to the two intersections of the curves. The difference of these values gives the logic voltage swing ΔU = 0.6 V. The inverter threshold voltage is U_{inv} = 0.45 V and the static inverter gain is $G = \Delta U_{out}/\Delta U_{in} = -6$. The "zero" noise margin (NM0) and "one" noise margin (NM1) are defined as the maximum input excursion such that the output is on the verge of erroneously switching to a logic zero or one, respectively, when the noise signal is applied to one set of alternate stages. From Fig. 5 one obtains for NM0 = 0.29 V and NM1 = 0.31 V.

6. DISCUSSION

From DC-characterizations and the time resolved pulse measurements follow that normally-off InGaAs JFETs are very attractive condidates for DCFL inverter applications. In particular the absence of gate-lag and drain-lag indicates the good quality of the surface and the interface between InGaAs channel and InP:Fe substrate. It is believed that these well behaved characteristics are mainly caused by the incorporation of an InGaAs buffer layer. The JFETs show a very high transconductance g_m which results in a high inverter gain G. The measured value of G = -6 from Fig. 5 results from the product of $g_m \approx 40$ mS and the drain load resistor of 120 Ω.

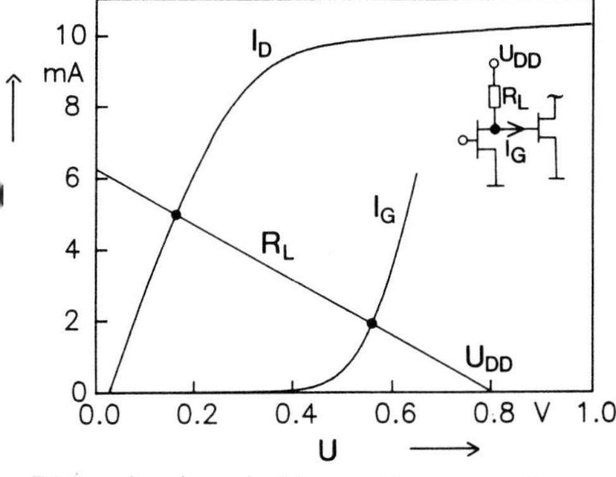

Fig. 4. Load line diagram for a normally-off InGaAs DCFL inverter.

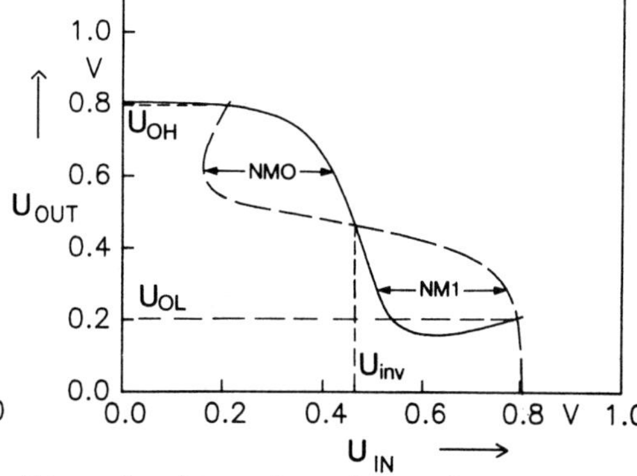

Fig. 5. Inverter transfer characteristic of a normally-off InGaAs DCFL inverter.

7. ACKNOWLEDGMENT

The technical assistance of H. Huber in growing the epitaxial material, W. Kunkel in SiN_x deposition and Ch. Lauterbach for device fabrication are gratefully acknowledged, as are helpful discussions with M. Plihal. This work was supported by the Ministerium für Forschung und Technologie of the Federal Republic of Germany.
The author alone is responsibel for the contents.

8. REFERENCES

Albrecht H. and Lauterbach Ch. 1987 IEEE Electr. Dev. Letters EDL-8 pp. 353-4

Chai, Y. G., Yuen C. and Zdasiuk G. A. 1985 IEEE Trans.Electr, Dev. EC-32 pp. 972-7

Hyun C. H., Shur M.S. and Peczalski A. 1986 IEEE Trans. Electr. Dev. ED-33 pp. 1421-26

Ketterson A. A. and Morkoc H. 1986 IEEE Trans. Electr. Dev. ED-33 pp. 1626-33

Lehovec K. and Zuleeg R. 1980 IEEE Trans. Electr. Dev. ED-27 pp. 1074-91

Pearsall T. P. 1982 GaInAsP Alloy Semiconductors (New York: Wiley) pp. 295-312

Rocchi M. 1985 Proc. of ESSDERC (Amsterdam: North-Holland) pp. 119-138

Wake D, Scott E.G. and Henning I.D. 1986 Electr. Letters 22 pp. 719-721

Paper presented at Int. Symp. GaAs and Related Compounds, Heraklion, Greece, 1987

Thermal wave measurements on ion-implanted GaAs: a comparison with device results

R.T. Blunt[1] and A.R. Lane[2]

[1] Plessey Research and Technology Ltd., Caswell, Towcester, Northants, NN12 8EQ, UK.
[2] Chell Instruments Ltd., North Walsham, Norfolk, NR28 9JH, UK.

ABSTRACT: Thermal waves have been used to map the uniformity of implants into GaAs wafers and the results compared to pinch-off voltage maps of processed devices. Excellent correlations were observed, both for 'bare wafer' and 'through nitride' implants. Implants made 'through nitride' gave better device uniformities than 'bare wafer' implants with the best results being obtained using implants through 500Å of nitride.

1. INTRODUCTION

Thermal wave techniques allow the rapid measurement of surface damage levels in semiconductor wafers in an entirely non-destructive manner. Up to the present time the major use of the technique has been the assessment of implant dose and uniformity (Smith et al 1986), residual polishing damage (Lee et al 1987), and damage caused by reactive ion etching (Geraghty and Smith 1986) in silicon wafers. Besides measurements on silicon the technique is also applicable to other materials including the III-V compounds (Lane and Blunt 1987). In this paper we present a comparison of thermal wave results with device measurements in ion implanted GaAs MESFETs. We believe that there is no other satisfactory method of measuring the uniformity of such low dose implants into GaAs and that this is the first time such an investigation has been made.

Device uniformity (both within the area of an individual IC and across a complete wafer) is vital to the successful manufacture of digital GaAs ICs. One of the factors which control device uniformity is implant uniformity itself since, in the absence of other variables, the value of pinch-off voltage in ion implanted GaAs MESFET devices is directly proportional to implant dose. Implantation through a layer of silicon nitride on the surface of the GaAs is often used to minimise ion channelling effects (Blunt and Davies 1986), thus improving device uniformity, and to reduce wafer contamination. However the uniformity of the implant into the GaAs (and thus the uniformity of devices) will be adversely affected by inevitable non-uniformities in the silicon nitride layer and will never be as good as for an implant into bare GaAs. The optimum thickness of the nitride layer is thus a compromise between these factors and one of the aims of this work was to determine this optimum thickness.

2. EXPERIMENTAL

Implants were made into bare wafers and also through silicon nitride film thicknesses of 300Å and 500Å using a ^{29}Si implant at constant energy (60 keV). The implant dose was varied to give a pinch-off voltage of approximately -1 volt for all the wafers. Following implantation the implant dose and nitride thickness uniformities were mapped using a Thermoprobe 200 (Smith et al 1986). After processing pinch-off voltage maps were obtained and compared to both thermal wave and, where appropriate, nitride thickness maps.

3. THERMAL WAVE MEASUREMENTS

Measurement of implant dose uniformity using thermal waves has previously been described in detail (Smith et al 1986, Lane and Blunt 1987) and only a brief description will be given here. Two low power laser beams are directed collinearly and focused to a single one-micron diameter spot on the wafer surface. Absorption of light from the acousto-optically modulated Ar-ion 'pump' laser generates thermal and plasma waves within the surface region down to a depth of about three microns. These waves are then detected by the He-Ne 'probe' laser through the pump-induced modulation of the sample reflectivity at the 'probe' wavelength. An increase in lattice disorder, as occurs near the surface in an ion implanted semiconductor, results in a sizeable increase in this modulated reflectance signal.

The thermal wave signal is a complex, but monotonic, function of the implant dose. Thus an unknown implant dose at a known energy can be determined by reference to a previously determined calibration curve as shown in Figure 1. By repeating the measurement at various points across a wafer a map of implant dose across the wafer can be quickly obtained (typically a 137 point map can be made in approximately 4 minutes). Since the purpose of this paper is primarily to demonstrate trends, the uniformity maps have been left as thermal wave maps rather than being converted into dose maps.

Since two lasers are used to obtain the thermal wave measurement it is a simple matter to determine the relative thickness of a dielectric layer on a wafer from absolute reflectance measurements at the two different laser wavelengths. This can be converted to an absolute thickness map if the refractive index of the film is known. Although not shown here we have compared the results to those obtained by ellipsometry and found excellent agreement. As thermal wave maps can be made on dielectric coated wafers as easily as on bare wafers it is thus possible to make essentially simultaneous measurements of thermal wave signal and dielectric thickness across complete wafers.

4. RESULTS

Figure 2 shows a thermal wave map of one of the Si implants (2.5×10^{12} cm^{-2} at 60 keV) into bare GaAs measured directly after implantation. Note that all thermal wave and nitride thickness maps shown in this paper use the same notation. Thus +, □ , and - symbols represent thermal wave signals or thicknesses greater than, within ± 0.25%, or less than the median value respectively. Contour lines have been drawn at 1% intervals with the median denoted by a bolder line. A map of FET pinch-off voltage

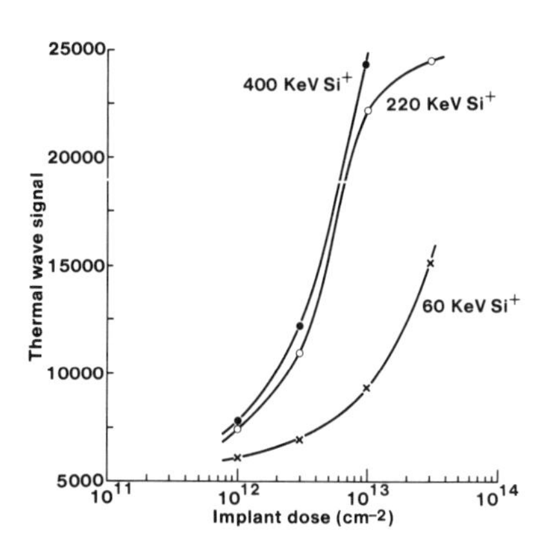

Figure 1

Thermal Wave Signal vs. Implant Dose into GaAs

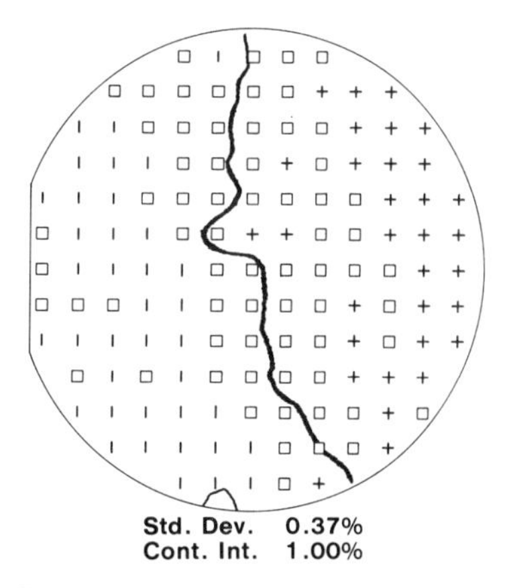

Figure 2

Thermal Wave Map (Bare Implant)

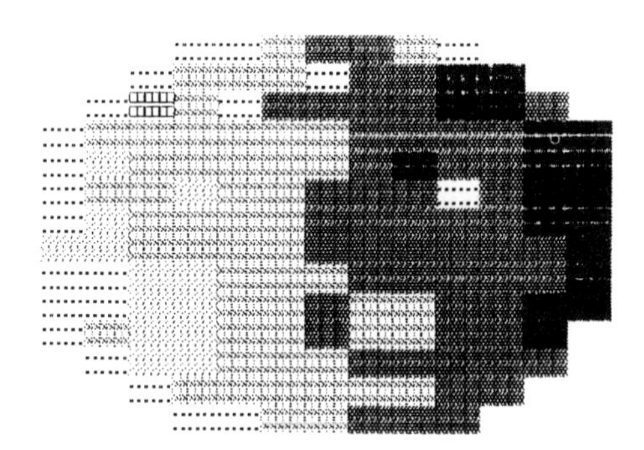

Figure 3

Device Pinch-off Voltage Map (Bare Implant)

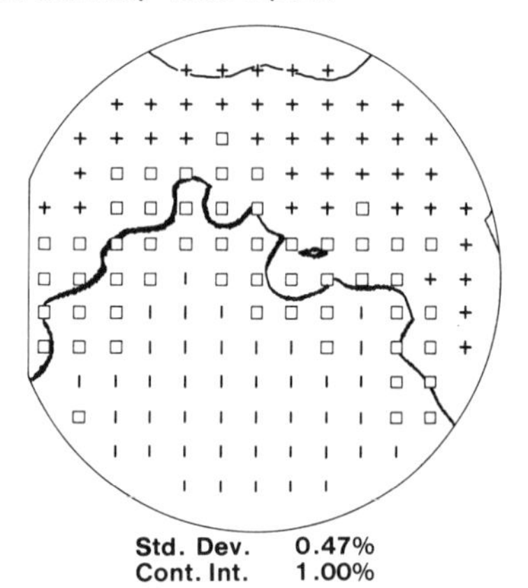

Figure 4

Thermal Wave Map (Nitride Coated Wafer)

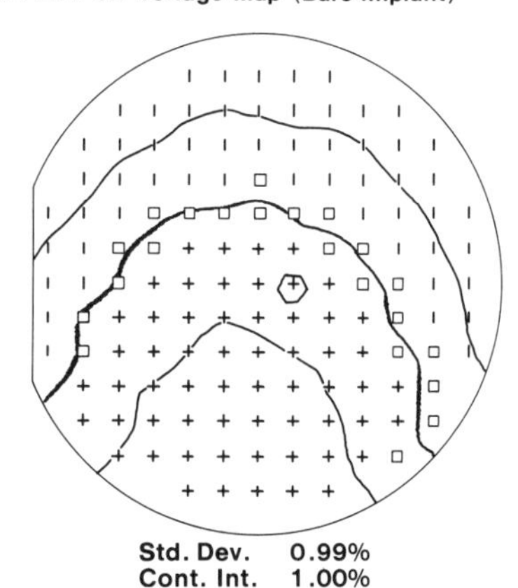

Figure 5

Film Thickness Map (Nitride Coated Wafer)

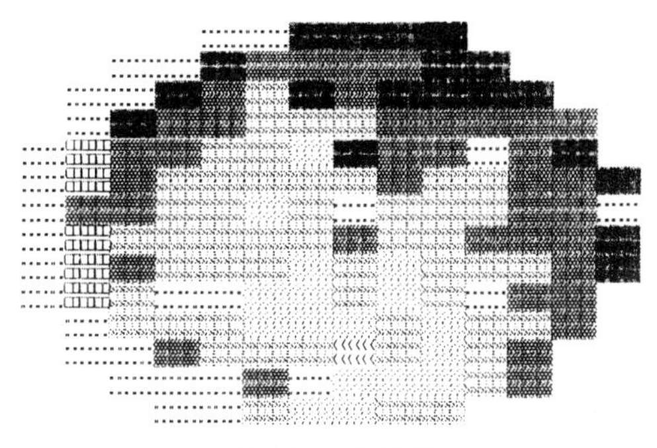

Figure 6

Device Pinch-off Voltage Map (Nitride Coated Wafer)

(higher pinch-off voltage corresponding to darker shading) obtained on the same wafer after processing is shown in Figure 3. There is excellent correlation between the maps, indicating the value of the technique. The standard deviation of pinch-off voltage on this wafer was 70mV.

Figures 4, 5, and 6 show plots for one of the 300Å coated wafers implanted using a slightly higher Si implant dose ($5x10^{12}$ cm^{-2} at 60 keV). The thermal wave map (figure 4) obviously correlates well with the nitride thickness map (figure 5) and suggests that the implant dose uniformity is being degraded by the non-uniformities in the nitride layer. The pinch-off voltage map for the same wafer (figure 6) also correlates very well with both the thermal wave and nitride maps. However it is important to note that the standard deviation in pinch-off voltage on this wafer was 49mV (ie lower than for the bare implant). Although not shown here the maps for the 500Å coated wafers also showed good correlations and gave a standard deviation in pinch-off voltage of around 32mV - the lowest value seen during this series of experiments.

5. CONCLUSIONS

Excellent correlations were obtained between thermal wave, pinch-off voltage, and, where appropriate, nitride thickness maps for ion implanted GaAs MESFETs. Thus clear correlations between thermal wave measurements made at an early stage in device processing and device parameters obtained on fully processed wafers have been observed. We believe this to be the first time such a correlation has been reported in GaAs. The speed and non-destructive nature of the thermal wave technique should ensure that it becomes a widely used 'on-line' tool in both manufacture and research into implanted GaAs ICs. Despite the fact that an obvious correlation was obtained between nitride thickness and threshold voltage for the 'through nitride' implanted wafers, better overall uniformities were obtained on these wafers than on the bare implants. This is believed to be due to the occurrence of slight planar channelling in the bare implant wafers, despite the use of both wafer tilt and rotation during implantation to prevent this effect.

6. ACKNOWLEDGEMENTS

We should like to thank P. Davies, T. Giddings, Z. Jackson, and D.J. Warner for implantation, processing and measurement of these wafers. This work was supported, in part, by the European Economic Community under ESPRIT Project 843.

7. REFERENCES

Blunt, R and Davies P, 1986 J. Appl. Phys. 60 1015.
Geraghty P and Smith W 1986 Mat. Res. Soc. Symp. Proc. 68 387.
Lane A and Blunt R 1987, to be published.
Lee J, Wong C, Tung C, Smith W, Hahn S and Arst M 1987, to be published.
Smith W, Rosencwaig A, Willenborg D, Opsal J, and Taylor M January 1986
Solid State Technology p. 85.

Inst. Phys. Conf. Ser. No. 91: Chapter 7
Paper presented at Int. Symp. GaAs and Related Compounds, Heraklion, Greece, 1987

A correlation between optical backgating and photogeneration mechanism in GaAs MESFETs

G.J. Papaioannou, J. Kaliakatsos, J.R. Forrest* and P.C. Euthymiou

University of Athens, Solid State Physics Section, Athens 106 80, Greece
*IBA Crawley Court, Winchester, Hants SQ21 2QA, England

ABSTRACT : The backgating effect of a GaAs MESFET induced by infrared illumination is investigated. The spectral dependence of carrier trapping in Cr (HL1) and EL2 levels at the semi-insulating Cr doped substrate is examined and related to the channel saturation current and backgate photocurrent response. Also the infrared photogeneration mechanism is confirmed by the diffusion photocurrent that is generated in an illuminated slab of semi-insulating Cr doped material.

1. INTRODUCTION

The variation of drain current when a bias is applied to the substrate is termed backgating. MESFET's are sensitive to light and illumination penetrating to the substrate can have backgating effects. In order to obtain a better knowledge of the contribution of various levels to the backgating, the spectral response of MESFET drain current and the photocurrent that flows through the substrate towards the channel were measured. The generation of excess electrons and holes in the substrate under illumination with photon energies lower than band-gap was estimated; the ratio of the excess electrons to the excess holes as a function of the illumination wavelength was compared in form to the variation of the diffusion photocurrent that is induced in a Cr-doped semi-insulating bulk material. The interpretation of the backgating effect is based on the recombination of the injected channel electrons with the substate excess holes.

2. EXPERIMENT

The measurements were carried out on a half a cell of Plessey GAT-6 MESFET chips. The gate contact was connected directly to the source, in order to avoid the generation of any photovoltage. An estimation of the device substrate contribution was obtained by measuring the spectral response of the diffusion photocurrent of a GaAs Cr-doped bulk material. In all cases the illumination spectrum extended over an interval of 0.9 - 1.6 μm and the measurements were carried out at room temperature.

3. RESULTS AND DISCUSSION

The mechanisms that contribute to the increase of the drain current with illumination are two: the channel conductivity due to excess carriers, and the channel opening due to the backgating effect. The gate is short-circuit and does not contribute. At wavelengths about 1 μm the material absorption coefficient decreases significantly ($\alpha<10$ cm^{-1}), Martin et al (1980 a). The light is practically

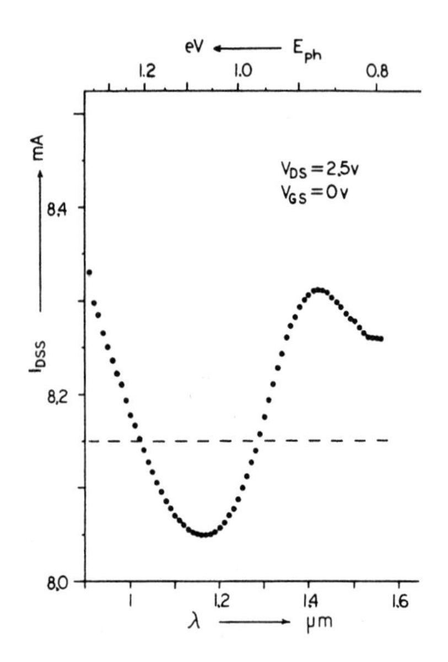

Fig.1 The drain saturation current vs. wavelength. The dashed line shows the level of I_{DSS} under dark.

Fig. 2 The backgate photocurrent vs. wavelength.

absorbed in the substrate, which in turn is almost uniformly illuminated, and no channel photoconductivity exists. Hence the spectral dependence of the backgating effect is monitored from the drain saturation current I_{DSS} (Fig. 1). Starting from short wavelengths, I_{DSS} presents a deep trough with a minimum lower than the dark saturation current, located at 1.17 μm (1.06 eV). Features in the curve are present also at 1.43 μm (0.87 eV). Upon illumination with light of appropriate energy, electrons are excited from deep traps into the conduction band. Similarly holes can be generated from the transition of valence band electrons into deep traps. The excess electrons and holes that are generated in the substrate and channel respectively, may diffuse across the channel-substrate barrier giving rise to a photovoltage, Fortin et al (1984), that forward biases the junction, thus increasing the channel opening hence the channel conductivity.

The origin of the drain current minimum at 1.17 μm (1.06 eV), with a value lower than the dark level is attributed, according to photoconductivity and photo-Hall spectra of Cr-doped GaAs, Look (1977), to a Cr^{+2} electron trap that has been also reported in optical absorption measurements, Martin et al (1979). On the other hand the structure centred at about 1.43 μm (0.87 eV) can be attributed to a Cr^{+2} resonance (0.87 eV) and a Cr hole trap (0.9 eV). Since the main electron trap in GaAs is the EL2 level, with a binding energy varying from 0.78 to 0.86 eV, depending on the growth conditions, the above structure can be attributed also to electron detraping from the EL2 level. Further in order to obtain more information on the photogeneration rate in the substrate, we measured the spectral dependence of the current (I_{SUB}) that flows through the backgate electrode (Fig. 2). This presents a broad peak at about 1.13 μm (1.10 eV), that is a significant absorption takes place due to electron excitation from the valence band into the Cr level which lies above Fermi level in the

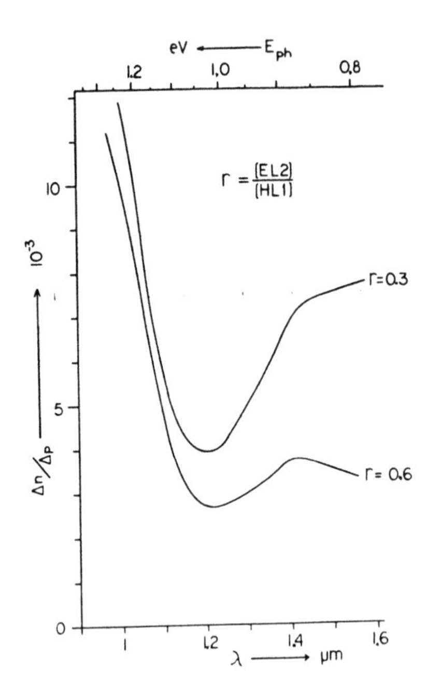

Fig. 3 The excess electron to excess hole ratio, vs. wavelength.

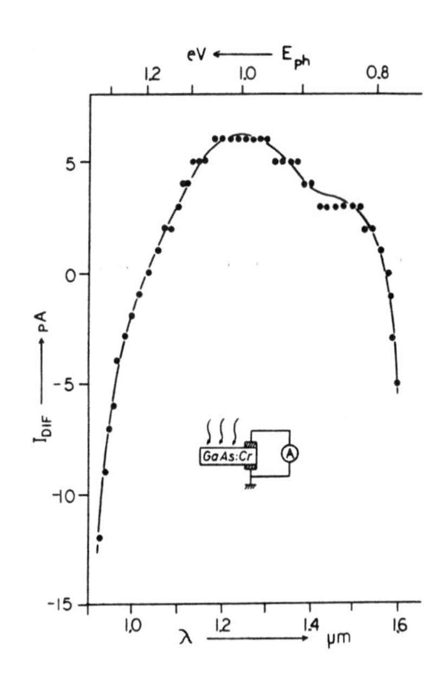

Fig. 4 The diffusion photocurrent vs. wavelength for a semi-insulating Cr doped slab.

intrinsic case at room temperature (Martin et al, 1980 b). Another contribution can be detected also above 1.43 μm although it is almost overlaped by the broad maximum. Although the drain current minimum at 1.17 μm has been attributed to electron traps, its origin is not well defined. This mechanism gives rise to photovoltage variations (Fortin et al, 1984) similar to those of the backgate current. Consequently, the minimum is ascribed to a decrease of the device channel opening. This is caused from accumulation of negative charge in the substrate due to the recombination, of channel electrons, which are injected into substrate with excess holes in the substrate. The result of this negative charge accumulation is the reverse bias of the active layer-substrate junction and the increase of the space charge region width, hence the decrease of the drain current. Since the backgating effect is related to photogeneration in the substrate, modelling of the photogeneration is further examined and results compared to the observed drain current variations.

4. PHOTOGENERATION MODEL AND COMPARISON

In order to obtain an estimation on the relative generation rates for electrons and holes under extrinsic illumination, the cross sections σ^o for optical transitions must be taken into account. Assuming low injeciton rates, Look (1983), it can be shown that the ratio of excess carrier concentrations $\Delta n/\Delta p$ is written :

$$\frac{\Delta n}{\Delta p} = K \frac{\sigma^o_{n1}\, r\, f_1 + \sigma^o_{n2}\, f_2}{\sigma^o_{p1}\, r\, (1-f_1) + \sigma^o_{p2}\, (1-f_2)}$$

where r is the ratio of EL2 to HL1 concentrations, the subscripts 1 and 2 are applied for EL2 and HL1 respectively. The factor K is a constant containing the thermal capture process for electrons and holes and depending on the temperature and the position of Fermi level only. It must be noted that the ratio in Eq. 2 is independent on illumination intensity. The plot of ratio $\Delta n/\Delta p$ vs. wavelength for two values of r (E_f = 0.7 eV) is given in Fig. 3. The data of cross sections for optical transitions were obtained from Martin et al. (1980 a).

In order to test the validity of the above model, the spectrum of the diffusion photocurrent of a bulk LEC Cr-doped GaAs sample was measured and plotted in Fig. 4. The positive values correspond to the domination of photo hole diffusion current while the negative ones correspond to photoelectron current.

The diffusion photocurrent reaches a maximum at about 1.24 μm (1.00 eV), where the ratio $\Delta n/\Delta p$ is a minimum, and shows additional features around 1.45 μm (0.85 eV). So an estimation of the value of EL2-to-HL1 ratio is obtained assuming that the ratio $\Delta n/\Delta p$ has to be the same at the wavelengths where the diffusion current vanishes (1.04 μm and 1.57 μm). The excess carrier ratio shows the same value at those wavelengths, assuming that E_f=E_c-0.70 eV, for r=0.28 which leads to a concentration (EL2)=2.8 x 10^{16} cm^{-3} for a typical value of (HL1)=10^{17} cm^{-3}.

In the case of the MESFET drain current it is assumed that the same condition has to take place with regard to excess carrier generation, carrier injection and channel substrate bias at the wavelengths where the drain photocurrent crosses the dark level (1.02 μm and 1.28 μm respectively). Again assuming the same position for the Fermi level, a value r=0.15 is obtained which leads to a concentration (EL2)=1.5 x 10^{16} cm^{-3} for (HL1)=10^{17} cm^{-3} which are values within reported limits (Martin, 1980).

REFERENCES

Fortin E, Charbonneau and Meikle S 1984 J. Appl. Phys. **56** 1141.

Look D C 1977 Sol. St. Comm. **24** 825.

Look D C 1983 Semicond. and Semimetals (New York, Academic Press) pp. 75-170.

Martin G M 1980 1st Int. Conf. in Semi-insulating III-V materials (London: Shira) pp. 13-28.

Martin G M, Verheijke M L, Jansen J and Poibland C 1979 J. Appl. Phys. **50** 467.

Martin G M, Jacob G, Poibland G, Goltzene A and Schwab C 1980 a 11th Int. Conf. on Defects and Radiation Effects in Semicond. p. 281.

Martin G M, Forges J P, Jacob G and Holtars J P 1980 b J. Appl. Phys. **51** 2840.

Inst. Phys. Conf. Ser. No. 91: Chapter 7
Paper presented at Int. Symp. GaAs and Related Compounds, Heraklion, Greece, 1987

A high-gain short-gate AlGaAs/InGaAs MODFET with 1 amp/mm current density

G.W. Wang, Y.K. Chen, D.C. Radulescu, P.J. Tasker and L.F. Eastman
School of Electrical Engineering and National Naofabrication Facility
Cornell University, Phillips Hall, Ithaca, NY 14853

ABSTRACT

Quadruple heterojunction MODFETs (Modulation Doped Field Effect Transitor) with a planar-doped lattice-strained AlGaAs/InGaAs structure have been fabricated and characterized at DC and microwave frequencies. At 300 K the 0.3-μ gate devices show a full channel current of 1100 mA/mm with a constant extrinsic transconductance of 350 mS/mm over a broad gate voltage range of 1.6 volts. Excellent microwave performance is also achieved with a maximum available gain cut-off frequency (f_{mag}) of 110 GHz and a current gain cut-off frequency (f_T) of 52 GHz. This is the highest current density ever reported for either GaAs MESFETs or MODFETs along with excellent high frequency performance.

Single heterojunction AlGaAs/GaAs modulation doped field-effect transistors (MODFETs) have shown excellent microwave performance especially for low noise application [Berenz 1984]. Because of a limited sheet charge concentration of less than 10^{12}cm^{-2} per heterojunction, multiple heterojunction MODFETs [Saunier, et al. 1986, Gupta 1985] were then investigated to increase the current driving capability and consequently improve the power performance and switching speed. Recently the AlGaAs/InGaAs pseudomorphic MODFET has shown better performance than the AlGaAs/GaAs MODFET [Chen, et al 1987]. The larger band discontinuity and better transport characteristics in the AlGaAs/InGaAs system have resulted in higher two-dimensional electron gas (2DEG) concentrations and higher current density. Among the highest current densities reported to date are 600 mA/mm from six-fold AlGaAs/GaAs heterojunctions [Saunier et al 1986] and 610 mA/mm from strained-layer AlGaAs/InGaAs double heterojunctions [Chen et al 1987].

We reported double heterojunction AlGaAs/GaAs/AlGaAs MODFET structures with two silicon planar-doped layers [Chen et al. 1986a]. This planar doping technique provides good charge control of the 2DEG sheet density. Little light sensitivity at 77K is observed due to the much reduced heavily doped AlGaAs region. In this paper, we report the fabrication and characterization of planar-doped AlGaAs/InGaAs pseudomorphic quadruple heterojunction MODFETs with high current density and high cut-off frequencies suitable for microwave and millimeter-wave power applications.

As shown in Fig. 1, the device layer structure is grown by MBE on a semi-insulating undoped LEC GaAs substrate. The structure consists of two AlGaAs/InGaAs/AlGaAs quantum wells with three silicon planar-doped AlGaAs layers. The silicon sheet densities are 6x10^{12} cm^{-2}, 3x10^{12} cm^{-2} and 1x10^{12} cm^{-2} respectively. The details of the growth conditions were similar to those reported in [Chen et al 1987].

The grown wafers were fabricated with a recess-gate FET process. Mesa isolation was performed by wet chemical etching. the source and drain of the device were defined by optical lithogrpahy with a mid-UV contact aligner. Ni/AuGe/Ag/Au contacts were subsequenty evaporated and alloyed at 475oC for 15 seconds with a rapid thermal annealer. A specific contact resistivity of 0.05 ohm-mm was obtained from the transmission line measurement. T-shape gates with 0.3-µm footprints were defined by electron beam lithography and followed by Ti/Pd/Au metalization. A PMMA/P(MMA-MAA)/PMMA triple layer electron beam resist system has been developed to obtain fine line width, high aspect ratio, and good lift-off profile [Wang, et al 1987]. These lead to a T-shaped gate with small gate length and low gate resistance which improves the high frequency performance of the device. The measured DC end-to-end resistance is typically 120 ohm-mm.

The room temperature current-voltage characteristics of a 0.3 x 100 µm device is shown in Fig. 2. A drain saturation current of 1100 mA/mm is obtained at a gate voltage of 1V. The high current density is the result of the high 2DEG concentration in the InGaAs channels. The MODFET of 0.3-µm gate length shows very good output conductance owing to good carrier confinement in the pseudomorphic quantum-well structure. A constant DC transconductance of 350 mS/mm is obtained over the gate voltage from -1V to 0.6V as shown in Fig. 3. The low transconductance variation is essential for low intermodulation distortion in power applications as reported in [Gupta et al 1985]. The gate-drain breakdown voltage is 6 to 7 V. Further optimization of the layer structure is required to achieve high current and high breakdown voltage simultaneously.

400 Å	GaAs:Si (2×10^{18} cm^{-3})
300 Å	undoped $Al_{.3}Ga_{.7}As$
	Si planar–doped (6×10^{12} cm^{-2})
30 Å	undoped $Al_{.3}Ga_{.7}As$ spacer
175 Å	$In_{.15}Ga_{.85}As$ Quantum Well
50 Å	undoped $Al_{.15}Ga_{.85}As$ spacer
	Si planar–doped (3×10^{12} cm^{-2})
30 Å	undoped $Al_{.15}Ga_{.85}As$ spacer
175 Å	$In_{.15}Ga_{.85}As$ Quantum Well
50 Å	undoped $Al_{.15}Ga_{.85}As$ spacer
	Si planar–doped (1×10^{12} cm^{-2})
100 Å	undoped $Al_{.15}Ga_{.85}As$ spacer
10,000 Å	$Al_{.3}Ga_{.7}As$/GaAs superlattice
	GaAs S.I. substrate

Fig. 1. Layer structure of the quadruple heterojunction AlGaAs/InGaAs MODFET.

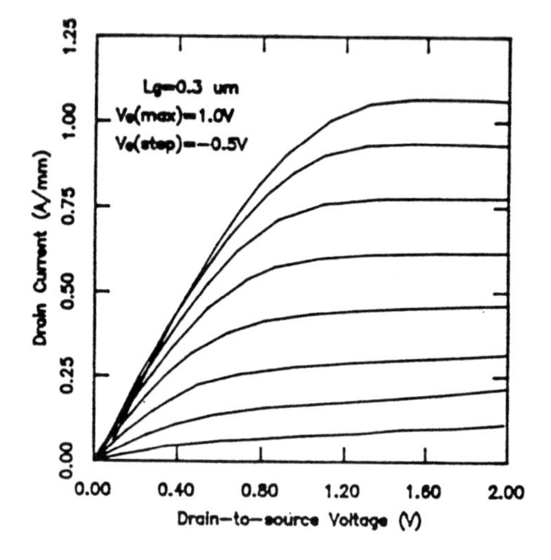

Fig. 2. Current-voltage characteristics of a 0.3-µm gate MODFET.

Microwave S-parameter measurements have been performed for various bias conditions from 0.5 to 26.5 GHz with microwave wafer probes and an HP 8510 automatic network analyzer. Fig. 4 shows the close agremeent between measured and modeled S-parameters at V_{gs} = 0 V and V_{ds} = 2 V. Power gains from measured S-parameters are depicted in Fig. 5. A current gain cut-off frequency (f_{MAG}) of 110 GHz can be extrapolated. Intrinsic f_T calculated by using the equivalent cirucit and $f_T = g_m/2\pi C_{gs}$ is 57 GHz, which is reasonably close to the one by extrapolation. The stability K factor is less than unity over the whole measurement frequency range. A carrier density as high as $6.5 \times 10^{12} cm^{-2}$ can be derived form a drain current of 870 mA/mm at a gate bias of 0.65V where an f_T of 44 GHz is measured.

In summary, we have demonstrated that extremely high current as well as excellent microwave characteristics can be achieved from a short-gate quadruple heterojunction pseudomorphic MODFETs. The 0.3 μm gate device shows a full channel current of 1100 mA/mm with an f_T of 52 GHz and an f_{MAG} of 110 GHz. This is the highest current density ever reported for microwave FETS. It indicates that short-gate multiple-heterojunction pseudomorphic MODFETs may be of great importance in millimeter-wave power applications.

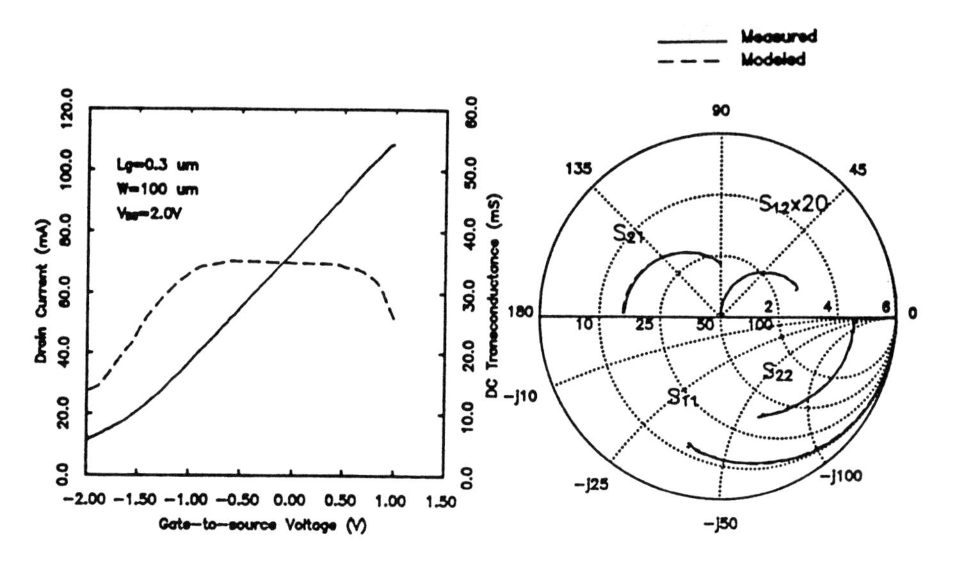

FIG. 3. Transconductance and current vs. V_{gs} at V_{ds} = 2V.

Fig. 4. Measured and modeled S-parameters at V_{gs} = 0V and V_{ds} = 2V.

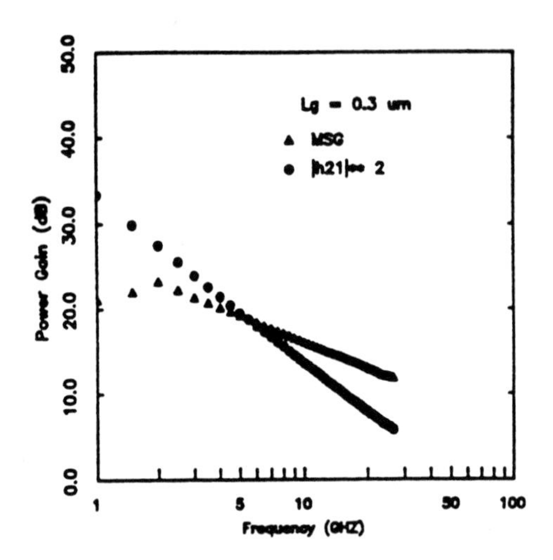

Fig. 5. Power gains vs. frequency at V_{gs} = 0V and V_{ds} = 2V.

We acknowledge Spectrum Technology Inc. (Holliston, Ma) for donation of the high-quality GaAs substrate used for this study. This work was supported in part by General Electric Company (Syracuse, NY), Martin Marietta Corp., the Office of Naval Research, Varian, Amoco, McDonnell Douglas and the Joint Services Electronics Program.

REFERENCES

Berenz J J, "Low Noise High Electron Mobility Transistor", Proc 1984 IEEE Microwave and Millimeter-Wave Monolithic Circuits Symp., p. 93, May 1984.

Chen Y K, Radulescu D C, Lepore A N, Foisy M C, Wang G C, Eastman L F "Enancement of 2DEG Density in GaAs/InGaAs/AlGaAs Double Heterojunction Power MODFET Structures by Buried Superlattice and Buried p^+-GaAs Buffer Layers", Proc. 45th IEEE Dev. Res. Conf., Santa Barbara, June 1987.

Chen Y K, Radulescu D C, Tasker P J, Wang G W, Eastman L F, "DC and RF Characteristics of a Planar-Doped Double Heterojunction MODFET", Int. GaAs and Related Comp. Symp., Las Vegas, Sept. 1986, Inst. Phys. Conf. Series 83 1987.

Gupta A K, Chen R T, Sovero E A, Higgins J A, "Power Saturation Characteristics of GaAs/AlGaAs HEMTs", Proc. 1985 IEEE Microwave and Millimeter-Wave Monolithic Circuits Symp., 50-53 June 1985.

Saunier P, Lee J W, "High-Efficiency Millimeter-Wave GaAs/GaAlAs Power HEMT's", IEEE Electron Device Lett EDL-7 (9) 505-505 Sept. 1986.

Wang G W, Chen Y K, Radulescu D C, Schaff W J, Tasker P J and Eastman L F, "Mutltiple Layer Electron beam Resist for Fabrication of Ultra-Short Gate MODFETs", accepted for presentation SOTAPOCS VII, 172 Electrochemical Soc. Meet. Hawaii, Oct. 1987.

Paper presented at Int. Symp. GaAs and Related Compounds, Heraklion, Greece, 1987

A quantum well tunnel triode

A. Kastalsky, M. Milshtein
Bell Communications Research
Red Bank, New Jersey 07701

Abstract: We demonstrate a novel three-terminal device, the tunnel triode, in which the current within the quantum well is a part of the tunnel current through the $p^+ - n^+$ junction. A tunnel-diode-like negative differential resistance effect with peak-to-valley ratio as high as 20 was observed, the tunnel current being controlled by the gate voltage. We show that tunneling occurs not in the quantum well but in the heavily doped n^+ $Al_{0.3}Ga_{0.7}As$ layer, and it is preceded by a real-space hot electron transfer from the quantum well into this layer. Logic operation of a bistable switch was obtained in a circuit comprising a tunnel triode and a series resistance.

The tunnel diode has long been considered as a promising device both for microwave generation and for logic applications due to its intrinsic negative differential resistance (NDR) (see, e.g., Chow 1964). Two tunnel diodes in series form a bistable element, since at certain bias conditions they have two stable configurations in which one of the devices takes almost all of the applied voltage. Logic operation consists in switching such a high-field "domain" from one device to another. In the case of two-terminal NDR devices, specially designed external circuitry is required which complicates the logic circuit and impedes the speed of operation. We describe here a novel three-terminal device called the Tunnel Triode (TT) in which the tunnel current of a highly doped p-n junction flows as lateral source-to-drain current within a single quantum well (QW) and is controlled by a third, gate, electrode. The latter feature allows a simple design for bistable switch circuits.

The device structure and contact geometry are illustrated in Fig. 1. The conducting layer is a 70Å GaAs QW sandwiched between the AlGaAs layers. The top $Al_{0.3}Ga_{0.7}As$ layer is n^+ heavily doped ($N_D \approx 3-4\times10^{18}cm^{-3}$) and provides electrons in the QW. The bottom undoped $Al_{0.45}Ga_{0.55}As$ layer separates the QW from an n^+ GaAs substrate which serves as a gate. The 100Å n^+ GaAs cap layer is needed to implement a high quality ohmic contact to the source. The device structure is designed to provide a complete depletion of the n^+ AlGaAs layer with electrons transferred to the QW. A key new element in our device is a p^+ contact at the drain (Be/Au alloyed at 420°C during 5 sec). This means that the QW is a part of p-n junction and the only medium containing free electrons. The concentrations of donors and acceptors are sufficiently high to expect the appearance of a tunnel current through the QW. The gate voltage V_G is considered to affect the tunnel current by altering the electron density in the channel.

The room temperature dependencies of the drain current I_D on the drain voltage V_D are presented in Fig. 2 for two different circuit arrangements. In Fig. 2a the source is kept grounded and $I_D - V_D$ characteristics are measured for fixed positive gate voltages, whereas in Fig. 2b the gate is grounded and negative source voltage V_S is a variable parameter. In Fig. 2a at $V_G = 0$ the QW is depleted and the rising I-V curve corresponds to a forward biased drain-to-gate p-n junction. As V_G increases the onset of the current rise shifts to higher V_D (as expected for such a junction) and new diode-like curves with subsequent current saturation arises at $V_D \geq 0.75V$, the magnitude of saturated current being controlled by the gate voltage. Finally, the current saturation is followed by a sharp current peak and a strongly pronounced negative differential resistance (NDR) with the amplitude of current peak as well as its position depending on the gate voltage. The highest peak-to-valley ratio was measured to be 20. The peak position progressively shifts to higher V_D as the gate voltage increases.

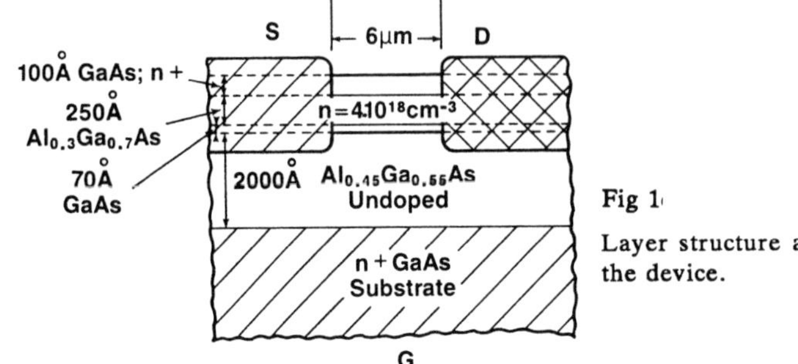

Fig 1

Layer structure and contact geometry of the device.

One can, however, use another circuit configuration (shown in Fig. 2b) at which the variations of peak position is small. Such a situation may be preferable for logic applications. Below we use this arrangement for demonstrating the logic operation of a bistable switch. Comparison of the characteristics shown in Fig. 2a and 2b suggests that the fixed gate voltage applied in the first case (Fig. 2a) affects the lateral built-in potential and changes the conditions for appearance of the current peak and the NDR.

We also measured the gate current versus V_D for both circuit configurations. These curves normally consist of two components, namely: the source-to-gate leakage current of negative polarity, independent of V_D, and at higher V_D the positive rising current of the forward biased p-i-n drain-to-gate junction. The important result of these measurements is that gate current does not contain the most essential features seen in $I_D - V_D$ characteristics. Therefore, the current saturation discussed above as well as the current peak and the NDR effect (see Fig. 2) belong to the source-to-drain component of the drain current.

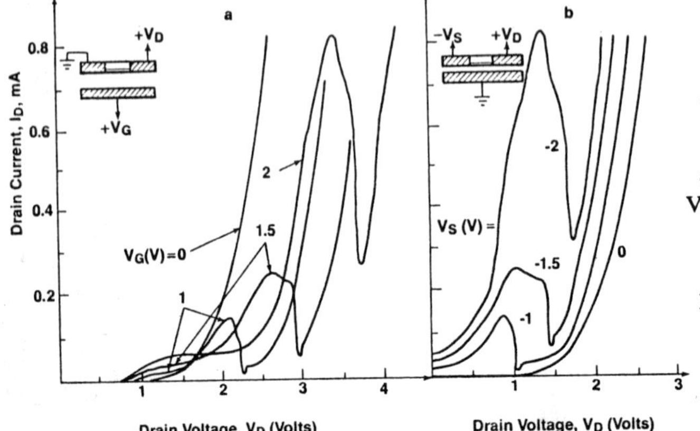

Fig. 2

Drain current vs drain voltage for two different circuit arrangements:

a. source grounded with V_G as a variable parameter.

b. gate grounded with $-V_S$ as a variable parameter.

We are going to discuss now the physical model accounting for the key features observed in the experiment. The diode-like characteristics seen at $V_D \geq 0.75V$ (Fig. 2a) with a subsequent current saturation can be reasonably attributed to the gate induced electron channel in the QW connected in series with the forward biased p-n junction. At drain voltages sufficient to open this p-n junction the source-drain current begins to be controlled by the channel resistance and current saturation results from a pinch-off effect in the field effect transistor formed by the channel and the n^+ substrate. Such a behavior of the channel current clearly indicates that the QW and p^+ contact do not create a tunnel diode in spite of high density of donors ($\sim 3-4 \times 10^{18} cm^{-3}$) and acceptors ($>10^{19} cm^{-5}$) at this junction. The obvious reason for that is that the p-n junction electric field, strong at this AlGaAs n-p transition, weakens in the vicinity of the QW. The remoteness of the channel electrons from the donors by 20−30Å is crucial for tunneling (Kane). Furthermore, the buried gate utilized in our experiments creates electrons at the bottom heterojunction interface in the QW thereby increasing the distance of the electron wave function from the donors. Therefore, only electrons entering n^+ AlGaAs layer will be capable of tunneling through the p-n junction. We believe that the **current peak and the NDR result from the process of electron transfer from the QW to the n^+ AlGaAs with a subsequent tunneling into p^+ region of this layer.** The source-drain electric field enhances the effect of hot-electron real-space transfer to the n^+ AlGaAs layer. This explains why the highest peak of the tunnel current is achieved at large source-drain voltages.

However, using illumination one can generate free electrons in AlGaAs layer and thereby provide conditions for tunneling even at low V_D. The results of such an experiment, allowing to identify location of the tunnel diode, are presented in Fig. 3 where we compare $I_D - V_D$ characteristics of one of the devices in the dark and under the light (probe-station illuminator). In both cases the value of saturated channel current at a given V_G remains almost the same (i.e., no light induced electron accumulation in the QW), whereas the difference in amplitude of the peak is clearly seen, especially at low gate voltages: at $V_G \leq 0.75V$ the peaks are well pronounced under illumination and do not appear in the dark at all. As V_G (and hence, the drain voltage corresponding to the peak) increases this difference gradually disappears. According to our model, this results from a development of hot-electron transfer from the QW to the n^+ AlGaAs which at high V_D becomes dominant. At low V_D the electron generation by light controls the conductivity of the n^+ AlGaAs layer and we see light induced tunnel-diode characteristics.

An interesting peculiarity of the observed tunnel current is the dependence of its peak position on the gate bias (see Fig. 2a): the peak shifts to the higher V_D as V_G increases. Such a sensitivity to the gate voltage seems unusual since the gate potential must be shielded by a corresponding charge in the channel. However, we notice that the tunnel peak always occurs in the regime of channel current saturation which is accompanied by a channel depletion near the drain (pinch-off-effect in the field-effect transistor, see e.g., Sze 1981) and therefore, by a

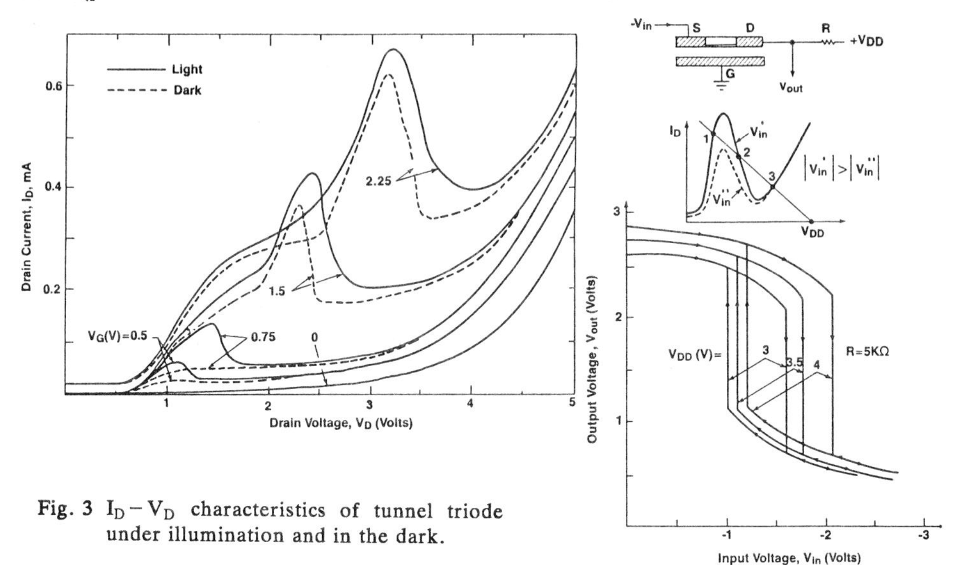

Fig. 3 $I_D - V_D$ characteristics of tunnel triode under illumination and in the dark.

Fig. 4 The circuit, the driver-load construct and output characteristics of the bistable switch.

penetration of the gate field beyond the channel. In the area of p^+ contact however, the gate electric field will be screened by free holes and will not reach the AlGaAs layer. This means that the gate voltage creates a potential difference at the p-n junction. Therefore, as V_G increases, in order to maintain identical conditions for tunneling one has to increase V_D to compensate variations of V_G. There is no need for such an adjustment, however, in the circuit where gate voltage is kept grounded. This is why we see rather small shift of the peak in the I-V characteristics shown in Fig. 2b.

Thus, according to our model, the tunneling and the resultant NDR effect originate from the AlGaAs layer in spite of larger energy gap and heavier electron effective mass in this material (as compared to the GaAs). The higher electric field at the p-n junction in AlGaAs is a prevaling factor for the observation of tunneling.

As is was mentioned in the introduction, the gate controlled NDR effect can be used for logic applications. The circuit (shown in Fig. 4) comprising the TT and a properly chosen series resistor can function as a bistable switch. For this purpose we use the regime of grounded gate (Fig. 2b) with the input voltage applied to the negatively biased source. The load line, as illustrated in the "driver-load" graphical construct in Fig. 4, provides two stable (1 and 3) and one unstable (2) operating points of this circuit. The latter becomes unstable when the drain voltage is kept within the NDR region. In this case the output voltage V_{out} will correspond to one of the stable points. Modulation of the input voltage V_{in} will switch circuit from one operating point to another by varying the amplitude of tunnel current peak (see dashed line in the construct).

Experimental realization of the bistable element is presented in Fig. 4 for the circuit comprising the TT (whose I-V characteristics are shown in Fig. 2) and the resistor of 5KΩ. The dependencies V_{out} (V_{in}) are obtained when different fixed biases V_{DD} are applied to the circuit and the negative voltage at the source (V_{in}) is varied. As $|V_{in}|$ increases the circuit experiences a switch from high to low V_{out} with a strongly pronounced hysteretic behavior as we reduce V_{in}. If one now keeps the input voltage within the hysteretic loop the subsequent positive (negative) pulse of V_{in} will snap the output voltage to its high (low) persistent level, thus providing the logic operation of bistable switch. As one can see in Fig. 4, the modulation of the output voltage reaches V_{out} (high) $-$ V_{out}(low)$\approx$1.5V. Finally, it is obvious that a similar bistable switch effect can be achieved in a circuit of two TT's in series. The advantage of such a circuit would be low currents for both stable operating points. It should be mentioned that a circuit consisting of two negative resistance field-effect transistors (NERFET) in which NDR is also controlled by the third electrode (collector), was demonstrated to function as a bistable switch [4]. In the NERFET the NDR results from the effect of real-space hot-electron transfer to the collector and is inevitably accompanied by a strong collector current, which causes undesirable power dissipation. In the bistable circuit of two TT in series the small currents corresponding to the stable operating points are the only source of energy consumption.

In conclusion, we demonstrate for the first time a novel three-terminal device, tunnel triode, in which characteristic features of a tunnel diode, such as the peak of the tunnel current and the NDR, are controlled by the gate voltage. It was found that the tunneling takes place in the heavily doped AlGaAs layer and is preceded by hot electron transfer from the GaAs quantum well. The logic operation of a bistable switch is demonstrated in a circuit including tunnel triode and series resistor.

We would like to thank E. O. Kane for many stimulating discussions.

References:

Chow W. F. (1964) "Principles of Tunnel Diode Circuits", NY, Wiley.
Kane E. O. (unpublished)
Kastalsky A., Luryi S., Gossard A. C. and Chan W. K. (1985) IEEE Electr. Lett. EDL-6, 347.
Sze S. M. (1981) "Physics of Semiconductor Devices", 2nd edition, J. Wiley & Sons, NY.

Paper presented at Int. Symp. GaAs and Related Compounds, Heraklion, Greece, 1987

Surface effect in submicronic GaAs MESFETs: two dimensional modelling and optimisation of recessed gate structures

F. Heliodore, M. Lefebvre and G. Salmer

Centre Hyperfrequences et Semiconducteurs U. A. C.N.R.S. N° 287 - U.S.T.L.F.A. 59655 Villeneuve d'Ascq cedex - France.

ABSTRACT : Due to surface potential effect in GaAs MESFET's, a depletion layer extends towards the substrate not only under the gate electrode but also on both sides of the gate. The purpose of this paper is to present its influence on device behaviour : DC, AC characteristics and expected microwave performances of submicronic normally on GaAs MESFET's. Thus, a two dimensional resolution of the basic semiconductor equations, including relaxation effects, is used. Some interesting conclusions concerning optimum values of device technological parameters for recessed gate structures, are drawn.

1. INTRODUCTION.

It has been admitted that surface potential of a normal free GaAs surface is close to 0.5 - 0.6 V, due to the fact that the surface Fermi level is pinned by a high density of surface states (Spicer 1980). Therefore, in GaAs MESFET's, a depletion layer extends between respectively source and gate and gate and drain. As a consequence, microwave performances of submicronic gate MESFET's can be modified, specially in millimeter wave range, where noise figure and associated gain are strongly dependent on the elements values of the equivalent circuit.

In the past, some theoritical and experimental analysis of this effect have been performed (Hariu 1982, Makkram-Ebeid 1984, Graffeuil 1986, Barton 1986). However, the available models which take into account surface effect don't allow to give accurate predictions for submicronic gate MESFET's at classical DC bias, because they don't take into account non stationary electron dynamic effects.

2. THEORITICAL MODEL.

The theoritical model, described elsewhere (Salmer 1984, El-Sayed 1987), is based on the two dimensional resolution of the fondamental semiconductor equations:

Poisson Equation $$\nabla^2 V = \frac{q}{\varepsilon}(n - N_D)$$

Current continuity $$\frac{\partial n}{\partial t} = \frac{1}{q}\nabla \vec{J}$$

The non stationary electron dynamic effects are taken into account by introducing the average energy relaxation equation,-derived from Boltzmann transport equation (Ibrahim 1984):

$$\frac{\partial}{\partial t}(n.w) - \frac{1}{q}\nabla(\vec{J}.w) = -(\vec{J}.\vec{E} - \nabla(\vec{J}(k_B T(w))) - n\frac{w - w_0}{\tau_w(w)}$$

where w is the total average energy of the carriers.
The major assumption is to assume that inertial terms can be neglected in the momentum relaxation equation (Ibrahim 1984), and consequently current density is defined by :

$$\vec{J} = q\mu(w)[n\vec{E} + \vec{\nabla}(n\,kB_B T(w))]$$

On the other hand, we postulate that the carrier behaviour is only dependent on the electron average energy w.

3. SURFACE EFFECT.

Surface potential is introduced by assuming a given value for the normal component of the electric field at the interface, depending on the doping level. This value is adjusted in order to obtain the typical surface potential difference in a zone where the component of the electric field parallel to the interface is negligible. We consider a typical submicronic gate MESFET's, which is represented on figure 1. In practical devices, an overdoped N^{++} layer is implanted on the epilayer and the gate is recessed. As a consequence, we can consider that the source and drain equivalent contacts are very close to the edges of the recessed zone, and that surface potential occurs only in this part of the device.

As a first assumption, we postulate that it isn't necessary to account for the exact form of the recessed structure, and we can treat such devices as planar ones, where surface potential effects are taken into account only on a distance R_i (figure 1b), which is the recessed zone width. In order to show the influence of surface potential effect, carrier concentration, equipotential contour and average energy are given on figure 2 for a recess width $R = R_1 = R_2 = 0.4\mu m$.

We can observe the following :
- The depletion region extends on both sides of the gate;
- The hot electron and high electric field domain have been moved towards the edge of the drain, by comparison with the corresponding contours obtained without taking into account surface effects (El-Sayed 1987);

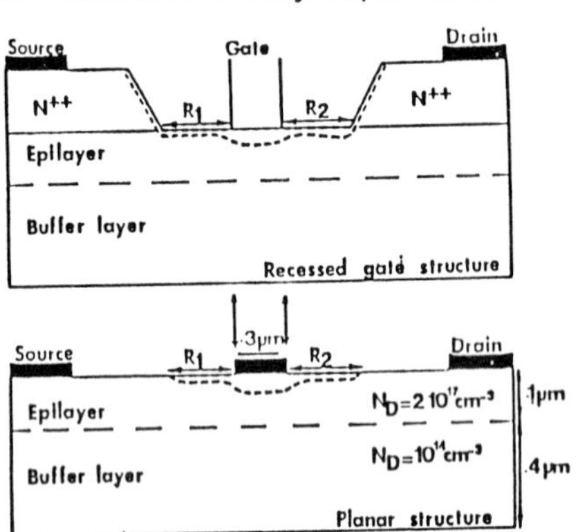

Fig.1 Recessed gate device : real and modellized structure.

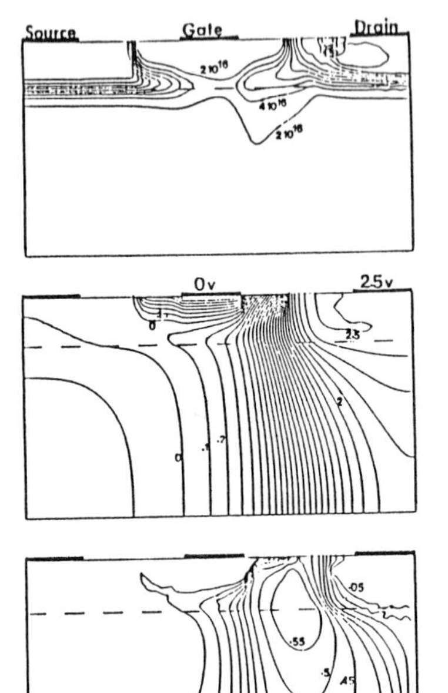

Fig.2 Equiconcentration, equipotential and equienergy contours. Influence of surface potential effect.

- Consequently, the carrier injection into the buffer layer is reduced.

These kinds of behaviour may have some important influence on the elements of the FET equivalent circuit.

The influence of recessed gate width R has been systematically studied in the symetrical case ($R = R_1 = R_2$). When R increases, the following variations have been observed (see also Heliodore 1986) :

- The decrease of the transconductance g_m, due to the increase of the source access resistance and the decrease of charge control by the gate potential, that causes also the decrease of the internal gate to source capacitance ;
- The decrease of the output conductance, due to the reduction of carriers injection into the buffer layer (figure 2);
- The decrease of the gate to drain capacitance resulting from the modifications of the potential distribution between gate and drain (figure 2).

In this symetrical case, the dependance of g_m, C_{gs}, g_d and C_{gd} upon R may then cause contradictory effects on the expected noise and gain performances. As a consequence, it seems better to consider an unsymetrical structure ($R_1 \neq R_2$), that may be realized by electron beam lithography technique and to optimize separately R_1 and R_2.

The choice of an optimum value for R_1 may result from a compromise between the increase with R_1 of both the access surce resistance (and consequently a decrease of g_m) and the parasitic gate to source air capacitance. By considering also technological requirements, an optimum value of R_1 may appear close to 0.2µm.

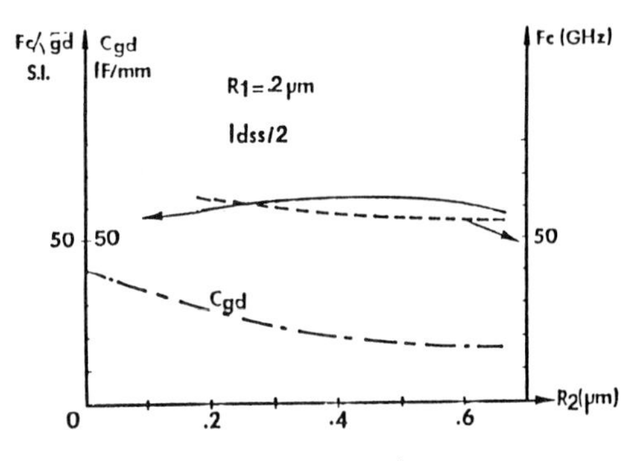

Fig.3 Variations of F_c, $F_c/\sqrt{g_d}$, C_{gd} versus R_2 (high power conditions).

In order to point out the influence of parameter R_2, typical evolutions of the main device characteristics are presented (for $R_1 = 0.2$µm) :

- The ratio $F_c / \sqrt{G_d}$, which is proportionnal to MAG and constitutes a merit factor for the gain device capability;
- The cut-off frequency F_c of current gain ($F_c = G_m / 2\pi C_{gs}$) and the ratio $F_0 = g_d / 2\pi C_{gd}$, that determine mainly noise performance (Cappy 1986).

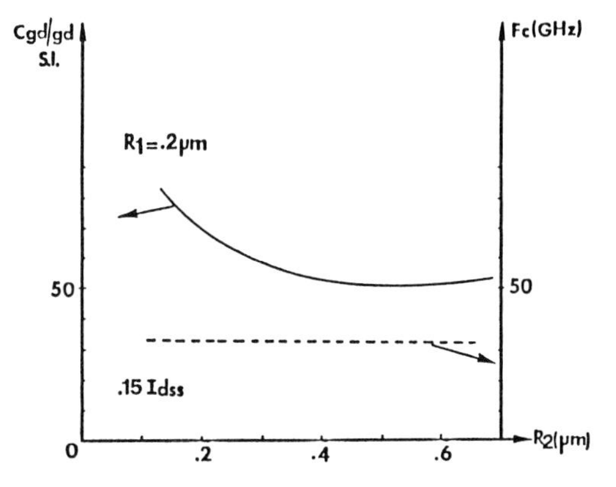

Fig.4 Variations of F_c, C_{gd}/g_d versus R_2 (low noise conditions).

From the variations observed on these figures 3 and 4, we can draw some interesting conclusions :

- Under large signal operation ($I_d \approx I_{dss}/2$, the merit factor $F_c / \sqrt{g_d}$ becomes maximum for R_2 values close to 0.5µm, then the cut-off frequency don't significally decreases. By taking into account the decrease of C_{gd} with R_2 (and then the corresponding improvement of the stability behaviour), it seems that optimum values of R_2 are close to 0.5µm. It is in qualitative good agreement with experimental results (Macksey 1986) obtained for power device

- Under low noise conditions ($I_d \approx 0.15 I_{dss}$), as the cut-off frequency remains constant with R2 and C_{gd}/g_d is minimum for $R_2 \approx$ 0.5µm, it seems that optimum value of this parameter is close to that obtained under large signal.

Using approximate formulae, taking into account the influence of F_0 on the noise drain current's amplitude :

$$N_F = 1 + \sqrt{8\pi}.F.[\frac{L_g}{F_c}(\alpha Z + \beta I_{ds})(R_g + R_s)(1 + (\frac{F}{F_0})^2]^{1/2}$$

(where a = 2.10-3S.I., b = 1.25 10-5 S.I., Lg gate length, Z gate width, Rs (Rg) parasitic source (gate) resistance)

It can be shown (figure 5) that an improvement of noise factor of about 0.7dB can be obtained at 40 GHz, if an optimum value of R_2 is choosen.

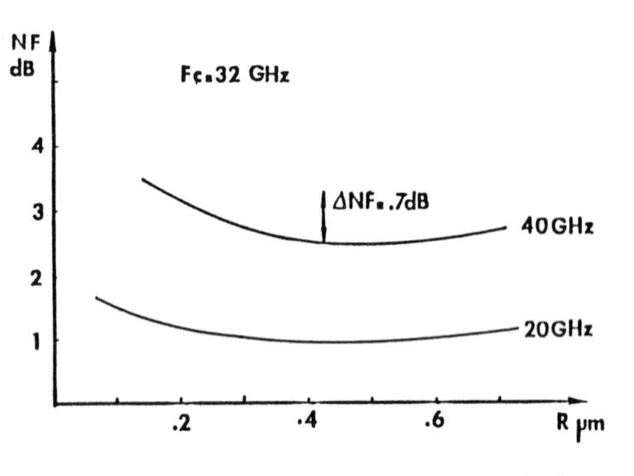

Fig.5 Noise figure variations versus R_2 ($I_{ds} = 0.15 I_{dss}$.)

4 CONCLUSION.

By using a two dimensional resolution of the basic semiconductor equations, including relaxation phenomena, it has been shown that surface effect may have important consequences influences on GaAs MESFET's behaviour.

With respect to surface effects in recessed gate structure, the recessed gate width constitues an important design parameter. Noise figure and available gain values are shown as strongly dependent on this technological, for which an optimum value close to 0.5µm has been determinated.

Work supported by CNET DAII under contract.

Barton J M Ladbrooke P H 1986 *Solid State Electr.* **20** 807
Cappy A 1986 *These doctorat d'etat* Lille
El-Sayed and al. 1987 *Solid State Electr.* **30** 643
Graffeuil J and al. 1986 *Solid State Electr.* **29** 1087
Hariu T 1982 *Jap. J. Appl. Phys.* **21** 77
Heliodore F 1987 *Ph D thesis* Lille
Ibrahim M 1984 *Ph. D. thesis* Cairo
Macksy H M 1986 *I.E.E.E. on Electr.Devices* **ED-33** 818
Makkram-Ebeid S Minondo P 1984 *Acta Electronica* **25** 241
Salmer and al. 1984 *11th Intern. symp GaAs and Relat. Compounds* Biarritz
Spicer W E and al. 1980 *J. Vac. Sci. Tech.* **17** 1019

Inst. Phys. Conf. Ser. No. 91: Chapter 7
Paper presented at Int. Symp. GaAs and Related Compounds, Heraklion, Greece, 1987

Two dimensional transient simulation of submicron-gate MODFETs

Tarek A. SHAWKI, G. SALMER
Centre Hyperfréquences et Semiconducteurs, U.A. CNRS n° 287
USTLFA, 59655 Villeneuve d'Ascq
Osman L. EL SAYED, Faculty of Engineering, Department of Electronics and Communications, Cairo University, Egypt

ABSTRACT : A novel 2-D numerical model featuring transient simulation of non stationary electron transport, hot electron and deep donor effects in submicron gate MODFET's is presented and used to investigate the large and small signal performances of 0.3 μm gate GaAs/AlGaAs MODFET's. The results show that real space transfer occurs and parasitic MESFET effects are only deleterious in the presence of deep donor levels.

1. INTRODUCTION

MODFET's are now holding the record for fastest switching speeds as well as for low noise microwave amplification. Nevertheless there is room for further optimization. This is hampered by the large number of optimization parameters together with the large variety of structures proposed and the large cost of the technological means used, thus stimulating the need for accurate device models. In general the models used for the simulation of MODFET's fall into two broad categories : particle (Monte Carlo) models and hydrodynamic models with these last ones divided into two subcategories : local and energy or temperature models. Hydrodynamic may be further categorized as quantum mechanical models in which the sub-band structure is evaluated in the absence of longitudinal field and classical models relying on 3D Fermi-Dirac or Boltzmann statistics.

In the following we present a rigorous model which retains the fundamental physical features of MODFET operation and discuss typical results obtained with it.

2. MODEL DESCRIPTION

This model is based on a previous one developed for the simulation of submicron gate MESFET's [SALMER 1984, EL-SAYED 1987]. It is basically a 2D hydrodynamic energy model that accounts for the non-stationary electron dynamics by using energy dependent transport parameters (mobility μ, electron temperature T and energy relaxation time τ_W) ; The average total electron energy w is obtained through the solution of an energy relaxation equation derived from the second moment of Boltzmann transport equation : the dependance of the transport parameters on energy is obtained from M.C. steady state simulations. The modifications introduced involve the simulation of GaAs/AlGaAs heterojunction and the introduction of the effects of deep donor levels.

The GaAs/AlGaAs heterojunction is modeled by assuming an equivalent effective electric field over a distance of 20Å, such that a conduction band discontinuity ΔE_c is obtained (here ΔE_c = 0,23eV). The shallow and deep donor levels (located at about 150 meV under the L valley) are taken into consideration by assuming a single equivalent donor level located at about 40 meV below the conduction band as it as been proposed by several authors, SHUBERT and PLOOG for instance. Thus the donor ionization equation and Poisson equation are solved self-consistently and the continuity equation modified to include time-varying ionization of the deep donor levels.

It should be noted that we assume in this model that the carrier dynamics of the 2 DEG are similar to electron transport in the bulk material. This assumption, although controversial, has been confirmed by M.C. and hydrodynamic simulation. Carrier transport is thus governed by the following set of equations :

$$\frac{\delta n}{\delta t} = \frac{1}{q} \nabla \vec{J}_n + G_n - R_n \qquad (1)$$

$$\vec{J}_n = -q\,\mu(w)\left[n(\vec{\nabla}\psi + \frac{1}{q}\vec{\nabla}\chi) - \vec{\nabla}(nkT)\right] \qquad (2)$$

$$\frac{\delta(nw)}{\delta t} = \vec{J}_n.(\vec{\nabla}\psi + \frac{1}{q}\vec{\nabla}\chi) + \vec{\nabla}.\left[w + kT(w)\right].\frac{\vec{J}_n}{q} - n\frac{(w - w_0)}{\tau_w(w)} \qquad (3)$$

$$\nabla^2\psi = \frac{q}{\varepsilon}(n - N^+_d) \quad (4) \qquad N^+_d = \frac{N_D}{1 + 2\exp\left[(E_{fn} - E_d - E_c)/kT\right]} \quad (5)$$

where χ is the electron affinity
N^+_d is the ionized donor concentration
E_d is the "equivalent" donor energy level

The other terms have their usual meaning. The above system of equations is solved using a finite difference scheme on a two-dimensionally non-uniform mesh structure. In the 2 DEG the mesh size reachs a minimum of 20 Å and increases gradually away from it. The structural features of the device modeled are shown in fig.1. In the simulations performed two cases are considered.

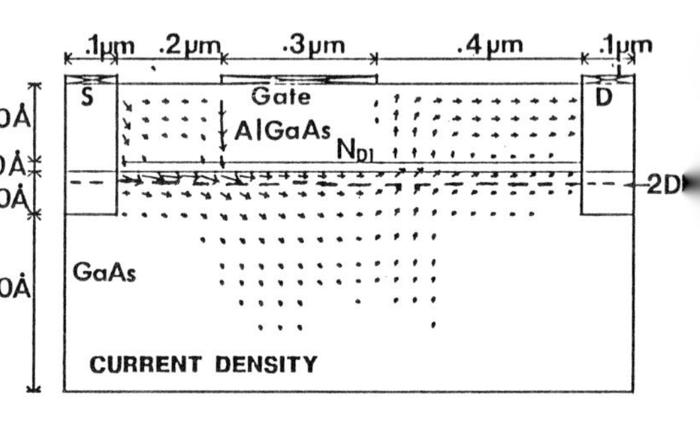

Fig.1 Device Structure showing current distribution

In the first, only shallow donors are considered thus N_{+d} = N_d. In the second, the effects of deep donor levels are also taken into consideration.

SIMULATION RESULTS

Fig. 1-4 illustrate the main features of electron transport in submicron gate GaAs/AlGaAs MODFET's namely :

(a) Current injection in the GaAs bulk together with real space transfer of energetic electrons back into the AlGaAs layer (fig. 1). It should be noted when electrons cross the interface from GaAs to AlGaAs, their average energy is reduced and they become colder (fig. 2) : consequently, in spite of the low mobility of AlGaAs, the electrons velocity in AlGaAs can be close to that reached in GaAs (fig. 3). This explains the importance of the current in the AlGaAs layer.

(b) Non stationary electron dynamics (fig. 2) confirming the above remarks and hot electron transport (fig. 3) which show electron energies under the gate reaching 0.5 eV thus legitimating our assumption on transport in the 2 DEG.

(c) Electron trapping by deep donor levels resulting in the neutralization of a large percentage of donors and consequently the reduction of the 2 DEG sheet carrier density (fig. 4).

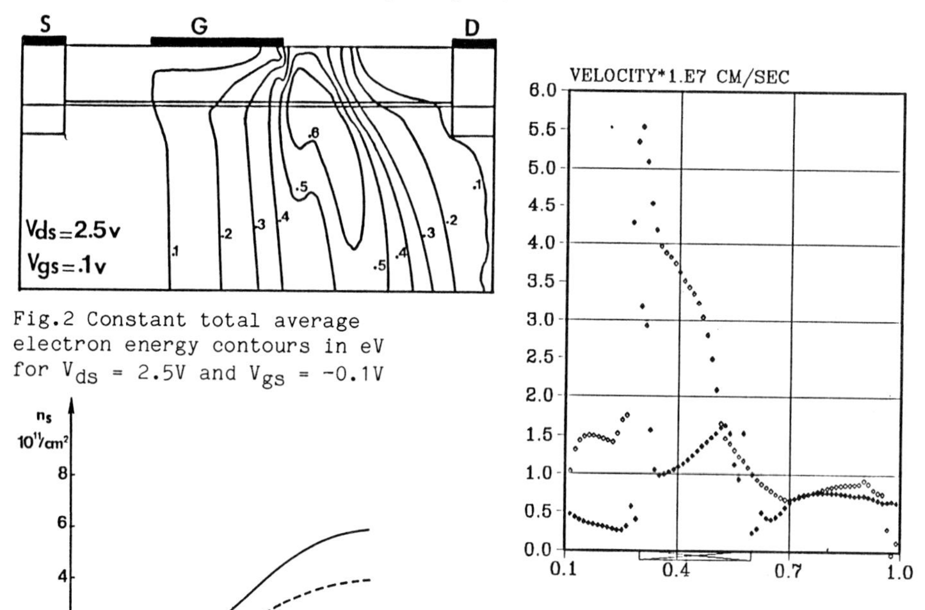

Fig.2 Constant total average electron energy contours in eV for V_{ds} = 2.5V and V_{gs} = -0.1V

Fig.3 Longitudinal distribution of x-directed carrier velocity in the 2DEG and midway in the AlGaAs layer

Fig.4 2-DEG sheet carrier density (n_s) versus gate voltage in the presence (---) and absence (——) of deep donor levels.

The device characteristics are shown in fig. 5-7. In the absence of deep donor levels, the evolutions of I_{ds}, g_m and Cgs with V_{gs} are quite normal. The values of g_m vary from 200 mS/mm at the low noise operating point to 600 mS/mm at Vgs = 0.3 V. The deleterious effects of the "parasitic MESFET" are only felt beyond this point where the current in the AlGaAs exceeds 50% of the total current.

The presence of deep donor levels results in a reduction of I_{ds} by over 30% at Vgs = 0.2 V and 55% at Vgs = 0.4 V. The effect on g_m is much more pronounced. The rapid drop above V_{gs} = 0.1V is visibly due to the combined effect of reduced 2 DEG sheet carrier density and increased source access resistance resulting from increased neutralization of donor atoms. However at large negative values of V_{gs}, g_m is not affected. As in general C_{gs} is slightly affected, the value of f_c at the low noise operating point is sensibly the same in both cases. However the values of g_m and f_c for large signal operation are drastically reduced.

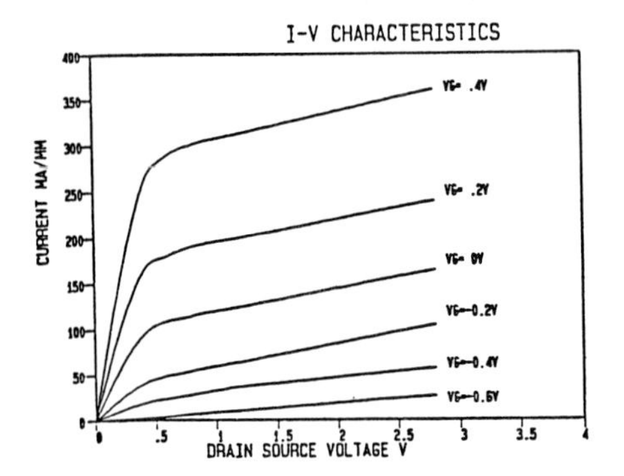

Fig.5 I-V characteristics of modeled device

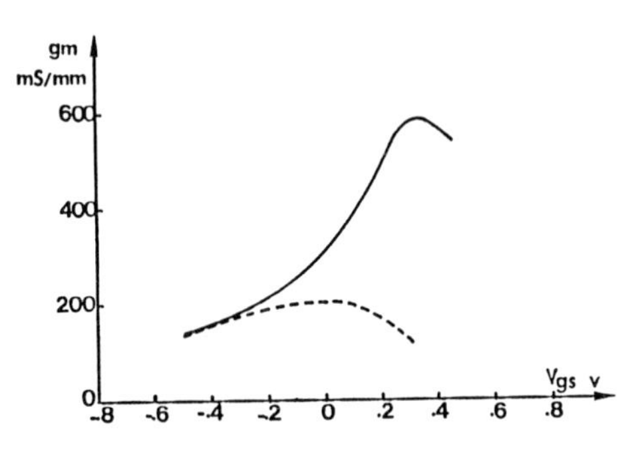

Fig.6 Transconductance versus gate voltage for V_{ds} = 2V in the presence (---) and absence (——) of deep donor levels.

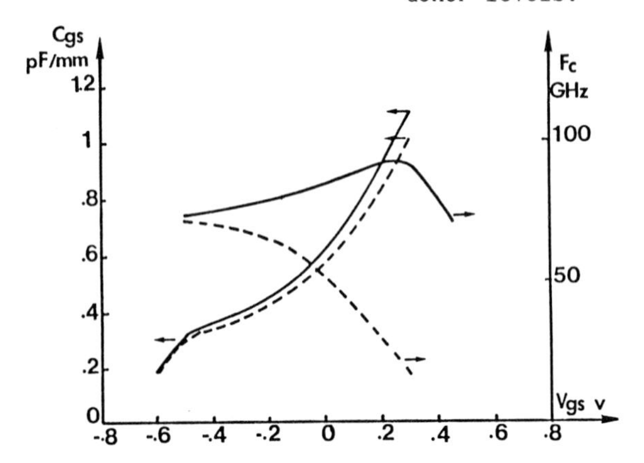

Fig. 7 Gate-to-source capacitance versus gate voltage and cut-off frequency in the same conditions as before.

CONCLUSION

A novel two-dimensional energy model incorporating non-stationary electron transport effects as well as parasitic MESFET and deep donor level effects has been presented. It demonstrates the importance of the effects of transport in the AlGaAs layer and deep donors on the performance of sub-micron gate MODFET's.

REFERENCES

(1) Osman L. EL-SAYED et al., Solid State Electronics, Vol. 30, n°6, pp. 643-654, 1987.

(2) G. SALMER et al., Proceedings of the 11th intern. Symp. on GaAs and Related Compounds, Biarritz, France, Sept. 1984, pp. 503-508.

(3) E.F. SCHUBERT and K. PLOOG, Physical Review B, Vol. 30, n°12, 15 déc. 1984, pp. 7021-29.

Paper presented at Int. Symp. GaAs and Related Compounds, Heraklion, Greece, 1987

The GaAs submicronic recessed gate MESFET: A Monte-Carlo study

R. Fauquembergue, M. Pernisek, J.L. Thobel, P. Bourel

Centre Hyperfrequences et Semiconducteurs, L.A. CNRS 287

Universite des Sciences et Techniques de LILLE – FLANDRES – ARTOIS

59655 VILLENEUVE D'ASCQ , FRANCE

ABSTRACT: We report on the results of Monte – Carlo simulations of recessed structure MESFETs. It is shown that, in order to obtain high performances, the recess – gate distance has to be as short as possible. The use of a very thin active layer increases the transconductance and K – value but this effect is reduced if the doping level is increased, and the use of a p – type buffer layer improves the output conductance and reduces the real pinch – off voltage . These computed results are compared to experimental ones and a good agreement is found.

1. INTRODUCTION

In this paper we present results related to the simulation of recessed structure MESFETs, the main characteristics of which are as follow : the recessed deepth is 0.1μm and in most of the cases, the gate length is 0.3μm. The simulation method uses an ensemble Monte – Carlo technique associated with a two – dimensional Poisson solver (R.W. Hockney, 1985) in order to take into account non stationnary electron dynamics and spatial two – dimensional effects which are very important in such short sized devices. We will study more specifically the influence of the recess, the surface potential, the thickness and doping of the active layer or the use of a p – type buffer layer on the performances of MESFETs.

2. INFLUENCE OF THE RECESS – GATE DISTANCE : Lrg

We simulated a structure with a 0.1μm active layer thickness doped to 2 10^{17} at/cm^3 and we varied the recess – gate distance Lrg from 250Å to 4000Å. The lower Lrg, the higher is the electric field under the gate and, the more important is the carrier's velocity overshoot phenomena. As an example, when Lrg is lowered from 4000Å to 250Å, the maximum electron velocity increases from 2.2 10^5 m/s to 3 10^5 m/s. As a result, the transconductance Gm and the cut – off frequency Fc = Gm/2πCgs increase with decreasing Lrg and, as shown in figures 1a and 1b, this effect is much more pronounced for low values of the internal gate potential Vgs – Vbi (which takes into account the Schottky built – in potential). These results show the interest of structures with very low recess to gate distances.

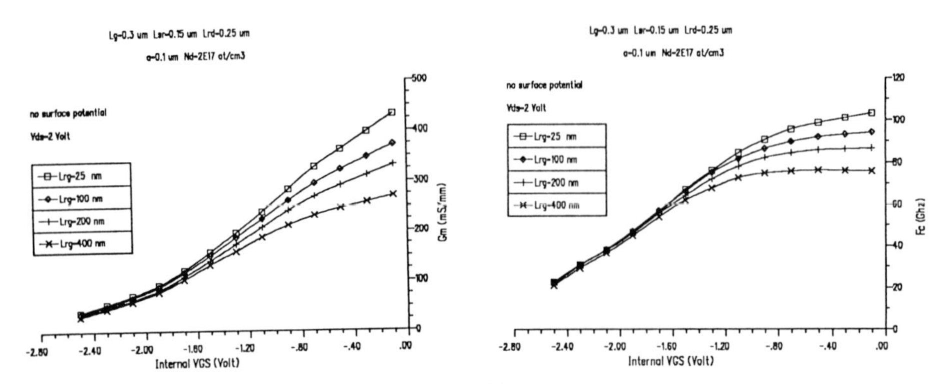

Fig. 1a: Gm versus Vgs for different Lrg

Fig. 1b: Fc versus Vgs for different Lrg

3. INFLUENCE OF THE SURFACE POTENTIAL

For GaAs, the surface density of charges induces a surface potential close to 0.5 V. This surface potential give rise to a transverse electric field which push the carriers away from the surface and create depleted regions. This effect is specially important under the recess as it increases the energy of the carriers which cannot reach anymore very high velocities. As an example, for Lrg = 4000Å , the maximum carrier's velocity is 1.2 10^5 m/s when surface potential is taken into account and 2.2 10^5 m/s without surface potential effect. In figures 2a and 2b are shown the evolution of Gm and Fc versus the recess – gate distance Lrg with and without surface potential effect. It can be seen that surface potential effect reduces the transconductance and cut – off frequency values and this effect is much more pronounced for large recess – gate distances. This give another reason to realise recessed MESFET with very low Lrg.

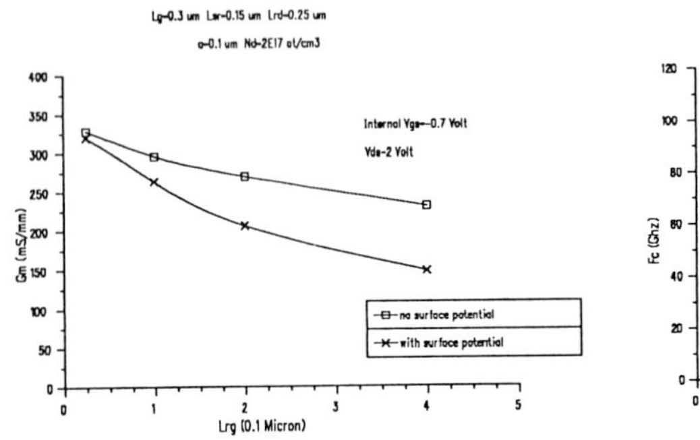

Fig. 2a: Gm versus Lrg with and without surface potential

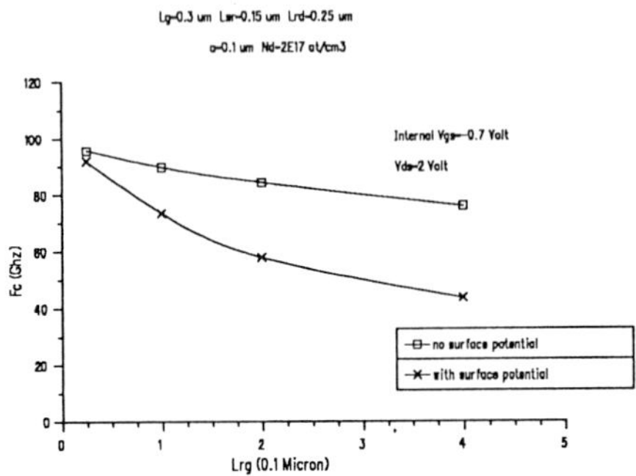

Fig. 2b: Fc versus Lrg with and without surface potential

4. INFLUENCE OF THE ACTIVE LAYER THICKNESS AND DOPING LEVEL

In this study, we varied simultaneously the thickness a and doping Nd of the active layer in order to keep unchanged the theoretical pinch – off voltage $Vp = qN_d a^2/2\varepsilon$ (Sze, 1981).

In figure 3 is shown the variation of the transconductance Gm versus the internal gate voltage (Vgs – Vbi) for Vp = 0.35 V and Vds = 2 V and for different active layer thickness. It can be seen that Gm increases from Gm = 175 mS/mm for a = 1000Å to Gm = 675 mS/mm for a = 250Å, whereas the cut-off frequency $Fc = Gm/2\pi Cgs$ changes only from 70 Ghz to 95 Ghz, the gain in Gm being partially compensated by the increase of the gate-source capacitance Cgs. So, Gm varies almost linearly with the reverse of the active layer thickness, which agrees with the relation Gm~Vm/a (Sze, 1981), where Vm is the carrier's mean velocity in the MESFET's channel. This augmentation of Gm with 1/a is reduced for large pinch-off voltages as shown in figure 4. This can be interpreted for large Vp (i.e. large doping level Nd), by a reduction of the mean velocity Vm due to increased Nd.

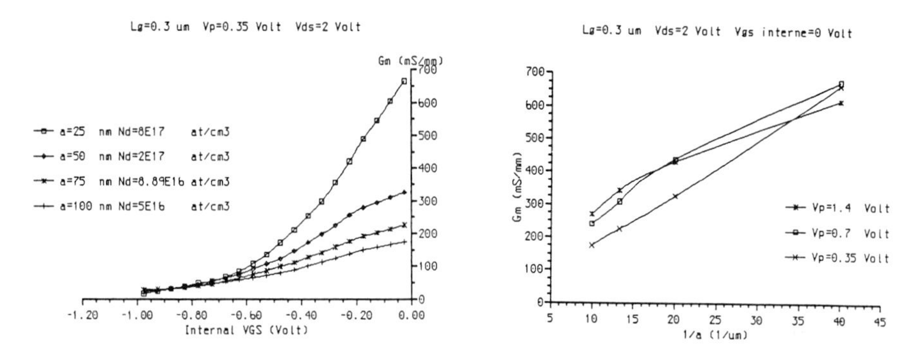

Fig. 3 : Gm versus Vgs for different a and for Vp=0.35 V

Fig. 4 : Gm versus 1/a for different Vp

The same phenomena is observed on the variation of the K-value versus 1/a for various pinch-off voltages Vp (fig. 5). The K-value, which is equal to half the derivative of the function Gm(Vgs), is shown to be proportional to the reverse of the logic gate delay of the device and high K-value are needed for D.C.F.L. applications. Now, following Lehovec (1980), K is to a first approximation proportional to μ/a, where μ is the carrier's mobility which is reduced due to increased doping level for large Vp.

We also found that the K-value is quite insensitive to gate length Lg. As an example, for a = 250Å and Vp = 0.4 volt, K changes from 560 mS/mm/V to 550 mS/mm/V as Lg increases from 0.3μm to 0.5μm. This result is in complete agreement with Ueno experimental results (Ueno, 1985) for Lg≦ 0.6μm.

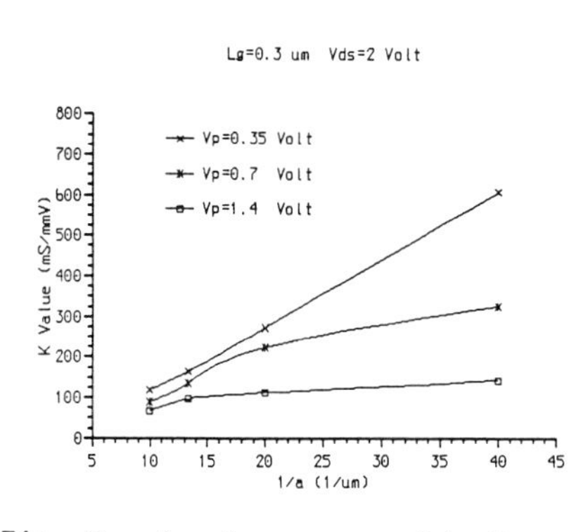

Fig. 5 : K-value versus 1/a for different Vp

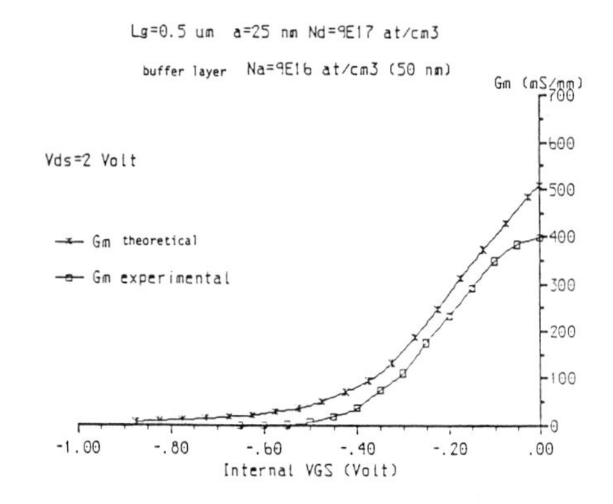

Fig. 6 : Comparison between the theoretical and experimental Gm versus Vgs

5. INFLUENCE OF P-TYPE BUFFER LAYER

We simulated a recessed structure MESFET with Lg=0.5μm, and a 250Å active layer thickness doped to 9 10^{17} at/cm^3. Introducing a p-doped layer (Na) between the active layer and the substrate reduces the injection of carriers into the substrate. As a result, the obtained pinch-off voltage value is closer to the theoretical one and the output conductance Gd is improved and changes from Gd=25 mS/mm with no p-type buffer layer to Gd=16 mS/mm when a p-type buffer layer doped to Na= 9 10^{16} at/cm^3 is used. On another hand the transconductance Gm decreases from 680 mS/mm (Na=0) to 500 mS/mm (Na=9 10^{16} at/cm^3). The K-value and cut-off frequency are almost insensitive to Na and we obtain respectively K=590 mS/mm/V and Fc=45 Ghz. These theoretical results are to be compared with experimental ones (B.J. Van Zegbroeck, 1987) obtained for a MESFET structure identical to the simulated one with a p-type buffer layer doped to Na=9 10^{16} at/cm^3. As an example, in figure 6, is presented the comparison between the computed and experimental transconductance versus Vgs curves. A fairly good agreement is obtained except for low internal gate voltages where the experimental structure begin to exhibit gate leakage current which is not taken into account in the simulation. The other experimental results are K=580 mS/mm/V, Gd=13 mS/mm and Fc=35 Ghz which are very close to the computed ones, respectively K=590 mS/mm/V, Gd=16 mS/mm and Fc=45 Ghz.

6. CONCLUSION

We used a 2-D Monte-Carlo simulation to analyse the influence of various technological parameters on the performances of recessed structure MESFETs. We have shown that we must use thin active layer with a reasonable doping level in order to obtain acceptables values for the mobility, transconductance, K-value and cut-off frequency. Similarly, the recess to gate distance has to be as short as possible. Last, the use of a p-type buffer layer does not change K-value and cutt-off frequency but allows to obtain a pinch-off voltage value closer to the theoretical one and improved output conductance.

7. REFERENCES

Hockney R.W. 1985 Comp. Phys. Com. 36 25-27

Lehovec K., Zuleeg R. 1980 IEEE Trans. Elec. Dev. ED27 1074-1091

Sze S.M. 1981 Physics of Semiconductor Devices

Ueno K., Furutsuka T., Toyoshima H., Kanamori M., Higashisaka A. 1985 Proc. IEDM Washington 82-85

Van Zegbroeck B.J., Patrick W., Meier H., Vettiger P. 1987 IEEE Elec. Dev. Let. EDL 8 N3 118-120

Monte Carlo simulation of impact ionization by electrons in $Al_xGa_{1-x}As$

D. LIPPENS and O. VANBESIEN

Centre Hyperfréquences et Semiconducteurs, Université de Lille Flandres Artois - FRANCE.

ABSTRACT : Monte Carlo simulations of Impact Ionization by electrons in AlGaAs for x = 0.25 and x = 0.45 is presented. Calculations show a drastic increase of ionization rates with electric field despite the large energy bandgap. This is explained by the fact that the electric field becomes strong enough to elevate the electron energy to a high energy state. For devices which involve GaAs-AlGaAs heterojunctions, this result shows the importance of maintaining the field values to values low enough to keep the advantage of the use of large bandgap materials.

1. INTRODUCTION

AlGaAs material is increasingly used in several devices such as TEGFET-MISlikeFET (Drummond 1986) and modulated semiconductor structures (Capasso 1985). As device dimensions are reduced below the 1μm range and applied voltage are increased, extremely high electric field values (in excess of 10^6 V/cm) may be found. These conditions, it is expected that carriers gain sufficient energy in the electric field to impact ionize. At the moment, impact ionization in AlGaAs material system has attracted relatively little attention. Theoretically, apart the recent work of Brennan (1986) on theory of high field transport of holes in $Al_{0.45}Ga_{0.55}As$, we are not aware of any published work in this area. Experimentally, very scarce measurements of AlGaAs impact ionization rates have been published often restricted to small values of Al content (Shabde 1970) or applied electric field (David 1985). Despite the rather low values measured in the field range investigated, these experimental studies have shown that the electron and hole rates vary very rapidly with electric field. Therefore, in the high electric field limit, Schabde (1970) has predicted an electron impact ionization rate of 4,3 10^5 cm^{-1} for an aluminium concentration of x = 0.25. In addition, calculations of Brennan (1985) of high field properties of holes show that the hole impact ionization rate is quite large at fields above 400 kV/cm for x = 0.45.

We report here a theoretical analysis of impact ionization by electrons based on Monte Carlo simulation of high field carrier transport. Even though Monte Carlo method applied to high field transport requires severe approximations in the calculation, they have the merit of providing a first physical insight into the problem of impact ionization. We will focus on the two Al concentrations encountered in practical applications x = 0.25 and x = 0.45. To be realistic, the simulation has to depict the high energy points in the k space. Therefore, in a first step, the band structures have been calculated by the pseudopotential method. Then, the impact ionization rates of electron were calculated in a large applied electric field range.

2. SIMULATION METHOD

The Monte Carlo method has been used extensively in the study of carrier transport in semiconductors. Let us recall that the method keeps track of an electron **k** vector in the Brillouin zone until it reaches the threshold energy for impact ionization. Thus, the simulation procedure requires the knowledge of the band structure and of the different scattering mechanisms that a carrier undergoes during its motion.

The rather complex conduction band structure was calculated by the empirical pseudopotential method. Figure 1 illustrates the isoenergy lines in the ΓXK cross section for $Al_{0.45}Ga_{0.55}As$. The pseudopotential forms factors, used in the calculation are given in table 1. The high energy states are given in shaded zones. One can note that these zones are located outside the main symetry lines only described in simple dispersion relations. The cross over between the Γ and X valley located at same energy level is also depicted for this concentration

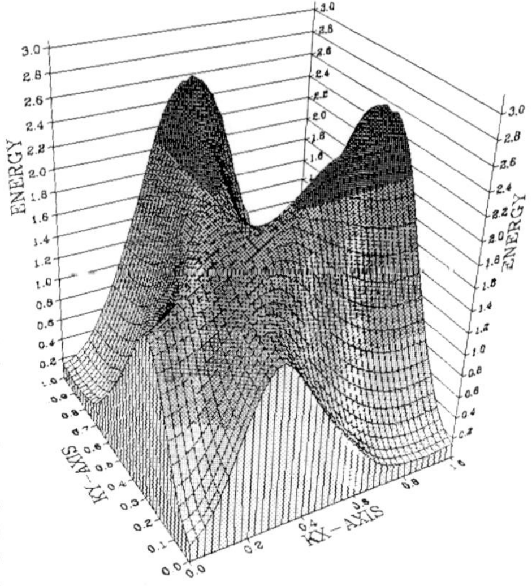

Figure 1 : Isoenergy lines in the ΓXK cross section for $Al_{0.45}$ $Ga_{0.55}$ A_s.

The scattering mechanisms described are acoustic phonon, polar optical phonon and intervally scatterings. Most of the parameters used to calculate the scatteringrates were taken from Adachi (1985). Due to the lack of extensive data on intervalley deformation potentials we have used the data given by Saxena (1982) for Ga_{1-x} Al_x A_s alloys. In the high energy range, the scattering rate has been simply extrapolated in view of the complexity of a more exact analysis as shown by Capasso (1985) and to limit the number of adjustable parameters. Figure 2 shows the energy dependence of total scattering rate in the Γ valley for the two Al concentrations studied in this work. It is worthwhile mentioning that rather high scattering rate values are calculated at high energies. This will randomize very often the orientation of the carrier momentum and will allow a rapid and complete excursion of the Brillouin zone.

$Al_{0.45}Ga_{0.55}As$ material lattice constant 5.6568 10^{-8}cm		
V_3^S = −0.2322	V_8^S = +0.0107	V_{11}^S = +0.0682
V_3^A = +0.0678	V_4^A = +0.0491	V_{11}^A = +0.0044

Table 1 : pseudopotential form factors, in rydbergs.

For the treatment of impact ionization we have assumed that the ionization cross section increases very rapidly with energy and consequently that impact ionization is immediate once the threshold energy is reached. In addition, we have assumed an isotropic threshold energy. Higher conduction bands have not been included in the calculations because of the lack of knowledge of the deformation potential constants. In spite of these restrictions we will see that the agreement between theory and experiment is quite good.

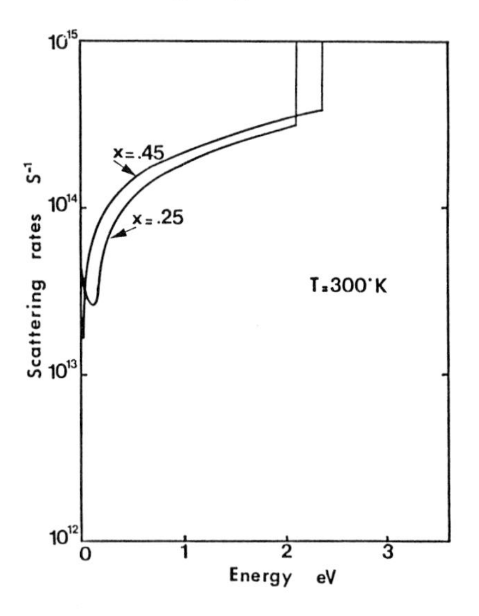

Figure 2 : Scattering rate as a function of electron energy.

3. RESULTS

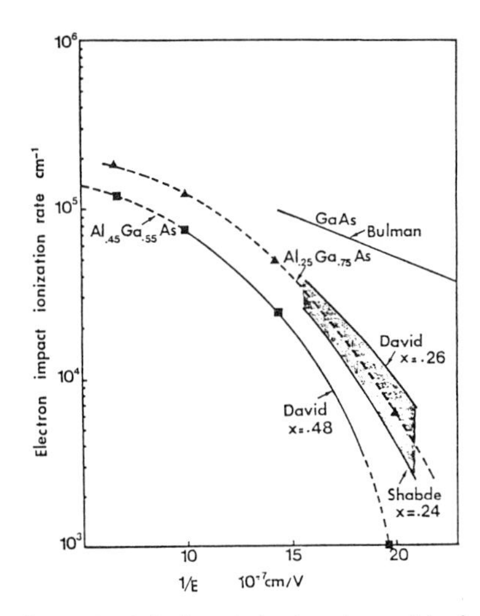

Figure 3 : Calculated electron impact ionization as a function of inverse electric field in AlGaAs.

Figure 3 shows the electric field dependence of the calculated electron ionization rates for the two Al concentrations.The isotropic ionization thereshold energy has been chosen equal to 2.10 for x = 0.25. The theoretical curve (dashed lines) falls within the range of experimental data (Shabde 1970) (David 1985). A value of 2.35 has been chosen for x = 0.45 to fit the results of David (1985). As a comparison, the ionization rate versus electric field strength measured by Bulman (1985) in GaAs is also reported. For electric field values below 500 kV/cm, it is found that no significant ionization is observed in AlGaAs. By contrast, above this value, the ionization rate increases drastically, despite the large energy band gap. In the high energy electric field limit, the calculated values become comparable to those in GaAs. This is in agreement with the trend revealed by the experimental AlGaAs ionization rates.
In order to interpret this result, the electric field dependence of the average energy for each composition is given in figure 4. For low electric field values, a smooth increase of carrier energy is obtained due to frequent intervalley scatterings. The effect is particularly pronounced for x = 0.45. This result may be correlated with the low mobility values measured and/or calculated by Saxena (1985) and Hill (1981). For extremely high electric fields, the field strengh becomes strong enough to defeat the scattering interactions. We can note that in this case the average energy is elevated to a high energy state (typically 1ev). Thus, the electron has the ability to impact ionize when its momentum vector is located near the very high energy sites.Or equivalently, the tail of the energy distribution exceeds significantly the impact ionization threshold energy. For a better understanding of how the electrons impact ionize we show in figure 5 the variation of electron energy after each scattering event at electric field strengh of 10^6 V/cm. The result was obtained for x = 0.45. It can be noted that typically electrons experience a large number of phonon scatterings. Occasionally electrons move up very rapidly to higher energy. Some reach ionization threshold. This feature is revealed in figure 5 during the time of observation.

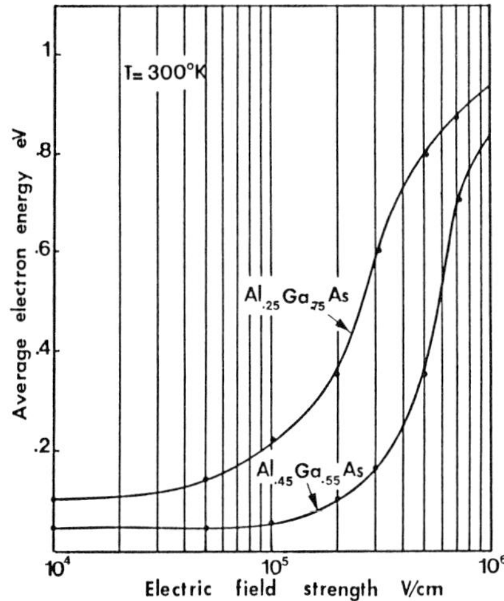

Figure 4 : Average electron energy as a function of electric field.

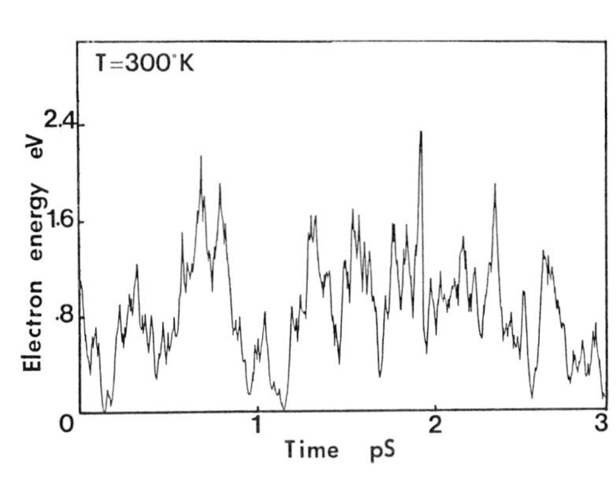

Figure 5 : Variation of electron energy after each scattering event.

4. DISCUSSION

The calculated ionization data in figure 3 can be parameterized as a function of electric field in volts per centimeter

$$\alpha(E) = 2.5 \times 10^5 \exp(-(8.6 \times 10^5/E)^{2.5})\ \text{cm}^{-1} \quad \text{for } x = 0.25$$

$$\alpha(E) = 1.4 \times 10^5 \exp(-(8.3 \times 10^5/E)^3)\ \text{cm}^{-1} \quad \text{for } x = 0.45$$

Thus, the numerical solution shows that the electron rate has an upper limit near 2×10^5 cm^{-1} in the Al concentration range considered in this work. This values is roughly two times lower than that given by Shabde (1970) for x = 0.24. (The discrepancy probably arises from a fit to rate data restricted in electric field range in Shabe experimental work) and is quite comparable with GaAs value.

The electric field dependence of electron rate found above can have a significant effect on the breakdown voltage of devices which involve GaAs/GaAlAs heterojonctions. Therefore, in a MISlikeFET structure which consists of an undoped GaAlAs layer inserted between the gate and the channel layer, most of the ionization integral comes from the region near the gate edge. In this region, the electric field values are generally larger than those where ionization rate has been measured by direct method (photomultiplication measurement essentially). The rapid increase in the electron rate can make high breakdown voltage difficult to achieve in practice. This conclusion is in contradiction with first analysis for which it is often thought that the undoped layer would increase the breakdown voltage for any electric field values. Concerning multilayer GaAs/AlGaAs avalanche photodiodes one of the requirements of band gap engineering concept for low excess noise is that there is no possibility of impact ionization in AlGaAs layers. The simulation shows that this requirement is no longer valid in very high electric field range.

5. CONCLUSIONS

Monte Carlo calculations of electron high field properties in $Al_xGa_{1-x}As$ for x = 0.25 and x = 0.45 have been presented. The main result is the drastic increase of ionization rates with electric field not expected in a first analysis based on band gap considerations. It is clearly of interest to take this fact into account in the design of devices for which one takes advantages of the use of GaAlAs layers in order to improve breakdown voltage or noise properties.

6. ACKNOWLEGDMENT

The authors would like to thank K. Hess and J.P. Leburton (University of Illinois) for their help in the calculation of band structures and J.L. Thobel (University of Lille) for hepful discussions. The technical assistance of Y. Tinel and S. Jennequin of the University computer Center is also greatly appreciated.

7. REFERENCES

Adachi S. 1985, J. Appl. Phys. **58** R1.
Brennan K. and Hess K. 1986, J. Appl. Phys. **59** 964.
Bulman G.E., Robbins V.M. and G. Stillman. IEEE Trans. Electron Device ED-22, 2454.
Capasso F. 1985 Semiconductors and semimetals (New York : Academic Press) Vol. 22, Part. D
David J.P.R., Marsland J.S., Hall H.Y., Hill G., Mason N.J., Pate M.A. Roberts J.S., Robson P.N., Sitch J.E., Woods R.C. 1985 Inst. Phys. Conf. Ser. **74** 247.
Drummond T.J., Masselink W.T. and Morkoc M. 1986 Proceeding of the IEE **74**, 773.
Hill G. and Robson P.N. 1981. Journal de Phys. **42**, 33
Saxena A.K. and Gurumurthy K.S. 1982 J. Phys. Chem. Solids 43 801.
Saxena A.K. and Mudares M.A.L. 1985, J. Appl. Phys. **58**, 2795.
Shabde S.N. and Yeh C., 1970, J. Appl. Phys. **41**, 4743.

Inst. Phys. Conf. Ser. No. 91: Chapter 7
Paper presented at Int. Symp. GaAs and Related Compounds, Heraklion, Greece, 1987

Theory of hot carrier transport in GaAs–$Ga_xAl_{1-x}As$ superlattices

D C Herbert, J H Jefferson and M A Gell*

Royal Signals and Radar Establishment, Malvern, UK
* British Telecom Research Labs, Ipswich, UK

ABSTRACT: Electron-phonon scattering rates in superlattices are computed from effective mass wave functions and used to discuss hot carrier transport. A new direct solution technique for solving the transport equations is used and compared with ensemble Monte-Carlo results.

1. INTRODUCTION

The physics of hot carrier transport in superlattices is only just beginning to be explored. By varying the alloy composition and unit cell parameters a wide range of electronic structure is possible (Gell and Herbert 1987, Gell et al 1986) and is expected to yield a rich variety of hot carrier transport properties with potential device applications. Due to the large range of possibilities for the physical parameters, a reliable theory is essential for selecting structures of most device interest.

To obtain useful transport in the superlattice growth direction it is desirable either to use ultra-thin barriers which have a high tunneling probability for the lowest sub-band, or to inject directly into the higher sub-bands. In the former case, application of an electric field leads to several interesting effects. The existence of the first mini-gap suppresses carrier heating, modifies scattering rates and leads to Bloch oscillation effects. For increasing field strength a threshold is reached where Zener tunneling to the higher mini-bands sets in and at sufficiently high fields the mini-gaps become transparent and the band structure description changes. In this paper we concentrate on a specific example with unit cell consisting of 100 Å GaAs, 10 Å AlGaAs with composition chosen to yield a barrier height of 0.25 eV. In this case the first sub-band width and mini-gap are 27 meV and 37 meV respectively. These values are obtained using the average mass description (Gell and Herbert 1987). Detailed comparison with pseudo-potential band structure calculations suggests that this approximation should be adequate for ultra-thin barriers. The transport calculations in this paper are performed at 100°K when population of the higher sub-bands can be neglected and effects associated with the lowest sub-band can be studied in isolation.

2. ELECTRON-PHONON SCATTERING RATES

The effective mass wave functions in a mini-band can vary rapidly with wave vector and this changes scattering rates through mass anisotropy, non-parabolicity, wave-function overlap and umklapp. To include umklapp the Frohlich interaction H_I is taken in the form

$$H_I(q) = V \sum_G |\underline{q}+\underline{G}|^{-1} \exp[i(\underline{q}+\underline{G}).\underline{r}], \qquad |V|^2 = 2\pi\omega_0 (\varepsilon_\infty^{-1} - \varepsilon_0^{-1})$$

where $\underline{G}$ are superlattice reciprocal lattice vectors, ω_0 is the optic phonon energy taken to be 36 meV and ε_0, ε_∞ are the usual low and high frequency dielectric constants. The overlap and umklapp matrix elements are obtained from the band structure wave functions and used in a numerical calculation of scattering rates.

In Fig 1 S_2 denotes bulk GaAs phonon scattering rates corrected for the presence of the mini-gap. S_0, S_1 are superlattice scattering rates allowing for mass anisotropy, wave function overlap and umklapp with k_x (the wave vector in the growth direction) taking the minimum and maximum allowed values respectively on the initial state energy shell. Above the threshold for optic phonon emission, suppression of scattering occurs due to the wave function overlap. The peak in the scattering at ~ 62 meV is due to umklapp scattering and an enhanced density of states when the final state energy is close to the first mini-band edge. Similar structure is observed in the higher mini-bands, and when a mini-band width exceeds the optic phonon emission threshold, a peak in the scattering is also found for an initial state lying close to the mini-band edge (Herbert to be published). Similar effects are present in the phonon absorption but due to the very low scattering rates they are not so evident in Fig 1. They do however influence the low temperature transport which is sensitive to the acoustic and phonon absorption scattering since a large fraction of the electrons are below the optic phonon emission threshold in this case. For the 100°K transport considered in this work, the first mini-gap inhibits electron heating in the applied field and the results are not sensitive to the phonon emission structure at energies greater than ~ 60 meV.

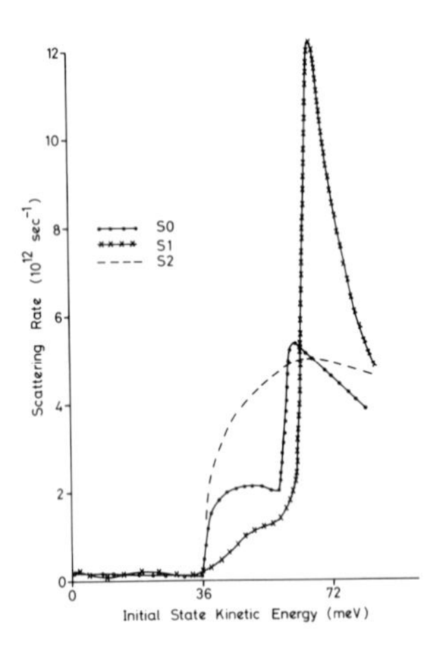

Fig 1. Superlattice scattering rates.

3. HOT CARRIER TRANSPORT

We are currently developing new approximate semi-analytic techniques for studying transport in microstructures. The methods are numerically fast, allowing an interactive approach to device modelling. The complexity of the hot carrier distribution in superlattices provides a severe test for these techniques and the results are compared with accurate semi-classical Monte-Carlo simulations (Fig 2) for a simplified scattering model. The essence of the method is to expand the distribution on constant energy shells in low order polynomials of the form

$$f = \sum_{s=0}^{N-1} a_s(E)(k_x/K)^s$$

where K is the maximum allowed value for k_x on the shell with total energy E. By using the Boltzmann equation or quantum kinetic equations (Herbert et al 1985, 1984), the Nth order term couples to a term of order N+1, and as a decoupling approximation we replace this by an equivalent coupling to the N-1 term. For a superlattice, Bragg reflection is included as a boundary condition that the antisymmetric part of f vanish when k_x coincides with the superlattice Brillouin zone edge. Scattering out and scattering in terms are readily computed on the polynomial sub-space, thus allowing for scattering anisotropy. The discretised transport equations can be formulated in matrix form as simultaneous equations and solved for the coefficient $a_s(E)$. In practice for the higher order polynomial approximations it was necessary to include a weak relaxation term for numerical stability. For bulk GaAs, N=2 gives a good description of transport, but for the superlattice, N=10 is necessary for convergence when excellent agreement with Monte-Carlo simulation is achieved (Fig 2). The Monte-Carlo simulation used bulk GaAs scattering rates but rejected scattering events leading to a forbidden final state. Since the energy gained between scattering events can never exceed the minibandwidth E_B, the simulation used the scattering rate (including self-scattering) at energy $E_\perp + E_B$ where $E_\perp$ is the energy perpendicular to the field. The results show a large negative differential mobility related to Bloch oscillation in this structure. The Bloch oscillations occur explicitly as a transient in the Monte-Carlo simulation (Fig 3) which used up to 20000 electrons in an initial thermal ensemble.

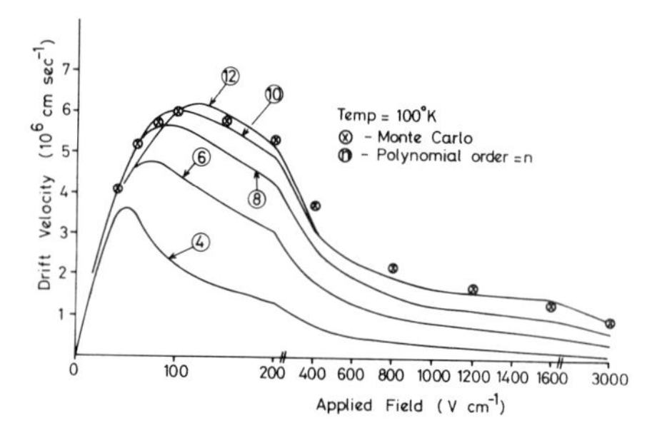

Fig 2. Comparison of Monte-Carlo with semi-analytic approximations.

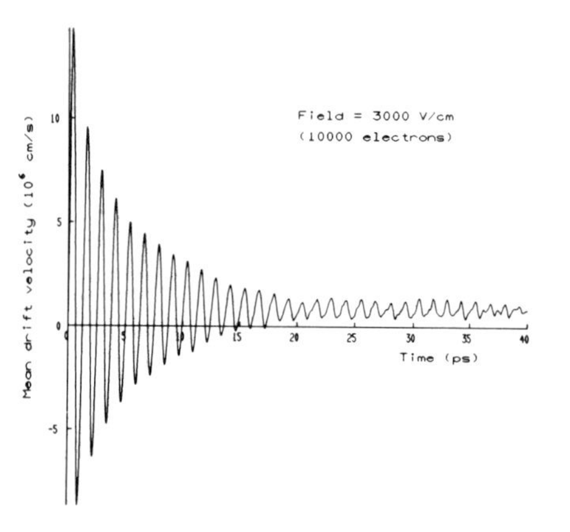

Fig 3. Transient Bloch oscillations.

In Fig 4 we compare results obtained for N=10 using the superlattice scattering rates (S_0,S_1) and the bulk scattering rate S_2 taken from Fig 1. We also show results obtained using the quantum kinetic equations to allow for the wave-packet character of the electrons. For this particular example, wave function overlap suppresses the optic phonon scattering at low energies leading to enhanced mobility and an increase in peak velocity. The suppression of phonon absorption scattering by the mini-gap for electrons with energy below the band edge, increases the importance of acoustic scattering in this energy range. To show this, the acoustic deformation potential has been reduced from 7 eV to 3.5 eV for the results in Fig 4 and the peak velocity using bulk scattering is increased by ~ 10% compared with the N=10 result in Fig 2 which used a 7 eV potential. The quantum calculation which allows for the wave-packet character of electrons suggests that corrections to the Boltzmann equation become significant (~ 50%) at high fields where Bloch oscillation effects dominate. It is expected that these quantum effects will greatly reduce the time dependent transient oscillations of Fig 3.

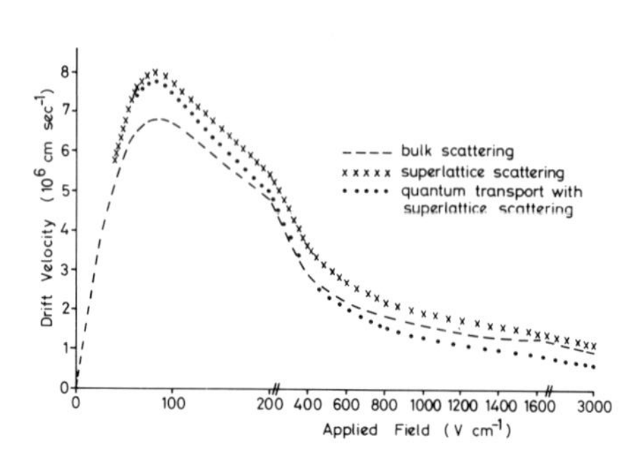

Fig 4. Semi-classical and quantum transport using superlattice scattering rates.

4. CONCLUSION

Computed polar phonon scattering rates in superlattices deviate considerably from the bulk GaAs form and show a great deal of structure which must be allowed for when calculating transport properties. Below the threshold for Zener tunneling in the specific case considered, negative differential mobility with a very high peak to valley ratio is predicted. Interesting structure also occurs in the high field current response, related to inter-mini-band Zener tunneling and transport in the higher mini-bands. We conclude that hot carrier transport in superlattices shows considerable potential for a range of applications in high speed electronics. Our new techniques for modelling transport greatly facilitate the detailed analysis of these effects.

REFERENCES

Gell M A, Ninno D, Jaros M and Herbert D C, 1986, Phys Rev B 34, 2416.
Gell M A and Herbert D C, 1987, Phys Rev B. 35, 9591
Herbert D C, 1984, J Phys C 17 6749-67.
Herbert D C and Kirton M J, 1985, Physica 129B, 537-41.
Herbert D C, Jefferson J H and Kirton M J, 1985, Physica 134B, 82-86.

Effect of current crowding on the performance of a GaAs/AlGaAs heterojunction phototransistor

J K Twynam*, R C Woods*, J C H Birbeck, D R Wight, J C Heaton & G R Pryce

* The University of Sheffield, Department of Electronic and Electrical Engineering, Mappin Street, Sheffield S1 3JD

Royal Signals and Radar Establishment, St Andrews Road, Great Malvern, Worcs. WR14 3PS

ABSTRACT: It is well known that the performance of a heterojunction phototransistor can be enhanced, in terms of gain, speed of response and signal to noise ratio, by the application of external bias either optically or electrically via a third (base) terminal. In this paper we show that crowding of the bias current and the signal current will limit the advantage given by electrical bias in 3-terminal devices but aid the performance of 2-terminal devices. Theoretical results are compared with experiment.

1. INTRODUCTION

The heterojunction phototransistor (HPT) is potentially useful as an optical receiver (Fritzsche et al. 1981, Campbell et al. 1981, Campbell and Ogawa 1982, Scavannec et al. 1983). An analysis by Alavi and Fonstad (1981) concluded that the sensitivity of an HPT in a pulse code modulated optical communications system could be superior to that of a PIN/FET combination, particularly at high bit rates, and that it could approach that of an APD/FET combination.

The response time of a heterojunction bipolar transistor is related to the emitter-collector signal delay time τ_{ec} (Ladd and Feucht 1970). For typical devices with base thickness $W_b \leq 0.5\mu m$ the base transit time is negligible, as is the transit time of minority carriers across the collector-base depletion region, compared with the charging time of the emitter-base capacitance. Hence τ_{ec} is determined primarily by the time constant τ_e for charging the emitter-base capacitance. This time constant can be written simply as $\tau_e = R_e C_e$, where R_e is the dynamic emitter-base junction resistance and C_e is the emitter-base junction capacitance, if the collector capacitance C_c is assumed to be much less than C_e. The gain-bandwidth product or cut-off frequency of a phototransistor is given by

$$f_c = \frac{1}{2\pi\tau_{ec}} \approx \frac{1}{2\pi R_e C_e} = \frac{q(I_{ph} + I_{dc})(1 + h_{fe})}{2\pi n\, kT\, C_e} \qquad (1)$$

where I_{ph} is the primary photocurrent, I_{dc} is the external base bias current and n is the ideality factor of the emitter-base heterojunction. However, this simple expression does not take into account base resistance.

2. CURRENT CROWDING IN HPTS

The current flow in an npn heterojunction phototrajsistor is shown schematically in Fig. 1. The optical signal beam is assumed to have finite diameter and to be incident at the centre of the phototransistor so that the positive charge generated spreads out giving rise to a lateral base current. If an external electrical bias is supplied then base current also flows into the device from the contacts. These lateral currents, flowing through the finite base resistance of the phototransistor, give rise to variations in the emitter-base junction voltage and the emitter current density across the device. This in turn leads to "fast" areas of the junction, where the RC time constant is small, and "slow" areas where the RC time constant is large. Using the transmission-line equivalent circuit model and the method described by Twynam and Woods (1987), we can compute the distribution of both dc currents and small signal ac currents.

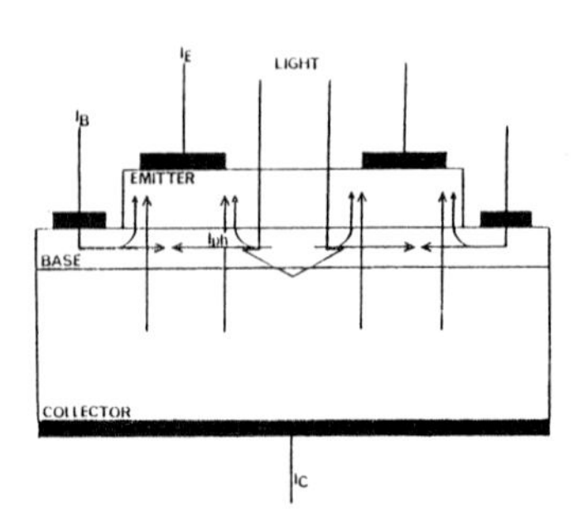

Fig. 1 Schematic diagram showing the current flow in a heterojunction phototransistor

We can therefore calculate the gain-frequency response of an HPT and relate the cut-off frequency of a device to such variables as the bias current level, the optical signal level and the optical signal beam diameter.

3. THEORETICAL RESULTS

Figs. 2a and 2b show theoretical plots of cut-off frequency against dc collector current for 2-terminal and 3-terminal HPTs with different base doping levels. For the 2-terminal devices in Fig. 2a the cut-off frequency is, in each case, an increasing function of the dc collector current which is determined entirely by the optical signal level. At higher current levels however the phototransistor with high base resistance gives a greater improvement in f_c. The explanation for this is that the signal current has been confined to the "fast" central region of the transistor, where the forward bias is high and it is effectively decoupled from the "slow" region of the transistor by the base resistance.

The cut-off frequencies of the 3-terminal devices represented in Fig. 2b are also increasing functions of dc collector current. In this case the optical signal level is fixed and the dc current is determined by the electrically applied bias. At higher current levels the base resistance tends to reduce the effect of the bias at the centre of the transistor, away from the ring contact. So, in contrast to the case of the 2-terminal devices, at higher current levels the 3-terminal HPT has better high frequency response when the base resistance is low.

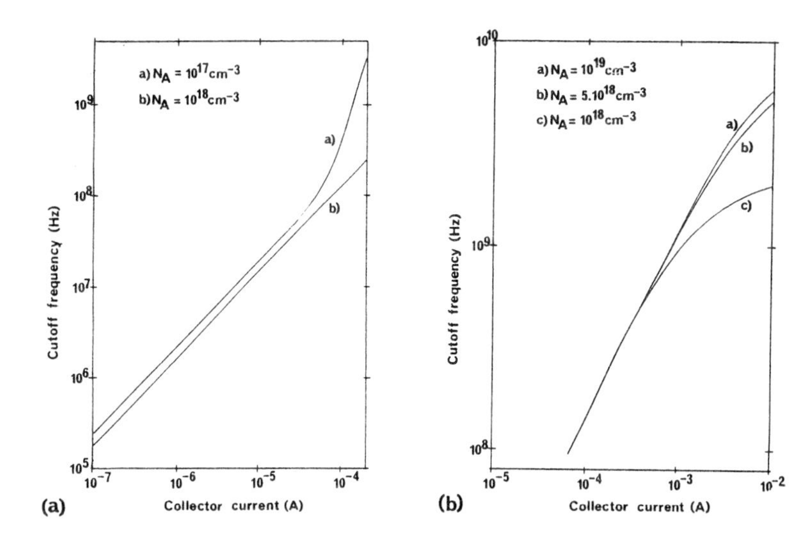

Fig. 2 Theoretical plot of cut-off frequency against dc collector current for HPTs with different base doping levels N_A: (a) 2-terminal devices; (b) 3-terminal devices with fixed optical power level P_o = 10μW. Other parameters: Device diameter = 50μm, optical signal beam diameter = 2μm, β_o = 10, $N_D = 10^{17}$ cm^{-3}, w = 0.1 μm

4. EXPERIMENTAL RESULTS

A schematic diagram of the GaAs/AlGaAs HPT structure is shown in Fig. 3. The devices were grown by MOCVD and were fabricated using conventional lift-off and selective etching techniques. High frequency measurements were made using a vector network analyser, both to modulate an AlGaAs/GaAs/AlGaAs laser and to measure the phototransistor output power. Fig. 4 shows a plot of small signal optical gain vs frequency for a large device (diameter = 200μm) under different illumination conditions with the base contact left unbonded. In each case the optical power to the device was 250μW but the diameter of the optical beam was approximately (a) 20μm (b) 40μm (c) 80μm (d) 160μm. The dashed curves (e), (f), (g) and (h) are the corresponding theoretical results. The theory predicts that the phototransistor should operate faster with a narrower optical beam because of the current crowding effect. This is confirmed by the experimental results.

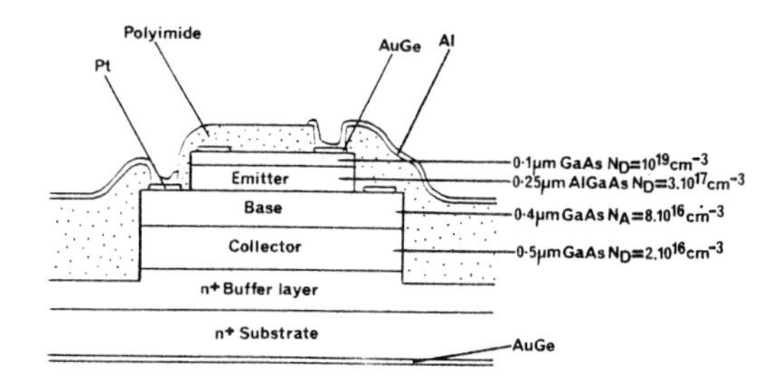

Fig. 3 Schematic cross section of HPT structure

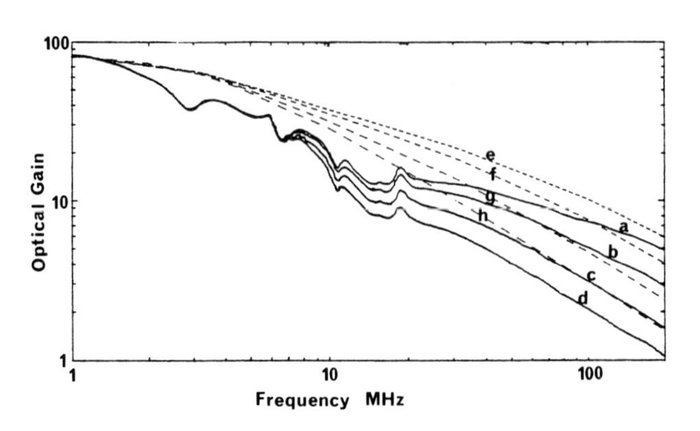

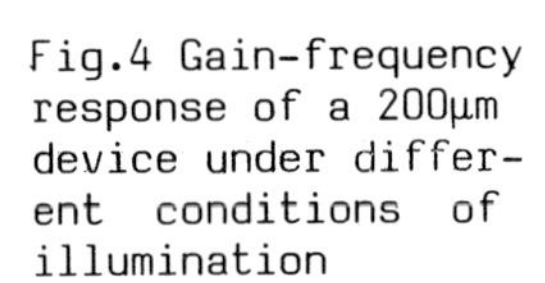
Fig.4 Gain-frequency response of a 200µm device under different conditions of illumination

5. CONCLUSIONS

A transmission-line equivalent circuit model has been used to determine the two-dimensional dc and ac current flow in a phototransistor with cylindrical geometry. The gain-frequency response of a device can be calculated using this method and we have presented theoretical results showing that whereas a 3-terminal HPT would perform better with low base resistance, a 2-terminal HPT should benefit from high base sheet resistance. The effect of current crowding on a 2-terminal device has been verified experimentally.

J K Twynam wishes to thank SERC and RSRE (Malvern) for the award of his CASE studentship.

REFERENCES

Campbell J C, Burrus C A, Dentai A G and Ogawa K 1981 Apply. Phys. Lett. 39 (10) pp 820-821
Campbell J C and Ogawa K 1982 J. Appl. Phys. 53 (2) pp 1203-1208
Fritzsche D, Kuphal E and Aulbach R 1981 Electron. Lett. 1981 17 (5) pp 178-179
Ladd G O and Feucht D L 1970 IEEE Trans. Electron Devices ED-17 (5) pp 413-420
Scavennec A, Ankri D, Besombes C, Courbet C, Riou J and Heliot F 1983 Electron. Lett. 19 (10) pp 394-395
Tabatabaie-Alavi K and Fonstad C G 1981 IEEE J Quantum Electron. QE-17 (12) pp 2259-2261
Twynam J K and Woods R C 1987 (Accepted for publication IEE proc-J)

Inst. Phys. Conf. Ser. No. 91: Chapter 7
Paper presented at Int. Symp. GaAs and Related Compounds, Heraklion, Greece, 1987

GaAs MESFET Schottky barrier height dependence on device scaling

S W Bland, J E Puleston Jones and J Mun

STC Technology Ltd., London Road, Harlow, Essex, CM17 9NA, UK

ABSTRACT: The effect of reductions in device contact geometry and active layer thickness on the GaAs MESFET Schottky barrier height and ideality factor have been investigated. Barrier height reductions are observed as either the gate length or gate-ohmic separation are reduced. The barrier height was also significantly reduced for both short and long gate length devices as the channel implantation energy was reduced from 120 keV to 25 keV. The possible cause of Schottky diode degradation is discussed.

1. INTRODUCTION

The MESFET is one of the most important devices for the realisation of high performance GaAs integrated circuits. The MESFET relies for its operation on a good quality rectifying Schottky barrier gate contact. Contact quality depends on a number of factors including the method of GaAs surface preparation, the type of metal employed, its method of deposition and any subsequent heat cycles. This paper describes additional factors which must be taken into account as the device is scaled to achieve higher speed performance and increased levels of integration. For example, the current trend in device fabrication, especially for self-aligned FETs, is toward thinner active layers in order to achieve improved device transconductance and alleviate short channel effects. The effect of such scaling on the Schottky diode characteristics is investigated.

2. EXPERIMENTAL

GaAs MESFETs were fabricated on semi-insulating substrates using Si^{29} ion-implantation and optical transient annealing (1050°C for 2 seconds using a 300 A SiNx cap). Contact metals were AuGe/Ni and Cr/Au for ohmics and gates respectively. FETs were fabricated with varying gate lengths (1.3 to 100 μm) and gate-ohmic separations (0.8 to 7.6 μm) for a fixed implantation energy of 100 keV through 300 A of SiNx. Long (101 μm) and short (1.2 μm) gate length FETs were also fabricated using direct implanation at energies of 25, 40, 80 and 120 keV to achieve variations in the active layer thickness. The implant dose was scaled in order to obtain an approximately constant device threshold voltage.

Schottky barrier height ϕ_b and ideality factor n were calculated from the forward IV characteristics assuming (Padovani and Stratton 1966)

$$I = I_s \, (e^{V/V_o} - 1) \qquad (1)$$

and

$$I_s = A^{**}T^2 \, e^{-\Phi_b/kT} \qquad (2)$$

where $V_o = nkT/q$, A^{**} is the effective Richardson constant and the rest of the symbols have their usual meanings.

3. RESULTS

Figure 1(a) shows the effect of variations in gate-length on the Schottky barrier height and ideality factor. There is a sharp reduction in barrier height and a corresponding increase in ideality factor as the gate length is reduced below about 20 µm (gate-ohmic separation of 0.8 µm and gate width of 100 µm). The reverse breakdown voltage was also measured and remained relatively constant for all gate lengths at around -30 V. A similar degradation in Schottky diode characteristics is observed for reductions in the gate-ohmic separation for a constant gate length of 1.3 µm (Figure 1(b)). In this instance the reverse breakdown voltage increases linearly with separation with a gradient of around 1.6×10^5 V/cm.

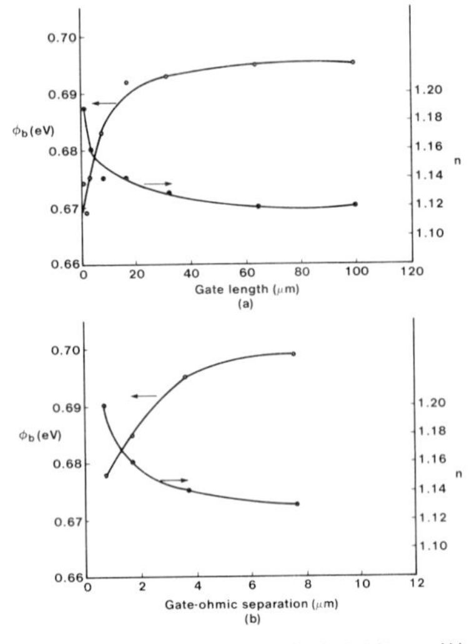

Figure 1. The variation of Schottky barrier height ϕ_b and ideality factor n with (a) gate length (gate-ohmic separation of 0.8µm) and (b) gate-ohmic separation (gate length of 1.3µm)

Figure 2 shows the effect of variation in implant energy on the Schottky diode barrier height. There is a marked reduction in barrier height for both long (curve a) and short (curve b) gate length devices as the implant energy is reduced. Similarly there is a corresponding increase in the ideality factor although the reverse breakdown voltage remains relatively constant at around -30 V. Equivalent results were obtained if the implants were performed through a dielectric layer.

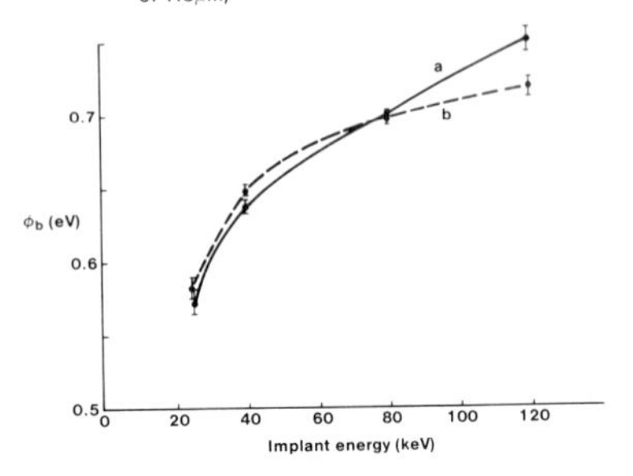

Figure 2. The variation of Schottky barrier height ϕ_b with implant energy for (a) 101 µm and (b) 1.2 µm gate length devices.

4. DISCUSSION

The results obtained illustrate the importance of considering the electric field distribution within the FET. Increased electric fields occur at the edges of the gate contact and this results in increased thermionic-field (T-F) emission and a reduction in the effective Schottky barrier height. Consequently as the ratio of diode periphery to diode area increases, the diode parameters are progressively degraded. This effect is clearly observed in Figure 1(a). The importance of considering the electric field distribution is further illustrated in Figure 1(b) by a similar degree of diode degradation due to reductions in the gate-ohmic separation.

The trend described in Figure 1(a) is also apparent in Figure 2 for implant energies in excess of about 80 keV. For energies below this value the barrier height difference between long and short gate length devices is relatively small. This later observation implies a reduced contribution from fringing fields at the gate edges which correlates well with improvements in short channel behaviour which we have observed in parallel studies on self-aligned devices. Nevertheless the overall Schottky barrier height is reduced as the implant energy is lowered and a number of additional experiments were performed to investigate this phenomenon.

The presence of T-F emission is characterised by a temperature independence of the IV characteristics at low temperature where field emission dominates. In this region

$$V_o = V_{oo} = \tfrac{1}{2}\hbar\,(N/m^*\varepsilon\varepsilon_o)^{\frac{1}{2}} \tag{3}$$

where N is the impurity concentration, ε is the dielectric constant of the semiconductor, m* is the effective mass of the electrons and the remaining symbols have their usual meanings (Saxena 1969). The absolute value of V_{oo} is a useful guide to the degree of T-F emission. V_{oo} is plotted in Figure 3 as a function of implant energy. Also included in Figure 3 is a theoretical curve based on the assumption of a Gaussian implant profile and 100% activation (in reality an activation efficiency of around 60% is more realistic). Good qualitative agreement is obtained although quantitatively there is a large discrepancy.

In common with Saxena (1969) it was found that the diode characteristics were best described using the empirical relation $V_o = k(T+T_o)/q$. T_o was found to increase toward lower temperatures which is an indication of T-F emission. If T is replaced by $T+T_o$ in equation 2 a roughly linear dependence of the form $\phi_b = \phi_{bo} - aT$ is obtained for all implant energies, where ϕ_{bo} = 0.965 eV and a = 3.1×10^{-4} eV/°K. Such a dependence is in reasonable agreement with the variation of energy gap with temperature. Plots of $\ell n(I_s/T^2)$ against $1/(T+T_o)$ were also obtained and found to be linear. A value for ϕ_{bo} of approximately 0.95 eV was calculated from these curves regardless of implantation energy.

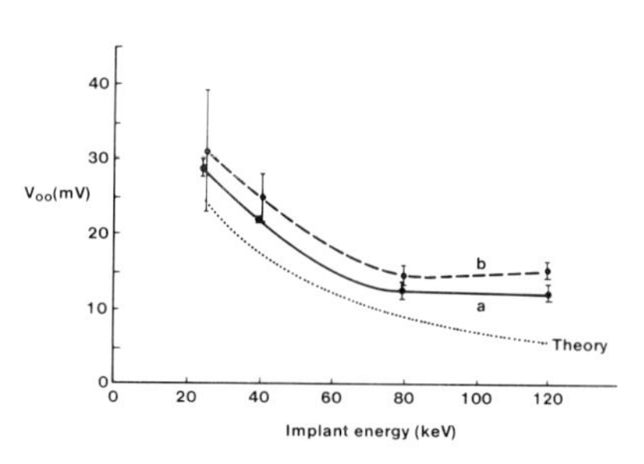

Figure 3. The variation of V_{oo} with implant energy for (a) 101 μm and (b) 1.2 μm gate length devices. A theoretical curve (see text) is included for comparison.

Consequently the barrier height variations are all empirically accounted for by the introduction of a T_o term. It should be noted that the image force barrier lowering is estimated to only contribute between 0.02 and 0.04 eV for the implantation conditions employed. Padovani (1971), Levine (1971) and Crowell (1977) have considered the physical significance of the excess temperature T_o and have to some extent linked it with the presence of traps at the interface. For example a high concentration of mid-gap trapping centres would produce a high and non-uniform interfacial space-charge density giving rise to a large electric field at the interface (Padovani 1971). An enhanced field at the interface would be a possible cause of the discrepancy between theory and experiment in Figure 3 and would also contribute to a larger image force lowering. A possible cause of the high interface state density is residual implantation damage near to the GaAs surface which is likely to become more pronounced as the implant energy is reduced.

5. CONCLUSIONS

It has been shown that the barrier height and ideality factor of a GaAs MESFET Schottky gate contact are degraded as the device is scaled. The principal cause of degradation is thermionic-field emission caused by high fields occuring at the gate edges. For implant energies greater than around 80 keV the barrier height is reduced as the ratio of diode periphery to diode area is increased. For low energy implants, The Schottky barrier height is reduced but appears roughly equivalent for both short and long gate length devices. These reductions are quite significant resulting in a barrier height of only 0.57 eV for a 25 keV implantation energy. The implications of this observation are quite serious for digital circuits based on E/D logic (e.g. DCFL) since a reduced EFET barrier height results in a reduced noise margin and hence a reduced yield. This latter effect may be associated with a high density of interface states and could represent a more fundamental limitation to the scaling of ion-implanted MESFETs.

6. ACKNOWLEDGEMENTS

The authors wish to thank the directors of STC Technology Ltd. for permission to publish this work. This work has in part been supported by the Procurement Executive, Ministry of Defence (Directorate of Components, Valves and Devices) and sponsored from the Royal Signals and Radar Establishment.

7. REFERENCES

Crowell C R 1977 Solid-State Electronics 20 pp 171-5
Levine J D 1971 Journal of Applied Physics 42(10) pp 3991-9
Padovani F A 1971 in "Semiconductors and Semimetals" Ed. Willardson and Beer (Academic Press, N.Y.) Chap. 2
Padovani F A and Stratton R 1966 Solid-State Electronics 9 pp 695-707
Saxena A N 1969 Surface Science 13 pp 151-171

Low frequency GaAs substrate phenomena and their effects on precision baseband analogue integrated circuits

W.S. Lee
STC Technology Limited
London Road, Harlow, Essex, CM17 9NA, England

Abstract The mechanisms of substrate phenomena occurring at low frequencies and their impact on the performance of precision GaAs baseband analogue ICs have been studied. The origin of the frequency dependent characteristic of the basic MESFET drain conductance which had hampered the design of simple high gain amplifiers and transient-free switching circuits has been identified, and a simple circuit solution successfully demonstrated. The effects of back-gating and the conditions for the generation of oscillatory substrate currents on two basic analogue circuit building blocks have been established. Simple guidelines have been established to minimise these effects in ICs.

Introduction

Recent developments of high precision baseband analogue integrated circuits based on GaAs MESFETS have revealed several performance-limiting effects occurring at low frequencies. These effects appeared to be unique to the present MESFET technologies based on semi-insulating (SI) substrates. A dominant feature is the frequency dependent characteristic of the basic FET drain conductance which has been shown to limit the performance accuracy of analogue circuits such as switched-capacitor filters, (Harrold et al 1985) and high speed analogue-to-digital converters (Duncourant et al 1984). Circuit instability resulting from back-gating via the SI substrate was also found to be important in complex ICs. A third anomalous behaviour which could limit the usefulness of GaAs analogue ICs at low frequencies is the generation of oscillatory currents in the substrate and its subsequent coupling into the active devices.

The objective of this paper is to highlight the sensitivity of two typical analogue IC building blocks (i.e. single stage inverter amplifier and source follower) to these effects and to identify the critical components. Solutions by which these effects may be minimised are also discussed.

Frequency dependent drain conductance

The frequency dependent behaviour of the FET drain conductance has been reported previously (Camacho-Penalosa and Aitchison 1985). Our study on a variety of FETs of different (i) gate metallisation, (ii) surface passivation, and (iii) channel material (i.e. epitaxial or implanted layer) has shown that the response characteristics in both frequency and time domains were consistent with the electrostatic feedback effect involving the substrate and trapping centres residing near the substrate channel interface as proposed by Makram-Ebeid and Minondo (1985). Channel current transient versus temperature measurements made on implanted FETs revealed a dominant activation energy of 0.71 eV closely related to residual implantation damage, regardless of gate metallisation and surface passivation. Epitaxial devices exhibited different activation energies suggesting that this effect is dominated by process-induced defects, rather than intrinsic defects associated with the bulk materials.

A variety of approaches have been explored to counter this effect. Modifications to the basic device structure by isolating the channel with p layers (Canfield and Forbes 1986) have been reported with some success. The use of circuit techniques, some of which are well exploited in Si ICs have also led to some encouraging results. In the latter approach, the so-called self-bootstrapping negative feedback technique is particularly suitable for the present IC technologies based on MESFETS. By using a dual-threshold arrangement (Lee and Mun 1987) in the basic bootstrapping pair of FETs (Figure 1) a reduction in drain conductance (and hence drain current transient) by a factor of 11 has been demonstrated for low threshold ion-implanted devices ($V_{th} < -1V$). The impact of this approach in circuit design is exemplified by the gain characteristic of a simple inverter amplifier, Figure 2. By using the dual-threshold bootstrapping arrangement for both the driver and load, an increase in mid-band gain by 20-21 dB was realised. The roll-off at high frequencies was due to capacitive loading of the measurement apparatus (10.8 pF//10MΩ). Further, by choosing the correct threshold voltage combination (i.e. $V_{th2} \sim 2\ V_{th1}$), the composite arrangement may be replaced by a dual-gate structure, thus simplifying circuit layout and minimising the extra area incurred.

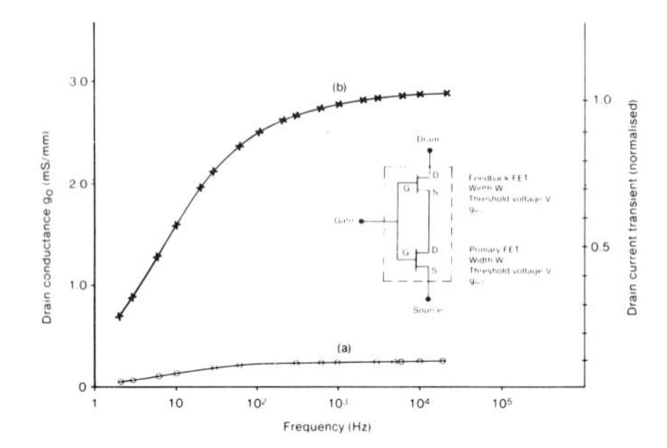

Fig. 1 Frequency variation of small signal drain conductance, (a) with and (b) without feedback

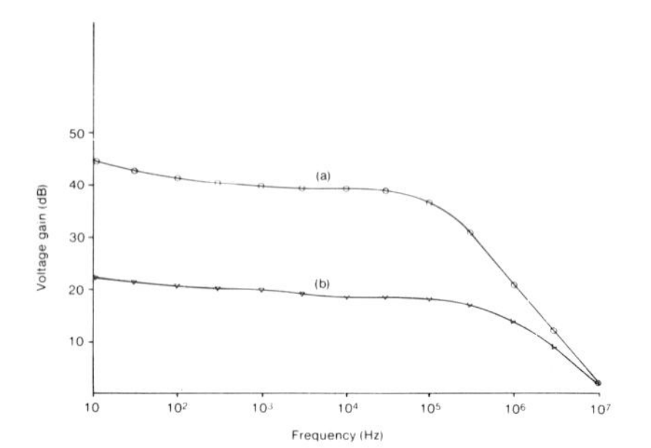

Fig. 2 Voltage gain charactertistic of simple inverter amplifiers, (a) with and (b) without local negative feedback

Back-gating instability

The mechanism of back-gating and its influence on the electrical characteristics of discrete FETs and the performance of digital integrated circuits have been studied extensively. Figure 3 shows a photograph and the circuit diagram of the two analogue building blocks investigated. External voltage bias was applied to side-contacts (CrAu/AuGeNi/n-implant) parallel to individual FETs to initiate back-gating. Details of circuit layout and the fabrication process were reported previously (Lee 1987).

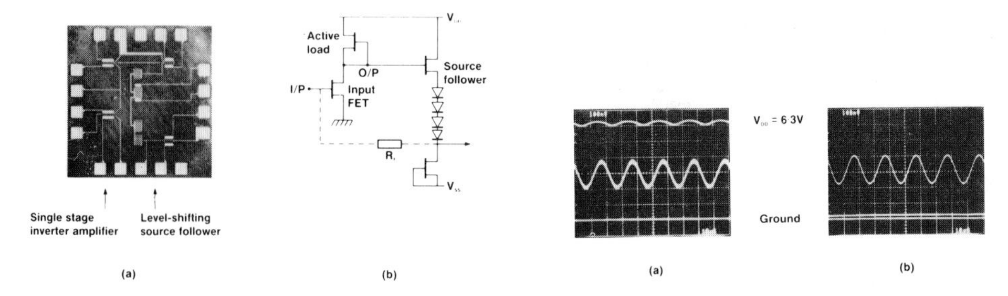

Fig. 3 Photomicrograph of single stage inverter amplifier/source follower IC

Fig. 4 Effects of side-contact potentials on the DC stability of the inverter amplifier: (a) +5V bias and (b) -5V bias. Side-contact separation = 5μm from drain

Figure 4 shows the inverter amplifier DC instability resulting from the application of side-contact potentials to the active load. Depending on the polarity of the bias, the FET was either saturated or pinched-off, driving the amplifier output to the supply rail or ground respectively. The bias levels required to saturate or pinch-off the FET were typically +8 and -5V respectively, for the contact 5 μm adjacent to the drain and was generally 2 to 3V higher for the contact 10 μm next to the source. The effect of negative side-contact bias adjacent to the amplifier input FET was opposite to that displayed by the active load. The input FET was pinched-off driving the amplifier output to the supply level. The voltage level required was similar to above. However, a bias in excess of +30V was needed to saturate the FET. In contrast, the effect of side-contact potentials on the source follower FETs was relatively small. Less than 200 mV shift in output DC voltage was induced by the application of ± 30V on the side-contacts.

This behaviour is consistent with the back-gating effect invoking the field-induced modification of the depletion region existing at the channel/substrate interface region. In the case of the inverter amplifier, this effect was aggravated by the internal gain mechanism, since the field-induced changes in the threshold voltage are indistinguishable from the input signal. The source follower is much less susceptible since the internal gain is always less than unity.

Low frequency oscillations

The DC stability of the single stage amplifier against the influence of side-contact potentials was improved significantly by negative feedback via the level-shifting source follower output and an external resistor to the gate of the input FET. However, when the bias applied to the side-contacts adjacent to the inverter amplifier components was sufficiently negative (typically - 15 volts) low frequency oscillations at the amplifier output were observed. Figure 5 shows an oscillatory waveform superimposed on the high frequency (100 kHz) output signal and the corresponding oscillation spectrum. The dominant oscillation frequency varied between 2 to 4 hertz as the bias potential increased towards -30 V. The oscillation waveforms were generally more complex at higher bias voltages. No oscillations were observed for positive biases up to at least 30 V.

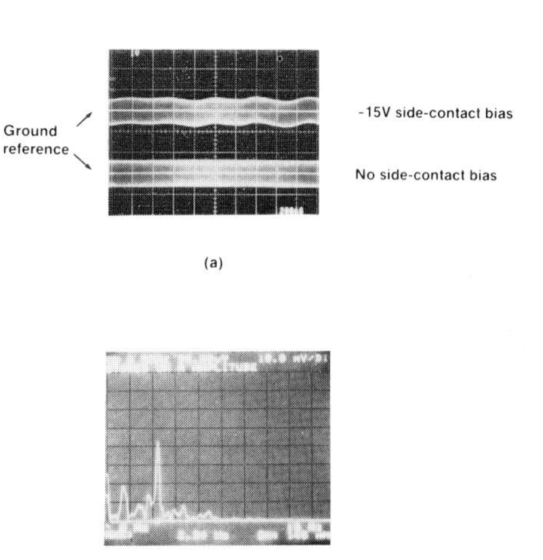

Fig. 5 Back-gating-induced oscillations in an inverter amplifier with negative feedback: (a) oscillation waveform and (b) oscillation frequency spectrum

The dependence of the oscillation phenomenon on the bias magnitude and polarity is consistent with the mechanism of electron injection into the substrate and subsequent transport via trapping centres. Under high field bias, the complexity of the oscillatory waveform suggests that several mechanisms including subsurface conduction may be involved.

Conclusions

From the results of this work, it is apparent that the dominant performance-limiting effects at low frequencies could be minimised by a combination of proper circuit design and layout. By replacing the single FET with a dual-threshold self-bootstrapping pair of devices, the magnitude of the drain conductance and hence drain current transient could be reduced. An improvement by a factor of 11 has been demonstrated. This circuit technique is expected to have a major impact in the design of GaAs MESFET-based baseband analogue ICs requiring low power/high gain amplifiers and transient-tolerant circuits.

It was also clearly demonstrated that an adequate separation between devices within any high gain stage and from other circuit components must be maintained to avoid back-gating. This is particularly important when negative feedback can not be implemented to ensure DC stability. Considerations should also be given to the location and magnitude of negative potential differences to avoid the generation of low frequency oscillations.

Acknowledgements

The author wishes to thank the directors of STC Technology Limited for permission to publish this work. This work has been supported by the Procurement Executive, Ministry of Defence (Directorate of Components, Valves and Devices) and sponsored from the Admiralty Research Establishment and the Royal Signals & Radar Establishment.

References

Camcho-Penalosa C., and Aitchison, C.S.: Electronics Letters, 1985, vol. 21, pp. 528-529.

Canfield P., and Forbes L.: IEEE Tran. on Electron Devices, Vol.ED-33, pp. 925-928. 1986

Duncourant T., Meignant D., and Binet M.: Int. Symp. GaAs and Related Compounds, Biarritz, 1984, pp. 659-664.

Harrold, S.J., Vance, I.A.W. and Haigh, D.G.: Electronics Letters, 1985, Vol. 21, pp 494-496.

Lee, W.S.: Electronics Letters, 1987, Vol. 23. pp. 587-589.

Lee, W.S. and Mun, J.: Electronic Letters, 1987. Vol.23, pp. 705-707.

Makram-Ebeid S., and Minondo P.: IEEE Tran. on Electron Devices, Vol. ED-32, 1985, pp. 632-640.

High efficiency 650nm aluminium gallium arsenide light emitting diodes

L W Cook, M D Camras, S L Rudaz, and F M Steranka

Hewlett Packard, Optoelectronics Division, 370 West Trimble Rd., San Jose, CA 95131, USA

ABSTRACT: An LPE grown double heterostructure AlGaAs red (650nm) LED with an AlGaAs transparent substrate has been demonstrated with an external quantum efficiency of 18% at 300 K and 50% at 90 K. The luminous efficiency at 300 K of this device is 30 lm/A and the optical output power is 7mW at 20mA. This paper will discuss the dependence of efficiency on both current and temperature. The degradation performance of these devices will be shown.

1. INTRODUCTION

AlGaAs red LEDs are now being produced with efficiencies significantly higher than other visible wavelength LEDs. This is primarily due to the ability to make high quality heterojunctions by LPE in this nearly lattice matched system. However, at these high Al compositions the AlGaAs bandgap is near the direct-indirect crossover, which makes the material quality very important (Varon et al 1981, Nishizawa et al 1983, and Kaliski et al 1985). Nishizawa et al (1977) demonstrated that material of high quality could be grown by temperature difference LPE on a production scale to manufacture efficient AlGaAs red LEDs. Single heterostructure AlGaAs devices exhibit encapsulated external quantum efficiencies of 1-3% (Nishizawa et al 1977 and Varon et al 1981). Double heterostructure devices increase the efficiency to 3-7% (Nishizawa et al 1983). Ishiguro et al (1983) obtained efficiencies of 8% by growth of the double heterostructure on a transparent AlGaAs substrate. We report on transparent substrate double heterostructure LEDs grown by slow-cooling LPE with efficiencies as high as 18%. This is the highest reported efficiency for a visible wavelength incoherent emitting diode to our knowledge. The efficiency of this device will be examined as a function of current, temperature, and time under stress conditions.

2. DEVICE STRUCTURE

Figure 1 depicts a transparent substrate double heterostructure AlGaAs LED. The transparent substrate is a very thick (100-150μm), high Al content ($x>0.4$) AlGaAs layer. The absorbing GaAs substrate is removed by selective chemical etch. This technique allows much of the light reflected internally to escape eventually from the chip and it increases the external efficiency by a factor of 2-3 over double heterostructure devices with GaAs substrates. Slow-cooling LPE is used to grow the entire structure. The composition of the transparent substrate varies from $x=0.75$ to 0.55 along the growth direction as determined by electron microprobe. The doping level of the transparent substrate is less than $1x10^{18}cm^{-3}$. These chips are about 265μm on a side and typically 125μm thick.

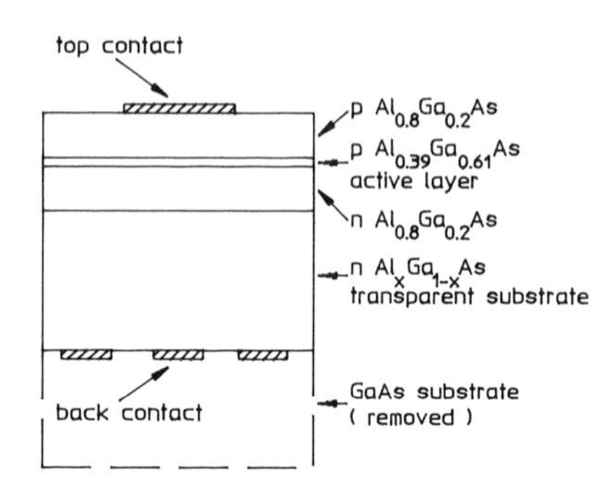

Fig. 1. Transparent substrate double heterostructure AlGaAs LED.

3. DEVICE CHARACTERISTICS

An electroluminescence spectrum for one of these LEDs is shown in Figure 2. The composition of the active layer is chosen to give a peak in intensity at 650nm. This wavelength is the optimal trade off between increasing eye response and decreasing internal quantum efficiency as the Al composition of the active layer is increased towards where the bandgap soon becomes indirect. As a result, the luminous efficiency is maximized, but the quantum efficiency of these devices is not as high as longer wavelength LEDs. Spectra of these devices typically have a FWHM of 18nm. The integrated luminous efficacy is 85 lm/W.

The dependence of room temperature luminous efficiency on forward current for plastic encapsulated LEDs is illustrated in Figure 3. The efficiency is observed to increase up to a current of 30-40mA where it reaches a maximum. The best device reaches a peak luminous efficiency of 30 lm/A and the optical power out of this diode is 7mW at 20mA. The measured external quantum efficiency of this device is 18%. Typically these devices reach a peak luminous efficiency of 20 lm/A. The relationship between the measured values of quantum efficiency and luminous efficiency for the best diode is consistent with the calculated luminous efficacy given above. Again, it

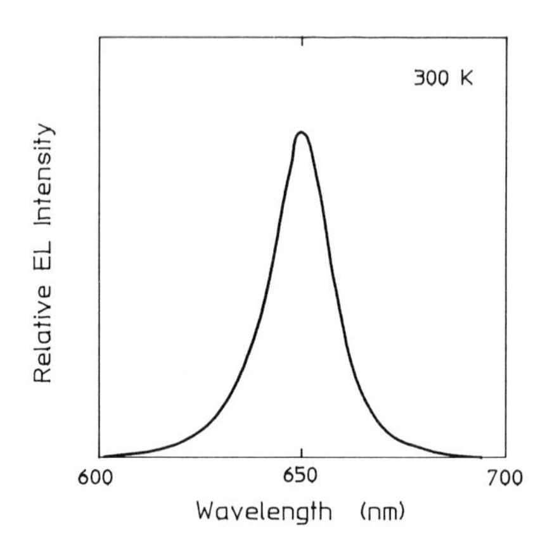

Fig. 2. Typical 300 K electroluminescence spectrum.

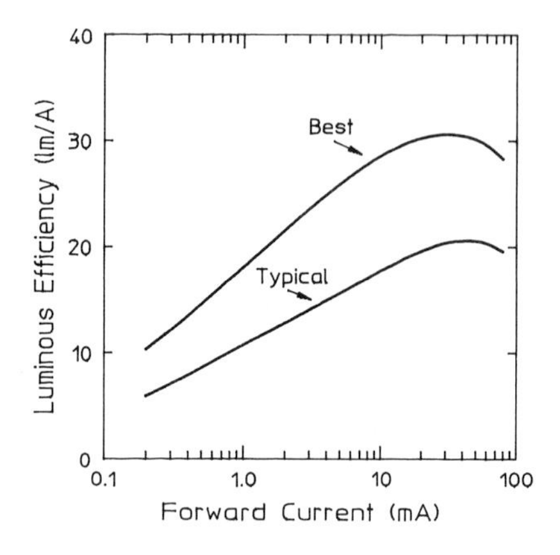

Fig. 3. Luminous efficiency at 300 K versus forward current for the best and a typical LED.

should be pointed out that the composition of the active layer was chosen to maximize the luminous efficiency. This composition is very close to the direct-indirect crossover. Therefore, an external quantum efficiency of 18% must indicate very high material quality.

The temperature dependence of the external quantum efficiency for the best diode is shown in Figure 4. It is observed that as the temperature decreases the difference between the high and low current efficiencies becomes smaller and the low current efficiency becomes slightly higher at the lowest temperature (90 K) where it reaches a value of 50%. This result shows that the internal quantum efficiency must be at least 50% at low temperatures. The internal quantum efficiency may be much higher than 50% (perhaps closer to 100%), but this cannot be concluded without first determining the extraction efficiency (fraction of generated photons that escape from the chip) which is not easily done. However, it can be concluded that the extraction efficiency is also at least 50% at low temperatures, since the external quantum efficiency is the product of the internal quantum efficiency and the extraction efficiency and since neither can exceed 100%.

The typical degradation characteristics of these LEDs are presented in Figure 5. In the test shown here, the units were stressed at a constant forward current of 30mA and at an ambient temperature of 55°C. The efficiency decreases monotonically with time and after 1000 hours it decreases by 17% from its initial efficiency. This is typical of most conventional LEDs. There was no evidence of dark-line-defect formation in any devices stressed at this current since this current density ($43A/cm^2$) is significantly below the current densities at which these defects usually form in AlGaAs LEDs or lasers.

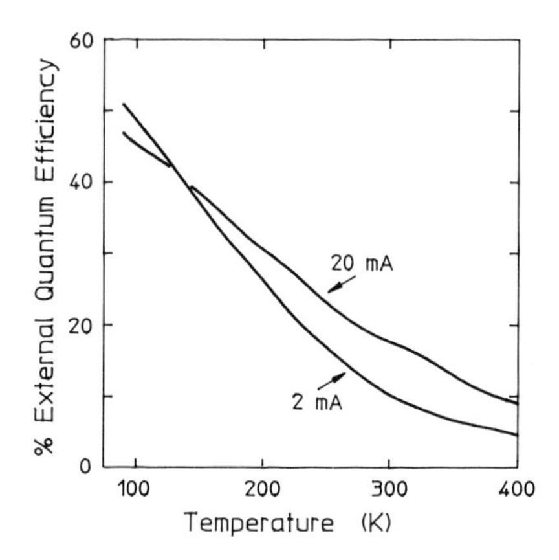

Fig. 4. Temperature dependence of the external quantum efficiency for the best diode.

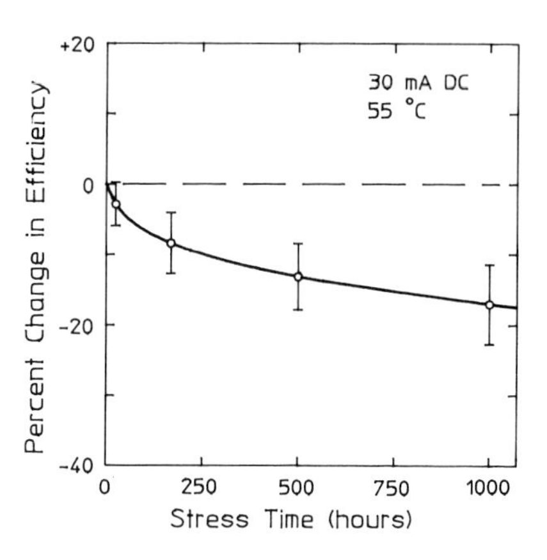

Fig. 5. Typical variation of efficiency with stress time at 30mA DC and 55°C.

4. CONCLUSION

Transparent substrate double heterostructure 650nm AlGaAs LEDs have been fabricated with external quantum efficiencies as high as 18% at room temperature. At low temperatures, the external efficiency reaches 50%. We believe this is the highest reported efficiency for a visible wavelength LED. It is accomplished despite the use of a material with a bandgap that is nearly indirect. These high brightness devices exhibit well behaved degradation characteristics and are expected to open up many new applications for LEDs.

ACKNOWLEDGMENTS

The authors wish to thank Drs M G Craford, R H Haitz, and W L Snyder for useful discussions and support, and A Davis, T Patterakis, and R L Pettit for technical assistance.

REFERENCES

Ishiguro H, Sawa K, Nagao S, Yamanaka H, and Koike S 1983 Appl. Phys. Lett. 43 1034

Kaliski R W, Epler J E, Holonyak N Jr, Peanasky M J, Herrmannsfeldt, G A, Drickamer H G, Tsai M J, Camras M D, Kellert F G, Wu C H, and Craford M G 1985 J. Appl. Phys. 57 1734

Nishizawa J, Suto K, and Teshima T 1977 J. Appl. Phys. 48 3484

Nishizawa J, Koike M, and Jin C C 1983 J. Appl. Phys. 54 2807

Varon J, Mahieu M, Vandenburg P, Boissy M, and Lebailly J 1981 IEEE Trans. Electr. Dev. ED-28 416

Monolithic integration of a GaInAs/GaAs photoconductor with a GaAs FET for 1.3–1.55 μm wavelength applications

*Razeghi M., **Ramdani J.,**Legry P.,**Vilcot J.P.,**Decoster D.

*LCR, Thomson-CSF, Domaine de Corbeville 91401 Orsay FRANCE ** CHS, Université des Sciences et Techniques de Lille Flandres Artois, 59655 Villeneuve d'Ascq Cedex FRANCE.

ABSTRACT: We present a monolithic integrated photoreceiver for long wavelength optical systems wich associates a GaInAs/GaAs strained photoconductor with a GaAs F.E.T. In order to reduce the dark current of the photoconductive detector, the GaInAs photosensitive layer has been deposited, by L.P.M.O.C.V.D., on the undoped GaAs layer. Static, dynamic and noise properties of the photoconductor and the integrated circuit are presented and discussed, taking into account the special structure of the material and integrated circuit. As an example, an amplification factor of 5 has been achieved for a 200 Ω bias resistor; the gain bandwith product of the I.C. is close to 1 GHz, which value is connected to electrode spacing of the photoconductor.

1. INTRODUCTION

Monolithic integrated photoreceivers are desirable devices for fiber-optic communication systems. Most photodetectors sensitive to the 1.3-1.55 micron wavelengths have been fabricated either on InP or Ge. But these detectors have the disadvantage that neither of these substrates have a well developed electronic integration technology. The purpose of this paper is to present an optoelectronic integrated circuit (O.E.I.C.) suitable for long wavelength optical communication systems (1.3-1.55μm) using $Ga_{0.47}In_{0.53}As$/ GaAs strained heteroepitaxies. It consists of a planar $Ga_{0.47}In_{0.53}As$ photoconductive detector associated in a classical common-source amplification (figure 2) with a GaAs field-effect

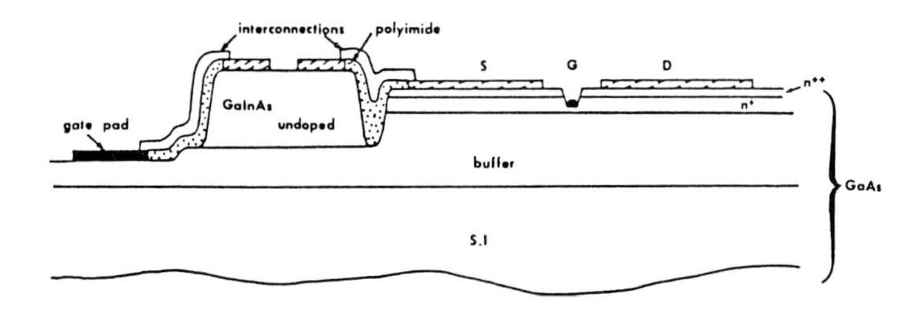

Fig. 1. Schematic cross section of the O.E.I.C.

transistor (FET). A first device has been previously proposed and fabricated (Razeghi 1986), with a $Ga_{0.47}In_{0.53}As$ layer grown on a classical GaAs FET epitaxy. But, for this device, the n^+ GaAs layer reduces the dark resistance; therefore, the GaInAs heteroepitaxy has been grown by L.P.M.O.C.V.D. in boxes previously etched in classical GaAs FET epitaxy (figure 1), in order to deposite the GaInAs active layer of the photoconductor on the GaAs buffer layer.

2. DEVICE FABRICATION

First, boxes are etched in the classical GaAs FET epitaxy and then the GaInAs (undoped, 1μm thick) layer is deposited by L.P.M.O.C.V.D. After these two steps, an ion milling of the GaInAs layer, except for the area corresponding to the photoconductor, delimits the photoconductor. A

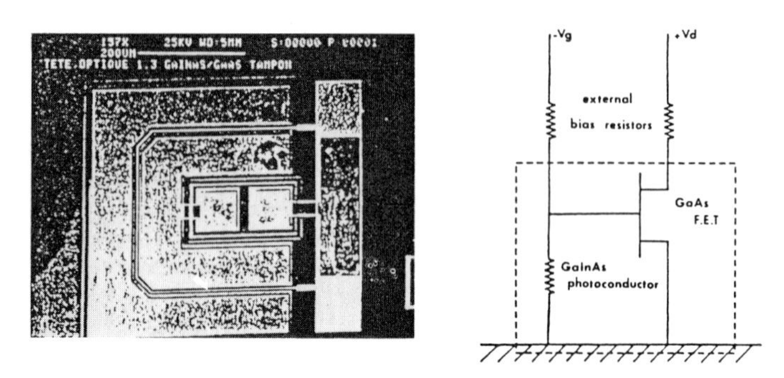

Fig. 2. Photograph and electrical circuit of the O.E.I.C.

planar structure is then obtained. A schematic cross section of the device is shown in figure 1. The two ohmic contacts (AuGeNi 2000 A) of the photoconductor are separated from each other by 20μm, leading to a 20μmx80μm photosensitive area. The FET has been fabricated on the GaAs layers. To obtain good amplification behaviour at frequencies up to several gigahertz, the FET has a 2μmx900μm gate (Ti 500 A; Pt 300 A; Ti 300 A; Au 2000 A). To make a very compact design integrated circuit, the FET surrounds the photoconductor. The interconnection between the photoconductor and the FET is achieved using a polyimide bridge. A scanning electron microscope of the integrated circuit is given in figure 2. On the same ship a photoconductor alone was also processed in order to perform experimental photoconduction evaluation.

3. EXPERIMENTAL RESULTS

We have verified that the dark resistance of the photoconductor is given by the GaInAs epitaxy characteristics. Typical photoconductor steady state gain (defined as the number of charges collected in the external circuit per incident photon) are given in figure 3. We observe that the gain increases when the bias voltage increases, corresponding to a reduction of the transit time τ_t, according to the expression of the gain: $G= \tau_v / \tau_t$ where τ_v is the electron-hole pair lifetime. The value of the gain is rather low and is related to a short response time (figure 4), approximatively given by τ_v. This short response time could be explained by trapping effects at the GaInAs/GaAs interface. The dynamic

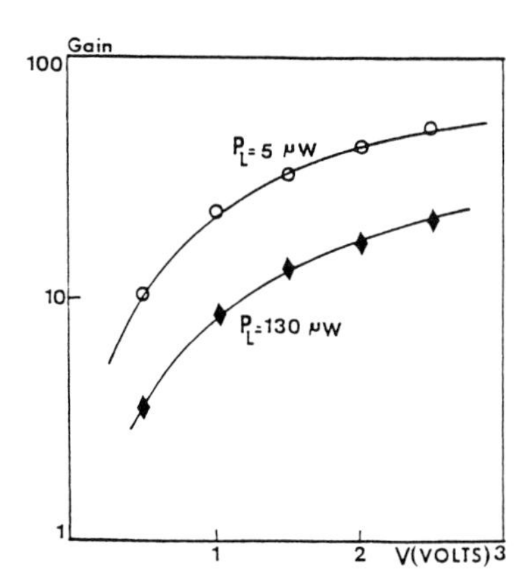

Fig. 3. Photoconductor steady state gain versus bias voltage λ = 1.3μm

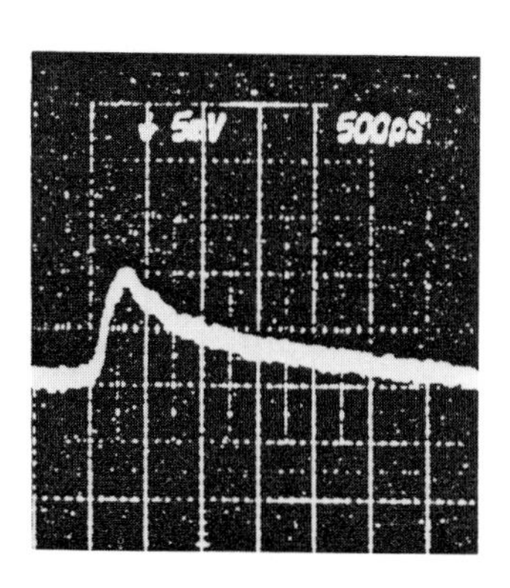

Fig. 4. Photoconductor picosecond response. λ = 1.3μm, P_L = 10μW V = 2V

gain is reported in figure 5. The FET transistor has also been characterized alone and its transconductance is approximatively equal to 60mS. Its amplification capabilities can be observed on the dynamic gain of the O.E.I.C.; as an example, an amplification factor of 5 has been achieved with a 200 Ω gate bias resistor; the gain bandwidth product of the O.E.I.C. is close to 1GHz (figure 5), which value is related to the

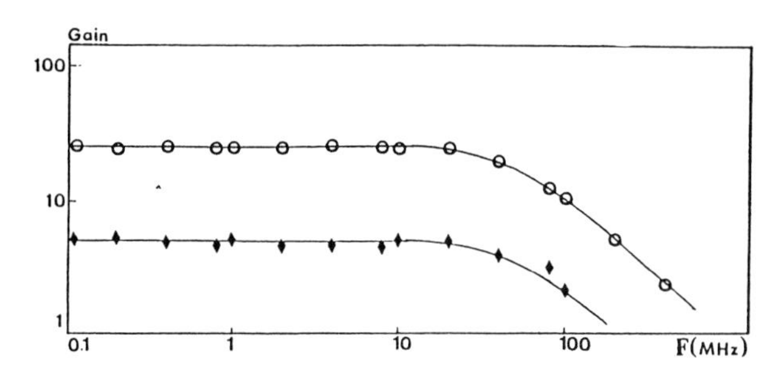

Fig. 5. Dynamic gains vs frequency. = 1.3μm, P_L = 100μW. (a) Photoconductor V = 2V; (b) O.E.I.C. Vds = 3V, Vgs = -2V.

electrode spacing of the photoconductor. Obviously, an increasing of the dark resistance leads to a better amplification factor (gm x R product); another advantage of growing the GaInAs layer in boxes, on the buffer GaAs epilayer, has been noticed on the noise level of the photoconductor which has been found lower than for a photoconductor which GaInAs layer has been deposited directly on the n^+ GaAs layer. From noise level measurements performed on the O.E.I.C. in the 10MHz - 1.5GHz frequency range and the dynamic responsivity; the optical sensitivity has been evaluated versus the data rate for a 10^{-9} bit error rate (figure 6); a comparison with results obtained elsewhere for other integrated photoreceivers (Ohnara, 1985; Renaud, 1987) shows that the performance of our O.E.I.C. is similar.

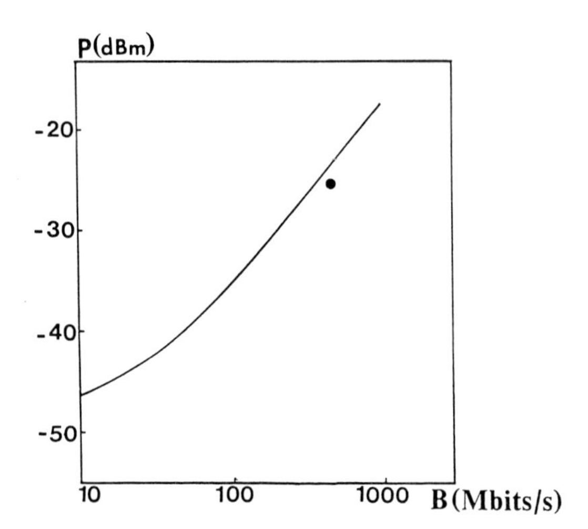

Fig. 6. O.E.I.C. minimum detectable power vs data rate (10^{-9} BER). Results for PINJFET (• Ohnaka)

4. CONCLUSION

We have shown that it is possible to fabricate planar monolithic integrated photoreceiver for 1.3 - 1.55μm wavelength applications, using strained GaInAs/GaAs heteroepitaxies, whose performances are comparable to those obtained elsewhere. Moreover, several improvements can be proposed to obtain better performance: for example using interdigitated structure for the design of the photoconductor to reduce the transit time of the carriers and by coupling the light with an optical waveguide (Mallecot, 1987)

5. REFERENCES

Razeghi M., Ramdani J., Verriele H., Decoster D., Constant M., and Vanbremeersch J., 1986 Appl.Phys.Lett. 49, pp.215-217

Ohnalsa K., Inoue K., Uno T., Hasegawa K., Hasen and Serizawa H., 1985 IEEE J.QUE, 8, 21

Renaud JC., Nguyen L., Allown M., Blanconnier P., Lugiez F. and Scavennec A., 1987 Paper presented at this conference, DO (1) 6

Mallecot F., Vilcot J.P., Decoster D. and Razeghi M., 1987 Proceedings of the 17th ESSDERC, Bologna Ed.G.Soncini.

LP-MOCVD multiplexed SWIR $In_xGa_{1-x}As$ photodiode arrays for spot IV satellite

M.A. di Forte-Poisson, C. Brylinski, and P. Poulain
Laboratoire Central de Recherches, Thomson-CSF, 91401 Orsay, France
and
J.P. Moy, M. Villard, S. Chaussat, J. Decachard and B. Vilotitch
Département Tubes Electroniques, Thomson-CSF, 38120 Saint-Egrève, France

ABSTRACT: A linear array of 3000 multiplexed GaInAs photodiodes is being developped for earth remote sensing in the 1.55µm - 1.72µm band. We report the performances of each elementary module. The individual photosensors are planar pin double heterojunction, prepared by LP-MOCVD growth. The structure is composed of an absorbing GaInAs layer between InP buffer and window layers. Typical dark current density at a reverse bias of 5V is lower than 5.10^{-8}A cm^{-2} at 300°K. Spectral response covers the whole 1.55 µm - 1.72µm range with an external quantum efficiency still equal to 0.8 at 1.70µm.

1. INTRODUCTION

Earth imaging in the short wavelength infrared (SWIR) band 1.55µm - 1.72µm presents growing interest because it will provide information upon the water content of vegetation and soil .

Consequently, the incorporation of a SWIR channel in the high resolution imaging instrument spot IV satellite (JP Moy, 1985) has been decided by the French Space Administration (CNES).

The main features of the sensor are :

- 3000 $In_{0.55}Ga_{0.45}As$ detectors 30x30µm^2 multiplexed in the focal plane, staggered and aligned at a pitch of 26µm, corresponding to a ground resolution of 20 m.
- Signal to noise ratio allowing the measurement of ground albedo with an accuracy of about 1%.
- Operation close to 300K.

As a monolithic chip array of 3000 detectors is not achievable, the complete sensor was decomposed into ten modules each of 300 photodiodes.

LP-MOCVD is used to prepare GaInAs/InP photodiode heterostructure sensitive to wavelength as long as 1.72 µm. In that case the lattice mismatch between the active InGaAs layer and the InP layers is 2.10^{-3}.

High quality GaInAs epitaxial layers both lattice matched and lattice mismatched on InP are obtained by LP-MOCVD.

Performances of a multiplexed linear array of 3000 $Ga_{1-x}In_xAs$ photodiodes are presented. Our results illustrate the potential of LP-MOCVD technique for the large yield production of high performance optoelectronic components.

2. SHORT WAVE INFRARED DETECTOR

2.1. Crystal growth

Various kind of sensitive detectors in the 1.55μm - 1.72μm band have been investigated. Among different materials that could be considered, $Ga_{1-x}In_xAs$ grown on InP by LP-MOCVD appeared to be the most promising (P. Poulain, 1984 - N.D. Scott, 1986). However, the bandgap of the InP-matched alloy $Ga_{1-x}In_xAs$ (x = 0.53) is too large to cover the above mentioned wavelength range and an alloy with an In mole fraction of 0.55 must be used. This results in a lattice mismatch of $\Delta a/a = 2.10^{-3}$.

The LP-MOCVD growth parameters have been optimized to obtain high quality GaInAs latice mismatched on InP and to minimize the density of interfacial cristallographic defects.

Some critical points for LP-MOCVD process are the quality of sources (TEI, TEG, A_sH_3, InP substrate) and appropriate growth conditions (T_{growth}, P_{AsH_3}, $\Delta a/a$, H_2-N_2 main flow).

By controlling these critical parameters in epitaxial GaInAs/InP heterojunction, no significant enhancement of the dislocation density has been observed if compared with the substrate.

Lattice matched GaInAs/InP heterojunctions exhibit background doping levels as low as $10^{14}cm^{-3}$ for InP and $8.10^{13}cm^{-3}$ for GaInAs. High resolution photoluminescence measurements performed on these GaInAs layers at 4.9K have shown a near bandgap emission line width of 2 meV : Figure 1. The Hall mobility is as high as $80.000 cm^2V^{-1}s^{-1}$ at 77K and $12000\ cm^2V^{-1}s^{-1}$ at 300K.

The composition of $Ga_{1-x}In_xAs$ lattice-mismatched on InP checked by optical measurement is very homogeneous over the whole wafer ($\Delta\lambda < 0.01\mu m$ on $10 cm^2$).

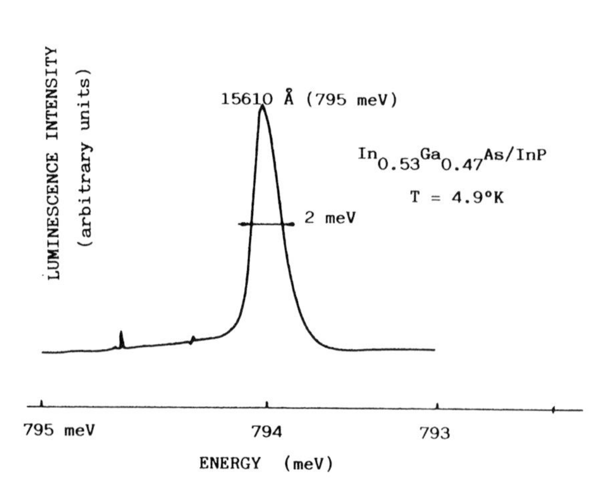

Such an homogeneity and a low concentration of interfacial defects allow the fabrication of linear arrays of more than one thousand photodiodes with an acceptable yield.

Fig. 1. 4.9°K Photoluminescence spectrum.

2.2. Planar pin photodiode

The $In_{0.55}Ga_{0.45}As$ photodiodes are of the planar double heterostructure type. The epitaxial heterostructure consists of an epitaxial Si doped $In_{0.55}Ga_{0.45}As$ absorbing layer (3 μm thick) followed by an InP Si doped window layer (1 μm thick) on a InP Si doped buffer layer (2μm thick) grown on a (100) InP S or Sn doped substrate. A planar process has been developped to fabricate the photodiodes (P. Poulain, 1987).

The p^+n junction is formed using a zinc diffusion through windows opened in SiO_2 layer by a standard photolithographic technique. A schematic cross section of the planar device structure is shown in fig. 2.

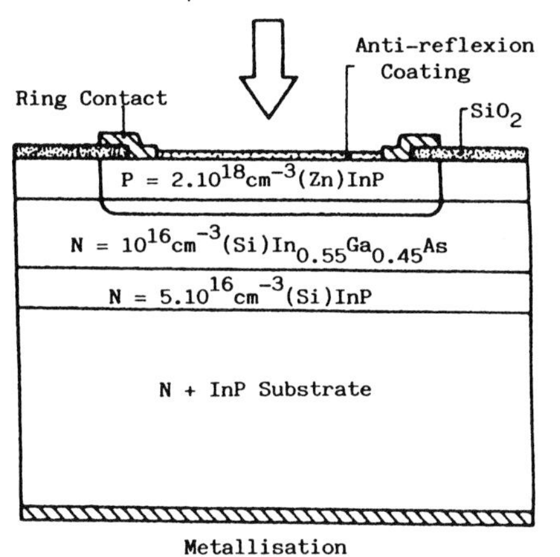

Fig. 2. Schematic representation of the photodiode heterostructure

The dark current density versus reverse voltage of a typical planar $In_{0.55}Ga_{0.45}As$ pin is shown figure 3. At a reverse bias of 5V, it is about $5.10^{-8}Acm^{-2}$ at 300K. As these photodiodes operate in a very low frequency range : 0 - 167 Hz, the low frequency noise is a critical parameter and the dark current must be kept very low to minimize shot noise. The frequency dependence of the noise has been investigated. No appreciable 1/F noise is observed down to few Hertz. The noise was found to be essentially the shot noise induced by the dark current.

The spectral response, covers the whole 1.55μm - 1.72μm range with an external quantum efficiency still equal to 0.8 at 1.70μm.

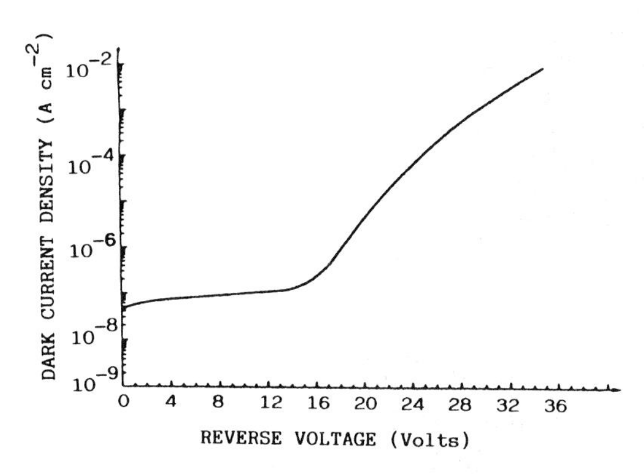

Fig. 3. Reverse I-V characteristic of a photodiode

These photodiodes have passed through aging tests (1000 hrs - 125°C) without significant degradation.

2.3. Detector arrangement

The independent elementary modules of 300 detectors consist of :

. a sapphire substrate
. a linear array of 300 InGaAs photodiodes
. two silicon CCD_S for the electronic readout of the two staggered lines of photodiodes.

The photodiodes are individually connected by wire bonding to each corresponding cell of the CCD.

These elementary modules are fully tested (spectral response, noise, dark current) before building the complete detector.

Figure 4 shows an histogram of the dark current along six such lines and gives typical example of the homogeneity. There is a very good correlation between pixels having visible fabrication defects and those whose dark current exceed the specification of 100 pA.

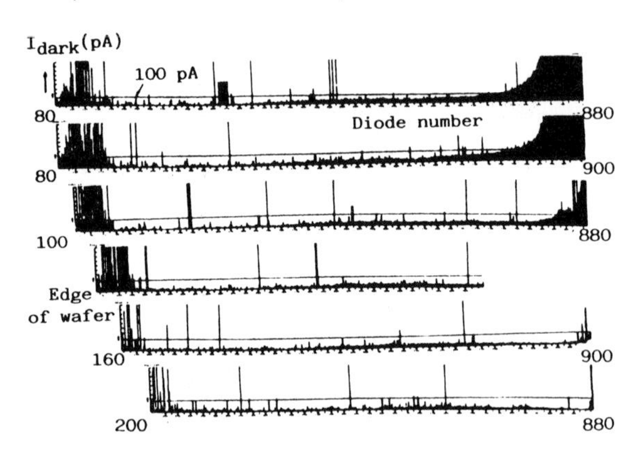

Fig. 4. Histogram of dark current along six lines of 1000 photodiodes

3. CONCLUSION

We have demonstrated the feasibility of a large GaInAs SWIR detector array operating at room temperature.

Prototypes of elementary modules and of the full working line of 3000 detectors have been already delivered. Spot IV flight models are scheduled for 1989.

ACKNOWLEDGMENT

We would like to thank M. Tordjman and A. Romann for technical assistance and D. Pons and N. Proust for helpful discussions. This work was supported by the French Space Administration (CNES).

REFERENCES

P. Poulain, MA di Forte-Poisson, K. Kazmierski and B. de Cremoux, 1984, Presented at Int. Symp. GaAs and Related Compounds, Biarritz.

JP. Moy, J. Rebondy, JP. Reboul, M. Villard, K. Kazmiersky, MA. Poisson, P. Poulain, 1985, Presented at 36th Int. Astronautical Congress, Stockholm.

M.D. Scott, AH. Moore, AJ. Moseley and RH. Wallis, 1986 Journal of Crystal Growth 77, 606-612.

P. Poulain, MA. di Forte-Poisson, K. Kazmiersky, B. de Cremoux, 1987, presented at Opto.87.

Paper presented at Int. Symp. GaAs and Related Compounds, Heraklion, Greece, 1987

A new low capacitance transverse junction stripe AlGaAs/GaAs laser for planar laser-MESFET integration

F. Brillouet, E.V.K. Rao, J. Beerens, Y. Gao

Centre National d'Etudes des Têlécommunications
Laboratoire de Bagneux
196 avenue Henri Ravera - 92220 BAGNEUX - FRANCE

ABSTRACT : A new transverse junction stripe laser structure compatible with a simple entirely planar laser-MESFET integration is presented. The laser, obtained by two successive P and N diffusions across undoped AlGaAs/GaAs double heterostructure layers, has a threshold current of Ith=55mA and a very low parasitic capacitance Cs=0.6 pF.

1. INTRODUCTION

One of the main benefits of a laser integrated with a driving circuit is to overcome the limits of modulation bandwidth due to parasitic inductance of a hybrid connection. However, as a prerequisite, high speed transistor and laser structures must be defined, each other being compatible with an integration process.

Up to now, most of integrated structures consisted of a short gate length MESFET (L = 1μm) connected side by side to a laser where carriers are vertically injected into the active region. In this case a rather sophisticated technological process must be elaborated to realize a quasi-planar integrated structure (Brillouet et al, 1985).

In this paper we describe the technological process and characteristics of a Double Diffused Transverse Junction Stripe (DDTJS) laser with a horizontal carrier injection and a very low parasitic capacitance, directly compatible with an entirely planar structure for Opto-Electronic Integrated Circuit (OEIC). The same undoped epitaxial layers, which lead to the DDTJS laser by successive P and N diffusions, can also be employed as the MESFET structure (Brillouet et al, 1986).

2. THE DDTJS LASER STRUCTURE

The DDTJS structure is a derivative of the standard TJS structure (Kumabe et al, 1979) where the P-N junction is obtained by a localized diffusion of Zn in a uniformly N-doped double-heterostructure.

In the DDTJS laser, the epitaxial layers are nominally undoped (except the active layer which can be N-doped) and the laser stripe is defined by localized diffusion of two separate P and N regions so that a stripe -s- is kept undiffused between P and N regions (Furuya et al, 1987) (fig. 1). The main advantages which can be expected from the undiffused stripe -s- are linked to the undoped cladding layers which provide a higher leakage resistance and a lower parasitic capacitance in comparison with a standard TJS laser.

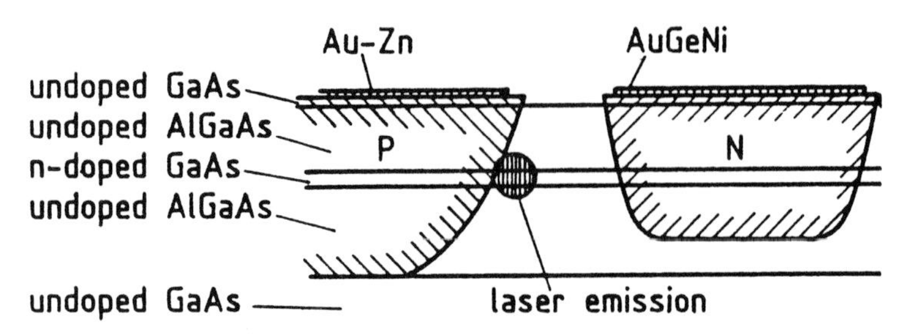

Fig. 1 : DDTJS Laser structure.

3. P AND N OPEN-TUBE DIFFUSION : APPLICATION TO DDTJS LASER

In previous experiments, transverse junction lasers with undoped cladding layers have been realized by two liquide phase epitaxy (P and N successively) selective regrowth (Suzuki et al, 1987) or by double diffusion (Si and Zn) in a vacuum sealed ampoule (Furuya et al, 1987). Here an open-tube diffusion technique is developed both for P-type (Zn) and N-type (S) diffusion. The same experimental configuration is used for Zn and S diffusion : the sample is positionned in the proximity of a solid diffusion source inside a graphite box which is closed by sliding a graphite lid after oxygen evacuation ; all diffusions are performed under Ar-H_2 flowing gas.

3.1 Zn diffusion in GaAs and AlGaAs

Diffusion of Zn at T = 650°C, using a powdered mixture of $ZnAs_2$ and GaAs as source, yielded typically a surface carrier concentration $p=10^{20}cm^{-3}$ and a diffusion velocity $D_{Zn}=1.2\ 10^{-11}cm^{-3}/s$. For a DDTJS laser, Zn must diffuse down to the N-type active layer, accross the GaAs undoped cap and $Al_{0.6}Ga_{0.4}As$ undoped cladding layers. Examples of Zn and Al depth profile, obtained by Secondary Ion Mass Spectroscopy (SIMS) after a .25 hour Zn diffusion at T=650°C are shown on fig. 2. The lower solutibility limit and enhanced diffusivity of Zn in AlGaAs with respect to GaAs, can explain the abrupt change of (Zn) at the first GaAs/$Al_{0.6}Ga_{0.4}As$ and $Al_{0.6}Ga_{0.4}As$/GaAs interfaces. The steep decrease of (Zn) at the bottom of the N-doped active layer is probably a consequence of a retarded Zn diffusion by N-type background impurity.

3.2 S diffusion

Using the same technique previously described, S is diffused in GaAs and AlGaAs with a mixture of Ga_2S_3 and GaAs at T=850°C as reported by Rao et al (1985). The carrier concentration profile of S diffusion in GaAs (fig. 3) is close to an error function complement curve with a diffusion coefficient $D_S=5.10^{-13}cm^{-3}$ and a surface concentration $n_S=10^{18}$ cm^{-3} (Rao et al, 1987).For the DDTJS technology, unlike D_{Zn}, since D_S is little dependent on S concentration and Al fraction in $Al_xGa_{1-x}As$, it leads to a steeper lateral profile on the N-side than on the P-side.

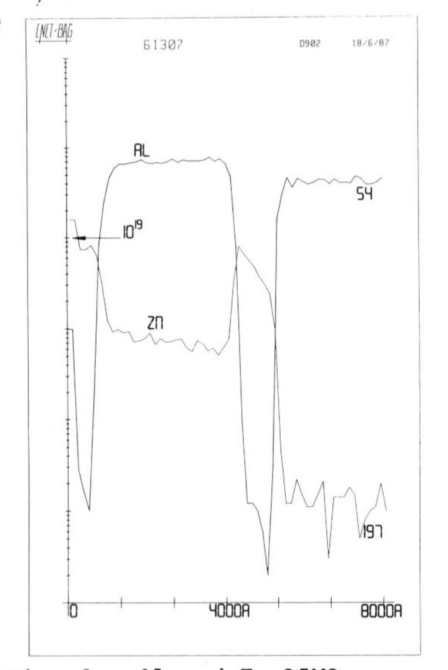

Fig. 2 : Al and Zn SIMS profile after Zn diffusion.

4. TECHNOLOGICAL PROCESS

Molecular Beam Epitaxy is used to grow the following layers : 1.5µm undoped GaAs buffer layer, 1.5µm undoped $Al_{0.4}Ga_{0.6}As$ first cladding layer, 0.2µm N-doped active layer, 0.5µm undoped $Al_{0.6}Ga_{0.4}As$ second cladding layer and 0.1µm undoped GaAs cap layer. Then the N-side of the laser junction is realized by a localized sulfur diffusion at T=850°C at atmospheric pressure. In these conditions, a 2000Å thick layer acts as an efficient mask. Si itself does not penetrate in the semiconductor as verified by SIMS and in the same time prevents sulfur diffusion.
After a two hours sulfur diffusion, the Si-mask is removed by plasma-etching and a new Si-mask is patterned so that Zn is diffused through openings adjacent to the N-diffused region.

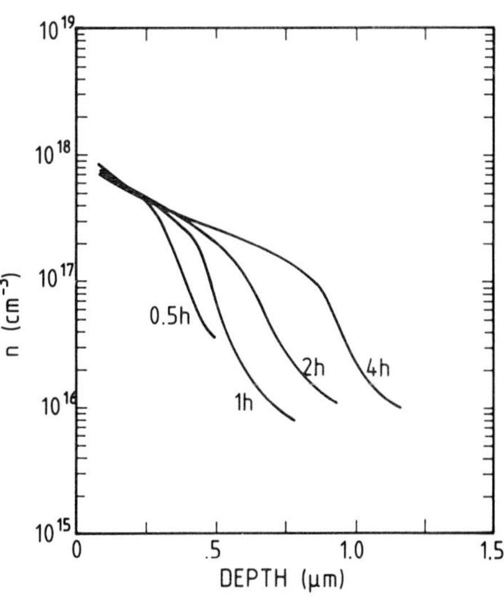

Fig. 3 : Donor concentration profile after S diffusion.

The Zn diffusion occurs at T=650°C for half of an hour in the same conditions as mentioned.

In this structure, the laser emission is located at the interface between the Zn-diffused region and the N-doped active layer. In order to obtain a low parasitic capacitance structure, an undiffused stripe -s- must be kept between P and N regions. In our experiment, the distance between the two masks for P and N diffusion is fixed at 4µm, and the amount of lateral P and N diffusion under the Si-mask is experimentally kept within 1µm. As a result, and taking account of the P and N depth diffusion profile, a 3µm undiffused stripe is obtained at the level of the active layer.

Au-Zn and Au-Ge-Ni are evaporated respectively on P and N diffused regions, and alloyed for ohmic contacts ; then the cap layer between the two contact pads is selectively etched.

5. LASER CHARACTERISTICS

A plot of the optical output power versus injected current is shown on fig. 4, for a cavity length of L=200µm : a threshold current I_{th}=55 mA with a differential quantum efficiency η_D=14 % per face up to P=10 mW is measured. These characteristics can be compared with those obtained by Suzuki et al. (1987) in the LPE buried transverse junction laser, and a lower threshold current can be expected with a quantum-well active layer.

The near-field pattern of the laser is single-transverse mode with a slight lateral asymetry due to a reduced optical loss on the N-side of the homojunction.

One of the main advantages of the DDTJS lies in the undoped cladding layers so that a very low parasitic capacitance (Cs = 0.6 pF) is measured. This capacitance value, combined with a series resistance of Rs = 15 Ohms results in a very low roll-off time constant : $\tau r = R_S C_S$ = 9 pS, to be compared with τr = 60 pS for a ridge laser structure (Tucker et al. 1984).

6 . CONCLUSION

A DDTJS laser has been realized by two successive local Zn and S open-tube diffusion in a double heterostructure with undoped cladding layers. These lasers, with a threshold current of I_{th}=55 mA, present a very low roll-off time constant (τr=9 pS). In addition, the DDTJS structure can be directly integrated with a FET where the same active layer can be used as the channel of the transistor, and Drain and Source obtained by the same S diffusion as used for the laser junction (Brillouet et al. 1986).

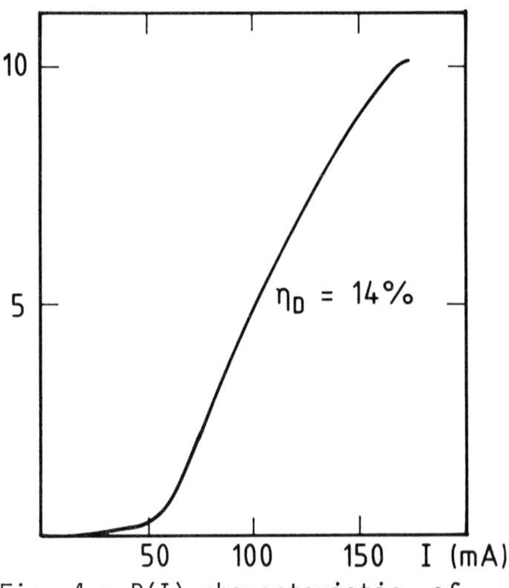

Fig. 4 : P(I) characteristic of a DDTJS laser.

REFERENCES

Brillouet F, Clei A, Bouadma N, Lefevre R, Azoulay R, Alexandre F and Duhamel N, 1985 SPIE, Vol. 587, p. 164.

Brillouet F, Rao E.V.K. and Alexandre F, 1986 French Patent n° 8600089.

Furuya A, Makiuchi M, Wada O, Fujii T and Nobuhara H, 1987 Jpn. J. Appl. Phys. Vol. 26, n°2, p. L 134.

Kumabe H, Tanaka T, Namizaki H, Takamiya S, Ishii M and Susaki W, 1979 Jpn. J. Appl. Phys. Vol 18, p. 371.

Rao E.V.K, Thibierge H, Brillouet F, Alexandre F and Azoulay R, 1985 Appl. Phys. Lett. 46(9), p. 867.

Rao E.V.K and Brillouet F, to be published.

Suzuki Y, Mukai S, Yajima H and Sato T, 1987 Electron. Lett., Vol. 23, n°8, p. 385.

Tucker R, and Kaminow I, 1984, J. of Light. Tech., Vol. LT-2, n°4, p. 385.

Inst. Phys. Conf. Ser. No. 91: Chapter 7
Paper presented at Int. Symp. GaAs and Related Compounds, Heraklion, Greece, 1987

Phase-locked index-guided semiconductor laser arrays

J.Opschoor, R.R.Drenten, C.J.Reinhoudt and C.J. van der Poel
Philips Research Labs, P.O. Box 80.000, 5600 JA, Eindhoven,
The Netherlands.

Abstract: Based on VSIS AlGaAs laser elements, as grown by LPE techniques, the performance of several phase-locked laser array geometries has been measured. The relation between array performance and growth parameters and tolerances is discussed. In an array of 5 parallel stripes, the near-field intensity distribution is manipulated by the spacing between the emitters. A uniform near-field distribution is achieved.

Catastrophic damage at the mirror facet limits the maximum available light output of an injection laser. In phase-locked laser arrays, the light power is distributed over several, closely spaced, emitters to form a coherently radiating, high brightness light source. Several distinct radiation patterns, so called supermodes [Yariv,1985] , are possible in a laser array. Whether or not such a supermode will reach threshold for lasing depends on its modal gain, and thus on the array geometry. For optical pumping purposes [Cross,1987] , the overall radiation pattern is not of great importance. However, applications like optical recording and laser beam printing require ,in addition to high power output, light beams of high optical quality so that in this case the laser-array has to operate in a single supermode. With gain-guided arrays [Scifres,1982] stability and beam quality cannot be guaranteed and appropriate index-guided array geometries have to be selected.

In parallel-stripe array configurations the distances between the subsequent emitters can be chosen such that favourable array properties result [Streifer,1986] . In such geometries it appears that precise control over the waveguide definition of the array elements is required. In this paper we study LPE grown chirped 5 emitter arrays, as composed with the VSIS laser [Hayakawa,1982] as basic element. The impact of LPE growth parameters on array performance is emphasized.

1. Growth parameters and tolerances.

The basic, well-known,element in building our arrays is the V-grooved Substrate Inner Stripe laser (VSIS), as grown by Liquid Phase Epitaxy (LPE). On a p-type GaAs substrate a n-GaAs current blocking layer (CBL) is grown. Stripe openings corresponding to the array geometry are etched into the CBL. Subsequently, a p-$Al_{53}Ga_{47}As$ confinement layer,a $Al_{14}Ga_{86}As$ active layer and n-type $Al_{53}Ga_{47}As$ confinement and GaAs contact layers are grown. A single emitter VSIS laser of this type has ,typically, a threshold current of 50 mA ,an emission wavelength of 780 nm and an external differential efficiency per facet of 0.3 W/A. A cross-section of a regular 5-emitter array with 3 um wide emitters on 5 um spaced centers is shown in figure 1(a).

Figure 1. Cross section of laser arrays.: (a) Regular 5 emitter array : spacing 5 um
(b) Chirped array : Center emitter 3.5 um

The thickness of the p-Al cladding layer (0.3-0.4 um) determines the lateral index guiding of the subsequent emitters in the array. With strong index guiding, care has to be taken in growing each emitter in the array identically. To illustrate this, figure 2 shows the near- and far-fields as measured from a chirped array. The spacing between the outer emitters is 4.3 um and the inner emitters are 4.9 um apart. Although the total near-field distribution looks acceptable, a spectrally resolved measurement (at 300 mA) shows that the total field is actually built up from 3 "decoupled" modes, separated in wavelength by about 1 A, with emission patterns corresponding to, respectively, antiphase locking of the two outer emitters and independent lasing of the central emitter. Careful inspection of the near field intensity pattern reveals that the term "decoupled" is somewhat misleading, since the observed radiation fields should still be interpreted as deformed supermodes of the array. The strong distortion with respect to the expected array supermodes are induced by small additional geometry changes in the array elements caused by the LPE growth.

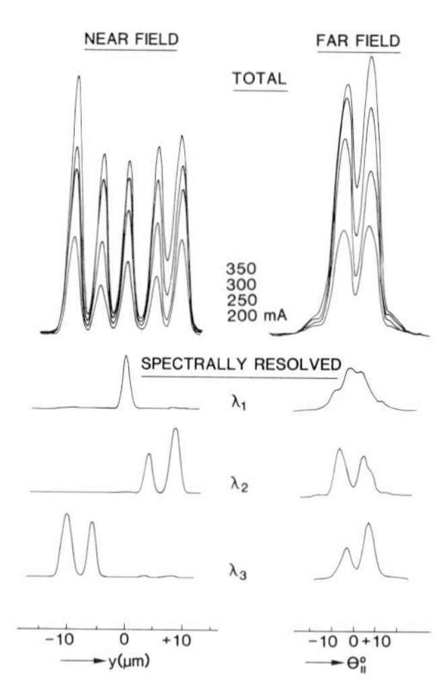

Figure 2. Decoupled chirped array: (a) Total NF and FF (b) Spectrally resolved NF and FF

Calculations using the effective index approach indicate that cladding layer thickness variation in the order of 0.03 um , or, equivalently, active layer thickness changes of 0.01 um can indeed induce such drastic changes in the supermode intensity patterns. Obviously,the tolerances on layer thickness control during LPE growth have to be critically reviewed. It should be noted that the current injection and temperature profiles can also lead to supermode distortion.

LPE growth over channeled substrates is strongly influenced by substrate misorientation [Botez,1983] . A misorientation of 0.3° off the (100) plane in a direction parallel to the V-grooves already lowers the ratio between the growth rate inside the groove and the growth rate on the plane outside the channeled area. Thus, crescent type layers are grown and, in particular, the active layer thickness varies smoothly in the lateral direction from a low value at the edges of the array to a higher value in the array center. As observed experimentally, this results in array structures with a tendency to "decouple" the central emitter from the two double-stripe outer arrays. At small misorientation in the direction perpendicular to the stripes, it appears that growth steps are located preferentially in the V-grooved region of the structure , so that steps in the p-type cladding layer and abrupt active layer thickness variation occur. Experimentally, it is found that in these cases the 5-stripe array effectively breaks up into two "decoupled" arrays, spatially divided by the growth step position. Obviously, care has to be taken to minimize the misorientation of the substrate (facet growth). A small misorientation ($< 0.05^0$) parallel to the channels can be allowed for.

During LPE-growth of the p-type cladding layer on the grooved n-GaAs CBL, melt back of the sharp edges of the V-groove occurs. The edge shape strongly influences the effective index step in the lateral direction. To minimize melt-etching effects the cladding layer should be grown at an as high as possible supersaturation, while still avoiding uncontrolled, spontaneous nucleation in the melt. However, at high supersaturation the ratio between the growth rates inside the groove and on the plane again is reduced. This competition between melt-etching rate and the growth rate ratio limits the supersaturation (to 3.5 K) and the minimum p-Al cladding thickness (to about 0.3 um). Furthermore, due to the geometry, the melt-etching rate depends on the position of the edge in the array: the small mesas in the center of the array are etched at a higher rate than the outer edges of the array.

2. Near-field shaping.

A regular array preferentially lases in the highest-order supermode (i.e. with neighbouring emitters in antiphase) and with the highest intensity at the central emitter. Therefore,the power density at the central emitter limits the maximum power output of the array. By manipulation of the relative distances between the emitters ("chirping") a uniform near field intensity distribution can be obtained. Figure 3 shows experimental results for a regular array with 5 emitters, each 3 um wide, on 5.0 um spacings as compared to a chirped array with an outer emitter spacing of 4.7 um and an inner emitter spacing of 5.3 um. Typically, the threshold current for these arrays is 220 mA. The external differential efficiency per facet is 0.3 W/A. Coated arrays (32% reflectivity) can emit more than 400 mW of pulsed light output power per facet. From the near field distribution of the regular array it is observed that the ratio between the central and outer emitter is about 1.5, whereas, theoretically, a ratio of 4.0 is expected. This discrepancy is caused by the higher melt-etch rate of the mesas with respect to the array outer edges. In practice, the thickness of the p-Al cladding layer at the edges of the array is 0.15 um less than at the mesa positions. The large index step at the array edges results in the experimentally observed near field pattern.

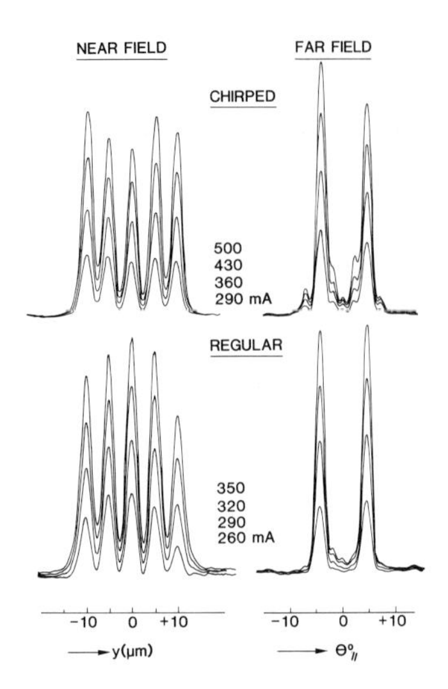

Figure 3. NF and FF of regular and chirped 5-emitter arrays.: (a) Regular array: spacing 5.0 um. (b) Chirped array: outer spacing 4.7 um , inner spacing 5.3 um.

Further equivalence between the emitters is obtained by shifting the outer emitters 0.3 um towards the center, corresponding to an increase in coupling constant by a factor of 1.13. The resulting near field pattern is almost uniform. From the far field measurements, it follows that the array operates in the highest order supermode.

3. Conclusions.

In applying LPE growth over channeled substrates to fabricate phase-locked index-guided laser arrays care has to be taken in choosing the growth parameters and misorientation of the substrate. As observed in spectrally resolved near- and far-field measurements and in accordance with calculations, small perturbation of the array structure induces large supermode distortions and "delocking". Substrate misorientation strongly affects the layer uniformity over the array. Furthermore, high supersaturation is needed to minimize local meltback. If these parameters are properly adjusted, high quality arrays can be grown. Finally, it is shown experimentally that a uniform light distribution can be achieved in a chirped array geometry.

References

D.Botez and J.C.Connolly,RCA Rev.,44,64,(1983).
P.S.Cross,Proc. CLEO'87,Baltimore,206.
T.Hayakawa,N.Miyauchi,S,Yamamoto,H.Hayashi,S.Yano and T.Hijikata, J.Appl.Phys.,53,7224,(1982).
D.R.Scifres,R.D.Burnham and W.Streifer,Appl.Phys.Lett.,41,118,(1982).
W.Streifer,M.Osinski,D.R.Scifres,D.F.Welch and P.S.Cross, Appl. Phys.Lett.,49,1496,(1986).
A.Yariv,Optical Electronics,ed.Holt-Saunders,(1985).

Ultraviolet-assisted growth of GaAs in LP-MOVPE

D.Grundmann, J. Wisser, R. Lückerath, W. Richter, H. Lüth and P. Balk
Institute of Semiconductor Electronics/Institute of Physics
Aachen Technical University, D-5100 Aachen, FRG

Abstract

Excimer laser stimulation (λ = 193 nm) LP-MOVPE growth at 0.1 bar was studied from 900 K down to 370 K. Localized growth could be realized below 700 K. The growth rate is determined by the photochemistry of surface reactions, but photolysis of reactants in the gas phase near the surface also plays a role. Absorption of radiation by reactants away from the substrate reduces the rate of deposition.

Introduction

Photostimulated growth offers considerable attraction for the localized deposition of materials for electronic devices. For this reason efforts have been made recently to extend the use of this technique from the growth of metal films to that of III-V semiconductor layers (Donnelly et al (1987) and Zinck et al (1986)). In an earlier paper we reported on the metal organic vapor phase epitaxy (MOVPE) of GaAs at temperatures of 700 K and higher using ArF excimer laser excitation (λ = 193 nm) and compared its effect with that of other laser wavelengths and of a mercury lamp for a range of growth temperatures (Balk et al (1987)). We showed that the absorption of radiation in the gas phase may considerably reduce the stimulation effect. The present report focusses at low substrate temperatures (< 700 K). This condition is favorable for localized growth since without UV irradiation no detectable deposition is observed.

Experimental

The growth studies were carried out in a cylindrical horizontal quartz reactor provided with a flat (70 x 10 mm^2) rectangular liner tube. The sample was placed on a graphite susceptor, which was heated using three 1000 W IR strip heaters (Balk et al (1987)). By providing UV transmitting windows on the top and at the sides of the outer tube and corresponding slots in the liner laser radiation could be introduced perpendicular and parallel to the substrate surface. The small height of the liner tube (10 mm) was chosen to reduce optical absorption by the reactants AsH_3 and TMG $(Ga(CH_3))_3$ in the gas phase in the case of perpendicular incidence, thereby obtaining a maximum intensity at the surface. This point is important because of the high absorbance of these compounds at 193 nm. The radiation was generated by an ArF excimer laser (Lambda Physik EMG 101 MSC) at 17 ns pulses of 2.8 MW/cm^2 at 10 Hz for λ = 193 nm.

Reactants were AsH_3 (UCAR, Phoenix Plus) and TMG (Alfa-Ventron). As a carrier Pd-diffused H_2 at a total pressure of 0.1 bar was used. Substrates were semi-insulating (100) GaAs wafers misoriented 2^0 towards the (110) plane. The thicknesses of the deposited films were determined optically.

Results and discussion

Photostimulation of the deposition process is successfully applied only for conditions where the growth rate is not mass transport limited but determined by a chemically activated chemical reaction. In our experiments this is the case below 900 K, where the rate drops off exponentially with temperature (fig. 1). The steep decrease is indicative of a thermally activated reaction. Increasing the linear flow velocity and thereby the rate of supply of reactants hardly affects the growth rate at these conditions. Only at very low gas velocities the rate becomes supply limited, as shown for 723 K in fig. 2. This indicates that for reasonable flow velocities significant depletion of the reactants in the gas phase does not take place. Consequently, the film thickness is uniform for these depostion conditions.

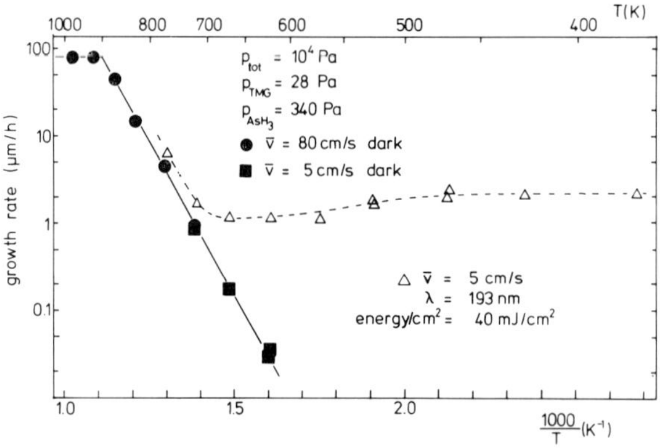

fig.1 Temperature dependence of growth rate with and without irradiation

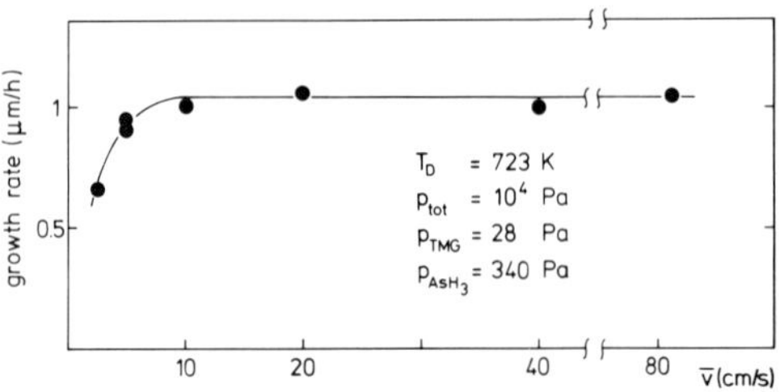

fig.2 Dependence of growth rate on gas velocity at 723 K

Upon entering the reaction rate limited region by lowering the growth temperature the magnitude of the UV stimulation effect (perpendicular incidence) is rather modest and amounts to less than a factor of two. However, below 700 K the rate observed with laser irradiation remains constant in first approximation at a value slightly above 1 μm/h for the experimental conditions used in our experiment (fig. 1). Upon lowering the temperature still further the rate exhibits a slight increase (to approximately 2 μm/h). We propose that this increase is caused by the combination of temperature dependencies, for example, of the hydrodynamics and the flow patterns in the system, of the concentrations in the gas phase (the partial pressures were held constant) and of the adsorption at the surface. Deposition was observed all the way down to 370 K.

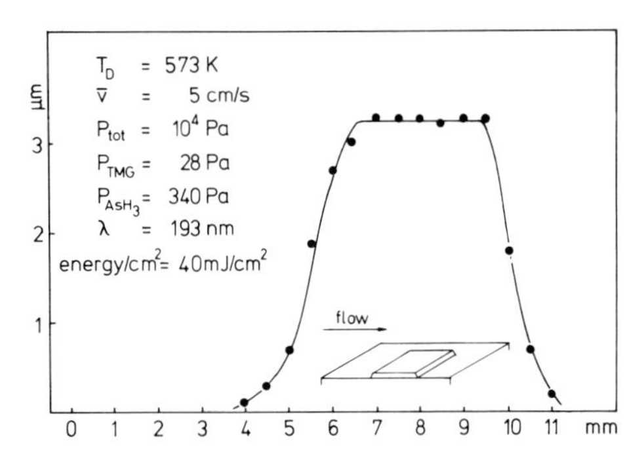

fig.3 Thickness profile of GaAs film grown with laser stimulation

The rectangular shape of the laser beam used in our study is rather distinctly imaged in the growth pattern. A profile of a layer deposited at 473 K is depicted in fig. 3. At least part of the sloped edges is caused by the flanks of the energy profile of the beam; these flanks extend over a comparable width. However, the diffusion of photolysed reactants out of the illuminated gas volume may also play a role in determining the shape of the flanks.

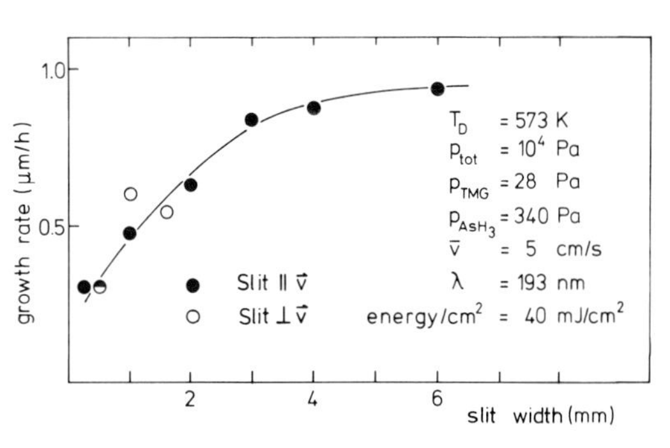

fig.4 Growth rate vs. width of slit used in defining laser beam

To determine the role of gas phase photolysis in the growth process experiments were carried out with the laser beam passing at close distance parallel to the surface. At 573 K, the other experimental parameters being the same as in the experiment of fig. 1, the growth rate was nearly two orders of magnitude lower than that found for perpendicular incidence. This finding documents that photolysis in the gas phase may be a necessary but not a sufficient condition for UV-stimulated growth. It also suggests that surface photochemical processes are essential for the UV stimulation effect.

When both processes are required for growth diffusion, of photolyzed reactants from the irradiated gas volume in the case of perpendicular incidence of the laser beam should reduce the amount of precursors available for growth on the irradiated area of the surface. To check for such an effect growth experiments were made at the same temperature using an external slit to vary the cross-section of the beam. Indeed, the overall growth rate drops off for smaller slit widths but saturates for wide slits (fig. 4). These data demonstrate the importance of gas phase reactions in UV-assisted MOVPE. The experiments were carried out with the slit pattern parallel and perpendicular to the direction of gas flow. The orientation of the slit is apparently not of major importance as shown by the data in fig. 4. This indicates that the gas phase photolysis affecting the growth apparently takes place in a region close to the surface, where the gas velocity is small.

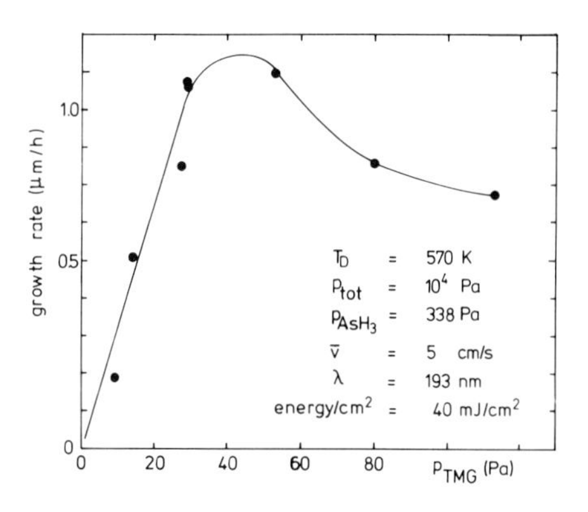

fig.5 Growth rate vs. TMG pressure

The effect of the concentration of the reactants on the rate was also studied at 573 K. Fig. 5 presents the data for TMG. It may be seen that at low partial pressures the rate increases roughly linearly with pressure. Here apparently the amount of photolyzed material at or near the surface determines the growth. However, the rate reaches a maximum and falls off again. The latter part of the curve is probably caused by the decreasing density of photons reaching the surface due to absorption in the gas phase. A qualitatively similar behavior is found in the dependence of the rate on the AsH_3 pressure. Here the same explanation appears to apply. Note that also the photolysis of AsH_3 is essential for the growth process. The decay of the growth rate for higher pressures may be modelled using Lambert-Beer's law of optical absorption by means of the approach employed in our earlier paper (Balk et al (1987)). The different position of the maximum in figs 5 and 6 reflects the different extinction coefficients for TMG and AsH_3 at 193 nm.

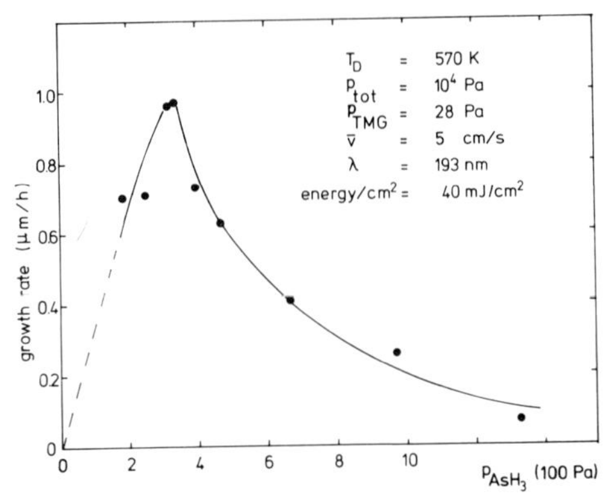

fig.6 Growth rate vs. AsH_3 pressure

SIMS studies showed that the carbon level in samples grown at 573 K was below the level of detection for our apparatus ($10^{17}cm^{-3}$). This material is polycrystaline with a grain structure of aprox. 2nm. The high carrier concentration (mostly p ≃ 10^{18} cm^{-3}; $\mu_{p(300)}$ in the range 10 -100 $cm^2V^{-1}s^{-1}$) suggests that the defects may be situated at the grain boundaries. These characteristics cannot be explained from the relatively low C concentration level.

Conclusions

Our data indicate the feasibility of localized growth of GaAs in the MOVPE system using ArF excimer radiation at 193 nm. To obtain this effect the substrate temperature should be below 700 K. Photolytic processes both at and near the surface in the gas phase play a role in the growth. The rate of deposition is determined by the amounts of the reactants reaching the area near the surface and by the density of photons in that region.

Acknowledgement

The authors like to thank the VW Foundation for financial support. They are indebted to Dr. B. Stritzker, KFA, Jülich for RBS studies, Dr. A. Vere of RSRE, Great Malvern, for HR-TEM data on the films, and to U. Breuer from our labratory for the SIMS study.

References

Balk P., Fischer M., Grundmann D., Lückerath R., Lüth H. and Richter W.
J.Vac.Sci.Technol., in press (1987)

Donnelly V.M., McCray V.R., Appelbaum A., Brasen, D. and Low W.
J.Appl.Phys., 61, 1410 (1987)

Karlicek R., Long J.A. and Donnelly V.M.
J.Cryst.Growth, 68, 123 (1984)

Zinck J.J., Brewer P.D., Jensen J.E., Olson G.L. and Tutt L.W.
presented at MRS fall meeting, Boston Dec. 1986

Inst. Phys. Conf. Ser. No. 91: Chapter 8
Paper presented at Int. Symp. GaAs and Related Compounds, Heraklion, Greece, 1987

Low temperature electron transport properties of exceptionally high purity InP

J M Boud, M A Fisher, D Lancefield, A R Adams, E J Thrush+ and C G Cureton+

Department of Physics, University of Surrey, Guildford, Surrey, UK.
+ STC Technology, Harlow, Essex, UK.

ABSTRACT: We report results of transport measurements on very high purity MOCVD InP with 77K mobilities of $3\times10^5 cm^2/Vs$ and peak mobilities $>4\times10^5 cm^2/Vs$. These are significantly higher than any previously reported. Good agreement with theory is obtained for a background impurity level in the low $10^{13} cm^{-3}$ range and assuming a deformation potential of 6.7eV. Above band gap persistent photoconductivity has also been observed. This is tentatively associated with band bending at the surface and substrate depletion regions.

In this paper we discuss results of low temperature transport measurements on ultra-high purity InP. The paper is divided into two sections. In the first part we discuss the use of such material to study acoustic phonon scattering. The second part is devoted to a description of the persistent photoconductive effect which is observed in the highest purity samples at low temperature.

The material in question was grown by Metal Organic Chemical Vapour Deposition using a low pressure (150 Torr) system operated at a low growth temperature (570°C) with a "diphos" purified Trimethyl indium source and 10% phosphine. Details of the growth system have been published elsewhere (Thrush et al (1987)).

Standard van der Pauw clover leaf samples were prepared with alloyed indium contacts. The epilayer thickness was estimated as 8μm from electrochemical C-V profiling. Samples were mounted in a small beryllium copper cell in a continuous flow helium cryostat through which an optical fibre was passed to allow optical experiments to be carried out. Temperature dependent Hall effect measurements were made at 0.035T down to 10K.

The temperature dependence of the electron mobility is shown in Fig.1. The mobility at 77K is $\sim3\times10^5 cm^2/Vs$ and reaches a peak of $4\times10^5 cm^2/Vs$ at 45K. To the authors' knowledge these values exceed those previously reported by a factor of two (Di Forte-Poisson et al (1985)). To obtain a fit to these experimental results we have used an iterative solution of the Boltzmann equation which allows for finite magnetic fields and includes electron scattering by ionized impurities, polar optic and acoustic phonons, piezoelectric scattering and scattering by neutral impurities using the model of Erginsoy (1950). These are combined at the matrix element level using the numerical technique described by Rode (1970) which eliminates approximations introduced in a Matthiessen's rule analysis. The parameters are as used previously by Lancefield et al (1987). The solid curve in Fig.1 shows the best fit using this technique. Clearly the agreement is very

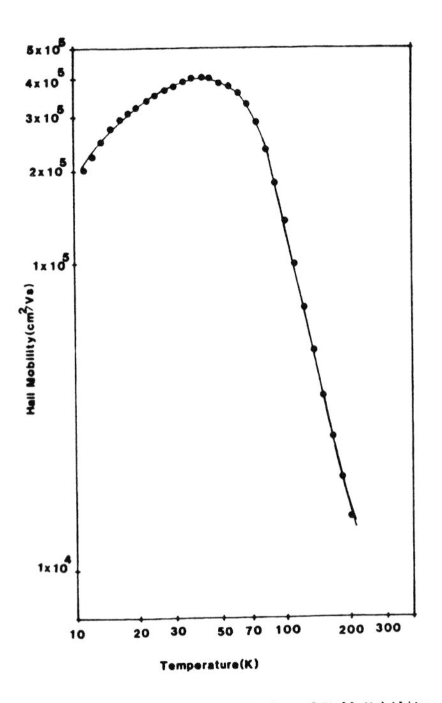

Fig.1 Temperature variation of Hall Mobility. • - experimental points; continuous curve - fit using iterative solution of the Boltzman equation.

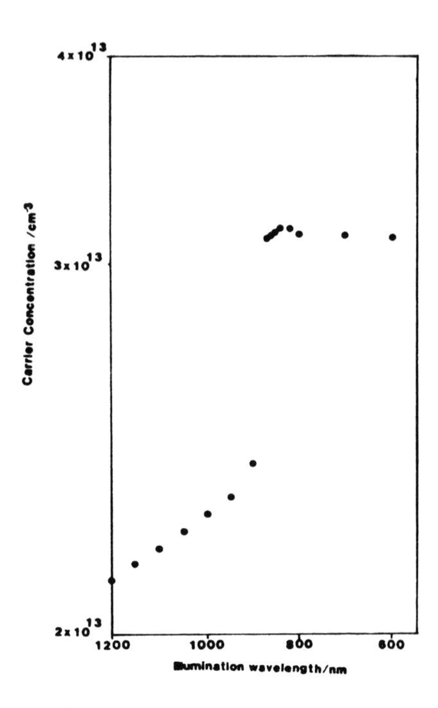

Fig.2 Carrier concentration measured in the dark against wavelength of prior illumination.

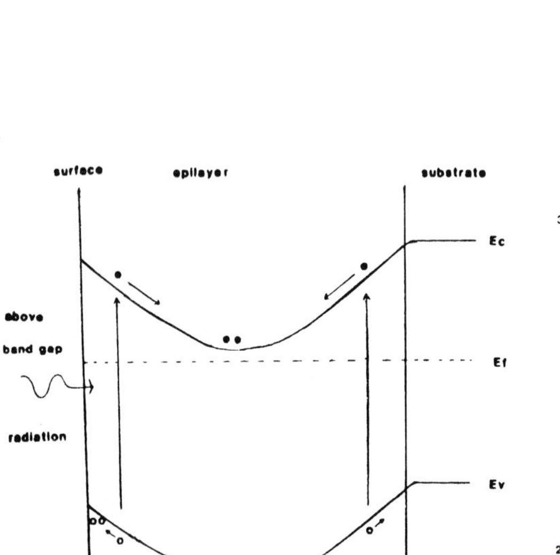

Fig.3 Band bending model : vertical axis -electron energy; horizontal axis -distance into sample from surface.

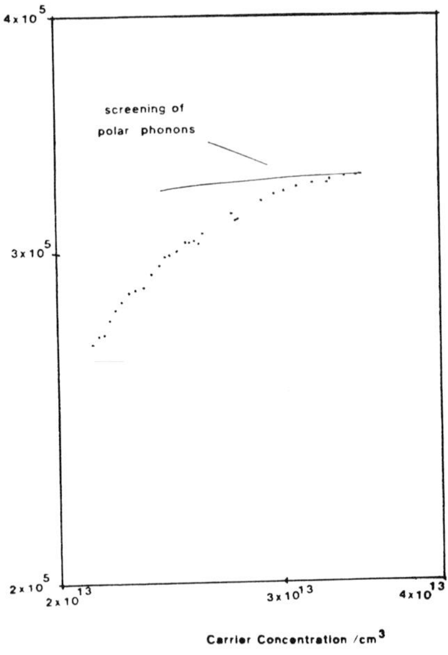

Fig.4 Hall mobility variation with carrier concentration using the persistent effect: • - experimental points; continuous curve - expected variation due to phonon screening

good over the whole temperature range considered.

The extremely high mobilities observed at low temperatures indicate a very low background impurity density, $\sim 3 \times 10^{13} cm^{-3}$ in this sample. Such low impurity densities make this an ideal sample for studying phonon scattering mechanisms and in particular we have used it to study acoustic phonon scattering.

The interaction of the electrons with acoustic phonons can be described in terms of a deformation potential which represents the change in the band edge energy under the strain associated with the passage of a phonon. Consequently it can be related to the effect of hydrostatic pressure on the optical band gap. However, it is only the conduction band deformation potential that scatters electrons, and this cannot be measured directly so that workers have often resorted to fitting theoretical expressions to the transport measurements using the deformation potential as an adjustable parameter. This has however proved generally unreliable because samples have ben dominated by polar phonon scattering at high temperatures and dominated by impurity scattering below this. This has led to considerable controversy with values quoted in the literature ranging from 3.4eV (Takeda and Sasaki (1984)) to 18eV (Nag and Dutta (1976)). Most of these are in excess of the direct measurements for the whole band gap. Our sample is ideal for this type of fit because there is an appreciable temperature range where the mobility peaks where the calculated mobility is highly sensitive to changes in the deformation potential used. Fitting in this temperature region we obtain a conduction band deformation potential of 6.7eV. This is considerably lower than a number of previously reported values and is close to values obtained from direct measurements of the total band deformation potential of 6.3eV Muller et al (1980). This implies that the valence band contribution is small which is in agreement with recent theoretical calculations, (Cardona and Christensen (1987)). The agreement with the direct measurement is also the case in GaAs, see for example Takeda (1984).

It should be noted that the above measurements have been achieved using the persistent photoconductive effect which is observed in these samples. When cooled to below 100K in the dark it is found that the material has a very high resistance and that it is only possible to drive current through a sample by first illuminating it.

The wavelength dependence of the persistent effect is shown in Fig.2. These measurements were carried out by cooling samples to 77K, illuminating for a fixed time (starting at long wavelength) and then using the Hall effect to measure carrier concentration in the dark. The figure shows a sharp rise in carrier density at photon energies near the band gap. The carrier concentration saturates at this point and a further increase can only be achieved under constant illumination. The non-zero slope below the band gap energy is not yet understood. The rapid change at the band gap would suggest that direct inter-band excitation is the dominant mechanism for the generation of additional carriers and that meta-stable impurities such as DX centres are not involved. If this is the case then a means of charge separation is necessary to achieve the persistent effect.

Noting that a potential of 0.5V at either the surface or interface would cause depletion to a depth of $\sim 4\mu m$ in such pure material the energy band diagram of Fig.3 applies. In this diagram conduction and valence band energies are plotted against distance into the epilayer. The 0.5V potential is consistent with near mid gap Fermi-level pinning at both surface and interface.

Under these conditions electron-hole pairs generated within the field regions would be separated with electrons remaining near the centre while the holes drift either to the surface or substrate where they are trapped. These effects combined with screening of the potentials by the additional electrons will result in a reduction of the barriers. This would explain the observed saturation in carrier density.

Finally the persistent photoconductive effect has been used to plot the variation of Hall mobility with carrier concentration at 77K. These results are shown in Fig.4. This curve shows the saturation of electron density observed previously.

So far it has not been possible to explain this effect quantitatively. Several effects need to be considered. At concentrations just below the peak mobility it will be dominated by polar phonons at 77K and the decreasing mobility with decreasing electron density may be due to decreasing screening of the phonon field (solid line in Fig.4). Initial calculations show that this effect cannot explain the steeper decrease in mobility at carrier concentrations below $3\times10^{13}cm^{-3}$ where it is possible that the non-uniformity of the very low doping may influence the conductivity which should then be described in terms of percolation theory as presented by Adkins (1979).

In conclusion, we have presented Hall effect and conductivity measurements on ultra pure InP which show an electron mobility more than twice that previously reported. Analysis of the mobility variation with temperature reveals acoustic phonon scattering with a deformation potential of 6.7eV consistent with the hydrostatic pressure dependence of the direct band gap. We have also observed persistent photoconductivity which has been interpreted in terms of charge separation by the electric fields in the surface depletion regions.

REFERENCES

Adkins J 1979 *J Phys C: Solid State Physics* **12** 3395
Cardona M A and Christensen N E 1987 **35** 12 6182
Di Forte-Poisson M A, Brylinski C and Duchemin J P 1985 *Appl Phys Letts* **46** 476
Erginsoy C 1950 *Phys Rev* **79** 1013
Lancefield D, Adams A R and Fisher M A 1987 *J Appl Phys* to be published
Muller H, Trommer R, Cardona M and Vogl P 1980 *Phys Rev B* **21** 4879
Nag B R and Dutta G M 1978 *J Phys C: Solid State Physics* **11** 1191
Rode D L 1970 *Phys Rev* **B2** 1012
Takeda Y and Sasaki A 1984 *Solid State Elect* **27** no.12 1127
Thrush E J, Cureton C G, Trigg J M, Stagg J P and Butler B R 1987 *J Chemtronics* **63** 62

Persistent current in 2D GaAs/GaAlAs rings

A. Raymond, J.L. Robert, C. Bousquet, J.Y. Mulot

G.E.S. Université des Sciences et techniques du Languedoc
34060 - Montpellier-Cédex, France

J.P. André, G.M. Martin

Laboratoire d'électronique et de physique appliquée
94451 - Limeil-Brévannes-Cédex, France

ABSTRACT : We report studies on persistent current in 2D GaAs/GaAlAs rings, related to the achievement of ultra-low resistance in the quantized Hall regime. The dependence of the conductivity σ_{xx} with the filling factor υ is determined and values of σ_{xx} lower than $10^{-21}(\Omega/\square)^{-1}$ are deduced at T = 1.4 K for υ = 2. The corresponding 3D resistivity is larger than $2\ 10^{13}$ Ω.m. This leads to a time constant for the test-sample itself larger than 1500 s. The longitudinal resisitivty ρ_{xx} is lower than $7\ 10^{-14}$ Ω/□, i.e. more than three orders of magnitude lower than the previously determined values.
Such values of σ_{xx} are correlated with the observation of persistent current, circulating along the ring path.

1. INTRODUCTION

In a quasi 2D system like a heterojunction, the electron motion is practically confined in a plane parallel to the interface. In presence of high magnetic fields B, perpendicular to the interface, the electron system is completely quantized into discrete Landau levels. Each level has a degeneracy N = eB/h. If n_s is the 2D electron density, the filling factor is given by $\upsilon = n_s/N$. Each Landau level, broadened by disorder, scattering etc, into a band, contains extended states, separated from the localized states by a mobility edge.
As a consequence of the total quantization of the density of states, the Hall voltage exhibits a series of flat plateaus when the magnetic field is increased. This is the well-known quantum Hall effect. In the quantum Hall regime, the Hall conductivity σ_{xy} is equal to ie^2/h (i is an integer equal to the number of filled Landau levels υ).
Such quantized values of σ_{xy}, corresponding to vanishing values of the longitudinal resistivity ρ_{xx}, can be observed whenever the Fermi level lies in localized states, or in real gaps in the absence of localized states (Laughlin 1981).

$$\rho_{xx} = \frac{\sigma_{xx}}{\sigma_{xx}^2 + \sigma_{xy}^2} \simeq 0.$$

In this study, we report experiments on persistent current in 2D GaAs-GaAlAs rings, related to the achievement of ultra-low resistance in the

quantum Hall regime.
The circulating Hall current resulting from the ρ_{xx} = 0 nature of the system does not vanish when the d.c current injected across two opposite probes is switched off. In the zero resistance state, the sample behaves like a capacitance C shunted by a large resistance R_t. This resistance presented a sharp maximum for each integer value of the filling factor. This maximum is higher for the lowest values of the filling factor and increases when the temperature is lowered.
Direct measurements of the decay time of the circulating Hall current is possible when R_t is lower than the impedance of the external circuit. In the absence of an external circuit, the time constant τ of the test-sample itself is equivalent to $\rho\ \epsilon$ (ϵ is the permittivity of the material and ρ is the 3D resistivity associated to the value of the resistance R_t). The time constant τ can be very large (~ 1500 s) since values of ρ larger than 10^{13} Ω.m are deduced from the υ and T dependence of R_t.

II. EXPERIMENTS

Our experiments in the zero-resistance state have been performed on modulation doped heterojunctions.
The characteristics of the samples, grown by MOCVD techniques are given in Table 1.

sample	Pressure (MPa)	n_s (m^{-2}) T= 4.2 K	μ (m^2/V.s) T= 4.2 K	d (Å) spacer
(†) 518	0.1	6.1 10^{15}	18.3	100
	400	3.9 10^{15}	13.8	
784	0.1	6.7 10^{15}	24.9	70
514	0.1	5.0 10^{15}	23.3	100

The geometry of the samples, corresponding to the so-called shunted-bridge configuration (S.B.C.),was defined in order to observe, on the same sample, the effects of the non-vanishing Hall current and the classical quantized magnetoresistance (Fig. 1).
This geometry is equivalent to a generalized Corbino device and, in the zero-resistance state, the application of a constant voltage across two opposite probes (3-4) of the S.B.C. sample creates a circulating Hall current practically without dissipation. The transverse resistance R_t between these probes is proportional to σ_{xx}^{-1} and so is very large.
To determine σ_{xx}, a d.c current was injected across the contacts 3-4 and the voltage was measured on the probes 5-6 with a high input impedance electrometer (10^{14} Ω) Direct measurements of R_t were possible only when its value was lower 10^{14} Ω.
For magnetic fields corresponding to the lowest values of the filling factor, the transverse resistance R_t is so high that the voltage V_{56} occurs a long time after the application of the d.c current I_{34}.
Moreover, the Hall current continues to circulate along the ring path after switching the d.c current off.

(†) Pressure can be used as an external parameter to change n_s (Mercy et al, 1984).

III. RESULTS

Figure 1 shows the variation of the voltage V_{56} ($\sim R_t$) as a function of the magnetic field B for sample 518.
Since there is no voltage drop along the ring path in the zero resistance state, the voltage V_{56} passes through a maximum when the longitudinal resistivity ρ_{xx} tends towards zero.
It must be noticed that the magnetic field range in which σ_{xx} tends towards zero, is very narrow.

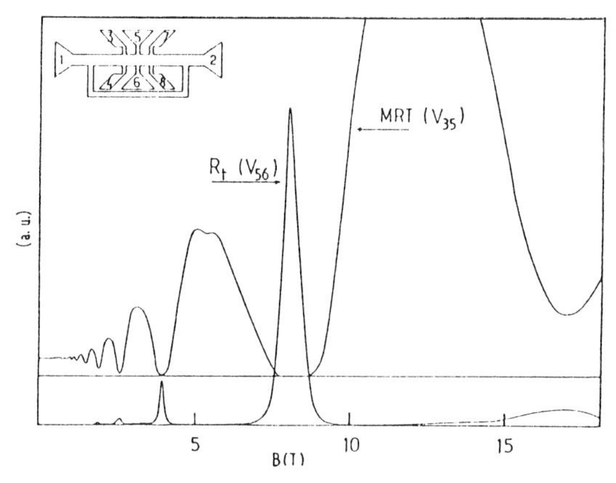

Fig. 1 : Magnetic field dependence of the transverse magnetoresistance (MRT) and of the resistance $R_t \sim \sigma_{xx}^{-1}$.

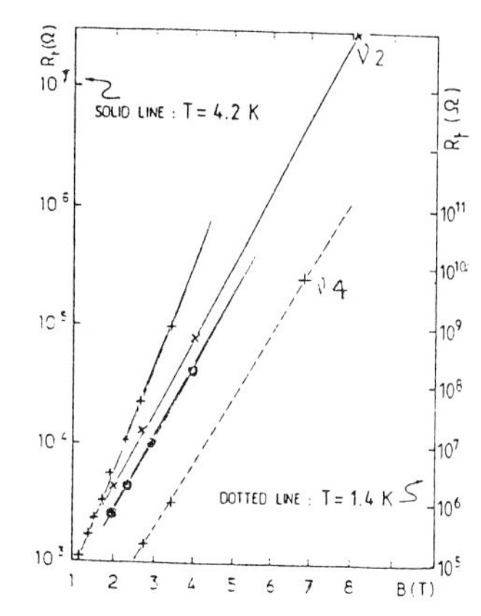

Fig. 2 : R_T versus B
+ sample 784 (P = 0.1 MPa)
o sample 518 (P = 0.1 MPa)
x sample 518 (P = 400 MPa)

In the same figure, the classical magnetoresistance voltage V_{35} is measured conventionally, using probes 1-2 as the source and drain contacts.
R_t presents a maximum R_T for each integer value of ν. A quasi-linear dependance of log R_T versus B is obtained, which shows that σ_{xx} varies over several orders of magnitude when ν decreases (Figure 2). Our results show that one deals with an activated conduction process : for a given value of ν, R_T increases strongly when the temperature is lowered (Robert et al 1987).
A typical curve, representing the voltage V_{56} after switching the d.c current off is given in Figure 3. It is interesting to notice that, when a magnetic field pulse is applied, the Hall current vanishes steeply, as a consequence of the strong decrease of R_T.
The variation of R_t around integer values of ν gives rise to a strong decrease of the time-constant τ of the Hall current : this clearly appears in Figure 4.
In order to find the equivalent circuit of the test-sample in the zero resistance state, we have determined the time constants with different capacitances in parallel with the sample. The values of R_T deduced from these experiments are the same than those directly measured.
This clearly demonstrates that the test-sample behaves like a capacitance C shunted by a large resistance R_T.
Then, we can write that the time constant τ' of the test-sample itself is equal to $\rho\ \epsilon$: its value can be, in particular, determined for the highest values of R_T.

When R_T cannot be directly measured, we extrapolate its value from its magnetic field dependence (Figure 2).
Values of σ_{xx} lower than 10^{-21} ($\Omega/\square$)$^{-1}$ have been determined at T = 1,4K for υ = 2 on sample 784.

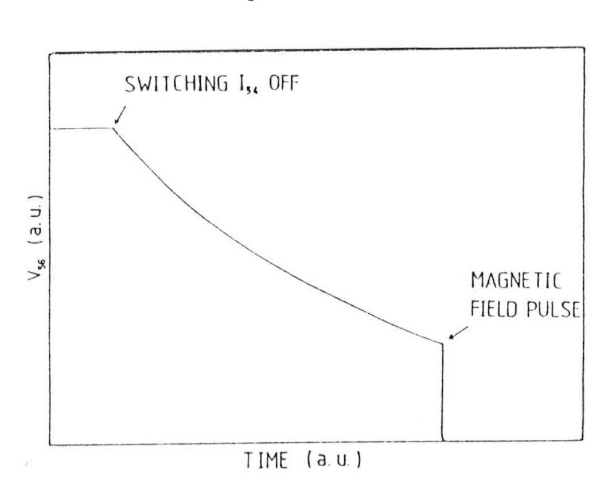

Fig. 3 : V_{56} voltage after switching the current I_{34} off, and after a magnetic pulse.

Fig. 4 : Magnetic field dependance of τ
T= 4.2 K, υ= 2, sample 514.

We have shown (Robert et al 1987) that the dependence of σ_{xx} on B and T is well described by considering that the conduction in the zero-resistance state is governed by a neighbor-hopping process.
Such a value of σ_{xx} corresponds to a value of ρ_{xx} lower than 7.10^{-14} ($\Omega/\square$), i.e. more than three orders of magnitude lower than the previously determined values (Tsui et al 1982, Haavasoja et al 1984).
Considering that the 2D channel has a thickness of 10^{-8} m, we get $\rho > 2.10^{13}$ Ω.m. This gives a time constant for the test-sample itself longer than 1500 s.
In summary, we have investigated the persistent Hall current circurlating in 2D GaAs/GaAlAs rings in the zero-resistance state.
The measured decay time, which can be very large, depends not only on the transverse resistance R_t but also on the capacitance of the external circuit.
Because of the strong variation of R_t around their maxima values when B is varied, a magnetic field pulse reduces drastically the time constant. The same result can be achieved, at a fixed magnetic field, by changing the 2D electron gas density in a gate-controlled device.

REFERENCES

- Laughlin R.B., 1981, Phys. Rev. B.23.
- Robert J.L., Raymond A, Mulot J.Y., Bouscuqet C, 1987, This conference.
- Tsui D.G., Stoermer H.L., Gossard A.C., 1982, Phys. Rev. B-25-2, 1405.
- Haavasoja T., Störmer H.L., Bishop D., Narrayanamurti V., Gossard A.C., Wiegmann W., 1984, Surface Sci. 142, 294.
- Mercy J.M., Bousquet C, Robert J.L., Raymond A, Grégoris G, Beerens J, Portal J.C., Frijlink P.M., Delescluse P, Chevrier J, Linh N.T., 1984 Surface Sci. 142, 298.

Paper presented at Int. Symp. GaAs and Related Compounds, Heraklion, Greece, 1987

Ranges, straggles and shape factors of 20 keV through 6 MeV random and channelled Si implants into GaAs, unannealed and annealed

R.G. Wilson, D.M. Jamba, D.C. Ingram,* P.P. Pronko,* P.E. Thompson,** and S.W. Novak***

Hughes Research Laboratories, Malibu, CA 90265
*Universal Energy Systems, Dayton, OH 45432
**Naval Research Laboratory, Washington, DC 20375-5000
***Charles Evans and Associates, Redwood City, CA 94063

ABSTRACT: We implanted Si into GaAs in a random orientation and in the <110> channeling direction from 20 keV to 6 MeV, annealed many of these implants using both furnace and flash lamp annealing with different caps, and measured the resulting depth distributions using secondary ion mass spectrometry (SIMS), and in some cases, capacitance-voltage electrical measurements. The ^{30}Si isotope was used to improve the SIMS background and profile dynamic range. SIMS profiles were processed using a Pearson IV computer fitting routine to determine the moments of the random depth distributions. We also report for the <110> channeled profiles, the maximum channeling range, the depth of the channeled peak, and the Si atom density achieved in the deep channeled region. These data allow a GaAs device designer to know what depths and densities are available for Si implantation, and the appropriate ion energies and fluences to use.

1. INTRODUCTION

Low energy (20 to 75 keV) Si implants into GaAs are important for the fabrication of shallow channels for devices. High energy (0.6 to 6 MeV) Si implants are of interest for the fabrication of fully implanted monolithic GaAs integrated circuits or devices where deep n regions are desired. Channeled implants of Si into the <110> direction of GaAs offer the potential to achieve greater doping depths for lower energy implants, or different profiles, but with some limitation in achievable doping density.

2. EXPERIMENTS AND TECHNIQUES

We implanted ^{30}Si ions into GaAs in a (100) random orientation and in the <110> direction of LEC and HB GaAs crystals at fluences between 5×10^{12} and 2×10^{13} cm^{-2} and energies from 10 keV to 6 MeV. For some of the energies, including 10, 20, and 40 keV, we also implanted GaAs predamaged by 5×10^{14} cm^{-2} implants of Ga plus As. Pieces of some of these implants, including both low and high energies, were annealed at 700°C/20 min or 800°C/15 min in a furnace, or at 800°C/10s or 1000°C/10s in a flash lamp annealer with proximity, SiO_2, and/or Si_3N_4 caps. ^{30}Si depth profiles were measured using Cs positive secondary ion mass spectrometry (SIMS). The Cs primary ion energy was 14.5 keV. Sputtering rates varied from about 1 nm/s for the lowest energies to about 10, 20, and 40 nm/s (each) for the MeV implants. Capacitance-voltage (CV) profiling was performed using an electrolytic, repetitive etch, low voltage profiler (Polaron).

3. RESULTS

Our experimental results are summarized in Table I and plotted in Figure 1. Values of the four moments of the depth distributions, R_p or μ, ΔR_p or σ, γ_1, and β_2, were obtained from a computer fitting routine that uses the Pearson IV algorithm. Corresponding LSS calculations using the Surface Alloys program are given in Table II and also plotted in Figure 1. Representative random and <110> channeled depth distributions for Si implanted into GaAs measured using SIMS are shown in Figures 2, 3, and 4, for 40, 700, and 6000 keV, respectively. Depth distributions were measured for selected annealed samples using SIMS for atom distributions, and using CV profiling for donor distributions (MeV implants). The results of the SIMS and CV measurements were in agreement within experimental accuracy, as was reported by Thompson et al.(1987).

Table I. Range and profile shape parameters for Si implants into GaAs

					For <110> Channeling			
Energy keV	R_m μm	R_p μm	ΔR_p μm	γ_1	N_{max}	R_c cm^{-3}	R_{max} μm	R_{mx}/R_p μm
10	0.012	0.012	0.011		-	-	-	-
20	0.026	0.018	0.033	-1.6	-	0.87	1.55	40
40	0.046	0.037	0.047	-1.3	2[16]	1.40	1.60	35
75	0.065	0.060	0.078	-0.7	2[16]	1.61	1.97	30
100	0.106	0.095	0.090	-0.7		2.07	2.25	25
150	0.162	0.157	0.125	-0.7	3[16]	2.35	2.65	17
200	0.21	0.20	0.15	-1.8	1[16]	2.8	3.2	15
300	0.32	0.27	0.17	-1.5	3[16]	3.4	3.7	12
350	0.39	0.33	0.22	-1.24	1.3[16]	3.6	4.1	10.5
450	0.47	0.40	0.19	-1.0	-	-	-	-
600	0.61	0.56	0.23	-0.6	-	-	-	-
700	0.75	0.64	0.28	-1.08	1.8[16]	4.9	5.7	7.6
1000	1.06	1.05	0.26	0.10	1.0[16]	5.6	6.5	6.1
2000	1.67	1.64	0.30	-0.07	-	-	-	-
4000	2.52	2.52	0.36	-0.06	1.3[16]	8.8	10.2	4.0
6000	3.19	3.11	0.35	-0.51	1.4[16]	10.5	12.5	3.9

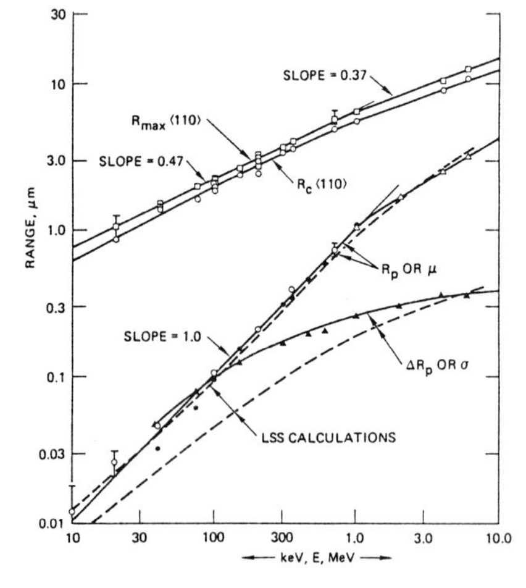

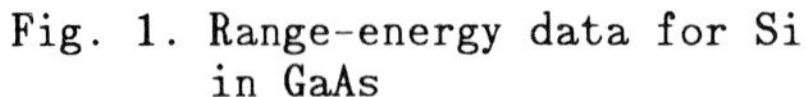
Fig. 1. Range-energy data for Si in GaAs

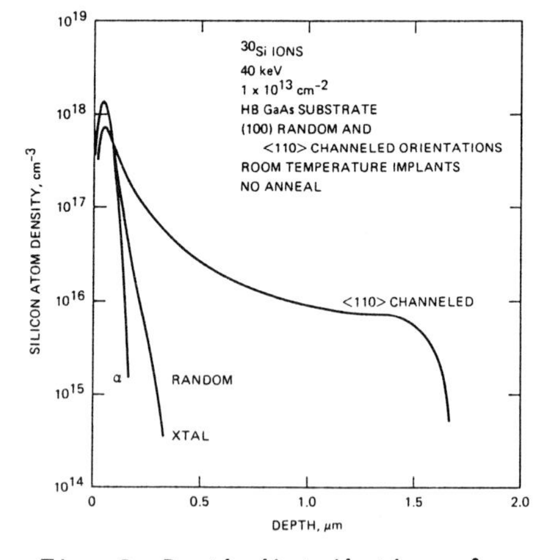

Fig. 2. Depth distributions for 40 keV Si in GaAs

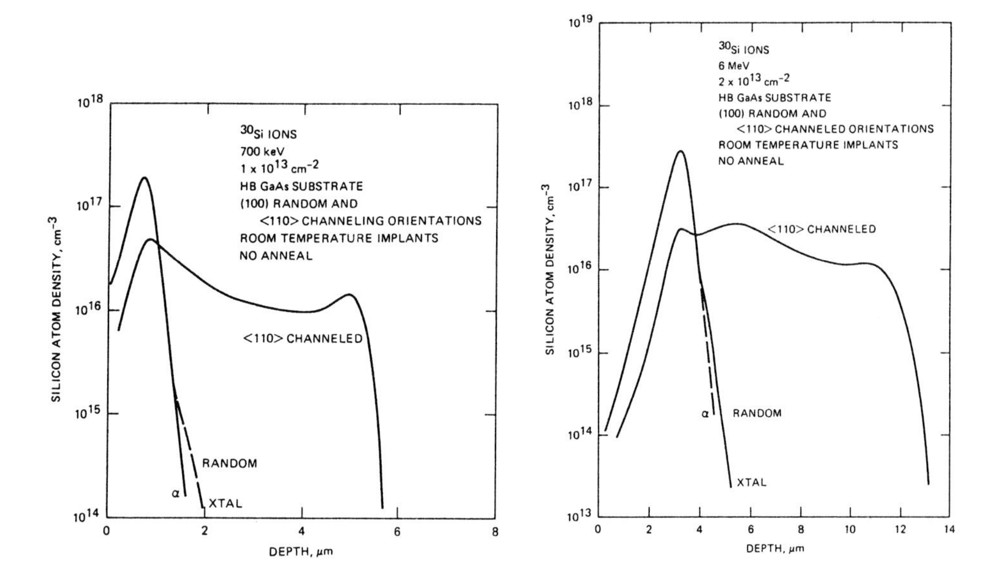

Fig. 3. Depth distributions for 700 keV Si in GaAs

Fig. 4. Depth distributions for 6 MeV Si in GaAs

Table II. LSS Calculations of Range Parameters for Si in GaAs

Energy keV	R_m μm	R_p, μ μm	$\Delta R_p, \sigma$ μm	γ_1	β_2	β_2'
10	0.008	0.011	0.008	0.54	3.66	0.00
20	0.016	0.020	0.013	0.44	3.51	0.01
40	0.032	0.037	0.022	0.31	3.38	0.18
75	0.064	0.067	0.035	0.14	3.31	0.24
100	0.088	0.090	0.044	0.06	3.30	0.28
150	0.136	0.135	0.060	-0.07	3.30	0.27
200	0.184	0.180	0.074	-0.16	3.31	0.23
300	0.28	0.271	0.097	-0.31	3.38	0.18
350	0.33	0.317	0.108	-0.36	3.43	0.16
450	0.43	0.40	0.125	-0.46	3.53	0.03
600	0.56	0.53	0.15	-0.58	3.7	0.0
700	0.65	0.61	0.16	-0.65	3.9	0.0
1000	0.90	0.84	0.18	-0.81	4.3	0.0
2000	1.60	1.49	0.25	-1.60	9.3	1.7
3000	2.16	2.04	0.29	-2.07	14.9	4.5
4000	2.64	2.50	0.33	-2.40	20.3	7.3
5000	3.08	2.91	0.36	-2.65	25.4	9.7

4. DISCUSSION AND CONCLUSIONS

No redistribution of implanted Si profiles was observed for any furnace or flashlamp annealing condition studied here, within the measurement accuracy

of SIMS, the estimated error for which is ±7% in depth. This observation applies to implant fluences of 10^{12} to 10^{14} cm^{-2} (Si density of ~10^{18} cm^{-3}). The maximum furnace and flash lamp temperatures were 800 and 1000°C, respectively. The maximum Si density achieved in the deep regions of the <110> channeling profiles is 1 or $2x10^{16}$ cm^{-3} for $1x10^{13}$ cm^{-2} fluence, and about 2 or $3x10^{16}$ cm^{-3} for $3x10^{13}$ or $1x10^{14}$ cm^{-2} fluences, so $2x10^{16}$ cm^{-3} is about the maximum achievable Si density for <110> channeled profiles in HB GaAs at depth from 2 to 12 μm (for implant energies from 0.1 to 6 MeV). This Si density is approximately constant (within a factor of 2) over 2.5 decades of ion energy. The maximum <110> channeling depths vary from about 1 μm for 20 or 40 keV ion energy, to about 12 μm for 6 MeV Si energy. The ratio of maximum <110> channeling range to the value of R_p, a measure of how much deeper <110> channeled depth distributions are than random ones, varies from 30 or 40 at 10-75 keV to 4 at 4 and 6 MeV. The SIMS background for Si (^{30}Si) profiling in this work was $2x10^{14}$ cm^{-3}, and the corresponding detection limit for Si in GaAs is ~$1x10^{14}$ cm^{-3} (3 times the background-subtracted density). The following observations are made regarding comparison of calculated (LSS) and experimental measurements (SIMS and CV). For R_m and R_p: As seen in Figure 1, these agree fairly well, varying from 0 to 20% difference, the experimental values being larger (except for 4 and 6 MeV). For ΔR_p or σ: The calculated values are less than the experimental ones, but approach them as energy increases, until they converge at the highest energy studied. The calculated values are about 50% of the measured values for the lower energies (<150 keV). Experimental moments are affected by channeling tails in random profiles in crystalline material and experimental values are not always accurate. The LSS calculated values are also not very accurate at their present stage of development. However, the experimental measurements and Pearson IV fits are probably good enough to say that the agreement should be better and that improved calculations are needed for moments 2, 3, and 4.
Finally, we note the shapes of the <110> channeled depth distributions as a function of implantation energy in Figures 2, 3, and 4: Low energies have a relatively high density pseudo-random peak, a gradual decrease through the dechanneling region, and no peak in the deep channeled region. Depth distributions for intermediate energies have clearly defined peaks in the deep <110> channeled distribution, and are constant within a factor of 10 (for these 10^{13} cm^{-2} fluences). For high energies (4 and 6 MeV), three peaks occur; one is the pseudo random peak, one, barely defined, is the deep channeling peak, and an intermediate peak is seen in the dechanneling region. For the $2x10^{13}$ cm^{-2} fluences, these profiles also vary within less than a decade from 3 to 10 or 12 μm in depth.

5. ACKNOWLEDGMENTS

This work was partially supported by the Army Research Office and the Naval Research Laboratory.

6. REFERENCES

Thompson P E, Wilson R G, Ingram D C, and Pronko PP 1987 Mat Res Soc Spring Meeting

MOCVD InGaAs photodiodes with extremely low dark current

M. Gallant, N. Puetz, A. Zemel*, and F.R. Shepherd

Bell Northern Research, P.O. Box 3511, Station C,
Ottawa, Ontario, Canada K1Y 4H7

ABSTRACT: Planar Zn-diffused MOCVD InGaAs photodiodes with extremely low dark current have been fabricated and evaluated. The typical dark current measured for 100μm diameter devices was 10pA at -10V bias. To the best of our knowledge, this is lower than the lowest dark current reported for InGaAs MOCVD photodiodes to date and is comparable to the lowest dark current demonstrated by VPE grown InGaAs detectors. Excellent uniformity of diode electrical characteristics was found over 2 inch wafers. A low capacitance of 0.45pF, a responsivity at 1.3μm of 0.90 A/W and rise/fall times of less than 150ps were measured at -5V bias.

1. INTRODUCTION

Extensive development of InGaAs p-i-n photodiodes has been carried out in recent years for use in lightwave communication systems operating in the 1.0-1.65μm wavelength region. It is well known that excellent material quality and thickness uniformity over a large substrate area can be achieved by the MOCVD growth technique (Okamoto et al 1984). However, to date MOCVD grown p-i-n photodiodes have not achieved very low dark current, characteristic of the best InGaAs LPE and VPE grown detectors (Trommer and Hoffmann 1986, Mikawa et al 1984). In this work we report, for the first time, extremely low dark current in planar Zn-diffused InP/InGaAs p-i-n photodiodes grown by MOCVD.

2. FABRICATION

A schematic cross-section of the photodiode is shown in Figure 1. The diode has a planar structure with a high bandgap InP cap layer which is expected to show advantages in low leakage current and reliability (Kim et al 1985). The MOCVD multilayer structure was grown on a 2" sulfur doped n^+-InP substrate with the following layer sequence: a 1μm thick undoped n-InP buffer layer, 2.7μm thick undoped n^--InGaAs absorbing layer lattice matched to InP, and a 1μm thick Si-doped n-InP cap layer. The p^+-n junction was formed in the InGaAs layer by diffusing Zn through the InP cap layer. A 1200Å thick CVD SiO_2 layer with various etched windows between 35 and 250μm in diameter was used as a diffusion mask. The depth of the p-n junction was approximately 1.5μm from the diode

* Permanent address: Soreq Nuclear Research Center
Yavne, 70600, Israel

surface. A 2000Å thick CVD SiN_x layer was subsequently deposited for passivation and antireflection. Contacts were defined on the p and n-side of the diode by standard lift-off processing of evaporated Cr/Au.

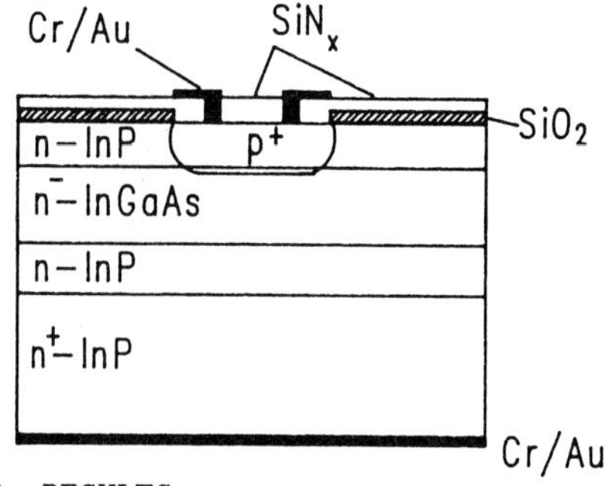

Fig. 1 Schematic cross-section of the planar InP/InGaAs p-i-n photodiode

3. RESULTS

The typical dark current measured for 100μm diameter devices was 10pA at -10V bias, with some devices having values as low as 3pA ($3.8x10^{-8}A/cm2$). To the best of our knowledge, these values are lower by more than an order of magnitude than the lowest dark current reported for InGaAs MOCVD grown photodiodes to date, (Ebbinghaus and Trommer 1987) and are comparable to the lowest dark current which has been demonstrated by LPE and VPE grown InGaAs detectors (Trommer and Hoffmann 1986, Mikawa et al 1984). Excellent uniformity (≤20% variation) of the dark current was found for devices within any "tile" of the wafer (corresponding to an area of 5x5mm2), and a slow variation in dark current, from 3-50pA, was found across a full 2" diameter wafer.

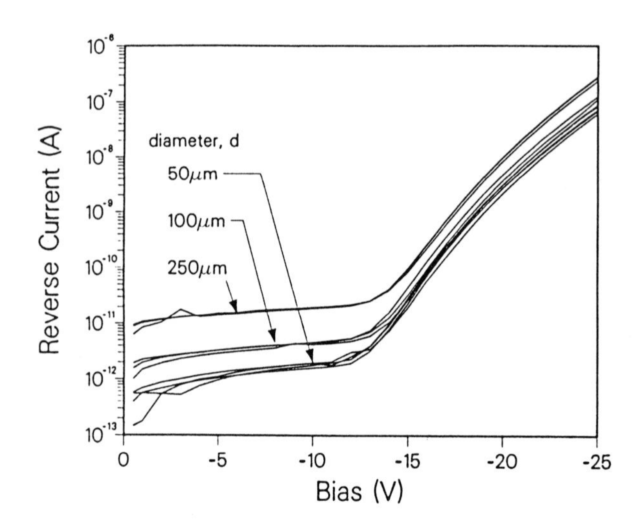

Fig. 2 Dark current characteristics measured on a number of diodes of various diameter d

Figure 2 shows the reverse I-V characteristics measured on a number of diodes with diameters of 50, 100 and 250μm. Curves for several diodes of each size from one tile are shown. From low to mid reverse bias, the plots show a relatively flat I-V characteristic indicating a nearly diffusion limited current transport at room temperature. This is supported by the temperature dependence of the dark current characteristics which will be published elsewhere. The current values for the 100μm diameter diodes in this tile are around 4.5pA at -10V bias. The very good uniformity in the dark current characteristics of the diodes over the entire bias range reveals the dependence of the dark current on diode dimensions at low and high reverse bias as shown in Figure 3. A clear area dependence ($I \alpha d2$) of the diode dark current at low to mid bias and a perimeter dependence ($I \alpha d$) at high bias are seen in the figure. The dependence of the dark current on diode area in the low to mid bias range indicates that the dark current is limited by the bulk material. The source for the current at high bias could be surface leakage, leakage through the p^+-n InP cylindrical homojunction or leakage through the p^+-n^- InP/InGaAs heterojunction. Further work is under way to study the origin and nature of this leakage current.

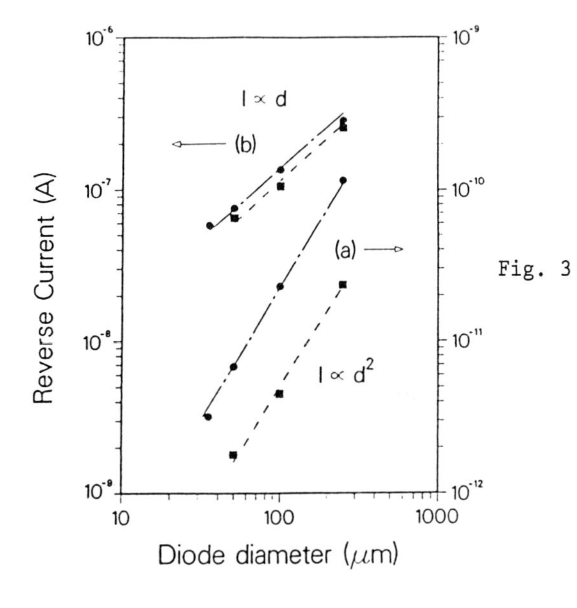

Fig. 3 Reverse current I as a function of diode diameter d, measured at -10V(a) and -25V(b) bias. The circles and triangles refer to results obtained in different areas of the wafer

A low junction capacitance of 0.45pF sufficient for high-speed operation is obtained at -5V bias. The high-speed response of the photodiodes at -5V bias is illustrated in Figure 4. The top-entry 80μm diameter diode was illuminated by 200ps wide 100MHz repetition rate pulses, using a 1.3μm wavelength InGaAsP diode laser. The rise/fall times of the response are less than 150ps and the full width at half maximum (FWHM) is 160ps. A DC responsivity, at 1.3μm, of 0.90A/W was measured at -5V bias.

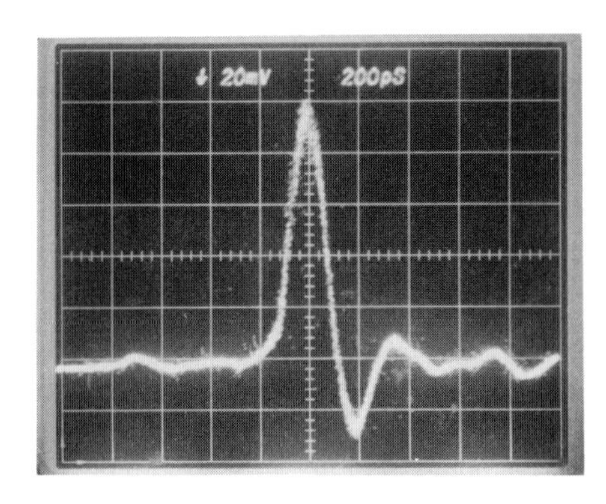

Fig. 4 1.3μm pulse response of the photodetector measured at -5V bias

4. CONCLUSIONS

In conclusion, planar Zn-diffused InGaAs photodiodes with extremely low dark current, comparable to the lowest dark current reported to date for LPE and VPE grown InGaAs detectors, were fabricated using MOCVD. Excellent uniformity of the dark current was found over 2" wafers. The electrical and optical characteristics of the photodiodes show that they are suitable for use in high sensitivity receivers operating up to 1Gb/s. This work demonstrates the feasibility of fabricating state-of-the-art InGaAs photodiodes from MOCVD grown material, with very high yield.

ACKNOWLEDGEMENTS

The technical assistance of J.P.D. Cook, M. Svilans, H. Postolek, and L. Tarof is gratefully acknowledged.

REFERENCES

Ebbinghaus G and Trommer R 1987 First European Workshop on MOVPE Aachen
Kim O K, Dutt B.V., McCoy R J and Zuber J R 1985 IEEE J. Quantum Electron. QE-21 138
Mikawa T, Kagawa S and Kaneda T 1984 Fujitsu Sci. Tech. J. 20 201
Okamoto A, Sumahawa H, Terao H and Watanabe H 1984 J. Crystal Growth 70 140
Trommer R and Hoffmann L 1986 El. Lett 22 360

Inst. Phys. Conf. Ser. No. 91: Chapter 8
Paper presented at Int. Symp. GaAs and Related Compounds, Heraklion, Greece, 1987

A planar 10 mW 30 GHz MIC compatible GaAs MESFET-like oscillator

H. Scheiber, K. Lübke, C. Diskus and H. Thim

University of Linz, Microelectronics Institute
Altenbergerstr. 69, A-4040 Linz, Austria

Abstract: Millimeter wave microstrip oscillators have been built which utilize the frequency independent negative resistance of the stationary charge dipole domain that forms in the channel of GaAs MESFET devices. The length of the gate was chosen 1.5 μm as the gate transit time is uncritical in this mode of operation. 10 mW fundamental output power at about 30 GHz has been obtained with 1.2 % efficiency with an ac-wise grounded gate biased with approx. -4 Volts.

1. Introduction

During the last few years GaAs FET performance has advanced to the point that FETs are considered as viable devices for millimeter wave systems. Both low noise and high power amplifiers have been built and operated at frequencies ranging from 30 GHz to 110 GHz. Power amplifiers have, for example, produced output power levels of a few tenths of a watt at ka-band frequencies (Hikosaka et al. 1986) with power added efficiencies near 10 %.

Interestingly, these high power levels cannot be obtained with FET oscillators even when mounted in waveguide or finline resonator circuits. The best results are 10 mW at 36 GHz (Tserng and Kim 1984) obtained with a single FET mounted in a finline resonator and 5 mW at 35.3 GHz obtained with two FETs in parallel mounted in a strip line circuit and loaded with a dielectric resonator (Talwar 1985). Already in 1981 Yamasaki et al. obtained 5 mW at 67.5 GHz with 5 % efficiency with a 0.25 μm FET mounted in an impatt-like waveguide circuit.

It is not so clear why FET oscillators produce much lower power levels than FET amplifiers. The difficulties in designing efficient feedback circuitry for a two port device might explain this discrepancy as it is well known that one port devices such as Gunn diodes are capable of producing tenths of a watt at ka-band frequencies. Unfortunately, presently available Gunn diodes are sandwich structures and are therefore not IC compatible. On the other hand planar Gunn diodes have not yet been developed to the point that they can produce sufficient power at ka-band frequencies or higher (Kuch et al. 1983, Binari et al. 1985).

The purpose of this paper is to present new encouraging results obtained with planar transferred electron devices with a MESFET cathode contact first described in 1982 (Kuch et al. 1982). A sufficiently high negative bias is applied to the Schottky gate so that normal Gunn oscillations due to travelling dipole domains are suppressed. Instead, a stationary dipole layer developes in the gate-drain region. The important difference to normal MESFET operation is that both gate and source are ac-wise grounded. An ac signal applied to the drain-gate/source contacts will be amplified by the differential negative resistance of the stationary dipole domain. This negative resistance is frequency independent allowing operation at any frequency below the frequency limit of the transferred electron (Gunn) effect. The important significance of this device is that it is not subject to the usual transit time limitation MESFETs and Gunn diodes suffer from. In other words, both gate length and drift length need not be made small in accordance with the transit time relation

$$f_T \leqslant \frac{v_D}{L} \qquad (1)$$

where L corresponds to either gate length or drift length and v_D to the drift velocity of electrons. This implies that sophisticated technological processes such as submicrometer gate technology are not necessary for fabrication of this device.

The advantages can be summarized as follows:
- easy fabrication due to simple technology
- it is fully planar and hence, IC-compatible
- it is not transit-time limited and therefore a broad-band device
- contact resistances are not as critical as those in MESFETs reducing transconductance
- circuit matching is easy as it is a two terminal device.

2. Device Structure

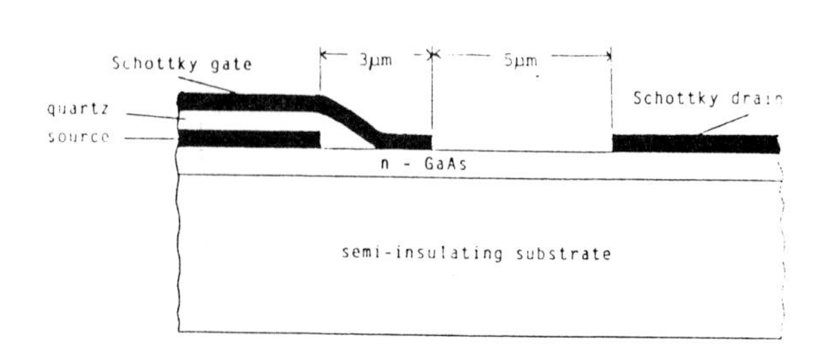

Fig. 1 Cross sectional view of the device

Fig. 1 shows a cross sectional view of the device. It consists of a 0.8 μm thick active n-layer ($N_D \sim 3.10^{16} cm^{-3}$), an ohmic source contact, a Schottky drain contact, a 1.5 μm long 400 μm wide Schottky gate and an integrated capacitor between gate and source with SiO_2 as the insulator. This capacitor connects the gate ac-wise to source. The MESFET thus acts as a constant current injector for the high field ("drift") region of this so called "field effect controlled transferred electron device" or "FECTED" (Kuch et al. 1983).

3. Device Technology

The epitaxial layers used for device fabrication were grown on Cr-doped substrates by chloride VPE. To form the ohmic source contact a Cr-Au-Ge-Cr layer is evaporated and alloyed at 400° C for two minutes after defining its pattern by a three level lithographical process including plasma etching. Next a Cr-Au layer is evaporated to form both the Schottky anode contact and a source metallization layer. After a mesa-etch a 500 nm thick SiO_2 layer is deposited at 325° C by chemical vapour deposition. Then the insulator is patterned by a careful plasma etch in order to get a 45° slope at the overlapping gate edge. Finally, the overlapping gate metal (Cr-Au) is deposited and patterned by a plasma etch process. Figure 2 shows a SEM micrograph of a device bonded to a microstrip circuit.

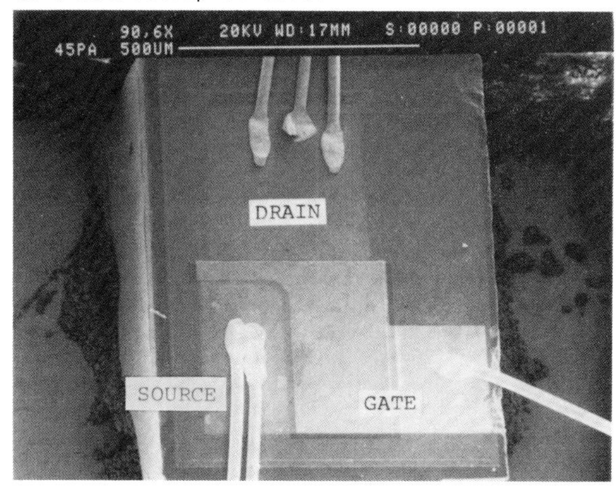

Fig. 2 SEM-micrograph of a mounted device

4. Experimental results

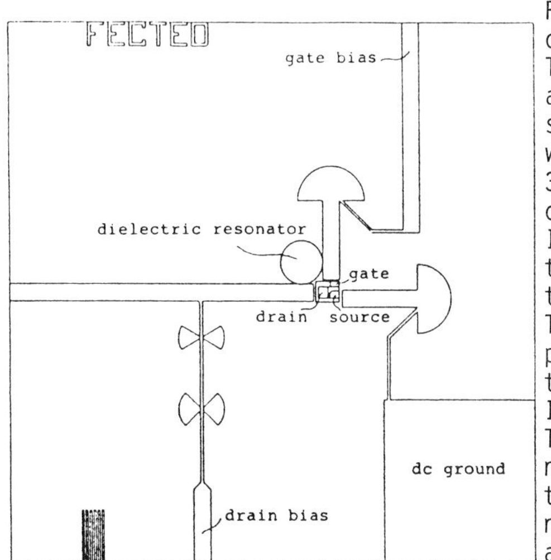

Fig. 3 Microstrip circuit

Figure 3 shows the microstrip circuit configuration of a FECTED oscillator. The drain voltage is applied to the 50 Ω -stripline via a bias filter which blocks the ac signal at 30 GHz. Both gate and source contact are connected to radial line stubs via $\lambda/2$ line sections providing ground potential to both gate and source. The dielectric resonator was positioned near these two electrodes yielding maximum oscillatory output power at 30 GHz. The microstrip circuit was connected to a ka-band waveguide testsystem via a Wiltron K-connector and a waveguide-coaxial adapter.

Devices operated at zero gate bias produced 40 mW at 13 GHz with 4 % efficiency in the conventional Gunn-effect transit-time mode. The frequency of operation corresponds to the transit-time frequency (equ.(1)) with a drift length of 8 µm which is the drain to source distance. When a negative gate bias of approx. -4 V was applied to the gate the same device produced 10 mW at 29.8 GHz with 1.2 % efficiency. Lower frequency probes had been inserted into the microstrip circuit and connected to a spectrum analyzer in order to make sure that no other signals at lower frequencies were present. Since the frequency of oscillation could not be influenced by replacing the 30 GHz dielectric resonator by other resonators it is speculated that the oscillation frequency was primarily determined by bond wire inductances and the depletion layer capacitance underneath the gate. The observed variation of frequency with gate bias voltage as shown in Figure 4 supports this speculation.

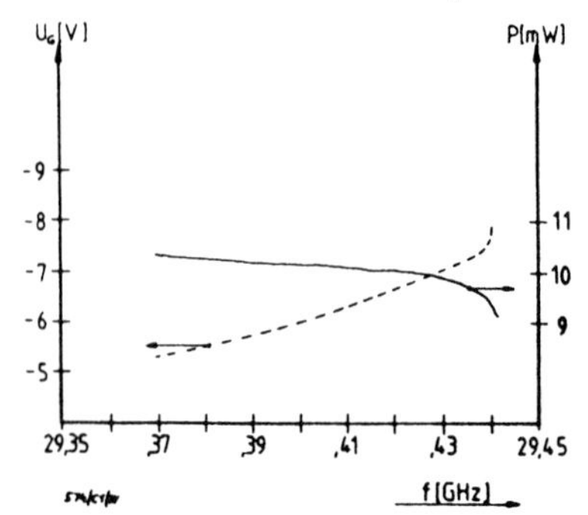

Fig. 4 Gate Voltage and RF power versus frequency

Some devices have also been operated at 38 GHz with 1.5 mW with a tuning range of ± 300 MHz.

5. Conclusions

It has been shown that planar transferred electron devices with MESFET cathode contacts ("FECTEDs") can be operated at ka-band frequencies and presumably at higher frequencies as these devices are not subject to the transit time limitations FETs and TEDs suffer from. Oscillatory output power levels in the 10 mW range have been obtained which are comparable to those obtained with 0.2 µm gate FET oscillators. Since measured efficiencies (1.2 %) are still a factor of 5 lower than calculated values higher power levels should be attainable with these devices making them attractive candidates as local oscillators in MMICs.

Acknowledgements:
This work was sponsored in part by the US Army through its European Research Office. The authors are grateful to Dr. Noll of AMS Unterpremstaetten for supplying the masks.

Binari S.C., Thompson P.E. and Grubin H.L. 1985 IEEE Electron Device Letters EDL-6 22
Hikosaka K., Hirachi Y. and Abe M. 1986 IEEE Transactions ED-33 583
Kuch R., Lübke K., Lindner G. and Thim H. 1982 Inst. Phys. Conf. Ser. 63 293
Kuch R., Lübke K., Thim H., Chabicovsky R., Lindner G. and Haydl W. 1983 Inst. Phys. Conf. Ser. 65 439
Talwar A.K. 1985 IEEE Trans. MTT-33 731
Tserng H.Q. and Kim B. 1984 Electronic Letters 20 297
Yamasaki H. and Schellenberg J.M. 1981 Inst. Phys. Conf. Ser. 63 431

Inst. Phys. Conf. Ser. No. 91: Chapter 8
Paper presented at Int. Symp. GaAs and Related Compounds, Heraklion, Greece, 1987

A high-gain superlattice bipolar transistor with controlled carrier multiplication

Albert Chin and Pallab Bhattacharya

Department of Electrical Engineering and Computer Science, The University of Michigan, Ann Arbor, Michigan 48109-2122

ABSTRACT: A novel n-p-n bipolar transistor with controlled avalanche multiplication and large current output over a significant bias region is demonstrated by incorporating in the collector junction a few periods of a symmetric or asymmetric multiquantum well in which only electrons predominantly multiply. The theory of operation, materials growth by molecular beam epitaxy and device performance are described. Optical gains as high as 140 are measured in phototransistors utilizing such structures.

1. INTRODUCTION

In the past, transistors have been designed with enhanced gain due to avalanching in the collector-base region at biases close to the collector breakdown voltage.(Tantraporn et al 1978) In such devices, the gain can be high, but is largely uncontrolled. We demonstrate here a controlled avalanche superlattice transistor (CAST) with III-V semiconductors which can lead to better performance, since these compounds usually have higher electron mobility and saturation velocity than those on Si. In the proposed device a superlattice region is incorporated in the collector-base junction to induce an enhancement of α_n/α_p. Such enhancement has been proposed and demonstrated in superlattice avalanche photodiodes.(Capasso 1985, 1983) In the proposed superlattice a large α_n/α_n can be obtained at fairly low fields.

2. PRINCIPLE OF OPERATION

We present the principle of operation with a staircase superlattice in the collector base junction, as shown in Fig. 1. When the conduction-band discontinuity E_c is comparable to the electron ionization threshold energy E_i in the low bandgap material, each electron will impact ionize a new electron-hole pair when it crosses the conduction-band step. A simplified analysis of an N-stage staircase superlattice structure gives the electron multiplication factor $M_e=(1+\gamma)^N$ when the hole ionization rate is negligible.[3] Here γ is the ionization yield. A proper choice of materials and

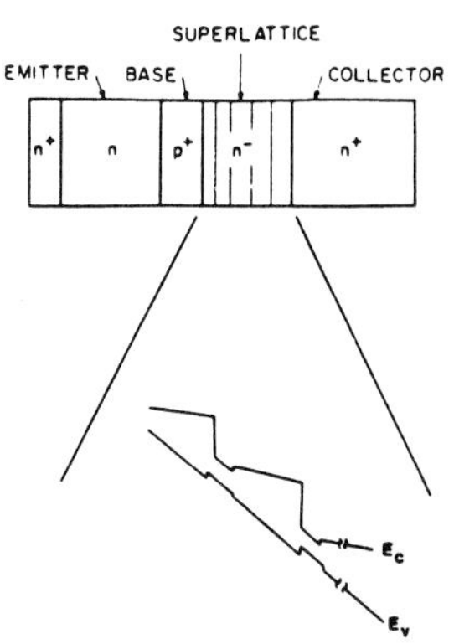

Figure 1 Controlled avalanche transistor with a staircase multiplier in the collector. The band diagram of the staircase superlattice under bias is also shown.

structural design can yield a negligible hole ionization rate for an applied bias which is large enough to create electron impact ionization at the conduction-band steps. The current gain of the CAST, with controlled multiplication M_e is expressed as

$$\alpha = \alpha_F M_e = \left[1 + \frac{D_{hE} N_A x_B}{D_{eB} N_D x_E} \left(\frac{m^*_{eE} m^*_{hE}}{m^*_{eB} m^*_{hB}}\right)^{3/2} .exp(-\Delta E_g/kT)\right] \left(1 - \frac{x_B^2}{2L_n^2}\right) M_e. \quad (1)$$

The multiplication can lead to a higher small-signal transconductance $M_e g_m$. The current-voltage characteristics can be expressed as

$$J_c = q \frac{n_{iB}^2 D_{eB}}{N_A x_B} (exp[qV_{BE}/kT] - 1)\, M_e \quad (2)$$

where $M_e=(1+\gamma)^N$. The computed current-voltage characteristics for the transistor structure shown in Fig. 2 for different values of N are shown in Fig. 3.

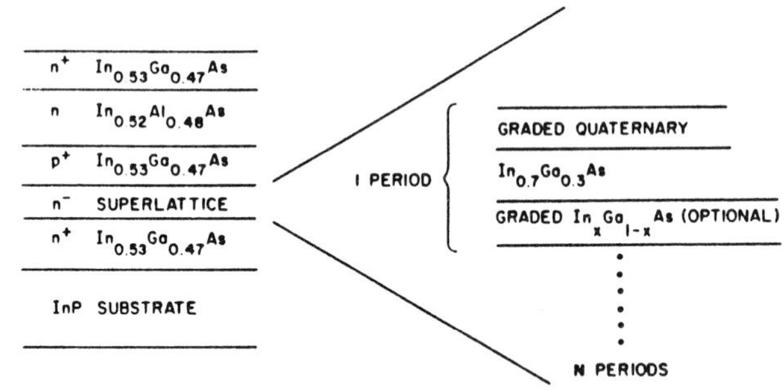

Figure 2 Schematic of proposed InP-based avalanche transistor showing the various materials used.

3. EXPERIMENTAL TECHNIQUES

The ideal device structure proposed and modeled in the previous section was difficult to realize because the graded InGaAlAs region is extremely difficult to grow reliably. We have, therefore, grown by molecular beam epitaxy an alternate structure, onve version of which is shown schematically in Fig. 4. The 0.465 μm depleted multiplication region includes 3 periods of 350Å graded $Al_xGa_{1-x}As$ (x=0.35-0), 250Å GaAs, 100Å $In_{0.3}Ga_{0.7}As$ and 450Å GaAs. Other variations of this structure, which were grown are, (a) a planar doped ($1x10^{13}cm^{-2}$) base with a collector depletion region consisting of 4 periods of the multiplying superlattice to enhance the absorption, and (b) a thick base (2000Å) structure with two $In_{0.3}Ga_{0.7}As$ wells (100Å) in the GaAs region of the multiplying structure.

4. DEVICE CHARACTERISTICS

Floating-base n-p-n phototransistor structures were made with area of 7.9×10^{-5} cm^2 from the grown layers by standard photolithography techniques. Typical dark currents are ~ 1-10 nA near collector breakdown voltages. Fig. 5 shows the output characteristics of the phototransistor UM784 as a function of 818 nm illumination intensity. The significant features are the steps in the photocurrent at 1.4 and 4.5 V and eventually a breakdown at higher voltages. The first threshold is due to the electrons from the base surmounting the first AlGaAs barrier. Assuming $E_c = 0.6\ Eg$, and some scattering, we calculate a threshold voltage of 1.6 V. This is very close to the experimentally observed value.

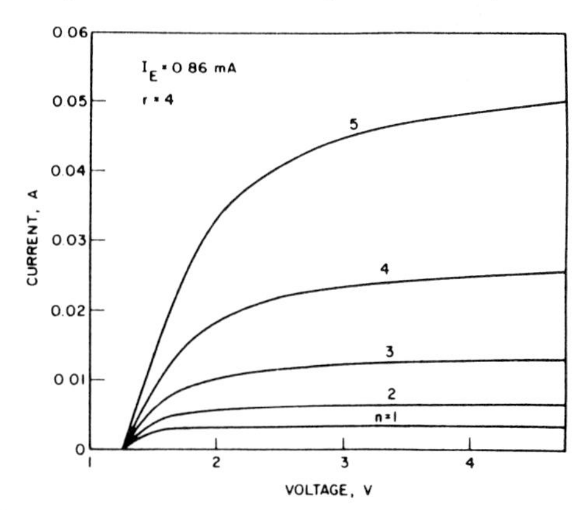

Figure 3 Calculated current-voltage characteristics for different number of multiplication steps.

The ternary wells were included in the design of the multiquantum wells in order to ensure electron impact ionization in that region. For this to happen, we must have $E_{ie} \approx E_g\ (In_{0.3}Ga_{0.7}As) = 0.9$ eV. Assuming near-ballistic motion of the electrons in the 600Å region before the ternary wells, the threshold voltage is calculated to be ≈ 4.5 V. The optical gain of the transistor UM784 is 23 at a bias of 9.6 V, which is lower than expected. Also, the maximum current is low. This is partly because in this structure the emitter-base heterojunction was graded to improve the high-speed response of the device. As a result the emitter injection efficiency is low.

In another device with a planar doped base, a similar set of characteristics were recorded, with a higher current output. This is due to the greater modulation of the emitter barrier possible in these structures. Another structure was grown with two ternary wells (L_z = 100Å) in the impact ionizing region and 3 periods of MQW, to increase the probability of impact ionizing. The currents are higher than the first device by a factor of 6. Also, the optical gain at the maximum incident power is 142. This is the highest optical gain for a phototransistor with a base width of 2000Å.

It should be noted that the devices presented here exhibit high currents at much lower biases than in conventional bipolar phototransistors exhibiting avalanche gain near breakdown voltages. (Chand et al 1985) With optimized device design it is expected that high gains and much higher current outputs will be obtained from these devices. Again, since base widths can be considerably reduced, and with hot electron injection from the emitter to the base, the devices are expected to operate at high speeds.

5 REFERENCES

Capasso F in Semi-conductors and Semimetals W T Tsang, Ed, Vol. 22. New York: Academic, 1985, Part D, ch. 1

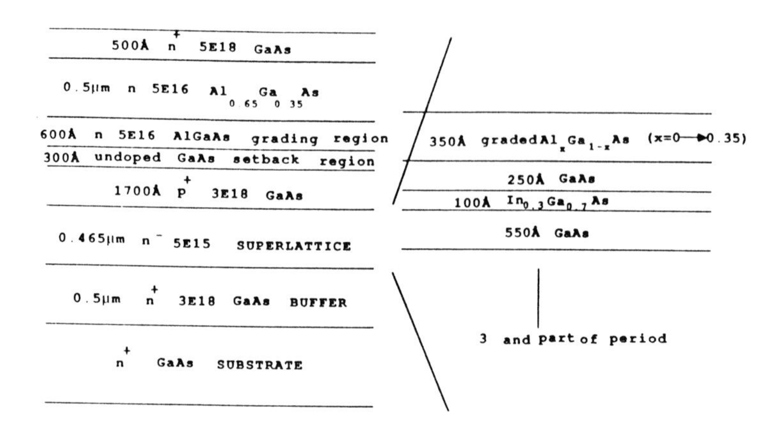

Figure 4 Avalanche bipolar transistor grown on GaAs substrates by molecular beam epitaxy.

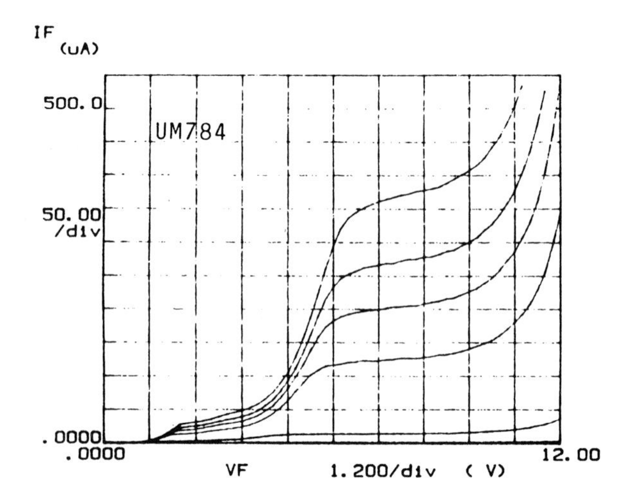

Figure 5 Output characteristics of phototransistor with one ternary InGaAs well in the MQW for different 818 nm light intensity levels.

Capasso F, Tsang W T and Williams G F 1983 IEEE Trans. Electron Devices, ED-30 pp 381

Chand N, Houston P A and Robson P N 1985 IEEE Trans. Electron Devices ED-32 pp 622

Tantraporn W, Yu S P and Cady W R 1978 IEEE Trans. Electron Devices ED-25 pp 520

Electrical characterization of gold–tantalum GaAs Schottky diodes using I–V and DLTS measurements

G. Pananakakis[1], N.C. Bacalis[2], P. Panayotatos[3], G. Kiriakidis[2] and A. Christou[4]

[1] LPCS/INPG (UA CNRS 840), Grenoble, France.
[2] Research Center of Crete, Iraklion, Greece.
[3] New Jersey University, USA.
[4] Naval Research Laboratory, Washington D.C., USA.
Research Center of Crete, Iraklion, Greece.

ABSTRACT : The energy location and density of "surface-like" states in Au-Ta-nGaAs Schottky barriers have been determined using I-V and DLTS measurements. The experimental data have been fitted to a numerical model for a metal-semiconductor structure.

1. INTRODUCTION

Gold-refractory type Schottky barriers are important to the GaAs technology from the point of view of self aligned gates and the high temperature stability of circuits. At the present time, there is lack of information concerning the location in the band gap and the density of surface states which are introduced by gold-refractory contacts such as Au-Ta. There is also a requirement to model the metal-semiconductor interface using the experimental IV and CV measurements. In the present investigation, using deep level transient spectroscopy (DLTS), we have measured the energy level of surface states in gold-tantalum Schottky barriers to n-GaAs and have fitted the experimental data to a static classical model of the interface. Excellent agreement between the theoretical calculations and the experimental results were obtained.

2. SAMPLE PREPARATION

The devices under test were fabricated on n^+GaAs substrate. 1 μm thick n-GaAs, silicon doped ($N_D \sim 1.2 \times 10^{16}$ cm^{-3}) epitaxial material was grown by Molecular Beam Epitaxy (MBE). Gold-tantalum dots were deposited on the epitaxial material with a contact area equal to 0.5 mm^2. The electrical back contact was ensured using silver paint.

3. EXPERIMENTAL RESULTS AND MODELLING

DLTS measurement allowed us to obtain the energy location E_t of the states at the metal-semiconductor interface. According to Lang (1974) the thermal emission rate of the states were plotted versus 1/T and the energy difference E_c-E_t was found equal to 0.44 eV, on several devices.

In order to fit the experimental I-V characteristics of the studied devices we have used a self consistent one dimensional numerical model essentially based on the thermionic emission theory (Rhoderick (1978)) developed by Pananakakis et al (1979). According to this model we assumed that the Au-Ta-GaAs contact can be described as a metal-insulator(air)-semiconductor structure with single energy level generation-recombination centers located at the insulator-semiconductor interface.

The expression of the currents between the interface states and the 3 reservoirs of free barriers (metal, semiconductor conduction and valence band) can be obtained by using SCHOCKLEY HALL READ statistics.

The total collection electron and hole currents at the metal-semiconductor interface are written by assuming the validity of Boltzmann statistics and that the carriers cross the insulator by tunnel effect.

Finally the I-V characteristics of the diode can be numerically calculated using several input parameters. The most important are the insulator thickness (taken equal to 10 A of air) the Schottky barrier height (taken equal to 0.9 eV for the gold-tantalum-n GaAs system) and the energy location of the of the states E_c-E_t = 0.44 eV, given by the DLTS measurements. A very good fitting of the experimental I-V characteristics was obtained (Figure 1). The interface states density was also determined in this way and was found equal to 1.2×10^{12} cm^{-2}. Besides the value of the series resistance of the probe and back contact circuit was evaluated in the order of several kΩ.

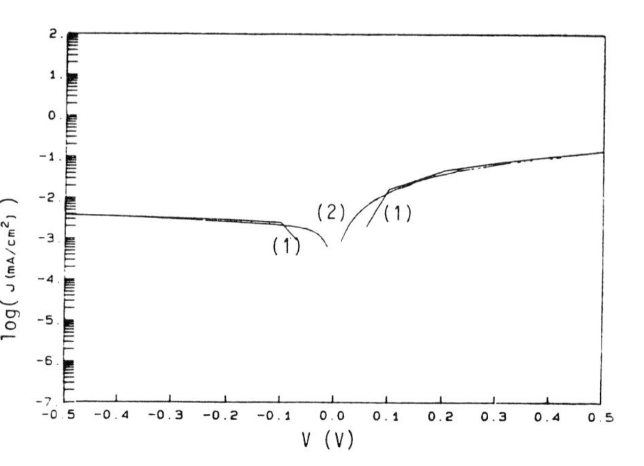

Figure1: Comparison between the experimental and the theoretical I-V characteristics of the Schottky diode (1) : experiment, (2) : theory.

4. CONCLUSION

We have determined the energy location and the density of surface states of Schottky barriers to GaAs using I-V and DLTS measurements as well as a numerical model of the working of a metal-semiconductor structure. We have now a theoretical tool for calculating the surface state location and density for Schottky barriers.

5. REFERENCES

Lang D V, *Journal of Appl. Physics*, 45, (1974), pp. 3023-3032.
Rhoderick E H, *Metal Semiconductor Contacts*, Clarendon Press, Oxford (1978).
Pananakakis G, Kamarinos G, Viktorovitch P, *Revue de Phys. Appliquee*, 14, (1979), pp. 639-647.

Author Index